General Advisors / Contributors

JOHN ALCOCK
Arizona State University

AARON BAUER
Villanova University

ROBERT COLWELL
University of Connecticut

JERRY COYNE
University of Chicago

GEORGE COX
San Diego State University

KATHERINE DENNISTON
Towson State University

DANIEL FAIRBANKS
Brigham Young University

PAUL HERTZ
Barnard College

JOHN JACKSON
North Hennepin Community College

EUGENE KOZLOFF
University of Washington

ROBERT LAPEN
Central Washington University

WILLIAM PARSON
University of Washington

CLEON ROSS
Colorado State University

SAMUEL SWEET
University of California, Santa Barbara

STEPHEN WOLFE
University of California, Davis

Primary Biological Illustrator

RAYCHEL CIEMMA

BIOLOGY
concepts and applications

THIRD EDITION

CECIE STARR

BELMONT, CALIFORNIA

WADSWORTH PUBLISHING COMPANY

I(T)P® AN INTERNATIONAL THOMSON PUBLISHING COMPANY

Belmont, CA • Albany, NY • Bonn • Boston
Cincinnati • Detroit • Johannesburg • London • Madrid
Melbourne • Mexico City • New York
Paris • San Francisco • Singapore
Tokyo • Toronto • Washington

BIOLOGY PUBLISHER: Jack C. Carey

ASSISTANT EDITOR: Kristin Milotich

EDITORIAL ASSISTANT: Kerri Abdinoor

MARKETING MANAGER: David Garrison

PRINT BUYER: Karen Hunt

PROJECT EDITOR: Sandra Craig

PRODUCTION: Mary Douglas, Rogue Valley Publications

TEXT AND COVER DESIGN, ART DIRECTION: Gary Head,
Gary Head Design

ART COORDINATOR: Myrna Engler-Forkner

EDITORIAL PRODUCTION: Rosaleen Bertolino, Karen Stough,
Mary Roybal

PRIMARY ARTISTS: Raychel Ciemma; Precision Graphics
(Jan Flessner, J.C. Morgan)

ADDITIONAL ARTISTS: Robert Demarest, Darwen Hennings,
Vally Hennings, Betsy Palay, Nadine Sokol, Kevin Somerville,
Lloyd Townsend

PHOTO RESEARCH AND PERMISSIONS: Marion Hansen

COVER PHOTOGRAPH: Art Wolfe/Tony Stone Images

COMPOSITION: AMERICAN COMPOSITION & GRAPHICS, INC.
(Jim Jeschke, Jody Ward)

COLOR PROCESSING: H&S GRAPHICS, INC. (Tom Anderson,
Nancy Dean, and John Deady)

PRINTING AND BINDING: Von Hoffmann Press, Inc.

COPYRIGHT © 1997 by Wadsworth Publishing Company
A Division of International Thomson Publishing Inc.

I(T)P

The ITP logo is a registered trademark under license.

Printed in the United States of America
5 6 7 8 9 10

Library of Congress Cataloging-in-Publication Data
Starr, Cecie.
    Biology : concepts and applications / Cecie Starr. — 3rd ed.
        p.   cm.
    Includes bibliographical references and index.
    ISBN 0-534-50440-X
    1. Biology.   I. Title
QH307.2.S73   1996
574—dc20   96-17245

For more information, contact Wadsworth Publishing Company, 10 Davis
Drive, Belmont, California 94002, or electronically at
http://www.thomson.com/wadsworth.html

International Thomson Publishing Europe
Berkshire House 168-173, High Holborn
London, WC1V7AA, England

Thomas Nelson Australia
102 Dodds Street
South Melbourne 3205, Victoria, Australia

Nelson Canada
1120 Birchmount Road
Scarborough, Ontario, Canada M1K 5G4

International Thomson Editores
Campos Eliseos 385, Piso 7
Col. Polanco, 11560 México D.F. México

International Thomson Publishing GmbH
Königswinterer Strasse 418
53227 Bonn, Germany

International Thomson Publishing Asia
221 Henderson Road, #05-10 Henderson Building
Singapore 0315

International Thomson Publishing Japan
Hirakawacho Kyowa Building, 3F
2-2-1 Hirakawacho, Chiyoda-ku, Tokyo 102, Japan

International Thomson Publishing Southern Africa
Building 18, Constantia Park
240 Old Pretoria Road
Halfway House, 1685 South Africa

CONTENTS IN BRIEF

DETAILED CONTENTS

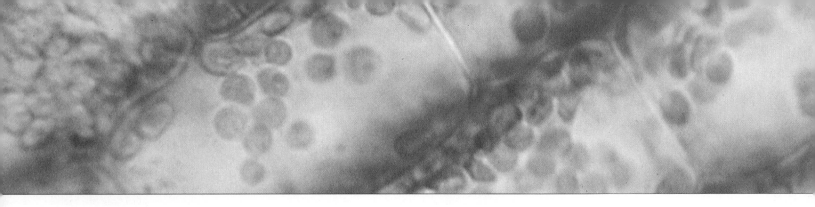

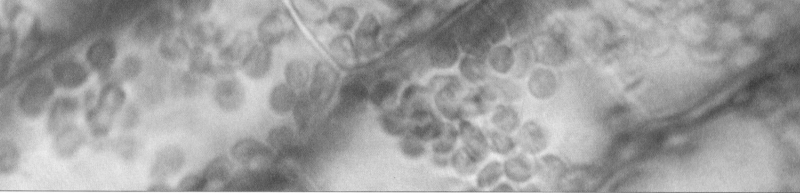

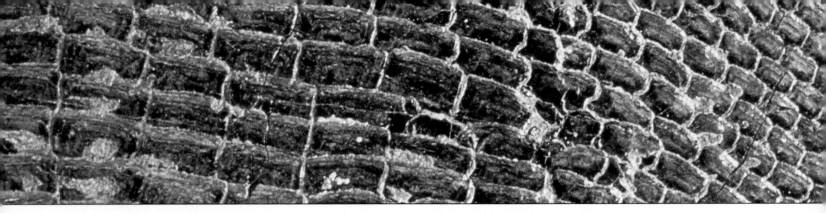

FOSSILIZED REMAINS OF AN ANCIENT FISH

SOLDIER TERMITES DEFENDING THEIR COLONY

FLOWERING PLUM BUSILY ENTICING POLLINATORS

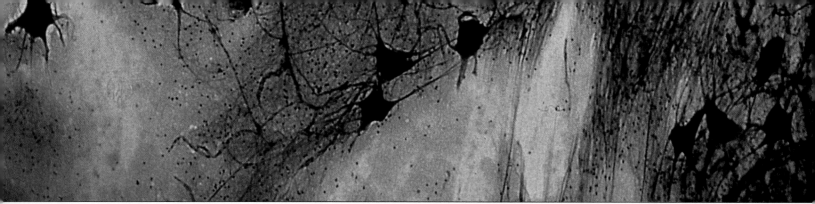

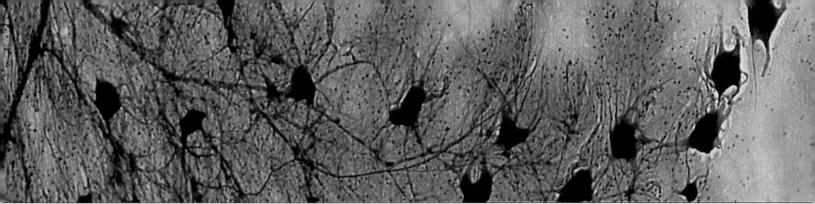

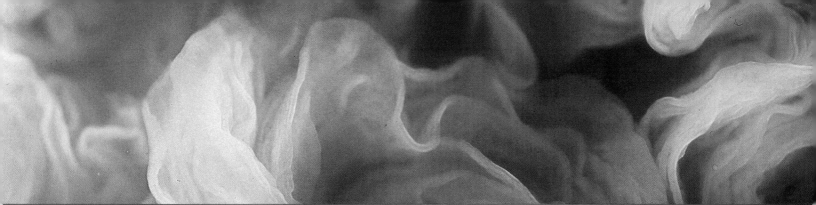

PREFACE

The book you are about to read probably costs the same as a night on the town or a single top-of-the-line athletic shoe. Is it worth it? That depends on what you expect to get out of a course in general biology. More than this, it depends on what you expect to get out of life.

Reading *Biology: Concepts and Applications* is like having your own private audience with biologists from colleges, universities, and research laboratories the world over. The book contains insights from about 2,000 biologists who contributed time and energy to its development over the past two decades—insights that can sharpen your view of life. The biological perspective happens to be one of the most useful tools for interpreting the world. With it, you can cut your own intellectual paths through social, medical, and environmental thickets. With it, you can come to understand the past and to predict possible futures for ourselves and all other organisms.

Writers who mean to do justice to the biological perspective know they must use the principles of biochemistry, inheritance, and evolution as the framework for a book. They also should convey the great power of evolutionary theory to explain life's unity and diversity by incorporating examples of problem solving and experiments. They should identify key concepts for all major fields of inquiry and select topics to give students a sense of current understandings and research trends. They should explain the structure and functioning of a broad sampling of organisms in enough detail so that students can develop a working vocabulary about life's parts and processes. Besides all of this, they must write in a way that is logical, clear, engaging, and never patronizing.

We met four more objectives for this latest revision of *Biology*. We recast the book in a modular format. We refined and updated its core material and applications. We started each chapter with a short story that leads to the chapter's key concepts in an informative yet entertaining, nonthreatening way. Finally, we created many more visual summary illustrations, which have a track record of helping several million of our student readers work their way through difficult concepts.

CONCEPTUAL FLOW THROUGH A MODULAR FORMAT

The book's *numbered, modular format* is the most significant departure from preceding editions. These modules give instructors far more flexibility in assigning or skipping different sections of each chapter. The advantages are so great that most introductory textbooks probably would be structured this way—if it were easy to modularize chapters without chopping up their story lines. Even though we have more than twenty years of experience in writing and illustrating biology textbooks, the structuring turned out to be an astonishing challenge. Yet we believe that the results move the book to a new, more effective level of pedagogy.

3.1 BASIC ASPECTS OF CELL STRUCTURE AND FUNCTION

Inside your body and at its moist surfaces, trillions of cells live in interdependency. In scummy pondwater, a single-celled amoeba moves about freely, thriving on its own. For humans, amoebas, and all other organisms, the **cell** is the smallest entity that still retains the properties of life. A cell either can survive on its own or has the potential to do so. Its structure is highly organized, and it engages in metabolism. A cell senses and responds to specific changes in the environment. And it has the potential to reproduce itself, based on inherited instructions in its DNA.

Structural Organization of Cells

Cells differ enormously in size, shape, and activities, as you might gather by comparing a bacterium with one of your liver cells. Yet they are alike in three respects. All cells start out life with a plasma membrane, a region of DNA, and a region of cytoplasm:

1. **Plasma membrane**. This thin, outermost membrane maintains the cell as a distinct entity. By doing so, it allows metabolic events to proceed apart from random events in the environment. A plasma membrane does not *isolate* the cell interior. Substances and signals continually move across it, in highly controlled ways.

2. **DNA-containing region**. DNA occupies part of the cell interior, along with molecules that can copy or read its hereditary instructions.

3. **Cytoplasm**. The cytoplasm is everything enclosed by the plasma membrane, *except* for the region of DNA. It is a semifluid substance in which particles, filaments, and often compartments are organized.

This chapter introduces two fundamentally different kinds of cells. **Eukaryotic cells** contain distinct arrays of organelles—tiny, internal sacs and other compartments formed from membranes. One organelle, the nucleus, houses the DNA. It is the defining feature of eukaryotic cells. By contrast, **prokaryotic cells** have no nucleus; no membranes intervene between the region of DNA and the cytoplasm that surrounds it. Bacteria are the only prokaryotic cells. Beyond the realm of bacteria, all other organisms—from amoebas to peach trees and puffball mushrooms to zebras—are eukaryotes.

The Lipid Bilayer of Cell Membranes

A cell's organization and activities depend on its membranes. If any membrane is to function properly, fluid must bathe its two surfaces. How, then, does it remain structurally intact? Is it like a solid wall separating two fluid regions? Not at all. Push a fine needle into and out of a cell, and cytoplasm will not ooze out. Instead, the

cell surface will flow over the puncture and seal it. The plasma membrane apparently shows fluid behavior!

How do "fluid" cells remain distinct from their fluid surroundings? For the answer, let's start by looking at phospholipids—the most abundant component of any cell membrane. A **phospholipid**, recall, is a molecule having a hydrophilic (water-loving) head and two fatty acid tails, which are hydrophobic (water-dreading). If you immerse many phospholipid molecules in water, they will interact with the water molecules and with one another until they spontaneously cluster together. Collectively, the jostlings might even force them to become arranged in two layers, with the fatty acid tails sandwiched between the hydrophilic heads. This arrangement, which is called a **lipid bilayer**, is the structural basis of cell membranes:

hydrophilic head
hydrophobic tails

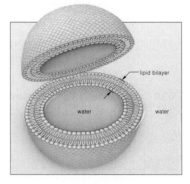
lipid bilayer
water   water

So why does a punctured membrane seal itself? A puncture is energetically unfavorable; it exposes many hydrophobic groups to water. Mutual repulsion forces the phospholipids back to the more stable arrangement.

Fluid Mosaic Model of Membrane Structure

Figure 3.2 shows a bit of membrane that corresponds to the **fluid mosaic model**. By this model, cell membranes consist of lipids *and* proteins. They incorporate a variety of phospholipids, glycolipids, and sterols (such as cholesterol in animals and phytosterols in plants). Many proteins are embedded in the bilayer or positioned at

FIGURE A   *A module from this edition.*

As the sample in Figure *A* shows, each module starts with a clearly numbered heading, and it ends with boldface summary statements to reinforce the key concepts. Each is self-contained on one or two pages—*yet careful transitions allow all the modules of a chapter to flow smoothly, one into the other, so students won't lose sight of the big picture.*

Each module contains all the text and illustrations for a particular topic, so students can work their way through one topic at a time without those annoying page-flipping searches for information.

Unless writers know what they are doing, they run the risk of having a modular format dictate content, like a tail wagging the proverbial dog. Fortunately, we had a history of being full-time fanatics about crafting the text, art, and captions for each chapter into an integrated story. When we decided to learn Quark for this edition, it was with the understanding that page makeup should not supersede our craft but rather should be an extension of it.

We quickly learned that you can't write good modules without clearing the clutter from your mind, for there simply is no room for superfluous detail. Where details are useful as expansions of key concepts, we often integrated them into suitable illustrations rather than allowing them to disrupt the text flow. We blocked out section headings and subheadings in order to rank the importance of bits of related information. Any good story has such a hierarchy of information, with background settings, major and minor characters, and high points and an ending where everything comes together. Without a hierarchy, information has all the excitement, flow, and drama of an encyclopedia. Additionally, we refined the paragraphs and carefully adjusted the size of photographs and line art. As part of the creative fine-tuning, we also introduced variations in page layouts to avoid the monotonous, encyclopedic look that tend to make student readers wonder whether they are brain dead.

Finally, we understand fully that many students approach biology textbooks with apprehension. If the words don't engage them, they sometimes end up hating both the book *and* the subject. It comes down to line-by-line judgment calls. Having devoted years to writing precisely and objectively, we have a strong sense of when to leave core material alone and when to loosen it up to give students breathing room. Interrupting, say, the description of mitotic cell division with a distracting anecdote does no good. Plunking a humorous aside into a chapter that ties together the evolution of the Earth and life trivializes a magnificent story. By contrast, it certainly is appropriate to follow modules on mitotic cell division with a brief look at the cell cycle and cancer. It is fine to entertain students *if* doing so reinforces key concepts, as with the story of the misguided species introduction that allowed rabbits to rampage through much of Australia or the story of how artificial selection produced bulldogs, which sometimes pass out because the overly wide, flabby roof of their mouth does not favor easy breathing.

The result is that students get the best of both worlds—appropriate, engaging content *and* cleanly served-up content— one module at a time.

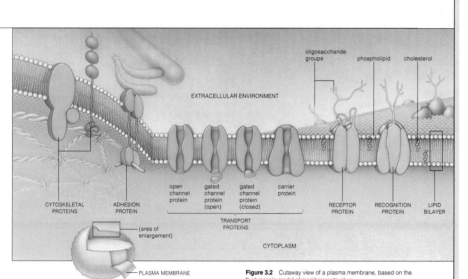

Figure 3.2 Cutaway view of a plasma membrane, based on the fluid mosaic model of membrane structure.

one of its surfaces. Different events proceed at the two surfaces, which differ in the number, kind, and arrangement of their lipids and proteins.

The term "mosaic" refers to the mixed composition of lipids and proteins. "Fluid" refers to the motions of lipids and their interactions. Phospholipid molecules spin about their long axis, they move sideways, and their tails flex—all of which helps keep neighboring molecules from packing into a solid layer. Lipids that have short tails or unsaturated (kinked) tails contribute to the membrane's fluidity.

Proteins associated with the bilayer carry out most membrane functions. Various **transport proteins** allow or encourage water-soluble substances to move through their interior, which opens on both sides of the bilayer. These proteins bind molecules or ions on one side of the membrane, then release them on the other.

Receptor proteins bind hormones and other extracellular substances that trigger changes in cell activities. For example, enzymes that crank up machinery for cell growth and division are switched on when a hormone (such as somatotropin) binds to receptors for it. Different cells have different combinations of receptors.

Diverse **recognition proteins** at the cell surface are like molecular fingerprints; they identify a cell as being of a specific type. For example, "self" proteins pepper the plasma membrane of your body's own cells. Certain white blood cells chemically recognize these proteins and leave your cells alone. But they will launch attacks against bacterial cells, viruses, and other invaders that have "nonself" proteins at their surfaces.

Finally, in multicelled organisms, **adhesion proteins** help particular kinds of cells stick together and thereby remain positioned in their proper tissues.

All cells have an outermost, plasma membrane, an internal region of cytoplasm, and an internal region of DNA.

Besides the plasma membrane, eukaryotic cells have internal, membrane-bound compartments, including the nucleus.

Each membrane has a lipid bilayer. The hydrophobic parts of its lipid molecules are sandwiched between the hydrophilic parts, which are dissolved in the fluid surroundings.

The lipid bilayer gives the membrane its basic structure and serves as a barrier to water-soluble substances.

Proteins associated with the bilayer carry out most membrane functions. Many types transport substances across the bilayer or serve as receptors for extracellular substances. Other types function in cell-to-cell recognition or adhesion.

Chapter 3 Cell Structure and Function **43**

VISUAL SUMMARIES

Over the years, many instructors and students who have used *Biology: Concepts and Applications* have expressed appreciation of the book's illustrations, which we develop simultaneously with the text, as inseparable parts of the same story. This integrative approach was the starting point for further refinements to this edition.

Today, for example, many students are visual learners. If they can visualize a process first, then reading the text becomes less intimidating. Students tell us how much they learn from our hundreds of summary illustrations, which contain step-by-step, written descriptions of biological parts and processes. We break down information into a series of illustrated steps, which are far less threatening than a complicated, "wordless" diagram. Figure *B* is a sample. Notice how simple captions, integrated with the art, walk students through the stages by which mRNA transcripts are translated into proteins, one step at a time.

Similarly, we continue to create visual summaries for anatomical drawings. These summaries integrate structure and function. Students need not jump back and forth from the text, to tables, to illustrations, and back again in order to comprehend how an organ system is put together and what its assorted parts do. Even the individual descriptions of parts are hierarchically arranged to reflect the structural and functional organization of that system.

COLOR CODING

Also consider something as basic as the line illustrations. In these, we consistently use the *same* colors for the *same* types of molecules and cell structures through the entire book. Such visual consistency makes it easier for students to track complex parts and processes. Figure C is our color coding chart. Figures *A* through *D* show how we use it.

ZOOM SEQUENCES

Many of our illustrations progress from macroscopic to microscopic views of the same subject. Figure 5.2 is a fine example; this zoom sequence shows where the reactions of photosynthesis proceed, starting with a plant growing by a roadside. As another example, Figure 26.17 moves down through levels of skeletal muscle contraction, starting with a muscle in the human arm.

ICONS

These small diagrams or sketches help relate the topic of an illustration to the big picture. For instance, in Figure *A*, a representation of a cell subtly reminds students of the location of the plasma membrane relative to the cytoplasm. Other icons serve as reminders of the location of reactions and processes in cells and how they interrelate to one another. Still others remind students of the evolutionary relationships among groups of organisms (Figure *E*).

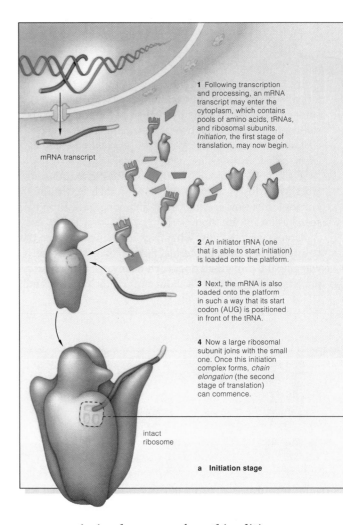

mRNA transcript

1 Following transcription and processing, an mRNA transcript may enter the cytoplasm, which contains pools of amino acids, tRNAs, and ribosomal subunits. *Initiation*, the first stage of translation, may now begin.

2 An initiator tRNA (one that is able to start initiation) is loaded onto the platform.

3 Next, the mRNA is also loaded onto the platform in such a way that its start codon (AUG) is positioned in front of the tRNA.

4 Now a large ribosomal subunit joins with the small one. Once this initiation complex forms, *chain elongation* (the second stage of translation) can commence.

intact ribosome

a   Initiation stage

FIGURE B   *A visual summary from this edition.*

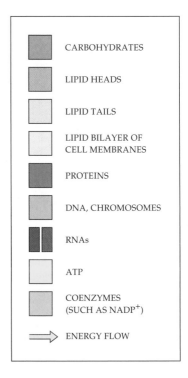

FIGURE C
Color coding chart for the diagrams of biological molecules and cell structures.

CARBOHYDRATES

LIPID HEADS

LIPID TAILS

LIPID BILAYER OF CELL MEMBRANES

PROTEINS

DNA, CHROMOSOMES

RNAs

ATP

COENZYMES (SUCH AS NADP⁺)

ENERGY FLOW

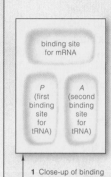

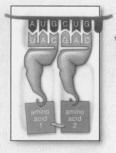

1 Close-up of binding sites on the platform of the small ribosomal subunit. It shows the relative positions of the binding sites for an mRNA transcript and for tRNAs that deliver amino acids to the intact ribosome.

2 The initiator tRNA is already positioned in the first tRNA binding site (*P*). Its anticodon matches up with the start codon (AUG) of the mRNA strand, which already is in position also. Another tRNA is about to move into the second binding site (*A*). This tRNA is one that can bind with the codon that follows the start signal.

3 Through enzyme action, the bond between the initiator tRNA and the amino acid hooked to it is broken. At the same time, an enzyme catalyzes the formation of a peptide bond between the two amino acids. After these bonding events are over, the initiator tRNA will be released from the ribosome.

4 Now the first amino acid is attached only to the second one—which is still hooked to the second tRNA. This tRNA will move into the *P* site on the ribosomal platform, sliding the mRNA with it by one codon. When it does so, the third codon will become aligned above the *A* site.

5 A third tRNA is about to move into the *A* site. Its anticodon is capable of base-pairing with the third codon of the mRNA transcript. Next, through enzyme action, a peptide bond will form between amino acids 2 and 3.

6 Steps 3 through 5 are repeated again and again. The polypeptide chain continues to grow this way until enzymes reach a stop codon in the mRNA transcript. Now *termination*, the last stage of transcription, can begin, as shown in (**c**).

b Chain elongation stage

1 Once a stop codon is reached, the mRNA transcript is released from the ribosome.

2 The newly formed polypeptide chain also is released.

c Chain termination stage

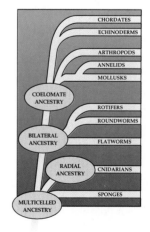

3 The ribosomal subunits separate.

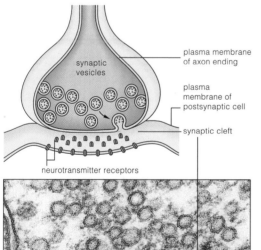

synaptic vesicles

plasma membrane of axon ending

plasma membrane of postsynaptic cell

synaptic cleft

neurotransmitter receptors

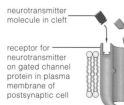

neurotransmitter molecule in cleft

ions

receptor for neurotransmitter on gated channel protein in plasma membrane of postsynaptic cell

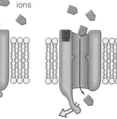

DNA

protein coat

sheath

baseplate

tail fiber

CHORDATES
ECHINODERMS
ARTHROPODS
ANNELIDS
MOLLUSKS
COELOMATE ANCESTRY
ROTIFERS
ROUNDWORMS
BILATERAL ANCESTRY
FLATWORMS
RADIAL ANCESTRY
CNIDARIANS
SPONGES
MULTICELLED ANCESTRY

FIGURE D *Examples of the consistent use of color for structures and processes in this book's diagrams.*

FIGURE E *Example of an icon. This one starts the text on each animal phylum to remind students of evolutionary relationships.*

FOUNDATIONS FOR CRITICAL THINKING

To help students develop their own capacity for critical thinking, we walk them through experiments that yielded evidence in favor of or against hypotheses being discussed. The book's Index will give you an idea of the number and types of experiments used (see the entry *Experiments*).

We also selected certain chapter introductions as well as entire chapters to show students how biologists apply critical thinking to solving problems. Among these are the introductions to Chapters 15 (Figure *F*) and 40, as well as the chapters on Mendelian genetics, DNA structure and function, and animal behavior. Our *Focus on Science* essays provide more detailed, optional examples. For example, one of these describes RFLP analysis and a few of its rather amazing applications (Section 13.3). Another essay (Section 17.5) helps convey to students that biology is not a closed book. Even when new research brings a sweeping story into sharp focus—in this case, the origin of the great prokaryotic and eukaryotic kingdoms—it also opens up new roads of inquiry.

New to this edition are the *Critical Thinking* questions at the end of chapters. Katherine Denniston of Towson State University developed these thought-provoking questions. Chapters 9 and 10 also include many *Genetics Problems* that help students grasp the principles of inheritance.

VIGNETTES LEADING INTO KEY CONCEPTS

Taking up valuable reading time with interesting stories is pointless, unless they lead into or reinforce key concepts. Our short stories (vignettes) that start each chapter do this. For example, the one shown in Figure *F* is a lively account of research that leads into the issue of how we go about defining speciation. A crisp list of key concepts follows this, as an advance organizer for the chapter as a whole.

BALANCING CONCEPTS AND APPLICATIONS

At strategic points, examples of applications parallel the core material—not so many as to be distracting, but enough to keep minds perking along with the conceptual development. Many brief accounts of applications are integrated with the text itself, to lend tangible support to concepts. Other applications are set apart as *Focus* essays for more detail on medical, environmental, and social issues. These provide more information for interested students but do not interrupt the text flow. Figure *G* is only one example. Another *Focus* essay (Section 13.7) asks students to look beyond the clinical trials for human gene therapy to the question of whether we as a species are ready to engage in gene tinkerings. All applications in this edition are listed separately, on the back end papers, for easy reference.

FIGURE F   *Example of the two-page chapter openings.*

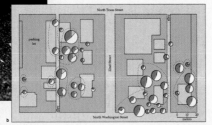

15 SPECIATION

The Case of the Road-Killed Snails

If you happen to be a snail living in a garden in Bryan, Texas, it doesn't take much to keep your genes away from snails in a backyard across the street. By day, the sunbaked asphalt would be about as inviting to a snail as a desert would be to a catfish. Besides, day or night, a street-traversing snail is vulnerable to cars, trucks, skateboards, and bicycles (Figure 15.1a). Whatever else it might be, that strip of asphalt is a formidable barrier to gene flow between populations. For snails, that is.

Whether or not a physical barrier deters gene flow depends partly on an organism's mode of locomotion or dispersal. It also depends on how fast and how long an organism *can* move in response to environmental factors or its own hormones.

A snail is not swift, and it does not roam far from its home population. Compare it to the wild black duck, banded in Virginia in 1969, that turned up eight years later in Korea. Compare it to the wandering albatross, one of the supreme barrier busters. After lifting off from Kerguelen Island in the Indian Ocean, one of these birds soared westward past Africa, across the Atlantic, and around Cape Horn. Thirteen thousand kilometers later, it landed in Chile. With a lightweight body and a wingspan of 3.65 meters (12 feet across), the albatross was able to exploit the great prevailing winds of the Southern Hemisphere.

Figure 15.1   (a) A snail (*Helix aspersa*) about to encounter a barrier to gene flow.

(b) Results from a study of neighboring snail populations, all descended from founders that arrived in a small town in Texas in the 1930s. For each population, a circle represents the relative abundances of three alleles (coded yellow, gold, or brown) for an enzyme, leucine aminopeptidase. Greater genetic variation exists between populations living on opposite sides of 22nd Street. Notice, for example, the higher frequencies of the allele coded gold in the block to the west of the street.

And yet, in 1859, snails did cross an ocean. Humans transported garden-variety snails (*Helix aspersa*) from France to California, then turned them loose. The idea was that the snails would multiply and so meet the demand for that French delicacy, *escargots aux fines herbes*. It was bad enough that the importers mistakenly brought over a less tasty species. Worse than that, the snails exceeded expectations and became an absolute nuisance in gardens and nurseries throughout the southwestern United States.

By the 1930s, *H. aspersa* had hitched rides, possibly as eggs in the soil of plant containers, to Bryan, Texas. They founded small colonies in the local vegetation. Forty years later, Robert Selander, now of Pennsylvania State University, was down on his hands and knees with a few graduate students, scouring patches of vegetation on two adjacent city blocks. Why? Selander wanted to determine the effect of genetic drift on the introduced populations. He and his students collected every single snail—2,218 of them—from fourteen local populations. For each population, they determined the allele frequencies at five different gene locations.

In all cases, the results from their analysis pointed to some genetic variation among the colonies on the same block—and to major differences in allele frequencies *between* blocks. Figure 15.1b shows the results for one of the genes studied.

Assume the genetic differences between populations continue to increase, as through natural selection and genetic drift. Will the time come when snails from opposite sides of the street can no longer interbreed successfully, even if they do manage to get together? In other words, *will they become members of separate species?* Or will selection work against increases in genetic differences by eliminating extreme phenotypes from the populations? After all, how much can a workable package of *H. aspersa* genes evolve in adjacent patches of vegetation—and under very similar environmental pressures—in a small town in Texas?

With this chapter, we turn to **speciation**—to changes in allele frequencies that are significant enough to mark the formation of daughter species from a parent species. Obviously, no one was around to watch the formation of species in the past. No one lives long enough to know whether many existing populations are at some intermediate stage leading to speciation. Evolutionary biologists are still working out theories about speciation processes, and what you are about to read may change in the near or distant future.

KEY CONCEPTS

1. A species consists of one or more populations of individuals that can interbreed under natural conditions and produce fertile offspring, and that are reproductively isolated from other such populations. This definition is restricted to sexually reproducing species.

2. The populations of a species have a shared genetic history, they are maintaining genetic contact over time, and they are evolving independently of other species.

3. Speciation is the process by which daughter species evolve from a parent species.

4. Speciation begins when gene flow is prevented between populations (or subpopulations) of a species. Thereafter, mutation, natural selection, and genetic drift operate independently in each population and lead to irreversible genetic divergence of one from the other.

5. By chance, two populations may come to differ in certain alleles that govern morphological, physiological, and behavioral traits associated with reproduction. When the differences become great enough, they are reproductively isolated from each other—and with reproductive isolation, speciation is completed.

6. Three models of speciation guide current research. With allopatric speciation, species form in separated populations or subpopulations. With parapatric speciation, they form in a region where populations share a border. With sympatric speciation, they form within the home range of the parent species.

7. The timing, rate, and direction of speciation vary within and between lineages. The extinction of some number of species is inevitable for all lineages.

239

END-OF-CHAPTER STUDY AIDS AND APPENDIXES

Figure *H* shows a sampling of our end-of-chapter study aids, which reinforce the key concepts. Each chapter ends with a summary in list form, review questions, a self-quiz, critical thinking questions, selected key terms, and a list of readings. Italicized page numbers tie the review questions and key terms to relevant text pages.

At the book's end, a Glossary includes the boldfaced terms in the text, with pronunciation guides and word origins when such information can make formidable words less so. Our Index is detailed enough to help the reader find a door to the text more quickly.

Students can use the detailed classification scheme in Appendix I for reference purposes. Appendix II includes metric-English conversion charts. The third appendix has detailed answers to genetics problems, and the fourth has answers to self-quizzes. For those interested students and professors who prefer the added detail, Appendix V gives structural formulas for the major metabolic pathways, and Appendix VI shows the periodic table of the elements.

Appendixes and the Glossary are printed on paper that is tinted different colors to preclude frustrating searches for where one ends and the next starts.

FIGURE G *One of the* **Focus** *essays from this edition.*

FIGURE H *Example of end-of-chapter study aids.*

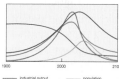
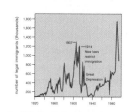

SIGNIFICANT CONTENT REVISIONS

Instructors who have used preceding editions of *Biology: Concepts and Applications* may wish to evaluate the following overview of significant modifications to the content.

Overall, each chapter's *conceptual development* is more focused. An overriding objective for the revision was to make the writing crisp but not too brief, because certain potentially confusing topics become even more so when they are not treated in enough detail. Great thought went into the selective expansions and condensations of every chapter in the new edition.

Also, *we revised all chapters wherever new research findings demanded shifts in presentation*. For example, the chapter on immunity (28) incorporates the current models of immune responses and the current theory of the HIV replication cycle. One *Focus* essay (Section 27.8) includes an update on atherosclerotic plaque formation. Another traces how molecular detectives investigated obese mice to isolate the gene for leptin, one of the hormones that contribute to appetite suppression and higher metabolic rates (Section 30.11). Chapter 35 includes new data on human population growth. Even end-of-chapter *Critical Thinking* questions incorporate current material—for example, on population growth models (Chapter 35) and on prions and the 1996 outbreak of mad cow disease in Great Britain (Chapter 30).

UNIT I. PRINCIPLES OF CELLULAR LIFE   Our advisor Steve Wolfe worked closely with us to refine and update the text and illustrations for many of the complex topics in this unit, including mechanisms of membrane transport and the reactions of aerobic respiration. For example, the text and accompanying Figures 4.9 and 4.10 provide updated models of active transport. As another example, Figure 6.3 is a new, integrative summary illustration of the energy yield of the aerobic pathway.

UNIT II. PRINCIPLES OF INHERITANCE   Rapid advances in genetics research demanded line-by-line evaluations for adjustments in these chapters. Intensive rewriting made the fascinating yet potentially heart-stopping sections—including protein synthesis, mutation, gene regulation, and recombinant DNA technology—more accessible to the introductory student. With our judicious sampling of diverse organisms, including the often-neglected plants, we implicitly remind students of the underlying molecular and biochemical unity of life. An example is Figure 12.19, which shows one outcome of transcriptional controls that affect photosynthesis.

UNIT III. PRINCIPLES OF EVOLUTION   We reorganized and extensively rewrote all the chapters in this unit. Instructors can expect cleaner text discussions, a better selection of graphs and photographs, and other improvements. New discussions and illustrations for stabilizing selection and genetic drift are examples. At the more inclusive level, we now have a new short chapter on speciation (15). We have a separate chapter (16) on patterns of macroevolution and on the functions of classification schemes.

UNIT IV. EVOLUTION AND DIVERSITY   The sweeping story of the origin and evolution of life, including its apparent endosymbiotic elaborations, is now the first chapter (17) of the diversity unit. This crucial survey can help students gain perspective on nature and their place in it. It also provides a conceptual and chronological framework for the subsequent chapters on diversity (18–21). Rather than being just another "march through the phyla," all of the diversity chapters now have crisp evolutionary story lines and plenty of applications, including one on *Ebola* and other emerging pathogens. Overall we have much better coverage of bacteria and viruses. In keeping with the consensus, we now classify algae as protistans. We now consider human evolution at a more logical juncture— at the end of the chapter on vertebrates.

UNIT V. PLANT STRUCTURE AND FUNCTION   You may wish to evaluate the refinements to the text as well as the new illustrations and engaging applications in this unit. For example, Figure 22.13 is a better illustration of leaf internal structure. Section 22.2 is an easy-to-follow overview of plant tissues. Chapter 24 opens with a suitable vignette on the coevolution of plants and pollinators.

UNIT VI. ANIMAL STRUCTURE AND FUNCTION   The chapters in this unit are prime examples of *how modularization can help move students step by step through individually complex yet interrelated topics*. Without such apparent structuring, topics might seem to tumble on top of one another. The chapters also have a *better balance between invertebrate and vertebrate systems*. They have many new illustrations, such as the ones for cell junctions, muscular system, skeletal systems, heart structure, fertilization, and the placenta. We have a new brief treatment of temperature regulation (32). In the chapter on endocrine controls (33), we clarified and updated the text on the molecular basis of hormonal signals and positioned it earlier in the chapter.

UNIT VII. ECOLOGY AND BEHAVIOR   The updated chapter on population ecology (35) is indicative of the extent of the revision. Its vignette extracts a lesson from Angel Island's fecund but starving deer, then raises questions concerning human population growth as a lead-in to the key concepts. Thanks to George Cox, Chapter 36 (Community Interactions) has crisper definitions of the niche and other topics, an update on Charles Krebs's ongoing studies of Canada's lynx/arctic hare population cycles, and updated text on succession, species introductions, and biodiversity patterns. We added a sedimentary cycle (the phosphorus cycle). In Chapter 37, we added material on wetlands and on coral reefs and coral banks, as well as updated material on ENSOs. G. Tyler Miller, Jr., helped us update Chapter 39 (Human Impact on the Biosphere). John Alcock did the same for his fine Chapter 40 (Animal Behavior), which now includes a thought-provoking example of research into a human social behavior (adoption).

SUPPLEMENTAL COURSE MATERIAL

Instructors who are evaluating the suitability of *Biology: Concepts and Applications* for their students may also wish to consider the following supplemental course material.

MULTIMEDIA

1. *Interactive Concepts in Biology*. Packaged free with all student copies, this is the first CD-ROM to address the entire field of biology. This cross-platform CD-ROM has a concept module for almost every numbered module in the book. Because students can learn by doing, they are encouraged to manipulate illustrations from the text to further their understanding of biology. Every chapter of the book is enhanced through a combination of text, graphics, photographs, animations, video, and audio. The CD-ROM also can be used as a presentation tool.

2. *An Introduction to the Internet*. This 100-page booklet introduces students to the Internet. It describes how to get around when using a browser (such as Netscape), search engines, e-mail, setting up home pages, and related topics. The booklet has a listing of useful biology sites on the net that correspond to chapters in the book. It also contains exercises for the net that can be assigned to students.

3. *Wadsworth's Web Page for Biology*. This site has critical thinking questions designed around the web. It has an instructor's forum for sharing ideas on teaching courses, presentation art, instructor's home page template, and Bio Updates. A cool event of the quarter will have an on-going experiment in which students and instructors can participate, a student feedback site, descriptions of degrees and careers in biology, cool clip art, and ideas for teaching on the web. *http://www.wadsworth.com/biology*.

4. *The Internet Student Survival Guide*. This kit contains Earthlink TotalAccess software, which includes Netscape Navigator and Eudora Light e-mail for communicating around the world. It includes a 56-page booklet that helps beginners become cyber savvy.

5. *Cycles of Life: Exploring Biology*. The twenty-six programs of this telecourse feature compelling footage from around the world, spectacular 3D animation, and original micro-videography. A student study guide and faculty manual are included. For information on course licensing and pricing, fax queries to Coast Telecourses at 714-421-6286.

6. *Liquid Assets—The Ecology of Water*. Two double-sided level-3 videodisks compare the ecologies of the Everglades and San Francisco Bay. A study guide is included.

7. *Overheads or 35mm Slides*. All of the book's 724 diagrams and micrographs are available and are reproduced with vivid colors and large, bold-lettered labels. Most diagrams are on *CD-ROM*, delivered in a Kudo run-time catalog format to facilitate access. The catalog has an image drag-and-drop feature for presentation tools.

8. In collaboration with Carolina Biological Supply, Bill Surver of Clemson University developed a *Videodisc* with new animations and films. All line art has large, boldface labels, often step-by-step or in-full-motion animation. Art is available with fill-in-the-blank labels for tests. There are 3,5000 photographic stills, a correlation directory, bar code guide, and hypercard and toolbook software. All items are organized together by book chapters.

9. Adopters can receive a version of *STELLA II*, a software tool to develop critical thinking skills, and a word book. This modeling software provides 23 simulations. Beyond the book's critical thinking questions, 150 more are arranged by chapter in *Critical Thinking Exercises*.

10. An *Electronic Study Guide* has over fifty multiple-choice questions per book chapter that differ from the test-item booklet. After students respond to each question, an on-screen prompt allows them to review their answers and learn why they are correct or incorrect.

ADDITIONAL SUPPLEMENTS

Respected test writers created a *Test Items* booklet with over 4,000 questions in electronic form for IBM and Macintosh in a test-generating data manager. Questions also are available in WordPerfect for DOS and Microsoft Word for Mac.

An *Instructor's Resource Manual* has these items for each chapter: outline, objectives, list of boldface or italic terms, and detailed lecture outline. It has hundreds of suggestions for lecture presentations, classroom and laboratory demonstrations, suggested questions for discussion, research paper topics, and annotations for filmstrips and videos. Lecture outlines in the *Instructor's Resource Manual* exist on a Mac or IBM disk for those who wish to modify the material.

A new, active-oriented *Study Guide and Workbook* allows students to write answers for most of the questions given. Its questions are arranged by chapter module for ease of assignment. For those who may wish to modify or select portions of the material, *chapter objectives* are available on disk as part of the testing file.

A 100-page *Answer Booklet* has answers to the book's review questions. Self-quiz questions are answered in the book itself.

Flashcards show 1,000 glossary items. A booklet, *Building Your Life Science Vocabulary*, helps students learn biological terms by explaining root words and their applications.

Jim Perry and David Morton's *Laboratory Manual* has 38 experiments and exercises, with 600+ full-color labeled photographs and diagrams. Many experiments are divided in parts that can be assigned individually, depending on time available. Each experiment has objectives, discussion (introduction, background, and relevance), list of materials for each part, procedural steps, prelab questions, and post-lab questions. An *Instructor's Manual* for the lab manual lists quantities, procedure for preparing reagents, time requirements for each part of an exercise, hints to make the lab a success, and vendors of materials with item numbers. It has more investigative exercises that can be copied for laboratory use.

Customized Laboratory Manuals by Phillip Shelp and its accompanying instructor's manual can be tailored for individual courses. A new *Photo Atlas* has 700 full-color, labeled photographs and micrographs of the cells and organisms that students typically deal with in the lab.

Seven additional-readings supplements are available:
• *Contemporary Readings in Biology* is a collection of articles on applications of interest to students. • *A Beginner's Guide to Scientific Method* is a supplement for those who wish to treat this topic in detail. • *The Game of Science* gives students a realistic view of what science is and what scientists do.
• *Environment: Problems and Solutions* provides a brief 120-page introduction to environmental concerns. • *Green Lives, Green Campuses* is a hands-on workbook to help students evaluate the environmental impact of their own life-styles.• *Watersheds: Classic Cases in Environmental Ethics* is a collection of important case studies.• *Environmental Ethics: An Introduction to Environment Philosophy* is a survey of environmental ethics and recent philosophical positions.

A COMMUNITY EFFORT

It's true that one, two, or a smattering of authors can write accurately and often well about their field of interest, but it takes more than this to deal with the full breadth of the biological sciences. For us, it takes an educational network that includes more than 2,000 teachers, researchers, and photographers in the United States, Canada, England, Germany, France, Australia, Sweden, and elsewhere. We list those reviewers whose recent contributions continue to shape our thinking. There simply is no way to describe the thoughtful effort that these individuals and other reviewers before them gave to our books. We can only salute their commitment to quality in education.

Biology has become widely respected largely because it reflects the understandings of our general advisors and contributors and their abiding concern for students. Steve Wolfe, author of acclaimed books on cell biology, works out alternative phrasings with us, sometimes line by line over the phone, to make the chapters as clear as possible. Dan Fairbanks always manages to find time to ferret out errors that creep into the genetics manuscripts. Katherine Denniston, another long-time genetics reviewer, wrote good critical thinking questions, itself no small feat. Aaron Bauer, Paul Hertz, Samuel Sweet, and more recently Jerry Coyne helped chisel major parts of the evolution unit. Jerry also created a great computer simulation of genetic drift. Our unit on plant structure and function is strong, thanks to initial resource manuscripts from Cleon Ross. And what would we do if our abiding friends Gene Kozloff and John Jackson didn't stop us from inventing biology? What would we do without Robert Lapen, who wrote the definitive manuscript on immunology and lived to tell about it?

We similarly recognize how Rob Colwell, George Cox, and Tyler Miller strengthened the ecology unit. The animal behavior chapter, with its evolutionary approach, began with a resource manuscript from John Alcock, himself the author of a highly respected book on animal behavior.

Also over the years, Nancy Dengler, Ron Hoham, Bruce Holmes, David Morton, and Frank Salisbury have been fine guardians of accuracy and teachability. So has Tom Garrison, himself an author, who dispenses sympathy with wit better than just about anybody.

Mary Douglas has now become the finest production manager in the business. She is the only one between us and extinction as publishing lurches into the electronics age. We simply cannot imagine these complex revisions moving through computerized design and production without the talented, flexible, even-tempered Mary. With this new edition, Myrna Engler-Forkner, art coordinator, proved herself as indispensable as Mary, no small feat.

Of all the artists with whom we have worked over the years, Raychel Ciemma unquestionably is the best and the brightest. She works directly with us to turn our rough sketches into works of art. This is amazing, given that we are so fanatically compulsive about the finished product.

Gary Head taught us Quark. He designed the book and its cover. He even went out on photographic assignment and came back with such treats as Fred-and-Ginger, the streetwise snail in Figure 15.1b. Once again we thank Marion Hansen. This fine photo researcher, permissions editor, and friend keeps on forgetting that she vows never to get sucked into another edition. We thank Mary Arbogast, a superb developmental editor, and Sandra Craig, who shepherds book development through its convoluted route. Kristin Milotich and Kerri Abdinoor are editorial professionals with endurance, a big heart, and no attitude problem. The amazing Kristin oversees the book's great supplements program. The amazing Pat Waldo almost singlehandedly brought us to the world of multimedia.

Kathie Head, Pat Brewer, Stephen Rapley, Carolyn Deacy, Bill Ralph, and Peggy Meehan have bestowed steady guidance over the years. We are glad that Gary Carlson and Dave Garrison are part of the team effort. We are grateful for the support of Rosaleen Bertolino, Karen Stough, Alexandra Nickerson, Mary Roybal, Jennie Redwitz, Bob Demarest, Natalie Hill, Carol Lawson, John Douglas, and the other individuals listed on the copyright page. The talented Jan Flessner and her colleagues at Precision Graphics, and Tom Anderson and Nancy Dean at H&S Graphics, accommodated our pursuit of excellence. So did Jim Jeschke and Jody Ward at American Composition and Graphics, even though our early fits and starts with Quark submissions must have sent them up the walls and clinging to the ceiling.

Twenty years ago, Jack Carey convinced us to do the first edition of this book. Ever since, he has remained our closest counseler, our abiding friend. And nothing, in all that time, has shaken our shared belief in the intrinsic capacity of the biological perspective to enrich the lives of each new generation of students.

REVIEWERS

Current configurations of the Earth's oceans and land masses—the geologic stage upon which life's drama continues to unfold. Thousands of separate images were pieced together to create this remarkable cloud-free view of our planet.

1 METHODS AND CONCEPTS IN BIOLOGY

Biology Revisited

Buried somewhere in that mass of tissue just above and behind your eyes are memories of first encounters with the living world. Still in residence are memories of discovering your hands and feet, your family, friends, the change of seasons, the smell of rain-drenched earth and grass. In that brain are memories of your early introductions to a great disorganized parade of

spiders, flowers, frogs, and furred things, mostly living, sometimes dead. There, too, are memories of questions—*"What is life?"* and, inevitably, *"What is death?"* There are memories of answers, some satisfying, others less so.

By observing, asking questions, and accumulating answers, you have built up a store of knowledge about

Figure 1.1 Think back on all you have ever known and seen. This is a foundation for your deeper probes into life.

the world of life. Experience and education have been refining your questions, and no doubt some answers are difficult to come by. Think of a young man whose brain is functionally dead as a result of a motorcycle accident. If his breathing and other basic functions proceed only as long as he remains hooked up to mechanical support systems, is he no longer "alive"? Or think of a recently fertilized egg growing inside a woman's body. At this early stage, it is no more than a cluster of a few dozen tiny cells. At what point in its development would you define it as "human" life? If questions like this have crossed your mind, your thoughts about life obviously run deep.

The point is, this book isn't your introduction to biology—"the study of life"—for you have been studying life ever since information began penetrating your brain. This book simply is biology *revisited*, in ways that may help carry your thoughts to more organized levels of understanding.

Return to the question, *What is life?* Offhandedly, you might respond that you know it when you see it. To biologists, however, the question opens up a story that has been unfolding in countless directions for several billion years! "Life" is an outcome of ancient events by which nonliving materials became assembled into the first living cells. "Life" is a way of capturing and using energy and raw materials. "Life" is a way of sensing and responding to changes in the environment. "Life" is a capacity to reproduce, grow, and develop. And "life" evolves, meaning that details in the body plan and functions of organisms can change through successive generations.

Yet this short description only hints at the meaning of life. Deeper insight requires wide-ranging study of life's characteristics.

Throughout this book you will come across many diverse examples of how organisms are constructed, how they function, where they live, and what they do. The examples support certain concepts which, taken together, will give you a sense of what "life" is. This chapter provides an overview of the basic concepts. As you continue reading the book, you may find it useful to return to this simple overview to reinforce your grasp of details.

KEY CONCEPTS

1. There is an underlying unity in the world of life, for all organisms are alike in key respects. They consist of the same kinds of substances, put together according to the same laws that govern matter and energy. Their activities depend on inputs of energy, which they must obtain from their environment. All organisms sense and respond to changing conditions in their environment. And they all grow and reproduce, based on instructions contained in their DNA.

2. There also is immense diversity in the world of life. Millions of different organisms inhabit the Earth, and many millions more lived and became extinct over the past 3.8 billion years. And each kind of organism is unique in some of its traits—that is, in some aspects of its body plan, body functions, and behavior.

3. Theories of evolution, especially a theory of natural selection as first formulated by Charles Darwin, help explain the meaning of life's diversity.

4. Biology, like other branches of science, is based on systematic observations, hypotheses, predictions, and relentless tests. The external world, not internal conviction, is the testing ground for scientific theories.

1.1 ORGANIZATION IN NATURE

Levels of Biological Organization

Picture a frog on a rock, busily croaking. Without even thinking about it, you know that the frog is alive and the rock is not. And yet, at a deeper level, the difference between them blurs. Both consist of the same particles (protons, electrons, and neutrons), organized in atoms according to the same physical laws. At the heart of those laws is something called **energy**—a capacity to make things happen, to do work. Energetic interactions bind atom to atom in orderly patterns, giving rise to the structured bits of matter we call molecules. Energetic interactions among molecules hold rocks together. And they hold frogs—and all other organisms—together.

A special type of molecule—deoxyribonucleic acid, or **DNA**—sets living things apart from the nonliving world. No chunk of granite or quartz has it. DNA has instructions for assembling organisms from "lifeless" molecules that contain carbon and a few other kinds of atoms. By analogy, with proper instructions and a little effort, you can turn a heap of just two kinds of ceramic tiles into ordered patterns such as these:

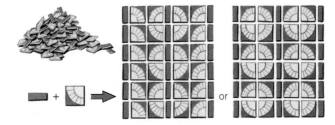

By itself, DNA—"the molecule of life"—is not alive. As Figure 1.2 indicates, the properties of life emerge in individual cells. The **cell** is an organized unit that can survive and reproduce on its own, given DNA instructions, raw materials, and inputs of energy. Clearly this definition fits a free-living, single-celled organism such as an amoeba. Does it fit a **multicelled organism**, which has specialized cells organized as tissues and organs? Yes. You may find this a strange answer. After all, your own cells could never live all by themselves in nature; body fluids must continuously bathe them. Yet even isolated human cells can be kept alive under controlled laboratory conditions. In fact, researchers throughout the world routinely maintain human cells for use in important experiments, including cancer studies.

The next level of organization is the **population**, a group of organisms of the same kind, such as a colony of Emperor penguins. Next is the **community**—all of the populations of all species living in the same area (such as Antarctica's penguins, seals, birds, and so on). The next level, the **ecosystem**, includes the community

Biosphere
Those regions of the earth's waters, crust, and atmosphere in which organisms can exist

↑

Ecosystem
A community and its physical environment

↑

Community
The populations of *all* species occupying the same area

↑

Population
A group of individuals of the same kind (that is, the same species) occupying a given area at the same time

↑

Multicellular Organism
An individual composed of specialized, interdependent cells arrayed in tissues, organs, and often organ systems

↑

Organ System
Two or more organs interacting chemically, physically, or both in ways that contribute to the survival of the whole organism

↑

Organ
A structural unit in which tissues are combined in specific amounts and patterns that allow them to perform a common task

↑

Tissue
A group of cells and surrounding substances, functioning together in a specialized activity

↑

Cell
Smallest *living* unit; may live independently or may be part of a multicellular organism

↑

Organelle
Sacs or other compartments that separate different activities inside the cell

↑

Molecule
A unit of two or more atoms of the same or different elements bonded together

↑

Atom
Smallest unit of an element that still retains the properties of that element

↑

Subatomic Particle
An electron, proton, or neutron; one of the three major particles of which atoms are composed

Figure 1.2 Levels of organization in nature.

and its physical and chemical environment. Finally, the **biosphere** includes all parts of the earth's waters, crust, and atmosphere in which organisms live. Astoundingly, *this globe-spanning organization starts with the convergence of energy, materials, and DNA in tiny, individual cells.*

Producers trap, convert, and use or store some energy from the sun.

PRODUCERS

NUTRIENT CYCLING

CONSUMERS, DECOMPOSERS

ONE-WAY FLOW OF ENERGY

Energy is transferred from one organism to another; in time, all flows back to the environment.

a

b   c

Figure 1.3 An example of the one-way energy flow and the cycling of materials through the biosphere. (**a**) Plants of a warm, dry grassland called the African savanna capture energy from the sun and use it to build plant parts. Some of the energy ends up in plant-eating organisms, including this adult male elephant. He eats huge quantities of plants to maintain his eight-ton self and produces huge piles of solid wastes—dung—that still contain some unused nutrients. Thus, although most organisms would not recognize it as such, elephant dung is an exploitable food source.

(**b**) And so we next have little dung beetles rushing to the scene almost simultaneously with the uplifting of an elephant tail. Working rapidly, they carve fragments of moist dung into round balls, which they roll off and bury in burrows. In these balls the beetles lay eggs—a reproductive behavior that helps assure forthcoming offspring (**c**) of a compact food supply. Thanks to beetles, dung does not pile up and dry out into rock-hard mounds in the intense heat of the day. Instead, the surface of the land is tidied up, beetle offspring get fed, and the leftover dung accumulates in beetle burrows—there to enrich the soil that nourishes the plants that sustain (among others) the elephants.

新陳代謝

Metabolism: Life's Energy Transfers

The metabolic activities of single cells and multicelled organisms maintain the great pattern of organization in nature. **Metabolism** refers to the capacity of the cell to (1) extract and convert energy from its surroundings and (2) use energy to maintain itself, grow, and make more cells. Simply put, it means *energy transfers*.

Think of a food-producing cell inside a leaf of a tree. By the process of **photosynthesis**, it traps energy from the sun and uses it to produce molecules of an energy carrier, called **ATP**. Then ATP transfers energy to sites inside the cell where "metabolic workers" (enzymes) put together sugars, starch, and other substances. The cell even stores some energy. By the process of **aerobic respiration**, it can later release some stored energy and produce ATP molecules—which transfer the energy that drives hundreds of different cellular activities.

Interdependencies Among Organisms

Plants and other photosynthetic organisms are the main entry point for an immense flow of energy into the world of life. They are the food **producers**. Animals are **consumers**. Directly or indirectly, they feed on energy that has been stored in tissues of the photosynthesizers. For example, some energy is transferred to elephants after they eat leaves. Energy is transferred again when lions eat a baby elephant that wandered away from its herd. And it is transferred again to fungi and bacteria

that are **decomposers**. When decomposers feed on the tissues or remains of elephants, lions, or any other organism, they break down sugars and other biological molecules to simple materials—which may be cycled back to the producers. In time, all of the energy that the producers initially captured from the sun flows back to the environment, but that's another story.

For now, keep in mind that interdependencies link organisms together, owing to a one-way flow of energy *through* them and a cycling of materials *among* them (Figure 1.3). Such interactions influence the structure, size, and composition of populations and communities. They influence ecosystems, even the biosphere. Understand the extent of these interactions and you will gain insight into amplification of the greenhouse effect, acid rain, and other modern-day problems.

Levels of organization exist in nature. The characteristics of life emerge at the level of single cells and extend through populations, communities, ecosystems, and the biosphere.

Organisms engage in metabolism. Their cells acquire and use energy to assemble, break down, stockpile, and dispose of materials in ways that promote survival and reproduction.

Interdependencies exist among nearly all organisms, based on a one-way flow of energy through them and a cycling of materials among them.

1.2 SENSING AND RESPONDING TO THE ENVIRONMENT

It is often said that only organisms can respond to the environment. Yet even a rock shows responsiveness, as when it yields to the force of gravity and tumbles down a hill or changes shape slowly under the battering of wind, rain, or tides. The difference is this: *Organisms can sense changes in their surroundings, then make controlled, compensating responses to them.* How? Each organism has **receptors**, which are certain molecules and structures that can detect specific kinds of information about the environment. When cells receive signals from receptors, their metabolic activities shift in ways that bring about a suitable response. When your brain receives signals from receptors inside your body or near its surface, it rapidly sends out commands to muscles or glands that can execute a suitable response.

For example, an organism's body can withstand only so much heat or cold. It must rid itself of harmful substances. It requires certain foods, in certain amounts. Yet temperatures do shift, harmful substances might be encountered, and food is sometimes plentiful or scarce.

Think about what happens after you finish a snack. Simple molecules, including sugars, leave your gut and enter your bloodstream. Blood is part of the body's "internal environment" (the other part is tissue fluid that bathes your cells). The sugar absorbed from the gut raises the concentration of sugar in the blood. Now cells in a glandular organ called the pancreas step up their secretion of insulin. As it happens, most cells in your body have receptors for insulin—a hormone that stimulates them to take up sugar molecules from the internal environment. When an enormous number of cells do this, the blood concentration of sugar returns to normal.

Suppose you skip breakfast, then lunch. The blood concentration of sugar starts to decrease. Now a different hormone goes into action. It stimulates liver cells to dig into their stores of energy-rich molecules and break them down to simple sugars. The cells release sugars, which enter the bloodstream. There they help return the blood concentration of sugar to normal.

Usually, the range of physical and chemical conditions inside an organism remains fairly constant. When the internal operating conditions stay within tolerable limits, we call this a state of **homeostasis**.

Single cells and multicelled organisms detect and respond to specific conditions in the environment. They do so with the help of molecules or structures called receptors.

Responsiveness to changing conditions helps organisms maintain physical and chemical conditions inside the body.

Homeostasis is a state in which internal operating conditions are being held within tolerable limits.

1.3 CONTINUITY AND CHANGE— THE NATURE OF INHERITANCE

Perpetuating Heritable Traits

We humans tend to think we enter the world rather abruptly and leave it the same way. Yet we and all other organisms are more than this. *We are part of an immense journey that began billions of years ago.*

Think of the first cell of a new individual, produced when a human sperm fertilizes an egg. It would not even exist if the sperm and egg had not formed earlier, according to DNA instructions that were passed down through countless human generations. With those time-tested instructions, a new individual may develop, grow, and then eventually engage in **reproduction**—that is, the production of offspring. By this process, life's journey continues.

Or think about a moth. Is it just a winged insect? What of the fertilized egg deposited upon a leaf by a female moth (Figure 1.4)? Inside that small egg are instructions for becoming an adult. They guide the egg's development into a caterpillar, a larval stage adapted for rapid feeding

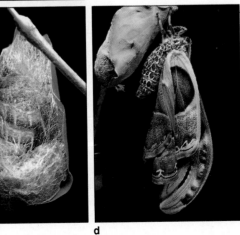

Figure 1.4 "The insect"— a continuous series of stages of development. Different adaptive properties emerge at each stage. Shown here, a silkworm moth, from the egg (**a**), to a larval stage (**b**), to a pupal stage (**c**), and on to the winged form of the adult (**d,e**).

and growth. The caterpillar eats and grows until an internal "alarm clock" goes off. Then its body enters a pupal stage, when tissues undergo drastic remodeling. Many cells die; others multiply and become organized in different patterns. In time an adult emerges that is adapted for reproduction. It has organs that house eggs or sperm. Its wings are brightly colored and flutter at a frequency suitable for attracting a mate.

None of these stages is "the insect." The insect is a series of organized stages, from one fertilized egg to the next. Each stage is vital for the ultimate production of new moths. Instructions for each stage were written into moth DNA long before each moment of reproduction—and so the ancient moth story continues.

Mutations—Source of Variations in Heritable Traits

Reproduction is possible because of a process known as **inheritance**. The word means that parent organisms transmit specific DNA instructions for duplicating their traits to offspring. Why does a baby stork look like its parents and not like pelicans? It inherited stork DNA—which is not exactly the same as pelican DNA.

DNA has two striking qualities. Its instructions assure that offspring will resemble their parents—yet they also permit *variations* in the details of most traits. For example, having five fingers on each hand is a human trait. Yet some humans are born with six fingers on each hand instead of five! This is an outcome of a **mutation**—a molecular change in the DNA. Mutations are the original source of variations in heritable traits.

Many mutations are harmful. A change in even a tiny bit of DNA may be enough to sabotage the body's growth, development, or functioning. One such mutation causes *hemophilia A*, a blood-clotting disorder. After even a small cut or bruise, an abnormally long time passes before a clot forms and stops the bleeding. Yet some variations are harmless or even beneficial under

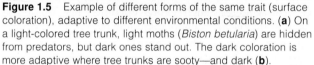

Figure 1.5 Example of different forms of the same trait (surface coloration), adaptive to different environmental conditions. (**a**) On a light-colored tree trunk, light moths (*Biston betularia*) are hidden from predators, but dark ones stand out. The dark coloration is more adaptive where tree trunks are sooty—and dark (**b**).

conditions that happen to prevail in the environment. A classic case is a mutation in lightly colored moths that results in dark offspring. Moths fly at night and rest during the day, when birds that eat them are active. When a light-colored moth rests on a light tree trunk, it is camouflaged—it "hides in the open"—so birds tend not to see it (Figure 1.5). Suppose people build coal-burning factories nearby. Over time, soot-laden smoke darkens the tree trunks. Now the dark moths are less conspicuous to predators—so they have a better chance of living long enough to reproduce. Under sooty conditions, the variant form of the trait is more adaptive.

An **adaptive trait** simply is any trait that helps an organism survive and reproduce under a given set of environmental conditions.

Each organism arises through reproduction. It is part of a reproductive continuum that extends back in time, through countless generations.

DNA is the molecule of inheritance. Parents transmit its instructions for reproducing traits to their offspring.

Mutations in DNA introduce variations in heritable traits.

Although many mutations are harmful, some give rise to variations in form, function, or behavior that are adaptive under conditions that prevail in a given environment.

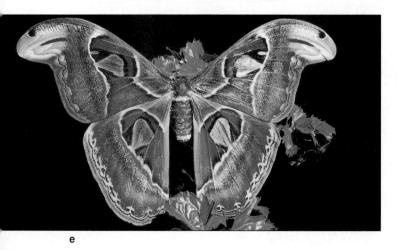

e

1.4 SO MUCH UNITY, YET SO MANY SPECIES

So far, we have focused on the unity of life—on the characteristics that all living things have in common. Think of it! All organisms are put together from the same "lifeless" materials, according to the same laws of energy. They remain alive by metabolism—by ongoing energy transfers at the cellular level. They interact with one another in their requirements for energy and raw materials. They have the capacity to sense and respond to specific conditions in the environment. They have a capacity to reproduce, based on heritable instructions that are encoded in their DNA. And their DNA instructions can change, as a result of mutations.

Superimposed on the shared heritage is immense diversity. Many millions of different kinds of organisms, or **species**, inhabit the earth. Many millions more lived at some time during the past 3.8 billion years, but they became extinct. Attempts to make sense of this confounding diversity led to a classification scheme that assigns each newly identified species a two-part name. The first part of the name designates the **genus** (plural, genera). Each genus encompasses all of the species that seem to be related by way of descent from a common ancestor. The second part of the name designates a particular species within that genus.

For instance, *Quercus alba* is the scientific name of the white oak. *Q. rubra* is the name of the red oak. As this example suggests, once you spell out the genus name in a document, you may abbreviate the name wherever else it appears in the document.

We further classify life's diversity by assigning species to groups at more encompassing levels. Genera that seem to share a common ancestor are grouped in the same *family*, related families are grouped in the same *order*, and then related orders in the same *class*. Related classes are grouped in a *phylum* (plural, phyla), which is assigned to one of five *kingdoms*.

Throughout this book, you will be reading about representatives from all five kingdoms—the Monera, Protista, Fungi, Plantae, and Animalia (Figure 1.6). At this point, it is enough simply to become familiar with a few of the defining features of their members.

All of the world's bacteria (singular, bacterium) are **monerans**. Monerans are single cells, of a type called *prokaryotic*. The word means they don't have a nucleus, a membranous sac that otherwise would keep their DNA separated from the rest of the cell interior. Different species of bacteria are producers, consumers, or decomposers. Of all kingdoms, theirs is the one that shows the greatest metabolic diversity.

Protistans include single-celled species as well as some multicelled forms. Most are larger than bacteria, and they have far greater internal complexity. All are *eukaryotic*, meaning that their DNA is enclosed inside a

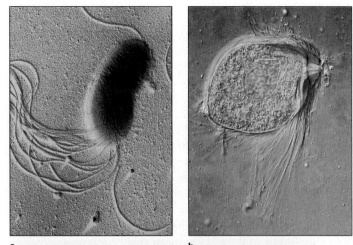

a

b

c

Figure 1.6 Representatives of life's diversity.

KINGDOM MONERA. (**a**) A bacterium, a microscopically small single cell. Bacteria live nearly everywhere, including in or on other organisms. The ones living in your gut and on your skin outnumber the many trillions of cells of your body.

KINGDOM PROTISTA. (**b**) A trichomonad, a type of protistan living in a termite's gut. Compared to bacteria, most protistans are much larger and have greater internal complexity. They range from single cells to giant seaweeds.

KINGDOM FUNGI. (**c**) A stinkhorn fungus. Some fungi are parasites, some cause diseases, but the vast majority are decomposers. Without decomposers, communities would gradually become buried in their own garbage.

KINGDOM PLANTAE. (**d**) A grove of redwoods near the coast of California. Like nearly all plants, redwoods produce their own food by photosynthesis. (**e**) Flower of a plant called a composite. Its colors and patterning guide bees to nectar. The bees get food, and the plant gets help reproducing. Like many other organisms, they interact in a mutually beneficial way.

KINGDOM ANIMALIA. (**f,g**) Male bighorn sheep competing for females and so displaying a characteristic of the kingdom—they move actively and often in highly conspicuous ways through their environment.

d

f

g

e

藻類

nucleus. A dazzling variety of microscopically small, single-celled producers and consumers are protistans. So are multicelled brown algae and other "seaweeds."

菌類 Most **fungi**, including the familiar mushrooms sold in grocery stores, are multicelled. These eukaryotic decomposers and consumers feed in a distinctive way. They secrete substances that digest food outside of the fungal body, then their cells absorb the digested bits.

Plants and **animals** include many familiar species and an astounding number of rather obscure ones. All are eukaryotic. Plants are multicelled, photosynthetic producers. Animals are diverse multicelled consumers that include plant eaters, parasites, and meat eaters. Unlike the plants, animals actively move about during at least some stage of their life.

Pulling this all together, you start to get a sense of what it means when someone says that life shows unity *and* diversity.

Unity threads through the world of life, for all organisms are alike in the following respects:

- **They consist of the same materials, which are put together in the same ways.**
- **They engage in metabolic activities, and they sense and respond to their environment.**
- **They have a capacity to reproduce, based on heritable instructions encoded in their DNA.**

Immense diversity exists in nature. As an outcome of mutations in DNA, organisms differ enormously in body form, in the functions of body parts, and in behavior.

To make the study of life's diversity more manageable, we group organisms into five great kingdoms—the monerans (bacteria), protistans, fungi, plants, and animals.

Given that organisms are so much alike, what could account for their great diversity? One key explanation is called evolution by means of natural selection. A few simple examples will be enough to introduce you to the main premises of this explanation.

Evolution Defined

Think back on the population of light moths in a sooty forest. At some point, a DNA mutation arose in the population and resulted in a moth of a different color. Some offspring of the mutated individual inherited the trait. Birds saw and ate many light moths, but most of the dark ones escaped detection and lived to reproduce. So did their dark offspring—and so did *their* offspring.

Thus the frequency of the dark form of the trait increased with the passing of generations, and the frequency of the light form decreased. Over time, the dark form may even become the more common, and we might end up referring to "the population of dark-colored moths."

Evolution is proceeding. The word simply refers to change in a line of descent over time. The frequencies of different forms of a trait that helps characterize the moth population are changing through successive generations.

Figure 1.7 Some of the 300+ varieties of domesticated pigeons produced by artificial selection practices. Breeders began with variant forms of traits in captive populations of wild rock doves, one of which is shown above.

Natural Selection Defined

Long ago, the naturalist Charles Darwin used pigeons to explain the connection between variation in traits and evolution. Domesticated pigeons show splendid variation in size, feather color, and other traits (Figure 1.7). As Darwin knew, pigeon breeders select certain forms of traits. For example, if breeders prefer black tail feathers that have curly edges, they allow only the individual pigeons having the most black and the most curl in their tail feathers to mate. Over time, "black" and "curly" will become the most common forms of tail feathers in the captive population, and other forms will gradually be eliminated.

Pigeon breeding is a case of <u>artificial selection</u>, because the selection of a specific form of a trait takes place in an artificial environment. Even so, Darwin saw the practice as a way to explain evolution by <u>natural selection</u>—that is, the selection of adaptive traits in nature. Later in the book, we will consider the actual mechanisms by which populations evolve. Meanwhile, keep the following points of Darwin's explanation in

mind. They are central to biological inquiry, and we will be using them to explain topics throughout the book. We express them here in modern terms:

Members of a population differ from one another in form, function, and behavior. Much of the variation is heritable.

Some forms of heritable traits are more *adaptive* than others. This simply means that they improve chances of surviving and reproducing. As a result, they become more common in the population.

Natural selection is the outcome of differences in survival and reproduction that have occurred among individuals that differ in one or more traits.

A population evolves as some forms of traits become more or less common, or even disappear, over the generations— as by natural selection.

Evolutionary processes help explain life's diversity—that is, the sum total of variations in traits that have accumulated in different lines of organisms.

On Scientific Methods

Biology, like science generally, is a methodical, ongoing search for information that helps reveal the secrets of the natural world. Since Darwin's time, the searchers have branched out and established a great number of specialized subdivisions. Biologists now study topics ranging from the molecular structure of a virus that causes AIDS to the formation of ozone holes in the stratosphere. So much remains to be learned about each topic that few now claim the whole of nature as their research interest. And no one seriously claims that one method alone is enough to study nature's complexity.

Despite all the specialization, scientists everywhere have practices in common. They ask questions, make educated guesses about possible answers, then devise ways to test predictions that may hold true—if their guesses are good ones.

Here is a more formal description of how scientists generally proceed with an investigation:

1. Ask a question (or identify a problem) about some aspect of nature. Develop one or more **hypotheses**, or educated guesses, about what the answer (or the solution) might be. This might require sorting through existing information about related phenomena.

2. Using hypotheses as a guide, make a **prediction**—that is, a statement of what you should observe in nature, if you were to go looking for it. This is often called the "if-then" process. (*If* gravity does not pull objects toward the Earth, *then* it should be possible to observe apples falling up, not down, from a tree.)

3. Devise ways to **test** the accuracy of predictions. This typically involves making observations, developing models, and performing experiments.

4. If the tests do not turn out as expected, check to see what might have gone wrong. (Maybe you overlooked something that influenced the results. Or maybe the hypothesis isn't a good one.)

5. Repeat the tests or devise new ones—the more the better, because hypotheses supported by many different tests are more likely to be correct. Then objectively report the results and conclusions drawn from them.

In broad outline, a scientific approach to studying the natural world is that simple. You can use this approach to advantage. You can use it to satisfy curiosity about mammoths or moth wings. You can use it to pick your way logically through environmental, medical, and social land mines of the sort described later in the book. Finally, you can use it to understand the past and predict possible futures for life on this planet.

About the Word "Theory"

How does a hypothesis differ from a theory? By way of example, go back to Darwin's ideas about evolution. When Darwin proposed them more than a century ago, he ushered in one of the most dramatic of all scientific revolutions. The core of his thinking became popularly known as "the theory of evolution."

In science, a **theory** is a related set of hypotheses that, taken together, form a broad-ranging, testable explanation about some fundamental aspect of the natural world. A scientific theory differs from a scientific hypothesis in its *breadth of application*. Darwin's theory fits this description. It is a big, encompassing, *"Aha!"* explanation. In a few intellectual strokes, it helps us make sense of an astounding number of observable phenomena. Think of it! Darwin's theory explains how most of the diversity among many millions of different living things came about! Section 1.7 provides you with one example of how biologists can use this theory to formulate and test ideas about the natural world.

Yet is any theory an "absolute truth" in science? Ultimately, no. Why? It simply would be impossible to perform the infinite number of tests required to show that a theory holds true under all possible conditions! Objective scientists say only that there is a high or low probability that a theory is (or is not) correct.

Even so, "high probability" can be truly impressive. Especially after many researchers perform exhaustive tests, a theory may be as close to the truth as we can get with the evidence at hand. After more than a century's worth of thousands of different tests, Darwin's theory still stands, with only minor modification. Most biologists accept the modified theory—although they still keep their eyes open for contradictory evidence.

Scientists must keep asking themselves: "Will some other evidence show my idea is not a good one?" They are expected to put aside pride or bias by testing ideas, even in ways that might prove them wrong. Even if an individual doesn't (or won't) do this, *others will*—for science proceeds as a community that is both cooperative and competitive. Its practitioners share ideas, with the understanding that it is just as important to expose errors as it is to applaud insights. Individuals can change their mind when presented with new evidence —and this is a strength of science, not a weakness.

A scientific approach to studying the natural world is based on asking questions, formulating hypotheses, making predictions, devising tests, and objectively reporting the results.

A scientific theory is a *testable* explanation about the cause or causes of a broad range of related phenomena. It remains open to tests, revision, and tentative acceptance or rejection.

DARWIN'S THEORY AND DOING SCIENCE

A time-tested theory serves as a general frame of reference for studying nature. Consider Charles Darwin's theory of evolution by natural selection. How might you use it to explain the patterned wings of the moth shown in Figure 1.4? According to this theory, the traits of moths and all other organisms exist because they have contributed to reproductive success. So your question might be this: "I wonder how the moth's wing pattern helps the moth leave descendants." Then you hypothesize about the possible answers.

"Maybe a wing pattern is a mating flag that helps males and females of a species identify each other." This may be a good guess, but you don't limit yourself to one hypothesis. Why? Nearly always, there's more than one possible answer to a question about some aspect of nature. So you also think up an alternative: "Maybe the pattern camouflages moths during the day." *In science, alternative hypotheses are the rule, not the exception.*

Testing Hypotheses Of any number of alternative hypotheses, how do you identify the most plausible one? The trick is to let each one guide you in making *testable* predictions. If wing patterns help moths identify mates (hypothesis), then it follows that moths should mate only when the patterns are visible (the prediction). Moths mate at night. If wing pattern is a mating flag, then on moonless nights, moths shouldn't be able to see it and you won't see moths mating. To test the prediction, you watch moths on a moonlit and then on a moonless night. You notice they mate with or without help from the light of the moon. Here is evidence that the prediction—and, by extension, the hypothesis—might be wrong. Then again, maybe you overlooked something important. For example, maybe moths (like cats) see better than you do in the dark.

The Role of Experiments Now you decide to test the same hypothesis by experimentation. An **experiment** is a test in which nature is manipulated to reveal one of its secrets. It requires careful design of a set of controls to evaluate possible side effects of the manipulation.

If wing pattern is a mating flag, then the moths with altered patterns might have a tough time attracting mates. To test this new prediction, you capture new moths and paint an altered pattern on their wings (Figure 1.8). After this, you put them in a cage with unaltered moths to see what happens.

You also set up a **control group**. Control groups are used to evaluate possible side effects of a test involving an experimental group. Ideally, members of a control group should be identical to members of an experimental group in all respects *except* for a key **variable** (the factor being investigated). The number of individuals in both groups also must be large enough so results won't be due to chance alone (compare Sections 9.2 and 14.9).

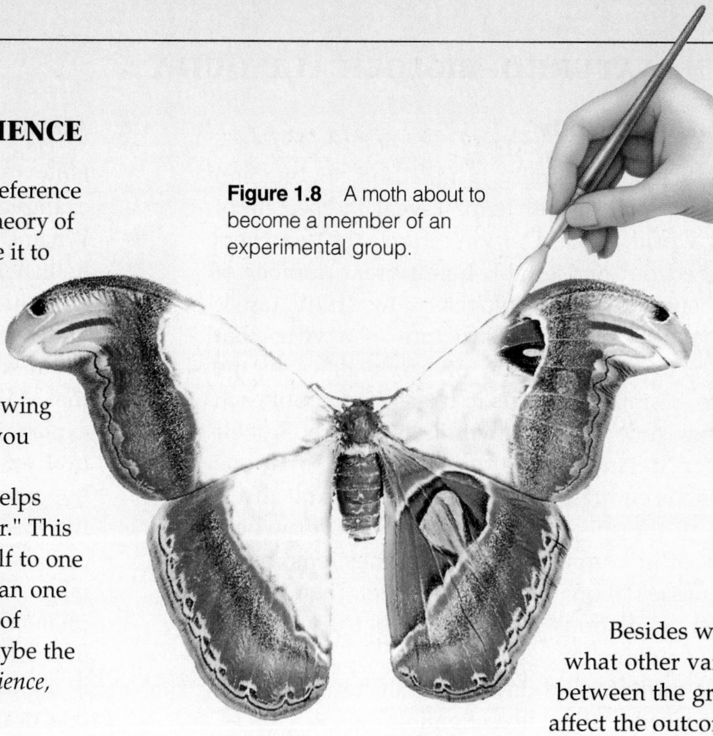

Figure 1.8 A moth about to become a member of an experimental group.

Besides wing pattern, what other variables between the groups might affect the outcome of your experiment? Maybe paint fumes are as repulsive to a potential mate as a painted-on pattern. Maybe when you paint the moths you somehow rough them up and make them less desirable than moths in the control group. Maybe the paint weighs enough to change the flutter frequency of the wings.

So you decide the control group also must be painted with the same kind of paint, using the same brushes, and must be handled the same way. For this group, however, you duplicate the natural wing color pattern as you paint. Now your experimental and control groups are identical *except* for the variable under study. If only those moths with altered wing patterns turn out to be unlucky in love, then your control group will help support the hypothesis.

Generally, experiments are devised to disprove a hypothesis. Why is this so? It would be impossible to prove beyond a shadow of a doubt that a hypothesis is correct. It would take an infinite number of experiments to demonstrate it holds under all possible conditions.

Have you been thinking that painting moths is a rather fanciful example of a scientific approach? As reported in *Nature* in 1993, Karen Marchetti, a graduate student of the University of California at Davis, wielded a paintbrush on birds in a forest in Kashmir, India. She found evidence for her hypothesis that bright feather color, not patterning, gives male yellow-browed leaf warblers a competitive edge in mating.

In early tests, Marchetti captured male birds and put a patternless dab of yellow paint on their head feathers. In later tests, she painted larger-than-normal yellow bands on the wings of one group of male birds and painted out part of the bands of another group. She used transparent paint for a third group. (Can you guess why?) As Marchetti discovered, color-enhanced birds secured larger territories and produced more offspring, compared to the toned-down birds used in her experiments.

1.8 THE LIMITS OF SCIENCE

The call for objective testing strengthens the theories that emerge from scientific studies. It also puts limits on the kinds of studies that can be carried out. Beyond the realm of science, some events remain unexplained. Why do we exist, for what purpose? Why does any one of us have to die at a particular moment? Answers to such questions are *subjective*—they come from within us, as an outcome of all the experiences and mental connections shaping our consciousness. Because people differ vastly in this regard, subjective answers do not readily lend themselves to scientific analysis.

This is not to say subjective answers are without value. No human society can function for long unless its members share a commitment to standards for making judgments, even subjective ones. Moral, aesthetic, philosophical, and economic standards vary from one society to the next. But all guide their members in deciding what is important and good, and what is not. All attempt to give meaning to what we do.

Every so often, scientists stir up controversy when they happen to explain some part of the world that was considered to be beyond natural explanation—that is, belonging to the "supernatural." This is sometimes true when moral codes are interwoven with religious narratives. Exploring some longstanding view of the world from a scientific perspective may be misinterpreted as questioning morality, even though the two are not the same thing.

For example, centuries ago in Europe, Nicolaus Copernicus studied the planets and concluded that the Earth circles the sun. Today this seems obvious enough. Back then it was heresy. The prevailing belief was that the Creator had made the Earth (and, by extension, humanity) the immovable center of the universe. Not long afterward a respected professor, Galileo Galilei, studied the Copernican model of the solar system. He thought it was a good one and said so. He was forced to retract his statement publicly, on his knees, and to put the Earth back as the fixed center of things. (Word has it that when Galileo stood up he still muttered, "Even so, it *does* move.")

Today, as then, society has sets of standards. Those standards may be questioned when a new, natural explanation runs counter to supernatural belief. This doesn't mean that the scientists who raise questions are less moral, less lawful, less sensitive, or less caring than anyone else. It simply means one more standard guides their work: *The external world, not internal conviction, must be the testing ground for scientific beliefs.*

Systematic observations, hypotheses, predictions, tests—in all these ways, science differs from systems of belief that are based on faith, force, or simple consensus.

SUMMARY

1. There is unity in the living world, for all organisms have these characteristics in common:

   a. They are assembled from the same kinds of atoms and molecules, according to the same laws of energy.

   b. They survive by metabolism, and by sensing and responding to specific conditions in the environment.

   c. They have the capacity for growth, development, and reproduction, based on heritable instructions encoded in the molecular structure of their DNA.

2. The characteristics of life emerge at the level of cells, and they extend through multicelled organisms, populations, communities, ecosystems, and the biosphere.

3. Many millions of species (kinds of organisms) exist; many millions more lived in the past and became extinct. In classification schemes, species are placed in ever more inclusive groupings, from genus on up through family, order, class, phylum, and kingdom.

4. The diversity among organisms arises through mutations—changes in the structure of DNA molecules. The changes may lead to variation in heritable traits (that is, traits that parents bestow on offspring, including most details of the body's form and functioning).

5. Darwin's theory of evolution by natural selection is a cornerstone of biological inquiry. Its key premises are:

   a. Individuals of a population show different versions of the same heritable trait. Variant forms of traits may affect the ability to survive and reproduce.

   b. Natural selection is the outcome of differences in survival and reproduction among individuals that differ in one or more traits. Adaptive forms of traits may become more common; less adaptive ones may become less common or disappear. Thus the traits that define a population change over time; the population evolves.

6. There are many diverse methods of scientific inquiry. The following terms are important to all of them:

   a. Theory: An explanation of a broad range of related phenomena, supported by many tests. An example is Darwin's theory of evolution by natural selection.

   b. Hypothesis: A possible explanation of a specific phenomenon. Sometimes called an educated guess.

   c. Prediction: A claim about what can be expected in nature, based on the premises of a theory or hypothesis.

   d. Test: An attempt to produce actual observations that match predicted or expected observations.

   e. Conclusion: A statement about whether a theory or hypothesis should be accepted, modified, or rejected, based on tests of predictions derived from it.

7. Scientific theories are based on systematic observations, hypotheses, predictions, and relentless tests. The external world, not internal conviction, is the testing ground for those theories.

Review Questions

1. Why is it difficult to give a simple definition of life? *3* (For this and subsequent chapters, *italic numbers* following review questions indicate the pages on which the answers may be found.)

2. What characteristics do all organisms have in common? *3, 8*

3. What is energy? What is DNA? *3*

4. Study Figure 1.2. Then, on your own, arrange and define the levels of biological organization. *4*

5. Define metabolic activity. *4–5*

6. Make a sketch of the one-way flow of energy and the cycling of materials through the biosphere. *5*

7. What is mutation? Explain its functional relationship to the diversity of life. *7*

Self-Quiz *(Answers in Appendix IV)*

1. The _____ is the smallest unit of life.

2. _____ is the ability of cells to extract and transform energy from the environment and use it to maintain themselves, grow, and reproduce.

3. _____ is a state in which the body's internal environment is being maintained within tolerable limits.

4. If a given form of a trait improves chances for surviving and reproducing in a particular environment, it is a(n) _____ trait.

5. The capacity to evolve is based on variations in heritable traits, which originally arise through _____ .

6. You have some number of traits that also were present in your great-great-great-great-grandmothers and -grandfathers. This is an example of _____ .
   a. metabolism   c. a control group
   b. homeostasis   d. inheritance

7. DNA molecules _____ .
   a. contain instructions for traits
   b. undergo mutation
   c. are transmitted from parents to offspring
   d. all of the above

8. For many years in a row, a dairy farmer allowed his best milk-producing cows but not the poor producers to mate. Over many generations, milk production increased. This outcome is an example of _____ .
   a. natural selection   c. evolution
   b. artificial selection d. both b and c

9. Match the terms with the most suitable descriptions.
   ____ adaptive trait   a. statement of what you should observe
   ____ natural   in nature, if you were to go looking for it
         selection   b. educated guess
   ____ theory   c. improves chance of surviving and
   ____ hypothesis   reproducing in prevailing environment
   ____ prediction   d. related set of hypotheses that together
                              form a broad-reaching, testable explana-
                              tion about a basic aspect of nature
                           e. outcome of differences in survival and
                              reproduction that occurred among indi-
                              viduals that differ in one or more traits

Critical Thinking

1. Witnesses in a court of law are asked to "swear to tell the truth, the whole truth, and nothing but the truth." What are some of the problems inherent in the question? Can you think of a better alternative?

2. Design a test to support or refute the following hypothesis: Body fat appears yellow in certain rabbits—but only when those rabbits also eat leafy plants that contain an abundance of a yellow pigment molecule called xanthophyll.

3. A group of scientists devised the following experiments to shed light on whether different species of fishes of the same genus compete with one another in their natural habitat. They set up twelve ponds that were identical in chemical composition and physical characteristics. In each pond, they released the following:

Ponds 1, 2, 3:	species A	(300 individuals each pond)
Ponds 4, 5, 6:	species B	(300 individuals each pond)
Ponds 7, 8, 9:	species C	(300 individuals each pond)
Ponds 10, 11, 12:	species A, B, C	(300 of each in each pond)

Does this experimental design take into consideration all factors that can affect the outcome? If not, how would you change it?

4. Many popular magazines publish articles on diet, exercise, and other health-related topics. Often the authors will recommend a specific diet or dietary supplement. What kinds of evidence do you think the articles should describe so you can decide whether you should accept their recommendations?

Selected Key Terms

For this and subsequent chapters, these are the **boldface** terms that occur in the text on the pages indicated by *italic* numbers. Make a list of these terms, write a definition next to each, then check it against the one in the text. (You will be using these terms later on.)

adaptive trait *7*	ecosystem *4*	natural
aerobic	energy *4*	selection *10*
respiration *5*	evolution *10*	photosynthesis *4*
animal *9*	experiment *12*	plant *9*
artificial	fungus *9*	population *4*
selection *10*	genus *8*	prediction *11*
ATP *4*	homeostasis *6*	producer *5*
biosphere *4*	hypothesis *11*	protistan *9*
cell *4*	inheritance *7*	receptor *6*
community *4*	metabolism *4*	reproduction *6*
consumer *5*	moneran *8*	species *8*
control group *12*	multicelled	test *11*
decomposer *5*	organism *4*	theory *11*
DNA *4*	mutation *7*	variable *12*

Readings

Committee on the Conduct of Science. 1989. *On Being a Scientist.* Washington, D.C.: National Academy of Sciences. Paperback.

Larkin, T. June 1985. "Evidence versus Nonsense: A Guide to the Scientific Method." *FDA Consumer* 19: 26–29.

Morris, I. December 1966. "Is Science Really Scientific?" *Science Journal*, p. 76.

FACING PAGE: *Living cells of a green plant* (Elodea), *as seen with the aid of a microscope. Each rectangular cell contains efficient chemical factories called chloroplasts (the green spheres).*

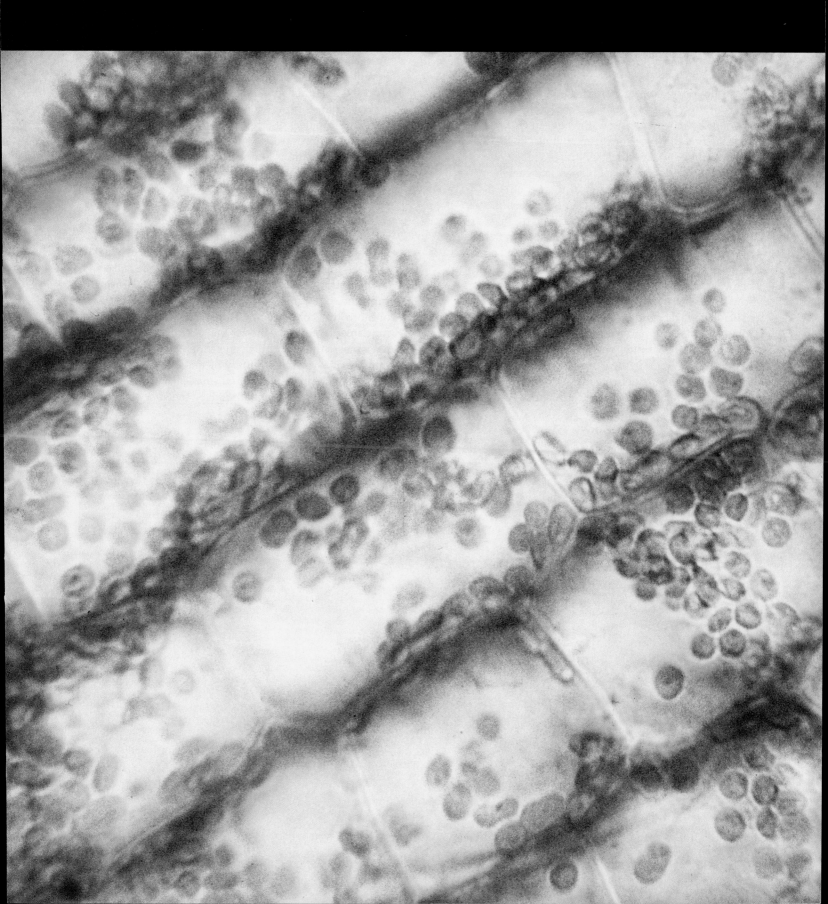

2 CHEMICAL FOUNDATIONS FOR CELLS

The Chemistry In and Around You

Right now you are breathing in oxygen. You would die without it. Two centuries ago, no one had a clue to what oxygen is, where it comes from, and how it helps keep people alive. Then researchers started unlocking the secrets of this chemical substance and others. As their knowledge of chemistry grew, they began to develop such amazing things as fertilizers, nylons, lipsticks, aspirin, antibiotics, and plastic parts of refrigerators, computers, television sets, jet planes, and cars.

Today, our chemical "magic" brings benefits *and* problems. For example, without synthetic fertilizers to boost crop yields, more humans than you might ever imagine would starve to death. Yet weeds don't know that fertilizers are for crop plants. Certain pests don't know that crops aren't meant to be their salad bars. Each year they ruin or gobble up nearly half of what we grow. That is why, in 1945, we began to use synthetic pesticides. We designed these substances to kill weeds, moths, worms, rats, and other organisms that threaten our food supplies, health, pets, and ornamental plants. In 1988 alone, people in the United States spread more than a billion pounds of pesticides through homes and gardens, offices, industries, and farmlands (Figure 2.1).

Among the pesticides, we find organophosphates (including malathion), carbamates, and halogenated compounds (including chlordane). Most cause death by blocking the brain's vital messages. Some remain active for days, others for weeks or years. Besides pests, they kill great numbers of birds and other predators that normally help control population sizes of pests. Worse, targeted populations become resistant to pesticides, for reasons that will become apparent in later chapters.

We, too, inhale pesticides, ingest them with food, or absorb them through skin. Some cause headaches and rashes. Some trigger hives, asthma, joint pain, even life-threatening allergic reactions in millions of people.

Pesticides, people, pumpkins, the solid earth beneath our feet, the very air we breathe—*everything in and around you is "chemistry."* Every solid, liquid, or gaseous substance you care to think about is a collection of one or more kinds of elements. Think of the **elements** as fundamental forms of matter that occupy space and have mass. You can't break an element apart into something else, at least by ordinary means. Break a chunk of some element into smaller and smaller bits, and the smallest bit you end up with will be the *same* element.

Figure 2.1 A low-flying cropduster with its rain of pesticides.

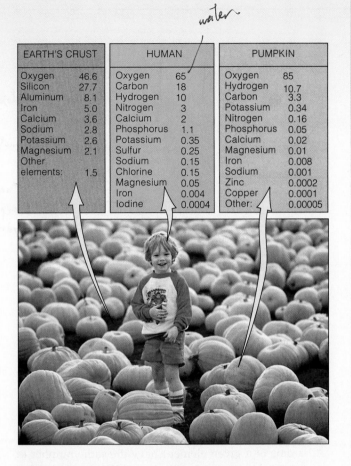

EARTH'S CRUST		HUMAN		PUMPKIN	
Oxygen	46.6	Oxygen	65	Oxygen	85
Silicon	27.7	Carbon	18	Hydrogen	10.7
Aluminum	8.1	Hydrogen	10	Carbon	3.3
Iron	5.0	Nitrogen	3	Potassium	0.34
Calcium	3.6	Calcium	2	Nitrogen	0.16
Sodium	2.8	Phosphorus	1.1	Phosphorus	0.05
Potassium	2.6	Potassium	0.35	Calcium	0.02
Magnesium	2.1	Sulfur	0.25	Magnesium	0.01
Other		Sodium	0.15	Iron	0.008
elements:	1.5	Chlorine	0.15	Sodium	0.001
		Magnesium	0.05	Zinc	0.0002
		Iron	0.004	Copper	0.0001
		Iodine	0.0004	Other:	0.00005

Figure 2.2 Proportions of elements that make up a human body and the fruit of pumpkin plants, compared to the proportions of elements in materials of the earth's crust. In what respects are the proportions similar? In what respects do they differ?

About ninety-two elements occur naturally on Earth. As is true of all organisms, most of your own body consists of only four kinds—oxygen, carbon, hydrogen, and nitrogen (Figure 2.2). It also incorporates some calcium, phosphorus, potassium, sulfur, sodium, and chlorine, as well as seemingly insignificant amounts of other, trace, elements. Normal functioning depends on all of them. Without daily intakes of magnesium, for instance, your muscles will become sore and weak, and your brain won't work properly. Without magnesium, older leaves of a tree will droop, turn yellow, then die. When a weed absorbs 2,4-D, a pesticide, it is stimulated to grow excessively. At the same time, it can't absorb vital elements fast enough to sustain its growing tissues. Quite literally, the weed grows itself to death.

In short, maintaining crops, industries, and health depends on knowledge of chemistry. So do efforts to minimize side effects of its applications. You owe it to yourself and others to gain understanding of chemical substances. By demystifying chemistry's "magic," you will be better equipped to assess its benefits and risks.

KEY CONCEPTS

1. All substances consist of one or more elements, such as hydrogen, oxygen, and carbon. Atoms of each element have some number of protons, electrons, and usually neutrons, and these component parts are arranged in particular ways. Atoms have no overall electric charge unless they become ionized—that is, unless they lose electrons or acquire more of them.

2. Whether an atom will interact with other atoms—as through ionic, hydrogen, and covalent bonds—depends on its electron structure. Such interactions are the basis of the molecular organization and activities of living things.

3. Life originated in water and is adapted to its properties. Water has temperature-stabilizing effects, cohesiveness, and a capacity to dissolve or repel a variety of substances.

4. Organic compounds have a backbone of one or more carbon atoms to which hydrogen, oxygen, nitrogen, and other atoms are attached. Cells are defined partly by their capacity to assemble the organic compounds known as carbohydrates, lipids, proteins, and nucleic acids.

5. Carbohydrates and lipids are the cell's main sources of energy and building blocks. Two nucleic acids, DNA and RNA, are the basis of inheritance and cell reproduction.

6. Many proteins are structural materials or enzymes, which speed up specific metabolic reactions. Many other proteins transport substances, contribute to movements, trigger changes in cell activities, or help the body defend itself against disease.

REGARDING THE ATOMS

The organization and activities of living things start with certain elements, especially those listed in Table 2.1. By international agreement, a one- or two-letter chemical symbol stands for each element, regardless of the element's name in different languages. What we call nitrogen is *azoto* in Italian and *stickstoff* in German—but the symbol for nitrogen remains N. Similarly, the symbol for sodium is Na (from the Latin *natrium*).

Structure of Atoms

Atoms are the smallest units that retain the properties of a given element. Each atom is composed of subatomic particles called protons, electrons, and (except for hydrogen) neutrons. **Protons** carry a positive electric charge (p^+). Together with **neutrons**, which have no charge, protons make up the atom's core region, the atomic nucleus. **Electrons** carry a negative charge (e^-). They move rapidly around the nucleus and they occupy most of the atom's volume (Figure 2.3).

Regardless of the element, atoms have just as many electrons as protons. This means that they carry no *net* charge, overall.

Each element has a unique **atomic number**, which refers to the number of protons in its atoms. For example, as Table 2.1 indicates, that number is 1 for the hydrogen atom (with one proton) and 6 for the carbon atom (with six protons). Each element also has a **mass number**, which refers to the number of protons *and* neutrons in the atomic nucleus. A carbon atom with six protons and six neutrons has a mass number of 12. You may have heard of atomic weights. It is an imprecise way to refer to the relative masses of atoms (mass is not quite the same thing as weight), but its use continues.

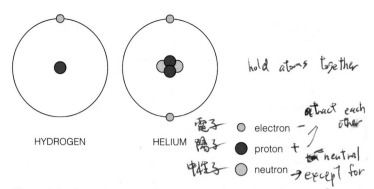

Figure 2.3 One model of atomic structure. Bear in mind, this model is extremely simplified. At this scale, the nucleus of these two representative atoms (hydrogen and helium) actually would be an invisible speck.

Why bother with atomic numbers and mass numbers? As you will see, they give us an idea of whether and how substances will interact. *And this knowledge can help us predict how the substances of life will behave in cells, in multicelled organisms, and in the environment, under a variety of conditions.*

Isotopes—Variant Forms of Atoms

All atoms of a given element have the same number of protons and electrons, but they can differ in the number of neutrons. Such atoms are **isotopes**. "A carbon atom," for example, might be carbon 12 (with six protons and six neutrons), carbon 13 (six protons, seven neutrons), or carbon 14 (six protons, eight neutrons). You can write these as ^{12}C, ^{13}C, and ^{14}C. All isotopes of an element interact with other atoms the same way. Thus cells can use any carbon isotope for a metabolic reaction.

You have probably heard of radioactive isotopes, or **radioisotopes**. Henri Becquerel, a physicist, discovered them in 1896, after he put a heavily wrapped rock into a desk drawer, on top of an unexposed photographic plate. The rock contained isotopes of uranium. A few days later, Becquerel noticed that the plate bore a faint image of the rock, which apparently had been emitting energy. His coworker, Marie Curie, named the phenomenon "radioactivity." As we now know, radioisotopes are unstable atoms (with dissimilar numbers of protons and neutrons) that emit electrons and energy. In this spontaneous process, called radioactive decay, they are transformed into a different isotope. The *Focus* essay on the facing page describes some uses of radioisotopes.

Elements are forms of matter that occupy space and have mass. Atoms, the smallest units unique to each element, have one or more positively charged protons, negatively charged electrons, and (except for hydrogen) neutrons.

Table 2.1	Atomic Number and Mass Number of Elements Common in Living Things		
Element	Symbol	Atomic Number	Most Common Mass Number
Hydrogen	H	1	1
Carbon	C	6	12
Nitrogen	N	7	14
Oxygen	O	8	16
Sodium	Na	11	23
Magnesium	Mg	12	24
Phosphorus	P	15	31
Sulfur	S	16	32
Chlorine	Cl	17	35
Potassium	K	19	39
Calcium	Ca	20	40
Iron	Fe	26	56
Iodine	I	53	127

USING RADIOISOTOPES TO DATE FOSSILS, TRACK CHEMICALS, AND SAVE LIVES

放射性测定

Radioactive Dating Each type of radioisotope has a certain number of protons and neutrons, and it decays spontaneously at a certain rate into a more stable form. *"Half-life"* is the time it takes for half of a given quantity of a radioactive element to decay into a different element. It cannot be modified by temperature, pressure, chemical reactions, or any other environmental factor.

Consider the half-life of ^{40}potassium. It takes 1.3 billion years for this radioisotope to decay to ^{40}argon, a stable isotope. Thus researchers can discern the age of anything that contains ^{40}potassium by measuring its ratio of ^{40}argon to ^{40}potassium. The following table lists the useful ranges of some isotopes that researchers use in dating methods:

Radioisotope (unstable)	More Stable Product	Half-Life (years)	Useful Range (years)
^{87}rubidium →	^{87}strontium	49 billion	100 million
^{232}thorium →	^{208}lead	14 billion	200 million
^{238}uranium →	^{206}lead	4.5 billion	100 million
^{40}potassium →	^{40}argon	1.3 billion	100 million
^{235}uranium →	^{207}lead	704 million	100,000
^{14}carbon →	^{14}nitrogen	5,730	0–60,000

✓estimate how old they are.

Radioactive dating is a reliable way to discern the age of bands of sediments, volcanic ash, and other layers in the Earth. By using ^{238}uranium (with a half-life of 4.5 billion years), researchers discovered that the Earth formed more than 4.6 billion years ago. Similarly, they have calculated the ages of many fossils, including this fossilized frond of a tree fern that lived more than 250 million years ago:

Tracking Chemicals Radioisotopes can be used as **tracers**. That is, if used in conjunction with scintillation counters and other devices, they may reveal a pathway or destination of a substance that has entered a cell, a body, an ecosystem, or some other "system." Carbon provides an example. All isotopes of an element have the same number of electrons, so they all interact with other atoms the same way. Therefore, cells can use any isotope of carbon in reactions that require carbon atoms—such as certain reactions of photosynthesis. By putting plant cells in a medium enriched in ^{14}carbon, researchers identified

the steps by which plants take up carbon and incorporate it into carbohydrates during photosynthesis. Researchers also use tracers to identify how plants use synthetic fertilizers as well as natural nutrients. The findings of some studies may lead to improved crop yields.

What about medical applications? Consider the human thyroid, the only gland of ours that takes up iodine. If a trace amount of the radioisotope ^{123}iodine is injected into a patient's blood, a photographic imaging device can scan the thyroid. The scans that follow are examples of (1) the normal gland, (2) the enlarged gland of a patient with a thyroid disorder, and (3) a cancerous thyroid gland:

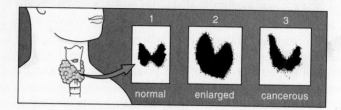

Saving Lives Medical practitioners use radioisotopes to diagnose and treat diseases. For example, patients with irregular heartbeats can receive artificial pacemakers, powered by energy from ^{238}plutonium. This dangerous radioisotope is sealed inside a case so its emissions won't damage tissues. As another example, radioisotopes are used in *radiation therapy*, which destroys or impairs living cells. In some therapies, localized cancers are bombarded with energy from a source of ^{226}radium or ^{60}cobalt.

As a final example, *PET* (short for *Positron-Emission Tomography*) yields images of metabolically active and inactive tissues. Clinicians attach radioisotopes to glucose or some other molecule. They inject the labeled glucose into a patient, who is moved into a PET scanner. When cells in certain tissues absorb the glucose, radioisotope emissions are intercepted and used to produce a vivid image of variations or abnormalities in metabolic activity:

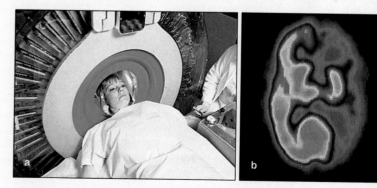

(a) PET scanner. (b) Brain scan of a child who has a severe neurological disorder. Normally, different colors in a brain scan signify differences in metabolic activity. The right half of this injured child's brain shows little activity.

2.3 WHAT IS A CHEMICAL BOND?

We turn now to **chemical bonds**, which are unions between the electron structures of atoms. Take a moment to review Figure 2.4, which summarizes a few of the conventions that are used to describe these events.

Electrons and Energy Levels

In a given chemical reaction, atoms may acquire extra electrons, share them, or donate them to another atom. The atoms of certain elements do this rather easily, but others do not. What determines whether one kind will interact with another in such ways? *The outcome depends on the number and arrangement of their electrons.*

By way of analogy, picture three preschoolers who are circling around a cookie jar, not yet expert in the fine art of sharing. Each is drawn to the cookies but dreads being shoved by the others. Two may maneuver themselves on opposite sides of the jar to avoid a direct hit. But all three never, ever position themselves in the same place at the same time.

Electrons behave roughly the same way. Because all of an atom's electrons carry a negative charge, they repel each other but are attracted to the positive charge of the protons. The electrons spend as much time as possible near the protons and far away from each other by moving about in different orbitals. Think of **orbitals** as *volumes of space* around the atomic nucleus in which electrons are likely to be at any instant.

Just as the number of electrons differs among atoms, so does the number of orbitals. Only one or *two electrons at most* can occupy any one of these orbitals.

Consider hydrogen, the simplest atom of all. Its lone electron occupies a spherical orbital that is closest to the nucleus, which corresponds to the *lowest available energy level*. In all other atoms, two electrons fill this orbital:

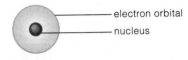

HYDROGEN ATOM

In larger atoms, the next two electrons occupy a second spherical orbital that surrounds the first. As many as six more electrons may be distributed in three dumbbell-shaped orbitals, and so on. These electrons spend more of the time farther away from the nucleus; they are said to be at *higher energy levels.*

The **shell model**, shown in Figure 2.5, is a simple although not quite accurate way to think about how electrons are distributed in a given atom. By this model,

---

Figure 2.4 Chemical bookkeeping.

We use symbols for elements when writing *formulas*, which identify the composition of compounds. For example, water has the formula H_2O. The subscript indicates that two hydrogen (H) atoms are present for every oxygen (O) atom. We use symbols and formulas in *chemical equations*—representations of reactions among atoms and molecules. An arrow in such an equation means "yields." Substances that enter a reaction (the reactants) are to the left of the arrow, and products are to the right, as shown by this equation for photosynthesis:

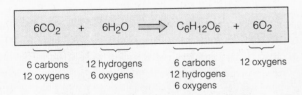

6 carbons	12 hydrogens	6 carbons	12 oxygens
12 oxygens	6 oxygens	12 hydrogens	
		6 oxygens	

Notice that there are as many atoms of each element to the right of the arrow as there are to the left, even though they are combined in different forms. Atoms taking part in chemical reactions may get rearranged, but they are never destroyed. According to the *law of conservation of mass*, the total mass of all materials entering a reaction equals the total mass of all the products. Bear in mind, the equations that you use to represent cellular reactions must be balanced this way, because no atoms are lost.

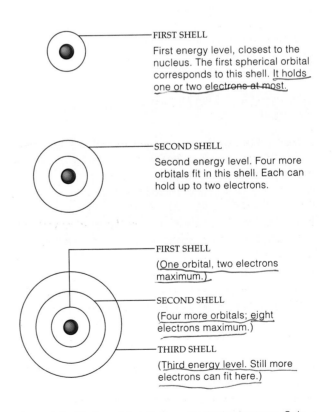

Figure 2.5 Shell model of electron distribution in atoms. Only three of the many possible energy levels (shells) are shown.

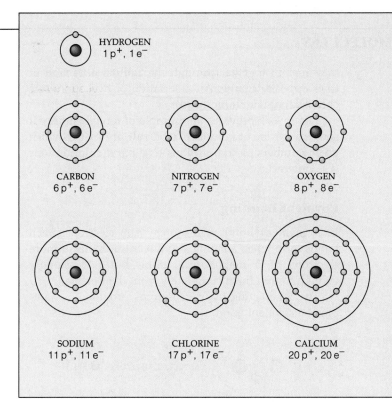

HYDROGEN
1 p⁺, 1 e⁻

CARBON
6 p⁺, 6 e⁻

NITROGEN
7 p⁺, 7 e⁻

OXYGEN
8 p⁺, 8 e⁻

SODIUM
11 p⁺, 11 e⁻

CHLORINE
17 p⁺, 17 e⁻

CALCIUM
20 p⁺, 20 e⁻

Element	Symbol	Atomic Number*	Distribution of Electrons			
			First Shell	Second Shell	Third Shell	Fourth Shell
Hydrogen	H	1	1	—	—	—
Helium	He	2	2	—	—	—
Carbon	C	6	2	4	—	—
Nitrogen	N	7	2	5	—	—
Oxygen	O	8	2	6	—	—
Neon	Ne	10	2	8	—	—
Sodium	Na	11	2	8	1	—
Magnesium	Mg	12	2	8	2	—
Phosphorus	P	15	2	8	5	—
Sulfur	S	16	2	8	6	—
Chlorine	Cl	17	2	8	7	—
Calcium	Ca	20	2	8	8	2

* The number of protons in the nucleus.

Figure 2.6 Examples of the shell model of the electron distribution in atoms. Helium, neon, and other atoms with no electron vacancies (no unfilled orbitals) in their outermost shell are inert; they tend not to enter into chemical reactions. Hydrogen, carbon, and other atoms with electron vacancies in their outermost shell tend to interact with other atoms.

a series of shells encompasses all of the orbitals that are available to electrons. The first spherical orbital fits in the first shell. The next four orbitals fit in a second shell (at a higher energy level) that encloses the first. More orbitals fit in a third shell, still more fit in a fourth shell, and so on up for large, complex atoms of the heavier elements, as listed in Appendix VI.

Electrons and the Bonding Behavior of Atoms

Figure 2.6 shows how the shell model can be used to represent electron distributions. Each electron, signified here by a *gold* dot, is assigned to a circle that represents one of the shells (energy levels). Count the electrons to find out which atoms have electron "vacancies"—that is, one or more unfilled orbitals in their outermost shell. As you can see, helium and neon have no such vacancy. Like many other atoms, they are inert, meaning they tend not to interact with other atoms.

Now look again at Figure 2.2, which lists the most abundant elements of a typical organism. The elements include hydrogen, oxygen, carbon, and nitrogen—the atoms of which do have electron vacancies. *These tend to form bonds with other atoms and so fill the vacancies*.

In short, whether any two atoms will attract or repel each other depends on their electron structures. And when an atom does give up, gain, or share electrons, *the number or the distribution of its electrons changes*—in ways that will be described shortly.

From Atoms to Molecules

When two or more atoms bond together, the outcome is a **molecule**. Many molecules are composed of only one element. Molecular nitrogen (N_2), with its two nitrogen atoms, is like this.

Many other kinds of molecules are "compounds," meaning that they consist of two or more elements, in proportions that never vary. Water is an example. Every water molecule has one oxygen atom bonded with two hydrogen atoms. Molecules of water in rainclouds or in a Siberian lake or in your bathtub *always* have twice as many hydrogen as oxygen atoms.

By contrast, in a "mixture," two or more elements are simply intermingling in proportions that can vary. Consider the sugar sucrose, which is a compound of carbon, hydrogen, and oxygen. Swirl sucrose molecules and water molecules together, and you get a mixture.

In atoms, electrons occupy orbitals (volumes of space) around the nucleus. Think of orbitals as being arranged inside a series of shells. The shells correspond to levels of energy that become greater with distance from the nucleus.

One or two electrons at most occupy any orbital. Atoms with lone electrons (unfilled orbitals) in their outermost shell tend to interact with other atoms.

In molecules of an element, all of the atoms are of the same kind. In molecules of a compound, atoms of two or more elements are bonded together, in unvarying proportions.

Eat your peas! Drink your milk! Probably for longer than you care to remember, somebody has been telling you to eat certain foods that are rich in carbohydrates, proteins, and other "biological molecules." Only living organisms put together and use these molecules, which consist of only a few kinds of atoms held together by a few kinds of bonds—especially the ones known as ionic, covalent, and hydrogen bonds.

Ionic Bonding

An atom, recall, has just as many electrons as protons, so it carries no *net* charge. Suppose a chlorine atom, which has an unfilled orbital in its outermost shell, latches onto an extra electron and thereby fills the vacancy (Figure 2.7a). Or suppose an electron is knocked out or pulled away from an orbital in an atom's outermost shell. Whenever an atom loses or gains one or more electrons, *the balance between its protons and its electrons shifts*—and the atom becomes positively or negatively charged. In this state, it is called an **ion.**

In cells, neighboring atoms typically accept or donate electrons among one another. When one atom loses an electron and one gains, *both* become ionized. Depending on cellular conditions, the ions may be separated, or they

may remain together through the mutual attraction of their opposite charge. An association of two oppositely charged ions is an **ionic bond**.

Figure 2.7b provides an example of ionic bonding. It shows a portion of a crystal of table salt, or NaCl, in which great numbers of sodium ions (Na^+) and chloride ions (Cl^-) interact.

Covalent Bonding

In a **covalent bond**, two atoms *share* electrons. Such bonds form when an atom strongly attracts one or more electrons of a neighboring atom—but not strongly enough to pull them completely from their orbital.

Two interacting hydrogen atoms provide an example of a covalent bond:

MOLECULAR HYDROGEN (H_2)

In structural formulas, a single line between two atoms represents a single covalent bond. In a double covalent bond, two atoms share two pairs of electrons. In a triple

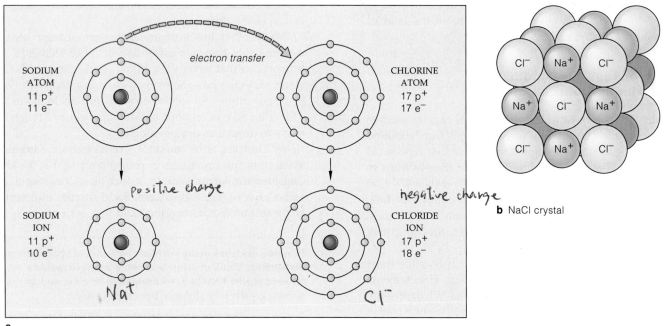

a

Figure 2.7 **(a)** Ionization by way of an electron transfer. In this case, a sodium atom donates the lone electron in its outermost shell to a chlorine atom, which has an unfilled orbital in *its* outermost shell. A sodium ion (Na^+) and a chloride ion (Cl^-) are the outcome of this interaction.

(b) In each crystal of table salt, or NaCl, many sodium and chloride ions remain together because of the mutual attraction of their opposite charge. Their interaction is an example of ionic bonding.

covalent bond, two atoms share three pairs of electrons. Here are a few simple examples of covalent bonding:

H—H   single covalent bond
           in molecular hydrogen (H₂)

O═O   double covalent bond
           in molecular oxygen (O₂)

N≡N   triple covalent bond
           in molecular nitrogen (N₂)

All three of these are gaseous molecules. Every time you breathe in, you draw some number of H_2, O_2, and N_2 molecules into your nose.

Covalent bonds are nonpolar or polar. In a *nonpolar* covalent bond, the participating atoms exert the same pull on the electrons and share them equally. The term "nonpolar" implies no difference at the ends of the bond (that is, at its two poles). Molecular hydrogen is a simple example of this. Its two H atoms, each with one proton, attract the shared electrons equally.

In a *polar* covalent bond, atoms of different elements (which have different numbers of protons) don't exert the same pull on shared electrons. The more attractive atom ends up with a slight negative charge; the atom is "electronegative." Its effect is balanced out by the other atom, which ends up with a slight positive charge. In other words, taken together, atoms interacting in a polar covalent bond have no *net* charge—but the charge is distributed unevenly between the bond's two ends.

Consider the water molecule, which has two polar covalent bonds (H—O—H). In this case, the electrons are less attracted to the hydrogens than to the oxygen, which has more protons. A water molecule carries no *net* charge, but its polarity makes it weakly attractive to neighboring polar molecules and ions.

Patterns of electron sharing in covalent bonds hold atoms together in specific arrangements in molecules. Some of these patterns also give rise to weak attractions and repulsions *between* molecules. These bonding forces have important roles in the structure and functioning of biological molecules.

Hydrogen Bonding

In a **hydrogen bond**, a small, highly electronegative atom of a molecule interacts weakly with a hydrogen atom that is already participating in a polar covalent bond. When taking part in such a bond, hydrogen has a slight positive charge, and this attracts other atoms that carry a slight negative charge (Figure 2.8).

Hydrogen bonds may form between two or more molecules. They also may form between different parts of the *same* molecule that is twisting back on itself. For

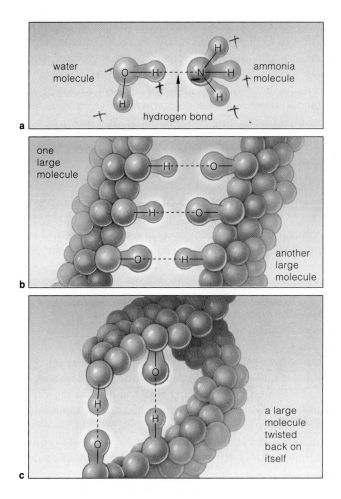

Figure 2.8 Examples of hydrogen bonds. Compared to covalent bonds, a hydrogen bond is easier to break. Collectively, however, such bonds are important in water and many other substances.

example, great numbers of the bonds form between the two strands of a DNA molecule. Individually, hydrogen bonds break easily. Collectively, they stabilize DNA's structure. Similarly, hydrogen bonds form between the molecules that make up water. As you will read next, they contribute to water's life-sustaining properties.

In an ionic bond, two ions of opposite charge attract each other and remain together. Ions form when atoms gain or lose electrons and so acquire a net positive or negative charge.

In a covalent bond, atoms share electrons. If the atoms share electrons equally, the bond is nonpolar. If the sharing is not equal, the bond itself is polar (slightly positive at one end, slightly negative at the other).

In a hydrogen bond, a small, highly electronegative atom or molecule interacts weakly with a neighboring hydrogen atom that is already taking part in a polar covalent bond.

PROPERTIES OF WATER

Life originated in water. Many organisms still live in it, and the ones that don't live in water carry it around with them, in cells and tissue spaces. Many metabolic reactions require water as a reactant. The very shape and internal structure of cells depends on it. We will consider these topics in many chapters, so you may find it useful to study the following points about water's properties.

Polarity of the Water Molecule

A water molecule, remember, has no *net* charge—but the charges that it does carry are unevenly distributed. As a result of its electron arrangements and bond angles, the water molecule's oxygen "end" is a bit negative and its other end is a bit positive. Figure 2.9a shows a simple way to think about this charge distribution.

Because of their built-in polarity, water molecules form hydrogen bonds with one another (Figure 2.9b). They also form hydrogen bonds with other polar substances, such as sugars. All polar molecules are attracted to water; they are **hydrophilic** (water loving). By contrast, water repels nonpolar substances, such as oil. All nonpolar molecules are **hydrophobic** (water dreading).

Shake a bottle containing water and salad oil, then put it on a counter. In time, hydrogen bonds reunite the water molecules. Why? They replace bonds that were broken when you shook the bottle. As water molecules hydrogen-bond with each other, they push the oil aside. Oil molecules are forced to cluster together in droplets or as a film at the water's surface. Among other things, these hydrophobic interactions help organize the rather oily, sheetlike layers of cell membranes.

Water's Temperature-Stabilizing Effects

The first forms of life on earth originated in *liquid* water, not water in a frozen or gaseous form. And most of the Earth's water tends to stay liquid because of hydrogen bonds between water molecules.

Consider that molecules of water or any other substance are in constant motion, and that energy inputs make them move faster. **Temperature** is a measure of this molecular motion. Compared to most other fluids, water requires a greater input of heat energy before its temperature increases measurably. Why? Hydrogen bonds in water absorb much of the incoming energy, so the motion of individual molecules does not increase as fast. This property helps stabilize temperatures in cells, which are mostly water. It helps cells resist temperature changes that might otherwise disrupt enzyme action and other vital activities.

Hydrogen bonds are breaking and forming again constantly in liquid water. But what happens during **evaporation**? Then, an input of heat energy converts

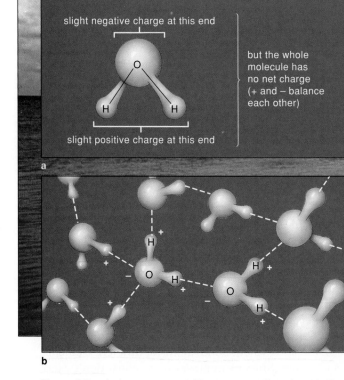

a

b

Figure 2.9 A view you never will see on any other planet in the solar system—an ocean of water, a substance that is vital for life. (**a**) Polarity of a water molecule. (**b**) Hydrogen bonding between water molecules in liquid water. Dashed lines signify hydrogen bonds.

(**c**) Hydrogen bonding between water molecules in ice. Below 0°C, each molecule becomes locked by four hydrogen bonds into a crystal lattice. In this bonding pattern, molecules are spaced farther apart than in liquid water at room temperature. When water is liquid, constant molecular motion usually prevents the maximum number of hydrogen bonds from forming.

the liquid water to the gaseous state. The extra energy increases the molecular motion so much that hydrogen bonds *stay* broken and molecules at the water's surface escape into the air. As molecules break free and depart in large numbers, they carry away some energy and lower the water's surface temperature.

Evaporative water loss also can help cool you (and some other mammals) when you work up a sweat on hot, dry days. Under such environmental conditions, sweat—which is 99 percent water—evaporates from your skin.

Below 0°C, hydrogen bonds resist breaking and lock water molecules in the latticelike bonding pattern of ice (Figure 2.9c). Ice is less dense than water. During winter freezes, ice sheets may form near the surface of ponds, lakes, and streams. Like a blanket, the ice "insulates" the liquid water beneath it—and so helps protect many fishes, frogs, turtles, and other water-dwelling organisms against freezing.

Figure 2.10 A water strider (*Gerris*). This long-legged bug feeds on lightweight insects that land or fall on the water's surface. Because of its fine, water-resistant leg hairs (and water's high surface tension), the bug scoots easily across the surface.

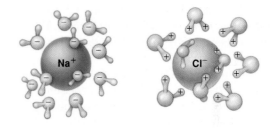

Figure 2.11 Spheres of hydration around two charged ions.

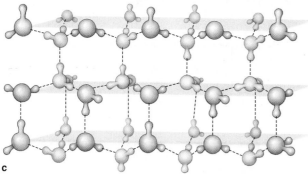

c

Water's Cohesion

Swimming on a hot summer night can be refreshing, if you don't mind all of the night-flying insects that hit the water's surface and float about on it. Because of the great numbers of hydrogen bonds among its molecules, liquid water shows cohesive behavior. **Cohesion** means something has the capacity to resist rupturing when it is placed under tension—that is, stretched—as by the weight of a bug's legs (Figure 2.10).

For any body of liquid water, the water molecules at or near the surface are continually being pulled inward by hydrogen bonding. This ongoing, collective bonding produces a high surface tension.

Like all organisms, land plants use water and the nutrients dissolved in it during growth and metabolism. Largely because of cohesion, water moves through pipelines that extend from roots to leaves, even in the tallest trees. As water evaporates from leaves, hydrogen bonds "pull" more water molecules up into leaf cells as replacements, in ways described in Chapter 23.

Water's Solvent Properties

Water is a terrific solvent, in that ions and polar molecules can readily dissolve in it. Dissolved substances are known as **solutes**. But what does "dissolved" mean? By way of example, pour some table salt into a glass of water. In time the salt crystals disappear; they separate into Na^+ and Cl^- ions. Each Na^+ attracts the negative end of water molecules, even as each Cl^- attracts the positive end of other water molecules. Many molecules of water cluster as "spheres of hydration" around ions and keep them dispersed in the fluid (Figure 2.11). *Any substance is "dissolved" when spheres of hydration form around its individual ions or molecules.* This happens to solutes in cells, body fluids, maple tree sap, and so on.

A water molecule has no net charge, yet it shows polarity.

Polarity allows water molecules to form hydrogen bonds with one another and with other polar (hydrophilic) substances. Water molecules repel nonpolar (hydrophobic) substances.

Water has temperature-stabilizing effects, internal cohesion, and a capacity to dissolve many substances. These properties influence the structure and functioning of organisms.

The pH Scale

The fluids inside and outside of cells contain a great variety of dissolved ions, which influence cell structure and function. Among the most influential are **hydrogen ions** (H^+), which are the same thing as free (unbound) protons. Hydrogen ions have amazingly widespread effects, largely because they are chemically active and there are so many of them.

At any given time in liquid water, a few molecules are breaking apart into hydrogen ions and **hydroxide ions** (OH^-). This ionization of water is the basis of the **pH scale**, shown in Figure 2.12. Biologists use the pH scale to measure the concentration of H^+ in blood, tree sap, or any other aqueous solution.

At pH 7, pure water always has just as many H^+ as OH^- ions. Whether for water or any other fluid, this condition represents *neutrality* on the pH scale, and it is assigned a value of 7. It is the midpoint of the pH scale, which ranges from 0 (highest H^+ concentration) to 14 (lowest concentration). *The greater the H^+ concentration, the lower the pH value.*

Starting at neutrality, each change by one unit of the pH scale corresponds to a tenfold increase or decrease in H^+ concentration. To get a sense of this logarithmic difference between units, dissolve baking soda (pH 9) on your tongue or taste egg white (pH 8). Next, sip pure water (pH 7), then vinegar (3) or lemon juice (2.3).

The fluid inside most of your cells hovers around pH 7. The pH of your blood and most tissue fluids ranges between 7.3 and 7.5. As another example, the pH of an aquatic habitat may be higher or lower than the pH inside the organisms living there.

Acids and Bases

Acids are any substances that release H^+ when they dissolve in water. **Bases** are substances that combine with those ions. *Acidic* solutions such as lemon juice have more H^+ than OH^- ions; their pH is below 7. *Basic* (or alkaline) solutions, such as egg whites, have fewer H^+ than OH^- ions; their pH is above 7.

Think of what happens when you sniff, chew, then swallow fried chicken and so send it on its way to the gastric fluid in your saclike stomach. As an outcome of your feeding activities, cells in the stomach's lining are stimulated to secrete hydrochloric acid (HCl), which separates into H^+ and Cl^-. These ions make the gastric fluid more acidic, and a good thing, too. Increased acidity switches on enzymes that can digest the chicken proteins. It also helps kill bacteria that may be lurking in or on the chicken.

If you eat too much of the fried chicken, you may end up with an "acid stomach." And you might reach

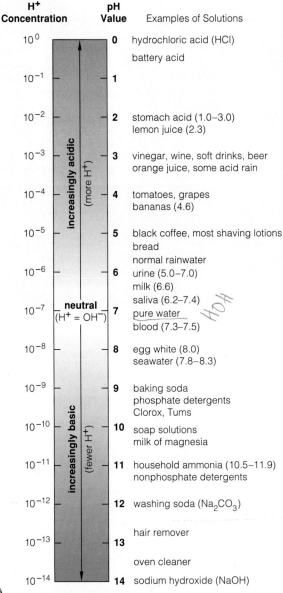

Figure 2.12 The pH scale, in which a liter of fluid is assigned a number according to the number of hydrogen ions in it. The scale ranges from 0 (most acidic) to 14 (most basic). A change of 1 on the scale means a tenfold change in H^+ concentration.

for an antacid tablet. Milk of magnesia is one kind of antacid. When dissolved, it releases magnesium ions and hydroxide ions. The hydroxide ions may *combine with* some of the excess hydrogen ions in your gastric fluid and raise its pH—and so help settle things down.

Certain airborne industrial wastes are acidic (Figure 2.13), and they alter the pH of rain. In turn, the pH of acid rainwater can alter the environments of organisms, in ways that will be considered in Chapter 39.

Figure 2.13 Sulfur dioxide emissions from a coal-burning power plant. Special camera filters revealed the otherwise invisible emissions. Along with other airborne pollutants, sulfur dioxides dissolve in atmospheric water to form acidic solutions. They are a major component of acid rain.

Salts

Acids commonly combine with bases. The results are ionic compounds called **salts**. Salts easily dissolve and form again, depending on pH. Consider how sodium chloride forms, then dissolves:

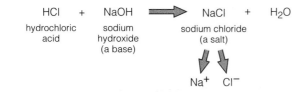

$$HCl + NaOH \longrightarrow NaCl + H_2O$$

hydrochloric acid | sodium hydroxide (a base) | sodium chloride (a salt)

$$Na^+ \quad Cl^-$$

Many other salts also dissolve into ions that serve key functions in cells. For example, many help maintain the body's acid-base balance, described next.

Buffers Against Shifts in pH

Metabolic reactions are sensitive to even slight shifts in pH—yet hydrogen ions continually enter or leave fluids in or around cells. Control mechanisms counter these potential disruptions and help maintain cellular pH.

In multicelled organisms, some controls maintain the pH of blood and tissue fluids, which together constitute the body's internal fluid environment. Later in the book, you will see how lungs and kidneys help maintain the acid-base balance of this internal environment, at levels suitable for life. For now, simply keep in mind that many control mechanisms involve buffer molecules.

A **buffer** is any molecule that can combine with hydrogen ions, release them, or both, and so help stabilize pH. Some weak acids and bases work as a buffer system. Consider how bicarbonate (HCO_3^-) helps restore pH when blood becomes too acidic. It combines with excess H^+ to form carbonic acid:

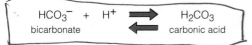

$$HCO_3^- + H^+ \rightleftharpoons H_2CO_3$$

bicarbonate | carbonic acid

When blood is not acidic enough, bicarbonate releases H^+. Its buffer action is vital. In one form of *acidosis*, the body can't eliminate carbon dioxide, so carbonic acid (and H^+) increases in blood. This abnormal condition makes breathing difficult and weakens the body.

Metabolism and the Cell's Fluid Environment

As a final example of how the composition of fluids can influence the structure and function of biological molecules, think about a protein dissolved in cellular fluid. Specific molecular regions at the protein's surface carry positive or negative charges, some strong, some weak. Those charged regions attract oppositely charged regions of neighboring water molecules. They attract neighboring dissolved ions as well, and these in turn attract *more* water molecules to the vicinity. In this way, an electrically charged "cushion" of ions and water molecules forms around the protein:

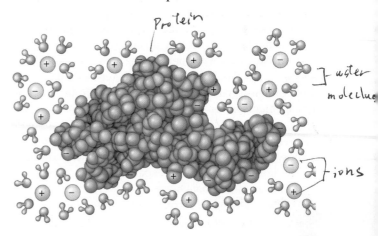

Through interactions with water and ions, the protein stays dispersed in the cellular fluid rather than settling any which way against some cell structure. Why is it important to prevent random settling? Many chemical reactions must proceed on specific molecular regions of proteins. *Cells must have access to those surfaces, for they are the stages for life-sustaining tasks.*

Ions dissolved in fluids inside and outside of cells affect the structure and behavior of biological molecules. The pH value of such a fluid refers to its concentration of hydrogen ions.

Buffer mechanisms counter slight shifts in pH by releasing hydrogen ions when their concentration is too low or by combining with them when the concentration is too high.

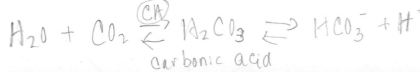

$$H_2O + CO_2 \underset{CA}{\rightleftharpoons} H_2CO_3 \rightleftharpoons HCO_3^- + H^+$$

carbonic acid

2.7 PROPERTIES OF ORGANIC COMPOUNDS

The molecules of life are **organic compounds**, meaning that they consist of atoms of carbon and one or more other elements, held together by covalent bonds. The term is a holdover from a time when chemists thought "organic" substances were the ones they obtained from animals and vegetables, as opposed to the "inorganic" substances obtained from minerals. The term persists even though researchers now synthesize organic compounds in laboratories. And it persists even though there are reasons to believe that organic compounds were present on the Earth *before* organisms were.

Effects of Carbon's Bonding Behavior

By far, organisms consist mainly of oxygen, hydrogen, and carbon (Figure 2.2). Much of the oxygen and the hydrogen is in the form of water. Remove the water, and carbon makes up more than half of what's left.

Carbon's importance in life arises from its versatile bonding behavior. *Each carbon atom can share pairs of electrons with as many as four other atoms.* Each covalent bond formed this way is quite stable. Such bonds link carbon atoms together in chains and rings, and these serve as a backbone to which hydrogen, oxygen, and other elements become attached (Figure 2.14).

The flattened structural formulas shown in Figure 2.14 might lead you to believe that the molecules they represent are flat, like paper dolls. Yet molecules have three-dimensional shapes, which arise from bonding arrangements in the carbon backbone. In that backbone, some carbon atoms may rotate freely around a single covalent bond. Where a *double* covalent bond joins two carbon atoms, rotation is restricted, so the two atoms rigidly hold their position in space.

Imagine a long backbone, with some atoms locked in position and others rotating at different angles in space. The resulting orientations of atoms attached to the backbone may promote or discourage interactions with other substances in the vicinity. As you will see shortly, *such interactions give rise to the three-dimensional shapes and functions of biological molecules.*

Hydrocarbons and Functional Groups

A "hydrocarbon" has only hydrogen atoms attached to a carbon backbone. Hydrocarbons do not break apart very easily, and they form the stable portions of most biological molecules.

Biological molecules also incorporate some number of **functional groups**. These are single atoms or clusters of atoms covalently bonded to the carbon backbone. To get a sense of their importance, consider a few of the functional groups shown in Figure 2.15. Sugars and other organic compounds that are classified as **alcohols**

Figure 2.14 Carbon compounds. There is a Tinkertoy quality to carbon compounds, for a single atom can be the start of diverse molecules assembled from "straight-stick" covalent bonds. Start with the simplest hydrocarbon, methane (CH_4). Stripping one hydrogen from methane leaves a methyl group, which occurs in fats, oils, and waxes:

Imagine stripping a hydrogen atom from each of two methane molecules, then bonding the two molecules together. If the resulting structure were to lose a hydrogen atom, you would end up with an ethyl group:

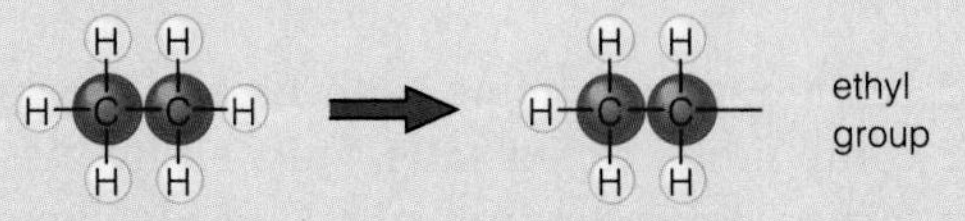

You could go on building a continuous chain:

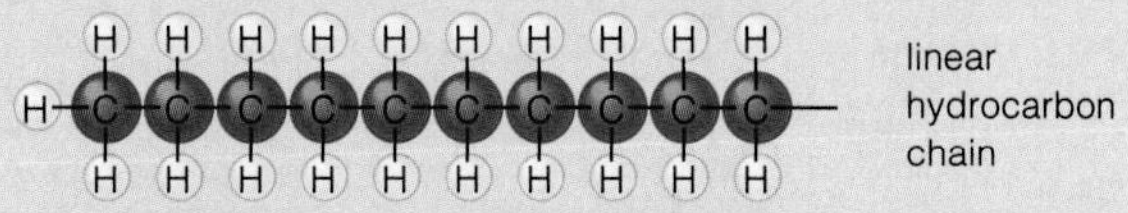

And you could add branches to the chain:

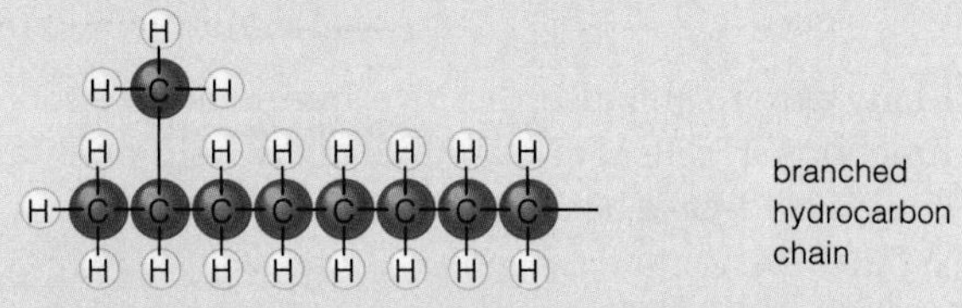

You might have the chain coil back on itself into a ring, which you could diagram in various ways:

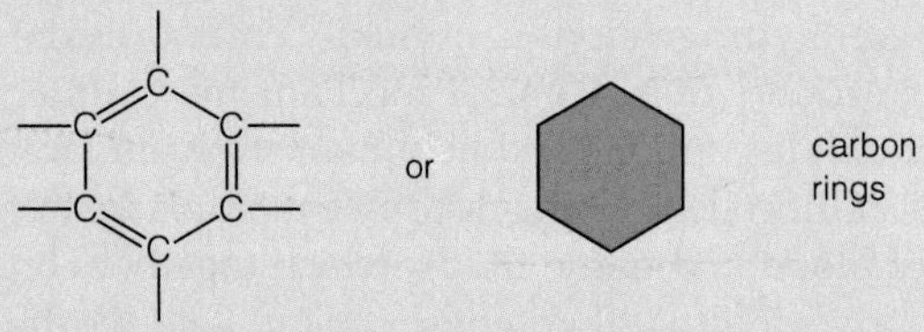

You might even join many carbon rings together and so produce larger molecules:

have one or more hydroxyl groups (—OH). Alcohols are quick to dissolve in water, because water molecules can readily form hydrogen bonds with —OH groups. Or consider the proteins. They have a backbone that forms by repeated reactions between many amino groups and carboxyl groups. And the particular bonding patterns associated with this backbone contribute to the three-dimensional structure of proteins. Amino groups also can combine with H^+ and so act as buffers against decreases in pH.

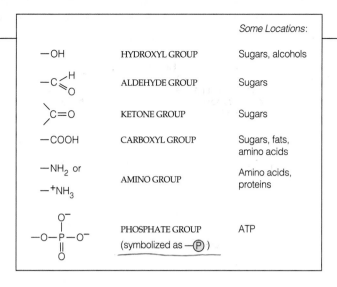

		Some Locations:
—OH	HYDROXYL GROUP	Sugars, alcohols
—C⟨H,O (ALDEHYDE)	ALDEHYDE GROUP	Sugars
>C=O	KETONE GROUP	Sugars
—COOH	CARBOXYL GROUP	Sugars, fats, amino acids
—NH₂ or —⁺NH₃	AMINO GROUP	Amino acids, proteins
—O—P—O⁻ (with O⁻ and O)	PHOSPHATE GROUP (symbolized as —Ⓟ)	ATP

Figure 2.15 A few examples of functional groups.

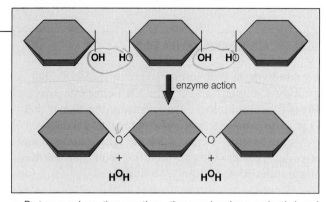

a By two condensation reactions, three molecules covalently bond into a larger molecule; and two water molecules typically form.

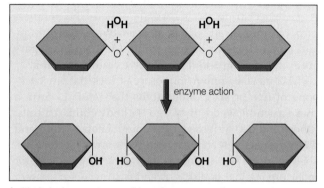

b Hydrolysis, a water-requiring cleavage reaction.

Figure 2.16 Examples of reactions by which most biological molecules are put together, rearranged, and broken apart.

How Cells Use Organic Compounds— Five Classes of Reactions

Enzymes, a class of proteins, speed specific metabolic reactions. We will study enzymes in later chapters. For now, it is enough to know they mediate five categories of reactions by which most biological molecules are assembled, rearranged, and broken apart:

1. **Functional-group transfer.** One molecule gives up a functional group, which another molecule accepts.

2. **Electron transfer.** One or more electrons stripped from one molecule are donated to another molecule.

3. **Rearrangement.** A juggling of internal bonds converts one type of organic compound into another.

4. **Condensation.** Through covalent bonding, two molecules combine to form a larger molecule.

5. **Cleavage.** A molecule splits into two smaller ones.

To get a sense of what goes on, consider just two of these events. In many **condensation** reactions, enzymes remove a hydroxyl group from one molecule and an H atom from another, then a covalent bond forms between the two molecules at the exposed sites (Figure 2.16a). The discarded atoms (now H^+ and OH^- ions) may combine to form a water molecule. Cells assemble starch and other polymers by repeated condensation reactions. A "polymer" is a large molecule composed of three to millions of subunits, which may or may not be identical. The individual subunits are sometimes called "monomers," as in the sugar monomers of starch.

One type of cleavage reaction, **hydrolysis**, is like condensation in reverse (Figure 2.16b). Enzyme action breaks covalent bonds at functional groups and splits a molecule into two or more parts. At the same time, —H and —OH derived from a water molecule are attached to the exposed sites. Cells commonly hydrolyze starch and other polymers, then use the released subunits as building blocks or energy sources.

The Molecules of Life

Under present-day conditions on Earth, only living cells can synthesize carbohydrates, lipids, proteins, and nucleic acids. *These are the molecules characteristic of life.* Different kinds function as packets of instant energy, energy stores, structural materials, metabolic workers, and libraries of hereditary information.

Some biological molecules are rather small organic compounds. The simple sugars, fatty acids, amino acids, and nucleotides are like this. As you will see next, cells use some assortment of these compounds as subunits for the synthesis of complex carbohydrates and other large biological molecules.

The diverse three-dimensional shapes and functions of organic compounds begin with flexible and rigid bonding arrangements in their carbon backbones. Functional groups covalently bonded to the backbone add to the diversity.

Carbohydrates, lipids, proteins, and nucleic acids are the organic compounds that are characteristic of life.

2.8 CARBOHYDRATES

A **carbohydrate** is a simple sugar or molecule that is composed of many sugar units. Carbohydrates are the most abundant biological molecules. All cells use them as structural materials, transportable packets of energy, and stored forms of energy. We recognize three classes of carbohydrates. We call them the **monosaccharides**, **oligosaccharides**, and **polysaccharides**.

Monosaccharides

"Saccharide" comes from a Greek word meaning sugar. A *monosaccharide*, or one sugar unit, is the simplest carbohydrate of all. It has at least two —OH groups and an aldehyde or ketone group. Most simple sugars are sweet tasting, and they dissolve readily in water.

The most common monosaccharides have a backbone of five or six carbon atoms that tends to form a ring structure when dissolved in body fluids or cells. Ribose and deoxyribose (sugar components of RNA and DNA, respectively) have five carbon atoms. Glucose has six (Figure 2.17a). Glucose is the main energy source for most organisms. It also serves as a precursor (parent molecule) of many compounds and as a building block for larger carbohydrates. Vitamin C (a sugar acid) and glycerol (an alcohol with three hydroxyl groups) are other examples of compounds that are derived from sugar monomers.

Oligosaccharides

An *oligo*saccharide is a short chain of two or more covalently bonded sugar units. (*Oligo-* means a few.) The simplest kinds, the *disaccharides*, have two sugar units. Lactose, sucrose, and maltose are examples.

Lactose (a glucose and a galactose unit) is present in milk. Sucrose, the most plentiful sugar in nature, consists of a glucose and a fructose unit (Figure 2.17b). Plants continually convert their carbohydrate stores to sucrose, which is easily transported throughout leaves, stems, and roots. Crystallizing sucrose extracts from sugarcane and sugar beets gives us our table sugar. Proteins and other large molecules often have oligosaccharides attached to their carbon backbone, as side chains. Some chains have roles in membrane function, others have roles in the responses to agents of disease.

Polysaccharides

A *poly*saccharide is a straight or branched chain of many sugar units—often hundreds or thousands of the same or different kinds. The most common—starch, cellulose, and glycogen—consist only of glucose.

Plant cells store sugars in the form of large starch molecules, which enzymes readily hydrolyze back to

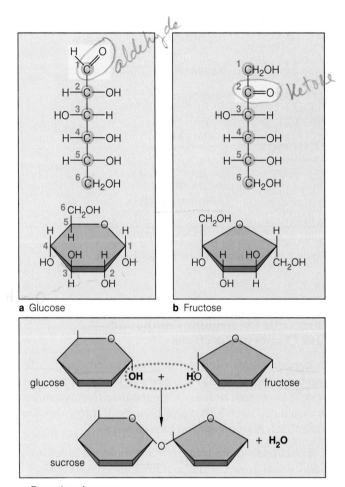

a Glucose **b** Fructose

c Formation of sucrose

Figure 2.17 Straight-chain and ring forms of glucose (**a**) and fructose (**b**). For reference purposes, the carbon atoms of sugar molecules are often numbered in sequence, starting at the end closest to the aldehyde or ketone group. (**c**) Condensation of two monosaccharides into a disaccharide.

sugar units. They use cellulose as a structural material in cell walls. Fibers of cellulose molecules are tough and insoluble. Like steel rods in reinforced concrete, the fibers withstand considerable weight and stress.

If both cellulose and starch consist of glucose, why do they have such different properties? The answer lies with differences in the patterns of covalent bonding between adjacent glucose units. Both substances have chains of glucose units. In starch, a flexible bonding arrangement allows the chains to twist into a coil, with many —OH groups facing outward. In this pattern, bonds are oriented in positions that are accessible to enzymes (Figure 2.18a). In cellulose, the chains stretch out, side by side, and hydrogen-bond to one another at —OH groups (Figure 2.18c). In this way, the chains become stabilized in a tight, bundled bonding pattern that resists digestion, at least by most enzymes.

Figure 2.18 Starch and cellulose compared. (**a**) Structure of amylose, a readily soluble form of starch that is composed of glucose units. (**b**) Bonding pattern between adjacent glucose units. These particular linkages cause chains of glucose to coil. Coiling orients the linkages in such a way that they are accessible to the enzymes of hydrolysis.

(**c**) Structure of cellulose, which is an insoluble structural material composed of glucose units. Very few organisms have enzymes that are able to digest it. (**d**) Bonding pattern between the adjacent glucose units. (**e**) In fibers of cellulose, chains of glucose units stretch out, side by side, and hydrogen-bond at —OH groups. The bonds stabilize the chains in tight bundles that are organized into increasingly larger fibers. (**f**) Diagram showing the thick cellulose fibers in the cell wall of *Cladophora*, a green alga.

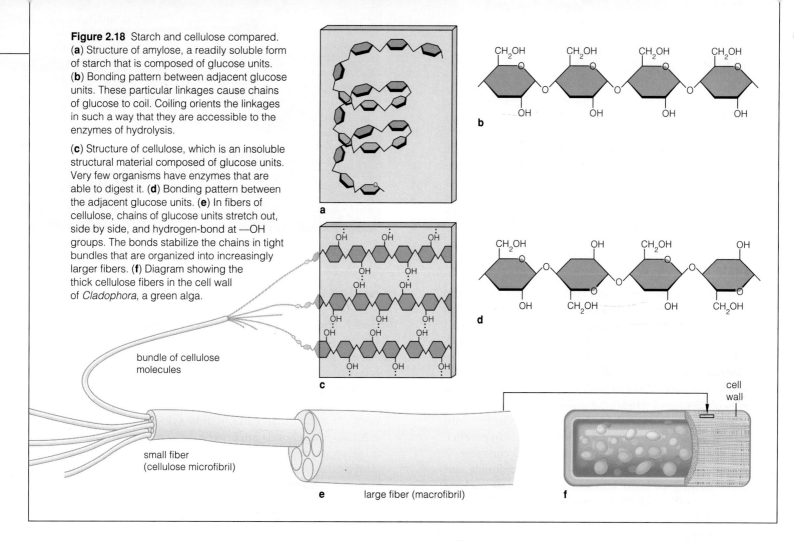

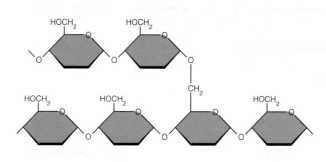

Figure 2.19 Branching structure of glycogen. Like cellulose and starch, glycogen is composed of glucose units.

Figure 2.20 Scanning electron micrograph of a tick. A protective, chitin-reinforced cuticle serves as its body covering.

Like starch in plants, glycogen is a sugar storage form in animals—notably so in liver and muscle tissues. When blood sugar levels fall, liver cells break down glycogen and release glucose units to the blood. During exercise, muscle cells tap into their glycogen stores for quick access to energy. Figure 2.19 shows a few of glycogen's numerous branchings.

Chitin, another polysaccharide, has nitrogen atoms attached to the backbone. Chitin is the main structural material in the external skeletons and other hard body parts of many animals, including crabs, insects, and ticks (Figure 2.20). It also is the main structural material in the cell walls of many species of fungi.

Carbohydrates are either simple sugars (such as glucose) or molecules composed of many sugar units (such as starch).

All cells require carbohydrates as structural materials, stored forms of energy, and transportable packets of energy.

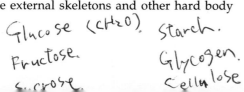

2.9 LIPIDS

The greasy or oily compounds called **lipids** dissolve readily in one another, but they show very little tendency to dissolve in water. In nearly all organisms, certain lipids are the main reservoirs of stored energy. Other lipids are structural materials in cell components (such as membranes) and products (such as surface coatings). Here, we consider the neutral fats, phospholipids, and waxes, all of which have fatty acid components. We also consider the sterols, each with a backbone of four carbon rings.

Fatty Acids

The lipids called **fatty acids** have a backbone of up to thirty-six carbon atoms, a carboxyl group (—COOH) at one end, and hydrogen atoms occupying most or all of the remaining bonding sites. When combined with other molecules, fatty acids typically stretch out, like flexible tails. Tails with one or more double bonds in their backbone are *unsaturated*. Those with single bonds only are *saturated*. Figure 2.21a shows examples.

When many saturated fatty acid tails occur in a substance, weak attractions make them snuggle in parallel, and this gives the substance a rather solid consistency. By contrast, in unsaturated fatty acids, the double and triple bonds put rigid kinks in the tails. The packing arrays are less stable and impart fluidity to substances.

Figure 2.22 Triglyceride-protected penguins taking the plunge.

Neutral Fats (Triglycerides)

Butter, lard, and oils are examples of **triglycerides**, or neutral fats—the body's most abundant lipids and its richest source of energy. These lipids have fatty acid tails attached to a backbone of glycerol (Figure 2.21b). Gram for gram, triglycerides yield more than twice as much energy as carbohydrates. They have far more covalent bonds than carbohydrates do—and energy is released when bonds are broken. In vertebrates, cells of adipose tissue store quantities of triglycerides as fat droplets. A thick layer of triglycerides also insulates penguins and some other animals against near-freezing environmental temperatures (Figure 2.22).

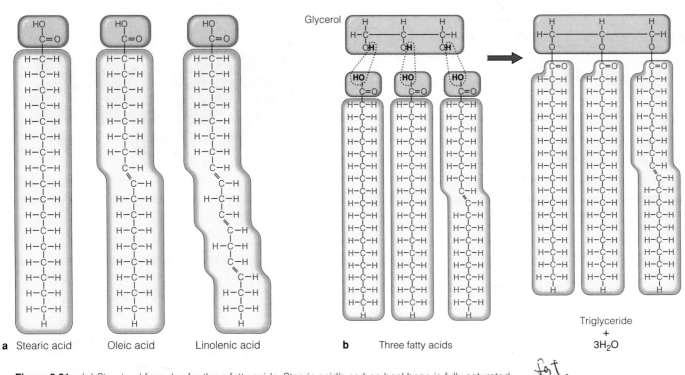

a Stearic acid   Oleic acid   Linolenic acid

Glycerol

b Three fatty acids

Triglyceride
+
3H₂O

Figure 2.21 (**a**) Structural formulas for three fatty acids. Stearic acid's carbon backbone is fully saturated with hydrogen atoms. Oleic acid, with a double bond in its backbone, is unsaturated. Linolenic acid, with three double bonds, is a "polyunsaturated" fatty acid. (**b**) Condensation of fatty acids into a triglyceride.

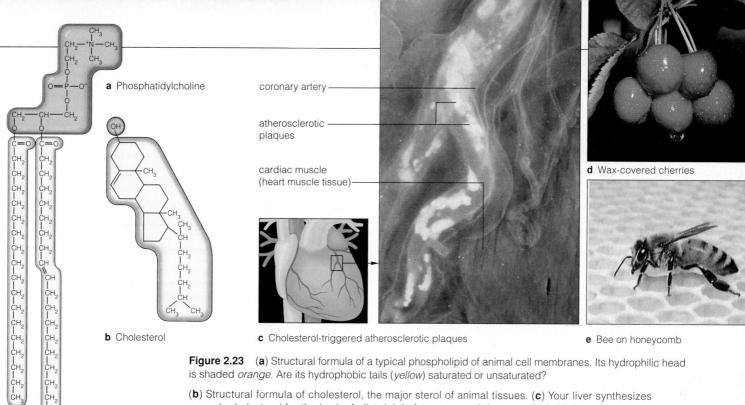

a Phosphatidylcholine

b Cholesterol

coronary artery

atherosclerotic plaques

cardiac muscle (heart muscle tissue)

c Cholesterol-triggered atherosclerotic plaques

d Wax-covered cherries

e Bee on honeycomb

Figure 2.23 (**a**) Structural formula of a typical phospholipid of animal cell membranes. Its hydrophilic head is shaded *orange*. Are its hydrophobic tails (*yellow*) saturated or unsaturated?

(**b**) Structural formula of cholesterol, the major sterol of animal tissues. (**c**) Your liver synthesizes enough cholesterol for the body. A diet rich in fats may result in excessively high levels of cholesterol in the blood and the formation of abnormal masses of material in blood vessels called arteries. These *atherosclerotic plaques* may clog arteries that deliver blood to the heart. We return to this topic in the introduction to Chapter 13.

(**d**) Demonstration of the water-repelling attribute of the cherry cuticle. (**e**) Honeycomb—a structural material constructed of a firm, water-repellent, waxy secretion called beeswax.

Phospholipids

— *cell membranes.*

A **phospholipid** has a backbone of glycerol, two fatty acid tails, and a hydrophilic "head" that includes a phosphate group (Figure 2.23*a*). Phospholipids are the main components of cell membranes, which have two lipid layers. The heads of one layer are dissolved in the cell's fluid interior. The heads of the other layer are dissolved in the surroundings. All fatty acid tails, which are hydrophobic, are sandwiched between the two.

Sterols and Their Derivatives

Sterols are among the lipids that have no fatty acid tails. Sterols differ in the number, position, and type of their functional groups, but they all have a rigid backbone of four fused-together carbon rings:

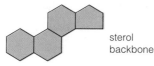

sterol backbone

Sterols occur in eukaryotic cell membranes. Cholesterol is the major type in animal tissues (Figure 2.23*b* and *c*). Cells also remodel cholesterol. Molecules derived from cholesterol include vitamin D (required for good bones and teeth), steroid hormones (such as estrogen and testosterone, which govern sexual traits and gamete formation), and bile salts (which assist fat digestion in the small intestine).

Waxes

The lipids called **waxes** have long-chain fatty acids, tightly packed and linked to long-chain alcohols or to carbon rings. Waxes have a firm consistency and repel water. Waxes and another lipid, cutin, make up plant cuticles. These coverings on aboveground parts help the plants conserve water and serve as barriers to some parasites. A waxy cherry cuticle is an example (Figure 2.23*d*). In many animals, waxy secretions from cells become incorporated into coatings that keep skin or hair protected, lubricated, and pliable. In waterfowl and other birds, wax secretions help keep feathers dry. Bees construct honeycombs of beeswax (Figure 2.23*e*).

Lipids (hydrophobic, greasy or oily compounds) include:

- **Neutral fats (triglycerides), major reservoirs of energy.**
- **Phospholipids, the main components of cell membranes.**
- **Sterols (such as cholesterol), membrane components and precursors of steroid hormones and other vital molecules.**
- **Waxes, firm yet pliable components of water-repelling and lubricating substances.**

2.10 AMINO ACIDS AND PROTEINS

Of all the large biological molecules, **proteins** are by far the most diverse. The ones called enzymes make metabolic reactions proceed much faster than they otherwise would. Structural types are the stuff of bone and cartilage, feathers and webs, and a dizzying array of other biological parts and products. Transport proteins help move substances across cell membranes or through body fluids. Nutritious proteins abound in milk, eggs, and a variety of seeds. Regulatory types, including the protein hormones, function as signals for change in cell activities. Many proteins function as chemical weapons against disease-causing bacteria and other invaders. Amazingly, cells build all of these diverse proteins from their pools of only twenty or so kinds of amino acids.

Each **amino acid** is a small organic compound with an amino group, a carboxyl group (an acid), a hydrogen atom, and one or more atoms called its *R* group. These parts are covalently bonded to the same carbon atom:

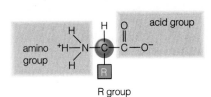

R group

Figure 2.24 shows some of the amino acids that we will consider later in the book.

Primary Structure of Proteins

When cells synthesize proteins, amino acids become linked, one after the other, by *peptide* bonds. As Figure 2.25 shows, this type of covalent bond forms between the amino group of one amino acid and the acid group (carboxyl) of another. Three or more amino acids joined this way form a **polypeptide chain.** The backbone of such chains incorporates nitrogen atoms in this regular pattern: —N—C—C—N—C—C—.

For each particular kind of protein, different amino acid units are selected in a prescribed order, one at a time, from the twenty kinds available. Overall, the resulting sequence of amino acids is unique for each kind of protein and represents its *primary* structure.

Three-Dimensional Structure of Proteins

Cells make thousands of different proteins. Many are fibrous, with polypeptide chains organized as strands or sheets. Collectively, many such molecules contribute to the shape, internal organization, and movement of cells. Many other proteins are globular, with one or more polypeptide chains folded into compact, rounded shapes. Most enzymes are globular proteins.

Each protein's shape and function arise from its primary structure—that is, from information built into its amino acid sequence. That information dictates which

Figure 2.24 Structural formulas for nine of the twenty common amino acids. The *green* boxes highlight the functional groups (which are designated R groups) that give each type of amino acid its distinctive properties.

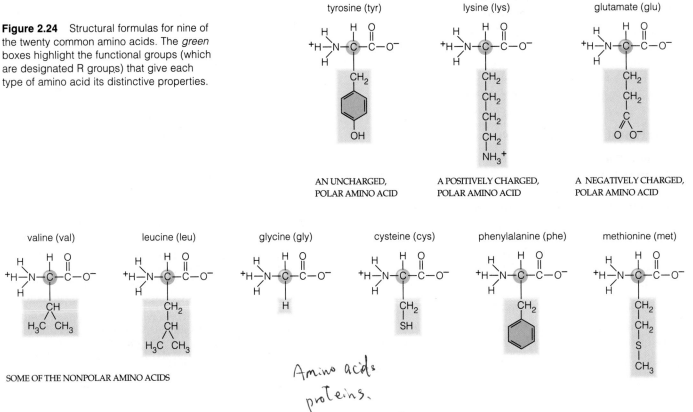

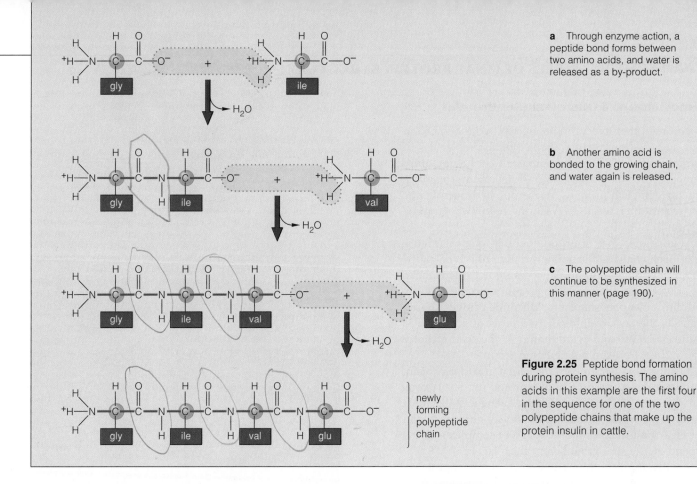

a Through enzyme action, a peptide bond forms between two amino acids, and water is released as a by-product.

b Another amino acid is bonded to the growing chain, and water again is released.

c The polypeptide chain will continue to be synthesized in this manner (page 190).

Figure 2.25 Peptide bond formation during protein synthesis. The amino acids in this example are the first four in the sequence for one of the two polypeptide chains that make up the protein insulin in cattle.

parts of a polypeptide chain will coil, bend, or even interact with other chains nearby. And the type and arrangement of atoms in the coiled, stretched-out, or folded regions determine whether a protein will act as, say, an enzyme, a receptor, or an inadvertent target for some bacterium or virus. In such ways, *a protein's primary structure dictates its final structure and chemical behavior.*

How does primary structure give rise to a protein's shape? *First*, it allows hydrogen bonds to form between different amino acids in a polypeptide chain. *Second*, it puts R groups in positions that force them to interact. The interactions force the chain to bend and twist.

Think of a chain as a set of dominoes joined by links that can swivel a bit. Each "domino" is a peptide group (Figure 2.26*a*). Atoms on either side of such groups can rotate somewhat around their covalent bonds *and form bonds with neighboring atoms*. In many chains, hydrogen bonds form between every third amino acid. The bonding pattern forces the peptide groups to coil helically—that is, like a spiral staircase (Figure 2.26*b*). In other cases, the hydrogen-bonding pattern holds two or more chains side by side, in sheetlike array (Figure 2.26*c*).

Thus, proteins have *secondary* structure—a coiled or extended pattern, as brought about by hydrogen bonds at regular intervals along their polypeptide chains. Most coiled chains undergo further folding. They do so when certain R groups interact with other R groups

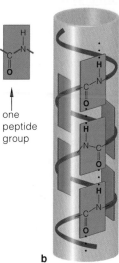

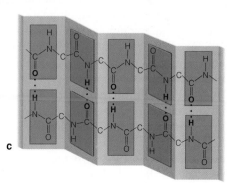

a

one peptide group

b

c

Figure 2.26 Major bonding patterns between the peptide groups (**a**) of polypeptide chains. Hydrogen bonding (*dotted lines*) can bring about coiling (**b**) or sheetlike formations (**c**).

some distance away, with the backbone of the chain, or with cell substances. The folding that arises through interactions among R groups of a polypeptide chain is the *tertiary* structure of proteins. The next section gives examples of the resulting three-dimensional shapes.

A protein consists of one or more chains of amino acids. The amino acid sequence (which kind follows another) is unique for each kind of protein and gives rise to its unique structure and function.

2.11 SOME EXAMPLES OF FINAL PROTEIN STRUCTURE

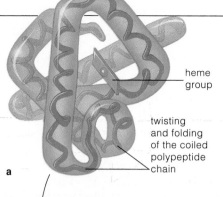

heme group

twisting and folding of the coiled polypeptide chain

a

Hemoglobin and Other Complex Proteins

As you read this, every single, mature red blood cell in your body is transporting about a billion molecules of oxygen, bound to 250 million molecules of **hemoglobin**. Hemoglobin is an example of a protein with *quaternary* structure. In such proteins, numerous hydrogen bonds and some other weak interactions hold two or more polypeptide chains together tightly.

Hemoglobin has four polypeptide chains (Figure 2.27). Each chain is a separate protein, of a type called globin. Hydrogen bonding and disulfide bridges make the polypeptide chain of each globin molecule coil and fold into a compact shape. (A "disulfide bridge" simply connects the sulfur atoms of two units of a particular amino acid, cysteine, in the chain.) At the chain's center is a ringlike, iron-containing heme group.

Hemoglobin is a fine example of a globular protein. Keratin, a structural protein of hair and fur, is a fibrous one (Section 26.1). So is collagen, which is the most common protein in animals. Skin, bones, corneas, blood vessels—these and other body parts of complex animals depend on the strength inherent in collagen.

Finally, think of the lipoproteins and glycoproteins, which we consider later in the book. **Lipoproteins** form when certain proteins that circulate freely in the blood encounter and combine with cholesterol, triglycerides, and phospholipids that were absorbed from the gut, after a meal. Lipoproteins differ from one another in their combinations of protein and lipid components.

Most **glycoproteins** form when new polypeptide chains are undergoing modification into mature proteins. These proteins have oligosaccharides covalently bonded to them. Some of the attached oligosaccharides are linear chains; others are branched. Nearly all of the proteins on the outer surface of animal cells are glycoproteins. So are most protein secretions from cells and most proteins in blood.

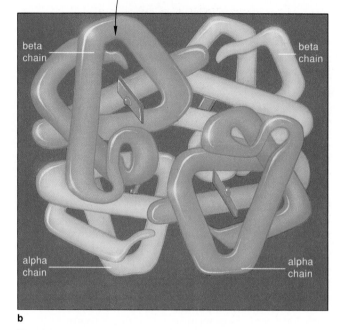

beta chain

beta chain

alpha chain

alpha chain

b

Figure 2.27 (**a**) A globin molecule. It is a single polypeptide chain (the *green* "ribbon") with a heme group. Oxygen is strongly attracted to this iron-containing group. There is a whole family of globin molecules. They are identical in critical regions of their amino acid sequences but they differ in other regions.

(**b**) Hemoglobin, an oxygen-transporting protein in blood, is composed of four globin molecules. To keep the diagram from looking like tangled noodles, the two chains in the foreground are shaded differently from the two in the background.

Denaturation and Protein Structure

Breaking the weak bonds of a protein or any other large molecule disrupts its three-dimensional shape. We call this event **denaturation**. For example, individually, hydrogen bonds are weak—and they are sensitive to increases or decreases in temperature and pH. If the temperature or pH exceeds a given protein's range of tolerance, then its polypeptide chains will unwind or change shape, and the protein will no longer function.

Consider the protein albumin, concentrated in the "egg white" of uncooked chicken eggs. When you cook eggs, the heat doesn't disrupt the strong covalent bonds of albumin's primary structure. But it destroys weaker bonds contributing to the three-dimensional shape. For

some proteins, denaturation can be reversed when normal conditions are restored—but albumin isn't one of them. There is no way to uncook a cooked egg.

The amino acid sequence of a polypeptide chain represents the primary structure of proteins.

Hydrogen bonds at regular intervals in the sequence force polypeptide chains to coil or extend, sheetlike. In many chains, R group interactions bring about even more folding.

Many proteins incorporate two or more polypeptide chains, held together by numerous hydrogen bonds.

2.12 NUCLEOTIDES AND NUCLEIC ACIDS

Nucleotides With Roles in Metabolism

核酸：細胞中の重要な
生体高分子

The small organic compounds called **nucleotides** have three components: a five-carbon sugar, a phosphate group, and a nitrogen-containing base. The sugar is either ribose or deoxyribose. The base has a single- or double-carbon ring structure (Figure 2.28).

One nucleotide is central to metabolism. We call it **ATP** (adenosine triphosphate). ATP can deliver energy from one reaction site to virtually any other reaction site in cells. Other nucleotides are subunits of enzyme helpers called **coenzymes**. They accept hydrogen atoms and electrons that are being stripped from molecules and transfer them elsewhere. Two major coenzymes are NAD^+ (nicotinamide adenine dinucleotide) and FAD (flavin adenine dinucleotide). Still other nucleotides function as chemical messengers within and between cells. You will encounter one of these (cyclic adenosine monophosphate, or cAMP) later in the book.

Nucleic Acids—DNA and RNA

In **nucleic acids**, four different kinds of nucleotides are bonded together in large single- or double-stranded molecules. Each strand's backbone consists of sugars covalently bonded to the phosphate groups of adjacent nucleotides. The nucleotide bases stick out to the side (Figure 2.28). In at least some regions, the sequence of particular bases is unique to each kind of nucleic acid.

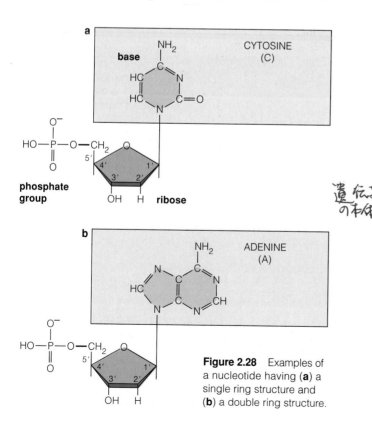

Figure 2.28 Examples of a nucleotide having (**a**) a single ring structure and (**b**) a double ring structure.

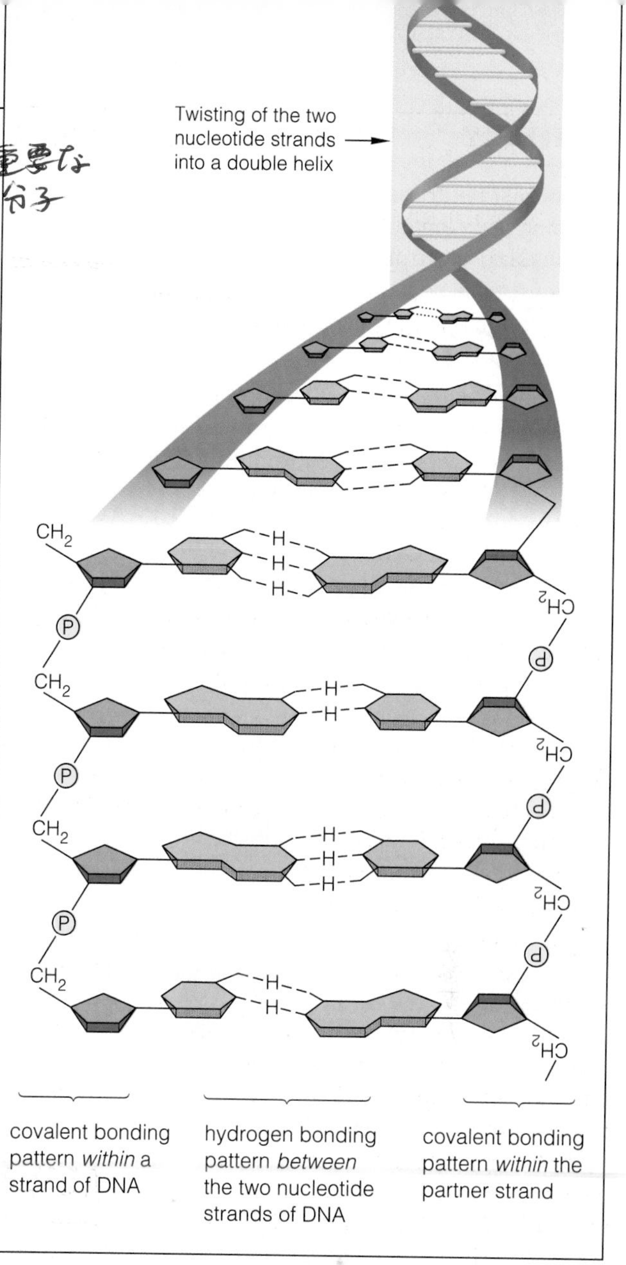

遺伝子
の本体

Twisting of the two nucleotide strands into a double helix →

covalent bonding pattern *within* a strand of DNA

hydrogen bonding pattern *between* the two nucleotide strands of DNA

covalent bonding pattern *within* the partner strand

Figure 2.29 Bonding patterns in a DNA molecule. Hydrogen bonds connect bases of one strand with bases of the other strand.

Heritable (genetic) information is encoded in such base sequences.

リボ核酸：タンパク質の合成

You have probably heard of **RNA** (ribonucleic acid) as well as **DNA** (deoxyribonucleic acid). RNA is a single nucleotide strand, most often. DNA is usually a double-stranded molecule that twists helically, like a spiral staircase, with hydrogen bonds holding the two strands together. Figure 2.29 shows this helical configuration. You will be reading more about RNA and DNA in later chapters. For now, it is enough to know this:

Genetic information is encoded in the particular order of nucleotide bases that follow one another in DNA.

RNA molecules function in the processes by which the genetic information in DNA is used to build proteins.

SUMMARY

1. Protons, neutrons, and electrons are building blocks of atoms. An element's atoms have the same number of protons and electrons but may vary in the number of neutrons. (The variant forms are isotopes.) Whether an atom interacts with others depends on the number and arrangement of its electrons.

2. By one model, electrons occupy orbitals (volumes of space) inside a series of shells around the nucleus of an atom. Atoms with one or more unfilled orbitals in their outermost shell tend to bond with other elements.

3. Atoms (which have no net charge) may gain or lose one or more electrons and become ions, with an overall positive or negative charge.

4. Hydrogen, oxygen, carbon, and nitrogen are the most abundant elements in organisms.

5. In ionic bonds, a positive ion and a negative ion remain together simply by the mutual attraction of their opposite charges. In covalent bonds, atoms share one or more electrons. In hydrogen bonds, a small, electronegative atom weakly interacts with a hydrogen atom that is already taking part in a polar covalent bond in the same molecule or in a different one.

Table 2.2 Summary of the Main Carbon Compounds in Living Things

Category	Main Subcategories	Some Examples and Their Functions	
CARBOHYDRATES *contain an aldehyde or a ketone group, and one or more hydroxyl groups*	**Monosaccharides** (simple sugars)	Glucose	Energy source
	Oligosaccharides	Sucrose (a disaccharide)	Form of sugar transported in plants
	Polysaccharides (complex carbohydrates)	Starch Cellulose	Energy storage Structural roles
LIPIDS *are largely hydrocarbon; generally do not dissolve in water but dissolve in nonpolar substances*	**Lipids with fatty acids:**		
	Glycerides: one, two, or three fatty acid tails attached to glycerol backbone	Fats (e.g., butter) Oils (e.g., corn oil)	Energy storage
	Phospholipids: phosphate group, another polar group, and (often) two fatty acids attached to glycerol backbone	Phosphatidylcholine	Key component of cell membranes
	Waxes: long-chain fatty acid tails attached to alcohol	Waxes in cutin	Water retention by plants
	Lipids with no fatty acids:		
	Sterols: four carbon rings; the number, position, and type of functional groups vary	Cholesterol	Component of animal cell membranes; can be converted into a variety of steroid molecules; precursor of vitamin D
PROTEINS *are polypeptides (up to several thousand amino acids, covalently linked)*	**Fibrous proteins:**		
	Individual polypeptide chains, often linked into tough, water-insoluble molecules	Keratin Collagen	Structural element of hair, nails Structural element of bones and cartilage
	Globular proteins:		
	One or more polypeptide chains folded and linked into globular shapes; many roles in cell activities	Enzymes Hemoglobin Insulin Antibodies	Increase in rates of reactions Oxygen transport Control of glucose metabolism Tissue defense
NUCLEIC ACIDS (AND NUCLEOTIDES) *are chains of units (or individual units) that each consist of a five-carbon sugar, phosphate, and a nitrogen-containing base*	**Adenosine phosphates**	ATP	Energy carrier
	Nucleotide coenzymes	NAD^+, $NADP^+$	Transport of protons (H^+) and electrons from one reaction site to another
	Nucleic acids: Chains of thousands to millions of nucleotides	DNA, RNAs	Storage, transmission, translation of genetic information

6. A pH value refers to the concentration of hydrogen ions in some fluid. Acidic substances release H⁺, bases combine with them. At pH 7, H⁺ and OH⁻ concentrations are equal. Buffering mechanisms maintain pH values of blood, tissue fluids, and the fluid inside cells.

7. Water is crucial for the metabolic activity, shape, and internal organization of cells. Because of hydrogen bonds between its polar molecules, water resists temperature changes, shows internal cohesion, and shows a remarkable capacity to dissolve other polar substances.

8. A carbon atom forms up to four covalent bonds with other atoms. Carbon atoms bonded together in linear or ring structures are the backbone of organic compounds.

9. Cells assemble, rearrange, and break apart most organic compounds by five kinds of enzyme-mediated reactions: functional-group transfers, electron transfers, internal rearrangements, condensation reactions, and cleavages (including hydrolysis).

10. Cells have pools of dissolved sugars, fatty acids, amino acids, and nucleotides, all of which are small organic compounds that include no more than twenty or so carbon atoms. They are building blocks for the large biological molecules—the complex carbohydrates, lipids, proteins, and nucleic acids. Table 2.2 summarizes the main features of these molecules.

Review Questions

1. Define element, atom, molecule, and compound. What are the six main elements (and their symbols) in most organisms? *16, 18, 21*

2. Define covalent, ionic, and hydrogen bonds. Explain the difference between a hydrophilic and a hydrophobic interaction. *22–24*

3. Explain how the amino acid sequence of one or more polypeptide chains gives rise to the final structure of proteins. *35, 36*

4. Identify the carbohydrate, fatty acid, amino acid, and polypeptide in the following list: *30, 32, 34*

  a. $^+NH_3$—CHR—COO⁻   c. (glycine)$_{20}$

  b. $C_6H_{12}O_6$ d. $CH_3(CH_2)_{16}COOH$

Self-Quiz *(Answers in Appendix IV)*

1. An _____ has a net charge of zero; an _____ has gained or lost one or more electrons, so becoming negatively or positively charged.
  a. ion; ion   c. atom; atom
  b. ion; atom   d. atom; ion

2. Orbitals within shells around the nucleus of an atom can each hold no more than _____ electrons.
  a. one   b. two   c. three   d. four

3. Electrons are shared unequally in a(n) _____ bond.
  a. nonpolar covalent   c. hydrogen
  b. ionic   d. polar covalent

4. Polar substances are _____ .
  a. hydrophilic   b. hydrophobic

5. A(n) _____ is any molecule that can combine with hydrogen ions or release them in response to changes in cellular pH.
  a. acid   b. salt   c. base   d. buffer

6. Match each molecule with the most suitable description.
  _____ long sequence of amino acids   a. carbohydrate
  _____ the main energy carrier   b. phospholipid
  _____ glycerol, fatty acids, phosphate   c. protein
  _____ two strands of nucleotides   d. DNA
  _____ one or more sugar monomers   e. ATP

Critical Thinking

1. A clerk in a health-food store tells you certain "natural" vitamin C tablets extracted from rose hips are better for you than synthetic vitamin C tablets. Given your understanding of the structure of organic compounds, what would be your response? Design an experiment to test whether these vitamins differ.

2. After looking at these functional groups of pantothenic acid, a vitamin, explain why this is a water-soluble or fat-soluble vitamin.

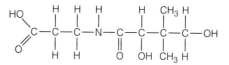

3. Rabbits that eat green, leafy vegetables containing xanthophyll accumulate this yellow pigment in body fat but not in muscles. What properties of the pigment cause this selective accumulation?

4. Cows, but not people, can digest grasses. Their four-chambered stomach houses large populations of certain microorganisms that are not normal inhabitants of our stomach. What kind of metabolic reactions do you think these microorganisms carry out? What do you think would happen to a cow undergoing treatment with an antibiotic that killed all these microorganisms?

Selected Key Terms

acid *26*
alcohol *28*
amino acid *34*
atom *18*
atomic
  number *18*
ATP *37*
base *26*
buffer *27*
carbohydrate *30*
chemical bond *20*
coenzyme *37*
cohesion *25*
condensation *29*
covalent bond *22*
denaturation *36*
DNA *37*
electron *18*
element *16*
enzyme *29*
evaporation *24*
fatty acid *32*

functional
  group *28*
glycoprotein *36*
hemoglobin *36*
hydrogen bond *23*
hydrogen
  ion (H⁺) *26*
hydrolysis *29*
hydrophilic *24*
hydrophobic *24*
hydroxide
  ion (OH⁻) *26*
ion *22*
ionic bond *22*
isotope *18*
lipid *32*
lipoprotein *36*
mass number *18*
molecule *21*
monosaccharide *30*
neutron *18*
nucleic acid *37*

nucleotide *37*
oligosaccharide *30*
orbital *20*
organic
  compound *28*
pH scale *26*
phospholipid *33*
polypeptide
  chain *34*
polysaccharide *30*
protein *34*
proton *18*
radioisotope *18*
RNA *37*
salt *27*
shell model *20*
solute *25*
sterol *33*
temperature *24*
tracer *19*
triglyceride *32*
wax *33*

Reading

Wolfe, S. 1995. *Introduction to Molecular and Cellular Biology.* Belmont, California: Wadsworth.

3 CELL STRUCTURE AND FUNCTION

Animalcules and Cells Fill'd With Juices

Early in the seventeenth century, a scholar by the name of Galileo Galilei arranged two glass lenses within a cylinder. With this instrument he happened to look at an insect, and later he described the stunning geometric patterns of its tiny eyes. Thus Galileo, who was not a biologist, was the first to record a biological observation made through a microscope. The study of the cellular basis of life was about to begin. First in Italy, then in France and England, scholars set out to explore a world whose existence had not even been suspected.

At midcentury Robert Hooke, a Curator of Instruments for the Royal Society of England, was at the forefront of these studies. When Hooke first turned a microscope to thinly sliced cork from a mature tree, he observed tiny compartments (Figure 3.1*a*). He gave them the Latin name *cellulae*, meaning small rooms—hence the origin of the biological term "cell." They were actually the walls of dead cells, which is what cork is made of, but Hooke did not think of them as being dead because he did not know cells could be alive. In other plant tissues, he discovered cells "fill'd with juices" but could not imagine what they represented.

Given the simplicity of their instruments, it is truly amazing that the pioneers in microscopy saw as much

as they did. Antony van Leeuwenhoek, a Dutch shopkeeper, had great skill in constructing lenses and quite possibly the keenest vision (Figure 3.1*b*). By the late 1600s he was observing wonders everywhere, including "many very small animalcules, the motions of which were very pleasing to behold," in scrapings of tartar from his teeth. He observed diverse protistans, sperm, even a bacterium—an organism so small it would not be seen again for another two centuries!

In the 1820s, improvements in lenses brought cells into sharper focus. Robert Brown, a botanist, noticed an opaque spot in a variety of cells. He called it a nucleus. In 1838 the botanist Matthias Schleiden wondered if the nucleus has something to do with a cell's development. He decided that each plant cell develops as an independent unit, even though it is an integral part of the plant.

By 1839, after years of studying animal tissues, the zoologist Theodor Schwann had this to say: Animals as well as plants consist of cells and cell products—and

Figure 3.1 Early glimpses into the world of cells. (**a**) Robert Hooke's compound microscope and his drawing of cell walls from cork tissue. (**b**) Antony van Leeuwenhoek, microscope in hand, and (**c**) one of his early sketches of sperm cells. (**d**) Cartoon evidence of the startling impact of microscopic observations on nineteenth-century London.

a   b

even though the c̲e̲l̲l̲s̲ are part of a whole organism, they have, to some extent, an individual life of their own.

[handwritten: 細胞.]

A question remained: Where do cells come from? A decade later Rudolf Virchow, a physiologist, completed his own studies of cell growth and reproduction—that is, their division into two cells. Every cell, he decided, comes from a cell that already exists.

And so, by the middle of the nineteenth century, microscopic analysis yielded three generalizations, which together constitute the **cell theory**: *First, all organisms are composed of one or more cells. Second, the cell is the smallest unit having the properties of life. Third, the continuity of life arises directly from the growth and division of single cells.* These insights still hold true.

This chapter provides an overview of our current understanding of the structure and functions of cells. It also introduces some of the modern microscopes that transport us more deeply into the spectacular worlds of juice-fill'd cells and animalcules.

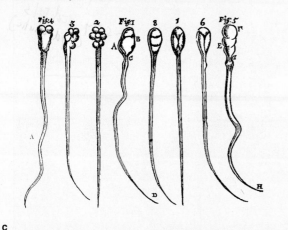

c

d

1. All organisms are composed of one or more cells. The cell is the smallest unit that still retains the characteristics of life. And each new cell arises from a preexisting cell. These are the three generalizations of the cell theory.

2. All cells have an outermost, plasma membrane that helps their interior remain distinct from the surroundings. Cells contain cytoplasm, a region that is organized for energy conversions, protein synthesis, movements, and other activities necessary for survival. Cells contain a nucleus or a comparable region in which DNA is located:

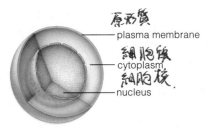

[handwritten: 原形質]
plasma membrane
[handwritten: 細胞膜]
cytoplasm
[handwritten: 細胞核.]
nucleus

3. The nucleus is one of many organelles. Organelles are membrane-bound compartments inside e̲u̲k̲a̲r̲y̲o̲t̲i̲c̲ ̲c̲e̲l̲l̲s̲. They physically separate different metabolic reactions and allow them to proceed in orderly fashion. Prokaryotic cells (bacteria) do not have comparable organelles.

4. Cell membranes consist mostly of [handwritten: two fatty acids] p̲h̲o̲s̲p̲h̲o̲l̲i̲p̲i̲d̲ and protein molecules. The phospholipids form two adjacent layers that give the membrane its basic structure and prevent water-soluble substances from freely crossing it. The proteins carry out most membrane functions.

BASIC ASPECTS OF CELL STRUCTURE AND FUNCTION

Inside your body and at its moist surfaces, trillions of cells live in interdependency. In scummy pondwater, a single-celled amoeba moves about freely, thriving on its own. For humans, amoebas, and all other organisms, the **cell** is the smallest entity that still retains the properties of life. A cell either can survive on its own or has the potential to do so. Its structure is highly organized, and it engages in metabolism. A cell senses and responds to specific changes in the environment. And it has the potential to reproduce itself, based on inherited instructions in its DNA.

Structural Organization of Cells

Cells differ enormously in size, shape, and activities, as you might gather by comparing a bacterium with one of your liver cells. Yet they are alike in three respects. All cells start out life with a plasma membrane, a region of DNA, and a region of cytoplasm:

1. **Plasma membrane.** This thin, outermost membrane maintains the cell as a distinct entity. By doing so, it allows metabolic events to proceed apart from random events in the environment. A plasma membrane does not *isolate* the cell interior. Substances and signals continually move across it, in highly controlled ways.

2. **DNA-containing region.** DNA occupies part of the cell interior, along with molecules that can copy or read its hereditary instructions.

3. **Cytoplasm.** The cytoplasm is everything enclosed by the plasma membrane, *except* for the region of DNA. It is a semifluid substance in which particles, filaments, and often compartments are organized.

This chapter introduces two fundamentally different kinds of cells. **Eukaryotic cells** contain distinct arrays of organelles—tiny, internal sacs and other compartments formed from membranes. One organelle, the nucleus, houses the DNA. It is the defining feature of eukaryotic cells. By contrast, **prokaryotic cells** have no nucleus; no membranes intervene between the region of DNA and the cytoplasm that surrounds it. Bacteria are the only prokaryotic cells. Beyond the realm of bacteria, all other organisms—from amoebas to peach trees and puffball mushrooms to zebras—are eukaryotes.

The Lipid Bilayer of Cell Membranes

A cell's organization and activities depend on its membranes. If any membrane is to function properly, fluid must bathe its two surfaces. How, then, does it remain structurally intact? Is it like a solid wall separating two fluid regions? Not at all. Push a fine needle into and out of a cell, and cytoplasm will not ooze out. Instead, the

cell surface will flow over the puncture and seal it. The plasma membrane apparently shows fluid behavior!

How do "fluid" cells remain distinct from their fluid surroundings? For the answer, let's start by looking at phospholipids—the most abundant component of any cell membrane. A **phospholipid**, recall, is a molecule having a hydrophilic (water-loving) head and two fatty acid tails, which are hydrophobic (water-dreading). If you immerse many phospholipid molecules in water, they will interact with the water molecules and with one another until they spontaneously cluster together. Collectively, the jostlings might even force

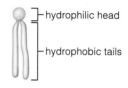

hydrophilic head

hydrophobic tails

them to become arranged in two layers, with the fatty acid tails sandwiched between the hydrophilic heads. This arrangement, which is called a **lipid bilayer**, is the structural basis of cell membranes:

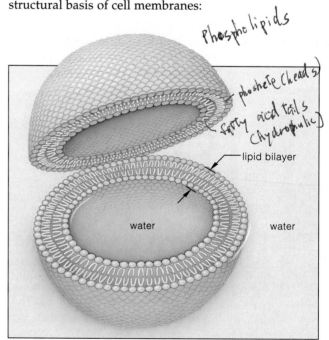

Phospholipids

phosphate (heads)

fatty acid tails (hydrophilic)

lipid bilayer

water

water

So why does a punctured membrane seal itself? A puncture is energetically unfavorable; it exposes many hydrophobic groups to water. Mutual repulsion forces the phospholipids back to the more stable arrangement.

Fluid Mosaic Model of Membrane Structure

Figure 3.2 shows a bit of membrane that corresponds to the **fluid mosaic model**. By this model, cell membranes consist of lipids *and* proteins. They incorporate a variety of phospholipids, glycolipids, and sterols (such as cholesterol in animals and phytosterols in plants). Many proteins are embedded in the bilayer or positioned at

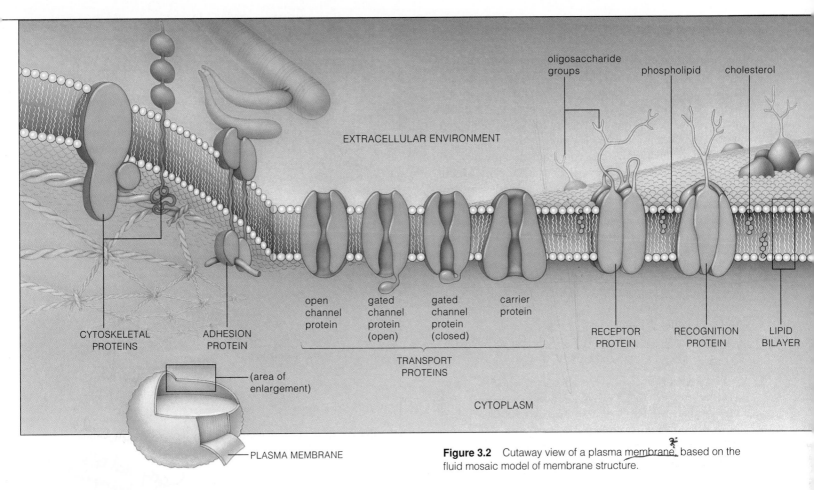

oligosaccharide groups   phospholipid   cholesterol

EXTRACELLULAR ENVIRONMENT

open channel protein   gated channel protein (open)   gated channel protein (closed)   carrier protein

CYTOSKELETAL PROTEINS   ADHESION PROTEIN

TRANSPORT PROTEINS

RECEPTOR PROTEIN   RECOGNITION PROTEIN   LIPID BILAYER

CYTOPLASM

(area of enlargement)

PLASMA MEMBRANE

Figure 3.2 Cutaway view of a plasma membrane, based on the fluid mosaic model of membrane structure.

one of its surfaces. Different events proceed at the two surfaces, which differ in the number, kind, and arrangement of their lipids and proteins.

The term "mosaic" refers to the mixed composition of lipids and proteins. "Fluid" refers to the motions of lipids and their interactions. Phospholipid molecules spin about their long axis, they move sideways, and their tails flex—all of which helps keep neighboring molecules from packing into a solid layer. Lipids that have short tails or unsaturated (kinked) tails contribute to the membrane's fluidity.

Proteins associated with the bilayer carry out most membrane functions. Various **transport proteins** allow or encourage water-soluble substances to move through their interior, which opens on both sides of the bilayer. These proteins bind molecules or ions on one side of the membrane, then release them on the other.

Receptor proteins bind hormones and other extracellular substances that trigger changes in cell activities. For example, enzymes that crank up machinery for cell growth and division are switched on when a hormone (such as somatotropin) binds to receptors for it. Different cells have different combinations of receptors.

Diverse **recognition proteins** at the cell surface are like molecular fingerprints; they identify a cell as being

of a specific type. For example, "self" proteins pepper the plasma membrane of your body's own cells. Certain white blood cells chemically recognize these proteins and leave your cells alone. But they will launch attacks against bacterial cells, viruses, and other invaders that have "nonself" proteins at their surfaces.

Finally, in multicelled organisms, **adhesion proteins** help particular kinds of cells stick together and thereby remain positioned in their proper tissues.

All cells have an outermost, plasma membrane, an internal region of cytoplasm, and an internal region of DNA.

Besides the plasma membrane, eukaryotic cells have internal, membrane-bound compartments, including the nucleus.

Each membrane has a lipid bilayer. The hydrophobic parts of its lipid molecules are sandwiched between the hydrophilic parts, which are dissolved in the fluid surroundings.

The lipid bilayer gives the membrane its basic structure and serves as a barrier to water-soluble substances.

Proteins associated with the bilayer carry out most membrane functions. Many types transport substances across the bilayer or serve as receptors for extracellular substances. Other types function in cell-to-cell recognition or adhesion.

3.2 CELL SIZE AND CELL SHAPE

Can any cell be observed with the unaided human eye? There are a few, including the "yolks" of bird eggs, cells in the red part of watermelons, and the fish eggs we call caviar. However, most cells cannot be observed without microscopes. To give you a sense of cell sizes, red blood cells are about 8 millionths of a meter across. You could fit a string of 2,000 of them across your thumbnail!

Why are most cells so small? The **surface-to-volume ratio** constrains increases in cell size. By this physical relationship, an object's volume increases with the cube of the diameter, but its surface area increases only with the square. Simply put, *if a cell expands in diameter during growth, its volume will increase more rapidly than its surface area will.* The following example illustrates this point:

diameter (cm):	0.5	1.0	1.5
surface area (cm²):	0.79	3.14	7.07
volume (cm³):	0.06	0.52	1.77
surface-to-volume ratio:	13.17:1	6.04:1	3.99:1

Suppose we figure out a way to make a round cell grow four times wider. Its volume increases 64 times (4^3) and its surface area increases 16 times (4^2). Unfortunately, each unit of plasma membrane must now serve four times as much cytoplasm as before! Past a certain point, the inward flow of nutrients and outward flow of wastes will not be fast enough, and the cell will die.

A large, round cell also would have trouble moving materials *through* its cytoplasm. In small or skinny cells, the random, tiny motions of molecules easily distribute materials. If a cell isn't small, it probably is long and thin or has outfoldings and infoldings that increase its surface relative to its volume. *The smaller or narrower or more frilly-surfaced the cell, the more efficiently materials cross its surface and become distributed through the interior.*

Multicelled body plans show evidence of surface-to-volume constraints, also. For example, cells attach end to end in strandlike alga, and each one interacts directly with the surroundings. Your circulatory system delivers materials to trillions of cells and sweeps away wastes. Its "highways" cut through the volume of tissue and so shrink the distance to and from individual cells.

During their growth, cells increase faster in volume than they do in surface area. This physical constraint on increases in size influences the size and shape of cells. It also influences the body plans of multicelled organisms.

3.3 *Focus on Science*

MICROSCOPES—GATEWAYS TO THE CELL

Modern microscopes are gateways to astounding worlds. Some kinds even afford glimpses into the structure of molecules. Figure 3.3 shows the range of magnifications possible. The micrographs in Figures 3.3 and 3.4 hint at the details now being observed. A **micrograph** simply is a photograph of an image that was formed with the aid of a microscope.

Light Microscopes   Light microscopes work by bending (refracting) light rays. Such rays naturally pass straight through the center of a curved lens. The farther they are from the center of the lens, the more they will bend:

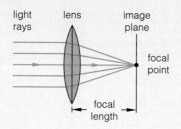

The angle at which rays of light enter a glass lens and the molecular structure of the glass dictate the extent to which the ray will bend.

In a *compound light microscope*, glass lenses bend incoming light rays to form an enlarged image of a cell or some other specimen. A living cell must be small or thin enough for light to pass through. It would help if cell parts differed in color and density from their surroundings, but most are nearly colorless and appear uniform in density. That is why microscopists stain cells (expose them to dyes that react with some parts but not others). Staining may alter and kill cells. Dead cells break down very fast, so they typically are pickled or preserved before being stained.

Suppose you are using the best glass lens system. When you magnify the diameter of a specimen by 2,000 times or more, you discover that cell parts appear larger but are not clearer. What limits the resolution of smaller details? The answer lies with a certain property of visible light.

Picture a train of waves moving across an ocean. Each **wavelength** is the distance from one wave's peak to the peak of the wave behind it. Light also travels as waves. In fact, different colors of light correspond to waves of certain, unvarying lengths. The wavelength is about 750 nanometers for red light and 400 nanometers for violet; all other colors fall in between. *And if a cell structure is less than one-half of a wavelength long, rays of light streaming by it will overlap so much that the object won't be visible.*

Electron Microscopes   Electrons are particles, but they, too, behave like waves. Electron microscopy is based on the use of accelerated streams of electrons with wavelengths of about 0.005 nanometer—*about 100,000 times shorter than those of visible light.* Ordinary glass lenses will scatter accelerated electrons. But an electron's charge responds to a magnetic field. The "lens" of a *transmission*

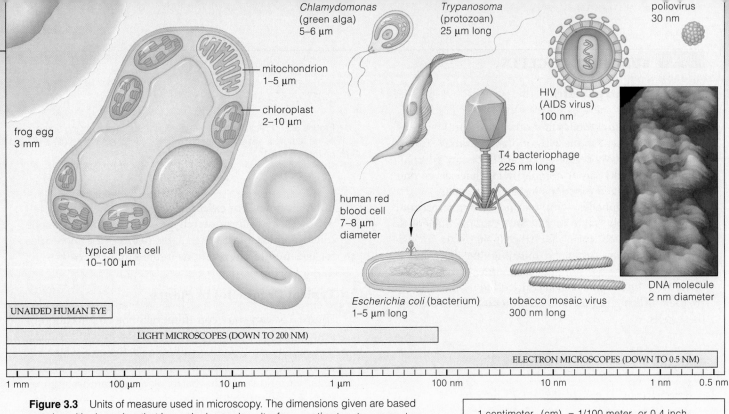

Chlamydomonas
(green alga)
5–6 μm

Trypanosoma
(protozoan)
25 μm long

poliovirus
30 nm

mitochondrion
1–5 μm

chloroplast
2–10 μm

frog egg
3 mm

HIV
(AIDS virus)
100 nm

T4 bacteriophage
225 nm long

human red
blood cell
7–8 μm
diameter

typical plant cell
10–100 μm

DNA molecule
2 nm diameter

Escherichia coli (bacterium)
1–5 μm long

tobacco mosaic virus
300 nm long

UNAIDED HUMAN EYE

LIGHT MICROSCOPES (DOWN TO 200 NM)

ELECTRON MICROSCOPES (DOWN TO 0.5 NM)

1 mm   100 μm   10 μm   1 μm   100 nm   10 nm   1 nm   0.5 nm

Figure 3.3 Units of measure used in microscopy. The dimensions given are based on a logarithmic scale—that is, each change in units, from centimeters to nanometers, represents a tenfold decrease in size. A *scanning tunneling microscope*, which provides magnifications up to 100 million, gave us the photomicrograph of part of a DNA molecule. Its needlelike probe has a single atom at its tip. Voltage applied between the tip and an atom at a specimen's surface causes a detectable tunnel to form in the electron orbitals. As the tip moves over a specimen's contours, a computer analyzes the motion and creates a three-dimensional view of the surface atoms.

1 centimeter (cm) = 1/100 meter, or 0.4 inch
1 millimeter (mm) = 1/1,000 meter
1 micrometer (μm) = 1/1,000,000 meter
1 nanometer (nm) = 1/1,000,000,000 meter

$1 \text{ meter} = 10^2 \text{ cm} = 10^3 \text{ mm} = 10^6 \text{ μm} = 10^9 \text{ nm}$

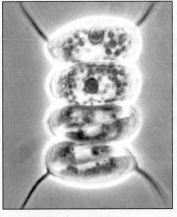

a Light micrograph
(phase-contrast process)

b Light micrograph
(Nomarski process)

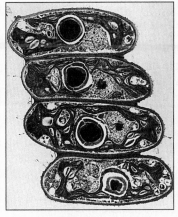

c Transmission electron
micrograph, thin section

d Scanning electron
micrograph

10 μm

electron microscope is a magnetic field. This microscope channels accelerated electrons through a specimen, focuses them into an image, and magnifies them. With *scanning electron microscopes*, a narrow beam of electrons moves back and forth across a specimen coated with a thin metal layer. The metal responds by emitting some of its own electrons. Camera equipment detects the emission patterns and forms an image that reveals surface features. Most of the images have fantastic depth of field.

Figure 3.4 How different microscopes can reveal different aspects of the same organism—in this case, a green alga (*Scenedesmus*). The micrographs show all four specimens at the same magnification. The phase-contrast and Nomarski processes mentioned in (**a**) and (**b**) create optical contrasts without staining the cells. Both processes enhance the usefulness of light micrographs.

As for other micrographs in the book, the short bar below the micrograph in (**d**) provides a visual reference for size. A micrometer (μm) is 1/1,000,000 of a meter.

Functions of Organelles

We turn now to organelles and other structural features that typically occur in the cells of plants, animals, fungi, and protistans. We define an **organelle** as an internal, membrane-bound sac or compartment that serves one or more specialized functions inside these cells.

Observe organelles with the aid of a microscope, and you probably will notice at once that the nucleus is one of the most conspicuous. All cells that start out life with a nucleus are *eukaryotic*, meaning that they have a "true nucleus." The numbers and kinds of organelles differ from one type of eukaryotic cell to the next. Even so, the following organelles and structures are typical of these cells:

Organelle or Structure	Main Function
Nucleus	*Localization of the cell's DNA*
Ribosomes	*Structures that organize and speed up the assembly of polypeptide chains*
Endoplasmic reticulum	*Initial routing and modification of newly formed polypeptide chains; also lipid synthesis*
Golgi body	*Further modification of polypeptide chains into mature proteins; sorting and shipping of proteins and lipids for secretion or for use inside cell*
Vesicles	*Different types function in transport or storage of substances; digestion inside the cell; other functions*
Mitochondria	*Efficient factories of ATP production*
Cytoskeleton	*Cell shape, internal organization, and movements*

What is the advantage of partitioning the cell interior into so many organelles? *The compartmentalization allows a large number of activities to proceed simultaneously in very limited space.*

Consider a photosynthetic cell inside a leaf. It can put together starch molecules by one set of reactions *and* break them apart by another set of reactions. Yet the cell would gain absolutely nothing if the synthesis and breakdown reactions proceeded at the same time on the same starch molecule. Without organelle membranes, the balance of diverse chemical activities within the cell would spiral out of control.

Organelle membranes do more than physically separate incompatible reactions. They also allow reactions that *are* compatible and interconnected to proceed at

different times. For example, a photosynthetic plant cell produces and stores starch molecules in an organelle called a chloroplast—then later releases the molecules for use in different reactions in the same organelle.

Typical Organelles in Plants

Figure 3.5*a* can start you thinking about the location of organelles in a typical plant cell. Bear in mind, calling this cell "typical" is like calling a cactus or a water lily or an elm tree a "typical" plant. As is true of animal cells, variations on the basic plan are mind-boggling. With this qualification in mind, also take a close look at Figure 3.6 (page 48). The micrograph in that figure shows the locations of organelles that you are likely to find in many kinds of specialized plant cells.

Typical Organelles in Animals

Next, start thinking about the organelles of a typical animal cell, as shown in Figures 3.5*b* and 3.7 (page 49). Like the plant cell, it contains a nucleus, a number of mitochondria, and the other components listed earlier. *These structural similarities point to basic functions that are necessary for survival, regardless of cell type.* We will return to this concept throughout the book.

Comparing Figures 3.5 through 3.7 also gives you an initial idea of how plant and animal cells *differ* in their structure. For example, you won't ever observe an animal cell surrounded by a cell wall. (You might see assorted fungal and protistan cells with one, however.) What other differences can you identify?

Eukaryotic cells contain a profusion of organelles—internal, membrane-bound sacs and compartments that serve specific metabolic functions.

Organelles physically separate chemical reactions, many of which are incompatible.

Organelles also separate different reactions in time, as when certain molecules are put together, stored, then used later in other reaction sequences.

Many organelles, including the nucleus, are common to all eukaryotic cells. They carry out functions that are essential for cell survival. Specialized cells may also contain other types of organelles.

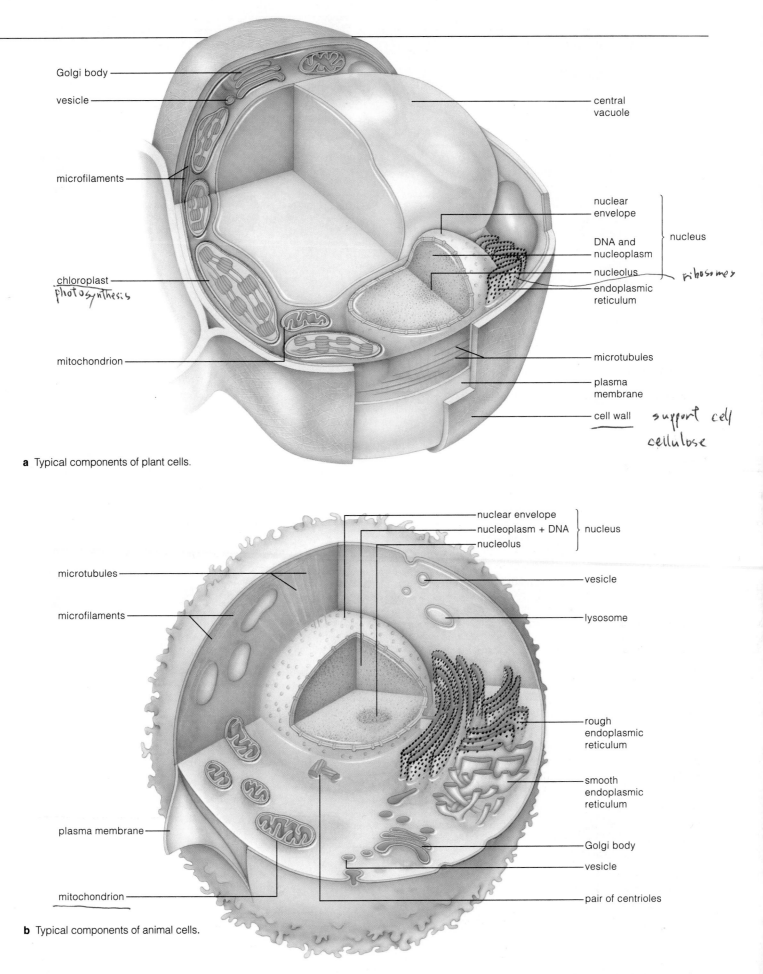

Golgi body

vesicle

microfilaments

chloroplast
photosynthesis

mitochondrion

central
vacuole

nuclear
envelope

DNA and
nucleoplasm

nucleolus

endoplasmic
reticulum

} nucleus

ribosomes

microtubules

plasma
membrane

cell wall

support cell
cellulose

a Typical components of plant cells.

nuclear envelope

nucleoplasm + DNA

nucleolus

} nucleus

microtubules

microfilaments

vesicle

lysosome

rough
endoplasmic
reticulum

smooth
endoplasmic
reticulum

plasma membrane

Golgi body

vesicle

mitochondrion

pair of centrioles

b Typical components of animal cells.

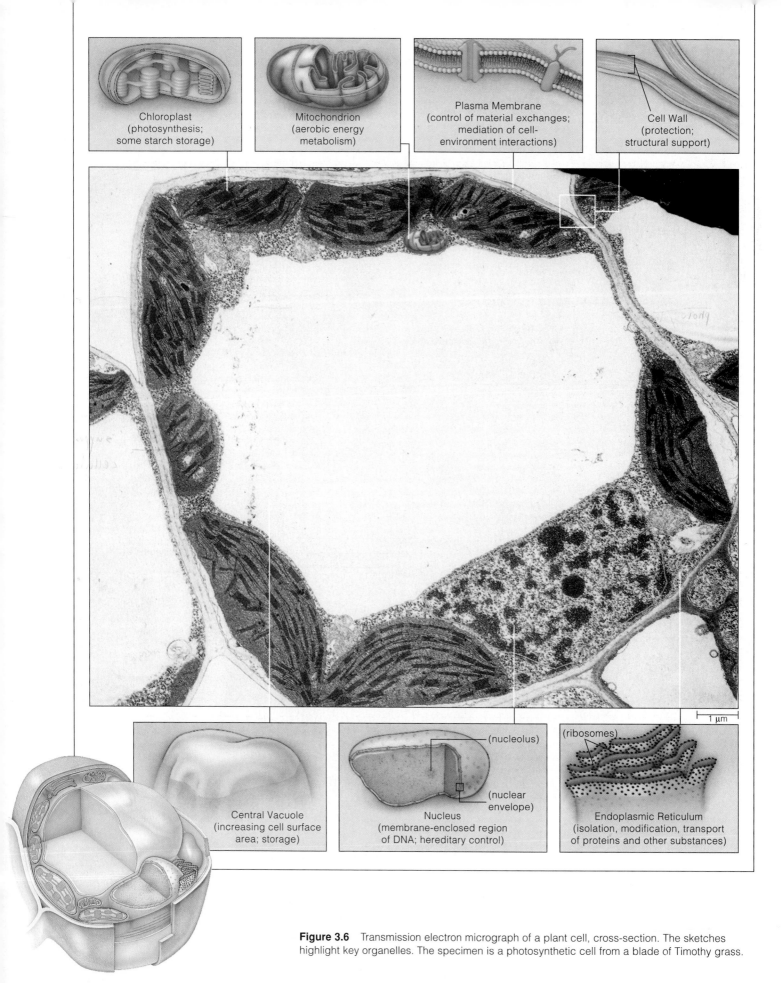

Chloroplast
(photosynthesis;
some starch storage)

Mitochondrion
(aerobic energy
metabolism)

Plasma Membrane
(control of material exchanges;
mediation of cell-
environment interactions)

Cell Wall
(protection;
structural support)

1 μm

Central Vacuole
(increasing cell surface
area; storage)

(nucleolus)

(nuclear
envelope)

Nucleus
(membrane-enclosed region
of DNA; hereditary control)

(ribosomes)

Endoplasmic Reticulum
(isolation, modification, transport
of proteins and other substances)

Figure 3.6 Transmission electron micrograph of a plant cell, cross-section. The sketches highlight key organelles. The specimen is a photosynthetic cell from a blade of Timothy grass.

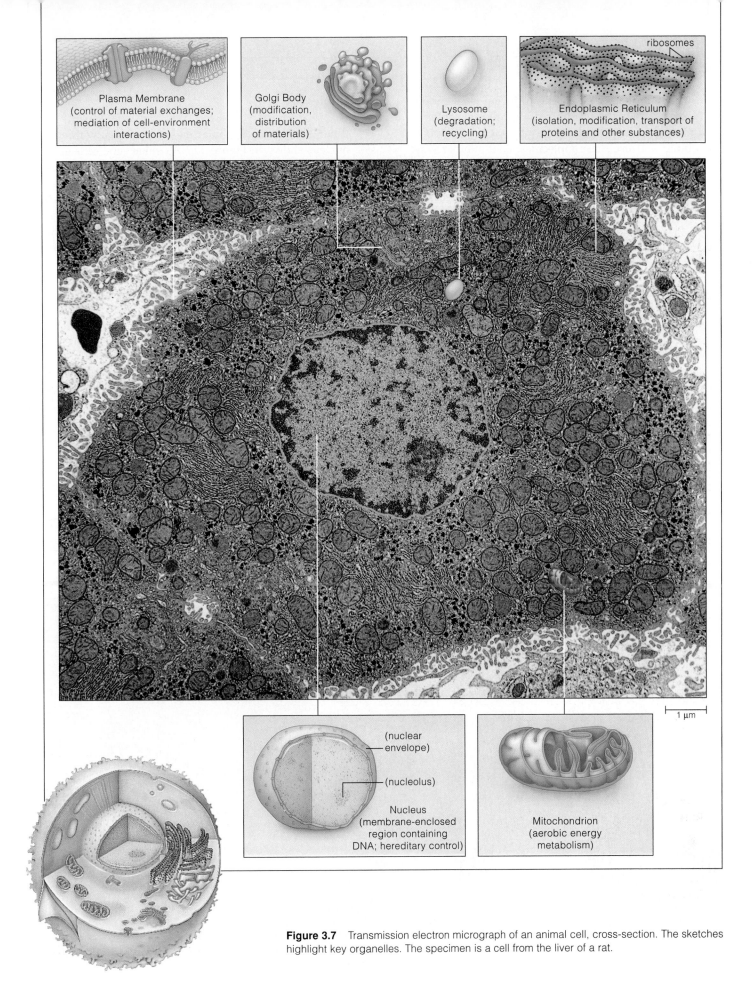

Plasma Membrane
(control of material exchanges;
mediation of cell-environment
interactions)

Golgi Body
(modification,
distribution
of materials)

Lysosome
(degradation;
recycling)

ribosomes

Endoplasmic Reticulum
(isolation, modification, transport of
proteins and other substances)

1 μm

(nuclear
envelope)

(nucleolus)

Nucleus
(membrane-enclosed
region containing
DNA; hereditary control)

Mitochondrion
(aerobic energy
metabolism)

Figure 3.7 Transmission electron micrograph of an animal cell, cross-section. The sketches highlight key organelles. The specimen is a cell from the liver of a rat.

3.5 THE NUCLEUS

Constructing, operating, and reproducing cells simply can't be done without carbohydrates, lipids, proteins, and nucleic acids. It takes a class of proteins—enzymes—to build and use these molecules. Said another way, a cell's structure and function begin with proteins. *And instructions for building proteins are contained in DNA.*

Unlike bacteria, eukaryotic cells have the hereditary instructions distributed in several to many DNA molecules of various lengths. For example, each of your body cells contains forty-six DNA molecules. Stretched end to end, they would be about 1 meter long. DNA in frog cells would be 10 meters, end to end. Compared to the single molecule in bacteria, that's a lot of DNA!

The **nucleus** houses eukaryotic DNA. This type of organelle has a distinctive structure, as shown by the example in Figure 3.8, and it serves two key functions. *First,* the nucleus physically tucks away all of the DNA molecules, apart from the intricate metabolic machinery of the cytoplasm. Localization of DNA makes it easier to sort out hereditary instructions when the time comes for a cell to divide. The DNA molecules can be assorted into parcels—one parcel for each new cell that forms. *Second,* the membranous boundary of the nucleus helps control the exchange of signals and substances between the nucleus and the cytoplasm.

Nucleolus

What is the dense, globular mass of material within the nucleus shown in Figure 3.8? As eukaryotic cells grow, one or more of these appear in the nucleus. Each mass is a **nucleolus** (plural, nucleoli). It is a site where the protein and RNA subunits of ribosomes are assembled. Later, the subunits are shipped from the nucleus, into the cytoplasm. When proteins are about to be synthesized, they join together as intact, functional ribosomes.

Table 3.1	Components of the Nucleus
Nuclear envelope	Pore-riddled double membrane system that selectively controls passage of substances into and out of the nucleus
Nucleolus	Dense cluster of RNA and proteins used in the assembly of ribosomal subunits
Nucleoplasm	Fluid portion of interior of nucleus
Chromosome	One DNA molecule and the proteins that are associated with it
Chromatin	Total collection of all DNA molecules and their associated proteins in the nucleus

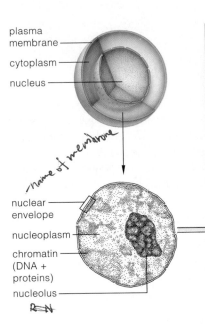

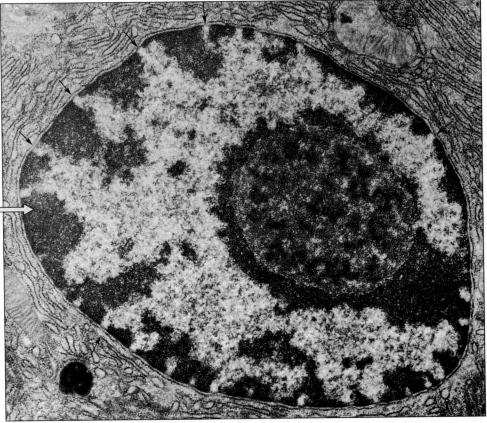

Figure 3.8 Transmission electron micrograph of the nucleus from a pancreatic cell, thin section. The small arrows inside the micrograph point to pores where controls operate to restrict or permit the passage of specific substances across the nuclear envelope.

Nuclear Envelope

Notice, in Figure 3.9, that the outermost component of the nucleus has *two* lipid bilayers, one wrapped around the other. They completely surround the fluid portion of the nucleus (nucleoplasm). This double-membrane system is the **nuclear envelope**. As is the case for all cell membranes, its lipid bilayers act as a barrier to water-soluble substances. Clusters of proteins form pores that span both bilayers. They allow ions and small, water-soluble molecules to move freely across the nuclear envelope, but they control the passage of ribosomal subunits, proteins, and other large molecules.

The innermost surface of the nuclear envelope has attachment sites for protein filaments. These anchor the DNA molecules to the membrane and help keep them organized. At the outermost surface are many of the cell's ribosomes, which take part in protein synthesis.

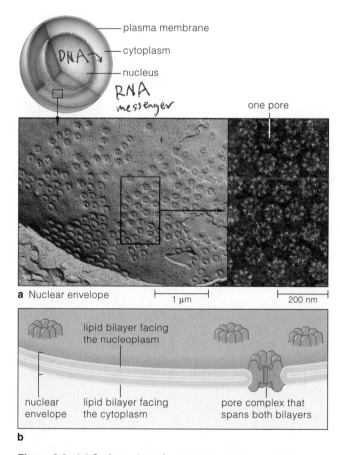

RNA messenger

a Nuclear envelope   1 μm   200 nm

lipid bilayer facing the nucleoplasm

nuclear envelope   lipid bilayer facing the cytoplasm   pore complex that spans both bilayers

b

Figure 3.9 (**a**) Surface view of a nuclear envelope, which consists of two pore-studded lipid bilayers. The specimen in the photograph at left was deliberately fractured for microscopy. The photograph at right is a closer view of the pores. Each pore across the envelope is an organized array of membrane proteins. It permits the selective transport of larger molecules into and out of the nucleus. (**b**) Diagram of the nuclear envelope.

Chromosomes

Between cell divisions, eukaryotic DNA is threadlike, with many enzymes and other proteins attached to it like beads on a string. Except at extreme magnification, the beaded threads look grainy, as in Figure 3.8. Before a cell divides, however, it duplicates its DNA molecules (so each new cell will get all of the required hereditary instructions). Besides this, the molecules are folded and twisted into condensed structures, proteins and all.

Early microscopists named the seemingly grainy substance chromatin. They also named the condensed structures chromosomes. We now define **chromatin** as a cell's total collection of DNA, together with all of the proteins associated with it. Each **chromosome** is an individual DNA molecule and its associated proteins, whether it is in threadlike or condensed form:

| unduplicated, not condensed chromosome (one DNA double helix + proteins) | duplicated but not condensed chromosome (two DNA double helices + proteins) | duplicated and now condensed chromosome |

In other words, the "chromosome" doesn't always look the same during the life of a eukaryotic cell. We will consider different aspects of chromosomes in chapters to come, so it will help to keep this point in mind.

What Happens to the Proteins Specified by DNA?

Enzymes assemble polypeptide chains on ribosomes in the cytoplasm. What happens to the chains? Many are used or stockpiled in the cytoplasm. Many others pass through the **cytomembrane system**. As you will read next, this system is a series of organelles, including endoplasmic reticulum, Golgi bodies, and vesicles.

Many proteins take on their final form and become packaged in vesicles in the system. Also, lipids are assembled and packaged here. Some vesicles deliver proteins and lipids to regions of the cell where new membrane must be built. Other vesicles store proteins or lipids for specific uses. Others move to the plasma membrane and release their contents outside the cell.

The nucleus keeps the cell's DNA molecules separated from cytoplasmic machinery, and this serves two functions:

• When a cell divides, the separation makes it easier to organize and parcel out hereditary instructions to new cells.

• Pores across the nuclear envelope help control the passage of larger molecules between the nucleus and the cytoplasm.

THE CYTOMEMBRANE SYSTEM

Figure 3.10 shows how organelles of the cytomembrane system are functionally related to one another. Again, new proteins undergo final modification in this system, and lipids are assembled within it. Later, these products are packaged and shipped to various destinations.

Endoplasmic Reticulum

Functionally speaking, the cytomembrane system starts with **endoplasmic reticulum**, or **ER**. In animal cells, the ER membranes begin at the nucleus and curve through the cytoplasm. Different regions of these membranes have a rough or smooth appearance. The difference arises largely from the presence or absence of ribosomes on the side of the membrane facing the cytoplasm.

Rough ER is often arranged as stacked, flattened sacs which have many ribosomes attached (Figure 3.11*a*). Polypeptide chains are assembled on the ribosomes. But only newly forming chains that have a built-in signal can enter the spaces within rough ER or become incorporated into ER membranes. (The "signal" is a string of about fifteen to twenty amino acids.) Once chains are inside rough ER, enzymes may attach oligosaccharides and other side chains to them. Many specialized cells secrete proteins, and rough ER is abundant in them. For example, some ER-rich cells of your pancreas produce and secrete enzymes that end up in the small intestine, where they have roles in digesting your meals.

Smooth ER is free of ribosomes. It curves through the cytoplasm like connecting pipes (Figure 3.11*b*). In many cells, smooth ER is the main site of lipid synthesis. It is especially developed in seeds. Smooth ER in liver cells even inactivates certain drugs and harmful by-products of metabolism. A type of smooth ER in skeletal muscle cells has a key role in muscle contraction.

Golgi Bodies

In **Golgi bodies**, enzymes put the finishing touches on lipids and proteins, sort them out, and package them in vesicles, which will move them to specific locations. For example, an enzyme in one Golgi region might attach a phosphate group to a protein and so give it a mailing tag to its proper destination.

The flattened membrane sacs of a Golgi body vaguely resemble a stack of pancakes (Figure 3.12). Vesicles form at the final region of a Golgi body (the uppermost pancakes) when parts of the membrane bulge, then break away. In animal cells, this is the region closest to the plasma membrane.

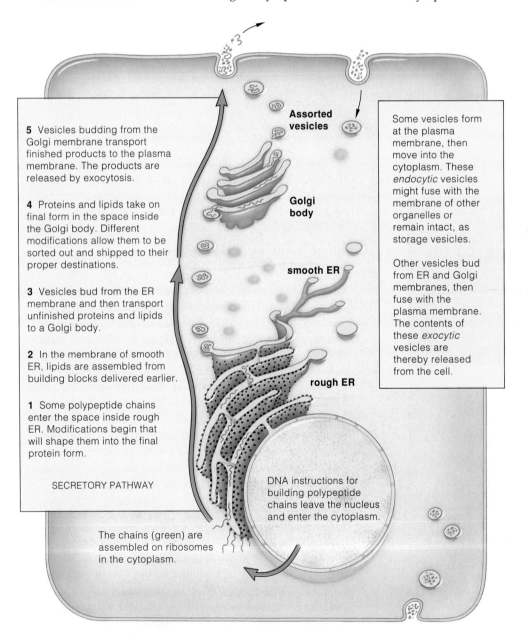

5 Vesicles budding from the Golgi membrane transport finished products to the plasma membrane. The products are released by exocytosis.

4 Proteins and lipids take on final form in the space inside the Golgi body. Different modifications allow them to be sorted out and shipped to their proper destinations.

3 Vesicles bud from the ER membrane and then transport unfinished proteins and lipids to a Golgi body.

2 In the membrane of smooth ER, lipids are assembled from building blocks delivered earlier.

1 Some polypeptide chains enter the space inside rough ER. Modifications begin that will shape them into the final protein form.

SECRETORY PATHWAY

Assorted vesicles

Golgi body

smooth ER

rough ER

DNA instructions for building polypeptide chains leave the nucleus and enter the cytoplasm.

The chains (green) are assembled on ribosomes in the cytoplasm.

Some vesicles form at the plasma membrane, then move into the cytoplasm. These *endocytic* vesicles might fuse with the membrane of other organelles or remain intact, as storage vesicles.

Other vesicles bud from ER and Golgi membranes, then fuse with the plasma membrane. The contents of these *exocytic* vesicles are thereby released from the cell.

Figure 3.10 Cytomembrane system. This membrane system in the cytoplasm functions in the assembly, modification, packaging, and shipment of proteins and lipids. The *green* arrows highlight a secretory pathway by which certain proteins and lipids are packaged and then released from many cells.

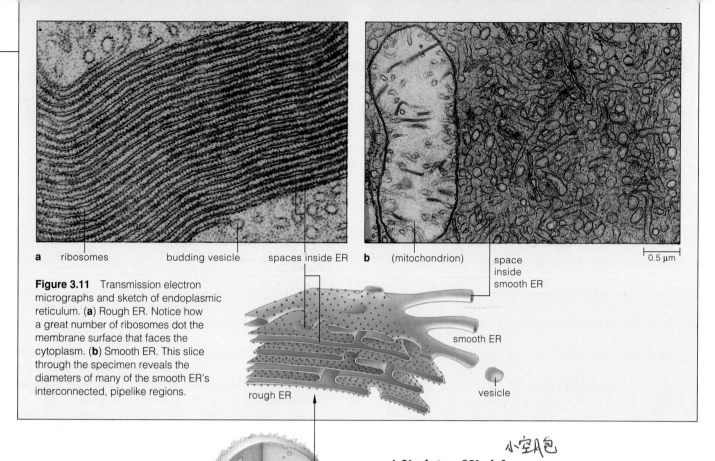

a ribosomes   budding vesicle   spaces inside ER   **b** (mitochondrion)   space inside smooth ER

0.5 µm

Figure 3.11 Transmission electron micrographs and sketch of endoplasmic reticulum. (**a**) Rough ER. Notice how a great number of ribosomes dot the membrane surface that faces the cytoplasm. (**b**) Smooth ER. This slice through the specimen reveals the diameters of many of the smooth ER's interconnected, pipelike regions.

smooth ER

rough ER

vesicle

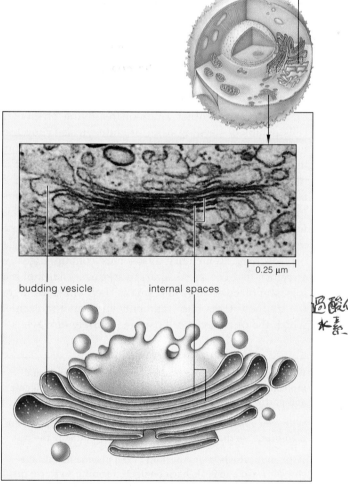

0.25 µm

budding vesicle   internal spaces

Figure 3.12 Micrograph and sketch of a Golgi body.

A Variety of Vesicles

Assorted vesicles move through the cytoplasm or take up positions within it. Consider the **lysosome**, a type of vesicle that buds from the Golgi membranes of animal cells and some fungal cells. Lysosomes are organelles of intracellular digestion. They contain a potent brew, rich with enzymes that speed the breakdown of complex carbohydrates, proteins, and nucleic acids, and some lipids. Often, lysosomes fuse with vesicles that formed at the plasma membrane. As described in the next section, such vesicles contain particles, bacteria, or other items that docked at the plasma membrane. Lysosomes may even help digest some or all of the cell's own parts. For example, as a tadpole develops into an adult frog, its tail disappears. Lysosomal enzymes destroy tail cells in a controlled, genetically programmed way.

Or consider the **peroxisomes**. These sacs of enzymes break down fatty acids and amino acids. The reactions produce hydrogen peroxide, a potentially harmful substance. Before hydrogen peroxide can do harm, another enzyme converts it to water and oxygen or uses it to break down alcohol. If you drink alcohol, nearly half of it is degraded in peroxisomes of liver and kidney cells.

In the ER and Golgi bodies of the cytomembrane system, many proteins take on final form and lipids are synthesized.

Lipids, proteins (including enzymes), and other items become packaged in vesicles destined for export, storage, membrane building, intracellular digestion, and other cell activities.

3.7 VESICLES THAT MOVE SUBSTANCES OUT OF AND INTO CELLS

Transport proteins can move ions and small molecules into or out of cells. However, when it comes to taking in or expelling large molecules or particles, cells rely on vesicles that form through exocytosis and endocytosis.

By **exocytosis**, a small, membrane-bound sac in the cytoplasm moves to the plasma membrane and fuses with it. Such sacs are *exocytic* vesicles. As the vesicle membrane becomes incorporated into the plasma membrane, its contents become exposed to the surroundings (Figure 3.13a). By **endocytosis**, part of the plasma mem-

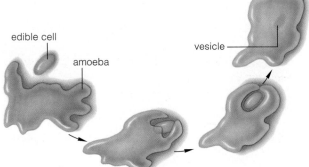

Figure 3.14 Phagocytosis, a process by which amoebas, macrophages (a type of white blood cell), and some other cells engulf their targets. A membrane-bound sac forms around the target and moves into the cytoplasm, then fuses with lysosomes.

brane sinks inward and then balloons around particles, liquid droplets, even tiny prey. It seals back on itself to form an *endocytic* vesicle, which transports its contents or stores them within the cytoplasm (Figure 3.13b).

Many times, receptors mediate endocytosis. In this case, molecules bind to membrane receptors that are specific for them. (For example, any cell that stores or uses cholesterol has receptors for lipoprotein particles.) The receptors are distributed throughout the plasma membrane. When they bind a molecule, the region of plasma membrane to which they are attached forms a shallow depression, or pit. The pit continues to sink into the cytoplasm and so forms an endocytic vesicle.

Amoebas and other kinds of free-living phagocytic cells exhibit a startling form of endocytosis. They trap bacterial cells and other foreign items by extending temporary lobes of cytoplasm around them. (Phagocyte literally means "cell eater.") The lobes are known as pseudopods. They wrap around the trapped item and then seal together, so that part of the plasma membrane forms a vesicle (Figure 3.14). The vesicle moves deeper into the cytoplasm, where it fuses with other vesicles —lysosomes. Remember, lysosomes contain digestive enzymes that can break apart nearly every kind of biological molecule, large and small. Their enzymes digest the trapped items into fragments and small molecules, which the phagocytic cell uses in a variety of ways.

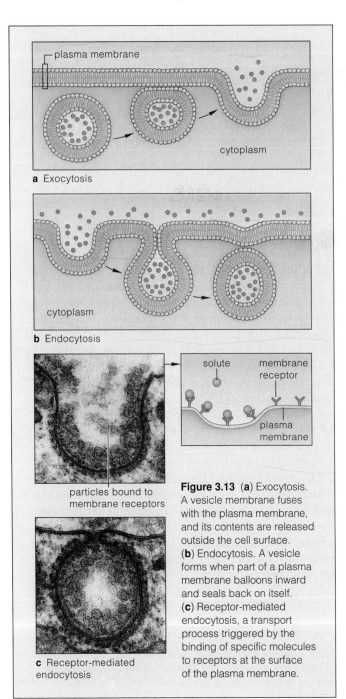

a Exocytosis

b Endocytosis

particles bound to membrane receptors

solute · membrane receptor · plasma membrane

c Receptor-mediated endocytosis

Figure 3.13 (a) Exocytosis. A vesicle membrane fuses with the plasma membrane, and its contents are released outside the cell surface. (b) Endocytosis. A vesicle forms when part of a plasma membrane balloons inward and seals back on itself. (c) Receptor-mediated endocytosis, a transport process triggered by the binding of specific molecules to receptors at the surface of the plasma membrane.

Whereas transport proteins deal only with ions and small molecules, exocytosis and endocytosis move large molecules and particles across the plasma membrane.

With exocytosis, a cytoplasmic vesicle fuses with the plasma membrane, so that its contents are released outside the cell.

With endocytosis, a tiny portion of the plasma membrane sinks inward and seals back on itself, forming a vesicle inside the cytoplasm. Receptors often mediate this process.

3.8 MITOCHONDRIA

Energy that ATP molecules carry from one reaction site to another drives nearly all cell activities. In organelles called **mitochondria** (singular, mitochondrion), energy released when organic compounds are broken apart is used to form *many* ATP molecules. These reactions, which require oxygen, extract far more energy than can be done by any other means. When you breathe in, you are taking in oxygen primarily for your mitochondria.

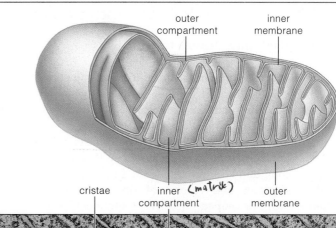

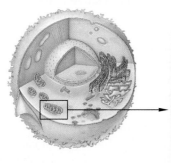

Figure 3.15 Cutaway sketch and transmission electron micrograph of a thin slice through a typical mitochondrion. Reactions within this organelle produce quantities of ATP, the major energy carrier between different reaction sites in cells. The reactions cannot proceed without oxygen.

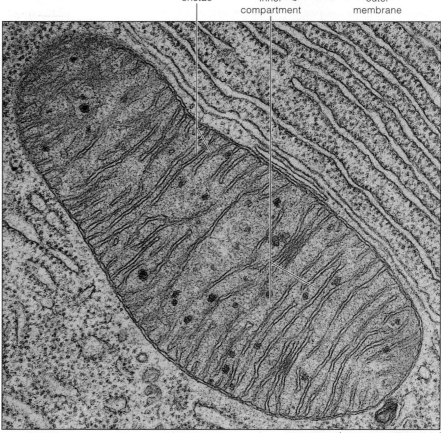

0.5 μm

The mitochondrion, as Figure 3.15 shows, has two membranes. Its outermost membrane faces the cytoplasm. In most cases, its inner membrane folds back on itself repeatedly. Each fold is a crista (plural, cristae).

This two-membrane system creates two distinct compartments within the mitochondrion. Enzymes and other proteins that are associated with the inner membrane are machinery for ATP formation; oxygen helps keep the machinery running. Chapter 6 gives details of these reactions.

All eukaryotic cells contain one or more mitochondria. Maybe you will find only one in a single-celled yeast. But you might find a thousand or more in energy-demanding cells, such as those of muscles. Take another look at the profusion of mitochondria in Figure 3.7—which is a micrograph of merely one thin slice from a liver cell. This micrograph alone tells you that your liver must be an exceptionally active, energy-demanding organ.

In their size and biochemistry, mitochondria resemble bacteria. They have their own DNA and some ribosomes. They even divide on their own. Possibly they evolved from ancient bacteria that were engulfed by a predatory, amoebalike cell, yet they managed to escape digestion. Perhaps they were able to reproduce inside the cell and its descendants. If they became permanent, protected residents, they may have lost structures and functions required for independent life while becoming mitochondria. We return to this intriguing idea later in the book, in Section 17.5.

Mitochondria are the powerhouses of eukaryotic cells.

Energy-releasing reactions proceed at their compartmented, internal membrane system. The reactions, which require oxygen, produce far more ATP than can be made by any other cellular means.

3.9 SPECIALIZED PLANT ORGANELLES

Chloroplasts and Other Plastids

Many plant cells contain plastids, a general category of organelles that function in photosynthesis or storage. Three types are common: chloroplasts, chromoplasts, and amyloplasts.

Only photosynthetic cells have **chloroplasts**. Within these organelles, sunlight is captured, ATP forms, and organic compounds are synthesized from simple raw materials (water and carbon dioxide). Chloroplasts often are oval or shaped like disks. Outermost are two membrane layers, one wrapped around the other. These surround a semifluid interior, the stroma. *Another* membrane weaves through the stroma and forms an inner system of interconnecting compartments (Figure 3.16). Commonly, the disk-shaped compartments are stacked together. Each stack is a granum (plural, grana).

The first stage of photosynthesis proceeds at light-trapping pigments, enzymes, and other proteins of the inner membrane system. That is where light energy is absorbed and ATP forms. The synthesis stage proceeds in the stroma. Here, sugars, starch, and other organic compounds are put together. Here also, starch grains (clusters of new starch molecules) may be stored temporarily.

Of the photosynthetic pigments, chlorophylls (green) are the most abundant, followed by carotenoids (yellow, orange, and red). The relative abundances of different pigments influence the coloration of plant parts. Some parts may be green, others golden brown, and so on.

In many ways, chloroplasts resemble certain photosynthetic bacteria. Like mitochondria, they may have evolved from bacteria that were engulfed by predatory cells, yet escaped digestion and became permanent residents in them. We return to this idea in Section 17.5.

Unlike chloroplasts, the chromoplasts have an abundance of carotenoids but no chlorophylls. They are the source of red-to-yellow colors of many flowers, autumn leaves, ripening fruits, and carrots and other roots. Their pigments also may visually attract animals that pollinate plants and disperse seeds. Amyloplasts have no pigments. They often store starch grains and are abundant in cells of stems, potato tubers, and many seeds.

Central Vacuole

Many mature, living plant cells contain a **central vacuole**, as shown in Figure 3.6. This organelle stores amino acids, sugars, ions, and toxic wastes. And it increases cell size and surface area. While a vacuole grows, fluid pressure builds up inside it and so forces the still-pliable cell wall to enlarge. The cell itself enlarges under this force. The resulting increased surface area enhances the rate at which substances are able to move into and out of the cell. In most cases, this fluid-filled organelle increases so much in volume that it takes up 50 to 90 percent of the cell's interior. The cytoplasm ends up as a very narrow zone between the central vacuole and the plasma membrane.

Photosynthetic eukaryotic cells contain chloroplasts and other plastids that function in food production and storage.

Many plant cells contain a central vacuole. Enlargement of this organelle during growth enlarges the cell itself and so increases the cell surface area available for absorption.

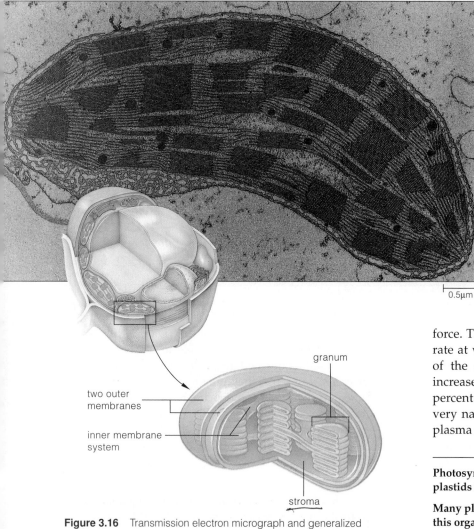

0.5μm

granum

two outer membranes

inner membrane system

stroma

Figure 3.16 Transmission electron micrograph and generalized sketch of a chloroplast, thin section. Chloroplasts are a key defining feature of photosynthetic eukaryotic cells.

CELL SURFACE SPECIALIZATIONS

Eukaryotic Cell Walls

Single-celled eukaryotes interact directly with the environment. Many have a **cell wall**, a continuous structure that surrounds the plasma membrane. The walls help support and protect the cell. They are porous, so water and solutes can easily move to and from the plasma membrane. In multicelled eukaryotic species, walls or other surface features allow adjacent cells to interact with one another and with their physical surroundings, as a few examples will illustrate.

Cells of leafy plants stick together, wall to wall. In growing plant parts, bundles of cellulose strands form a *primary* cell wall, of the sort shown in Figure 3.17. The primary walls are pliable, so that the surface area of cells can enlarge under the pressure of incoming water.

Later, in many cell types, more layers are deposited inside the first wall. They form a rigid, *secondary* wall that helps maintain cell shape. Pectin deposits cement the adjacent walls together. (They also thicken jams and jellies.) Other kinds of deposits, such as waxes, protect and reduce water loss from cells at the plant surfaces exposed to the air.

In plants, numerous channels cross adjacent walls and connect the cytoplasm of neighboring cells. They are called plasmodesmata (singular, plasmodesma). The number of channels affects how fast substances are transported through the plant.

Intercellular Material in Animals

Cell secretions and other kinds of materials intervene between the cells of many animal tissues. For example, think of the cartilage at the knobby ends of your leg bones. Cartilage consists of cells scattered in a "ground substance" (of modified polysaccharides) and protein fibers rich in collagen or elastin. Through this material, nutrients and other substances diffuse from cell to cell. The accumulated deposits of intercellular material in mature bone and other tissues account for much of your body weight.

Cell Junctions in Animals

Three cell-to-cell junctions are common in the tissues of most kinds of animals. *Tight* junctions link the cells of epithelial tissues, which line the body's outer surface, inner cavities, and organs. Tight junctions form seals that prevent molecules from freely crossing the tissue lining, as when they keep acids from leaking out of the stomach.

Adhering junctions are like spot welds. They keep cells together in tissues of the skin, heart, and other organs that are subject to stretching. *Gap* junctions link

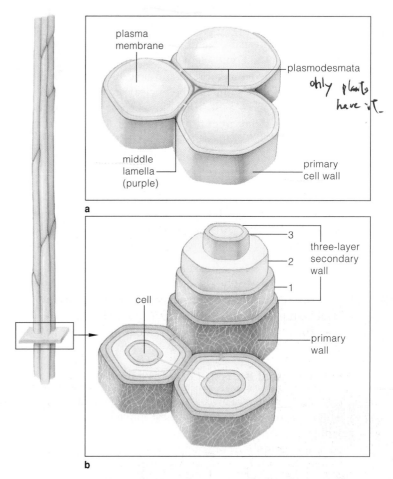

Figure 3.17 (**a**) Primary cell wall of cells in living plant tissues. Between the walls of adjoining cells, material is deposited to form a layer called the middle lamella (*purple*). The layer is thickest in adjoining corners. Channels called plasmodesmata extend across adjacent walls and connect the cytoplasm of neighboring cells.

(**b**) In many types of cells, such as the long fiber cells shown here, more layers become deposited inside the primary wall. These stiffen the wall and help it maintain its shape. Later, the cell itself may die, leaving the stiffened walls behind. This also is true of the water-conducting pipelines that thread through most plant tissues. They are "tubes" of many interconnected, stiffened cell walls.

the cytoplasm of adjacent cells. They are open channels for the rapid flow of signals and substances.

We will return to these cell-to-cell interactions and others in later chapters. For now, the point to remember is this:

In multicelled organisms, coordinated cell activities depend on specialized forms of communication and physical links between cells.

3.11 THE CYTOSKELETON

An interconnected system of fibers, threads, and lattices extends between the nucleus and plasma membrane of eukaryotic cells. This system, the **cytoskeleton**, gives the cells their internal organization, overall shape, and capacity to move. Some elements of the cytoskeleton reinforce the plasma membrane and hold its proteins in place; others do the same for the nuclear envelope. Still other elements serve as scaffolds for specific regions of the cytoplasm. They might even stabilize the enzyme systems of protein synthesis and other activities.

A cytoskeleton consists mainly of **microtubules** and **microfilaments**. Animal cells may contain **intermediate filaments** as well. Microtubules have parallel rows of tubulin subunits (Figure 3.18). Different microfilaments occur in different cells. For example, in muscle cells they consist of myosin and actin subunits; in skin cells and hair cells, they have keratin subunits.

Many parts of the cytoskeleton are permanent, but other parts appear only at certain times in a cell's life. Before the cell divides, for instance, new microtubules assemble into a "spindle" structure that moves chromosomes, then they disassemble when the task is done.

The Structural Basis of Cell Movements

If a cell lives, it moves. This is true even of cells with fixed positions in tissues. By the simplest movements, a cell rearranges its internal structures. With intricate movements, it might shunt organelles or chromosomes to a new location, bulge forward, or propel itself through fluid environments. Some cells even put out temporary lobes called pseudopods ("false feet," as in Figure 3.14). Microfilaments, microtubules, or both take part in most of these movements. They do so by three mechanisms:

1. *Through the controlled assembly and disassembly of their subunits, microtubules and microfilaments grow or shrink in length.* As they do, any structures attached to them are dragged or pushed through the cytoplasm. This is how microtubules move chromosomes about.

2. *Microfilaments or microtubules actively slide past one another.* As an example, muscle cells contain parallel, permanent arrays of microfilaments that bring about contraction. Briefly, myosin molecules repeatedly bind and release their actin neighbors, so that the actin slides on past. The cumulative, directional sliding shortens the arrays of microfilaments (they contract).

A similar sliding mechanism produces amoeboid motion. Here, contractions in one part of a free-living cell or cancer cell squeeze some cytoplasm into other parts. The cell body bulges forward or pseudopods form. Another sliding mechanism underlies the beating of flagella and cilia, as described shortly.

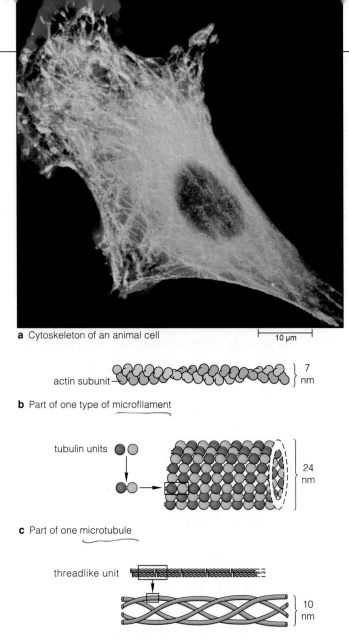

a Cytoskeleton of an animal cell |—— 10 µm ——|

actin subunit ⎯ } 7 nm

b Part of one type of microfilament

tubulin units ⬤⚪ } 24 nm

c Part of one microtubule

threadlike unit } 10 nm

d Part of one type of intermediate filament

Figure 3.18 (**a**) Cytoskeleton of a fibroblast, a type of animal cell that gives rise to certain animal tissues. In this composite of three images, *green* identifies microtubules. *Blue* and *red* identify two different kinds of microfilaments. (**b**) Structural organization of a microfilament, (**c**) microtubule, and (**d**) an intermediate filament.

3. *Microtubules or microfilaments shunt organelles from one location to another.* For example, between dawn and dusk, the sun's overhead position changes. In response to the changing angle of incoming rays, chloroplasts in photosynthetic cells move to new light-intercepting positions. Myosin molecules that are attached to them are "walking" over bundles of actin microfilaments, carrying the chloroplasts with them. This mechanism is the basis of "cytoplasmic streaming." The term refers to an active flowing of cytoplasmic components that is especially pronounced in plant cells.

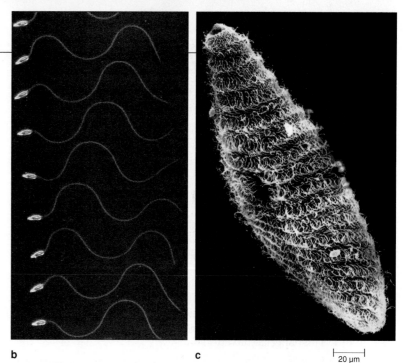

one of nine pairs
of microtubules
of the outer ring

dynein arm

two central
microtubules

central sheath

spokes and
links of the
connective
system

plasma
membrane

9 + 2 Array

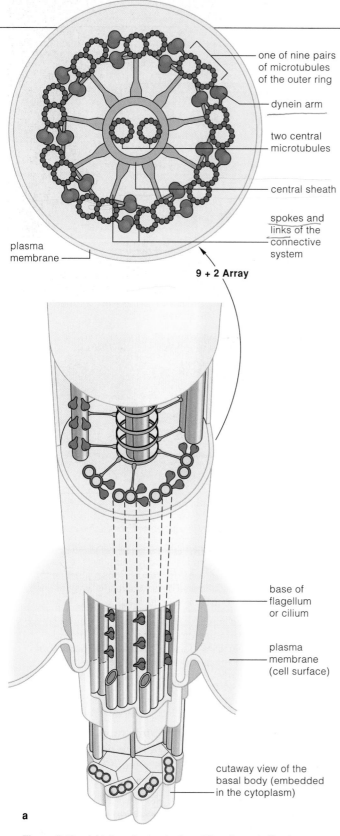

base of
flagellum
or cilium

plasma
membrane
(cell surface)

cutaway view of the
basal body (embedded
in the cytoplasm)

a

b

c

$\vdash$ 20 µm $\dashv$

Figure 3.19 (**a**) Internal organization of flagella and cilia. A system of microtubules and a connective system of spokes and linking elements are the basis of beating movements. (**b**) Composite micrographs of the beating of a sperm flagellum. Waves start at the flagellum's base and move toward its tip. (**c**) Scanning electron micrograph of *Paramecium*. The coordinated beating of rows of cilia propels this protistan through its fluid surroundings.

Flagella and Cilia

A **flagellum** (plural, flagella) or **cilium** (plural, cilia) is an example of cytoskeletal organization. As Figure 3.19*a* shows, nine pairs of microtubules ring a central pair in these motile structures. A system of spokes and links holds this "9 + 2 array" together. The system bends the motile structure when a sliding mechanism operates.

Like whiplike tails, flagella propel sperm and many other free-living cells through fluid environments (Figure 3.19*b*). Cilia are shorter and usually more profuse than flagella (Figure 3.19*c*). In multicelled organisms, some ciliated epithelial cells stir their surroundings. For example, thousands line airways of your respiratory tract, and their beating cilia can direct mucus-trapped bacteria and other particles away from the lungs.

Centrioles and MTOCs

The microtubules of a flagellum or cilium arise from centrioles. These remain at the base of the completed structure, as a basal body (Figure 3.19*a*). In many cells, we find the centrioles located in MTOCs (microtubule organizing centers). These sites of dense material give rise to large numbers of microtubules with different roles in the cytoskeleton. For example, one type of MTOC near the cell nucleus gives rise to microtubules that move chromosomes about before a cell divides.

Microtubules and other elements of the cytoskeleton are the basis of a eukaryotic cell's shape, its internal structure, and its capacity for movements.

Cilia and flagella are motile structures that project from the surface of many cells and beat in distinctive patterns. Their function arises from their cytoskeletal organization.

3.12 PROKARYOTIC CELLS—THE BACTERIA

We turn now to the bacteria. Unlike the cells we have considered so far, all species of bacteria are prokaryotic; their DNA is *not* enclosed within a nucleus. *Prokaryotic* means "before the nucleus." The word implies that bacteria existed on Earth before the nucleus evolved in forerunners of all other types of cells.

Bacteria are the smallest of all cells. They usually are not much more than one micrometer wide; even rod-shaped species are only about a few micrometers long.

In structural terms, bacteria are the simplest kinds of cells to think about. Most species have a semi-rigid or rigid cell wall that surrounds the plasma membrane (Figure 3.20*a*). Although the wall is porous, it supports and imparts shape to the cell. Polysaccharides cover many cell walls. They help a bacterium attach to rocks, teeth, the vagina, and other interesting surfaces. In many disease-causing species, polysaccharides form a jellylike capsule that surrounds and protects the wall.

As is true of eukaryotic cells, bacteria have a plasma membrane that controls the movements of substances to and from the cytoplasm. Their plasma membrane, too, has a variety of receptors and built-in machinery for metabolic reactions, such as the breakdown of energy-rich molecules. Clusters of proteins in the plasma membrane of photosynthetic bacteria harness light energy and convert it to the chemical energy of ATP.

Bacterial cells are too small to contain more than a small volume of cytoplasm, but they have many ribosomes, where polypeptide chains are assembled. These cells also are small enough and so internally simple that they apparently do not require a cytoskeleton. Unlike eukaryotic cells, a bacterium's cytoplasm is continuous with an irregularly shaped region of DNA. No membranes surround this region, which often is called a nucleoid. As Figure 3.20*c* indicates, a single, circular molecule of DNA occupies this region.

One or more **bacterial flagella** (singular, flagellum) extends from the surface of many bacterial cells. These

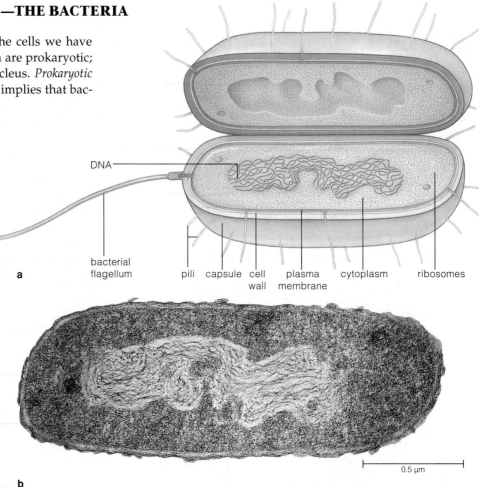

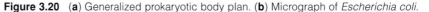

a   bacterial flagellum   pili   capsule   cell wall   plasma membrane   cytoplasm   ribosomes

DNA

0.5 μm

b

Figure 3.20 (**a**) Generalized prokaryotic body plan. (**b**) Micrograph of *Escherichia coli*.

Your own gut is home to a large population of a normally harmless strain of *E. coli*. In 1993, a dangerous strain contaminated meat that was sold to some fast-food restaurants. The same strain also contaminated hard apple cider sold at a few roadside stands. Cooking the meat thoroughly (or boiling the cider) would have killed the bacterial cells. Where this was not done, people who ate the meat or drank the cider became quite sick. Some died.

Facing page: (**c**) This *E. coli* cell has been shocked into releasing its circular molecule of DNA. (**d**) Cells of different bacterial species are shaped like balls, rods, or corkscrews. The ball-shaped cells of *Nostoc*, a photosynthetic bacterium, stick together inside a thick, gelatinlike sheath of their own secretions. Chapter 18 gives other splendid examples. (**e**) Like this *Pseudomonas marginalis* cell, many species have one or more bacterial flagella—motile structures that propel the cell through fluid environments.

threadlike motile structures permit rapid movements through fluid environments. They are not the same as eukaryotic flagella; they do not have a 9 + 2 array of microtubules. Other surface projections include protein filaments called pili (singular, pilus), which help many bacteria attach to various surfaces, even to one another.

Taken as a group, bacteria are certainly the most metabolically diverse organisms. They have managed to exploit energy and raw materials in just about every kind of environment. Besides this, ancient members of their kingdom gave rise to all the protistans, plants, fungi, and animals ever to appear on earth. The evolu-

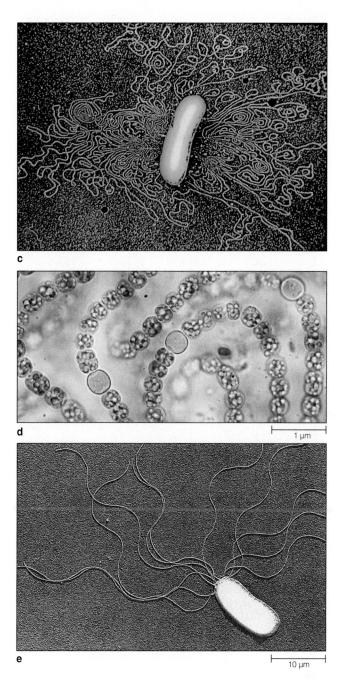

c

d

1 μm

e

10 μm

tion, structure, and functioning of bacteria are topics of later chapters.

Bacteria alone are prokaryotic cells; their cytoplasm does not contain a nucleus. Most species have a cell wall around the plasma membrane. Generally, their cytoplasm does not have any organelles comparable to those of eukaryotic cells.

Although bacteria are the simplest cells, when considered as a group, they are by far the most metabolically diverse. Most metabolic events proceed at the plasma membrane and at numerous ribosomes in the cytoplasm.

SUMMARY

1. Three generalizations constitute the cell theory:

   a. All living things are composed of one or more cells.

   b. The cell is the smallest unit that retains the properties of life. That is, it either lives independently or has a built-in, genetic capacity to do so.

   c. A new cell arises only from cells that already exist.

2. At the minimum, a newly formed cell has a plasma membrane, a region of cytoplasm, and a region of DNA.

3. Membranes are crucial to cell structure and function. Cell membranes are composed of lipids and proteins, for the most part. The lipids are mainly phospholipids, arranged as two layers. This lipid bilayer imparts structure to the membrane and bars passage of water-soluble substances across it. Diverse proteins are embedded in the bilayer or attached to its surfaces.

4. The proteins of a cell membrane carry out most of its functions. They include the following:

   a. All cell membranes incorporate transport proteins, which allow or promote the passage of water-soluble substances across the lipid bilayer.

   b. All plasma membranes also incorporate receptor proteins, which bind extracellular substances (such as hormones) that trigger changes in cell activities.

   c. In multicelled organisms, the plasma membrane incorporates recognition proteins, which are like molecular fingerprints that identify the cell as being of a certain type. It also incorporates adhesion proteins, which help cells stick together in their proper tissues.

5. Cell membranes divide the cytoplasm of eukaryotic cells into functional compartments called organelles. Prokaryotic cells do not have comparable organelles.

6. Organelle membranes separate metabolic reactions in the space of the cytoplasm and allow different kinds to proceed in orderly fashion. (In bacteria, many similar reactions proceed at the plasma membrane.)

   a. The nuclear envelope functionally separates the DNA from the metabolic machinery of the cytoplasm.

   b. The cytomembrane system includes the ER, Golgi bodies, and various vesicles. New proteins are modified into final form and lipids are assembled in this system. The finished products are packaged and then shipped off for destinations inside or outside the cell.

   c. Mitochondria specialize in oxygen-requiring reactions that produce many ATP molecules.

   d. Chloroplasts trap sunlight energy and produce organic compounds in photosynthetic eukaryotic cells.

7. Eukaryotic cells have a cytoskeleton, the components of which function in cell shape, internal organization, and movements.

8. Table 3.2 on the next page summarizes the defining features of both prokaryotic and eukaryotic cells.

Table 3.2 Summary of Typical Components of Prokaryotic and Eukaryotic Cells

Cell Component	Function	Prokaryotic Moneran	Eukaryotic Protistan	Fungus	Plant	Animal
Cell wall	Protection, structural support	✓*	✓*	✓	✓	none
Plasma membrane	Control of substances moving into and out of cell	✓	✓	✓	✓	✓
Nucleus	Physical separation and organization of DNA	none	✓	✓	✓	✓
DNA	Encoding of hereditary information	✓	✓	✓	✓	✓
RNA	Transcription, translation of DNA messages into specific proteins	✓	✓	✓	✓	✓
Nucleolus	Assembly of ribosomal subunits	none	✓	✓	✓	✓
Ribosome	Protein synthesis	✓	✓	✓	✓	✓
Endoplasmic reticulum (ER)	Initial modification of many newly forming proteins; lipid synthesis	none	✓	✓	✓	✓
Golgi body	Final modification of proteins, lipids; sorting and packaging them for use inside cell or for export	none	✓	✓	✓	✓
Lysosome	Intracellular digestion	none	✓	✓*	✓*	✓
Mitochondrion	ATP formation	**	✓	✓	✓	✓
Photosynthetic pigment	Light–energy conversion	✓*	✓*	none	✓	none
Chloroplast	Photosynthesis, some starch storage	none	✓*	none	✓	none
Central vacuole	Increasing cell surface area, storage	none	none	✓*	✓	none
Cytoskeleton	Cell shape, internal organization, basis of cell motion	none	✓*	✓*	✓*	✓
9+2 flagellum, cilium	Movement	none	✓*	✓*	✓*	✓

*Known to occur in at least some groups.
**Oxygen-requiring (aerobic) pathways of ATP formation do occur in many groups, but mitochondria are not involved.

Review Questions

1. Label the organelles in this diagram of a plant cell. *47*

2. Label the organelles in this diagram of an animal cell. *47*

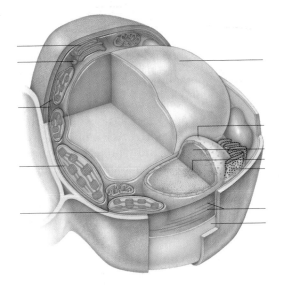

3. State the three key points of the cell theory. *41*

4. Suppose you want to observe the three-dimensional surface of an insect's eye. Would you benefit most from a compound light microscope, transmission electron microscope, or scanning electron microscope? *44–45*

5. Describe the three features that all cells have in common. Then, after reviewing Table 3.2 and the appropriate text sections, write a paragraph describing the key differences between prokaryotic and eukaryotic cells. *42, 46–47, 60–61*

6. Which organelles are part of the cytomembrane system? Sketch their arrangement in an animal cell, from the nuclear envelope to the plasma membrane. *52–53*

7. Is this statement true or false? Plant cells contain chloroplasts, but not mitochondria. *55*

8. What are the functions of the central vacuole in mature, living plant cells? *56*

9. Define cytoskeleton. How does it aid in cell functioning? *58*

10. Are all components of the cytoskeleton permanent? *58*

11. What gives rise to the 9 + 2 array of cilia and flagella? *59*

12. Cell walls are features of which organisms: bacteria, protistans, plants, fungi, or animals? Are cell walls solid or porous? *57*

13. In certain plant cells, is a secondary wall deposited inside or outside the surface of the primary wall? *57*

14. In multicelled organisms, coordinated interactions depend on linkages and communications between adjacent cells. What types of junctions occur between adjacent animal cells? Plant cells? *57*

Self-Quiz *(Answers in Appendix IV)*

1. Cell membranes consist mainly of a _____ .
   a. carbohydrate bilayer and proteins
   b. protein bilayer and phospholipids
   c. phospholipid bilayer and proteins
   d. none of the above

2. Organelles _____ .
   a. are membrane-bound compartments
   b. are typical of eukaryotic cells, not prokaryotic cells
   c. separate chemical reactions in time and space
   d. all of the above are functions of organelles

3. Plant cells but not animal cells have _____ .
   a. mitochondria
   b. a plasma membrane
   c. ribosomes
   d. a cell wall

4. Unlike eukaryotic cells, prokaryotic cells _____ .
   a. lack a plasma membrane   c. do not have a nucleus
   b. have RNA, not DNA   d. all of the above

5. Match each cell component with its function.
   ____ mitochondrion   a. synthesis of polypeptide chains
   ____ chloroplast   b. initial modification of new
   ____ ribosome   polypeptide chains
   ____ rough ER   c. final modification, sorting, shipping
   ____ Golgi body   of proteins, lipids
   d. photosynthesis
   e. ATP formation

Critical Thinking

1. Why is it likely that you will never encounter a predatory two-ton living cell on the sidewalk?

2. Many compound light microscopes have blue filters. Think about the spectrum of visible light (Figure 5.3 shows one diagram of this) and explain why blue light might be most efficient when viewing objects at high magnification.

3. Your biology professor shows you an electron micrograph of a cell that contains very large numbers of mitochondria and Golgi bodies. You notice that this particular cell also has a great deal of rough endoplasmic reticulum. What kinds of cellular activities would require such an abundance of the three kinds of organelles?

4. *Cystic fibrosis* is a disabling and, in time, fatal genetic disorder. A defective form of a certain protein is implicated in the disorder. The protein is a component of the plasma membrane of cells of the glands that secrete mucus, digestive enzymes, and sweat.

The glands of affected individuals secrete far too much, with far-reaching effects on the body's functioning. For example, in time, digestive enzymes clog a duct leading from the pancreas to the small intestine, food cannot be digested properly, and malnutrition results even if food intake increases. Cysts form in the pancreas, which degenerates and becomes fibrous (hence the name of this disorder). Also, thick mucus accumulates in the respiratory tract, so affected individuals have great difficulty expelling airborne bacteria and particles that enter the lungs. Finally, too much sweat forms. Midwives used to lick the forehead of newborns. If they tasted far too much salt in the sweat, they predicted that the new individual would develop lung congestion.

Using your understanding of cell structure and function, track this disorder back from the affected protein to its original source.

Selected Key Terms

adhesion protein *43*	endocytosis *54*	nuclear envelope *51*
bacterial flagella *60*	ER (endoplasmic	nucleolus *50*
cell *42*	reticulum) *52*	nucleus *50*
cell theory *41*	eukaryotic cell *42*	organelle *46*
cell wall *57*	exocytosis *54*	peroxisome *53*
central vacuole *56*	flagellum *59*	phospholipid *42*
centriole *59*	fluid mosaic	plasma
chloroplast *56*	model *42*	membrane *42*
chromatin *51*	Golgi body *52*	prokaryotic cell *42*
chromosome *51*	intermediate	receptor protein *43*
cilium (cilia) *59*	filament *58*	recognition
cytomembrane	lipid bilayer *42*	protein *43*
system *51*	lysosome *53*	surface-to-volume
cytoplasm *42*	microfilament *58*	ratio *44*
cytoskeleton *58*	micrograph *44*	transport protein *43*
DNA-containing	microtubule *58*	wavelength *44*
region *42*	mitochondria *55*	

Readings

deDuve, C. 1985. *A Guided Tour of the Living Cell.* New York: Freeman. Beautifully illustrated introduction to the cell; two short volumes.

Wolfe, S. 1995. *An Introduction to Molecular and Cellular Biology.* Belmont, California: Wadsworth. Outstanding reference text.

4 GROUND RULES OF METABOLISM

Growing Old With Molecular Mayhem

Somewhere in those slender strands of DNA in your cells are snippets of instructions for constructing two truly wonderful proteins. The proteins go by the names superoxide dismutase and catalase. Both are enzymes—a type of molecule that makes metabolic reactions proceed much faster than they would spontaneously, on their own. And both help keep you from growing old before your time.

The two enzymes help your cells clean house, so to speak. Together, they produce and then disassemble hydrogen peroxide (H_2O_2), a normal but potentially toxic by-product of certain oxygen-requiring reactions. Oxygen (O_2) is supposed to pick up electrons from the reactions. Sometimes it picks up only one—which is not enough to complete the reactions but is enough to give the oxygen a negative charge (O_2^-).

Like other unbound, molecular fragments with the wrong number of electrons, O_2^- is a **free radical**. Free radicals are *so* reactive, they even attach to molecules that usually will not take part in just any reaction—molecules like DNA. An attack by free radicals can disrupt DNA's structure and destroy its function.

a

b

Figure 4.1 (**a**) Owner of a good supply of functional enzymes that help keep free radicals in check.

(**b**) Owner of skin with a spattering of age spots—visible evidence of free radicals on the loose.

Enter <u>superoxide dismutase</u>. Under its chemical prodding, two of the rogue oxygen molecules combine with hydrogen ions. H_2O_2 and O_2 are the outcome.

Enter <u>catalase</u>. Under *its* prodding, two molecules of hydrogen peroxide react and split into ordinary water and ordinary oxygen: $2H_2O_2 \longrightarrow 2H_2O + O_2$.

As people age, their capacity to produce functional proteins—including enzymes—begins to falter. Among those enzymes are superoxide dismutase and catalase. Cells produce them in diminishing numbers, crippled form, or both.

When that happens, free radicals and hydrogen peroxide accumulate. Like loose cannons, they careen through cells with tiny blasts at the structural integrity of proteins, DNA, membranes, and other vital parts.

Cells suffer or die outright. Those brown "age spots" you may have noticed on an older person's skin are evidence of assaults by free radicals. The prominent spot on the forehead of the gentleman shown in Figure 4.1*b* is an example. Each age spot is a mass of brownish-black pigment molecules that build up in cells when free radicals finally take over—all for the want of two enzymes.

With this chapter, we start to examine activities that keep cells alive and functioning smoothly. At times the topics may seem remote from the world of your interests. But they help define who *you* are and who you will become, age spots and all.

KEY CONCEPTS

1. Cells engage in metabolism—that is, they use energy to build, stockpile, break apart, and eliminate substances in ways that help them survive and reproduce.

2. With each metabolic reaction, energy escapes into the environment. To stay alive, cells must balance their energy losses with energy gains. But they cannot create energy from scratch. They can only draw upon existing sources, such as light energy from the sun or chemical energy that has become tucked away in glucose and other substances.

3. Compared to their environment, cells maintain greater or lesser amounts of certain substances, as required for metabolism. They do so even though the molecules or ions of any substance have a natural tendency to diffuse into regions where they are less concentrated. Membrane transport proteins work with or against this tendency.

4. Metabolic pathways maintain, increase, or decrease the relative amounts of various substances in cells. Typically, they couple reactions that release usable energy from substances to other reactions that require energy.

5. Chemical reactions proceed far too slowly on their own to sustain life. In cells, enzymes greatly increase the rate of specific reactions.

6. ATP transports usable energy in chemical form from one reaction site to another. When it donates a phosphate group to a substance, the substance becomes activated—that is, primed for chemical change. Metabolic pathways depend on such group transfers.

Find a microscope and watch a living cell, suspended in a water droplet. The image might jar you—something that small pulsates with movement. Even as you watch it, the cell takes in energy-rich solutes from the water. It builds membranes, stores things, checks out the DNA, and replenishes enzymes. It is alive; it is growing; it may divide in two. Multiply these activities by *trillions* of cells and you have an idea of what goes on in your own body, even when you do nothing more than sit quietly and watch a single cell!

These dynamic activities are signs of **metabolism**—of a cell's capacity to acquire energy and use it to build, break apart, store, and release substances in controlled ways. Metabolism is the means by which cells survive and reproduce—and it all begins with energy.

Energy is a capacity to make things happen, to do work. We can measure it, as in kilocalories. A *kilocalorie* is the same thing as a thousand calories—the amount of energy that will heat 1,000 grams of water from 14.5°C to 15.5°C at standard pressure. You use energy when you wash a car, even when you sleep. In both cases, your muscles, brain, and other body parts are being put to work. Energy drives the muscle cell contractions that move bones or hold them in various positions. Energy allows brain cells to chatter among themselves and guide what you do. Gain insight into how cells secure, use, and lose energy, and you gain insight into life.

How Much Energy Is Available?

Like single cells, you cannot create energy from scratch; you must get it from someplace else. Why? By the **first law of thermodynamics**, the total amount of energy in the universe remains constant. More energy cannot be created, and existing energy cannot be destroyed. It can only be converted from one form to another.

Think about what this law means. The universe has only so much energy, distributed in a variety of forms. One form can be converted to another, as when corn plants absorb energy from the sun and convert it to the chemical energy of starch. After you eat corn, your cells extract energy from the starch and convert it to other forms—such as mechanical energy for movements. With each conversion, a bit of energy escapes to the surroundings, as heat. Even when you are "doing nothing," your body gives off about as much heat as a 100-watt light bulb, because conversions are proceeding in your cells. The released energy "heats up" the surroundings as it is transferred to the disorderly array of atoms and molecules that make up the air (Figure 4.2). The added energy increases the number of ongoing, random collisions among molecules—and with every collision, a bit more of the energy spreads outward. However, none of that energy *vanishes*. It simply keeps on spreading out in a random, increasingly disorderly way, among molecules in the air.

The One-Way Flow of Energy

Most of the energy available for conversions in cells resides in covalent bonds. Glucose, glycogen, starches, fatty acids, and other organic compounds with many of these bonds are said to have a high energy content. When they enter metabolic reactions, the bonds may break or parts may become rearranged. At such times, atoms move. Their movements add to the molecular commotion (heat energy) of the surroundings. In general, cells cannot recapture energy lost as heat. Their enzymes can only direct atoms into different bonds.

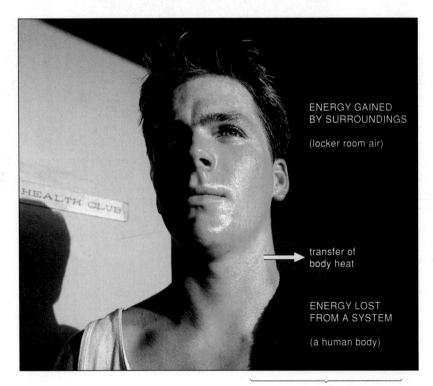

ENERGY GAINED
BY SURROUNDINGS

(locker room air)

transfer of
body heat

ENERGY LOST
FROM A SYSTEM

(a human body)

net energy change = 0

Figure 4.2 Example of how the total energy content of a system *together with its surroundings* remains constant.

"System" means all matter in a specific region, such as a human body, a plant, a DNA molecule, or a galaxy. "Surroundings" can be a small region in contact with the system or as vast as the entire universe. The system shown is giving off heat to the surroundings by evaporative water loss from sweat. What one region loses, the other gains, so the total energy content of both doesn't change.

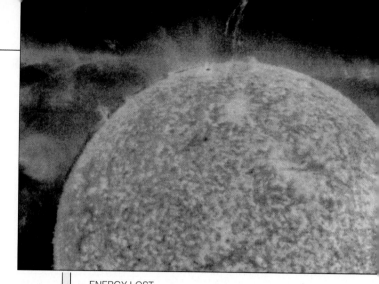

Figure 4.3 Example of the one-way flow of energy into the world of life that compensates for the one-way flow out of it. The sun continuously loses energy, much of it in the form of wavelengths of light (Section 3.3). Living cells intercept some of that energy and convert it to useful forms of energy, stored in bonds of organic compounds. Each time a metabolic reaction proceeds in cells, stored energy is released—and some inevitably is lost to the surroundings, mostly as heat.

In the lower photograph, the *green* "dots" are tiny, spherical colonies of water-dwelling, photosynthetic cells (*Volvox*). *Orange* dots are cells set aside for reproduction. They form new colonies inside the parent sphere.

ENERGY LOST
one-way flow of energy away from the sun

ENERGY GAINED
one-way flow of energy from the sun into organisms

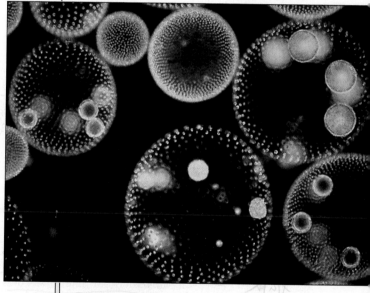

Think of what happens when your cells break all the covalent bonds in glucose that they can. After many reaction steps, six molecules of carbon dioxide and six of water remain. Compared to glucose, these leftovers have more stable arrangements of atoms. However, the energy remaining in all of their bonds is much less than the total bond energy of glucose. Why? *Some energy was lost as heat at each step leading to their formation.*

Said another way, as a source of usable energy, the carbon dioxide is lower in quality than glucose. And the small amount of heat energy that was transferred to the surroundings during its formation is very low quality, because it doesn't lend itself to conversions in cells.

Bad news for cells of the remote future: The amount of "low-quality" energy in the universe is increasing. Because no energy conversion can ever be 100 percent efficient, the total amount of energy in the universe is spontaneously flowing from forms of higher to lower energy content. That, basically, is the point to remember about the **second law of thermodynamics**.

Without energy to maintain it, any organized system tends to become disorganized over time. **Entropy** is a measure of the degree of a system's disorder. Think of the Egyptian pyramids—originally organized, presently crumbling, and many thousands of years from now, dust. The ultimate destination of the pyramids and everything else in the universe is a state of maximum entropy. Billions of years from now, all of the energy available for conversions will be dissipated. And nothing we know of will pull it together again.

Can it be that life is one glorious pocket of resistance to the depressing flow toward maximum entropy? After all, every time a new organism grows, new bonds form and hold atoms together in precise arrays—so energy becomes *more* concentrated and organized, not less so! Yet a simple example will show that the second law does indeed apply to life on Earth.

The primary energy source for life on Earth is the sun—which is steadily losing energy. Plants capture some sunlight energy, convert it in various ways, then lose energy to other organisms that feed, directly or indirectly, on plants. At each energy transfer along the way, some energy is lost, usually as heat that joins the

ENERGY LOST
one-way flow of energy away from organisms
to the surroundings

universal pool. *Overall, energy still flows in one direction.* The world of life maintains a high degree of organization only because it is being resupplied with energy lost from someplace else (Figure 4.3).

The amount of energy in the universe remains constant. Energy can undergo conversions from one form to another, but it cannot be created from scratch or destroyed.

The total amount of energy in the universe is spontaneously flowing from forms of higher to lower energy content.

A steady flow of sunlight energy into the interconnected web of life compensates for the steady flow of energy leaving it.

The substances within a cell differ in kind and amounts from those of its surroundings. Metabolism depends on the differences—which membranes help establish and maintain. Cell membranes, recall, are protein-studded lipid bilayers (Section 3.1). An outer, plasma membrane encloses the cytoplasm. In eukaryotic cells, internal membranes form compartments that can hold different kinds and amounts of substances.

Every cell membrane shows **selective permeability**. That is, *some substances but not others can cross them in certain ways, at certain times*. Oxygen, carbon dioxide, and other small molecules with no net charge readily cross the bilayer itself. Ions and large, water-soluble molecules such as glucose only cross the membrane if and when transport proteins assist them. Such movements raise an interesting question. Regardless of whether its movement is unassisted or assisted, what compels any substance to move in a given direction in the first place?

Following the Gradients

Two concepts help explain why substances move in certain directions. First, molecules (and ions) have internal energy that keeps them in constant motion. Second, **concentration gradients** can exist for each kind of molecule. "Concentration" refers to the number of molecules of a substance in a specified volume of fluid. Add the word "gradient," and this means one region of the fluid contains more molecules than a neighboring region.

In the absence of other forces, molecules move down their concentration gradient. They do so because they collide constantly, millions of times a second. The randomly colliding molecules move back and forth, but their *net* movement is away from the place where they are most concentrated (Figure 4.4). **Diffusion** is the name for this net movement of like molecules down their concentration gradient.

Suppose a fluid contains more than one substance. This makes no difference. *Each* substance will diffuse in the direction set by its *own* concentration gradient.

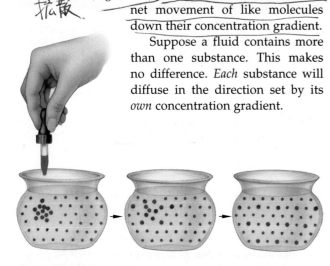

Figure 4.4 Diffusion. Dye molecules *and* water molecules get evenly dispersed through a bowl of water as each substance shows a net movement down its own concentration gradient.

Diffusion rates are faster when a gradient is steep. (Far more molecules are moving outward, compared to the number moving the other way.) As the gradient decreases, the difference in the number moving one way or the other becomes less pronounced. When the gradient disappears, individual molecules are still in motion. But the total number moving one way or the other is just about the same at any given time.

Temperature also affects diffusion rates (heat causes molecules to move faster and collide more frequently). Besides this, smaller molecules move faster than large ones do. Finally, an *electric gradient* (a difference in charge between two adjoining regions) can modify the rate and direction of diffusion. So can a *pressure gradient* (a difference in pressure between two regions).

Osmosis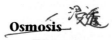

By itself, water cannot get more or less concentrated. Why? Hydrogen bonds keep its molecules from crowding closer together or drifting apart. Liquid water gets "less concentrated" only when substances are dissolved in it (Figure 4.5). Therefore, the number of molecules of all solutes on either side of a membrane affects the water concentration gradient across that membrane.

Tonicity is the name for the relative concentrations of solutes in two fluids. Solute concentrations are the same in *isotonic* fluids. A *hypotonic* fluid has fewer solutes than a *hypertonic* one. If a cell membrane separates two isotonic fluids, water shows no net movement either way. But water tends to move into regions of higher solute concentrations. Usually, solute concentrations on the two sides of a membrane are not the same, but most cells have built-in means to adjust to this.

water molecule   semipermeable membrane   protein molecule

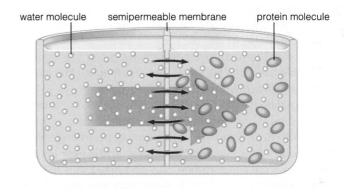

Figure 4.5 Osmosis in a container divided in two by a membrane that water molecules can cross. First pour pure water into the left compartment, then the same volume of a protein-rich solution into the right one. Proteins cannot move across the membrane. They always occupy some of the available space in the compartment at the right—which therefore has fewer water molecules than the one at left. Water molecules follow the water concentration gradient; the net osmotic movement is from left to right (*large blue arrow*).

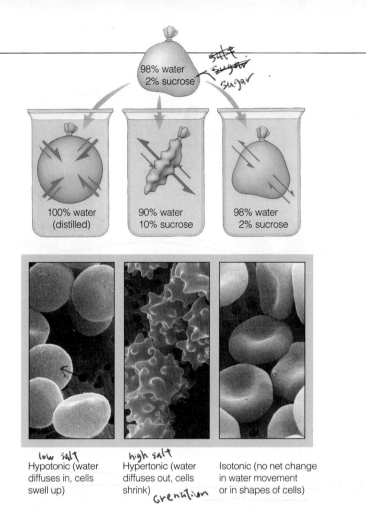

98% water
2% sucrose

salt
sugar
sugar

100% water
(distilled)

90% water
10% sucrose

98% water
2% sucrose

low salt
Hypotonic (water
diffuses in, cells
swell up)

high salt
Hypertonic (water
diffuses out, cells
shrink)
Crenation

Isotonic (no net change
in water movement
or in shapes of cells)

Figure 4.6 Effect of tonicity on water movement. Picture a sac made of a membrane that water but not sucrose can move across. Fill three such bags with a 2% sucrose solution. Immerse one in water, one in a solution of higher sucrose concentration, and the third in the same solution. The direction and relative amounts of water movement will differ among them (indicated by arrow widths).

The micrographs correspond to the sketches. They show the kinds of shapes that human red blood cells will assume when placed in fluids of higher, lower, and equal solute concentrations. Normally, the solutions inside and outside of red blood cells are in balance. These particular cells do not have built-in means to adjust to drastic changes in solute levels in their fluid surroundings.

Cells that do not will shrivel or burst when the solute concentration of their surroundings rises or falls past their range of tolerance. Figure 4.6 gives examples of this.

Pressure may modify the direction of water movement. Think about the fluid pressure generated by your beating heart. In your kidneys, it forces some water and small solutes to leave the blood, enter the surrounding tissue fluid, then cross cell membranes. Tissue fluid is more dilute than plasma (the fluid portion of blood), yet pressure forces water to move *against* the gradient. When a pressure gradient compels water and solutes to move in the same direction, we call this "bulk flow."

Osmosis is the name for the movement of water across a selectively permeable membrane in response to

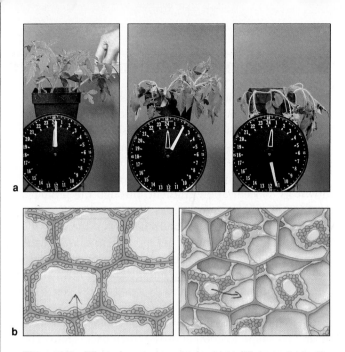

a

b

Figure 4.7 Effect of osmosis on leafy plants. When a plant with soft green leaves is growing well, you can safely bet that water in the surrounding soil is dilute (hypotonic), compared to fluids in the plant's living cells. Plant cells have fairly stiff walls. As water responds to solute concentration gradients and moves into cells, fluid pressure increases against the wall.

Internal fluid pressure on a cell wall is called _turgor pressure._ It builds up against any walled cell when water moves inward by osmosis. But water also is squeezed out when the pressure is great enough to counter the attractive force of cytoplasmic fluid—which usually has more solutes than soil water does. Both forces have the potential to cause directional movement of water. The sum of these two opposing forces is the "water potential."

When soft plant parts are erect, as much water is moving into cells as is moving out, and the constant turgor pressure keeps cell walls plump. Suppose the soil dries or has too many solutes. Water stops moving in, and cells lose water. *Wilting* occurs; soft plant parts droop with the loss of turgor pressure.

(**a**) A simple experiment shows the wilting effect. Put 10 grams of table salt (NaCl) in 60 milliliters of water. Pour the salty solution into the soil around a tomato plant. The plant starts to collapse after 5 minutes. In less than 30 minutes, wilting is severe.

(**b**) What happens to a plant cell when water leaves by osmosis? Plasmolysis occurs—the cytoplasm and central vacuole shrink, causing the plasma membrane to pull away from the cell wall.

concentration gradients, fluid pressure, or both. Figure 4.7 is a fine example of how solute concentrations and pressure influence the direction of water movement.

Diffusion is the net movement of like molecules or ions down a concentration gradient. Osmosis is the movement of water across a selectively permeable membrane in response to concentration gradients, fluid pressure, or both.

Cells work with or against concentration gradients. By doing so, they establish and maintain internal stores of substances at levels that are required for metabolism.

MOVEMENT THROUGH TRANSPORT PROTEINS

Let's now take a look at how water-soluble substances diffuse into and out of cells or organelles. Molecules or ions of such substances must be assisted by transport proteins, which span the lipid bilayer of every cell membrane (Figure 4.8). By a mechanism called **passive transport**, a specific solute is simply allowed to diffuse down its concentration gradient, through the protein's interior. (This also is known as "facilitated" diffusion.) By **active transport**, a solute still moves through the interior of a protein, but its net movement is *against* the concentration gradient. This mechanism operates only when the protein receives an energy boost.

To understand how these proteins work, you have to know they are not rigid blobs of atoms. When the protein interacts with a particular solute, it changes from one shape to another shape, then back again. The changes start when a solute becomes weakly bound at a specific site on the protein surface. Part of the protein closes in behind the bound solute—and part opens up to the opposite side of the membrane. There, the solute dissociates (separates) from the site. Think of the solute as hopping onto the transport protein on one side of the membrane, then hopping off on the other side.

Passive Transport

Transport proteins can move solutes both ways across a membrane. With passive transport, the *net* direction of movement at a given time depends on how many molecules or ions of the solute are making random contact with vacant binding sites on those proteins (Figure 4.9). Binding and transport proceed more often on the side of the membrane where the solute is more concentrated. With more molecules around, random encounters with the binding sites are more frequent than they are on the other side of the membrane.

By itself, the passive two-way transport would continue until solute concentrations became equal on both sides of the membrane. In most cases, however, other

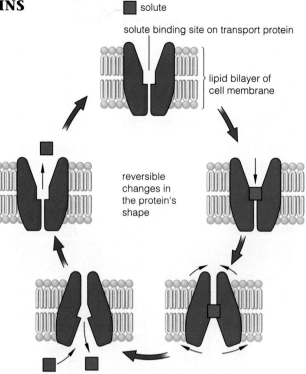

Figure 4.9 Passive transport across a cell membrane. Transport proteins can move a solute in both directions. In passive transport, the *net* movement will be down the solute's concentration gradient (from higher to lower concentration). This will continue until the concentrations become the same on both sides of the membrane.

processes affect the outcome. For example, your bloodstream delivers glucose to all of your body's tissues. Nearly all cells require glucose as an energy source and as building blocks. When the glucose concentration in blood (and tissues) is high, cells take up glucose. But as fast as molecules of glucose diffuse into cells, metabolic reactions usually are withdrawing other glucose molecules from the cytoplasm. Thus, when cells rapidly use glucose, they are actually maintaining a concentration gradient that favors the uptake of *more* glucose.

Figure 4.8 Examples of transport proteins that span the lipid bilayer of a plasma membrane. Transport proteins passively or actively permit ions as well as glucose and other large molecules having no net charge to move through their interior, from one side of the membrane to the other.

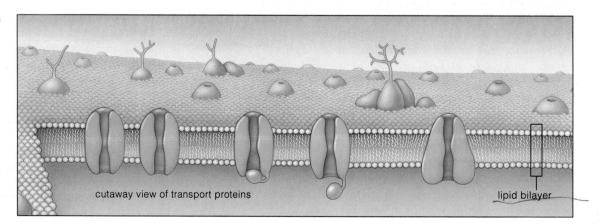

cutaway view of transport proteins

lipid bilayer

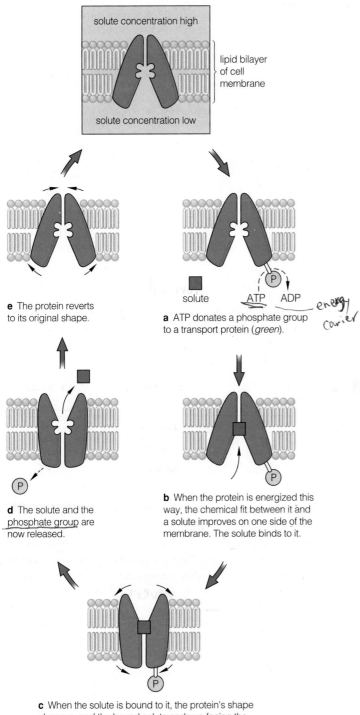

solute concentration high

lipid bilayer of cell membrane

solute concentration low

e The protein reverts to its original shape.

solute   ATP   ADP   *energy carrier*

a ATP donates a phosphate group to a transport protein (*green*).

d The solute and the phosphate group are now released.

b When the protein is energized this way, the chemical fit between it and a solute improves on one side of the membrane. The solute binds to it.

c When the solute is bound to it, the protein's shape changes, and the bound solute ends up facing the opposite side of the membrane. There, the binding site reverts to the less attractive configuration.

Figure 4.10 Active transport across a cell membrane. When a transport protein receives an energy boost from ATP, the binding site for its solute becomes altered—and so becomes more attractive to the solute. After the solute is moved to the opposite side of the membrane, the site reverts to its less attractive shape. Even at low concentrations, the solute occupies the altered sites more often and moves across the membrane faster. Thus, there is a greater *net* movement against the concentration gradient.

Active Transport

Transport proteins with roles in active transport across cell membranes move a solute against its concentration gradient. Unlike passive transport, this mechanism can continue until the solute becomes *more* concentrated on the side to which it is being pumped.

Active transport does not proceed spontaneously. It requires an energy input, most often from ATP. In brief, ATP donates a phosphate group to the protein. When that happens, the chemical fit between the binding site and the solute improves on one side of the membrane. After a solute molecule or ion binds at the site, the folded shape of the protein changes. With this change, the bound solute becomes exposed to the fluid bathing the opposite side of the membrane (Figure 4.10). Now the binding site reverts to its less attractive state and the solute is released. When the site is less attractive, fewer molecules or ions of the solute make the return trip. So the *net* movement is to the side of the membrane where the solute is more concentrated. By analogy, picture hordes of skiers rapidly hopping onto the chairs of a ski lift that moves them up a mountain. At the top, the chairs tilt in a way that is not very inviting for skiers who want to hop on for a downhill ride. Thus, more skiers will be transported uphill than downhill.

One active transport system, known as the calcium pump, helps keep the calcium concentration inside a cell at least a thousand times lower than outside.

The sodium-potassium pump is a major *co*transport system. In this case, an activated protein entices sodium ions to bind with it. Bound sodium causes alteration in another binding site, making it more easily occupied by potassium—the solute. When a cell membrane contains many sodium-potassium pumps, their operation can set up or maintain the concentration and electric gradients across that membrane. And energy inherent in those gradients can be used to drive cell activities.

We will return to mechanisms of active transport in chapters dealing with muscle contraction, information flow through nervous systems, and other physiological processes. For now, keep these points in mind:

Transport proteins bind molecules or ions of water-soluble substances. When they do, they undergo a reversible change in shape that shunts the solute across a cell membrane.

In passive transport, diffusion proceeds through the interior of a transport protein that spans a cell membrane. The net movement of the solute is down its concentration gradient.

In active transport, a solute crosses a cell membrane only after a transport protein receives an energy boost, as from ATP. The solute moves through a protein's interior—but in this case the net movement is against the concentration gradient.

4.4 CHARACTERISTICS OF METABOLIC REACTIONS

Cells never stop building and tearing down molecules and moving them in specific directions until they die. What does their great juggling act accomplish? Think it through. Cells can use only so many molecules at a given time, and they have only so much internal space to hold any excess. If they produce more of a substance than they can use, store, or secrete, the extra amounts might cause problems.

Consider *phenylketonuria*, or PKU. People affected by this heritable disorder produce a defective enzyme that prevents cells from using phenylalanine, which is one of the amino acids. When this amino acid accumulates, the excess enters reactions that produce phenylketones. An accumulation of *these* molecules damages the brain in less than a few months. In most developed countries, affected newborns are detected through routine screening programs. They can grow up symptom-free if they are placed on a phenylalanine-restricted diet.

The Direction of Metabolic Reactions

Concentration gradients help drive metabolic reactions in certain directions. Recall that molecules or ions are in constant random motion, which puts them on collision courses. The more concentrated they are, the more often they collide. Energy released during the collisions may be enough to cause a chemical reaction—that is, make a molecule change shape, combine with something else, or split into smaller parts.

Most metabolic reactions are reversible. They may proceed in the forward direction, from reactants (the starting substances) to products. They also may proceed in reverse, with products being converted back to the reactants. If you come across a chemical equation with two opposing arrows, as in Figure 4.11, this signifies that the reaction is reversible.

Which way a reaction runs depends partly on the ratio of reactant to product molecules. With a high reactant concentration, it runs strongly in the forward direction. When the product concentration is high enough, more products are available to revert to reactants.

Any reversible reaction tends to run spontaneously toward **chemical equilibrium**—at which time it will be proceeding at about the same pace in *both* directions (Figure 4.12). Bear in mind, the amounts of the reactant and product molecules may or may not be the same at equilibrium. Picture a party, with just as many people wandering in as wandering out of two adjoining rooms. The total number in each room remains the same—say, thirty in one and ten in the other even when the mix of people in each room changes.

Each reaction has a particular ratio of reactant to product molecules at equilibrium. Consider the conversion of glucose-1-phosphate into glucose-6-phosphate.

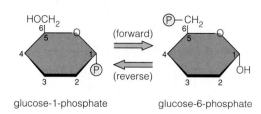

Figure 4.11 A reversible reaction. Glucose is primed to enter reactions when a phosphate group becomes attached to it. With a high concentration of glucose-1-phosphate, the reaction tends to run in the forward direction. With a high concentration of glucose-6-phosphate, it runs in reverse. (The 1 and 6 of these names simply identify which particular carbon atom of the glucose ring has a phosphate group attached to it.)

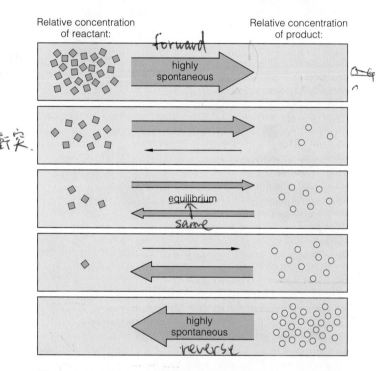

Figure 4.12 Chemical equilibrium. When the concentration of reactant molecules is high, a reaction runs most strongly in the forward direction (to products). When the concentration of product molecules is high, it runs most strongly in reverse. At equilibrium, the rates of the forward and reverse reactions are the same.

(As Figure 4.11 shows, these are glucose molecules with a phosphate group attached to one of their six carbon atoms.) Depending on the starting ratio of reactants and products, the reaction initially will proceed in the forward or the reverse direction. Eventually, the forward and reverse reactions proceed at the same rate, but only when there are nineteen glucose-6-phosphate molecules for every glucose-1-phosphate molecule. In this case, the ratio at equilibrium is 19:1.

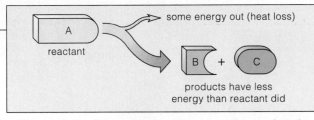

some energy out (heat loss)

A
reactant

B + C

products have less
energy than reactant did

a The reactants of some reactions have *more* energy than the products. Many such reactions release energy that cells can use, as happens during aerobic respiration.

we need energy 有酸素 呼吸

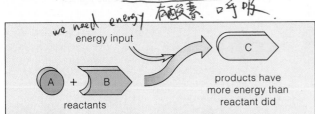

energy input

A + B

reactants

C

products have
more energy than
reactant did

b The reactants of other reactions have *less* energy than the products. Energy inputs drive such reactions, as when an input of sunlight energy drives photosynthesis.

c Coupled reactions. Some reactions (such as glucose breakdown) release usable energy that can be channeled to other reactions (such as synthesis reactions, which require energy).

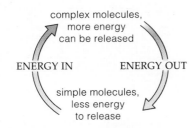

complex molecules, more energy can be released

ENERGY IN   ENERGY OUT

simple molecules, less energy to release

Figure 4.13 Energy changes in metabolic reactions.

Energy Flow and Coupled Reactions

Metabolic reactions release usable energy, as when cells break down glucose to carbon dioxide and water. These products have more stable structures (it takes more energy to break them apart). But the released energy was channeled elsewhere or lost as heat, so their total bond energies are lower than they were in glucose. Any reaction that ends with a net loss in energy is *exergonic*, meaning "energy out" (Figure 4.13*a*). In adult humans, the daily breakdown of glucose and other molecules releases an average of 1,200–2,800 kilocalories of energy.

When cells release energy, they channel much of it into specific reactions, including those by which ATP forms. With inputs of ATP energy, other reactions run in energetically unfavorable directions—that is, in ways that would not happen on their own. This is how cells build starch, fatty acids, and other "energy-rich" molecules from smaller molecules of lower energy content. Said another way, the synthesis reactions end with a net gain in energy (Figure 4.13*b*). They are *endergonic*, which means "energy in." The next two chapters especially will show how life depends on the coupling of *energy-releasing* reactions with *energy-requiring* reactions.

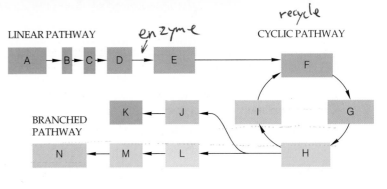

LINEAR PATHWAY   *enzyme*   *recycle*   CYCLIC PATHWAY

A → B → C → D → E → F

BRANCHED PATHWAY

K ← J   I   G

N ← M ← L ←   H

Figure 4.14 Types of reaction sequences of metabolic pathways.

Metabolic Pathways

At any instant, thousands of reactions are transforming thousands of substances in the confines of a cell. The reactions don't proceed willy-nilly. Most are organized as **metabolic pathways**, in which reactions proceed one after another, in orderly steps that enzymes mediate. Often the steps proceed in linear sequence. Often they proceed in a circle, with the end product becoming the starting substance. Many times, the end products or substances produced at earlier steps of one pathway become reactants elsewhere, so that two or more pathways become coupled in linear or branching fashion. Figure 4.14 shows some of the possibilities.

The main metabolic pathways are biosynthetic or degradative, overall. In the *biosynthetic* pathways, small molecules are assembled into complex carbohydrates, proteins, and other large molecules of higher energy content. In the *degradative* pathways, large molecules are broken down to products of lower energy content. Later in the book, you will consider examples of both kinds of pathways. You may find it useful to start thinking about their participants, as introduced next.

Substrates (also called reactants or precursors) are substances that enter a reaction. **Intermediates** are substances that form between the start and conclusion of a metabolic pathway. **End products** are the substances present at the end of a metabolic reaction or pathway.

Nearly all **enzymes** are proteins that catalyze (speed up) specific reactions. **Cofactors** are organic molecules or metal ions that assist enzymes or transport electrons or atoms. **Energy carriers** activate (donate energy) to substances by transferring functional groups to them. ATP is the main type. Finally, **transport proteins** adjust concentration gradients at cell membranes in ways that influence the direction of metabolic reactions.

A cell can simultaneously increase, decrease, and maintain the concentrations of thousands of different substances. It accomplishes this juggling act by coordinating thousands of reactions, organized into metabolic pathways.

The reactions proceed from substrates, to intermediates, to end products. Enzymes, cofactors, energy carriers, and transport proteins have roles in metabolic reactions.

4.5 ENZYMES

Characteristics of Enzymes

Even when you are about to fall asleep, when you think your body is shutting down for the night, its metabolic machinery is synthesizing and degrading uncountable numbers of molecules. Much of the glucose from your latest meal was dismantled earlier for a quick energy fix in your heart, brain, and muscle cells. More glucose is being converted to fat or something else. Hemoglobin molecules built two months ago are being digested to bits even as new hemoglobin molecules replace them.

Without enzymes, the dynamic, steady state called "you" would quickly cease to exist. Reactions simply would not proceed fast enough for the body to process food, build and tear down hemoglobin and other vital molecules, send signals among brain cells, make muscles contract, and do everything else to stay alive.

Enzymes are *catalytic* molecules; they enhance the rate at which reactions approach equilibrium. Nearly all of them are proteins, although some RNA molecules also show catalytic activity. Enzymes have four features in common. *First*, they do not make anything happen that couldn't happen on its own. But they usually make it happen at least a million times faster. *Second*, they are not permanently altered or used up in a reaction; the same one may act over and over again. *Third*, the same enzyme usually works for both the forward and reverse directions of a reaction. *Fourth*, each type of enzyme is highly selective about its substrates.

An enzyme's substrates are specific molecules that it can chemically recognize, bind, and modify in certain ways. For example, thrombin, an enzyme involved in blood clotting, recognizes a side-by-side arrangement of two amino acids, arginine and glycine, and cleaves a peptide bond between them:

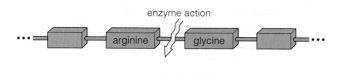

enzyme action

arginine   glycine

Enzyme-Substrate Interactions

A reaction occurs only when the participating atoms collide with some minimum amount of energy, called the **activation energy**. This is true regardless of whether it occurs spontaneously or with the help of enzymes. Think of the activation energy as an "energy hill" that must be surmounted before a reaction will proceed.

Enzymes make the energy hill smaller, so to speak. How do they do this? As Figures 4.15*a* and *b* indicate, an enzyme has one or more **active sites**. These are crevices in its surface where substrates interact with the enzyme and where a specific reaction is catalyzed.

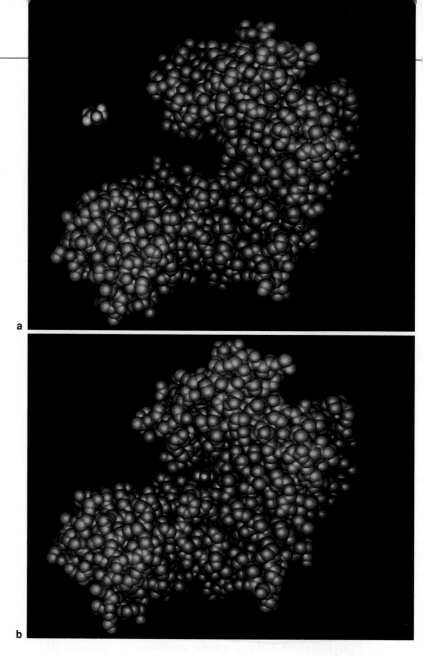

a

b

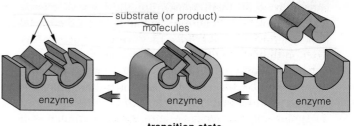

substrate (or product) molecules

enzyme   enzyme   enzyme

transition state
(tightest binding but least stable)

c

Figure 4.15 An enzyme at work. (**a**) Model of hexokinase (*green shading*). Its substrate, a glucose molecule (*red*), is heading toward the active site, a cleft in the enzyme. (**b**) When glucose contacts the site, parts of the enzyme temporarily close in around it and prod the molecule to enter a specific reaction.

(**c**) Induced-fit model of enzyme-substrate interactions. Only when the substrate is bound in place is an enzyme's active site complementary to it. The fit is most precise during a transition state of a reaction. The enzyme-substrate complex is short-lived, partly because the bonds holding it together are usually weak.

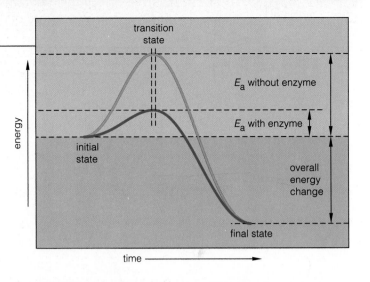

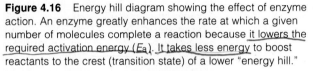

Figure 4.16 Energy hill diagram showing the effect of enzyme action. An enzyme greatly enhances the rate at which a given number of molecules complete a reaction because it lowers the required activation energy (E_a). It takes less energy to boost reactants to the crest (transition state) of a lower "energy hill."

According to Daniel Koshland's **induced-fit model**, each substrate has a surface region that almost *but not quite* matches chemical groups in an active site. When substrates first settle in the site, the contact strains some of their bonds. Strained bonds are easier to break, and this helps pave the way for new bonds (in products). In addition, interactions among charged or polar groups in the site favor a redistribution of electric charge that primes substrates for conversion to an activated state.

When substrates fit most precisely in the active site of an enzyme, they reach an activated, *transition* state (Figure 4.15c). In that state, they react spontaneously, just as a boulder pushed up and over the crest of a hill rolls down on its own (Figure 4.16).

Enzymes induce the fit in several ways. For instance, weak but extensive bonding at the active site puts substrates in positions that make them collide, and thereby promotes reaction. If those same molecules were colliding on their own, they would do so from random directions. The reaction rate would not be very impressive, because the mutually attractive chemical groups would meet up chancily and far less frequently. Enzymes enhance reaction rates by putting substrates on precise collision courses.

Temperature, pH, and Enzyme Activity

Rates of reactions also depend on temperature and pH. Each type of enzyme functions best within a certain temperature range. As Figure 4.17a indicates, reaction rates decrease sharply when temperatures become too high. The increase in heat energy disrupts weak bonds holding an enzyme in its three-dimensional shape. This alters the active site, and substrates cannot bind to it.

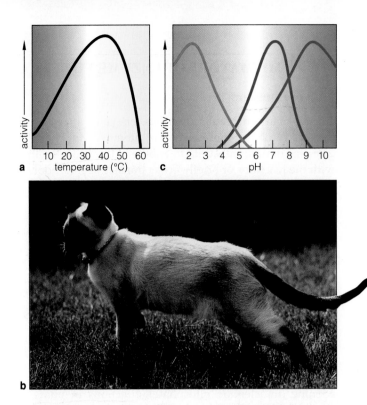

Figure 4.17 Examples of how changes in temperature and pH affect enzyme activity. (**a**) Graph of changes in the activity of one type of enzyme that was subjected to increases in temperature. (**b**) Siamese cats show observable effects of such changes. Fur on the ears and paws contains more dark brown pigment (melanin) than the rest of the body. A heat-sensitive enzyme controlling melanin production is less active in warmer body regions, and this results in lighter fur in those regions.

(**c**) Diagram showing how the activity of three different enzymes is influenced by pH. One enzyme (*brown* line) functions best in neutral solutions. Another (*red* line) functions best in basic solutions, and another (*purple* line) in acidic solutions.

Exposure to temperatures that are much higher than an organism usually encounters may destroy enzymes and disrupt metabolism. For example, this happens during a dangerously high *fever*. Humans usually die when their internal temperature reaches 44°C (112°F).

Each enzyme also functions best within a certain pH range. Higher or lower pH values generally disrupt its structure and function (Figure 4.17c). Most enzymes function best in neutral solutions (pH 7). Pepsin is one of the exceptions. This protein-digesting enzyme works in gastric fluid, which is extremely acidic.

An enzyme enhances the rate at which a specific reaction reaches equilibrium. It does so by lowering the amount of activation energy necessary to make substrates react.

Enzymes change the *rate*, not the outcome, of a reaction. They only act on specific substrates. And they may catalyze the same reaction repeatedly, as long as substrates are available.

Enzymes function best when the cellular environment stays within limited ranges of temperature and pH.

4.6 MEDIATORS OF ENZYME FUNCTION

Enzyme Helpers

During many metabolic reactions, enzymes speed the transfer of one or more electrons, atoms, or functional groups from one substrate to another. "Cofactors" help with the reactions or briefly act as transfer agents. Some cofactors are specific organic compounds. Others are metal ions that associate with the enzyme.

The enzyme helpers called **coenzymes** are complex organic molecules. Many are derived from vitamins. NAD$^+$ (nicotinamide adenine dinucleotide) and FAD (flavin adenine dinucleotide) are examples. Both accept electrons and hydrogen atoms that are liberated during glucose breakdown. They transfer the electrons to other reaction sites. Being attracted to the opposite charge of electrons, the unbound protons (H$^+$) go along for the ride. When loaded with electrons and hydrogen, these coenzymes are abbreviated NADH and FADH$_2$, respectively. Another coenzyme, NADP$^+$ (for nicotinamide adenine dinucleotide phosphate), functions in photosynthesis. Like NAD$^+$, it is derived from the vitamin niacin. When loaded down, it is abbreviated NADPH.

Ferrous iron (Fe^{++}), one of the metal ions that act as cofactors, is a component of cytochrome molecules. Cytochromes are transport proteins that show enzyme activity. They are embedded in cell membranes, such as the membranes of chloroplasts and mitochondria.

Control of Enzyme Function

Cells maintain, increase, and decrease concentrations of substances mostly by controlling enzymes. Some controls govern enzyme synthesis and so cut down or beef up the number of enzyme molecules that are operating at any time at a key step in a pathway. Other controls stimulate or inhibit enzymes that are already formed.

Internal controls come into play when concentrations of substances change within the cell. Picture a bacterium, busily synthesizing tryptophan and other amino acids necessary to construct its proteins. After a bit, protein synthesis slows and no more tryptophan is necessary. But the tryptophan pathway is still in full swing, so the cellular concentration of its end product rises.

Now **feedback inhibition** operates. By this mechanism, when production of a substance triggers a cellular change, the substance itself shuts down its further production (Figure 4.18).

In this case, unused molecules of tryptophan inhibit a key enzyme in the pathway. This enzyme happens to be governed by *allosteric* control. Besides the active site, allosteric enzymes have control sites to which specific substances bind and alter enzyme activity. When the pathway is blocked, fewer tryptophan molecules are around to inhibit the key enzyme—so production rises.

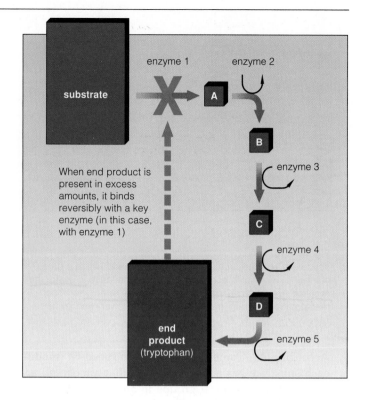

Figure 4.18 Example of feedback inhibition of a metabolic pathway. Five enzymes act in sequence to convert a substrate to an end product (tryptophan). When end product accumulates, some of the excess molecules bind to the first enzyme and thereby block the entire pathway.

In such ways, feedback inhibition of enzyme activity quickly adjusts concentrations of substances in cells.

In humans and other multicelled organisms, control of enzyme activity is just amazing. Cells not only work to keep themselves alive, they work with other cells in coordinated ways that benefit the whole body! Consider the hormones, which are signaling agents in this vast enterprise. Specialized cells release hormones into the bloodstream. Any cell having receptors for a given hormone will take it up, then its program for constructing a particular protein or some other activity will be altered. The hormone trips internal control agents into action—and the activities of specific enzymes change.

Enzyme function depends on cofactors (coenzymes and metal ions), which help catalyze reactions or transfer electrons, atoms, and functional groups from one substrate to another.

Enzyme function is subject to control mechanisms that influence the synthesis of new enzymes and that stimulate or inhibit existing enzymes.

By controlling enzymes, cells control the concentrations and kinds of substances required for survival.

4.7 ATP—THE MAIN ENERGY CARRIER

Structure and Function of ATP

Photosynthetic cells don't run directly on sunlight. First they convert light energy to the chemical energy of **ATP**. (This is an abbreviation for adenosine triphosphate, one of the nucleotides.) And no cell whatsoever *directly* uses energy that is released when glucose or other molecules are degraded. First the energy is converted to the chemical energy of ATP.

In every living cell, *molecules of ATP couple energy-releasing reactions with energy-requiring ones.*

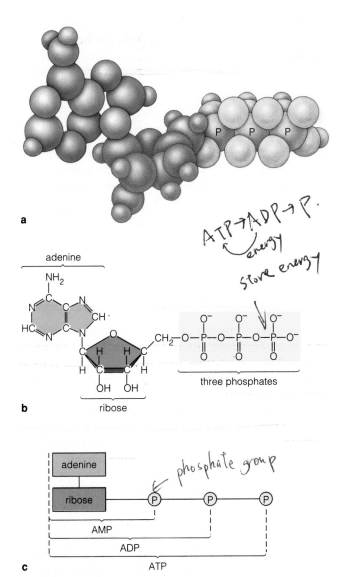

a

adenine

ribose

three phosphates

b

ATP→ADP→P.
energy
store energy

adenine

ribose

phosphate group

AMP

ADP

c ATP

Figure 4.19 ATP—adenosine triphosphate, the main energy carrier in all cells. (**a**) Three-dimensional model showing ATP's component atoms. (**b**) The structural formula for ATP.

(**c**) Adenosine diphosphate (ADP) forms when ATP gives up one phosphate group to another molecule. A second phosphate group transfer leaves adenosine monophosphate (AMP).

As Figure 4.19*b* shows, an ATP molecule is composed of adenine (a nitrogen-containing compound), the five-carbon sugar ribose, and a string of three phosphate groups. Covalent bonds hold its components together, but the bonding arrangement is not that stable. Many hundreds of different enzymes can easily split off the outermost phosphate group of ATP and attach it to a substrate.

The transfer of a phosphate group to a molecule of any sort is called **phosphorylation**. During these phosphate group transfers, a great deal of usable energy is released. It can activate hundreds of different molecules and drive hundreds of activities—such as synthesizing or degrading organic compounds, actively transporting substances across membranes against concentration gradients, and making muscle cells contract. Thus ATP molecules are like the coins of a nation—they are the cell's common currency of energy. Often you will see a cartoon "coin" being used to symbolize them:

ATP

The ATP/ADP Cycle

The **ATP/ADP cycle** is the cell's way of renewing its supply of ATP. During photosynthesis, aerobic respiration, and many other metabolic pathways, an energy input drives the attachment of unbound phosphate (P_i) りんさん塩 or a phosphate group to ADP (adenosine diphosphate). The result is an ATP molecule. Later on, when the ATP molecule donates a phosphate group somewhere, it reverts back to ADP:

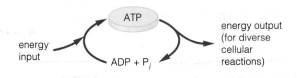

energy input

ATP

ADP + P_i

energy output (for diverse cellular reactions)

Given the central role of ATP in metabolism, it comes as no surprise that cells have a mechanism for renewing the supply of ATP.

ATP is the main carrier of energy from one reaction site to another in all cells. This molecule is the key intermediate between energy-releasing and energy-requiring reactions.

When a molecule gets phosphorylated—as by a phosphate group transfer from ATP—its store of energy increases and it becomes primed to enter a reaction.

The ATP/ADP cycle is a renewable means of conserving energy and transferring it to specific reactions.

4.8 ENERGY AND THE FLOW OF ELECTRONS

The Nature of Electron Transfers

Much of the flow of energy through cells involves the transfer of electrons from one molecule to another. Occasionally you might hear someone refer to this as an *oxidation-reduction* reaction. The name simply means an electron transfer has taken place. A donor molecule that gives up electrons is said to be oxidized. A molecule that accepts electrons is reduced.

Such transfers proceed after an atom or molecule absorbs enough energy to boost one or more electrons farther from the nucleus—or completely out of the atom. An electron excited this way gives off energy as it returns to the lowest energy level available to it, either in the same atom or in a new one. The *Focus* essay describes an example of this.

Certain electron transfers help cells save energy. Imagine tossing a bit of glucose into a wood fire. Its atoms would quickly let go of one another and combine with oxygen in the air, forming CO_2 and H_2O. In this case, all of the released energy would be lost as heat to the surroundings. By contrast,

a

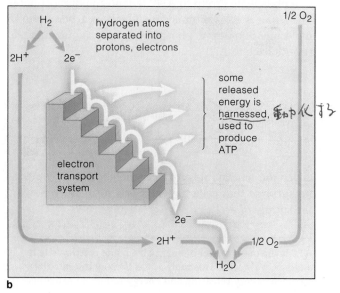

b

Figure 4.20 Example of the controlled release of energy during metabolic reactions. (**a**) Hydrogen and oxygen exposed to an electric spark will react and release energy all at once. (**b**) In a cell, the same type of reaction proceeds in many small, enzyme-mediated steps that allow some released energy to be harnessed. These steps involve electron transfers, often between molecules that operate together as an electron transport system.

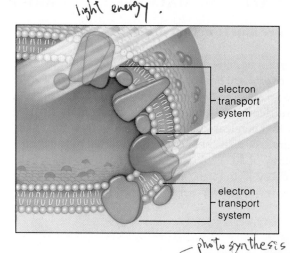

Figure 4.21 Example of electron transport systems embedded in the internal membrane system of chloroplasts. Each is positioned to receive electrons that sunlight energy knocks out of a complex of pigment molecules that also are embedded in the membrane.

cells do not "burn" glucose all at once, so they do not waste much energy. Instead, during glucose break-down, enzymes pluck atoms from intermediates that form at different reaction steps. Each step releases only some of the energy stored in chemical bonds. Such bonds, remember, are unions between the electron structures of atoms. Breaking the bonds puts electrons up for grabs, so to speak. And in cells, electrons are made to flow in ways that do useful work.

Electron Transport Systems

For example, consider an **electron transport system**. This is an organized array of enzymes and coenzymes that transfer electrons, in sequence. One molecule donates electrons, and the next molecule in line accepts them. You find these systems in cell membranes, such as the ones inside chloroplasts and mitochondria.

An electron transport system operates by accepting electrons at higher energy levels, then releasing them at lower energy levels. Think of it as a staircase (Figure 4.20). Excited electrons at the top step have the most energy. They drop down, one step at a time, and release a bit of energy as they do. At certain steps, some energy is harnessed to do work—for instance, to move ions into a membrane-bound compartment and thereby set up concentration and electric gradients across the membrane. Such gradients are central to the formation of ATP. Figure 4.21 shows two of the electron transport systems you will encounter in the next chapter.

Energy conversions in cells involve a flow of electrons among molecules, as in electron transport systems.

YOU LIGHT UP MY LIFE—VISIBLE EFFECTS OF METABOLIC ACTIVITY

At night, as an outcome of certain metabolic reactions, fireflies, certain beetles, and some other organisms show **bioluminescence**—they flash with light. These flashes take place when enzymes called luciferases convert chemical energy to light energy. The reactions begin when ATP transfers a phosphate group to luciferin, a molecule that belongs to a class of highly fluorescent substances. Luciferases, with the assistance of oxygen, convert the activated luciferin molecule to a different chemical form. Their action excites electrons of their substrate to a higher energy level. Very quickly, the excited electrons return to a lower energy level. When they do, they release energy—in the form of light.

In the tropical forests of Jamaica, click beetles fly about at night and give startling displays of bioluminescence (Figure 4.22*a–c*). These particular beetles, known locally as kittyboos, belong to the genus *Pyrophorus*. Different varieties emit green, greenish yellow, yellow, or orange flashes.

Molecular biologists have identified the kittyboo genes responsible for the flashes. They even have managed to insert copies of the genes into other organisms, including bacteria (Figure 4.22*d*).

Besides being fun to think about, transfers of the genes for bioluminescence have practical applications. Each year, for example, 3 million people die from a lung disease that follows infection by *Myobacterium tuberculosis*. Different strains of this bacterium are resistant to different kinds of antibiotics. A patient cannot be treated effectively until the strain causing her or his particular infection is identified.

Today, bacterial cells from a patient can be cultured and exposed to luciferase genes. Some of the cells take up the genes, which may become inserted into the bacterial DNA. After identifying these cells, clinicians can expose colonies of the genetically modified bacteria to different antibiotics. If an antibiotic has no effect, the cells churn out gene products—*including luciferase*—and the colony glows. If a colony doesn't glow, this is visible evidence that the antibiotic has stopped the particular strain of infectious cells in its metabolic tracks.

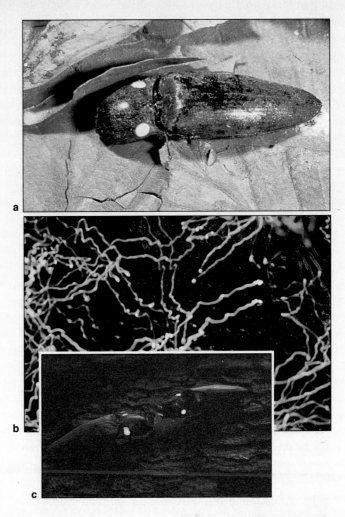

a

b

c

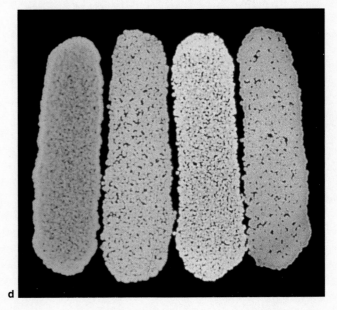

d

Figure 4.22 (**a**) One of the Jamaican kittyboo beetles (*Pyrophorus noctilucus*). The time-lapse photograph in (**b**) reveals the random paths of kittyboos on the wing, as they light up the night with their splendid bioluminescent flashes. The flashes help potential mates find each other in the dark (**c**).

The micrograph in (**d**) shows four colonies of bacterial cells, each grown from a parent bacterium that took up a kittyboo gene for a particular bioluminescent color.

SUMMARY

1. Cells acquire and use energy to build, store, break down, and rid themselves of substances. Collectively, these energy-driven activities are called metabolism. They underlie the survival of all living things.

2. Two laws of thermodynamics affect life. First, energy undergoes conversion from one form to another, but its amount never increases or decreases during a conversion. So the total amount of energy in the universe holds constant. Second, energy flows spontaneously in one direction, from forms of higher to lower energy content.

3. Like all organized systems, a cell tends to become disorganized without energy. It loses some energy with each metabolic reaction. It stays organized and alive by countering energy outputs with energy inputs.

4. The sun is life's primary energy source. In plants and other photosynthesizers, cells trap sunlight energy and convert it to the chemical energy of organic compounds. The plants (and organisms that feed on plants and one another) tap energy that became stored in these compounds to do cellular work.

5. Like all substances, the molecules and ions required for metabolism have energy that keeps them in constant motion. Where they have become more concentrated, they spontaneously tend to move down their concentration gradient (to adjoining regions where they are less concentrated). This tendency is called diffusion.

6. The steepness of concentration gradients (and electric or pressure gradients), temperature changes, and the size of molecules influence diffusion rates.

7. In osmosis, water moves across a selectively permeable membrane in response to concentration gradients, a pressure gradient, or both.

8. Table 4.1 lists the mechanisms that move substances into and out of cells, as described in this chapter and the preceding one. Lipid-soluble substances diffuse across a membrane's lipid bilayer. Membrane proteins bind solutes, change shape, and shunt them across.

a. In passive transport, a solute simply moves down its concentration gradient, through the interior of a transport protein spanning the membrane.

b. In active transport, an energy boost (as from ATP) changes the protein's shape in a way that promotes the net movement of the solute *against* its gradient.

9. Most metabolic reactions are reversible and tend to run on their own in the forward and reverse directions until they reach equilibrium. Cellular mechanisms work with or against this tendency. At any given time, they increase, decrease, or maintain the concentrations of substances required for specific metabolic reactions.

10. In metabolic pathways, reactions run in orderly, enzyme-mediated steps. The *biosynthetic* pathways

Table 4.1 Summary of Membrane Transport Mechanisms

Feature	Simple Diffusion	Facilitated Diffusion	Active Transport	Endocytosis, Exocytosis
Part involved in movement	Lipid bilayer or membrane proteins	Membrane proteins	Membrane proteins	Small portion of the plasma membrane
Binding of transported substance?	No	Yes	Yes	In many cases
Energy source for transport	Concentration gradients	Concentration gradients	ATP energy boost or concentration gradient	Varied
Direction of net movement	With the gradient	With the gradient	Against the gradient	Endocytic vesicle forms from plasma membrane; exocytic vesicle forms in cytoplasm, fuses with plasma membrane

After S. Wolfe, 1995, *Introduction to Molecular and Cellular Biology.*

assemble large, energy-rich organic compounds from smaller molecules of lower energy content. *Degradative* pathways break down molecules to smaller ones of lower energy content. Cells commonly couple energy-releasing reactions with energy-requiring reactions.

11. Substrates (reactants) are the substances that enter a reaction or pathway. Intermediates are substances that form between the reactants and end products.

12. Enzymes are catalysts; they greatly enhance the rate (not the outcome) of a reaction involving specific substrates. Nearly all enzymes are proteins.

a. Enzymes lower the activation energy required to start a reaction. They bind substrates at an active site, strain its bonds, and make the bonds easier to break.

b. Each kind of enzyme functions best within limited ranges of temperature and pH.

c. Cofactors assist enzymes in speeding a reaction, or they carry electrons, hydrogen, or functional groups stripped from substrate to other sites. They include coenzymes (such as NAD^+) and metal ions.

d. Controls stimulate or inhibit enzyme activity at key steps in metabolic pathways. They help coordinate the kinds and amounts of substances available.

13. Energy carriers couple energy-releasing reactions with energy-requiring ones. They activate molecules, as by phosphate group transfers, and so prime them for reaction. ATP is the main energy carrier in all cells; it delivers energy (inherent in its phosphate groups) to hundreds of different reaction sites.

14. Many energy conversions in cells involve a flow of electrons, as through electron transport systems.

Review Questions

1. What is a free radical? How does it relate to aging? *64–65*

2. State the first and second law of thermodynamics. Does life violate the second law? *66, 67*

3. Define diffusion and explain how it works. *68*

4. In what ways are membrane proteins that engage in passive and active transport similar? In what way do they differ? *70–71*

5. Name some substances that diffuse across the lipid bilayer of cell membranes. What kind of substances must be transported across the membrane? *68*

Self-Quiz *(Answers in Appendix IV)*

1. _____ is the primary source of energy for life on earth.
   a. Food   c. The sun
   b. Water   d. ATP

2. Place a cell in a hypotonic fluid and water will _____ .
   a. move into the cell   c. show no net movement
   b. move out of the cell   d. move by endocytosis

3. Diffusion of a solute in one direction or another will *not* be influenced by _____ .
   a. how steep the concentration gradient is
   b. an electric gradient
   c. a pressure gradient
   d. all of these factors can influence diffusion

4. _____ readily diffuses across the lipid bilayer of cell membranes.
   a. Glucose   c. Oxygen
   b. Carbon dioxide   d. b and c are correct

5. Certain transport proteins that receive an energy boost move sodium ions across a membrane. This is an example of _____ .
   a. passive transport   c. facilitated diffusion
   b. active transport   d. a and c are correct

6. Which is *not* true of chemical equilibrium?
   a. Product and reactant concentrations are always equal.
   b. The rates of the forward and reverse reactions are the same.
   c. There is no further net change in product and reactant concentrations.

7. Enzymes _____ .
   a. enhance reaction rates   c. act on specific substrates
   b. are affected by pH   d. all are correct

8. NAD^+ and $NADP^+$ are _____ of a type called _____ .
   a. cofactors; coenzymes   c. organic compounds; enzymes
   b. cofactors; metal ions   d. energy carriers; enzymes

9. The main energy carriers in cells are _____ .
   a. NAD^+ and $NADP^+$ c. ATP molecules
   b. metal ions   d. enzyme molecules

10. Electron transport systems involve _____ .
    a. enzymes and cofactors   c. cell membranes
    b. electron transfers   d. all are correct

11. Match each substance with the most suitable description.
    ____ coenzyme or metal ion   a. reactant or substrate
    ____ mainly ATP   b. enzyme
    ____ substance entering a reaction c. cofactor
    ____ substance formed while   d. intermediate
         a reaction is proceeding   e. product
    ____ substance at end of reaction f. energy carrier
    ____ enhances reaction rate

Critical Thinking

1. When Siamese cats are born, their coloration is very light over their entire body. After many months, fur on their ears, paws, tail, and nose darkens. From what you have learned about the effect of temperature on enzyme activity, propose a hypothesis that might account for this phenomenon.

2. The bacterium *Vibrio cholerae* causes the disease *cholera*. Infected people have such severe diarrhea, they may lose as much as twenty liters of fluid in a day's time. The bacterium enters the body when a person drinks contaminated water, then adheres to the intestinal lining. During its metabolic activities, it secretes a substance that is toxic to cells of the lining, and they start secreting chloride ions (Cl^-). Sodium ions (Na^+) follow the chloride ions into the fluid in the intestines. Explain how this sequence of events causes the massive fluid loss.

3. When a toxic organic compound called cyanide binds to a type of enzyme that is a component of electron transport systems, *cyanide poisoning* results. The binding prevents the enzyme from donating electrons to its neighboring acceptor molecule in the system. What effect will this have on ATP production? From what you know of the function of ATP, what effect will this have on a person's health?

4. AZT (azidothymidine) is a drug used to alleviate symptoms of *AIDS* (acquired immunodeficiency syndrome). It is very similar in molecular structure to thymidine, one of the nucleotides that occur in DNA. It also can fit into the active site of an enzyme of HIV, the virus that causes AIDS. Infection puts the viral enzyme and the viral genetic material (an RNA molecule) into a cell. There, the RNA serves as a template (structural pattern) upon which nucleotides are joined together to form a strand of DNA. The viral enzyme takes part in the assembly reactions. Propose a model to explain how AZT might inhibit replication of the virus inside cells.

Selected Key Terms

activation energy *74*	end product *73*	passive transport *70*
active site *74*	energy *66*	phosphorylation *77*
active transport *70*	energy carrier *73*	
ATP *77*	entropy *67*	second law of thermodynamics *67*
ATP/ADP cycle *77*	enzyme *73*	
bioluminescence *79*	feedback inhibition *76*	selective permeability *68*
chemical equilibrium *72*	first law of thermodynamics *66*	substrate *73*
coenzyme *76*	free radical *64*	tonicity *68*
cofactor *73*	induced-fit model *75*	transport protein *73*
concentration gradient *68*	intermediate *73*	
diffusion *68*	metabolic pathway *73*	
electron transport system *78*	metabolism *66*	
	osmosis *69*	

Readings

Fenn, J. *Engines, Energy, and Entropy.* New York: Freeman. Deceptively simple paperback.

Rusting, R. December 1992. "Why Do We Age?" *Scientific American* 267(6): 131–141.

Wolfe, S. 1995. *Introduction to Molecular and Cellular Biology.* Belmont, California: Wadsworth.

5 ENERGY-ACQUIRING PATHWAYS

Sun, Rain, and Survival

Just before dawn in the Midwest, the air is dry and motionless. The heat that has scorched the land for weeks still rises from the earth and hangs in the air of a new day. There are no clouds in sight. There is no promise of rain. For hundreds of miles, crops stretch out, withered and nearly dead. All the marvels of modern agriculture can't save them. In the absence of one vital substance—water—life in each cell of those many hundreds of thousands of plants has ceased.

In Los Angeles, a student wonders if the Midwest drought will bump up food prices. In Washington, D.C., economists analyze crop failures in terms of tonnage available for domestic consumption and export.

Thousands of kilometers away, in the vast Sahel Desert of Africa, grasses and cattle are dying after a similar unrelenting drought. Children with bloated bellies and spindly legs wait passively for death. Deprived of nourishment for too long, cells of their bodies will never function normally again.

You are about to explore pathways by which cells trap and use energy. At first these pathways may seem to be far removed from your everyday world. *Yet the food that nourishes you and nearly all other organisms cannot be produced or used without them.*

We will return to this point in later chapters, when we address the major issues of human population

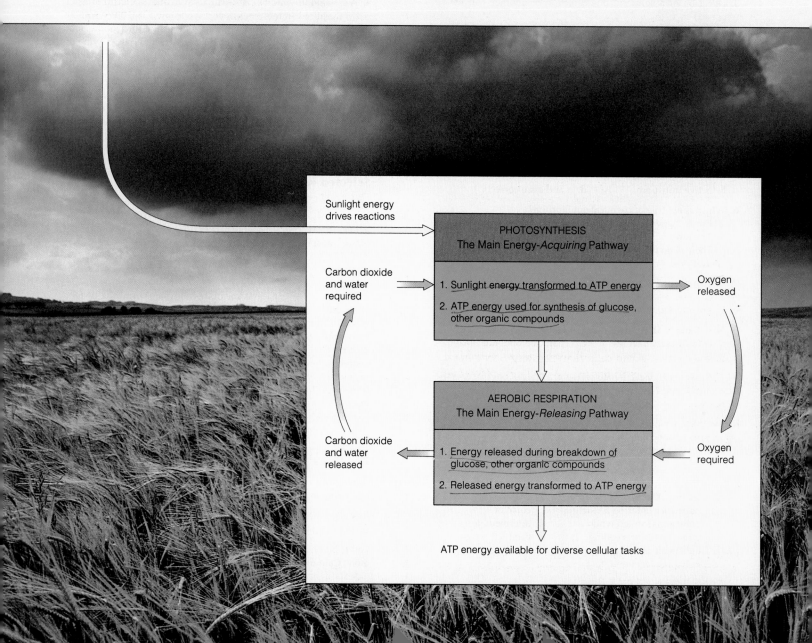

Sunlight energy drives reactions

PHOTOSYNTHESIS
The Main Energy-*Acquiring* Pathway

Carbon dioxide and water required

1. Sunlight energy transformed to ATP energy

2. ATP energy used for synthesis of glucose, other organic compounds

Oxygen released

AEROBIC RESPIRATION
The Main Energy-*Releasing* Pathway

1. Energy released during breakdown of glucose, other organic compounds

2. Released energy transformed to ATP energy

Carbon dioxide and water released

Oxygen required

ATP energy available for diverse cellular tasks

growth, nutrition, limits on agriculture and genetic engineering, and the impact of pollution on crops. Here, our point of departure is the *source* of food—which isn't a farm or a supermarket or a refrigerator. What we call "food" was put together somewhere in the world by living cells from glucose and other organic compounds. Such compounds are built on a framework of carbon atoms, so the questions become these:

1. Where does the carbon come from in the first place?

2. Where does the energy come from to drive the synthesis of carbon-based compounds?

Answers to these questions depend on an organism's mode of nutrition.

Many organisms are classified as **autotrophs**. They are "self-nourishing," which is what autotroph means. Their carbon source is carbon dioxide (CO_2), a gaseous substance all around us in the air and dissolved in water. The world's plants, many protistans, and certain bacteria are *photo*autotrophs; sunlight is their energy source. A few bacteria are *chemo*autotrophs; they extract energy from an inorganic substance, such as sulfur.

Other organisms are classified as **heterotrophs**. They get their carbon and energy from organic compounds that other organisms have already put together. They feed on autotrophs, each other, and organic wastes. (*Hetero-* means other, as in "being nourished by other organisms.") That is how animals, fungi, many of the protistans, and most bacteria stay alive.

When you think about this grand pattern of who eats whom, one thing becomes clear. *Survival of nearly all organisms ultimately depends on photosynthesis, the main pathway by which carbon and energy enter the world of life.* Once glucose and other organic compounds are assembled, cells put them in storage or use them as building blocks. When cells require energy, they break such compounds apart by several pathways. However, *the predominant energy-releasing pathway is called aerobic respiration*. Figure 5.1 provides you with a preview of the chemical links between photosynthesis and aerobic respiration—the focus of this chapter and the next.

Figure 5.1 Links between photosynthesis and aerobic respiration, the main energy-acquiring and energy-releasing pathways in the world of life.

KEY CONCEPTS

1. Carbon-based compounds are the building blocks and energy stores of life. Plants assemble these compounds by photosynthesis. First the plants trap sunlight energy and convert it to chemical energy (in the form of certain bonds in ATP molecules). Then ATP delivers energy to reactions in which glucose is put together from carbon dioxide and water. Finally, glucose subunits are combined to form starch and other molecules.

2. Photosynthesis is the biosynthetic pathway by which most carbon and energy enter the web of life.

3. In plant cells, photosynthesis proceeds in organelles called chloroplasts. The pathway starts at a membrane system inside the chloroplast. Its machinery includes organized arrays of light-absorbing pigments, enzymes, and a coenzyme ($NADP^+$), which delivers hydrogen and electrons to the synthesis reactions.

4. Photosynthesis is often summarized this way:

$$12H_2O + 6CO_2 \xrightarrow{\text{sunlight}} 6O_2 + C_6H_{12}O_6 + 6H_2O$$

glucose. water

5.1 PHOTOSYNTHESIS: AN OVERVIEW

Energy and Materials for the Reactions

Photosynthesis is an ancient pathway, and it evolved in distinct ways in different organisms. To keep things simple, let's focus on what goes on in lettuce, weeds, and other leafy plants.

The pathway has two stages, each with its own set of reactions. In the light-dependent reactions, energy from sunlight is absorbed and converted to ATP energy. Water molecules are split, and the coenzyme $NADP^+$ picks up the liberated hydrogen and electrons, thereby becoming NADPH. In the *light-independent* reactions, ATP donates energy to sites where glucose ($C_6H_{12}O_6$) is put together from carbon, hydrogen, and oxygen. Water provides hydrogen, as delivered by NADPH. Carbon dioxide (CO_2) provides the carbon and oxygen.

Photosynthesis is often summarized in this manner:

$$12H_2O + 6CO_2 \xrightarrow{\text{sunlight}} 6O_2 + C_6H_{12}O_6 + 6H_2O$$

In order to keep the chemical bookkeeping simple, this summary equation shows glucose as an end product.

a

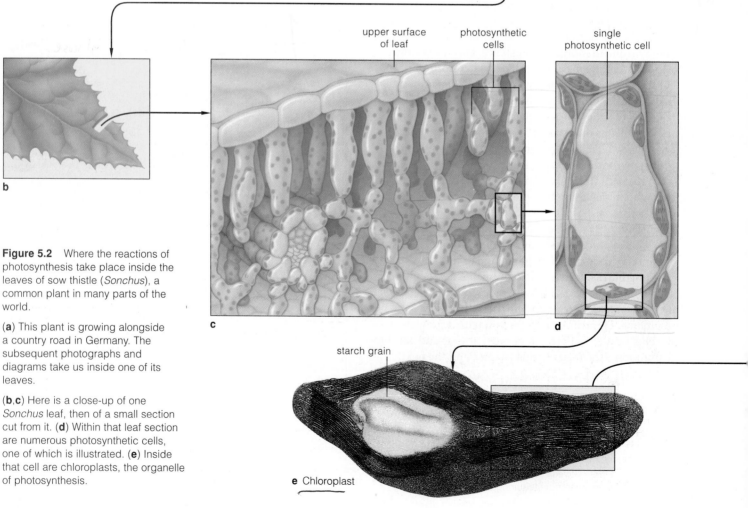

upper surface of leaf

photosynthetic cells

single photosynthetic cell

b

c

d

starch grain

e Chloroplast

Figure 5.2 Where the reactions of photosynthesis take place inside the leaves of sow thistle (*Sonchus*), a common plant in many parts of the world.

(**a**) This plant is growing alongside a country road in Germany. The subsequent photographs and diagrams take us inside one of its leaves.

(**b,c**) Here is a close-up of one *Sonchus* leaf, then of a small section cut from it. (**d**) Within that leaf section are numerous photosynthetic cells, one of which is illustrated. (**e**) Inside that cell are chloroplasts, the organelle of photosynthesis.

However, the reactions don't really stop with glucose. Glucose and other simple sugars combine at once to form sucrose, starch, and other carbohydrates—the true end products of photosynthesis.

Where the Reactions Take Place

The two stages of photosynthesis proceed at different sites inside each **chloroplast**. Only the photosynthetic cells of plants and some protistans contain this type of organelle (Section 3.9). Each chloroplast has two outer membranes that are wrapped around a largely fluid interior, called the **stroma**. An inner membrane extends through the stroma. Often this membrane has the form of interconnected channels and disks, which are arranged in stacks called grana (singular, granum). The first stage of photosynthesis proceeds at this inner membrane, which is the **thylakoid membrane system**. The spaces inside the disks and channels connect as a single compartment that collects hydrogen ions, which are used in ATP production. The second stage of photo-

synthesis—the set of reactions by which sugars are assembled—takes place in the stroma.

Figure 5.2 marches down through a chloroplast from sow thistle (*Sonchus*), a common weed. Two thousand of these chloroplasts, lined up single file, would be no wider than a dime. Imagine all the chloroplasts in just one weed or lettuce leaf—each a tiny factory for producing sugars and starch—and you get an idea of the magnitude of metabolic events required to feed you and all other organisms living together on this planet.

The chloroplasts of plants are organelles of photosynthesis.

The first stage of photosynthesis (in which energy from the sun is absorbed and converted to ATP energy) takes place at a membrane system inside the chloroplast.

The second stage (in which hydrogen, carbon, and oxygen combine to form carbohydrates) takes place in the stroma, which is located between the chloroplast's outer and inner membrane systems.

---

light energy absorbed

thylakoid membrane system within stroma chloroplast

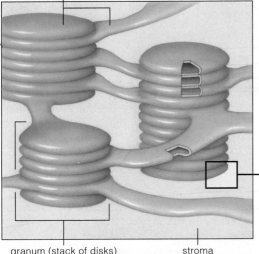

granum (stack of disks)   stroma

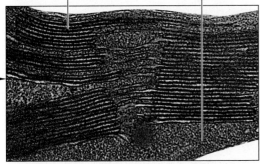

f Inside the photosynthetic cell, light-dependent reactions proceed at the thylakoid membrane system. Light-independent reactions proceed in the stroma.

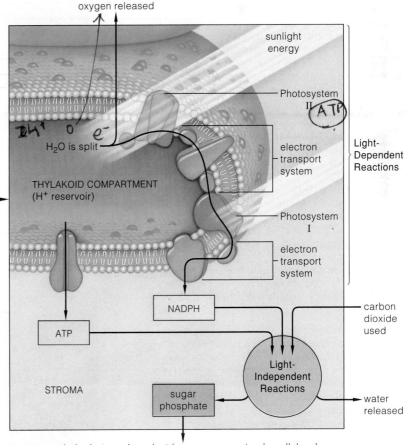

oxygen released

sunlight energy

Photosystem II

ATP

H^+ O e^-

H_2O is split

electron transport system

THYLAKOID COMPARTMENT (H^+ reservoir)

Photosystem I

electron transport system

Light-Dependent Reactions

NADPH

carbon dioxide used

ATP

STROMA

Light-Independent Reactions

sugar phosphate

water released

carbohydrate end product (e.g., sucrose, starch, cellulose)

g This diagram provides an overview of the sites where the key steps of both stages of reactions proceed.

5.2 LIGHT-TRAPPING PIGMENTS

Recall, from Section 3.3, that light from the sun travels through space in undulating motion, rather like ocean waves. Light has another interesting property. It travels as distinct packets of energy, called **photons**. The wavelength of a photon (the distance from one wave peak to the next) is related to its energy. The most energetic photons travel as very short wavelengths, and the least energetic photons as long wavelengths. Some animals, including humans, bees, and birds, can perceive part of the spectrum of wavelengths as colors. Rainbows especially give us spectacular displays of the visible spectrum of light (Figure 5.3a).

The molecules called **pigments** absorb photon energy. In animals, melanin and other pigments have roles in the coloration of body surfaces, vision, and other functions. In plants and other photoautotrophs, certain pigments absorb energy that drives photosynthesis (Figures 5.3b and 5.4).

Figure 5.3 (**a**) Electromagnetic spectrum, which includes wavelengths of visible light. (**b**) Range of wavelengths that photosynthetic pigments absorb. Here we focus on three classes of pigments: the chlorophylls (*green* lines), carotenoids (*yellow-orange*), and phycobilins (*purple* and *blue*). Colors within the graph line for each pigment correspond to wavelengths that it absorbs. Peaks in the ranges of absorption correspond to the measured amount of energy absorbed and used in photosynthesis.

The dashed line shows what would happen if we were to *combine* the amounts of energy that all of the different classes of photosynthetic pigments absorb. Taken together, they can absorb most wavelengths of visible light.

Figure 5.4 T. Englemann's 1882 experiment that correlated photosynthetic activity of a strandlike green alga (*Spirogyra*) with certain wavelengths of light. Oxygen is a by-product of photosynthesis. Many species use oxygen (in aerobic respiration). Among them are certain free-living bacterial cells. By tumbling about, these cells move into places where conditions favor their activities and away from places where conditions are less favorable.

Englemann reasoned that the oxygen-requiring bacteria living in the same habitats as the green alga would stop tumbling and congregate where oxygen was being produced. He put a strand of the alga in a drop of water containing the bacteria, then mounted it on a microscope slide. He used a crystal prism to cast a spectrum of colors across the slide. The bacterial cells congregated mostly where violet and red wavelengths fell across the algal strand. Englemann concluded that these wavelengths are most effective for photosynthesis (and oxygen production).

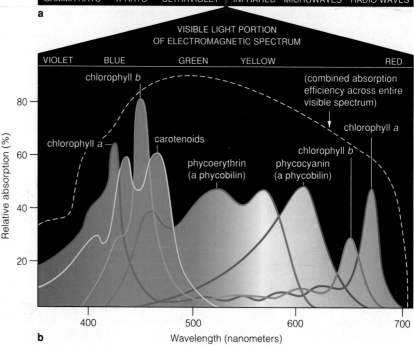

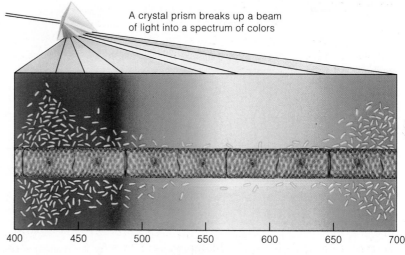

A crystal prism breaks up a beam of light into a spectrum of colors

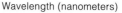

maple leaf
in autumn

Figure 5.5 Changes in leaf color in autumn.

The chloroplasts of mature leaves contain chlorophylls, carotenoids, and other pigment molecules. (The carotenoids include the yellow carotenes and xanthophylls.) Intensely green leaves have an abundance of chlorophyll pigments that are masking the presence of other pigments. In many species, the gradual decline of daylength in autumn and other factors trigger the breakdown of chlorophyll—and more colors show through the leaf surface.

Also in autumn, water-soluble anthocyanins accumulate in the central vacuoles of leaf cells. The pigments appear red when fluids being transported through plants are slightly acidic, blue when the fluids are basic (alkaline), or colors in between when they are at intermediate pH levels. Soil conditions contribute to the differences in pH.

Birch, aspen, and other tree species always have the same color in autumn. The coloration of maple, ash, sumac, and other species varies around the country. Color even varies from one leaf to the next, depending on the pigment combinations.

maple leaf
in summer

The wavelengths that various organisms use for light-requiring processes range between about 400 and 750 nanometers. Wavelengths that are shorter than this are energetic enough to break the chemical bonds of organic compounds. Among them are wavelengths of ultraviolet light and x-rays, which can destroy cells. (Hundreds of millions of years ago, a protective ozone layer formed in the atmosphere and shielded the Earth from ultraviolet radiation in the sun's rays. Before then, the lethal bombardment kept photosynthetic bacteria and other forms of life confined to the seas. We return to this fascinating story in Section 17.4.)

The thylakoid membranes of chloroplasts contain clusters of different pigments, including chlorophylls and carotenoids. Just as some ocean waves are more exciting than others to surfers, certain wavelengths of light are more exciting to each kind of pigment.

The **chlorophylls** absorb wavelengths of violet-to-blue and red light (Figure 5.3b). They reflect green wavelengths, which is why plant parts having an abundance of chlorophyll molecules appear green to us. In all green algae and plants, the chlorophylls are the main photosynthetic pigments. The one designated chloro-

phyll *a* is the main photosynthetic pigment in plants. Chlorophyll *b* and other pigments are present as well. They are better at absorbing energy of different wavelengths. They do not retain the energy; they transfer it to the main chlorophyll and enhance its effectiveness.

Carotenoids absorb violet and blue wavelengths but transmit red, orange, and yellow ones. They are far less abundant than chlorophylls in green leaves. Often they become visible in autumn, when many plants stop producing chlorophyll (Figure 5.5).

Several other pigments contribute to the coloration of certain photosynthetic species. Among them are the red and blue **phycobilins**, the signature pigments of red algae and cyanobacteria.

Photons are packets of light energy that differ in wavelength.

The wavelengths of photons of visible light correspond to different colors.

Taken together, chlorophylls and other photosynthetic pigment molecules can absorb most of the wavelengths of visible light.

5.3 LIGHT-DEPENDENT REACTIONS

During the **light-dependent reactions**, the first stage of photosynthesis, three events unfold. *First*, pigments absorb sunlight energy and give up electrons. *Second*, transfers of electrons and hydrogen lead to the formation of ATP and NADPH. *Third*, the pigments that gave up electrons in the first place get electron replacements.

Photosystems

Embedded in thylakoid membranes are light-trapping clusters called **photosystems**. A membrane may have many thousands of them, and each includes 200 to 300 pigment molecules. Most of the pigments do no more than harvest energy. When they absorb a photon, one of their electrons gets boosted to a higher energy level. When the electron returns to a lower level, it quickly gives up its extra energy:

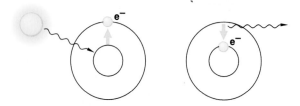

As the released energy bounces randomly among the system's pigments, a little is lost as heat at each bounce. In less than a billionth of a second, the energy still left corresponds to a wavelength that only a chlorophyll *a* molecule can trap. Such a molecule is the photosystem's reaction center; *it traps excitation energy and then gives up electrons for photosynthesis* (Figure 5.6). As you will see, these electrons get transferred to an acceptor molecule poised at the start of a neighboring transport system.

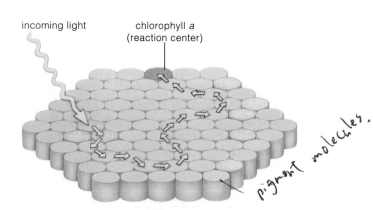

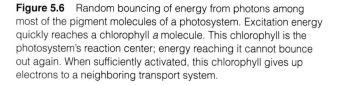

Figure 5.6 Random bouncing of energy from photons among most of the pigment molecules of a photosystem. Excitation energy quickly reaches a chlorophyll *a* molecule. This chlorophyll is the photosystem's reaction center; energy reaching it cannot bounce out again. When sufficiently activated, this chlorophyll gives up electrons to a neighboring transport system.

ATP and NADPH: Loading Up Energy, Hydrogen, and Electrons

Electron transport systems, recall, are organized arrays of enzymes, coenzymes, and other proteins in all cell membranes (Section 4.8). Electrons are transferred from one component to another, in orderly sequence. At each transfer, some energy is released and is harnessed to drive certain reactions.

In thylakoid membranes, transport systems adjoin two kinds of photosystems and accept electrons from them. They allow plants to produce ATP by two kinds of pathways—one cyclic and the other noncyclic.

Cyclic Pathway One pathway starts at an electron-donating chlorophyll of a *type I* photosystem. In the **cyclic pathway of ATP formation**, the electrons "cycle" from a reaction center designated chlorophyll P700, through a transport system, then back to P700. Energy associated with the electron flow drives the formation of ATP from ADP and unbound phosphate:

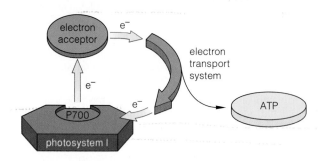

The cyclic pathway is probably the oldest means of ATP production. The first cells to use it were as tiny as existing bacteria, so their body-building programs were scarcely enormous. ATP alone would have provided enough energy to build enough organic compounds. Building larger organisms requires far greater amounts of organic compounds—and vast amounts of hydrogen atoms and electrons. Long ago, in the forerunners of multicelled plants, the cyclic pathway's machinery underwent expansion and became the basis of a more efficient ATP-forming pathway. Amazingly, this bit of modified machinery changed the course of evolution.

Noncyclic Pathway Today, the cyclic pathway still operates in trees, weeds, and other leafy members of the plant kingdom. But the **noncyclic pathway of ATP formation** dominates. Electrons are not cycled through this pathway. They depart (in NADPH)—and electrons from water molecules replace them.

The pathway goes into operation when the sun's rays bombard a *type II* photosystem. Photon energy makes this photosystem's reaction center (chlorophyll

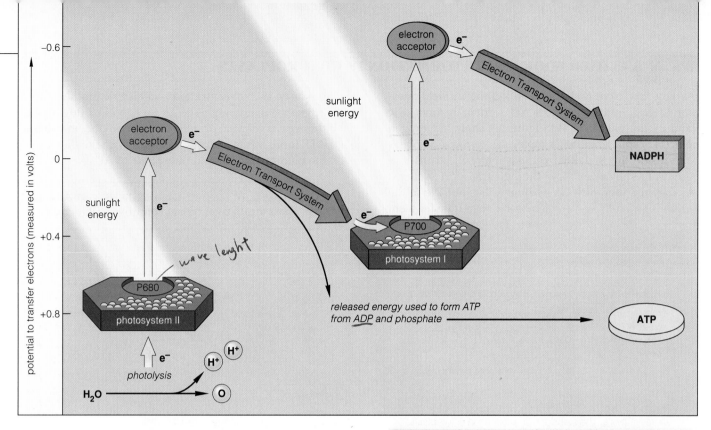

Figure 5.7 Noncyclic pathway of photosynthesis, which yields ATP *and* NADPH. Electrons derived from the splitting of water molecules (photolysis) travel through two kinds of photosystems. By working together, the photosystems boost electrons to an energy level that is high enough to drive NADPH formation.

P680) give up electrons. The energy input also triggers **photolysis**, a reaction sequence in which molecules of water split into oxygen, hydrogen ions, and electrons. P680 attracts the rather unexcited electrons as replacements for the excited ones that got away (Figure 5.7).

Meanwhile, excited electrons are moving through a transport system, then to P700 of *photosystem I*. They still have some extra energy left, and they get even more upon their arrival—because photons also happen to be bombarding P700. The energy boost puts electrons at a higher energy level that allows them to enter a second transport system. A coenzyme, NADP⁺, assists one of the system's enzymes. It becomes NADPH by picking up two of the electrons and a hydrogen ion—*which it then transports to sites where organic compounds are built.*

The Legacy—A New Atmosphere

On sunny days, on the surfaces of aquatic plants, you can see bubbles of oxygen. Figure 5.8 gives an example of this. The oxygen is a by-product of the noncyclic pathway of photosynthesis. As described in Chapter 17, this pathway may have evolved more than 2 billion years ago. At first, the oxygen simply dissolved in seas, lakes, wet mud, and other bacterial habitats. By about 1.5 billion years ago, however, significant amounts of

Figure 5.8 Visible evidence of photosynthesis—bubbles of oxygen emerging from leaves of *Elodea*, an aquatic plant.

dissolved oxygen were escaping into what had been an oxygen-free atmosphere. The accumulation of oxygen changed the atmosphere forever. And it made possible aerobic respiration—the most efficient pathway for releasing energy from organic compounds. The emergence of the noncyclic pathway ultimately allowed you and all other animals to be around today, breathing the oxygen that helps keep your cells alive.

In the light-dependent reactions, energy from the sun drives the formation of ATP (which carries energy) and NADPH (which transports hydrogen and electrons).

Oxygen, a by-product of photosynthesis, profoundly changed the early atmosphere and made aerobic respiration possible.

5.4 A CLOSER LOOK AT ATP FORMATION IN CHLOROPLASTS

As you may have noticed, we saved the trickiest question for last. Exactly *how* does ATP form in the noncyclic (and cyclic) pathway? As Figure 5.9 shows, hydrogen ions (H^+) split from water molecules during photolysis stay inside the thylakoid compartment. Electrons that also were split away flow through transport systems. As they do, they pick up hydrogen ions outside of the membrane, then transfer them into the compartment. In this way, H^+ concentration and electric gradients form across the membrane. The ions respond by flowing out, through transport proteins that span the membrane. ATP forms when energy associated with the flow drives the binding of unbound phosphate to ADP.

In chloroplasts, H^+ concentration and electric gradients that form across the thylakoid membrane drive ATP formation.

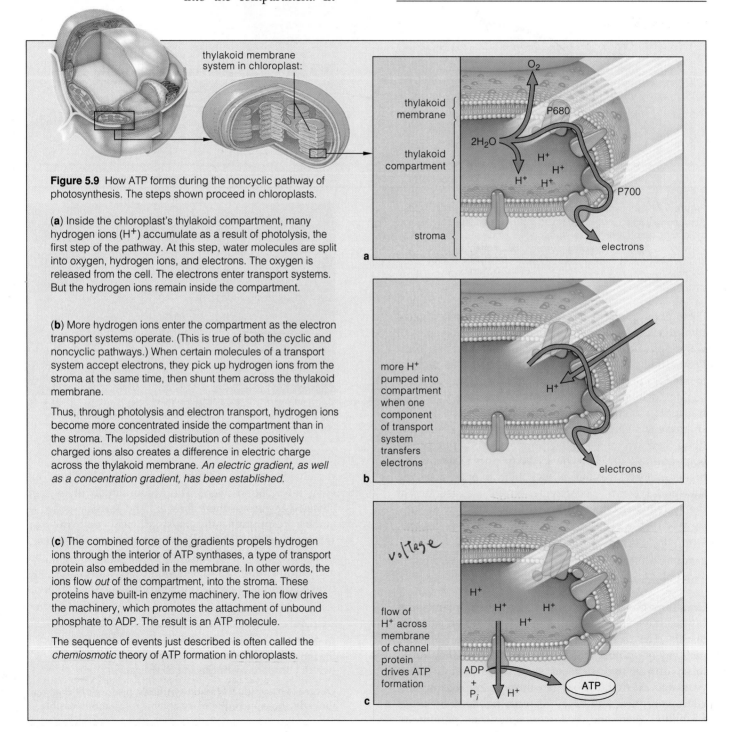

Figure 5.9 How ATP forms during the noncyclic pathway of photosynthesis. The steps shown proceed in chloroplasts.

(**a**) Inside the chloroplast's thylakoid compartment, many hydrogen ions (H^+) accumulate as a result of photolysis, the first step of the pathway. At this step, water molecules are split into oxygen, hydrogen ions, and electrons. The oxygen is released from the cell. The electrons enter transport systems. But the hydrogen ions remain inside the compartment.

(**b**) More hydrogen ions enter the compartment as the electron transport systems operate. (This is true of both the cyclic and noncyclic pathways.) When certain molecules of a transport system accept electrons, they pick up hydrogen ions from the stroma at the same time, then shunt them across the thylakoid membrane.

Thus, through photolysis and electron transport, hydrogen ions become more concentrated inside the compartment than in the stroma. The lopsided distribution of these positively charged ions also creates a difference in electric charge across the thylakoid membrane. *An electric gradient, as well as a concentration gradient, has been established.*

(**c**) The combined force of the gradients propels hydrogen ions through the interior of ATP synthases, a type of transport protein also embedded in the membrane. In other words, the ions flow *out* of the compartment, into the stroma. These proteins have built-in enzyme machinery. The ion flow drives the machinery, which promotes the attachment of unbound phosphate to ADP. The result is an ATP molecule.

The sequence of events just described is often called the *chemiosmotic* theory of ATP formation in chloroplasts.

5.5 LIGHT-INDEPENDENT REACTIONS

Light-independent reactions are the "synthesis" part of photosynthesis. ATP delivers the required energy for the reactions. NADPH delivers the required hydrogen and electrons. Carbon dioxide (CO_2) in the air around photosynthetic cells provides the carbon and oxygen.

We say the reactions are light-independent because they don't depend directly on sunlight. They proceed even in the dark, as long as ATP and NADPH hold out.

Capturing Carbon

Let's track a CO_2 molecule that diffuses into air spaces inside a leaf and ends up next to a photosynthetic cell (Figure 5.2d). From there, it diffuses across the plasma membrane and on into the stroma of a chloroplast. The light-independent reactions start when the carbon atom of the CO_2 becomes attached to **RuBP** (ribulose bisphosphate), a molecule with a backbone of five carbon atoms. This event is called **carbon dioxide fixation**.

Building the Glucose Subunits

Attaching carbon to RuBP is the first step of the **Calvin-Benson cycle**, a cyclic pathway that produces a sugar phosphate molecule *and* regenerates RuBP. A specific enzyme catalyzes each step. For our purposes, we can simply focus upon the carbon atoms of the substrates, intermediates, and end products, as Figure 5.10 shows.

The attachment of a carbon atom to RuBP produces an unstable intermediate with six carbon atoms. This splits into two molecules of **PGA** (phosphoglycerate), each with a three-carbon backbone. Now ATP donates a phosphate group to each PGA. Also, NADPH donates hydrogen and electrons to the resulting intermediate, thereby forming **PGAL** (phosphoglyceraldehyde).

The steps just outlined proceed *six* times. In other words, six CO_2 molecules are fixed and twelve PGAL molecules form in the cycle. Most of the PGAL becomes rearranged to form new RuBP, which can be used to fix more carbon. But *two* PGAL combine, forming a six-carbon sugar phosphate that can enter further reactions.

The Calvin-Benson cycle produces enough RuBP molecules to replace the ones used in carbon dioxide fixation. The ADP, $NADP^+$, and phosphate leftovers diffuse through the stroma. Upon reaching the sites of the light-dependent reactions, they are converted once more to NADPH and ATP. The sugar phosphate formed in the cycle serves as a building block for sucrose, starch, or cellulose—the plant's main carbohydrates. *And synthesis of these organic compounds by other pathways marks the end of the light-independent reactions.*

During daylight hours, photosynthetic cells convert their newly formed sugar phosphates to sucrose or starch. Of all plant carbohydrates, sucrose is the most

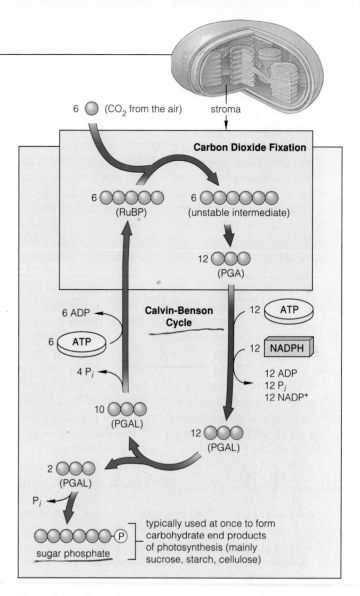

Figure 5.10 Summary of the light-independent reactions of photosynthesis. *Red* circles signify the carbon atoms of the key molecules. All of the intermediates have one or two phosphate groups attached. However, for simplicity, this diagram shows only the phosphate group on the resulting sugar phosphate.

easily transportable. Starch is the most common storage form. The cells also can convert excess PGAL to starch. They briefly store starch as grains in the stroma. (You can see one of these grains in Figure 5.2e.) After the sun goes down, cells convert starch to sucrose, for export to other living cells in leaves, stems, and roots. Ultimately, the products and intermediates of photosynthesis end up as energy sources and building blocks for all of the lipids, amino acids, and other organic compounds that are required for growth, survival, and reproduction.

During the Calvin-Benson cycle, carbon is "captured" from carbon dioxide, a sugar phosphate forms in reactions that require ATP and NADPH, and RuBP (necessary to capture the carbon) is regenerated.

5.6 FIXING CARBON—SO NEAR, YET SO FAR

If light intensity, air temperature, rainfall, and the composition of soil were uniform all year long, everywhere in the world, maybe photosynthetic reactions would proceed the same way in all plants. But environments obviously differ—and so do details of photosynthesis. A brief comparison of two carbon-fixing adaptations to stressful environments will illustrate this point.

Think of a Kentucky bluegrass plant. Like all plants, it depends on CO_2 uptake for growth. CO_2 is plentiful in the air but isn't always plentiful inside leaves. Leaves have a waxy cover that restricts water loss—except at stomata (singular, stoma). These tiny openings span the leaf surface. Nearly all of the inward diffusion of CO_2 and outward diffusion of O_2 can occur only at stomata.

On hot, dry days, stomata close and plants conserve water—but CO_2 cannot diffuse into leaves. Meanwhile, photosynthetic cells are busy, and oxygen builds up in leaves. The stage is set for "photorespiration." By this wasteful process, *oxygen, not CO_2, attaches to the RuBP used in the Calvin-Benson cycle.* Remember, the production of sugar phosphates in that cycle requires PGA. When photorespiration wins out, less PGA forms—and the bluegrass plant's capacity for growth suffers.

Kentucky bluegrass is one of many **C3 plants**. The name refers to the first intermediate (three-carbon PGA) of their carbon-fixing pathway. By contrast, the first intermediate formed when crabgrass, corn, and many other plants fix carbon is the four-carbon oxaloacetate (Figure 5.11). Hence *their* name, **C4 plants**.

C4 plants maintain adequate amounts of CO_2 inside leaves even though they, too, close stomata on hot, dry days. *They fix CO_2 not once but twice*, in two types of photosynthetic cells. Mesophyll cells have first crack at the CO_2. They use it to produce oxaloacetate, which is quickly transferred to bundle-sheath cells around leaf veins. In these cells, CO_2 is released and fixed again—in the Calvin-Benson cycle. With their more efficient way of fixing carbon, C4 species get by with tinier stomata—and therefore lose less water than C3 species.

Generally, C4 plants are better adapted where temperatures are highest during the growing season. Thus, of all the plants that evolved in Florida, 80 percent are C4 plants, compared to 0 percent in Manitoba, Canada. Kentucky bluegrass and other C3 species have greater advantage where temperatures drop below 25°C; they are less sensitive to cold. Mix C3 and C4 species from different regions in a garden, and one or the other kind will do better during at least part of the year. That is why a Kentucky bluegrass lawn does well during cool spring weather in San Diego, only to be overwhelmed during a hot summer by a C4 plant—crabgrass.

We find another carbon-fixing adaptation in deserts and other taxingly dry environments. Think of a cactus plant—a succulent with juicy, water-storing tissues and

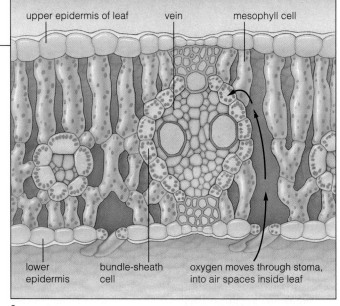

upper epidermis of leaf   vein   mesophyll cell

lower epidermis   bundle-sheath cell   oxygen moves through stoma, into air spaces inside leaf

a

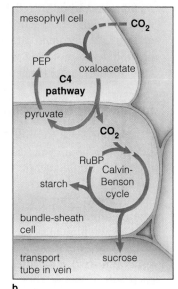

Figure 5.11 (**a**) The internal structure of a leaf from corn (*Zea mays*), a typical C4 plant. Photosynthetic mesophyll cells (*light green*) surround bundle-sheath cells (*light blue*) that surround tubes in veins. The tubes transport water into leaves and photosynthetic products from them. (**b**) The carbon-fixing system that precedes the Calvin-Benson cycle in C4 plants.

mesophyll cell CO_2
PEP   oxaloacetate
C4 pathway
pyruvate
CO_2
RuBP   Calvin-Benson cycle
starch
bundle-sheath cell
transport tube in vein   sucrose

b

thick surface layers that restrict water loss. It can't open its stomata during the day without losing precious water. Instead, it opens them and fixes CO_2 *at night*. Its cells store the resulting intermediate in central vacuoles and use it in photosynthesis the next day, when stomata close. Many plants are adapted this way. They are known as **CAM plants** (short for *Crassulacean Acid Metabolism*). *Unlike C4 species, CAM plants don't fix carbon twice, in different types of cells. They fix it in the same cells, at different times.*

During prolonged droughts, when many plants die, some CAM plants survive by keeping their stomata closed even at night. They repeatedly fix the CO_2 that forms during aerobic respiration. Not that much forms, but it is enough to allow these plants to maintain very low rates of metabolism. As you may have deduced, CAM plants grow slowly. Try growing a cactus plant in Seattle or some other place with a mild climate and it will compete poorly with C3 and C4 plants.

C4 plants and CAM plants both have modified ways of fixing carbon for photosynthesis. The modifications counter the stress imposed by hot, dry conditions in their environments.

AUTOTROPHS, HUMANS, AND THE BIOSPHERE

As we conclude this chapter, remind yourself of how photosynthesis "fits" in the world of living things. Think of the mind-boggling numbers of single-celled and multi-celled photosynthetic organisms in which the reactions are proceeding at this very moment, on land and in the sunlit waters of the earth. The sheer volume of reactant and product molecules that the photosynthesizers deal with might surprise you.

For example, drifting through the surface waters of the world ocean are uncountable numbers of single cells. You can't see them without a microscope—a row of 7 million cells of one species would be less than a quarter-inch long. Yet are they abundant! In some parts of the world, a cup of seawater may hold 24 million cells of one species, and that doesn't even include all the cells of *other* aquatic species.

Many of the drifters are photosynthetic bacteria, plants, and protistans. Together they are the pastures of the seas, the food base for diverse consumers. The pastures "bloom" in spring, when waters become warmer and enriched with nutrients that winter currents churn up from the deep. Then, populations burgeon, by cell divisions. Until NASA gathered data from satellites in space, biologists had no idea that the number of cells and their distribution were so mind-boggling. For example, the satellite image in Figure 5.12*b* shows a springtime bloom in the North Atlantic. It stretches from North Carolina all the way to Spain!

Collectively, these single cells have enormous impact on the world's climate. As they photosynthesize, they sponge up nearly half the carbon dioxide we humans release each year, as when we burn fossil fuels or burn forests. Without them, carbon dioxide in the air would accumulate more rapidly and possibly accelerate global warming, as described in Section 37.7. If our planet warms too much, vast coastal regions may become submerged, and nations that are now the greatest food producers may be hit hard.

Amazingly, every day, humans dump industrial wastes, fertilizers, and raw sewage into the ocean and thereby alter the living conditions for those drifting cells. How much of that noxious chemical brew can they tolerate?

One final point: Photosynthesis is so pervasive, it is sometimes easy to overlook other, less common energy-acquiring routes. Deep in the sea, in hot springs, even in waste heaps of coal mines, we find diverse bacteria that are classified as **chemoautotrophs**. They obtain energy not from sunlight but rather by pulling away hydrogen and electrons from ammonium ions, iron or sulfur compounds, and other inorganic substances. They, too, exist in monumental numbers. And they influence the global cycling of nitrogen and other vital elements through the biosphere. We will return later to their environmental effects. In this unit, we turn next to pathways by which energy is released from glucose and other biological molecules—the chemical legacy of autotrophs everywhere.

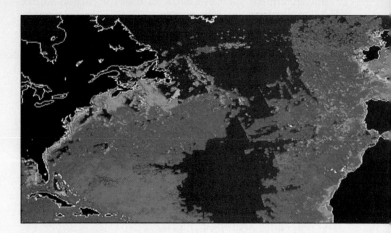

a Photosynthetic activity in winter. In this color-enhanced image, *red-orange* shows where chlorophyll concentrations are greatest.

b Photosynthetic activity in spring.

Figure 5.12 Two satellite images that help convey the astounding magnitude of photosynthetic activity during the spring in the North Atlantic portion of the world ocean.

SUMMARY

1. Plants and other photoautotrophs use sunlight as an energy source and carbon dioxide as the carbon source for building organic compounds. Animals and other heterotrophs get carbon and energy from organic compounds synthesized by plants and other autotrophs.

2. Photosynthesis is the main biosynthetic pathway by which carbon and energy enter the web of life. Its two sets of reactions start by trapping light energy ("photo") and end with assembly reactions ("synthesis").

    a. In plants, the light-dependent reactions take place at the thylakoid membrane system inside chloroplasts. The reactions produce ATP and NADPH (Figure 5.13).

    b. The light-independent reactions proceed outside the membrane system, in the chloroplast's stroma. They produce the sugar phosphates used in building sucrose, starch, and other end products of photosynthesis.

3. Regarding the light-dependent reactions:

    a. Photosystems (clusters of pigments in thylakoid membranes) absorb photon energy from the sun. With sufficient excitation energy, the photosystem's reaction center (a chlorophyll *a* molecule) transfers electrons to an acceptor molecule that donates them to an electron transport system also embedded in the membrane.

    b. In the cyclic pathway of ATP formation, excited electrons leave the P700 chlorophyll of photosystem I, give up energy in a transport system, and return to that photosystem.

    c. In the noncyclic pathway of ATP formation, the electrons from the P680 chlorophyll of photosystem II pass on through a transport system, then enter photosystem I, pass through a second transport system, and finally end up in NADPH. Also, at the start of the pathway, photolysis occurs: water molecules are split into oxygen, hydrogen ions (H^+), and electrons (which replace the electrons that P680 gives up).

    d. H^+ accumulates in the inner compartment of the membrane system as H^+ is split from water and as electron transport systems run. This sets up concentration and electric gradients that drive ATP formation.

4. The key points concerning the light-independent reactions:

    a. ATP delivers energy and NADPH delivers hydrogen and electrons to the stroma of chloroplasts, where sugar phosphates form and are combined into starch, cellulose, and other end products (Figure 5.13).

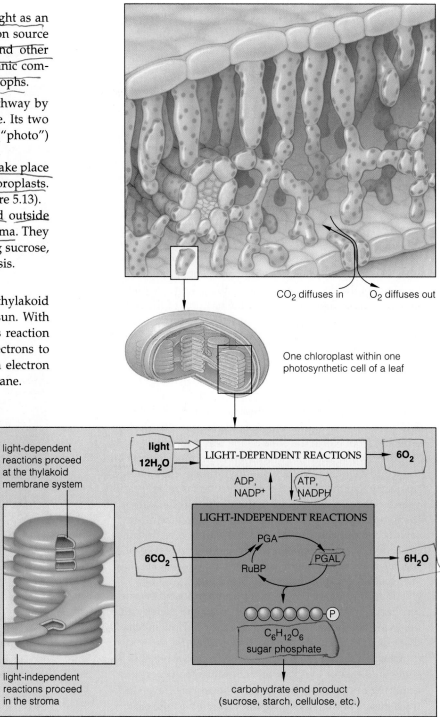

Figure 5.13 Summary of the main reactants, intermediates, and products of photosynthesis, corresponding to the equation:

$$12H_2O + 6CO_2 \xrightarrow{\text{sunlight}} 6O_2 + C_6H_{12}O_6 + 6H_2O$$

Starting with the light-dependent reactions, splitting twelve water molecules yields twenty-four electrons and six O_2. For every four electrons, three ATP and two NADPH form. During the light-independent reactions, *each turn* of the Calvin-Benson cycle requires one CO_2, three ATP, and two NADPH. Each sugar phosphate that forms has a backbone of six carbon atoms—so its formation requires six turns of the cycle.

b. The sugar phosphates form in the Calvin-Benson cycle. This pathway begins when CO_2 from the air is affixed to RuBP, making an unstable intermediate that splits into two PGA. ATP donates a phosphate group to each PGA. The resulting molecule receives H^+ and electrons from NADPH to form PGAL.

c. For every six CO_2 molecules that enter the cycle, twelve PGAL are produced. Two of those are used to produce a six-carbon sugar phosphate. The remainder are used to regenerate RuBP for the cycle.

5. C3 plants fix carbon by the Calvin-Benson cycle only. C4 plants can fix carbon twice, in two cell types. CAM plants fix and store carbon at night, then use it by day.

Review Questions

1. A cat eats a bird, which earlier speared and ate a caterpillar that had been chewing on a weed. Which of these organisms are the autotrophs? The heterotrophs? *83*

2. Summarize the photosynthesis reactions as an equation. Then state the main events of the light-dependent reactions, and next, of the light-independent reactions. *84, 88, 91*

3. Identify which one of the following substances accumulates in the thylakoid compartment of chloroplasts: glucose, chlorophyll, carotenoids, hydrogen ions, or fatty acids. *90*

4. Which is *not* required for the light-independent reactions: ATP, NADPH, RuBP, carotenoids, free oxygen, CO_2, or enzymes? *91*

5. How many CO_2 molecules must enter the Calvin-Benson cycle to produce one sugar phosphate? Why? *91*

6. Suppose a plant busily photosynthesizing is exposed to CO_2 molecules that contain radioactively labeled carbon atoms ($^{14}CO_2$). Identify the compound in which the labeled carbon will first appear: NADPH, PGAL, pyruvate, or PGA. *91*

Self-Quiz *(Answers in Appendix IV)*

1. Photosynthetic autotrophs use _____ from the air as a carbon source and _____ as their energy source.

2. In plants, light-*dependent* reactions proceed at the _____ .
   a. cytoplasm
   b. plasma membrane
   c. stroma
   d. thylakoid membrane

3. The light-*independent* reactions proceed in the _____ .
   a. cytoplasm
   b. plasma membrane
   c. stroma
   d. grana

4. In the light-dependent reactions, _____ .
   a. carbon dioxide is fixed
   b. ATP and NADPH form
   c. carbon dioxide accepts electrons
   d. sugar phosphates form

5. When a photosystem absorbs light, _____ .
   a. sugar phosphates are produced
   b. electrons are transferred to ATP
   c. RuBP accepts electrons
   d. light-dependent reactions begin

6. The Calvin-Benson cycle starts when _____ .
   a. light is available
   b. light is not available
   c. carbon dioxide is attached to RuBP
   d. electrons leave a photosystem

7. In the light-independent reactions, ATP donates phosphate to _____ .
   a. RuBP
   b. $NADP^+$
   c. PGA
   d. PGAL

8. Match each event with its suitable description.
   ____ RuBP used; PGA forms
   ____ ATP and NADPH used
   ____ NADPH forms
   ____ ATP and NADPH form
   ____ only ATP forms

   a. cyclic pathway
   b. noncyclic pathway
   c. CO_2 fixation
   d. PGAL forms
   e. transfer of H^+ and electrons to $NADP^+$

Critical Thinking Questions

1. Imagine a garden filled with red, white, and blue petunias. Explain the colors of these flowers in terms of which wavelengths of light they are absorbing and reflecting.

2. About 200 years ago, Jan Baptista van Helmont performed an experiment on the nature of photosynthesis. He wanted to know where a growing plant acquired the raw materials necessary to increase in size. For his experiment, he planted a tree seedling weighing 5 pounds in a barrel filled with 200 pounds of soil. He watered the tree regularly. After five years passed, van Helmont again weighed the tree and the soil. At that time the tree weighed 169 pounds, 3 ounces. The soil weighed 199 pounds, 14 ounces. Because the tree's weight had increased so much and the soil's weight had decreased so very little, he concluded the tree gained weight as a result of the water he had added to the barrel.

   Given your knowledge of the composition of biological molecules, why was van Helmont's conclusion misguided? On the basis of the current model of photosynthesis, provide a more plausible explanation of his results.

3. Captain Kirk and Spock land on a planet in a distant galaxy, where they find populations of a vibrantly purple, carbon-based life form. Spock suspects the form secures the carbon and energy necessary for survival through a process similar to photosynthesis. How would he go about determining whether this is the case?

Selected Key Terms

autotroph *83*
C3 plant *92*
C4 plant *92*
Calvin-Benson cycle *91*
CAM plant *92*
carotenoid *87*
chemoautotroph *93*
chlorophyll *87*
chloroplast *85*
cyclic pathway of ATP formation *88*
electron transport system *88*
heterotroph *83*
light-dependent reaction *88*
light-independent reaction *91*
noncyclic pathway of ATP formation *88*
PGA *91*
PGAL *91*
photolysis *89*
photon *86*
photosystem *88*
phycobilin *87*
pigment *86*
RuBP *91*
stroma *85*
thylakoid membrane system *85*

Readings

Davis, B. February 1992. "Going for the Green." *Discover* 13: 20. Describes work on artificial systems for photosynthesis.

Hendry, George. May 1990. "Making, Breaking, and Remaking Chlorophyll." *Natural History* 36–41.

Youvan, D., and B. Marrs. 1987. "Molecular Mechanisms of Photosynthesis." *Scientific American* 256: 42–50.

6 ENERGY-RELEASING PATHWAYS

The Killers Are Coming!

In 1990, descendants of "killer" bees that left South America a few decades earlier buzzed across the border between Mexico and Texas. By 1995, they had invaded 13,287 square kilometers of Southern California and were busily setting up colonies.

When provoked, the bees behave in a terrifying way. For instance, thousands of bees flew into action simply because a construction worker started up a tractor a few hundred yards away from their hive. The agitated bees entered a nearby subway station and started stinging passengers on the platform and inside the trains. They killed one person and injured a hundred others. Just recently, they put two tree trimmers in Indio, California, in the hospital.

Where did the bees come from? In the 1950s, some queen bees were shipped from Africa to Brazil for selective breeding experiments. Why? Honeybees are big business. Besides serving as honey producers, bees can be rented out to pollinate commercial orchards—and their activities make a large difference in fruit production. Position a screened cage around an orchard tree in bloom and less than 1 percent of the flowers will set fruit. Put a hive of honeybees in the same cage and 40 percent of the flowers will set fruit.

Compared with their African relatives, bees in Brazil are sluggish pollinators and honey producers. By cross-breeding the two varieties, researchers thought they might be able to come up with a strain of mild-mannered but zippier bees. They put local bees and the imported bees together inside netted enclosures, complete with artificial hives. Then they let nature take its course.

Figure 6.1 One of the mild-mannered honeybees buzzing in for a landing on a flower, wings beating with energy provided by ATP. If this were one of its Africanized relatives protecting a hive, possibly you would not stay around to watch the landing. Both kinds of bees look alike. How can we tell them apart? From our own biased perspective, Africanized bees are the ones with an attitude problem.

Twenty-six African queen bees escaped. That was bad enough. Then beekeepers got wind of preliminary experimental results. After learning that the first few generations of offspring were more energetic but not overly aggressive, they imported hundreds of African queens and encouraged them to mate with the locals. And they set off a genetic time bomb.

Before long, African bees became established in commercial hives—and in wild bee populations. And their traits became dominant. The "Africanized" bees do everything other bees do, but they do more of it faster. Their eggs develop into adults more quickly. Adults fly more rapidly, outcompete other bees for nectar, and even die sooner.

When something disturbs their hives or swarms, Africanized bees become extremely agitated. They can remain that way for as long as eight hours. Whereas a mild-mannered honeybee might chase an intruding animal fifty yards or so, a squadron of Africanized bees will chase it a quarter of a mile. If they catch up to it, they collectively can sting it to death.

Doing things faster means having a continuous supply of energy and efficient ways of using it. An Africanized bee's stomach can hold thirty milligrams of sugar-rich nectar—which is enough fuel to fly sixty kilometers. That's more than thirty-five miles! Besides this, compared to other kinds of bees, the flight muscle cells of an Africanized bee have larger mitochondria. These organelles specialize in releasing a great deal of energy from sugars and other organic compounds, then converting it to the energy of ATP.

Whenever they tap into the stored energy of organic compounds, Africanized bees reveal their biochemical connection with other organisms. Study a primrose or puppy, a mold growing on stale bread, an amoeba in pond water or a bacterium living on your skin, and you will discover that their energy-releasing pathways differ in some details. But all of the pathways require characteristic starting materials. They yield predictable products and by-products. And they yield the universal energy currency of life—ATP.

In fact, throughout the biosphere, organisms put energy and raw materials to use in amazingly similar ways. *At the biochemical level, we find undeniable unity among all forms of life.* We will return to this idea in the concluding section of the chapter.

KEY CONCEPTS

1. All organisms can release energy stored in glucose and other organic compounds, then use it in ATP production. The energy-releasing pathways differ from one another. But the main types all start with the breakdown of glucose to pyruvate.

2. The initial breakdown reactions, known as glycolysis, can proceed in the presence of oxygen or in its absence. Said another way, the reactions can be the first stage of either aerobic or anaerobic pathways.

3. Two kinds of energy-releasing pathways are completely anaerobic, from start to finish. We call them fermentation and anaerobic electron transport. They proceed only in the cytoplasm, and none yields more than a small amount of ATP for each glucose molecule metabolized.

4. Another pathway, aerobic respiration, also starts in the cytoplasm. But it alone runs to completion in organelles called mitochondria. Compared with the other pathways, aerobic respiration releases far more energy from glucose.

5. Aerobic respiration has three stages. First, pyruvate forms from glucose (through glycolysis). Second, different reactions break down the pyruvate to carbon dioxide. These reactions liberate electrons and hydrogen, which coenzymes deliver to a transport system. Third, stepwise electron transfers through the system help set up the conditions that favor ATP formation. Free oxygen accepts the electrons at the end of the line and combines with hydrogen, thereby forming water.

6. Over evolutionary time, photosynthesis and aerobic respiration became linked on a global scale. The oxygen-rich atmosphere, a long-term outcome of photosynthetic activity, sustains aerobic respiration—which has become the dominant energy-releasing pathway. And most kinds of photosynthesizers use carbon dioxide and water from aerobic respiration as raw materials when they synthesize organic compounds:

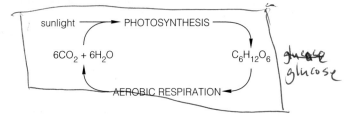

6.1 HOW CELLS MAKE ATP

Organisms stay alive by taking in energy. Plants and all other organisms that engage in photosynthesis get energy from the sun. Animals get energy secondhand, thirdhand, and so on, by eating plants and one another. Regardless of its source, however, energy must be put into a form that can drive life-sustaining reactions. The energy carried from one reaction site to another by adenosine triphosphate—ATP—serves that function.

Plants make ATP during photosynthesis. They and all other organisms also make ATP by breaking down carbohydrates (glucose especially), fats, and proteins. During the breakdown reactions, electrons get stripped from intermediates—then energy that is associated with the liberated electrons drives the formation of ATP. Electron transfers of the sort described in Section 4.8 are central to these energy-releasing pathways.

Comparison of the Main Types of Energy-Releasing Pathways

The first energy-releasing pathways evolved about 3.8 billion years ago, when conditions were very different on Earth. Because the atmosphere had little free oxygen, the pathways must have been anaerobic, which means they could run to completion without utilizing oxygen. Many bacteria and protistans still live in places where oxygen is absent or not always available, and they must make ATP by anaerobic routes. (Fermentation pathways and anaerobic electron transport are the most common of these.) Some of your own cells can use an anaerobic pathway for short periods—but only when they are not receiving enough oxygen. Your cells, and most others, depend primarily on **aerobic respiration**, an oxygen-requiring pathway of ATP formation. With each breath you take, you provide your actively respiring cells with a fresh supply of oxygen.

Make note of this point: *The main energy-releasing pathways all start with the same reactions in the cytoplasm.* During this initial stage of reactions, called glycolysis, enzymes cleave and rearrange each glucose molecule into two pyruvate molecules.

Once that first stage is over, however, the energy-releasing pathways differ. Most importantly, the aerobic pathway continues inside a mitochondrion:

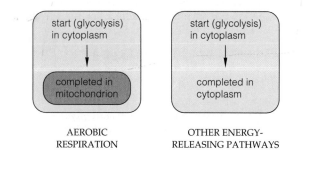

AEROBIC RESPIRATION OTHER ENERGY-RELEASING PATHWAYS

a

In mitochondria, oxygen serves as the final acceptor of electrons that were released at different reaction steps. By contrast, the anaerobic pathways start *and end* in the cytoplasm—and a substance other than oxygen serves as the final acceptor of released electrons.

As you examine these energy-releasing pathways in sections to follow, keep in mind that the reaction steps do not proceed by themselves. Enzymes catalyze each step, and intermediate molecules formed at one step serve as substrates for the next enzyme in the pathway.

Overview of Aerobic Respiration

Of all the energy-releasing pathways, aerobic respiration gets the most ATP for each glucose molecule. Whereas the anaerobic routes typically have a net yield of two ATP molecules, the aerobic pathway commonly yields thirty-six or more.

If you were a bacterium, you wouldn't require much ATP. Being far larger, more complex, and highly active, you depend absolutely on the high yield of the aerobic route. When a glucose molecule is the starting material, aerobic respiration can be summarized this way:

$$C_6H_{12}O_6 + 6O_2 \rightarrow 6CO_2 + 6H_2O$$
glucose   oxygen   carbon dioxide water

$$C_6H_{12}O_6 + 6O_2 \rightarrow 6CO_2 + 6H_2O$$

b

CYTOPLASM

glucose

energy input → GLYCOLYSIS → 2 ATP (net)

electron

coenzymes — 2 NADH   2 pyruvate

MITOCHONDRION

2 NADH ← → 2 CO₂

6 NADH ← KREBS CYCLE → 4 CO₂

2 FADH₂ → 2 ATP

release

ELECTRON TRANSPORT PHOSPHORYLATION → water

→ 32 ATP

oxygen

Typical Energy Yield: 36 ATP

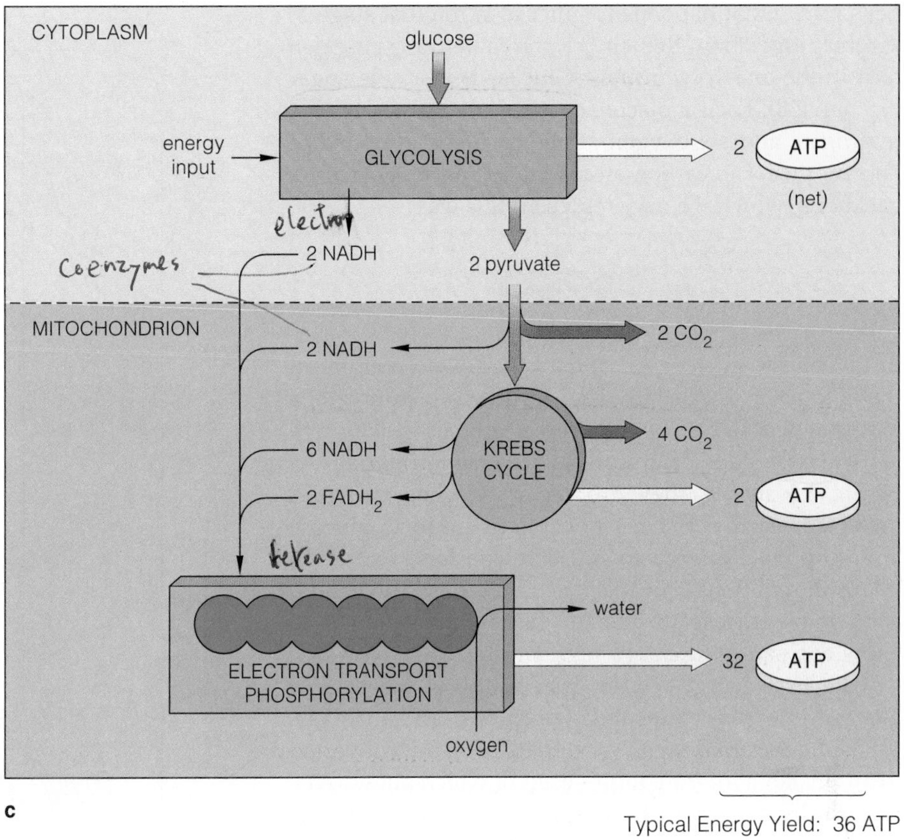

c

Figure 6.2 Overview of aerobic respiration, which proceeds through three stages. (**a,b**) Only this pathway delivers enough ATP to build and maintain giant redwoods and other large, multicelled organisms. Only great amounts of ATP can sustain birds, bees, humans, and other highly active animals.

(**c**) Glucose partially breaks down to pyruvate in the first stage of reactions (glycolysis). In the second stage, which includes the Krebs cycle, pyruvate breaks down to carbon dioxide. Coenzymes (NAD⁺ and FAD) pick up electrons and hydrogen stripped from intermediates at both stages.

In the final stage (electron transport phosphorylation), the loaded-down coenzymes (NADH and FADH₂) give up the electrons and hydrogen to a transport system. Energy released during the flow of electrons through the system drives ATP formation. Oxygen accepts the electrons at the end of the third stage.

From start (glycolysis) to finish, the typical net energy yield from each glucose molecule is thirty-six ATP.

However, the summary equation only tells us what the substances are at the start and finish of the pathway. In between are three stages of reactions (Figure 6.2).

The initial stage, again, is glycolysis. In the second stage, which includes the **Krebs cycle**, pyruvate breaks down to carbon dioxide and water. During both stages, coenzymes pick up electrons and hydrogen stripped from the intermediates. Not much ATP forms in either stage. The large energy harvest comes *after* coenzymes deliver their cargo to an electron transport system.

In the third stage, the transport system functions as the machinery for **electron transport phosphorylation**. It is during this stage that so many ATP molecules are produced. As it ends, oxygen inside the mitochondrion accepts the "spent" electrons from the last component

of the transport system. Oxygen picks up hydrogen at the same time and thereby forms water.

Cells can release energy stored in glucose and other organic compounds and convert it to ATP energy, which drives nearly all metabolic activities. The main kinds of energy-releasing pathways all start in the cytoplasm with glycolysis, a stage of reactions that break down glucose to pyruvate.

The most common anaerobic pathways (fermentation routes and anaerobic electron transport) end in the cytoplasm. Each has a net energy yield of two ATP.

Aerobic respiration, an oxygen-dependent pathway, runs to completion in the mitochondrion. From start (glycolysis) to finish, it commonly has a net energy yield of thirty-six ATP.

6.2 GLYCOLYSIS: FIRST STAGE OF THE ENERGY-RELEASING PATHWAYS

Let's track what happens to glucose in the first stage of aerobic respiration. Remember, the same things happen to it in the anaerobic routes. Glucose is a simple sugar (Section 2.8). Each molecule of it has six carbon, twelve hydrogen, and six oxygen atoms covalently bonded to one another. The carbon atoms make up the molecule's backbone, which we may represent this way:

In glycolysis, glucose (or some other carbohydrate) in the cytoplasm is partially broken down to **pyruvate**, a molecule with a backbone of three carbon atoms.

The first steps of glycolysis are *energy-requiring*. As Figure 6.3 shows, they proceed only when two ATP molecules each transfer a phosphate group to glucose and so donate energy to it. Such transfers, recall, are "phosphorylations." In this case, they raise the energy content of glucose to a level that is high enough to allow entry into the *energy-releasing* steps of glycolysis.

The first energy-releasing step cleaves the primed glucose into two molecules, which we can call PGAL (phosphoglyceraldehyde). Each PGAL gets converted to an unstable intermediate—each of which allows ATP to form by giving up a phosphate group to ADP. The next intermediate in the sequence does the same thing.

Thus, a total of four ATP form by **substrate-level phosphorylation**. By definition, this metabolic event is the direct transfer of phosphate group from a substrate of a reaction to another molecule, such as ADP.

Remember, though, two ATP were invested to start the reactions. So the *net* energy yield is only two ATP.

Meanwhile, the coenzyme **NAD⁺** (for nicotinamide adenine dinucleotide) picks up electrons and hydrogen liberated from each PGAL, thereby becoming NADH. When the NADH gives up its cargo at another reaction site, it becomes NAD⁺ once more. Like other enzyme helpers, then, the NAD⁺ is reusable (Section 4.5).

In sum, glycolysis converts a bit of the energy stored in glucose to a transportable form of energy, in ATP. A coenzyme picks up electrons and hydrogen stripped from glucose. These have key roles in the next stage of reactions. So do the end products of glycolysis—the two molecules of pyruvate.

Cells obtain some usable energy from glycolysis, a stage of reactions that partially break down glucose or some other carbohydrate to two pyruvate molecules. Two NADH and four ATP molecules form at certain steps in the reaction sequence.

Subtracting the two ATP required to start the reactions puts the net energy yield of glycolysis at two ATP.

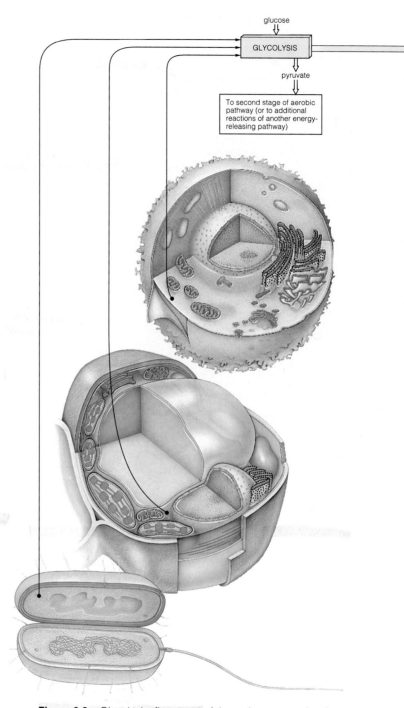

Figure 6.3 Glycolysis, first stage of the main energy-releasing pathways. The reaction steps proceed inside the cytoplasm of every living prokaryotic and eukaryotic cell.

In this example, glucose is the starting material. By the time the reactions end, two pyruvate, two NADH, and four ATP have been produced. Cells invest two ATP to start glycolysis, so the net energy yield of glycolysis is two ATP.

Depending on the type of cell and on prevailing conditions in its environment, the pyruvate molecules may be used in the second set of reactions of the aerobic pathway, which includes the Krebs cycle. Or they may be used in different reactions, such as those of fermentation pathways.

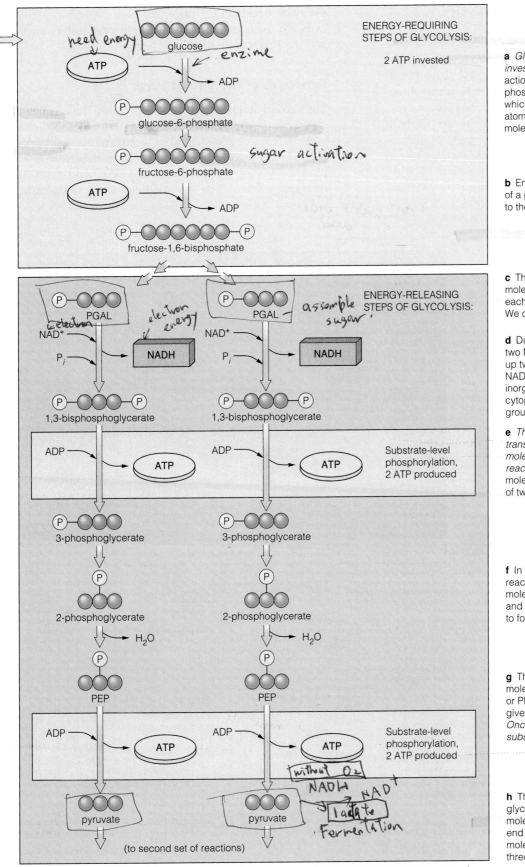

need energy

← enzime

glucose

ENERGY-REQUIRING
STEPS OF GLYCOLYSIS:

2 ATP invested

a *Glycolysis starts with an energy investment of two ATP.* First, enzyme action promotes the transfer of a phosphate group from ATP to glucose, which has a backbone of six carbon atoms. With this transfer, the glucose molecule becomes slightly rearranged.

glucose-6-phosphate

sugar activation

fructose-6-phosphate

b Enzyme action promotes the transfer of a phosphate group from another ATP to the rearranged molecule.

fructose-1,6-bisphosphate

electron energy

assemble sugar

ENERGY-RELEASING
STEPS OF GLYCOLYSIS:

PGAL   PGAL

electron

NAD^+ NAD^+

NADH   NADH

P_i P_i

1,3-bisphosphoglycerate   1,3-bisphosphoglycerate

c The resulting fructose-1,6-bisphosphate molecule splits at once into two molecules, each with a three-carbon backbone. We can call these two PGAL.

d During enzyme-mediated reactions, two NADH form after each PGAL gives up two electrons and a hydrogen atom to NAD^+. Each PGAL also combines with inorganic phosphate (P_i) present in the cytoplasm, then donates a phosphate group to ADP.

ADP → ATP   ADP → ATP

Substrate-level phosphorylation, 2 ATP produced

e *Thus two ATP have formed by the direct transfer of phosphate from two intermediate molecules that serve as substrates in the reactions.* With this formation of two ATP molecules, the original energy investment of two ATP is paid off.

3-phosphoglycerate   3-phosphoglycerate

2-phosphoglycerate   2-phosphoglycerate

→ H_2O → H_2O

f In the next two enzyme-mediated reactions, each of the two intermediate molecules releases a hydrogen atom and an —OH group, which combine to form water.

PEP   PEP

ADP → ATP   ADP → ATP

Substrate-level phosphorylation, 2 ATP produced

g The resulting intermediates (two molecules of 3-phosphoenolpyruvate, or PEP) are rather unstable. Each gives up a phosphate group to ADP. *Once again, two ATP have formed by substrate-level phosphorylation.*

without O₂
NADH → NAD^+
lactate
Fermentation

pyruvate   pyruvate

(to second set of reactions)

h Thus the net energy yield from glycolysis is two ATP for each glucose molecule entering the reactions. The end products of glycolysis are two molecules of pyruvate, each with a three-carbon backbone.

NET ENERGY YIELD: 2 ATP

SECOND STAGE OF THE AEROBIC PATHWAY

Preparatory Steps and the Krebs Cycle

Suppose two pyruvate molecules, formed by glycolysis, leave the cytoplasm and enter a **mitochondrion** (plural, mitochondria). In this organelle alone, the second and third stages of the aerobic pathway run to completion. Figure 6.4 shows its structure and functional zones.

In the second stage, a bit more ATP forms. Carbon atoms depart from the pyruvate, in the form of carbon dioxide. And coenzymes latch onto the electrons and hydrogen from intermediates of the reactions. Figure 6.5 illustrates key steps of this stage.

In a few preparatory steps, an enzyme strips a carbon atom from each pyruvate molecule. Coenzyme A, an enzyme helper, becomes **acetyl-CoA** when it joins with the two-carbon fragment that remains. It transfers this fragment to **oxaloacetate**—the entry point of the Krebs cycle. The name of this cyclic pathway honors Hans Krebs, who began working out its details in the 1930s. Notice that *six* carbon atoms enter this stage of reactions (three in each

pyruvate backbone). Notice also that *six* leave (in six carbon dioxide molecules) during the preparatory steps and the Krebs cycle proper.

Functions of the Second Stage

The second stage serves three functions. First, it loads electrons and hydrogen onto NAD$^+$ and **FAD** (flavin adenine dinucleotide, another coenzyme) to produce NADH and FADH$_2$. Second, through substrate-level phosphorylations, it produces two ATP. And third, it rearranges Krebs cycle intermediates into oxaloacetate.

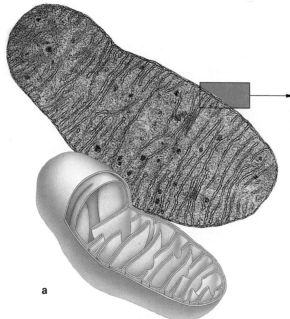

a

Figure 6.4 (**a**) Mitochondrion, in thin section. (**b,c**) Functional zones inside the mitochondrion. An inner membrane system divides the interior into two compartments. The second and third stages of aerobic respiration proceed here. Many coenzymes pick up electrons and hydrogen from intermediates of the second-stage reactions, then deliver them to transport systems embedded in the inner membrane. Operation of the systems during the third stage drives ATP formation at nearby proteins (ATP synthases) in the membrane.

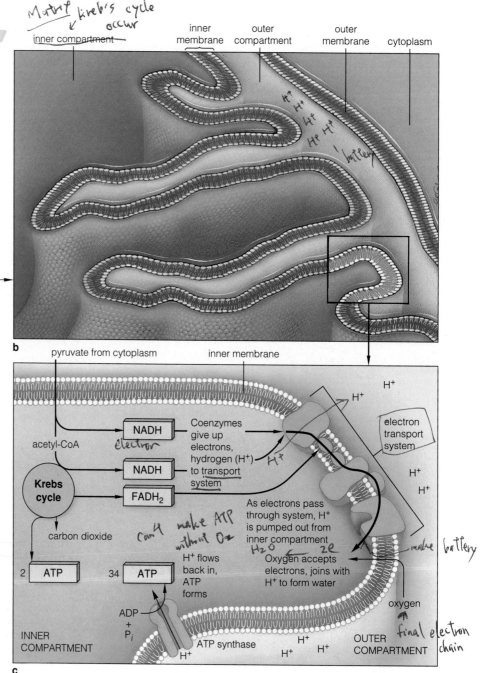

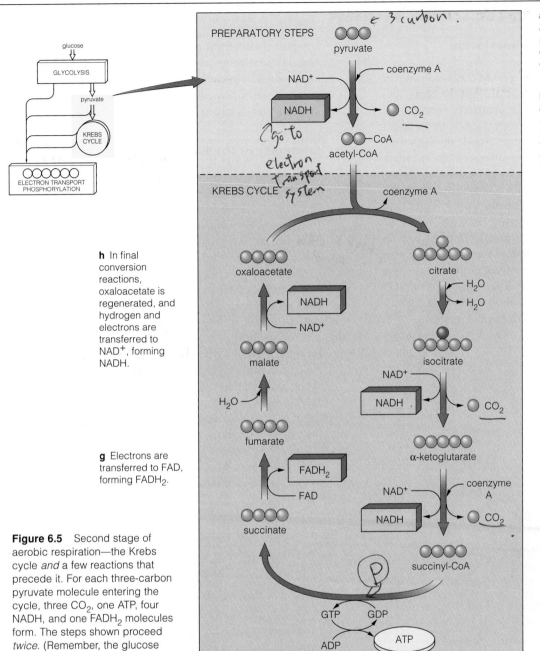

PREPARATORY STEPS

← 3 carbon. *(handwritten)*

a A pyruvate molecule enters a mitochondrion. It undergoes preparatory conversions before entering cyclic reactions (Krebs cycle).

b First, the pyruvate is stripped of a functional group (COO⁻), which departs as CO_2. Next, it gives up hydrogen and electrons to NAD⁺, forming NADH. A coenzyme joins with the remaining two-carbon fragment, forming acetyl-CoA.

pyruvate

NAD⁺ — coenzyme A

NADH

CO_2

go to *(handwritten)*

acetyl-CoA —CoA

electron transport system *(handwritten)*

KREBS CYCLE   coenzyme A

c The acetyl-CoA is transferred to oxaloacetate, a four-carbon compound that is the point of entry into the Krebs cycle. The result is citrate, with a six-carbon backbone.

h In final conversion reactions, oxaloacetate is regenerated, and hydrogen and electrons are transferred to NAD⁺, forming NADH.

oxaloacetate

citrate

H_2O
H_2O

NADH

NAD⁺

malate

isocitrate

NAD⁺

H_2O

NADH

CO_2

fumarate

α-ketoglutarate

d Citrate enters conversion reactions in which a COO⁻ group departs (as CO_2). Also, hydrogen and electrons are transferred to NAD⁺, forming NADH.

g Electrons are transferred to FAD, forming FADH₂.

FADH₂

FAD

NAD⁺ — coenzyme A

NADH

CO_2

succinate

e Another COO⁻ group departs (as CO_2) and another NADH forms. *At this point, three carbon atoms have been released, balancing out the three that entered the mitochondrion (in pyruvate).*

Figure 6.5 Second stage of aerobic respiration—the Krebs cycle *and* a few reactions that precede it. For each three-carbon pyruvate molecule entering the cycle, three CO_2, one ATP, four NADH, and one FADH₂ molecules form. The steps shown proceed *twice*. (Remember, the glucose molecule was broken down earlier to *two* pyruvate molecules.)

succinyl-CoA

P

GTP   GDP

ADP   ATP

f After further reactions, ATP forms by substrate-level phosphorylation.

Cells have only so much oxaloacetate, and it must be regenerated to keep the cyclic reactions going.

The two ATP that form don't add much to the small yield from glycolysis. However, *many* coenzymes pick up electrons and hydrogen for transport to the sites of the third and final stage of the aerobic pathway:

Glycolysis:	2 NADH
Pyruvate conversion before Krebs cycle:	2 NADH
Krebs cycle:	2 FADH₂ + 6 NADH
Coenzymes sent to third stage:	2 FADH₂ + 10 NADH

Overall, these are the key points to remember about the second stage of aerobic respiration:

During the second stage of the aerobic pathway, the two pyruvate molecules from glycolysis enter a mitochondrion. Each gives up a carbon atom, and its remnant enters the Krebs cycle. All of its carbon atoms end up in carbon dioxide.

The preparatory steps and the cycle proper yield two ATP. Oxaloacetate, the entry point for the cycle, is regenerated. Many coenzymes pick up electrons and hydrogen stripped from substrates, for delivery to the final stage of the pathway.

---

glucose

GLYCOLYSIS

pyruvate

KREBS CYCLE

ELECTRON TRANSPORT PHOSPHORYLATION

6.4 THIRD STAGE OF THE AEROBIC PATHWAY

Electron Transport Phosphorylation

ATP production goes into high gear in the third stage of the aerobic pathway. Electron transport systems and neighboring proteins called ATP synthases serve as the production machinery. They are embedded in the inner membrane that divides the mitochondrion into two compartments (Figure 6.6). They interact with electrons and unbound hydrogen—that is, H^+ ions. Remember, coenzymes deliver this bounty from reaction sites of the first two stages of the aerobic pathway.

Briefly, electrons get transferred from one molecule of each transport system to the next in line. As certain molecules accept and then donate electrons, they also pick up H^+ ions in the inner compartment, then release them to the outer compartment. Their action sets up H^+ concentration and electric gradients across the inner mitochondrial membrane. Nearby in the membrane, H^+ ions follow the gradients and flow back to the inner compartment—through the interior of ATP synthases. *This H^+ flow drives the formation of ATP from ADP and unbound phosphate.* Free oxygen keeps ATP production going. It withdraws electrons at the end of the transport systems and combines with H^+. Water is the result.

Summary of the Energy Harvest

In many types of cells, thirty-two ATP form during the third stage of aerobic respiration. Add these to the net yield from the preceding stages, and the total harvest is thirty-six ATP from one glucose molecule (Figure 6.7). That's a lot! An anaerobic pathway may use eighteen glucose molecules to produce the same amount of ATP.

Think of thirty-six ATP as a typical yield only. The actual amount depends on cellular conditions, as when cells require a given intermediate elsewhere and pull it out of the reaction sequence.

The yield also depends on how particular cells use the NADH that formed during glycolysis. Any NADH produced in the cytoplasm can't enter a mitochondrion. It can only transfer electrons and hydrogen to proteins in the outer mitochondrial membrane—which shuttle them to NAD^+ or to FAD molecules already inside the mitochondrion. Both coenzymes deliver the electrons to transport systems of the inner membrane. However, FAD puts them to a *lower* entry point in the transport system, so *its* deliveries produce less ATP (Figure 6.7).

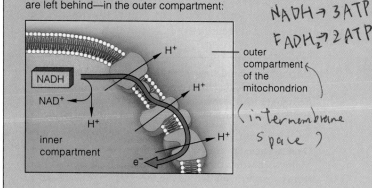

Figure 6.6 Electron transport phosphorylation, the third and final stage of aerobic respiration. The reactions proceed at transport systems and at channel proteins (ATP synthases) that are embedded in the inner mitochondrial membrane. Each transport system consists of enzymes, cytochromes, and other proteins that act in sequence.

The inner membrane divides the mitochondrion into two compartments. The third-stage reactions start in the inner compartment, when NADH and $FADH_2$ give up hydrogen (as H^+ ions) and electrons to the transport system. The electrons are transferred through the system, but the ions are left behind—in the outer compartment:

NADH → 3 ATP
FADH₂ → 2 ATP

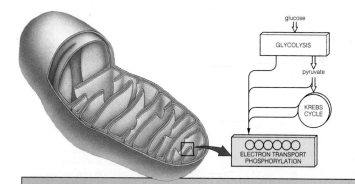

(intermembrane space)

Soon there is a higher concentration of H^+ in the outer compartment than in the inner one. Concentration and electric gradients now exist across the membrane. The ions follow the gradients and flow across the membrane, through the interior of the transport proteins. Energy associated with the flow drives the formation of ATP from ADP and unbound phosphate (hence the name, electron transport *phosphorylation*):

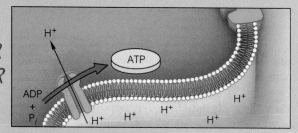

Do these events sound familiar? They should. ATP forms in much the same way in chloroplasts (Figure 5.9). By the chemiosmotic theory, recall, concentration and electric gradients across a membrane drive ATP formation. The theory also applies to mitochondria (even though ions flow in the opposite direction compared to chloroplasts).

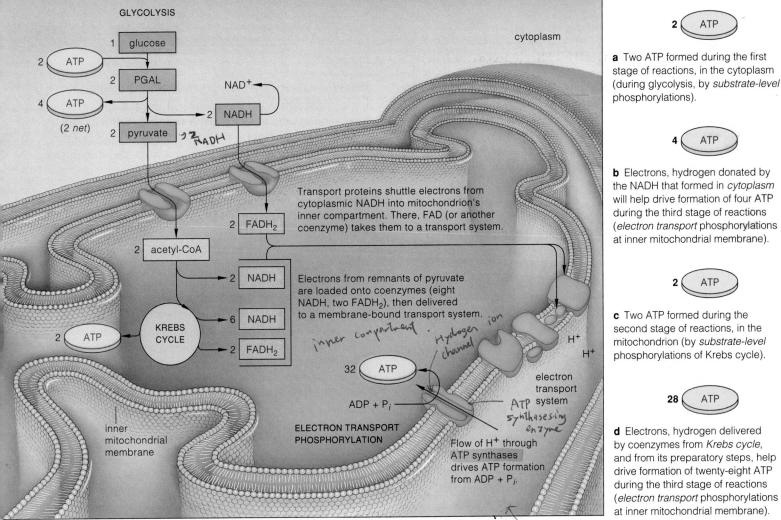

Figure 6.7 Summary of the harvest from the energy-releasing pathway of aerobic respiration. Commonly, thirty-six ATP form for each glucose molecule that enters the pathway. However, the net yield varies according to shifting concentrations of reactants, intermediates, and end products of the reactions. It also varies among different types of cells.

For example, cells differ in how they use the NADH from glycolysis. These NADH cannot enter mitochondria. They only deliver their cargo of electrons and hydrogen to certain transport proteins of the outer mitochondrial membrane. The proteins shuttle the electrons and hydrogen across the membrane, to NAD^+ or to FAD already inside the mitochondrion, to form NADH or $FADH_2$.

Any NADH inside the mitochondrion delivers electrons to the highest possible point of entry into a transport system. When it does, enough H^+ can be pumped across the inner membrane to produce *three* ATP. By contrast, any $FADH_2$ delivers them to a lower entry point in the transport system. Fewer hydrogen ions can be pumped, so only *two* ATP can be produced.

In liver, heart, and kidney cells, for example, electrons and hydrogen enter the highest entry point of transport systems, so the overall energy harvest is thirty-eight ATP. More commonly, as in skeletal muscle and brain cells, they get transferred to FAD—so the overall harvest is thirty-six ATP.

In figure area (handwritten annotations):
- 2 ATP
- 6 (4)   2 NADH × 3(2)
- 2 ATP
- 24   8 NADH × 3
- 4   2 FADH₂ × 2
- 38 (36)

a Two ATP formed during the first stage of reactions, in the cytoplasm (during glycolysis, by *substrate-level* phosphorylations).

b Electrons, hydrogen donated by the NADH that formed in *cytoplasm* will help drive formation of four ATP during the third stage of reactions (*electron transport* phosphorylations at inner mitochondrial membrane).

c Two ATP formed during the second stage of reactions, in the mitochondrion (by *substrate-level* phosphorylations of Krebs cycle).

d Electrons, hydrogen delivered by coenzymes from *Krebs cycle*, and from its preparatory steps, help drive formation of twenty-eight ATP during the third stage of reactions (*electron transport* phosphorylations at inner mitochondrial membrane).

36 ATP
TYPICAL NET
ENERGY YIELD

One final point. Glucose, recall, has more energy (stored in more covalent bonds) than carbon dioxide or water does. When it breaks down to those more stable end products, about 686 kilocalories are released. Most of this energy escapes (as heat), but 7.5 kilocalories or so are conserved in each ATP molecule. Therefore, when 36 ATP form through the breakdown of a glucose molecule, the energy-conserving efficiency of aerobic respiration is $(36)(7.5)/(686) \times 100$, or 39 percent.

In the final stage of the aerobic pathway, coenzymes deliver electrons to transport systems of the inner mitochondrial membrane. As electrons move through the system, they set up H^+ gradients that drive ATP formation at nearby proteins in the membrane. Oxygen is the final electron acceptor.

Again, from start (glycolysis in the cytoplasm) to finish (in mitochondria), the pathway commonly has a net yield of thirty-six ATP for every glucose molecule metabolized.

So far, we have tracked the fate of a glucose molecule through the pathway of aerobic respiration. We turn now to its use as a substrate for fermentation pathways. Remember, these are anaerobic pathways; they do *not* use oxygen as the final acceptor of the electrons that ultimately drive the ATP-forming machinery.

Fermentation Pathways

Diverse kinds of organisms use fermentation pathways. Many are bacteria and protistans that make their homes in marshes, bogs, mud, deep-sea sediments, the animal gut, canned foods, sewage treatment ponds, and other oxygen-free settings. Some kinds of fermenters actually die if exposed to oxygen. The bacteria responsible for many diseases, including botulism and tetanus, are like this. Other kinds of fermenters, including the bacterial "employees" of yogurt manufacturers, are indifferent to the presence of oxygen. Still others can use oxygen, but they also can use a fermentation pathway when oxygen becomes scarce. Even your muscle cells do this.

As is true of aerobic respiration, glycolysis serves as the first stage of the fermentation pathways. Here also, enzymes split glucose and rearrange the fragments into two pyruvate molecules. Here again, two NADH form, and the net energy yield is two ATP. However, as Figure 6.8 indicates, the reactions do not completely break down glucose to carbon dioxide and water, and they produce no more ATP beyond the yield from glycolysis. *The final steps serve only to regenerate NAD$^+$—a coenzyme with absolutely central roles in the breakdown reactions.*

Fermentation yields enough energy to sustain many single-celled anaerobic organisms. It even helps carry some aerobic cells through times of stress. But it is not enough to sustain large, active, multicelled organisms —this being one reason why you never will meet up with an anaerobic elephant.

Lactate Fermentation With these points in mind, take a look at Figure 6.9, which tracks the main steps of **lactate fermentation**. During this anaerobic pathway, the *pyruvate* molecules from the first stage of reactions (glycolysis) accept the hydrogen and electrons from NADH. This transfer regenerates NAD$^+$ and, at the same time, converts each pyruvate to a three-carbon compound called lactate. You may hear people refer to this compound as "lactic acid." However, its ionized form (lactate) is far more common in cells.

Some bacteria, such as *Lactobacillus*, rely exclusively on this anaerobic pathway. Left to their own devices, their fermentation activities often spoil food. Yet certain fermenters have commercial uses, as when they break down glucose in huge vats where cheeses, yogurt, and sauerkraut are produced.

In humans, rabbits, and many other animals, some types of cells also can switch to lactate fermentation for a quick fix of ATP. When your own demands for energy are intense but brief—say, during a short race—muscle cells use this pathway. They cannot do so for long; they would throw away too much of glucose's stored energy for too little ATP. When glucose stores are depleted, muscles fatigue and lose their ability to contract.

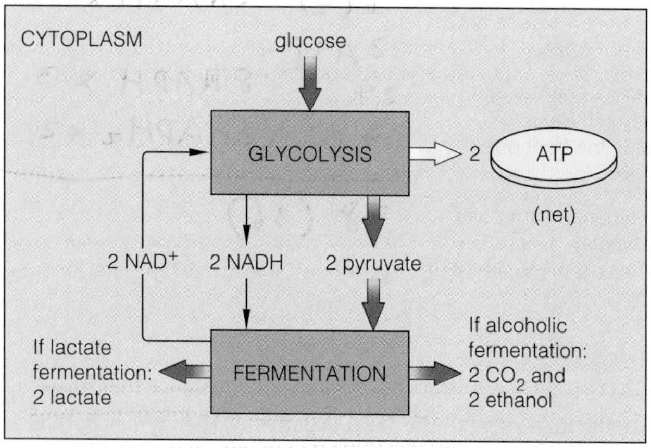

Figure 6.8 Overview of two fermentation routes. In this type of energy-releasing pathway, only the initial stage (glycolysis) has a net energy harvest, in the form of two ATP. The remaining reactions serve to regenerate NAD$^+$.

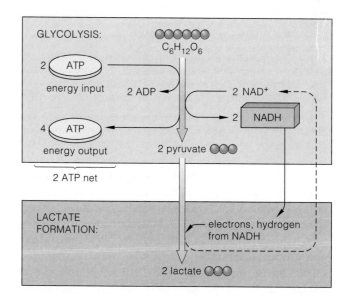

Figure 6.9 Lactate fermentation. In this anaerobic pathway, electrons end up in lactate, the reaction product.

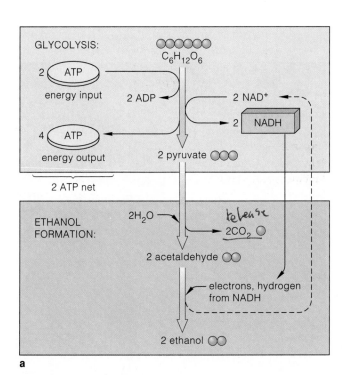

GLYCOLYSIS:

$C_6H_{12}O_6$

2 ATP
energy input

2 ADP

2 NAD^+

2 NADH

4 ATP
energy output

2 pyruvate

2 ATP net

ETHANOL
FORMATION:

$2H_2O$

release

$2CO_2$

2 acetaldehyde

electrons, hydrogen
from NADH

2 ethanol

a

b   c

Figure 6.10 (**a**) Alcoholic fermentation. In this anaerobic pathway, acetaldehyde, an intermediate of the reactions, is the final acceptor of electrons. Ethanol is the end product. Yeasts, single-celled organisms, use this pathway. (**b**) One species of *Saccharomyces* makes bread dough rise. Another (**c**) lives on sugar-rich tissues of ripened grapes.

Alcoholic Fermentation In **alcoholic fermentation**, each pyruvate molecule from glycolysis is converted to an intermediate form, acetaldehyde. NADH transfers electrons and hydrogen to this form and so converts it to an alcoholic end product—ethanol (Figure 6.10).

Certain species of single-celled fungi called yeasts are renowned for their use of this anaerobic pathway. One type, *Saccharomyces cerevisiae*, makes bread dough rise. Bakers mix the yeast with sugar, then blend the mixture into dough. When yeast cells break down the sugar, they release carbon dioxide. Bubbles of the gas expand the dough (make it "rise"). Oven heat forces the gas out of the dough, and a porous product remains.

Beer and wine producers also use yeasts on a large scale. Vintners use wild yeasts that live on grapes. They also employ cultivated strains of *S. ellipsoideus*, which remain active until the alcohol concentration in wine vats exceeds 14 percent. In the wild, yeasts die when the concentration exceeds 4 percent—but 4 percent still packs a punch. Landscapers no longer plant pyracantha along highways; robins getting drunk on the naturally fermenting berries of these shrubs kept weaving into windshields. Wild turkeys get similarly tipsy when they gobble fermenting apples in untended orchards.

Anaerobic Electron Transport

Especially among the bacteria, we find less common energy-releasing pathways, some of which are topics of later chapters in the book. For example, many bacterial species have key roles in the global cycling of sulfur, nitrogen, and other crucial elements. Collectively, their metabolic activities influence nutrient availability for organisms everywhere.

For example, certain bacteria use **anaerobic electron transport**. Electrons stripped from some type of organic compound move through transport systems embedded in the plasma membrane. Often an inorganic compound in the environment serves as the final electron acceptor. The net energy yield varies, but it is always small.

Even as you read this, some anaerobic bacteria that live in waterlogged soil are stripping electrons from a variety of compounds. They dump electrons on sulfate. Hydrogen sulfide, a foul-smelling gas, is the result. These sulfate-reducing bacteria also live in aquatic habitats that are enriched with decomposed organic material. They even live on the deep ocean floor, near hydrothermal vents. As described in Section 38.12, they form the food production base for unique communities.

In fermentation pathways, an organic substance that forms during the reactions serves as the final acceptor of electrons from glycolysis. The reactions regenerate NAD^+, which is required to keep the pathway operational.

In anaerobic electron transport, an inorganic substance (but not oxygen) usually serves as the final electron acceptor.

For each glucose molecule metabolized, anaerobic pathways typically have a net energy yield of two ATP, which form during glycolysis.

Carbohydrate Breakdown in Perspective

So far, you have looked at what happens after a lone glucose molecule enters an energy-releasing pathway. Now you can start thinking about what cells do when they have too many or too few molecules of glucose.

For example, consider what happens after you or any other mammal finishes a meal. Your body absorbs a great deal of glucose and other small organic molecules, which your bloodstream delivers to tissues throughout the body. There, cells rapidly take up excess glucose, then "trap" it by converting it to glucose-6-phosphate. (In its phosphorylated state, glucose can't be transported back out, across the plasma membrane.) Look again at Figure 6.3a and you see that glucose-6-phosphate is the first activated intermediate of glycolysis.

When your food intake exceeds cellular demands for energy, ATP-producing machinery goes into high gear. In fact, the ATP concentration may rise to such an extent that it inhibits glycolysis. Glucose-6-phosphate is diverted into a biosynthesis pathway at such times. The end product of that pathway is glycogen—a storage polysaccharide composed of glucose units (Figure 2.19).

Between meals, the amount of free glucose entering the internal environment dwindles. At such times, cells break down glycogen to glucose-6-phosphate, which enters glycolysis. Liver cells do more. They convert the glucose-6-phosphate back to free glucose, which they release. The bloodstream quickly delivers glucose to energy-demanding cells of muscles and other organs that have depleted their own glycogen stores.

A word of caution: Don't let the preceding examples lead you to believe that *all* cells squirrel away large amounts of glycogen. Liver and muscle cells maintain the largest stores. Even then, the glycogen represents a mere 1 percent or so of the total energy conserved in the body of an adult human. On the average, 78 percent is conserved in fats and another 21 percent in proteins.

Energy From Fats

Maybe you avoid butter, ice cream, and other fatty foods, thinking it is better to fill up on carbohydrates and proteins. This is a good idea, as long as you don't stuff yourself with these organic compounds—because your body will convert excess amounts to fats. A fat molecule, recall, has a glycerol head and one, two, or three fatty acid tails. Most of the fats that become stored in your body are in the form of triglycerides, with three tails each. Triglycerides accumulate inside the fat cells of certain tissues (adipose tissues), which develop at the buttocks and other strategic locations beneath the skin.

Between meals or during exercise, the body taps triglycerides as energy alternatives to glucose. At such

Figure 6.11 Points at which a variety of organic compounds can enter the reaction stages of aerobic respiration.

Complex carbohydrates, fats, and proteins cannot enter the aerobic pathway. In humans and other mammals, the digestive system as well as individual cells must first break apart these large molecules into simpler, degradable subunits.

times, enzymes in fat cells cleave the bonds holding glycerol and fatty acids together. Then the breakdown products enter the bloodstream. When glycerol reaches the liver, enzymes convert it to PGAL—an intermediate of glycolysis. Most cells can take up the circulating fatty acids. Enzymes cleave the carbon backbone of fatty acid tails and convert the fragments to acetyl-CoA—which can enter the Krebs cycle (Figures 6.5 and 6.11).

Each fatty acid tail has many more carbon-bound hydrogen atoms than glucose—so its breakdown yields much more ATP. In between meals or during sustained exercise, fatty acid conversions supply about half of the ATP that your muscle, liver, and kidney cells require.

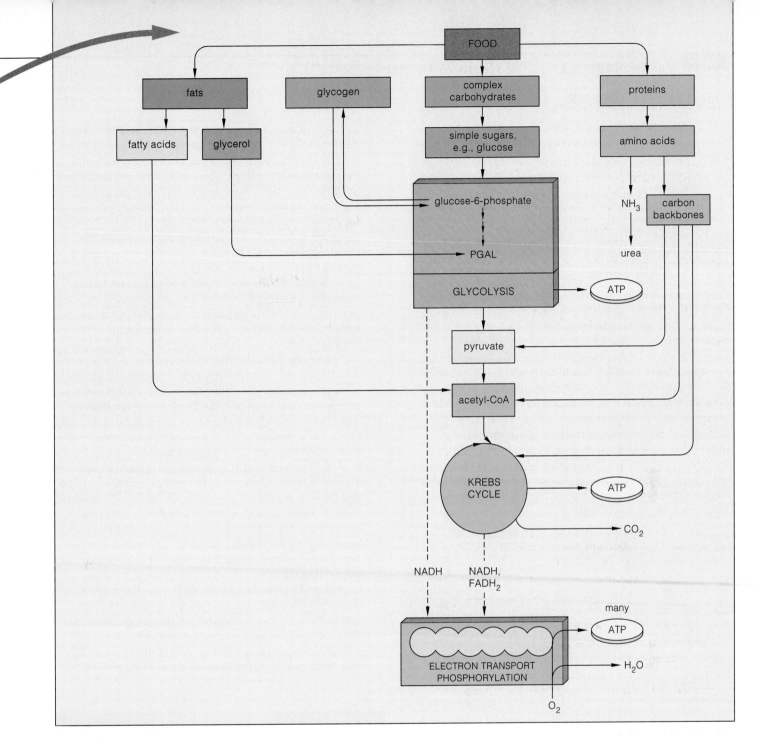

Energy From Proteins

Eat more proteins than your body requires to grow and maintain itself, and your cells will not store them. Enzymes split the proteins into amino acid units. Then they remove the amino group ($—NH_3^+$) from each unit, and ammonia (NH_3) forms. What happens to the left-over carbon backbones? Depending on conditions in the cell, the outcome varies. These backbones can be converted to carbohydrates or fats. Or they may enter the Krebs cycle, as in Figure 6.11, where coenzymes can pick up the hydrogen and electrons stripped away from the carbon atoms. The ammonia that forms undergoes conversions to become urea. This nitrogen-containing waste product would be toxic if allowed to accumulate to high concentrations. The body excretes it, in urine.

This concludes our look at aerobic respiration and other energy-releasing pathways. The essay to follow may help you get a sense of how they fit into the larger picture of life's evolution and interconnectedness.

In humans and other mammals, the entrance of glucose or some other organic compound into an energy-releasing pathway depends on its concentrations inside and outside the cell, and on the type of cell.

PERSPECTIVE ON LIFE

In this unit, you read about photosynthesis and aerobic respiration—the main pathways by which cells trap, store, and release energy. What you may not know is that the two pathways became linked, on a grand scale, over evolutionary time.

When life originated about 3.8 billion years ago, the Earth's atmosphere had little free oxygen. Probably the earliest single-celled organisms used reactions similar to glycolysis to make ATP. Without oxygen, fermentation pathways must have dominated. About 1.5 billion years later, oxygen-producing photosynthetic cells had emerged. They irrevocably changed the course of evolution. Oxygen, a by-product of the noncyclic pathway of photosynthesis, began to accumulate in the atmosphere. Probably through mutations in the proteins of electron transport systems, some cells started using oxygen as an electron acceptor. At some point in the past, descendants of those fledgling aerobic cells abandoned photosynthesis. Among them were the forerunners of animals and other organisms that engage in aerobic respiration.

With aerobic respiration, the flow of carbon, hydrogen, and oxygen through the metabolic pathways of living organisms came full circle. For the final products of this key aerobic pathway—carbon dioxide and water—are precisely the materials that are used to build organic compounds in photosynthesis:

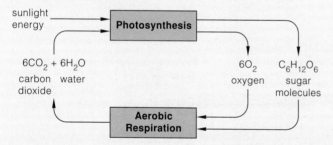

Photosynthesis

sunlight energy

$6CO_2 + 6H_2O$
carbon   water
dioxide

$6O_2$ $C_6H_{12}O_6$
oxygen   sugar
             molecules

Aerobic Respiration

Perhaps you have difficulty fathoming the connection between yourself—an intelligent being—and such remote-sounding events as energy flow and the cycling of carbon, hydrogen, and oxygen. Is this really the stuff of humanity?

Think back, for a moment, on the structure of a water molecule. Two hydrogen atoms sharing electrons with an oxygen atom may not seem close to your daily life. And yet, through that sharing, water molecules show polarity—and they hydrogen-bond with one another. Their chemical behavior is a beginning for the organization of lifeless matter that leads to the organization of all living things.

For now you can imagine other molecules interspersed through water. The nonpolar kinds resist interaction with water; the polar kinds dissolve in it. On their own, the phospholipids among them assemble into a two-layered film. Such lipid bilayers, remember, serve as the very framework of all cell membranes, hence all cells. From the beginning, the cell has been the fundamental *living* unit.

The essence of life is not some mysterious force. It is metabolic control. With a cell membrane to contain them, reactions *can* be controlled. With mechanisms built into their membranes, cells can respond to energy changes and shifting concentrations of substances in the environment. The response mechanisms operate by "telling" proteins—enzymes—when and what to build or tear down.

And it is not some mysterious force that creates the proteins themselves. DNA, the slender double-stranded treasurehouse of heredity, has the chemical structure—*the chemical message*—that allows molecule to reproduce molecule, one generation after the next. In your own body, DNA strands tell trillions of cells how countless molecules must be built or torn apart for their stored energy.

So yes, carbon, hydrogen, oxygen, and other organic molecules represent the stuff of you, and us, and all of life. But it takes more than molecules to complete the picture. Life exists as long as an unbroken flow of energy sustains its organization. Molecules are assembled into cells, cells into organisms, organisms into communities, and so on up through the biosphere. It takes energy inputs from the sun to maintain these levels of organization. And energy flows through time in one direction—from organized to less organized forms. Only as long as energy continues to flow into the web of life can life continue in all its rich diversity.

In short, life is no more *and no less* than a marvelously complex system of prolonging order. Sustained by energy transfusions from the sun, life continues on, through its capacity for self-reproduction. For with the hereditary instructions contained in DNA, energy and materials can be organized, generation after generation. Even with the death of individuals, life elsewhere is prolonged. With death, molecules are released. They may be cycled once more, as raw materials for new generations.

In this flow of energy and cycling of material through time, each birth is affirmation of our ongoing capacity for organization, each death a renewal.

SUMMARY

1. Metabolic reactions run on the energy, inherent in phosphate groups, that ATP molecules deliver to them. Aerobic respiration, fermentation, and other pathways that release chemical energy from organic compounds, such as glucose, produce ATP.

2. After glucose enters these pathways, enzymes strip electrons and hydrogen from intermediates that form along the way. Coenzymes pick these up and deliver them to other reaction sites at which the pathway is completed. NAD+ is the main coenzyme; the aerobic route also uses FAD. When loaded with electrons and hydrogen, they are designated NADH and FADH₂.

3. All of the main energy-releasing pathways begin with glycolysis, a stage of reactions that start and end in the cytoplasm. These reactions can be completed either in the presence of oxygen or in its absence.

   a. During glycolysis, enzymes break down a glucose molecule to two pyruvate molecules. Two NADH and four ATP form during the reactions.

   b. The net energy yield is two ATP (because two ATP had to be invested up front to get the reactions going).

4. Aerobic respiration continues on through two more stages: the Krebs cycle and a few steps preceding it, and electron transport phosphorylation. The stages proceed only inside the organelles called mitochondria.

5. The second stage of the aerobic pathway starts when an enzyme strips a carbon atom from each pyruvate. Coenzyme A binds the remaining two-carbon fragment (to form acetyl-CoA), then transfers it to oxaloacetate, the entry point of the Krebs cycle. The cyclic reactions, along with the steps immediately preceding them, load ten coenzymes with electrons and hydrogen (eight NADH and two FADH₂). Two ATP form. Three carbon dioxide molecules are released for each pyruvate that entered this second stage.

6. The third stage of the aerobic pathway proceeds at a membrane that divides the interior of a mitochondrion into two compartments. Electron transport systems and ATP synthases are embedded in this inner membrane.

   a. Coenzymes deliver electrons from the first two stages to transport systems. In the outer compartment, hydrogen ions accumulate so that concentration and electric gradients form across the membrane.

   b. The hydrogen ions follow the gradients and flow from the outer to the inner compartment, through the interior of ATP synthases. Energy released during the ion flow drives the formation of ATP from ADP and unbound phosphate.

   c. Oxygen withdraws electrons from the transport system and at the same time combines with hydrogen ions to form water. This oxygen is the final acceptor of electrons that initially resided in glucose.

7. Aerobic respiration has a typical net energy yield of thirty-six ATP for each glucose molecule metabolized. Yields vary, according to cell type and cell conditions.

8. Fermentation pathways as well as anaerobic electron transport also start with glycolysis, but they don't use oxygen; they are anaerobic, start to finish.

   a. Lactate fermentation has a net energy yield of two ATP, which form in glycolysis. The remaining reactions regenerate NAD+. The two NADH from glycolysis transfer electrons and hydrogen to two pyruvate from glycolysis. Two lactate molecules are the end products.

   b. Similarly, alcoholic fermentation has a net energy yield of two ATP from glycolysis, and its remaining reactions regenerate NAD+. Enzymes convert pyruvate from glycolysis to acetaldehyde, and carbon dioxide is released. The NADH from glycolysis transfer electrons and hydrogen to the two acetaldehyde molecules, and so form two ethanol molecules—the end products.

   c. Certain bacteria use anaerobic electron transport. Electrons are stripped from various organic compounds and travel through transport systems in the bacterium's plasma membrane. An inorganic compound in the environment often serves as the final electron acceptor.

9. In humans and other mammals, simple sugars such as glucose (from carbohydrates), glycerol and fatty acids (from fats), and the carbon backbone of amino acids also enter the ATP-producing pathways. Their entrance depends on their concentrations inside and outside the cell, and on cell type.

Review Questions

1. For this diagram of the aerobic pathway, fill in the blanks and write in the number of molecules of pyruvate, coenzymes, and end products. Also write in the net ATP produced in each stage, as well as the net ATP produced from start (glycolysis) to finish. *99, 105*

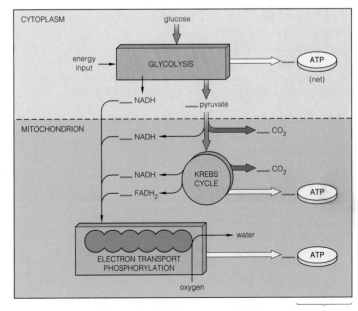

Typical Energy Yield: ___ATF

2. Is glycolysis energy *requiring* or energy *releasing*? Or do both kinds of reactions occur during glycolysis? *100, 101*

3. What is the difference between *electron transport* phosphorylation and *substrate-level* phosphorylation? *99, 100 (see also Figure 6.6)*

4. Sketch the double-membrane system of the mitochondrion and show where transport systems and ATP synthases are located. *102*

5. Which compound serves as the entry point for the Krebs cycle? Does it directly accept the pyruvate from glycolysis? *102, 103*

6. For each glucose molecule, how many carbon atoms enter the Krebs cycle? How many depart from it, and in what form? *102*

7. Is this statement true or false: Aerobic respiration occurs in animals but not plants, which make ATP only by photosynthesis. *98*

8. Is this statement true or false: Muscle cells cannot contract at all when deprived of oxygen. If true, explain why. If false, name the alternative(s) available to them. *106*

Self-Quiz *(Answers in Appendix IV)*

1. Glycolysis starts and ends in the _____ .
   a. nucleus   c. plasma membrane
   b. mitochondrion   d. cytoplasm

2. Which of the following does *not* form during glycolysis?
   a. NADH   c. FAD
   b. pyruvate   d. ATP

3. The pathway of aerobic respiration is completed in the _____ .
   a. nucleus   c. plasma membrane
   b. mitochondrion   d. cytoplasm

4. In the last stage of aerobic respiration, _____ is the final acceptor of electrons that originally resided in glucose.
   a. water   c. oxygen
   b. hydrogen   d. NADH

5. In the last stage of aerobic respiration, a flow of _____ in response to concentration and electric gradients drives the formation of ATP from ADP and phosphate.
   a. electrons   c. NADH
   b. hydrogen ions   d. $FADH_2$

6. _____ engage in lactate fermentation.
   a. *Lactobacillus* cells   c. Sulfate-reducing bacteria
   b. Muscle cells   d. a and b are correct

7. In alcoholic fermentation, _____ is the final acceptor of electrons stripped from glucose.
   a. oxygen   c. acetaldehyde
   b. pyruvate   d. sulfate

8. The fermentation pathways produce no more ATP beyond the small yield from glycolysis, but the remaining reactions _____ .
   a. regenerate ADP   c. dump electrons on an
   b. regenerate NAD^+ inorganic substance (not oxygen)

9. In certain organisms and under certain conditions, _____ can be used as an energy alternative to glucose.
   a. fatty acids   c. amino acids
   b. glycerol   d. all are correct

10. Match the event with its suitable metabolic description.
    ____ glycolysis   a. ATP, NADH, $FADH_2$, CO_2, and
    ____ fermentation   water form
    ____ Krebs cycle   b. glucose to 2 pyruvate
    ____ electron transport c. NAD^+ regenerated, 2 ATP net
         phosphorylation   d. H^+ flows through ATP synthases

Critical Thinking

1. Diana suspects that a visit to her family doctor is in order. After eating carbohydrate-rich food, she always experiences sensations of being intoxicated and becomes nearly incapacitated—as if she had been drinking alcohol. She even wakes up with a hangover the next day. Having completed a course in freshman biology, Diana has an idea that something is affecting the way her body is metabolizing glucose. Explain why.

2. The cells of your body absolutely do not use nucleic acids as alternative energy sources. Suggest why.

3. The body's energy needs and its programs for growth depend on balancing the levels of amino acids in blood with proteins in cells. Cells of the liver, kidneys, and intestinal lining are especially important in this balancing act. When the levels of amino acids in blood decline, lysosomes in cells can rapidly digest some of their proteins (structural and contractile proteins are spared, except in cases of malnutrition). The amino acids released this way enter the blood and thereby help maintain the required levels.

   Suppose you embark on a body-building program. You already eat plenty of carbohydrates, but a nutritionist advises a protein-rich diet that includes protein supplements. Speculate on how the extra dietary proteins will be put to use, and in which tissues.

4. Each year, Canada geese lift off in precise formation from their northern breeding grounds. They head south to spend the winter months in warmer climates, and then they make the return trip in spring. As is true of other migratory birds, their flight muscle cells are efficient at using fatty acids as an energy source. (Remember, the carbon backbone of fatty acids can be cleaved into fragments that can be converted to acetyl-CoA for entry into the Krebs cycle.)

   Suppose a lesser Canada goose from Alaska's Point Barrow has been steadily flapping along for three thousand kilometers and is nearing Klamath Falls, Oregon. It looks down and notices a rabbit sprinting like the wind from a coyote with a taste for rabbit. With a stunning burst of speed, the rabbit reaches the safety of its burrow.

   Which energy-releasing pathway predominated in the rabbit's leg muscle cells? Why was the Canada goose relying on a different pathway for most of its journey? And why wouldn't the pathway of choice in goose flight muscle cells be much good for a rabbit making a mad dash from its enemy?

5. Reflect on this chapter's introduction, then on Question 4. Now speculate on which energy-releasing pathway is predominating in an agitated Africanized bee chasing a farmer across a cornfield.

Selected Key Terms

acetyl-CoA *102*   Krebs cycle *99*
aerobic respiration *98*   lactate fermentation *106*
alcoholic fermentation *107*   mitochondrion *102*
anaerobic electron transport *107* NAD^+ *100*
electron transport   oxaloacetate *102*
  phosphorylation *99*   pyruvate *100*
FAD *102*   substrate-level
glycolysis *98*   phosphorylation *100*

Readings

Levi, P. October 1984. "Travels with C." *The Sciences.* Journey of a carbon atom through the world of life.

Wolfe, S. 1995. *An Introduction to Molecular and Cellular Biology.* Belmont, California: Wadsworth. Exceptional reference text.

FACING PAGE: *Human sperm, one of which will penetrate this mature egg and so set the stage for the development of a new individual in the image of its parents.*

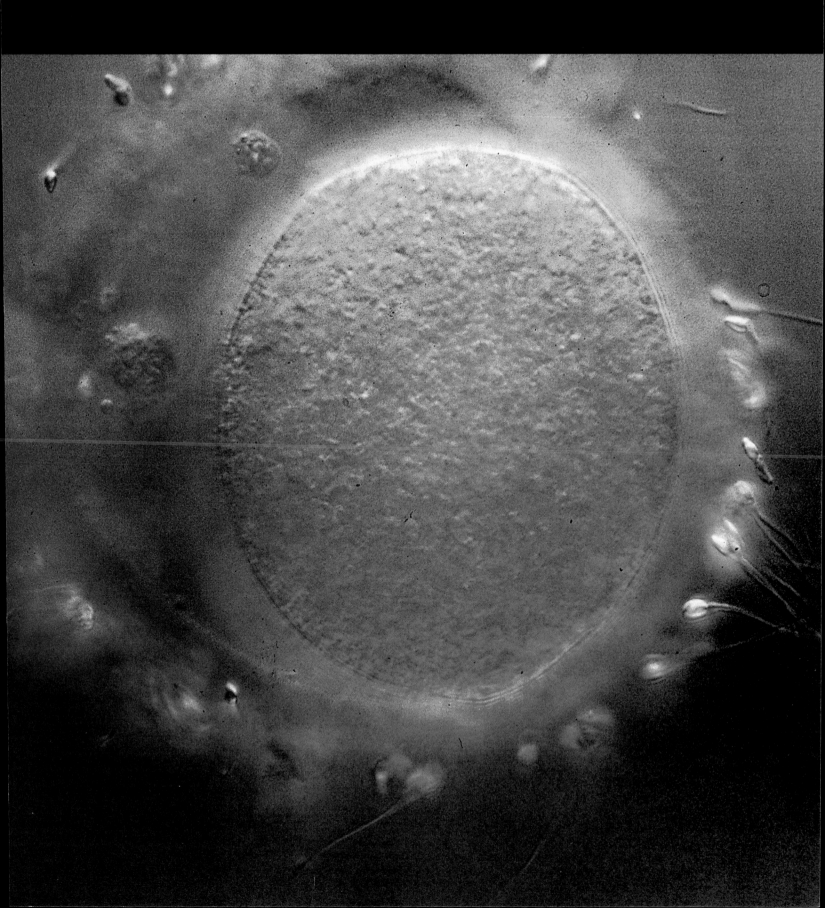

7 CELL DIVISION AND MITOSIS

Silver In the Stream of Time

Five o'clock, and the first rays of the sun dance over the wild Alagnak River of the Alaskan tundra. It is September, and life is both ending and beginning in the clear, frigid waters. By the thousands, mature silver salmon have returned from the open ocean to spawn in their shallow native home. The females are tinged with red, the color of spawners, and they are dying.

This morning, a female salmon releases translucent pink eggs into a "nest," hollowed out by her fins in the gravel riverbed (Figure 7.1). Within moments, a male sheds a cloud of sperm, and fertilization follows. Trout and other predators eat most of the eggs—but some eggs survive and give rise to a new generation.

Within three years, the pea-size eggs have become streamlined salmon, fashioned from billions of cells. A few of the cells will develop into eggs or sperm. In time, on some September morning, they will take part in an ongoing story of birth, growth, death, and rebirth.

For you, as for the salmon and all other multicelled species, growth as well as reproduction depends on *cell division*. In your mother's body, a fertilized egg divided in two, then the two into four, and so on until billions of cells were growing, developing in specialized ways, and dividing at different times to produce all of your genetically prescribed body parts. Your body now has roughly 65 trillion living cells—and many of the cells are still dividing. Every five days, for instance, cell divisions replace the lining of your small intestine.

Understanding cell division—and, ultimately, how new individuals are put together in the image of their parents—begins with answers to three questions. *First*, what instructions are necessary for inheritance? *Second*, how are those instructions duplicated for distribution into daughter cells? *Third*, by what mechanisms are those instructions divided into daughter cells? We will

Figure 7.1 The last of one generation and the first of the next in the Alagnak River of Alaska.

require more than one chapter to consider the nature of cell reproduction and other mechanisms of inheritance. Even so, the points made early in this chapter can help you keep the overall picture in focus.

Begin with the word **reproduction**. In biology, this means producing a new generation of cells or multi-celled individuals. The process starts with the division of single cells. And the ground rule for cell division is this: *Parent cells must provide their daughter cells with hereditary instructions, encoded in DNA, and enough cytoplasmic machinery to start up their own operation.*

DNA, recall, contains instructions for synthesizing proteins. Some proteins are structural materials. Many of the proteins are enzymes that speed the assembly of specific sugars, lipids, and other building blocks of cells. Unless a daughter cell receives the necessary instructions for making proteins, it simply will not be able to grow or function properly.

Also, the parental cell's cytoplasm already has operating machinery—enzymes, organelles, and so on. When a daughter cell inherits what looks like a blob of cytoplasm, it really is getting start-up machinery for its operation—until it has time to use its inherited DNA for growing and developing on its own.

1. When a cell divides, its two daughter cells must each receive a required number of DNA molecules as well as cytoplasm. In eukaryotes, a division mechanism called mitosis sorts out the DNA into two new nuclei. A separate mechanism divides the cytoplasm in two.

2. Mitotic cell division is the basis of growth and tissue repair in multicelled eukaryotes. It also is the means by which single-celled eukaryotes and many multicelled eukaryotes reproduce asexually.

3. In eukaryotic cells, DNA has many proteins attached to it. Each DNA molecule, with its attached proteins, is a chromosome.

4. Members of the same species have the same number of chromosomes in their reproductive cells (sperm or eggs). The chromosomes differ in length and shape, and they carry different portions of the hereditary instructions.

5. A reproductive cell has a haploid chromosome number; it contains *one* of each type of chromosome characteristic of the species. By contrast, body cells of humans and many other organisms are diploid. They contain not one but *two* of each chromosome characteristic of the species.

6. When a cell prepares for mitosis, its chromosomes are duplicated. Now each chromosome consists of two DNA molecules, which remain attached until they are separated for distribution to forthcoming daughter cells.

7. Mitosis keeps the chromosome number constant, from one cell generation to the next. Thus, if a parent cell is diploid, so will be the daughter cells.

Overview of Division Mechanisms

Before a cell of a plant, animal, or another eukaryotic organism reproduces, it undergoes **mitosis** or **meiosis**. Both are *nuclear* division mechanisms; they sort out and package a parent cell's DNA molecules into new nuclei, for the forthcoming daughter cells. Then some other mechanism splits the cytoplasm into daughter cells.

Multicelled organisms grow by way of mitosis and the cytoplasmic division of body cells, which are called **somatic cells**. They also repair tissues that way. Nick yourself peeling a potato, and mitotic cell divisions will replace the cells that the knife sliced away. Besides this, many protistans, fungi, plants, and even some animals reproduce asexually by mitotic cell division.

By contrast, meiosis occurs only in **germ cells**, a cell lineage set aside for the formation of gametes (such as sperm and eggs) and sexual reproduction. As you will read in the next chapter, meiosis has much in common with mitosis, but the end result is different.

What about prokaryotic cells—the bacteria? They reproduce asexually by way of a different mechanism, called prokaryotic fission. We will consider the bacteria later, in Chapter 18.

Some Key Points About Chromosomes

Before a cell starts preparing for mitosis, its DNA molecules are stretched out like threads, with many proteins attached to them. Each DNA molecule, with its attached proteins, is a **chromosome**.

Chromosomes undergo duplication while they are in the threadlike form. Afterward, each consists of two DNA molecules, which will stay together until late in mitosis. For as long as they remain attached, the two are called **sister chromatids** of the chromosome:

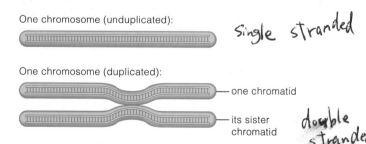

One chromosome (unduplicated):   Single stranded

One chromosome (duplicated):
— one chromatid
— its sister chromatid   double stranded

Notice how the chromosome narrows down in a small region. This is the **centromere**, a region with attachment sites for the microtubules that move the chromosome during nuclear division. Bear in mind, the preceding sketch is simplified. As you might deduce from Figure 7.2, the centromere location is not the same in all chromosomes. And the two strands of a DNA molecule don't look like a ladder—they twist rather like a spiral staircase and are much longer than can be shown here.

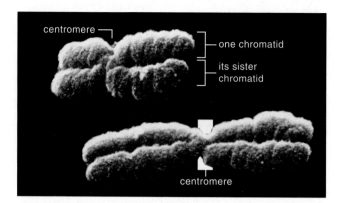

Figure 7.2 Scanning electron micrograph of two human chromosomes, each in the duplicated state.

Mitosis and the Chromosome Number

Each species has a characteristic **chromosome number**, which refers to the sum total of chromosomes in cells of a given type. Human somatic cells have 46, those of gorillas have 48, and those of pea plants have 14.

Actually, your 46 chromosomes are like volumes of two sets of books. Each set is numbered 1 to 23. For example, you have two "volumes" of chromosome 22—that is, *a pair of them*. Generally, both members of each pair have the same length and shape, and they carry instructions for the same traits. In these respects, they are not like the other pairs. Think of them as two sets of books on how to build a house. Your father gave you one set. Your mother had her own ideas about storage, plumbing, and so on, so she gave you a revised edition. Her set covers the same topics but has slightly different things to say about many of them.

We say the chromosome number is **diploid**, or 2*n*, if a cell has two of each type of chromosome characteristic of the species. The body cells of humans, gorillas, pea plants, and a great many other organisms are like this. (By contrast, as explained in the following chapter, their sperm and eggs have a *haploid* chromosome number, *n*, with only one of each type characteristic of the species.)

With mitosis, a diploid parent cell can produce two diploid daughter cells. This doesn't mean each merely gets forty-six or forty-eight or fourteen chromosomes. If only the total mattered, one cell might get, say, two pairs of chromosome 22 and no pairs whatsoever of chromosome 9. However, neither cell would function properly *without two of each type of chromosome*.

Mitosis keeps the chromosome number constant, division after division, from one cell generation to the next. Thus, if a parent cell is diploid, its daughter cells will be diploid also.

7.2 MITOSIS AND THE CELL CYCLE

Mitosis is only one phase of the **cell cycle**. Such cycles start each time new cells are produced and end when those cells complete their own division. The cycle starts again for each new daughter cell (Figure 7.3). Three other phases occur during **interphase**, which is usually the longest part of a cell cycle. At this time, a cell increases its mass, roughly doubles its number of cytoplasmic components, and duplicates its DNA. These abbreviations signify the four phases of the cell cycle:

M   *Mitosis; nuclear division only, usually followed by cytoplasmic division*

G_1 *Of interphase, a "Gap" (interval) of cell growth before the onset of DNA replication*

S   *Of interphase, the time of "Synthesis" (replication) of DNA*

G_2 *Of interphase, a second Gap (interval) following DNA replication, when the cell prepares for division*

The cell cycle lasts about the same length of time for cells of a given type. Its duration differs among cells of different types. For example, all of the neurons (nerve cells) in your brain are arrested at interphase, and they usually will not divide again. By contrast, cells of a new sea urchin may double in number every two hours.

Adverse conditions may disrupt a cell cycle. When deprived of a vital nutrient, for instance, the free-living cells called amoebas do not leave interphase. Even so, if any cell proceeds past a certain point in interphase, the cycle normally will continue regardless of outside conditions, owing to built-in controls over its duration.

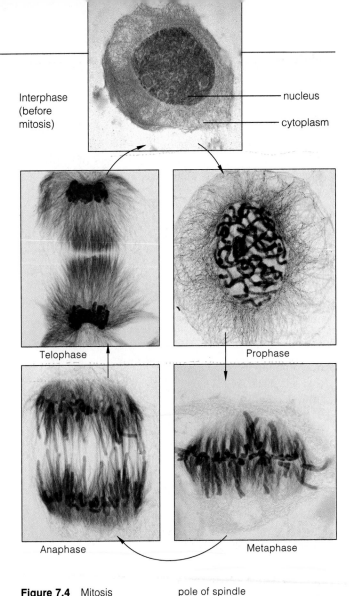

Interphase (before mitosis) — nucleus — cytoplasm

Telophase

Prophase

Anaphase

Metaphase

Figure 7.4 Mitosis in a cell from the African blood lily. Chromosomes are stained *blue*, and microtubules that move them about are stained *red*. Before reading further, take a moment to become familiar with the labels on the micrographs.

pole of spindle — microtubules organized as a spindle apparatus — equator of spindle — condensed chromosome — pole of spindle

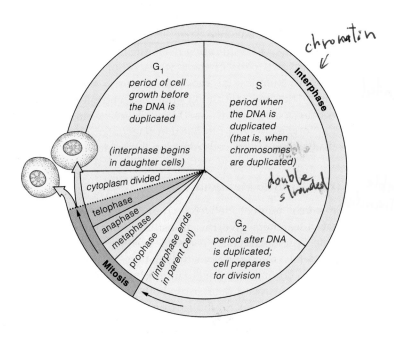

chromatin

G_1 period of cell growth before the DNA is duplicated

(interphase begins in daughter cells)

S period when the DNA is duplicated (that is, when chromosomes are duplicated)

double stranded

Interphase

cytoplasm divided

telophase

anaphase

metaphase

prophase

(interphase ends in parent cell)

G_2 period after DNA is duplicated; cell prepares for division

Mitosis

Figure 7.3 Generalized eukaryotic cell cycle. The length of different stages differs among cells.

We turn now to mitosis and how it maintains the chromosome number through turn after turn of the cell cycle. Figure 7.4 only hints at the precise division mechanisms that take over as a cell leaves interphase.

A cell cycle begins at interphase, when a new cell (formed by mitosis and cytoplasmic division) increases its mass, doubles the number of its cytoplasmic components, then duplicates its chromosomes. The cycle ends when the cell divides.

When a cell makes the transition from interphase to mitosis, it has stopped constructing new cell parts, and its DNA has been replicated. Within that cell, profound changes now proceed smoothly, one after the other, through four stages. The sequential stages of mitosis are **prophase**, **metaphase**, **anaphase**, and **telophase**.

Figure 7.5 shows an animal cell undergoing mitosis. Compare it with the plant cell in Figure 7.4, and it becomes clear that the chromosomes in both cells are moving about. They don't do so on their own. A **spindle apparatus** moves them.

A spindle consists of microtubules, arranged as two sets. Each set extends from one pole (end point) of the spindle. It overlaps the other set a bit at the spindle equator, midway between the poles. The formation of this bipolar spindle establishes what the ultimate destinations of the chromosomes will be during mitosis, as you will see shortly.

Figure 7.5 Mitosis. This nuclear division mechanism assures that daughter cells will have the same chromosome number as the parent cell. For clarity, the diagram shows only two pairs of chromosomes from a diploid (*2n*) animal cell. With rare exceptions, the picture is more involved than this, as indicated by the micrographs (*facing page*) of mitosis in a whitefish cell.

MITOSIS

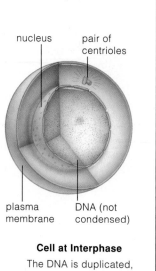

Cell at Interphase
The DNA is duplicated, then the cell prepares for nuclear division.

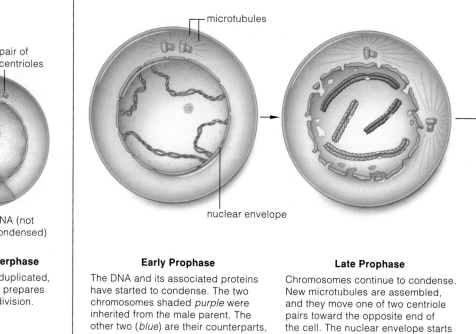

Early Prophase
The DNA and its associated proteins have started to condense. The two chromosomes shaded *purple* were inherited from the male parent. The other two (*blue*) are their counterparts, inherited from the female parent.

Late Prophase
Chromosomes continue to condense. New microtubules are assembled, and they move one of two centriole pairs toward the opposite end of the cell. The nuclear envelope starts to break up.

Prophase: Mitosis Begins

Prophase is evident when chromosomes become visible in the light microscope as threadlike forms. ("Mitosis" comes from the Greek *mitos*, meaning thread.) Each was duplicated earlier, at interphase. In other words, each chromosome now consists of two sister chromatids, joined together at the centromere. Early on, the sister chromatids twist and fold into a more compact form. By late prophase, all the chromosomes will be condensed into thicker, rod-shaped forms.

Meanwhile, in the cytoplasm, most microtubules of the cytoskeleton are breaking apart, into their tubulin subunits (page 58). The subunits reassemble near the nucleus, as *new* microtubules. Many of the microtubules will extend from one spindle pole or the other to the centromere of a chromosome. The remainder will not interact at all with the chromosomes. Rather, they will extend from the poles and overlap each other.

While new microtubules are assembling, the nuclear envelope prevents them from interacting with chromosomes inside the nucleus. However, as prophase ends, the nuclear envelope starts to break up.

Many cells have two barrel-shaped **centrioles**. Each centriole started duplicating itself during interphase, so there are two pairs of them when prophase is under way. Microtubules start moving one pair to the opposite pole of the newly forming spindle. Centrioles, recall, give rise to flagella or cilia. If you observe them in cells of an organism, you can bet that flagellated or ciliated cells (such as sperm cells) develop during its life cycle.

Transition to Metaphase

So much happens between prophase and metaphase that researchers give this transitional period its own name, "prometaphase." The nuclear envelope breaks up completely, into numerous tiny, flattened vesicles. Now the chromosomes are free to interact with microtubules that extend toward them, from both poles of the developing spindle. Microtubules from both poles harness each chromosome and start pulling on it. The two-way pulling orients the chromosome's two sister chromatids toward opposite poles. Meanwhile, overlapping spindle

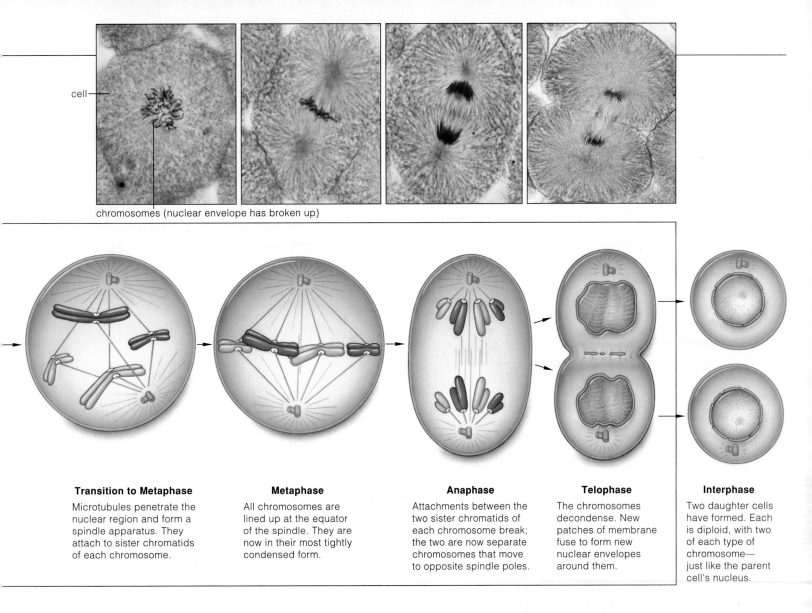

cell

chromosomes (nuclear envelope has broken up)

Transition to Metaphase

Microtubules penetrate the nuclear region and form a spindle apparatus. They attach to sister chromatids of each chromosome.

Metaphase

All chromosomes are lined up at the equator of the spindle. They are now in their most tightly condensed form.

Anaphase

Attachments between the two sister chromatids of each chromosome break; the two are now separate chromosomes that move to opposite spindle poles.

Telophase

The chromosomes decondense. New patches of membrane fuse to form new nuclear envelopes around them.

Interphase

Two daughter cells have formed. Each is diploid, with two of each type of chromosome— just like the parent cell's nucleus.

microtubules ratchet past each other and push the poles farther apart. All of the push–pull forces are balanced when the chromosomes reach the spindle's midpoint.

When all the duplicated chromosomes are aligned midway between the poles of a completed spindle, we call this metaphase (*meta-* means midway between). The alignment is crucial for the next stage of mitosis.

From Anaphase Through Telophase

At anaphase, sister chromatids of each chromosome separate and move to opposite poles. Two mechanisms account for this. First, the microtubules attached to the centromere regions shorten, pulling the chromosomes to the poles. Second, the spindle elongates when overlapping microtubules ratchet past each other and push the two spindle poles farther apart. Once separated, the sister chromatids have a new name. Each is now a separate chromosome in its own right.

Telophase begins once the chromosomes arrive at opposite spindle poles. The chromosomes are no longer

harnessed to microtubules, and they return to thread-like form. Vesicles of the old nuclear envelope fuse to form patches of membrane around the chromosomes. Patch joins with patch, and soon a new nuclear envelope separates each cluster of chromosomes from the cytoplasm. If the parent cell was diploid, each cluster will contain two of each type of chromosome. With mitosis, remember, each new nucleus has the same chromosome number as the parent nucleus. Once the two nuclei form, telophase is over—and so is mitosis.

Prior to mitosis, each chromosome in a cell's nucleus is duplicated, so that it consists of two sister chromatids.

Mitosis proceeds through four consecutive stages, called prophase, metaphase, anaphase, and telophase.

A microtubular spindle moves sister chromatids of every chromosome apart, to opposite spindle poles. A new nuclear envelope forms around the two clusters of chromosomes. Both of the daughter nuclei formed this way have the same chromosome number as the parent cell's nucleus.

7.4 DIVISION OF THE CYTOPLASM

The cytoplasm usually divides at some time between late anaphase and the end of telophase. As you might gather from Figures 7.6 and 7.7, the mechanism of this **cytoplasmic division** (or cytokinesis, as it is commonly called) differs among organisms.

Most plant cells have walls, which means their cytoplasm cannot be merely pinched in two. In this case, the cytoplasm divides by a mechanism that involves **cell plate formation**. As Figure 7.6 shows, vesicles filled with wall-building materials fuse with remnants from the microtubular spindle. Together, they form a disklike

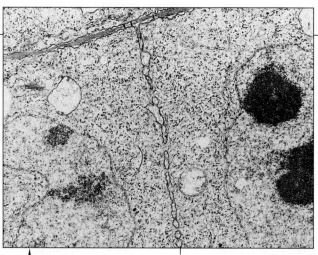

cell plate

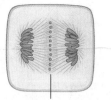

spindle equator

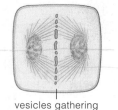

vesicles gathering

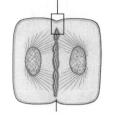

cell plate growing

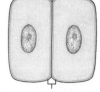

two new primary walls

a As mitosis draws to a close, vesicles converge at the equator of the microtubular spindle. They contain cementing materials and "starting materials" for a new primary cell wall.

b A cell plate starts to form as membranes of the vesicles fuse. Their contents become sandwiched between two membranes that form along the plane of the cell plate.

c Inside the sandwich, two walls will form as cellulose becomes deposited over both membranes. Other substances inside will form the middle lamella that cements new cell walls together.

d The cell plate grows at its margins until it fuses with the plasma membrane of the parent cell. During plant growth, when cells expand and walls are still thin, new material also is deposited over the old primary wall.

Figure 7.6 Cytoplasmic division of a plant cell, as brought about by cell plate formation.

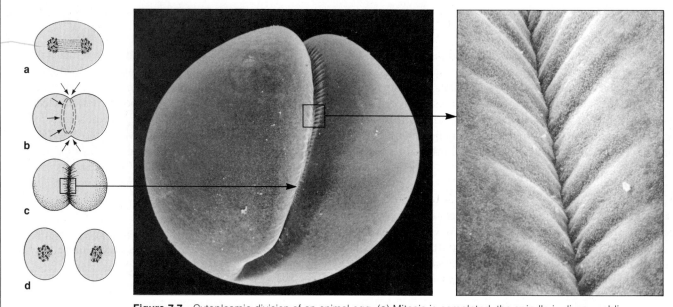

Figure 7.7 Cytoplasmic division of an animal egg. (**a**) Mitosis is completed; the spindle is disassembling. (**b**) Just beneath the plasma membrane of the parent cell, rings of microfilaments at the former spindle equator contract, like a purse string. (**c**,**d**) Contractions continue and in time will divide the cell in two. The micrographs show how the surface of the plasma membrane sinks inward. The depression defines the cleavage plane.

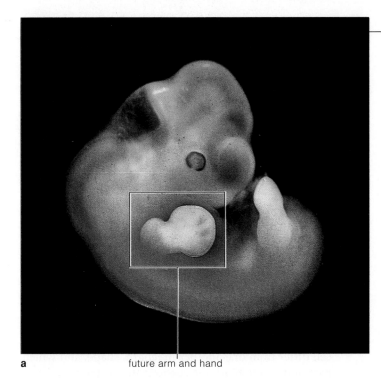

a   future arm and hand

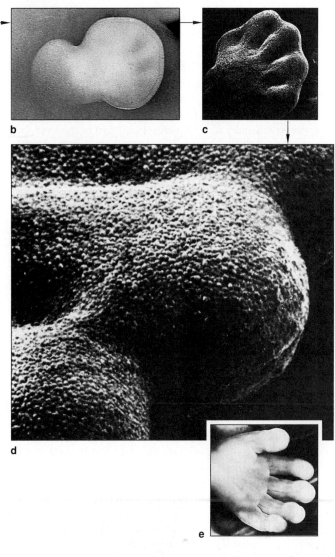

b   c

d

e

Figure 7.8 Development of the human hand by way of mitosis, cytoplasmic divisions, and other processes. Individual cells resulting from mitotic cell divisions are visible in (**d**).

structure (the "cell plate"). At this location, deposits of cellulose accumulate and form a crosswall that divides the parent cell into two daughter cells.

By contrast, animal cells do not have a cell wall, and the cytoplasm of most of them does "pinch in two." Consider a newly fertilized egg. Through a mechanism called <u>cleavage</u>, a layer of material becomes deposited around microtubules at the cell midsection. After this, a shallow, ringlike depression forms above the layer, at the cell surface (Figure 7.7). The depression is called the cleavage furrow. It is evidence of parallel arrays of microfilaments in the cytoplasm that connect with the plasma membrane. Microfilaments, recall, are thread-like cytoskeletal elements. These particular ones are organized so that they slide past one another (Section 3.11). As they do so, they pull the plasma membrane inward—and thereby cut the cell in two.

This concludes our picture of mitotic cell division. Look now at your hands—and try to envision all the cells in your palms, thumbs, and fingers. Imagine all the divisions of all the cells that preceded them when you were developing early on, inside your mother's body (Figure 7.8). And be grateful for the astonishing precision of the mechanisms that led to their formation, for the alternatives can be terrible indeed. Why? Good health, even survival itself, depends absolutely on the proper timing and completion of mitosis and other events of the cell cycle. Mistakes in the duplication or distribution of even one chromosome may result in a genetic disorder. As you will read in Section 12.9, cancer may rapidly destroy a mature tissue if controls are lost that otherwise prevent cells from dividing. The *Focus* essay that concludes this chapter describes a landmark case of such unchecked cell divisions.

With mitotic cell division, two daughter nuclei are produced (mitosis proper), and another mechanism cuts the cytoplasm into two daughter cells.

In most plants, a cell plate and a crosswall form in between the developing plasma membranes of the daughter cells.

In most animals, microfilaments arranged in a ring around the parent cell's midsection slide past one another in a way that pinches the cytoplasm in two.

HENRIETTA'S IMMORTAL CELLS

Each human starts out as a single fertilized egg. By the time of birth, mitotic cell divisions and other processes have resulted in a human body of about a trillion cells. Even in an adult, billions of cells are still dividing. For example, cells of the stomach's lining divide every day. Liver cells usually don't divide—but if part of the liver becomes injured or diseased, many will divide repeatedly and produce new cells until the damaged part is replaced.

In 1951, George and Margaret Gey of Johns Hopkins University were trying to develop a way to keep human cells dividing *outside* the body. With such isolated cells, these researchers and others could investigate basic life processes. They also could conduct studies of cancer and other diseases without having to experiment directly on humans and thereby gamble with lives.

The Geys had samples of normal and diseased human cells, which local physicians had provided to them. But they just couldn't stop the descendants of these cells from dying out within a few weeks.

Mary Kubicek, a laboratory assistant, worked with the Geys in their efforts to start a self-perpetuating lineage of cultured human cells. She was about to give up after dozens of failed attempts. Even so, in 1951 she prepared yet another sample of cancer cells for culture. The sample was code-named HeLa, for the first two letters of the patient's first and last names.

The HeLa cells began to divide. And divide. And divide again. By the fourth day there were so many cells that they had to be subdivided into more tubes. As months passed, the culture continued to thrive!

Unfortunately, the tumor cells inside the patient's body were just as vigorous. Six months after the patient was first diagnosed as having cancer, tumor cells had spread to tissues throughout her body. Two months later, Henrietta Lacks, a young woman from Baltimore, was dead.

Although Henrietta passed away, some of her cells lived on in the Geys' laboratory as the first successful human cell culture. HeLa cells were soon shipped to other researchers, who passed cells on to others, and so on. HeLa cells came to live in laboratories all over the world.

Some of the cultured cells even journeyed into space, aboard the *Discoverer XVII* satellite. Every year, hundreds of scientific papers describe research that is based on work with HeLa cells.

Henrietta was only thirty-one years old when runaway cell divisions killed her. Now, many decades later, her legacy is still benefiting humans everywhere—in cells that are still alive and dividing, day after day after day.

SUMMARY

1. Through certain division mechanisms, summarized in Table 7.1, parent cells provide each daughter cell with the hereditary instructions (DNA) and cytoplasmic machinery necessary to start up its own operation.

   a. In eukaryotic cells, the nucleus divides by mitosis or meiosis. Cytoplasmic division typically follows.

   b. Prokaryotic cells divide by binary fission.

2. Mitosis divides the parental DNA into two equivalent parcels. It is the basis of growth and tissue repair in multicelled eukaryotes and of asexual reproduction in many single-celled and multicelled species. (Meiosis proceeds only in germ cells that have been set aside for sexual reproduction.)

3. A different mechanism divides the cytoplasm near the end of nuclear division or at some point thereafter. Plant cells divide after a cell plate develops and gives rise to a crosswall and new plasma membrane. Animal cells are pinched in two, as by cleavage.

4. A eukaryotic chromosome is a single DNA molecule with many proteins attached. The chromosomes in a cell differ from one another in length, shape, and which part of the hereditary instructions they carry.

5. The chromosome number of a given cell is the total number of chromosomes characteristic of the species. Cells with a *diploid* chromosome number (2*n*) have two of each type of chromosome.

6. Mitosis keeps the chromosome number constant from one cell generation to the next. If a cell is 2*n*, its daughter cells resulting from mitosis will be 2*n* also.

7. The cell cycle starts when a new cell forms, proceeds through interphase, and ends when the cell reproduces by mitosis and cytoplasmic division. At interphase, a cell increases in mass, doubles its number of cytoplasmic components, then duplicates its chromosomes.

8. After duplication, each chromosome consists of two DNA molecules attached at the centromere. For as long as the two remain attached, they are known as sister chromatids of the chromosome.

Table 7.1 Cell Division Mechanisms	
Mechanisms	Functions
Mitosis, cytoplasmic division	In multicelled eukaryotes, the basis of bodily growth. In single-celled and many multicelled eukaryotes, the basis of asexual reproduction
Meiosis, cytoplasmic division	In single-celled and multicelled eukaryotes, the basis of gamete formation and of sexual reproduction
Prokaryotic fission	In bacterial cells, this mechanism is the basis of asexual reproduction

9. Mitosis proceeds through four continuous stages:

a. Prophase. Duplicated chromosomes, which are in threadlike form, start to condense. Microtubules of the cytoskeleton disassemble, and then new ones start to assemble near the nucleus. They will form a spindle apparatus. The nuclear envelope starts to break up.

b. Metaphase. During the *transition* to metaphase, the nuclear envelope breaks up completely into tiny vesicles. Microtubules of the forming spindle attach to the sister chromatids of each chromosome and orient them toward opposite spindle poles. *At* metaphase, all chromosomes are aligned at the spindle equator.

c. Anaphase. Many microtubules pull the two sister chromatids of each chromosome away from each other, toward opposite spindle poles. Every chromosome that was present in the parent cell is now represented by a daughter chromosome at both poles.

d. Telophase. The chromosomes decondense to the threadlike form. A new nuclear envelope forms around them. Each nucleus has the same chromosome number as the parent cell. Mitosis is completed.

Review Questions

1. Define the two types of division mechanisms of eukaryotic species. Does either one divide the cytoplasm? *116*

2. Define somatic cell and germ cell. *116*

3. What is a chromosome called when it is in the unduplicated state? In the duplicated state (with two sister chromatids)? *116*

4. Describe the microtubular spindle and its functions, then name and briefly describe the stages of mitosis. *118–119*

5. How does cytoplasmic division differ in plant and animal cells? *120–121*

Self-Quiz *(Answers in Appendix IV)*

1. A eukaryotic chromosome consists of _____ .
   a. DNA only   c. DNA plus membrane
   b. DNA plus proteins   d. DNA plus lipids

2. A cell that has two of each type of chromosome characteristic of the species is a(n) _____ cell.
   a. diploid   c. abnormal
   b. mitotic   d. a and c

3. A duplicated chromosome has _____ chromatid(s).
   a. one   c. three
   b. two   d. four

4. Of a chromosome, a _____ is a constricted region with attachment sites for microtubules.
   a. chromatid   c. cell plate
   b. centromere   d. cleavage furrow

5. Interphase is the part of the cell cycle when _____ .
   a. a cell ceases to function
   b. a germ cell forms its spindle apparatus
   c. a cell grows and duplicates its DNA
   d. mitosis proceeds

6. After mitosis, the chromosome number of a daughter cell is _____ the parent cell's.
   a. the same as   c. rearranged compared to
   b. one-half   d. doubled compared to

7. Mitosis and cytoplasmic division function in _____ .
   a. asexual reproduction of single-celled eukaryotes
   b. growth, tissue repair, and sometimes asexual reproduction in many multicelled eukaryotes
   c. gamete formation in prokaryotes
   d. both a and b are correct
   e. all are correct

8. Only _____ is not a stage of mitosis.
   a. prophase   c. metaphase
   b. interphase   d. anaphase

9. Match each stage with its key events.
   ____ metaphase   a. sister chromatids of each
   ____ prophase   chromosome are moved apart
   ____ telophase   b. chromosomes start to condense
   ____ anaphase   c. chromosomes decondense,
                          daughter nuclei form
                       d. all chromosomes are aligned
                          at the spindle equator

Critical Thinking

1. Suppose you have a means of measuring the amount of DNA in a single cell during the cell cycle. You first measure the amount during the G_1 phase. At what points during the remainder of the cycle would you predict changes in the amount of DNA per cell?

2. A cell from a tissue culture has 38 chromosomes. After mitosis and cytoplasmic division, one daughter cell has 39 chromosomes, and the other has 37. Speculate on what might have occurred to cause the abnormal chromosome numbers. Generally speaking, how might the abnormality affect cell structure, function, or both?

3. The Pacific yew (*Taxus brevifolius*) faces extinction. People started stripping its bark—and killing the trees—when they heard that taxol, a chemical extract from the bark, may be effective in treating cancer of the breast and ovaries. (Synthesizing taxol in the laboratory may save the species.) Taxol prevents the disassembly of microtubules. What does this tell you about its potential as an anti-cancer drug?

4. X rays and gamma rays emitted from some radioisotopes cause chemical damage to DNA, especially in cells engaged in DNA replication. High-level exposure can result in *radiation poisoning*. Hair loss and damage to the lining of the stomach and intestines are two early symptoms. Speculate why. Also speculate on why highly focused radiation therapy is used against some cancers.

Selected Key Terms

anaphase *118*	interphase *117*
cell cycle *117*	meiosis *116*
cell plate formation *120*	metaphase *118*
centriole *118*	mitosis *116*
centromere *116*	prophase *118*
chromosome *116*	reproduction *115*
chromosome number *116*	sister chromatid *116*
cleavage *121*	somatic cell *116*
cytoplasmic division *120*	spindle apparatus *118*
diploid cell *116*	telophase *118*
germ cell *116*	

Reading

Murray, A., and M. Kirschner. March 1991. "What Controls the Cell Cycle?" *Scientific American* 264(3): 56–63.

MEIOSIS

Octopus Sex and Other Stories

The couple clearly are interested in each other. First he caresses her with one tentacle, then another—and then another and another. She reciprocates with a hug here, a squeeze there. This goes on for hours. Finally the male reaches under his mantle (a fold of tissue that drapes around most of his body). He removes a packet of sperm and inserts it into an egg chamber under the female's mantle. For every sperm that fertilizes an egg, a new octopus may develop.

Unlike the coupling between a male and female octopus, sex for the slipper limpet is a group activity. Slipper limpets are marine animals, relatives of land snails. Before one of them becomes transformed into a sexually mature adult, it passes through a free-living stage called a larva. When the time comes for a larva to undergo transformation, it settles down on a pebble or shell or rock. If it settles down alone, it will develop into a female. If a second larva settles and develops on the first one, *it* will function right off as a male. But if another male develops on top of it, that first male will

gradually become a female. Then the third male also will become a female if a fourth male develops on top of it, and so on amongst ten or more limpets.

Slipper limpets typically live in such piles, with the bottom one always being the oldest female, and the uppermost one being the youngest male (Figure 8.1*a*). Until they make the gender switch, the males release sperm, these fertilize a female's eggs, which then grow to become males and then, most likely, females—and so it goes, from one limpet generation to the next.

Even within a single species, we may come across variations in the mode of reproduction. For example, the life cycle of many sexually reproducing organisms also includes asexual episodes that are based on mitotic cell divisions. Consider certain orchids or dandelions or many other species of plants, which reproduce without engaging in sex. Consider the flatworms, which can split lengthwise into two roughly equivalent parts that each grow into a new flatworm. Or consider the aphids. In summer, nearly all aphids are females, which busily

Figure 8.1 Variations in reproductive modes among eukaryotic organisms.

(**a**) Slipper limpets busily perpetuating the species by group participation in sexual reproduction. The tiny crab in the foreground is merely a passerby, not a voyeur. (**b**) Live birth of an aphid, a type of insect that reproduces sexually in autumn but switches to an asexual mode in summer.

With this chapter, we turn to the kinds of reproductive cells that bridge the generations for eukaryotic organisms. For many species, specialized phases of reproduction and development—including asexual episodes—loop out from the basic life cycle.

Regardless of the specializations, all of these cycles turn on two basic events: *gamete formation* and *fertilization*.

a

produce more females from numbers of unfertilized egg cells (Figure 8.1b). When autumn approaches, male aphids finally develop and do their part in the sexual phase of the life cycle. Even so, females that survive the winter can do without the males. Come summer, the females begin another round of producing offspring all by themselves.

These examples only hint at the immense variation in reproductive modes among eukaryotic organisms. And yet, despite the variation, sexual reproduction dominates their life cycles—and it inevitably involves certain events. Before cell division, chromosomes are duplicated in germ cells. As the germ cells mature, they undergo meiosis and cytoplasmic division. And then gametes develop. When gametes join at fertilization, they form the first cell of a new individual.

Meiosis, the formation of gametes, and fertilization are the hallmarks of sexual reproduction. As you will see in this chapter, these processes contribute to the splendid diversity of life.

b

KEY CONCEPTS

1. Sexual reproduction proceeds through three events: meiosis, formation of gametes—such as sperm and eggs—and fertilization.

2. Meiosis, a nuclear division mechanism, occurs only in cells set aside for sexual reproduction. Meiosis sorts out the cell's chromosomes into four new nuclei. Following meiosis, gametes form by way of cytoplasmic division and other events.

3. Cells with a diploid chromosome number have two of each type of chromosome characteristic of the species. The two function as a pair during meiosis. Typically, one chromosome of the pair is maternal, with hereditary instructions from a female parent. The other is paternal, with comparable instructions from a male parent.

4. Meiosis divides the chromosome number by half for each forthcoming gamete. Thus, if both parents are diploid ($2n$), the union of two gametes at fertilization will restore the diploid number in the new individual ($n + n = 2n$).

5. During meiosis, each pair of chromosomes may swap segments. Each time they do so, they exchange hereditary instructions about certain traits. Also, meiosis assigns one of every pair of chromosomes to a forthcoming gamete—but *which* gamete is its destination is a matter of chance. Hereditary instructions are further shuffled at fertilization. All three of these reproductive events lead to variations in traits among offspring.

6. In most plants, spore formation and other events intervene between meiosis and gamete formation.

8.1 COMPARISON OF ASEXUAL AND SEXUAL REPRODUCTION

When an orchid, flatworm, or aphid reproduces all by itself, what sort of offspring does it get? By the process of **asexual reproduction**, one parent alone produces offspring, and each new individual gets the same number and kinds of genes as its parent. **Genes** are specific stretches of chromosomes—that is, of DNA molecules. Taken together, the genes for each species contain all of the heritable bits of information that are necessary to produce new individuals. Rare mutations aside, this means that asexually produced offspring can only be clones—genetically identical copies of the parent.

Inheritance gets much more interesting with **sexual reproduction**. This process involves meiosis, gamete formation, and fertilization (the union of two gametes). In humans and many other species, two parents—each with pairs of genes—use this process. And the first cell of a new individual ends up with pairs of genes.

If instructions encoded in every pair of genes were identical down to the last detail, sexual reproduction would produce clones, also. Just imagine—you, everyone you know, the entire human population might be a clone, with everybody looking alike. But the two genes of a pair may *not* be identical. Why not? The molecular structure of genes can change; this is what we mean by mutation. Depending on their structure, the two genes that happen to be paired in a person's cells may "say" slightly different things about the trait. Each unique molecular form of the same gene is called an **allele**.

Such tiny differences affect thousands of traits. For example, whether you have a chin dimple depends on which pair of alleles you inherited at a certain chromosome location. One kind of allele says "put a dimple in the chin," another kind says "no dimple." And this leads us to a key reason why members of sexually reproducing species don't all look alike. *Through sexual reproduction, offspring get new combinations of alleles—and these lead to variations in physical and behavioral traits.*

This chapter describes the cellular basis of sexual reproduction. More importantly, it starts us thinking about far-reaching effects of gene shufflings at different stages of the process. The process introduces variations in traits among offspring that may be acted upon by agents of natural selection. Thus, *variation in traits is a foundation for evolutionary change.*

Asexual reproduction produces genetically identical copies of the parent. Sexual reproduction introduces variations in the details of traits among offspring.

Sexual reproduction dominates the life cycle of eukaryotic species, even though many cycles also include asexual phases. Meiosis, formation of gametes, and fertilization are the basic events of this process.

8.2 MEIOSIS AND THE CHROMOSOME NUMBER

Think "Homologues"

Think back on the preceding chapter, which focused on mitotic cell division. Unlike mitosis, **meiosis** divides chromosomes into separate parcels not once *but twice* prior to cell division. Unlike mitosis, it is the first step leading to the formation of gametes.

Gametes are sex cells, such as sperm and eggs. In multicelled eukaryotic organisms, including plants and animals, gametes arise only from germ cells, which are produced in reproductive structures and organs. Figure 8.2 shows examples of where gametes form.

Recall that the sum total of chromosomes in cells of a given type is the chromosome number. Germ cells

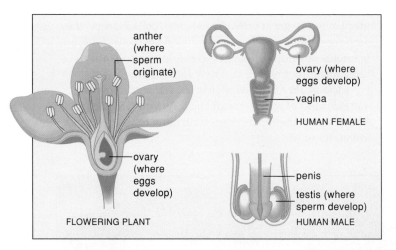

anther (where sperm originate)

ovary (where eggs develop)

FLOWERING PLANT

ovary (where eggs develop)

vagina

HUMAN FEMALE

penis

testis (where sperm develop)

HUMAN MALE

Figure 8.2 Examples of gamete-producing structures.

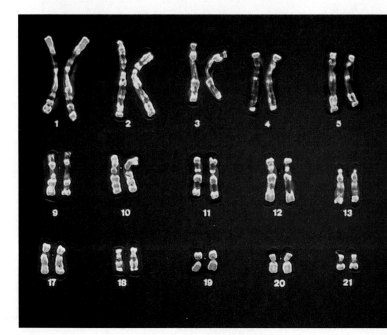

have the same chromosome number as the rest of the body's cells. The ones that are **diploid** (2*n*) have _two of each type of chromosome_, often from two parents. In general, the two members of each pair of chromosomes have the same length and shape. Their genes deal with the same traits. And they line up with each other at meiosis. Think of them as **homologous chromosomes** (*hom-* means alike).

As you can probably deduce from Figure 8.3, your germ cells contain 23 + 23 homologous chromosomes. After meiosis, 23 chromosomes—one of each type—end up in gametes. Said another way, meiosis halves the chromosome number, so the gametes are **haploid** (*n*).

Two Divisions, Not One

Meiosis resembles mitosis in some respects, even though the outcome is different. Before interphase gives way to meiosis, a germ cell duplicates its DNA. Each duplicated chromosome now consists of two DNA molecules. These remain attached at a narrowed-down region, called the centromere. For as long as the two remain attached, they are known as **sister chromatids** of the chromosome:

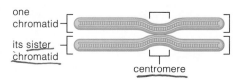

one chromatid
its sister chromatid
centromere

As in mitosis, the microtubules of a spindle apparatus move the chromosomes in prescribed directions.

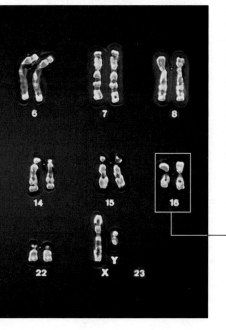

Figure 8.3 From a diploid cell of a human male, twenty-three pairs of homologous chromosomes. The sketch corresponding to the two boxed chromosomes indicates that the chromosomes are in the duplicated state (each consists of two sister chromatids).

one pair of duplicated chromosome

22 pairs somatic
1 pair sex.

With meiosis alone, however, *chromosomes proceed through two consecutive divisions, which end with the formation of four haploid nuclei*. The two divisions are called meiosis I and II:

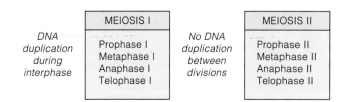

DNA duplication during interphase

MEIOSIS I
Prophase I
Metaphase I
Anaphase I
Telophase I

No DNA duplication between divisions

MEIOSIS II
Prophase II
Metaphase II
Anaphase II
Telophase II

During meiosis I, each duplicated chromosome lines up with its partner, *homologue to homologue*; then the partners are separated from each other:

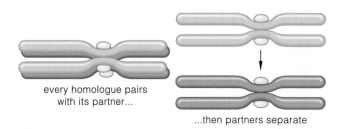

every homologue pairs with its partner...

...then partners separate

The cytoplasm typically divides after the separation of homologues. The two daughter cells are haploid, with only one of each type of chromosome. But remember, those chromosomes are still duplicated.

During meiosis II, *the two sister chromatids of each chromosome are separated from each other*:

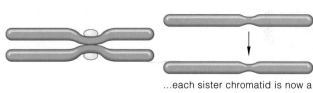

...each sister chromatid is now a chromosome in its own right

After the four nuclei form, the cytoplasm often divides once more, the outcome being four haploid cells.

On the next two pages, Figure 8.4 illustrates the key events of meiosis I and II.

Meiosis is a type of nuclear division that reduces the parental chromosome number by half—to the haploid number (*n*).

Meiosis proceeds only in germ cells, a cell lineage set aside for sexual reproduction. It is the first step leading to the formation of gametes.

A VISUAL TOUR OF THE STAGES OF MEIOSIS

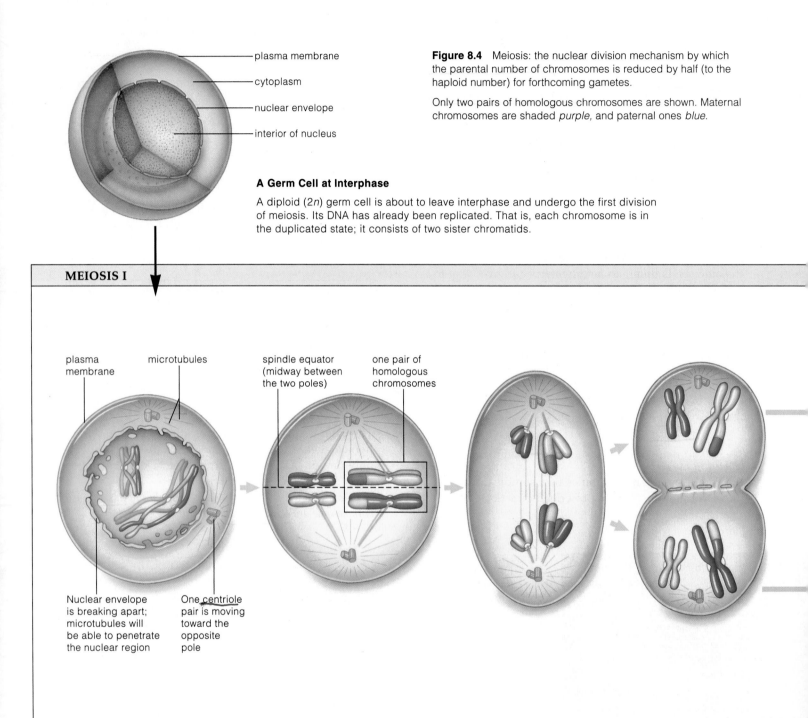

plasma membrane

cytoplasm

nuclear envelope

interior of nucleus

Figure 8.4 Meiosis: the nuclear division mechanism by which the parental number of chromosomes is reduced by half (to the haploid number) for forthcoming gametes.

Only two pairs of homologous chromosomes are shown. Maternal chromosomes are shaded *purple,* and paternal ones *blue.*

A Germ Cell at Interphase

A diploid (2*n*) germ cell is about to leave interphase and undergo the first division of meiosis. Its DNA has already been replicated. That is, each chromosome is in the duplicated state; it consists of two sister chromatids.

MEIOSIS I

plasma membrane

microtubules

spindle equator (midway between the two poles)

one pair of homologous chromosomes

Nuclear envelope is breaking apart; microtubules will be able to penetrate the nuclear region

One centriole pair is moving toward the opposite pole

Prophase I

Each chromosome starts to twist and fold into more condensed form. It pairs up with its homologue, and the two typically swap segments. The swapping, called crossing over, is indicated by the break in color of the pair of larger chromosomes. As in mitosis, chromosomes become attached to microtubules of a newly forming spindle.

Metaphase I

Microtubules have pushed and pulled all the chromosomes into position, midway between the two poles of the spindle. At this stage, the spindle is fully formed, owing to the interactions between its microtubules and the chromosomes.

Anaphase I

Each chromosome is now separated from its homologue. The two are moved toward opposite poles of the spindle apparatus.

Telophase I

When the cytoplasm divides, two haploid (*n*) cells are the result. Each cell has one chromosome of each type, although these are still in the duplicated state.

Of the four haploid cells that may form by way of meiosis and cytoplasmic divisions, one or all may develop into gametes.

(In plants, the cells that form after meiosis has been completed may develop into spores, which precede the formation of gametes.)

MEIOSIS II

There is no DNA replication 複製 between the two divisions.

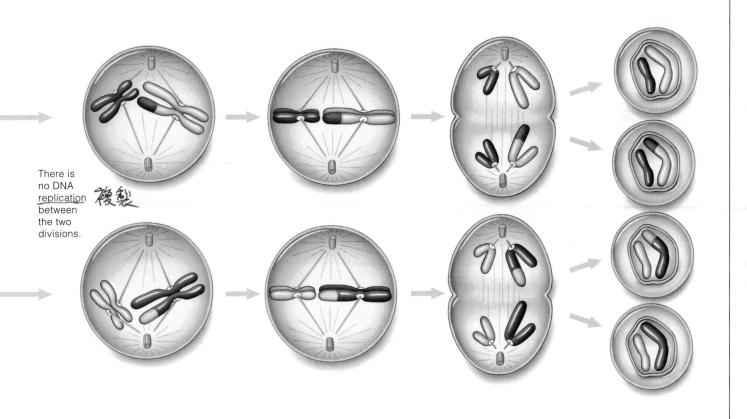

Prophase II

During the transition to prophase II, microtubules moved one member of each pair of centrioles to the opposite pole of the spindle. Now, at prophase II, microtubules become attached to the chromosomes and start moving them toward the equator of the spindle apparatus.

Metaphase II

All of the duplicated chromosomes are now positioned midway between the two poles of the spindle.

Anaphase II

The attachment between the two chromatids of each chromosome breaks. The former "sister chromatids" are now chromosomes in their own right and are moved to opposite poles of the spindle.

Telophase II

Four daughter nuclei now form. When the cytoplasm divides, each new cell has a haploid (*n*) number of chromosomes, all in the unduplicated state.

The preceding overview, in Sections 8.2 and 8.3, is enough to convey the overriding function of meiosis—that is, *the reduction of the chromosome number by half for forthcoming gametes.*

However, as you will now read, two other events that take place during prophase I and metaphase I of meiosis contribute greatly to the adaptive advantage of sexual reproduction—*the production of offspring with new combinations of alleles, and therefore individuals that vary in the details of their traits.*

Prophase I Activities

Prophase I of meiosis is a time of major gene shufflings. Consider Figure 8.5a, which shows two chromosomes that have condensed to threadlike form. All chromosomes in a germ cell condense this way. When they do, each is drawn close to its homologue by a process called synapsis. It is as if homologues become stitched together point by point along their length, with little space in between. The intimate, parallel array favors **crossing over**, a type of molecular interaction between two "nonsister" chromatids of a pair of homologous chromosomes. The nonsister chromatids break at the same places along their length and exchange corresponding segments—that is, genes—at the break points. Figure 8.5c is a simplified picture of this interaction.

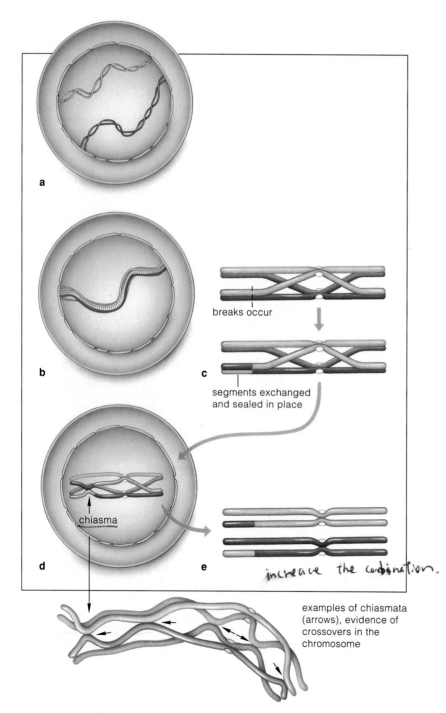

breaks occur

segments exchanged and sealed in place

chiasma

increase the combination.

examples of chiasmata (arrows), evidence of crossovers in the chromosome

Figure 8.5 Key events of prophase I, the first stage of meiosis. For clarity, the diagram shows only a single pair of homologous chromosomes and one crossover event. *Blue* signifies the paternal chromosome, and *purple* signifies its maternal homologue.

a The chromosomes were duplicated previously, during interphase. Early in prophase I, each duplicated chromosome is in threadlike form, attached at both ends to the nuclear envelope. Its two sister chromatids are so close together, they look like a single thread.

b The two homologous chromosomes become zippered together, so all four chromatids are intimately aligned. (X and Y chromosomes, which are two forms of sex chromosomes, also become zippered together for a small part of their length.)

c Typically, one or more crossovers occur at intervals along the chromosomes. Each time, two nonsister chromatids break at identical sites. Then they swap segments at the breaks, and enzymes seal the broken ends. For clarity, the two chromosomes are shown in condensed form and pulled apart. Crossing over may seem more plausible when you realize that it proceeds while the chromosomes are extended like threads and tightly aligned.

d As prophase I ends, the chromosomes continue to condense, to become thicker, rodlike forms. They detach from the nuclear envelope and from each other—except at "chiasmata." Each chiasma is indirect evidence that a crossover occurred at some point in the chromosomes.

e Crossing over breaks up old combinations of alleles and puts new ones together in pairs of homologous chromosomes.

Gene swapping would be rather pointless if each type of gene never varied from one chromosome to the next. But remember, a gene can have slightly different forms—alleles. You can safely bet that some alleles on one chromosome will *not* be identical to those on the homologue. Therefore, every crossover represents a chance to swap a *slightly different version* of hereditary instructions for a particular trait.

We will look at the mechanism of crossing over in later chapters. For now, it is enough to remember this: *Crossing over leads to genetic recombination, which in turn leads to variation in the traits of offspring.*

Metaphase I Alignments

Major shufflings of whole chromosomes begin during the transition from prophase I to metaphase I, which is the second stage of meiosis. Suppose the shufflings are proceeding at this very moment in one of your germ cells. By now, crossovers have made genetic mosaics out of the chromosomes, but let's put this aside in order to simplify tracking. Just call the twenty-three chromosomes that you inherited from your mother the *maternal* chromosomes, and call their twenty-three homologues from your father the *paternal* chromosomes.

Microtubules have already harnessed and oriented one chromosome of each pair toward one spindle pole and its homologue toward the other pole. Now they are moving all the chromosomes, which soon will become positioned at the spindle's equator.

Are all the maternal chromosomes attached to one pole and all the paternal ones attached to the other? Maybe, but probably not. Remember, the first contacts between spindle microtubules and chromosomes are random. Because of this random harnessing, *the eventual positioning of a maternal or paternal chromosome at the spindle equator at metaphase I follows no particular pattern.* Carrying this one step further, either one of each pair of homologous chromosomes might end up at either pole of the spindle after they move apart at anaphase I.

Think of the possibilities when merely three pairs of homologues are being tracked. As Figure 8.6 shows, by metaphase I, these may be arranged in any one of four possible positions. In this case, 2^3 or eight combinations of maternal and paternal chromosomes are possible for the forthcoming gametes.

Of course, a human germ cell has twenty-three pairs of homologous chromosomes, not just three. So 2^{23} or *8,388,608 combinations* of maternal and paternal chromosomes are possible every time a germ cell gives rise to sperm or eggs!

Moreover, in each sperm or egg, many hundreds of alleles inherited from the mother might not "say" the exact same thing about particular traits as the alleles

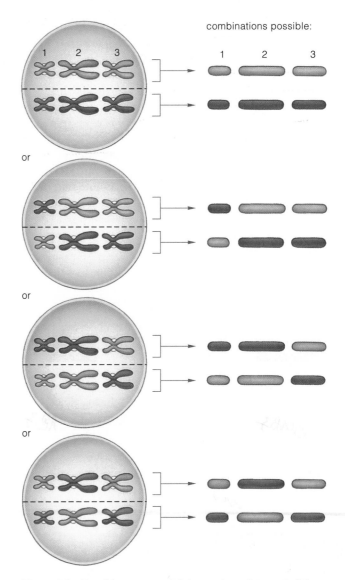

combinations possible:

Figure 8.6 Possible outcomes of the random alignment of three pairs of homologous chromosomes at metaphase I of meiosis. Three types of chromosomes are labeled 1, 2, and 3. Maternal chromosomes are *purple;* paternal ones are *blue.* With merely four possible alignments, eight combinations of maternal and paternal chromosomes are possible in forthcoming gametes.

inherited from the father. Are you beginning to get an idea of why such splendid mixes of traits show up even in the same family?

Crossing over is an interaction between a pair of homologous chromosomes. It breaks up old combinations of alleles and puts new ones together during prophase I of meiosis.

The random attachment and subsequent positioning of each pair of maternal and paternal chromosomes at metaphase I lead to different combinations of maternal and paternal traits in offspring.

8.5 FROM GAMETES TO OFFSPRING

The gametes that form after meiosis are not all the same in their details. For example, human sperm have one tail, opossum sperm have two, and roundworm sperm have none. Crayfish sperm look like pinwheels. Most eggs are microscopic, yet ostrich eggs tucked inside a shell are as large as a baseball. From appearance alone, you might not believe a plant gamete is even remotely like an animal's.

Later chapters contain details of how gametes form in the life cycles of representative organisms, including humans. Figure 8.7 and the following points may help you keep the details in perspective.

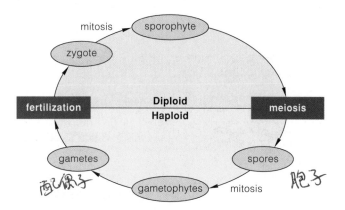

a Key events in the life cycle of plants.

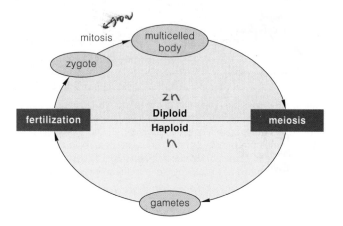

b Key events in the life cycle of animals.

Figure 8.7 Generalized life cycles for (**a**) most plants and (**b**) animals. The zygote is the first cell that forms when the nuclei of two gametes fuse together at fertilization.

For plants, a sporophyte (spore-producing body) develops, by way of mitotic cell divisions, from the zygote. Following meiosis, gametophytes (gamete-producing bodies) form. A pine tree is a sporophyte. Gametophytes develop in its cones.

Chapters 19, 20, and 34 especially incorporate illustrated examples of the life cycles of representative plants and animals.

Gamete Formation in Plants

For pine trees, roses, dandelions, and other familiar plants, certain events intervene between the time of meiosis and gamete formation. Among other things, spores form.

Spores are haploid cells, often with walls that allow them to resist episodes of drought or other adverse conditions in the environment. Under favorable conditions, spores germinate and then develop into a haploid body or structure that will produce gametes. Thus, *gamete*-producing bodies and *spore*-producing bodies develop during the life cycle of plants. Figure 8.7*a* provides a generalized diagram of these events.

Gamete Formation in Animals

In male animals, gametes form by a process that is known as spermatogenesis. Inside the male reproductive system, a diploid germ cell increases in size. It becomes a large, immature cell (the primary spermatocyte) that undergoes meiosis and cytoplasmic divisions. The outcome is four haploid daughter cells, which develop into immature cells called spermatids (Figure 8.8). The spermatids change in form, and each develops a tail. In this way, each becomes a **sperm**, a type of mature male gamete.

In female animals, gametes form by a process that is known as oogenesis. Each diploid germ cell develops into an oocyte, which is an immature egg. Compared to a sperm, an oocyte accumulates far more cytoplasmic components. Also, as Figure 8.9 indicates, its daughter cells differ in size and function. As the oocyte divides following meiosis I, one cell (the secondary oocyte) gets nearly all of the cytoplasm. Hence the other cell (the first polar body) is quite small. At some time after this, both cells may undergo meiosis II and then cytoplasmic division. Once again, one cell gets most of the cytoplasm. It develops into the gamete. A mature female gamete is called an ovum or, more commonly, an **egg**.

Division of the smaller cell means there are now three polar bodies. These do not function as gametes. They function as dumping grounds for chromosomes, so that the egg ends up with the appropriate (haploid) number of chromosomes at fertilization. Because polar bodies do not have much cytoplasm, they do not have much in the way of nutrients and metabolic machinery. In time they degenerate.

More Gene Shufflings at Fertilization

The chromosome number characteristic of the parents is restored at **fertilization**, when the nuclei of two haploid gametes fuse. Unless meiosis precedes it, fertilization would result in a doubling of the chromosome number in every new generation. The hereditary instructions

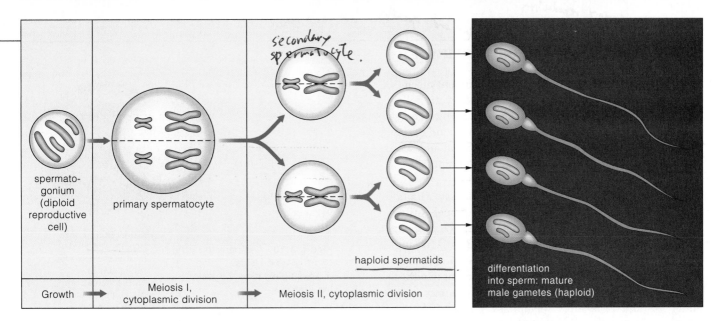

Figure 8.8 Generalized picture of sperm formation in male animals.

Handwritten annotation: secondary spermatocyte.

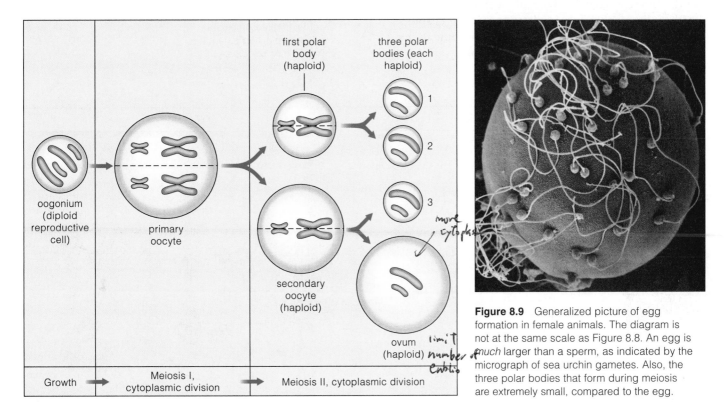

Figure 8.9 Generalized picture of egg formation in female animals. The diagram is not at the same scale as Figure 8.8. An egg is much larger than a sperm, as indicated by the micrograph of sea urchin gametes. Also, the three polar bodies that form during meiosis are extremely small, compared to the egg.

Handwritten annotations: more cytoplasm; limit number of eggs.

encoded in chromosomes operate as a fine-tuned package for each individual. Changes in the chromosome number disrupt the instructions, usually for the worse.

Fertilization contributes to variation in offspring. Reflect on the possibilities for humans alone. During prophase I of meiosis, an average of two or three crossovers take place in each human chromosome. Even without crossovers, the random positioning of pairs of paternal and maternal chromosomes at metaphase I results in one of millions of possible chromosome combinations in each gamete. And of all the male and female gametes that are produced, *which two* actually get together is a matter of chance. Thus the sheer number of combinations that can exist at fertilization is staggering!

Cumulatively, crossing over, the distribution of random mixes of homologous chromosomes into gametes, and fertilization contribute to variation in the traits of offspring.

---

sexual reproduction ↑

& repair and growth.

8.6 MEIOSIS AND MITOSIS COMPARED

In this unit, our focus has been on two different mechanisms that divide the nuclear DNA of eukaryotic cells. Single-celled species use mitosis in asexual reproduction; multicelled species use mitosis during growth and tissue repair. Meiosis is the basis of gamete formation and sexual reproduction. Figure 8.10 summarizes the similarities and differences between the two mechanisms.

The two mechanisms differ in a crucial way. *Mitotic cell division produces clones* (genetically identical copies of a parent cell). *Meiosis, in conjunction with fertilization, promotes variation in traits among offspring*. First, crossing over at prophase I of meiosis puts new combinations of alleles in chromosomes. Second, the movement of either member of each pair of homologous chromosomes to either spindle pole after metaphase I puts different mixes of maternal and paternal alleles into gametes. Third, different combinations of alleles are brought together by chance at fertilization. Later chapters describe how meiosis and fertilization contribute to the evolution of sexually reproducing organisms.

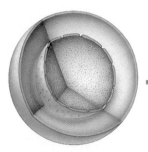

Figure 8.10 Comparison of mitosis and meiosis, using a diploid (2n) animal cell as the example.

This diagram is arranged to help you compare the similarities and differences between the two division mechanisms. Maternal chromosomes are shaded *purple*, and paternal chromosomes are *blue*.

A diploid (2n) *somatic* cell is at interphase. DNA is replicated (all chromosomes are duplicated) before nuclear division begins.

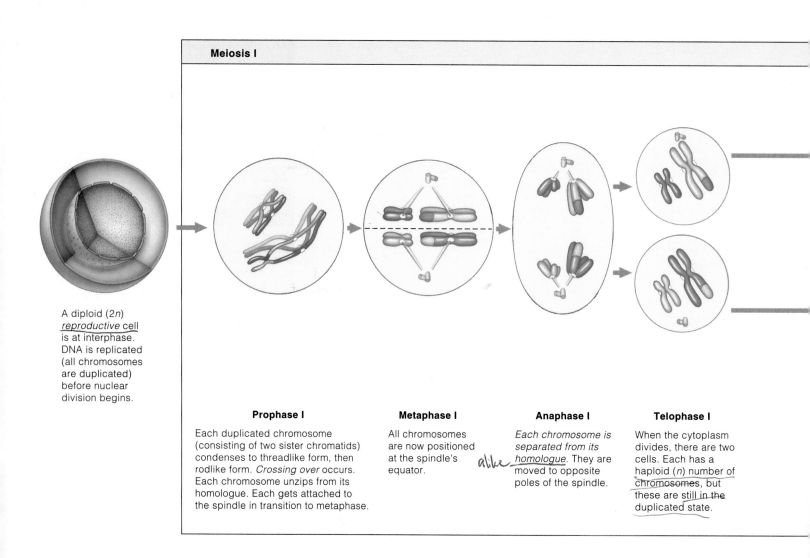

Meiosis I

A diploid (2n) *reproductive* cell is at interphase. DNA is replicated (all chromosomes are duplicated) before nuclear division begins.

Prophase I

Each duplicated chromosome (consisting of two sister chromatids) condenses to threadlike form, then rodlike form. *Crossing over* occurs. Each chromosome unzips from its homologue. Each gets attached to the spindle in transition to metaphase.

Metaphase I

All chromosomes are now positioned at the spindle's equator.

Anaphase I

alike *Each chromosome is separated from its homologue.* They are moved to opposite poles of the spindle.

Telophase I

When the cytoplasm divides, there are two cells. Each has a haploid (n) number of chromosomes, but these are still in the duplicated state.

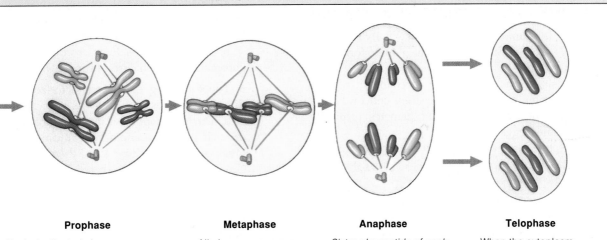

Prophase

Each duplicated chromosome (consisting of two sister chromatids) condenses from threadlike form to rodlike form. Each gets attached to the spindle during the transition to metaphase.

Metaphase

All chromosomes are now positioned at the spindle's equator.

Anaphase

Sister chromatids of each chromosome are separated from each other. These new, daughter chromosomes are moved to opposite poles of the spindle.

Telophase

When the cytoplasm divides, there are two cells. Each is diploid (2n)—*it has the same chromosome number as the parent cell.*

Meiosis II

There is no DNA replication between the two divisions

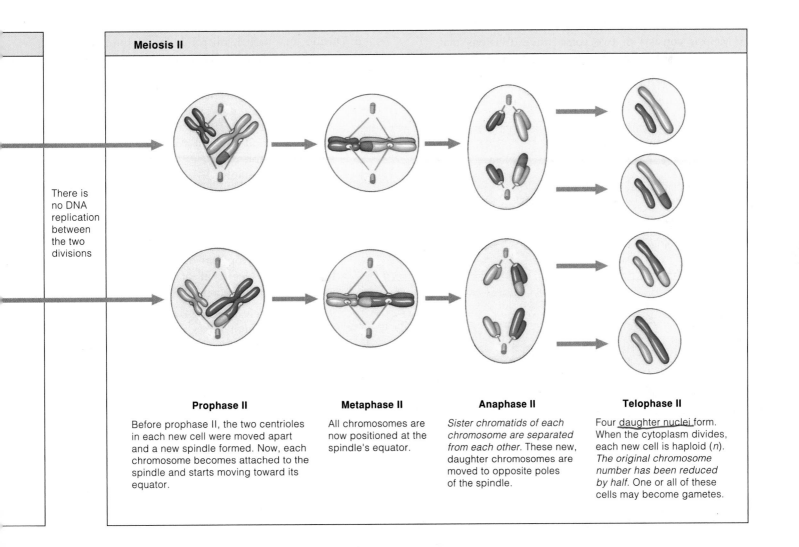

Prophase II

Before prophase II, the two centrioles in each new cell were moved apart and a new spindle formed. Now, each chromosome becomes attached to the spindle and starts moving toward its equator.

Metaphase II

All chromosomes are now positioned at the spindle's equator.

Anaphase II

Sister chromatids of each chromosome are separated from each other. These new, daughter chromosomes are moved to opposite poles of the spindle.

Telophase II

Four daughter nuclei form. When the cytoplasm divides, each new cell is haploid (n). *The original chromosome number has been reduced by half.* One or all of these cells may become gametes.

SUMMARY

1. At the minimum, the life cycles of sexually reproducing species proceed through meiosis, gamete formation, and fertilization.

  a. Meiosis, a nuclear division mechanism, reduces the chromosome number of the parent cell by half. It *precedes* the formation of haploid gametes (typically, sperm in males, eggs in females).

  b. At fertilization, a sperm nucleus fuses with an egg nucleus. As Figure 8.11 shows, this event restores the chromosome number.

2. If the germ cells of sexually reproducing organisms are diploid (2n), they have *two* of each type of chromosome characteristic of the species. Commonly, one of the two is maternal (inherited from a female parent), and the other is paternal (from a male parent).

3. Each pair of maternal and paternal chromosomes shows homology (the two are alike). Generally, the two have the same length, same shape, and same sequence of genes. They interact during meiosis.

4. Chromosomes are duplicated during interphase. So before meiosis even begins, each consists of two DNA molecules that remain attached (as sister chromatids).

5. Meiosis consists of two consecutive divisions that require a spindle apparatus, which is assembled from many microtubules.

  a. During meiosis I, microtubules of the spindle move each duplicated chromosome away from its homologue (which is also duplicated).

  b. During meiosis II, microtubules move the sister chromatids of each chromosome away from each other.

6. The following events characterize the first nuclear division (meiosis I):

  a. During prophase I, the two nonsister chromatids of each pair of homologous chromosomes commonly break at corresponding sites, then exchange segments. Crossing over puts new combinations of alleles together. Alleles, which are slightly different molecular forms of the same gene, code for different forms of the same trait. Different combinations of alleles lead to variation in the details of a given trait among offspring.

  b. Also during prophase I, a microtubular spindle starts forming outside the nucleus. The nuclear envelope starts to break up. If a pair of centrioles is present, one of the pair starts moving to the opposite pole of the newly forming spindle.

  c. At metaphase I, all of the pairs of homologous chromosomes have become positioned at the equator of the spindle apparatus. For each pair, either the maternal chromosome or its homologue can be oriented toward either pole.

  d. During anaphase I, microtubules of the spindle apparatus move each chromosome away from its homologue, toward opposite spindle poles.

7. The following events characterize the second nuclear division (meiosis II):

  a. At metaphase II, the chromosomes are still duplicated, and they all are positioned at the spindle equator.

  b. During anaphase II, the sister chromatids of each chromosome are moved apart. Each is now a separate, unduplicated chromosome.

  c. By the end of telophase II, four haploid nuclei have formed.

8. When the cytoplasm divides, there are four haploid cells. One or all of these may function as gametes (or as spores, in flowering plants).

9. Cumulatively, crossing over, the chance allocation of different mixes of pairs of maternal and paternal chromosomes to different gametes, and the chance of any two gametes meeting at fertilization contribute to immense variation in the details of traits among offspring.

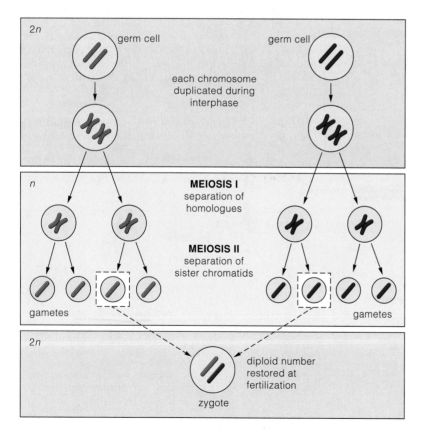

Figure 8.11 Summary of changes in the chromosome number at different stages of sexual reproduction, using diploid (2n) germ cells as the example. Meiosis reduces the chromosome number by half (n). Then the union of haploid nuclei of two gametes at fertilization restores the diploid number.

Review Questions

1. Genetically speaking, what is the key difference between the outcomes of sexual and asexual reproduction? *126*

2. Refer to the following numbers of chromosomes in the diploid body cells of a few organisms. In each case, how many chromosomes would end up in gametes? *127*

Fruit fly, *Drosophila melanogaster*	8
Garden pea, *Pisum sativum*	14
Corn, *Zea mays*	20
Frog, *Rana pipiens*	26
Earthworm, *Lumbricus terrestris*	36
Human, *Homo sapiens*	46
Chimpanzee, *Pan troglodytes*	48
Amoeba, *Amoeba*	50
Horsetail, *Equisetum*	216

3. Suppose a diploid germ cell contains four pairs of homologous chromosomes, designated AA, BB, CC, and DD. How would the chromosomes of the gametes be designated? *127*

4. Define meiosis and describe its main stages. In what respects is meiosis *not* like mitosis? *127, 134*

5. Is this cell at anaphase I or anaphase II? *128–129*

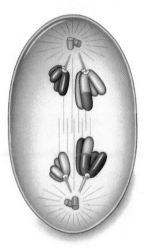

6. Outline the steps by which sperm and eggs form in animals. *132–133*

Self-Quiz *(Answers in Appendix IV)*

1. Sexual reproduction requires _____ .
   a. meiosis
   b. gamete formation
   c. fertilization
   d. all of the above

2. Meiosis is a division mechanism that produces _____ .
   a. two cells
   b. two nuclei
   c. four cells
   d. four nuclei

3. An animal cell with two of each type of chromosome characteristic of the species is _____ .
   a. diploid
   b. haploid
   c. a normal gamete
   d. both b and c

4. Meiosis _____ the parental chromosome number.
   a. doubles
   b. reduces
   c. maintains
   d. corrupts

5. Generally, a pair of homologous chromosomes _____ .
   a. carry the same genes
   b. are the same length, shape
   c. interact at meiosis
   d. all of the above

6. Before the onset of meiosis, all chromosomes are _____ .
   a. condensed
   b. released from protein
   c. duplicated
   d. b and c

7. Each chromosome moves away from its homologue and ends up at the opposite spindle pole during _____ .
   a. prophase I
   b. prophase II
   c. anaphase I
   d. anaphase II

8. Sister chromatids of each chromosome move apart and end up at opposite spindle poles during _____ .
   a. prophase I
   b. prophase II
   c. anaphase I
   d. anaphase II

9. Match each term and its description.
   ____ chromosome number
   ____ alleles
   ____ metaphase I
   ____ interphase

   a. different molecular forms of the same gene
   b. none between meiosis I, II
   c. pairs of homologues aligned at spindle equator
   d. the sum total of chromosomes in cells of a given type

Critical Thinking

1. Assume you can measure the amount of DNA in single cells. You measure the amount in a primary oocyte, then in a primary spermatocyte, which gives you a mass *m*. What mass of DNA would you expect to find in each mature gamete (egg and sperm) that forms following meiosis? What mass of DNA would you expect to find in (1) a cell resulting from fertilization of an egg by one of the sperm, and (2) in the fertilized egg after the first DNA replication?

2. Jeff has a pair of alleles (on a pair of homologous chromosomes) that influence whether a person is right- or left-handed. One allele says "left," and its partner says "right." Now visualize one of Jeff's germ cells, in which chromosomes are being duplicated prior to meiosis. Visualize what happens to the chromosomes at anaphase I and II. (It may help to use index cards as models of the sister chromatids of each chromosome.) What fraction of Jeff's sperm will carry the gene for right-handedness? For left-handedness?

3. Jeff also has one allele for long eyelashes, and a partner allele (on the homologous chromosome) for short eyelashes. What fraction of his sperm will have these gene combinations:
   right-handed, long eyelashes   left-handed, long eyelashes
   right-handed, short eyelashes   left-handed, short eyelashes

Selected Key Terms

Readings

Klug, W., and M. Cummings. 1994. *Concepts of Genetics.* Fourth edition. New York: Macmillan.

Wolfe, S. 1995. *Introduction to Molecular and Cellular Biology.* Belmont, California: Wadsworth.

9 OBSERVABLE PATTERNS OF INHERITANCE

A Smorgasbord of Ears and Other Traits

Basketball ace Charles Barkley has them. So does actor Tom Cruise. Actress Joan Chen doesn't, and neither did a monk named Gregor Mendel. To see how *you* fit in with these folks, use a mirror to check out your ears. Is the fleshy lobe at the base of each ear attached to the side of your head? If so, you and Barkley and Cruise have something in common. Or is the fleshy lobe not attached, so that you can flap it back and forth? If so, you are like Chen and Mendel (Figure 9.1).

Whether a person is born with attached or detached earlobes depends on a single kind of gene. That gene comes in slightly different molecular forms—alleles. Only one form has information about detached lobes. The information is put to use while a human body is developing inside the mother. It calls for a death signal, which is sent to all the cells positioned between the newly forming lobes and the head. Without the signal, the cells don't die, and earlobes don't detach.

We all have genes for thousands of traits, including earlobes, cheeks, lashes, and eyeballs. Most traits vary in their details from one person to the next. Remember,

we inherit pairs of genes, on pairs of chromosomes. In some pairings, one allele has strong effects and overwhelms the other allele's contribution. The outgunned allele is said to be recessive to the dominant one. If you have *detached* earlobes, *dimpled* cheeks, *long* lashes, or *large* eyeballs, you carry at least one and possibly two dominant alleles that affect the trait in a particular way.

When both alleles of a pair are recessive, nothing masks their effect on a trait. You get *attached* earlobes with one pair of recessive alleles (and *flat* feet with another, a *straight* nose with another, and so on).

How did we discover such remarkable things about our genes? It started with Gregor Mendel. By analyzing pea plants generation after generation, Mendel found indirect but *observable* evidence of how parents bestow units of hereditary information—genes—on offspring. This chapter focuses on the methods and results of Mendel's experiments. They remain a classic example of how a scientific approach can pry open important secrets about the natural world. And to this day, they serve as the foundation for modern genetics.

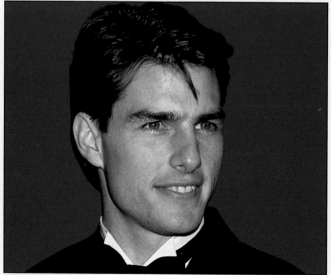

Tom Cruise

(*Right*) Charles Barkley

Figure 9.1 The attached and detached earlobes of a few representative humans. This sampling provides observable evidence of a trait that is governed by a single gene. Do you have one or the other version of this trait? It depends on whether you inherited a specific molecular form of that gene from your mother, your father, or both of your parents. As Gregor Mendel perceived, such easily observable traits can be used to identify patterns of inheritance that exist from one generation to the next.

Joan Chen

Gregor Mendel

1. Genes are units of information about heritable traits. Each gene has a specific location in the chromosomes of a species. But its molecular form may differ slightly from one chromosome to the next. Different molecular forms of a gene (alleles) specify different versions of the same trait.

2. Humans, pea plants, and other organisms with diploid body cells inherit pairs of genes, on pairs of homologous chromosomes.

3. The two genes of a pair segregate from each other during meiosis, and they end up in different gametes. Gregor Mendel found indirect evidence of this when he crossbred plants having different versions of the same trait, such as the color of their flowers.

4. Each pair of homologous chromosomes is assorted into gametes independently of the other pairs. Mendel found indirect evidence of this when he tracked plants having observable differences in *two* traits, such as flower color and height.

5. If the two genes of a pair specify different versions of a trait (if they are nonidentical alleles), one may have a more pronounced effect on the trait. Besides this, two or more gene pairs often influence the same trait, and some single genes influence many traits. Finally, environmental conditions may alter gene expression.

More than a century ago, people wondered about the basis of inheritance. It was common knowledge that sperm and eggs both transmit information about traits to offspring. But almost no one suspected the information is organized in units (genes). Instead, the idea was that a father's blob of information "blended" with a mother's blob at fertilization, like cream into coffee.

Carried to its logical conclusion, blending gradually would dilute a population's pool of hereditary information until there was only a single version left of each trait. Yet if that were so, why did, say, freckled children keep showing up among nonfreckled generations? Why weren't all the descendants of a herd of white stallions and black mares gray? The theory of blending scarcely explained the variation that people could see with their own eyes. Even so, few disputed the theory.

Charles Darwin was among the scholarly dissidents. According to a key premise of his theory of natural selection, individuals of a population show variation in heritable traits. Through the generations, variations that improve chances of surviving and reproducing occur with greater frequency than those that do not. The less advantageous variations may persist among fewer individuals or may even disappear. It is not that separate versions of a trait are "blended out" of the population. Rather, *each version of a trait may persist in a population, at frequencies that may change over time.*

Even before Darwin presented his theory, someone was gathering evidence that eventually would support his key premise. A monk, Gregor Mendel, was about to prove that sperm and eggs carry distinct "units" of information about heritable traits. By analyzing pea plants generation after generation, Mendel found indirect but *observable* evidence of how parents transmit genes to offspring.

Mendel's Experimental Approach

Mendel spent most of his life in a monastery in Brünn, an Austrian village that is now located in the Czech Republic. The monastery of St. Thomas was somewhat removed from the European capitals, which were then the centers of scientific inquiry. Yet Mendel was not a man of narrow interests who accidentally stumbled by chance onto principles of great import.

Having been raised on a farm, Mendel was aware of agricultural principles and their applications. He kept abreast of breeding experiments and developments described in the available literature. He was a member of the regional agricultural society. He also won several awards for developing improved varieties of vegetables and fruits. Shortly after entering the monastery, he spent two years studying mathematics at the University of Vienna. Few scholars of his time had similarly combined talents in plant breeding and mathematics.

b Pollen from a plant that breeds true for purple flowers is brushed onto a floral bud of a plant that breeds true for white flowers and that had its own stamens snipped off.

c The cross-fertilized plant produces seeds, each of which is allowed to grow into a new plant.

d The flower color of the new plants can be used as visible evidence of patterns in how hereditary material might be transmitted from each parent plant.

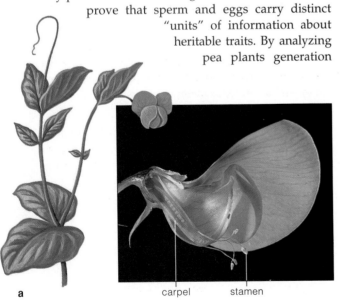

a

carpel   stamen

Figure 9.2 The garden pea plant (*Pisum sativum*), the focus of Mendel's experiments. (**a**) A flower has been sectioned to show the location of its stamens and carpel. Sperm-producing pollen grains form in stamens. Eggs develop, fertilization takes place, and seeds mature inside the carpel.

Shortly after his university training, Mendel began experiments with the garden pea plant, *Pisum sativum* (Figure 9.2). This plant is self-fertilizing. Its flowers produce sperm and eggs, which meet up inside the same flower. Nearly all the plants are **true-breeding**. In other words, successive generations are just like the parents in one or more traits, as when all plants grown from seeds of white-flowered plants have white flowers.

As Mendel knew, pea plants also will cross-fertilize when sperm and eggs from different plants are brought together under controlled conditions. For his studies, he could open flower buds of a plant that bred true for a trait—say, white flowers—and snip out the stamens. (Stamens bear pollen grains in which sperm develop.) Then he could brush the "castrated" buds with pollen from a plant that bred true for a *different* version of the same trait (purple flowers). As Mendel hypothesized, such clearly observable differences could be used to

track a given trait through many generations. If there were patterns to the trait's inheritance, *those patterns might tell him something about the hereditary material itself.*

Some Terms Used in Genetics

Having read the chapter on meiosis, you already have insight into the mechanisms of sexual reproduction— which is more than Mendel had. He did not know about chromosomes. So he could not have known that the chromosome number is reduced by half in gametes, then restored at fertilization. Yet Mendel sensed what was going on. As we follow his thinking, let's simplify things by substituting a few of the modern terms used in studies of inheritance (see also Figure 9.3):

1. **Genes** are units of information about specific traits, and they are passed from parents to offspring. Each gene has a specific location (locus) on a chromosome.

2. Diploid cells have a pair of genes for each trait, on a pair of homologous chromosomes.

3. Although both genes of a pair deal with the same trait, they may vary in their information about it. This happens when they have slight molecular differences, as when one gene for flower color specifies purple and another specifies white. The different molecular forms of a gene are called **alleles** of that gene.

4. If it turns out that the two alleles of a pair are the same, this is a *homozygous* condition. If the two alleles are different, this is a *heterozygous* condition.

5. An allele is *dominant* when its effect on a trait masks that of any *recessive* allele paired with it. We use capital letters for dominant alleles and lowercase letters for recessive ones (for instance, *A* and *a*).

6. Putting this together, a **homozygous dominant** individual has a pair of dominant alleles (*AA*) for the trait that is being studied. A **homozygous recessive** individual has a pair of recessive alleles (*aa*) for the trait. And a **heterozygous** individual has a pair of nonidentical alleles (*Aa*) for the trait.

7. Two terms help keep the distinction clear between genes and the traits they specify. **Genotype** refers to the particular genes that are present in an individual. **Phenotype** refers to an individual's observable traits.

8. When tracking the inheritance of traits through generations of offspring, these abbreviations apply:

P parental generation
F_1 first-generation offspring
F_2 second-generation offspring

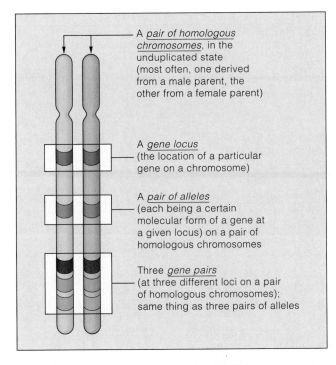

A *pair of homologous chromosomes*, in the unduplicated state (most often, one derived from a male parent, the other from a female parent)

A *gene locus* (the location of a particular gene on a chromosome)

A *pair of alleles* (each being a certain molecular form of a gene at a given locus) on a pair of homologous chromosomes

Three *gene pairs* (at three different loci on a pair of homologous chromosomes); same thing as three pairs of alleles

Figure 9.3 A few genetic terms illustrated. Diploid organisms have pairs of genes, on pairs of homologous chromosomes. For example, you inherited one chromosome of each pair from your mother, and the other, homologous chromosome from your father.

Genes may have different molecular forms, called alleles. Different alleles specify slightly different versions of the same trait. An allele at one location on a chromosome may or may not be identical to its partner on the homologous chromosome.

When offspring of genetic crosses inherit identical alleles for a trait, generation after generation, they are a *true-breeding* lineage. When they inherit nonidentical alleles for a trait under study, they are said to be *hybrid* offspring.

Predicting the Outcome of Monohybrid Crosses

Mendel had an idea that in every generation, a plant inherits two "units" (genes) of information for a trait, one from each parent. To test his idea, he performed what we now call **monohybrid crosses**. Offspring of such crosses are heterozygous for the one trait being studied (which is what monohybrid means). The two parents breed true for different versions of the trait, so their offspring inherit a pair of nonidentical alleles.

Mendel tracked many individual traits through two generations. For instance, in one series of experiments, he crossed true-breeding purple-flowered plants and true-breeding white-flowered ones. All plants grown from the seeds that resulted from this cross had purple flowers. Mendel allowed these plants to self-fertilize. Some plants grown from the seeds had white flowers!

If Mendel's hypothesis were correct—if each plant had inherited two units of information about flower color—then the unit for "purple" had to be dominant, because it had masked the unit for "white" in F_1 plants.

Let's rephrase his thinking. Germ cells of pea plants are diploid, with pairs of homologous chromosomes. Assume one parent is homozygous dominant (AA) and the other is homozygous recessive (aa) for flower color. After meiosis, a sperm or egg carries one allele for flower color (Figure 9.4). Thus, when a sperm fertilizes an egg, one outcome is possible: $A + a = Aa$.

Before continuing, you should know that Mendel crossed hundreds of plants and tracked thousands of offspring. Besides this, he counted and recorded the plants showing dominance or recessiveness. As you can see from Figure 9.5, an intriguing ratio emerged. On the average, three of every four F_2 plants had the dominant phenotype, and one had the recessive phenotype.

To Mendel, the ratio suggested that fertilization is a chance event, with a number of possible outcomes. And he had an understanding of **probability**, which applies

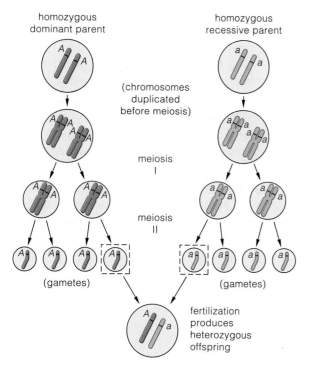

Figure 9.4 A monohybrid cross, showing how one gene of a pair segregates from the other. Two parents that breed true for two different versions of a trait can produce only heterozygous offspring.

Figure 9.5 Results from Mendel's monohybrid cross experiments with the garden pea. The numbers are his counts of F_2 plants that he assumed were carrying dominant or recessive hereditary "units" (alleles) for the trait. On the average, the dominant-to-recessive ratio was 3:1.

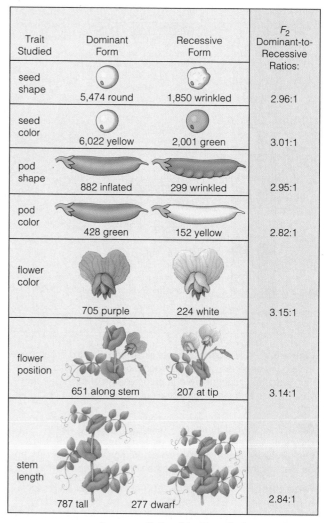

Trait Studied	Dominant Form	Recessive Form	F_2 Dominant-to-Recessive Ratios:
seed shape	5,474 round	1,850 wrinkled	2.96:1
seed color	6,022 yellow	2,001 green	3.01:1
pod shape	882 inflated	299 wrinkled	2.95:1
pod color	428 green	152 yellow	2.82:1
flower color	705 purple	224 white	3.15:1
flower position	651 along stem	207 at tip	3.14:1
stem length	787 tall	277 dwarf	2.84:1

Average ratio for all traits studied: 3:1

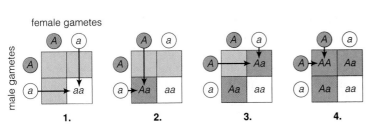

female gametes

male gametes

1.   2.   3.   4.

Figure 9.6 Punnett-square method of predicting the probable outcome of a genetic cross. Circles represent gametes. The *italic* letters on the gametes represent dominant or recessive alleles. The different squares show the different genotypes possible among offspring. In this example, the gametes are from a self-fertilizing plant that is heterozygous (*Aa*) for a trait.

to chance events *and so could help him predict the possible outcomes of crosses.* "Probability" simply means that the chance of each outcome occurring is proportional to the number of ways it can be reached.

The **Punnett-square method**, explained in Figure 9.6 and applied in Figure 9.7, may help you visualize the possibilities. As you can see, if half of a plant's sperm (or eggs) were *a* and half were *A*, then four outcomes were possible every time a sperm fertilized an egg:

Possible Event	Probable Outcome
sperm *A* meets egg *A*	1/4 *AA* offspring
sperm *A* meets egg *a*	1/4 *Aa*
sperm *a* meets egg *A*	1/4 *Aa*
sperm *a* meets egg *a*	1/4 *aa*

By this prediction, an F_2 plant had three chances in four of getting at least one dominant allele (purple flowers). It had one chance in four of getting two recessive alleles (white flowers). That is a probable phenotypic ratio of three purple to one white, or 3:1.

Mendel's observed ratios weren't *exactly* 3:1. You can see this for yourself by looking at the numerical results listed in Figure 9.5. Why did Mendel put aside the deviations? To understand why, flip a coin a couple of times. As we all know, a coin is just as likely to end up heads as tails. But often it ends up heads, or tails, several times in a row. So if you flip the coin only a few times, the observed ratio may differ greatly from the predicted ratio of 1:1. Flip the coin many, many times, and you are more likely to come close to the predicted ratio. Mendel understood the rules of probability—and he performed a large number of crosses. Almost certainly, this kept him from being confused by minor deviations from the predicted results of the experimental crosses.

Testcrosses

Mendel gained support for his prediction with the **test-cross**. In this type of experimental cross, an organism

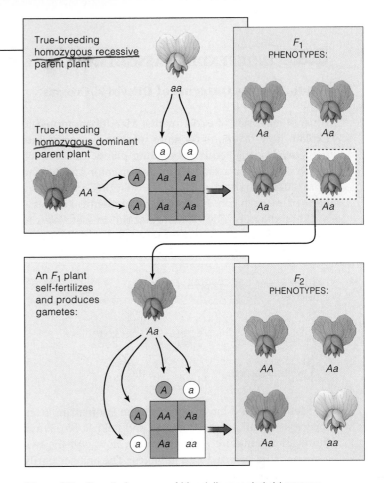

Figure 9.7 Results from one of Mendel's monohybrid crosses. On the average, the dominant-to-recessive ratio among the second-generation (F_2) plants was 3:1.

shows dominance for a specified trait but its genotype is unknown, so it is crossed to a known homozygous recessive individual. Results may reveal whether the organism is homozygous dominant or heterozygous.

Regarding the monohybrid crosses just described, Mendel tested his prediction that the purple-flowered F_1 offspring were heterozygous by crossing them with true-breeding, white-flowered plants. If they were all homozygous dominant, then all the F_2 offspring would show the dominant form of the trait. If heterozygous, there would be about as many dominant as recessive plants. Sure enough, about half of the F_2 plants had purple flowers (*Aa*) and half had white (*aa*).

Can you construct two Punnett squares that show the possible outcomes of this testcross?

On the basis of results from Mendel's monohybrid crosses and his testcrosses, it is possible to formulate a theory, which we state here in modern terms:

MENDEL'S THEORY OF SEGREGATION.   **Diploid cells have pairs of genes (on pairs of homologous chromosomes). During meiosis, the two genes of each pair segregate from each other. As a result, they end up in different gametes.**

Predicting the Outcome of Dihybrid Crosses

By another series of experiments, Mendel attempted to explain how *two* pairs of genes might be assorted into gametes. He selected true-breeding plants that differed in two traits—for example, in flower color and height. With such **dihybrid crosses**, F_1 offspring inherit two gene pairs, each consisting of two nonidentical alleles.

Let's diagram one of his dihybrid crosses. We can use A for flower color and B for height as the dominant alleles, and a and b as their recessive counterparts:

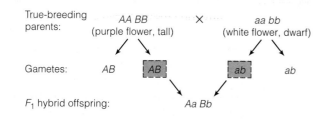

As Mendel would have predicted, the F_1 offspring from this cross are all purple-flowered and tall (*Aa Bb*). When the F_1 plants mature and reproduce, how will the two gene pairs be assorted into gametes? The answer partly depends on the chromosomal locations of the gene pairs. Assume one pair of homologous chromosomes carries the *Aa* alleles and a *different* pair carries the *Bb* alleles. Next, think of how all chromosomes become positioned at the spindle equator during metaphase I of meiosis (Figures 8.4 and 9.8). The chromosome with the *A* allele might be positioned to move to either spindle pole (and on into one of four gametes). The same is true of its homologue. And the same is true of the chromosomes with the *B* and *b* alleles. Thus, after meiosis and gamete formation, four combinations of alleles are possible in the sperm or eggs: 1/4 *AB*, 1/4 *Ab*, 1/4 *aB*, and 1/4 *ab*.

Given the alternative alignments of chromosomes at metaphase I, several allelic combinations are possible at fertilization. Simple multiplication (four kinds of sperm times four kinds of eggs) tells us sixteen combinations of gametes are possible in the F_2 offspring of a dihybrid cross. Use the Punnett-square method to diagram the probabilities (Figure 9.9). Now add up all the possible phenotypes and you get 9/16 tall purple-flowered, 3/16 dwarf purple-flowered, 3/16 tall white-flowered, and 1/16 dwarf white-flowered plants. That is a probable phenotypic ratio of 9:3:3:1. Results from one dihybrid cross that Mendel described were close to this ratio.

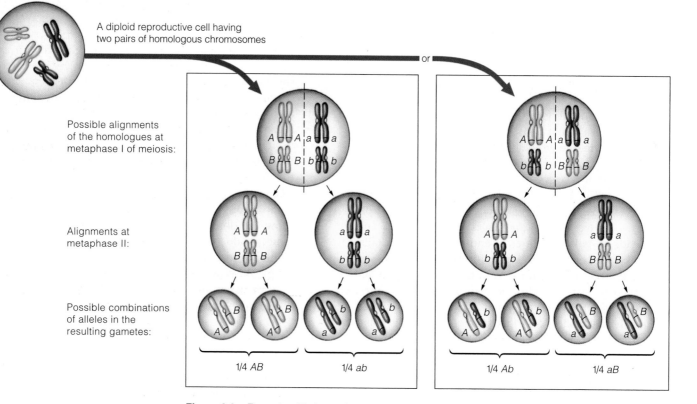

Figure 9.8 Example of independent assortment, using just two pairs of homologous chromosomes. An allele at one locus on a chromosome may or may not be identical to its partner allele on the homologous chromosome. At meiosis, either chromosome of a pair may get attached to either spindle pole. So two different lineups are possible at metaphase I.

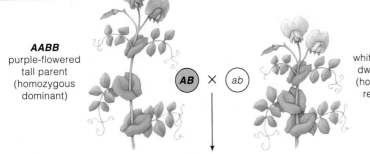

AABB
purple-flowered
tall parent
(homozygous
dominant)

AB × ab

aabb
white-flowered
dwarf parent
(homozygous
recessive)

F_1 OUTCOME: All F_1 plants purple-flowered, tall
(**A**a**B**b heterozygotes)

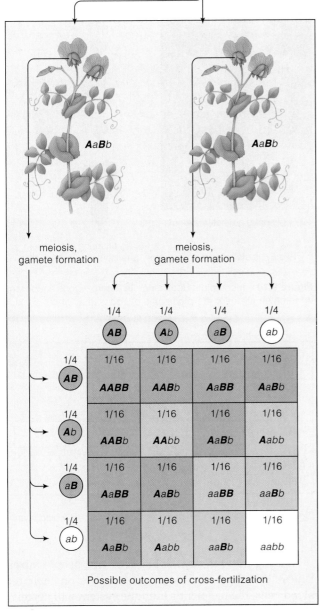

AaBb   AaBb

meiosis,   meiosis,
gamete formation   gamete formation

	1/4 **AB**	1/4 **A**b	1/4 a**B**	1/4 ab
1/4 **AB**	1/16 **AABB**	1/16 **AAB**b	1/16 **A**a**BB**	1/16 **A**a**B**b
1/4 **A**b	1/16 **AAB**b	1/16 **AA**bb	1/16 **A**a**B**b	1/16 **A**abb
1/4 a**B**	1/16 **A**a**BB**	1/16 **A**a**B**b	1/16 aa**BB**	1/16 aa**B**b
1/4 ab	1/16 **A**a**B**b	1/16 **A**abb	1/16 aa**B**b	1/16 aabb

Possible outcomes of cross-fertilization

ADDING UP THE F_2 COMBINATIONS POSSIBLE:

▢ 9/16 or 9 purple-flowered, tall

▢ 3/16 or 3 purple-flowered, dwarf

▢ 3/16 or 3 white-flowered, tall

▢ 1/16 or 1 white-flowered, dwarf

Figure 9.9 Results from Mendel's dihybrid cross between parent plants that bred true for different versions of two traits (flower color and plant height). *A* and *a* represent dominant and recessive alleles for flower color. *B* and *b* represent dominant and recessive alleles for height. As the Punnett square indicates, the probabilities of certain combinations of phenotypes among F_2 offspring occur in a 9:3:3:1 ratio, on the average.

The Theory in Modern Form

Mendel could do no more than analyze the numerical results from his dihybrid crosses, because he didn't know that a number of chromosomes carry the pea plant's "units" of inheritance. It simply seemed to him that the two units for the first trait he was tracking had been assorted into gametes independently of the two units for the second trait. In time, his interpretation became known as the theory of **independent assortment**, which we state here in modern terms: By the end of meiosis, each pair of homologous chromosomes—and the genes they carry—have been sorted for shipment into gametes independently of how the other pairs were sorted out.

Independent assortment and hybrid crossing lead to staggering variety among offspring. In a monohybrid cross involving only a single gene pair, three genotypes are possible: *AA*, *Aa*, and *aa*. We can represent this as 3^n, where *n* is the number of gene pairs. When we consider more gene pairs, the number of possible combinations increases dramatically. Even if parents differ in merely ten gene pairs, nearly 60,000 genotypes are possible among their offspring. If they differ in twenty gene pairs, the number approaches 3.5 billion!

In 1865 Mendel presented his ideas to the Brünn Natural History Society. His ideas had little impact. The next year his paper was published, and apparently it was read by few and understood by no one. In 1871 Mendel became an abbot of the monastery, and his experiments gave way to administrative tasks. He died in 1884, never to know his work would be the starting point for the development of modern genetics.

Today, Mendel's theory of segregation still stands. Hereditary material is indeed organized in units (genes) that retain their identity and are segregated from each other for distribution into different gametes. However, the theory of independent assortment has undergone some modification, as you will see in the next chapter.

MENDEL'S THEORY OF INDEPENDENT ASSORTMENT. **By the end of meiosis, the genes on pairs of homologous chromosomes have been sorted out for distribution into one gamete or another independently of gene pairs of other chromosomes.**

9.4 DOMINANCE RELATIONS

For the most part, Mendel studied traits having clearly dominant or recessive forms. As the remaining sections of this chapter will make clear, however, the expression of other traits is not as straightforward.

Incomplete Dominance

In **incomplete dominance**, one allele of a pair isn't fully dominant over its partner, so a heterozygous phenotype *somewhere in between* the two homozygous phenotypes emerges. Cross a true-breeding red snapdragon and a true-breeding white one. All F_1 offspring will have pink flowers. Cross two F_1 plants, and expect red, pink, and white snapdragons in a predictable ratio (Figure 9.10). What causes this inheritance pattern? Red snapdragons have two alleles that allow them to make an abundance of red pigment molecules. White ones are pigment-free; they have different versions of these alleles. Pink ones are heterozygous. Their one red allele specifies enough pigment to make flowers pink, but not red.

ABO Blood Types: A Case of Codominance

In **codominance**, a pair of nonidentical alleles specify two phenotypes, both of which are expressed at the same time in heterozygotes. As an example, consider how your body cells have markers—certain molecules at their surface that give them a unique identity. One kind of marker on red blood cells can have different molecular forms. A method of analysis called *ABO blood typing* reveals which form a person has.

In the human population, the gene for this marker has three alleles. Two, I^A and I^B, are codominant when paired. The third allele, i, is recessive; when paired with either I^A or I^B, its effects are masked. When three or more alleles of a gene are present among members of a population, we call this a **multiple allele system**.

Picture a cell that is making new molecules of this marker. Before each molecule becomes positioned at the cell surface, it takes on final form in the cytomembrane system (Section 3.6). First a carbohydrate chain becomes attached to a lipid. Then an enzyme attaches a sugar unit to the chain. Alleles I^A and I^B code for two versions of that enzyme. The two attach *different* sugar units, and this gives the marker a special identity—either A or B.

Which of those two alleles do you have? With either I^AI^A or I^Ai, your blood is type A. With either I^BI^B or I^Bi, it is type B. With codominant alleles I^AI^B, your blood type is AB—you have both versions of the enzyme, and both sugar units are attached to the marker molecules. If you are homozygous recessive (ii), the markers never did get a sugar unit attached to them. Therefore, your blood type is neither A nor B (that's what type "O" means). Figure 9.11 summarizes the possibilities.

homozygous parent × homozygous parent

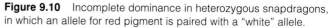

All F_1 offspring are heterozygous.

Two F_1 plants are crossed. ⟶ F_2 offspring show three phenotypes in a 1:2:1 ratio.

Figure 9.10 Incomplete dominance in heterozygous snapdragons, in which an allele for red pigment is paired with a "white" allele.

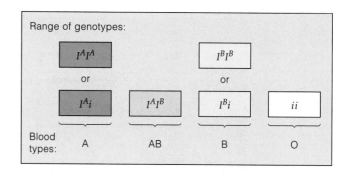

Range of genotypes:

I^AI^A		I^BI^B	
or		or	
I^Ai	I^AI^B	I^Bi	ii

Blood types: A   AB   B   O

Figure 9.11 Allelic combinations associated with ABO blood typing.

During *transfusions*, the blood of two people mixes. Unless they have the same self markers on their red blood cells, the recipient's immune system will perceive the donor's markers as "nonself." It will act against the transfused cells and may cause death (Chapter 29).

One allele may be fully dominant, incompletely dominant, or codominant with its partner on the homologous chromosome.

9.5 MULTIPLE EFFECTS OF SINGLE GENES

Expression of the alleles at a single gene location on a chromosome may influence two or more traits—and it may do so in positive or negative ways. This condition is known as **pleiotropy** (after the Greek *pleio-*, meaning more, and *-tropic*, meaning to change).

A genetic disorder called *sickle-cell anemia* is a classic example of pleiotropic effects. It arises from a mutated gene for hemoglobin (Hb^S instead of Hb^A). Hemoglobin is a protein that transports oxygen in red blood cells. Heterozygotes (Hb^A/Hb^S) typically show few symptoms. They are able to make enough normal hemoglobin molecules to compensate for their abnormal ones. But homozygotes (Hb^S/Hb^S) only produce abnormal hemoglobin molecules, and drastic changes in phenotype may follow. The changes start with disruptions to the concentration of oxygen in the bloodstream.

Like most organisms, humans depend on the intake of oxygen for aerobic respiration. Oxygen from the air enters the lungs. Then it diffuses into blood, which circulates through the body. The concentration gradient for oxygen is steepest in tissues that are serviced by capillaries, the blood vessels with the thinnest walls. The capillaries service living cells in all tissues. Most of the oxygen from the lungs moves out of capillaries, then into tissues, and on into cells. This cellular uptake lowers the oxygen concentration in blood. The drop is more pronounced during strenuous activity and at high altitudes.

A low oxygen concentration is harmful for people who have sickle cell anemia. Under such conditions, the abnormal hemoglobin molecules stick together into rodlike arrangements inside their red blood cells. The rods distort cells into a shape rather like a sickle (a short-handled farm tool with a crescent-shaped blade). The distorted blood cells rupture very easily, and their remains clog and rupture the blood capillaries. When that happens, the body's tissues become starved for oxygen. Besides this, carbon dioxide and other metabolic wastes accumulate.

Over time, the ongoing expression of the mutated gene can have multiple and damaging effects. Figure 9.12c tracks how successive changes in phenotype may

Figure 9.12 (**a**) A red blood cell from a person with the genetic disorder sickle-cell anemia. This scanning electron micrograph shows the surface appearance of the affected individual's red blood cells when the blood is adequately oxygenated. (**b**) This scanning electron micrograph shows the sickle shape that the red blood cells assume when the oxygen concentration in blood is low. (**c**) The range of symptoms that are characteristic of an individual who is homozygous recessive for the disorder.

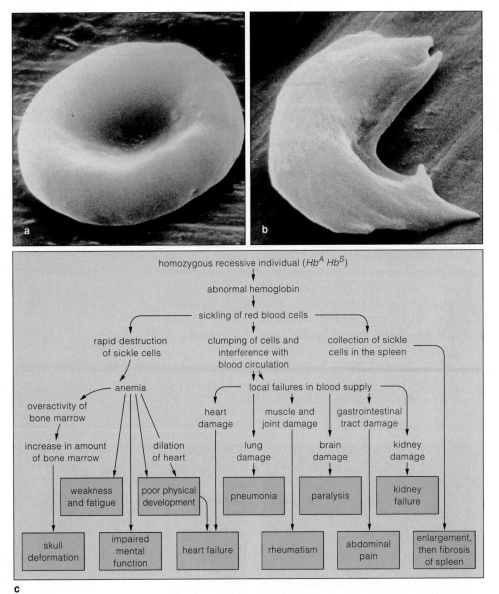

proceed. In chapters to come, you will read about other aspects of this genetic disorder.

The alleles at a single gene location may have positive or negative effects on two or more traits.

The effects may not be simultaneous. Rather, they may have repercussions over time. The gene may lead to an alteration in one trait, that change may alter another trait, and so on.

9.6 INTERACTIONS BETWEEN GENE PAIRS

Often a trait results from interactions among products of two or more gene pairs. For example, two alleles of one gene may mask expression of another gene's alleles, so some expected phenotypes may not appear. Such interactions are called **epistasis** (meaning the act of stopping).

Hair Color in Mammals

Epistasis is common among the gene pairs responsible for the coloration of fur or skin in mammals. Consider the black, brown, or yellow fur of Labrador retrievers (Figure 9.13). Variations in the amount and distribution of melanin, a brownish-black pigment, produce the differences in color. The enzymes and other products of many gene pairs affect different steps in the production of melanin and its deposition in certain body regions.

The alleles of one gene specify an enzyme required to produce melanin. Expression of allele *B* (black) has a more pronounced effect and is dominant to *b* (brown). Alleles of a different gene control the extent to which molecules of melanin will be deposited in a retriever's hairs. Allele *E* permits full deposition. Two recessive alleles (*ee*) reduce deposition, and the fur will be yellow.

The interactions between these two gene pairs may not even be possible if a certain allelic combination exists at still another gene locus. That gene (*C*) codes for tyrosinase, the first of several enzymes in the metabolic pathway that produces melanin. An individual with one or two dominant alleles (*CC* or *Cc*) can produce the functional enzyme. An individual with two recessive alleles (*cc*) cannot. When the biosynthetic pathway for melanin production is blocked, *albinism*—the absence of melanin—is the resulting phenotype (Figure 9.14).

Figure 9.13 Coat color of Labrador retrievers, as dictated by interactions among alleles of two gene pairs. Allele *B* (black) of one kind of gene involved in melanin production is dominant to allele *b* (brown). Allele *E* of a different gene allows melanin pigment to be deposited in individual hairs. Two recessive alleles (*ee*) of this gene block deposition, and a yellow coat results.

F₁ offspring of a dihybrid cross produce *F₂* offspring in a 9:3:4 ratio, as the Punnett-square diagram shows. The yellow Labrador in photograph (**b**) probably has genotype *BB ee*, because it can produce melanin but can't deposit pigment in hairs. Looking at that photograph, can you say why?

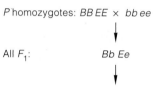

P homozygotes: *BB EE* × *bb ee*

All *F₁*: *Bb Ee*

F₂ combinations possible:

	BE	Be	bE	be
BE	BB EE	BB Ee	Bb EE	Bb Ee
Be	BB Ee	BB ee	Bb Ee	Bb ee
bE	Bb EE	Bb Ee	bb EE	bb Ee
be	Bb Ee	Bb ee	bb Ee	bb ee

RESULTING PHENOTYPES:
- 9/16 or 9 black
- 3/16 or 3 brown
- 4/16 or 4 yellow

a Black Labrador

b Yellow Labrador

c Chocolate Labrador

Figure 9.14 A rare albino rattlesnake. Like other animals that cannot produce melanin, it has pink eyes and its body is white, overall. (Eyes look pink because the absence of melanin from a tissue layer in the eyeball allows red light to be reflected from blood vessels in the eyes.) In birds and mammals, surface coloration results from pigments in feathers, fur, or skin. In fishes, amphibians, and reptiles, color-bearing cells give skin its surface coloration. Some of the cells contain melanin pigments or yellow-to-red pigments. Others contain crystals that reflect light and alter the effect of other pigments present.

The mutation affecting melanin production in the snake shown here had no effect on the production of yellow-to-red pigments and light-reflecting crystals. So the snake's skin appears iridescent yellow as well as white.

a Walnut comb   b Rose comb   c Pea comb   d Single comb

Figure 9.15 Interaction between two genes affecting the same trait in domestic breeds of chickens. The initial cross is between a Wyandotte (with a rose comb, **b**, on the crest of its head) and a brahma (pea comb, **c**). With complete dominance at the locus for pea comb and at the locus for rose comb, products of the two gene pairs interact and give rise to a walnut comb (**a**). With complete recessiveness at both loci, products interact and give rise to a single comb (**d**).

Comb Shape in Poultry

In some cases, interaction between two pairs of genes results in a phenotype that neither pair can produce alone. The geneticists W. Bateson and R. Punnett identified two interacting gene pairs (R and P) that produce comb shape in chickens. The allelic combinations of rr at one chromosome locus and pp at the other locus result in the least common phenotype, the single comb. The presence of dominant alleles (R, P, or both) results in varied phenotypes. The Figure 9.15 diagram shows the combinations of alleles that lead to rose, pea, or walnut combs.

Genes often interact, as when alleles of one gene mask the expression of another gene, and some expected phenotypes may not appear at all.

LESS PREDICTABLE VARIATIONS IN TRAITS

Regarding the Unexpected Phenotype

As Mendel demonstrated, the phenotypic effects of one or two pairs of certain genes show up in predictable ratios from one generation to the next. Certain interactions among gene pairs also may produce phenotypes in predictable ratios, as the example of Labrador coat color clearly demonstrated.

However, even when you are tracking a single gene through the generations, the resulting phenotypes may be not quite as expected. Consider *camptodactyly*, a rare, genetic abnormality that affects the shape and movement of fingers. Some people who carry the gene for this heritable trait have immobile, bent fingers on both hands. Other people have immobile, bent fingers on the left or right hand only. Still others carry the mutated gene—yet their fingers are not affected.

Why the confounding variation? Bear in mind, most pathways for synthesizing biological compounds are a series of small metabolic steps—*and an enzyme that is the product of a gene regulates each step*. Maybe one gene is mutated in one of possibly a number of different ways. Maybe the product of another gene blocks the pathway or causes it to run nonstop, or not long enough. Or maybe poor nutrition or some other variable condition related to the environment affects a crucial enzyme in the pathway. These are the sorts of factors that might introduce less predictable variations in the phenotypes that result from gene expression.

Continuous Variation in Populations

Generally, the individuals of a population show a range of small differences in most traits. We call this condition **continuous variation**. The greater the number of genes and environmental factors that can influence a trait, the more continuous will be the expected distribution of all the versions of that trait.

Think about the color of your eyes. As is true of all humans, the color is the cumulative outcome of many genes that are involved in the stepwise production and distribution of the light-absorbing pigment melanin. Black eyes have abundant deposits of melanin molecules in the iris, and these molecules absorb most of the incoming light. Dark brown eyes have smaller deposits of melanin, so some light is not absorbed but rather is reflected from the iris. Pale brown or hazel eyes have even less (Figure 9.16).

Green, gray, and blue eyes do not have green, gray, or blue pigments. In these cases, the iris contains different quantities of melanin, but not very much of it. As a result, many or most of the blue wavelengths of light that enter the eye are reflected out.

How can you describe the continuous variation of a trait within a group—say, the students in Figure 9.17? They range from very short to very tall, with average heights more common than either extreme. Start out by dividing the full range of different phenotypes into measurable categories. Next, count all the individual students in each category. This gives you the relative frequencies of phenotypes that are distributed across the range of measurable values.

The bar chart in Figure 9.17c plots the proportion of students in each category against the range of measured phenotypes. The vertical bars that are shortest represent categories with the least number of students. The bar that is tallest represents the category with the greatest

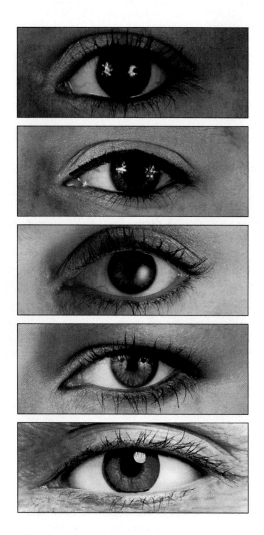

Figure 9.16 Samples from the range of continuous variation in human eye color. Different gene pairs interact to produce and deposit melanin. Among other things, this pigment helps color the eye's iris.

Different combinations of alleles result in small differences in eye color. So the frequency distribution for the eye-color trait appears to be continuous over the range from black to light blue.

1	4	8	10	16	16	15	15	14	13	13	11	9	8	8	5	1	2

number of individuals

60	61	62	63	64	65	66	67	68	69	70	71	72	73	74	75	76	77
(5 feet)																	

height (inches)

a Students at Brigham Young University in a splendid example of continuous variation in height.

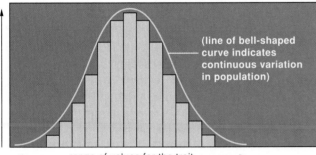

(line of bell-shaped curve indicates continuous variation in population)

range of values for the trait

b Idealized bell-shaped curve for a population that displays continuous variation in some trait.

number of individuals with some value of the trait

1	4	8	10	16	16	15	15	14	13	13	11	9	8	8	5	1	2

range of values for height

c Bell-shaped curve corresponding to the distribution of the trait (height) illustrated by the photograph in (**a**).

Figure 9.17 Continuous variation in a trait that is characteristic of the human population.

(**a**) Suppose you want to know the frequency distribution for height in a group of 169 biology students at Brigham Young University. You decide on how finely the range of possible heights should be divided. Then you measure each student and assign her or him to the appropriate category. Finally, you divide the number in each category by the total number of all students in all categories.

(**b**) Often a bar graph is used to depict continuous variation in a population. In such graphs, the proportion of individuals in each category is plotted against the range of measured phenotypes. Notice the curved line above the bars. It is an idealized example of the kind of "bell-shaped" curve that emerges for populations showing continuous variation in a trait. The bell-shaped curve in (**c**) is a specific example of this type of diagram.

number of students. Finally, draw a graph line around all of the bars and you end up with a "bell-shaped" curve. Such curves are typical of populations that show continuous variation in a trait.

Enzymes and other products of genes regulate each step of most metabolic pathways. Mutations, gene interactions, and environmental conditions may affect one or more of the steps, and this leads to variations in phenotypes.

In general, individuals of a population show continuous variation—that is, a range of small differences in most traits.

The greater the number of genes and environmental factors that can influence a trait, the more continuous will be the expected distribution of all the versions of that trait.

EXAMPLES OF ENVIRONMENTAL EFFECTS ON PHENOTYPE

We have mentioned in passing that environmental conditions may well contribute to variable gene expression among individuals of a population. Before leaving this chapter, consider just two examples of environmentally induced variations in phenotype.

Figure 9.19 Effect of different environmental conditions on gene expression in plants. Leaves growing from submerged stems of the water buttercup (*Ranunculus aquatilis*) are more finely divided than leaves growing in air. The variation appears even in the *same* leaf if its growth and development happen to proceed half in and half out of the water.

Figure 9.18 Effect of different environmental conditions on gene expression in animals. A Himalayan rabbit normally has black hair only on its long ears, nose, tail, and lower leg limbs. In one experiment, a patch of a rabbit's white fur was plucked clean, then an icepack was secured over the hairless patch. Where the colder temperature had been maintained, the hairs that grew back were black.

Himalayan rabbits are homozygous for the *ch* allele of a gene that codes for tyrosinase (an enzyme needed to produce melanin). The allele specifies a heat-sensitive version of the enzyme that functions only when temperatures are below about 33°C. When hairs grow under warmer conditions, they appear light (no melanin is produced). Light fur normally covers warm body regions. Ears and other slender extremities are cooler; they tend to lose metabolic heat more rapidly.

Possibly you have seen some Himalayan rabbits and probably some Siamese cats. Both of these furry mammals carry an allele that specifies a heat-sensitive version of one of the enzymes necessary for melanin production. At the surface of warm body regions, the enzyme is less active. There, fur grows in lighter than it does at cooler regions, which include the ears and other extremities. The experiment that is illustrated in Figure 9.18 provided observable evidence of the variation.

The environment also influences the genes that govern plant growth. You may have observed some leaves of water buttercups growing half in and half out of a pond. As Figure 9.19 indicates, the submerged leaves have a notably different appearance, compared to the leaves growing above the water's surface. In this type of plant, the genes responsible for leaf shape produce very different phenotypes under the two different sets of physical and chemical conditions imposed by air and water.

And so we conclude this chapter, which has dealt with heritable and environmental factors that give rise to variations in phenotype. What is the take-home lesson? Simply this:

Owing to gene mutations, cumulative gene interactions, and environmental effects on genes, individuals of a population show degrees of variation for many traits.

SUMMARY

1. A gene is a unit of information about a heritable trait. The alleles of a gene are different molecular versions of that information. Through many experimental crosses with pea plants, Mendel gathered evidence that diploid organisms have two genes for each trait and that genes retain their identity when transmitted to offspring.

2. Mendel performed monohybrid crosses (between two true-breeding plants showing different versions of a single trait). The crosses provided indirect evidence that some forms of a gene may be dominant over other, recessive forms.

3. Homozygous dominant individuals have two dominant alleles (AA) for the trait being studied. Homozygous recessives have two recessive alleles (aa). Heterozygotes have two nonidentical alleles (Aa).

4. In Mendel's monohybrid crosses ($AA \times aa$), all F_1 offspring were Aa. Crosses between F_1 plants resulted in these combinations of alleles in F_2 offspring:

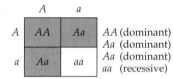

AA (dominant)
Aa (dominant)
Aa (dominant)
aa (recessive)

This produced the expected phenotypic ratio of 3:1.

5. The results from Mendel's monohybrid crosses led to the formulation of a theory of segregation. In modern terms, diploid organisms have pairs of genes, on pairs of homologous chromosomes. The two genes of each pair segregate from each other during meiosis, such that each gamete formed ends up with one or the other.

6. Mendel performed dihybrid crosses (between two true-breeding plants showing different versions of two traits). Results were close to a 9:3:3:1 phenotypic ratio:

9 dominant for both traits
3 dominant for *A*, recessive for *b*
3 dominant for *B*, recessive for *a*
1 recessive for both traits

7. Mendel's dihybrid crosses led to the formulation of a theory of independent assortment. In modern terms, by the end of meiosis, the gene pairs of two homologous chromosomes have been sorted for distribution into one gamete or another independently of how the gene pairs of other chromosomes were sorted out.

8. Four factors may influence gene expression:
a. Degrees of dominance may exist between some gene pairs.
b. Gene pairs can interact to influence the same trait.
c. A single gene can have positive or negative effects on two or more traits.
d. Environmental conditions affect gene expression.

Review Questions

1. Do segregation and independent assortment proceed during mitosis, meiosis, or both? *143, 145*

2. Define the difference between these terms. *141*
a. gene and allele
b. dominant allele and recessive allele
c. homozygote and heterozygote
d. genotype and phenotype

3. Define true-breeding. What is a hybrid? *141*

4. Distinguish between monohybrid and dihybrid crosses. What is a testcross, and why is it useful in genetic analysis? *142, 143, 144*

5. What do the vertical and horizontal arrows of this diagram represent? What do the bars and the curved line represent? *151*

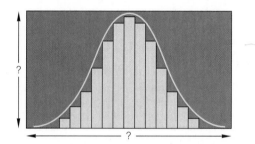

Self-Quiz *(Answers in Appendix IV)*

1. Alleles are _____ .
a. different molecular forms of a gene
b. different molecular forms of a chromosome
c. self-fertilizing, true-breeding homozygotes

2. A heterozygote has a _____ for the trait being studied.
a. pair of identical alleles
b. pair of nonidentical alleles
c. haploid condition, in genetic terms
d. a and c

3. The observable traits of an organism are its _____ .
a. phenotype   c. genotype
b. sociobiology   d. pedigree

4. Offspring of a monohybrid cross $AA \times aa$ are _____ .
a. all AA c. all Aa
b. all aa d. 1/2 AA and 1/2 aa

5. Second-generation offspring from a cross are the _____ .
a. F_1 generation   c. hybrid generation
b. F_2 generation   d. none of the above

6. Assuming complete dominance, offspring of the cross $Aa \times Aa$ will show a phenotypic ratio of _____ .
a. 3:1   c. 9:1
b. 1:2:1   d. 9:3:3:1

7. Crosses between F_1 individuals resulting from the cross $AABB \times aabb$ lead to F_2 phenotypic ratios close to _____ .
a. 1:2:1   c. 3:1
b. 1:1:1:1   d. 9:3:3:1

8. Match each example with its appropriate description.
____ dihybrid cross   a. *bb*
____ monohybrid cross   b. $AaBb \times AaBb$
____ homozygous condition c. *Aa*
____ heterozygous condition d. $Aa \times Aa$

Critical Thinking—Genetics Problems

(Answers in Appendix III)

1. One gene has alleles *A* and *a*. Another has alleles *B* and *b*. For each genotype listed, what type(s) of gametes will be produced? (Assume independent assortment occurs before gametes form.)

    a. *AA BB*   c. *Aa bb*
    b. *Aa BB*   d. *Aa Bb*

2. Still referring to Problem 1, what will be the genotypes of the offspring from the following matings? Indicate the frequencies of each genotype among them.

    a. *AA BB* × *aa BB*   c. *Aa Bb* × *aa bb*
    b. *Aa BB* × *AA Bb*   d. *Aa Bb* × *Aa Bb*

3. In one experiment, Mendel crossed a pea plant that bred true for green pods with one that bred true for yellow pods. All the F_1 plants had green pods. Which form of the trait (green or yellow pods) is recessive? Explain how you arrived at your conclusion.

4. Return to Problem 1, and assume you now study a third gene having alleles *C* and *c*. For each genotype listed, what type(s) of gametes will be produced?

    a. *AA BB CC*   c. *Aa BB Cc*
    b. *Aa BB cc*   d. *Aa Bb Cc*

5. Mendel crossed a true-breeding tall, purple-flowered pea plant with a true-breeding dwarf, white-flowered plant. All F_1 plants were tall and had purple flowers. If an F_1 plant self-fertilizes, what is the probability that a randomly selected F_2 plant is heterozygous for the genes specifying height and flower color?

6. At a certain gene location on a human chromosome, a dominant allele controls whether you can curl up the sides of your tongue (Figure 9.20). People who are homozygous for the recessive allele cannot roll their tongue. At a different gene location, a dominant allele controls whether earlobes are attached or detached (refer to Figure 9.1). These two gene pairs assort independently. Suppose a tongue-rolling, detached-earlobed woman marries a man who has attached earlobes and cannot roll his tongue. Their first child has the father's phenotype. Given this outcome,

    a. What are the genotypes of the mother, father, and child?
    b. What is the probability that a second child will have detached earlobes and won't be a tongue roller?

7. Bill and his wife, Marie, plan to have children. Both have flat feet and long eyelashes, and tend to sneeze a lot (the *achoo syndrome*). A dominant allele gives rise to each of these traits: *A* (foot arch), *E* (eyelash length), and *S* (chronic sneezing). With respect to all three traits, Bill is heterozygous and Marie is homozygous.

    a. What is Bill's genotype? Marie's genotype?
    b. Marie gets pregnant four times. What is the probability that each child will show the three phenotypes of Bill? Of Marie?
    c. What is the probability that each child will have short lashes, high arches, and no chronic tendency to sneeze?

8. Prior to DNA fingerprinting, described in Chapter 13, attorneys often relied on the ABO blood-typing system to settle disputes over paternity. Suppose, as a geneticist, you are called to testify during a case in which the mother has type A blood, the child has type O blood, and the alleged father has type B blood. How would you respond to the following statements?

    a. Attorney of the alleged father: The mother's blood is type A, so the child's type O blood must have come from the father. Because my client has type B blood, he could not be the father.
    b. Mother's attorney: Because further tests prove this man is heterozygous, he must be the father.

9. Suppose you identify a new gene in mice. One of its alleles specifies white fur color. A second allele specifies brown fur color. You are asked to determine whether the relationship between the two alleles is one of simple dominance or incomplete dominance. What sorts of genetic crosses would give you the answer? On what types of observations would you base your conclusions?

10. Your sister moves away and gives you her magnificent pure-bred Labrador retriever, a female named Dandelion. Suppose you decide to breed Dandelion and sell the puppies to help pay for your college tuition. But then you discover that two of her four brothers and sisters have developed a hip disorder. If Dandelion mates with a male Labrador that is known to be free of the harmful allele, can you guarantee to a purchaser that the puppies will not carry that allele? Explain your answer.

11. A dominant allele *W* confers black fur on guinea pigs. If a guinea pig is homozygous recessive (*ww*), it has white fur. Fred would like to know whether his pet black-furred guinea pig is homozygous dominant (*WW*) or heterozygous (*Ww*). How might he determine his pet's genotype?

12. Red-flowering snapdragons are homozygous for a certain allele (R^1R^1). White-flowering snapdragons are homozygous for a different allele (R^2R^2). The heterozygotes (R^1R^2) bear pink flowers. What phenotypes should appear among the F_1 offspring of the following crosses? What are the expected proportions for each phenotype?

    a. R^1R^1 × R^1R^2 c. R^1R^2 × R^1R^2
    b. R^1R^1 × R^2R^2 d. R^1R^2 × R^2R^2

(Note: in cases of incomplete dominance, it is inappropriate to refer to either allele of a pair as dominant or recessive. When the phenotype of a heterozygote is halfway between those of the two homozygotes, then there is no dominance. Such alleles are usually designated with superscript numerals, as shown above, rather than by uppercase letters for dominance and lowercase letters for recessiveness.)

Figure 9.20 A student at San Diego State University exhibiting the tongue-rolling trait for the benefit of a tongue-roll-challenged student.

13. In chickens, two pairs of genes affect the comb type (see Figure 9.15). When both genes are recessive, the chicken will have a single comb. The dominant allele of one gene, P, gives rise to a pea comb. The dominant allele of the other (R) gives rise to a rose comb. An epistatic interaction occurs when a chicken has at least one of both dominants, $P - R -$, producing a walnut comb. Predict the F_1 ratios resulting from a cross between two walnut-combed chickens that are heterozygous for both genes ($PpRr$).

14. An allele codes for a mutated form of hemoglobin (Hb^S instead of Hb^A). Homozygotes ($Hb^S Hb^S$) develop sickle cell anemia. But heterozygotes ($Hb^A Hb^S$) show few outward symptoms.

Suppose a female's mother is unaffected by this genetic disorder, but her father is homozygous for the Hb^S allele. She marries a male who is heterozygous for the allele, and they plan to have children. For *each* pregnancy, state the probability that this couple will have a child who is:

   a. homozygous for the Hb^S allele
   b. homozygous for the Hb^A allele
   c. heterozygous $Hb^A Hb^S$

15. Certain dominant alleles are so crucial for normal development that an individual who is homozygous recessive for a mutant recessive form of the allele cannot survive. Such recessive, *lethal alleles* can be perpetuated by heterozygotes. Consider the Manx allele (M^L) in cats. Homozygous cats ($M^L M^L$) die when they are still embryos inside the mother cat. In heterozygous cats ($M^L M$), the spine develops abnormally, and the cats end up with no tail whatsoever (Figure 9.21).

Suppose that two $M^L M$ cats mate. Among their *surviving* progeny, what is the probability that any individual kitten will be heterozygous?

Figure 9.21 Manx cat.

16. A recessive allele a is responsible for albinism, an inability to produce or deposit melanin in tissues. Humans and some other organisms can have this phenotype (Figure 9.22). In each of the following cases, what are the possible genotypes of each parent and of their children?

   a. Both parents have normal phenotypes; some of their children are albino and others are unaffected.
   b. Both parents are albino and have only albino children.

Figure 9.22 An albino male in India.

   c. The woman is unaffected, the man is albino, and they have one albino and three unaffected children. (What gives rise to this 3:1 ratio?)

17. Kernel color in wheat plants is determined by two pairs of genes. Alleles of one pair show incomplete dominance over alleles of the other pair. For the gene pair at one chromosome location (locus), allele A^1 gives one dose of red color to the kernel, whereas allele A^2 does not. At the second locus, allele B^1 gives one dose of red color to the kernel, whereas allele B^2 does not. A kernel with genotype $A^1 A^1 B^1 B^1$ is dark red. A kernel with genotype $A^2 A^2 B^2 B^2$ is white. All other genotypes have kernel colors in between these two extremes.

   a. If you cross a plant grown from a dark-red kernel with one grown from a white kernel, what genotypes and phenotypes would be expected among the offspring? In what proportions?

   b. If a plant with genotype $A^1 A^2 B^1 B^2$ self-pollinates, what genotypes and phenotypes would be expected among the offspring? In what proportions?

Selected Key Terms

allele *141*
codominance *146*
continuous variation *150*
dihybrid cross *144*
epistasis *148*
gene *141*
genotype *141*
heterozygous *141*
homozygous dominant *141*
homozygous recessive *141*
incomplete dominance *146*
independent assortment *145*
monohybrid cross *142*
multiple allele system *146*
phenotype *141*
pleiotropy *147*
probability *142*
Punnett-square method *143*
testcross *143*
true-breeding *141*

Readings

Cummings, M. 1994. *Human Heredity*. Third edition. St. Paul: West.

Mendel, G. 1966. "Experiments on Plant Hybrids." Translation in C. Stern and E. Sherwood, *The Origin of Genetics: A Mendel Source Book*. New York: Freeman.

Orel, V. 1984. *Mendel*. New York: Oxford University Press.

10 CHROMOSOMES AND HUMAN GENETICS

Too Young To Be Old

Imagine being ten years old, trapped in a body that with each passing day becomes a bit more shriveled, more frail, *old*. You are just tall enough to peer over the top of the kitchen counter, and you weigh less than thirty-five pounds. Already you are bald, and your nose has become crinkled and beaklike. Possibly you have only a few more years to live. Yet, like Mickey Hayes and Fransie Geringer, you still play, laugh, and hug your friends (Figure 10.1).

Of every 8 million newborns, one is destined to grow old far too soon. That rare individual possesses a mutated gene on just one of the forty-six chromosomes inherited from its mother or father. Through hundreds, thousands, then many billions of DNA replications and mitotic cell divisions, terrible information encoded in that gene was methodically distributed to every cell in the growing human embryo, then in the newborn. The outcome is accelerated aging and a greatly reduced life expectancy. These are defining features of *Hutchinson-Gilford progeria syndrome*. There is no cure.

The mutation is the start of severe disruptions in the gene interactions that bring about bodily growth and development. Outwardly apparent symptoms begin to emerge when the child is less than two years old. Skin that should be plump and resilient becomes thinned. Muscles become flabby. Limb bones that should lengthen and become stronger start to soften instead. Hair loss is pronounced and typically ends in extremely premature baldness.

Most progeriacs die in their early teens as a result of strokes or heart attacks. These final insults are brought on by a hardening of the arteries—a condition that is typical of advanced age.

There are no documented cases of progeria running in families, which suggests that the gene must mutate spontaneously, at random. The mutated gene apparently is dominant over its partner on the homologous chromosome.

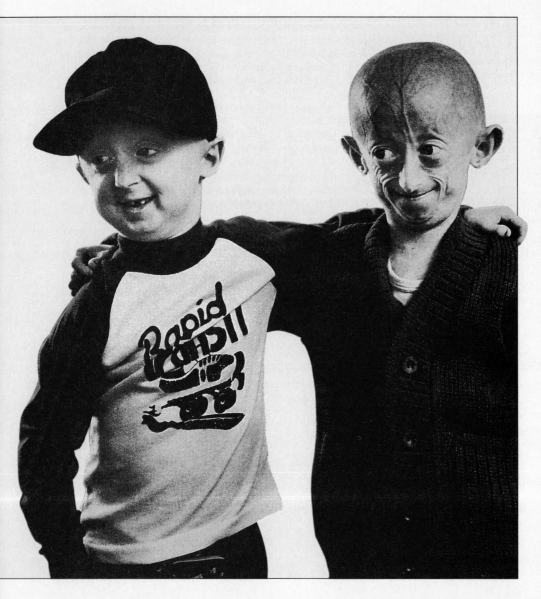

Figure 10.1 Two boys who met during a gathering of progeriacs at Disneyland, California, when they were not yet ten years old. Progeria, a heritable disorder, arises through a mutation in a dominant allele. It is characterized by accelerated aging and reduced life expectancy.

We began this unit of the book by looking at cell division, the starting point of inheritance. Then we started thinking about how chromosomes—and the genes they carry—are shuffled during meiosis and fertilization. In this chapter we delve more deeply into patterns of chromosomal inheritance. If at times the described methods of analysis seem remote from the world of your interests, remember this: For better or worse, *an inherited collection of specific bits of information gives rise to traits that help define each organism*—and this includes yourself and other human individuals. When Mickey Hayes turned all of eighteen, he was the oldest living progeriac. Fransie was seventeen when he died.

So in that sense, our story of Mickey and Fransie takes us back to the 1880s—and to the rediscovery of Gregor Mendel's studies of inheritance. In 1884 Mendel himself had just passed away, and his paper on pea plants had been gathering dust in a hundred libraries for nearly two decades. However, improvements in the resolving power of microscopes had rekindled efforts to locate the hereditary material within cells. By 1882, Walther Flemming had observed threadlike bodies—chromosomes—inside the nuclear region of dividing cells. By 1884, a question was taking shape: Could chromosomes be the hereditary material?

Then researchers realized each gamete has half the number of chromosomes of a fertilized egg. In 1887, August Weismann proposed that a special division process must reduce the chromosome number by half before gametes form. Sure enough, in that same year meiosis was discovered. Weismann began to promote his theory of heredity: The chromosome number is halved during meiosis, then restored at fertilization. Thus a cell's hereditary material is half paternal in origin and half maternal. His theory was hotly debated, and it prompted a flurry of experimental crosses—just like the ones Mendel had carried out.

Finally, in 1900, researchers came across Mendel's paper while checking literature related to their own genetic crosses. To their surprise, their experimental results confirmed what Mendel's results had already suggested: Diploid cells have two units (genes) for each heritable trait, and the two units are segregated before gametes form.

During the decades that followed, researchers learned a great deal more about chromosomes. Let's turn now to a few high points of their work, which will serve as background for our understanding of human inheritance.

KEY CONCEPTS

1. One gene follows another in sequence along the length of a chromosome. Each gene has its own position in that sequence.

2. The combination of alleles in a chromosome does not necessarily remain intact through meiosis and gamete formation. During an event called crossing over, some alleles in the sequence swap places with their partners on the homologous chromosome. The alleles that are swapped may or may not be identical.

3. The structure of a chromosome may change, as when a chromosome segment is deleted, duplicated, inverted, or moved to a new location. Also, the chromosome number may change as a result of improper separation of duplicated chromosomes during meiosis or mitosis.

4. Allele shufflings, changes in chromosome structure, and changes in chromosome number contribute to variation in traits. Often these events lead to genetic abnormalities or disorders.

THE CHROMOSOMAL BASIS OF INHERITANCE—AN OVERVIEW

Genes and Their Chromosome Locations

From the preceding chapters in this unit, you already are familiar with some concepts concerning the structure of chromosomes and their behavior at meiosis. Let's now integrate these concepts with a few points that help explain patterns of human inheritance:

1. **Genes** are units of information about heritable traits. Each gene has a particular location in a particular type of chromosome.

2. The genes of each eukaryotic species are distributed among a number of chromosomes.

3. In many species, cells that give rise to gametes are diploid ($2n$); they have chromosome pairs. Typically, one chromosome of each type is inherited from the mother, and its homologue from the father.

4. The pairs are **homologous chromosomes**. That is, one of each pair has the same length, shape, and gene sequence as its partner and interacts with it during meiosis. The two forms of sex chromosomes (X and Y) are homologous in one small region only.

5. A pair of homologous chromosomes may carry identical or nonidentical alleles at corresponding locations in their gene sequence. **Alleles** are slightly different molecular forms of the same gene, and they arise through mutation.

6. Genes close together on the same chromosome *tend* to stay together through meiosis. But chromosomes break and exchange corresponding segments with their homologous partners. We call this event **crossing over**. Such an event puts new combinations of alleles into chromosomes; in other words, it results in **genetic recombination**.

7. At metaphase I of meiosis, all the maternal and paternal chromosomes are positioned at the equator of the microtubular spindle—but not in any particular pattern. As a result of this random alignment, gametes and then offspring end up with mixes of maternal and paternal chromosomes—and more mixing of alleles.

8. A chromosome's structure may change, as when a segment is deleted, duplicated, inverted, or moved to a new location. Besides this, improper separation of chromosomes that have been duplicated in advance of meiosis or mitosis may alter the chromosome number in gametes and the new individual.

9. For any trait, the particular combination of alleles that the new individual inherits may have neutral, beneficial, or harmful effects.

Autosomes and Sex Chromosomes

Generally, the two members of each pair of homologous chromosomes are exactly alike in their length, shape, and gene sequence. However, in the late 1800s, microscopists discovered an exception to this. For humans and many other organisms, a distinctive chromosome is present in females *or* males, but not in both.

Figure 10.2 in the *Focus* essay shows a lineup of the twenty-three pairs of human chromosomes. Notice the last two in the series, which are the **X chromosome** and the **Y chromosome**. As you can see, they are physically different; one is much shorter than the other. Despite the difference, the two are able to synapse in a small region, and this allows them to function as homologues during meiosis. If you are male, you have one X and one Y chromosome (designated XY). If you are female, you are XX; you have two X chromosomes.

The X and Y chromosomes of humans are examples of **sex chromosomes**. The term means that a particular combination of these two types of chromosomes determines gender (whether a new individual will become a male or a female). All of the other chromosomes in an individual's cells are the same in both sexes; they are designated **autosomes**.

Karyotype Analysis

Today, microscopists can routinely analyze the physical appearance of a cell's sex chromosomes and autosomes. Chromosomes, recall, are in the most highly condensed state at metaphase of mitosis. At that time, their size, length, and centromere location are easiest to identify. In addition, after microscopists stain them in certain ways, the chromosomes of many species show distinct banding patterns. The most condensed chromosome regions often take up more stain and therefore appear as darker bands. Figure 8.3 gives an example.

The number of metaphase chromosomes and their defining characteristics are called the **karyotype** of an individual (or species). The *Focus* essay explains how to construct a karyotype diagram. Such a diagram is a cut-up, rearranged photograph in which all the autosomes are lined up, largest to smallest, and sex chromosomes are positioned last.

All but one pair of chromosomes are the same in both males and females of a species. They are designated autosomes.

One pair of chromosomes—the sex chromosomes—governs a new individual's gender.

A karyotype for an individual (or species) is a preparation of metaphase chromosomes, sorted by their defining features.

PREPARING A KARYOTYPE DIAGRAM

Karyotype diagrams can help answer questions about an individual's chromosomes. Chromosomes are in their most condensed form—and easiest to identify—in cells that are proceeding through metaphase of mitosis. Technicians don't count on finding a cell that happens to be dividing in the body when they go looking for it. Instead, they culture cells *in vitro* (literally, "in glass"). They put a small sample of blood, skin, bone marrow, or some other tissue in a glass container. In that container is a solution that stimulates cell growth and mitotic cell division for many generations.

Dividing cells can be arrested at metaphase by adding colchicine to the culture medium. Colchicine is an extract of the autumn crocus (*Colchicum autumnale*). Technicians and researchers use it to block formation of microtubular spindles. This way, they prevent duplicated chromosomes from separating during nuclear division. With suitable colchicine concentrations and exposure time, many metaphase cells can accumulate, and this increases the chances of finding candidates for karyotype diagrams.

Following colchicine treatment, the culture medium is transferred to the tubes of a centrifuge, a motor-driven spinning device (Figure 10.2*a*). Cells have greater mass and density than the surrounding solution. As a result, the spinning force moves them farthest away from the motor, to the bottom of the attached tubes. Separation in response to the spinning force is called *centrifugation*.

1 Add sample (e.g. blood) to culture medium that has stimulator for mitosis. Incubate at 37°C. Add colchicine to arrest mitosis at metaphase.

(blood cells)

2 Transfer to centrifuge tube and spin down.

3 Remove the culture medium.

4 Add dilute saline solution, then fixative.

5 Gently suspend cells.

6 Prepare and stain cells for microscopy.

7 Put cells on microscope slide; observe.

8 Photograph, enlarge image of chromosomes.

9 Cut out each chromosome; arrange as a set.

a

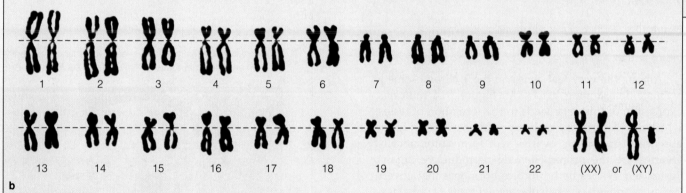

1 2 3 4 5 6 7 8 9 10 11 12

13 14 15 16 17 18 19 20 21 22 (XX) or (XY)

b

Afterward, the cells are transferred to a saline solution. When immersed in this hypotonic fluid, they swell up (by osmosis) and move apart. And so do the metaphase chromosomes. The cells are ready to be dropped onto a microscope slide, fixed (as by air-drying), and stained.

Chromosomes take up some stains uniformly throughout their length. This is enough to allow identification of chromosome size and shape. With other staining procedures, horizontal bands show up in the chromosomes of certain species. When ultraviolet light is directed at these chromosomes, the bands fluoresce. You can see an example of this in Figure 8.3, which is a human karyotype diagram.

Figure 10.2 (**a**) Karyotype preparation. (**b**) Human karyotype. Human somatic cells have 22 pairs of autosomes and 1 pair of sex chromosomes (XX or XY). That is a diploid number of 46. These are metaphase chromosomes; each is in the duplicated state.

In the last steps of karyotype preparation, metaphase chromosomes are photographed, then the microscope image is enlarged. The photographed chromosomes are cut apart, one at a time, then arranged according to their size, shape, and length of arms. Finally, the pairs of homologous chromosomes are horizontally aligned by their centromeres, as shown in Figure 10.2*b*.

10.3 SEX DETERMINATION IN HUMANS

Analyzing human gametes, as with karyotyping, yields evidence that each egg produced by a female carries one X chromosome. Half of the sperm cells produced by a male carry an X chromosome and half carry a Y.

If an X-bearing sperm fertilizes an X-bearing egg, the new individual will develop into a female. If the sperm happens to carry a Y chromosome, the individual will develop into a male (Figure 10.3).

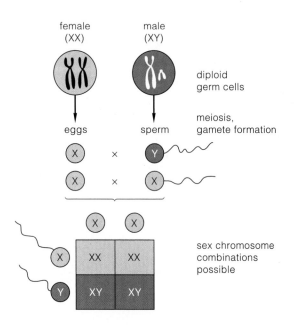

Figure 10.3   Pattern of sex determination in humans.

Among the very few genes on the Y chromosome is a "male-determining gene." As Figure 10.4 indicates, expression of this gene leads to the formation of testes, which are the primary male reproductive organs. In the absence of this gene, ovaries will form automatically. Ovaries are the primary female reproductive organs. Testes and ovaries both produce hormones that govern the development of particular sexual traits.

A human X chromosome carries more than 2,300 genes. Like other chromosomes, it carries some genes associated with sexual traits, such as the distribution of body fat and hair. However, most of its genes deal with *nonsexual* traits, such as blood-clotting functions. These genes can be expressed in males as well as females (males, remember, also carry one X chromosome).

A certain gene on the human Y chromosome dictates that a new individual will develop into a male. In the absence of the Y chromosome (and the gene), a female develops.

Figure 10.4   Boys, girls, and the Y chromosome.

For about the first four weeks of its existence, a human embryo has neither male nor female traits, even though it normally carries XY or XX chromosomes. However, ducts and other structures start forming that can go *either way*.

(**a–c**) In an XX embryo, the primary female reproductive organs (ovaries) start to form automatically—*in the absence of a Y chromosome*. By contrast, in an XY embryo, the primary male reproductive organs (testes) start to form during the next four to six weeks. Apparently, a gene region on the Y chromosome governs a fork in the developmental road that can lead to maleness.

The newly forming testes start to produce testosterone and other sex hormones. These hormones are crucial for the development of a male reproductive system. By contrast, in the XX embryo, the newly forming ovaries start to produce different kinds of sex hormones. These are crucial for the development of a female reproductive system.

The master gene for male sex determination is named SRY (short for *S*ex-determining *R*egion of the *Y* chromosome). So far, the same gene has been identified in DNA from male humans, chimpanzees, rabbits, pigs, horses, cattle, and tigers. None of the females tested had the gene. Tests with mice indicate that the gene region becomes active about the time that testes are starting to develop.

The SRY gene resembles regions of DNA that are known to specify regulatory proteins. As described in Chapter 12, such proteins bind to certain parts of DNA and so turn genes on and off. It seems that the SRY gene product regulates a whole cascade of reactions that are necessary for male sex determination.

umbilical cord (lifeline between embryo and maternal tissues)   embryo floating in protective, fluid-filled sac (the amnion)

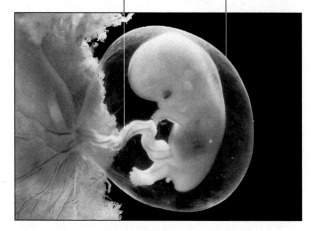

a   A human embryo, eight weeks old. Male reproductive organs have already started to develop.

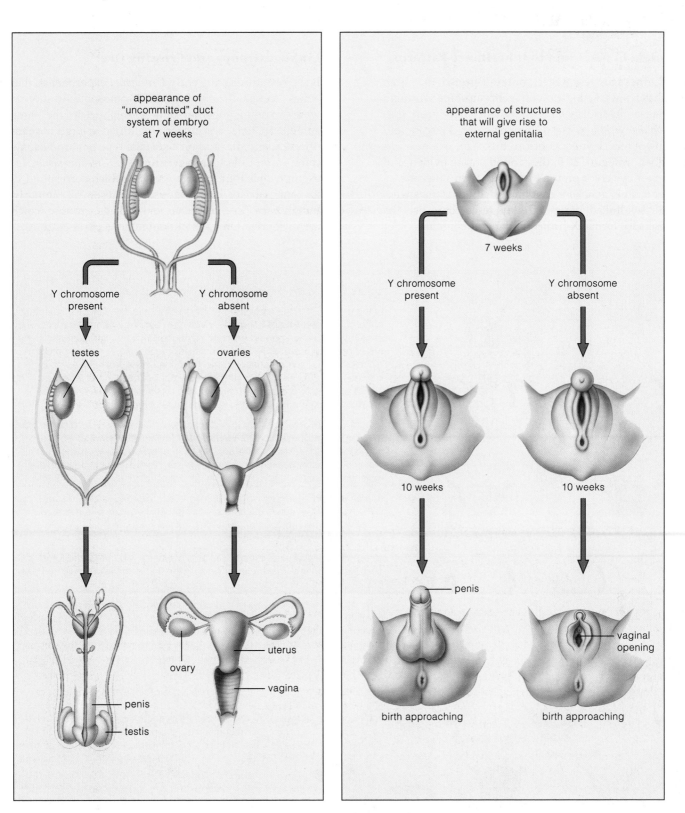

appearance of
"uncommitted" duct
system of embryo
at 7 weeks

Y chromosome
present

Y chromosome
absent

testes

ovaries

penis

testis

ovary

uterus

vagina

b The duct system in the early human embryo that may develop into the primary male reproductive organs *or* into the primary female reproductive organs.

appearance of structures
that will give rise to
external genitalia

7 weeks

Y chromosome
present

Y chromosome
absent

10 weeks

10 weeks

penis

vaginal
opening

birth approaching

birth approaching

c External appearance of the newly forming reproductive organs in human embryos.

X-Linked Genes: Clues to Inheritance Patterns

Some time ago, researchers only suspected that each gene has a specific location on a chromosome. Through a series of hybridization experiments with fruit flies (*Drosophila melanogaster*), Thomas Hunt Morgan and his coworkers helped confirm this. For example, as described in Figure 10.5, they found strong evidence of a gene for eye color on the *Drosophila* X chromosome.

For a time, genes located on sex chromosomes were called "sex-linked genes." Today researchers use the more precise terms, **X-linked** and **Y-linked genes**.

Linkage Groups and Crossing Over

It seemed, during the early *Drosophila* experiments, that genes "linked" to one type of chromosome or another were traveling together and ending up in the same gamete. In time, researchers identified a large number of genes, and the ones on a specific type of chromosome came to be called **linkage groups**. *D. melanogaster*, for example, has four linkage groups, which correspond to its four pairs of homologous chromosomes. Similarly, Indian corn (*Zea mays*) has ten linkage groups, which correspond to ten pairs of homologous chromosomes.

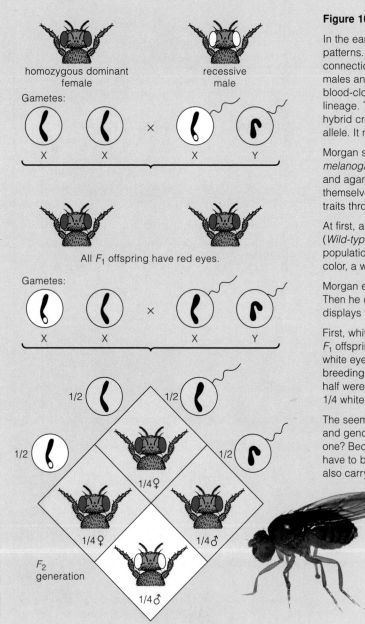

Figure 10.5   X-linked genes—clues to inheritance patterns.

In the early 1900s, the embryologist Thomas Morgan was studying inheritance patterns. He and his coworkers discovered an apparent genetic basis for the connection between gender and certain nonsexual traits. For example, human males and females both have blood-clotting mechanisms. Yet hemophilia (a blood-clotting disorder) shows up most often in males, not females, of a family lineage. This gender-specific outcome was not like anything Mendel saw in his hybrid crosses of pea plants. (Either parent plant could carry a recessive allele. It made no difference; the resulting phenotype was the same.)

Morgan studied eye color and other nonsexual traits of fruit flies (*Drosophila melanogaster).* These flies can be raised in bottles on cornmeal, molasses, and agar. A female lays hundreds of eggs in a few days, and offspring can themselves reproduce in less than two weeks. Morgan could track hereditary traits through nearly thirty generations of thousands of flies in a year's time.

At first, all the flies were wild-type for eye color; they had brick-red eyes. (*Wild-type* simply means the normal or most common form of a trait in a population.) Then, through an apparent mutation in a gene controlling eye color, a white-eyed male appeared.

Morgan established true-breeding strains of white-eyed males and females. Then he did paired, *reciprocal* crosses. In the first of these crosses, one parent displays the trait. In the second, the other parent displays it.

First, white-eyed males were mated with homozygous red-eyed females. All F_1 offspring had red eyes. But of the F_2 offspring, only some of the males had white eyes. In the second cross, white-eyed females were mated with true-breeding red-eyed males. Half of the F_1 offspring were red-eyed females and half were white-eyed males. Of the F_2 offspring, 1/4 were red-eyed females, 1/4 white-eyed females, 1/4 red-eyed males, and 1/4 white-eyed males!

The seemingly odd results implied a relationship between an eye-color gene and gender. Probably the gene was located on a sex chromosome. But which one? Because females (XX) could be white-eyed, the recessive allele would have to be on one of their X chromosomes. Suppose white-eyed males (XY) also carry the recessive allele on their X chromosome—and suppose there is no corresponding eye-color allele on the Y chromosome. The males would have white eyes, for they have no dominant allele to mask the effect of the recessive one.

The diagram shows the expected results when the idea of an X-linked gene is combined with Mendel's concept of segregation. By proposing that a specific gene is located on an X chromosome but not on the Y, Morgan was able to explain the outcome of his reciprocal crosses. His experimental results matched the predicted outcomes.

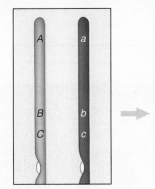
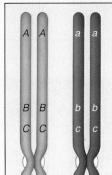

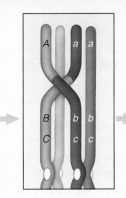
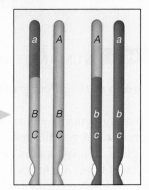

a Diploid (2n) cells have pairs of genes at corresponding locations, on pairs of homologous chromosomes. The two genes of each pair may or may not be identical. In this case, the alleles are nonidentical (A with a, B with b, and C with c).

b The same gene regions at interphase, after DNA replication. Both of the homologous chromosomes are in the duplicated state.

c At prophase I of meiosis, two of the nonsister chromatids break while tightly aligned together (see Figure 8.5).

d Nonsister chromatids swap segments, then enzymes seal the broken ends. The breakage and exchange represent one crossover event.

e The outcome of the crossover is genetic recombination between two of four chromatids. (They are shown after meiosis, as unduplicated, separate chromosomes.)

Figure 10.6 Crossing over. As shown in Figure 8.4, this event occurs in prophase I of meiosis. Here we see how it disrupts part of a linkage group (genes that otherwise would stay together on the same chromosome through meiosis and gamete formation).

As we now know, linked genes are vulnerable to crossing over (Figure 10.6). We also know that crossing over is not a rare event. In humans and most other eukaryotic species, meiosis cannot even be completed properly unless each pair of homologous chromosomes takes part in at least one crossover.

Imagine any two genes at two different locations on the same chromosome. *The probability that a crossover will disrupt their linkage is proportional to the distance that separates them.* Suppose genes A and B are twice as far apart as two other genes, C and D:

We would expect crossing over to disrupt the linkage between A and B much more often.

Two genes are very closely linked when the distance between them is small; they nearly always end up in the same gamete (Figure 10.7). Linkage is more vulnerable to crossover when the distance between genes is greater. When two genes are very far apart, crossing over disrupts linkage so often that those genes assort independently of each other into gametes.

The results of many experimental crosses tell us genes undergo recombination in fairly regular patterns. The patterns have been used to locate relative positions of genes along chromosomes, an activity called linkage mapping. For example, of the several thousand known genes in the four types of *Drosophila* chromosomes, the positions of about a thousand have been mapped. What about the 50,000 to 100,000 genes on the twenty-three types of human chromosomes? These must be mapped by methods described in Chapter 16. Humans, unlike

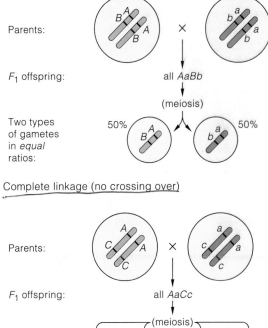

a Complete linkage (no crossing over)

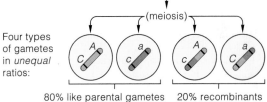

80% like parental gametes   20% recombinants

b Incomplete linkage owing to crossing over

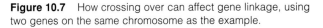

Figure 10.7 How crossing over can affect gene linkage, using two genes on the same chromosome as the example.

watermelon plants, do not lend themselves to experimental crosses. And with this jarring thought fresh in your mind, you are ready to begin the next section.

The farther apart two genes are on a chromosome, the greater will be the frequency of crossing over and recombination.

10.5 HUMAN GENETIC ANALYSIS

Constructing Pedigrees

Some organisms, including pea plants and fruit flies, are ideal for genetic analysis. They grow and reproduce rapidly in small spaces, under controlled conditions. It doesn't take long to track a trait through many generations. Humans are another story. We live under variable conditions in diverse environments. We select our own mates and reproduce if and when we want to. Human subjects live as long as the geneticists who study them, so tracking traits through generations can be tedious. Most human families are so small, there aren't enough offspring for easy inferences about inheritance.

To get around some of the problems associated with analyzing human inheritance, geneticists put together **pedigrees**. A pedigree is a chart that is constructed, by standardized methods, to show the genetic connections among individuals. Figure 10.8 shows examples, along with definitions of a few of the standardized symbols used to represent individuals.

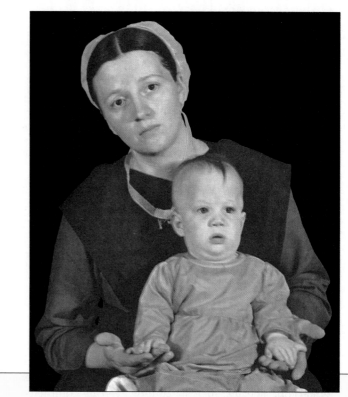

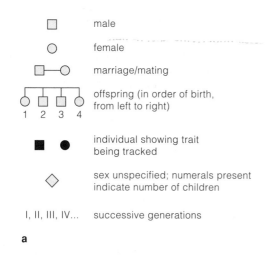

a

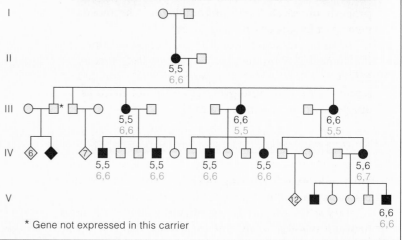

b

Figure 10.8 (**a**) Some symbols used in constructing pedigree diagrams. (**b**) Example of a pedigree for *polydactyly*, a condition in which a person has extra fingers, extra toes, or both. Expression of the gene for this trait can vary from one individual to the next. *Black* numerals signify the number of fingers on each hand where data were available; *blue* signify the number of toes on each foot.

(**c**) From the human genetic researcher Nancy Wexler, a pedigree for *Huntington disorder*, by which the nervous system progressively degenerates. Wexler's team constructed an extended family tree for nearly 10,000 people in Venezuela. Analysis of affected and unaffected individuals revealed that a dominant allele on human chromosome 4 is the genetic culprit. Wexler has a special interest in Huntington disorder; she herself has a fifty percent chance of developing it.

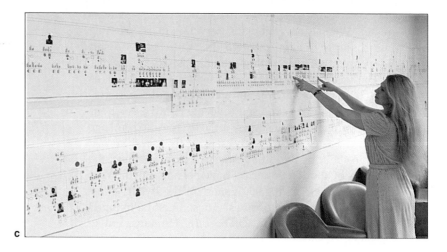

c

Table 10.1 Examples of Human Genetic Disorders and Genetic Abnormalities

Disorder or Abnormality*	Main Consequences
Autosomal Recessive Inheritance *non-sex chromosome.*	
Albinism *148*	Absence of pigmentation
Sickle-cell anemia *147*	Severe tissue, organ damage
Galactosemia *166*	Brain, liver, eye damage
Phenylketonuria *171*	Mental retardation
Autosomal Dominant Inheritance	
Achondroplasia *167*	One form of dwarfism
Achoo syndrome *154*	Chronic sneezing
Camptodactyly *150*	Rigid, bent little fingers
Familial hypercholesterolemia *202*	High cholesterol levels in blood, clogged arteries
Huntington disorder *164, 166*	Nervous system degenerates progressively, irreversibly
Polydactyly *164*	Extra fingers, toes, or both
Progeria *156, 166*	Drastic premature aging
X-Linked Dominant Inheritance *Sex chromosome.*	
Faulty enamel trait *167*	Problems with teeth
X-Linked Recessive Inheritance	
Hemophilia A *167*	Deficient blood-clotting
Duchenne muscular dystrophy *167*	Muscles waste away
Testicular feminization syndrome *576*	XY individual but female traits, sterility
Red-green color blindness *569*	Reds, greens appear to be the same color
Changes in Chromosome Structure	
Cri-du-chat *170*	Retardation, skewed larynx
Fragile X syndrome *170*	Mental retardation
Changes in Chromosome Number	
Down syndrome *168*	Mental retardation, heart defects
Turner syndrome *169*	Sterility, abnormal ovaries and sexual traits
Klinefelter syndrome *169*	Sterility, retardation
XYY condition *169*	Mild retardation in some; no symptoms in others

*Italic numbers indicate the pages on which the disorder is described.

[Handwritten annotations: "problem with processing galactose → sugar"; "can't make D-ピレッジ"; "change in number of chromosome"; "trisomy 21"; "XO"; "XXY"; "Edwards syndrome trisomy 18"; "XXX"; "no retardation"]

When analyzing pedigrees, geneticists rely on their knowledge of probability and Mendelian inheritance patterns, which might yield clues to the genetic basis for a trait. For instance, they might determine that the responsible allele is dominant or recessive, or that it is located on an autosome or a sex chromosome.

Gathering a great many family pedigrees increases the numerical base for analysis. When a trait clearly shows a simple Mendelian inheritance pattern, a geneticist may be justified in predicting the probability of its occurrence among the children of prospective parents. We will return to this topic later in the chapter.

Regarding Human Genetic Disorders

Table 10.1 lists some of the heritable traits that have been studied in detail. A few of these are abnormalities, or deviations from the average condition. Said another way, a **genetic abnormality** is nothing more than a rare, uncommon version of a trait, as when a person is born with six toes on each foot instead of five.

Whether we view an abnormal trait as disfiguring or merely interesting is subjective. There is nothing inherently life-threatening or even ugly about it.

By contrast, a **genetic disorder** is an inherited condition that results in mild to severe medical problems. The alleles underlying severe genetic disorders do not abound in populations, for they put individuals at great risk. They do not disappear entirely for two reasons. First, rare mutations can put new copies of the alleles in the population. Second, in heterozygotes, the harmful allele is paired with a normal one that might cover its functions, so it can be passed on to offspring.

You may hear someone refer to a genetic disorder as a disease, but the terms are not always interchangeable. "Diseases" result from infection by bacteria, viruses, or some other environmental agent that invades the body and multiplies inside its tissues. Illness follows only if the infection leads to tissue damage that interferes with normal body functions. When a person's genes increase susceptibility to infection or weaken the response to it, the resulting illness might be called a genetic disease.

With these qualifications in mind, we will turn next to some examples of patterns of inheritance in the human population.

For many genes, pedigree analysis might reveal simple Mendelian inheritance patterns that will allow inferences about the probability of their transmission to children.

A genetic abnormality simply is a rare or less common version of an inherited trait. A genetic disorder is an inherited condition that results in mild to severe medical problems.

10.6 PATTERNS OF INHERITANCE

Autosomal Recessive Inheritance

For some traits, inheritance patterns reveal two clues that point to a recessive allele on an autosome. *First*, if both parents are heterozygous, any child of theirs will have a 50 percent chance of being heterozygous and a 25 percent chance of being homozygous recessive (Figure 10.9). *Second*, if both parents are homozygous recessive, any child of theirs will be, also.

About 1 in 100,000 newborns are homozygous for the recessive allele that causes *galactosemia*. They cannot produce functional molecules of an enzyme that stops a product of lactose breakdown from accumulating to toxic levels. Lactose normally is converted to glucose and galactose, then to glucose-1-phosphate (which is broken down by glycolysis or converted to glycogen). In affected persons, the full conversion is blocked:

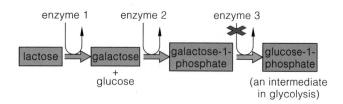

A high blood level of galactose can damage the eyes, liver, and brain. Malnutrition, diarrhea, and vomiting are early symptoms. Untreated galactosemics often die in childhood. The telling symptom—a high galactose level—can be detected in urine samples. If affected individuals are placed on a restricted diet that excludes dairy products, they can grow up symptom-free.

Autosomal Dominant Inheritance

Two clues of a different sort indicate that an autosomal dominant allele is responsible for a trait. *First*, the trait typically appears in each generation, for the allele is usually expressed even in heterozygotes. *Second*, if one parent is heterozygous and the other is homozygous recessive, there is a 50 percent chance that any child of theirs will be heterozygous (Figure 10.10).

A few dominant alleles cause serious genetic disorders. Even so, they persist in populations. Why? Some are perpetuated by spontaneous mutations. This is the case for progeria, the rare aging disorder described in the chapter introduction. In other cases, expression of the dominant allele may not interfere with reproduction or affected individuals have children before symptoms become severe.

For example, *Huntington disorder* is a progressive deterioration of the nervous system. Symptoms may not even begin until about age forty, and most people have already reproduced by then. As another example,

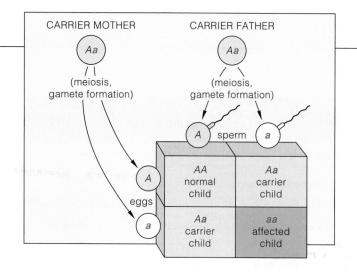

Figure 10.9 One pattern for autosomal recessive inheritance. In this example, both parents are heterozygous carriers of the recessive allele (shown in *red*).

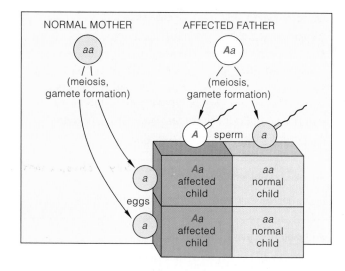

Figure 10.10 One pattern for autosomal dominant inheritance. Assume the dominant allele (*red*) is fully expressed in carriers.

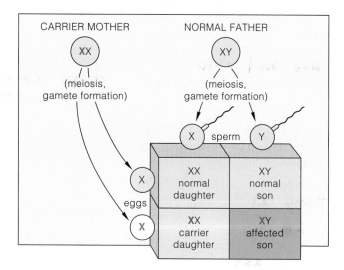

Figure 10.11 One pattern for X-linked inheritance. In this example, the mother carries a recessive allele on one of her X chromosomes (*red*).

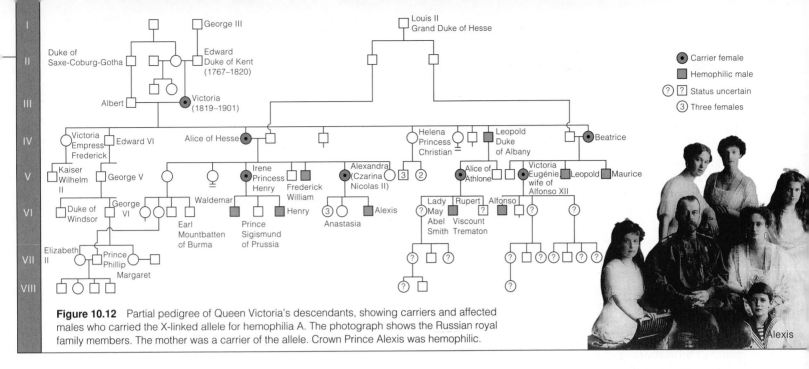

Figure 10.12 Partial pedigree of Queen Victoria's descendants, showing carriers and affected males who carried the X-linked allele for hemophilia A. The photograph shows the Russian royal family members. The mother was a carrier of the allele. Crown Prince Alexis was hemophilic.

achondroplasia affects about 1 in 10,000 humans. Usually, the homozygous dominant condition leads to stillbirth. Heterozygotes are able to reproduce. They just cannot form cartilage properly when limb bones are growing, so they end up with abnormally short arms and legs. Adults are less than 4 feet, 4 inches tall. The dominant allele often has no other phenotypic effects. In a few cases, severe symptoms do not appear until after the affected person reproduces and so transmits the allele.

X-Linked Recessive Inheritance

Distinctive clues often show up when a recessive allele on an X chromosome causes a trait. *First*, males show the recessive phenotype far more often than females do. A recessive allele can be masked in females, who may inherit a dominant allele on their other X chromosome. It can't be masked in males, who have one X chromosome only (Figure 10.11). *Second*, a son cannot inherit the recessive allele from his father. A daughter can. If a daughter is heterozygous, there is a 50 percent chance each son of hers will inherit the allele.

Hemophilia A, a blood-clotting disorder, is a case of X-linked recessive inheritance. In most people, a blood-clotting mechanism quickly stops bleeding from minor injuries. Reactions that result in clot formation require the products of several genes. If any of these genes is mutated, its defective product will allow bleeding to proceed for an abnormally long time. About 1 in 7,000 males inherit the gene for hemophilia A. Clotting time is close to normal in heterozygous females.

The frequency of hemophilia A was unusually high in royal families of nineteenth-century Europe. Queen Victoria of England was a carrier (Figure 10.12). At one time, the recessive allele was present in eighteen of her

sixty-nine descendants. A hemophilic great-grandchild, Crown Prince Alexis, was a focus of political intrigue that helped trigger the Russian revolution of 1917.

Another, rare case of X-linked recessive inheritance is *Duchenne muscular dystrophy*. Muscles in an affected person slowly enlarge with fat and connective tissue deposits, but the muscle cells atrophy (waste away). At first, children with this form of dystrophy appear to be normal. But between their second and tenth birthdays, their muscles start to weaken. They become progressively clumsier and fall frequently. Usually, after their twelfth birthday, they no longer can walk. In time, muscles of the chest and head deteriorate. Affected people typically die of respiratory failure in their early twenties. At present, there is no cure.

X-Linked Dominant Inheritance

The *faulty enamel trait* is one of the few known cases of X-linked dominant inheritance. With this disorder, the hard, thick enamel coating that normally protects teeth fails to develop properly. The inheritance pattern is like that for X-linked recessive alleles, except the allele is also expressed in heterozygous females. The phenotype tends to be less pronounced in females than in males. A son can't inherit the dominant allele for the trait from an affected father, but all daughters will. If a woman is heterozygous, she will transmit the allele to half of her offspring, regardless of their sex.

Genetic analyses of family pedigrees have revealed simple Mendelian inheritance patterns for certain traits.

Many of these traits have been traced to either dominant or recessive alleles on an autosome or X chromosome.

10.7 CHANGES IN CHROMOSOME NUMBER

Categories and Mechanisms of Change

Sometimes, owing to abnormal events, new individuals end up with the wrong chromosome number. The consequences range from minor to lethal physical changes.

With **aneuploidy**, individuals have one extra or one less chromosome. This condition is a major cause of human reproductive failure; possibly it affects half of all fertilized eggs. Autopsies show that most human embryos that were miscarried (spontaneously aborted before pregnancy reached full term) were aneuploids.

With **polyploidy**, individuals have three or more of each type of chromosome. About half of all species of flowering plants are polyploid (page 243). So are some insects, fishes, and other animals. Polyploidy is lethal for humans. It may disrupt interactions between the genes of autosomes and sex chromosomes at key steps in the complex pathways of development and reproduction. All but 1 percent of human polyploids die before birth. The rare newborns die within a month.

Chromosome numbers can change during mitotic or meiotic cell divisions. Suppose a cell cycle goes through DNA duplication and mitosis but is arrested before the cytoplasm divides. The cell is "tetraploid," with four of each type of chromosome. Suppose one or more pairs of chromosomes fail to separate in mitosis or meiosis—an event called **nondisjunction**. Some or all of the forthcoming cells have too many or too few chromosomes, as in Figure 10.13. Researchers routinely expose cells to colchicine to induce nondisjunction (Section 10.2).

The chromosome number of a normal gamete also may change at fertilization. Suppose that gamete unites by chance with an $n + 1$ gamete (one extra chromosome). The new individual will be "trisomic" ($2n + 1$), with three of one type of chromosome and two of every other type. What if the other gamete is $n - 1$? Then the new individual will be "monosomic" ($2n - 1$).

Change in the Number of Autosomes

Most changes in the number of autosomes arise through nondisjunction during gamete formation. Here we consider one of the most common of the resulting disorders.

A trisomic 21 newborn, with three chromosomes 21, will show the effects of *Down syndrome*. ("Syndrome" means a set of symptoms that characterize a disorder.) Symptoms vary greatly, but most affected individuals show moderate to severe mental impairment. About 40 percent develop heart defects. Because of abnormal skeletal development, older children have shortened body parts, loose joints, and poorly aligned bones of the hips, fingers, and toes. Muscles and muscle reflexes are weaker than normal, and development of speech and other motor functions is quite slow. Yet with special training, trisomic 21 individuals often take part in normal activities (Figure 10.14). As a group, they are cheerful, impish, affectionate people who typically derive great pleasure from music and dancing.

Down syndrome is one of many disorders that can be detected by prenatal diagnosis (Section 10.9). Before detection procedures were widespread, about 1 in 700 newborns of all ethnic groups were trisomic 21. Now the number is closer to 1 in 1,100. The risk is greater if pregnant women are over thirty-five (Figure 10.14b).

Figure 10.13 Nondisjunction. In this example, a pair of chromosomes fails to separate at anaphase I of meiosis. The chromosome number changes in gametes. Make a sketch of nondisjunction at anaphase II. What will the chromosome numbers be in gametes?

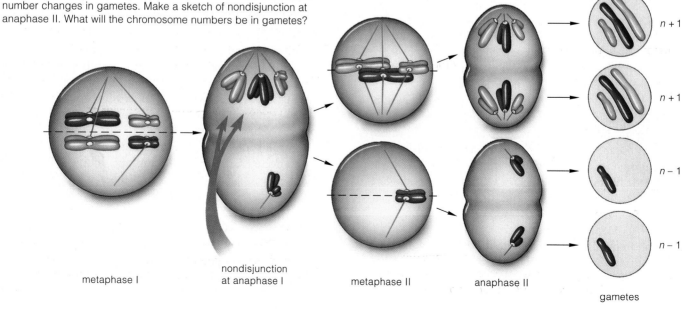

metaphase I | nondisjunction at anaphase I | metaphase II | anaphase II

gametes

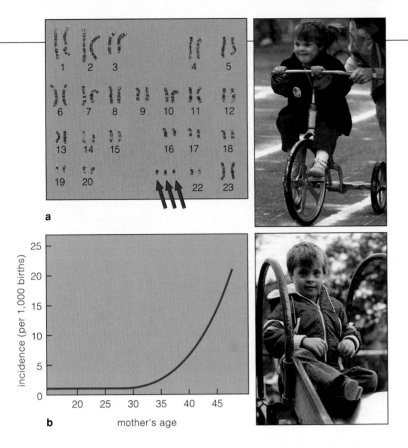

Figure 10.14 Down syndrome. (**a**) Karyotype showing a trisomic 21 condition (*red* arrows). (**b**) Relation between the frequency of Down syndrome and mother's age at the time of childbirth. Results are from a study of 1,119 children with the disorder who were born in Victoria, Australia, between 1942 and 1957. The young lady in the upper photograph was an exuberant participant in the Special Olympics, held annually in San Mateo, California.

Change in the Number of Sex Chromosomes

Most sex chromosome abnormalities arise as a result of nondisjunction during meiosis and gamete formation. Let's look at a few phenotypic outcomes.

Turner Syndrome Inheritance of one X chromosome without a partner X or Y chromosome gives rise to *Turner syndrome*, which affects 1 in 2,500 to 10,000 newborn girls. Nondisjunction affecting sperm accounts for 75 percent of the cases. We see fewer people with Turner syndrome compared to people with other sex chromosome abnormalities. The likely reason is that at least 98 percent of all X0 zygotes spontaneously abort early in pregnancy. About 20 percent of all spontaneously aborted embryos with chromosome abnormalities are X0.

Despite the near-lethality, X0 survivors are not as disadvantaged as other aneuploids. They grow up well proportioned, albeit short (4 feet, 8 inches tall, on the average). Generally, their behavior is normal during childhood. But most Turner females are infertile. They do not have functional ovaries and so cannot produce eggs or sex hormones. Without sex hormones, breast enlargement and the development of other secondary sexual traits are reduced. Possibly as a result of their arrested sexual development and small size, females often become passive and are easily intimidated by peers during their teens. Some patients have benefited from hormone therapy and corrective surgery.

Klinefelter Syndrome About 1 in 500 to 2,000 liveborn males has two X chromosomes *and* one Y chromosome. This XXY condition results mainly from nondisjunction in the mother (about 67 percent of the time, compared to 33 percent in the father). Symptoms of the resulting *Klinefelter syndrome* develop after the onset of puberty.

XXY males are taller than average and are sterile or nearly so. Their testes usually are much smaller than average; the penis and scrotum are not. Facial hair is often sparse, and there may be some breast enlargement. Injections of testosterone, a hormone, can reverse the feminized traits but not the low fertility. Some XXY males show mild mental impairment, although many fall within the normal range of intelligence. Except for low fertility, many show no outward symptoms at all.

XYY Condition About 1 in 1,000 males has one X and two Y chromosomes. XYY males tend to be taller than average. Some may be mildly retarded, but most are phenotypically normal. At one time, they were thought to be genetically predisposed to become criminals. This erroneous conclusion was based on small numbers of cases in highly selected groups, such as prison inmates. Investigators often knew who the XYY males were, and this may have biased their evaluations.

There were no *double-blind studies,* by which different investigators gather data independently of one another and match them up *after* both sets of data are completed. In this case, the same investigators gathered karyotypes and personal histories. Fanning the stereotype was a sensationalized report in 1968 that a mass-murderer of young nurses was XYY. He wasn't.

In 1976, a Danish geneticist issued a report on a large-scale study that was based on records of 4,139 tall males, twenty-six years old, who had reported to their draft board. Besides giving results of physical examinations and intelligence tests, the records provided clues to socioeconomic status, educational history, and any criminal convictions. Only twelve of the males were XYY, which left more than 4,000 for the control group. The only significant finding was that tall, mentally impaired males who engage in criminal activity are more likely to get caught—irrespective of karyotype.

Most changes in chromosome number arise as a result of nondisjunction during meiosis and gamete formation.

10.8 CHANGES IN CHROMOSOME STRUCTURE

Rare conditions also arise when a chromosome changes structurally through deletion, inversion, translocation, or duplication of one or more of its segments.

Deletions

A **deletion** is the loss of a chromosome region by viral attack, irradiation, chemical assaults, or other environmental factors. One or more genes may be lost, and this nearly always causes problems.

For example, one deletion from human chromosome 5 leads to mental retardation and an abnormally shaped larynx. When affected infants cry, they produce meowing sounds. Hence the name of the disorder, *cri-du-chat* (cat-cry). Figure 10.15*a* and *b* shows an affected child.

Inversions and Translocations

An **inversion** is a segment that separated from a chromosome and then was inserted at the same place—but in reverse. The reversal alters the position and order of the chromosome's genes:

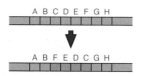

In most **translocations**, part of one chromosome exchanges places with a corresponding part of another chromosome that is *not* its homologous partner. As an example, chromosome 14 may end up with a segment of chromosome 8 (and chromosome 8 with a segment of 14). The normal controls over genes of that segment are lost at the new location, and a form of cancer results.

Duplications

Even normal chromosomes contain **duplications**, which are gene sequences that are repeated several to many times. Often the same gene sequence has been repeated thousands of times. Some duplications are built into the DNA of the species; others are not.

Consider *fragile X syndrome*, which affects 1 of 1,500 males and 1 of 2,500 females. The disorder is associated with many repeats of a sequence of three nucleotides in the FMR-1 gene on the X chromosome. Most people have about 29 of these repeats, but affected people have 700 or more. Apparently, when the number of repeats passes a certain threshold, gene expression is reduced. In cultured cells, most X chromosomes that carry the unstable array of repeats have an unusual constricted region (Figure 10.15*c*).

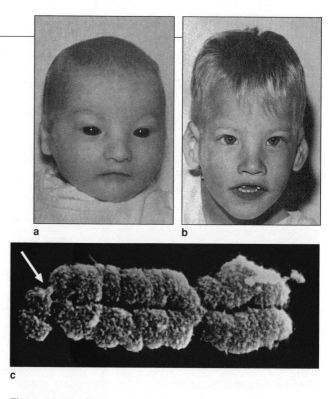

Figure 10.15 Cri-du-chat syndrome, a result of a deletion on the short arm of chromosome 5. (**a**) One affected boy just after birth. (**b**) The same boy four years later, showing how facial features change. By this age, affected individuals no longer make the mewing sounds typical of the cri-du-chat syndrome.

(**c**) Fragile X syndrome. A constricted region (*arrow*) on the long arm of the X chromosome is associated with the disorder.

Not all duplications are harmful. Many had pivotal roles in evolution. Cells require specific gene products, so mutations that change most genes are probably selected against. But *duplicates* of some gene sequences that do no harm could be retained and would be free to mutate—for the normal gene would still provide the required product. If one or more of those duplicated sequences did become slightly modified, their products could function in slightly different or new ways.

This apparently happened in gene regions coding for the polypeptide chains of the hemoglobin molecule. In humans and other primates, these regions contain copies of strikingly similar gene sequences, and they produce whole families of slightly different chains.

Together with duplications, certain inversions and translocations probably helped put the primate ancestors of humans on a unique evolutionary road. Of the twenty-three pairs of human chromosomes, eighteen of them are nearly identical to their counterparts in chimpanzees and gorillas. The other five pairs differ at inverted and translocated regions.

On rare occasions, a segment of a chromosome may be lost, inverted, moved to a new location, or duplicated.

PROSPECTS IN HUMAN GENETICS

With the arrival of their newborn, parents typically ask, "Is our baby normal?" Quite naturally, they want their baby to be free of genetic disorders, and most of the time it is. But what are the options when it is not?

We do not approach diseases and heritable disorders the same way. "Diseases" follow infection by bacteria and other agents from the outside, and we attack them with antibiotics and other weapons. But how do we attack an "enemy" within our genes? Do we institute regional, national, or global programs to identify people carrying harmful alleles? Do we tell them they are "defective" and run a risk of bestowing their disorder on their children? Who decides which alleles are "harmful"? Should society bear the cost of treating genetic disorders? If so, should it also have a say in whether affected embryos will be born at all, or aborted? These questions are the tip of an ethical iceberg. We don't have universally acceptable answers.

Phenotypic Treatments   Often, symptoms of genetic disorders can be minimized or suppressed by controlling the diet, adjusting to outside environmental conditions, or intervening surgically or with hormone therapy.

Diet control works in PKU (phenylketonuria). A certain gene specifies an enzyme that converts one amino acid to another (phenylalanine to tyrosine). If someone is homozygous recessive for a mutated form of the gene, the first of these amino acids accumulates in the body. If the excess is diverted to other pathways, phenylpyruvate and other compounds may form. High phenylpyruvate levels can impair brain function. By restricting their intake of phenylalanine, affected persons don't have to dispose of excess amounts and can lead normal lives. Among other things, they can avoid soft drinks and other products sweetened with aspartame—which contains phenylalanine.

Environmental adjustments help counter the symptoms of some disorders, as when albinos avoid direct sunlight. Surgical reconstructions can minimize problems such as a form of *cleft lip* in which a vertical fissure cuts through the lip and extends into the roof of the mouth.

Genetic Screening   Through screening programs in the population at large, affected persons or carriers can be detected early enough to start preventive measures before symptoms develop. For example, most hospitals in the United States routinely screen newborns for PKU, so it is less common to see people with symptoms of this disorder.

Genetic Counseling   If a first child or close relative has a severe problem, prospective parents may worry and request help. Counseling often includes diagnosis of parental genotypes, detailed pedigrees, and biochemical testing for many metabolic disorders. Geneticists may predict risks for disorders that follow simple Mendelian inheritance, but not all disorders do. Prospective parents must know the same risk applies to each pregnancy.

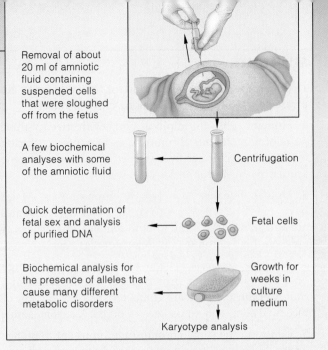

Removal of about 20 ml of amniotic fluid containing suspended cells that were sloughed off from the fetus

A few biochemical analyses with some of the amniotic fluid

Centrifugation

Quick determination of fetal sex and analysis of purified DNA

Fetal cells

Biochemical analysis for the presence of alleles that cause many different metabolic disorders

Growth for weeks in culture medium

Karyotype analysis

Figure 10.16   Steps in amniocentesis.

Prenatal Diagnosis   Suppose a pregnant woman forty-five years old worries that her fetus might be trisomic 21. Methods of *prenatal diagnosis* can detect this condition and more than a hundred others (prenatal means before birth).

In **amniocentesis**, a doctor draws a sample of fluid inside the amnion, a sac around the fetus (Figure 10.16). Fetal cells in the sample are cultured and then analyzed, as by karyotyping. In **chorionic villi sampling** (CVS), cells are drawn from the chorion, a sac around the amnion. CVS can be performed between the ninth and twelfth weeks of pregnancy. Like amniocentesis, it is risky. Either procedure may accidentally cause infection or puncture the fetus.

What happens if prenatal diagnosis reveals a serious problem? Do prospective parents opt for induced **abortion** —that is, expulsion of the embryo from the uterus? They must weigh their awareness of the crushing severity of the disorder against ethical and religious beliefs. Worse, they must play out their personal tragedy on a larger stage, dominated now by a nationwide battle between fiercely vocal "pro-life" and "pro-choice" factions.

Another procedure, *preimplantation diagnosis*, relies on **in-vitro fertilization**. Sperm and eggs from prospective parents are put in an enriched medium in a petri dish. One or more eggs may get fertilized. In two days, cell divisions may convert a fertilized egg into a ball of eight cells. In one view, the tiny, free-floating ball is a *prepregnancy* stage. Like the unfertilized eggs discarded monthly from a woman, it is not attached to the uterus. All cells in the ball have the same genes and are not yet committed to giving rise to specialized cells of a heart and other organs. Doctors take one of the undifferentiated cells and analyze its genes for suspected disorders. If the cell has no detectable genetic defects, the ball is reinserted into the uterus.

Several couples who are at risk of passing on muscular dystrophy, cystic fibrosis, and other disorders have opted for the procedure. Several "test-tube" babies have been born in good health—and are free of the harmful genes.

SUMMARY

1. Genes, the units of instruction for heritable traits, are arranged one after the other along chromosomes.

2. Human cells are diploid (2*n*), with twenty-three pairs of homologous chromosomes that interact during meiosis. Each pair has the same length, shape, and gene sequence (except for an XY pairing).

3. Human females have two X chromosomes. Males have one X paired with one Y. All other chromosomes are autosomes (the same in both females and males). A gene on the Y chromosome determines gender.

4. Pedigrees (charts of genetic connections through lines of descent) often provide clues to inheritance of a trait. Certain patterns are characteristic of dominant or recessive alleles on autosomes or on the X chromosome.

5. Genes on the same chromosome represent a linkage group. However, crossing over (breakage and exchange of segments between homologues) disrupts linkages. The farther apart two genes are, the greater will be the frequency of crossovers between them.

6. A chromosome's structure may become altered on rare occasions. A segment might be deleted, inverted, moved to a new location (translocated), or duplicated.

7. The chromosome number may change on rare occasions. New individuals may end up with one more or one less chromosome than the parents (aneuploidy) or with three or more of each type (polyploidy). Nondisjunction at meiosis (prior to gamete formation) accounts for most of these abnormalities. The following chart summarizes cases of nondisjunction involving an X chromosome, as described in the text (here, an abnormal egg is fertilized by a normal sperm):

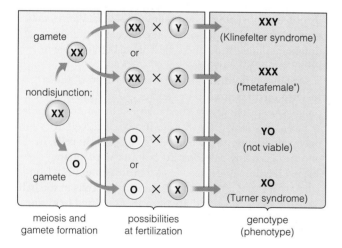

8. Crossing over and changes in chromosome number or structure may influence the course of evolution. The changes in genotype (genetic makeup) lead to variations in phenotype (observable traits) among members of a population, so that evolution is possible.

Review Questions

1. What is a gene? What are alleles? *158*

2. Distinguish between: *158*
   a. Homologous and nonhomologous chromosomes
   b. Sex chromosomes and autosomes
   c. Karyotype and karyotype diagram

3. Define genetic recombination, and describe how crossing over can bring it about. What are the possible consequences of genetic recombination for a given trait in a new individual? *158, 163*

4. Explain why a genetic abnormality is not the same as a genetic disorder or genetic disease. *165*

5. Contrast a typical pattern of autosomal recessive inheritance with that of autosomal dominant inheritance. *166*

6. Contrast a typical pattern of X-linked recessive inheritance with that of X-linked dominant inheritance. *167*

7. Define aneuploidy and polyploidy. Make a sketch of an example of nondisjunction. *168*

8. Explain how chromosomal deletions, duplications, inversions, and translocations differ from one another. *170*

Self-Quiz *(Answers in Appendix IV)*

1. _____ segregate during _____ .
   a. Homologues; mitosis
   b. Genes on one chromosome; meiosis
   c. Homologues; meiosis
   d. Genes on one chromosome; mitosis

2. The probability of a crossover occurring between two genes on the same chromosome is _____ .
   a. unrelated to the distance between them
   b. increased if they are closer together on the chromosome
   c. increased if they are farther apart on the chromosome
   d. impossible

3. Chromosome structure can be altered by _____ .
   a. deletions   d. translocations
   b. duplications   e. all of the above
   c. inversions

4. Nondisjunction can be caused by _____ .
   a. crossing over in mitosis
   b. segregation in meiosis
   c. failure of chromosomes to separate during meiosis
   d. multiple independent assortments

5. A gamete affected by nondisjunction would have _____ .
   a. a change from the normal chromosome number
   b. one extra or one missing chromosome
   c. the potential for a genetic disorder
   d. all of the above

6. Genetic disorders can be caused by _____ .
   a. gene mutations
   b. changes in chromosome structure
   c. changes in chromosome number
   d. all of the above

7. Which of the following events contributes to phenotypic variation in a population?
   a. independent assortment
   b. crossing over
   c. changes in chromosome structure and number
   d. all of the above

8. Match the chromosome terms appropriately.

_____ crossing over
_____ deletion
_____ nondisjunction
_____ translocation
_____ karyotype

a. number and defining features of individual's metaphase chromosomes
b. chromosome segment moves to a nonhomologous chromosome
c. disrupts gene linkages at meiosis
d. causes gametes to have abnormal chromosome numbers
e. loss of a chromosome segment

Critical Thinking—Genetics Problems

(Answers in Appendix III)

1. Human females are XX and males are XY.
   a. Does a male inherit the X from his mother or father?
   b. With respect to X-linked genes, how many different types of gametes can a male produce?
   c. If a female is homozygous for an X-linked gene, how many different types of gametes can she produce with respect to that gene?
   d. If a female is heterozygous for an X-linked gene, how many different gametes can she produce with respect to that gene?

2. Suppose one allele of a Y-linked gene results in nonhairy ears in males. Another allele results in rather long hairs, a condition called *hairy pinnae* (Figure 10.17).
   a. Why would you *not* expect females to have hairy pinnae?
   b. Any son of a hairy-eared male will also be hairy-eared, but no daughter will be. Explain why.

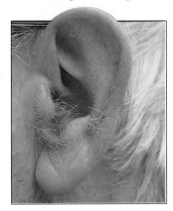

Figure 10.17   Hairy pinnae.

3. Suppose that you have two linked genes with alleles *Aa* and *Bb* respectively. An individual is heterozygous for both genes, as in the following:

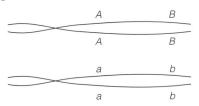

If the crossover frequency between these two genes is 0 percent, what genotypes would be expected among the gametes produced by this individual, and with what frequencies?

4. In *D. melanogaster*, a gene governs wing length. A dominant allele at this gene locus results in the formation of long wings. A recessive allele results in the formation of vestigial (short) wings, as shown in Figure 10.18.

Figure 10.18   A vestigial-winged fly (*Drosophila*).

Suppose you cross a homozygous dominant long-winged fruit fly with a homozygous recessive vestigial-winged fly. Shortly after the mating, fertilized eggs are exposed to a level of x-rays known to induce mutation and chromosomal deletions. Later, when the irradiated fertilized eggs develop into adults, most of the flies are heterozygous and have long wings. However, a few are vestigial-winged. Give possible explanations for the appearance of these vestigial-winged adults.

5. Suppose you have alleles for lefthandedness and straight hair on one chromosome and alleles for righthandedness and curly hair on the homologous chromosome. If the two gene loci for the alleles are very close together along the length of the chromosome, how likely is it that crossing over will occur between them? If they are distant from each other, is a crossover between them more or less likely to occur?

6. Individuals affected by *Down syndrome* typically have an extra chromosome 21. In other words, their body cells contain a total of 47 chromosomes.
   a. At which stages of meiosis I and II could a mistake occur that could result in the altered chromosome number?
   b. In a few cases of Down syndrome, 46 chromosomes are present. Included in this total are two normal-appearing chromosomes 21 and a longer-than-normal chromosome 14. Interpret this observation and indicate how these few individuals can have a normal chromosome number.

7. The mugwump, a type of tree-dwelling mammal, has a reversed sex-chromosome condition. The male is XX and the female is XY. However, perfectly good sex-linked genes are found to have the same effect as in humans. For example, a recessive, X-linked allele *c* produces *red-green color blindness*. If a normal female mugwump mates with a phenotypically normal male mugwump whose mother was color blind, what is the probability that a son from that mating will be color blind? A daughter?

8. One type of childhood *muscular dystrophy* is a recessive, X-linked trait. A slowly progressing loss of muscle function leads to death, usually by age twenty or so. Unlike color blindness, this disorder is restricted to males. Suggest why.

Selected Key Terms

abortion *171*	inversion *170*
allele *158*	in-vitro fertilization *171*
amniocentesis *171*	karyotype *158*
aneuploidy *168*	linkage group *162*
autosome *158*	nondisjunction *168*
chorionic villi sampling *171*	pedigree *164*
crossing over *158*	polyploidy *168*
deletion *170*	sex chromosome *158*
duplication *170*	translocation *170*
gene *158*	X chromosome *158*
genetic abnormality *165*	X-linked gene *162*
genetic disorder *165*	Y chromosome *158*
genetic recombination *158*	Y-linked gene *162*
homologous chromosome *158*	

Readings

Cummings, M. 1994. *Human Heredity: Principles and Issues.* Third edition. St. Paul, Minnesota: West.

Holden, C. 1987. "The Genetics of Personality." *Science* 237: 598–601. For students interested in human behavioral genetics.

Weiss, R. November 1989. "Genetic Testing Possible Before Conception." *Science News* 136(21): 326.

11 DNA STRUCTURE AND FUNCTION

Cardboard Atoms and Bent-Wire Bonds

One might have wondered, in the spring of 1868, why Johann Friedrich Miescher was collecting cells from the pus of open wounds and, later, from the sperm of a fish. Miescher, a physician, wanted to identify the chemical composition of the nucleus. These particular cells have very little cytoplasm, which makes it easier to isolate the nuclear material for analysis.

Miescher finally succeeded in isolating an organic compound with the properties of an acid. Unlike other substances in cells, it incorporated a notable amount of phosphorus. He called the substance nuclein. He had discovered what came to be known many years later as **deoxyribonucleic acid**, or **DNA**.

The discovery caused scarcely a ripple through the scientific community. At the time, no one knew much about the physical basis of inheritance—that is, *which chemical substance encodes the instructions for reproducing parental traits in offspring*. Only a few researchers even suspected that the nucleus might hold the answer.

And then, in 1951, Linus Pauling did something that no one had done before. Through his training in biochemistry, a talent for model building, and a few educated guesses, he deduced the three-dimensional structure of the protein collagen. Pauling's discovery was electrifying. If someone could pry open the secrets of proteins, then why not assume the same might be done for other biological molecules? Further, wouldn't structural details provide clues to biological functions? *And who would go down in history as having discovered the very secrets of inheritance?* Scientists around the world started scrambling after that ultimate prize.

Maybe hereditary instructions were encoded in the structure of some unknown class of proteins. After all, heritable traits are spectacularly diverse. Surely the molecules encoding information about those traits were structurally diverse also. Proteins are put together from potentially limitless combinations of twenty different amino acid subunits, so they almost certainly could function as the sentences (genes) in each cell's book of inheritance.

Yet there was something about another substance—DNA—that excited many researchers. Among them were James Watson, a young postdoctoral student from Indiana University, and Francis Crick, an energetic researcher at Cambridge University. How could DNA, a molecule consisting of only four kinds of subunits, hold genetic information? Watson and Crick spent long hours arguing over everything they had read about the size, shape, and bonding requirements of the subunits of DNA. They fiddled with cardboard cutouts of the subunits. They badgered chemists to identify potential bonds they might have overlooked. They assembled models from bits of metal, held together with wire "bonds" bent at seemingly suitable angles.

In 1953, they put together a model that fit all of the pertinent biochemical rules and all of the facts about DNA they had gleaned from other sources (Figure 11.1). They had discovered the structure of DNA. The breathtaking simplicity of that structure also enabled them to solve another long-standing riddle—*how life can show unity at the molecular level and yet give rise to so much diversity at the level of whole organisms.*

With this chapter, we turn to investigations that led to our current understanding of DNA. The story is more than a march through details of its structure and function. *It also is revealing of how ideas are generated in science.* On the one hand, having a shot at fame and fortune quickens the pulse of men and women in any profession, and scientists are no exception. On the other hand, science proceeds as a community effort, with individuals sharing not only what they can explain but also what they do not understand. Even when an experiment fails to produce the anticipated results, it might turn up information that others can use—or lead to questions that others can answer. Unexpected results, too, might be clues to something important about the natural world.

Figure 11.1 (*Lower photograph*) James Watson and Francis Crick posing in 1953 by their newly unveiled structural model of DNA. (*Upper photograph*) This is a more recent, computer-generated model of DNA. Although it is more sophisticated in appearance, it is basically the same as the prototype that Watson and Crick put together decades ago.

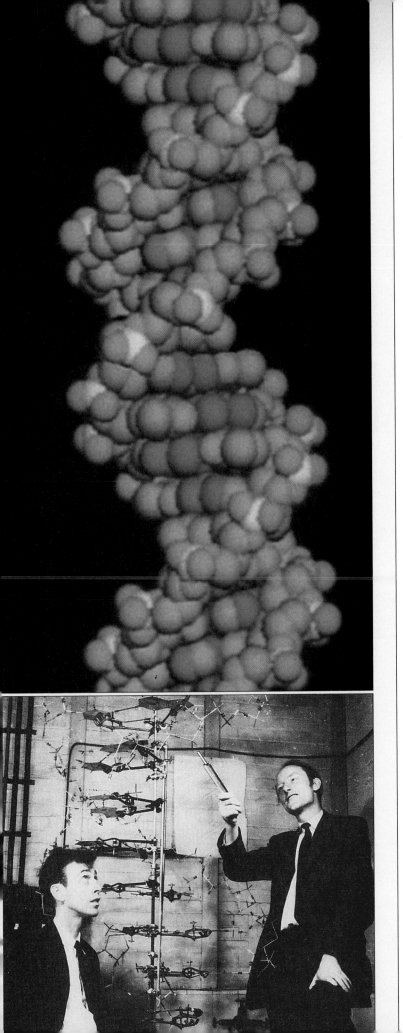

1. In all living cells, DNA is the storehouse of information about heritable traits.

2. In a DNA molecule, two strands of nucleotides twist together, like a spiral stairway. Each strand consists of four kinds of nucleotides, which are the same except for one component—a nitrogen-containing base. The four bases are adenine, guanine, thymine, and cytosine.

3. Great numbers of nucleotides are arranged one after another in each strand of the DNA molecule. In at least some regions, the order in which one kind of nucleotide follows another is unique for each species. Hereditary information is encoded in that particular sequence.

4. Hydrogen bonds connect the bases of one strand of the DNA molecule to bases of the other strand. As a rule, adenine pairs (hydrogen-bonds) with thymine, and guanine with cytosine.

5. Before a cell divides, its DNA is replicated with the help of enzymes and other proteins. Each double-stranded DNA molecule starts unwinding. A new, complementary strand is assembled on the exposed bases of each parent strand, according to the base-pairing rule.

6. There is only one DNA molecule in an unduplicated chromosome. Except in bacteria, numerous proteins are attached to the DNA, and these function in its structural organization.

11.1 DISCOVERY OF DNA FUNCTION

Early Clues

The year was 1928. Frederick Griffith, an army medical officer, was attempting to develop a vaccine against *Streptococcus pneumoniae*, a bacterium that causes a form of pneumonia. (When introduced into a person's body, vaccines can mobilize internal defenses against a real attack. Many are preparations of killed or weakened bacterial cells.) Griffith never did develop a vaccine. But his work unexpectedly opened a door to the molecular world of heredity.

Griffith isolated and cultured two different strains of the bacterium. He noticed that colonies of one strain had a rough surface appearance, but those of the other strain appeared smooth. He designated the strains *R* and *S* and used them in a series of four experiments:

1. Laboratory mice were injected with live R cells. As Figure 11.2 indicates, they did not develop pneumonia. *The R strain was harmless.*

2. Mice were injected with live S cells. The mice died. Blood samples taken from them teemed with live S cells. *The S strain was pathogenic* (disease-causing).

3. S cells were killed by exposure to high temperature. Mice injected with these cells did not die.

4. Live R cells were mixed with heat-killed S cells and injected into mice. The mice died—and blood samples from them teemed with *live* S cells!

What was going on in the fourth experiment? Maybe heat-killed S cells in the mixture weren't really dead. But if that were true, then mice injected with heat-killed S cells alone (experiment 3) would have died. Maybe harmless R cells in the mixture had mutated into a killer form. But if that were true, then mice injected with the R cells alone (experiment 1) would have died.

The simplest explanation was this: *Heat did kill the S cells but did not destroy their hereditary material—including the part that specified "how to cause infection."* Somehow, that material had been transferred from dead S cells to living R cells, which put it to use.

Further experiments showed that harmless cells had indeed picked up information about infection and had been permanently transformed into pathogens. After hundreds of generations, bacterial descendants of the transformed cells were still causing infections!

The unexpected results of Griffith's work intrigued the microbiologist Oswald Avery and his colleagues. In time they were able to transform harmless cells with extracts of killed pathogens. In 1944, they reported that the hereditary substance in the extracts was probably DNA—not proteins, as was widely believed. They had added protein-digesting enzymes to some extracts, but

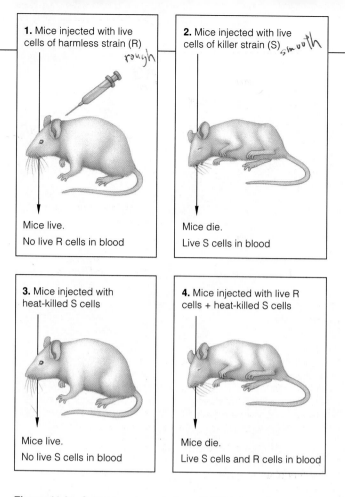

1. Mice injected with live cells of harmless strain (R) *rough*

Mice live.
No live R cells in blood

2. Mice injected with live cells of killer strain (S) *smooth*

Mice die.
Live S cells in blood

3. Mice injected with heat-killed S cells

Mice live.
No live S cells in blood

4. Mice injected with live R cells + heat-killed S cells

Mice die.
Live S cells and R cells in blood

Figure 11.2 Summary of results from Griffith's experiments with a harmless strain and a disease-causing strain of *Streptococcus pneumoniae*, as described in the text.

cells were transformed anyway. Then they added an enzyme that breaks apart DNA but not proteins—and the enzyme blocked hereditary transformation.

Despite these impressive experimental results, most biochemists refused to give up on the proteins. Avery's findings, they said, probably applied only to bacteria.

Confirmation of DNA Function

By the early 1950s, Max Delbrück, Alfred Hershey, Martha Chase, Salvador Luria, and other molecular detectives were using certain viruses as experimental subjects. These viruses, called **bacteriophages**, infect *Escherichia coli* and other bacteria.

Viruses are as biochemically simple as you can get. They are not alive, but they do contain hereditary information on building new virus particles. At some point after they infect a host cell, viral enzymes take over a portion of the cell's metabolic machinery—which starts churning out substances necessary to construct new virus particles.

By 1952, researchers knew that some bacteriophages consist only of DNA and a protein coat. Also, electron micrographs revealed that the main part of the viruses

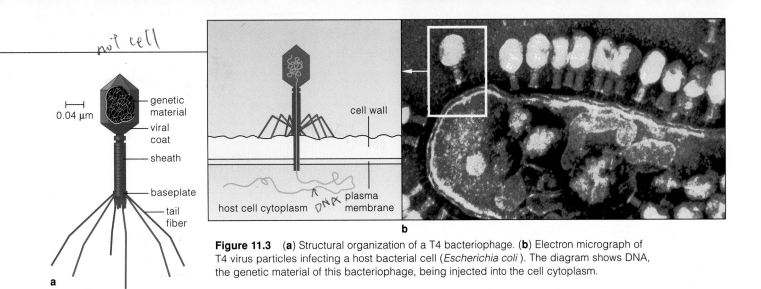

not cell

genetic
material

viral
coat

sheath

baseplate

tail
fiber

0.04 μm

a

cell wall

host cell cytoplasm — DNA — plasma
membrane

b

Figure 11.3 (**a**) Structural organization of a T4 bacteriophage. (**b**) Electron micrograph of T4 virus particles infecting a host bacterial cell (_Escherichia coli_). The diagram shows DNA, the genetic material of this bacteriophage, being injected into the cell cytoplasm.

Figure 11.4 Two examples of the landmark experiments that pointed to DNA as the substance of heredity. In the 1940s, Alfred Hershey and his colleague, Martha Chase, were studying the biochemical basis of inheritance. They were aware that certain bacteriophages were composed of proteins and DNA. Did the proteins, DNA, or both contain the viral genetic information?

To find out, Hershey and Chase designed two experiments, based on two known biochemical facts. First, the proteins of bacteriophages incorporate sulfur (S) but not phosphorus (P). Second, their DNA incorporates phosphorus but not sulfur.

(**a**) In one experiment, bacterial cells were grown on a culture medium that included the radioisotope ^{35}S and no other form of sulfur. To synthesize proteins, the bacterial cells would have to use the radioisotope—which would serve as a tracer. After the cells became labeled with the tracer, bacteriophages were allowed to infect them. Afterward, viral proteins were synthesized inside the host cells. These proteins also became labeled with ^{35}S. So did the new generation of virus particles.

Next, the labeled bacteriophages were allowed to infect a new batch of unlabeled bacteria that were suspended in a fluid culture medium. Afterward, Hershey and Chase whirred the fluid in a kitchen blender. Whirring dislodged the viral protein coats from the cells. The particles became suspended in the fluid medium. Analysis revealed the presence of labeled protein in the fluid—_not_ inside the bacterial cells.

(**b**) In the second experiment, new bacterial cells were cultured. The only phosphorus available for synthesizing DNA was the radioisotope ^{32}P. Bacteriophages were allowed to infect the cells. As predicted, viral DNA assembled inside the infected cells became labeled, as did the new generation of virus particles. Next, the labeled particles were allowed to infect bacteria that were suspended in a fluid medium. Then they were dislodged

sulfur label

phosphorus
label

bacterial cell (cutaway view)

label
outside cell

label inside cell

a FIRST EXPERIMENT

b SECOND EXPERIMENT

from the host cells. Analysis revealed that the labeled viral DNA was not in the fluid. It remained _inside_ the host cells, where its hereditary information had to be put to use to make more virus particles. Here was evidence that DNA is the genetic material of this type of virus.

remains _outside_ of the cells they are infecting (Figure 11.3). Quite possibly, such viruses were injecting genetic material alone _into_ host cells. If that were true, was the material DNA, protein, or both?

Through many ingenious experiments, researchers accumulated strong evidence that DNA, not proteins,

functions as the molecule of inheritance. Figure 11.4 describes two of these landmark experiments.

Information for producing the heritable traits of single-celled and multicelled organisms is encoded in DNA.

11.2 DNA STRUCTURE

Components of DNA

Long before the bacteriophage studies were under way, biochemists knew that DNA contains only four types of nucleotides that are the building blocks of nucleic acids. A **nucleotide** consists of a five-carbon sugar (which is deoxyribose in DNA), a phosphate group, and one of the following nitrogen-containing bases:

adenine	guanine	thymine	cytosine
(A)	(G)	(T)	(C)

As you can see from Figure 11.5, all four types of nucleotides in DNA have their component parts joined together in much the same way. However, T and C are pyrimidines, which are single-ring structures. A and G are purines, which are larger, bulkier molecules; they have double-ring structures.

By 1949 Erwin Chargaff, a biochemist, had shared two crucial insights into the composition of DNA with the scientific community. First, the amount of adenine relative to guanine differs from one species to the next. Second, the amount of adenine in DNA always equals that of thymine, and the amount of guanine always equals that of cytosine. We may show this as:

$$A = T \quad \text{and} \quad G = C$$

The relative proportions of the four kinds of nucleotides were tantalizing clues. In some way, those proportions surely were related to the arrangement of the two chains of nucleotides in a DNA molecule.

The first convincing evidence of that arrangement came from Maurice Wilkins's laboratory in England. One of Wilkins's colleagues, Rosalind Franklin, had obtained especially good **x-ray diffraction images** of DNA fibers. By this process, a beam of x-rays is directed at a molecule, which scatters the beam in patterns that can be captured on film. The pattern itself consists only of dots and streaks; in itself, it doesn't reveal molecular structure. However, the images of those patterns can be used to calculate the positions of the molecule's atoms.

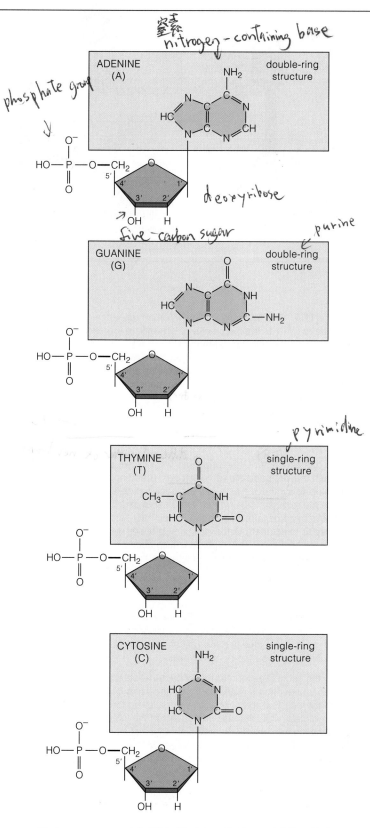

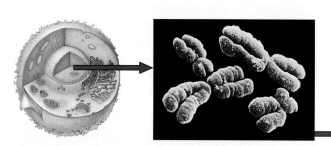

Figure 11.5 The four kinds of nucleotide subunits of DNA. Small numerals on the structural formulas identify the carbon atoms to which other parts of the molecule are attached.

All chromosomes contain DNA. What does DNA contain? Four kinds of nucleotides. Each nucleotide has a five-carbon sugar (shaded red). That sugar has a phosphate group attached to the fifth carbon atom of its ring structure. It also has one of four kinds of nitrogen-containing bases (shaded blue) attached to its first carbon atom. The nucleotides differ only in which base is attached to that atom.

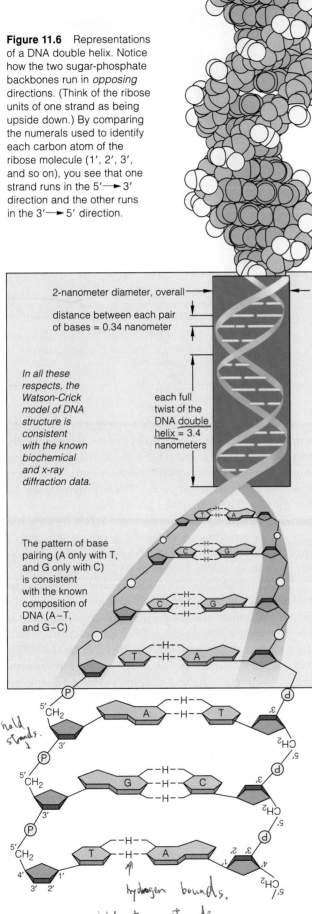

Figure 11.6 Representations of a DNA double helix. Notice how the two sugar-phosphate backbones run in *opposing* directions. (Think of the ribose units of one strand as being upside down.) By comparing the numerals used to identify each carbon atom of the ribose molecule (1′, 2′, 3′, and so on), you see that one strand runs in the 5′ ──► 3′ direction and the other runs in the 3′ ──► 5′ direction.

2-nanometer diameter, overall

distance between each pair of bases = 0.34 nanometer

In all these respects, the Watson-Crick model of DNA structure is consistent with the known biochemical and x-ray diffraction data.

each full twist of the DNA double helix = 3.4 nanometers

The pattern of base pairing (A only with T, and G only with C) is consistent with the known composition of DNA (A–T, and G–C)

DNA does not readily lend itself to x-ray diffraction. However, researchers could rapidly spin a suspension of DNA molecules, spool them onto a rod, and gently pull them into gossamer fibers, like cotton candy. If the atoms in DNA were arranged in a regular order, x-rays directed at a fiber should scatter in a regular pattern that could be captured on film.

Calculations based on Franklin's images strongly indicated that the DNA molecule had to be long and thin, with a 2-nanometer diameter. Some molecular configuration was being repeated every 0.34 nanometer along its length, and another one, every 3.4 nanometers.

Could the sequence of nucleotide bases be twisting, like a circular stairway? Certainly Pauling thought so. After all, he discovered the helical shape of collagen. He and everybody else—including Wilkins, Watson, and Crick—were thinking "helix." Watson later wrote, "We thought, why not try it on DNA? We were worried that *Pauling* would say, why not try it on DNA? Certainly he was a very clever man. He was a hero of mine. But we beat him at his own game. I still can't figure out why."

Pauling, it turned out, made a big chemical mistake. In his model, hydrogen bonds at phosphate groups held DNA's structure together. That does happen in highly acidic solutions. It doesn't happen in cells.

Patterns of Base Pairing

As Watson and Crick perceived, DNA consists of *two* strands of nucleotides, held together at their bases by hydrogen bonds. The bonds form when the two strands run in opposing directions and twist together into a double helix (Figure 11.6). Two kinds of base pairings form along the length of the molecule: A-T and G-C. This bonding pattern permits variation in the order of bases in any given strand. For example, in even a tiny stretch of DNA from a rose, gorilla, human, or any other organism, the sequence might be:

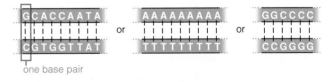

one base pair

In fact, even though all DNA molecules show the same bonding pattern, each species has unique base sequences in its DNA. *This molecular constancy and variation among species is the foundation for the unity and diversity of life.*

The *pattern* of base pairing between the two strands in DNA is constant for all species—A with T, and G with C. However, the DNA molecules of each species show unique differences in the *sequence* of base pairs along their length.

11.3 A CLOSER LOOK AT DNA

DNA Replication and Repair

The discovery of DNA structure was a turning point in studies of inheritance. Until then, no one could explain **DNA replication**—that is, how the molecule of inheritance is duplicated prior to cell division. The Watson-Crick model suggested at once how this might be done.

Enzymes can easily break hydrogen bonds between the two nucleotide strands of a DNA molecule. When they act at a given site, one strand unwinds from the other, thereby exposing some nucleotide bases. Cells have stockpiles of free nucleotides, and these pair with exposed bases. Each parent strand remains intact, and a companion strand is assembled on each one, according to this base-pairing rule: A to T, and G to C. As soon as a stretch of the parent strand has a stretch of its new, partner strand, the two twist into a double helix.

Because the parent strand is conserved, each "new" DNA molecule is really half old, half new (Figure 11.7). That is why biologists sometimes refer to this process as *semiconservative* replication.

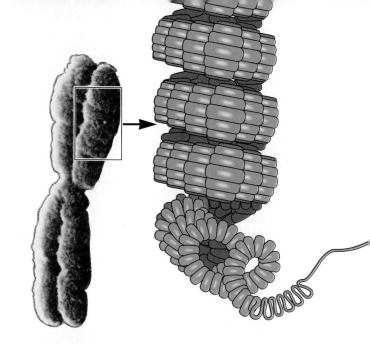

a Human chromosome at metaphase, in its most condensed form. Interactions among certain chromosomal proteins keep loops of DNA tightly packed in a "supercoiled" arrangement.

Figure 11.8 A model of the levels of organization of a eukaryotic chromosome.

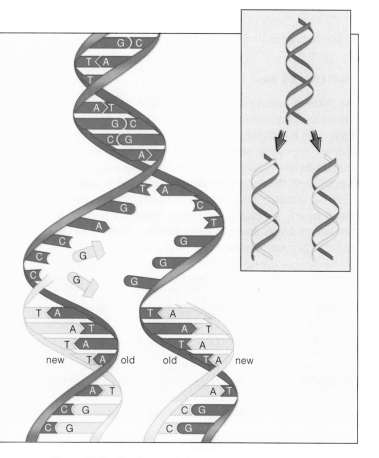

Figure 11.7 Semiconservative nature of DNA replication. The original two-stranded DNA molecule is shown in *blue*. Each parent strand remains intact, and a new strand (*yellow*) is assembled on each one.

DNA replication requires a large team of molecular workers. For example, one kind of enzyme unwinds the two nucleotide strands. At the same time, many proteins bind to the unwound portions and hold them apart as replication proceeds. **DNA polymerases** are key players. These enzymes attach free nucleotides to a growing strand. Other enzymes, the **DNA ligases**, seal new short stretches of nucleotides into a continuous strand. As you will read in Chapter 13, some of these enzymes have uses in recombinant DNA technology.

DNA polymerases, DNA ligases, and other enzymes also engage in **DNA repair**. By this process, enzymes excise and repair altered parts of the base sequence in one strand of a double helix. DNA polymerases "read" the complementary sequence on the other strand. Then, with the aid of other repair enzymes, they restore the original sequence. At this chapter's end, a *Focus* essay will give you a sense of what can happen if something impairs the excision-repair function.

Organization of DNA in Chromosomes

Each chromosome has one DNA molecule. If the DNA of all forty-six chromosomes in one of your somatic cells were stretched out end to end, the string would be about two meters long! What keeps all of that DNA from becoming a tangled mess? Proteins.

Eukaryotic DNA has many **histones** and other protein molecules bound tightly to it (Figure 11.8). Some histones are like spools for winding up small stretches

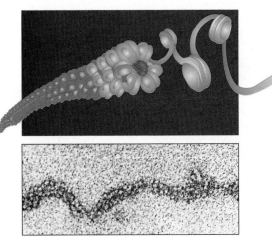

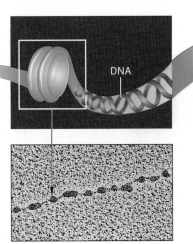

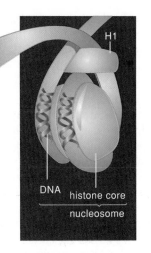

b At a deeper level of organization, the chromosomal proteins and DNA are arranged as a cylindrical fiber (solenoid), 30 nanometers in diameter.

c Immerse a chromosome in a salt solution, and it will loosen up to a beads-on-a-string organization. The "string" is one DNA molecule. Each "bead" is a nucleosome, the smallest organizational unit of the eukaryotic chromosome.

d Each nucleosome consists of a double loop of DNA around a core of eight proteins, of a class called histones. Another histone (H1) stabilizes the arrangement.

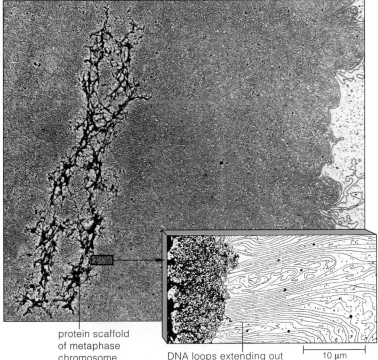

protein scaffold of metaphase chromosome

DNA loops extending out from and back to scaffold

10 μm

Figure 11.9 From a human cell, the scaffold of a metaphase chromosome after its DNA was experimentally "decondensed." The cell was treated with detergent and highly charged molecules that removed the chromosomal proteins. The boxed inset shows a small portion of the scaffold at higher magnification. Notice how the DNA has no disorganized, free ends. Each loop of the DNA begins and ends near the same scaffold region.

of DNA. Each histone-DNA spool is one **nucleosome**. Another histone stabilizes the spools.

Histone-DNA interactions can make a chromosome coil back on itself again and again. The coiling greatly increases its diameter. Further folding results in a series of loops. As Figure 11.9 shows, other proteins besides histones serve as a structural "scaffold" for the loops.

Apparently, some portions of the scaffold intervene between genes, not in regions that contain information for building proteins. Do scaffold proteins organize the chromosome into "domains" having distinct functions? Does the structural organization make it easier for enzymes to replicate the DNA, even to "read" the genes as the first step in protein synthesis? These are just two of the possibilities being probed by the new generation of molecular detectives.

DNA is replicated prior to cell division. Enzymes unwind its two strands. Each strand remains intact—it is conserved—and enzymes assemble a new, complementary strand on each one.

Some of the enzymes also repair the DNA where base-pairing errors have crept into the nucleotide sequence.

Eukaryotic DNA is organized on a scaffold of many proteins. The smallest unit of organization, the nucleosome, is a stretch of DNA spooled around a core of histone proteins.

Chromosomal proteins partially loosen their hold on DNA during interphase. They interact to put the DNA in its most condensed form in the metaphase chromosome.

WHEN DNA CAN'T BE FIXED

1992 was truly an unforgettable year for Laurie Campbell. She finally turned eighteen. And in that same year she happened to notice a peculiar mole on her skin. This one was suspiciously black. It had an odd lumpiness about it, a ragged border, and a crusty surface. Laurie quickly made an appointment with her family doctor, who ordered a biopsy. The mole was a *malignant melanoma*—the deadliest form of skin cancer.

Laurie was lucky. She detected the cancer in its earliest stage, before it could spread through her body. Ever since, she consistently checks out the appearance of other moles that pepper her skin. She is acutely aware of having become a statistic—one of 500,000 people in the United States alone who develop skin cancer in any given year, and one of the 23,000 with malignant melanoma. She knows now that 7,500 die each year from skin cancer, and that 5,600 of them die from malignant melanoma.

Laurie is smart. She plotted out the position of every mole on her body. Once a month, this body map is her guide for a quick but thorough self-examination. Figure 11.10 shows examples of what she looks for. Laurie also schedules a medical examination every six months.

Changes in DNA are the triggers for skin cancer. The ultraviolet wavelengths in light from the sun, tanning lamps, and other sources can cause the molecular changes. Among other things, these wavelengths can promote covalent bonding between two adjacent thymine bases in a nucleotide strand. In this way, the two nucleotides to which the bases belong are combined into an abnormal, bulky molecule called a "thymine dimer."

Normally, a DNA repair mechanism gets rid of such bulky lesions. The mechanism requires at least seven gene products. Mutation in one or more of the required genes can skew the repair machinery. Thymine dimers can accumulate in skin cells of individuals with this type of mutation. The accumulation triggers development of lesions, including skin cancer.

The risk of skin cancer is greater for some than others. At one extreme, a newborn destined to develop *xeroderma pigmentosum* literally faces a dim future. The individuals affected by this genetic disorder cannot be exposed to sunlight, even briefly, without risking disfiguring skin tumors and possible early death from cancer.

You are at risk if you have moles that are chronically irritated (as by shaving or abrasive clothing). You are at risk if your skin, including your lip surface, is chronically chapped, cracked, or sore. You are at risk if your family has a history of cancer or if you have undergone radiation therapy. And, like Laurie, you are at risk if you have pale skin and burn easily in the sun. Even on slightly overcast days, fair-skinned people must apply a high SPF (Sun Protection Factor) sunblock, especially between 10 A.M. and 2 P.M. They should wear wide-brimmed hats, long sleeves, and long pants or skirts. Damaged DNA and skin cancer are the reality, in spite of the ill-advised, socially promoted allure of a golden tan.

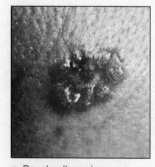

a Basal cell carcinoma

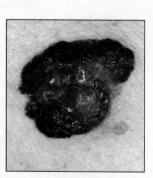

b Squamous cell carcinoma

c Malignant melanoma

Figure 11.10 Some examples of what happens when repair enzymes cannot fix changes in the nucleotide sequence of DNA. (**a**) *Basal cell carcinoma*, the most common form of skin cancer. This slow-growing, raised lump may be uncolored, reddish-brown, or black. (**b**) *Squamous cell carcinoma*, the second most common skin cancer. These pink growths, firm to the touch, grow rapidly under the surface of skin exposed to the sun. (**c**) *Malignant melanoma*, which spreads most rapidly. These malignant cells form very dark, encrusted lumps. They may itch like an insect bite or bleed easily. (**d**) Laurie Campbell, avoiding the sun—and melanoma.

SUMMARY

1. The hereditary information of cells and multicelled organisms is encoded in DNA (deoxyribonucleic acid).

2. DNA consists of nucleotide subunits. Each of these has a five-carbon sugar (deoxyribose), one phosphate group, and one of four kinds of nitrogen-containing bases (adenine, thymine, guanine, or cytosine).

3. A DNA molecule consists of two nucleotide strands twisted together into a double helix. The bases of one strand pair (hydrogen-bond) with bases of the other.

4. The bases of the two strands in a DNA double helix pair in constant fashion. Adenine pairs with thymine (A to T), and guanine with cytosine (G to C). *Which* base pair follows the next (A–T, T–A, G–C, or C–G) varies along the length of the strands.

5. Overall, the DNA of one species includes a number of unique stretches of base pairs that set it apart from the DNA of all other species.

6. During DNA replication, enzymes unwind the two strands of a double helix and assemble a new strand of complementary sequence on each parent strand. Two double-stranded molecules result. One strand of each molecule is "old" (it is conserved); the other is "new."

7. Numerous histones and other proteins are tightly bound to eukaryotic DNA. The structural organization of chromosomes arises from DNA–protein interactions.

Review Questions

1. Name the three molecular parts of a nucleotide in DNA. Name the four different bases that occur in these nucleotides. *178*

2. What kind of bond joins two DNA strands in a double helix? Which nucleotide base-pairs with adenine? With guanine? *179*

3. The relative amounts of the four nucleotide bases in DNA differ among species, yet they are the same among members of a single species. How does this base-pairing rule explain the constancy and variation we find among DNA molecules of different species? *179*

Self-Quiz *(Answers in Appendix IV)*

1. Which is *not* a nucleotide base in DNA?
   a. adenine   c. uracil   e. guanine
   b. thymine   d. cytosine

2. What are the base-pairing rules for DNA?
   a. A–G, T–C   c. A–U, C–G
   b. A–C, T–G   d. A–T, C–G

3. A DNA strand having the sequence C–G–A–T–T–G would be complementary to the sequence _____ .
   a. C–G–A–T–T–G   c. T–A–G–C–C–T
   b. G–C–T–A–A–G   d. G–C–T–A–A–C

4. One species' DNA differs from others in its _____ .
   a. sugars   c. base sequence
   b. phosphate groups   d. all of the above

5. When DNA replication begins, _____ .
   a. the two DNA strands unwind from each other
   b. the two DNA strands condense for base transfers
   c. two DNA molecules bond
   d. old strands move to find new strands

6. DNA replication requires _____ .
   a. free nucleotides   c. many enzymes
   b. new hydrogen bonds   d. all of the above

7. Match the DNA concepts appropriately.
   _____ base pairing
   _____ metaphase chromosome
   _____ constancy in base pairing
   _____ replication
   _____ DNA double helix

   a. two nucleotide strands twisted together
   b. A with T, G with C
   c. hereditary material duplicated
   d. accounts for life's diversity
   e. DNA with protein scaffold

Critical Thinking

1. Chargaff's data suggested that adenine pairs with thymine, and guanine pairs with cytosine. What other data available to Watson and Crick suggested that adenine-guanine and cytosine-thymine pairs normally do not form?

2. Matthew Meselson and Frank Stahl found experimental support for the semiconservative model of DNA replication. They made "heavy" DNA by growing *Escherichia coli* in a medium containing ^{13}N, the heavy isotope of nitrogen. They prepared "light" DNA by growing *E. coli* in the presence of ^{14}N, the more common isotope. An available technique helped them identify which replicated molecules were heavy, light, or hybrid (one heavy strand, one light).

   Use two pencils of different colors, one for heavy strands and one for light strands. Starting with a DNA molecule having two heavy strands, sketch the daughter molecules that would form after one round of replication in an ^{14}N-containing medium. Now sketch the four DNA molecules that would result if these daughter molecules were replicated a second time in the ^{14}N medium.

3. Mutations are the original source of genetic variation. This variation is the raw material of evolution. How can both statements be true, given that cells have efficient mechanisms to repair DNA before mutations (permanent changes in base sequences) arise?

4. As described in Figure 3.20, a new strain of *E. coli* has appeared. It has caused medical problems and fatalities, especially among children who ingested undercooked, contaminated beef. Develop a hypothesis to explain how a normally harmless bacterium can become a pathogen.

Selected Key Terms

bacteriophage *176*
DNA (deoxyribonucleic acid) *174*
DNA ligase *180*
DNA polymerase *180*
DNA repair *180*
DNA replication *180*
histone *180*
nucleosome *181*
nucleotide *178*
x-ray diffraction image *178*

Readings

Watson, J. 1978. *The Double Helix*. New York: Atheneum. Highly personal view of scientists and their methods, interwoven into an account of how DNA structure was discovered.

Wolfe, S. 1995. *Introduction to Molecular and Cellular Biology*. Belmont, California: Wadsworth. Comprehensive, current, and accessible.

12 FROM DNA TO PROTEINS

Beyond Byssus

Picture a mussel, of the sort shown in Figure 12.1. Hard-shelled but soft of body, it is using its muscular foot to probe a wave-scoured rock. At any moment, pounding waves can whack the mussel into the water, hurl it repeatedly against the rock with shell-shattering force, and so offer up a gooey lunch for gulls.

By chance, the mussel's foot comes across a crevice in the rock. The foot moves, broomlike, and sweeps the crevice clean. It presses down, forcing air out from underneath it, then arches up. The result is a vacuum-sealed chamber, rather like the one that forms when a plumber's rubber plunger is being squished down and up to unclog a drain. Into this vacuum chamber the mussel spews a fluid composed of keratin and other proteins, which bubbles into a sticky foam. And now, by curling its foot into a small tubular shape and pumping the foam through it, the mussel forms sticky threads about as wide as a human whisker. As a final touch, it varnishes the threads with another protein and thereby ends up with an adhesive called byssus—which anchors the mussel to the rock.

Byssus is the world's premier underwater adhesive. Nothing that humans have manufactured comes close. (Sooner or later, water chemically degrades or deforms synthetic adhesives.) Byssus fascinates biochemists, dentists, and surgeons looking for better ways to do tissue grafts and to rejoin severed nerves. Genetic

Figure 12.1 Mussels, busily demonstrating the importance of proteins for survival. When mussels come across a suitable anchoring site, they use their foot like a plumber's plunger to create a vacuum chamber. In this chamber they manufacture the world's best underwater adhesive from a mix of proteins. The adhesive anchors them to substrates in their wave-swept habitat.

engineers insert mussel DNA into yeast cells, which reproduce in large numbers and serve as "factories" for translating mussel genes into useful quantities of proteins. This exciting work, like the mussel's own byssus building, starts with one of life's universal concepts. *Every protein is synthesized in accordance with instructions contained in DNA.*

You are about to trace the steps leading from DNA to protein. Many enzymes are players in this pathway, and so is another kind of nucleic acid besides DNA. The same steps produce *all* proteins, from mussel-inspired adhesives to the keratin in your hair and fingernails to the insect-digesting enzymes of a Venus flytrap.

Start out by thinking of each cell's DNA as a book of protein-building instructions. The alphabet used to create the book is simple enough—A, T, G, and C (for the nucleotide bases adenine, thymine, guanine, and cytosine). How do you get from this alphabet to a protein? The answer starts with DNA's structure.

DNA, recall, is a double-stranded molecule. Which kind of nucleotide base follows the next along the length of a strand—that is, the **base sequence**—differs from one kind of organism to the next. As you read in the preceding chapter, before a cell divides, its DNA is replicated and the two strands unwind entirely from each other. However, at other times in a cell's life, the two strands unwind only in certain regions to expose particular base sequences—genes. *And those genes contain the instructions for building proteins.*

It takes two steps, **transcription** and **translation**, to carry out a gene's protein-building instructions. In eukaryotic cells, transcription proceeds in the nucleus. In this step, a selected base sequence in DNA serves as a structural pattern—as a template—for assembling a strand of **ribonucleic acid** (**RNA**) from the cell's pool of free nucleotides. Afterward, the RNA moves into the cytoplasm, where translation proceeds. In this second step, RNA directs the assembly of amino acids into polypeptide chains. Later, the chains become folded into the three-dimensional shapes of proteins.

In short, DNA guides the synthesis of RNA, then RNA guides the synthesis of proteins:

$$DNA \xrightarrow{\text{transcription}} RNA \xrightarrow{\text{translation}} protein$$

The new proteins will have structural and functional roles in cells. Some even will have roles in building more DNA, RNA, and proteins.

KEY CONCEPTS

1. Life cannot exist without enzymes and other proteins. Proteins consist of polypeptide chains, which consist of amino acids. The sequence of amino acids corresponds to a gene, which is a sequence of nucleotide bases in DNA.

2. The path leading from genes to proteins has two steps, called transcription and translation.

3. In transcription, the double-stranded DNA molecule is unwound at a gene region, then an RNA molecule is assembled on the exposed bases of one of the strands.

4. In translation, RNA directs the linkage of one amino acid after another, in the sequence required to produce a specific kind of polypeptide chain.

5. With few exceptions, the genetic "code words" by which DNA instructions are translated into proteins are the same in all species of organisms.

6. A gene's base sequence is vulnerable to permanent changes, called mutations. Such changes are the original source of genetic variation.

7. Mutations lead to alterations in protein structure, protein function, or both. These alterations may lead to small or large differences in traits among the individuals of a given population.

12.1 TRANSCRIPTION OF DNA INTO RNA

Classes of RNA

The introduction to this chapter may have left you with the impression that protein synthesis requires only one class of RNA molecules. Actually, it requires three. Transcription of most genes produces **messenger RNA (mRNA)**, the only class of RNA that carries protein-building instructions. Transcription of some other genes produces **ribosomal RNA (rRNA)**, the main component of ribosomes. Ribosomes, recall, are structures upon which polypeptide chains are assembled. Transcription of still other genes produces **transfer RNA (tRNA)**. As you will see, tRNA delivers amino acids one by one to a ribosome, in the order specified by mRNA.

How RNA Is Assembled

An RNA molecule is almost, but not quite, like a single strand of DNA. It, too, is composed of only four types of nucleotides, each consisting of a five-carbon sugar (ribose), a phosphate group, and a base. Three of its bases (adenine, cytosine, and guanine) are the same as those in DNA. But the fourth type of base in RNA is **uracil**, not thymine. Figure 12.2 shows its structure. Like thymine, uracil can pair with adenine. This means a new RNA strand can be put together on a DNA region according to base-pairing rules (Figure 12.3).

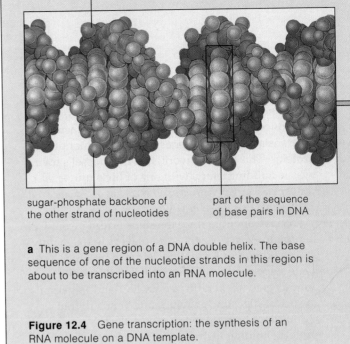

sugar-phosphate backbone of one strand of nucleotides

sugar-phosphate backbone of the other strand of nucleotides

part of the sequence of base pairs in DNA

a This is a gene region of a DNA double helix. The base sequence of one of the nucleotide strands in this region is about to be transcribed into an RNA molecule.

Figure 12.4 Gene transcription: the synthesis of an RNA molecule on a DNA template.

Transcription resembles DNA replication in another respect. Enzymes add the nucleotide units to a growing RNA strand one at a time, in the 5′ → 3′ direction. (Here you may wish to refer to Figures 11.6 and 11.7.)

Transcription *differs* from DNA replication in three key respects. First, only a certain stretch of one DNA strand—not the whole molecule—acts as the template. Second, different enzymes, called **RNA polymerases**, catalyze the addition of nucleotides to the 3′ end of a growing strand. Third, transcription results in a *single* strand of nucleotides in RNA.

Transcription starts at a **promoter**, a base sequence in DNA that signals the start of a gene. Proteins help position an RNA polymerase on the DNA so it binds with the promoter. The enzyme moves along the DNA strand, joining nucleotides one after another (Figure 12.4). When it reaches a base sequence that serves as a stop signal, the RNA is released as a free transcript.

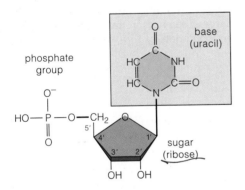

Figure 12.2 Structure of one of four types of nucleotides of RNA. The three other kinds have a different base (adenine, guanine, or cytosine instead of the uracil shown here). Compare Figure 11.5, which shows the nucleotides of DNA.

Finishing Touches on mRNA Transcripts

Newly formed mRNA is an unfinished molecule; it must be modified before its protein-building instructions can be put to use. Just as a dressmaker might snip off some threads or add bows on a dress before it leaves the shop, so do eukaryotic cells tailor this pre-mRNA.

Very quickly, enzymes attach a cap to the 5′ end of the pre-mRNA molecule. The cap is a nucleotide that has a methyl group and phosphate groups bonded to it.

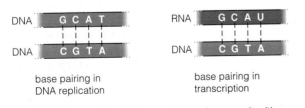
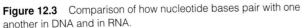

base pairing in DNA replication

base pairing in transcription

Figure 12.3 Comparison of how nucleotide bases pair with one another in DNA and in RNA.

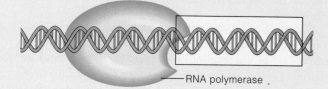

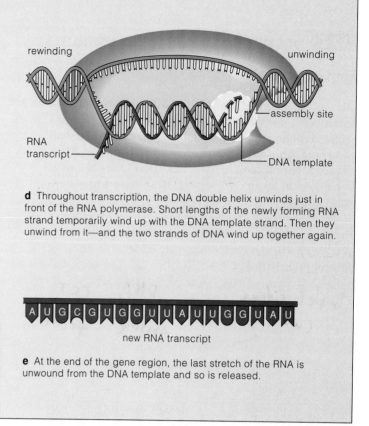

b An RNA polymerase molecule binds to a promoter in the DNA. It will use the base sequence positioned downstream from that site as a template for linking nucleotides into a strand of RNA.

d Throughout transcription, the DNA double helix unwinds just in front of the RNA polymerase. Short lengths of the newly forming RNA strand temporarily wind up with the DNA template strand. Then they unwind from it—and the two strands of DNA wind up together again.

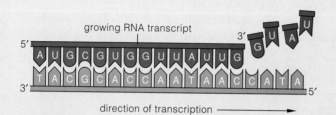

direction of transcription ⟶

c During transcription, RNA nucleotides are base-paired, one after another, with the exposed bases on the DNA template.

e At the end of the gene region, the last stretch of the RNA is unwound from the DNA template and so is released.

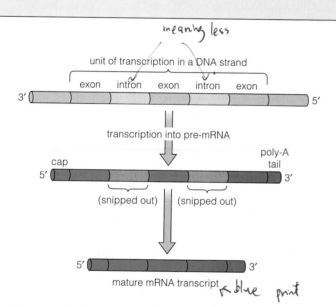

meaning less

↗ blue print

Figure 12.5 Transcription and modification of newly formed mRNA in the nucleus of eukaryotic cells. The cap simply is a nucleotide with functional groups attached. The poly-A tail is a string of adenine nucleotides.

Also, enzymes attach a tail of about 100 to 200 adenine-containing nucleotides to the 3′ end of most pre-mRNA transcripts. Hence the name, "poly-A tail" (for multiple adenine units). The tail becomes wound up with proteins. Later, in the cytoplasm, the cap will facilitate the binding of mRNA to a ribosome. Also, enzymes will gradually destroy the wound-up tail from the tip on back, then the mRNA. Such tails "pace" enzyme access

to mRNA. Apparently they help keep protein-building messages intact for as long as the cell requires them.

Besides these alterations, the mRNA message itself becomes modified. Most eukaryotic genes have one or more **introns**, which are base sequences that do not get translated into an amino acid sequence. The introns intervene between **exons**, the only parts of mRNA that become translated into protein. As Figure 12.5 indicates, the introns are transcribed right along with exons, but enzymes snip them out before the mRNA leaves the nucleus in mature form.

It could be that some introns are evolutionary junk, the leftovers of past mutations that led nowhere. Yet other introns are sites where instructions for building a particular protein can be snipped apart and spliced back together in different ways. The alternative splicing allows different cells in your body to use the same gene to make modified versions of a pre-mRNA transcript—and modified versions of the resulting protein. We will return to this topic in the next chapter.

In gene transcription, a sequence of exposed bases in one of the two strands of a DNA molecule serves as the template for assembling a single strand of RNA. The assembly follows base-pairing rules (adenine only with uracil, cytosine only with guanine).

Before leaving the nucleus, new RNA transcripts undergo modification into final form.

The Genetic Code

Like a DNA strand, mRNA is a linear sequence of nucleotides. But what are the protein-building "words" encoded in its sequence? Gobind Khorana, Marshall Nirenberg, and others came up with the answer. They deduced that nucleotide bases are "read" *three at a time*, as triplets. In an mRNA strand, these base triplets are called <u>codons</u>.

Figure 12.6 will give you an idea of how the order of different codons in an mRNA strand dictates the order in which particular amino acids are added to a growing polypeptide chain.

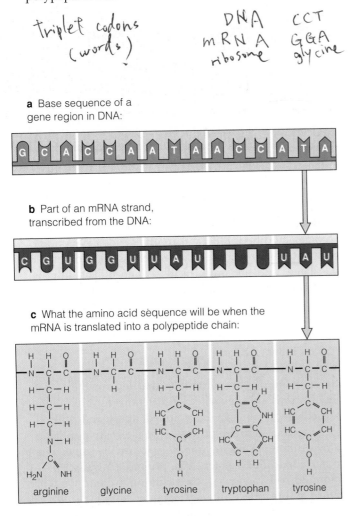

a Base sequence of a gene region in DNA:

b Part of an mRNA strand, transcribed from the DNA:

c What the amino acid sequence will be when the mRNA is translated into a polypeptide chain:

arginine   glycine   tyrosine   tryptophan   tyrosine

Figure 12.6 The steps from genes to proteins. (**a**) This diagram represents a region of a DNA double helix that was unwound during transcription.

(**b**) The exposed bases on one DNA strand served as a template for assembling an mRNA strand. In the new mRNA transcript, every three nucleotide bases equaled one codon. Each codon called for one amino acid in this polypeptide chain.

(**c**) Referring to Figure 12.7, can you fill in the blank codon for tryptophan in the chain?

Count the codons listed in Figure 12.7, and you see there are sixty-four kinds. Most of the twenty kinds of amino acids correspond to more than one kind of codon. (Glutamate corresponds to the code words GAA *or* GAG, for example.) The codon AUG also establishes the reading frame for translation. That is, at ribosomes, the "three-bases-at-a-time" selections start at an AUG that functions as the start signal in an mRNA strand. Three other codons (UAA, UAG, UGA) are stop signals. They prevent the further addition of more amino acids to a new polypeptide chain.

The set of sixty-four different codons is the **genetic code**. It is the basis of protein synthesis in all organisms.

First Letter	Second Letter				Third Letter
	U	C	A	G	
U	phenylalanine	serine	tyrosine	cysteine	U
	phenylalanine	serine	tyrosine	cysteine	C
	leucine	serine	stop	stop	A
	leucine	serine	stop	tryptophan	G
C	leucine	proline	histidine	arginine	U
	leucine	proline	histidine	arginine	C
	leucine	proline	glutamine	arginine	A
	leucine	proline	glutamine	arginine	G
A	isoleucine	threonine	asparagine	serine	U
	isoleucine	threonine	asparagine	serine	C
	isoleucine	threonine	lysine	arginine	A
	(start) methionine	threonine	lysine	arginine	G
G	valine	alanine	aspartate	glycine	U
	valine	alanine	aspartate	glycine	C
	valine	alanine	glutamate	glycine	A
	valine	alanine	glutamate	glycine	G

Figure 12.7 The genetic code. The codons in mRNA are nucleotide bases, read in blocks of three. Sixty-one of these base triplets correspond to specific amino acids. Three others serve as signals that stop translation. The left column of the diagram shows the first of the three nucleotides in each codon in mRNA. The middle columns show the second nucleotide. The right column shows the third. Reading from left to right, for instance, the triplet U G G corresponds to tryptophan. Both U U U and U U C correspond to phenylalanine.

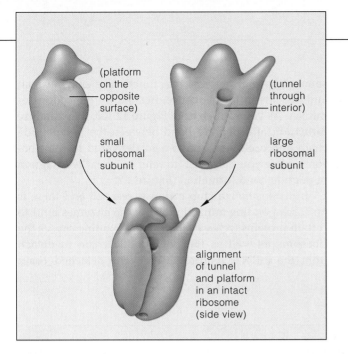

Figure 12.8 Model of eukaryotic ribosomes. Polypeptide chains are assembled on the small subunit's platform. Newly forming chains may move through the large subunit's tunnel.

Figure 12.9 *Right* (**a**) Computer-generated, three-dimensional model of one type of tRNA molecule. The tRNA (*reddish brown*) is shown attached to a bacterial enzyme (*green*), along with an ATP molecule (*gold*). This particular enzyme attaches amino acids to tRNAs. (**b**) Structural features common to all tRNAs. (**c**) Simplified model of tRNA that you will come across in illustrations that follow. The "hook" sketched at one end is the site to which a specific amino acid can become attached.

Roles of tRNA and rRNA

Before its message can be translated, mRNA must bind to specific sites on the surface of **ribosomes**. As shown in Figure 12.8, each ribosome has two subunits. These are assembled in the nucleus from rRNA and proteins, some of which show enzyme activity. The subunits are shipped separately to the cytoplasm. There they will combine as functional units only during translation.

In addition to the ribosomal subunits, the cytoplasm also contains pools of free amino acids and free tRNA molecules. The tRNAs each have a molecular "hook," an attachment site for amino acids. They also have an **anticodon**, a nucleotide triplet that can base-pair with codons (Figure 12.9). When tRNAs bind to codons, they position their attached amino acids automatically, in the order specified by mRNA.

A cell has more than sixty kinds of codons but can utilize fewer kinds of tRNAs. How do they match up? According to base-pairing rules, adenine must pair with uracil, and cytosine with guanine. However, for codon-anticodon interactions, the rules loosen up at the third base. For example, CCU, CCC, CCA, and CCG all

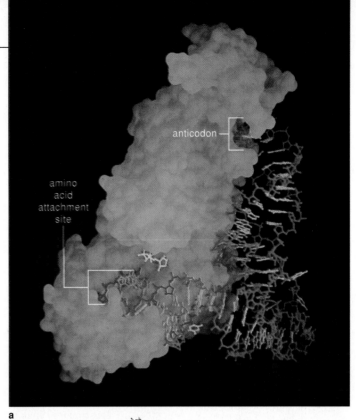

a

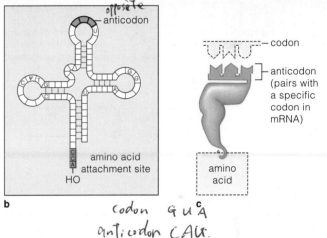

b

specify proline. Such freedom in codon-anticodon pairing at the third base is called the "wobble effect."

The nucleotide sequence of both DNA and mRNA encodes protein-building instructions. The genetic code is a set of sixty-four base triplets (nucleotide bases, read in blocks of three). A codon is a base triplet in mRNA.

Different combinations of codons specify the amino acid sequence of different polypeptide chains, start to finish.

mRNAs are the only molecules that carry protein-building instructions from DNA to the cytoplasm.

tRNAs bind to codons and position their attached amino acids in the order specified by mRNA. Their action translates mRNA into a corresponding sequence of amino acids.

rRNAs are components of ribosomes, the structures upon which amino acids are assembled into polypeptide chains.

Translation proceeds in the cytoplasm. It requires three stages, called initiation, elongation, and termination.

In *initiation*, a certain tRNA that can start transcription and an mRNA transcript become loaded onto a ribosome. First, the initiator tRNA binds with the small ribosomal subunit. So does the start codon (AUG) of the mRNA transcript. After this, a large ribosomal subunit binds with the small subunit, and so completes the initiation complex (Figure 12.10a). The next stage can begin.

In *elongation*, a new polypeptide chain forms as the mRNA passes between the ribosomal subunits, like a thread being moved through the eye of a needle. Again,

some ribosomal proteins that function as enzymes join amino acids in the sequence dictated by the codons of mRNA. As you can see from Figure 12.10b, they catalyze formation of a peptide bond between the polypeptide chain and each new amino acid delivered to the ribosome. Here you may wish to refer to an earlier diagram of peptide bond formation (Figure 2.25).

In *termination*, a stop codon is reached and there is no corresponding anticodon. Now the enzymes bind to certain proteins (release factors). The binding causes the ribosome as well as the polypeptide chain to detach from the mRNA (Figure 12.10c). The detached chain

Figure 12.10 Translation, the second step of protein synthesis.

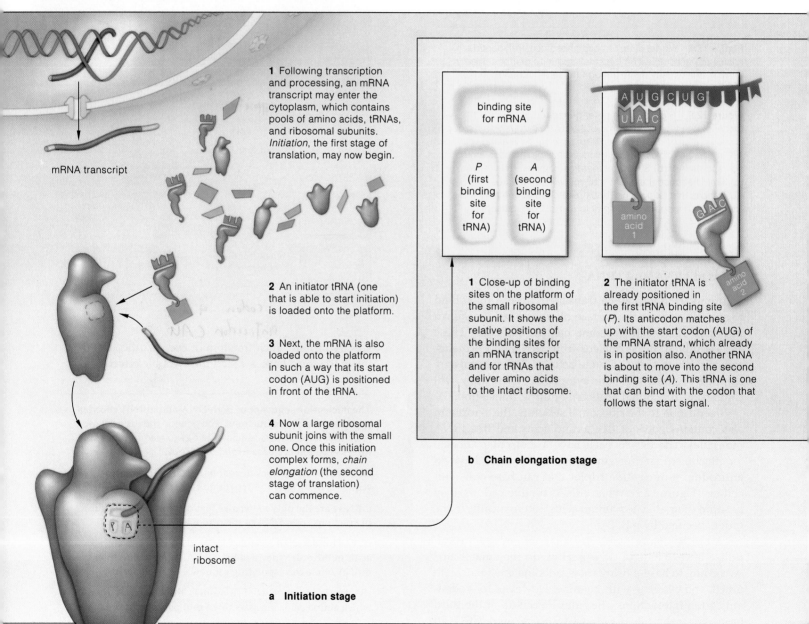

1 Following transcription and processing, an mRNA transcript may enter the cytoplasm, which contains pools of amino acids, tRNAs, and ribosomal subunits. *Initiation*, the first stage of translation, may now begin.

mRNA transcript

2 An initiator tRNA (one that is able to start initiation) is loaded onto the platform.

3 Next, the mRNA is also loaded onto the platform in such a way that its start codon (AUG) is positioned in front of the tRNA.

4 Now a large ribosomal subunit joins with the small one. Once this initiation complex forms, *chain elongation* (the second stage of translation) can commence.

intact ribosome

a   Initiation stage

binding site for mRNA

P (first binding site for tRNA)

A (second binding site for tRNA)

1 Close-up of binding sites on the platform of the small ribosomal subunit. It shows the relative positions of the binding sites for an mRNA transcript and for tRNAs that deliver amino acids to the intact ribosome.

AUGCUG
UAC

amino acid 1

GAC

amino acid 2

2 The initiator tRNA is already positioned in the first tRNA binding site (*P*). Its anticodon matches up with the start codon (AUG) of the mRNA strand, which already is in position also. Another tRNA is about to move into the second binding site (*A*). This tRNA is one that can bind with the codon that follows the start signal.

b   Chain elongation stage

may join the pool of free proteins in the cytoplasm. Or it may enter the cytomembrane system, starting with the compartments of rough ER, even before completion. Many of the newly formed chains take on final form in that system before they are shipped on to their destinations, either inside or outside the cell (Section 3.6).

Often, many ribosomes and tRNAs translate the same mRNA transcript simultaneously. The transcript threads through all the ribosomes, which are arranged one after another in assembly-line fashion. These close arrays of ribosomes on the transcript are sometimes called polysomes. Their presence indicates that a cell is producing many copies of a polypeptide chain from the same mRNA transcript.

Translation is initiated through the convergence of a small ribosomal unit, an initiator tRNA, an mRNA transcript, and then a large ribosomal subunit.

Next, anticodons of tRNAs base-pair with mRNA codons. A polypeptide chain grows as peptide bonds form between it and each new amino acid delivered to the ribosome.

Translation ends when a stop codon triggers events that cause the chain and the mRNA to detach from the ribosome.

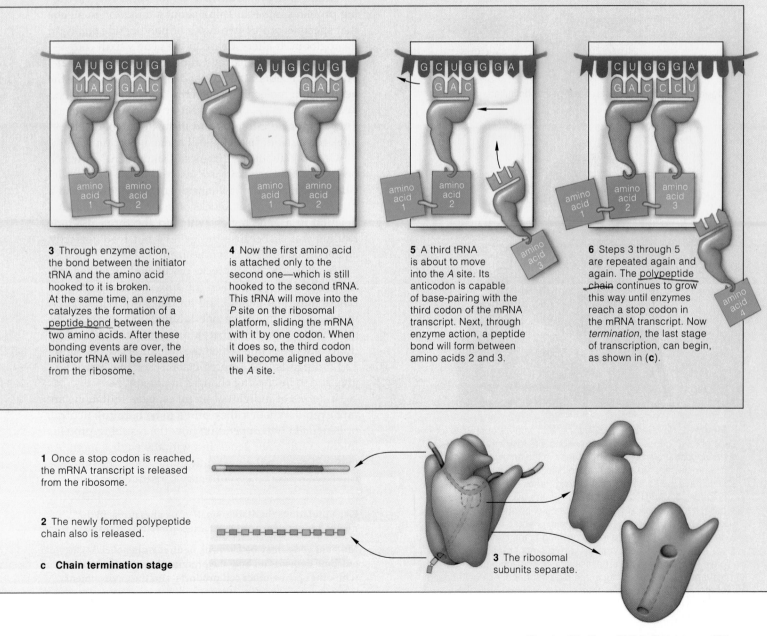

3 Through enzyme action, the bond between the initiator tRNA and the amino acid hooked to it is broken. At the same time, an enzyme catalyzes the formation of a peptide bond between the two amino acids. After these bonding events are over, the initiator tRNA will be released from the ribosome.

4 Now the first amino acid is attached only to the second one—which is still hooked to the second tRNA. This tRNA will move into the *P* site on the ribosomal platform, sliding the mRNA with it by one codon. When it does so, the third codon will become aligned above the *A* site.

5 A third tRNA is about to move into the *A* site. Its anticodon is capable of base-pairing with the third codon of the mRNA transcript. Next, through enzyme action, a peptide bond will form between amino acids 2 and 3.

6 Steps 3 through 5 are repeated again and again. The polypeptide chain continues to grow this way until enzymes reach a stop codon in the mRNA transcript. Now *termination*, the last stage of transcription, can begin, as shown in (**c**).

1 Once a stop codon is reached, the mRNA transcript is released from the ribosome.

2 The newly formed polypeptide chain also is released.

c Chain termination stage

3 The ribosomal subunits separate.

HOW MUTATIONS AFFECT PROTEIN SYNTHESIS

Whenever a cell puts its genetic code into action, it is making precisely those proteins that it requires for its structure and functions. If something changes a gene's code words, the resulting protein will change. If the protein is central to cell architecture or metabolism, we can expect the outcome to be a dead or damaged cell.

Every so often, genes do change. Maybe one base is substituted for another in the DNA sequence. Or maybe

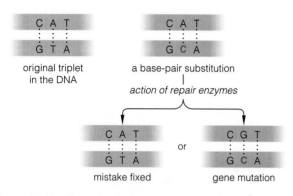

Figure 12.11 Example of a base-pair substitution.

Figure 12.12 Barbara McClintock, who won a Nobel Prize for her insight that genes can move from one site to another in DNA, as transposable elements. In her hands is Indian corn (*Zea mays*); its kernels sent her on the road to discovery. In the kernels, all cells have the same pigment-coding genes. Yet some kernels are colorless or spottily colored. In this plant's ancestor, a gene in a cell left its position in one DNA molecule, invaded another, and shut down a pigment gene. The plant inherited the mutation. As cell divisions proceeded in the growing plant, none of the mutated cell's descendants could synthesize pigment molecules. They gave rise to colorless kernel tissue. Later, in some cells, the movable gene slipped out of the pigment gene. All descendants of *those* cells produced pigment—and colored kernel tissue.

an extra base is inserted or a base is lost. These small-scale changes in the nucleotide sequence of the DNA molecule are **gene mutations**.

Some mutations result from exposure to **mutagens**, agents that heighten the risk of heritable alterations in DNA's structure. Ultraviolet light, some substances in tobacco smoke, and those rogue molecular fragments called free radicals (page 64) are common mutagens.

Mutations may arise in cells even in the absence of mutagens. For example, spontaneous mutations may follow a replication error, as when adenine is wrongly paired with a cytosine unit in a DNA template strand. DNA repair enzymes may detect the error—then "fix" the wrong base (Figure 12.11).

Whether it is spontaneous or induced by a mutagen, the outcome of a "base-pair substitution" may be the replacement of an amino acid with a different one during protein synthesis. Think about a mutation in a gene that specifies one of two types of polypeptide chains in hemoglobin. It alters a base in the gene's sixth codon—which thereafter calls for valine instead of glutamate. This change puts a sticky (hydrophobic) patch on the hemoglobin molecules. When the oxygen level in blood declines, the molecules interact at their sticky patches. They aggregate into rods and distort red blood cells. In time, the outcome is sickle-cell anemia (Section 9.5).

Or think about the "frameshift mutation," in which one to several base pairs are inserted into a DNA molecule or deleted from it. Remember, polymerases read a nucleotide sequence in blocks of three. An insertion or a deletion in a gene region will shift the three-at-a-time reading frame. Because the gene is not read correctly, an abnormal protein is synthesized.

As Barbara McClintock discovered, spontaneous mutations also result when transposable elements (once called "jumping genes") are on the move. These DNA regions can move from one location to another in the same DNA molecule or to a different one. Often they inactivate genes into which they are inserted. As Figure 12.12 suggests, the unpredictability of the jumps can give rise to interesting changes in phenotype.

In terms of individual lifetimes, gene mutations are rare events. Whether they prove to be harmful, neutral, or beneficial will depend on how the resulting proteins interact with other genes and with the environment. As you will read in the next unit of the book, the outcome can have powerful evolutionary consequences.

Gene mutations (heritable, small-scale alterations in the nucleotide sequence of DNA) affect protein synthesis.

Mutated genes may be harmful, neutral, or beneficial. The outcome depends on how the proteins they specify interact with other genes, other cell products, and the environment.

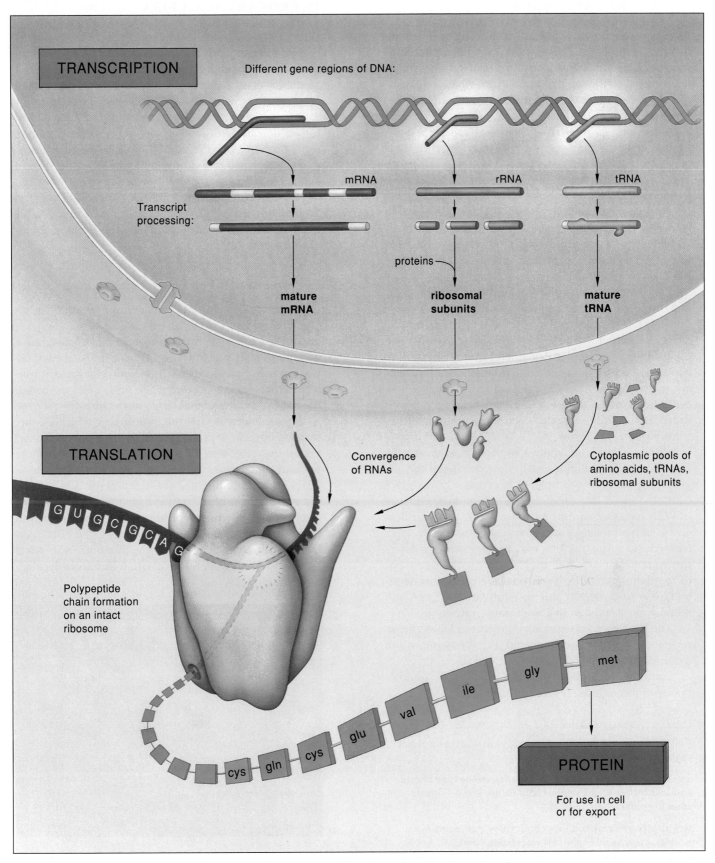

Figure 12.13 Summary of transcription and translation—the steps of protein synthesis in eukaryotic cells.

12.6 THE NATURE OF CONTROLS OVER GENE EXPRESSION

At this very moment, bacteria are feeding on nutrients inside your gut. Your own red blood cells are binding, transporting, or giving up oxygen, and great numbers of epithelial cells deep in your skin are busily synthesizing the protein keratin. Like cells everywhere, they are functioning by virtue of the protein products of genes.

Cells don't express all of their genes all of the time. Rather, cells use some genes and their products only once. They use other genes at certain times, all of the time, or not at all. *Which genes are being expressed depends on the type of cell, its moment-by-moment adjustments to changing chemical conditions, which signals from the outside it happens to be receiving—and its built-in control systems.*

The control systems consist of certain molecules, including **regulatory proteins**, that interact with DNA, RNA, or gene products.

For example, availability of nutrients shifts rapidly and often for enteric bacteria, which live in animal intestines. Like other prokaryotic cells, they can rapidly transcribe certain genes and synthesize many nutrient-digesting enzyme molecules when nutrients happen to be moving past—then restrict synthesis when nutrients are scarce. By contrast, the composition and solute concentrations of the fluid bathing your cells do not shift drastically. And few of your cells exhibit rapid shifts in transcription.

Different kinds of control systems operate during transcription, translation, and after translation (on the gene product). For example, cells commonly have two control systems that can block transcription or enhance it. With **negative control systems**, a regulatory protein is able to bind to DNA at a certain site and thereby *block* transcription. The blocking effect of negative regulators can be reversed, as when they interact with other types of regulatory proteins. With **positive control systems**, a regulatory protein is able to bind with the DNA and thereby *promote* the initiation of transcription.

The binding of a regulatory protein to DNA can be reversed, as when conditions that triggered synthesis of a particular protein change.

Summing up, these are the points to keep in mind as you read through the rest of this chapter:

Cells exert control over when, how, and to what extent each of their genes is expressed.

The expression of a given gene depends on the type of cell and its functions, on chemical conditions, and on signals from the outside environment.

Regulatory proteins and other molecules exert control over gene expression through their interactions with DNA, RNA, and gene products (polypeptide chains, and then proteins).

12.7 EXAMPLES OF GENE CONTROL IN PROKARYOTIC CELLS

When nutrients are plentiful and other environmental conditions also favor growth, bacteria tend to grow and divide indefinitely. Gene controls promote the rapid synthesis of enzymes with roles in nutrient digestion and other growth-related activities. Transcription is rapid, and translation begins even before mRNA transcripts are finished. (Bacteria have no nucleus; nothing separates their DNA from ribosomes in the cytoplasm.)

If a nutrient-degrading pathway requires several enzymes, all genes for those enzymes are transcribed, often into one continuous mRNA molecule. The genes are not transcribed when conditions turn unfavorable. Let's look at two examples of this all-or-nothing control of transcription.

Negative Control of Transcription

Escherichia coli is an enteric bacterium residing in the mammalian gut. It lives on glucose, lactose (a sugar in milk), and other ingested nutrients. Like other adult mammals, you probably don't drink milk around the clock. When you do drink it, *E. coli* cells rapidly transcribe three genes for enzymes with roles in certain breakdown reactions that begin with lactose.

A promoter precedes the three genes, which are next to one another. A promoter, recall, is a base sequence that signals the start of a gene. Another sequence, an **operator**, intervenes between a promoter and bacterial genes. It is a binding site for a **repressor**, a regulatory protein that can block transcription (Figure 12.14). An arrangement in which a promoter and operator service more than one gene is an **operon**. Elsewhere in *E. coli* DNA, a different gene codes for this repressor, which can bind with the operator *or* with a lactose molecule.

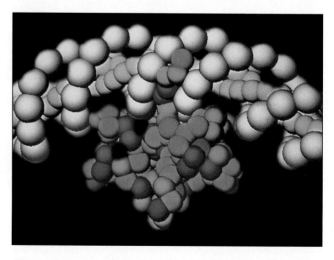

Figure 12.14 Model of a repressor protein (*green*) binding to an operator in bacterial DNA (*blue*).

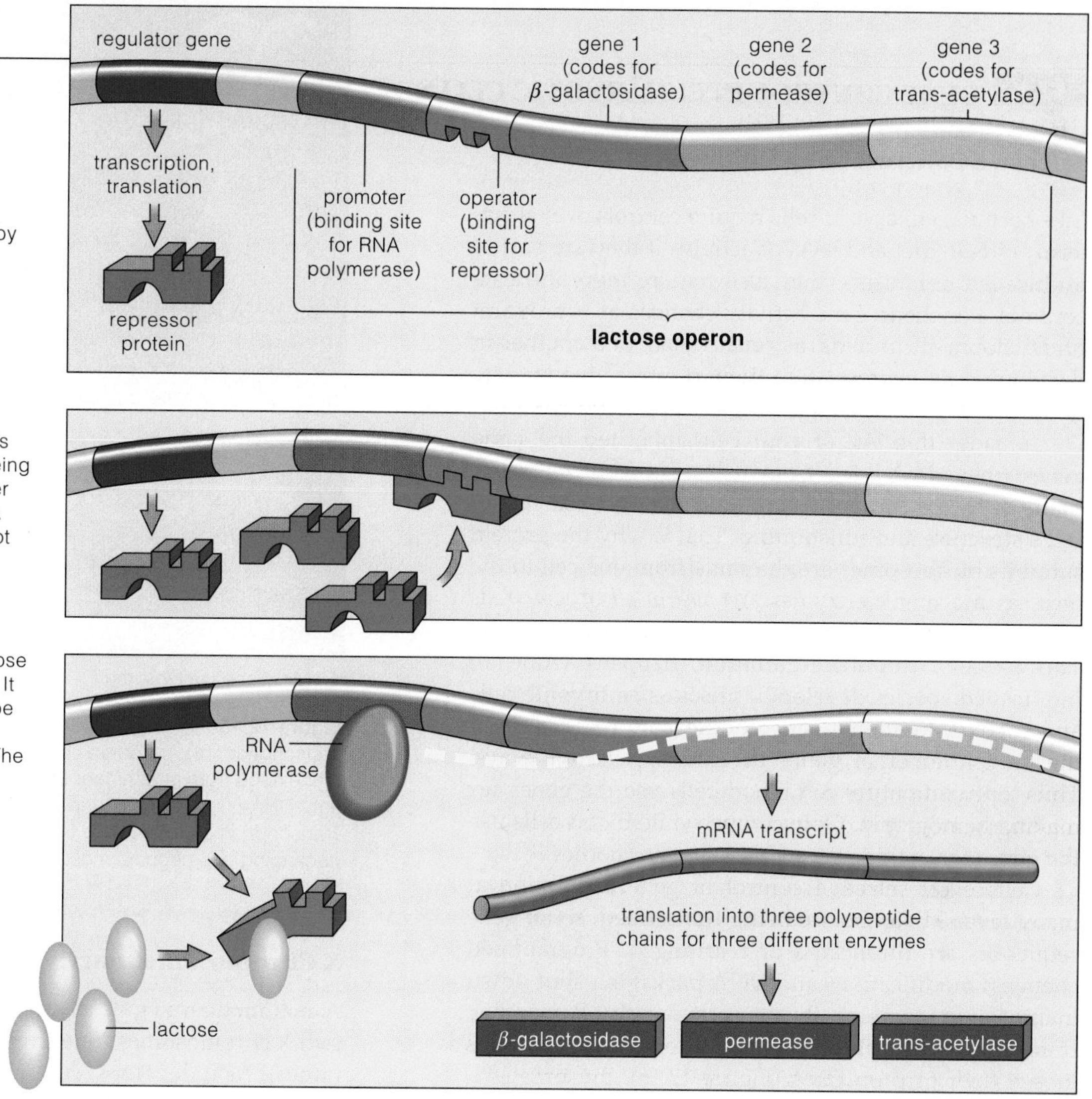

a A repressor protein exerts negative control over three genes of the lactose operon by binding to the operator and inhibiting transcription.

b When the concentration of lactose is low, the repressor is free to block transcription. Being bulky, it overlaps the promoter and prevents binding by RNA polymerase. The enzymes (not needed) are not produced.

c At high concentration, lactose is an inducer of transcription. It binds to and distorts the shape of the repressor—which now cannot bind to the operator. The promoter is exposed and the genes can be transcribed.

Figure 12.15 Negative control of the lactose operon. The first gene of this operon codes for an enzyme that splits lactose, a disaccharide, into two subunits (glucose and galactose). The second codes for an enzyme that transports lactose into cells. The third enzyme functions in metabolizing certain sugars.

When the lactose concentration is low, a repressor binds with the operator (Figure 12.15). Being a large molecule, it overlaps the promoter and blocks transcription. Thus, *lactose-degrading enzymes are not built when they are not required.* When the lactose concentration is high, the odds are greater that a lactose molecule will bind with the repressor. Binding alters the repressor's shape, so it cannot bind with the operator—and RNA polymerase can now transcribe the genes. Thus, *lactose-degrading enzymes are built only when required.*

Positive Control of Transcription

E. coli cells pay far more attention to glucose than to lactose. They transcribe genes for glucose breakdown continually, at faster rates. Even if lactose is present, the lactose operon isn't used much—*unless glucose is absent.*

At such times, a regulatory protein called CAP acts on the operon. CAP is an **activator protein**, a key player in a positive control system. To understand how it works, you have to know that the lactose operon's promoter is not good at binding RNA polymerase. It does a better job when CAP adheres to it first. But CAP won't do this unless it is activated by a small molecule called cAMP.

Among other things, cAMP is produced from ATP—which *E. coli* can produce by glucose breakdown (that is, by way of glycolysis). Not much cAMP is available in the cell when glucose is plentiful and glycolysis is proceeding full bore. When these conditions prevail, the activator protein does not become primed to adhere to the promoter, and transcription of the lactose operon genes slows almost to a standstill. Now suppose that glucose is scarce and lactose becomes available. cAMP can accumulate, CAP-cAMP complexes can form—and the lactose operon genes can receive suitable attention.

Prokaryotic cells, which must respond rapidly to changing conditions, commonly rely on a small number of regulatory proteins that exert rapid, on-off control of transcription.

12.8 GENE CONTROL IN EUKARYOTIC CELLS

Selective Controls in Specialized Cells

Like bacteria, eukaryotic cells require controls over short-term shifts in diet and level of activity. If they are part of multicelled organisms, they also require more intricate controls. For them, gene activity changes as a program of development unfolds, as cells contact one another in developing tissues, and as cells start interacting by way of hormones and other signaling molecules.

Consider this. All of your cells inherited the same genes (they descended from the same fertilized egg). Many of the genes specify proteins that are basic to any cell's structure and functioning. That is why the protein subunits of ribosomes are the same from one cell to the next, as are many enzymes. *Yet nearly all of your cells became specialized in composition, structure, and function.* This process, **cell differentiation**, happens while all multicelled species develop. It arises as embryonic cells and their descendants activate and suppress a few of the total number of genes in unique, selective ways. Thus, only immature red blood cells use the genes for making hemoglobin. Only certain white blood cells use the genes for making weapons called antibodies.

Cells exert selective control of gene expression at many levels. Even before being transcribed, some gene sequences are duplicated or rearranged. Programmed chemical modifications and DNA packaging shut down many genes, temporarily or permanently. Remember, transcription can proceed when chromosomal proteins loosen their grip on DNA (Figure 12.16). By some estimates, cells of a multicelled organism rarely use more than 5 to 10 percent of their genes at a given time. One way or another, most of the genes are being repressed. Which ones are used depends on the stage of growth and development that the organism is passing through.

We see less of the rapid, all-or-nothing control over transcription that is common among bacteria, for the "internal environment" of the multicelled body affords more stable operating conditions. Most often, transcription rates rise or fall by degrees because of slight shifts in the concentrations of signaling molecules, substrates, and products. Post-transcriptional controls deal with transcript processing, transport of mature RNAs from the nucleus, and translation rates. Other controls deal with modifying new polypeptides or with activating, inhibiting, and degrading the protein products.

Finally, many genes specify enzymes, so control of their transcription, translation, and expression extends to *all* metabolic reactions. Indirectly, they influence the type and number of molecules that will be synthesized and degraded at a given time. *Thus they govern all short-term and long-term aspects of cell structure and function.*

We will return to this topic in Chapter 34. For now, let's simply consider a few examples of gene control.

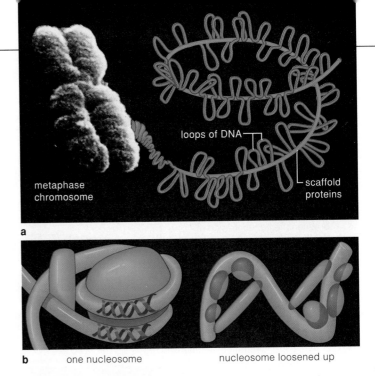

a

b   one nucleosome   nucleosome loosened up

Figure 12.16 Changes in DNA organization that may promote transcription. (**a**) At certain times, DNA decondenses into loops that extend from attachment sites on scaffold proteins. (**b**) The tight DNA-histone packing in nucleosomes also may loosen up. Transcription occurs mainly in regions where chromosome packaging has become most relaxed. Compare Figure 11.8.

X Chromosome Inactivation

A mammalian zygote destined to become a female has two X chromosomes (one from the mother, one from the father). As it develops, *one* of the two condenses in each cell. The condensation is a programmed event, but the outcome is random. *One or the other chromosome may be inactivated.* The condensed X chromosome is visible as a dark spot in the interphase nucleus (Figure 12.17a). It is called a **Barr body** after its discoverer, Murray Barr.

When the maternal (or paternal) X chromosome is inactivated in a cell, it also becomes inactivated in all of the cell's descendants. Thus, by adulthood, every adult female is a "mosaic" for the X chromosomes. That is, *she has patches of tissues in which maternal genes are being expressed—and patches of tissue in which paternal genes are being expressed.* Because any pair of alleles on her two X chromosomes may or may not be identical, the tissue patches may or may not have the same characteristics.

Mary Lyon discovered the mosaic tissue effect that arises from random X chromosome inactivation. A good example is *anhidrotic ectodermal dysplasia.* Human females with this heritable condition are heterozygous for a recessive allele that prevents sweat glands from forming. The allele is on the X chromosome. In such females, the X chromosome bearing the dominant allele has been condensed—and genes on the one bearing the mutated allele are being transcribed in some patches of skin that have no sweat glands (Figure 12.17b).

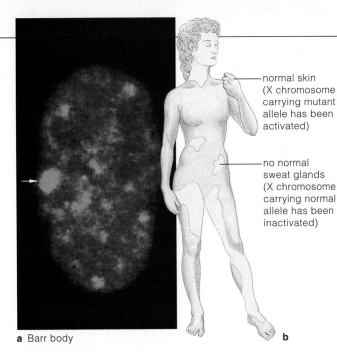

normal skin
(X chromosome
carrying mutant
allele has been
activated)

no normal
sweat glands
(X chromosome
carrying normal
allele has been
inactivated)

a Barr body

b

Figure 12.17 (**a**) An inactivated X chromosome, called a Barr body (*arrow*), as it appears in a human female's somatic cell at interphase. The X chromosome is not condensed this way in a human male's cells. (**b**) Anhidrotic ectodermal dysplasia, a mosaic pattern of gene expression. The condition arises from random X chromosome inactivation. It leads to patches of skin with and without normal sweat glands. The mutated allele is inactivated in the normal patches.

Figure 12.18 Why is this female calico cat "calico"? In her cells, one X chromosome carries a dominant allele for the brownish-black pigment melanin. The allele on her other X chromosome specifies yellow fur. At an early stage of the cat's embryonic development, one of the two X chromosomes was inactivated at random in each cell that had formed by then. In all descendants of those cells, the same chromosome also became inactivated, leaving only one functional allele for the coat-color trait. We see patches of different colors, depending on which allele was inactivated in cells that formed a given tissue region. (The white patches result from a gene interaction involving the "spotting gene," which blocks melanin synthesis entirely.)

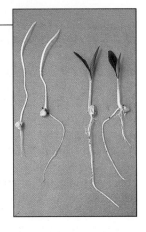

Figure 12.19 Effect of the absence of light on corn seedlings. The two seedlings to the right (the control group) were grown in a greenhouse. The two seedlings to the left of them were grown in darkness for eight days. The two dark-grown plants could not convert their stockpiled precursors of chlorophyll molecules to active form. They never did green up.

The mosaic effect is splendidly apparent in female calico cats. These cats are heterozygous for black and yellow coat-color alleles on their X chromosomes. The coat color in a given body region depends on which X chromosome's genes are transcribed (Figure 12.18).

Sunlight As a Signal

Plant a few corn seeds in moist, nutrient-rich soil but keep them in total darkness. After eight days, spindly, pale seedlings develop (Figure 12.19). They can't make chlorophyll, the key photosynthetic pigment. Expose the seedlings to a single burst of dim light from a flashlight. Within ten minutes, they start to convert some stockpiled molecules to active forms of chlorophyll!

Phytochrome, a blue-green pigment, helps plants adapt to changes in light. It becomes inactive at sunset, at night, or even in shade, where far-red wavelengths predominate. It becomes activated at sunrise, when red wavelengths dominate the sky. As days alternate with nights, as days grow shorter and then longer with the changing seasons, the quantity of incoming red or far-red wavelengths varies. The variations in light govern phytochrome activity and so influence transcription of certain genes at certain times of day and year.

Elaine Tobin and her coworkers conducted experiments with dark-grown seedlings of duckweed. They discovered a marked increase in the number of certain mRNA transcripts after they exposed the seedlings to one minute of red light. Apparently, exposure enhanced the transcription of genes for proteins that are required for the development and greening of chloroplasts.

All cells of a multicelled organism inherit the same genes, yet nearly all become specialized in composition, structure, and function. This process of cell differentiation arises because different populations of cells activate and suppress a few of their total number of genes in highly selective, unique ways.

Gene controls operate at many different levels. They underlie basic, short-term housekeeping tasks and intricate, long-term patterns of growth and development.

GENES, PROTEINS, AND CANCER

Every second, millions of cells in your skin, gut lining, liver, and other body regions divide and replace their worn-out, dead, and dying predecessors. They do not divide willy-nilly. Certain genes specify the enzymes and other proteins required for cell growth, DNA replication, spindle formation, chromosome movements, and division of the cytoplasm. Regulatory proteins control the synthesis and use of these gene products, and they control when the division machinery is put to rest. When controls are lost, cell divisions will not stop as long as conditions for growth remain favorable.

When cells are not responding to normal controls over growth and division, they form a tissue mass called a **tumor** (Figure 12.20). Cells of common skin warts and other *benign* tumors grow slowly, in an unprogrammed way, and they still have surface recognition proteins that can hold them together in their home tissue. Surgically remove a benign tumor, and you remove its potential threat to the surrounding tissues.

In a *malignant* tumor, abnormal cells grow and divide more rapidly, with destructive physical and metabolic effects on surrounding tissues. These are grossly disfigured cells. They cannot construct a normal cytoskeleton or plasma membrane, and they cannot synthesize normal versions of recognition proteins. Such cells can break loose from their home tissue. They can enter lymph or blood vessels, travel through the body—and become lodged where they don't belong. Figure 12.21 shows this process of migration and invasion. It is called **metastasis**.

Figure 11.10 shows merely 3 of more than 200 types of malignant tumors that have been identified so far. All malignant tumors are grouped into the general category of **cancer**. Each year in developed countries alone, 15 to 20 percent of all deaths result from cancer. And it is not just a human affliction. It has been observed in most of the

animals that have been studied. Similar abnormalities have even been observed in many kinds of plants.

At the minimum, all cancer cells show the following characteristics:

1. *Profound changes in the plasma membrane and cytoplasm.* Membrane permeability increases. Membrane proteins are lost or altered, and different ones form. The cytoskeleton becomes disorganized, shrinks, or both. Enzyme activity shifts, as in amplified reliance on glycolysis.

2. *Abnormal growth and division.* Controls that prevent overcrowding in tissues are lost. Cell populations reach high densities. New proteins trigger abnormal increases in small blood vessels that service the growing cell mass.

3. *Weakened capacity for adhesion.* Recognition proteins are altered or lost; cells can't stay anchored in proper tissues.

4. *Lethality.* Unless cancer cells are eradicated, they will kill the individual.

Any gene with the potential to induce the cancerous transformations is an **oncogene**. Oncogenes were first identified in retroviruses (a class of RNA viruses). However, certain gene sequences in many organisms are nearly identical to oncogenes, yet they rarely trigger cancer! The sequences, **proto-oncogenes**, specify certain proteins required for normal cell function. Some proto-oncogenes call for regulatory proteins. Others specify growth factors (transcriptional signals sent by one cell to trigger growth in other cells). Still others specify receptors for growth factors. Thus, the normal expression of proto-oncogenes is vital, which helps us understand why their *abnormal* expression is lethal.

Transformation may begin with mutations in proto-oncogenes or in certain control elements that govern them. For instance, this happens with certain insertions of viral DNA into cellular DNA. It can happen when **carcinogens** (cancer-inducing agents) cause changes in the DNA. Ultraviolet radiation, x-rays, and gamma rays are common carcinogens. So are many natural and synthetic compounds, including asbestos and certain components of tobacco smoke.

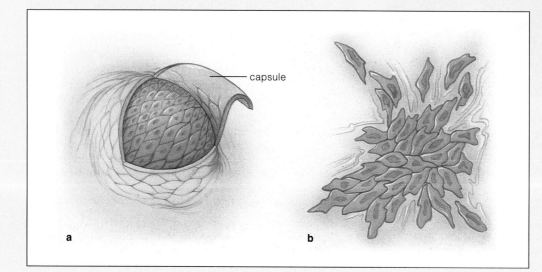

capsule

a b

Figure 12.20 (**a**) Benign tumor, a mass of cells that have an outwardly normal appearance and that are enclosed in a capsule of connective tissue. (**b**) Malignant tumor, a disorganized collection of cancer cells.

1 Cancer cells break away from tumor in tissue.

2 Migrating cells attach to the membrane located between the outside wall of all blood or lymph vessels and the surrounding tissue. They secrete enzymes that digest wall material, then enter the vessel.

3 Cancer cells creep or tumble through blood or lymph vessels. In time, possibly in response to molecular cues at particular sites, they leave the blood or lymph vessels the same way they got in.

a

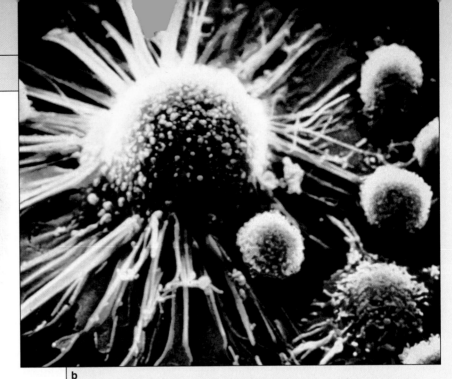

b

Figure 12.21 (**a**) Metastasis—the migration of cancer cells from the tissue in which they originated and their invasion of other tissues in the body. (**b**) Scanning electron micrograph that shows a few of the body's defenders—white blood cells—flanking a cancer cell. The defenders may or may not be able to destroy it. Cancer arises through loss of controls that govern specific genes, many of which influence cell growth and division.

Some cancers arise if base substitutions or deletions alter one of these genes or one of the controls over its transcription. Others arise if a gene becomes abnormally duplicated over and over again. Still others arise if a whole gene or part of it moves to a new location in the same chromosome or to an entirely different chromosome.

Consider the *myc* gene, located in human chromosome 8. It specifies a regulatory protein that helps control the cell cycle. Sometimes a break occurs in this gene region *and* in a region of chromosome 14. The two chromosomes swap segments. The *myc* gene ends up in a new location. There,

the gene may escape normal controls over its transcription and come under the influence of new ones. Whatever the case, transformations begin that convert *myc* from a proto-oncogene to an oncogene—and Burkitt's lymphoma may be the outcome.

Burkitt's lymphoma is a malignant tumor of the white blood cells called B lymphocytes. This is the only type of cell that produces antibody molecules—protein weapons against specific agents of disease (page 482). In activated B cells, a gene region that specifies antibody is transcribed at an exceptionally high rate. The translocation between chromosomes 8 and 14 puts the *myc* gene next to this gene region and makes it abnormally active.

Yet cancer seems to be a multistep process, involving more than one oncogene. Researchers have identified three such genes with roles in nearly all colon cancers. In 1995 they identified a key gene in ovarian and breast cancers. Through such discoveries, it may be possible to diagnose carriers early and frequently. A cancer detected early enough may still be curable by surgery.

Perhaps more than any other example, cancerous transformations bring home the extent to which you and all other organisms depend on gene controls. A dizzying variety of controls govern transcription and translation. They influence when and where gene products will be used. And they reinforce the profound importance of the topics of this chapter.

199

SUMMARY

1. A gene is a sequence of nucleotide bases in DNA. For most genes, the nucleotide sequence corresponds to a sequence of specific amino acids in a polypeptide chain. Proteins consist of one or more polypeptide chains.

2. Most genes contain instructions for building RNA molecules, which are necessary for protein synthesis. There are three classes of RNA molecules:

    a. Messenger RNA (mRNA) carries protein-building instructions from DNA to the cytoplasm.

    b. Ribosomal RNA (rRNA) is a component of ribosomes (sites where polypeptide chains are assembled).

    c. Transfer RNA (tRNA) is the vehicle of translation; it delivers amino acids to ribosomes in a sequence that corresponds to the sequential message of mRNA.

3. The path from genes to proteins has two steps. First, in *transcription*, enzymes assemble RNA on a DNA template. Second, in *translation*, RNAs interact in ways that lead to the synthesis of polypeptide chains, which later take on the three-dimensional shape of proteins:

4. Here are the key points concerning transcription:

    a. A double-stranded DNA molecule is unwound at a particular gene region. Then an RNA molecule is assembled on the exposed bases of one DNA strand.

    b. Base-pairing rules govern transcription of DNA into RNA. However, unlike DNA replication, uracil (not thymine) pairs with adenine in RNA:

    c. In eukaryotic cells, RNA transcripts are modified into final form before shipment from the nucleus. The noncoding portions (introns) of mRNAs are excised, and coding portions (exons) are spliced together. Mature mRNA transcripts are translated in the cytoplasm.

5. Here are the key points concerning translation:

    a. mRNA interacts with tRNAs and ribosomes in such a way that amino acids become linked one after another, in the sequence required to produce a specific kind of polypeptide chain.

    b. Translation is based on the genetic code, a set of sixty-four base triplets (nucleotide bases that certain ribosomal proteins read in blocks of three).

    c. In mRNA, a base triplet is a codon. An anticodon is a complementary triplet in tRNA. Some combination of codons specifies the amino acid sequence of a polypeptide chain, start to finish.

6. Translation proceeds through three stages:

    a. Initiation. A small ribosomal subunit, an initiator tRNA, and an mRNA transcript converge. The small subunit then binds with a large ribosomal subunit.

    b. Chain elongation. tRNAs deliver amino acids to the ribosome. Their anticodons base-pair with codons in mRNA. The amino acids become linked by peptide bonds to form a new polypeptide chain.

    c. Chain termination. A stop codon in mRNA triggers detachment of the chain and mRNA from the ribosome.

7. Gene mutations are heritable, small-scale alterations in the nucleotide sequence of DNA. Their effect on the individual depends on how the proteins they specify interact with other genes and with the environment. Mutations are induced by exposure to ultraviolet radiation and other mutagens. They also arise spontaneously, as by replication errors or transposable elements.

8. DNA is the source of life's unity. Mutations are the original source of life's diversity. The environment is the testing ground for gene products (proteins).

9. Cells control *gene expression* (which gene products appear, when, and in what amounts). When the control mechanisms come into play depends on the type of cell, prevailing chemical conditions, and outside signals that can change the cell's activities.

10. Regulatory proteins, enzymes, hormones, and other control elements interact with one another, with DNA and RNA, and with the gene products. They operate before, during, and after transcription and translation.

11. Two common types of control systems operate in cells to block or enhance transcription. Their effects are reversible.

    a. In negative control systems, a regulatory protein binds at a specific DNA sequence and prevents one or more genes from being transcribed.

    b. In positive control systems, a regulatory protein binds to DNA and promotes transcription.

12. Most prokaryotes (bacteria) do not require many genes to grow and reproduce. Control systems often enhance transcription rates when gene products are required. This is true of operons (groupings of related genes and control elements), for example.

13. Eukaryotes require complex gene controls. Gene activity changes rapidly in response to short-term shifts in the surroundings, and during long-term growth and development as cells multiply, make physical contact in tissues, and interact chemically.

14. The cells in a multicelled eukaryote inherit the same genes, but different cell types activate and suppress some fraction of the genes in different ways.

15. Selective gene control brings about cell differentiation. In other words, cell lineages become specialized in appearance, composition, and function.

Review Questions

1. Are proteins assembled on DNA? If so, state how. If not, tell where they are assembled, and on which molecules. *188–189*

2. Name three classes of RNA and define their functions. How do the base sequences of RNA and DNA differ? *186*

3. Describe the steps of gene transcription. How does this process resemble DNA replication? How does it differ? *186*

4. In the nucleus, are introns or exons snipped out of pre-mRNA transcripts? *187*

5. Define the genetic code, codon, and anticodon. How do the codons for the same amino acid usually differ? *188–189*

6. Briefly describe the steps of translation. *190–191*

7. Describe one kind of mutation. What determines whether its altered product will have helpful, neutral, or harmful effects? *192*

8. Define promoter, operator, and repressor protein. *194*

9. Compare gene control in prokaryotic and eukaryotic cells. As part of your comparison, define cell differentiation. *194–196*

10. What are the characteristics of all cancer cells? *198*

Self-Quiz *(Answers in Appendix IV)*

1. DNA contains different genes that are transcribed into _____ .
   a. proteins   c. tRNAs and rRNAs
   b. mRNAs   d. b and c

2. The RNA molecule is _____ .
   a. a double helix   c. always double-stranded
   b. usually single-stranded   d. usually double-stranded

3. mRNA is produced by _____ .
   a. replication   c. transcription
   b. duplication   d. translation

4. Each codon calls for a specific _____ .
   a. protein   c. amino acid
   b. polypeptide   d. carbohydrate

5. Anticodons pair with _____ .
   a. mRNA codons   c. tRNA anticodons
   b. DNA codons   d. amino acids

6. The expression of a given gene depends on _____ .
   a. cell type and functions   c. environmental signals
   b. chemical conditions   d. all of the above

7. Regulatory proteins interact with _____ .
   a. DNA   c. gene products
   b. RNA   d. all of the above

8. In prokaryotic but not eukaryotic cells, a(n) _____ precedes the genes of an operon.
   a. lactose molecule   c. operator
   b. promoter   d. both b and c

9. Cell differentiation _____ .
   a. occurs in all multicelled organisms
   b. requires different genes in different cells
   c. involves selective gene expression
   d. both a and c
   e. all of the above

10. X chromosome inactivation in mammalian females may result in a _____ for some traits.
    a. male phenotype   c. rise in transcription rates
    b. mosaic tissue effect d. rise in translation rates

11. Match the terms with the suitable description.
    ____ phytochrome   a. inhibits gene transcription
    ____ Barr body   b. base triplets coding for amino acids
    ____ oncogene   c. gene with the potential to induce
    ____ genetic code   cancerous transformation
    ____ repressor   d. inactivated X chromosome
                         e. helps plants adapt to changes in light

Critical Thinking

1. Refer to Figure 12.7. Use the genetic code to translate the mRNA sequence UAUCGCACCUCAGGAUGAGAU. The first codon in the frame is UAU. Which amino acid sequence is being specified?
   a. tyr-arg-thr-ser-gly-stop-asp . . .
   b. tyr-arg-thr-ser-gly . . .
   c. tyr-arg-tyr-ser-gly-stop-asp . . .
   d. none of the above

2. Josh and Noah have isolated a tRNA with a mutated codon (AAU instead of AUU). What effect will this have on the process of protein synthesis in cells having the mutated tRNA?

3. *Duchenne muscular dystrophy*, a genetic disorder, affects boys almost exclusively. Early in childhood, muscles begin to atrophy (waste away) in affected individuals, who typically die in their teens or early twenties as a result of respiratory failure. Muscle biopsies of women who carry a gene associated with the disorder reveal some regions of atrophied muscle tissue. Yet muscle tissue adjacent to these regions was normal or even larger and more chemically active, as if to compensate for the weakness of the adjoining region. How can you explain these observations?

4. Kristen has isolated a strain of *E. coli*. A mutation has affected the capacity of CAP to bind to a region of the lactose operon, as it would do normally. State how the mutation affects transcription of the lactose operon when cells of this bacterial strain are subjected to the following conditions:
   a. Lactose and glucose are available.
   b. Lactose is available but glucose is not.
   c. Both lactose and glucose are absent.

5. Some researchers suspect that the "three-letter" genetic code (64 base triplets) evolved from a "two-letter" code that specified fewer amino acids. Reflect on Figure 12.7, then speculate why.

Selected Key Terms

activator protein *195*	metastasis *198*	repressor *194*
anticodon *189*	mutagen *192*	ribonucleic acid
Barr body *196*	negative control	(RNA) *185*
base sequence *185*	system *194*	ribosomal RNA
cancer *198*	oncogene *198*	(rRNA) *186*
carcinogen *198*	operator *194*	ribosome *189*
cell differentiation *196*	operon *194*	RNA
codon *188*	phytochrome *197*	polymerase *186*
exon *187*	positive control	transcription *185*
gene mutation *192*	system *194*	transfer RNA
genetic code *188*	promoter *186*	(tRNA) *186*
intron *187*	proto-oncogene *198*	translation *185*
messenger RNA	regulatory	uracil *186*
(mRNA) *186*	protein *194*	

Readings

Grunstein, M. October 1992. "Histones As Regulators of Genes." *American Scientist* 267(4): 68–74B.

Liotta, L. February 1992. "Cancer Cell Invasion and Metastasis." *Scientific American* 266(2): 54–63.

13 RECOMBINANT DNA AND GENETIC ENGINEERING

Mom, Dad, and Clogged Arteries

Butter! Bacon! Eggs! Ice cream! Cheesecake! Possibly you think of such foods as enticing, off-limits, or both. After all, who among us doesn't know about animal fats and the dreaded cholesterol?

Soon after you feast on these fatty foods, cholesterol enters the bloodstream. Cholesterol *is* important. It is a structural component of animal cell membranes, and without membranes, there would be no cells. Cells also remodel cholesterol into various molecules, including the vitamin D that is necessary for the development of good bones and teeth. Normally, however, your liver synthesizes enough cholesterol for your cells.

Certain proteins that circulate in the blood combine with cholesterol and other substances, forming particles called lipoproteins. HDLs (high-density lipoproteins) collect cholesterol and transport it to the liver, where it can be metabolized. LDLs (low-density lipoproteins) normally end up in cells that store or use cholesterol.

Sometimes too many LDLs form, and the excess infiltrates the walls of arteries. There they promote formation of abnormal masses called athero-sclerotic plaques (Figure 13.1). These interfere with blood flow and narrow the arterial diameter. If the plaques clog one of the tiny, coronary arteries that deliver blood to the heart, the resulting symptoms can range from mild chest pains to a heart attack.

How your body handles dietary cholesterol depends on what you inherited from your parents. Consider the gene for a protein that serves as the cell's receptor for LDLs. Inherit two "good" alleles of that gene, and your blood level of cholesterol will tend to remain so low that your arteries will never get clogged, even with a high-fat diet. Inherit two copies of a certain mutated allele, however, and you are destined to develop a rare genetic disorder called *familial cholesterolemia*. With this disorder, cholesterol builds up to abnormally high levels. Many of the affected individuals die of heart attacks during childhood or their teens.

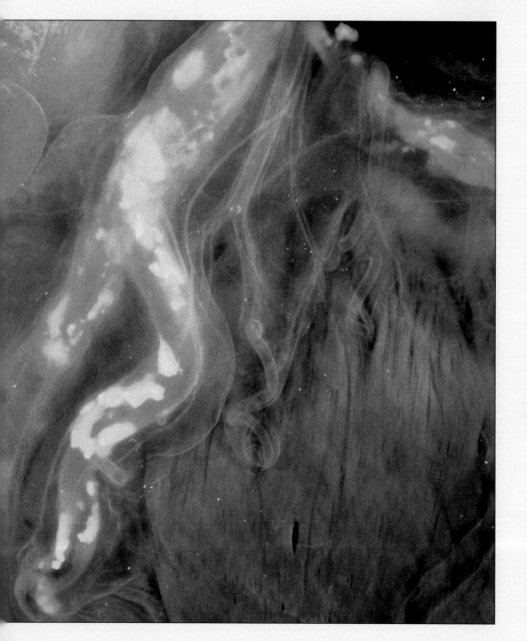

Figure 13.1 Potentially life-threatening plaques (*bright yellow*) inside one of the coronary arteries, which deliver blood to the heart. Such plaques are the legacy of abnormally high levels of cholesterol.

In 1992 a woman from Quebec, Canada, became a milestone in the history of genetics. She was thirty years old. Like two of her younger brothers, who had already died from heart attacks in their early twenties, she inherited the defective gene for the LDL receptor. She herself survived a heart attack when she was sixteen. At twenty-six, she had coronary bypass surgery.

At the time, people were hotly debating the risks and promise of **gene therapy**—the transfer of one or more normal or modified genes into an individual's body cells to correct a genetic defect or boost resistance to disease. Even so, the woman consented to undergo an untried, physically wrenching procedure designed to give her working copies of the good gene.

Medical researchers removed about 15 percent of the woman's liver. They placed liver cells in a nutrient-rich medium that promoted growth and division. *And they spliced the good gene into the genetic material of a harmless virus.* That modified virus served roughly the same function as a hypodermic needle. The researchers allowed it to infect the cultured liver cells—and thereby insert copies of the good gene into them.

Later, the researchers infused about a billion of the modified cells into the woman's portal vein, which leads directly to the liver. There, at least some cells took up residence—and they started to produce the missing cholesterol receptor.

Two years later, between 3 and 5 percent of the woman's liver cells were behaving normally and sponging up cholesterol from the bloodstream. Her blood levels of LDLs had declined nearly 20 percent. Scans of her arteries showed no evidence of the progressive clogging that had nearly killed her. At a recent press conference, the woman announced that she is active and doing quite well.

Her cholesterol levels do remain more than twice as high as normal, and it is too soon to know whether the gene therapy will prolong her life. Yet the intervention provides solid proof that the concept of gene therapy is sound, and hopes are high.

As you might gather from this pioneering clinical application, recombinant DNA technology has truly staggering potential for medicine. It also has staggering potential for agriculture and industry. It does not come without risks. With this chapter, we consider some basic aspects of the new technology. We also address ecological, social, and ethical questions related to its application.

KEY CONCEPTS

1. Genetic experiments have been going on in nature for billions of years, through gene mutations, crossing over and recombination, and other natural events.

2. Humans are now purposefully bringing about genetic changes by way of recombinant DNA technology. Such enterprises are called genetic engineering.

3. With this technology, researchers isolate, cut, and splice together gene regions from different species, then greatly amplify the number of copies of the genes that interest them. The genes, and in some cases their protein products, are produced in quantities that are large enough for research and for practical applications.

4. Three activities are at the heart of recombinant DNA technology. First, procedures based on specific enzymes are used to cut DNA molecules into fragments. Second, the fragments are inserted into cloning tools, such as plasmids. Third, fragments containing the genes of interest are identified, then copied rapidly and repeatedly.

5. Genetic engineering involves isolating, modifying, and inserting genes back into the same organism or into a different one. The goal is to beneficially modify traits that the genes influence. Human gene therapy, which focuses on controlling or curing genetic disorders, is an example.

6. The new technology raises social, legal, ecological, and ethical questions regarding its benefits and risks.

RECOMBINATION IN NATURE—AND IN THE LABORATORY

For at least 3 billion years, nature has been conducting countless genetic experiments, through mutation, crossing over, and other events that introduce changes in genetic messages. This is the source of life's diversity.

For many thousands of years, we humans have been changing many genetically based traits of species. Through artificial selection, we produced new crop plants and breeds of cattle, birds, dogs, and cats from wild ancestral stocks. We developed meatier turkeys and sweeter oranges, larger corn, seedless watermelons, flamboyant ornamental roses, and other useful plants. We produced amazing hybrids, including the tangelo (tangerine × grapefruit) and mule (horse × donkey).

Researchers now use **recombinant DNA technology** to analyze genetic changes. With this technology, they cut and splice DNA from different species, then insert the modified molecules into bacteria or other types of cells that engage in rapid replication and cell division. The cells copy the foreign DNA right along with their own. In short order, huge populations can produce useful quantities of recombinant DNA molecules. The new technology also is the basis of **genetic engineering**, by which genes are isolated, modified, and inserted back into the same organism or into a different one.

Plasmids, Restriction Enzymes, and the New Technology

Believe it or not, this astonishing technology originated with the innards of bacteria. Bacterial cells have a single chromosome, a circular DNA molecule that has all the genes they require to grow and reproduce. But many species also have **plasmids**—small, circular molecules of "extra" DNA that contain a few genes (Figure 13.2).

Usually, plasmids are not essential for survival, but some of the genes they carry may benefit the bacterium. For instance, some plasmid genes confer resistance to antibiotics. (*Antibiotics* are normal metabolic products of certain microorganisms, and they kill or inhibit the growth of other microbial species.) The bacterium's replication enzymes copy and reproduce plasmid DNA, just as they copy chromosomal DNA.

In nature, many bacteria are able to transfer plasmid genes to a bacterial neighbor of the same species or a different one (Section 18.2). Replication enzymes may even integrate the transferred plasmid into a recipient's chromosome. In such cases, a molecule of recombinant DNA is the result.

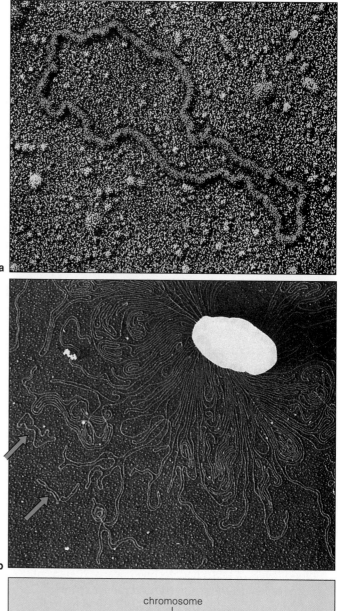

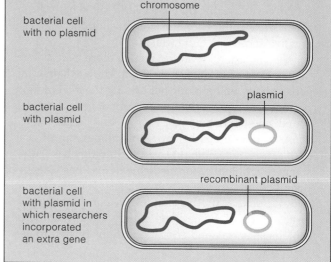

Figure 13.2 (**a**) A plasmid at high magnification. (**b**) Plasmids (*blue* arrows) released from a ruptured *Escherichia coli* cell. (**c**) Naturally occurring and modified plasmids.

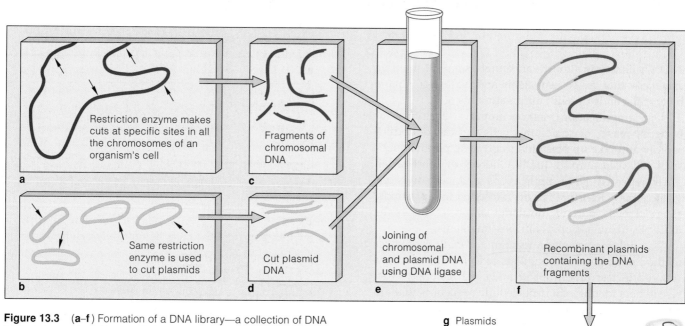

| a | c | e | f |

a Restriction enzyme makes cuts at specific sites in all the chromosomes of an organism's cell

c Fragments of chromosomal DNA

b Same restriction enzyme is used to cut plasmids

d Cut plasmid DNA

e Joining of chromosomal and plasmid DNA using DNA ligase

f Recombinant plasmids containing the DNA fragments

Figure 13.3 (**a–f**) Formation of a DNA library—a collection of DNA fragments, produced by restriction enzymes and inserted into plasmids or another cloning tool. (**g**) Plasmid insertion into host cells to form cloned DNA—multiple, identical copies of the DNA fragments.

g Plasmids inserted into host cells for amplification

Infectious particles called viruses as well as bacteria dabble in gene transfers and recombinations, as do most eukaryotic species. As you might imagine, viral infection does a bacterium no good. Over evolutionary time, bacteria developed an arsenal against invasion by harmful DNA. They became equipped with many types of **restriction enzymes**, which are able to recognize and cut apart foreign DNA that may enter a cell. Eventually, researchers learned how to use plasmids *and* restriction enzymes for genetic recombination in the laboratory.

Producing Restriction Fragments

Each type of restriction enzyme makes a cut wherever it recognizes a specific, very short nucleotide sequence in the DNA. Cuts at two identical sequences in the same DNA molecule produce a fragment. Because some types of enzymes make *staggered* cuts, the fragments may have single-stranded portions at both ends. These are sometimes referred to as sticky ends:

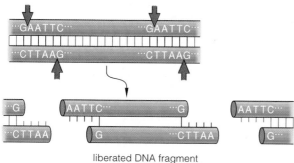

liberated DNA fragment with sticky ends

By "sticky," we mean that the short, single-stranded ends of a DNA fragment will be able to base-pair with any other DNA molecule that also has been cut by the same restriction enzyme.

For example, suppose you use the same restriction enzyme to cut plasmids *and* DNA molecules that you have isolated from a human cell. When you mix the cut-up molecules together, they base-pair at the cut sites. Afterward, you add **DNA ligase** to the mixture. This type of enzyme seals DNA's sugar-phosphate backbone at the cut sites, just as it does during DNA replication. In this way, you create "recombinant plasmids," which have pieces of DNA from another organism inserted into them (Figure 13.2c).

You now have a **DNA library**—a collection of DNA fragments, produced by restriction enzymes, that have been incorporated into plasmids (Figure 13.3).

With recombinant DNA technology, DNA from different species is cut and spliced together, then the recombinant molecules are amplified (by way of copying mechanisms) to produce useful quantities of the genes of interest.

Recombination is made possible by restriction enzymes that make specific cuts in DNA molecules and by DNA ligases that seal the cut ends.

Recombinant DNA technology has uses in basic research.

The new technology also has uses in genetic engineering—the deliberate modification of genes, followed by their insertion into the same individual or a different one.

13.2 WORKING WITH DNA FRAGMENTS

Amplification Procedures

A DNA library is almost vanishingly small. It must be *amplified*—that is, copied again and again—so that a biochemist ends up with useful amounts of it.

One amplification method employs "factories" of bacteria, yeasts, or some other cells that can reproduce rapidly and take up plasmids. A growing population of such cells can amplify a DNA library in short order. Their repeated DNA replications and cell divisions yield cloned DNA that has been inserted into plasmids. (The "cloned" part of the name simply refers to the multiple, identical copies of DNA fragments.)

The **polymerase chain reaction**, or **PCR**, is a newer method of amplifying fragments of DNA. The reactions proceed in test tubes, not in microbial factories. First, researchers identify short nucleotide sequences that are located just before and just after a region of DNA from an organism that interests them. Then they synthesize primers—nucleotide sequences that will base-pair with the sequences in the DNA. They mix the primers with enzyme molecules, free nucleotides, and all of the DNA from one of the organism's cells. Then they expose the mixture to precise temperature cycles.

During the cycles, the two strands of all the DNA molecules unwind from each other. Then, as Figure 13.4 indicates, the short sequences become positioned at the specific target sites. That is, they hydrogen-bond with the exposed, complementary sequence according to the base-pairing rules.

DNA polymerase, a replication enzyme, recognizes the synthesized primers as signals to get busy. The DNA polymerase used for PCR comes from a bacterium that lives in hot springs and even in hot water heaters. This enzyme can function at the elevated temperatures required to unwind the DNA and at the lower temperatures required for base-pairing.

With each round of reactions, the number of DNA molecules doubles. For example, if there are 10 such molecules in the test tube, there soon will be 20, then 40, 80, 160, 320, 640, 1,280, and so on. Very quickly, a target region from a single DNA molecule can be amplified to *billions* of molecules.

In short, *PCR can amplify samples that contain even tiny amounts of DNA.* As you will see in the next section, such samples can be obtained from fossils—even from a single hair or drop of blood left at the scene of a crime.

Sorting Out Fragments of DNA

When restriction enzymes cut DNA, the fragments they produce are not all the same length. Researchers can employ **gel electrophoresis** to separate the fragments from one another according to their length. This labora-

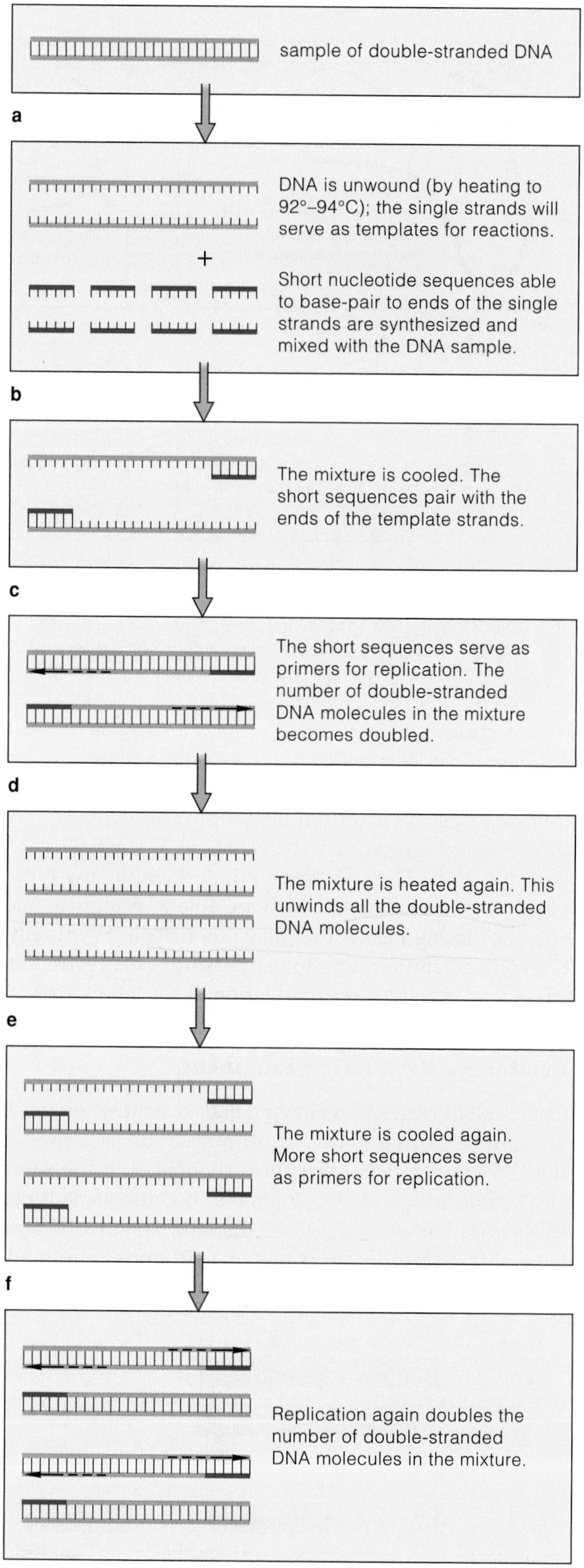

a sample of double-stranded DNA

b DNA is unwound (by heating to 92°–94°C); the single strands will serve as templates for reactions.

+ Short nucleotide sequences able to base-pair to ends of the single strands are synthesized and mixed with the DNA sample.

c The mixture is cooled. The short sequences pair with the ends of the template strands.

d The short sequences serve as primers for replication. The number of double-stranded DNA molecules in the mixture becomes doubled.

e The mixture is heated again. This unwinds all the double-stranded DNA molecules.

f The mixture is cooled again. More short sequences serve as primers for replication.

g Replication again doubles the number of double-stranded DNA molecules in the mixture.

Figure 13.4 The polymerase chain reaction (PCR).

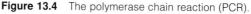

Figure 13.5 One method of sequencing DNA, as initially developed by Frederick Sanger. The method can be used to determine the nucleotide sequence of specific DNA fragments, as this example illustrates.

a Single-stranded DNA fragments are added to a solution in four different test tubes:

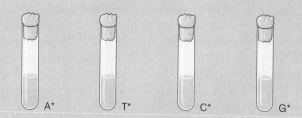

All four tubes contain DNA polymerases, short nucleotide sequences that can serve as primers for replication, and the nucleotide subunits of DNA (A, T, C, and G). Each tube contains a modified, labeled version of only one of the four kinds of nucleotides. We can show these as A*, T*, C*, and G*. The labeled form is present in low concentration, along with a generous supply of the unmodified form of the same nucleotide. Let's follow what happens in the tube with the A* subunits.

b As expected, the DNA polymerase recognizes a primer that has become attached to a fragment, which it uses as a template strand. The enzyme assembles a complementary strand according to base-pairing rules (A only to T, and C only to G). Sooner or later, the enzyme picks up an A* subunit for pairing with a T on the template strand. The modified nucleotide is a chemical roadblock—it prevents the enzyme from adding more nucleotides to the growing complementary strand. In time, the tube contains labeled strands of different lengths, as dictated by the location of each A* in the sequence:

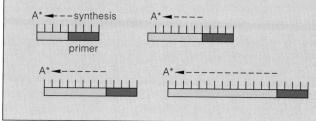

The same thing happens in the other three tubes, with strand lengths dictated by the location of T*, C*, and G*.

c DNA from each of the four tubes is placed in four parallel lanes in the same gel. Then the DNA can be subjected to electrophoresis. The resulting nucleotide sequence can be read off the resulting bands in the gel. Look at the numbers running down the side of this diagram:

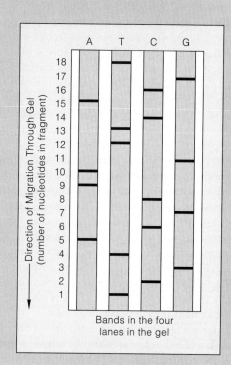

Bands in the four lanes in the gel

Start with "1" and read across the four lanes (A, T, C, G). As you can see, T is the closest to the start of the nucleotide sequence; it has migrated farthest through the gel. At "2," the next nucleotide is C, and so on. The entire sequence, read from the first nucleotide to the last, is

T C G T A C G C A A G T T C A C G T

Now, by using the rules of base-pairing, you can deduce the sequence of the DNA fragment that served as the template.

tory procedure uses an electric field to move molecules through a viscous gel and separate them according to certain properties. For DNA, size alone affects how far they move. (For proteins, size, shape, and net surface charge are determining factors.)

A gel that contains DNA fragments is immersed in a buffered solution, and electrodes connect the solution to a power source. Apply voltage, and the DNA fragments respond to it (because of their negatively charged phosphate groups). They move away from one another in the gel. Again, how far they actually migrate toward the positively charged pole depends only on their size—the larger ones cannot move as fast through it. Following a predetermined period of electrophoresis, fragments of different length can be identified by staining the gel.

DNA Sequencing

Once DNA fragments from a sample have been sorted out according to length, researchers can work out the nucleotide sequence of each type. The Sanger method of DNA sequencing will give you a sense of one of the ways this can be done (Figure 13.5).

Restriction fragments can be rapidly amplified by PCR and by large populations of bacterial cells or yeast cells.

Restriction fragments can be sorted out according to length, as by gel electrophoresis. Then sequencing methods can be employed to determine the nucleotide sequences of the DNA fragments.

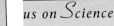

ND DNA FINGERPRINTS

...rcher has isolated a good-sized sample of ...cules. She cuts them up with restriction enzymes, then subjects the restriction fragments to gel electrophoresis. Large ones can't move as fast as small ones, so the fragments separate from one another in the gel. Now the researcher uses the *Southern blot method*. She transfers the electrophoretically separated fragments from the gel to absorbent paper, then immerses the paper in a solution containing a labeled probe that can bind to the fragment that interests her. Afterward, she does the same to DNA samples from several other people.

When she compares the banding patterns, she finds small variations among them. Why? Although the DNA molecules of any two people are alike in most respects, they differ in a few base sequences. These molecular differences shift the number and location of sites where restriction enzymes can make their cuts. As a result, some fragments from corresponding regions of DNA from two or more individuals differ in length. These differences in the DNA electrophoresis patterns among individuals are called **RFLPs** (pronounced RIFF-lips). The term stands for **restriction fragment length polymorphisms**.

As you know, every human has a unique set of fingerprints, a marker of his or her identity. As is true of other sexually reproducing species, every human also has a **DNA fingerprint**—a unique array of RFLPs, inherited from each parent in a Mendelian pattern.

RFLP analysis is a wonderful new procedure for basic research. It also has startling applications in society at large. Consider the human genome project. As described in Section 13.6, this is a monumental effort to establish the nucleotide sequence of human DNA, with its 3.2 billion base pairs. Or consider how evolutionary biologists are analyzing RFLPs to decipher DNA from mummies, from fossilized insects and plants, even from mammoths and humans that were preserved many thousands of years ago in glacial ice. Investigators even used RFLP analysis to confirm that bones exhumed from a shallow pit in Siberia belonged to five members of the Russian imperial family, all shot to death in secrecy in 1918.

RFLP analysis has spectacular potential for medicine. Researchers have already pinpointed unique restriction sites within several mutated genes, including the ones that cause sickle-cell anemia, cystic fibrosis, and other genetic disorders. They can use the sites to find out whether an individual carries a mutated gene. RFLP analysis has uses in prenatal diagnosis, the topic of Section 10.9.

What about social issues? Merely think about a woman who is presently claiming that a motion-picture superstar fathered her child, and who is demanding large support payments. Paternity *and* maternity cases brought against celebrities and ordinary folks alike can be resolved by comparing DNA fingerprints of the child and the disputed parent. Or think about forensic medicine. Often, a few

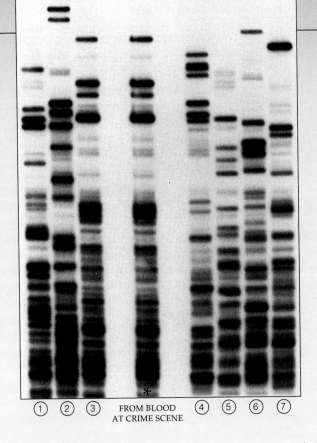

Figure 13.6 DNA fingerprint from a bloodstain discovered at a crime scene, with DNA fingerprints from blood samples of seven suspects (*circled numbers*). Which one of the seven was a match?

drops of blood, drops of semen, or even cells from a hair follicle at a crime scene or on a suspect's clothing yield enough DNA to identify the perpetrator (Figure 13.6).

Britain, which has the world's first DNA database, is at the forefront of solving crimes with DNA fingerprinting. Also, police in Britain can lawfully collect samples of hair or saliva from crime suspects, even if by force, for RFLP analysis. At this writing, detectives in Cardiff, Wales, are now engaged in a massive "blood drive" to locate the man who raped and killed a local teenager. They expect to contact as many as 5,000 "donors." The detectives have let it be known that anyone who declines to donate blood voluntarily risks calling attention to himself—and inviting exceedingly fine scrutiny. Mass DNA screening has already helped solve other murders in Britain and Germany.

Meanwhile, in the United States, defense lawyers in a high-profile and seemingly interminable criminal case are hammering away at DNA evidence. Someone nearly cut off the head of a celebrity's wife, slashed her friend to death, and left a trail of blood leading to his house. RFLP analysis seems to implicate him. His lawyers are pounding on the competence and character of just about everybody who found blood at the crime scene and at the suspect's house, collected samples, ran the PCRs, and performed the RFLP analyses. And one can only wonder if their sensationalized efforts alone will send DNA research back to the twelfth century.

13.4 MODIFIED HOST CELLS

Use of DNA Probes

Recombinant plasmids are not much use in themselves. They must be mixed with living cells that can take them up. So how do you find out which cells do this? You use **DNA probes**—short DNA sequences synthesized from radioactively labeled nucleotides. Part of a probe must be able to base-pair with part of the DNA of interest. Base-pairing between nucleotide sequences from different sources is called **nucleic acid hybridization**.

The challenge is to "select" cells that have taken up recombinant plasmids and the gene of interest. One way to do this is to use a plasmid containing a gene that confers resistance to a particular antibiotic. You put prospective host cells on a culture medium that has this antibiotic added to it. The antibiotic prevents growth of all cells *except* the ones that house plasmids with the antibiotic-resistance gene.

As the cells divide, they form colonies. Assume the colonies are growing on agar, a gel-like substance, in a petri dish. You blot the agar against a nylon filter. Some cells stick to the filter in locations that mirror the locations of the original colonies (Figure 13.7). You use solutions to rupture the cells, fix the released DNA onto the filter, and make the double-stranded DNA unwind. Then you add DNA probes. The probes hybridize only with the gene region having the proper base sequence. Probe-hybridized DNA emits radioactivity and allows you to tag the colonies that harbor the gene of interest.

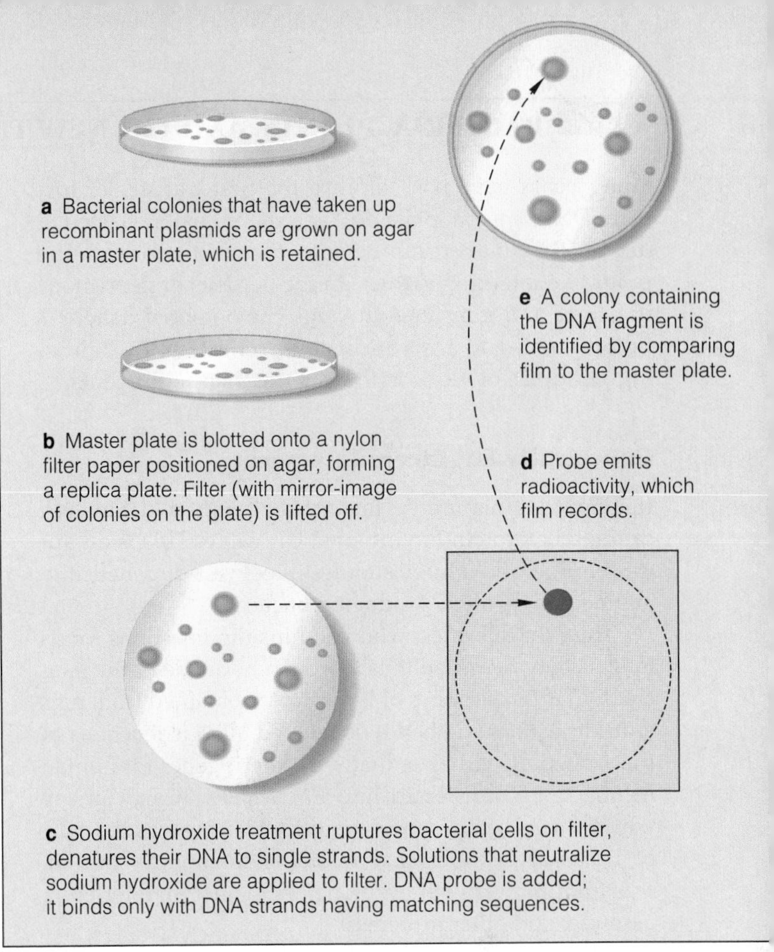

Use of cDNA

Researchers study genes to find out about the protein products and how they are put to use. However, even if a host cell takes up a gene, the cell might not be able to produce the protein. For example, recall that new mRNA transcripts of human genes contain noncoding sequences (introns). They cannot be translated until the introns are snipped out and coding regions (exons) are spliced together into mature form. Bacterial enzymes do not recognize the splice signals, so bacterial host cells cannot always properly translate human DNA.

Sometimes researchers get around this problem by using **cDNA**. This is a DNA strand "copied" from a *mature* mRNA transcript for the gene. By a backwards process called **reverse transcription**, a matching DNA strand is assembled on mRNA. The result is a hybrid mRNA-cDNA molecule (Figure 13.8). Enzymes remove the RNA, then they synthesize a complementary DNA strand to make double-stranded DNA.

Double-stranded cDNA can be further modified, as by adding signals for transcription and translation to it. The modified cDNA can be inserted into a plasmid for amplification. Recombinant plasmids can be inserted

a Bacterial colonies that have taken up recombinant plasmids are grown on agar in a master plate, which is retained.

e A colony containing the DNA fragment is identified by comparing film to the master plate.

b Master plate is blotted onto a nylon filter paper positioned on agar, forming a replica plate. Filter (with mirror-image of colonies on the plate) is lifted off.

d Probe emits radioactivity, which film records.

c Sodium hydroxide treatment ruptures bacterial cells on filter, denatures their DNA to single strands. Solutions that neutralize sodium hydroxide are applied to filter. DNA probe is added; it binds only with DNA strands having matching sequences.

Figure 13.7 Use of a DNA probe to identify a colony of bacterial cells that have taken up plasmids with a specific DNA fragment.

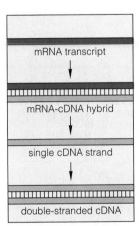

a An mRNA transcript of a desired gene is used as a template for assembling a DNA strand. An enzyme (reverse transcriptase) does the assembling.

b An mRNA-cDNA hybrid molecule results.

c Enzyme action removes the mRNA and assembles a second strand of DNA on the remaining DNA strand. The result is double-stranded cDNA, "copied" from an mRNA template.

mRNA transcript

mRNA-cDNA hybrid

single cDNA strand

double-stranded cDNA

Figure 13.8 Formation of cDNA from an mRNA transcript.

into bacteria, which may use the cDNA directions for making the protein.

Host cells that take up modified genes may be identified by procedures involving DNA probes.

Modified eukaryotic genes may not be expressed in host bacterial cells unless first introduced into the cells as cDNA.

Many years have passed since the first transfer of foreign DNA into a plasmid, yet the transfer started a debate that may continue into the next century. The point of contention is this: Do the benefits of gene modifications and transfers outweigh the potential dangers? Before coming to any conclusions, reflect on the following examples of work with bacteria, then with plants.

Genetically Engineered Bacteria

Imagine a miniaturized factory that churns out insulin or another protein having medical value. This is an apt description of the huge stainless steel vats of genetically engineered, protein-producing bacteria.

Think of diabetics, who need insulin injections for as long as they live. Insulin is a protein hormone of the pancreas. Medical supplies of it once were obtained from pigs and cattle. Some diabetics developed allergic reactions to the foreign insulin. Eventually, synthetic genes for human insulin were transferred into *E. coli* cells. (Can you say why the genes had to be synthesized?) This was the start of bacterial factories that manufacture human insulin, somatotropin (growth hormone), hemoglobin, interferon, albumin, and other proteins.

Also, in research laboratories throughout the world, certain strains of bacteria are being en-

Figure 13.9 Spraying an experimental strawberry patch in California with "ice-minus" bacteria. The sprayer used elaborate protective gear to meet government regulations in effect then.

gineered to degrade oil spills, manufacture alcohol and other chemicals, process minerals, or leave crop plants alone. The strains are harmless to begin with. The ones meant to be confined to the laboratory are designed to prevent escape. As an extra precaution, the foreign DNA usually includes "fail-safe" genes. Such genes are silent *unless* engineered bacteria become exposed to conditions that exist in the environment. Upon exposure, these genes become activated, with lethal results.

For example, foreign DNA may include a *hok* gene next to a promoter of the lactose operon (Section 12.7). If an engineered bacterium does manage to escape into the environment—where sugars are common—the *hok* gene trips into action. The protein specified by this gene will destroy membrane function and so destroy the cell.

Even so, some people worry about the possible risks of introducing genetically engineered bacteria into the environment. Consider how Steven Lindow engineered a bacterium that can make many crop plants resist frost. A protein at the bacterial surface promotes formation of ice crystals. Lindow excised the ice-forming gene from some cells. He hypothesized that spraying his "ice-minus bacteria" on strawberry plants in an isolated field before a frost would make plants less vulnerable to freezing. Even though he deleted a *harmful* gene from an organism, it triggered a bitter legal debate about releasing engineered bacteria into the environment. The courts finally ruled in favor of allowing the release. Researchers sprayed a small strawberry patch (Figure 13.9). Nothing bad happened. Since that time, rules governing the release of genetically engineered organisms have become much less restrictive.

Genetically Engineered Plants

A Hunt for Beneficial Genes   Botanists are combing the world for seeds and other living tissues from the wild ancestors of potatoes, corn, and other plants. They send their prizes—genes from a plant's lineage—to a safe storage laboratory in Colorado. Why? Farmers use only a few strains of high-yield crop plants—which feed most of the human population. The lack of genetic diversity means our food base is vulnerable to disease-causing viruses, bacteria, and fungi.

It is a race against time. The human population is increasing to astounding levels. People, saws, and bulldozers are rapidly encroaching on uninhabited areas, and hundreds of wild plant species disappear weekly.

The danger is real. Recently a new fungal strain that rots potatoes entered the United States from Mexico. In 1994, crop losses from the resulting *potato blight* topped $100 million. Pennsylvania farmers had little to harvest. Perhaps laboratory-stored strains will save potato crops in the future.

Plant Regeneration Botanists also search for good genes in the laboratory. Years ago, Frederick Steward and his coworkers cultured cells of carrot plants and induced the cells to grow into small embryos. Some embryos grew into whole plants. Other plant species, including major crop plants, are now regenerated from cultured cells. The methods increase mutation rates, so the cultures are a source of genetic modifications.

Researchers can pinpoint a useful mutation among millions of cells. Suppose a culture medium contains a toxic product of a disease agent. If a few cells have a mutated gene that confers resistance to the toxin, only they live in the culture. If the cells regenerate whole plants that can be hybridized with other varieties, the hybrids may end up with the good gene.

Methods of Gene Transfer Today, genetic engineers insert genes into cultured plant cells. For example, they put DNA fragments into the "Ti" plasmid from a bacterium (*Agrobacterium tumefaciens*) that infects many flowering plants. Some plasmid genes invade the plant DNA and induce the formation of crown gall tumors (Figure 13.10*a*). Before introducing the Ti plasmid into plants, researchers first remove the tumor-inducing genes and insert desired genes into it. They grow the modified bacterial cells with cultured plant cells, then regenerate plants from cells that take up the genes. In some cases, the foreign genes have been expressed in plant tissues, with observable effects (Figure 13.10*b*).

A. tumefaciens can only infect beans, peas, potatoes, and other dicots. However, wheat, corn, rice, and many major food crops are monocots—and the bacterium can't infect them. In some cases, genetic engineers can use electric shocks or chemicals to deliver modified DNA into protoplasts. (Protoplasts are plant cells stripped of their walls.) Regenerating some species of plants from protoplast cultures is not yet possible. However, researchers have had some success delivering genes into cultured plant cells by shooting at them with pistols. Instead of bullets, they use blanks to drive DNA-coated, microscopic particles into cells.

On the Horizon Despite many obstacles, improved varieties of crop plants have been developed or are in the works. For example, genetically engineered cotton plants now resist worm attacks (Figure 13.10*c*). Only the targets are killed. Unlike pesticides, the plants don't kill beneficial insects that eat aphids and other pests.

Also on the horizon are engineered plants that can serve as factories for pharmaceuticals. A few years ago, genetically engineered tobacco plants that make human hemoglobin, melanin, and other proteins were planted in a field in North Carolina. Ecologists later found no trace of foreign genes or proteins in the soil or in other

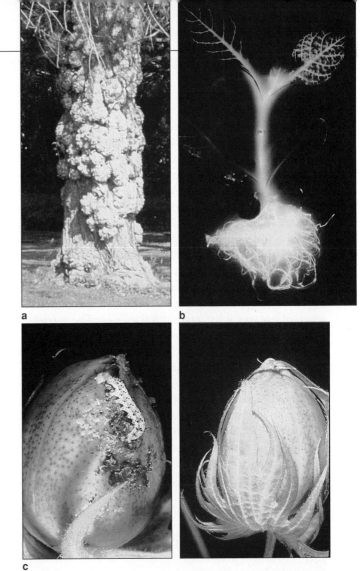

a   b

c

Figure 13.10 Gene transfers in plants. (**a**) Genes in a plasmid from a common bacterium (*Agrobacterium tumefaciens*) cause crown gall tumors on woody plants. Scientists use this plasmid to move desirable genes into cultured plant cells. (**b**) Evidence of a successful gene transfer: a modified tobacco plant that glows in the dark. A firefly gene became incorporated into the plant's DNA and is being translated. The protein product is luciferase, an enzyme required for bioluminescence (Section 4.9).

(**c**) Example of a valuable plant that benefits from genetic engineering. *Left:* Buds of cotton plants are vulnerable to worm attack. *Right:* Buds of a modified cotton plant resist attack.

plants or animals in the vicinity. Recently, mustard plant cells made beads of plastic in a Stanford University laboratory. ("Plastics" are long-chain polymers of identical subunits.) The plant DNA contains three bacterial enzymes that convert carbon dioxide, nutrients, and water into inexpensive, biodegradable plastic.

Fail-safe measures must be built into genetically engineered organisms that may end up where they shouldn't be.

Genetically engineered plants are one of the few options available for protecting our vulnerable food supply.

13.6 GENETIC ENGINEERING OF ANIMALS

Supermice and Biotech Barnyards

Mice were the first mammals that were used in genetic engineering experiments. Consider how R. Hammer, R. Palmiter, and R. Brinster corrected a hormone deficiency that leads to dwarfism in mice. Such mice have trouble producing somatotropin. The researchers used a microneedle to inject the gene for rat somatotropin into fertilized mouse eggs. Then they implanted the eggs into an adult female. The gene became integrated into the mouse DNA. Baby mice in which the foreign gene was expressed were 1-1/2 times larger than their dwarf littermates. In other experiments, the gene for human somatotropin was transferred into a mouse embryo, where it became integrated into the DNA. The modified embryo grew up to be "supermouse" (Figure 13.11).

Today, as part of research into the molecular basis of Alzheimer disease and other genetic disorders, several human genes are being inserted into mouse embryos. Besides microneedles, microscopic laser beams are used to open temporary holes in the plasma membrane of cultured cells, although such methods have varying degrees of success. Retroviruses also are used to insert genes into cultured cells, virtually all of which incorporate the foreign genes into their DNA. But the genetic

material of retroviruses can undergo rearrangements, deletions, and other alterations that shut down the introduced genes. And if the virus particles escape from laboratory isolation, they can infect individuals.

Soon, "biotech barnyards" may be competing with bacterial factories as genetically engineered sources of proteins. Farm animals produce the proteins in greater amounts, at less cost. Herman, a Holstein bull, received the human gene for lactoferrin, a milk protein, when he was just an embryo. His female offspring may mass-produce the protein, which can be used as a supplement for infant formulas. Similarly, goats provide CFTR protein (used in the treatment of cystic fibrosis) and TPA (which lessens the severity of heart attacks). Sheep produce alpha-1 antitrypsin, used in the treatment of emphysema. Cattle may soon produce human collagen, a key component of skin, cartilage, and bone.

Applying the New Technology to Humans

Researchers around the world are working their way through the 3.2 billion base pairs of the twenty-three pairs of human chromosomes. This ambitious effort is the **human genome project**. Some researchers focus on specific chromosomes; others are deciphering certain gene regions only. For example, rather than studying noncoding sequences (introns), J. Venter and Sidney Brenner isolate mRNAs from brain cells and use them to make cDNA. By sequencing only cDNAs, they have identified hundreds of previously unknown genes.

About 99.9 percent of the nucleotide sequence is the same in every human on earth. This means that, once the sequencing project is completed, we will have the ultimate reference book on human biology and genetic disorders. Or should we say reference *books*—the entire sequence will fill the equivalent of about 200 Manhattan telephone directories.

What will we do with the information? Certainly we will use it in the search for effective treatments and cures for genetic disorders. Of 2,000 or so genes studied so far, 400 have already been linked to genetic disorders. The knowledge opens doors to gene therapy—the transfer of one or more normal or modified genes into body cells of an individual to correct a genetic defect or boost resistance to disease.

But what about forms of human gene expression that are neither disabling nor life-threatening? Will we tinker with these, also? We leave the chapter with a *Focus* essay that looks at these questions.

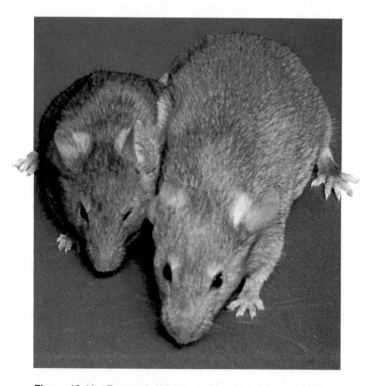

Figure 13.11 Ten-week-old mouse littermates, the one on the left weighing 29 grams, and the one on the right, 44 grams. The larger mouse grew from a fertilized egg into which the gene for human somatotropin (growth hormone) had been inserted.

As with any new technology, the potential benefits of recombinant DNA technology and genetic engineering should be weighed carefully against the potential risks.

REGARDING HUMAN GENE THERAPY

This chapter opened with a historic, inspiring case, the first proof that gene therapy can work to save lives. It closes now with a few questions that invite you to consider some issues associated with the application of recombinant DNA technology to our rapidly advancing knowledge of the human genome. We are only beginning to work our way through its social and ethical implications.

To most of us, human gene therapy to correct genetic abnormalities seems like a socially acceptable goal. Let's take this idea one step further. Is it also socially acceptable to insert genes into a normal human individual (or sperm or egg) to alter or enhance traits?

The idea of selecting desirable human traits is called *eugenic engineering*. Yet who decides which forms of a trait are most "desirable"? For example, what if prospective parents could pick the sex of a child by way of genetic engineering? In one survey group, three-fourths of the people who were asked this question said they would choose a boy. So what would be the long-term social implications of a drastic shortage of girls?

As another example, would it be okay to engineer taller or blue-eyed or fair-skinned or curlier-haired individuals? If so, would it be okay to engineer "superhuman" children with exceptional strength or breathtaking intelligence? Suppose a person of average intelligence moved into a town composed of 800 Einsteins. Would the response go beyond a few mutterings of "There goes the neighborhood"?

Some say that the DNA of any organism must never be altered. Put aside the fact that nature itself alters DNA much of the time, and has done so for nearly all of life's history. The concern is that *we* don't have the wisdom to bring about beneficial changes without causing great harm to ourselves or to the environment.

When it comes to manipulating human genes, one is reminded of our human tendency to leap before we look. When it comes to restricting genetic modifications of any sort, one also is reminded of an old saying: "If God had wanted us to fly, he would have given us wings."

And yet, something about the human experience gave us the capacity to imagine wings of our own making—and that capacity has carried us to the frontiers of space.

Where are we going from here with recombinant DNA technology, this new product of our imagination? To gain perspective on the question, spend some time reading the history of our species. It is a history of survival in the face of all manner of new challenges, threats, bumblings, and sometimes disasters on a grand scale. It is also a story of our connectedness with the environment and with one another.

The questions confronting you today are these: Should we be more cautious, believing that one day the risk takers may go too far? And what do we as a species stand to lose if the risks are not taken?

SUMMARY

1. Countless gene mutations and other forms of genetic "experiments" have been proceeding in nature for at least 3 billion years.

2. Through artificial selection practices, humans have been manipulating the genetic character of different species for many thousands of years. Today, recombinant DNA technology has enormously expanded our capacity to modify organisms genetically.

3. With recombinant DNA technology, researchers isolate, cut, and splice together gene regions from different species, then greatly amplify the number of copies of those regions into usefully large quantities. Enzymes make the cuts and do the splicing.

4. Researchers use various restriction enzymes and DNA ligase to cut and insert DNA into plasmids from bacterial cells.
   a. Plasmids are small, circular DNA molecules that contain extra genes, besides those genes of the circular bacterial chromosome.
   b. A DNA library is a collection of DNA fragments, produced by restriction enzymes and incorporated into plasmids.

5. Restriction fragments may be amplified by a population of rapidly dividing cells (such as bacteria or yeasts). Short DNA fragments may be amplified more rapidly in a test tube by the polymerase chain reaction (PCR).

6. Following amplification, the DNA fragments can be sorted out (according to length), and their nucleotide sequences can be determined.

7. Each individual of a species has a DNA fingerprint: a unique array of restriction fragment length polymorphisms (RFLPs). Small molecular variations in the base sequence of the DNA of different individuals lead to small, identifiable differences in restriction fragments cut from the DNA.

8. Cells that take up foreign genes can be identified by DNA probes, which base-pair with the genes of interest. Base-pairing of nucleotide sequences from different sources is called nucleic acid hybridization.

9. In genetic engineering, genes are isolated, modified, and inserted into the same organism or a different one. In gene therapy, one or more normal or modified genes are inserted into an individual to correct a genetic defect or boost resistance to disease.

10. Recombinant DNA technology and genetic engineering have enormous potential for research and applications in medicine, agriculture, the home, and industry. As with any new technology, potential benefits must be weighed against potential risks, including ecological and social repercussions.

Review Questions

1. Distinguish these terms from one another:
   a. recombinant DNA technology and genetic engineering *204*
   b. restriction enzyme and DNA ligase *205*
   c. cloned DNA and cDNA *206, 209*
   d. PCR and DNA sequencing *206–207*

2. Define RFLPs. Briefly explain one of the methods by which biotechnologists produce them, and then name some applications for RFLP analysis. *208*

3. Explain how DNA probes and nucleic acid hybridization can be used to identify genetically modified host cells. *209*

Self-Quiz *(Answers in Appendix IV)*

1. _____ are small circles of bacterial DNA that are separate from the circular bacterial chromosome.

2. DNA fragments result when _____ cut DNA molecules at specific sites.
   a. DNA polymerases
   b. DNA probes
   c. restriction enzymes
   d. RFLPs

3. PCR stands for _____ .
   a. polymerase chain reaction
   b. polyploid chromosome restrictions
   c. polygraphed criminal rating
   d. politically correct research

4. A _____ is a collection of DNA fragments, produced by restriction enzymes and incorporated into plasmids.
   a. DNA clone
   b. DNA library
   c. DNA probe
   d. gene map

5. A _____ is multiple, identical copies of a collection of DNA fragments inserted into plasmids.
   a. DNA clone
   b. DNA library
   c. DNA probe
   d. gene map

6. Gel electrophoresis, a standard laboratory procedure, can be used to separate DNA restriction fragments according to _____ .
   a. size and shape
   b. length
   c. net surface charge
   d. all of the above

7. In reverse transcription, _____ is assembled on _____ .
   a. mRNA; DNA
   b. cDNA; mRNA
   c. DNA; enzymes
   d. DNA; agar

8. _____ is the transfer of normal genes into body cells to correct a genetic defect.
   a. Reverse transcription
   b. Nucleic acid hybridization
   c. Gene mutation
   d. Gene therapy

9. Tobacco plant leaves that produce hemoglobin are a result of _____ .
   a. gene therapy
   b. genetic engineering
   c. pressure on tobacco growers
   d. a and b

10. Match the terms with the most suitable description.
    ____ DNA fingerprint
    ____ Ti plasmid
    ____ nature's genetic experiments
    ____ nucleic acid hybridization
    ____ human genome project
    ____ eugenic engineering

    a. selecting "desirable" traits
    b. deciphering 3.2 billion base pairs of 23 human chromosomes
    c. used in gene transfers
    d. unique array of RFLPs
    e. base-pairing of nucleotide sequences from different DNA or RNA sources
    f. mutations, crossovers

Critical Thinking

1. Besides this chapter's examples, list what you believe might be some potential benefits and risks of genetic engineering.

2. Ryan, a forensic scientist, has obtained a very small DNA sample from a crime scene. In order to examine the sample by DNA fingerprinting, he must first amplify the sample using the polymerase chain reaction. He estimates there are 50,000 copies of the DNA in his sample. Derive a simple formula and calculate the number of copies he will have after fifteen cycles of PCR.

3. A game warden in Africa has confiscated eight ivory tusks from elephants. Some tissue is still attached to the tusks. He must determine whether they were taken illegally, from northern populations of endangered elephants, or from elephants from populations to the south that can be hunted legally. How can he use DNA fingerprinting to find the answer?

4. A restriction enzyme dubbed EcoRI is specified by a gene on an R plasmid in *E. coli*. In nature, it cleaves the circular DNA of a small virus that infects mammals. Lisa, a biotechnologist, has cloned an EcoRI fragment into a plasmid at the one site where replication enzymes can insert it. The cloned EcoRI fragment is 850 base pairs (bp) long. And 300 bp from one end of the fragment is a site where a different restriction enzyme (BamHI) can cleave it. The plasmid is 1,900 bp long and has no BamHI sites.
   a. When EcoRI and the plasmid are mixed together in solution, what length does Lisa expect the resulting fragments to be? What can she expect when she uses BamHI alone to cut it? When she uses both EcoRI and BamHI?
   b. Draw a circular map of the recombination plasmid, showing the relative distance between the two restriction enzyme sites.

5. Last winter, Lunardi's Market put out a bin of tomatoes having splendid vine-ripened redness, flavor, and texture. The sign posted above the bin identified them as genetically engineered produce. Most shoppers selected the unmodified tomatoes in the adjacent bin, even though these tomatoes were pale pink, mealy-textured, and tasteless. Which ones would you pick? Why?

Selected Key Terms

Readings

Anderson, W. F. 1992. "Human Gene Therapy." *Science* 256: 808–813.

Gasser, C. S., and R. T. Fraley. 1989. "Genetically Engineering Plants for Crop Improvement." *Science* 244: 1293–1299.

Joyce, G. December 1992. "Directed Molecular Evolution." *Scientific American* 267(6): 90–97.

Pursel, V. G., et al. 1989. "Genetic Engineering of Livestock." *Science* 244: 1281–1288.

Watson, J. D. 1990. "The Human Genome Project: Past, Present, and Future." *Science* 248: 44–49.

FACING PAGE: *Millions of years ago, a bony fish died, and sediments gradually buried it. Today its fossilized remains are studied as one more piece of the evolutionary puzzle.*

14 MICROEVOLUTION

Designer Dogs

We humans have tinkered rather ruthlessly with the modern descendants of a long and distinguished lineage. That lineage originated some 40 million years ago with the appearance of small, weasel-shaped, tree-dwelling carnivores. They were the forerunners of bears, raccoons, pandas, badgers—and dogs. About 10,000 years ago, we began domesticating wild dogs.

No doubt the advantages of doing so were obvious. Times were tough in the days before supermarkets and police protection. Dogs could guard people and their possessions. They could corner, kill, eat, and thereby dispose of rats and other vermin that were not at all welcome in the home.

It didn't take us long to develop different varieties (breeds) through artificial selection. Individual dogs having the desired forms of traits were selected from each new litter and, later, encouraged to breed. Those having undesired forms of traits were passed over.

After favoring the pick of the litter over hundreds or thousands of generations, we ended up with sheep-herding collies, badger-hunting dachshunds, wily retrievers, and snow-traversing, sled-pulling huskies. And at some point we began to delight in the odd, extraordinary dog. Imagine! In practically no time at all, evolutionarily speaking, we picked our way through the pool of variant dog genes and came up with such extremes as Great Danes and Chihuahuas (Figure 14.1).

Of course, canine designs can exceed the limits of biological common sense. For example, how long would a tiny, finicky-eating, nearly hairless, nearly defenseless Chihuahua last in the wild? Not very long. Or what about the English bulldog, bred for a very short snout and a compressed face? Long ago, breeders thought these particular traits would allow the dogs to get a better grip on the nose of bulls. (Why they wanted dogs to bite bulls is a story in itself.) So now the roof of the bulldog mouth is ridiculously wide and it is often flabby, so bulldogs have trouble breathing. They sometimes get so short of air, they pass out.

Through our centuries-old fascination with artificial selection, we produced thousands of varieties of crop plants, cats, cattle, and birds as well as dogs. With the currently available technologies of genetic engineering, we are now mixing the genes of different species and producing incredible new varieties, including tobacco plants that produce hemoglobin and mustard plants that produce plastic.

So, when you hear someone wonder about whether "evolution" takes place, remind yourself that evolution simply means *genetic change through time*. Selective breeding practices provide abundant, tangible evidence that heritable changes do, indeed, occur. Just *how* those changes are brought about in a given line of descent is the subject of this chapter.

KEY CONCEPTS

1. We define biological evolution as heritable changes in lineages (lines of descent). Evidence of evolution comes from investigations that began nearly two centuries ago.

2. As Charles Darwin and Alfred Wallace perceived long ago, members of the same population share the same heritable traits—yet they don't look or act exactly the same, and they don't all survive and reproduce. Differences in survival and reproduction among individuals that differ in one or more traits are the basis of natural selection.

3. In the population as a whole, each gene may exist in two or more slightly different molecular forms, called alleles. Often these differences result in variations in the details of traits. Thus phenotypic variation may occur in the population, because its individuals may or may not inherit the same alleles for any given trait.

4. Any allele may become more or less common in the population, relative to the other kinds, or it may disappear. *Microevolution* refers to such changes in a population's allele frequencies over time.

5. Mutation, gene flow, genetic drift, and natural selection can change allele frequencies. Only mutation gives rise to new alleles.

6. Natural selection is not an "agent," purposefully searching for the "best" individuals in a population. It simply is the result of a difference in survival and reproduction among individuals that differ in one or more traits.

Figure 14.1 Two designer dogs. About 10,000 years ago, humans began domesticating wild dogs. From that ancestral stock, artificial selection produced diverse yet rather closely related breeds, such as the Great Dane (*legs, left*) and the Chihuahua (*possibly fearful of being stepped on, right*).

EARLY BELIEFS, CONFOUNDING DISCOVERIES

The Great Chain of Being

More than two thousand years ago, the seeds of biological inquiry were taking hold among the ancient Greeks. At the time, popular belief held that supernatural beings could intervene directly in the affairs of human beings. For example, the gods were said to cause a common ailment known as the sacred disease. And yet, from a physician of the school of Hippocrates, these thoughts come down to us:

It seems to me that the disease called sacred . . . has a natural cause, as other diseases have. Men think it divine merely because they do not understand it. But if they called everything divine that they did not understand, there would be no end of divine things! . . . If you watch these fellows treating the disease, you see them use all kinds of incantations and magic—but they are also very careful in regulating diet. Now if food makes the disease better or worse, how can they say it is the gods who do this? . . . It does not really matter whether you call such things divine or not. In Nature, all things are alike in this, in that they can be traced to preceding causes.

— ON THE SACRED DISEASE (400 B.C.)

Passages such as this reflect the early attempts to find natural explanations for observable events.

Aristotle was foremost among the early naturalists, and he described the world around him in excellent detail. He had no reference books or instruments to guide him, for biological science in the Western world began with the great thinkers of this age. Yet here was a man who was no mere collector of random bits of information. In his descriptions, we find evidence of a mind perceiving connections between diverse observations and attempting to explain the order of things.

Aristotle believed (as did others) that each kind of organism was distinct from all the rest. Nevertheless, he wondered about organisms that seemed to have rather blurred positions in nature. For example, even though some sponges look very much like plants, they don't make their own food; they capture and digest it—as animals do. In time, Aristotle came to view nature as a continuum of organization, from lifeless matter through complex forms of plant and animal life.

By the fourteenth century, this idea had been transformed into a rigid view of life. A great Chain of Being was seen to extend from the lowest forms of life to humans and on to spiritual beings. Each kind of being, or "species," as it was called, was a separate link in the chain. All the links were designed and forged at the same time, at the same center of creation, and had not changed in any way since. Scholars thought that once they had discovered, named, and described all of the links, the meaning of life would be revealed to them.

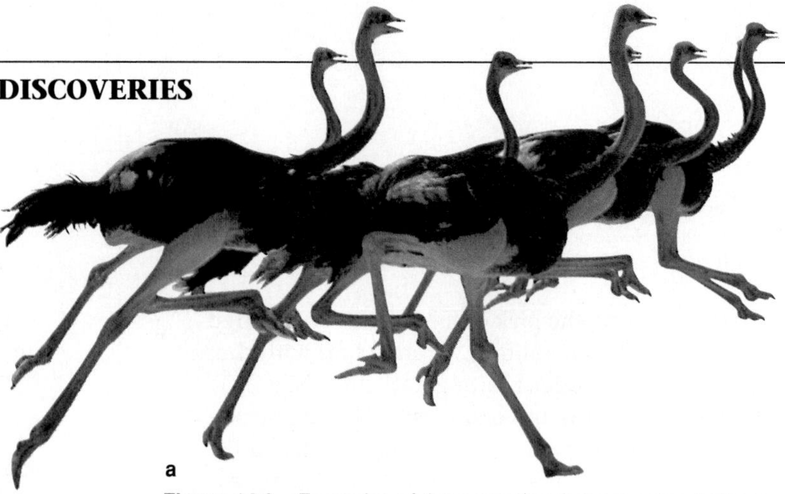

a

Figure 14.2 Examples of three species that are native to three geographically separate parts of the world. (**a**) *Above*, ostrich of Africa. *Right:* Rhea of South America (**b**) and emu of Australia (**c**). All three species are like one another and unlike most birds in several ways, including their inability to get airborne.

Questions From Biogeography

Until the fifteenth century, naturalists were unaware that the world is a great deal bigger than Europe, so the task of locating and describing all species seemed manageable. Then global explorations began and the known world expanded enormously. Naturalists were soon overwhelmed by descriptions of tens of thousands of plants and animals that explorers were discovering in Asia, Africa, the Pacific Islands, and the New World.

In 1590 the naturalist Thomas Moufet attempted to sort through the bewildering array. He simply gave up and wrote such gems as this description of locusts and grasshoppers: "Some are green, some black, some blue. Some fly with one pair of wings; others with more; those that have no wings they leap; those that cannot fly or leap they walk; some have long shanks, some shorter. Some there are that sing, others are silent." It was not exactly a work of subtle distinctions.

Even so, a few scholars began to examine the world distribution of plants and animals, a discipline now called **biogeography**. It became clear that many unique species live in isolated places, such as oceanic islands. It also became clear that some species look very similar to one another yet live in places that are separated by great oceans or impenetrable mountains (Figure 14.2).

How did so many species get from the center of creation to islands and other isolated places?

Questions From Comparative Anatomy

In the eighteenth century, scholars were noticing similarities and differences in the body plans of different mammals, reptiles, and other major groups, a discipline called **comparative anatomy**. Think of a human arm, whale flipper, and bat wing. These body parts differ in

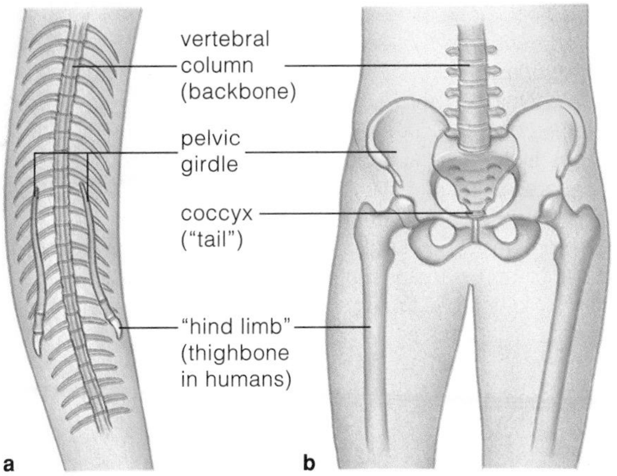

Figure 14.3 (a) Python bones corresponding to the pelvic girdle of other vertebrates, including humans (b). Small "hind limbs" protrude through the skin on the underside of the snake.

Labels in figure:
- vertebral column (backbone)
- pelvic girdle
- coccyx ("tail")
- "hind limb" (thighbone in humans)

a   b

size, shape, and function. Yet all have similar locations in the body. They consist of the same tissues, arranged in the same patterns. And they develop in similar ways in embryos. Comparative anatomists who discovered all of this wondered: Why are certain animals that are so different in some features so much alike in others?

According to one hypothesis, basic body plans were so perfect, there was no need to come up with a totally new one for each organism at the time of their creation. Yet if that were true, then how could there be parts with no function? For instance, some snakes have bones that correspond to a pelvic girdle—a set of bones to which hind legs attach (Figure 14.3). Snakes don't have legs. Why the bones? Humans have bones that correspond to a few of the tail bones of many other mammals, but they don't have a tail. Why do they have parts of one?

Questions About Fossils

Starting in the late 1600s, geologists added to the confusion. They began mapping **sedimentary beds** at sites where erosion or quarrying had cut deep into the earth. They found similar beds around the world. (Such beds consist of distinct, multiple layers of sedimentary rock. Figure 16.3 shows an example.) Most agreed that sand and other sediments were deposited at different times and gradually compacted into rock layers. As they soon realized, certain layers contained **fossils**, which were known to be evidence of life in the past. For example, deep layers contained fossils of very simple marine organisms. In layers above these, fossils of organisms were more complex in their structure. And in layers above these, the fossils closely resembled living marine organisms. Were these *sequences* of fossils, separated in time? If so, *what did the increasing structural complexity among fossils of a given type of organism represent?*

Taken as a whole, the findings from biogeography, comparative anatomy, and geology did not fit well with prevailing beliefs. Georges-Louis Leclerc de Buffon and a few others started formulating startling hypotheses. If dispersal of all species from one center of creation was not possible, given the vast oceans and other barriers, *then perhaps species had originated in more than one place.* If species were not created in a perfect state, as suggested by the layering of different fossils and the occurrence of "useless" body parts in certain organisms, *then perhaps species had been modified over time.* Awareness of change in lines of descent—**evolution**—was in the wind.

Awareness of biological evolution emerged over centuries, through the cumulative observations of many naturalists, biogeographers, comparative anatomists, and geologists.

A FLURRY OF NEW THEORIES

Squeezing New Evidence Into Old Beliefs

In the nineteenth century, naturalists tried to reconcile the growing evidence of change in lines of descent with a traditional conceptual framework that did not allow for change. Foremost among them was Georges Cuvier, a respected anatomist. For years, Cuvier had compared body plans of fossils and living organisms. He acknowledged there were abrupt changes in the fossil record, corresponding to discontinuities between certain layers of sedimentary beds. Were these evidence of changes in the populations of organisms that inhabited ancient environments? Cuvier thought so. And he was right, as you will see from the evolutionary story in Chapter 17.

Cuvier constructed a theory of **catastrophism** based on his inference. There was one time of creation, he said, that populated the world with all species. A global catastrophe destroyed many of them. A few survivors repopulated the world. It wasn't that survivors were *new* species; naturalists simply hadn't yet found earlier fossils of them that would date to the time of creation. Later catastrophes wiped out more species, and repopulation by the survivors followed, as recorded by fossils.

Many scholars accepted his theory. Others kept chipping away at the puzzle. Consider Jean-Baptiste Lamarck's **theory of inheritance of acquired characteristics**. During an individual's life, said Lamarck, environmental pressures and internal "needs" bring about permanent changes in the body, and offspring inherit the needed changes. And so life, created long ago in a simple state, gradually improved. The force for change was a drive for perfection, up the Chain of Being. The drive was centered in nerves that directed an unknown "fluida" to body parts in need of change.

For instance, suppose modern giraffes had a short-necked ancestor. Pressed by the need to find food, it kept stretching its neck to browse on leaves beyond the reach of other animals. Stretching directed fluida to its neck, which lengthened permanently. The longer neck was bestowed on offspring, which stretched their necks also. And so generations of animals desiring to reach higher leaves led to the modern giraffe.

As Lamarck correctly inferred, the environment *is* a factor in evolution. But his overall theory, like others proposed at this time, has not been supported by any investigation carried out since then. No one has found evidence that environmental modifications of existing traits can be passed to offspring.

Voyage of the *Beagle*

In 1831, in the midst of the confusion, Charles Darwin was twenty-two years old and wondering what to do with his life. Ever since he was eight, he had wanted to

Figure 14.4 (**a**) Charles Darwin and a blue-footed booby, one of the species he encountered during his five-year voyage around the world on H.M.S. *Beagle*. A replica of this ship is shown in (**b**), sailing off the coast of South America. During stops along Argentina's coast, Darwin explored parts of the Andes. He saw fossils of marine organisms embedded in rocks 3.6 kilometers above sea level. (**c**) The Galápagos Islands, isolated bits of land in the ocean, far to the west of Ecuador. We now know they arose through volcanic action about 5 million years ago, so organisms could not have originated there. Winds or ocean currents must have carried them to the new islands.

hunt, fish, collect shells, or simply watch insects and birds—anything but sit in school. Later, at his father's insistence, he did try to study medicine in college. The crude, painful procedures used on patients at that time sickened him. Then his father strongly urged him to become a clergyman, so Darwin packed for Cambridge. His grades were good enough to earn him a degree in theology. But he spent most of his time among faculty members with leanings toward natural history.

John Henslow, a botanist, perceived Darwin's real interests. He arranged for Darwin to function as ship's naturalist aboard H.M.S. *Beagle*. The *Beagle* was about to embark on a five-year voyage that would take Darwin around the world (Figure 14.4). The young man who hated schoolwork and who had no formal training as a naturalist suddenly began to work enthusiastically.

The *Beagle* sailed first to South America, to complete work on mapping the coastline. During the Atlantic crossing, Darwin collected and examined marine life. He also read Henslow's parting gift, the first volume of Charles Lyell's *Principles of Geology*. During stops along the coast and at various islands, he observed diverse

b

species in environments ranging from sandy shores to high mountains. And he started circling the question of evolving life, which was now on the minds of many respected individuals—including his own grandfather.

Darwin started mulling over a rather radical theory that Lyell was advancing in his book. By the **theory of uniformity**, the same geologic processes that are now slowly sculpting the Earth's surface also were at work in the past. Over millions of years, mountain ranges rose from ancient seafloors. Season after season of rains and winds eroded mountain flanks. The washed-down sediments slowly accumulated in layers, in valleys and on the seafloor. Violent earthquakes, major floods, and other catastrophes were like abrupt punctuation points in this history of gradual geologic change.

The theory bothered many scholars, who believed the Earth was less than 6,000 years old. They thought people had recorded everything that happened during those thousands of years, and in all that time, no one ever mentioned seeing a species evolve. Yet by Lyell's calculations, it must have taken millions of years to mold the present landscape. *Wasn't that enough time for species to evolve in diverse ways?* Darwin thought so. *But how did they evolve?* He would end up devoting the rest of his life to that burning question.

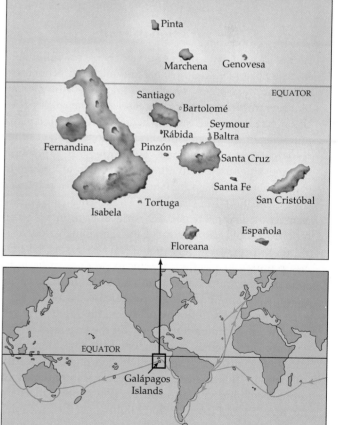

Prevailing beliefs can influence how we interpret clues to natural processes and their observable outcomes.

Darwin's observations during a global voyage helped him think about these processes in a novel way.

c

Old Bones and Armadillos

After he returned to England in 1836, Darwin talked with other naturalists about the possible evidence that life evolves. By carefully studying all of the notes from his journey, he came up with some possibilities.

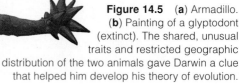

Figure 14.5 (**a**) Armadillo. (**b**) Painting of a glyptodont (extinct). The shared, unusual traits and restricted geographic distribution of the two animals gave Darwin a clue that helped him develop his theory of evolution.

For example, he had observed fossils of glyptodonts in Argentina. Of all the animals on Earth, only living armadillos are like glyptodonts, which are now extinct (Figure 14.5). Of all places on Earth, armadillos live only in the same regions where glyptodonts once lived. If the two kinds of animals had been created at the same time, lived in the same place, and were so much alike, why is only one kind still alive? Wouldn't it be reasonable to assume glyptodonts were early relatives of armadillos? Many of their shared traits might have been retained through countless generations. Other traits might have been modified in the armadillo branch of a family tree. *Descent with modification*—it seemed possible.

A Key Insight—Variation in Traits

While Darwin was assessing his notes, an influential essay by Thomas Malthus, a clergyman and economist, made him take a closer look at a topic of social interest. Malthus correlated famine, disease, and war with population size. Humans, he said, tend to produce offspring much faster than food supplies, living space, and other resources can be sustained. The larger the population becomes, the more individuals there are to reproduce in each generation. As population size burgeons, resources dwindle—and the struggle for existence intensifies. Many people starve, get sick, and engage in war and other forms of competition for the remaining resources.

From his own observations, Darwin suspected that any population has the capacity to produce more individuals than the environment can support. To give an example, a single sea star can release 2,500,000 eggs each year, but the seas obviously are not filled with sea stars. In each generation, nearly all the eggs and larvae end up in the bellies of predators. Still others starve or succumb to disease or some other environmental insult.

Assume that the environment keeps the number of reproducing individuals in check. *Which* individuals are the winners and losers? What might affect the outcome?

Darwin thought about the populations of animals and plants he had observed during his voyage. He recalled that individuals in those populations weren't alike down to the last detail. They showed variations in size, coloration, and other traits. *And it dawned on him that variations in traits might affect an individual's ability to secure resources—and to survive and reproduce in particular environments.*

Did the Galápagos Islands provide evidence of this? About 900 kilometers of open ocean separate these volcanic islands from the South American coast. They offer a variety of habitats along rocky shorelines, in deserts, and on mountain flanks. Most of their inhabitants live nowhere else—although they do resemble species on the mainland. Were the islands colonized by species that flew, floated, or were blown over from the mainland? If so, then island-hopping descendants of colonizers that took hold in different island habitats could have become adapted over time to local conditions.

Darwin learned, through conversations with other naturalists, that populations of thirteen finch species were distributed among the various Galápagos Islands. He himself had collected specimens of the birds, and now he attempted to correlate variations in their traits with environmental challenges.

For example, in one population, the finches have a large, strong bill, suitable for cracking seeds (Figure 14.6). Yet the bill of a few individuals is a bit stronger—and they can crack open seeds that other birds find too tough to deal with. If most seeds being produced in a particular environment have hard coats, a strong-billed bird will have a competitive edge. Assuming the trait has a heritable basis, the same will be true of its strong-billed descendants. Take this one step further. Over time, if factors in the natural environment continue to "select" the most adaptive version of the trait, then the population will become one of mostly strong-billed birds. *And a population is evolving when its heritable traits are changing through successive generations.*

Recall, from Chapter 1, that Darwin used pigeon breeding and other *artificial* selection practices as a way to explain *natural* selection of adaptive traits in the wild. Thus, when breeders favor pigeons with black tail feathers, they encourage every new black-tailed pigeon to mate—but they might put the white-tailed ones in the stewpot. It was an easy way to show how selection could cause an increase in the frequency of one version of a trait in the captive population, and how it might eliminate the other version.

Figure 14.6 Four species of finches from the Galápagos Islands. (**a**) *Geospiza conirostris* and *G. scandens*, both with a bill adapted for eating cactus flowers and fruits. Other finches have thick, strong bills that crush cactus seeds. (**b**) *Left*: *Certhidea olivacea*, a tree-dwelling finch, resembles warblers in song and behavior. It uses its slender beak to probe for insects. *Right*: *Camarhynchus pallidus* feeds on wood-boring insects such as termites. It does not have the woodpecker's long, probing tongue. It has learned to break cactus spines and twigs to suitable lengths, then hold the "tools" and use them to probe bark for insects hidden from its view.

Anticipating that his view of evolution by natural selection would generate controversy, Darwin waited to announce it and searched for flaws in his reasoning. He waited too long. More than a decade after he wrote but never published an essay on his theory, he received a paper from Alfred Wallace, another naturalist (Figure 14.7*a*). On his own, Wallace had developed the same theory, then quickly wrote up and circulated his ideas. Darwin's colleagues prevailed on him to formally present a paper with Wallace (who also believed Darwin deserved most of the credit). The next year, in 1859, Darwin's impressively detailed evidence in support of the theory was published in book form.

Darwin's theory faced a crucial test. If he were correct, then all existing species are related, by descent, to ancestral species. Given the rapid outcomes of artificial selection, Darwin believed one kind of organism could evolve slowly into another, over hundreds or thousands of generations. But where were the "missing links"? If his theory were correct, then there must be fossils of transitional forms between major groups of organisms.

In 1861, such a fossil was unearthed. *Archaeopteryx* appears to be a transitional form between reptiles and birds (Figure 14.7*b*). Like modern birds, it had a beak, wings, and feathers. Like fossils of small, two-legged reptiles, it had teeth and a long, bony tail. Other fossils

Figure 14.7 (**a**) Alfred Wallace. Darwin worked out a theory of natural selection long before Wallace did, but Wallace was first to circulate a letter outlining the process. (**b**) *Archaeopteryx*, a transitional form that lived more than 140 million years ago, was on or very near the evolutionary road from reptiles to birds. The fossil might have been classified as reptilian had it not been for clear imprints of feathers in the limestone tomb.

were found during Darwin's lifetime. Even so, nearly seventy years passed before advances in a new field, genetics, led to widespread acceptance of his theory of natural selection. In the meantime, Darwin's name was associated primarily with the idea that life evolves—something others had proposed before him. And with this bit of history behind us, we are ready to consider some current views of the mechanisms of evolution.

DARWIN'S THEORY OF EVOLUTION BY NATURAL SELECTION. **Any population can evolve (change over time) when individuals differ in one or more heritable traits that are responsible for differences in the ability to survive and reproduce.**

14.4 INDIVIDUALS DON'T EVOLVE; POPULATIONS DO

Examples of Variation in Populations

As Charles Darwin perceived, the individual doesn't evolve; populations do. By definition, a **population** is a group of individuals of the same species occupying a given area. To understand how a population evolves, start with the variation in traits among its individuals.

Certain features define a population. Its members have the same body plan, as when jays have wings, feathers, feet with three toes forward and one toe back, and so on. These are *morphological* traits (*morpho-* means form). Cells and body parts of the individuals operate much the same way during growth, day-to-day tasks, and reproduction. These are *physiological* traits (related to body function). Also, individuals respond the same way to basic stimuli, as when babies instinctively imitate adult facial expressions. These are *behavioral* traits.

Most traits differ in their details from one individual to the next, especially in sexually reproducing species. Pigeon feathers or snail shells differ in patterning and coloration in a given population (Figures 1.7 and 14.8). Some individuals of a frog population may be more sensitive to winter cold or more successful at attracting a mate than others. Humans differ in the color, texture, amount, and distribution of hair. And these examples only hint at the staggering variation in populations.

Many traits, such as those Gregor Mendel studied, show *qualitatively different* variation. They come in two or more distinct forms (morphs) in the population, such as white or yellow butterfly wings. This feature is called **polymorphism**. Other traits show *continuous* variation, with small, incremental differences among individuals that can be quantified, as described in Section 9.7.

Figure 14.8 Variation in shell color and banding patterns in populations of a single species of snail that lives on islands of the Caribbean. For most traits, different individuals carry different alleles. Alleles are alternative molecular forms of the same gene. They code for alternative forms of the same trait. Members of a population vary in traits because they carry different combinations of alleles at many gene locations along their chromosomes.

The "Gene Pool"

The information about heritable traits resides in genes, which are specific regions of DNA molecules. Generally, individuals of a population have the same number and kinds of genes. (We say "generally" because males and females differ in some genes on the sex chromosomes.)

Think of all the genes in the entire population as a **gene pool**—a pool of genetic resources that, in theory at least, is available to all members of a population. Each kind of gene in the pool usually exists in two or more slightly different molecular forms, called **alleles**.

Individuals inherit different combinations of alleles. This leads to variations in phenotype—to differences in the details of traits. For example, whether your hair is black, brown, red, or blond depends on which alleles of certain genes you inherited from your two parents. When studying the basis of evolution, remember this: *offspring inherit genes, not phenotypes*. What may appear to be a heritable trait may actually be the outcome of environmental effects on gene expression (Section 9.8).

Which alleles end up in a given gamete and, later, a new individual? The outcome depends on five events, described in earlier chapters and summarized here:

1. Gene mutation (produces new alleles)

2. Crossing over at meiosis (leads to new combinations of alleles in chromosomes)

3. Independent assortment at meiosis (leads to mixes of maternal and paternal chromosomes in gametes)

4. Fertilization (combining of alleles from two parents)

5. A change in chromosome number or structure (leads to the loss, duplication, or repositioning of genes)

Of the five events just listed, mutation alone *creates* new alleles. The other events can only shuffle *existing* alleles into different combinations. But what a shuffle! Consider this: A human gamete ends up with one of 10^{600} possible combinations of alleles. Not even 10^{10} humans are alive today. So unless you have an identical twin, it is extremely unlikely that another person with your exact genetic makeup has ever lived—or ever will.

Stability and Change in Allele Frequencies

Imagine yourself in a large flower garden in summer. Gradually you notice the butterflies flitting about. They appear to be all the same, except in wing coloration. A few wings are white, but most are blue. Most likely, you muse, the "blue" allele must be more common than the "white" allele. With genetic analysis, you could identify the **allele frequencies**—the abundance of each kind of allele in the population as a whole. By doing so, you could track the rate of genetic change over time.

You can start with the "Hardy-Weinberg rule," as given in Section 14.5, to set up a theoretical reference point for measuring patterns of change. At this point, called **genetic equilibrium**, the frequencies of alleles at a given gene locus remain stable, one generation after the next. The population is *not* evolving with respect to that gene, for five conditions are being met. First, there is no mutation. Second, the population is very, very large. Third, it is isolated from other populations of the species. Fourth, the gene has no effect at all on survival or reproduction. Finally, all mating is random.

Rarely, if ever, do all five conditions prevail at the same time in a natural population. Mutation is rare but inevitable over long spans of time. Besides this, three processes—*natural selection, gene flow,* and *genetic drift*—can drive a population away from genetic equilibrium, even in the span of a few generations.

The term **microevolution** refers to the small-scale changes in allele frequencies brought about by mutation, natural selection, gene flow, and genetic drift.

Mutations Revisited

Mutations, recall, are heritable changes in DNA that typically give rise to altered gene products. They are the only source of *new* alleles. We can't predict exactly when or in which individual mutations will appear. Yet each gene has a characteristic **mutation rate**, which is the probability of its mutating between or during DNA replications. On average, the rate is between 10^{-5} and 10^{-6} per gene locus per gamete, each generation. Thus, in a single reproductive season, only 1 gamete in 100,000 to 1,000,000 will have a new mutation at any locus.

Most mutations lead to changes in structure, function, or behavior that decrease an individual's chances of surviving and reproducing. Even one biochemical change sometimes has devastating effects throughout the body. For example, the growth and development of vertebrate skin, bones, tendons, and many other organs depends on collagen, a protein. If the gene specifying the molecular form of this one protein is mutated, then pronounced changes in the skeleton, arteries, lungs, liver, and other body parts may follow. Remember, the genes of any complex organism are expressed in the context of a truly intricate developmental program. A mutation that has drastic effects on phenotype usually causes the individual's death; it is a **lethal mutation**.

By contrast, **neutral mutations** neither help nor harm the individual. Natural selection cannot increase or decrease their frequency in a population, because these mutations simply don't influence the chances of survival or reproduction. For example, if you have the mutated gene that is responsible for attached rather than detached earlobes (page 138), this alone shouldn't stop you from surviving and reproducing just as well as anybody else.

Finally, every so often, a new mutation bestows an advantage on the individual. For example, a mutated growth-regulating gene might make a corn plant grow larger or faster, and so give it the best access to sunlight and nutrients. Or maybe a previously neutral mutation proves advantageous when environmental conditions change. Even if the advantage is small, natural selection or chance events may preserve the mutated gene and assure its representation in the next generation.

Mutations are so rare, they usually have little or no immediate effect on allele frequencies of a population. But beneficial mutations (and neutral ones) have been accumulating in different lineages for billions of years. Through all that time, they have been the source of new genetic variation—the foundation for the staggering range of biological diversity, past and present.

In the evolutionary view, then, the reason you don't look like a bacterium or a grass plant or an earthworm or even your neighbors down the street began with different mutations that originated at different times in the past, in different lines of descent.

A population shows genetic variation—differences in the combinations of alleles that its individuals carry.

Genetic variation gives rise to variations in phenotypes—differences in the details of structural, functional, and behavioral traits among individuals.

For sexually reproducing species, individuals of a population represent a pool of genetic resources—a gene pool.

Mutation alone creates *new* alleles. Natural selection, gene flow, and genetic drift change the *frequencies* of alleles in the gene pool. The evolutionary story begins with these changes.

WHEN IS A POPULATION *NOT* EVOLVING?

Take a look at Figure 9.1, which shows the earlobes of Tom Cruise, Joan Chen, and other celebrated individuals. Is the number of individuals with *detached* earlobes staying the same, generation after generation? What about the number of individuals with *attached* earlobes? Stated more broadly, how do we know whether a population is evolving with respect to earlobes or any other trait?

Almost a hundred years ago, a mathematician and a doctor came up with the answer. Working independently, they thought about the alleles of an idealized population and came up with what is now called the Hardy-Weinberg rule, in their honor.

They started with this formula: In a population at genetic equilibrium, the proportions of genotypes at one gene locus, for which there are two kinds of alleles, are

$$p^2 \; AA + 2pq \; Aa + q^2 \; aa = 1$$

where p is the frequency of allele A, and q is the frequency of allele a. The rule is this: The allele frequencies will stay the same over the generations *if* there is no mutation, *if* the population is infinitely large and isolated from other populations of the same species, *if* mating is random, and *if* all individuals survive and reproduce successfully.

To test whether a population at Hardy-Weinberg equilibrium will remain there, you decide to track two alleles of a gene through a population of butterflies. You start with the assumption that the population of these sexually reproducing organisms is currently at genetic equilibrium. The two alleles you are interested in govern wing color. Allele A specifies dark-blue wings. Allele a is associated with white wings. Medium-blue wings result from the heterozygous (Aa) condition.

In the population as a whole, the frequencies of A and a must add up to 1. For example, if A occupies half of all the loci for this gene in the population, then $0.5 + 0.5 = 1$. If A occupies 90 percent of all the loci, then a must occupy the remaining 10 percent ($0.9 + 0.1 = 1$). No matter what the proportions of the two kinds of alleles,

$$p + q = 1$$

The body cells of butterflies are diploid, with pairs of homologous chromosomes. During meiosis, each allele segregates from its partner, and the two end up in separate gametes. Therefore, p is also the proportion of gametes with the A allele, and q is the proportion with the a allele.

To find the expected frequencies of the three possible genotypes (AA, Aa, and aa) in the next generation, you construct a Punnett square:

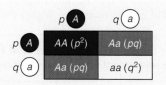

The frequencies of the genotypes add up to 1:

$$p^2 + 2pq + q^2 = 1$$

To see whether the allele frequencies and genotypic frequencies will remain the same through the generations, you work through an example. You decide the population has 1,000 butterflies, each of which produces two gametes:

490 AA individuals produce 980 A gametes
420 Aa individuals produce 420 A and 420 a gametes
 90 aa individuals produce 180 a gametes

You notice the frequency of A among the 2,000 gametes is

$$(980 + 420)/2{,}000 = 0.7$$

Also,

$$q = (420 + 180)/2{,}000 = 0.3$$

At fertilization, the gametes combine at random and produce the next generation, as given in the Punnett square. So now you have

$p^2 \; AA \; = 0.7 \times 0.7 = 0.49$ 490 AA individuals
$2pq \; Aa \; = 2 \times 0.7 \times 0.3 = 0.42$ *or*   420 Aa individuals
$q^2 \; aa \; = 0.3 \times 0.3 = 0.09$ 90 aa individuals

and

$$p^2 + 2pq + q^2 = 0.49 + 0.42 + 0.09 = 1$$

The allele frequencies have not changed:

$$A = \frac{2 \times 490 + 420}{2{,}000 \text{ alleles}} = \frac{1{,}400}{2{,}000} = 0.7 = p$$

$$a = \frac{2 \times 90 + 420}{2{,}000 \text{ alleles}} = \frac{600}{2{,}000} = 0.3 = q$$

You notice that genotype frequencies have not changed, either. And as long as the assumptions of the Hardy-Weinberg rule hold true, frequencies should stay the same through the generations. To test this, you calculate allele frequencies in the gametes of the *next* generation:

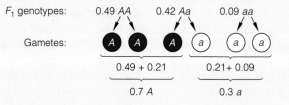

which is back where you started from. Because the allele frequencies for dark-blue, medium-blue, and white wings are the same as they were in the original gametes, they will yield the same phenotypic frequencies as you saw in the second generation.

You could do similar calculations for other wing colors. You could go on with your calculations until you ran out of paper (or patience). As long as the five assumptions hold true, allele frequencies and the range of values for the wing-color trait will not change, as you can see from Figure 14.9.

Thus, when genotypes and phenotypes do *not* show up in the proportions that you predicted on the basis of the Hardy-Weinberg rule, this tells you one or more conditions of the rule are being violated— and the hunt can begin for the specific evolutionary force, or forces, driving the change.

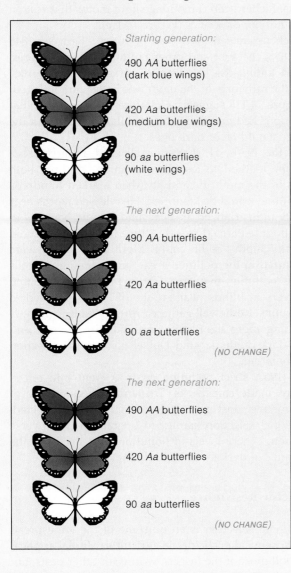

Starting generation:

490 *AA* butterflies
(dark blue wings)

420 *Aa* butterflies
(medium blue wings)

90 *aa* butterflies
(white wings)

The next generation:

490 *AA* butterflies

420 *Aa* butterflies

90 *aa* butterflies

(NO CHANGE)

The next generation:

490 *AA* butterflies

420 *Aa* butterflies

90 *aa* butterflies

(NO CHANGE)

Figure 14.9 A hypothetical population of butterflies at genetic equilibrium.

We now turn from our idealized population that never changes to real-world processes of change. Of these processes, natural selection probably accounts for most of the morphological and physiological changes that have occurred throughout the history of life.

Darwin, recall, was able to explain natural selection after correlating his understanding of inheritance with certain features of populations and the environment. Before we consider the modes of natural selection, let's restate his correlations in modern terms:

1. *Observation:* All populations in nature have the reproductive capacity to grow in numbers by ever larger amounts over the generations.

2. *Observation:* Population size cannot increase indefinitely, for its individuals will run out of food, living space, and other resources that sustain it.

3. *Inference:* Sooner or later, individuals of the population will end up competing for resources.

4. *Observation:* All of the individuals have the same genes, which specify the same assortment of traits. Collectively, their genes represent a pool of heritable information.

5. *Observation:* Most, if not all, kinds of genes occur in different molecular forms (alleles), which give rise to differences in phenotypic details.

6. *Inference:* Some phenotypes are better than others at helping the individual compete for resources, and therefore to survive and reproduce. Thus, the alleles for those phenotypes increase in the population, and other alleles don't. In time the genetic change leads to increased **fitness**—that is, an increase in adaptation to the environment.

7. *Conclusion:* **Natural selection** is the outcome of differences in survival and reproduction among individuals that vary in heritable traits. Adaptation is one outcome of this microevolutionary process.

Evolutionary biologists have been documenting the results of natural selection in thousands of field studies of populations from all five kingdoms. They find it has different outcomes. As you will see from the examples in the sections to follow, sometimes the outcome is a shift in the range of values for a trait in some direction. At other times, the outcome may be either stabilization or disruption of an existing range of values.

Over the generations, natural selection can lead to increased fitness—an increase in adaptation to the environment.

Directional Selection

With **directional selection**, allele frequencies underlying a range of phenotypic variation shift in a consistent direction. The shift is a response to directional change in the environment or to new environmental conditions. As Figure 14.10 shows, forms of traits at one end of the range become more common than midrange forms. Consider now a few documented cases of this outcome.

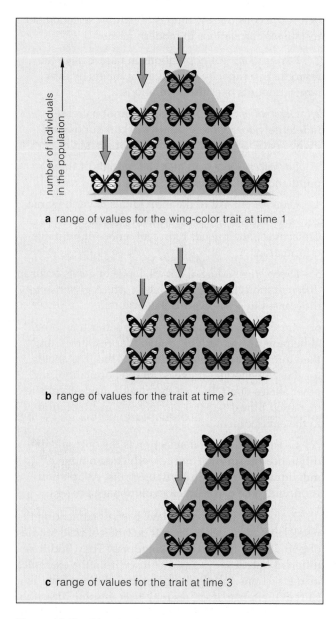

a range of values for the wing-color trait at time 1

b range of values for the trait at time 2

c range of values for the trait at time 3

Figure 14.10 Directional selection, using phenotypic variation within a population of butterflies as the example. A bell-shaped curve (*brown*) represents the range of continuous variation in wing color. The most common forms (powder blue) are between extreme forms of the trait (white at one end of the curve, deep purple at the other). The *orange* arrows signify which forms are being selected against over time.

The Case of the Peppered Moths

In England, biologists tracked directional selection in about a hundred moth species, including the peppered moth (*Biston betularia*). Peppered moths feed and mate at night. During daylight hours, they rest motionless on birches and other trees. Their wings and body have a mottled pattern, in shades ranging from light gray to nearly black. Apparently, their behavior, coloration, and wing patterns help camouflage them from moth-eating birds, which actively hunt for food during the day.

In the 1850s, an industrial revolution began. Outpourings of sooty smoke changed conditions in many parts of the surrounding countryside. Before then, the light moths were the most common form, and a dark form was rare. Also before living conditions changed, light-gray speckled lichens grew quite profusely on tree trunks. Lichens can camouflage light moths that rest on them, but not dark ones (Figure 14.11*a*).

Lichens are sensitive to air pollution. Between 1848 and 1898, the soot and other pollutants from factories started killing the lichens and darkening tree trunks. Now the rare form of the moth was better camouflaged, as Figure 14.11*b* suggests. Collectors hypothesized that conditions had previously favored light moths—but the change would favor dark ones.

In the 1950s, H. B. Kettlewell used a *mark-release-recapture* method to test the prediction. He bred both forms of the moth in captivity, then marked hundreds so that they could be identified. He released moths near heavily industrialized areas around Birmingham *and* in an unpolluted area (Dorset). After a time, he recaptured as many moths as he could. More dark moths were recaptured in the polluted area—and more light moths in the pollution-free area (Table 14.1). By stationing observers in blinds near groups of moths tethered to tree trunks, Kettlewell gathered direct evidence of birds capturing more light moths around Birmingham and more dark moths around Dorset. Directional selection *was* operating.

In 1952, strict pollution controls went into effect. Lichens made comebacks. Tree trunks became free of soot, for the most part. As you might have predicted, directional selection started to operate in the reverse direction. Where levels of pollution have declined, the frequency of dark moths is declining, also.

Pesticide Resistance

Directional selection is an outcome of the widespread use of chemical pesticides in agriculture. Initial applications kill most of the insects, worms, or other pests, but some usually manage to survive. Some aspect of their structure, physiology, or behavior allows them to resist

a

b

Figure 14.11 Individuals of a moth population that has undergone directional selection in response to changes in the environment. Light-winged and dark-winged peppered moths (*Biston betularia*) are resting on a lichen-covered tree trunk in (**a**) and on a soot-darkened tree trunk in (**b**).

Table 14.1	Marked Moths (*Biston betularia*) Recaptured in a Polluted Area and a Nonpolluted Area	
	Near Birmingham (pollution high)	Near Dorset (pollution low)
Light-gray moths:		
Released	64	393
Recaptured	16 (25%)	54 (13.7%)
Dark-gray moths:		
Released	154	406
Recaptured	82 (53%)	19 (4.7%)

Data after H. B. Kettlewell.

the chemical effects. If the resistance has a heritable basis, it becomes more common in the next generation, and the next—and the next. The chemicals are agents of selection; they favor the most resistant forms! Today, 450 different species are resistant to one or more pesticides. Worse yet, the pesticides also kill the natural predators of pests. Freed from natural constraints, the populations of resistant pests burgeon—and crop dam-

age is greater than ever. This outcome of directional selection is called **pest resurgence**.

Genetic engineering may reduce pesticide use. Even though plants engineered to resist pests won't escape the coevolutionary arms race, they may help keep our food supplies one step ahead of the pests. However, many consumers are leery of genetically engineered food. What about *biological control*? By this practice, the natural enemies of pests (such as parasitic wasps and predatory beetles) are raised in commercial insectaries in great numbers, then released at selected sites. The practice has one advantage, in that the control species can coevolve with the pests. But farmers must replace the ones that migrate from the fields or are destroyed at harvest time, and the cost is passed on to consumers.

Antibiotic Resistance

When your grandparents were children, the bacterial agents of tuberculosis, pneumonia, and scarlet fever may have caused up to one-fourth of all annual deaths in the United States. Bacteria that cause diphtheria, dysentery, and whooping cough were common killers. Bacterial infections at childbirth killed or maimed thousands of women. From the 1940s onward, we started treating these and other diseases with antibiotics.

An **antibiotic** is a normal metabolic by-product of certain microorganisms that kills or inhibits the growth of other microorganisms. For example, streptomycins block protein synthesis in target cells. The penicillins disrupt formation of covalent bonds that hold bacterial cell walls together. Penicillin derivatives cause the wall to weaken until it ruptures.

Antibiotics should be prescribed with restraint and care. Besides performing their intended function, some disrupt populations of bacteria that normally live in the intestines and of yeast cells in the vaginal canal. Such disruptions can lead to secondary infections.

Antibiotics have been overprescribed in the human population. Too frequently they are used for simple infections that the body often can fight successfully on its own. As a result, antibiotics have lost their punch. Over time, they did destroy the most susceptible cells of the target populations. They also favored their replacement by much more resistant ones. Today, antibiotic-resistant strains have made typhoid, tuberculosis, gonorrhea, "staph" infections, and other bacterial diseases much more difficult to treat. In a few cases, "superbugs" that cause tuberculosis cannot be treated successfully at all.

With directional selection, allele frequencies underlying a range of variation tend to shift in a consistent direction in response to directional change in the environment.

SELECTION AGAINST OR IN FAVOR OF EXTREME PHENOTYPES

As you have seen, natural selection can bring about a directional shift in a population's range of phenotypic variation. Depending on prevailing environmental conditions, natural selection also may favor either the most common or the most extreme phenotypes in that range.

Stabilizing Selection

In **stabilizing selection**, intermediate forms of a trait are favored, and alleles that specify extreme forms are eliminated from a population (Figure 14.12). This mode of selection tends to counter the effects of mutation, genetic drift, and gene flow and to preserve the most common phenotypes.

Consider the selection pressures on a gallmaking fly (*Eurosta solidaginus*). After a fly larva emerges from an egg, it bores into a stem of a tall goldenrod (*Solidago altissima*). In response, plant cells multiply rapidly and surround the invader with a tumorous mass (a gall). The larva feeds on juicy tissues and becomes an adult. As genetic analysis shows, flies of different phenotypes induce formation of galls of different sizes.

A wasp (*Eurytoma gigantea*) can only parasitize the larvae inside small galls (Figure 14.13), so it promotes selection in favor of individual flies that are associated with formation of *large* galls. But some insect-eating birds, including the downy woodpecker, preferentially assault large-diameter galls. As researchers discovered through a number of studies in Pennsylvania, flies that induced the formation of *intermediate-size* galls had the highest survival rate and fitness. Extreme phenotypes of the fly—which invited natural enemies with small galls and large galls of their own doing—were selected against. And so were the alleles for traits to which the goldenrod plants responded.

Thus, wasps work against one extreme in the range of phenotypes, birds work against the other—and the net result is an increase in intermediate phenotypes.

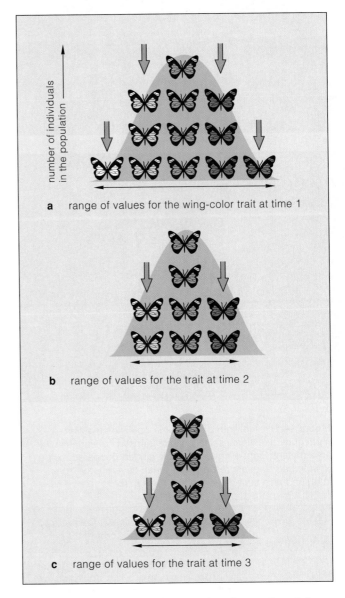

a   range of values for the wing-color trait at time 1

b   range of values for the trait at time 2

c   range of values for the trait at time 3

Figure 14.12 Stabilizing selection, using phenotypic variation within a population of butterflies as the example.

Figure 14.13 Example of stabilizing selection. (**a,b**) The larvae of a fly (*Eurosta solidaginus*) induce formation of tumors (galls) on goldenrod stems. (**c**) Downy woodpeckers (*Dendrocopus pubescens*) and other birds prefer to chisel into large-size galls and eat the larvae. (**d**) The egg-laying device of a wasp (*Eurytoma gigantea*) can only penetrate the thin wall of small galls. Its eggs develop into larvae, then *its* larvae eat fly larvae. Warren Abrahamson and his coworkers monitored twenty populations of *Eurosta* in Pennsylvania. They found that larvae in small and large galls have low relative fitnesses, and larvae in intermediate-size galls have relatively high fitnesses. Thus there is a stabilizing component to selection pressures created by the fly's natural enemies.

gall

a

Eurostra

b

c

d

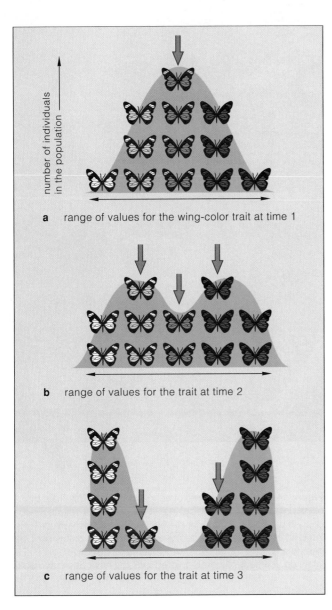

a range of values for the wing-color trait at time 1

b range of values for the trait at time 2

c range of values for the trait at time 3

Figure 14.14 Disruptive selection, using phenotypic variation within a population of butterflies as the example.

Disruptive Selection

In **disruptive selection**, forms at both ends of the range of variation are favored and intermediate forms are selected against (Figure 14.14). Thomas Smith discovered a clear example of this in a remote rain forest in Cameroon, West Africa. Smith had read about unusual variation in bill size in populations of the black-bellied seedcracker (*Pyrenestes ostrinus*). These African finches have large or small bills, but no sizes in between. The pattern holds for females and males, throughout their geographic range. (This is remarkable. Imagine every person in Texas being 4 feet or 6 feet tall.) If the pattern is unrelated to gender or geography, what causes it?

a

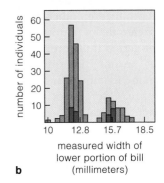

Figure 14.15 Disruptive selection among African finches. In feeding trials conducted by the biologist Thomas Smith, large-billed birds were more efficient at using hard seeds. Small-billed birds were most efficient at using soft seeds, not hard ones. (**a**) Two specimens displaying small and large bill sizes.

measured width of lower portion of bill (millimeters)

b

(**b**) Survival of juvenile birds during the dry season, when competition for resources is most intense. Individuals with *very* small, *very* large, or intermediate-size bills cannot feed efficiently on either type of seed; they survive poorly. For this graph, *tan* bars show the number of nestlings; *brown* bars show the only survivors among them. The findings are based on measurements of 2,700 netted individuals.

Smith hypothesized that if the persistence of only two bill sizes in the populations relates to seed-cracking ability (which directly affects survival), then disruptive selection may be eliminating birds with intermediate-size bills. What factors could affect seed hardness and so act as a selection pressure on feeding performance? Cameroon's swamp forests flood in the wet season. Lightning-sparked fires burn in the dry season. Two sedge species (grasslike, fire-resistant plants) dominate the region. One has hard seeds; the other has soft seeds. When finches reproduce, hard and soft seeds are abundant. All birds prefer soft seeds as long as they can get them. But food dwindles at the peak of the dry season. Then, the young birds especially are at a competitive disadvantage; many do not survive (Figure 14.15).

Smith also performed experimental crosses between large- and small-billed birds. All offspring had large *or* small bills. Together with other data, this suggests that bill size (and feeding performance) may be largely controlled by a single autosomal gene with two alleles.

With stabilizing selection, intermediate phenotypes are favored, and extremes at both ends of the range of variation are eliminated.

With disruptive selection, allele frequencies shift through selection against intermediate forms in the range of variation.

14.9 SPECIAL OUTCOMES OF SELECTION

Sexual Selection

As you may have noticed, individuals of most sexually reproducing species have a distinctively male or female phenotype. We call this **sexual dimorphism** (*dimorphos* means "having two forms"). How does this condition come about, and what maintains it? Here again, natural selection is at work. In this case, it is **sexual selection**, for it favors traits with no advantage for survival and reproduction, other than the fact that males or females prefer them. Through nonrandom mating, alleles for preferred traits prevail over the generations.

Sexual dimorphism is especially striking among many mammals and birds, including the pair in Figure 14.16. Often males are larger, have more vivid coloration and patterning, and are far more aggressive than the females. Remember the male bighorn sheep butting heads in Figure 1.6*g*? Fighting wastes time, wastes energy, and may cause injuries. Why, then, do alleles that contribute to aggressive behavior persist in the population? The increased chance of mating offsets the costs. Male bighorn sheep fight only to control areas where receptive females gather during a winter rutting season. Winners mate often, with many females; and losers will not challenge a stronger, larger male.

Most often, females are the agents of selection. By choosing mates, they exert direct control over reproductive success. We return to this topic in Chapter 40.

Balanced Polymorphism

Smith's work with African finches is a fine example of how selection can maintain two or more alleles for the same trait in fairly steady proportions, generation after generation. When selection has this outcome, a population is said to display **balanced polymorphism** for the trait (after the Greek *polymorphos*, which means "having many forms"). A population is in a state of balanced polymorphism when nonidentical alleles for a trait are maintained at frequencies greater than 1 percent. Even if the frequencies shift slightly, over time they often bounce back to the same values.

Sickle-Cell Anemia—Lesser of Two Evils?

We sometimes find cases of balanced polymorphism where environmental conditions favor heterozygotes (which carry nonidentical alleles for a specified trait), not the homozygotes (which carry identical alleles for the trait). Said another way, *one combination of alleles at a given locus may increase an individual's chance of surviving and reproducing—yet a different combination may lower it*.

Let's look at the environmental pressures that favor the pairing of an *HbA* and an *HbS* allele in humans. As

Figure 14.16 One outcome of sexual selection. This male bird of paradise (*Paradisaea raggiana*) is engaged in a spectacular courtship display. He caught the eye (and, perhaps, sexual interest) of the smaller, less colorful female. Males of this species compete fiercely for females, which serve as selective agents. (Why do you suppose drab-colored females have been favored?)

described in Section 9.5, *HbS* specifies a mutated form of hemoglobin, the oxygen-transporting molecule in blood. *HbS/HbS* homozygotes develop *sickle-cell anemia*, a genetic disorder with potentially severe phenotypic outcomes. The *HbS* frequency is high in tropical and subtropical regions of Africa and Asia. Often, *HbS/HbS* homozygotes die in their early teens or early twenties. But heterozygotes (*HbS/HbA*) make up nearly a third of the populations in these regions! Why is this allelic combination maintained at such a high frequency?

The balancing act, an outcome of natural selection, is most pronounced in areas with the highest incidence of *malaria* (Figure 14.17). There, a mosquito transmits *Plasmodium*, the parasite responsible for the disease, to humans. The parasite multiplies in the liver, then in red blood cells. The cells rupture and release new parasites during severe, recurring bouts of infection. People are far more likely to survive these recurrences *if* they are

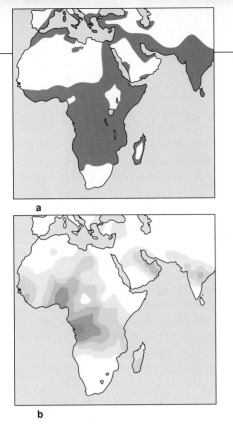

Figure 14.17
(**a**) Distribution of cases of malaria in parts of Africa, the Middle East, and Asia in the 1920s, before extensive mosquito control programs had been instituted.
(**b**) From the same areas, the distribution and frequency of individuals with the sickle-cell trait. Notice the close correlation between the color-tinted areas of the two maps.

☐ less than 1 in 1,600
☐ 1 in 400–1,600
☐ 1 in 180–400
☐ 1 in 100–180
☐ 1 in 64–100
☐ more than 1 in 64

HbS/HbA heterozygotes. The allelic combination gives them two forms of hemoglobin, with interesting results. Each has enough nonmutated molecules to maintain tissue functions, even if at reduced levels. But their *altered* molecules distort the shape of red blood cells, and this interferes with normal blood circulation. The slowdown in blood flow is the key. It hampers the parasite's ability to rapidly infect new cells during an infective cycle.

So the persistence of the "harmful" *HbS* allele is a matter of relative evils. Natural selection has favored the allelic combination *HbS/HbA* because its bearers show greatest fitness in environments where malaria is most prevalent. *HbA/HbA* is a workable combination also—but not as much as *HbS/HbA* in these regions.

In tropical and subtropical parts of Asia and the Middle East, malaria has been a selective force for more than 2,000 years. Although sickle-cell anemia occurs at high frequencies in such regions, its symptoms are much less severe than they are in Central Africa, where the *HbS* allele became established much later. Most likely, the products of other genes may be modifying some of the widespread effects of the *HbS* allele in ways that can minimize the symptoms of the disorder.

With sexual selection, a trait gives an individual an advantage in reproductive success. Sexual dimorphism is one outcome of sexual selection.

Balanced polymorphism is a state in which natural selection is maintaining two or more alleles over the generations at frequencies greater than 1 percent.

14.10 GENE FLOW

With emigration, individuals leave a population. With immigration, new individuals enter it. Either way, the frequencies of alleles can change. This physical flow of alleles, called **gene flow**, helps keep neighboring populations genetically similar. Over time, it tends to counter differences between populations that are brought about through mutation, genetic drift, and natural selection.

Think of the acorns that blue jays disperse when they store nuts for the winter. Each fall the jays may make hundreds of round trips from acorn-bearing oak trees to bury acorns in the soil of their home territories, which may be up to a mile away (Figure 14.18). The alleles flowing with "immigrant acorns" help reduce genetic differences that arise among neighboring stands of oaks.

Figure 14.18 Gene flow among oak populations, courtesy of feathered travel agents. Blue jays hoard acorns in their home territory, but they might shop at nut-bearing trees up to a mile away. Some acorns contribute to the allele pool of an oak population some distance away from the parent tree.

Or think of the millions of people from economically bankrupt, politically explosive countries who seek more stable homes. The scale is unprecedented, but hardly unique. Throughout human history, immigrations may have minimized many genetic differences that otherwise would have accumulated among geographically separated groups of people.

Gene flow is the physical movement of alleles into and out of a population, through immigration and emigration.

14.11 GENETIC DRIFT

Chance Events and Population Size

Genetic drift is a random change in allele frequencies over the generations, brought about by chance alone. By comparing the two charts in Figure 14.19, you can see that the magnitude of its effect on genetic diversity and the range of phenotypes relates to population size. The impact of genetic drift is minor or insignificant in very large populations but significant in small ones.

Genetic drift has nothing to do with *how* a small population got that way. The chance of any given allele becoming more or less prevalent in a population simply is more pronounced because of it. **Sampling error**, a rule of probability, helps explain why this is so. By this rule, the fewer times a chance event occurs, the greater will be the variance from the expected outcome of that occurrence. For example, flip a coin. You know there is a 50 percent chance it will turn up heads or tails. Flip it a thousand times and the odds are great that you will observe something close to this expected outcome. Flip it only ten times, and the odds diminish greatly.

Sampling error applies every time random mating and fertilization occur in a population. Consider an experiment on genetic drift. Researchers started with 1,320 beetles (*Tribolium*) that mate with one another at random. They grouped these into twelve populations of 10 beetles and twelve of 100. *Which* beetles ended up in which group was a matter of chance. At the start, the frequency of a wild-type allele (call it *A*) was 0.5. (Wild-type refers to the most common allele at a given locus

in a population.) The researchers tracked allele *A* for twenty generations. Each time, they randomly removed some offspring to maintain each population's original size. At the end of the experiment, *A* wasn't the only allele remaining in the *large* groups, but it had become fixed in seven of the *small* groups. **Fixation** means that only one kind of allele remains at a specified locus in a population; all individuals are homozygous for it.

This controlled experiment, along with Figure 14.19, reinforces a key point: *Over time, and in the absence of other forces, random change in allele frequencies will lead to the homozygous condition and a loss of genetic diversity.* This happens in all populations; it just happens more rapidly in small ones. Once the alleles inherited from the original population are fixed, their frequencies will not change unless mutation or gene flow introduces other kinds of alleles into the population.

Drift also operates in nature. For example, daisies and other perennials tend to produce many seeds, but only a few survive. Fixation often results when members of small populations reproduce among themselves.

Bottlenecks and the Founder Effect

Genetic drift is pronounced when very few individuals rebuild a population or found a new one. A **bottleneck** is a severe reduction in population size, as brought about by intense selection pressure or natural calamity. Suppose contagious disease, loss of habitat, hunting, or a massive volcanic blast destroys much of a population.

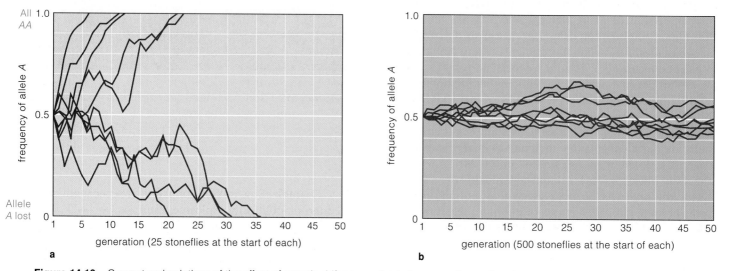

Figure 14.19 Computer simulations of the effect of genetic drift on an allele's frequency in small and large populations. In nine groups of stoneflies, population size was maintained at 25 breeding individuals every generation, for fifty generations. Population size of nine more groups was maintained at 500 for fifty generations. (**a**) Allele *A* became fixed in five of the small populations (the lines reaching the top of the graph) and was lost from four of them (the lines plummeting off the bottom of the graph). As this tells you, even the "best" allele doesn't always win. (**b**) The allele did not become fixed in any large population. Thus the magnitude of drift was much less in every generation than in the small populations.

Equal fitnesses assumed for these simulations:

AA = 1
Aa = 1
aa = 1

original population

island
subpopulation

Even if a moderate number do survive the bottleneck, allele frequencies will have been altered at random.

In the 1890s, hunters killed all but twenty of a large population of northern elephant seals (*Mirounga angustirostris*). The population recovered. When its numbers passed 30,000, electrophoretic analysis of a sample of twenty-four *M. angustirostris* genes revealed no variation. A number of alleles were lost in the bottleneck.

The genetic outcome can be similarly dicey after a few individuals leave a population and establish a new one elsewhere. This form of bottlenecking is called the **founder effect**. Allele frequencies of the founders may not be the same as those of the original population, and selection pressures in a new habitat may favor different phenotypes. The outcome can be stunning on isolated islands (Figure 14.20). Thus long ago, seabirds, winds, or ocean currents carried a few seeds from the Pacific Northwest to the Hawaiian Islands. New plant populations evolved rapidly, in spectacular ways.

Genetic Drift and Inbred Populations

Inbreeding refers to nonrandom mating among closely related individuals, which have many identical alleles in common. Inbreeding can be viewed as a form of drift in a small population (the group of relatives that are preferentially interbreeding). Like genetic drift, it leads to the homozygous condition. It also can lower fitness when the alleles that are increasing in frequency are recessive and have harmful effects.

Most human societies forbid or discourage incest (inbreeding between parents and children or between siblings). But inbreeding among other close relatives is common in small communities that are geographically or culturally isolated from a larger population. Consider Pennsylvania's Old Order Amish, a community with distinct genotypes. One outcome of inbreeding is the high frequency of a recessive allele that causes *Ellis-van Creveld syndrome*. Affected individuals have extra fingers or toes and short limbs (compare Figure 10.8). The allele probably was rare when the small number of founders immigrated to Pennsylvania. Now 1 in 8 are heterozygous and 1 in 200 are homozygous for it.

Bottlenecks and inbreeding are an especially awful combination for **endangered species**, the populations of which are very small and vulnerable to extinction. Consider the cheetah (Figure 14.21), which apparently went through a drastic bottleneck in the nineteenth century. The surviving parents mated with their own offspring when no other options were available.

Inbreeding among the survivors and their descendants left the 20,000 existing cats with strikingly similar alleles. One of these is a mutated allele with bad effects on fertility. Typically, a male cheetah's sperm count is

Figure 14.20 Example of the founder effect. A seabird carries a few seeds, stuck to its feathers, to an island. By chance, most of the seeds carry an allele for orange flowers that was uncommon in the original population. In the absence of further gene flow or selection for flower color, genetic drift will fix the allele in the island population.

Figure 14.21 A few of the remaining cheetahs, all of which have some very bad alleles that made it through a severe bottleneck.

low and 70 percent of the sperm are abnormal. Other shared alleles result in lower resistance to disease. Infections that are seldom life-threatening to other cat species can be devastating to cheetahs. In one outbreak of *feline infectious peritonitis* in a wild animal park, the pathogen (a type of coronavirus) had little effect on captive lions but killed a majority of cheetahs. This infection triggers an uncontrollable inflammatory response. Fluid fills the body cavity housing the heart and other internal organs, and the cat dies in agony. There is no vaccine.

Pressure from hunting and urban sprawl has put the Florida panther on the endangered species list, also. Only fifty highly inbred cats remain.

Genetic drift is the random change in allele frequencies over the generations, brought about by chance alone. The magnitude of its effect is greatest in small populations, such as the ones that make it through a bottleneck.

Barring mutation, selection, and gene flow, the chance losses and increases of the various alleles at a given locus lead to the homozygous condition and a loss of genetic diversity.

SUMMARY

1. Awareness of evolution—changes in lines of descent over time—emerged through comparisons of the body structure and patterning for major groups of animals (comparative anatomy), questions about the world distribution of plants and animals (biogeography), and observations of fossils of different types present in different layers of sedimentary rocks.

2. Generally, the individuals of a population have the same number and kind of genes. But genes come in different allelic forms, and this leads to variations in traits.

3. A population is evolving when some forms of a trait (and the alleles that specify them) are becoming more or less common, relative to the other kinds, over the generations. This happens as a result of mutation, gene flow, genetic drift, and natural selection (Table 14.2).

4. Genetic equilibrium, a state in which a population is not evolving, is used as a baseline to measure change. It occurs only if there is no mutation, if the population is very large and isolated from other populations of the same species, and if there is no selection (all members survive and reproduce by random mating).

5. Mutations, heritable changes in DNA molecules, are the only source of *new* alleles. Crossing over, independent assortment, and fertilization can only produce new combinations of *existing* alleles.

6. Natural selection is a difference in survival and reproduction among members of a population that differ in one or more heritable traits. It affects the relative abundances of alleles that are responsible for adaptive and maladaptive versions of traits over the generations.

   a. Selection pressure may shift the range of variation for a trait in one direction (directional selection), disrupt the range at or near the midpoint (disruptive selection), or favor intermediate forms and eliminate extreme forms from a population (stabilizing selection).

   b. Natural selection can result in balanced polymorphism. A population is in this state when nonidentical alleles for a given trait are being maintained over the generations at frequencies greater than 1 percent, even after slight perturbations in those frequencies.

   c. Sexual selection (by one sex or the other) leads to forms of traits that give the individual an advantage in reproductive success. Sexual dimorphism, the persistence of phenotypic differences between males and females of a species, is one outcome of sexual selection.

7. Gene flow is a change in allele frequencies brought about by the physical movement of alleles into and out of a population (by immigration and emigration).

8. Genetic drift is a change in allele frequencies over the generations due to chance events alone.

   a. Because of sampling error, the magnitude of its effect is greater in small populations than in large ones.

Table 14.2	Summary of Major Microevolutionary Processes
Mutation	A heritable change in DNA
Gene flow	Change in allele frequencies as individuals leave or enter a population
Genetic drift	Random fluctuation in allele frequencies over time, due to chance occurrences alone
Natural selection	Change or stabilization of allele frequencies as a result of differences in survival and reproduction among variant individuals of a population

In the absence of mutation, selection, and gene flow, sooner or later genetic drift can result in a homozygous condition and loss of genetic diversity.

   b. Genetic drift has great impact after a bottleneck, (a severe reduction in population size from which the population recovers). Alleles that make it through a bottleneck may give rise to a different range of phenotypes, compared to the original population. In one type of bottleneck, the founder effect, a few emigrants from one population successfully establish a small subpopulation in a new environment.

Review Questions

1. Define genetic equilibrium and explain how these processes can send allele frequencies out of equilibrium: *225*
   a. mutation   c. gene flow
   b. natural selection   d. genetic drift

2. Define fitness, as evolutionary biologists use the term. *227*

3. Define bottleneck and the founder effect. Are these cases of genetic drift, or do they set the stage for it? *234–235*

Figure 14.22 A male and a female sugarbird.

4. Figure 14.22 is a photograph of a brilliantly hued male and a subdued-hued female sugarbird. In the evolutionary view, why is the morphological difference between them being maintained over time? *232*

5. Long-term studies indicate that prospects for human newborns of very high or very low birthweight are not good. Being born too small increases the risk of stillbirth and early infant death (Figure 14.23). Similarly, pre-term rather than full-term pregnancies also increase the risks. What microevolutionary process is at work here? *230*

6. In Figure 14.24, identify the modes of selection (stabilizing, directional, and disruptive) in each diagram. *228, 230–231*

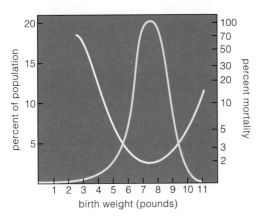

Figure 14.23 Weight distribution for 13,730 human newborns (*yellow* curve) correlated with mortality rate (*white* curve).

Figure 14.24 Bell-shaped curves characteristic of three modes of natural selection.

Self-Quiz *(Answers in Appendix IV)*

1. Individuals don't evolve; _____ do.

2. Sickle-cell anemia first appeared in Asia, the Middle East, and Africa. The causative allele entered the United States population when people were forcibly brought over from Africa prior to the Civil War. In microevolutionary terms, this is a case of _____ .
   a. mutation   c. gene flow
   b. genetic drift   d. natural selection

3. Directional selection _____ .
   a. eliminates uncommon forms of alleles
   b. shifts allele frequencies in a steady, consistent direction
   c. doesn't favor intermediate forms of a trait
   d. works against adaptive traits

4. Disruptive selection _____ .
   a. eliminates uncommon forms of alleles
   b. shifts allele frequencies in a steady, consistent direction
   c. doesn't favor intermediate forms of a trait
   d. works against adaptive traits

5. Match the following individuals and ideas or perceptions.
   ____ Cuvier   a. theory of natural selection
   ____ Darwin   b. populations outgrow resources
   ____ Lyell   c. catastrophism
   ____ Malthus   d. theory the earth is extremely ancient

6. Match the evolution concepts.
   ____ gene flow   a. source of new alleles
   ____ natural selection b. changes in a population's allele frequencies due to chance
   ____ mutation   c. immigration, emigration change allele frequencies
   ____ genetic drift   d. differences in survival and reproduction among variant members of a population

Critical Thinking

1. For centuries, *tuberculosis* (TB) has caused many deaths around the world. A bacterium (*Mycobacterium tuberculosis*) causes this contagious lung disease. TB declined steadily in the United States, where public health officials even predicted there would be no more cases by the year 2000. But caseloads are increasing. An AIDS epidemic, the influx of immigrants from regions where TB is still common, and overcrowding in tenements and homeless shelters contribute to the increase. At one time, antibiotics could cure TB within six to nine months. But now an antibiotic-resistant strain of *M. tuberculosis* has evolved. Currently, two or even three different antibiotics must be used to block different metabolic pathways of this bacterium. Why would this approach be more effective?

2. A few families that live in a remote region of Kentucky show a high frequency of *blue offspring*, an autosomal recessive disorder. The skin of affected individuals appears bright blue. Homozygous recessives lack the enzyme diaphorase. The enzyme catalyzes reactions that maintain hemoglobin (the oxygen-carrying red pigment in blood) in its normal molecular form. Without the enzyme, a blue form of hemoglobin accumulates in blood. Skin and the blood capillaries associated with it are transparent, so the color of pigments in blood shows through and contributes to skin color. This gives light skin a pinkish cast—and blue skin its color. Formulate a hypothesis to explain why this trait is rather common among a cluster of families but rare in the human population at large.

Selected Key Terms

allele 224	genetic equilibrium 225
allele frequency 225	inbreeding 235
antibiotic 229	lethal mutation 225
balanced polymorphism 232	microevolution 225
biogeography 218	mutation rate 225
bottleneck 234	natural selection 227
catastrophism 220	neutral mutation 225
comparative anatomy 218	pest resurgence 229
directional selection 228	polymorphism 224
disruptive selection 231	population 224
endangered species 235	sampling error 234
evolution 219	sedimentary bed 219
fitness 227	sexual dimorphism 232
fixation 234	sexual selection 232
fossil 219	stabilizing selection 230
founder effect 235	theory of inheritance of acquired characteristics 220
gene flow 233	theory of uniformity 221
gene pool 224	
genetic drift 234	

Readings

Darwin, C. 1957. *Voyage of the Beagle*. New York: Dutton. In his own words, what Darwin saw and thought about during his voyage.

Smith, T. B. January 1991. "A Double-Billed Dilemma." *Natural History*: 14–21.

15 SPECIATION

The Case of the Road-Killed Snails

If you happen to be a snail living in a garden in Bryan, Texas, it doesn't take much to keep your genes away from snails in a backyard across the street. By day, the sunbaked asphalt would be about as inviting to a snail as a desert would be to a catfish. Besides, day or night, a street-traversing snail is vulnerable to cars, trucks, skateboards, and bicycles (Figure 15.1a). Whatever else it might be, that strip of asphalt is a formidable barrier to gene flow between populations. For snails, that is.

Whether or not a physical barrier deters gene flow depends partly on an organism's mode of locomotion or dispersal. It also depends on how fast and how long an organism *can* move in response to environmental factors or its own hormones.

A snail is not swift, and it does not roam far from its home population. Compare it to the wild black duck, banded in Virginia in 1969, that turned up eight years later in Korea. Compare it to the wandering albatross, one of the supreme barrier busters. After lifting off from Kerguelen Island in the Indian Ocean, one of these birds soared westward past Africa, across the Atlantic, and around Cape Horn. Thirteen thousand kilometers later, it landed in Chile. With a lightweight body and a wingspan of 3.65 meters (12 feet across), the albatross was able to exploit the great prevailing winds of the Southern Hemisphere.

Figure 15.1 (**a**) A snail (*Helix aspersa*) about to encounter a barrier to gene flow.

(**b**) Results from a study of neighboring snail populations, all descended from founders that arrived in a small town in Texas in the 1930s. For each population, a circle represents the relative abundances of three alleles (coded *yellow*, *gold*, or *brown*) for an enzyme, leucine aminopeptidase. Greater genetic variation exists between populations living on opposite sides of 22nd Street. Notice, for example, the higher frequencies of the allele coded *gold* in the block to the west of the street.

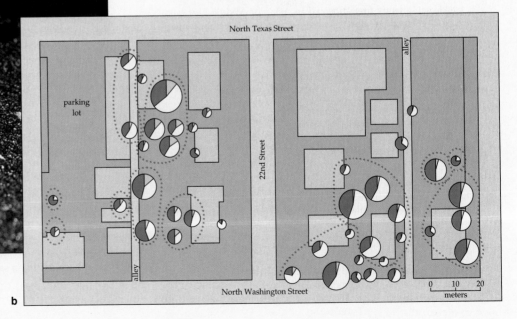

a

b

And yet, in 1859, snails did cross an ocean. Humans transported garden-variety snails (*Helix aspersa*) from France to California, then turned them loose. The idea was that the snails would multiply and so meet the demand for that French delicacy, *escargots aux fines herbes*. It was bad enough that the importers mistakenly brought over a less tasty species. Worse than that, the snails exceeded expectations and became an absolute nuisance in gardens and nurseries throughout the southwestern United States.

By the 1930s, *H. aspersa* had hitched rides, possibly as eggs in the soil of plant containers, to Bryan, Texas. They founded small colonies in the local vegetation. Forty years later, Robert Selander, now of Pennsylvania State University, was down on his hands and knees with a few graduate students, scouring patches of vegetation on two adjacent city blocks. Why? Selander wanted to determine the effect of genetic drift on the introduced populations. He and his students collected every single snail—2,218 of them—from fourteen local populations. For each population, they determined the allele frequencies at five different gene locations.

In all five cases, the results from their analysis pointed to some genetic variation among the colonies on the same block—and to major differences in allele frequencies *between* blocks. Figure 15.1*b* shows the results for one of the genes studied.

Assume the genetic differences between populations continue to increase, as through natural selection and genetic drift. Will the time come when snails from opposite sides of the street can no longer interbreed successfully, even if they do manage to get together? In other words, *will they become members of separate species?* Or will selection work against increases in genetic differences by eliminating extreme phenotypes from the populations? After all, how much can a workable package of *H. aspersa* genes evolve in adjacent patches of vegetation—and under very similar environmental pressures—in a small town in Texas?

With this chapter, we turn to **speciation**—to changes in allele frequencies that are significant enough to mark the formation of daughter species from a parent species. Obviously, no one was around to watch the formation of species in the past. No one lives long enough to know whether many existing populations are at some intermediate stage leading to speciation. Evolutionary biologists are still working out theories about speciation processes, and what you are about to read may change in the near or distant future.

KEY CONCEPTS

1. A species consists of one or more populations of individuals that can interbreed under natural conditions and produce fertile offspring, and that are reproductively isolated from other such populations. This definition is restricted to sexually reproducing species.

2. The populations of a species have a shared genetic history, they are maintaining genetic contact over time, and they are evolving independently of other species.

3. Speciation is the process by which daughter species evolve from a parent species.

4. Speciation begins when gene flow is prevented between populations (or subpopulations) of a species. Thereafter, mutation, natural selection, and genetic drift operate independently in each population and lead to irreversible genetic divergence of one from the other.

5. By chance, two populations may come to differ in certain alleles that govern morphological, physiological, and behavioral traits associated with reproduction. When the differences become great enough, they are reproductively isolated from each other—and with reproductive isolation, speciation is completed.

6. Three models of speciation guide current research. With allopatric speciation, species form in separated populations or subpopulations. With parapatric speciation, they form in a region where populations share a border. With sympatric speciation, they form within the home range of the parent species.

7. The timing, rate, and direction of speciation vary within and between lineages. The extinction of some number of species is inevitable for all lineages.

15.1 ON THE ROAD TO SPECIATION

What Is a Species?

"If it looks like a duck, walks like a duck, and quacks like a duck, probably it's a duck." Let's use this familiar saying as a starting point for defining what is and what is not a species. **Species** is a Latin word. Generally, it simply means "kind," as in "a particular kind of duck."

Few of us have trouble identifying a duck. But how do we go about distinguishing one kind from another? By outward appearance alone? Phenotype, recall, refers to observable aspects of an individual's morphology, physiology, and behavior. But phenotypic details can vary enormously, so perhaps we should move past the details, down to a basic function that unites members of a species and isolates them from all other species.

For example, using reproduction as a basic, defining function is central to the **biological species concept**. The evolutionary biologist Ernst Mayr phrases the concept in this way: "Species are groups of interbreeding natural populations that are reproductively isolated from other such groups." No matter how extensive the phenotypic variation, individuals will remain members of the same species as long as their form, physiology, and behavior permit them to interbreed and produce fertile offspring. Their capacity to contribute to a shared pool is qualification for membership. Mayr's concept doesn't work for bacteria and other kinds of organisms that reproduce asexually. But it is useful for research into factors that define sexually reproducing species.

By Mayr's concept, *we can identify a species mainly in terms of the portion of alleles that promote or maintain its reproductive isolation.* And it may not take many unique genes to maintain this isolation in nature. Consider the hybrid mammals called zebroids. Hybrid individuals, recall, are the offspring of parents of different genotypes—in this example, wild zebras and domesticated horses confined to the same pasture (Figure 15.2). The lineages that gave rise to horses and zebras diverged more than 3 million years ago. Existing members of this family of mammals range from true horses (*Equus*), to zebras (*Hippotigris* and *Dolichohippus*), to donkeys and asses (*Asinus*). The unnatural confinement of zebras with horses promoted a breach of reproductive barriers between them. The production of zebroids in captivity is evidence that considerable genetic compatibility still exists between individuals of the divergent lineages.

One final point: Reproductive isolation does not evolve purposefully to promote formation of a species or to maintain species identity. Rather, *any structural, functional, and behavioral differences that arise in modes of reproduction between two diverging populations are chance outcomes of genetic changes, as brought about by selection pressures of their environments, sexual selection, and genetic drift through the generations.*

Figure 15.2 A mixed herd of zebroids and horses. Zebroids are interspecies hybrids, from crosses between horses and zebras.

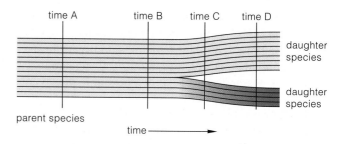
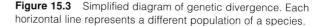

Figure 15.3 Simplified diagram of genetic divergence. Each horizontal line represents a different population of a species.

Genetic Divergence and Speciation

Individuals remain members of a species even if they belong to geographically distant populations, provided that gene flow persists among them. As defined in the preceding chapter, **gene flow** is the physical movement of alleles into and out of a population by immigration and emigration. Over time, it helps maintain a common reservoir of alleles and counters differences between populations that may arise through mutation, natural selection, and genetic drift.

If something *prevents* gene flow, the stage is set for **genetic divergence**—a buildup of differences between the allele pools of two or more populations or subpopulations. Now mutation, natural selection, and genetic drift are free to operate independently in each of the genetically isolated populations. Reproductive isolation may be an accidental outcome of their divergence. When reproductive isolation is complete, **speciation**—the process by which species form—is the outcome.

In most cases, we can't say precisely when daughter species will form. Consider Figure 15.3. Each horizontal line of this diagram represents a different population of a species. At time A, there is one species. At time D, there are two. At time B, divergence is under way but far from complete. We expect it to be difficult to tell species apart even by time C.

Figure 15.4 A sampling of courtship displays that precede copulation (sexual union) between a male and female albatross. Members of the same species recognize the visual, acoustical, and tactile components of such displays.

Reproductive Isolation

One way or another in nature, the mechanisms of reproductive isolation that evolve between genetically divergent populations take effect before, during, or after the time of fertilization. Some of the mechanisms prevent mating or pollination between individuals of the divergent populations, so hybrid zygotes cannot form. Other mechanisms take effect *after* zygotes have formed.

Prezygotic Isolation   The mechanisms of *prezygotic* isolation come into play before or during fertilization.

For example, *temporal* isolation results from inter-populational differences in the timing of reproduction. For most animals or plants, mating or pollination is over rather quickly, sometimes in less than a day. Even closely related species may be isolated when their reproductive cycles don't coincide. Consider cicadas, a type of insect that spends nearly all of its life underground in immature form, feeding on plant juices. One species matures, emerges, and reproduces every 13 years. The other does so every 17 years. Once every 221 years, they release gametes at the same time!

With *ecological* isolation, populations have become adapted to different microenvironments in the same habitat. For example, some copper-polluted slag heaps (residues of mining activities) have accumulated next to pastures. Within twenty meters of this environmental discontinuity, different populations of the same grass (*Agrostis tenuis*) differ in copper tolerance. Transplant a pasture plant to polluted soil a few meters away from its home range, and it dies. Transplant copper-tolerant plants to a pasture, and they show reduced fitness.

Behavioral isolation is a major barrier to gene flow among related species in the same territory. For example, before male and female birds copulate, they engage in courtship rituals (Figure 15.4). A female of the same species is equipped to recognize the male's singing, head bobbing, wing spreading, or prancing as an overture to sex. Females of another species usually do not.

Incompatibility between body parts of individuals is a form of *mechanical* isolation. Consider two species of sage (*Salvia melliflera* and *S. apiana*). Some petals of their flowers are arranged as a small landing platform

for visitors—small or medium-size pollinators. Other petals are tightly clustered, as a pollinator-attracting nectar cup. *S. apiana* has a larger landing platform and long, pollen-bearing stamens that extend some distance away from the nectar cup. Small bees that land on the large platform usually don't brush against the stamens. Larger pollinators do. Hence most small pollinators of *S. melliflera* are incapable of spreading pollen to flowers of *S. apiana*. Large pollinators of *S. apiana* can't land on and cross-pollinate *S. melliflera*.

With *gametic* isolation, gametes of different species have evolved in ways that make them incompatible at the molecular level. Thus, when pollen of one species lands on a plant of a different species, it usually does not recognize the molecular signals that can trigger its growth down through the plant's tissues, to the egg.

Postzygotic Isolation   The mechanisms of *postzygotic* isolation take effect after fertilization, while an embryo is developing. Unsuitable interactions among genes, gene controls, or gene products lead to early death, sterility, or hybrids of low fitness. Hybrid offspring commonly are weak, and their survival rates are not good. A few types are sturdy but sterile. Mules, resulting from a cross between a female horse and a male donkey, are like this. So are most zebroids, owing to chromosomal differences between horses and zebras.

A species consists of one or more populations of individuals that can interbreed under natural conditions and produce fertile offspring, and that are reproductively isolated from other such populations.

Speciation, the process by which daughter species form, begins with genetic divergence among populations of a species and ends with their reproductive isolation.

The prevention of gene flow between populations sets the stage for genetic divergence—a buildup of genetic differences between them. Then, mutation, selection, and genetic drift operate independently in each one.

By chance, reproductive isolating mechanisms evolve. These are heritable aspects of body form, physiology, or behavior that prevent interbreeding before or after fertilization.

15.2 MODELS OF SPECIATION

Speciation may unfold slowly or rapidly, and it varies in its details, depending on the distribution patterns and interactions among the affected populations. Three current models account for these differences: They are called allopatric, sympatric, and parapatric speciation.

Allopatric Speciation

By the model for **allopatric speciation**, some type of physical barrier forms and thereby prevents gene flow between populations or subpopulations of a species. (*Allo-* means different, and *patria* can be taken to mean homeland.) Over time, by genetic divergence and by chance, mechanisms of reproductive isolation evolve. Speciation is completed when individuals of the two populations no longer will interbreed, even if changing circumstances put them back together again.

Whether a geographic barrier proves to be effective at blocking gene flow between populations depends on an organism's means of travel, how fast it can travel, and whether it is inclined or compelled to disperse. You have only to think back on those garden snails and the wandering albatross described at the start of this chapter to conclude that this is so.

If we assume that the genetic changes required for speciation require physical separation, then allopatry probably is the main speciation route. Some distance separates the populations of most species, so that gene flow among them is more of an intermittent trickle than a steady stream. It wouldn't take much of a barrier to shut off the trickles. For example, this happened when a major earthquake changed the course of the Mississippi River in the 1800s. The abrupt change isolated some populations of insects that could not swim or fly.

Geographic isolation also can proceed slowly, over great spans of time. We find evidence of this in the fossil record, which provides glimpses into the breakup of formerly continuous environments.

For example, vast glaciers advanced down through North America and Europe many times in the past, and they slowly cut off populations from one another. When the glaciers retreated, groups descended from those populations came in contact. In some cases, they were no longer reproductively compatible; they had become separate species. In other cases, divergences had not proceeded far enough. Descendant populations can still interbreed; for them, speciation was not completed.

As another example, imperceptibly slow but colossal movements in the Earth's crust have resulted in the breakup and crunching together of huge land masses. Such movements uplifted the seafloor in the region we now call the Isthmus of Panama. The uplifting divided an ocean basin and set the stage for allopatric speciation among populations of fishes and other marine species.

Figure 15.5 (**a**) Blue-headed wrasse (*Thalassoma bifasciatum*) from the Atlantic side of the Isthmus of Panama and (**b**) Cortez rainbow wrasse (*T. lucasanum*) from the Pacific side. The fishes apparently are related by descent from a common ancestral population that split when geologic forces created the Isthmus. As is common among reef fishes, these individuals of the same species differ in body coloration and patterning.

In the early 1980s, John Graves compared four enzymes from the muscle cells of related Isthmus fishes (Figure 15.5). All of these fishes are strong swimmers. Graves knew that temperature affects the activity of different enzymes in different ways. He also knew that seawater on the Pacific side of the Isthmus is cooler by about 2°–3°C than it is on the Atlantic side, and that it varies more with the changing seasons.

As Graves discovered, all four "Pacific" enzymes function better than the four "Atlantic" enzymes do at lower temperatures. Electrophoresis studies revealed slight differences in electric charge between two of the four pairs of enzyme molecules, owing to slight differences in their amino acid sequences. Graves drew these tentative conclusions: First, even small differences in environmental conditions may impose enough selection pressure to promote divergences in molecular structure. Second, it appears that the closely related populations of fishes on the two sides of the Isthmus are beginning to diverge from each other. The alternative molecular forms of the enzymes that Graves studied are already showing detectable differences in activity.

How might we interpret the results from this study of geographically isolated populations? For these fishes, a gradual accumulation of mutations that are neutral or possibly adaptive might signify an evolutionary "foot in the door." In other words, *the genetic divergences may be evidence of speciation in progress.*

Less Prevalent Speciation Routes

By the model for **sympatric speciation**, species may form *within* the home range of an existing species—in the absence of a physical barrier. (*Sym-* means together with, as in "together with others in the homeland.")

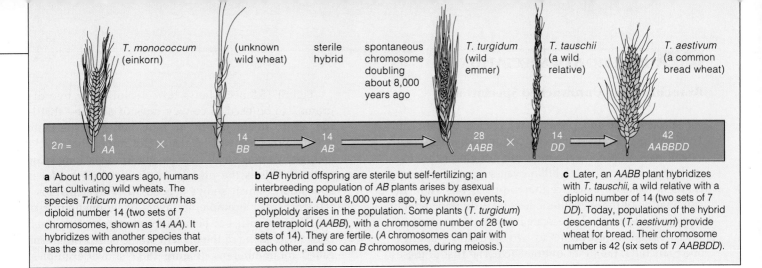

| | 14
AA | | 14
BB | | 14
AB | | 28
AABB | | 14
DD | | 42
AABBDD |

$2n =$ 14 AA   ×   14 BB → 14 AB → 28 AABB × 14 DD → 42 AABBDD

T. monococcum (einkorn)   (unknown wild wheat)   sterile hybrid   spontaneous chromosome doubling about 8,000 years ago   T. turgidum (wild emmer)   T. tauschii (a wild relative)   T. aestivum (a common bread wheat)

a About 11,000 years ago, humans start cultivating wild wheats. The species *Triticum monococcum* has diploid number 14 (two sets of 7 chromosomes, shown as 14 *AA*). It hybridizes with another species that has the same chromosome number.

b *AB* hybrid offspring are sterile but self-fertilizing; an interbreeding population of *AB* plants arises by asexual reproduction. About 8,000 years ago, by unknown events, polyploidy arises in the population. Some plants (*T. turgidum*) are tetraploid (*AABB*), with a chromosome number of 28 (two sets of 14). They are fertile. (*A* chromosomes can pair with each other, and so can *B* chromosomes, during meiosis.)

c Later, an *AABB* plant hybridizes with *T. tauschii*, a wild relative with a diploid number of 14 (two sets of 7 *DD*). Today, populations of the hybrid descendants (*T. aestivum*) provide wheat for bread. Their chromosome number is 42 (six sets of 7 *AABBDD*).

Figure 15.6 Presumed sympatric speciation in wheat through polyploidy and hybridizations. Wheat grains 11,000 years old have been found in the Near East. Diploid wild wheats still grow there.

In 1994, Ulrich Schliewen found evidence in favor of the model. He studied nine species of fishes (cichlids) in a small African lake that formed in a collapsed volcanic cone. Physical and chemical conditions are so uniform throughout the lake, even *micro*geographic separation is not possible. All species move about and breed near the bottom of the lake, in sympatry. They apparently descended from the same founder species. (In their mitochondrial DNA sequences, they are like each other and not like related species in nearby lakes and rivers.) Some species feed in the open waters and others on the lake bottom. Maybe the ecological separation promoted selective mating, then reproductive isolation.

Sympatric speciation might have occurred among flowering plants, about half of which are polyploid. **Polyploidy** refers to the inheritance of three or more of each type of chromosome characteristic of the parental stock. As described in Section 10.7, such a change in chromosome number can arise through the improper separation of chromosomes at meiosis or mitosis. It also arises when a germ cell duplicates its DNA but fails to divide, then goes on to function as a gamete.

Speciation could have been instantaneous for many kinds of flowering plants that engage in self-fertilization or asexual reproduction. Their offspring could inherit a novel chromosome number. Extra chromosomes could pair at meiosis, and the extra genes would do no harm. Some species also may have originated when polyploidy was followed by cross-fertilization (Figure 15.6).

However, most hybrids between species are sterile. This is especially true of the animal kingdom, where polyploidy is less common. It may upset the complex gene interactions required for sex determination, which is crucial for animal development. Usually, embryos that develop from animal gametes with mismatched chromosomes cannot survive. Those that do commonly have severe genetic disorders, shorter lives, or both.

Bullock's orioles   Baltimore orioles

NORTH DAKOTA

SOUTH DAKOTA

NEBRASKA

KANSAS

OKLAHOMA

hybrid zone

Figure 15.7 A hybrid zone where parapatric speciation among orioles (*Icterus*) may be proceeding along the borders. To the east of this zone of overlapping contact are Baltimore orioles; to the west are Bullock's orioles. Hybridization was once so common, they were called variants of a single species. However, the hybrids are becoming less frequent. Isolating mechanisms might be reinforcing divergence in this zone.

Evidence for **parapatric speciation** is sketchy. By this model, species form where two populations share a common border that is permeable to gene flow. (*Para*- means near, as in "near another homeland.") Initially, reproductive isolation is confined to a **hybrid zone**, where the adjoining populations meet, interbreed, and produce hybrid offspring. Such a zone extends through Nebraska and adjoining states (Figure 15.7). In time, reproductive isolation may spread through the range of the parent species.

ALLOPATRIC SPECIATION. **In the absence of gene flow between geographically separate populations, daughter species form gradually, by divergence.**

SYMPATRIC SPECIATION. **Daughter species arise from a group of individuals within an existing population.**

PARAPATRIC SPECIATION. **Daughter species form following reproductive isolation that starts in a group of individuals along a common border between two populations.**

Branching and Unbranched Speciation

All species, past and present, are related by descent. They are genetically connected through lineages that extend back in time to the molecular origin of the first prototypic cells, some 3.8 billion years ago. Subsequent chapters focus on evidence that supports this view. In anticipation of those chapters, let's start thinking about ways to interpret the large-scale histories of species.

The fossil record points to two speciation patterns, one branching, the other unbranched. The first of these patterns is called **cladogenesis** (from the Greek *klados*, meaning branch, and *genesis*, meaning origin). It applies to populations that become isolated from one another and then diverge genetically in different directions, as described earlier in this chapter.

Species also form through accumulated changes in allele frequencies and in phenotypes within a single, unbranched line of descent. This speciation pattern is called **anagenesis**. (In this context *ana-* means renewed.) Later in time, someone divides the lineage into two or more species, although deciding where to carve up the whole pattern of phenotypic change is rather subjective.

Evolutionary Trees and Rates of Change

You can summarize information about the continuity of relationship among species in the form of **evolutionary tree diagrams**. Take a look at Figure 15.8. It shows a simple way to start thinking about how the diagrams are constructed. Each branch in a tree diagram represents one line of descent from a common ancestor. A branch point is a time of divergence and speciation, as brought about by microevolutionary processes.

Figure 15.8 also shows how evolutionary tree diagrams can be used to convey rates of change—that is, the approximate length of time between speciation events. Branches with slight angles imply that species emerge by many small changes in form over long spans of time. This is the premise of the **gradual model of speciation**, which shows a good fit with many fossil sequences. For example, in layers of sedimentary rock we find sequences of beautifully preserved, intricately perforated shells of the ancient single-celled organisms called foraminiferans. (Figure 18.17 shows examples.) The sequences provide evidence of slow change.

Alternatively, evolutionary tree diagrams for some lineages have short, horizontal branches that abruptly make a 90-degree turn (Figure 15.8*b*). The diagrams are consistent with the **punctuation model of speciation**. By this model, most morphological changes are compressed into a brief period when populations first start to diverge—say, within merely hundreds or thousands of years. The idea is that bottlenecks, founder effects, strong directional selection, or some combination of these brings about rapid speciation. Daughter species recover quickly from the adaptive wrenching, and they change little over the next 2 million to 6 million years or so. Therefore, adaptive cohesion prevailed in about 99 percent of the history of most lineages. This pattern seems to pervade many parts of the fossil record.

Apparently, change can be gradual, abrupt, or both. Species originate at different times and differ in their persistence on the evolutionary stage. Remember, some lineages have endured without much change, producing a species here, losing a species there, over many millions of years. Still others have branched bushily and sometimes spectacularly, during times of adaptive radiation.

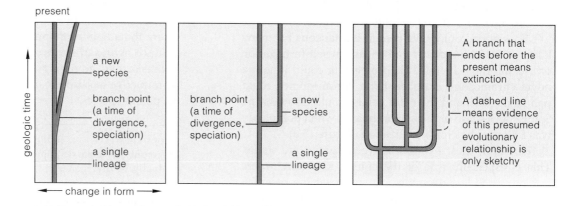

Figure 15.8 How to read an evolutionary tree diagram.

present

geologic time

a new species

branch point (a time of divergence, speciation)

a single lineage

← change in form →

a Softly angled branching means speciation occurred through gradual changes in traits over geologic time.

branch point (a time of divergence, speciation)

a new species

a single lineage

b Horizontal branching means traits changed rapidly around the time of speciation. Vertical continuation of a branch means traits of the new species did not change much thereafter.

A branch that ends before the present means extinction

A dashed line means evidence of this presumed evolutionary relationship is only sketchy

c Many branchings of the same lineage at or near the same point in geologic time means that an adaptive radiation occurred.

Adaptive Radiations

An **adaptive radiation** is a burst of microevolutionary activity within a lineage, one that results in the formation of many new species in a wide range of habitats. Figure 15.9 is a tree diagram of one adaptive radiation. Lineages often spread out this way when their members are presented with unfilled adaptive zones.

Adaptive zones are not easy to define. Think of them as ways of life, such as "burrowing in seafloor sediments" or "catching insects in the air at night." A lineage may radiate into such adaptive zones when it has physical, evolutionary, or ecological access to them.

Physical access means a lineage happens to be there when adaptive zones open up. For example, mammals were once distributed through uniform tropical regions of a single continent. That continent split into several land masses. Habitats and resources changed in different ways on those land masses and set the stage for independent radiations.

Evolutionary access means that modification of some structure or function will permit a lineage to exploit the environment in improved or novel ways. When the forelimbs of certain five-toed vertebrates evolved into wings, for example, this opened new adaptive zones to the ancestors of birds and bats.

Ecological access means that the lineage can enter an unoccupied adaptive zone or displace resident species. We will say more about this in Chapter 36.

Extinctions—End of the Line

Some number of species within a lineage inevitably disappear as local conditions change. This expected rate of disappearance over time is called **background extinction**. Table 15.1 lists examples. By contrast, an abrupt rise in extinction rates above the background level is called **mass extinction**. It is a catastrophic, global event in which entire families and other major groups are wiped out simultaneously.

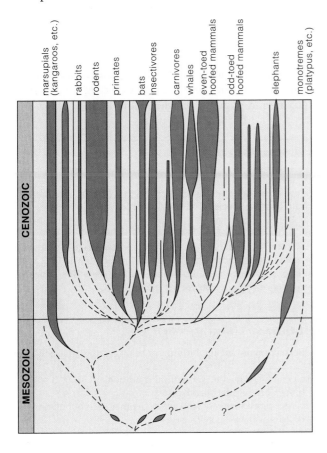

Figure 15.9 Evolutionary tree diagram of the great adaptive radiation of mammals, which began about 65 million years ago at the start of a geologic era called the Cenozoic. Variations in the width of a given branch correspond to the range of diversity within the lineage represented at different points in time. The wider the branch, the greater the diversity.

Table 15.1	Examples of the Estimated Average Durations of Species	
	Group	Duration (millions of years)
Protistans:	Foraminiferans	20–30
	Diatoms	25
Plants:	Bryophytes	20+
	Higher plants	8–20+
Animals:	Gastropods	10–13.5
	Ammonites	1–2, 6–15
	Trilobites	1+
	Beetles	2+
	Freshwater fishes	3
	Snakes	2+
	Mammals	1–2+

Lineages that are not widely dispersed tend to be hit hard in times of mass extinctions. This is especially true of the tropics. Global temperatures drop at such times. Species adapted to cool climates survive, but those in the tropics have nowhere to go. And luck has a lot to do with it, as when asteroids hit the Earth and level the playing field. Some survivors of one such impact radiated into adaptive zones occupied earlier by dinosaurs. Among them were mammals that had previously remained, wisely, hidden in the shrubbery.

Taken together, the persistence, branchings, and extinctions of species account for the full range of biological diversity at any point in geologic time.

SUMMARY

1. With respect to sexually reproducing organisms, a species consists of one or more populations of individuals that can interbreed under natural conditions and produce fertile offspring, and that are reproductively isolated from other such populations.

2. The populations of a species have a shared genetic history, they are maintaining genetic contact over time, and they are evolving independently of other species.

3. Speciation, the process by which daughter species form, requires an end to gene flow between populations (or subpopulations) of a species. In the absence of gene flow, mutation, natural selection, and genetic drift are free to operate independently in genetically separate populations, which may then irreversibly diverge.

4. Genetic divergence is a buildup of differences in allele frequencies between populations (or subpopulations) of a species that become reproductively isolated from one another.

5. Prezygotic isolating mechanisms (Table 15.2) prevent mating or pollination between populations. Differences in reproductive timing and behavior, incompatibilities in reproductive structures or gametes, and occupation of different microenvironments in the same area are examples. Postzygotic mechanisms lead to early death, sterility, or unfit hybrid offspring. They take effect after fertilization, while the embryo develops.

6. There are three current models of speciation:

a. Allopatric speciation. Gene flow is prevented between geographically separated populations, which diverge genetically into new species. Probably the most prevalent speciation route.

b. Sympatric speciation. Species form after there is reproductive isolation of individuals in the home range. Speciation by polyploidy is an example.

c. Parapatric speciation. Species form from a group of individuals along the common border between two populations, where gene flow is localized.

7. Lineages vary in the times, rates, and direction of speciation. Speciation may proceed gradually, rapidly, or both within the same lineage. Extensive branching of a lineage (divergence of one or more of its populations) in the same geologic time span is an adaptive radiation.

8. Evolutionary tree diagrams show the relationship among groups of species. Each branch of such trees represents a line of descent (lineage). Branch points are speciation events, as brought about by natural selection, genetic drift, and other microevolutionary processes.

9. All lineages inevitably lose some number of species over time. This loss is called background extinction. By contrast, in a mass extinction, major groups of species perish abruptly in a catastrophic, global event.

10. The persistences, branchings, and extinctions of species account for the full range of biological diversity at any period of geologic time (Table 15.3).

Table 15.2	Summary of Isolating Mechanisms
Prezygotic Isolation (*Mating or zygote formation is blocked*)	
Temporal isolation	Potential mates occupy overlapping ranges but reproduce at different times
Behavioral isolation	Potential mates meet but cannot figure out what to do about it
Mechanical isolation	Potential mates attempt engagement, but sperm cannot be successfully transferred
Gametic mortality	Sperm is transferred but egg is not fertilized (gametes die or gametes are incompatible)
Ecological isolation	Potential mates occupy different local habitats within the same area
Postzygotic Isolation (*Hybrids don't work*)	
Zygotic mortality	Egg is fertilized, but zygote or embryo dies
Hybrid inviability	First-generation hybrid forms but shows very low fitness
Hybrid infertility	Hybrid is sterile or partially so

After Alan Templeton, Jerry Coyne, and others.

Review Questions

1. Describe the biological species concept. *240*

2. How does the evolution of reproductive isolating mechanisms contribute to the speciation process? *241*

3. Define each of the three speciation models. *242–243*

4. Interpret these features of evolutionary tree diagrams: a single line, softly angled branching, horizontal branching, vertical continuation of a branch, many branchings of the same line, a dashed line, a branch that ends before the present. *244*

Self-Quiz (*Answers in Appendix IV*)

1. Sexually reproducing individuals belong to the same species if they _____ .
   a. can interbreed under natural conditions
   b. can produce fertile offspring
   c. are reproductively isolated from other such populations
   d. have a shared genetic history
   e. all of the above

2. Genetic divergence is an accumulation of differences in allele frequencies between _____ .
   a. species
   b. populations of a species
   c. subpopulations of a species
   d. both b and c may be correct

3. Isolating mechanisms _____ .
   a. prevent interbreeding
   b. prevent gene flow
   c. reinforce genetic divergence
   d. all of the above

Table 15.3 Summary of the Processes and Patterns of Evolution

Microevolutionary Processes

Mutation	Original source of alleles	Stability or change in a species is the outcome of balances or imbalances among all of these processes, the effects of which are influenced by population size and by prevailing environmental conditions.
Gene flow	Preserves species cohesion	
Genetic drift	Erodes species cohesion	
Natural selection	Preserves or erodes species cohesion, depending on environmental pressures	

Macroevolutionary Patterns

Genetic persistence	The basis of the unity of life. The biochemical and molecular basis of inheritance extends from the origin of the first cells through all subsequent lines of descent.
Genetic divergence	The basis of the diversity of life, brought about by adaptive shifts, branchings, and radiations. The rates and times of change have varied within and between lineages.
Genetic disconnect	The steady loss (background extinction) or abrupt, catastrophic loss (mass extinction) of lineages.

4. Speciation occurs through _____ .
   a. the evolution of reproductive isolating mechanisms
   b. genetic divergence
   c. mutation, genetic drift, and natural selection
   d. all of the above

5. When potential mates occupy overlapping ranges but reproduce at different times, this is a case of _____ isolation.
   a. postzygotic   c. temporal
   b. mechanical   d. gametic

6. Probably the most prevalent speciation route is _____ .
   a. allopatric speciation   c. parapatric speciation
   b. sympatric speciation   d. both a and c

7. Whether a geographic barrier deters gene flow between populations of the same species depends on an organism's _____ .
   a. built-in means of travel, speed, and endurance
   b. environmental proddings to disperse
   c. both a and b

8. In an evolutionary tree diagram, a branch point represents _____ , and a branch that ends represents _____ .
   a. a single species; incomplete data on lineage
   b. a single species; a time of extinction
   c. a time of divergence; extinction
   d. a time of divergence; speciation complete

9. An evolutionary tree diagram with horizontal branches that abruptly become vertical is consistent with _____ .
   a. the gradual model of speciation
   b. the punctuation model of speciation
   c. the idea of small changes in form over long spans of time
   d. both a and c
   e. both b and c

10. Match the terms with the suitable descriptions.
   ____ cladogenesis   a. unbranched lineage
   ____ anagenesis   b. burst of microevolutionary
   ____ adaptive radiation   activity within a lineage
   ____ background extinction c. catastrophic disappearance
   ____ mass extinction   of major groups of organisms
                              d. branching lineages
                              e. expected rate of species
                                 loss within a lineage

Critical Thinking

1. An archipelago is a chain of islands some distance from the mainland. In such isolated settings, organisms often undergo adaptive radiation, as when a group of animals or plants evolve into many related species. The Galápagos finches are one example. Fruit flies are another; although the Hawaiian Islands represent less than 2 percent of the world's land mass, they are home to 40 percent of all fruit fly species. Explain why allopatric speciation might produce adaptive radiations on archipelagoes.

2. Several duck species may occupy the same lake habitat. The females of different species may look very similar. But males of each species have distinctively patterned and colored feathers. What forms of reproductive isolation may be keeping the species distinct? How does the appearance of the male ducks provide you with a clue to the answer?

Selected Key Terms

adaptive radiation *245*
adaptive zone *245*
allopatric speciation *242*
anagenesis *244*
background extinction *245*
biological species concept *240*
cladogenesis *244*
evolutionary tree diagram *244*
gene flow *240*
genetic divergence *240*
gradual model of speciation *244*
hybrid zone *243*
mass extinction *245*
parapatric speciation *243*
polyploidy *243*
punctuation model of speciation *244*
speciation *239, 240*
species *240*
sympatric speciation *242*

Readings

Futuyma, D. 1986. *Evolutionary Biology.* Second edition. Sunderland, Massachusetts: Sinauer.

Mayr, E. 1976. *Evolution and the Diversity of Life.* Cambridge, Massachusetts: Belknap Press of Harvard University Press.

Otte, D., and J. Endler (editors). 1989. *Speciation and Its Consequences.* Sunderland, Massachusetts: Sinauer.

16 THE MACROEVOLUTIONARY PUZZLE

Of Floods and Fossils

About 500 years ago, Leonardo da Vinci was brooding about seashells entombed in the layered rocks of northern Italy's high mountains, hundreds of kilometers from the sea. How did they get there? If he accepted the traditional explanation, he would have to agree that stupendous floodwaters deposited those shells in the mountains during the Great Deluge (Figure 16.1). But many shells were thin, fragile—and intact. Surely they would have been battered to bits if they had been swept across such distances, then up the mountains.

Da Vinci also brooded about the rock layers. They were stratified (stacked like cake layers), and some had shells but others had none. Then he remembered how large rivers deposit silt during each spring flood. *Did all those layers slowly accumulate long ago, as a series of silt deposits where rivers emptied into the sea?* If so, then shells in the mountains would be evidence of a progression of communities of organisms that once lived in the seas!

Da Vinci did not announce his novel idea, perhaps knowing that it would have been met with deafening silence, imprisonment, or worse.

By the 1700s, fossils were being accepted as evidence of past life. They were still interpreted in a traditional way, as when a Swiss naturalist excitedly unveiled the remains of a giant salamander and announced that they were the skeleton of a man who drowned in the Deluge.

By midcentury, however, scholars began to question such interpretations. Extensive mining, quarrying, and canal excavations were under way. The diggers were finding similar rock layers and similar fossil sequences in distant places, such as the cliffs on both sides of the English Channel. More than a few scholars began to view the findings as evidence of connections between Earth history and the history of life.

Ever since, fossils have been analyzed in more and more refined ways. Together with biochemical studies and other modern sources of information, they provide good evidence of evolution through vast spans of time.

Evolution, recall, can only proceed by divergences from populations that already exist. Those forks in the evolutionary road began with the first species ever to appear on Earth. Thus *all species that have ever evolved*

are related, by way of descent. This principle of evolution guides efforts to make sense of scattered and often puzzling scraps of evidence of past life. It guides the task of identifying and sorting out lines of descent, or lineages, that connect all species, past and present.

Ultimately, then, life is a story of *species*—of how and when each originated, whether its defining traits persisted or changed, and whether it vanished or endured. It also is a story of **macroevolution**—of large-scale patterns, trends, and rates of change among families and other more inclusive groups of species. We turn to this larger picture in the next unit. Here, we begin with the nature of the evidence that supports it.

Figure 16.1 (*Left*) From the Sistine Chapel, Michelangelo's painting of the onset of a catastrophic flood, traditionally called the Great Deluge. Events of this magnitude actually punctuated geologic time, although they have been explained in different ways throughout recent human history. (*Below*) A modern-day photographer captures the intricate structural pattern of the fossilized shells of marine animals called ammonites. About 65 million years ago, all ammonites perished, along with many other groups of organisms. That mass extinction is but one intriguing piece of the evolutionary puzzle.

KEY CONCEPTS

1. In the evolutionary view, all species that have ever lived are related—some closely, others remotely so. This is because each new species evolved from variant individuals of species that *already existed*, starting with the first living cells to appear on Earth.

2. The term "macroevolution" refers to the patterns, trends, and rates of change among lineages over geologic time.

3. The fossil record, geologic record, and radioactive dating of rocks yield evidence of macroevolution.

4. Anatomical comparisons between major lineages help us understand and reconstruct patterns of change through time. Among the most revealing aspects of anatomy are homologous structures in different lineages. They signify descent from a common ancestor.

5. Biochemical comparisons within and between major lineages also provide evidence of macroevolution.

6. Stunning diversity characterizes the distribution of species through time and through the global environment. Biological systematics attempts to discern patterns in life's diversity through taxonomy, phylogenetic reconstruction, and classification.

7. With taxonomy, new species are identified and named. With phylogenetic reconstruction, evolutionary connections are worked out. With classification, the information about species is organized into retrieval systems.

Fossilization

"Fossil" comes from a Latin word for something that has been "dug up." In general, **fossils** are recognizable, physical evidence of ancient life. Figure 16.2 shows a few examples.

Most of the fossils discovered so far are bones, teeth, shells, seeds, capsules of spores, and other hard parts. (Usually, soft parts are the first to decompose when an organism dies.) Besides this, imprints of leaves, stems, tracks, trails, burrows, and other *trace* fossils provide us with indirect evidence of past life. Even fossilized feces (coprolites) indicate which species were being eaten—and therefore were present—in ancient environments.

Figure 16.2 Fossils—evidence of life in the distant past.

(**a**) A fossil hunter's dream—the complete skeleton of a bat that lived 50 million years ago. Erosion, movements in the Earth's crust, and other forces of nature have left few burial sites undisturbed, so intact fossils are extremely rare. Even the jumbled skeletal parts of several ducklike birds in (**b**) are a good find. It will take many hours of careful preparation and analysis to identify the species.

(**c**) Fossilized parts of *Cooksonia*, the oldest known land plant. Its stems were less than 7 centimeters (less than 3 inches) tall.

(**d**) One of the most magnificent, complete fossils ever found—skeletal remains of an ichthyosaur, a dolphin-like marine reptile that lived about 200 million years ago.

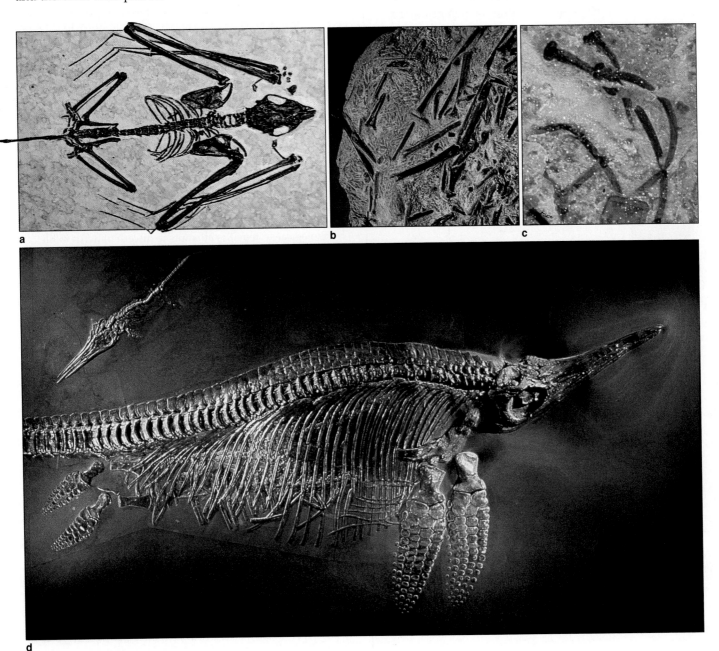

a b c

d

Fossilization is a gradual process that starts when an organism (or traces of it) becomes buried in volcanic ash or in sediments. Sooner or later, water infiltrates the organic remains, which become infused with dissolved metal ions and other inorganic compounds. More and more sediments accumulate above the burial site and exert increasing pressure on the remains. Over great spans of time, chemical changes and pressure transform the remains to stony hardness.

Preservation is favored when organisms are buried rapidly, in the absence of oxygen. Gentle entombment by volcanic ash or anaerobic mud is best. Preservation also is favored when a burial site is not disturbed. Most often, however, erosion and other geologic insults have crushed, deformed, broken, or scattered the fossils.

Interpreting the Geologic Tombs

We find similar fossil-containing layers of sedimentary rock over vast areas, even on different continents. Such layers formed long ago, by gradual deposits of volcanic ash, silt, and other materials, one above the other. This layering of sedimentary deposits is called **stratification**. Figure 16.3 shows an example. Generally, the deepest layers were the first to form. The layers closest to the surface formed last. Because particles tend to settle in response to gravity, most sedimentary layers must have formed horizontally. Where they are ruptured or tilted, this is evidence of subsequent geologic disturbance.

Understand how rock layers form, and you realize that fossils within a given layer are from a similar age in Earth history. Therefore, *the older the layer, the older the fossils.* Early naturalists had no way of determining the age of the fossils or the rocks. However, they divided the past into great eras, on the basis of abrupt changes in the fossil sequences. Much later, researchers assigned dates to this "geologic time scale." The scale serves as the chronological framework for the next chapter.

Interpreting the Fossil Record

At present, we have fossils for about 250,000 species, which give insights into evolutionary history. However, judging from the current range of diversity, there must have been many, many millions of ancient, now-extinct species. We never will be able to recover fossils for most of them, so our "record" of past life is incomplete, with built-in biases. Why is this so?

Most important, colossal movements in the Earth's crust obliterated evidence from crucial periods in the past. Besides this, most members of ancient communities simply weren't preserved. For example, bony fishes and hard-shelled mollusks are well represented in the fossil record. Soft-bodied worms and jellyfishes are not,

Figure 16.3 A splendid slice through time: the Grand Canyon of the American Southwest, once part of an ocean basin. Its layers of sedimentary rock formed gradually over hundreds of millions of years. Geologic forces lifted them above sea level. Later, the erosive force of rivers carved the deep canyon walls and so exposed the ancient stacked layers.

even though they may have been just as common or more so. Population density and body size further skew the record. For example, a plant population may have released millions of spores in a single growing season, whereas the earliest humans lived in very small groups and produced few offspring. What are your chances of finding a fossilized skeleton of an early human, compared to spores of plants that lived at the same time?

The fossil record also is heavily biased toward some environments. Most of the species we know about lived on land or in shallow seas that, through geologic uplifting, became part of continents. We have few fossils from sediments beneath the ocean—which extends across three-fourths of the earth's surface! Finally, most fossils have been found in the Northern Hemisphere, simply because that's where most fossil hunters have lived.

Fossils, the stone-hard physical evidence of ancient life, are present in layers of sedimentary rock. The deeper the layers, the older the fossils.

The completeness of the fossil record varies as a function of the kinds of organisms represented, where they lived, and the stability of their burial sites.

EVIDENCE FROM COMPARATIVE EMBRYOLOGY

Fine evidence of macroevolution comes from anatomical comparisons of major lineages. This field of inquiry is called **comparative morphology**. A few examples will give insight into one of its guiding principles. Namely, *when it comes to changes in the body's development, evolution tends to follow the line of least resistance.*

Developmental Program of Larkspurs

Most constraints on evolution take effect as embryos are developing. The sequence of a developmental program is so finely tuned, mutations that would alter the early steps especially tend to be selected against. Every so often, though, workable mutations allow evolution to proceed—but only in particular directions.

Flowering plants give evidence of evolution by way of mutations that changed the developmental program. Consider the flowers of *Delphinium decorum*, the most common larkspur. A ring of petals guides honeybees to the nectar-storing tube (Figure 16.4). Outward-bulging reproductive structures at the flower's center give the bees something to hang onto while they draw nectar—and pollinate the flowers. *D. nudicaule*, a larkspur of more recent origin, has tight flowers that discourage bees. Its pollinators are hummingbirds, which hover in front of a flower as they draw nectar from it.

As you can see from Figure 16.4*b–d*, the mature flowers of *D. nudicaule* strongly resemble the buds of *D. decorum*. The difference in morphology is an outcome of a gene mutation that slowed down the rate of floral development in *D. nudicaule*.

Developmental Program of Vertebrates

Now consider the vertebrates, which range from fishes to amphibians, reptiles, birds, and mammals. These are diverse animals. Yet comparisons of the ways in which their embryos develop provide compelling evidence of their evolutionary connection to one another.

Every vertebrate starts out life as a fertilized egg. It proceeds through specific stages of development before becoming an adult. Intriguingly, early in the program of development, the embryos of all the different lineages proceed through strikingly similar stages. Take a look at Figure 16.5. Without the labels, would you know which embryo is from a fish, lizard, chicken, or human?

a *D. decorum*

b *D. nudicaule*

Figure 16.4 From comparative embryology, evidence of evolutionary relationship among two species of larkspurs (*Delphinium*). Although the time required for flowers to develop and mature is much the same in both species, the *rate* of development is much slower for most of the floral structures of *D. nudicaule*, so the changes in petal shape are not nearly as great as they are in the more common species, *D. decorum*.

c Bud and mature flower (side and front views) of *D. decorum*

d Mature flower of *D. nudicaule*

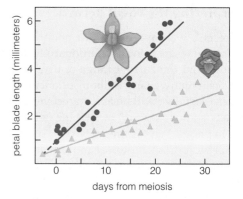

e Lengthening of petals of *D. decorum* (*dark blue* graph line) and of *D. nudicaule* (*light blue*), plotted against development of pollen and eggs in floral structures, starting with meiosis.

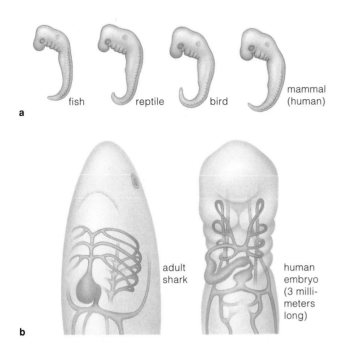

fish   reptile   bird   mammal (human)

a

adult shark

human embryo (3 milli-meters long)

b

Figure 16.5 From comparative embryology, some evidence of evolutionary relationship among vertebrates. (**a**) Adult vertebrates show great diversity, yet the very early embryos retain striking similarities. This is evidence of change in a common program of development. (**b**) Fishlike structures still form in early embryos of reptiles, birds, and mammals. For example, a two-chambered heart (*orange*), certain veins (*blue*), and portions of arteries called aortic arches (*red*) develop in fish embryos and persist in adult fishes. The same structures form in an early human embryo.

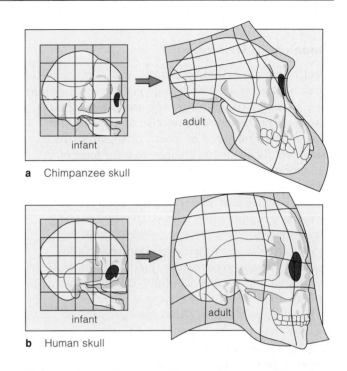

infant   adult

a Chimpanzee skull

infant   adult

b Human skull

Figure 16.6 Morphological differences between two lines of descent, presumably an outcome of changes in the timing of developmental steps. This example compares proportional changes in the skull bones of a chimpanzee and a human. Both skulls are quite similar in infants. Think of the representations of infant skulls as paintings on a blue rubber sheet divided into a grid. Stretching the sheet deforms the grid's squares. For the adult skulls, differences in size and shape within corresponding grid sections reflect differences in growth patterns.

The early embryos of vertebrates strongly resemble one another because they have inherited the same ancient plan for development. According to this plan, tissues can form only when cells divide in certain patterns and interact in prescribed ways. Later on, the heart, bones, skeletal muscles, gut, and other body parts grow in prescribed ways—but only if each developmental step is properly completed before the next step begins.

During vertebrate evolution, most mutations that disrupted early stages of development must have had lethal effects on the organized interactions required for later stages. The embryos of different groups remained similar because mutations that altered steps in early developmental stages were strongly selected against.

How, then, did adults of different groups get to be so different? At least some differences probably came about by mutations that altered the onset, rate, or time of completion of certain developmental steps. Such mutations would bring about changes in shape through increases or decreases in the relative size of body parts. (These are called allometric changes.) They also could lead to adult forms that retain some juvenile features.

For example, Figure 16.6 illustrates how changes in the growth rate at a key developmental step may have caused differences in the proportions of chimpanzee and human skull bones. Consider that the skull bones of both kinds of primates are alike at the time of birth. Thereafter, they change dramatically for chimps but only slightly for humans. Several lines of evidence imply that chimps and humans arose from the same ancestral stock. *Their genes are nearly identical.* However, somewhere along the separate evolutionary road that led to humans, certain regulatory genes apparently underwent mutation in ways that proved adaptive. Ever since that time, instead of promoting the rapid growth required for dramatic changes in skull bones (as in chimps), mutated human genes have blocked it.

Similar patterns of embryonic development may be a clue to evolutionary relationship among lineages.

Mutations that affect certain steps in a shared program of early development may be enough to bring about major differences in the adult forms of related lineages.

Homologous Structures

Recall, from the preceding chapter, that populations of the same species start to diverge genetically after they have become separated from one another. Over large spans of time, the populations also diverge in their appearance, functions, or both. This pattern of macroevolution—that is, change from the form of a common ancestor—is known as **morphological divergence**. (*Morpho-* means body form.)

Even with great morphological divergence, however, evolutionarily related species remain alike in many ways. They started out with the same body plan, and their evolution only could proceed through conservative modifications to the shared plan. Look carefully enough, and you can discover their underlying similarities.

For example, reptiles, birds, and mammals are land-dwelling vertebrates, presumably descended from amphibians that started to venture onto land. Our knowledge of the earliest land vertebrates is based on the fossilized, five-toed limb bones of a rather squat reptile (Figure 16.7a). Descendants of this reptile apparently expanded into new environments on land. A few of the later descendants even retreated from the land, through adaptations that allowed them to live in the seas.

And so, in the lineages of pterosaurs, birds, and bats, the five-toed limb served as evolutionary clay, which gradually became molded into wings (Figure 16.7b–d). In the lineages that led to porpoises and to penguins, the five-toed limb gradually evolved into flippers (Figure 16.7e–f). In still other lineages, it slowly became modified into the long, one-toed limbs of modern horses, the stubby limbs of moles and other burrowing mammals, and the pillarlike limbs of elephants. It also became modified into the human arm and five-fingered hand (Figure 16.7g).

What we have been describing is an example of **homology**—a similarity in one or more body parts in different organisms that we can attribute to descent from a common ancestor. (*Homo-* means the same.) Such similarities are probably the most important clues for researchers who are chipping away at the macroevolutionary puzzle.

Potential Confusion From Analogous Structures

Body parts that have similar form and functions in different lineages aren't *always* homologous. If the last shared connection between two lineages was in the remote past, major differences in form

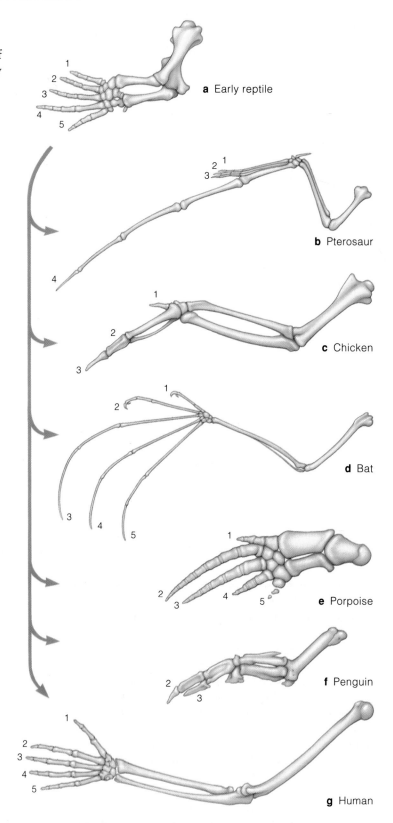

Figure 16.7 Morphological divergence in the vertebrate forelimb, starting with a generalized form of ancestral early reptiles. Diverse forms evolved even as similarities in the number and position of bones were preserved. The drawings are not to the same scale.

a Early reptile

b Pterosaur

c Chicken

d Bat

e Porpoise

f Penguin

g Human

Figure 16.8 Example of morphological convergence among three lineages that diverged considerably over evolutionary time. In the lineages leading to modern sharks, penguins, and porpoises, natural selection favored adaptations that promoted rapid swimming. As one outcome, sharks, penguins, and porpoises have similar-appearing body forms—even though they are only remotely related.

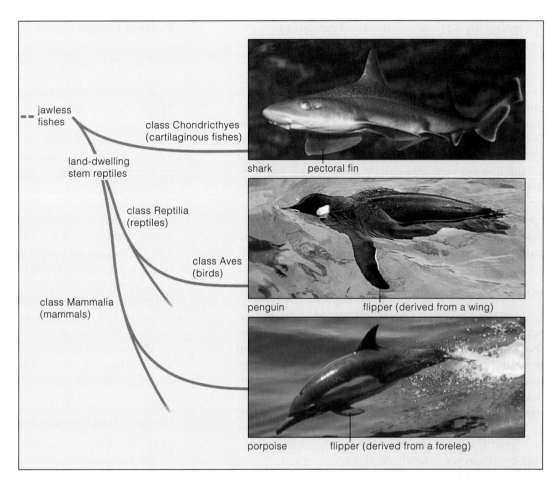

jawless fishes

class Chondricthyes (cartilaginous fishes)

land-dwelling stem reptiles

shark   pectoral fin

class Reptilia (reptiles)

class Aves (birds)

penguin   flipper (derived from a wing)

class Mammalia (mammals)

porpoise   flipper (derived from a foreleg)

may have emerged during their separate evolutionary journeys. Suppose, by chance, some of their descendants encountered similar kinds of environmental pressures. Suppose those only remotely related organisms started using comparable body parts in much the same way. Most likely, they would have become modified, as an outcome of natural selection, and would have ended up resembling one another in structure and function.

This macroevolutionary pattern, in which lineages that are only remotely related slowly evolve in similar directions, is called **morphological convergence**.

For example, sharks, penguins, and porpoises are all fast-swimming predators of the seas. As you can see from Figure 16.8, penguin and porpoise flippers are similar in shape to the shark pectoral fin. All three structures stabilize the body in water—but they are *not* homologous. The shark, penguin, and porpoise lineages diverged very early in vertebrate evolution, so they are related only in a very distant sense.

Sharks never left the water, and shark fins haven't changed much from the ancestral form. By contrast, penguins are descended from four-legged vertebrates that evolved into birds, which later became adapted to

life in water instead of in the air. Penguin flippers are modified wings—which are modified forelimbs. And porpoises descended from four-legged vertebrates that evolved into mammals, which later became adapted to life in the seas—their flippers are modified front legs. Thus, the penguin and porpoise flippers converged on the pectoral fins of sharks.

The body parts just mentioned are an example of **analogy**. The word refers to body parts that were once quite different in evolutionarily distant lineages, but that converged in structure and function because those lineages responded to similar environmental pressures. (*Analogos* means similar to one another.)

Homologous structures provide very strong evidence of morphological divergence. These are the same body parts that became modified in different ways in different lines of descent from a common ancestor.

Analogous structures are evidence of morphological convergence. At one time these body parts, now similar in form and function, differed in evolutionarily remote lineages, but then were used in comparable ways in similar environments.

All species are a mix of ancestral and derived traits. The kinds and numbers of traits they do or do not share are clues to how closely they are related. This is also true of their biochemical traits.

For example, on the basis of morphological studies alone, you might already have an idea that monkeys, humans, chimpanzees, and other primates are related to one another. You could test this idea by studying small differences in the amino acid sequences of proteins that all three primates synthesize. You also could test it by determining whether the nucleotide sequences in their DNA match up closely or not much at all. Logically, the species that are most closely related share the greatest biochemical similarities. This premise is central to comparative analysis at the molecular level.

Molecular Clocks

Like its close relatives among the primates, the human species has unique traits, the result of gene mutations that accumulated after a divergence from the ancestral primate stock. Yet humans and other species still have many genes in common. For example, the last shared ancestor of certain yeasts and humans lived many hundreds of millions of years ago—and yet, more than 90 percent of their known gene products (proteins) have remained the same. In other words, the overall structure of certain genes has been highly conserved.

Even so, neutral mutations have introduced slight structural differences in the highly conserved genes of different lineages. **Neutral mutations**, recall, have little or no effect on survival or reproduction. They can slip past agents of selection and accumulate in the DNA.

By some calculations, neutral mutations in highly conserved genes accumulated at a regular rate. Think of the accumulation of neutral mutations in a lineage as a series of predictable ticks of a **molecular clock**. Then turn the clock back—so that the total number of ticks "unwinds" down through the great geologic eras of the past. Where the last tick stops, that is roughly the time of origin for the lineage.

Figure 16.9 Primary structure of three versions of cytochrome *c* molecules. The amino acid sequence of this protein has not changed much, even in evolutionarily distant lineages. The three sequences shown below are from a representative yeast (*top row*), wheat (*middle row*), and primate (*bottom row*). The *gold* portions highlight the amino acids that are identical in all three species. The probability that this striking molecular resemblance resulted from chance alone is extremely low.

Protein Comparisons

Suppose two species have the same conserved gene. The amino acid sequences of the gene's product are the same or nearly so. The absence of mutation implies that the species are closely related. What if the sequences differ quite a bit? Then, many neutral mutations must have accumulated. A long time must have passed since the species shared a common ancestor.

Consider the highly conserved gene that specifies cytochrome *c*, a protein component of electron transport chains. Organisms ranging from aerobic bacteria to corn plants to humans synthesize this protein, which has a primary structure of 104 amino acids. Figure 16.9 shows how the amino acid sequences for cytochrome *c* from a fungus, plant, and animal are strikingly similar. Now think about this: Compared to human cytochrome *c*, the *entire* sequence is identical in chimps. It differs by 1 amino acid in rhesus monkeys, 18 in chickens, 19 in turtles, and 56 in yeasts.

On the basis of this information, would you assume that humans are more closely related to a chimpanzee or a rhesus monkey? A chicken or a turtle?

Nucleic Acid Comparisons

Typically, structural alterations that resulted from gene mutations pepper the nucleotide sequences of DNA and RNA molecules. Therefore, *the extent to which a strand of DNA or RNA isolated from one species will base-pair with a comparable strand from a different species is a rough measure of the evolutionary distance between them.*

As an example, some nucleic acid comparisons are based on **DNA-DNA hybridization**. By this method, two strands of a double-stranded DNA molecule are induced to unwind from each other, in the manner described in Section 13.4. A DNA molecule from another species is induced to unwind, also. Then the single strands are allowed to recombine into "hybrid" molecules. Next, the two double-stranded, hybridized molecules are subjected to heating, which breaks the hydrogen bonds holding them together.

The amount of heat energy required to pull apart a hybrid DNA molecule is a measure of the similarity between its two strands. It takes more heat energy to disrupt the hybrid DNA of closely related species. Figure 16.10 describes one application of this method.

Biochemical similarities are greatest among the most closely related species and weakest among the most distantly related.

$^+$NH$_3$-gly asp val glu lys gly lys lys ile phe ile met lys cys ser gln cys his thr val glu lys gly gly lys his lys thr gly pro asn leu his gly leu phe gly arg lys thr gly gln ala pro gly tyr ser

$^+$NH$_3$-ala ser phe ser glu ala pro pro gly asn pro asp ala gly ala lys ile phe lys thr lys cys ala gln cys his thr val asp ala gly ala gly his lys gln gly pro asn leu his gly leu phe gly arg gln ser gly thr thr ala gly tyr ser

$^+$NH$_3$-thr glu phe lys ala gly ser ala lys lys gly ala thr leu phe lys thr arg cys leu gln cys his thr val glu lys gly gly pro his lys val gly pro asn leu his gly ile phe gly arg his ser gly gln ala glu gly tyr ser

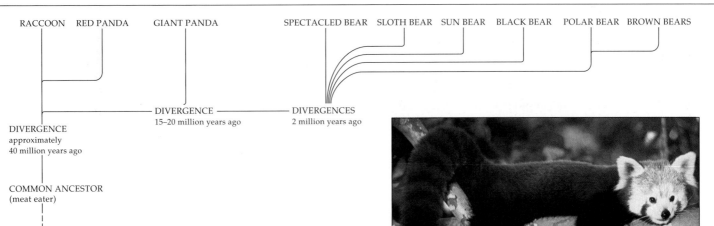

RACCOON RED PANDA GIANT PANDA SPECTACLED BEAR SLOTH BEAR SUN BEAR BLACK BEAR POLAR BEAR BROWN BEARS

DIVERGENCE
15–20 million years ago

DIVERGENCES
2 million years ago

DIVERGENCE
approximately
40 million years ago

COMMON ANCESTOR
(meat eater)

Figure 16.10 How the red panda, the giant panda, and bears are related, according to DNA-DNA hybridization studies.

A giant panda is a plant-eating mammal with paws. Unlike antelope, deer, cattle, and other ungulates, which are plant-eating mammals with hooves, the panda cannot efficiently digest the tough cellulose fibers in plants.

The ungulates pummel plant material inside an elaborately chambered stomach. They benefit from bacteria that reside permanently in their gut and that produce cellulose-digesting enzymes for them. By contrast, the panda has the gut of a *meat-eating* mammal. Its diet consists only of bamboo which, pound for pound, is not as nutrient-packed as meat. A panda must grasp and eat great quantities of bamboo to get enough carbohydrates, proteins, and fats. Its rounded, short-toed paws are not much good at holding and stripping leaves from bamboo stalks. But each front paw also has a thumblike bony digit. When a panda uses a "thumb" in opposition to the five toes on one of its paws, it gets a better grip on a bamboo meal.

The giant panda certainly looks like a bear. Yet it also resembles the red panda, which lives in the same general area in China and eats bamboo. Then again, neither bears nor the red panda has thumbs. Is the giant panda more closely related to red pandas or bears?

Researchers studied the extent to which single-stranded DNA from the giant panda would hybridize with DNA from the red panda and from bears. The more mismatched nucleotide bases, the greater the evolutionary distance between these mammals.

Results from DNA-DNA hybridization studies supported an earlier view that these three kinds of mammals are related by descent from a common ancestor that apparently lived more than 40 million years ago. The results indicate that a divergence took place among the descendants. One branch eventually led to modern raccoons and to the red panda. Another branch led to bears. Much later, between 20 million and 15 million years ago, a split from the bear lineage put the ancestors of the giant panda on a unique evolutionary road. Therefore, the giant panda seems to be much more closely related to bears than it is to the red panda.

RED PANDA

GIANT PANDA

BROWN BEAR

a ala asn lys asn lys gly ile ile trp gly glu asp thr leu met glu tyr leu glu asn pro lys lys tyr ile pro gly thr lys met ile phe val gly ile lys lys lys glu glu arg ala asp leu ile ala tyr leu lys lys ala thr asn glu-COO⁻
a ala asn lys asn lys ala val glu trp glu glu asn thr leu tyr asp tyr leu glu asn pro lys lys tyr ile pro gly thr lys met val phe pro gly leu lys lys pro gln asp arg ala asp leu ile ala tyr leu lys lys ala thr ser ser-COO⁻
a ala asn ile lys lys asn val leu trp asp glu asn asn met ser glu tyr leu thr asn pro lys lys tyr ile pro gly thr lys met ala phe gly gly leu lys lys glu lys asp arg asn asp leu ile thr tyr leu lys lys ala cys glu-COO⁻

16.5 ORGANIZING THE EVIDENCE—CLASSIFICATION SCHEMES

So far in this book, we have mentioned only a few of the many millions of species that originated and later vanished over the past 3.8 billion years. Yet even this sampling is enough to convey the challenges that face taxonomists, who identify, name, and classify species in ever more inclusive groups. Such groupings of species are known as **higher taxa** (singular, taxon). They are the organizing units of **classification schemes**, which are ways of retrieving information about particular species.

Assigning Names to Species

Whenever someone identifies a new organism, it gets a scientific name in accordance with a system that Carl von Linné devised in the eighteenth century. You may know this Swedish naturalist by his more prestigious, latinized name, Linnaeus.

Linnaeus was a naturalist whose enthusiasm knew no bounds. He sent ill-prepared students around the world to gather specimens of plants and animals, and is said to have lost a third of his collectors to the rigors of their expeditions. Although perhaps not commendable as a student adviser, Linnaeus did go on to develop a **binomial system** by which species are assigned a two-part Latin name. The first part is generic; it is descriptive of similar species that are thought to be the same type of organism and so are grouped together. Such a grouping is a **genus** (plural, genera). The second part is the specific name which, in combination with the generic name, refers to one kind of organism only. Such scientific names provide an unambiguous way for people anywhere to discuss the same organism.

For example, there is only one *Ursus maritimus*, or polar bear. Other bears include *Ursus arctos* (the brown bear) and *Ursus americanus* (the black bear). Notice the first letter of the generic name is capitalized and the second (specific) name is not. The specific name is never used without the full or abbreviated generic name preceding it, for it also can be the second name of a species in a different group. Thus, *U. americanus* does indeed mean black bear, but *Homarus americanus* means Atlantic lobster, and *Bufo americanus* means American toad. (Hence one would not order *americanus* for dinner unless one is willing to take what one gets.)

Shortly after Linnaeus devised it, the binomial system became the core of a classification scheme that was based only on perceived similarities or differences in physical features—the number of legs, body size, wings or no wings, coloration, and so forth. Later, biologists invented more inclusive groupings that show broader relationships. These broader taxa include family, order, class, and phylum, on up to kingdom.

Table 16.1 shows how taxa can help classify four species. This ranking alone does not yield information about patterns of relationships. The taxa reflect only *relative degrees* of relationship. A corn plant and vanilla orchid are closely related (both are monocots) but belong to separate orders. A human and a housefly are both animals but are distantly related; they are not even in the same phylum. Humans belong to the superfamily Hominoidea, not Ceboidea. Yet they are closer to Ceboidea than to members of the suborder Prosimii, as you can see from Figure 16.11.

A Five-Kingdom, Phylogenetic Scheme

Today, classification schemes are constructed to reflect **phylogeny**—the evolutionary relationships among species, starting with the most ancestral forms and including all the branches that lead to all of their descendants. In this book, we use a modified version of a five-kingdom, phylogenetic scheme, as devised by Robert Whittaker. In the broadest terms, the kingdoms may be defined in the manner shown in the paragraphs that follow.

Kingdom Monera. All of these single-celled prokaryotes are bacteria. They show little internal complexity but great biochemical diversity. They include producers (photoautotrophs and chemoautotrophs) and decomposers (heterotrophs).

Category (Taxon)	CORN	VANILLA ORCHID	HOUSEFLY	HUMAN
Kingdom	Plantae	Plantae	Animalia	Animalia
Phylum	Anthophyta (flowering plants)	Anthophyta	Arthropoda	Chordata
Class	Monocotyledonae (monocots)	Monocotyledonae	Insecta	Mammalia
Order	Commelinales	Orchidales	Diptera	Primates
Family	Poaceae	Orchidaceae	Muscidae	Hominidae
Genus	*Zea*	*Vanilla*	*Musca*	*Homo*
Species	*Z. mays*	*V. planifolia*	*M. domestica*	*H. sapiens*

Table 16.1 Classification of Four Organisms

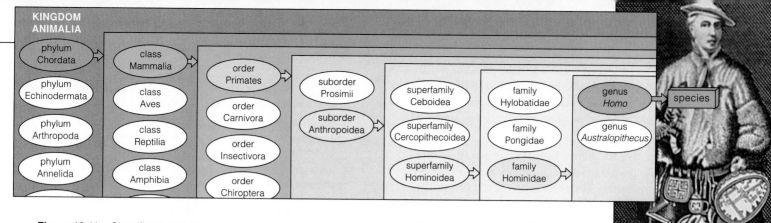

Figure 16.11 Classification of one species, *Homo sapiens* (modern humans) of the animal kingdom. The species itself is represented by Linnaeus, here outfitted with equipment suitable for exploration. Appendix I includes additional taxa.

Homo sapiens (only living species of this genus)

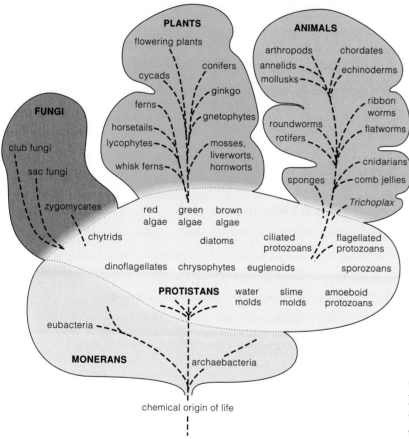

Figure 16.12 A modified version of Robert Whittaker's five-kingdom system of classification.

Kingdom Protista. Protistans. These single-celled or multicelled eukaryotes show great internal complexity, compared to the bacteria. They are photoautotrophs or heterotrophs. Many are major pathogens and parasites.

Kingdom Fungi. These multicelled eukaryotes are all heterotrophs. They feed by extracellular digestion and absorption. Many are major decomposers (and nutrient cyclers). Others are pathogens and parasites.

Kingdom Plantae. Plants. All are eukaryotes. Nearly all of the 295,000+ known plant species are multicelled producers (photoautotrophs).

Kingdom Animalia. Animals. There are more than a million known species. All are multicelled eukaryotes. All are heterotrophs, with predators and parasites predominating.

Existing members of all five kingdoms are the current representatives of long, independent lines of evolution. The evolutionary tree of Figure 16.12 only hints at their relationships to one another. It is important for you to understand that *species* are the only real entities in such family trees. The groupings of "twigs, branches, and limbs" are not real. They are *categories of relationship* that researchers have perceived and then superimposed on species.

As you read through the next unit, remember the underlying genetic continuity that threads through the history of life. The boundaries of kingdoms and other taxa cut across lines of descent that have continued unbroken, through time. Although the boundaries have not been assigned arbitrarily, they are not absolute. All classification schemes, including the one outlined here, are subject to ongoing modification as more pieces of the macroevolutionary puzzle come together.

Classification schemes organize information about species and simplify its retrieval. The newer, phylogenetic schemes attempt to reflect evolutionary relationships among species.

A widely accepted classification scheme groups organisms into five kingdoms, as determined partly by biochemistry, cell structure, and modes of nutrition. No classification scheme is etched in granite, including this one.

SUMMARY

1. A continuity of relationship exists among all species, past and present. The patterns, trends, and rates of change among groups of species over long spans of time are called macroevolution.

2. The fossil record, Earth history, comparative morphology, and comparative biochemistry yield extensive evidence of evolution. The evidence is based on similarities and differences in body form, functions, behavior, and biochemistry.

3. The completeness of the fossil record varies as a function of the species represented, where they lived, and the stability of a region since fossilization occurred.

4. Comparative morphology has revealed similarities in embryonic development that indicate evolutionary relationship. It also identifies homologous structures, shared as a result of descent from a common ancestor.

5. Comparative biochemistry relies on neutral mutations that accumulated in highly conserved genes of different species. The mutations serve as a molecular clock for dating divergences from a common ancestor. Nucleic acid hybridization and other methods reveal the degrees of similarity among species.

6. In the binomial system, developed by Linnaeus, each kind of organism is assigned a two-part Latin name. The first part, the genus, identifies all the species with shared similarities that are perceived to be derived from common ancestry. The second part is the name of the particular species.

7. Classification systems consist of a series of ever more inclusive taxa, from species and genera, through families, orders, classes, phyla, to kingdoms. Phylogenetic schemes base the groupings on presumed evolutionary relatedness. In this book , we use a five-kingdom phylogenetic scheme.

Review Questions

1. Will the fossil record ever be complete? Why or why not? *251*

2. Explain the difference between:
   a. microevolution and macroevolution *247, 249*
   b. homologous and analogous structures *254–255*
   c. morphological divergence and convergence *254–255*

3. What mutations are the basis of a molecular clock, and why? *256*

4. Name a protein specified by a gene that has been highly conserved in organisms ranging from bacteria to humans. *256*

5. Why do evolutionary biologists apply heat energy to hybrid molecules of DNA from two species? *256*

6. Give reasons why two organisms that are quite different in outward appearances may belong to the same lineages. *254*

7. Give reasons why two organisms that seem identical in outward appearance may not belong to the same lineages. *255*

Self-Quiz *(Answers in Appendix IV)*

1. Morphological divergences may lead to _____ .
   a. analogous structures   c. convergent structures
   b. homologous structures   d. both a and c

2. Morphological convergences may lead to _____ .
   a. analogous structures   c. divergent structures
   b. homologous structures   d. both a and c

3. A classification system that is _____ is based on presumed evolutionary relationship.
   a. epigenetic   c. credited to Linnaeus
   b. phylogenetic   d. both b and c

4. *Pinus banksiana, Pinus strobus,* and *Pinus radiata* are _____ .
   a. three families of pine trees
   b. three different names for the same organism
   c. several species belonging to the same genus
   d. both a and c

5. Increasingly inclusive taxa range from _____ to _____ .
   a. kingdom; species   c. genera; kingdom
   b. kingdom; genera   d. species; kingdom

6. Match these terms suitably.
   _____ phylogeny   a. accumulation of neutral mutations
   _____ fossil   b. stone-hard evidence of life in past
   _____ stratification   c. similar body parts in different
   _____ homology   lineages owing to common descent
   _____ biological clock   d. e.g., shark fins, penguin flippers
   _____ analogy   e. evolutionary relationship among
                               species, ancestors to descendants
                            f. layers of sedimentary rock

Critical Thinking

1. Protein comparisons and nucleic acid hybridization studies help us estimate evolutionary relationship and approximate times for divergences from ancestral stocks. *DNA sequence comparisons* yield even more accurate estimates. Reflect on the genetic code (Figure 12.7), then suggest why this may be a more accurate measure of mutations, mutation rates, and biochemical relatedness.

2. You probably have heard about "missing links"—undiscovered transitional forms between major groups of organisms. What are some possible reasons for apparent gaps in the fossil record?

Selected Key Terms

analogy *255*
Animalia *259*
binomial system *258*
classification scheme *258*
comparative morphology *252*
DNA-DNA hybridization *256*
fossil *250*
fossilization *251*
Fungi *259*
genus *258*
higher taxa *258*

homology *254*
macroevolution *249*
molecular clock *256*
Monera *258*
morphological convergence *255*
morphological divergence *254*
neutral mutation *256*
phylogeny *258*
Plantae *259*
Protista *259*
stratification *251*

Reading

Brooks, D. R., and D. A. McLennan. 1991. *Phylogeny, Ecology, and Behavior.* Chicago: University of Chicago Press.

FACING PAGE: *Patterns of diversity in nature, here represented by different species of plants and fungi.*

EVOLUTION AND DIVERSITY

17 THE ORIGIN AND EVOLUTION OF LIFE

In the Beginning . . .

Some clear evening, watch the moon as it rises from the horizon and think of the 380,000 kilometers between it and you. *Five billion trillion times* the distance between the moon and you are galaxies—systems of stars—at the boundary of the known universe. Wavelengths traveling through space move faster than anything else, millions of meters per second. Yet the long wavelengths that originated from faraway galaxies many billions of years ago are only now reaching the Earth.

By every known measure, all the near and distant galaxies suspended in the vast space of the universe are moving away from one another—which means the

universe is expanding. And the prevailing view of how the colossal expansion came about accounts for every bit of matter in the universe, in every living thing.

Think about how you rewind a videotape on a VCR, then imagine yourself "rewinding" the universe. As you do, the galaxies start moving back together. After 10 to 20 billion years of rewinding, all galaxies, all matter, and all of space have become compressed into a hot, dense volume about the size of the sun. You have arrived at time zero.

That incredibly hot, dense state lasted only for an instant. What happened next is known as the "big

Figure 17.1 Part of the great Eagle nebula, a hotbed of star formation 7,000 light-years away from Earth, in the constellation Serpens. (The Latin *nebula* means mist.) Above the pillars are a few huge, young stars (not visible in this image). For the past few million years, intense ultraviolet radiation from the stars has been eroding the less dense surface of the pillars. (By analogy, imagine a strong, prevailing wind blowing away sand in a desert and exposing large rocks.) Globules of denser gases and dust that have resisted erosion are visible at the surface. Each one is wider than our own solar system—more than 10 *billion* miles across! New stars are hatching from the protruding globules; some are shining brightly on the tips of gaseous streamers. The Hubble space telescope provided this remarkable image.

bang"—a truly inadequate name for the stupendous, nearly instantaneous distribution of all matter and energy everywhere, through all of the known universe. About a minute later, temperatures dropped to a billion degrees. Fusion reactions produced most of the light elements, including helium, which are still the most abundant elements throughout the universe. Radio telescopes have detected relics of the big bang—the cooled and diluted background radiation, left over from the beginning of time.

Over the next billion years, stars started forming as gaseous material contracted in response to the force of gravity. When stars became massive enough, nuclear reactions were ignited in their central region, and they gave off tremendous light and heat. As massive stars continued to contract, many became dense enough to promote the formation of heavier elements.

All stars have a life history, from birth to a sometimes spectacularly explosive death. In what might be called the original stardust memories, heavier elements released during the explosions became swept up during the gravitational contraction of new stars—and they became raw materials for the formation of even heavier elements. Even as you are reading this page, the Hubble space telescope is providing astounding glimpses of star-forming activity in the dust clouds of Orion, Serpens, and other constellations (Figure 17.1).

Now imagine a time long ago, when explosions of dying stars ripped through our own galaxy and left behind a dense cloud of dust and gas that extended trillions of kilometers in space. As the cloud cooled, countless bits of matter gravitated toward one another. By 4.6 billion years ago, the cloud had flattened out into a slowly rotating disk. At the dense, hot center of that disk, the shining star of our solar system—the sun—was born.

The remainder of this chapter is a sweeping slice through time, one that cuts back to the formation of the Earth and the chemical origins of life. It is the starting point for the next four chapters, which will take us along major lines of descent that led to the present range of species diversity. It's true that the story is far from complete. Even so, all the available evidence, from many avenues of research, points to a principle that can be used to organize separate bits of information about the past:

Life is a magnificent continuation of the physical and chemical evolution of the universe, of galaxies and stars, and of the planet Earth.

KEY CONCEPTS

1. A great body of evidence suggests that life originated more than 3.8 billion years ago. Its origin and subsequent evolution have been linked to the physical and chemical evolution of the universe, the stars, and the planet Earth.

2. All of the inorganic and organic compounds necessary for self-replication, membrane assembly, and metabolism —that is, for the structure and functioning of living cells— could have formed spontaneously under conditions that existed on the early Earth.

3. The history of life, from its chemical beginnings to the present, spans five intervals of geologic time. It extends through two great eons—the Archean and the Proterozoic— and the Paleozoic, Mesozoic, and Cenozoic eras.

4. Not long after life originated, divergences led to three great prokaryotic lineages—the archaebacteria, eubacteria, and the ancestors of eukaryotes.

5. Prokaryotes dominated the Archean and Proterozoic. Eukaryotes originated late in the Proterozoic, and they became spectacularly diverse. A theory of endosymbiosis helps explain the profusion of specialized organelles that we see in eukaryotic cells.

6. All five kingdoms of organisms are characterized by the persistences, extinctions, and radiations of many different lineages over time.

7. Throughout the history of life, asteroid impacts, drifting and colliding continents, and other environmental insults have had profound impact on the direction of evolution.

CONDITIONS ON THE EARLY EARTH

Origin of the Earth

Cloudlike remnants of stars, including those shown in Figure 17.1, are mostly hydrogen gas. They also contain water, iron, silicates, hydrogen cyanide, ammonia, methane, formaldehyde, and other simple inorganic and organic substances. Most likely, the contracting cloud from which our solar system evolved was similar in composition. Between 4.6 and 4.5 billion years ago, the cloud's outer regions cooled. Mineral grains and ice orbiting the sun started to clump together, through electrostatic attraction and the pull of gravity (Figure 17.2). In time, the larger, faster clumps started colliding and shattering. Some became more massive by sweeping up asteroids, meteorites, and other rocky remnants of collisions—and gradually they evolved into planets.

As the Earth was forming, much of its inner rocky material melted. Internal compression, huge asteroid impacts, and radioactive decay of minerals could have generated the required heat. As rocks melted, nickel, iron, and other heavy materials moved to the interior and lighter ones floated to the surface. This process of differentiation produced a crust of basalt, granite, and other low-density rocks, a rocky region of intermediate density (called the mantle), and a high-density, partially molten core of nickel and iron.

Four billion years ago, the Earth was a thin-crusted inferno (Figure 17.3a). Yet within 200 million years, life had originated on its surface! We have no record of the event; as far as we know, movements in the mantle and

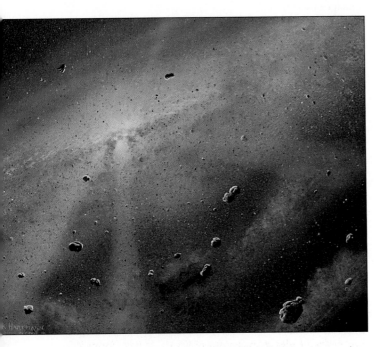

Figure 17.2 Representation of the cloud of dust, gases, and clumps of rock and ice around the early sun.

crust, volcanic activity, and erosion obliterated all traces of it. Still, we can put together a plausible explanation of how life originated by considering three questions:

1. What were the prevailing physical and chemical conditions on Earth at the time of life's origin?

2. Based on physical, chemical, and evolutionary principles, could large organic molecules have formed spontaneously and then evolved into molecular systems displaying the characteristics of life?

3. Can we devise experiments to test whether living systems could have emerged by chemical evolution?

The First Atmosphere

When the first patches of crust were forming, hot gases blanketed the Earth. This first atmosphere probably was a mix of gaseous hydrogen (H_2), nitrogen (N_2), and carbon monoxide (CO), as well as carbon dioxide (CO_2). Were gaseous oxygen (O_2) and water present? Probably not. Rocks release oxygen during volcanic eruptions, but not much. Besides, oxygen would have reacted at once with other elements. And any water that formed would have evaporated because of the intense atmospheric heat.

When the crust cooled and solidified, however, water condensed into clouds and the rains began. For millions of years, runoff from rains stripped mineral salts and other compounds from the parched rocks. Salt-laden waters collected in depressions in the crust and formed the early seas.

If the first atmosphere had not been free of oxygen, the organic compounds that became organized into the first cells could not have formed on their own. Oxygen would have attacked them (compare page 64). If liquid water had not accumulated, cell membranes could not have formed. Cells, recall, are the basic units of life. *Each has a capacity to survive and reproduce on its own.*

Synthesis of Organic Compounds

When we reduce cells to their lowest common denominator, we are left with proteins, complex carbohydrates, lipids, and nucleic acids. Today, cells assemble these molecules from small organic compounds—the simple sugars, fatty acids, amino acids, and nucleotides. Energy from the environment drives the synthesis reactions. Were small organic compounds also present on the early Earth? Were there sources of energy that drove their spontaneous assembly into the large molecules of life?

Mars, meteorites, the Earth's moon, and the Earth formed at the same time, from the same cosmic cloud. Rocks collected from Mars, meteorites, and the moon contain precursors of biological molecules, so the same precursors must have been present on the Earth. If this

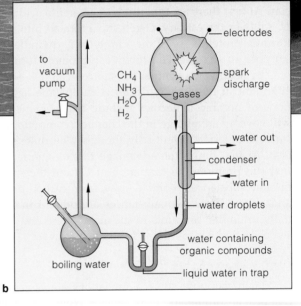

a

Figure 17.3 **(a)** Representation of the Earth during its formation, when the moon's orbit was much closer than it is today.

If the Earth had condensed into a smaller planet, its gravitational mass would not have been great enough to hold onto an atmosphere. If it had settled into an orbit closer to the sun, water would have evaporated from its hot surface. If Earth's orbit had been more distant from the sun, its surface would have been too cold, and water would have been locked up as ice. Without liquid water, life never would have originated on Earth.

(b) Stanley Miller's experimental apparatus, used to study the synthesis of organic compounds under conditions that presumably existed on the early Earth. The condenser cools circulating steam so that water droplets form.

b

were the case, *then energy from sunlight, lightning, or even heat escaping from the crust could have been enough to drive their combination into organic molecules.*

In the first of many tests of that prediction, Stanley Miller mixed hydrogen, methane, ammonia, and water in a reaction chamber (Figure 17.3*b*). He recirculated the mixture and then bombarded it with a spark discharge to simulate lightning. In less than a week, amino acids and other small organic compounds had formed.

In other experiments that simulated conditions on the early Earth, glucose, ribose, deoxyribose, and other sugars formed from formaldehyde. Adenine formed from hydrogen cyanide. Adenine and ribose are present in ATP, NAD, and other nucleotides vital to cells.

If *complex* organic compounds formed in the seas, they wouldn't have lasted long. In water, the spontaneous direction of the required reactions would have been toward hydrolysis, not condensation (Figure 2.16).

Maybe more lasting bonds formed at the margins of seas. By one scenario, clay in the rhythmically drained muck of tidal flats and estuaries served as templates

(structural patterns) for the spontaneous assembly of proteins and other complex organic compounds. Clay consists of thin, stacked layers of aluminosilicates with metal ions at its surface. Clay and metal ions attract amino acids. When clay is first warmed by sunlight, then alternately dried out and moistened, it actually promotes condensation reactions that yield complex organic compounds. We know this from experiments.

Suppose proteins that formed on clay templates had the shape and chemical behavior necessary to function as weak enzymes in hastening bonds between amino acids. If certain templates promoted such bonds, they would have selective advantage over other templates in the chemical competition for available amino acids.

Perhaps selection was at work even before the origin of cells, favoring the chemical evolution of enzymes.

Many experiments provide indirect evidence that the complex organic molecules characteristic of life could have formed under conditions that existed on the early Earth.

Origin of Agents of Metabolism

A defining characteristic of life is metabolism. The word refers to all the reactions by which cells harness energy and use it to drive their activities, such as biosynthesis. During the first 600 million years of Earth's history, enzymes, ATP, and other organic compounds may have assembled spontaneously in the same locations. If so, their close association would have promoted chemical interactions—and the beginning of metabolic pathways.

Imagine an ancient estuary, rich in clay deposits. Countless aggregations of organic molecules stick to the clay. At first there are quantities of an amino acid; call it *D*. All over the estuary, *D* gets incorporated into newly forming proteins—until the supply of *D* dwindles. But suppose an enzymelike protein can promote formation of *D* by acting on a simpler substance—call it *C*—that is abundant in the estuary. If, by chance, some aggregations of organic molecules include that enzyme, they will have an advantage in the chemical competition for starting materials. Eventually, though, *C* becomes scarce also. At that point, the advantage tilts to aggregations that can promote the formation of *C* from the simpler organic substances *B* and *A*—say, from carbon dioxide and water. These two substances are present in essentially unlimited amounts in the atmosphere and in the sea. Selection has favored a synthetic pathway:

$$A + B \longrightarrow C \longrightarrow D$$

Finally, suppose some aggregations are better than others at absorbing and using energy. What sort of molecules could bestow this advantage? Think of an energy-trapping pathway that now dominates the world of life: photosynthesis. It starts at chlorophylls, which are light-trapping pigment molecules. The portion of chlorophyll that absorbs sunlight energy and gives up electrons is a porphyrin ring structure. Porphyrins also are found in cytochromes—which happen to be part of the electron transport systems in all photosynthetic and aerobically respiring cells. Porphyrins can spontaneously assemble from formaldehyde—one of the molecular legacies of cosmic clouds (Figure 17.4). Was porphyrin an electron transporter of early metabolic pathways? Perhaps.

Origin of Self-Replicating Systems

Another defining characteristic of life is a capacity for reproduction—which starts with the protein-building instructions of DNA. The DNA molecule is fairly stable, and it is easily replicated before each cell division. As you know from earlier chapters, batteries of enzymes and RNA molecules carry out its encoded instructions.

Most existing enzymes are assisted by small organic molecules or metal ions called coenzymes. Intriguingly,

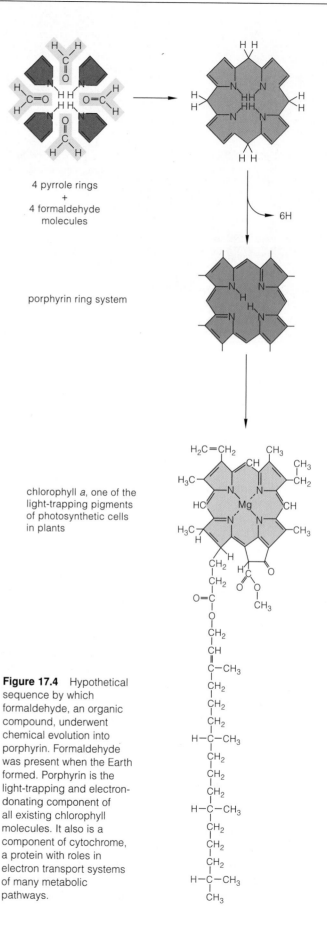

4 pyrrole rings
+
4 formaldehyde
molecules

6H

porphyrin ring system

chlorophyll *a*, one of the light-trapping pigments of photosynthetic cells in plants

Figure 17.4 Hypothetical sequence by which formaldehyde, an organic compound, underwent chemical evolution into porphyrin. Formaldehyde was present when the Earth formed. Porphyrin is the light-trapping and electron-donating component of all existing chlorophyll molecules. It also is a component of cytochrome, a protein with roles in electron transport systems of many metabolic pathways.

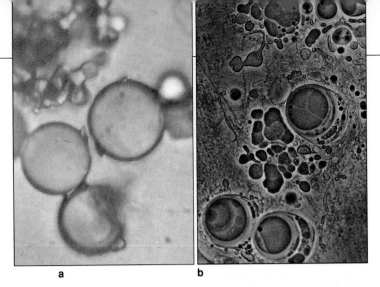

a   b

Figure 17.5 Microscopic spheres of (**a**) proteins and (**b**) lipids that self-assembled under abiotic conditions.

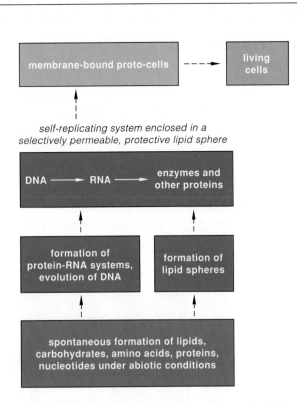

Figure 17.6 One possible sequence of events that led to the first self-replicating systems, then to the first living cells.

some of the coenzymes have the *same* structure as RNA nucleotides. Another clue: Heat nucleotide precursors and short chains of phosphate group together, and they assemble into RNA strands. On the early Earth, energy from sunlight alone could have been enough to drive the spontaneous formation of RNA molecules.

Simple self-replicating systems of RNA, enzymes, and coenzymes have been created in the laboratory. Experiments with these systems imply that RNA could have formed on clay templates. Did RNA later replace clay as information-storing templates for protein synthesis? Perhaps it did initially, although existing RNA is too chemically fragile for this role. Yet it may have set up an **RNA world** that preceded DNA's dominance as the main informational molecule. Eventually DNA replaced RNA for this function, probably because it can form long nucleotide chains in more stable fashion.

We still don't know how DNA entered the picture. Until we identify the likely chemical ancestors of RNA and DNA, the story of life's origin will be incomplete. Filling in the details will require imaginative sleuthing. For example, researchers ran a program on a rapid, advanced computer. The program included information on simple compounds, of the sort thought to have been present on the early Earth. The researchers asked the computer to subject the compounds selected to random chemical competition and natural selection. They ran the program repeatedly. And the outcome was always the same: Simple precursors inevitably evolved into interacting systems of large, complex molecules.

Origin of the First Plasma Membranes

Experiments are more revealing of the origin of the plasma membrane, which surrounds every living cell. The membrane is a lipid bilayer, studded with proteins that carry out diverse functions. Its primary role is to control which substances move into and out of the cell. Without this control, cells cannot exist. Before cells,

then, there must have been **proto-cells**. These may have been little more than membrane sacs, protecting information-storing templates and various metabolic agents from the environment. We know that simple membrane sacs can form spontaneously.

In one experiment, Sidney Fox heated amino acids until they formed protein chains, which he put in hot water. After cooling, the chains assembled into small, stable spheres (Figure 17.5*a*). Like cell membranes, the spheres were selectively permeable to different substances. They picked up free lipids, and a lipid-protein film formed at their surface. In other experiments, fatty acids and glycerol combined to form long-tail lipids under conditions that simulated evaporating tidepools. The lipids self-assembled into small, water-filled sacs. Many were like cell membranes (Figure 17.5*b*).

In short, there are major gaps in the story of life's origins. But there also is strong experimental evidence that chemical evolution probably led to the molecules and structures that are characteristic of life. Figure 17.6 summarizes the milestones in that chemical evolution, which preceded the first cells.

Although the story is not yet complete, many laboratory experiments and computer simulations indirectly show that chemical and molecular evolution gave rise to proto-cells.

17.3 LIFE ON A CHANGING GEOLOGIC STAGE

Drifting Continents and Changing Seas

By about 3.8 billion years ago, the earth's crust was becoming more stable—but things have never settled down completely. According to **plate tectonic theory**, plumes of molten rock rising from the mantle spread out beneath the crust or broke through it. The fracturing created the oceanic and continental plates—enormous slabs of crust that move apart and crunch together at their margins (Figure 17.7). The plates move no more than a few centimeters a year, on the average. Yet that displacement causes pressure changes along the plate margins. Earthquakes, volcanic eruptions, and lava flows are short-term consequences. Crustal uplifting and subsidence, mountain building, seafloor spreading, and continental drift are long-term consequences.

Take a look at Figure 17.8. In the remote past, plate movements put huge land masses on collision courses. In time they converged and formed supercontinents—which later split open at deep rifts that became new ocean basins. One early continent, Gondwana, drifted southward from the tropics, across the south polar region, then northward until it crunched together with other land masses to form Pangea. That single world continent extended from pole to pole, and an immense ocean spanned the rest of the globe!

Figure 17.7 Some forces of geologic change.

(**a**) The Earth's crust is fractured into rigid plates that gradually split apart, drift about, and collide.
(**b**) Great plumes of molten material drive the crustal movements. They well up from the Earth's interior and spread out laterally beneath the crust.

In the past, such plumes ruptured the crust at great ridges on the ocean floor. At these ridges, molten material seeps out, cools, and forces the seafloor on either side of them away from the rupture. This is *seafloor spreading*, and it displaces plates away from oceanic ridges. Often the displacement forces other margins of the plate under an adjacent plate, which is slowly uplifted as a result. Major mountain ranges that parallel the coasts of continents formed this way.

Plumes also cause deep rifting and splitting in the interior of continents. Such rifting is now proceeding in Missouri, at Lake Baikal in Russia, and in eastern Africa. In ancient times, *superplumes* violently ruptured the Earth's crust. Volcanoes still form at the ruptured "hot spots." The Hawaiian Islands formed

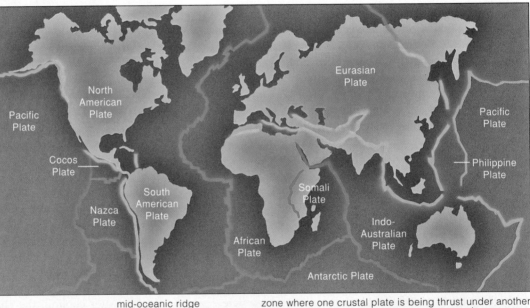

North American Plate · Pacific Plate · Cocos Plate · Nazca Plate · South American Plate · Eurasian Plate · Pacific Plate · Philippine Plate · Somali Plate · African Plate · Indo-Australian Plate · Antarctic Plate

a

mid-oceanic ridge · zone where one crustal plate is being thrust under another

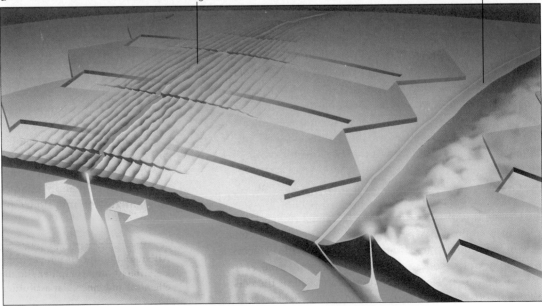

b

Figure 17.8 The geologic time scale. In this chart, the time spans of the different eras are not to scale. If they were, the Archean and Proterozoic portions would run off the page.

Think of the time spans as minutes on a clock. Life originated at midnight. The Paleozoic began at 10:04 A.M., the Mesozoic at 11:09 A.M., and the Cenozoic at 11:47 A.M. The recent epoch of the Cenozoic started during the last 0.1 second before noon.

The two maps below are an example of how dramatically the distribution of land masses has changed over time.

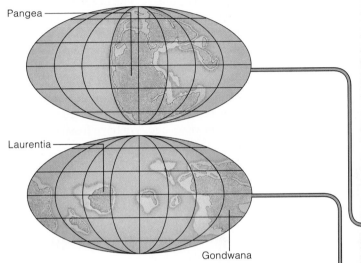

	Period	Epoch	Millions of Years Ago (mya)
CENOZOIC ERA	Quaternary	Recent	0.01–
		Pleistocene	1.65
	Tertiary	Pliocene	5
		Miocene	25
		Oligocene	38
		Eocene	54
		Paleocene	65
MESOZOIC ERA	Cretaceous	Late	100
		Early	138
	Jurassic		205
	Triassic		240
PALEOZOIC ERA	Permian		290
	Carboniferous		360
	Devonian		410
	Silurian		435
	Ordovician		505
	Cambrian		550
PROTEROZOIC			
			2,500
ARCHEAN			
			3,900

Besides this, the slow erosive forces of water and wind have continually resculpted the Earth's surface through time. Asteroids and meteorites have continued to bombard the crust, and the impacts have had long-term effects on global temperature and climate.

These changes in the land, oceans, and atmosphere profoundly influenced the evolution of life. Think of life flourishing in the warm, shallow waters along shorelines of early continents. When the continents collided, the shorelines vanished. These and other changes had dire consequences for many lineages. Yet, even as old habitats were lost, new ones opened for the survivors—and evolution took off in new directions.

The Geologic Time Scale

When early naturalists started discovering fossils in layers of sedimentary rock, they had no way of knowing the ages of either the fossils or their rocky tombs. However, they did notice four abrupt transitions in the kinds of fossils in different layers. They interpreted the transitions as boundaries between four great intervals in time. The oldest fossils were from the first geologic interval, the **Proterozoic**. They were followed by fossils from the **Paleozoic** era, the **Mesozoic** era, and finally the "modern" era, the **Cenozoic**.

We still use the boundaries between these intervals in a chronological chart of Earth history. Figure 17.8 shows this **geologic time scale**. We now know that the boundaries mark the times of **mass extinctions**. Thanks to radioisotope dating work (Section 2.2), the time scale was assigned actual dates. But the "Proterozoic" turned out to be so vast, it has been subdivided into eons, each of which lasted for hundreds of millions of years. One of these, the **Archean**, was the time of life's origin.

Over the past 3.8 billion years, changes in the Earth's crust, the atmosphere, and the oceans profoundly affected the evolution of life. Certain boundaries of the geologic time scale mark abrupt shifts in the direction of evolution.

ORIGIN OF PROKARYOTIC AND EUKARYOTIC CELLS

The first cells emerged as molecular extensions of the evolving universe, of our solar system and Earth. They may have originated in tidal flats or muddy sediments of an ancient seafloor, warmed by heated water from volcanic vents. (Such hydrothermal vents are described in Section 38.12.) Like existing bacteria, the first were **prokaryotic cells**, with no nucleus. Possibly they were little more than self-replicating, membranous sacs of DNA and other complex organic molecules. Given the absence of free oxygen, they must have secured energy by anaerobic routes—fermentation, most likely. Energy and carbon were plentiful; organic compounds had accumulated in the seas by natural geologic processes. So "food" was available, predators were absent, and biological molecules were free from oxygen attacks.

Early in the Archean, the first prokaryotic lineage diverged in three evolutionary directions. As shown in Figure 17.9, two branchings led to the **archaebacteria** and **eubacteria**. The third gave rise to the prokaryotic ancestors of **eukaryotic cells**.

Between 3.5 and 3.2 billion years ago, the cyclic pathway of photosynthesis evolved in certain anaerobic eubacteria. (Section 5.3 describes the cyclic pathway.) In those bacterial cells, components of electron transport systems had become modified in ways that allowed them to tap into sunlight—an unlimited source of energy. For nearly 2 billion years, the photosynthetic descendants of those bacterial cells dominated the world of life. Their populations formed mats, in which sediments collected. The mats

accumulated, one atop the other, becoming the large structures called **stromatolites** (Figure 17.10). Calcium deposits hardened and preserved these mats.

By the dawn of the Proterozoic, 2.5 billion years ago, the noncyclic pathway of photosynthesis had evolved in some eubacterial species. Oxygen, one of its by-products, started to accumulate. It had two irreversible effects. First, *an oxygen-rich atmosphere stopped the further chemical origin of living cells*. Except in a few anaerobic habitats, organic compounds could no longer form spontaneously and resist degradation. Second, *aerobic respiration became the dominant energy-releasing pathway*. In many prokaryotic lineages, selection favored metabolic equipment that "neutralized" oxygen by using it as an electron acceptor. This innovation contributed to the rise of multicelled eukaryotes and their invasion of far-flung environments.

Eukaryotic cells evolved in the Proterozoic, possibly before 1.2 billion years ago. We have fossils, 900 million

Figure 17.9 An evolutionary tree of life that reflects mainstream thinking about the connections among major lineages. The diagram incorporates ideas about the origins of some eukaryotic organelles.

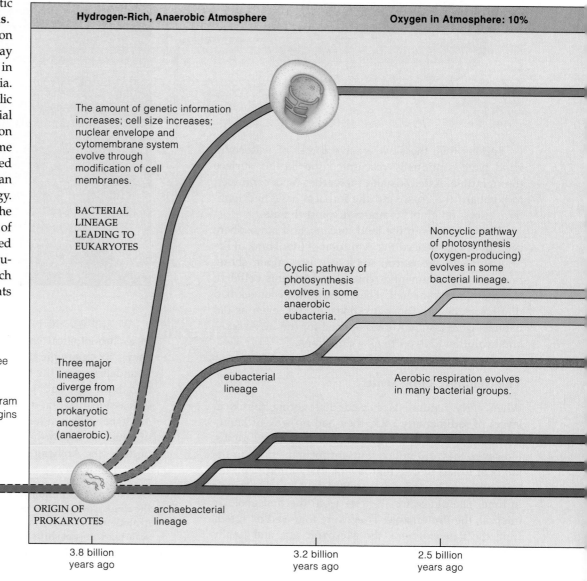

Hydrogen-Rich, Anaerobic Atmosphere

Oxygen in Atmosphere: 10%

The amount of genetic information increases; cell size increases; nuclear envelope and cytomembrane system evolve through modification of cell membranes.

BACTERIAL LINEAGE LEADING TO EUKARYOTES

Noncyclic pathway of photosynthesis (oxygen-producing) evolves in some bacterial lineage.

Cyclic pathway of photosynthesis evolves in some anaerobic eubacteria.

Three major lineages diverge from a common prokaryotic ancestor (anaerobic).

eubacterial lineage

Aerobic respiration evolves in many bacterial groups.

Chemical and molecular evolution into self-replicating systems and into membranes of proto-cells, some 4 billion years ago

ORIGIN OF PROKARYOTES

archaebacterial lineage

3.8 billion years ago

3.2 billion years ago

2.5 billion years ago

Figure 17.10 Stromatolites exposed at low tide in Shark Bay, Western Australia. These started forming 2,000 years ago. They are identical to stromatolites more than 3 billion years old.

years old, of well-developed red algae, green algae, fungi, and plant spores. Organelles, remember, are the hallmark of eukaryotic cells. *Where did they come from?* The next section describes a few plausible hypotheses.

About 800 million years ago, stromatolites began to decline dramatically. They had become a concentrated source of food for newly evolved, tiny bacteria-eating animals. About the same time, tiny soft-bodied animals were making tracks and digging burrows in seafloor sediments. They lived near the shores of Laurentia, an early supercontinent. About 570 million years ago, in "Precambrian" times, their descendants started the first adaptive radiation of animals.

The first cells evolved by about 3.8 billion years ago, during the Archean. Early on, divergences led to archaebacteria, eubacteria, and the ancestors of eukaryotic cells.

Oxygen released by certain photosynthetic bacteria changed the atmosphere. In time, the oxygen-rich atmosphere stopped the further spontaneous chemical evolution of life and was a key selection pressure in the evolution of eukaryotic cells.

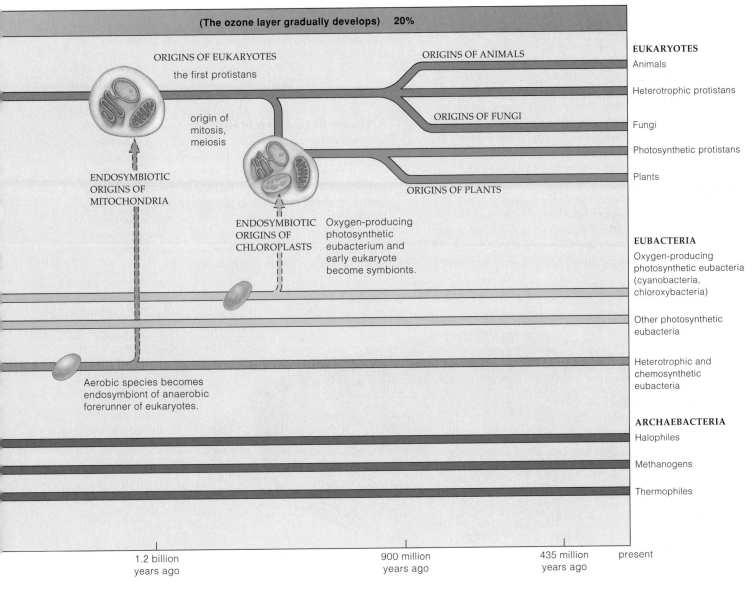

17.5 *Focus on Science*

WHERE DID ORGANELLES COME FROM?

Thanks to Andrew Knoll, William Schopf, Jr., and other globe-hopping fossil hunters, we have tantalizing evidence of early life. One fossil treasure is a strand of bacterial cells that lived 3.5 billion years ago, not long after the time that life originated. Others are from the Proterozoic. They were eukaryotic cells with a few membrane-bound organelles in the cytoplasm (Figure 17.11). Their living descendants have a profusion of them (Figure 17.12).

Where did the organelles come from? Speculations abound. Some organelles probably evolved through gene mutations and natural selection. For others, researchers make a good case for evolution by way of endosymbiosis.

Origin of the Nucleus and ER Prokaryotic cells don't have an abundance of organelles, but some species have interesting infoldings of the plasma membrane (Figure 17.13). Enzymes and other metabolic agents are embedded in that membrane. In the forerunners of eukaryotic cells, similar infoldings may have extended into the cell and served as channels to the surface. They may have evolved into ER channels and into an envelope around the DNA.

What would be the advantage of such membranous enclosures? Maybe they protected the genes and protein products from "foreigners." Remember, bacterial species can transfer plasmid DNA among themselves. So can yeasts, which are simple eukaryotic cells. Initially, a nuclear envelope may have been favored because it got the cell's genes, replication enzymes, and transcription enzymes out of the cytoplasm. It would have allowed vital genetic messages to be produced free of metabolic competition from an unmanageable hodgepodge of foreign genes. Similarly, ER channels might have kept vital proteins and other organic compounds away from metabolically hungry "guests"—foreign cells that one way or another became permanent residents inside the host cell, as described next.

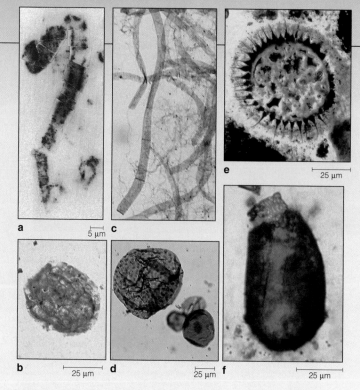

Figure 17.11 From Australia, (**a**) a strand of walled prokaryotic cells 3.5 billion years old and (**b**) one of the oldest known eukaryotes—a protistan 1.4 billion years old. From Siberia, (**c**) a multicelled alga 900 million to 1 billion years old and (**d**) eukaryotic microplankton 850 million years old. (**e**) From China, a splendid eukaryotic cell that lived 560 million years ago. (**f**) From Spitsbergen, Norway, a protistan that was alive 50 million years before the dawn of the Cambrian.

A Theory of Endosymbiosis Accidental partnerships between different prokaryotic species must have formed countless times on the evolutionary road to eukaryotic cells. Some partnerships possibly resulted in the origin of mitochondria, chloroplasts, and other organelles. This is a **theory of endosymbiosis**, developed in greatest detail by Lynn Margulis. *Endo-* means within, and *symbiosis* means living together. In endosymbiosis, one cell (a guest species) lives permanently inside another kind of cell (the host species), and the interaction benefits both.

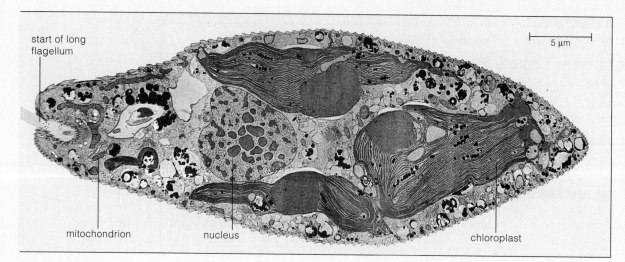

start of long flagellum

mitochondrion   nucleus   chloroplast

Figure 17.12
A fine example of the profusion of diverse organelles that are hallmarks of eukaryotic cells—*Euglena*, a single-celled protistan, sliced lengthwise for this micrograph. It also has a long flagellum, which could not fit in the image area at this magnification.

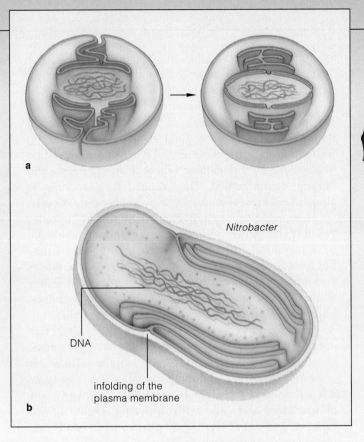

Nitrobacter

DNA

infolding of the
plasma membrane

b

a

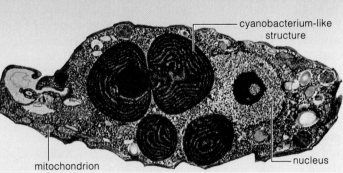

cyanobacterium-like
structure

nucleus

mitochondrion

Figure 17.14 *Cyanophora paradoxa*, a protistan. Its mitochondria resemble bacteria. Its photosynthetic structures look like spherical cyanobacteria (which are photosynthetic) without the cell wall.

Figure 17.13 (**a**) One idea concerning the origin of endoplasmic reticulum and the nuclear envelope. In the prokaryotic ancestors of eukaryotic cells, infoldings of the plasma membrane may have given rise to these structures. (**b**) Such infoldings are present in the cytoplasm of many kinds of existing bacteria, including *Nitrobacter*, sketched here in cutaway view.

According to this theory of endosymbiosis, eukaryotic cells arose after the noncyclic pathway of photosynthesis emerged and after oxygen accumulated to significant levels in the atmosphere. In some groups of bacteria, electron transport systems had already become expanded to include "extra" cytochromes. As it happened, those cytochromes were able to donate electrons to oxygen. In other words, those bacteria could extract energy from organic compounds by way of aerobic respiration. By 1.2 billion years ago, and possibly much earlier than this, the forerunners of eukaryotes were engulfing aerobic bacteria. Maybe they were like existing amoebalike cells that weakly tolerate free oxygen. If so, they trapped food by sending out cytoplasmic extensions of the cell body. Endocytic vesicles formed around the food and delivered it to the cytoplasm for digestion.

Some aerobic bacteria resisted digestion. They actually thrived in the new, protected, nutrient-rich environment. In time they were releasing extra ATP—which the hosts came to depend on for growth, greater activity, and assembly of more structures, such as hard body parts. The guests were no longer duplicating metabolic functions that the hosts performed for them. The anaerobic and aerobic cells were now incapable of independent existence. The guests had become mitochondria, supreme suppliers of ATP.

Evidence of Endosymbiosis Strong evidence favors Margulis's theory. There are plenty of examples that nature continues to tinker with endosymbionts, including the cell shown in Figure 17.14. Its mitochondria are like bacteria in size and structure. The inner mitochondrial membrane is like a bacterial plasma membrane. Each mitochondrion replicates its own DNA and divides independently of the host cell's division. A few genetic code words in its DNA and mRNA have unique meanings. They are translated into a few proteins required for specialized mitochondrial tasks. Thus, compared to the genetic code of cells, the "mitochondrial code" has distinct differences.

Also consider chloroplasts, which may be the stripped-down descendants of oxygen-evolving, photosynthetic bacteria. Perhaps predatory aerobic bacteria engulfed such photosynthetic cells, which escaped digestion, absorbed nutrients from their host's cytoplasm, and continued to function. By providing their respiring host with oxygen, their endosymbiotic existence would have been favored.

In metabolism and overall nucleic acid sequence, chloroplasts resemble some eubacteria. Their DNA is self-replicating, and they divide independently of the cell's division. Chloroplasts vary in shape and in their array of light-absorbing pigments, just as different photosynthetic eubacteria do. They may have originated a number of times, in a number of different lineages. Adding to the intrigue are species of ciliated protistans, even marine slugs, that "enslave" chloroplasts! The slugs eat algae but retain the algal chloroplasts in their gut. The chloroplasts draw nutrients from the host tissues, and they continue to photosynthesize and release oxygen for weeks.

However they arose, new kinds of cells did appear on the evolutionary stage. They had become equipped with a nucleus, ER membranes, and mitochondria or chloroplasts (or both). They were the eukaryotes, the first **protistans**. With their efficient metabolic strategies, early protistans underwent rapid divergences and adaptive radiations. In no time at all, evolutionarily speaking, some of their descendants gave rise to the great kingdoms of plants, fungi, and animals, as sketched earlier in Figure 17.9.

LIFE IN THE PALEOZOIC ERA

We subdivide the Paleozoic era into six periods. These are named the Cambrian, Ordovician, Silurian, Devonian, Carboniferous, and Permian. Before the dawn of this era, gradual rifting had split a supercontinent, Laurentia (Figure 17.8). By Cambrian times, the fragmented land masses were straddling the equator, and warm, shallow seas lapped their margins. Global conditions restricted pronounced seasonal changes in winds, ocean currents, and upward churning of nutrients from deep waters.

Thus, along shorelines at or near the equator, nutrient supplies were stable but limited.

Most of the major animal phyla had evolved in the Precambrian seas. Animals of the early Cambrian had flattened bodies, with a good surface-to-volume ratio for taking up nutrients (Figure 17.15a,b). Most lived on or just beneath the seafloor, where dead organisms and organic debris settled. They ranged from sponges to simple vertebrates—and they were exuberantly diverse.

How could so much diversity arise so early in time? Possibly, the genes governing embryonic development were less intertwined than they are today, so there may not have been as much selection against mutated alleles and novel traits among the animals. Also, warm waters along the new coastlines offered vacant adaptive zones, with opportunity for new ways to secure food.

Entombed in sedimentary beds from the Cambrian are plenty of fossilized organisms that have punctures, missing chunks, and healed-over wounds. These are not artifacts of fossilization; the animals were damaged while they were still alive. In short order, a great variety of predators and prey with armorlike shells, spines, mouths, and novel feeding structures evolved. Things were starting to get lively!

Later in the Cambrian, temperatures in the shallow seas changed drastically. Trilobites (Figure 17.15b), one of the most common animals, nearly vanished. In the Ordovician, the supercontinent called Gondwana had been drifting south, and parts became submerged in shallow seas. Vast marine environments opened up and promoted adaptive radiations. Many diverse reef organisms appeared. Among them were swift, shelled predators called nautiloids. Their modern descendants include the chambered nautiluses, as shown in Figure 38.26.

Later on, Gondwana straddled the South Pole and vast glaciers formed across its surface. When huge volumes of water were locked up as ice, shallow seas throughout the world were drained. This was the first ice age that we know about. It may have triggered the first global mass extinction. At the Ordovician-Silurian boundary, reef life everywhere collapsed.

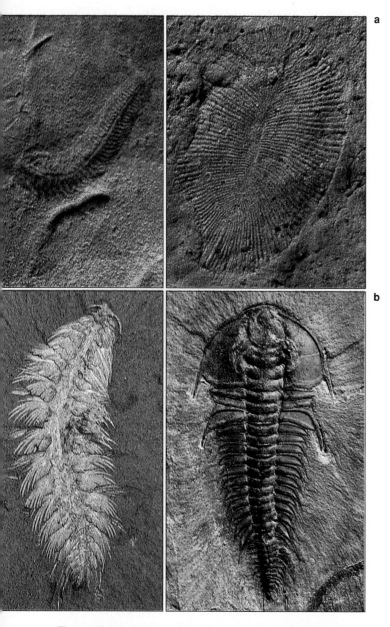

Figure 17.15 Representative Cambrian animals. (**a**) From the Ediacara Hills of Australia, fossils of two animals, about 600 million years old. (**b**) *Left:* From the Burgess Shale of British Columbia, a fossilized marine worm. *Right:* A beautifully preserved fossil of one of the earliest trilobites.

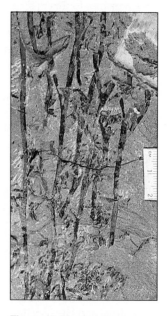

Figure 17.16 Fossils of a Devonian plant (*Psilophyton*), possibly one of the earliest ancestors of seed-bearing plants. Compare Figure 16.2c.

a

b

Gondwana drifted north during the Silurian and on into the Devonian. This was a pivotal time of evolution. Reef organisms recovered. Vertebrates—armor-plated, massive-jawed fishes—diversified. In the wet lowlands, small stalked plants were taking hold (Figures 17.16 and 17.17b). So were fungi and invertebrates, including the segmented worms. Later in the Devonian, some fishes that were ancestral to amphibians invaded the land. These fishes had lobed fins, the forerunners of legs and other limbs. And they had simple lungs. As described in Section 21.6, both adaptations would prove advantageous for life out of water, in dry land habitats.

Then, as they say, something bad happened. At the Devonian-Carboniferous boundary, sea levels swung catastrophically for unknown reasons and triggered another mass extinction. Afterward, plants and insects embarked on adaptive radiations on land.

Throughout the Carboniferous, land masses were gradually submerged and drained many times. Organic debris accumulated, became compacted, and was converted to coal, in the manner described in Section 19.3.

Insects, amphibians, and early reptiles flourished in vast swamp forests of Permian times (Figure 17.17c). The ancestors of the seed-producing plants called cycads, ginkgos, and conifers dominated these forests.

As the Permian drew to a close, the greatest of all mass extinctions took place. Nearly all known species perished. At that time, all land masses were colliding to form Pangea. This vast supercontinent extended from pole to pole. A single world ocean lapped its margins:

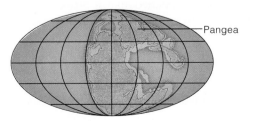

Pangea

c

Figure 17.17 (**a**) Life in the Silurian seas. Some of these shelled animals (nautiloids) were twelve feet long. (**b**) A Silurian swamp, dominated by the forerunners of modern ferns and club mosses. (**c**) Reptiles (*Dimetrodon*) of a largely hotter and drier time, the Permian. Fossils of these carnivores were found in Texas. Giant club mosses and horsetails had declined, and conifers, cycads, and other gymnosperms replaced them.

As you will see, the new distribution of oceans, land masses, and land elevations had catastrophic effects on global climates—and on the course of life's evolution.

Early in the Paleozoic era, organisms of all five kingdoms were flourishing in the seas. By the end of the era, many lineages had successfully invaded land environments, including the wet lowlands of the supercontinent Pangea.

17.7 LIFE IN THE MESOZOIC ERA

Speciation on a Grand Scale

The Mesozoic is divided into the Triassic, Jurassic, and Cretaceous periods. It lasted about 175 million years.

Early in the Cretaceous, the supercontinent Pangea started to break up. Its huge fragments slowly drifted apart, and we can well assume that the resulting geographic isolation favored divergences and speciation:

This was an era of spectacular expansion in the range of global diversity. Marine invertebrates and fishes underwent major adaptive radiations in the seas. Diverse angiosperms (flowering plants), insects, and reptiles became the most visibly dominant lineages on land.

Flowering plants originated about 127 million years ago. Within a mere 30 to 40 million years, they would displace conifers and related plants that were already declining in nearly all environments (Figure 17.18).

Rise of the Ruling Reptiles

Early in the Triassic, the first dinosaurs evolved from a reptilian lineage. They weren't much larger than a wild turkey. Many sprinted about on two legs. Most early dinosaurs may have had high metabolic rates and may have been warm-blooded. They weren't the dominant land animals. Center stage belonged to *Lystrosaurus* and other plant-eating, mammal-like reptiles that were too large to be bothered by most predators.

In time, adaptive zones did open up for dinosaurs, possibly when an asteroid struck the Earth. In central Quebec, a crater about the size of Rhode Island shows how huge such an impact could have been. The blast wave and global firestorm, earthquakes, and lava flows would have been stupendous. Most of the animals that survived this time of mass extinction (and later ones) were smaller, with higher rates of metabolism, and less vulnerable than others to drastic changes in outside temperatures. Descendants of the surviving dinosaurs became the ruling reptiles; they endured for 140 million years. Some, including ultrasaurs, reached monstrous proportions. These were 15 meters tall.

Many dinosaurs perished in another mass extinction at the end of the Jurassic, then in a pulse of extinctions in the Cretaceous. Perhaps plumes ruptured the crust and triggered changes in the global climate. Or perhaps

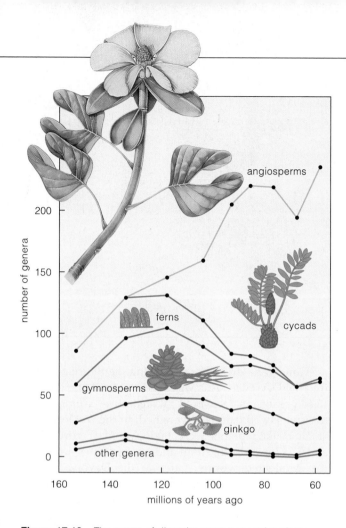

Figure 17.18 The range of diversity among vascular plants during the Mesozoic era. Accompanying the spectacular adaptive radiation of flowering plants were significant declines in the other groups shown. The painting depicts a floral shoot of *Archaeanthus linnenbergeri*, a flowering plant that lived during the Cretaceous. This now-extinct species resembled living magnolias in many respects.

a swarm of asteroid bombardments inflicted the blows. Whatever the causes, conditions changed for dinosaurs, and not all of them lived through it. Yet some lineages recovered and new ones evolved. Duckbilled dinosaurs appeared in forests and swamps. Tanklike *Triceratops* and other plant eaters flourished in open regions. They were prey for the swift, agile, and fearsomely toothed *Velociraptor* of motion picture fame.

About 120 million years ago, global temperatures shot up by 25 degrees. By one theory, plumes spread out beneath the crust and "greased" the crustal plates into moving twice as fast. A superplume (or a rash of them) broke through the crust. The crust in what is now the South Atlantic opened like a zipper. Basalt and lava poured out from fissures; volcanoes spewed nutrient-rich ashes. Simultaneously, the plumes released great amounts of carbon dioxide. This is one of the "greenhouse" gases that absorb heat radiating from the Earth before it can escape into space. The nutrient-enriched planet warmed—and stayed warm for 20 million years.

Figure 17.19 If dinosaurs of this sort had not disappeared at the end of the Cretaceous, would the then-tiny mammals ever have ventured out from under the shrubbery? Would you even be here today?

On land and in the shallow seas, photosynthetic organisms flourished. Their remains were slowly buried and converted into the world's oil reserves.

About 65 million years ago, the last dinosaurs and many marine organisms vanished in a mass extinction (Figure 17.19). As described in the next section, their disappearance apparently coincided with an asteroid impact. The impact site eventually drifted into a position that we now call the northern Yucatán peninsula.

The Mesozoic was a time of major adaptive radiations and of a mass extinction in which the last dinosaurs and many marine organisms disappeared.

HORRENDOUS END TO DOMINANCE

The first modern humans walked the Earth about 50,000 years ago. Since then, how many times have people puffed up with self-importance and set out to conquer neighbors, the land, and the seas? Think about that—then think about the dinosaurs. Were they good at reigning supreme? No question about it; their lineage dominated the land for 140 million years. In the end, did it matter? Not a bit. Sixty-five million years ago, at the Cretaceous-Tertiary (K-T) boundary, nearly all of the remaining members of their most excellent lineage perished. Why? Bad luck.

A thin layer of iridium-rich rock distributed around the world dates precisely to the K-T boundary. Iridium is rare on the earth's surface but common in asteroids.

Asteroids are rocky, metallic bodies, 1,000 kilometers to a few meters in diameter, that are hurtling through space. When planets were forming, their gravitational pull swept most asteroids from the sky. At least 6,000 still orbit the sun in a belt between Mars and Jupiter (Figure 17.20*a*). The orbits of many dozens of others take them across Earth's orbit. Think of it as Russian roulette on a cosmic scale.

By analyzing gravity maps, iridium levels in soils, and other evidence, researchers identified what appears to be a site where the K-T asteroid slammed into Earth. Massive movements in the crust transported the site to what is now the northern Yucatán peninsula of Mexico (Figure 17.20*b*). The crater is 9.6 kilometers deep and 300 kilometers across —wider than the state of Connecticut.

To make a crater this large, the asteroid had to hit the Earth at 160,000+ kilometers (100,000 miles) per hour. At least 200,000 cubic kilometers of debris and dense gases were blasted skyward. The crust heaved with earthquakes, and monstrous, 120-meter waves raced across the ocean. Researchers long thought that atmospheric debris blocked out sunlight for months. If so, plants and other producers that sustained the web of life would have withered and died; animals and other consumers would have starved to death. The theory has problems. By some calculations, the volume of debris that was blasted aloft would not have been enough to have such significant consequences.

Then, in 1994, the comet Shoemaker-Levy 9 slammed into Jupiter. Particles blasted into the Jovian atmosphere caused intense heating over an area larger than the Earth. This outcome favors a *global broiling theory*, first proposed by H. J. Melosh and his colleagues and expanded in 1995 by geologists at the University of California at Berkeley. Energy released at the K-T impact site was equivalent to detonating 100 million nuclear bombs. Trillions of tons of vaporized debris rose in a colossal fireball, then rapidly condensed into particles the size of sand grains. Seconds later, a cooler fireball of steam, unmelted rock, and carbon dioxide formed. When the debris fell to the Earth, it raised the atmospheric temperature by thousands of degrees. The sky above the entire planet glowed with heat ten times more intense than the noonday sun above Death Valley in

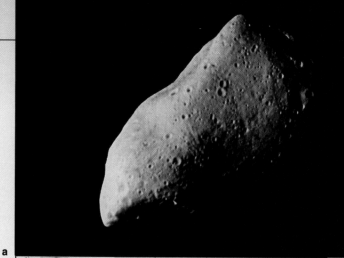

a

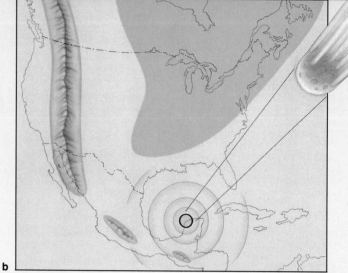

b

Figure 17.20 (**a**) What one asteroid looks like, courtesy of the Galileo spacecraft that flew past on its way to Jupiter. This is one of the smaller asteroids; it would only extend from Washington, D.C., halfway to Baltimore. (**b**) Artist's interpretation of what might have happened in the last few minutes of the Cretaceous.

summer. In one horrific hour, nearly all plants erupted in flames and every animal out in the open was broiled alive.

Things haven't settled down much since the demise of the dinosaurs. About 2.3 million years ago, for example, a huge object from space hit the Pacific Ocean. About the same time, vast ice sheets started forming abruptly in the Northern Hemisphere. Long-term shifts in climate may have been ushering in this most recent ice age, but a global impact would have accelerated it. Water vaporized by the impact could have contributed to a global cloud cover that limited the amount of sunlight reaching the surface. The ancestors of humans were around when this happened. The extreme climate shift surely tested their adaptability.

Almost certainly, severe environmental tests await all existing lineages. When we become too smitten with our importance in the world of life, we would do well to step back, from time to time, and reflect on what is going on above and beneath the Earth's surface. Asteroids and superplumes do have a way of leveling the playing field, as they did for the dinosaurs.

17.9 LIFE IN THE CENOZOIC ERA

The breakup of Pangea in the Mesozoic set events in motion that have continued to the present. At the dawn of the Cenozoic, land masses were on collision courses:

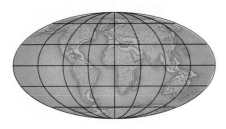

Coastlines fractured. The Cascades, Andes, Himalayas, and Alps rose through volcanic activity and uplifting along the margins of rifts and plate boundaries. These geologic changes brought about major shifts in climate that influenced the further evolution of life.

During the Paleocene epoch, climates were warmer and wetter. Tropical and subtropical forests extended farther north and south than they do today. Woodlands and forests spread even into polar regions. Although their key traits developed well before the dinosaurs disappeared, mammals now began their major radiation.

The global climate warmed even more during the Eocene epoch, and subtropical forests extended to the polar regions. Diverse mammals, including primates, bats, rhinos, hippos, elephants, horses, and assorted carnivores, emerged. By the late Eocene, however, the climates became cooler and drier, and seasons became pronounced. Now woodlands and semiarid grasslands dominated. Patterns of vegetational growth changed, and this drove many mammalian species to extinction.

From the Oligocene through the Pliocene, climate and vegetation favored an abundance of grazers and browsers, such as camels and the "giraffe rhinoceros"—and the carnivores that stalked them (Figure 17.21).

Today the distribution of land masses favors species diversity. Its richest ecosystems are the tropical forests of South America, Madagascar, and Southeast Asia, as well as the marine ecosystems of island chains in the tropical Pacific. Yet we are in the midst of what may be one of the greatest mass extinctions. About 50,000 years ago, nomadic humans followed migrating herds of wild animals around the Northern Hemisphere. Within a few thousand years, major groups of mammals were extinct. The pace of extinction has since accelerated as humans hunt for food, fur, feathers, or fun, and as they destroy habitats to clear land for farm animals or crops. Chapter 39 focuses on the global repercussions.

Major geologic changes during the Cenozoic triggered shifts in climate. The great adaptive radiation of mammals began, first in tropical forests, then in woodlands and grasslands.

Figure 17.21 Representatives from Cenozoic times. (**a**) In the early Paleocene, in what is now Wyoming, diverse mammals lived in dense forests of sequoia trees, laurel, and other plants. On the ground is the raccoonlike *Chriacus*, facing a tree-climbing rodent (*Ptilodus*). Higher up in the tree is *Peradectes*, an early marsupial. (**b**) From the late Eocene to early Miocene, *Indricotherium* browsed on trees in woodlands. This "giraffe rhinoceros," the largest land mammal ever discovered, weighed about 15 tons and measured 5.5 meters (18 feet) at shoulder. (**c**) From the Pleistocene, a small horse and the saber-tooth cat (*Smilodon*), a formidable predator. Both kinds of mammals are extinct. This reconstruction is based on fossils recovered from a pitch pool at Rancho LaBrea, California.

17.10 SUMMARY OF EARTH AND LIFE HISTORY

We conclude this overview chapter with an illustration that correlates milestones in the evolution of life with the evolution of the Earth. As you study Figure 17.22, keep in mind that it is only a generalized summary. For example, it charts five of the greatest mass extinctions, but there were many others in between. The diagram of global diversity conveys the shrinking and expanding range of species over time—but the range is for *all* of the major groups combined. Remember, each major group has its own distinct history of persisting lineages, of radiations, and of extinctions.

And now, with these qualifications in mind, we are ready to turn to the next chapters in this unit. They will provide you with richer detail of the history and current range of diversity for all five kingdoms of organisms.

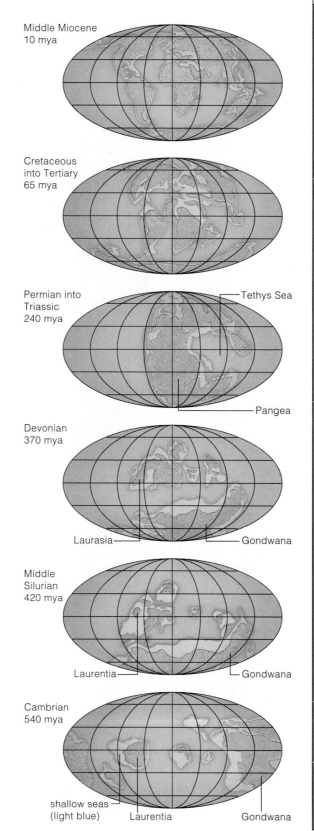

Figure 17.22 Summary of major events in the evolution of the Earth and of life. As you read through the chapters to follow, you may wish to return to this illustration now and then to remind yourself of how the details fit into the greater evolutionary picture.

	Period	Epoch
CENOZOIC ERA	Quaternary	Recent
		Pleistocene
	Tertiary	Pliocene
		Miocene
		Oligocene
		Eocene
		Paleocene
MESOZOIC ERA	Cretaceous	Late
		Early
	Jurassic	
	Triassic	
PALEOZOIC ERA	Permian	
	Carboniferous	
	Devonian	
	Silurian	
	Ordovician	
	Cambrian	
PROTEROZOIC EON		
ARCHEAN EON AND EARLIER		

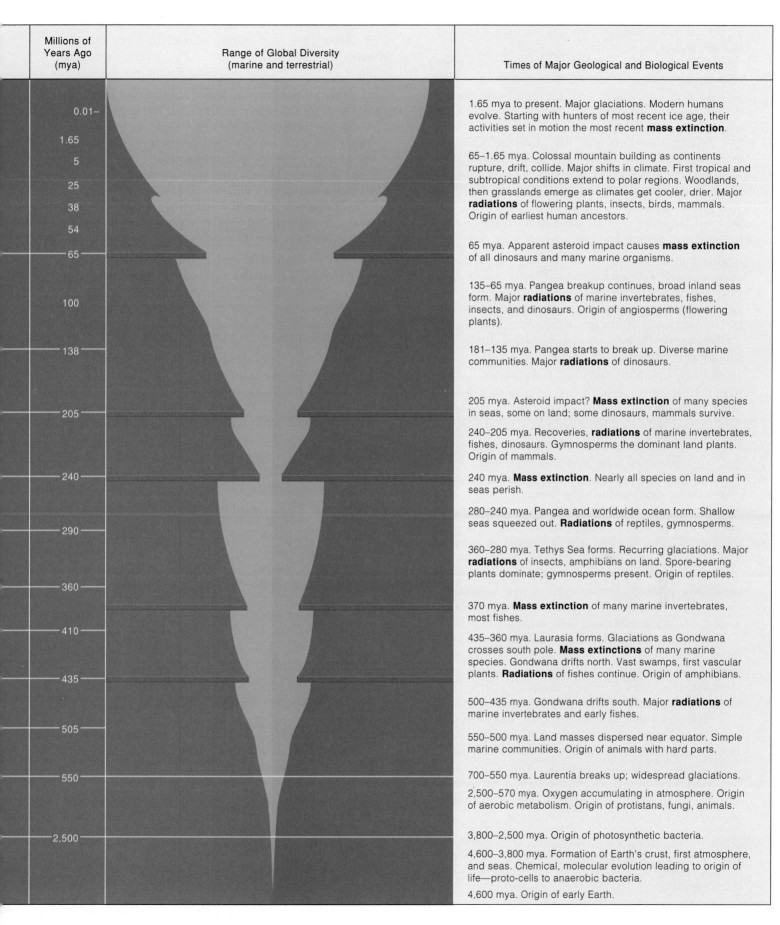

Millions of Years Ago (mya)	Range of Global Diversity (marine and terrestrial)	Times of Major Geological and Biological Events
0.01–		1.65 mya to present. Major glaciations. Modern humans evolve. Starting with hunters of most recent ice age, their activities set in motion the most recent **mass extinction**.
1.65		
5		65–1.65 mya. Colossal mountain building as continents rupture, drift, collide. Major shifts in climate. First tropical and subtropical conditions extend to polar regions. Woodlands, then grasslands emerge as climates get cooler, drier. Major **radiations** of flowering plants, insects, birds, mammals. Origin of earliest human ancestors.
25		
38		
54		
65		65 mya. Apparent asteroid impact causes **mass extinction** of all dinosaurs and many marine organisms.
100		135–65 mya. Pangea breakup continues, broad inland seas form. Major **radiations** of marine invertebrates, fishes, insects, and dinosaurs. Origin of angiosperms (flowering plants).
138		181–135 mya. Pangea starts to break up. Diverse marine communities. Major **radiations** of dinosaurs.
205		205 mya. Asteroid impact? **Mass extinction** of many species in seas, some on land; some dinosaurs, mammals survive.
		240–205 mya. Recoveries, **radiations** of marine invertebrates, fishes, dinosaurs. Gymnosperms the dominant land plants. Origin of mammals.
240		240 mya. **Mass extinction**. Nearly all species on land and in seas perish.
290		280–240 mya. Pangea and worldwide ocean form. Shallow seas squeezed out. **Radiations** of reptiles, gymnosperms.
360		360–280 mya. Tethys Sea forms. Recurring glaciations. Major **radiations** of insects, amphibians on land. Spore-bearing plants dominate; gymnosperms present. Origin of reptiles.
410		370 mya. **Mass extinction** of many marine invertebrates, most fishes.
435		435–360 mya. Laurasia forms. Glaciations as Gondwana crosses south pole. **Mass extinctions** of many marine species. Gondwana drifts north. Vast swamps, first vascular plants. **Radiations** of fishes continue. Origin of amphibians.
505		500–435 mya. Gondwana drifts south. Major **radiations** of marine invertebrates and early fishes.
550		550–500 mya. Land masses dispersed near equator. Simple marine communities. Origin of animals with hard parts.
		700–550 mya. Laurentia breaks up; widespread glaciations.
		2,500–570 mya. Oxygen accumulating in atmosphere. Origin of aerobic metabolism. Origin of protistans, fungi, animals.
2,500		3,800–2,500 mya. Origin of photosynthetic bacteria.
		4,600–3,800 mya. Formation of Earth's crust, first atmosphere, and seas. Chemical, molecular evolution leading to origin of life—proto-cells to anaerobic bacteria.
		4,600 mya. Origin of early Earth.

SUMMARY

1. The story of life begins with the "big bang," a model of the origin of the universe.

   a. By this model, all matter and all of space were once compressed in a fleeting state of enormous heat and density. Time began with the near-instantaneous distribution of all matter and energy throughout the known universe, which has been expanding ever since.

   b. Nearly all helium and other light elements, the most abundant elements of the universe, formed right after the big bang. Heavier elements originated during the formation, evolution, and death of stars.

   c. Every element of the solar system, the Earth, and life itself are products of the physical and chemical evolution of the universe and its stars.

2. Four billion years ago, the Earth had a high-density core, a mantle of intermediate density, and a thin, extremely unstable crust of low-density rocks. Probably gaseous hydrogen, nitrogen, carbon monoxide, and carbon dioxide made up the first atmosphere. Free oxygen and water could not have accumulated under the prevailing conditions.

3. Water accumulated after the Earth's crust cooled. Runoff from rains carried dissolved mineral salts and other compounds to crustal depressions, where early seas formed. Life could not have originated without this salty liquid water.

4. Many studies and experiments have yielded indirect evidence that life originated under the conditions that presumably existed on the early Earth.

   a. Comparative analysis of the composition of cosmic clouds, rocks from other planets, and rocks from the Earth's moon suggest that precursors of complex molecules associated with life were available.

   b. In laboratory tests that simulated the primordial conditions, including the absence of free oxygen, the precursors spontaneously assembled into sugars (such as glucose), amino acids, and other organic compounds.

   c. Known chemical principles and advanced computer simulations indicate that metabolic pathways could have evolved through chemical competition for the limited supplies of organic molecules (which had accumulated by natural geologic processes in the seas).

   d. Self-replicating systems of RNA, enzymes, and coenzymes have been synthesized in the laboratory. How DNA entered the picture is not yet understood.

   e. Lipid and lipid-protein membranes with some of the properties of cell membranes will form spontaneously under conditions thought to have existed on the early Earth.

5. Life originated by 3.8 billion years ago. Since then, it has been influenced by major changes in the Earth's crust, atmosphere, and oceans. Forces of change have included plate movements, asteroid impacts, and the activities of organisms (including oxygen-producing photosynthesizers and, currently, the human species).

6. Abrupt discontinuities in the fossil record mark the times of global mass extinctions. We use them as boundary markers for five great intervals in the geologic time scale. Radiometric dating has allowed us to assign absolute dates to this time scale:

   a. Archean: 3.9 to 2.5 billion years ago

   b. Proterozoic: 2.5 billion to 550 million years ago

   c. Paleozoic: 550 to 240 million years ago

   d. Mesozoic: 240 to 65 million years ago

   e. Cenozoic: 65 million years ago to the present

7. The first living cells were prokaryotic (bacteria). Not long after they appeared, divergences began that led to three great lineages: the archaebacteria, the eubacteria, and the prokaryotic ancestors of all eukaryotes. Among the eubacteria were species that used a cyclic pathway of photosynthesis.

8. During the Proterozoic, the noncyclic pathway of photosynthesis had evolved in some lineages. Oxygen, a by-product of the pathway, started accumulating in the atmosphere.

   a. In time, the atmospheric concentration of free oxygen prevented the further spontaneous formation of organic molecules. Thus the spontaneous origin of life was no longer possible on Earth.

   b. The abundance of atmospheric oxygen was a selective pressure that brought about the evolution of aerobic respiration. Aerobic respiration was a key step toward the origin of the first eukaryotic cells.

   c. Mitochondria and chloroplasts, two important eukaryotic organelles, probably evolved as an outcome of endosymbiosis between aerobic bacteria and the anaerobic forerunners of eukaryotes.

   d. The oxygen-rich atmosphere promoted formation of a layer of ozone (O_3). In time, that shield against destructive ultraviolet radiation allowed some lineages to move out of the seas, into low wetlands.

9. Early in the Paleozoic, diverse organisms of all five lineages had become established in the seas. By the end of the era, the invasion of land was under way. From that time on, there have been pulses of mass extinctions and adaptive radiations. Asteroid bombardments and other catastrophes triggered many of these events. So did plate movements that changed the distribution of oceans and land as well as the prevailing global and regional climates.

Review Questions

1. Describe the presumed chemical and physical conditions of the Earth four billion years ago. How do we know what it may have been like? *264*

2. Describe some of the experimental evidence for the spontaneous origin of large organic molecules, the self-assembly of proteins, and the formation of organic membranes and spheres, under conditions similar to those of the early Earth. *265–267*

3. How does continental drift occur, and in what ways has this almost imperceptibly slow occurrence influenced life on land and in the seas? *268–269*

4. Summarize the theory of endosymbiosis being advanced by Lynn Margulis. Cite evidence in favor of this theory. *272–273*

5. When did plants, fungi, and insects invade the land? What kind of vertebrates first invaded the land, and when? *275*

Self-Quiz *(Answers in Appendix IV)*

1. Through study of the geologic record, we know the evolution of life is linked to _____ .
   a. tectonic movements of the Earth's crust
   b. bombardment of the Earth by celestial objects
   c. profound shifts in land masses, shorelines, and oceans
   d. physical and chemical evolution of the Earth
   e. all of the above

2. The first cells emerged in the _____ .
   a. Paleozoic   d. Proterozoic
   b. Mesozoic   e. Cenozoic
   c. Archean

3. _____ was the first to obtain indirect evidence that organic molecules could have been formed on the early Earth.
   a. Darwin   c. Fox
   b. Miller   d. Margulis

4. An abundance of _____ was conspicuously absent from the Earth's atmosphere 4 billion years ago.
   a. hydrogen   c. carbon monoxide
   b. nitrogen   d. free oxygen

5. Life originated by _____ .
   a. 4.6 billion years ago   c. 3.8 billion years ago
   b. 2.8 million years ago   d. 3.8 million years ago

6. Abrupt discontinuities in the fossil record indicate _____ .
   a. the death of all organisms when the atmospheric oxygen levels dropped
   b. the times of global mass extinctions
   c. boundary markers for five great intervals of the geologic time scale
   d. both b and c

7. Which of the following statements is false?
   a. The first living cells were prokaryotes.
   b. The cyclic pathway of photosynthesis first appeared in some eubacterial species.
   c. Oxygen began accumulating in the atmosphere after the noncyclic pathway of photosynthesis evolved.
   d. During the Proterozoic, an increasing amount of atmospheric oxygen promoted further spontaneous generation of organic molecules.

8. Match the geologic time interval with the events listed.
   ____ Archean
   ____ Proterozoic
   ____ Paleozoic
   ____ Mesozoic
   ____ Cenozoic

   a. major radiations of dinosaurs, origin of flowering plants and mammals
   b. chemical evolution, origin of life
   c. major radiations of flowering plants, insects, birds, mammals, emergence of human forms
   d. oxygen present, origin of aerobic metabolism, protistans, algae, fungi, animals
   e. rise of early plants, origin of amphibians, origin of reptiles

Critical Thinking

1. Explain, in terms of hydrophilic and hydrophobic interactions, how proto-cells might have formed in water from aggregations of lipids, proteins, and nucleic acids.

2. The Atlantic Ocean is gradually widening, and the Pacific Ocean and Indian Ocean are closing. Many millions of years from now, the continents will collide and form a second Pangea. Write a short essay on what environmental conditions might be like on that future supercontinent and on what types of species might survive there.

3. By some estimates, there is a chance that about 10^{20} planets have formed in the universe that are capable of sustaining life—but only one chance at intelligent life per planet. Given your knowledge of molecular biology and evolutionary processes, speculate on why the odds are so low.

Selected Key Terms

archaebacterium *270*   plate tectonic theory *268*
Archean *269*   prokaryotic cell *270*
asteroid *278*   Proterozoic *269*
Cenozoic *269*   protistan *273*
eubacterium *270*   proto-cell *267*
eukaryotic cell *270*   radiation *281*
geologic time scale *269*   RNA world *267*
mass extinction *269*   stromatolite *270*
Mesozoic *269*   theory of endosymbiosis *272*
Paleozoic *269*

Readings

Bambach, R., C. Scotese, and A. Ziegler. 1980. "Before Pangea: The Geographies of the Paleozoic World." *American Scientist* 68(1): 26–38.

Dobb, E. February 1992. "Hot Times in the Cretaceous." *Discover* 13: 11–13.

Dott, R., Jr., and R. Batten. 1988. *Evolution of the Earth.* Fourth edition. New York: McGraw-Hill.

Hartman, W., and Chris Impey. 1994. *Astronomy: The Cosmic Journey.* Fifth edition. Belmont, California: Wadsworth.

Horgan, J. February 1991. "Trends in Evolution: In the Beginning . . . " *Scientific American* 264(2): 116–125.

18 BACTERIA, VIRUSES, AND PROTISTANS

The Unseen Multitudes

Did a friend ever mention that you are nearly 1/1,000 of a mile tall? Probably not. What would be the point of measuring people in units as big as miles? Yet we think this way, in reverse, when we measure **microorganisms**. For the most part, these are single-celled organisms too small to be seen without a microscope. The bacteria in Figure 18.1 are a case in point. To measure them, you would have to divide a meter into a thousand units (millimeters). Then you'd have to divide one millimeter into a thousand smaller units (micrometers). A single millimeter would be about as small as the dot of this "i." A *thousand* bacteria would fit side by side on the dot!

To be sure, viruses are smaller still. We measure them in nanometers (billionths of a meter). But viruses are not living things. We consider them here because one kind or another infects just about every organism.

Bacteria generally are the smallest microorganisms, but they vastly outnumber the individuals in all other kingdoms combined. Their reproductive potential is staggering. Under ideal conditions, some divide about every twenty minutes. If that rate of reproduction were to hold constant, a single bacterium would have nearly a billion descendants in ten hours! Many protistans also reproduce at impressive rates.

So why don't the unseen multitudes take over the world? Sooner or later, their burgeoning populations use up nutrients and pollute the surroundings with

their own metabolic wastes. In other words, they alter the very conditions that favored their reproduction. Besides this, viral attacks help keep their populations in check. So do seasonal changes in living conditions. So do the microbial species that eat or otherwise attack other species, as Figure 18.1*e* so vividly demonstrates.

Of course, you probably don't find much comfort in this when you serve as host for one of the pathogenic types. **Pathogens** are infectious, disease-causing agents that invade target organisms and multiply inside or on them. Disease follows when their activities damage tissues and interfere with normal body functions.

Certain pathogens can indeed make you suffer. But they shouldn't give every microorganism a bad name. For example, think back on the uncountable numbers of photosynthetic bacteria and protistans in the seas

Figure 18.1 (**a–c**) How small are bacteria? Shown here, *Bacillus* cells on the tip of a pin. The cells in (**c**) are magnified 14,000 times. (**d**) How small are viruses? Shown here, bacteriophage particles, each about 225 nanometers tall, infecting a bacterial cell.

(**e**) Mealtime for *Didinium*, a single-celled, predatory protistan with a big mouth. *Paramecium*, another protistan, is poised at the mouth (*left*) and swallowed (*right*). Generally, protistans are much larger than bacteria, as you can tell by comparing the scale bars at the lower right corner of all five micrographs.

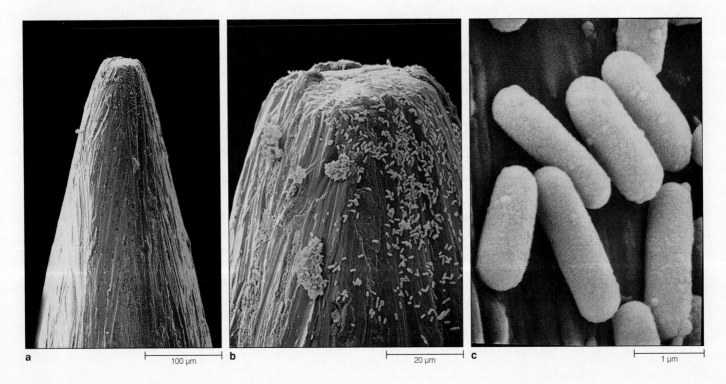

a   100 μm   b   20 μm   c   1 μm

(Section 5.7). Collectively, they provide food and oxygen for entire communities, and they have major roles in the global cycling of carbon. Or think about microorganisms that decompose organic debris. Collectively, they help cycle nutrients that sustain entire communities.

From the human perspective, microorganisms are good, bad, or dangerous. Basically, however, they simply are busy at surviving and reproducing like the rest of us, in ways that are the topics of this chapter.

infected bacterial cell — one virus particle

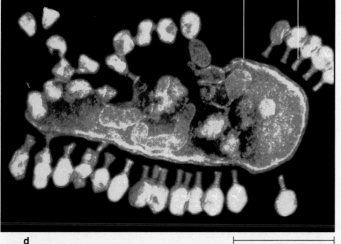

d 1.5 μm

Paramecium

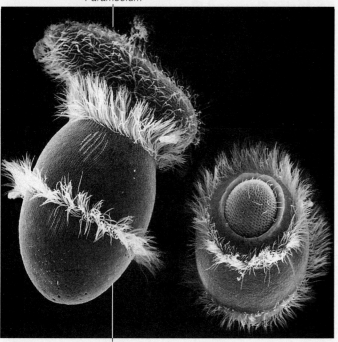

e *Didinium* 50 μm

KEY CONCEPTS

1. In structural terms, we find the simplest forms of life among the bacteria and protistans. Most of these are microscopically small. Smaller still are the viruses.

2. Bacteria alone are prokaryotic cells. They do not have a profusion of internal, membrane-bound organelles in the cytoplasm, as eukaryotic cells do. Yet bacteria show great metabolic diversity, and many show complex behavior.

3. Most bacteria reproduce by binary fission. This cell division mechanism follows DNA replication and divides a cell into two genetically equivalent daughter cells.

4. Bacteria were the first cells on Earth. Not long after they originated, they diverged into three great lineages. Today, archaebacteria, which may resemble the first cells, are living representatives of one lineage, and eubacteria represent another. Most bacterial species belong to this second lineage. A third lineage gave rise to the ancestors of eukaryotic cells.

5. A virus is a noncellular infectious particle that consists of nucleic acid and a protein coat, and sometimes an outer envelope. Viruses cannot be replicated without pirating the metabolic machinery of a specific type of host cell.

6. Nearly all viral multiplication cycles include five steps: attachment to a host cell, penetration, replication of viral DNA or RNA and synthesis of viral proteins, assembly of new viral particles, and release from the infected cell.

7. All protistans are eukaryotic and easily distinguishable from bacteria. Membrane-bound organelles, including a true nucleus, are a distinguishing feature. Some groups have close evolutionary ties to other kingdoms.

8. Most protistans are single celled, yet nearly every lineage also includes multicelled species. Protistans differ enormously from one another in body form and life-styles.

9. We tend to judge microorganisms through the prism of human interests. Yet their lineages are the most ancient, their adaptations are breathtakingly diverse, and they are simply surviving and reproducing like the rest of us.

CHARACTERISTICS OF BACTERIA

Of all organisms, bacteria are the most abundant and far-flung. Many thousands of species live in places that range from deserts to hot springs, glaciers, and seas. Billions of bacterial cells may occupy a handful of rich soil. The ones in your gut and on your skin outnumber your body cells! (Your cells are larger, so you are only "a few percent bacterial" by weight.)

Bacteria also have the longest evolutionary histories. Two great lineages, the archaebacteria and eubacteria, have endured almost since the time life originated. A third lineage gave rise to eukaryotic cells (Section 17.4). Trace the lineage of any organism back far enough, and you will discover bacterial ancestors. From *Escherichia coli* to amoebas, elephants, clams, and coast redwoods, *all* organisms interconnect—regardless of differences in size, numbers, and evolutionary distance. Table 18.1 and Figure 18.2 introduce features that help characterize these remarkable organisms.

Splendid Metabolic Diversity

All organisms take in energy and carbon to meet their nutritional requirements. Yet compared to other species, bacteria show the greatest diversity in the means of securing these resources.

Like plants, *photoautotrophic* bacteria are self-feeders that make organic compounds by photosynthesis. Sunlight is their source of energy; carbon dioxide is their source of carbon. Their photosynthetic machinery is embedded in the plasma membrane. Some species use a noncyclic pathway. Electrons and hydrogen from water molecules are used for the synthesis reactions, and the oxygen is released as a by-product. Other species are anaerobic; they cannot use oxygen or die in its presence. They rely on a cyclic pathway in which electrons and hydrogen are stripped from gaseous hydrogen, hydrogen sulfide, and other inorganic compounds.

By contrast, *photoheterotrophic* bacteria are not self-feeders. They can use sunlight for photosynthesis, but they can't harness carbon dioxide. They get carbon from organic compounds, including fatty acids and carbohydrates, that other organisms have already produced.

Chemoautotrophic bacteria make organic compounds by using carbon dioxide as the main source of carbon. These self-feeders strip electrons and hydrogen from various inorganic substances, including gaseous hydrogen, sulfur, nitrogen compounds, and a form of iron.

Then there are the *chemoheterotrophic* bacteria. These are parasites or saprobes, not self-feeders. The parasitic types live on or in a living host and draw glucose and other nutrients from it. The saprobic types get nutrients from the organic products, wastes, or remains of other organisms.

Table 18.1 Characteristics of Bacterial Cells

1. Bacterial cells are prokaryotic (they have no membrane-bound nucleus or other organelles in the cytoplasm).

2. Bacterial cells in general have a single chromosome (a circular DNA molecule); many species also have plasmids.

3. Most bacteria have a cell wall composed of peptidoglycan.

4. Most bacteria reproduce by binary fission.

5. Collectively, bacteria show great diversity in their modes of metabolism.

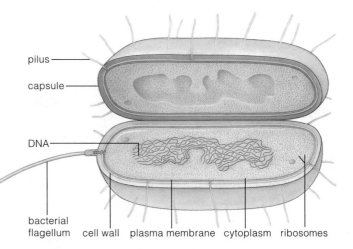

pilus —
capsule —
DNA —
bacterial flagellum   cell wall   plasma membrane   cytoplasm   ribosomes

Figure 18.2 Generalized body plan of a bacterium.

Bacterial Sizes and Shapes

So far, you have a sense of the microscopically small sizes of bacteria. Typically, the width or length of these cells falls between 1 and 10 micrometers.

Three basic shapes are common among bacteria. A spherical shape is a coccus (plural, cocci; from a word that really means berries). A rod shape is a bacillus (plural, bacilli, meaning small staffs). A bacterium with one or more twists in the cell body is a spiral.

coccus   bacillus   spiral

Don't let these simple categories fool you. Cocci may also be oval or a bit flattened, and bacilli may be skinny (like straws) or tapered (like cigars). After they divide, daughter cells may stay stuck together in chains, sheets, and other aggregations. Some spiral species are curved, like a comma; others are flexible or like stiff corkscrews. You may even observe square or star-shaped bacteria.

Figure 18.3 Gram staining. Cocci and rods smeared on a slide are stained with a purple dye (such as crystal violet), washed off, and then stained with iodine. All the cells are now purple. The slide is washed with alcohol, which renders the Gram-negative cells colorless. Now the slide is counterstained (with safranin), washed, and dried. Gram-positive cells (in this case, *Staphylococcus aureus*) remain purple. But the counterstain colors Gram-negative cells (in this case, *Escherichia coli*) pink.

■ stain with purple dye
□ stain with iodine
□ wash with alcohol
■ counterstain with safranin

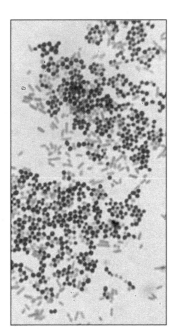

Structural Features

Bacteria alone are **prokaryotic cells**, meaning that they were around before cells with a nucleus evolved (*pro-*, before; *karyon*, nucleus). The interior of most bacteria is not subdivided into membrane-bound compartments. Reactions take place in the cytoplasm or at the plasma membrane. For example, proteins are synthesized at ribosomes that are distributed through the cytoplasm or attached to the plasma membrane.

This doesn't mean that bacteria are in some way inferior to eukaryotic cells. Being tiny, fast reproducers, they do very well without great internal complexity.

In most cases, the plasma membrane is surrounded by a semirigid, permeable **cell wall**, which helps a cell maintain its shape and resist rupturing when internal fluid pressure increases. Eubacterial cell walls consist of a type of molecule (peptidoglycan) in which peptides crosslink many polysaccharide strands to one another.

Differences in the cell wall's structure and composition help clinicians identify many bacterial species. In one staining reaction, the Gram stain, they expose cells to a purple dye, then to iodine, an alcohol wash, and a counterstain. Walls of *Gram-positive* species stay purple. Walls of *Gram-negative* species lose color after the wash but turn pink with the counterstain (Figure 18.3).

A sticky mesh (glycocalyx) often surrounds the cell wall. It consists of polysaccharides, polypeptides, or both. When highly organized and attached firmly to the wall, we call the mesh a capsule. When less organized and loosely attached to the wall, we call it a slime layer. The mesh helps a bacterial cell attach to teeth, mucous membranes, rocks in streambeds, and other interesting

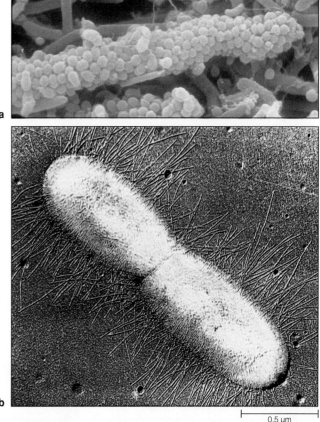

Figure 18.4 (**a**) Surface view of numerous bacteria that have become attached, by means of a sticky mesh, to the surface of a human tooth—and doesn't this micrograph make you want to grab a toothbrush? (**b**) Example of pili, filamentous structures that project from the surface of many bacterial cells. This particular *Escherichia coli* cell is dividing in two.

surfaces (Figure 18.4*a*). It even helps some encapsulated species avoid being engulfed by phagocytic, infection-fighting cells of their host organism.

Some bacteria have one or more **bacterial flagella** (singular, flagellum), which are used in motility. These don't have the same structure as a eukaryotic flagellum, and they don't operate the same way. They move the cell by rotating like a propeller.

Finally, many bacterial species have pili (singular, pilus). These short, filamentous proteins project above the cell wall. Figure 18.4*b* shows an example. Some pili help cells adhere to surfaces. Others help them attach to one another as a prelude to conjugation, described next.

Bacteria are microscopic, prokaryotic cells. In general, they have one circular bacterial chromosome and, often, extra DNA in the form of plasmids.

Nearly all bacteria have a wall around the plasma membrane, and many have a capsule or slime layer around the cell wall.

18.2 BACTERIAL REPRODUCTION

Most bacteria reproduce by a cell division mechanism called **binary fission**, although some types simply bud from a parent cell. Daughter cells inherit one **bacterial chromosome**, which is a circular, double-stranded DNA molecule with only a few proteins attached to it.

With binary fission, a parent cell replicates its DNA and the two DNA molecules start moving apart. Then the cytoplasm divides into two genetically equivalent daughter cells. As shown in Figure 18.5, the division requires suitable membrane growth and deposition of wall material at the midsection of the dividing cell.

Figure 18.5 (*Below*) Bacterial reproduction by binary fission, a cell division mechanism. The micrograph shows the division of the cytoplasm of *Bacillus cereus*, as brought about by the formation of new membrane and wall material. This bacterial cell has been magnified 13,000 times its actual size.

Figure 18.6 A simplified sketch of bacterial conjugation —the transfer of plasmid DNA from a donor cell to a recipient cell. A long pilus brings the two bacterial cells into close contact. Then a conjugation tube develops between them.

For the sake of clarity, the bacterial chromosome is not included in this sketch, and the size of the plasmid is greatly enlarged (for example, compare Figure 13.2).

plasmid (nick) └conjugation tube

a A conjugation tube forms and unites a donor cell and a recipient cell.

b Replication starts on plasmid DNA in donor cell; one of the two DNA strands is displaced and enters recipient cell.

c Replication starts on transferred DNA.

d Cells separate; plasmids circularize.

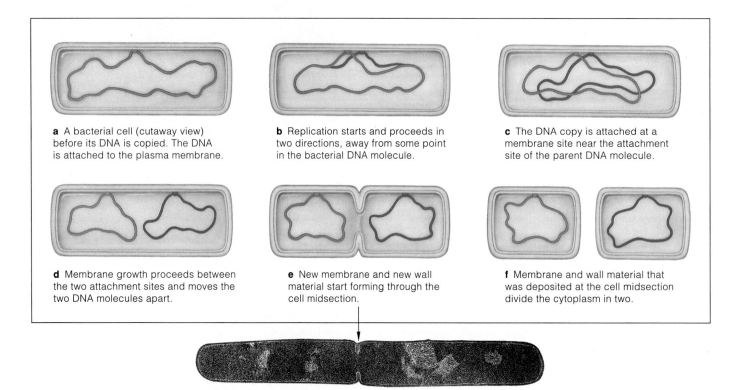

a A bacterial cell (cutaway view) before its DNA is copied. The DNA is attached to the plasma membrane.

b Replication starts and proceeds in two directions, away from some point in the bacterial DNA molecule.

c The DNA copy is attached at a membrane site near the attachment site of the parent DNA molecule.

d Membrane growth proceeds between the two attachment sites and moves the two DNA molecules apart.

e New membrane and new wall material start forming through the cell midsection.

f Membrane and wall material that was deposited at the cell midsection divide the cytoplasm in two.

In many species, daughter cells also inherit one or more plasmids. A **plasmid**, recall, is a small circle of extra DNA that has only a few genes and that is self-replicating (Section 13.1). Usually the genes confer a survival advantage, as when they specify an enzyme that allows a cell to synthesize or use some nutrient. Some genes confer resistance to antibiotics, substances that kill or inhibit growth of bacteria. Others confer the

means to engage in **bacterial conjugation**. By this mechanism, one bacterial cell transfers plasmid DNA to another cell, even of a different species (Figure 18.6).

Most bacteria use binary fission, a cell division mechanism that follows DNA replication and divides the parent cell into two genetically equivalent daughter cells.

BACTERIAL CLASSIFICATION

A few decades ago, reconstructing the evolutionary history of bacteria seemed an impossible task. Except for stromatolites (Section 17.4), early bacteria are not well represented in the fossil record. Most bacterial groups are not represented at all. Given their elusive histories, the many thousands of known species traditionally have been grouped together on the basis of cell shape, staining attributes of their cell wall, modes of nutrition, metabolic patterns, and other traits.

Table 18.2 lists representatives from some of the major groupings. Comparative biochemistry, including nucleic acid hybridization studies, is now providing insight into bacterial phylogeny. Such studies provide impressive evidence that three great bacterial lineages diverged from a common ancestor not long after living cells originated. These gave rise to the archaebacteria, the eubacteria, and the ancestors of eukaryotes.

Until researchers gather more data on evolutionary histories, bacterial species will continue to be classified largely in terms of cell shape, wall staining patterns, and other traits.

Table 18.2	Some Major Groups of Bacteria		
Group	Main Habitats	Characteristics	Representatives
Archaebacteria			
Methanogens	Anaerobic sediments of lakes, swamps; also animal gut	Chemosynthetic; methane producers; used in sewage treatment facilities	*Methanobacterium*
Halophiles	Brines (extremely salty water)	Heterotrophic; some also have unique photosynthetic machinery	*Halobacterium*
Extreme thermophiles	Acidic soil, hot springs, hydrothermal vents on seafloor	Heterotrophic or chemosynthetic; use inorganic substances as electron donors	*Sulfolobus, Thermoplasma*
Photoautotrophic eubacteria			
Cyanobacteria, green sulfur bacteria, and purple sulfur bacteria	Mostly lakes, ponds; some marine, terrestrial	Photosynthetic; use sunlight energy and carbon dioxide (carbon source); cyanobacteria produce oxygen as by-product	*Anabaena, Nostoc, Chloroflexus, Rhodopseudomonas*
Photoheterotrophic eubacteria			
Purple nonsulfur and green nonsulfur bacteria	Anaerobic, organically rich muddy soils, and sediments of aquatic habitats	Use sunlight, organic compounds (electron donors); some purple nonsulfur may also grow chemotrophically	*Rhodospirillum, Chlorobium*
Chemoautotrophic eubacteria			
Nitrifying, sulfur-oxidizing, and iron-oxidizing bacteria	Soil; freshwater, marine habitats	Use carbon dioxide, electrons from inorganic compounds; some have roles in agriculture, cycling of nutrients in ecosystems	*Nitrosomonas, Nitrobacter, Thiobacillus*
Chemoheterotrophic eubacteria			
Spirochetes	Aquatic habitats; parasites of animals	Helically coiled, motile; free-living and parasitic species; some major pathogens	*Spirochaeta, Treponema*
Gram-negative, aerobic rods and cocci	Soil, aquatic habitats; parasites of animals, plants	Some major pathogens; some (e.g., *Rhizobium*) fix nitrogen	*Pseudomonas, Neisseria, Rhizobium, Agrobacterium*
Gram-negative, facultative anaerobic rods	Soil, plants, animal gut	Many are major pathogens; one (*Photobacterium*) is bioluminescent	*Salmonella, Proteus, Escherichia, Photobacterium*
Rickettsias and chlamydias	Host cells of animals	Intracellular parasites; many pathogens	*Rickettsia, Chlamydia*
Myxobacteria	Decaying plant, animal matter; bark of living trees	Gliding, rod-shaped; cells aggregate and migrate together	*Myxococcus*
Gram-positive cocci	Soil; skin and mucous membranes of animals	Some major pathogens	*Staphylococcus, Streptococcus*
Endospore-forming rods and cocci	Soil; animal gut	Some major pathogens	*Bacillus, Clostridium*
Gram-positive nonsporulating rods	Fermenting plant, animal material; gut, vaginal tract	Some important in dairy industry, others major contaminators of milk, cheese	*Lactobacillus, Listeria*
Actinomycetes	Soil; some aquatic habitats	Include anaerobes and strict aerobes; major producers of antibiotics	*Actinomyces, Streptomyces*

MAJOR GROUPS OF BACTERIA

Archaebacteria

We divide archaebacteria into three major groups—the methanogens, halophiles, and extreme thermophiles. In many ways, their composition, structure, metabolism, and nucleic acid sequences are unique; they differ as much from other bacteria as they do from eukaryotes. Archaebacteria may even resemble the first cells, which apparently originated in exceptionally hot, acidic, and salty habitats. Existing species live in similar habitats— hence the name (*archae-* means beginning).

Methanogens ("methane-makers") live in swamps, sewage, the animal gut, stockyards, and other anaerobic places. They make ATP by converting hydrogen gas (H_2) and carbon dioxide to methane (CH_4). As a group, methanogens produce about 2 billion tons of methane every year. They influence atmospheric levels of carbon dioxide and the global carbon cycle (Section 37.6).

Halophiles, or "salt-lovers," live in brackish ponds, salt lakes, near hydrothermal vents (volcanic fissures on the seafloor), and other high-salinity settings (Figure 18.7a). They sometimes spoil salted fish, animal hides, and commercially produced sea salt. Most halophiles make ATP by aerobic pathways. At low oxygen levels, some also produce ATP by photosynthesis. Pigments (bacteriorhodopsin) form in the plasma membrane and absorb light energy. Absorption triggers an increase in an H^+ gradient across the membrane. ATP forms when ions flow down the gradient, through the interior of membrane proteins (compare Section 5.5).

Extreme thermophiles ("heat lovers") live in such places as highly acidic soils, hot springs, even coal mine wastes. Some species are the basis of remarkable food webs in sediments around hydrothermal vents. Water temperatures above the sediments may be as high as 110°C. Extreme thermophiles use the hydrogen sulfide spewing from the vents as a source of electrons for ATP formation. They are cited as evidence that life could have originated deep in the oceans.

Eubacteria

The *eu-* in eubacteria implies "typical." Eubacteria are far more common than the archaebacteria.

Photoautotrophic Eubacteria The most common photoautotrophic bacteria are the cyanobacteria (once called blue-green algae). These species are aerobic; they produce oxygen when photosynthesizing. Most live in ponds and other freshwater habitats. They may grow as mucus-sheathed chains of cells, which can form dense, slimy mats near the surface of nutrient-enriched water.

Anabaena and other types also convert nitrogen gas (N_2) to ammonia, which they use in biosynthesis. When

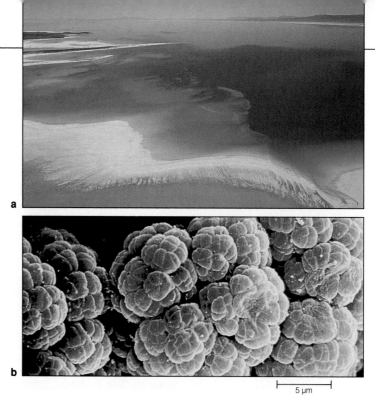

Figure 18.7 Archaebacteria. (**a**) Pinkish, saline water in Great Salt Lake, Utah—a sign of colonies of halophilic bacteria and certain algae that contain pinkish to red-orange carotenoid pigments. (**b**) A dense population of methanogenic cells (*Methanosarcina*).

nitrogen compounds are scarce, some cells develop into **heterocysts**, modified cells that make a nitrogen-fixing enzyme (Figure 18.8). These modified cells produce and share nitrogen compounds with the photosynthetic cells and receive carbohydrates in return. Substances move freely through cytoplasmic junctions between the cells.

Anaerobic photoautotrophs, such as green bacteria, get electrons from hydrogen sulfide or hydrogen gas, not water. They may resemble anaerobic bacteria in which the cyclic pathway of photosynthesis evolved.

Chemoautotrophic Eubacteria Many eubacteria in this category influence the global cycling of nitrogen, sulfur, and other nutrients. For example, as you know, nitrogen is a key building block for amino acids and proteins. Without it, there would be no life. Nitrifying bacteria in soil strip electrons from ammonia. Plants use the end product, nitrate, as a nitrogen source.

Chemoheterotrophic Eubacteria Most bacteria fall in this category. Many, including pseudomonads, are decomposers; their enzymes break down organic compounds and pesticides in soil. Other "good" species, by human standards, include *Lactobacillus* (employed in the manufacture of pickles, sauerkraut, buttermilk, and yogurt) and actinomycetes (sources of antibiotics). They include *Escherichia coli*, which makes vitamin K and compounds useful in fat digestion, and helps newborns digest milk. Its activities help keep many food-borne

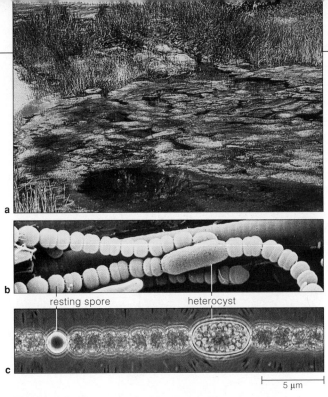

a

b

resting spore | heterocyst

c

5 µm

Figure 18.8 Some premier photoautotrophs—cyanobacteria. (**a**) A cyanobacterial population near the surface of a nutrient-enriched pond. (**b**,**c**) Resting spores form when conditions do not favor growth. A nitrogen-fixing heterocyst is also shown.

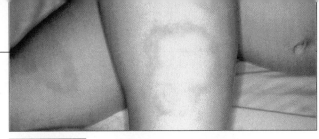

actual size: •

Figure 18.9 An extreme reaction to an infection by *Borrelia burgdorferi*, a spirochete. The rash is one symptom of what may now be the most common tick-borne disease in the United States: Lyme disease. Tick bites deliver the spirochete from one host to another. Statewide incidences in 1989 are mapped (*tan*, less than 10; *yellow*, between 15 and 90; *gold*, more than 90).

pathogens from colonizing the gut. Sugarcane and corn benefit from a symbiont, the nitrogen-fixing spirochete *Azospirillum*. These plants use some of the nitrogen and give up some sugars to the bacterium. Beans and other legumes rely on *Rhizobium*, a symbiont in their roots.

Also in this category are most pathogenic bacteria. We admire pseudomonads as decomposers in soil, not when they grow on "our" antiseptics, soaps, and other carbon-rich materials. They can transmit plasmids with antibiotic-resistance genes, so infections can be serious.

Some *E. coli* strains cause a form of diarrhea that is the main cause of infant death in developing countries. *Clostridium botulinum* can taint fermented grain as well as food in improperly sterilized or sealed cans and jars. Its toxins cause *botulism*, a form of poisoning that can disrupt breathing and lead to death. *C. tetani*, one of its relatives, causes the disease *tetanus* (Section 32.6).

Like many bacteria, *C. tetani* can form an **endospore** around a copy of its DNA and some cytoplasm. This structure can resist heat, drying out, radiation, and boiling. New bacterial cells form after it germinates.

Finally, like many bacteria, *Borrelia burgdorferi* taxis from host to host inside the gut of blood-sucking ticks. Tick bites transmit this spirochete from deer, field mice, and some other wild animals to humans, who develop *Lyme disease*. A pronounced rash, often shaped like a bull's-eye, develops around the tick bite (Figure 18.9). Severe headaches, backaches, chills, and fatigue follow. Without prompt treatment, the condition worsens.

Regarding the "Simple" Bacteria

Bacteria are small. Their insides are not elaborate. *But bacteria are not simple.* A brief look at their behavior will reinforce this point. Bacteria move toward nutrient-rich regions. Aerobes move toward oxygen; anaerobes avoid it. Photosynthetic types move into light (or away from intense light). Many species tumble away from toxins. Such behaviors often start with membrane receptors that change shape when stimulated by light or chemical compounds. Receptors are stimulated differently when a bacterium alters its direction. This triggers a fleeting "memory," a changing biochemical condition that can be compared against that of the immediate past.

Some species even show *collective* behavior, as when millions of *Myxococcus xanthus* cells form a "predatory" colony. The cells secrete enzymes that digest "prey"—cyanobacteria and other microorganisms that get stuck to the colony—then they absorb breakdown products. What's more, the cells migrate, change direction, and move as a single unit toward what may be food!

Many myxobacteria produce **fruiting bodies** (spore-bearing structures). Under suitable conditions, some cells in a colony differentiate and form a slime stalk, others form branches, and still others form clusters of spores. Spores are dispersed when a cluster bursts open, and each may form a new colony. As you will see, certain eukaryotic organisms also form such structures.

The first prokaryotes diverged into three great lineages: archaebacteria, eubacteria, and the bacterial ancestors of eukaryotes. Archaebacteria are confined to extreme habitats and may resemble the first cells. Eubacteria are the most common prokaryotes. They occur in nearly all environments.

18.5 THE VIRUSES

Defining Characteristics

In ancient Rome, *virus* meant "poison" or "venomous secretion." In the late 1800s, this rather nasty word was bestowed on newly discovered pathogens, smaller than the bacteria being studied by Louis Pasteur and others. Many viruses deserve the name. They attack humans, cats, cattle, birds, insects, plants, fungi, protistans, and bacteria. You name it, there are viruses that can infect it.

Today we define a **virus** as a noncellular infectious agent that has two characteristics. First, a viral particle consists of a protein coat around a nucleic acid core—that is, around its genetic material. Second, a virus cannot reproduce itself. It can be reproduced only after its genetic material and maybe a few enzymes enter a host cell and subvert the cell's biosynthetic machinery.

The genetic material of a virus is DNA *or* RNA. The viral coat consists of one or more types of protein subunits arranged in a rodlike or many-sided shape (Figure 18.10). The coat protects the genetic material during the journey from one host cell to the next. It also contains proteins that can bind with specific receptors on host cells. Some coats are enclosed in an envelope of mostly membrane remnants from the previously infected cell. Protein-carbohydrate structures spike out from these envelopes. Coats of complex viruses have sheaths, tail fibers, and other accessory structures attached.

The immune system of vertebrates can detect certain viral proteins. The problem is, the genes for many viral proteins mutate so frequently that a virus may elude the immune fighters. For example, people susceptible to lung infections get new "flu shots" each year because envelope spikes on influenza viruses keep changing.

Examples of Viruses

Each kind of virus can multiply only in certain hosts. It cannot be studied easily unless researchers culture living host cells. This is why much of our knowledge of viruses comes from **bacteriophages**, a group of viruses that infect bacterial cells. Unlike cells of humans and other complex, multicelled species, bacterial cells can be cultured easily and rapidly. This also is why bacteria and bacteriophages were used in early experiments to determine DNA function (Section 11.1). They are still used as research tools in genetic engineering.

Table 18.3 lists some major groups of animal viruses. They contain double- or single-stranded DNA or RNA, which is replicated in various ways. Animal viruses range in size from parvoviruses (18 nanometers) to the brick-shaped poxviruses (350 nanometers). Many cause diseases, including the common cold, certain cancers, warts, herpes, and influenza (Figure 18.11). One, HIV, is the trigger for *AIDS*. By attacking certain white blood

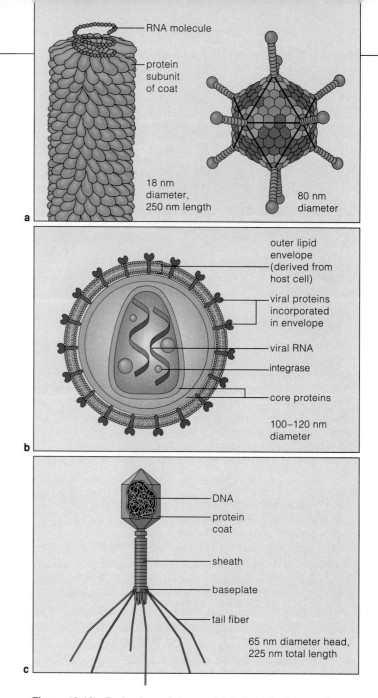

Figure 18.10 Body plans of viruses. (**a**) *Left: Helical* viruses have a rod-shaped coat of protein subunits, coiled helically around the nucleic acid. The upper subunits have been removed from this portion of a tobacco mosaic virus to reveal the RNA. *Right: Polyhedral* viruses, such as this adenovirus, have a many-sided coat. (**b**) *Enveloped* viruses, such as HIV, have an envelope around a helical or polyhedral coat. (**c**) *Complex* viruses, such as T-even bacteriophages, have additional structures attached to the coat.

cells, it weakens the immune system's ability to fight infections that might not otherwise be life threatening. Researchers attempting to develop drugs against HIV and other diseases use HeLa cells and other immortal cell lines for early experiments (Section 7.5). Later they must use laboratory animals, then human volunteers, to test a new drug for toxicity and effectiveness. Why? It takes a functioning immune system to test responses.

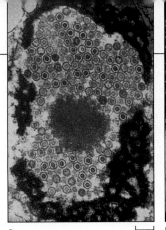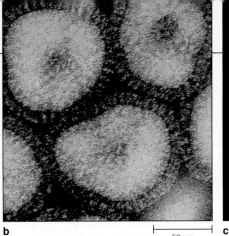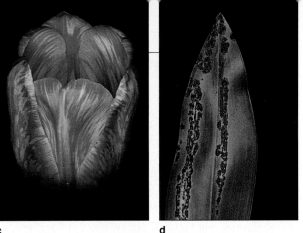

a
300 nm
b
50 nm
c
d

Figure 18.11 Some viruses and their effects. (**a**) Particles of a DNA virus that causes a herpes infection in humans. (**b**) Particles of an enveloped RNA virus that causes influenza in humans. Spikes project from the lipid envelope. (**c**) Streaking of a tulip blossom. A harmless virus infected pigment-forming cells in the colorless parts. (**d**) An orchid leaf infected by a rhabdovirus.

Table 18.3	Classification of Some Major Animal Viruses
DNA Viruses	**Some Diseases**
Parvoviruses	Gastroenteritis; roseola (fever, rash) in small children; aggravation of symptoms of sickle-cell anemia
Adenoviruses	Respiratory infections; some cause tumors in animals
Papovaviruses	Benign and malignant warts
Poxviruses	Smallpox, cowpox
Herpesviruses:	
H. simplex type I	Oral herpes, cold sores
H. simplex type II	Genital herpes
Varicella-zoster	Chicken pox, shingles
Epstein-Barr	Infectious mononucleosis, some forms of cancer
Hepadnavirus	Hepatitis B, severe liver damage
RNA Viruses	**Some Diseases**
Picornaviruses:	
Enteroviruses	Polio, hemorrhagic eye disease, hepatitis A (infectious hepatitis)
Rhinoviruses	Common cold
Hepatitis A virus	Inflammation of liver, kidneys, spleen
Togaviruses	Forms of encephalitis, rubella
Flaviviruses	Yellow fever (fever, chills, jaundice), dengue (fever, severe muscle pain)
Coronaviruses	Upper respiratory infections, colds
Rhabdoviruses	Rabies, other animal diseases
Filoviruses	Hemorrhagic fevers, as by Ebola virus
Paramyxoviruses	Measles, mumps
Orthomyxoviruses	Influenza
Bunyaviruses	Hemorrhagic fevers, as by hantaviruses
Arenaviruses	Hemorrhagic fevers
Retroviruses:	
HTLV I, II	Some cancers (leukemia)
HIV	AIDS
Reoviruses	Respiratory and intestinal infections

Plant viruses must breach plant cell walls to cause diseases. They typically hitch rides on the piercing or sucking devices of insects that feed on plant juices. Some RNA viruses infect tobacco plants (the tobacco mosaic virus), barley, potatoes, and other major crop plants. Certain DNA viruses infect such valuable crops as cauliflower and corn. Figure 18.11*c,d* shows the effects of other viral infections.

Infectious Agents Tinier Than Viruses

Some infectious agents are more stripped down than viruses. **Prions** are small proteins linked to eight rare, fatal degenerative diseases of the nervous system. They are altered products of a gene that is present in normal and infected individuals. Both unaltered and altered forms of the protein are found at the surface of neurons (communication cells of the nervous system). You may have heard of *kuru* and *Creutzfeldt-Jakob* diseases. They slowly destroy muscle coordination and brain function in humans. *Scrapie*, a prion-linked disease of sheep, is so named because infected animals rub against trees or posts until they scrape off most of their wool.

Viroids are tightly folded strands or circles of RNA, smaller than anything in viruses. They resemble introns (noncoding portions of eukaryotic DNA) and may have evolved from them. Viroids have no protein coat, but their tight folding may help protect them from host enzymes. These bits of "naked" RNA are known to cause plant diseases. Each year, they destroy millions of dollars' worth of potatoes, citrus, and other crop plants.

A virus is a nonliving infectious particle that consists of nucleic acid enclosed in a protein coat and sometimes an outer envelope. It cannot be replicated without pirating the metabolic machinery of a specific type of host cell.

18.6 VIRAL MULTIPLICATION CYCLES

Viruses multiply in a variety of ways. Even so, nearly all of the multiplication cycles include five basic steps:

1. *Attachment.* The virus attaches to a host cell. Any cell is a suitable host if a virus can chemically recognize and lock onto specific molecular groups at the cell surface.

2. *Penetration.* Either the whole virus or its genetic material alone penetrates the cell's cytoplasm.

3. *Replication and Synthesis.* In an act of molecular piracy, the viral DNA or RNA directs the host cell into producing many copies of viral nucleic acids and proteins, including enzymes.

4. *Assembly.* The viral nucleic acids and viral proteins are put together to form new infectious particles.

5. *Release.* New virus particles are released from the cell.

Consider this list with respect to some bacteriophages. Lytic and lysogenic pathways are common among their replication cycles. In a *lytic* pathway, steps 1 through 4 proceed rapidly, and lysis releases new particles (Figure 18.12). **Lysis** means a plasma membrane, cell wall, or both are damaged, so cytoplasm leaks out and the cell dies. Late into most lytic pathways, a viral enzyme is synthesized that swiftly destroys the bacterial cell wall.

In a *lysogenic* pathway, a latent period extends the cycle. The virus does not kill its host outright. Instead, a viral enzyme cuts a host chromosome, then integrates viral genes into it. When the infected cell prepares to divide, the recombinant molecule gets replicated. As a result of a single instance of genetic recombination, miniature time bombs get passed on to all of that cell's descendants. Later on, a molecular signal or some other stimulus may reactivate the cycle.

Latency occurs in the multiplication cycles of many viruses, not just among bacteriophages. Type I *Herpes simplex*, which causes *cold sores*, is an example. Nearly everybody harbors this virus. It remains latent inside a ganglion (a cluster of cell bodies of neurons) in the face. Stress factors, such as sunburn, can reactivate the virus. Then, virus particles move down the neurons, to their tips near the skin. There they infect epithelial cells and cause painful skin eruptions.

Like other enveloped viruses, the herpesviruses can enter a host cell by their own version of endocytosis, then leave it by budding from the plasma membrane. Figure 18.13 shows how they accomplish this.

The multiplication cycle of the RNA viruses has an interesting twist to it. In the host cell's cytoplasm, their

Figure 18.12 Generalized multiplication cycle for some of the bacteriophages. New viral particles may be produced and released by a lytic pathway alone. For certain viruses, the lytic pathway may be expanded to include a lysogenic pathway.

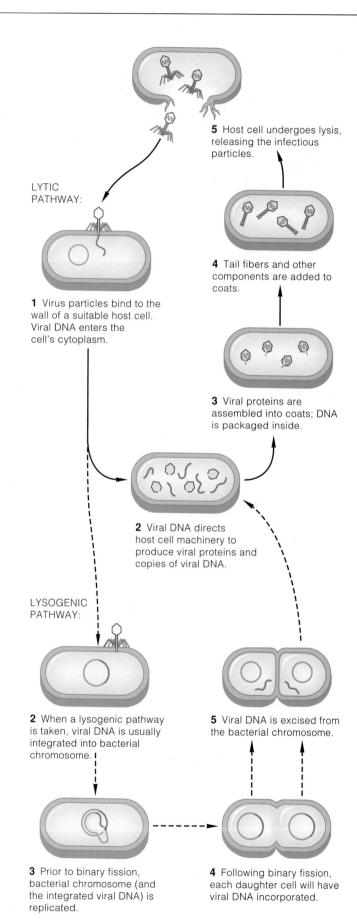

LYTIC PATHWAY:

1 Virus particles bind to the wall of a suitable host cell. Viral DNA enters the cell's cytoplasm.

5 Host cell undergoes lysis, releasing the infectious particles.

4 Tail fibers and other components are added to coats.

3 Viral proteins are assembled into coats; DNA is packaged inside.

2 Viral DNA directs host cell machinery to produce viral proteins and copies of viral DNA.

LYSOGENIC PATHWAY:

2 When a lysogenic pathway is taken, viral DNA is usually integrated into bacterial chromosome.

5 Viral DNA is excised from the bacterial chromosome.

3 Prior to binary fission, bacterial chromosome (and the integrated viral DNA) is replicated.

4 Following binary fission, each daughter cell will have viral DNA incorporated.

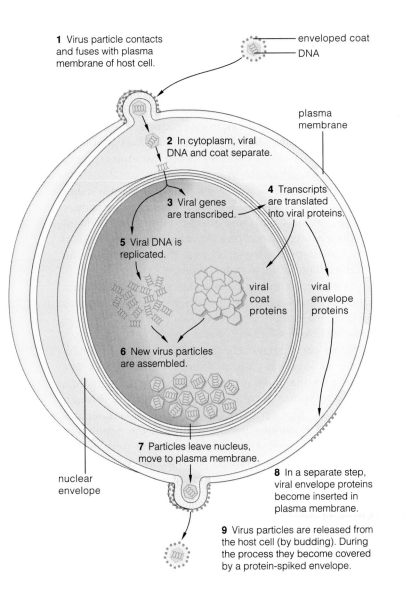

1 Virus particle contacts and fuses with plasma membrane of host cell.

enveloped coat
DNA

plasma membrane

2 In cytoplasm, viral DNA and coat separate.

3 Viral genes are transcribed.

4 Transcripts are translated into viral proteins.

5 Viral DNA is replicated.

viral coat proteins

viral envelope proteins

6 New virus particles are assembled.

7 Particles leave nucleus, move to plasma membrane.

nuclear envelope

8 In a separate step, viral envelope proteins become inserted in plasma membrane.

9 Virus particles are released from the host cell (by budding). During the process they become covered by a protein-spiked envelope.

Figure 18.13 Multiplication cycle of one animal virus. This is an enveloped DNA virus. When its envelope fuses with the plasma membrane of a host cell, its nucleic acid is dumped into the cytoplasm. The cell's metabolic machinery churns out viral DNA and proteins. Some proteins get embedded in cell membranes; others are used in the assembly of new virus particles. The particles acquire an envelope, spiked with viral proteins, either from the nuclear envelope or from the plasma membrane.

RNA serves as a template for synthesizing either DNA or mRNA. For example, HIV, a retrovirus, carts its own enzymes into cells. These assemble DNA on viral RNA by a process of reverse transcription (Section 13.4).

Nearly all viral multiplication cycles include five steps: attachment to a suitable host cell, penetration, viral DNA or RNA replication and protein synthesis, assembly of new viral particles, and release from the infected cell.

EBOLA AND OTHER EMERGING PATHOGENS

You are a potential host for many pathogenic bacteria, viruses, fungi, protozoans, and parasitic worms. **Infection** means a pathogen has invaded the body and is multiplying in cells and tissues. Its outcome, **disease**, results if defenses cannot be mobilized fast enough to prevent the pathogen's activities from interfering with normal body functions.

The good news is, most pathogens have less-than-fatal effects on a host. After all, an infected individual that lives longer can spread more of the next generation and so contribute to the pathogen's reproductive or replicative success. Generally, death results only when a pathogen attacks an individual in great numbers or when it infects a novel host—one with no coevolved defenses against it.

This brings us to the bad news. Thanks to planes, trains, and automobiles, great numbers of humans now travel to different or remote places, such as virgin tropical forests. There, a strange and often deadly pathogen may infect them. In a matter of hours, infected individuals can taxi the pathogen far away from its home range.

We know little about the deadly, **emerging pathogens**. Maybe they have been around for a long time and are taking advantage of the increased presence of a novel host. Maybe they are newly mutated strains of existing species.

Consider *Ebola*, one of the deadliest viruses that causes hemorrhagic fever. *Ebola* may have coevolved with monkeys in tropical forests in Africa. By 1976 and possibly earlier, it was infecting humans. It kills 70 to 90 percent of its victims. There is no treatment or vaccine, and it is not a pretty death. The disease starts abruptly, with fever and flu-like aches. The virus destroys cells making up the lining of blood vessels. Blood seeps out of the circulatory system. The liver and kidneys may quickly turn to mush. Victims become deranged before dying of circulatory shock.

Since 1976, there have been only four *Ebola* epidemics. During an **epidemic**, a disease abruptly spreads through part of a population for a limited period. This particular pathogen spreads mainly through direct contact with the body fluids or secretions of infected individuals, although there is evidence that infections may also occur through airborne transmission.

What happens when epidemics break out in several countries around the world in a given period? This is a **pandemic**. AIDS, an incurable disease caused by the human immunodeficiency virus (HIV) is an example. Millions of people around the world are infected.

By contrast, *sporadic* diseases such as whooping cough break out irregularly and affect few people. *Endemic* diseases occur more or less continuously, but they don't spread far in large populations. (Tuberculosis is like this. So is impetigo, a highly contagious bacterial infection that is often confined to individual day-care centers.) These less alarming patterns of infection may change in the near future. Increasingly virulent strains of the causative agents are emerging in countries throughout the world.

18.8 PROTISTAN CLASSIFICATION

Structurally, **protistans** are the simplest eukaryotes. And through no fault of their own, they are viewed as the ragbag of classification schemes. It's been said that they are most easily classified by what they are *not*. Later chapters describe structural and functional traits that distinguish protistans from plants, fungi, and animals. By a process of elimination, "protistans" are organisms that do not show these traits and that are not bacteria.

Like all other eukaryotes, protistans differ from the bacteria in several respects. At the least, they have a nucleus, large ribosomes, endoplasmic reticulum, Golgi bodies, and mitochondria. The nucleus has two or more chromosomes, each consisting of a DNA molecule with many histones and other proteins. Protistans assemble microtubules for use in a cytoskeleton, as spindles for moving chromosomes, and in the 9 + 2 core of flagella and cilia. Many protistans have chloroplasts. All divide by way of mitosis, and many by way of meiosis.

Table 18.4 lists the major groups. Most are single celled, yet nearly every lineage also has multicelled forms, and some have close evolutionary ties to other kingdoms. The ones called chytrids, water molds, slime molds, protozoans, and sporozoans are heterotrophs. The euglenoids and dinoflagellates are photosynthetic, heterotrophic, or both. Nearly all chrysophytes and the red, brown, and green algae are photosynthesizers.

Table 18.4 Classification of Major Groups of Protistans	
Phylum	Common Name
Heterotrophs (decomposers, predators, grazers, parasites):	
Chytridiomycota	Chytrids
Oomycota	Water molds
Acrasiomycota	Cellular slime molds
Myxomycota	Plasmodial slime molds
Sarcodina	Amoeboid protozoans: Amoebas Foraminiferans Heliozoans Radiolarians
Ciliophora	Ciliated protozoans
Mastigophora	Flagellated protozoans Sporozoans*
Autotrophs and Heterotrophs:	
Euglenophyta	Euglenoids
Pyrrhophyta	Dinoflagellates
Autotrophs (mostly photosynthesizers):	
Chrysophyta	Chrysophytes (yellow-green algae) Diatoms
Rhodophyta	Red algae
Phaeophyta	Brown algae
Chlorophyta	Green algae

*The name has no formal taxonomic status.

18.9 PREDATORY AND PARASITIC MOLDS

Chytrids and Water Molds

The **chytrids** live mainly in freshwater and marine habitats. Most of the 575 species are decomposers; some are parasites. Like fungi, they secrete enzymes that digest organic matter of other organisms, such as marsh grasses, then absorb breakdown products. Single-celled types produce flagellated asexual spores that germinate on host cells. Complex species form a mycelium (plural, mycelia), a mesh of absorptive filaments.

Most of the 580 species of **water molds** are major decomposers of aquatic habitats. Some types parasitize aquatic animals and land plants (Figure 18.14). Most produce a mycelium. Some mycelial filaments develop into gamete-producing structures. Water molds also can produce flagellated, asexual spores.

Some water molds have influenced human affairs. Over a century ago, Irish peasants cultivated potatoes as their main food crop. Between 1845 and 1860, the growing seasons were cool and damp, year after year. The cool conditions encouraged the rapid spread of a water mold (*Phytophthora infestans*) that causes *late blight*, a rotting of potato and tomato plants. The mold's abundant spores were dispersed unimpeded through the watery film on plants. Destruction was rampant. During a fifteen-year period, a third of the population in Ireland starved to death, died in the outbreak of typhoid fever that followed as a secondary effect, or fled to the United States and other countries.

Slime Molds

During the life cycle of heterotrophic protistans called **slime molds**, free-living, amoebalike cells form. These crawl on rotting plant parts, such as decaying leaves and bark. Like true amoebas, they are phagocytes; they engulf bacteria, spores, and organic compounds. When cells are starving, they aggregate into a slimy mass that may migrate to a new place. It moves by contractions. Later, the amoebalike cells develop into a few cell types

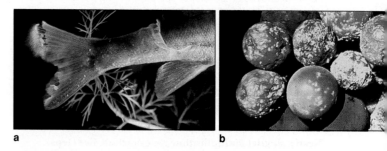

a   b

Figure 18.14 Effects of two water molds. (**a**) A parasitic water mold (*Saprolegnia*) has destroyed tissues of this aquarium fish. (**b**) *Plasmopara viticola* causes downy mildew in grapes. At times it has threatened large vineyards in Europe and North America.

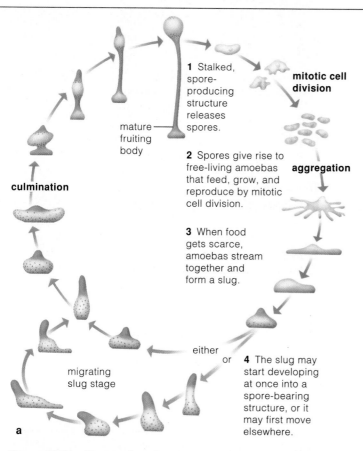

1 Stalked, spore-producing structure releases spores.

mitotic cell division

2 Spores give rise to free-living amoebas that feed, grow, and reproduce by mitotic cell division.

aggregation

3 When food gets scarce, amoebas stream together and form a slug.

mature fruiting body

culmination

either or

4 The slug may start developing at once into a spore-bearing structure, or it may first move elsewhere.

migrating slug stage

a

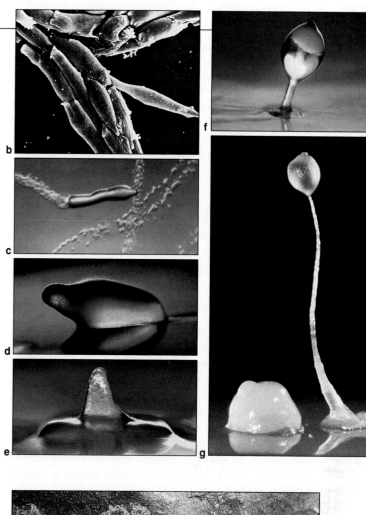

b

c

d

e

f

g

Figure 18.15 *Dictyostelium discoideum*, a cellular slime mold. (**a**) The life cycle includes a spore-producing stage. Spores give rise to free-living amoebas, which grow and divide until food (soil bacteria) dwindles. In response to a chemical signal (cyclic AMP) that they secrete, amoebas stream toward one another. As the amoebas aggregate, their plasma membranes become sticky, and they adhere to one another. A cellulose sheath forms around the aggregation, which starts crawling like a slug (**b–d**). Some slugs may consist of 100,000 or more amoebas.

As a slug migrates, the amoebas develop into different types of cells: prestalk (*red*), prespore (*white*), and anteriorlike cells (*brown dots*). Prestalk cells secrete ammonia, in amounts that vary with temperature and light intensity. The slug moves most rapidly in response to intermediate amounts of ammonia (not too little, not too much), which correspond to warm, moist conditions (not too cold or hot, not soppy or dry). Where such conditions exist at the surface of a substrate, prestalk cells and prespore cells differentiate and form a stalked, spore-bearing structure (**e–g**). Anteriorlike cells sort into two groups at each end of the posterior tissue. Possibly they function in elevating the nonmotile spores for dispersal from the top of the spore-bearing structure.

Figure 18.16 *Physarum*, one of the plasmodial slime molds. This aggregation of cells is migrating along a rotting log.

that differentiate and form a spore-bearing structure. After a spore germinates on a warm, damp surface, it gives rise to an amoebalike cell. Sexual reproduction (by gametes) also is common among slime molds.

Figure 18.15 illustrates the life cycle of *Dictyostelium discoideum*, one of 70 species of *cellular* slime molds.

There are approximately 500 species of *plasmodial* slime molds (Figure 18.16). When *their* amoebalike cells aggregate, the plasma membrane of each cell breaks

down. The cytoplasm flows and distributes nutrients and oxygen through the whole mass. The streaming mass (plasmodium) may grow several square meters and migrate if food runs out. You may have seen one crossing lawns or roads, even climbing trees.

Most chytrids and water molds are decomposers of aquatic habitats. Some are single cells. Like slime molds, many are free-living predators some of the time and simple experiments in multicellularity at other times.

During a slime mold life cycle, amoeboid cells aggregate to form a migrating mass. Cells in the mass differentiate, then form reproductive structures and spores or gametes.

18.10 ANIMAL-LIKE PROTISTANS

Collectively, about 65,000 single-celled protistans are called **protozoans** ("first animals") because they may resemble the single-celled, heterotrophic protistans that gave rise to animals. These are predators, parasites, or grazers. Asexual reproduction by budding or fission is most common, but protozoans also reproduce sexually. Some undergo multiple fission. (More than two nuclei form, then each nucleus and a bit of cytoplasm separate as a daughter cell.) Many parasitic types form **cysts**, structures that help them survive adverse conditions.

Fewer than two dozen protozoans cause diseases in humans but they are infecting hundreds of millions each year. There are no effective vaccines against them.

Amoeboid Protozoans

Amoebas, foraminiferans, heliozoans, and radiolarians are **amoeboid protozoans**. Adults move or capture prey by sending out pseudopods ("false feet"), which are temporary, cytoplasmic extensions of the cell body. Most feed on algae, bacteria, and other protozoans.

The soft-bodied amoebas may resemble the earliest eukaryotes. They engulf prey in soil, freshwater, and seawater. *Amoeba proteus* is a favorite for laboratory experiments in biology (Figure 18.17*a*). One parasitic species, *Entamoeba histolytica,* causes a severe intestinal disorder, *amoebic dysentery.* It travels in encysted form in feces, which may contaminate water and soil in places that do not have adequate sewage treatment facilities.

Most foraminiferans live in the ocean. They have hardened shells with tiny holes through which sticky, threadlike pseudopods extend (Figure 18.17*b,c*). Many bear stiffened spines. Radiolarians have silica skeletons (Figure 18.17*d*). Most are members of marine plankton; some form colonies in which many cells are cemented together. In many ocean sediments, staggering numbers of foraminiferan and radiolarian shells attest to the abundance of these organisms in the past. Heliozoans, or "sun animals," have fine pseudopods that radiate from the body like sun rays (Figure 18.17*e*). Most are free floating or bottom-dwellers in freshwater habitats.

Ciliated Protozoans

About 8,000 species of **ciliated protozoans** have notable arrays of cilia, which they use to swim through water (Figure 18.18). The ciliates abound in freshwater and marine habitats, where they feed on bacteria, algae, and one another. *Paramecium* is typical of the group. Cilia in its gullet sweep water laden with bacteria and food particles into the cell body. There, the food gets enclosed in enzyme-filled vesicles and is digested. Like amoeboid protozoans, *Paramecium* also has **contractile vacuoles**, organelles for collecting excess water that moves into a

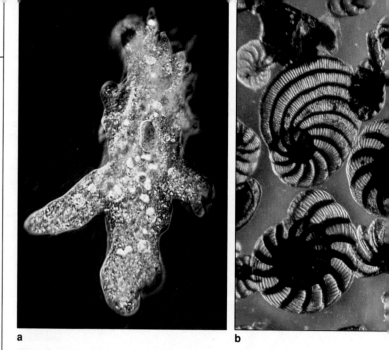

a   b

Figure 18.17 Amoeboid protozoans. (**a**) *Amoeba proteus.* (**b,c**) Shells and body plan of foraminiferans. (**d**) Radiolarian shell. (**e**) Micrograph and body plan of a living heliozoan. Its pseudopods have a core of supportive microtubules. Fluid-filled vacuoles make it neutrally buoyant so that it can remain suspended in water.

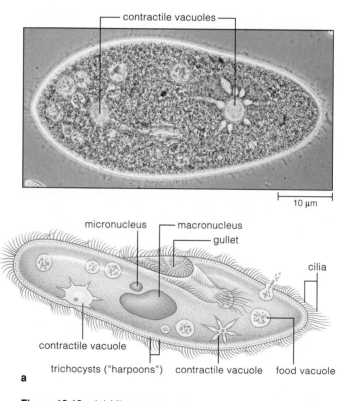

10 µm

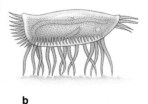

a

b

Figure 18.18 (**a**) Micrograph and body plan of *Paramecium*, a ciliated protozoan. Compare Figure 3.19*c*, which shows its ciliated surface. (**b**) From the Bahamas, a hypotrich. Hypotrichs are the most animal-like ciliates. They run around on leglike tufts of cilia. Some have a "head" end with modified sensory cilia.

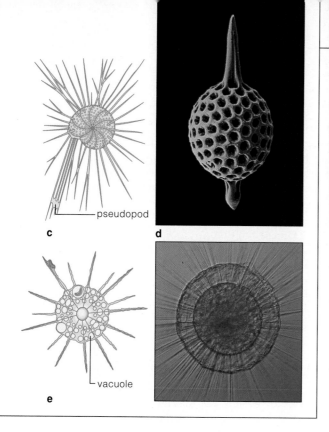

pseudopod

c

d

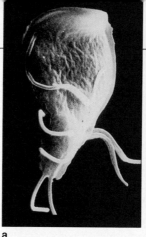

a

b

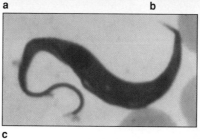

c

vacuole

e

Figure 18.19 A few flagellated protozoans. (**a**) *Giardia lamblia* causes intestinal disturbances. (**b**) *Trichomonas vaginalis* causes trichomoniasis, a sexually transmitted disease. (**c**) *Trypanosoma brucei* causes African sleeping sickness.

cell by osmosis. A filled vacuole contracts and forces water through a small pore to the outside.

Ciliates reproduce sexually and asexually. Each has a large macronucleus and one or more smaller, haploid micronuclei. During **protozoan conjugation**, two cells exchange micronuclei, which fuse and thereby form a diploid macronucleus in each cell. When cells divide by fission, each daughter cell is diploid.

Flagellated Protozoans

The **flagellated protozoans** bear one to several flagella. Many live in freshwater or marine habitats; others are internal parasites. *Giardia lamblia* (Figure 18.19*a*) may infect as many as 10 percent of the people in the United States. It usually causes mild intestinal upsets, such as diarrhea, but it has severe effects on a few susceptible people. Wild animals and foraging cattle often contain the parasite. It leaves the body encysted in feces, then infects people who drink water or ingest food that was contaminated with the feces. Even remote streams may harbor cysts, so water should be boiled before drinking.

Many trichomonads are among the parasitic types. *Trichomonas vaginalis*, a worldwide nuisance, can infect new human hosts during sexual intercourse (Figure 18.19*b*). Unless trichomonad infections are treated, they can damage the urinary and reproductive tracts.

Trypanosomes also are parasites. One, *Trypanosoma brucei*, causes *African sleeping sickness* (Figure 18.19*c*). Bites of the tsetse fly transmit it to hosts, where it may damage the nervous system. *T. cruzi*, prevalent in South

America and Mexico, causes *Chagas disease*. Bugs pick up the parasite when they feed on infected humans and other animals. The parasite multiplies in the insect gut, then it may be excreted onto a host. Scratches in skin invite infection. The liver and spleen enlarge, eyelids and the face swell, then the brain and heart become severely damaged. There is no treatment or cure.

Sporozoans

Sporozoan is an informal name for parasitic protistans that must complete part of their life cycle *inside* specific cells of their hosts. These intracellular parasites produce infective, motile stages known as sporozoites. Many become encysted during the life cycle.

Section 18.11 details the life cycle of *Plasmodium*, a sporozoan that causes malaria. *Toxoplasma*, another dangerous sporozoan, completes the sexual phase of its life cycle in cats and asexual phases in humans, cattle, pigs, and other animals. Humans can get infected when they ingest contaminated meat that is raw or undercooked. Sporozoite-containing cysts in the feces of infected cats are spread by houseflies, cockroaches, and other insects. The resulting disease, *toxoplasmosis*, is not prevalent in the population but is a major cause of birth defects. A woman who becomes infected during pregnancy may transmit the disease to her fetus, which may suffer brain damage or die. Pregnant women should never empty litterboxes or otherwise clean up after any cat.

Protozoans are single-celled predators, grazers, and parasites, some with animal-like traits. Fewer than two dozen parasitic species of protozoans are infecting hundreds of millions of people in a given year.

MALARIA AND THE NIGHT-FEEDING MOSQUITOES

Each year in Africa alone, about a million people die of *malaria*, a long-lasting disease caused by four species of *Plasmodium*. By some estimates, these sporozoans have infected more than 100 million people.

Female mosquitoes (*Anopheles*) transmit the parasite to human (or bird) hosts. There it must complete part of its life cycle inside liver cells and red blood cells. Sporozoites (an infective, motile stage) repeatedly engage in multiple fission, an asexual process that produces thousands of progeny called merozoites. These infect and multiply inside more and more red blood cells (Figure 18.20).

Symptoms start when infected cells suddenly rupture and release merozoites, metabolic wastes, and cellular debris into the bloodstream. Shaking, chills, a burning fever, and then drenching sweats are classic symptoms of malaria. After one such episode, symptoms subside for a few weeks or months. Infected individuals may even feel normal, but relapses inevitably recur. In time, anemia and gross enlargement of the liver and spleen may result.

The *Plasmodium* life cycle is attuned to the host's body temperature, which normally rises and falls rhythmically every twenty-four hours. Gametocytes, the infective stage that is transmitted to new hosts, are ready for travel at night—when mosquitoes feed.

Malaria has been around for a long time. More than 2,000 years ago, people were describing its symptoms. It got its name in the seventeenth century, when Italians made a connection between the disease and noxious gases from swamps near Rome—where mosquitoes flourished (*mal*, bad; *aria*, air). By severely incapacitating so many people, malaria contributed to the decline of the ancient empires of Greece and Rome. It incapacitated soldiers during the United States Civil War, World War II, and the Korean and Viet Nam conflicts.

Throughout history, malaria has been most prevalent in tropical and subtropical parts of Africa. Today, however, the number of cases in North America and elsewhere is increasing dramatically, owing to globe-hopping travelers and unprecedented levels of human immigration.

Travelers in countries with high rates of malaria are advised to use antimalarial drugs such as chloroquine. Drug resistance is common, however, and a vaccine has been difficult to develop. Vaccines induce the body to build up resistance to a specific pathogen. Experimental vaccines for malaria have not been equally effective against all the different stages that develop during sporozoan life cycles. This is true of most parasites with complex life cycles.

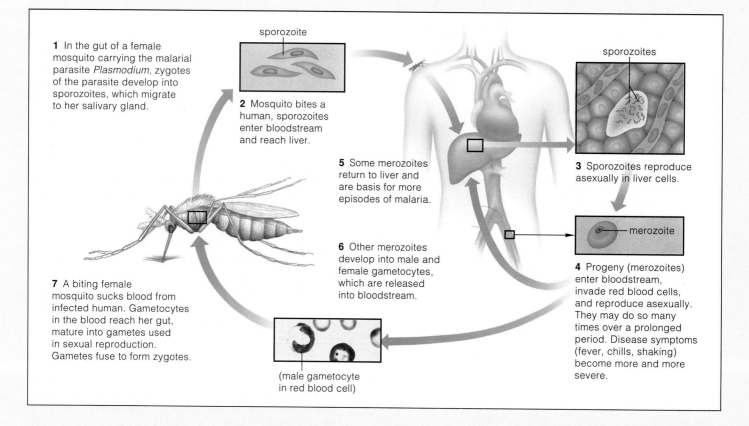

1 In the gut of a female mosquito carrying the malarial parasite *Plasmodium*, zygotes of the parasite develop into sporozoites, which migrate to her salivary gland.

sporozoite

2 Mosquito bites a human, sporozoites enter bloodstream and reach liver.

5 Some merozoites return to liver and are basis for more episodes of malaria.

6 Other merozoites develop into male and female gametocytes, which are released into bloodstream.

7 A biting female mosquito sucks blood from infected human. Gametocytes in the blood reach her gut, mature into gametes used in sexual reproduction. Gametes fuse to form zygotes.

(male gametocyte in red blood cell)

sporozoites

3 Sporozoites reproduce asexually in liver cells.

merozoite

4 Progeny (merozoites) enter bloodstream, invade red blood cells, and reproduce asexually. They may do so many times over a prolonged period. Disease symptoms (fever, chills, shaking) become more and more severe.

Figure 18.20 Life cycle of one of the sporozoans (*Plasmodium*) that causes malaria.

18.12 THE (MOSTLY) PHOTOSYNTHETIC SINGLE-CELLED PROTISTANS

In the open ocean, seas, lakes, and smaller bodies of water, we find most of the single-celled photosynthetic protistans, often in vast numbers. They are important members of **phytoplankton**, the food producers that are the basis for nearly all food webs in aquatic habitats.

Euglenoids

Euglenoids, which abound in freshwater or stagnant ponds and lakes, are a classic example of evolutionary experimentation. The *Euglena* cell shown in Figure 18.21 corresponds to Figure 17.21. Its many organelles include chloroplasts and a contractile vacuole. Its flexible outer layer, called a pellicle, has spiral strips of a translucent material. Pigment granules (an "eyespot") partly shield a light-sensitive receptor. By moving a long flagellum, the cell keeps the receptor exposed to light and so keeps itself where the light is most suitable for its activities.

Members of most euglenoid genera are heterotrophs that subsist on organic compounds dissolved in water. Yet at least 1,000 species are photosynthetic. Were the ancestors of these species *also* heterotrophs that long ago acquired chloroplasts by endosymbiosis, as described in Section 17.5? Probably.

Chrysophytes

Most **chrysophytes** are single photosynthetic cells with chlorophylls a, c_1, and c_2 and hardened walls, shells, or scales. Silica-hardened parts are common among the 500 species of **golden algae**. Fucoxanthin, a golden-brown pigment, masks the chlorophylls in photosynthetic types (Figure 18.22*a*). Golden algae are important in freshwater and marine habitats. So are 5,600 species of **diatoms** (Figure 18.22*b*). Most diatoms are photosynthetic. Their shells are two perforated structures of silica that overlap like a pillbox. Substances move to and from the cell's plasma membrane through the shell's numerous perforations. For 100 million

Figure 18.22 (**a**) *Synura*, a golden alga that grows as colonies in phytoplankton. (**b**) Diatom shells. (**c**) One dinoflagellate, *Gymnodinium breve*, that causes red tides along Florida's coast. (**d**) Part of a fish kill that resulted from a dinoflagellate "bloom."

years, the silica shells of at least 35,000 species of now-extinct diatoms accumulated at the bottom of lakes and seas. Many sediments contain deposits of their finely crumbled shells, which we use in abrasives, filters, and insulation. More than 270,000 metric tons are quarried annually near Lompoc, California.

Dinoflagellates

Dinoflagellates occur in marine phytoplankton. Most of the 1,200 species are photosynthetic; but many are heterotrophs. Some have flagella and cellulose plates (Figure 18.22*c*). Dinoflagellates are yellow-green, green, brown, blue, or red, depending on their pigments and endosymbiotic history. Some types undergo population explosions and color the seas red or brown. Because a few species produce a toxin, the resulting **red tides** can be devastating. Great numbers of plankton-eating fish may be poisoned and wash up along coasts. The toxin also accumulates in tissues of clams, oysters, and other mollusks. Humans who eat tainted mollusks may die.

Nearly all single-celled photosynthetic protistans, including most euglenoids, chrysophytes, and dinoflagellates, belong to phytoplankton—the food-producing base of aquatic habitats.

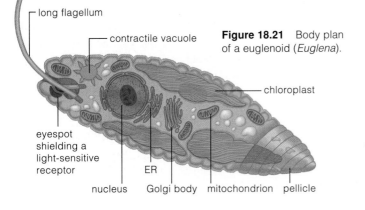

long flagellum

contractile vacuole

Figure 18.21 Body plan of a euglenoid (*Euglena*).

chloroplast

eyespot shielding a light-sensitive receptor

ER

nucleus   Golgi body   mitochondrion   pellicle

18.13 THE (MOSTLY) MULTICELLED PHOTOSYNTHETIC PROTISTANS

Red Algae

Nearly all of the 4,000 known species of **red algae** live in the seas; fewer than 3 percent live in lakes, streams, and springs. Red algae are especially abundant in warm currents and tropical seas, often at surprising depths (past 265 meters) in clear water. A few are members of plankton. The accumulation of some encrusting types contributes to the formation of coral reefs and banks.

Cells of red algae have walls hardened with calcium carbonate deposits. Different species appear red, green, purple, or nearly black, depending on which accessory pigments (phycobilins, mostly) mask their chlorophyll *a*. Phycobilins are good at absorbing the green and blue-green wavelengths, which can penetrate deep waters. (Chlorophylls are more efficient at absorbing red and blue wavelengths, which may not reach as far below the water's surface.) The chloroplasts of red algae resemble cyanobacteria, which suggests endosymbiotic origins.

As is true of brown and green algae, the modes of reproduction are diverse, with complex asexual and sexual phases. Most species become multicelled during the life cycle. Filamentous, often branching bodies form (Figure 18.23*a*), but these don't have tissues or organs. Red algae have a flexible, slippery texture owing to mucous material in their cell walls. Agar is made from extracts of wall material of several species. This inert, gelatinous substance is used as a moisture-preserving agent in baked goods and cosmetics, as a setting agent for jellies, and as culture gels. We also shape it into soft capsules (for drugs and food supplements). And we use carrageenan, extracted from *Euchema*, as a stabilizer in paints, dairy products, and many other emulsions.

Brown Algae

Walk along a rocky coast at low tide and you may come across brown-tinged seaweeds (Figure 18.23*b*). They are among the 1,500 species of **brown algae**. Nearly all inhabit cool or temperate marine waters. Depending on their pigment arrays, brown algae appear olive-green, golden, or dark brown. Like the chrysophytes to which they may be related, they contain fucoxanthin and other carotenoids as well as chlorophylls *a*, *c*₁, and *c*₂.

Brown algae range from microscopic, filamentous *Ectocarpus* to giant kelps twenty to thirty meters long. The giant kelps, including *Macrocystis* and *Laminaria*, are the largest and most complex protistans. During the life cycles, gamete-producing bodies (gametophytes) and spore-producing bodies (sporophytes) form. The multicelled sporophytes have holdfasts (anchoring structures), stipes (stemlike parts), and blades (leaflike parts). Hollow, gas-filled bladders impart buoyancy to

a

Figure 18.23 (**a**) A red alga (*Bonnemaisonia hamifera*) showing the most common growth pattern: filamentous and branched. (**b**) Diagram of *Macrocystis*, which shows the structural organization characteristic of many large brown algae.

bladder

blade

stipe

holdfast

b

stipes and blades and keep them upright. Stipes have tubelike arrangements of elongated cells. Like the tissues that conduct food in plants, the tubes rapidly carry dissolved sugars and other products of photosynthesis to all living cells throughout the alga.

Giant kelps function as productive ecosystems. Think of them as underwater forests within which great numbers of diverse bacteria and protistans, as well as fishes and other animals, carry out their lives. Extensive masses of *Sargassum*, another brown alga, are floating ecosystems in the vast Sargasso Sea, which lies between the Azores and the Bahamas.

If you enjoy ice cream, pudding, salad dressings, canned and frozen foods, jellybeans, or beer, if you use cough syrup, toothpaste, cosmetics, floor polish, or paper, thank the brown algae. The cell walls of some species contain algin, which can serve as thickening, emulsifying, and suspension agents. Especially in the Far East, people also harvest various kelps as sources of food and mineral salts, and as a fertilizer for crops.

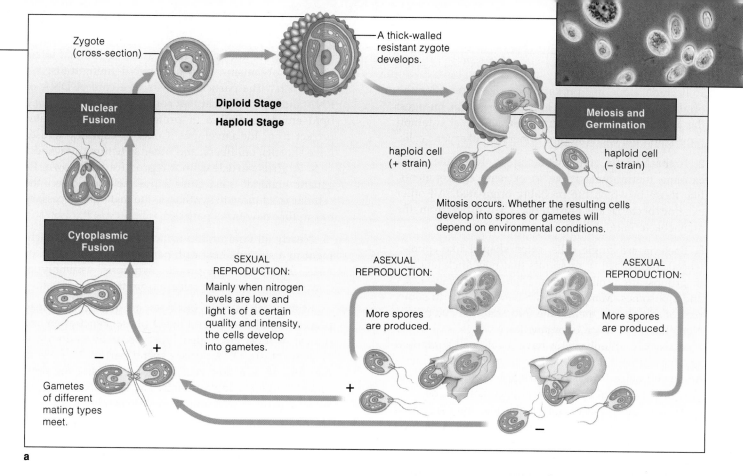

Zygote
(cross-section)

A thick-walled
resistant zygote
develops.

Nuclear Fusion

Diploid Stage

Haploid Stage

Meiosis and Germination

haploid cell
(+ strain)

haploid cell
(– strain)

Mitosis occurs. Whether the resulting cells develop into spores or gametes will depend on environmental conditions.

Cytoplasmic Fusion

SEXUAL REPRODUCTION:

Mainly when nitrogen levels are low and light is of a certain quality and intensity, the cells develop into gametes.

ASEXUAL REPRODUCTION:

More spores are produced.

ASEXUAL REPRODUCTION:

More spores are produced.

−   +

+

−

Gametes of different mating types meet.

a

Figure 18.24 (a) Life cycle of a species of *Chlamydomonas*, one of the most common green algae of freshwater habitats. This single-celled species reproduces asexually most of the time and sexually under certain environmental conditions. (b) One mode of sexual reproduction in *Spirogyra*, or watersilk. (*1*) This green alga has spiral, ribbonlike chloroplasts. (*2*) A conjugation tube forms between cells of adjacent haploid filaments of different mating strains. (*3,4*) The cellular contents of one strain pass through the tubes into cells of the other strain, where zygotes form. The zygotes develop thick walls. Later, as they germinate, they will undergo meiosis and give rise to new haploid filaments.

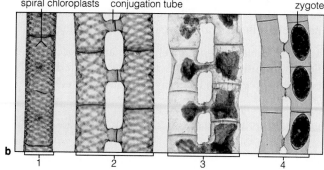

spiral chloroplasts   conjugation tube   zygote

b   1   2   3   4

Green Algae

Of all protistans, **green algae** bear the greatest structural and biochemical resemblances to plants and may be their closest relatives. For example, like plants, their chlorophylls are the molecular types designated *a* and *b*. Their chloroplasts, too, contain starch grains. And the cell walls of some species contain cellulose, pectins, and other polysaccharides typical of plants.

With at least 7,000 species, green algae show more diversity than other algae. Most live in freshwater. You also can find them at the ocean surface, just below the surface of soil and marine sediments, and on rocks, tree bark, other organisms, and snow. Some are symbionts with fungi, protozoans, and a few marine animals. A colonial form (*Volvox*) is a hollow whirling sphere of 500 to 60,000 flagellated cells. White, powdery beaches in the tropics are largely the work of countless green algal cells (*Halimeda*) that formed calcified cell walls, then died and disintegrated. Some day, green algae

might accompany astronauts in space. They can grow in small spaces, give off vital oxygen, and take up carbon dioxide exhaled by the aerobically respiring crew.

Chlamydomonas, a tiny green alga, is a food producer in nutrient-poor peat bogs and ponds (Figure 18.24*a*). Another green alga, *Codium magnum,* is taller than you are. In 1956, *Codium* from Washington's Puget Sound was accidentally introduced to the Connecticut River—where predators had not coevolved with them. The introduced alga spread along the Atlantic coast from Maine to North Carolina in just over two decades.

Red, brown, and green algae are photosynthesizers, mainly in aquatic habitats. They show great diversity in their size, morphology, life-styles, reproductive modes, and habitats.

Structurally and biochemically, the green algae strongly resemble plants and may be their closest relatives.

SUMMARY

1. Bacteria alone are prokaryotic cells. Soon after the origin of life, divergences led to three great lineages: the archaebacteria, eubacteria (now the most common types), and the bacterial forerunners of eukaryotic cells.

2. Archaebacteria (the methanogens, halophiles, and extreme thermophiles) live in extreme environments, like those in which life probably originated. All other existing bacteria are eubacteria. They differ from the archaebacteria in wall structure and other features.

3. Bacteria are single cells, often shaped like spheres (cocci), rods (bacilli), or spirals that may stick together in groups after cell division. Each cell contains DNA and ribosomes. Many species have plasmids. In some, part of the plasma membrane folds into the cytoplasm. None has a profusion of organelles.

   a. Nearly all eubacteria have a cell wall composed of a type of peptidoglycan. It protects the plasma membrane and helps it resist rupturing. Its composition and structure help identify particular bacterial species.

   b. A sticky mesh of polysaccharides may surround the wall as a capsule or slime layer. It helps bacteria attach to substrates and sometimes to resist a host organism's infection-fighting mechanisms.

   c. Bacterial flagella, structures used for motility, rotate like a propeller. Pili are filamentous proteins that help cells adhere to a surface or facilitate conjugation.

4. Most bacteria reproduce by binary fission. After DNA replication, this cell division mechanism divides a cell into two genetically equivalent daughter cells.

5. Collectively, bacteria show great metabolic diversity.

   a. Cyanobacteria and other photoautotrophs use sunlight energy and carbon dioxide in photosynthesis. Photoheterotrophs use light and organic compounds (instead of carbon dioxide) as carbon sources.

   b. Nitrifying bacteria and other chemoautotrophs use carbon dioxide but not sunlight. They strip electrons from inorganic substances.

   c. Chemoheterotrophs get carbon and energy from living hosts or organic wastes and remains. Most bacteria are like this. They include important decomposers and pathogens.

6. Many bacterial types make complex behavioral responses to their environment.

7. Viruses are nonliving, noncellular agents that infect nearly all organisms. They have two defining traits:

   a. Each virus particle consists of a core of DNA or RNA and a protein coat that sometimes is enclosed in a lipid envelope. Spikes of proteins and carbohydrates project from the envelopes. Coats of complex viruses have sheaths, tail fibers, and other structures attached.

   b. A virus particle cannot reproduce on its own. Its genetic material must enter a host cell and direct the cellular machinery to synthesize the materials necessary to produce new virus particles.

8. Nearly all viral multiplication cycles require attachment to a suitable host cell, penetration, viral DNA or RNA replication and protein synthesis, assembly of new viral particles, and release from the infected cell.

9. Multiplication cycles of viruses are diverse. They may proceed rapidly or enter a latent phase. Penetration and release of most enveloped types occur by endocytosis and exocytosis. DNA viruses spend part of the cycle in the nucleus of a host cell. For RNA viruses, the cycle proceeds in the cytoplasm only. The viral RNA serves as a template for mRNA and for protein synthesis.

10. In this chapter and others, we have considered characteristics of the simplest eukaryotes—the protistans—and speculated on their evolutionary links with other kingdoms. Table 18.5 pulls together the key similarities and differences among prokaryotes and eukaryotes.

11. Many protistans are heterotrophs. These include the water molds, chytrids, slime molds (both cellular and plasmodial), protozoans (the amoeboid, flagellated, and ciliated species), and sporozoans.

Table 18.5	Comparison of Prokaryotes With Eukaryotes	
	Prokaryotes	Eukaryotes
Organisms represented:	Bacteria only	Protistans, fungi, plants, and animals
Ancestry:	Two major lineages (archaebacteria and eubacteria) that began more than 3.5 billion years ago	Equally ancient prokaryotic ancestors gave rise to forerunners of eukaryotes, which emerged before 1.2 billion years ago
Level of organization:	Single celled	Protistans, single celled or multicelled. Nearly all others are multicelled, with a division of labor among differentiated cells, tissues, and often organs.
Typical cell size:	Small (1–10 micrometers)	Large (10–100 micrometers)
Cell wall:	Most have distinctive walls	Cellulose or chitin; none in animal cells
Membrane-bound organelles:	Very rarely	Typically profuse
Modes of metabolism:	Both anaerobic and aerobic	Aerobic modes predominate
Genetic material:	Bacterial chromosome (and sometimes plasmids)	Complex chromosomes (DNA, many associated proteins) within a nucleus
Mode of cell division:	Binary fission, mostly; also budding	Nuclear division (mitosis, meiosis, or both), followed by cytoplasmic division

12. Like fungi, chytrids and water molds are decomposers or parasites. They secrete enzymes that digest organic matter, then absorb breakdown products. Some form a mycelium (a mesh of absorptive filaments).

13. Like animals, slime molds are predatory. For part of their life cycles, they are phagocytic, amoebalike cells. Like fungi, they produce spore-producing structures.

14. Like animals, protozoans are predators, grazers, or parasites. The amoebas, foraminiferans, heliozoans, and radiolarians are amoeboid protozoans. *Paramecium* and other ciliated protozoans use arrays of cilia for motility. The flagellated protozoans are internal parasites or free-living in aquatic habitats. Many cause human diseases.

15. Sporozoans are intracellular parasites that produce infective, motile stages and that may form cysts. An example is *Plasmodium*, which causes malaria.

16. Like plants, most species of euglenoids, chrysophytes (golden algae, diatoms, and yellow-green algae), dinoflagellates, and the red, brown, and green algae are photoautotrophs. Many are members of phytoplankton.

17. Red, brown, and green algae differ from one another in body plan, size, reproductive modes, and habitats. Most are multicelled photosynthesizers. Some species of brown algae are the largest protistans. Plants may be descended from early green algae.

Review Questions

1. Label the structures on this generalized bacterial cell: *286*

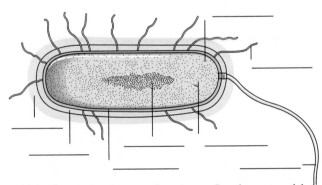

2. Name the major categories of protistans. Correlate some of their structural features with conditions in their environments. *296–303*

Self-Quiz *(Answers in Appendix IV)*

1. _____ live in habitats much like those of the early Earth.
   a. Cyanobacteria
   b. Eubacteria
   c. Archaebacteria
   d. Protozoans

2. Viruses have a _____ and a _____ .
   a. DNA core; carbohydrate coat
   b. DNA or RNA core; plasma membrane
   c. DNA-containing nucleus; lipid envelope
   d. DNA or RNA core; protein coat

3. Protistans differ from bacteria in which way?
   a. The protistan cell body contains cytoplasm.
   b. Some protistans have cell walls.
   c. Protistans have DNA.
   d. Protistans engage in mitosis and meiosis.

4. Match the organisms with the most suitable descriptions.
   ____ archaebacteria   a. bacteria other than archaebacteria
   ____ eubacteria   b. nonliving infectious particle, nucleic acid core, protein coat
   ____ viruses
   ____ *Plasmodium*   c. small circles of bacterial DNA
   ____ plasmids   d. methanogens, halophiles, extreme thermophiles
                          e. agent of malaria

Critical Thinking

1. *Salmonella* bacteria cause a dangerous form of food poisoning of the same name. The bacteria often live in poultry and eggs, but not in newly hatched chicks—until chicks eat the bacteria-harboring feces of healthy adult chickens. Harmless bacteria ingested this way colonize the surface of intestinal cells, leaving no place for pathogenic bacteria to take hold and cause infection. Some farmers raise thousands of chicks in confined quarters with no adult chickens. Should they consider feeding the chicks a known mixture of bacteria from a laboratory or a mixture of unknown bacteria from healthy adult chickens? Devise an experiment to find out which approach might be more effective.

2. In 1852 in the Coorong district of Australia, a rubbery substance was discovered near a lake that had formed after a period of heavy rainfall. The substance, later named coorongite, consisted of the same kinds of hydrocarbons as crude oil. The discovery sparked sixty years of feverish oil drilling even though geologists knew the place wasn't likely to yield oil. Later, biologists wondered if microorganisms had formed the substance. Drilling continued. Then someone found a green, scummy mat of bacteria (*Botyrococcus braunii*) in a natural lagoon. After water around the mat evaporated, the mat congealed into coorongite. Chemical analysis suggests it's an early stage in the formation of shale oil. Formulate a hypothesis to explain how the Earth's great oil reserves formed.

Selected Key Terms

amoeboid	diatom *301*	methanogen *290*
protozoan *298*	dinoflagellate *301*	microorganism *284*
bacterial	disease *295*	pandemic *295*
chromosome *288*	emerging	pathogen *284*
bacterial	pathogen *295*	phytoplankton *301*
conjugation *288*	endospore *291*	plasmid *288*
bacterial	epidemic *295*	prion *293*
flagellum *287*	euglenoid *301*	prokaryotic cell *287*
bacteriophage *292*	extreme	protistan *296*
binary fission *288*	thermophile *290*	protozoan *298*
brown alga *302*	flagellated	protozoan
cell wall *287*	protozoan *299*	conjugation *299*
chrysophyte *301*	fruiting body *291*	red alga *302*
chytrid *296*	golden alga *301*	red tide *301*
ciliated	green alga *303*	slime mold *296*
protozoan *298*	halophile *290*	sporozoan *299*
contractile	heterocyst *290*	viroid *293*
vacuole *298*	infection *295*	virus *292*
cyst *298*	lysis *294*	water mold *296*

Readings

Geunno, B. October 1995. "Emerging Viruses." *Scientific American* 273(4): 56-62.

Ingraham, J., and C. Ingraham. 1995. *Introduction to Microbiology.* Belmont, California: Wadsworth.

19 PLANTS AND FUNGI

Pioneers In a New World

Seven hundred million years ago, no shorebirds stirred and noisily announced the dawn of a new day. No crabs clacked their tiny claws together and skittered off to burrows. The only sounds were the rhythmic muffled thuds of waves in the distance, at the outer limits of another low tide. About 3 billion years before, life had its beginnings in the waters of the Earth—and now, quietly, the invasion of the land was under way.

Astronomical numbers of photosynthetic cells had come and gone, and oxygen-producing ones had slowly changed the atmosphere. High above the Earth, the sun's energy had converted much of the oxygen into a dense ozone layer. This became a shield against lethal doses of ultraviolet radiation—which had kept early organisms below the water's surface.

Were cyanobacteria the first to adapt to intertidal zones, where mud dried out with each retreating tide? Were they the first to spread into shallow, freshwater streams meandering down to the coasts? Probably so. We know that later in time, green algae and fungi made the same journey together.

Every plant around you is a descendant of green algae that lived near the water's edge or made it onto the land. Diverse fungi still associate with nearly all of them. Together, plants and fungi became the basis of communities in coastal lowlands, near the snowline of high mountains, and just about all places in between.

We have some tantalizing fossils of the pioneers. We also are learning about them through comparative biochemistry and studies of existing species. Today, as in Precambrian times, cyanobacteria and green algae grow in mats in nearshore waters and on the banks of freshwater streams (Figure 19.1*a*). After a volcano erupts or a glacier retreats, cyanobacteria are the first to colonize the barren rocks. Then symbiotic associations between green algae and fungi follow. Gradually the products and remains of these organisms accumulate and create pockets of soil. Mosses and other plants take hold in the newly forming soil and further enrich it.

With this chapter we turn to plants and their fellow travelers through time, the fungi. Nearly all plants are multicelled photoautotrophs. They conjure up organic compounds, using energy from the sun, carbon dioxide from the air, and minerals dissolved in water. And these metabolic wizards split water molecules. In doing so, they obtain the stupendous numbers of electrons and hydrogen atoms required for growth into multicellular forms as tall as giant redwoods, as extensive as an aspen forest that is one continuous clone.

Fungi, all heterotrophs, break down the discards, the remains, and sometimes the tissues of plants and other organisms. Fungi absorb the released nutrients—and so do plants.

We know of at least 295,000 kinds of plants and 56,000 kinds of fungi. Be glad their ancient ancestors left the water. Without them, we humans and all other land-dwelling animals never would have made it onto the evolutionary stage.

a

b

c

Figure 19.1 (**a**) Filaments of a green alga, massed in a shallow stream. More than 400 million years ago, green algae that may have been ancestral to plants lived in similar streams that meandered down to the shore of early continents. (**b**) Land-dwelling descendants of those ancestral forms—the conifers of Silver Star Mountain in Washington. (**c**) A denuded mountain in Alaska, symbolic of the astoundingly rapid deforestation now under way throughout the world. With this chapter, we turn to the beginning—and ends of the line—of some truly ancient lineages.

KEY CONCEPTS

1. All but a few plants are multicelled photosynthesizers. Although their earliest ancestors lived in water, most are adapted to land. Most have a waxy cuticle, root and shoot systems, and internal tissues for conducting water and solutes. Plants also nourish, protect, and disperse their gametes and offspring in ways that are responsive to the conditions of specific habitats.

2. Early divergences gave rise to the bryophytes, then seedless vascular plants, and then seed-bearing vascular plants. Of these categories, the seed producers were most successful in radiating into drier environments.

3. Fungi are heterotrophs. Their enzyme secretions digest food outside their body, then their cells absorb breakdown products. Saprobic types obtain nutrients from nonliving organic matter. Parasitic types get them from living hosts.

4. Together with heterotrophic bacteria, fungi are major decomposers. Their activities release carbon dioxide to the atmosphere and return many nutrients to the soil or water, where they become available to producer organisms.

5. Certain fungi are symbionts with cyanobacteria and algae, in structures called lichens. Others are symbionts with the young roots of plants, forming mycorrhizae.

6. We tend to assign value to plants and fungi in terms of their direct effect on our lives. Our battles with "bad" ones and reliance on "good" ones should start from a solid understanding of their long-established roles in nature.

19.1 EVOLUTIONARY TRENDS AMONG PLANTS

Overview of the Plant Kingdom

The plant kingdom includes at least 295,000 species of photoautotrophs and a few heterotrophs. Most kinds are **vascular plants**, with internal tissues that conduct and distribute water and solutes throughout the plant body. These plants have roots, stems, and leaves, which are defined in part by the presence of vascular tissues. Fewer than 19,000 species are *nonvascular* plants called **bryophytes**. Together with the photosynthetic bacteria and protistans, plants are the key producers of organic compounds for nearly all communities of life.

Liverworts, hornworts, and mosses are bryophytes. The whisk ferns, lycophytes, horsetails, and ferns are *seedless* vascular plants. Cycads, ginkgo, gnetophytes, and conifers belong to a group of *seed-bearing* vascular plants called **gymnosperms**. The **angiosperms**, another group of vascular plants, bear flowers as well as seeds. There are two classes of flowering plants, which we informally call the dicots and monocots.

The ancestors of plants had evolved in the seas by 700 million years ago. About 265 million more years passed before simple stalked species appeared along coasts and streams. Evolutionarily speaking, the pace picked up after that. Within a mere 60 million years, plants had radiated through much of the land. Some long-term changes in structure and reproductive events help explain how the diversity came about.

Evolution of Roots, Stems, and Leaves

Underground structures started evolving among the first colonizers of land. In lineages leading to vascular plants, these developed into **root systems**. Most root systems consist of underground, cylindrical absorptive structures with a large surface area. They rapidly take up soil water and dissolved mineral ions, and often they anchor the plant. Aboveground, other structures evolved into **shoot systems**. Shoot systems have stems

and leaves, which function in the absorption of energy from the sun and carbon dioxide from the air. Stems grew and branched extensively after plants developed a biochemical capacity to form and deposit **lignin**, an organic compound, in cell walls. Great numbers of cells with lignified walls structurally support the shoots and so increase the light-intercepting surface area of leaves.

In time, many plants were equipped with cellular pipelines for water and solutes. The pipelines were key factors in the evolution of roots, stems, and leaves. They evolved as components of two vascular tissues—xylem and phloem. **Xylem** distributes both water and dissolved ions through plant parts. **Phloem** distributes dissolved sugars and other photosynthetic products.

Life on land also depended on water conservation, which wasn't a problem in most aquatic habitats. Stems and leaves became protected by a **cuticle**, a waxy coat that helps reduce water loss on hot, dry days. **Stomata** (singular, stoma), tiny openings across leaf surfaces, became strategic in the controlled absorption of carbon dioxide and restriction of evaporative water loss. Later chapters describe these tissue specializations.

From Haploid to Diploid Dominance

As early plants moved into higher, drier places, their life cycles underwent modification. Think of the green algae, which quickly form and release gametes into the water. Actually, their gametes can't get together *except* in liquid water. A haploid (*n*) phase begins at meiosis and dominates the life cycle (Figure 19.2a). A diploid (2*n*) phase starts at fertilization, when gametes fuse.

By contrast, a diploid phase dominates the life cycle of most plants (Figure 19.2c). The phase begins when gametes (sperm and egg) fuse, thus forming a diploid zygote. Mitotic cell divisions transform the zygote into a multicelled diploid body called a **sporophyte**. Later, in parts of a mature sporophyte, haploid cells of a type known as **spores** form by way of meiosis. (Hence the

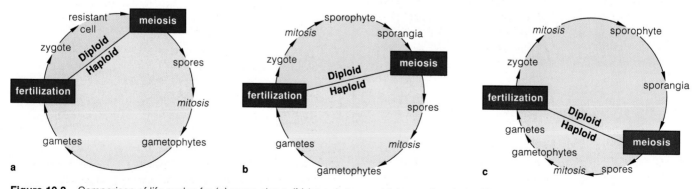

Figure 19.2 Comparison of life cycles for (**a**) some algae, (**b**) bryophytes, and (**c**) vascular plants. The diagrams highlight an evolutionary trend from haploid to diploid dominance during the colonization of land.

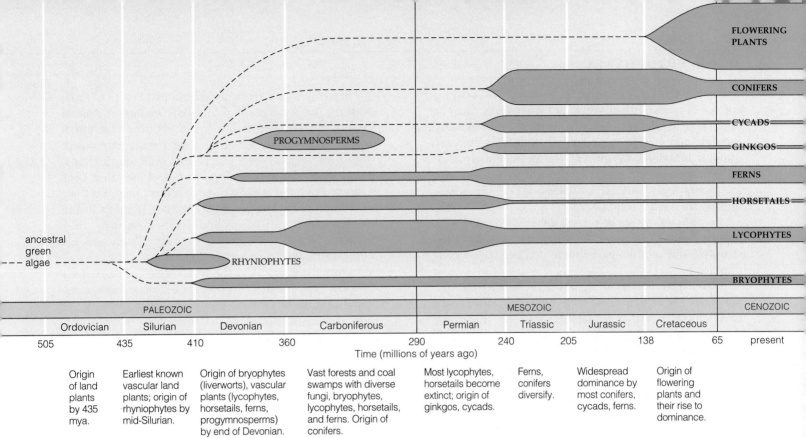

	PALEOZOIC				MESOZOIC				CENOZOIC
	Ordovician	Silurian	Devonian	Carboniferous	Permian	Triassic	Jurassic	Cretaceous	
505	435	410	360		290	240	205	138	65 present

Time (millions of years ago)

Origin of land plants by 435 mya.	Earliest known vascular land plants; origin of rhyniophytes by mid-Silurian.	Origin of bryophytes (liverworts), vascular plants (lycophytes, horsetails, ferns, progymnosperms) by end of Devonian.	Vast forests and coal swamps with diverse fungi, bryophytes, lycophytes, horsetails, and ferns. Origin of conifers.	Most lycophytes, horsetails become extinct; origin of ginkgos, cycads.	Ferns, conifers diversify.	Widespread dominance by most conifers, cycads, ferns.	Origin of flowering plants and their rise to dominance.

Figure 19.3 Milestones in plant evolution.

name sporophyte, for "spore-producing body." Pine trees and rose bushes are examples.) In certain sporophyte tissues, the spores divide by mitosis and give rise to **gametophytes**—multicelled, haploid plant parts that produce, nourish, and protect the forthcoming gametes. (You can find such parts in pine cones and flowers.)

The shift to diploid dominance was an adaptation to habitats on land, most of which show seasonal changes in the availability of water and dissolved nutrients. Long ago in these challenging habitats, natural selection must have favored sporophytes with well-developed root systems. Young roots of these systems interact with fungal symbionts, as described later in the chapter. The association (a mycorrhiza) enhances the plant's uptake of water and scarce minerals, even during dry seasons.

Unlike algae, then, vascular plants retain eggs within gametophytes; and after eggs are fertilized, gametophyte tissues protect and nourish the embryo sporophytes. *And they do so until environmental conditions are most suitable for dispersal and fertilization.*

Evolution of Pollen and Seeds

The evolution of pollen grains and seeds underlies the successful radiation of gymnosperms and angiosperms into high and dry habitats. Both kinds of seed-bearing plants produce not one but two types of spores. (This condition is *hetero*spory, as opposed to *homo*spory.) One type is the start of **pollen grains**, which become mature, sperm-bearing male gametophytes. The other type of

spore develops into female gametophytes, the parts where eggs form and get fertilized. Pollen grains hitch rides on insects, birds, air currents, and so on. Unlike algae, they don't need liquid water to reach eggs.

Also, the parent sporophyte of seed-bearing plants retains the female gametophytes. Its tissues nourish and protect fertilized eggs as they grow to young embryos. Actually, what we call a **seed** is a package of nutritive and protective tissues, together with the embryo. Seeds can withstand hostile conditions. And it is probably no coincidence that dominant seed plants arose during Permian times, when shifts in climate were extreme.

Before turning to the spectrum of diversity among plants, take a look at Figure 19.3. You can use it as a map for tracking the branching evolutionary roads.

The plant kingdom includes multicelled, photosynthetic species called bryophytes, seedless vascular plants, and seed-bearing vascular plants. Most of these live on land.

In most lineages, structural adaptations to dry conditions included root and shoot systems, waxy cuticles, stomata, vascular tissues, and lignin-reinforced tissues.

Sporophytes with well-developed roots, stems, and leaves came to dominate the life cycle of most land plants. These complex sporophytes can nourish and protect their fertilized eggs and embryos through unfavorable conditions.

Some plants started producing two types of spores, not one. This led to the evolution of male gametes adapted for dispersal without liquid water and to the evolution of seeds.

The bryophyte lineage includes about 18,600 existing species of **mosses**, **liverworts**, and **hornworts**. Most of these nonvascular plants grow in fully or seasonally moist habitats, although you will find some mosses growing in deserts and even on the excruciatingly cold, windswept plateaus of Antarctica. Mosses especially are highly sensitive to air pollution. Where air is bad, mosses are often few or absent.

Bryophytes are small plants, generally less than 20 centimeters (8 inches) tall. Although they have leaflike, stemlike, and rootlike parts, these don't contain xylem or phloem.

Like lichens and some algae, the bryophytes can dry out, then revive after absorbing moisture. Most kinds have rhizoids—either elongated cells or threadlike structures that attach the gametophytes to the soil and serve as absorptive structures.

Bryophytes are the simplest plants to display three features that emerged early in plant evolution. *First*, a cuticle prevents water loss from aboveground parts. *Second*, a cellular jacket around the parts that produce sperm and eggs holds in moisture. *Third*, of all plants, bryophytes alone have large gametophytes that hold onto sporophytes and do not depend on them for their nutrition. Embryo sporophytes start to develop inside the gametophyte tissues—and even at maturity, they remain *attached* to the gamete-producing body and get some nutritional support from it.

With 10,000 species, the true mosses are the most common bryophytes. Gametophytes of some species grow in clusters, forming low, cushiony mounds of the sort shown in Figure 19.4. Others have a branched, feathery growth pattern. In humid regions, masses of these often grow on trunks and tree branches.

Figure 19.4 Life cycle of a moss (*Polytrichum*), a representative bryophyte. The sporophyte remains attached to the gametophyte and depends upon it for water and nutrients.

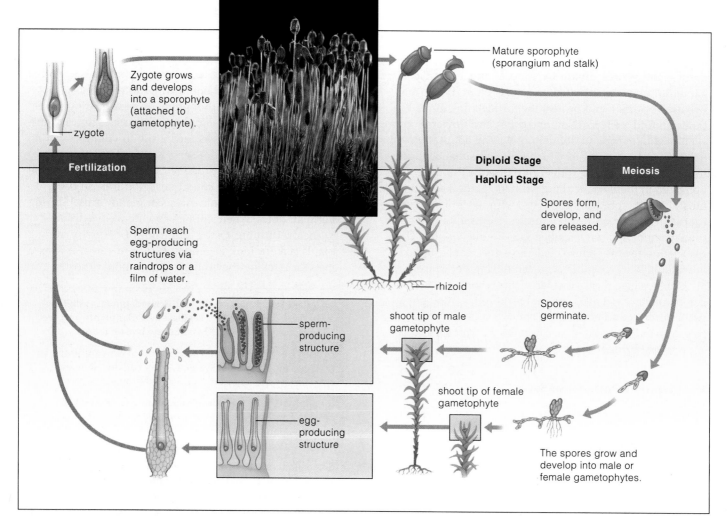

Zygote grows and develops into a sporophyte (attached to gametophyte).

zygote

Fertilization

Sperm reach egg-producing structures via raindrops or a film of water.

sperm-producing structure

egg-producing structure

rhizoid

Mature sporophyte (sporangium and stalk)

Diploid Stage

Haploid Stage

Meiosis

Spores form, develop, and are released.

Spores germinate.

shoot tip of male gametophyte

shoot tip of female gametophyte

The spores grow and develop into male or female gametophytes.

a

Figure 19.5 (**a**) From Alaska, a forest floor cloaked with a thick mat of moss. Over time, such mats contribute to the formation of peat—excessively moist, compressed organic matter that resists decomposition. (**b**) Gametophytes of a peat moss (*Sphagnum*). Attached to them are several sporophytes (the rust-colored, jacketed structures on the white stalks). Along with grasses and other plants, peat mosses contributed to the formation of extensive peat bogs in cold and temperate regions. In Ireland and elsewhere, dried peat is a major source of fuel.

Eggs and sperm develop inside tiny, jacketed vessels at the shoot tips of the gametophytes. Sperm reach eggs by swimming through a film of water on plant parts. After fertilization, the zygotes give rise to sporophytes. Each sporophyte has a stalk and a jacketed structure in which the spores develop.

Figure 19.5 shows one of 350 kinds of peat mosses (*Sphagnum*). These plants live in bogs, where turgor pressure keeps their parts erect. They have impressive absorptive properties. They soak up five times as much water as cotton does, owing to large, dead cells in their leaflike parts. Gardeners mix peat mosses into soil to increase its water-holding capacity and acidity.

Bryophytes are nonvascular plants with flagellated sperm that require liquid water to reach and fertilize the eggs.

A sporophyte of these plants develops within gametophyte tissues. It remains attached to the gametophyte and receives some nutritional support from it.

ANCIENT PLANTS, CARBON TREASURES

Between 360 and 280 million years ago, during the Carboniferous, swamp forests carpeted the wet lowlands of continents. Among the diverse species were the ancient ancestors of existing bryophytes, lycophytes, horsetails, ferns, and possibly seed-bearing plants. This was a period when sea levels rose and fell fifty times. When the seas receded, swamp forests flourished. When the seas moved back in, forest plants were submerged and became buried in sediments that protected them from decay. Gradually the sediments compressed the saturated, undecayed remains into what we now call **peat**. As more sediments accumulated, increased heat and pressure made the peat even more compact. It became **coal** (Figure 19.6).

Coal has a high percentage of carbon; it is energy-rich. It is one of our premier "fossil fuels." It took a fantastic amount of photosynthesis, burial, and compaction to form each major seam of coal in the Earth. It has taken us only a few centuries to deplete much of the world's known coal deposits.

Often you will hear about annual "production rates" for coal or some other fossil fuel. But how much do we really produce each year? None. We simply *extract* it from the Earth. Coal is a nonrenewable source of energy.

Figure 19.6 Reconstruction of a Carboniferous forest. The boxed inset shows part of a seam of coal.

EXISTING SEEDLESS VASCULAR PLANTS

The descendants of certain lineages of seedless vascular plants are with us today. We call them the **whisk ferns**, **lycophytes**, **horsetails**, and **ferns**. Like their ancestors, they differ from bryophytes in some important respects. First, the sporophyte does not remain attached to the gametophyte. Second, it has complex vascular tissues. Third, it is the larger, longer lived phase of the life cycle.

Most seedless vascular plants live in wet, humid places, and their gametophytes lack vascular tissues. Water droplets clinging to the plants are the only means by which flagellated sperm can reach the eggs. The few species living in dry habitats reproduce sexually during brief, seasonal pulses of heavy rains. Thus, whisk ferns, lycophytes, horsetails, and ferns are the "amphibians" of the plant kingdom. They haven't fully escaped the aquatic habitats of their ancestors.

Whisk Ferns

Whisk ferns (Psilophyta), which are not ferns, resemble whisk brooms. Florist suppliers commonly cultivate them in Hawaii, Texas, Louisiana, Florida, Puerto Rico, and other tropical or subtropical places. One genus, *Psilotum*, is a unique vascular plant, for its sporophytes have no roots or leaves. The photosynthetic, branched stems have xylem, phloem, and scalelike projections (Figure 19.7*a*). Belowground are **rhizomes**, branching, short, mostly horizontal stems that serve in absorption.

Lycophytes

About 350 million years ago, lycophytes (Lycophyta) included tree-sized members of swamp forests. About 1,100 far tinier species exist today. The most familiar are club mosses—members of communities in the Arctic, tropics, and regions in between. Many types form mats on forest floors. One type in Texas, New Mexico, and Mexico is commonly known as the resurrection plant.

Sporophytes of most club mosses have leaves and a branching rhizome that gives rise to vascularized roots and stems. Some have nonphotosynthetic, cone-shaped clusters of leaves that bear spore sacs (Figure 19.7*b*). Each cluster is a **strobilus** (plural, strobili). After spores disperse, they germinate and then develop into small, free-living gametophytes. In one genus (*Selaginella*), two kinds of spores develop in the same strobilus.

Horsetails

Tree-sized sphenophytes (Sphenophyta) flourished in ancient swamp forests, but only fifteen smaller species of one genus (*Equisetum*) made it to the present. These are the horsetails (Figure 19.17*c*). Of all existing plants, these may be the oldest, and their body plan changed

Figure 19.7 Representative seedless vascular plants. (**a**) The sporophytes of a whisk fern (*Psilotum*). You can see the spore-producing structures, shaped a bit like pumpkins, at the ends of stubby branchlets. (**b**) Sporophytes of a lycophyte (*Lycopodium*). (**c**) Nonphotosynthetic, fertile shoots of *Equisetum*. The cone-shaped clusters at the stem tips are spore-bearing structures. (**d**) Vegetative shoots of *Equisetum*, the shape of which provides you with a clue to why someone thought "horsetails" would be a suitable name for these plants.

little over the past 300 million years. Today they grow in streambank mud and vacant lots, roadsides, and other disrupted habitats. Figure 19.7*c* and *d* shows fertile stems and vegetative, photosynthetic stems of one species. Its spores give rise to free-living, pinhead-sized gametophytes. Most of the horsetail sporophytes have rhizomes, a hollow photosynthetic stem, and scalelike leaves. Clusters of xylem and phloem are arrayed as a ring inside the stems. Silica-reinforced ribs support the

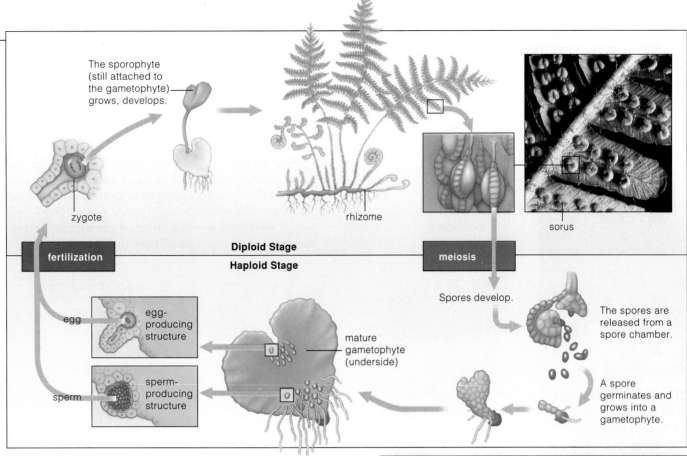

The sporophyte (still attached to the gametophyte) grows, develops.

zygote

rhizome

sorus

fertilization

Diploid Stage

Haploid Stage

meiosis

Spores develop.

The spores are released from a spore chamber.

egg

egg-producing structure

mature gametophyte (underside)

sperm

sperm-producing structure

A spore germinates and grows into a gametophyte.

Figure 19.8 Life cycle of a fern. The photograph shows a moist habitat of ferns in Indiana.

stems and give them a texture like sandpaper. Pioneers of the American West, who had to travel light, gathered horsetails along the way to use as pot scrubbers.

Ferns

With 12,000 or so species, the ferns (Pterophyta) are the largest and most diverse group of seedless vascular plants. All but about 380 are native to the tropics, but they are popular houseplants all over the world. Their size range is stunning. Some floating species are less than 1 centimeter across. Some tropical tree ferns are 25 meters (82 feet) tall. One climbing fern has a modified leaf stalk about 30 meters long.

Most ferns have vascularized rhizomes that give rise to roots and leaves. Exceptions include tropical tree ferns and epiphytes. ("Epiphyte" refers to any aerial plant that grows on tree trunks or branches.) While they develop, young fern leaves are coiled, rather like a fiddlehead. At maturity, the leaves (fronds) commonly are divided into leaflets.

You may have noticed rust-colored patches on the lower surface of many fern fronds. Each patch, a cluster of spore chambers, is called a sorus (plural, sori). At dispersal time, the chambers snap open and cause the spores to catapult through the air. Each germinating

spore develops into a small gametophyte, such as the green, heart-shaped type shown in Figure 19.8.

Seedless vascular plants (whisk ferns, lycophytes, horsetails, and ferns) have sporophytes adapted to conditions on land. Yet they have not entirely escaped their aquatic ancestry. When they reproduce sexually, their flagellated sperm cannot reach the eggs unless liquid water is clinging to the plant.

19.5 GYMNOSPERMS—PLANTS WITH "NAKED" SEEDS

Gymnosperms, such as conifers, produce their seeds on exposed parts of the sporophyte. (*Gymnos* means naked; *sperma* is taken to mean seed.) Like other seed-bearing plants, they have an advantage over seedless ones. Air currents, water currents, or insects can transport their sperm, packaged in pollen grains, to the eggs. *Thus, gymnosperms do not depend on free water for fertilization.*

Seed formation starts at an **ovule**—a structure that contains a female gametophyte with egg cell, nutritive tissue, and a jacket of cell layers (Figure 19.9). As an embryo sporophyte develops from a fertilized egg, the jacket's outer layers mature into a seed coat. The coat protects the embryo sporophyte during its dispersal. Nutrients stored inside the female gametophyte will help it through the critical time of germination, before its roots and shoots become fully functional.

Figure 19.9 Life cycle of a gymnosperm (ponderosa pine).

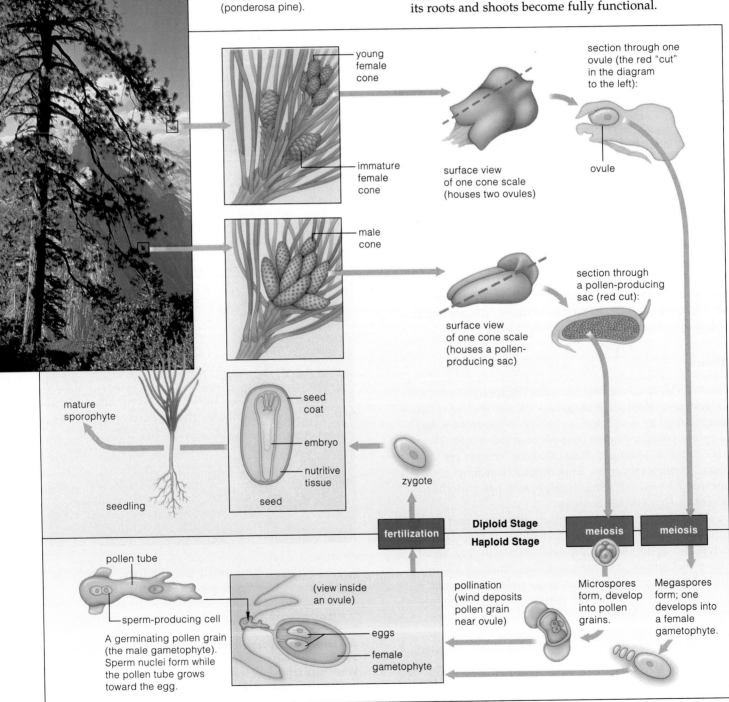

a

b

c

d

e

Figure 19.10 (**a**) Massive female cone of a cycad. (**b**) Ginkgo trees. (**c**) Fleshy-coated ginkgo seeds. (**d**) From California, *Ephedra viridis*. (**e**) *Welwitschia mirabilis*. The inset shows its female cones.

Conifers

Conifers (Coniferophyta) are woody trees and shrubs with needlelike or scalelike leaves. Most are *evergreen*; they remain leafy even though they shed old leaves throughout the year. A few are *deciduous*; they shed all leaves in autumn. Conifers include redwoods, pines, spruces, firs, and junipers. They have **cones**: clusters of modified leaves where spore-producing parts develop. They produce two kinds of spores (Figure 19.9). In conifers, **microspores** form in male cones and develop into pollen grains. **Megaspores** form in shelflike scales of female cones and develop into female gametophytes.

Each spring, millions of pollen grains drift off male cones and some land on ovules. The arrival of pollen on female reproductive parts is called **pollination**. After a pollen grain germinates, it develops into a tubelike structure that grows through the female tissues. This pollen tube transports sperm to female gametophytes. Fertilization occurs months or a year after pollination.

Conifers dominated many land habitats during the Mesozoic, but their slow reproductive pace may have put them at a competitive disadvantage when flowering plants began their spectacular adaptive radiations (Section 17.7). Coniferous forests still flourish at higher elevations, in the far north, and in parts of the Southern Hemisphere. However, besides having to compete with flowering plants for resources, conifers are vulnerable to **deforestation**—the removal of all trees from large tracts of land, as by clear cutting (Figure 19.1c). Conifers happen to be premier sources of lumber, paper, and other products. We return to this topic in Chapter 39.

Lesser Known Gymnosperms

Figure 19.10 shows a few relatives of the conifers. The **cycads** (Cycadophyta) superficially resemble palm trees (which are flowering plants). They flourished with the dinosaurs in the Mesozoic. About 100 species survived to the present in tropical and subtropical places. Cycads have massive cones that bear pollen or ovules. Insects or air currents transfer pollen from "male" to "female" plants. Cycad seeds and a flour made from the trunks are edible after their poisonous alkaloids are removed.

Ginkgos (Ginkgophyta) were diverse in dinosaur times. The maidenhair tree, *Ginkgo biloba*, is the only surviving species. Several thousand years ago, ginkgo trees were planted around temples in China. Then the natural populations nearly became extinct, even though ginkgos seem hardier than many other trees. Perhaps they became targets for firewood. Today, male ginkgo trees are widely planted. Besides having attractive, fan-shaped leaves, they are resistant to insects, disease, and air pollutants. Female trees are not favored; their fleshy-coated seeds give off an awful stench when stepped on.

Gnetophytes (Gnetophyta) are grouped into three genera. Trees and leathery leafed vines (*Gnetum*) grow in the humid tropics. Shrubs (*Ephedra*) grow in deserts and other arid places, including parts of California. Truly bizarre plants (*Welwitschia*) grow in hot deserts of south and west Africa. The sporophyte is mainly a deep-reaching taproot. The only exposed part, a woody disk-shaped stem, has cones and one or two strap-shaped leaves that split lengthwise repeatedly as the plant ages (Figure 19.10e).

Conifers, cycads, ginkgo, and gnetophytes are gymnosperms. They produce ovules and seeds, which they bear on exposed surfaces of cones and other spore-producing structures.

ANGIOSPERMS—FLOWERING, SEED-BEARING PLANTS

Angiosperms alone produce flowers. *Angeion*, meaning vessel, refers to female reproductive parts at the center of flowers. The enlarged base of the "vessel" is the ovary, where ovules and seeds develop (Figure 19.11*a*).

Most flowering plants coevolved with **pollinators**—insects, bats, birds, and other animals that withdraw nectar or pollen from a flower and, in so doing, transfer pollen to its female reproductive parts. The recruitment of animals as assistants in reproduction probably has

In flowers (reproductive structures), seeds develop from ovules and tissues of parent sporophyte after fertilization.

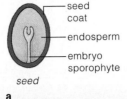

contributed to the success of flowering plants, which have dominated the land for 100 million years. At least 260,000 existing species live in a great variety of habitats. They range in size from the tiny duckweeds (about a millimeter long) to towering *Eucalyptus* trees, some of which are more than 100 meters tall. Mistletoe, Indian pipe, and a few other rare species are not even photosynthetic; they directly or indirectly feed on other plants (Figure 19.11*d*).

Figure 19.11 (**a**) Unique trademark of angiosperms—the flower, a reproductive structure with roles in pollination and seed formation. (**b**) This passion flower beckons a hovering hummingbird. (**c**) A few flowering plants, such as water lilies (*Nymphea*), live in water. (**d**) Indian pipe (*Monotropa uniflora*), one of the few nonphotosynthetic species, draws nutrients from fungi that are symbionts with the young roots of photosynthetic plants.

Dicots and Monocots

There are two classes of flowering plants, called the dicots and monocots (formally, the Dicotyledonae and Monocotyledonae). Among the 180,000 dicots are most nonwoody plants, shrubs and trees (such as oaks and apples), water lilies, and cacti. Among the 80,000 or so species of monocots are orchids, palms, lilies, and grasses, such as rye, sugarcane, corn, rice, wheat, and other valued crop plants.

Key Aspects of the Life Cycles

The next unit deals with the structure and function of flowering plants. For now, simply start thinking about the large sporophyte that dominates their life cycles (Figure 19.12). It retains and nourishes gametophytes; its sperm are dispersed within pollen grains. Inside flowering plant seeds, a special nutritive tissue called endosperm surrounds the embryo. Also, while seeds develop, the ovaries mature into **fruits**, which protect the embryo sporophytes and aid in their dispersal.

Angiosperms are the most successful plants in terms of their diversity, numbers, and distribution. They alone produce flowers. Most species coevolved with animal pollinators.

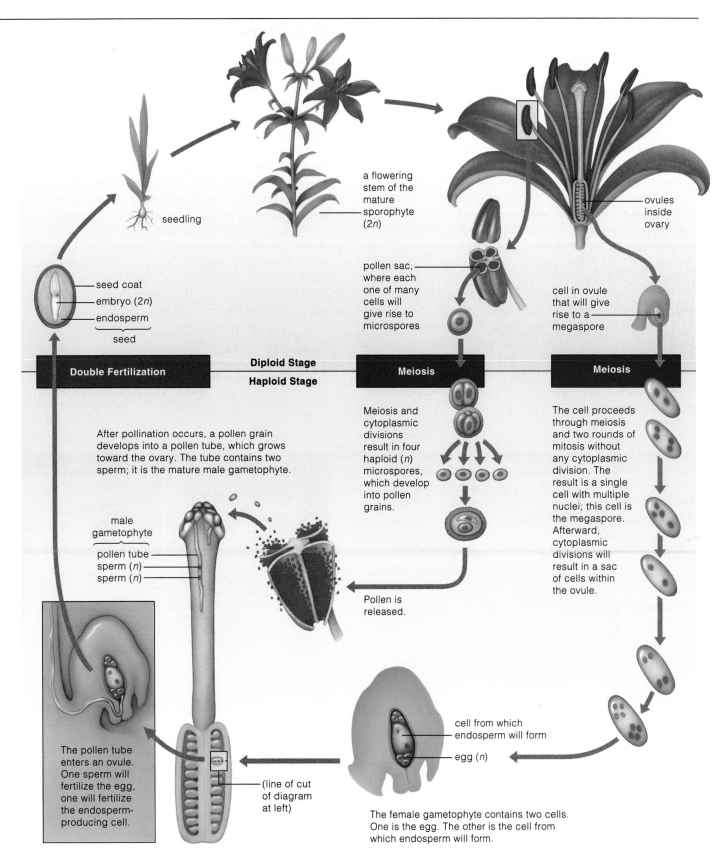

seedling

a flowering
stem of the
mature
sporophyte
(2n)

ovules
inside
ovary

seed coat
embryo (2n)
endosperm
seed

pollen sac,
where each
one of many
cells will
give rise to
microspores

cell in ovule
that will give
rise to a
megaspore

Double Fertilization

Diploid Stage
Haploid Stage

Meiosis

Meiosis

After pollination occurs, a pollen grain
develops into a pollen tube, which grows
toward the ovary. The tube contains two
sperm; it is the mature male gametophyte.

Meiosis and
cytoplasmic
divisions
result in four
haploid (n)
microspores,
which develop
into pollen
grains.

The cell proceeds
through meiosis
and two rounds of
mitosis without
any cytoplasmic
division. The
result is a single
cell with multiple
nuclei; this cell is
the megaspore.
Afterward,
cytoplasmic
divisions will
result in a sac
of cells within
the ovule.

male
gametophyte
pollen tube
sperm (n)
sperm (n)

Pollen is
released.

The pollen tube
enters an ovule.
One sperm will
fertilize the egg,
one will fertilize
the endosperm-
producing cell.

(line of cut
of diagram
at left)

cell from which
endosperm will form

egg (n)

The female gametophyte contains two cells.
One is the egg. The other is the cell from
which endosperm will form.

Figure 19.12 Life cycle of a lily (*Lilium*), a monocot. "Double" fertilization is a distinctive feature of
flowering plant life cycles. A male gametophyte delivers *two* sperm to an ovule. One sperm fertilizes the
egg, and the other fertilizes a cell that gives rise to endosperm, a tissue that will nourish the forthcoming
embryo. Figure 24.6 provides a closer look at flowering plant life cycles, using a dicot as the example.

19.7 CHARACTERISTICS OF FUNGI

Major Groups of Fungi

For many of us, "fungi" are mushrooms sold in grocery stores. Yet those drab mushrooms are produced by a fungus with stunningly diverse relatives. The few representatives in Figure 19.13 simply don't do justice to the 56,000 fungal species we know about—and there may be at least a million more we don't know about!

We know, from the fossil record, that fungi evolved before 900 million years ago. Some accompanied plants onto the land 430 million years ago. About 100 million years later, three major lineages were well established. We call them the zygomycetes (Zygomycota), sac fungi (Ascomycota), and club fungi (Basidiomycota). Other, puzzling kinds known as "imperfect fungi" are lumped together but are not a formal taxonomic group. The vast majority of species in all these groups are multicelled.

Nutritional Modes

Fungi are heterotrophs, meaning they require organic compounds that other organisms synthesize. Most are **saprobes**; they obtain nutrients from nonliving organic matter and so cause its decay. Others are **parasites**; they extract nutrients from tissues of a living host. The cells of all species grow in or on organic matter. As they do, they secrete digestive enzymes, then absorb breakdown products. The "extracellular digestion" benefits plants, which absorb some of the released nutrients as well as carbon dioxide. Together with heterotrophic bacteria, fungi are nature's grand decomposers. Without them, communities would be buried in their own garbage, nutrients would not be cycled, and life could not go on.

Key Features of Fungal Life Cycles

Fungi reproduce asexually most often but, given the opportunity, also reproduce sexually. They produce great numbers of nonmotile spores. As in plants, their spores are reproductive cells or multicelled structures, often walled, that germinate after dispersal from the parent. In multicelled species, spores give rise to a mesh of branched filaments. The mesh, a **mycelium** (plural, mycelia), rapidly grows over or into organic matter and has a good surface-to-volume ratio for food absorption (Figure 19.14). Each filament in a mycelium is a **hypha** (plural, hyphae). Hyphal cells commonly have chitin-reinforced walls. Their cytoplasm interconnects, so that nutrients flow unimpeded throughout the mycelium.

Fungi are major decomposers that engage in extracellular digestion and absorption of organic matter. Multicelled types form absorptive mycelia and spore-producing structures.

19.8 CONSIDER THE CLUB FUNGI

A Sampling of Spectacular Diversity

Fungal life cycles and life-styles show dizzying variety. The most we can do here is to sample a few species, starting with the **club fungi**. The 25,000 or so club fungi include mushrooms, shelf fungi, coral fungi, puffballs, and stinkhorns. Figures 19.13 and 1.6c show splendid examples. Many species are symbionts with the young roots of forest trees. Some saprobic types decompose plant debris. The ones called rusts and smuts can destroy fields of wheat, corn, and other crop plants.

Or consider *Armillaria bulbosa*, one of the oldest and largest of all organisms. One individual's mycelium extends about 15 hectares through a forest in northern Michigan. (A hectare is 10,000 square meters.) By some estimates, it weighs more than 10,000 kilograms and has been spreading through the soil for 1,500 years!

Figure 19.13 Club fungi. (**a**) Trumpet chanterelle (*Craterelius*). (**b**) The fly agaric mushroom (*Amanita muscaria*) contains toxins and intoxicants that cause hallucinations when ingested. It was used in ancient rituals in Central America, Russia, and India. (**c**) Light-red coral fungus (*Ramaria*). (**d**) Big laughing mushroom (*Gymnophilus*). (**e**) Shelf fungus (*Polyporus*) on a rotting log.

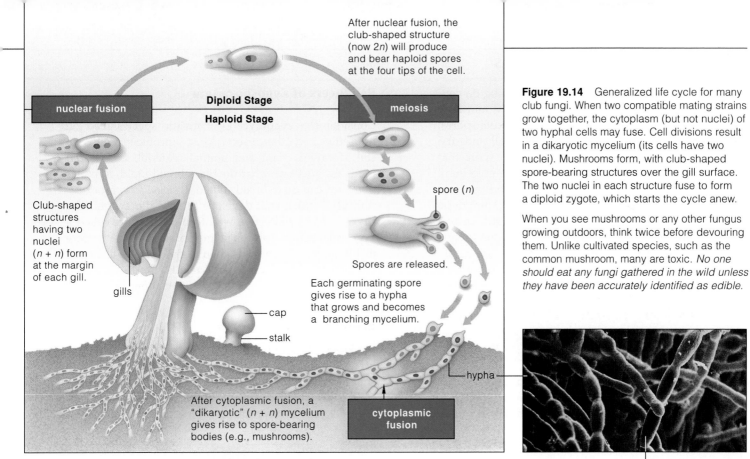

After nuclear fusion, the club-shaped structure (now 2n) will produce and bear haploid spores at the four tips of the cell.

nuclear fusion

Diploid Stage

Haploid Stage

meiosis

Club-shaped structures having two nuclei (n + n) form at the margin of each gill.

gills

spore (n)

Spores are released.

cap

stalk

Each germinating spore gives rise to a hypha that grows and becomes a branching mycelium.

After cytoplasmic fusion, a "dikaryotic" (n + n) mycelium gives rise to spore-bearing bodies (e.g., mushrooms).

cytoplasmic fusion

hypha

Figure 19.14 Generalized life cycle for many club fungi. When two compatible mating strains grow together, the cytoplasm (but not nuclei) of two hyphal cells may fuse. Cell divisions result in a dikaryotic mycelium (its cells have two nuclei). Mushrooms form, with club-shaped spore-bearing structures over the gill surface. The two nuclei in each structure fuse to form a diploid zygote, which starts the cycle anew.

When you see mushrooms or any other fungus growing outdoors, think twice before devouring them. Unlike cultivated species, such as the common mushroom, many are toxic. *No one should eat any fungi gathered in the wild unless they have been accurately identified as edible.*

hypha in a mycelium

d

e

Example of a Fungal Life Cycle

No doubt you have an idea of what the common mushroom (*Agaricus brunnescens*) looks like, so let's use it as our example of a fungal life cycle. As for most other club fungi, this species forms short-lived reproductive bodies—mushrooms—that are merely its aboveground parts; its living mycelium is buried in soil or decaying wood. A mushroom has a stalk and a cap. Club-shaped structures that bear spores on their outer surface form on the sides of gills, or sheets of tissue, in the cap.

When a spore dispersed from a particular strain of mushroom lands on a suitable site, it germinates and gives rise to a haploid mycelium. Suppose hyphae of two compatible mating strains make contact. They may undergo cytoplasmic fusion, but their nuclei do not fuse at once. As Figure 19.14 shows, the fused part may be the start of a *dikaryotic* mycelium, in which hyphal cells have one nucleus of each mating type. After an extensive mycelium develops and when conditions are favorable, mushrooms will form. Each spore-producing structure of the mushroom is dikaryotic at first, but then its two nuclei fuse to form a short-lived zygote. The zygote undergoes meiosis, haploid spores develop —and then air currents disperse them.

Club fungi, the fungal group with the greatest diversity, produce spores in distinctive club-shaped structures.

19.9 SPORES AND MORE SPORES

A fungus has a thing about spores. Depending on contact with a suitable hypha, food availability, and how cool or damp conditions are, it produces sexual spores, asexual spores, or both. Its spores are small and dry, and air currents easily disperse them. Each spore that germinates can be the start of a hypha and a mycelium. Stalked reproductive structures may form on many of the hyphae and produce asexual spores. After these spores germinate, each may be the start of still *another* extensive mycelium. In no time at all, that one fungus and staggering numbers of its descendants are busily decomposing organic stuff or pirating nutrients from a host! Look at what can happen to a slice of stale bread:

Each fungal class produces unique sexual spores. Club fungi form basidiospores, zygomycetes form spores by way of zygosporangia, and sac fungi form ascospores.

Producers of Zygosporangia

Consider the **zygomycetes**. Parasitic species feed on houseflies and other insects. Most saprobic types live in soil, decaying plant and animal material, and stored food. You just saw what the black bread mold, *Rhizopus stolonifer*, can do to bread. When *R. stolonifer* reproduces sexually, a thick wall develops around the zygote. The zygote *and* its protective wall form a structure called a zygosporangium (plural, zygosporangia). The zygote inside undergoes meiosis, then it gives rise to a stalked reproductive structure, as shown in Figure 19.15. Sexual spores form inside this, then germinate after release.

Producers of Ascospores

We know of more than 30,000 kinds of **sac fungi**, most of which form sexual spores (ascospores) inside sac-shaped cells. They alone produce these cells, which are called asci (singular, ascus). In multicelled species, asci are enclosed in reproductive structures consisting of

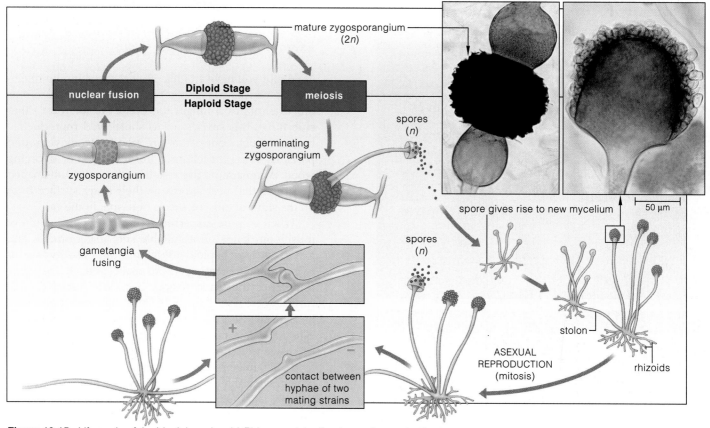

Figure 19.15 Life cycle of the black bread mold *Rhizopus stolonifer.* Asexual reproduction is common, but different mating strains (+ and −) also reproduce sexually. Either way, haploid spores form and give rise to new mycelia. Chemical attraction between a + hypha and a − hypha causes them to fuse. Two gametangia form, each with several haploid nuclei inside. Later their nuclei fuse to form a zygote. The zygote develops a thick wall, thereby becoming a zygosporangium, and may remain dormant for several months. Meiosis occurs as the zygosporangium germinates, and sexual spores form.

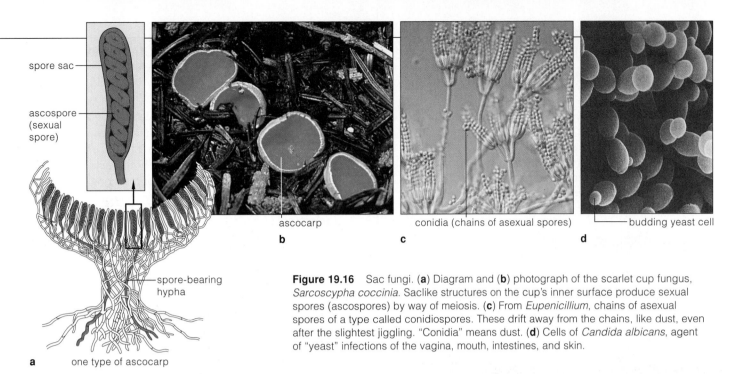

spore sac

ascospore (sexual spore)

spore-bearing hypha

a one type of ascocarp

ascocarp
b

conidia (chains of asexual spores)
c

budding yeast cell
d

Figure 19.16 Sac fungi. (**a**) Diagram and (**b**) photograph of the scarlet cup fungus, *Sarcoscypha coccinia*. Saclike structures on the cup's inner surface produce sexual spores (ascospores) by way of meiosis. (**c**) From *Eupenicillium*, chains of asexual spores of a type called conidiospores. These drift away from the chains, like dust, even after the slightest jiggling. "Conidia" means dust. (**d**) Cells of *Candida albicans*, agent of "yeast" infections of the vagina, mouth, intestines, and skin.

tightly interwoven hyphae. These structures resemble flasks, globes, and shallow cups (Figure 19.16*a*).

The vast majority of sac fungi are multicelled. They include morels and truffles. Certain morels are prized edibles. So are truffles (underground symbionts with oak and hazelnut tree roots). Trained pigs and dogs snuffle out truffles in the wild. In France, truffles also are being cultivated on roots of inoculated seedlings.

Other multicelled sac fungi include certain species of *Penicillium* that "flavor" Camembert and Roquefort cheeses and species that make penicillins, which we use as antibiotics. *Aspergillus* produces citric acid for candies and soft drinks, and it ferments soybeans for soy sauce. Most of the red, bluish-green, and brown molds that spoil stored food also are multicelled. One species, the salmon-colored *Neurospora sitophila*, wreaks havoc in bakeries or research laboratories. It produces so many spores, it is extremely difficult to eradicate. Yet one of its relatives, *N. crassa*, is an important organism in genetic research.

Sac fungi also include about 500 species of single-celled yeasts (although certain yeasts are classified as club fungi). Yeasts reproduce sexually when two cells fuse and become a spore-producing sac. Some types live in the nectar of flowers and on fruits and leaves. Bakers and vintners put fermenting by-products of vast populations of yeasts to good use. For example, carbon dioxide by-products of *Saccharomyces cerevisiae* leaven bread. The commercial production of wine and beer depends on its ethanol end product. Many yeast strains with desirable properties have been developed through artificial selection and genetic engineering. Then again, *Candida albicans*, a notorious relative of "good" yeasts, causes vexing infections in humans (Figure 19.16*d*).

roundworm   noose formed by hypha

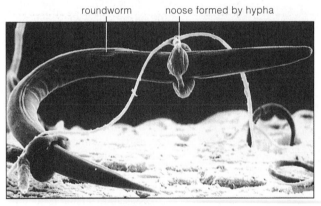

Figure 19.17 *Arthropbotrys dactyloides*, an imperfect fungus. This predatory species forms a nooselike ring that swells rapidly with turgor pressure when stimulated. The "hole" in the noose shrinks and captures a worm, into which hyphae will grow.

Elusive Spores of the Imperfect Fungi

Imperfect fungi are set aside in a taxonomic holding station not because they are somehow defective but mainly because no one has yet discovered what kind of sexual spores they produce (if any). Figure 19.17 shows one of the species, a puzzling predatory fungus, that is awaiting formal classification. Investigators recently reunited the previously orphaned *Aspergillus*, *Candida*, and *Penicillium* with their kin—other sac fungi.

Through their exuberant and rapid production of asexual and sexual spores, fungi take quick advantage of available organic matter, whether it has been discarded or is part of a living or dead organism. Their penchant for spore production is central to their success as decomposers and parasites.

A LOOK AT THE UNLOVED FEW

You know you are a serious student of biology when you view organisms objectively in terms of their place in nature, not in terms of their impact on humans generally and you in particular. As a student you can salute saprobic fungi as vital decomposers and praise parasitic fungi that keep populations of harmful insects and weeds in check. The true test is when you open the fridge for a dish of high-priced raspberries and discover that a fungus beat you to them. The true test is when a fungus starts feeding on warm, damp tissues between your toes and turns skin scaly, reddened, and cracked (Figure 19.18).

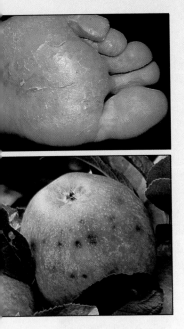

And which home gardeners wax poetic about black spot or powdery mildew on their roses? Which farmers happily give up millions of dollars each year to rusts and smuts? Who willingly inhales the spores of *Histoplasma capsulatum*? When the airborne spores land on moist lung tissues instead of soil, they still grow—and they cause *histoplasmosis*, a respiratory infection that, in rare cases, ends up damaging most organs. The fungus, common worldwide, is especially at home in regions drained by the Ohio and Mississippi rivers. It thrives in the nitrogen-rich droppings of bats, opossum, and chickens, pigeons, and other birds.

Figure 19.18 Love those fungi! *Above:* Athlete's foot, courtesy of *Epidermophyton fluccosum. Below:* Apple scab, trademark of *Venturia inequalis.*

Fungi even thread through human history. One notorious species, *Claviceps purpurea*, is a parasite of rye and other grains. We use some of its by-products (alkaloids) to treat migraine headaches and, following childbirth, to shrink the uterus to prevent hemorrhaging. Yet the alkaloids are toxic in large amounts. Eat a lot of bread made with tainted rye flour and you end up with *ergotism*. Disease symptoms include vomiting, diarrhea, hallucinations, hysteria, and convulsions. Untreated, the fungal infection can turn limbs gangrenous and it can end in death.

Ergotism epidemics were common in Europe during the Middle Ages, when rye was a key crop. Ergotism thwarted Peter the Great, the Russian czar who was obsessed with conquering ports along the Black Sea for his nearly land-locked empire. Soldiers laying siege to the ports ate mostly rye bread and fed rye to their horses. The former went into convulsions and the latter into "blind staggers." Possibly, outbreaks of ergotism were used as an excuse to launch witch-hunts in colonial Massachusetts and elsewhere.

19.11 BENEFICIAL ASSOCIATIONS BETWEEN FUNGI AND PLANTS

Symbiosis refers to species that live in close association. (The word literally means "living together.") In many cases, one species is a victim, not a partner, of a parasite. In other cases, called **mutualism**, both partners benefit. We find classic examples of symbionts among the fungi.

Lichens

A **lichen** is commonly called a mutualistic interaction between a fungus and a photosynthetic species. But it may well be a form of controlled parasitism, in which the fungus holds a cyanobacterium, green alga, or both captive. Many sac fungi enter into these interactions. A lichen forms after a hypha penetrates a host cell and starts absorbing carbohydrates from it. If the captive cell survives, both it and the fungus multiply together, so that crusty structures form. Figures 19.19 and 19.20 give examples. The fungus benefits from a long-term source of nutrients from the cell's descendants. The captive suffers in terms of its own growth, although it might benefit a bit from the lichen's sheltering effect.

Sheltering is useful in many environments. Lichens can colonize places too hostile for other organisms. They grow (slowly) on rocks in deserts and mountains, on bark and fence posts. They become activated when moistened with raindrops, fog, or dew. Lichens at the South Pole suspend activities when it gets too cold or bright. Their crusty part thickens and blocks sunlight. Photosynthesis stops until favorable conditions return.

Lichens absorb mineral ions from rocks and nitrogen from the air, and their metabolic activities slowly change the composition of their substrates. They can contribute to soil formation, thus setting the stage for colonization by different species. In fact, ancient lichens may have been the first invaders of land.

Lichens also serve as early warnings of deteriorating environmental conditions. They absorb but cannot rid

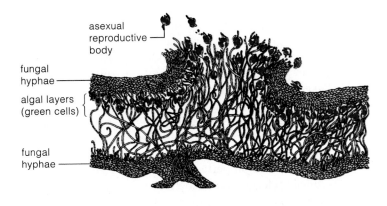

Figure 19.19 Diagram of one type of lichen, cross-section.

Figure 19.20 Lichens. (**a**) *Usnea*, old man's beard. (**b**) *Cladonia rangiferina*—sometimes called reindeer moss, but not a moss.

themselves of toxins. From extensive studies in New York City and in England, we know that when lichens die around cities, air pollution is getting bad.

Mycorrhizae

Fungi enter into mutually beneficial associations with young roots of most vascular plants. We call these associations **mycorrhizae**, or "fungus-roots." The fungus benefits by absorbing carbohydrates from the plant, which benefits by absorbing minerals from the fungus. Collectively, the fungal hyphae have a huge surface area for absorbing water and dissolved mineral ions. The fungus can take up ions when they are abundant in soil and release them to the plant when ions are scarce. Many plants simply cannot grow as efficiently when mycorrhizae are not present to help them absorb crucial ions, such as phosphorus (Figure 19.21).

The example in Figure 19.21*a* is an *exo*mycorrhiza, in which hyphae form a dense net around living cells in roots but do not penetrate them. Other hyphae form a velvety wrapping around the root, and the mycelium radiates outward from it. Exomycorrhizae are common in temperate forests, and they help the trees withstand adverse seasonal changes in temperature and rainfall.

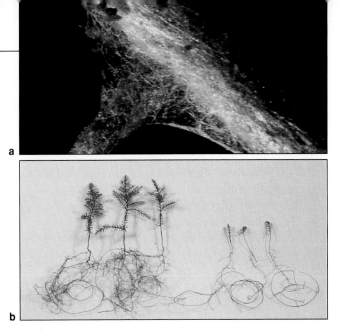

Figure 19.21 (**a**) Mycorrhiza from a hemlock tree. The white "threads" are hyphal strands wrapped around a small, young root. (**b**) Effects of the presence or absence of mycorrhizae on plant growth. The juniper seedlings at the left are six months old. They were grown in sterilized, phosphorus-poor soil with a mycorrhizal fungus. The seedlings at the right were grown without the fungus.

About 5,000 fungal species enter the associations. Most are club fungi, including those truffles described earlier.

The more common *endo*mycorrhizae form in about 80 percent of all vascular plants. These fungal hyphae penetrate plant cells, as they do in lichens. Fewer than 200 species of zygomycetes serve as the fungal partner. Their hyphae branch extensively, forming tree-shaped absorptive structures within cells. They also extend for several centimeters into the surrounding soil. Chapter 23 provides a closer look at these beneficial species.

As Fungi Go, So Go the Forests

Since the early 1900s, collectors have been recording data on the wild mushroom populations in the forests of Europe. A look at the records tells us that the number and diversity of fungi are now declining at alarming rates. Certainly mushroom gatherers aren't to blame; toxic as well as edible species are vanishing. But the decline does correlate with rising air pollution. Vehicle exhaust, smoke from coal burning, and emissions from nitrogen fertilizers pump ozone, nitrogen oxides, and sulfur oxides into the air. Normally as a tree ages, one species of mycorrhizal fungus gives way to another in predictable patterns. When fungi die, the tree loses its support system and becomes vulnerable to severe frost and drought. Are the North American forests at risk? Conditions there are deteriorating in comparable ways.

Lichens and mycorrhizae are symbiotic interactions between fungi and other organisms, with mutual benefits.

SUMMARY

1. Ancient green algae probably gave rise to plants, nearly all of which are multicelled photoautotrophs. Plants and fungi moved onto land about 430 million years ago. Table 19.1 summarizes and compares the major plant phyla.

2. Several trends in plant evolution can be identified by comparing different lineages (see also Table 19.2):

   a. Structural adaptations to dry conditions, such as vascular tissues (xylem and phloem).

   b. A shift from haploid to diploid dominance in the life cycle. Complex sporophytes evolved that hold onto, nourish, and protect spores and gametophytes.

   c. A shift from one to two kinds of spores. Among gymnosperms and angiosperms (flowering plants), this led to the evolution of pollen grains and seeds.

3. Mosses, liverworts, and hornworts are nonvascular plants called bryophytes. None has well-developed xylem and phloem.

4. Vascular land plants have a cuticle and stomata that help control water loss. Their root and shoot systems are well adapted to conditions on land. Tissues enclose and protect their spores and gametes. Their embryo sporophytes start development *within* the gametophyte tissues.

5. Whisk ferns, lycophytes, horsetails, and ferns are seedless vascular plants. Their flagellated sperm require ample water to swim to the eggs.

6. Gymnosperms and angiosperms are vascular plants that produce pollen grains (protected, immature male gametophytes) and seeds (mature ovules).

   a. Ovules are reproductive structures that contain the egg-producing female gametophytes, the precursor of nutritive tissue, and a jacket of cell layers, the outer portion of which develops into the seed coat.

   b. The evolution of pollen grains freed these plants from dependence on water for fertilization. Their seeds are efficient means of dispersing new generations even during hostile conditions. Pollen grains and seeds were key adaptations in the move to high, dry habitats.

7. Only angiosperms produce flowers. Most coevolved with pollinators, which enhance the transfer of pollen grains to female reproductive parts. Their seeds have a unique nutritive tissue (endosperm) and are usually surrounded by fruits, which aid in dispersal.

8. Fungi are heterotrophs and major decomposers. The saprobes feed on nonliving organic matter; parasites feed on living organisms.

   a. Some species of fungi are symbiotic partners with other organisms.

   b. The cells of all species of fungi secrete digestive enzymes that break down food into small molecules, which the cells absorb.

9. Nearly all fungi are multicelled. The food-absorbing portion (mycelium) consists of a mesh of filaments (hyphae). Aboveground reproductive structures, such as mushrooms, form from tightly interwoven hyphae. Fungi form asexual and sexual spores.

10. The major fungal classes (zygomycetes, sac fungi, and club fungi) form distinctive spores. When a sexual phase cannot be detected or is absent from the life cycle, a fungus is assigned to an informal category called the imperfect fungi.

Table 19.1   Comparison of Major Plant Groups

Nonvascular land plants. Fertilization requires free water. Haploid dominance. Cuticle, stomata present in some.

Bryophytes	18,600 species. Moist, humid habitats.

Seedless vascular plants. Fertilization requires free water. Diploid dominance. Cuticle, stomata present.

Lycophytes	1,100 species with simple leaves. Mostly wet or shady habitats.
Horsetails	15 species of single genus. Swamps, disturbed habitats.
Ferns	12,000 species. Wet, humid habitats in mostly tropical, temperate regions.

Vascular plants with "naked seeds" (gymnosperms). Diploid dominance. Cuticle, stomata present.

Conifers	550 species, mostly evergreen, woody trees and shrubs having pollen- and seed-bearing cones. Widespread distribution.
Cycads	185 slow-growing species. Tropics, subtropics.
Ginkgo	1 species, a tree with fleshy-coated seeds.
Gnetophytes	70 species, limited distribution in deserts and tropics.

Vascular plants with flowers and protected seeds (angiosperms). Diploid dominance. Cuticle, stomata present.

Flowering plants:

Monocots	80,000 species. Floral parts often arranged in threes or multiples of three; one seed leaf; parallel leaf veins common.
Dicots	At least 180,000 species. Floral parts often arranged in fours, fives, or multiples of these; two seed leaves; net-veined leaves common.

Table 19.2   Summary of Trends Among Plants

Bryophytes	Ferns	Gymnosperms	Angiosperms
Nonvascular ———→	Vascular ———————————————→		
Haploid ———————→ dominance	Diploid ———————→ dominance		
Spores of ——————→ one type	Spores of ——————→ two types		
Motile gametes ————————————→		Nonmotile gametes*	
Seedless ———————————→	Seeds ————————————→		

*Require pollination by wind, insects, etc.

11. A lichen is a symbiotic association of a fungus with a photosynthetic partner, such as a cyanobacterium, a green alga, or both. A mycorrhiza is a mutualistic association between a fungus and the young roots of plants. Fungal hyphae provide nutrients for their symbiont, which provides the fungus with carbohydrates.

Review Questions

1. Which evolutionary trends helped plants invade the land? *308–309*

2. How do vascular plants differ from bryophytes? *310, 312*

3. Describe the fungal mode of nutrition. *318*

4. Of what is a mycelium constructed and what is its function? *318*

5. How does a lichen differ from a mycorrhiza? *322–323*

Self-Quiz *(Answers in Appendix IV)*

1. Of all land plants, bryophytes alone have independent _____ and attached, dependent _____ .
   a. sporophytes; gametophytes   c. rhizoids; zygotes
   b. gametophytes; sporophytes   d. rhizoids, ascospores

2. A seed is _____ .
   a. a female gametophyte   c. a mature pollen tube
   b. a mature ovule   d. an immature embryo

3. New mycelia form following the germination of _____ .
   a. hyphae   c. mycelia
   b. spores   d. mushrooms

4. A "mushroom" is _____ .
   a. the food-absorbing part of a fungal body
   b. the part of the fungal body not constructed of hyphae
   c. a reproductive structure
   d. a nonessential part of the fungus

5. A mycorrhiza is a _____ .
   a. fungal disease of the foot   c. parasitic water mold
   b. fungus-plant relationship   d. fungus endemic to barnyards

6. Parasitic fungi obtain nutrients from _____ .
   a. tissues of living host   c. only living animals
   b. nonliving organic matter d. none of the above

7. Saprobic fungi derive nutrients from _____ .
   a. nonliving organic matter   c. root hairs
   b. living organisms   d. both b and c

8. Match the terms appropriately.
   _____ gymnosperm   a. produces haploid gametes
   _____ sporophyte   b. control water loss
   _____ seedless vascular   c. "naked" seeds
         plants   d. protect, disperse embryo
   _____ ovary   sporophyte
   _____ bryophytes   e. produces haploid spores
   _____ gametophyte   f. nonvascular land plants
   _____ stomata   g. lycophytes
   _____ angiosperm seed   h. usually a fruit at maturity

9. Match the terms appropriately.
   _____ zygomycetes   a. interaction between species in which
   _____ dikaryotic   positive benefits flow both ways
         mycelium   b. some yeasts and morels
   _____ hypha   c. each filament in a mycelium
   _____ basidiomycetes   d. black bread mold
   _____ mutualism   e. mushrooms, shelf fungi, stinkhorns
   _____ ascomycetes   f. cell has a nucleus of each mating type

Critical Thinking

1. Elliot Meyerowitz of the California Institute of Technology has studied the genetic basis of flower formation in *Arabidopsis thaliana*. By inducing mutations in seeds of this small weed, he eventually discovered three genes (call them *A*, *B*, and *C*) that work together in different parts of a developing flower. Interactions among these genes lead to formation of very different structures (sepals, petals, stamens, and carpels)—all from the same mass of undifferentiated tissue. From what you know of gene regulation, suggest ways that the *A*, *B*, and *C* genes might be controlling flower development.

2. Genes nearly identical to the *A*, *B*, and *C* genes of *A. thaliana* also have been isolated from snapdragons and other flowering plants. About 150 million years ago, these plants originated and rose to dominance rather abruptly, in nearly all land environments. Suggest how their rapid, spectacular radiation may have come about.

3. Diana discovers in the laboratory that a fungus (*Trichoderma*) can grow very well in distilled water. It continues to do so even after she rigorously treats the water and glassware to remove all traces of organic carbon. This type of fungus is *not* a photoautotroph. Suggest a metabolic life-style that would allow it to grow under these conditions.

4. Some strains of *Trichoderma* are being tested as a natural pest control agent. Laboratory experiments have demonstrated that the fungal strains combat other fungi that cause diseases in plants. Some even can promote seed germination and plant growth. In one set of twenty trials, workers increased lettuce yields by 54 percent. What concerns must be addressed before *Trichoderma* can be released into the environment for commercial applications?

Selected Key Terms

angiosperm *308*	hypha *318*	root system *308*
bryophyte *308*	lichen *322*	sac fungus
club fungus	lignin *308*	(ascomycete) *320*
(basidiomycete) *318*	liverwort *310*	saprobe *318*
coal *311*	lycophyte *312*	seed *309*
cone *315*	megaspore *315*	shoot system *308*
conifer *315*	microspore *315*	spore *308*
cuticle *308*	moss *310*	sporophyte *308*
cycad *315*	mutualism *322*	stoma (plural,
deforestation *315*	mycelium *318*	stomata) *308*
fern *312*	mycorrhiza *323*	strobilus *312*
fruit *316*	ovule *314*	symbiosis *322*
gametophyte *309*	parasite *318*	vascular plant *308*
ginkgo *315*	peat *311*	whisk fern *312*
gnetophyte *315*	phloem *308*	xylem *308*
gymnosperm *308*	pollen grain *309*	zygomycete *320*
hornwort *310*	pollinator *316*	
horsetail *312*	rhizome *312*	

Readings

Gensel, P., and H. Andrews. 1987. "The Evolution of Early Land Plants." *American Scientist* 75: 478–489.

Moore, R., W. D. Clark, and K. Stern. 1995. *Botany*. Dubuque, Iowa: W. C. Brown.

Moore-Landecker, E. 1990. *Fundamentals of the Fungi*. Third edition. Englewood Cliffs, New Jersey: Prentice-Hall.

Raven, P., R. Evert, and S. Eichhorn. 1986. *Biology of Plants*. Fourth edition. New York: Worth.

20 ANIMALS: THE INVERTEBRATES

Madeleine's Limbs

In August of 1994, about 900 million years after the first animals appeared on Earth, Madeleine made *her* entrance. As they are wont to do, grandmothers and aunts made a quick count on the sly—arms, legs, ears, and eyes, two of each; fully formed mouth and nose—just to be sure these were present and accounted for.

One grandmother, having been too long in the company of biologists, experienced an epiphany as she witnessed Madeleine's birth. In that profound instant she sensed ancestral connections, emerging from the distant past and through her, into the future.

Madeleine's body plan didn't emerge out of thin air. Thirty-five thousand years ago, people just like us were having children just like Madeleine. If we are interpreting the fossil record correctly, then five million years ago the offspring of individuals on the road to modern humans resembled her in some respects but not others. Sixty million years ago, primate ancestors of those individuals were giving birth precariously, up in the trees. Two hundred and fifty million years ago, mammalian ancestors of those primates were giving

birth—and so on back in time to the very first animals, which had no limbs or eyes or noses at all.

We have very few clues to what the first animals looked like, but one thing is clear. By the dawn of the Cambrian, they had given rise to all major groups of **invertebrates**—animals without backbones—even to Madeleine's backboned but limbless ancestors.

And what stories those Cambrian animals tell! One bunch flourished 530 million years ago, in a submerged basin that had formed between a reef and the coast of an early continent. Protected from ocean currents, the sediments had piled against the steep reef. About 500 feet below the surface, the water was oxygenated and clear—and tiny, well-developed animals lived in, on, and above the dimly lit, muddy sediments (Figure 20.1a).

Like castles built from wet sand along a seashore, their living quarters were unstable. Part of the bank above the community slumped abruptly and wiped it out. Sediments from that underwater avalanche kept scavengers from reaching and obliterating all traces of the dead. Over time, muddy silt rained down on the

a

Figure 20.1 **(a)** Reconstruction of a few Cambrian animals, known from the Burgess Shale fossils. **(b)** Presumed evolutionary relationships among major groups of animals. Take a moment to study this family tree diagram. We will use it repeatedly as our road map through discussions of each group. **(c)** Madeleine.

single-celled, protistanlike ancestors

b

c

tomb. The increased pressure and chemical changes gradually transformed the sediments into finely stratified shale; and soft parts of the flattened animals became shimmering mineralized films.

Sixty-five million years ago, part of the seafloor was plowing under the North American plate, and western Canada's mountain ranges were rising. By 1909, the fossils were high in the eastern mountains of British Columbia. There, a fossil hunter tripped over a chunk of shale, which split apart into fine layers—and the Burgess Shale story came to light.

In this chapter and the next, you will be comparing the body plans of different groups of animals. Such comparisons give insight into evolutionary relatedness and help us construct family trees, such as the one in Figure 20.1b. Don't assume that the structurally simple animals of the most ancient lineages are somehow primitive or evolutionarily stunted. As you will see, they, too, are exquisitely adapted to their environment.

As you poke about the branches of the animal family tree, keep the greater evolutionary story in mind. At each branch point, microevolutionary processes gave rise to workable changes in body plans. Madeleine's uniquely human traits, and yours, emerged through modification of certain traits that had evolved earlier, in countless generations of vertebrates and, before them, in ancient invertebrate forms.

KEY CONCEPTS

1. Animals are multicelled, aerobic heterotrophs that ingest or parasitize other organisms. Nearly all kinds have tissues, organs, and organ systems, and most are motile during at least part of the life cycle. Animals reproduce sexually and often asexually, and new individuals go through embryonic stages of development.

2. Animals originated about 900 million years ago. We know there are well over 2 million species alive today. Of these, more than 1,950,000 are invertebrates (animals with no backbone). Fewer than 50,000 species are vertebrates (animals with a backbone).

3. Comparisons of the body plans of existing animals, in conjunction with the fossil record, reveal several trends in animal evolution. The most revealing aspects of an animal's body plan are its type of symmetry, gut, and cavity (if any) between the gut and body wall; whether it has a distinct head end; and whether it is divided into a series of segments.

4. Placozoans and sponges are structurally simple, with no body symmetry. Both are at the cellular level of construction. Jellyfishes and other cnidarians have radial symmetry. They are at the tissue level of construction.

5. Flatworms, roundworms, rotifers, and nearly all other animals more complex than cnidarians show bilateral symmetry, and they have tissues, organs, and organ systems.

6. Apparently, two major lineages diverged shortly after ancestral flatworms evolved. One gave rise to mollusks, annelids, and arthropods. The other gave rise to echinoderms and chordates.

7. By biological measures, including diversity, sheer numbers, and distribution, insects and other arthropods have been the most successful group.

General Characteristics of Animals

What, exactly, are **animals**? We can only define them by a list of general characteristics, not with just a sentence or two. *First*, animals are multicelled, and in most cases their body cells form tissues that are arranged as organs and organ systems. The body cells are diploid in nearly all species. *Second*, animals are heterotrophs that obtain carbon and energy by ingesting other organisms or by absorbing nutrients from them. *Third*, animals require oxygen, for use in aerobic respiration. *Fourth*, animals reproduce sexually and, in many cases, asexually. *Fifth*, most animals are motile during at least part of the life cycle. *Sixth*, their life cycles proceed through a period of embryonic development. In brief, mitotic cell divisions transform the animal zygote into a multicelled embryo. The embryonic cells are the forerunners of **ectoderm**, **endoderm**, and, in most species, **mesoderm**. These are the primary tissue layers that give rise to all tissues and organs of the adult, as described in Section 34.2.

Diversity in Body Plans

Mammals, birds, reptiles, amphibians, and fishes are the most familiar animals. All are **vertebrates**—the only animals with a "backbone." And yet, of probably more than *2 million* species of animals, fewer than 50,000 are vertebrates! What we call **invertebrates** are animals with diverse features, but not a backbone.

We group animals into more than thirty phyla. Table 20.1 lists the ones described in this book. The characteristics they share with one another arose early in time, before divergences from a common ancestor gave rise to separate lineages. Later, morphological differences accumulated among lineages and were the foundation for dazzling diversity. How can we get a conceptual handle on their modern-day descendants—on animals as different as flatworms, hummingbirds, platypuses, humans, and giraffes? We can compare similarities and differences with respect to five basic features. These are body symmetry, cephalization, type of gut, type of body cavity, and segmentation.

Body Symmetry and Cephalization With very few exceptions, animals are radial *or* bilateral. Those with **radial symmetry** have body parts arranged regularly around a central axis, like spokes of a bike wheel. Thus a cut down the center of a hydra (Figure 20.2a) divides it into equal halves; another cut at right angles to the first divides it into equal quarters. All radial animals live in water. Their body plan is adapted to intercepting food that is coming toward them from any direction.

Animals with **bilateral symmetry** have right and left halves that are mirror images of each other. Most

Phylum	Some Representatives	Number of Known Species
Placozoa (*Trichoplax*)	Simplest animal; like a tiny plate, but with tissue layers	1
Porifera (poriferans)	Sponges	8,000
Cnidaria (cnidarians)	Hydrozoans, jellyfishes, corals, sea anemones	11,000
Platyhelminthes (flatworms)	Turbellarians, flukes, tapeworms	15,000
Nematoda (roundworms)	Pinworms, hookworms	20,000
Rotifera (rotifers)	Tiny body with crown of cilia, great internal complexity	1,800
Mollusca (mollusks)	Snails, slugs, clams, squids, octopuses	110,000
Annelida (segmented worms)	Leeches, earthworms, polychaetes	15,000
Arthropoda (arthropods)	Crustaceans, spiders, insects	1,000,000+
Echinodermata (echinoderms)	Sea stars, sea urchins	6,000
Chordata (chordates)	Invertebrate chordates: tunicates, lancelets	2,100
	Vertebrates: Fishes	21,000
	Amphibians	3,900
	Reptiles	7,000
	Birds	8,600
	Mammals	4,500

Table 20.1 Animal Phyla Described in This Book

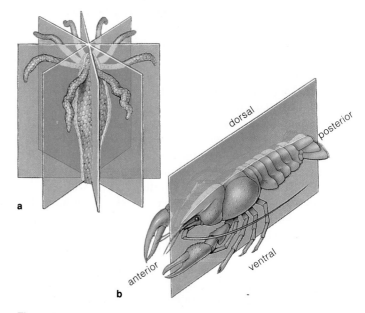

Figure 20.2 (**a**) Radial symmetry of a hydra. (**b**) Bilateral symmetry of a crayfish.

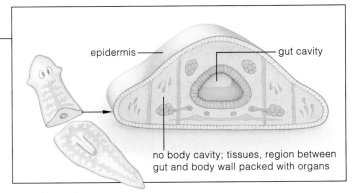

a No coelom

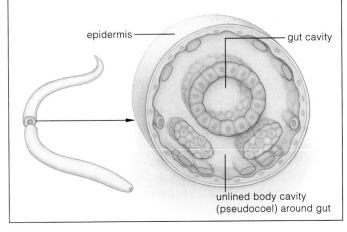

b False coelom

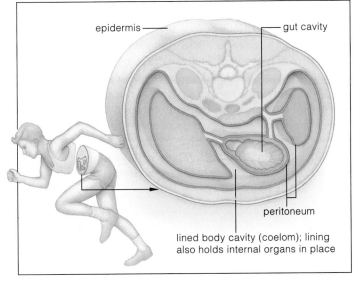

c Coelom

Figure 20.3 Type of body cavity (if any) in animals.

have an *anterior* end (head) and opposite, *posterior* end. They also have a *dorsal* surface (a back) and an opposite, *ventral* surface (Figure 20.2b). This body plan evolved after forward-creeping species emerged. Mostly, their forward end must have been the first to encounter food and other stimuli. We can imagine there was selection for **cephalization**. By this evolutionary process, sensory

structures and nerve cells became concentrated in the head. The joint evolution of bilateral body plans and cephalization resulted in *pairs* of muscles and *pairs* of sensory structures, nerves, and brain regions.

Type of Gut   The **gut** is a region inside the body in which food is digested, then absorbed into the internal environment. Saclike guts have one opening (a mouth) for taking in food and expelling residues. Other guts are part of a tubelike system with a mouth and an anus—that is, openings at both ends. Such "complete" digestive systems are very efficient. Different parts of them have specialized functions, such as preparing, digesting, and storing material. The evolution of such systems helped pave the way for increases in body size and activity.

Body Cavities   A body cavity separates the gut and the body wall of most bilateral animals (Figure 20.3). One type of cavity, a **coelom**, has a unique lining, the peritoneum. This lining also encloses the organs in the coelom and helps hold them in place. You, for example, have a coelom. A sheetlike muscle divides it into an upper, *thoracic* cavity and a lower, *abdominal* cavity. Your thoracic cavity holds a heart and lungs; your abdominal cavity holds a stomach, intestines, and other organs.

Some invertebrates don't have a body cavity. Tissues fill the region between the gut and body wall. Others have a pseudocoel ("false coelom"), a kind of body cavity with no peritoneum. A true coelom was a major step in the evolution of animals that were larger and more complex than worms. By cushioning and protecting organs, a coelom favored increases in size and activity.

Segmentation   Segmented animals have a series of body units, which may or may not be similar to one another. Think of earthworms. Their segments appear the same on the outside. Insect segments are grouped in three regions (head, thorax, and abdomen), and they differ greatly from one another. Insects especially show how a huge variety of head parts, legs, wings, and other appendages evolved from less specialized segments.

Animals are multicelled, and most have tissues, organs, and organ systems. Body cells of most species are diploid.

Animals are aerobically respiring heterotrophs that ingest other organisms or absorb nutrients from them.

Animals reproduce sexually and, in many cases, asexually. They go through a period of embryonic development, and most are motile during at least part of the life cycle.

The body plans of animals differ with respect to five features: body symmetry, cephalization, type of gut, type of body cavity, and segmentation.

20.2 PUZZLES ABOUT ORIGINS

Eight hundred million years ago, the first multicelled animals were making tracks and burrows in marine sediments (Section 17.4). *Where did these animals come from?* They probably originated from ancient protistans, but we don't know which ones.

By one hypothesis, the forerunners of animals were ciliates, much like *Paramecium*, that had multiple nuclei in a single-celled body. In time, each nucleus became compartmentalized in a separate cell. Yet no existing animal develops by compartmentalization.

By another hypothesis, multicelled animals arose from flagellated cells that lived in hollow, spherical colonies, rather like the *Volvox* colonies shown in Figure 4.3. In time, some cells in the colony became modified for reproduction and other specialized tasks. So began the division of labor that characterizes multicellularity.

Suppose some colonies became flattened and started creeping about on the seafloor. This may have led to the evolution of layers of cells—such as those of *Trichoplax adhaerens*. This soft-bodied marine animal, shaped a bit like pita bread, has no symmetry and no mouth. It is the only known **placozoan** (Placozoa, after *plax*, meaning plate; and *zoon*, meaning animal). Its several thousand cells are arrayed in two distinctly different layers. When it glides over food, its body briefly humps up (Figure 20.4). Glandular cells present in the lower layer secrete digestive enzymes on the food, then individual cells absorb the breakdown products. Reproduction might be asexual (by budding or fission) or it might be sexual, by mechanisms not yet understood. In sum, structurally and functionally, *Trichoplax* is as simple as animals get.

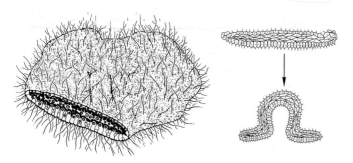

Figure 20.4 Cutaway view of *Trichoplax adhaerens*, an animal with a two-layer body that measures about three millimeters across.

Possibly the question of origins requires more than one answer. It may be that different lineages descended from more than one group of protistanlike ancestors.

Multicelled animals arose from protistanlike ancestors that may have resembled ciliates, colonial flagellates, or both.

20.3 SPONGES—SUCCESS IN SIMPLICITY

Sponges (Porifera) are one of nature's true success stories. Since Cambrian times, they have been abundant in the seas, especially in the waters off coasts and along reefs. Of 8,000 or so known species, about 100 live in freshwater. Worms, shrimps, tiny fishes, barnacles, and other animals make their home inside or on the sponge body.

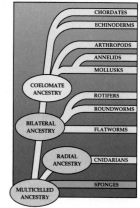

Some sponges are as tiny as a fingernail. Yet others are big enough to sit in. Figures 20.5 and 20.6 show a few of their sprawling, flat, lobed, compact, tubular, cuplike, or vaselike shapes. Regardless of its shape, the body has no symmetry or organs. Flattened cells line the outer surface and inner cavities. But the linings are not much more than the cell layers of *Trichoplax*, and they differ from the tissues of other animals. Amoeboid cells reside in a gelatinlike substance between the inner and outer linings (Figure 20.6b). Tough protein fibers (spongin), glasslike spicules of silica or calcium carbonate, or both, stiffen the sponge body and impart structure to it.

The skeletal elements may be a reason why sponges as a group have endured so long. Cleveland Hickman put it this way: Most potential predators discover that sampling a sponge is about as pleasant as eating a mouthful of glass splinters embedded in fibrous gelatin. Besides, chemically speaking, many sponges stink.

Water flows into a sponge body through microscopic pores and chambers, then out through one or more openings (oscula; singular, osculum). It does so mainly in response to the collective beating of thousands or millions of flagellated cells. These **collar cells** help make up the inner lining of the sponge body. Besides flagella,

Figure 20.5 A sprawling, red-orange sponge, one of many types that encrust underwater ledges in temperate seas.

Figure 20.6 (**a,b**) Body plan of a simple sponge. The outer lining consists of flattened cells. It also has some contractile cells, arranged as bands around the tiny openings across the body. These cells contract slowly, independently of the other cells, and so influence water flow. Amoeboid cells inside the gelatin-like matrix between the inner and outer linings secrete materials for spicules and fibers. Other amoeboid cells digest and transport food. They retain the capacity to divide, and their cellular descendants can differentiate into any other type of sponge cell. They also have roles in asexual processes, such as gemmule formation.

(**c**) Flagellated cells line the inner canals and chambers of the sponge body. Each has a collar of food-trapping structures called microvilli. Fine filaments connect the microvilli to each other, forming a "sieve" that strains food particles from the water. At the base of the collar, the cell engulfs the trapped food (by phagocytosis).

(**d**) A splendid example of skeletal elements—Venus's flower basket (*Euplectella*). This marine sponge has six-rayed spicules of silica, fused as a rigid, elaborate network. A thin cell layer stretches over the interconnecting spicules. At the base of the body is an anchoring tuft of spicules.

(**e**) A basket sponge of the Caribbean, releasing a cloud of sperm into the water.

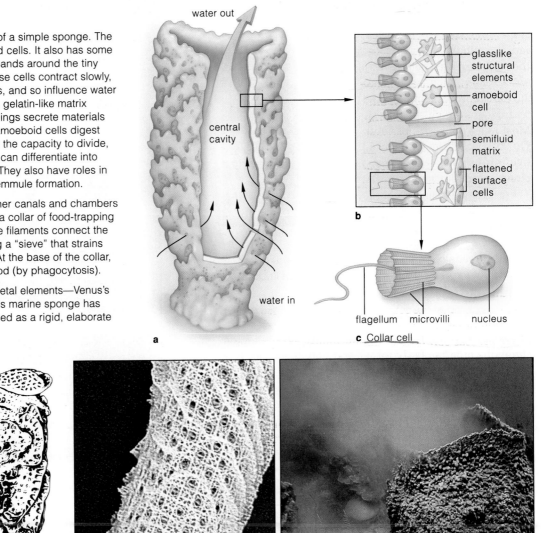

water out

central cavity

water in

a

glasslike structural elements

amoeboid cell

pore

semifluid matrix

flattened surface cells

b

flagellum   microvilli   nucleus

c Collar cell

d

e

the cells have "collars" of absorptive structures called microvilli (Figure 20.6*c*). Bacteria and other bits of food dissolved in the water become trapped in the collars. Some food gets transferred to the amoebalike cells for further breakdown, storage, and distribution.

Sponges reproduce sexually when some cells give rise to eggs and to sperm. Most release sperm into the water (Figure 20.6*e*), but they release or retain the eggs until those eggs have been fertilized and embryos start developing. The young, developing sponges proceed through a microscopic, swimming larval stage. Each **larva** (plural, larvae) is a sexually immature form that precedes the adult stage of many animal life cycles. Some sponges reproduce asexually by fragmentation

(small fragments break from the parent and grow into new sponges). Most freshwater species also can reproduce asexually by way of gemmules. These are clusters of sponge cells, some of which form a hard covering around others. The ones inside are protected from extreme cold or drying out. When conditions are favorable, the gemmules germinate and establish a new colony of sponges.

Sponges have no symmetry, tissues, or organs; they are at the *cellular* level of construction. Yet they have successfully endured through time, possibly because most predators find their spicule-rich and often stinky bodies unappetizing.

20.4 CNIDARIANS—TISSUES EMERGE

All **cnidarians** are tentacled, radial animals. They include the scyphozoans (jellyfishes), anthozoans (such as the sea anemones and corals), and hydrozoans (such as *Hydra*). Most live in the seas; of about 11,000 species, fewer than 50 live in freshwater. Cnidarians alone produce **nematocysts**. These capsules can discharge threads that entangle prey and that may fend off predators. As humans know, the toxin-tipped threads of some species cause stings. Hence the phylum name Cnidaria, after the Greek word for nettle.

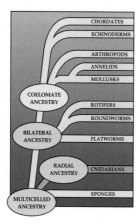

Cnidarian Body Plans

The **medusa** (plural, medusae) and **polyp** are the most common cnidarian body forms, and both have a saclike gut (Figure 20.7a). Medusae float. Some look like bells; others look like upside-down saucers. The mouth, centered under the bell, may have armlike extensions that assist in prey capture and feeding. Polyps have a tubelike body with a tentacle-fringed mouth at one end. The other end is usually attached to a firm substrate. When *Trichoplax* is draping over and digesting a blob of food, it is rather like a gut on the run. By contrast, the saclike cnidarian gut is a permanent food-processing chamber. It has a gastrodermis, a sheetlike lining with glandular cells that secrete digestive enzymes. Another lining, the epidermis, covers the rest of the body's surfaces (Figure 20.7d). Each lining is an **epithelium** (plural, epithelia), a

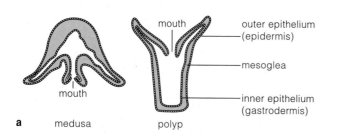

a medusa   polyp

- mouth
- outer epithelium (epidermis)
- mesoglea
- inner epithelium (gastrodermis)

Figure 20.7 Cnidarian body plans. (**a**) Diagram of a medusa and a polyp, sliced through the midsection. (**b**) Medusa of the sea nettle (*Chrysaora*), a jellyfish. (**c**) Hydrozoan polyp (*Hydra*), attached to a substrate, as it captures and digests its prey. (**d**) Tissue organization of a sea anemone. Cnidarian tentacles bristle with nematocysts. These are capsules with an inverted tubular thread. The one shown has a bristlelike trigger. When prey touch the trigger, the capsule becomes more "leaky" to water. Water diffuses inward, pressure inside the capsule builds up, and the thread is forced to turn inside out. The thread's tip may penetrate prey and release a toxin.

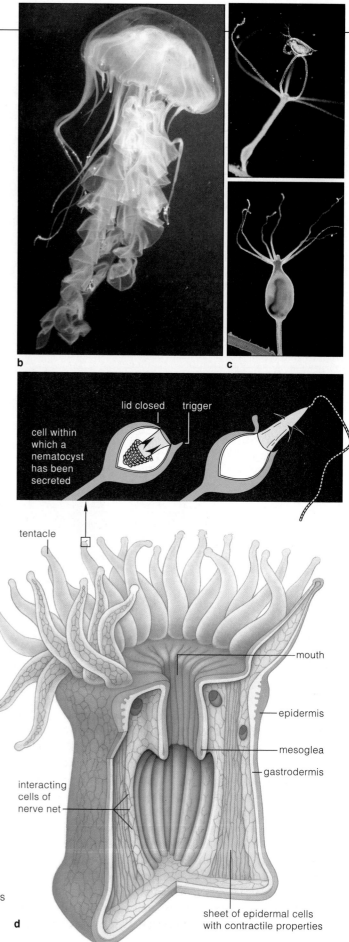

lid closed   trigger

cell within which a nematocyst has been secreted

tentacle

mouth

epidermis

mesoglea

gastrodermis

interacting cells of nerve net

sheet of epidermal cells with contractile properties

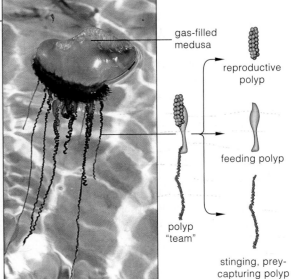

a land   barrier reef   open ocean

b

c

gas-filled
medusa

reproductive
polyp

feeding polyp

polyp
"team"

stinging, prey-
capturing polyp

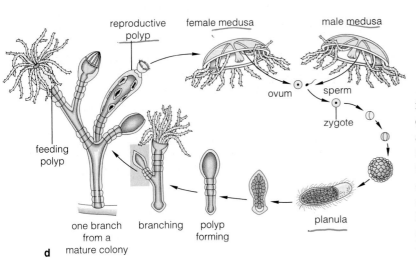

reproductive
polyp

female medusa

male medusa

ovum   sperm

zygote

feeding
polyp

one branch
from a
mature colony

branching

polyp
forming

planula

d

Figure 20.8 (**a**) Aerial view of a barrier reef, formed mainly of accumulated skeletons of certain corals, such as the one in (**b**). This colonial form has interconnected polyps with external skeletons. (**c**) Portuguese man-of-war (*Physalia*). (**d**) Life cycle of *Obelia*, which includes both medusa and polyp stages.

type of tissue that lines the free surfaces of all animals more complex than sponges. **Nerve cells** threading through the epithelia form a "nerve net," a very simple nervous tissue that coordinates responses to stimulation. Together with **sensory cells** and **contractile cells**, the nerve net controls movement and changes in shape.

Between the epidermis and gastrodermis is a layer of gelatinous secreted material known as the mesoglea ("middle jelly"). Jellyfishes contain enough mesoglea to impart buoyancy and to serve as a firm yet deformable skeleton against which the contractile cells can act. The contractions narrow the bell, forcing a jet of water from it and so propelling the animal forward. Most polyps, which have little mesoglea, use water in their gut as a hydrostatic skeleton. A **hydrostatic skeleton** is a fluid-filled cavity or cell mass. When certain cells contract, the volume of the cavity or mass doesn't change but rather is shunted about, so the shape of the body changes.

Colonial anthozoans show interesting variations on the basic body plan. For example, reef-forming corals live in clear, warm water (at least 20°C). They consist of

interconnected polyps that secrete calcium-reinforced external skeletons. Over time, the skeletons accumulate and thus become the main building material for reefs (Figure 20.8a and b). The most extensive accumulation, the Great Barrier Reef, parallels the eastern Australian coast for about 1,600 kilometers. *Physalia*, informally called the Portuguese man-of-war, is one of the colonial hydrozoans. In its colony, polyps and medusae interact as "teams" that engage in feeding, defense, reproduction, and other specialized tasks (Figure 20.8c).

Cnidarian Life Cycles

Many cnidarians have only a polyp stage *or* a medusa stage in the life cycle. *Physalia, Obelia,* and some other types have both, with the medusa being the sexual form (Figure 20.8d). They have **gonads** (primary reproductive organs) that are associated with the epidermis or gastrodermis. These simple gonads release gametes by rupturing. Most zygotes formed at fertilization develop into **planulas**, a kind of swimming or creeping larva that usually has ciliated epidermal cells. In time a mouth opens at one end, the larva is transformed into a polyp or medusa, and the cycle begins anew.

Cnidarians are radial animals with tentacles, a saclike gut, epithelia, and a nerve net. They alone form nematocysts.

Cnidarians are at the *tissue* level of construction; compared with sponges, their cells interact in much more coordinated fashion in the performance of specific tasks.

tissue = group of cell function together in specialized task.

FLATWORMS, ROUNDWORMS, ROTIFERS—AND SIMPLE ORGAN SYSTEMS

When we move beyond the cnidarians in our survey, we find animals that range from flatworms to humans. All of these animals have simple or complex organs. An **organ** is an association of one or more kinds of tissues, arranged in certain proportions and patterns. Most often, one organ interacts with others to carry out an activity that helps the body function. By definition, two or more organs that are interacting efficiently in the performance of some task represent an **organ system**. We turn now to the simplest animals at the *organ-system* level of construction. These are not breathtakingly complex animals, but a few can make our lives breathtakingly miserable (Section 20.6).

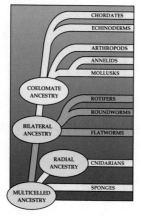

CHORDATES
ECHINODERMS
ARTHROPODS
ANNELIDS
MOLLUSKS
COELOMATE ANCESTRY
ROTIFERS
ROUNDWORMS
BILATERAL ANCESTRY
FLATWORMS
RADIAL ANCESTRY
CNIDARIANS
SPONGES
MULTICELLED ANCESTRY

Flatworms

Among the 15,000 or so known species of **flatworms** (phylum Platyhelminthes) are turbellarians, flukes, and tapeworms. Most of these bilateral, cephalized animals have a flattened body (hence their name) and simple organ systems. For example, their digestive system has a pharynx (a muscular tube used for feeding) and a saclike, often branched gut. Reproductive systems vary, but in most cases a flatworm is a **hermaphrodite**, with female and male gonads. Two such worms reproduce sexually by mutual transfer of sperm. Each has a penis (a sperm-delivery structure) and glands that produce a protective capsule around fertilized eggs.

Most turbellarians live in the seas; planarians and a few others live in freshwater. Some eat tiny animals or suck tissues from dead or wounded ones. Commonly, planarians reproduce asexually by *transverse* fission: They divide in half at their midsection, then each half regenerates the missing parts. Like you, a planarian is able to adjust the composition and volume of its body fluids. Its water-regulating system has one or more tiny, branched tubes. These tubes are called protonephridia (singular, protonephridium). They extend from pores at the body surface to bulb-shaped flame cells in body tissues. When excess water moves into the cells, a tuft of cilia "flickering" in the bulb drives it through the tubes, to the outside (Figure 20.9d).

Flukes are parasites. (A parasite, recall, lives in or on a living host and feeds on its tissues.) Their life cycles have sexual and asexual phases and at least two kinds of hosts. As you will see from the examples in Section 20.6, flukes reach sexual maturity in a vertebrate that serves as their *primary* host. The larval stages develop or become encysted in an *intermediate* host.

Tapeworms thrive on predigested food in vertebrate intestines. These parasites attach to the intestinal wall by a scolex, a structure with suckers, hooks, or both (Figure 20.10). Proglottids (tapeworm body units) bud

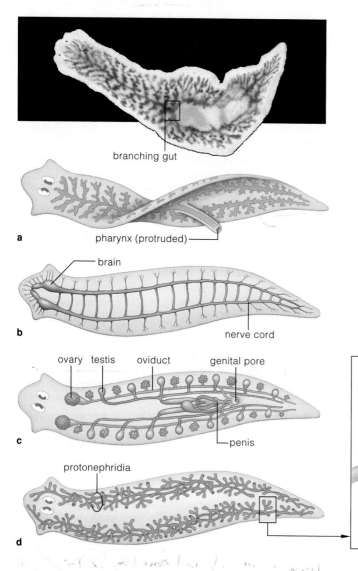

branching gut

a pharynx (protruded)

b brain

nerve cord

ovary testis oviduct genital pore

c penis

protonephridia

d

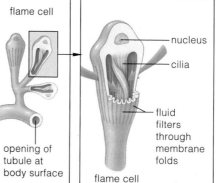

flame cell

opening of tubule at body surface

nucleus

cilia

fluid filters through membrane folds

flame cell

Figure 20.9 Organ systems of one type of flatworm (planarian). (**a**) A pharynx opens to the gut. It projects to the outside when the animal feeds. Between feedings the pharynx retracts into a chamber. (**b**) Nervous system, (**c**) reproductive system, and (**d**) water-regulating system.

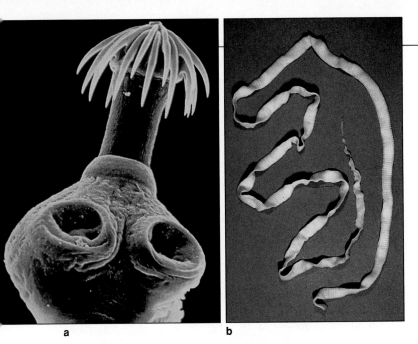

a　　　　　　　　　　　　　　**b**

Figure 20.10 (**a**) Tapeworm scolex. This one attaches to the gut lining of a shorebird, its primary host. (**b**) Sheep tapeworm.

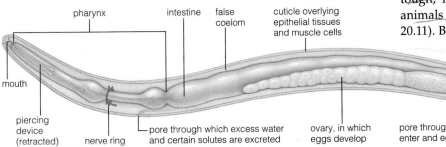

pharynx　　　　intestine　　false
　　　　　　　　　　　　　coelom

cuticle overlying
epithelial tissues
and muscle cells

mouth

piercing
device
(retracted)

nerve ring

pore through which excess water
and certain solutes are excreted

ovary, in which
eggs develop

pore through which sperm
enter and eggs are released

anus

Figure 20.11 Body plan of a roundworm (*Paratylenchus*) that parasitizes the roots of certain plants.

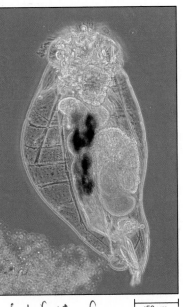

Figure 20.12 Lateral view of a rotifer, busily laying eggs. Males are unknown in many species; the females produce diploid eggs that develop into diploid females. The females of other species do the same, but they also produce haploid eggs that develop into haploid males. If a haploid egg happens to be fertilized by a male, it develops into a female. The male rotifers appear only occasionally, and are dwarfed and short-lived.

50 µm

just behind the scolex. Proglottids are hermaphroditic; they mate and transfer sperm. Older ones (farthest from the scolex) store fertilized eggs. They break off and then leave the body in feces. Afterward, an intermediate host may meet up with the eggs.

Some turbellarians and larval flukes and tapeworms resemble cnidarian planulas. This inspires speculation that bilateral animals may have evolved from planula-like ancestors, by way of increased cephalization and the development of tissues derived from mesoderm.

Roundworms

Roundworms live nearly everywhere, from snowfields to deserts and hot springs. A handful of rich soil has thousands of them; scavenging species make fast work of dead earthworms or rotting fruit. Most are harmless, even beneficial. But hookworms, pinworms, and other types parasitize animals or plants. All roundworms have a bilateral, cylindrical body, usually tapered at both ends. They have a **cuticle**, which in animals is a tough, flexible body covering. They are the simplest animals that have a complete digestive system (Figure 20.11). Between the gut and body wall is a false coelom, often packed with reproductive organs. Cells in all tissues absorb nutrients from this fluid and give up wastes to it.

Rotifers

Of **rotifers** (Rotifera), we might say this: Seldom have so many organs been packed in so little space. Rotifers are abundant in plankton (communities of floating or weakly swimming organisms), although some crawl on algae and wet mosses. They eat bacteria and tiny algae. They show the complexity possible in animals with a false coelom. Most species are less than a millimeter long, yet they have a pharynx, an esophagus, digestive glands, a stomach, usually an intestine and anus, and protonephridia (Figure 20.12). The head has nerve cell clusters. Some species have "eyes" (absorptive pigments). Two "toes" exude substances that attach many species to substrates at feeding time. All rotifers have a crown of cilia used in swimming and wafting food to the mouth. Its motions reminded early microscopists of a turning wheel (rotifer means wheel-bearer).

Flatworms, roundworms, and rotifers are simple bilateral, cephalized animals at the *organ-system* level of construction.

Chapter 20　Animals: The Invertebrates　**335**

A ROGUE'S GALLERY OF PARASITIC WORMS

Many parasitic flatworms and roundworms call the human body home. In any given year, about 200 million people house blood flukes responsible for *schistosomiasis*. Figure 20.13 shows the life cycle of a Southeast Asian blood fluke (*Schistosoma japonicum*). The cycle requires a human primary host, standing water in which larvae can swim, and an aquatic snail (an intermediate host). The flukes reproduce sexually (*1*). After being fertilized, eggs leave the human body in feces and hatch into ciliated, swimming larvae (*2*) that burrow into a snail and multiply asexually (*3*). In time, many fork-tailed larvae develop (*4*). These leave the snail (*5*) and swim until they contact human skin (*6*). They bore inward and migrate to thin-walled intestinal veins, and the cycle begins anew. In infected humans, white blood cells that defend the body attack the masses of fluke eggs, and grainy masses form in tissues. In time, the liver, spleen, bladder, and kidneys deteriorate.

Some tapeworms parasitize humans. Different species use pigs, freshwater fish, or cattle as intermediate hosts. Humans become infected when they eat pork, fish, or beef that is raw, improperly pickled, or insufficiently cooked—and contaminated with tapeworm larvae (Figure 20.14).

Or consider the parasitic roundworm that causes thin, serpentlike ridges at the surface of human skin. For several thousand years, healers have removed the "serpents" by winding them out slowly, painfully, around a stick.

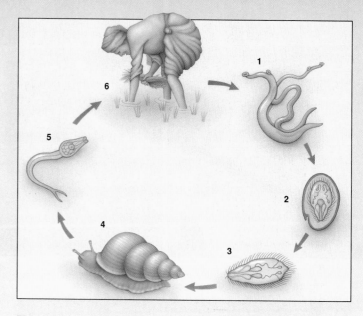

Figure 20.13 Life cycle of a blood fluke, *Schistosoma japonicum*.

Other roundworms, called pinworms and hookworms, also cause problems. The pinworm *Enterobius vermicularis* parasitizes humans in temperate regions. It lives in the large intestine, but at night the centimeter-long females migrate to the anal area and lay eggs. Their presence causes itching, and scratchings made in response transfer

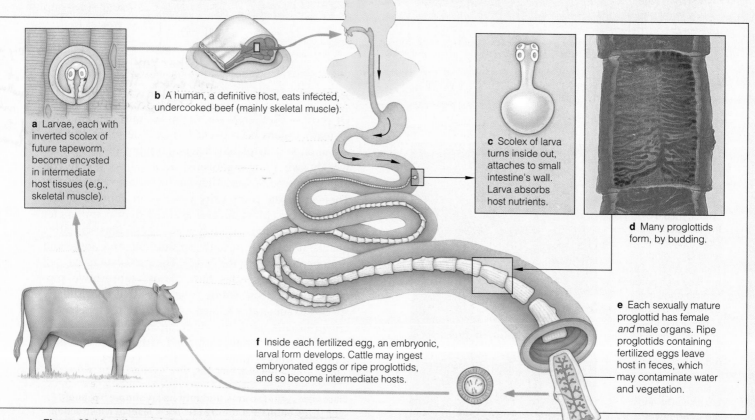

b A human, a definitive host, eats infected, undercooked beef (mainly skeletal muscle).

a Larvae, each with inverted scolex of future tapeworm, become encysted in intermediate host tissues (e.g., skeletal muscle).

c Scolex of larva turns inside out, attaches to small intestine's wall. Larva absorbs host nutrients.

d Many proglottids form, by budding.

e Each sexually mature proglottid has female *and* male organs. Ripe proglottids containing fertilized eggs leave host in feces, which may contaminate water and vegetation.

f Inside each fertilized egg, an embryonic, larval form develops. Cattle may ingest embryonated eggs or ripe proglottids, and so become intermediate hosts.

Figure 20.14 Life cycle of a beef tapeworm, *Taenia saginata*.

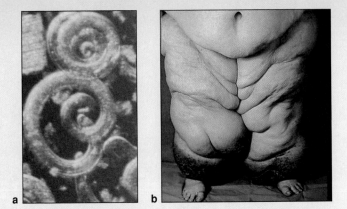

Figure 20.15 (**a**) Juveniles of a roundworm, *Trichinella spiralis*, living inside the muscle tissue of a host animal. (**b**) Legs of a woman infected by *Wuchereria bancrofti*, a roundworm.

some eggs to hands, then to other objects. Newly laid eggs contain embryos, but within a few hours they are juveniles and ready to hatch if another human ingests them.

Hookworms are especially serious in impoverished parts of the tropics and subtropics. Adult hookworms live in the small intestine. They feed on blood and other tissues after teeth or sharp ridges bordering their mouth cut into the intestinal wall. Adult females, about a centimeter long, can release a thousand eggs daily. These leave the body in feces, then hatch into juveniles. Walk barefoot, and a juvenile may penetrate the skin. Inside a host, the parasite travels the bloodstream to the lungs. There it works its way into the air spaces. After moving up the windpipe, the parasite is swallowed. Soon it is in the small intestine, where it may mature and live for several years.

The roundworm *Trichinella spiralis* causes painful, sometimes fatal symptoms. Adults live in the lining of the small intestine. Females release juveniles (Figure 20.15*a*), which work their way into blood vessels and travel to muscles. There they become encysted (they produce a covering around themselves and enter a resting stage). Humans get infected mostly by eating insufficiently cooked meat from pigs or certain game animals. The presence of encysted juveniles cannot easily be detected when fresh meat is examined, even in a slaughterhouse.

Figure 20.15*b* shows the results of prolonged, repeated infections by *Wuchereria bancrofti*, a roundworm. Adult worms become lodged in the body's lymph nodes, where they can obstruct the flow of a fluid (lymph) that normally trickles back to the bloodstream. When the obstruction causes fluid to accumulate, legs and other body regions undergo grotesque enlargement—a condition called *elephantiasis*. A mosquito is the intermediate host. Female *Wuchereria* produce active young that travel the bloodstream at night. If a mosquito sucks blood from an infected human, the juveniles may enter the insect's tissues. In time they move near the insect's sucking device—ready to enter a new host when the mosquito draws blood again.

20.7 A MAJOR DIVERGENCE

Bilateral animals not much more complex than modern flatworms evolved in Cambrian times. Not long afterward, some of these gave rise to two lineages of animals equipped with a coelom (Figure 20.1*b*). We call these great lineages the **protostomes** and the **deuterostomes**. Mollusks, annelids, and arthropods are protostomes. Echinoderms and chordates are deuterostomes.

As an outcome of mutations, developmental pathways diverged in these two lineages. Their embryonic stages developed differently, and this led to differences in body plans. For example, mitotic cell divisions that transform a zygote into an early embryo proceed relative to a body axis (like the axis that runs between the Earth's poles). Protostomes undergo **spiral cleavage**, a pattern in which the early cell divisions are at oblique angles relative to the original body axis. Deuterostomes undergo **radial cleavage**, a pattern in which the early cell divisions are parallel and perpendicular to the axis:

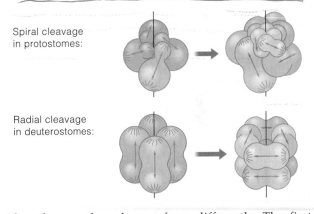

Spiral cleavage
in protostomes:

Radial cleavage
in deuterostomes:

Also, the mouth and anus form differently. The first external opening in protostome embryos becomes the mouth, and an anus forms elsewhere. The first opening in deuterostome embryos becomes the anus; the second becomes the mouth. As a last example, a protostome coelom arises from spaces in the mesoderm. A deuterostome coelom forms from outpouchings of the gut wall:

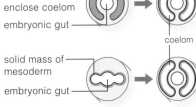

How the coelom forms in protostomes:

pouch that will form mesoderm, enclose coelom

embryonic gut

coelom

How the coelom forms in deuterostomes:

solid mass of mesoderm

embryonic gut

Animals with a coelom evolved in Cambrian times. Through an accumulation of mutations, they diverged into two great lineages—the protostomes and deuterostomes. The crucial mutations affected how the early embryos develop, and this led to major differences in basic body plans.

20.8 MOLLUSKS—A WINNING BODY PLAN

Maybe it was their fleshy, soft bodies that gave the ancestors of **mollusks** the potential to diversify and radiate into so many habitats. (The name of the phylum, Mollusca, means soft bodied.) Among 110,000 or so species are tiny snails in treetops and giant predators of the seas. All mollusks are bilateral, and they alone have a mantle, a tissue fold that hangs skirtlike over the body. Most have a shell of calcium carbonate or an evolutionary remnant of one. They have gills, of a kind called ctenidia (respiratory organs with thin-walled leaflets for gas exchange). Most have a fleshy foot; those with a well-developed head have eyes and tentacles. Many have a tonguelike, toothed organ (radula) that shreds food destined for the gut. Figure 20.16 shows a few species. Here we focus on three classes, the gastropods, bivalves, and cephalopods.

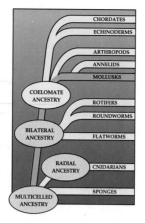

Gastropods

With 90,000 species, gastropods ("belly foots") are the largest class. These are shelled snails or slugs with no shell (or a remnant of one). Different kinds are grazers, predators, and parasites that creep, swim, or float. One, the garden snail, is a nuisance (page 238).

Evolution put an unusual twist in gastropods. The embryos undergo torsion during development. By this process, the visceral mass (gut, heart, gills, kidneys, and other internal organs) twists and realigns, so wastes end up being discharged above the head (Figure 20.16a).

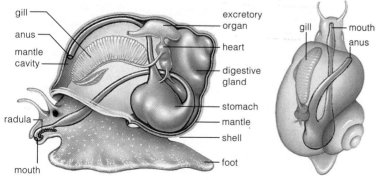

Figure 20.16 (a) Features of the molluscan body plan, evident in an aquatic snail. The radula of snails and some other mollusks is protracted and retracted in a rhythmic way. Food is rasped off and drawn to the gut on the retraction stroke. (b) Two "Mexican dancers," sea slugs belonging to the group of gastropods called nudibranchs. (c) From Monterey Bay, California, a chiton, a mollusk with a dorsal shell divided into eight plates. (d) An octopus, one of the cephalopods. (e) A garden snail.

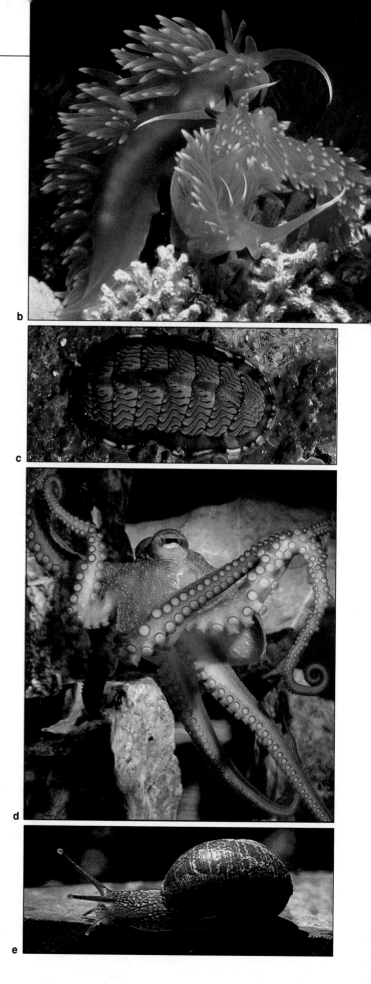

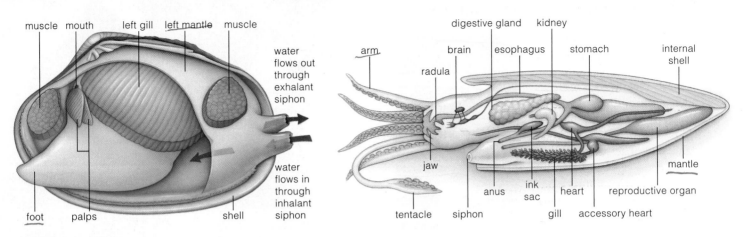

Figure 20.17 Body plan of a clam, with half of its shell (the left valve) removed for this diagram.

Figure 20.18 General body plan of a cuttlefish, one of the cephalopods. The tentacles, more slender than the arms, are specialized for prey capture.

Bivalves

Clams, scallops, oysters, and mussels are well-known bivalves (animals with a "two-valved shell"). Humans have been eating one type or another since prehistoric times. A few bivalves produce pearls; many have an inner lining of iridescent mother-of-pearl. Some species are only a millimeter across. A few giant clams of the South Pacific are more than a meter across and weigh 225 kilograms. The bivalve head is not much to speak of, but the foot is usually large and good for burrowing. In nearly all bivalves, gills function in collecting food and in respiration. As water moves through the mantle cavity, mucus on the gills traps bits of food (Figure 20.17). Cilia move the mucus to palps, which sort out the food before acceptable bits are driven to the mouth.

Bivalves hunkered in mud or sand have a pair of siphons (extensions of mantle edges, fused into tubes). Water is drawn into the mantle cavity through one siphon and leaves through the other, carrying wastes from the anus and kidneys. Siphons of the giant geoduck (pronounced "gooey duck") of the Pacific Northwest may be more than a meter long.

Cephalopods

Squids, cuttlefish, and octopuses are among the active marine predators called cephalopods. Long ago, the foot of ancestral forms became divided into arms and prey-capturing tentacles, which became strategically arranged around the mouth. Hence the name cephalopod ("head-foot"). Among the descendants of those ancestral forms are the largest of all known invertebrates, such as giant squid that may measure eighteen meters from tip to tip.

Take a look at Figure 20.18, which is a body plan of a typical cephalopod. These predators have a radula that draws prey into the mouth, and a beaklike pair of jaws that bite or crush it. They move through their habitat by jet propulsion. By this mode of locomotion, they force water out of the mantle cavity, through a funnel-shaped siphon. First, water moves into the cavity as muscles in the mantle relax. Other muscles contract and squeeze down on the water. Meanwhile, the mantle's free edge closes down on the head and siphon, so water shoots out in a jet through the siphon. By adjusting the siphon, a cephalopod partly controls which way it moves.

Being so active, cephalopods demand a good supply of oxygen. These are the only mollusks with a *closed* circulatory system; their blood stays inside the walls of hearts and blood vessels. Blood pumped from a main heart circulates rapidly to all body regions, then to two gills. Each gill has a booster heart that hastens blood flow and thereby enhances the rates of gas exchange. A cephalopod rhythmically circulates water through the mantle cavity even when it is resting. This behavior moves gases to and from the body surface; it also helps eliminate wastes from excretory organs and the anus.

Compared with some other mollusks, cephalopods are not hermaphrodites; they have separate sexes. They have the largest brain relative to body size, along with well-developed nervous and sensory systems. Giant nerve fibers connect the brain with the muscles used in jet propulsion, making swift responses to food or danger possible. A cephalopod's eyes resemble yours, although they form in a different way in the embryo. Finally, in terms of memory and learning, the cephalopods are the most complex invertebrates. Give an octopus a mild electric shock after showing it an object with a distinct shape, for example, and it thereafter avoids the object.

Mollusks are fleshy, soft-bodied animals with bilateral symmetry. With more than 100,000 species, mollusks vary enormously in body details, size, and life-styles.

20.9 ANNELIDS—SEGMENTS GALORE

Those earthworms you see at the soil surface at night or after a rain are examples of bilateral, segmented worms called **annelids**. (Annelida means ringed forms, but the "rings" are a series of body segments.) The phylum has 15,000 or so species of earthworms, diverse polychaetes, and leeches (Figure 20.19). Except for the leeches, nearly every segment has pairs or clusterings of chitin-stiffened bristles. These provide traction for crawling or burrowing; in swimming species, they are often broadened into paddles. Earthworms have a few bristles per segment. Marine polychaetes typically have a lot (*poly-*, many).

Advantages of Segmentation

When the body is segmented, different parts can evolve separately and become specialized for different tasks. Most segments in an annelid are much the same. But some species made evolutionary forays into specialization. Leeches have suckers at both ends. Polychaetes have an elaborate head and fleshy-lobed appendages called parapods (a word taken to mean "closely resembling feet"). The annelid lineage diverged early from their protostome ancestors. Even so, as you will see next, annelids remind us of developments that led to increases in size and to more complex internal organs.

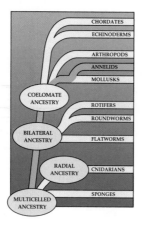

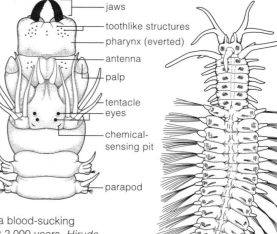

Figure 20.19 Representative annelids. (**a**) Most leeches have sharp jaws and a blood-sucking device. This one is shown before and after gorging on human blood. For at least 2,000 years, *Hirudo medicinalis*, a freshwater leech, has been used as a blood-letting tool to "cure" problems ranging from nosebleeds to obesity. Nowadays, leeches are still used, but more selectively. Thus, after surgeons reattach a severed ear, lip, or fingertip, leeches may be used to draw off pooled blood. A patient's body cannot do this on its own until the severed blood circulation routes are reestablished.

(**b**) A less scary annelid—an earthworm. (**c,d**) From the polychaetes, three examples of the dizzying variety of modifications that have evolved, starting with a segmented, coelomic body plan. The photograph shows a tube-dweller. The featherlike structures at its head end are coated with mucus. After the mucus has trapped bacteria and other bits of food, the coordinated beating of cilia sweeps them to the mouth. Most polychaetes live in marine habitats. They actually are one of the most common types of animals along coasts. Many are predators or scavengers; others dine on algae.

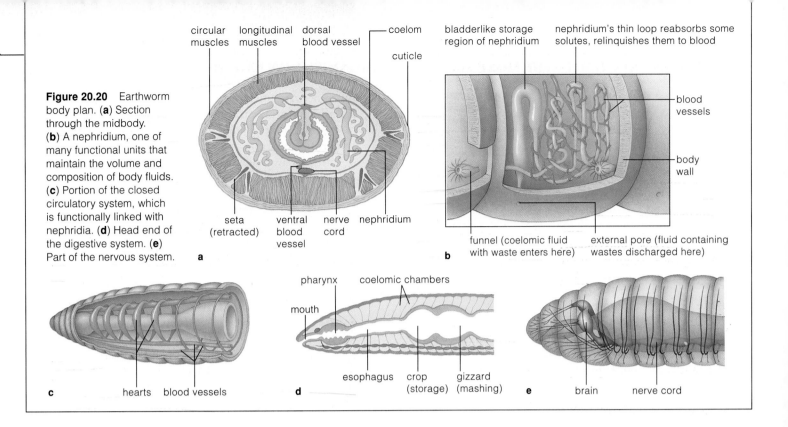

Figure 20.20 Earthworm body plan. (**a**) Section through the midbody. (**b**) A nephridium, one of many functional units that maintain the volume and composition of body fluids. (**c**) Portion of the closed circulatory system, which is functionally linked with nephridia. (**d**) Head end of the digestive system. (**e**) Part of the nervous system.

a — circular muscles; longitudinal muscles; dorsal blood vessel; coelom; cuticle; seta (retracted); ventral blood vessel; nerve cord; nephridium

b — bladderlike storage region of nephridium; nephridium's thin loop reabsorbs some solutes, relinquishes them to blood; blood vessels; body wall; funnel (coelomic fluid with waste enters here); external pore (fluid containing wastes discharged here)

c — hearts; blood vessels

d — pharynx; mouth; coelomic chambers; esophagus; crop (storage); gizzard (mashing)

e — brain; nerve cord

Annelid Adaptations—A Case Study

Earthworms, the usual textbook examples of annelids, are oligochaetes (which have land-dwelling and aquatic members). These scavengers burrow in moist soil and mud, feeding on decomposing plant material and other organic matter. A worm ingests its own weight in soil every twenty-four hours. Its burrowing and feeding activities aerate the soil and lift nutrients to the surface, to the benefit of many plants and other organisms.

Like most annelids, an earthworm's body segments have a cuticle, a layer of secreted material that wraps around the body's outer surface (Figure 20.20). Its thin, flexible cuticle bends easily and permits respiratory gases to move to and from the body surface. Partitions inside the body separate segments from one another, so there is a series of coelomic chambers. The gut extends through all the chambers, from mouth to anus.

As for other annelids, the fluid-cushioned coelomic chambers serve as a *hydrostatic* skeleton against which muscles operate. The wall of each segment has circular muscles (Figure 20.20a). When longitudinal muscles that span several segments contract, the circular ones relax—so the segments shorten and fatten. When the pattern reverses, the segments lengthen.

Muscle contractions alone do not move the earthworm forward. Locomotion also requires protraction and retraction of the bristles on different segments. The body moves forward when its first few segments elongate. The segments behind them plunge their bristles into the ground and hold their position. Then the first segments expand and plunge *their* bristles into the

ground. As they do, the segments behind them retract their bristles and are pulled forward. The alternating contractions and elongations proceed along the length of the body and move the worm forward.

Figure 20.20*b* shows part of a system of **nephridia** (singular, nephridium), which regulates the volume and composition of body fluids. In many annelids, the units of this system have cells similar to flame cells, which may imply an evolutionary link between flatworms and annelids. More often, the beginning of each nephridium is a funnel that collects excess fluid from one coelomic chamber. The funnel connects with a tubular part of the nephridium, which delivers fluid to a surface pore in the body wall of the next coelomic chamber in line.

A rudimentary brain at the head end can integrate sensory input for the whole worm. Leading away from the brain is a double nerve cord that extends the length of the body. A **nerve cord** is a bundle of slender extensions of nerve cell bodies. In each segment, the cord broadens into a **ganglion** (plural, ganglia), a cluster of nerve cell bodies that controls local activity.

Finally, as is true of most annelids, earthworms have a closed circulatory system, with blood confined in hearts and muscularized blood vessels. Contractions keep blood circulating in one direction. Smaller blood vessels service the gut, nerve cord, and body wall.

Annelids (earthworms, polychaetes, and leeches) are bilateral, segmented worms that have complex organ systems. Some species show the amazing degree of specialization possible with a segmented body plan.

Arthropod Diversity

Biologically, "success" means having the most species, the greatest number of offspring, occupancy of the most habitats, terrific defenses against predators and competitors, and a capacity to exploit the greatest amounts and kinds of food. These are the features that characterize **arthropods** (Arthropoda). We know of more than a million species (mostly insects), and new ones are being discovered weekly. Ancient annelids or animals like them probably gave rise to arthropod lineages. The trilobites, one of the four major lineages, are extinct (Section 17.6). The other three lineages are the chelicerates (spiders and their kin), crustaceans (such as crabs and barnacles), and uniramians (centipedes, millipedes, and insects).

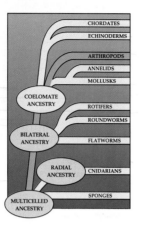

CHORDATES
ECHINODERMS
ARTHROPODS
ANNELIDS
MOLLUSKS
COELOMATE ANCESTRY
ROTIFERS
ROUNDWORMS
BILATERAL ANCESTRY
FLATWORMS
RADIAL ANCESTRY
CNIDARIANS
SPONGES
MULTICELLED ANCESTRY

Adaptations of Insects and Other Arthropods

Six adaptations contributed to the success of arthropods in general and insects in particular:

1. A hardened exoskeleton
2. Specialized segments
3. Jointed appendages
4. Specialized respiratory structures
5. Efficient nervous system and sensory organs
6. Division of labor in the life cycle

Hardened Exoskeletons Arthropods have a cuticle that serves as an external skeleton, or **exoskeleton**. Although hardened, its protein and chitin components make it light, flexible—and protective. For example, crabs and lobsters are armor-plated with a calcium-stiffened exoskeleton. Perhaps exoskeletons evolved as defenses against predators. When certain arthropods invaded the land, exoskeletons took on more functions. They restrict evaporative water loss, and they can support a body deprived of water's

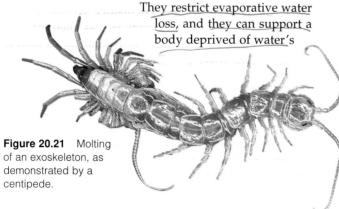

Figure 20.21 Molting of an exoskeleton, as demonstrated by a centipede.

buoyancy. A cuticle does restrict increases in size. But arthropods grow in spurts, by **molting**. At different stages in the life cycle, they grow a new exoskeleton under the old one, which they shed (Figure 20.21). Before the new one hardens, aquatic arthropods enlarge by swelling with water; land-dwellers swell with air.

Specialized Segments As arthropods evolved, body segments became fewer in number, grouped or fused in various ways—and more specialized in function. For example, in insects, a head, a thorax, and an abdomen evolved from different fused segments. In spiders, segments became combined into a forebody and hindbody.

Jointed Appendages Arthropods have specialized, jointed appendages. (Arthropod means "jointed foot.") For example, various paired appendages are adapted for the tasks of feeding, detecting stimuli, locomotion, transferring sperm to females, or spinning silk.

Respiratory Structures Many aquatic arthropods rely on gills for gas exchange. Air-conducting tubes evolved among insects and other land-dwellers. Insect **tracheas** begin as pores on the body surface and branch into tubes that deliver oxygen directly to tissues. They support energy-consuming activities, such as flight.

Specialized Sensory Structures Intricate eyes and other sensory organs contributed to arthropod success. Numerous species have a wide angle of vision and can process visual information from many directions.

Division of Labor Many insects have a division of labor among different stages of their life cycle. They first develop as sexually immature larvae that molt and change as they grow. Then larvae enter a pupal stage, which involves massive reorganization and remodeling of body tissues. Growth and major transformation of a larva into the adult form is called **metamorphosis**.

Moths, butterflies, beetles, and flies are examples of metamorphosing insects (Figure 1.4). Their larval stages specialize in *feeding and growth*, whereas the adult is concerned mostly with *dispersal and reproduction*. This division of labor is adaptive to seasonal changes in the environment, such as variations in food sources.

As a group, arthropods are exceptionally abundant, they engage in enormously different life-styles, and they have been evolutionary successes in nearly all habitats.

Hardened exoskeletons, specialized segments, jointed appendages, specialized respiratory, nervous, and sensory organs, and often a division of labor in the life cycle have contributed to this group's success.

20.11 A LOOK AT SPIDERS AND THEIR KIN

Chelicerates originated in shallow seas in the early Paleozoic. Except for a few mites, the only marine survivors are the horseshoe crabs (Figure 20.22) and sea spiders. The familiar chelicerates—scorpions, spiders, ticks, and chigger mites—are classified as arachnids, and most live on land. Of the arachnids, we might say this: Never have so many been loved by so few.

a

b

c
forebody

hindbody

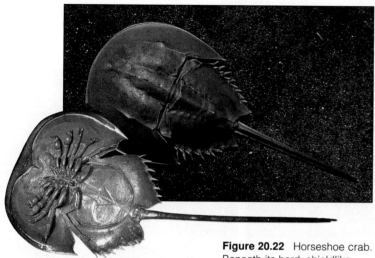

Figure 20.22 Horseshoe crab. Beneath its hard, shieldlike cover are *five* pairs of legs, one of its defining features.

Figure 20.23 *Right:* (**a**) Wolf spider. Like most spiders, it helps keep insect populations in check and is harmless to humans. (**b**) Female black widow. Its bite can be painful and sometimes dangerous. (**c**) Brown recluse, with a violin-shaped mark on its forebody. Its bite can be severe to fatal for humans. (**d**) Internal organization of the spider body.

Scorpions and spiders are predators; they sting or bite and may dispense venom to subdue prey. When they sting or bite us, the reaction can be painful but is rarely serious. But pain makes some of us forget about the splendid tendency of spiders especially to kill great numbers of pestiferous insects. Bites of ticks (bloodsucking parasites of deer, mice, and other vertebrates) cause maddening itches and often serious diseases. For example, some ticks transmit bacterial agents of Rocky Mountain spotted fever or Lyme disease to humans (Section 18.4). Most mites are free-living scavengers.

Arachnids have a splendidly appendaged forebody and hindbody. For example, four pairs of legs, a pair of pedipalps that serve mainly sensory functions, and a pair of chelicerae that can inflict wounds and discharge venom extend from a spider's forebody. Its hindbody appendages spin out silk threads for webs and for egg cases. Most webs are netlike. One type of spider spins a vertical thread with a ball of sticky material at the end. It uses a leg to swing the ball at insects passing by!

d

eye   brain   heart — digestive gland
poison gland
malpighian tubule
anus
pedipalp
mouth   book lung   ovary   spinnerets
chelicera   sperm receptacle   silk gland

Inside the body is an *open* circulatory system, with a heart that pumps blood into body tissues, then receives blood through openings in its wall. Some blood travels through moist folds of book lungs. These respiratory organs resemble book pages, and they greatly increase the surface area available for gas exchange with the air. Figure 20.23*d* shows the arrangement of book lungs and other major organs inside the spider body.

Spiders, scorpions, and their relatives have a variety of appendages specialized for predatory or parasitic life-styles.

20.12 A LOOK AT THE CRUSTACEANS

Shrimps, lobsters, crabs, barnacles, pillbugs, and other crustaceans got their name because they have a hard yet flexible "crust" (exoskeleton), but nearly all arthropods have this feature. Some of the 35,000 species live in freshwater or on land. But the vast majority live in the seas. They are so abundant, they have been called "the insects of the seas." The lobsters and crabs are the "giants" of this subphylum; most crustaceans are less than a few centimeters long. All have important roles in food webs, and humans harvest many edible types.

The simplest crustaceans have many pairs of similar appendages for most of their length. Some may resemble their ancient ancestors. In other lineages, however, the unspecialized appendages evolved into diverse structures. Figure 20.24 shows examples. Lobsters and crabs have strong claws, used to collect food, intimidate other animals, and sometimes dig burrows. Barnacles have feathery appendages that comb microscopic bits of food from the water (Figure 20.24d,e). Many crustaceans have sixteen to twenty segments; some have more than

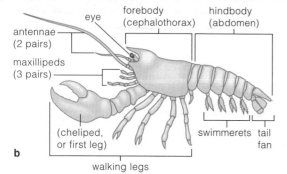

Figure 20.24 A sampling of crustaceans and the diversity in their life-styles. (**a**) Crab, (**b**) lobster, (**c**) copepod, and (**d,e**) stalked barnacles. Lobsters are secretive for most of their lives, whereas crabs skitter actively and openly across sand and rocks. Copepods are free-living filter feeders, predators, or parasites. This female has a pair of long antennae and is carrying an egg sac around with her. Adult barnacles cement themselves to one spot. As the goose barnacles in (**d**) suggest, they have no problem attaching to boat hulls, floating logs or bottles, and other available objects in the water. You might mistake barnacles for mollusks, but as soon as they open their hinged shell to filter feed, you see jointed appendages that are the hallmark of arthropods.

Labels in figure b: antennae (2 pairs); eye; forebody (cephalothorax); hindbody (abdomen); maxillipeds (3 pairs); (cheliped, or first leg); walking legs; swimmerets; tail fan

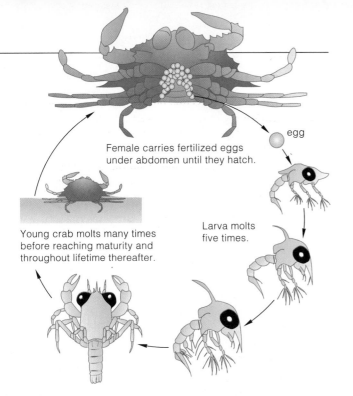

Female carries fertilized eggs under abdomen until they hatch.

egg

Larva molts five times.

Young crab molts many times before reaching maturity and throughout lifetime thereafter.

Figure 20.25 Life cycle of a crab. The larval and juvenile stages molt repeatedly during their growth.

sixty. In crabs, lobsters, and some other crustaceans, the dorsal cuticle extends back from the head as a shield-like cover (carapace) over some or all body segments. The head incorporates two pairs of antennae, a pair of mandibles (jawlike appendages), and a pair of maxillae (food-handling appendages). Crayfish, crabs, lobsters, shrimps, and their kin have five pairs of walking legs.

Figure 20.24*c* shows a copepod. Copepods are less than 2 millimeters long. Yet they are the most numerous animals in aquatic habitats, maybe even in the world. About 1,500 kinds parasitize various invertebrates and fishes. The vast majority (8,000 species) are premier consumers of phytoplankton, the pastures of the seas. Most also eat larval or tiny adult invertebrates, fish eggs, and fish larvae, which they grab with a pair of food-handling appendages. In turn, copepods are food for different invertebrates, fishes, and baleen whales.

Of all arthropods, only barnacles have a calcified "shell." Their modified exoskeleton protects them from predators, drying out, and the force of currents and waves. Adult barnacles cement themselves to wharf pilings, rocks, and similar surfaces (Figure 20.24*d,e*). A few kinds attach themselves only to the skin of whales.

Like other arthropods, crustaceans molt repeatedly to shed the exoskeleton during their life cycle. Figure 20.25 shows a crab's larval stages and increases in size.

Crustaceans differ greatly in the number and kind of their appendages. As for arthropods generally, they repeatedly replace their exoskeletons (by molting) during growth.

VARIABLE NUMBERS OF MANY LEGS

Millipedes and **centipedes** have a long, segmented body with many legs. Of course, millipedes don't have "a thousand," as their name implies. Most have about 100, although one exuberant individual grew 752. And centipedes have between 15 and 177 pairs of legs, not a nicely rounded number of "one hundred."

Millipedes have a rounded body (Figure 20.26*a*). As they develop, pairs of segments fuse, so what looks like a single segment in the adult has *two* pairs of legs. They typically bulldoze through soil and forest litter, mostly as scavengers of decaying vegetation. Centipedes have a flattened body, and all but two segments have a pair of walking legs. All species are fast-moving, aggressive predators, outfitted with fangs and venom glands. They prey on insects, earthworms, and snails. The one shown in Figure 20.26*b* subdues small lizards, toads, and frogs. The house centipede (*Scutigera coleoptrata*) often scurries about in buildings, where it preys on cockroaches, flies, and other pests. Athough helpful, its vaguely terrifying appearance keeps it from being welcomed.

Figure 20.26 (**a**) A millipede and (**b**) a Southeast Asian centipede.

The mild-mannered, scavenging millipedes and aggressive, predatory centipedes do not lend themselves to leg counts as they walk by.

20.14 A LOOK AT INSECT DIVERSITY

As a group, insects share the adaptations listed earlier in Section 20.10. Here we expand the list a bit. Insects have a head, a thorax, and an abdomen. The head has a pair of sensory antennae and paired mouthparts for biting, chewing, sucking, or puncturing (Figure 20.27). The thorax has three pairs of legs and usually two pairs of wings. Abdominal appendages are mostly reproductive structures, such as egg-laying devices. An insect has a foregut, midgut (where most digestion proceeds), and hindgut (where water is reabsorbed). Insects get rid of waste material by **Malpighian tubules**, small tubes that connect with the midgut. Nitrogen-containing wastes from protein breakdown diffuse from blood into the tubules and are converted into harmless crystals of uric acid. The crystals are eliminated with feces. This system allows land-dwelling insects to get rid of potentially toxic wastes without losing precious water.

We've already catalogued more than 800,000 species of insects. If we use sheer numbers and distribution as the yardstick, the most successful insects are small in size and have a staggering reproductive capacity. For example, you may find some of these species growing and reproducing in great numbers on a single plant that might be only an appetizer for another animal. By one estimate, if all the progeny of a single female fly were to survive and reproduce through six more generations, that fly would have more than 5 trillion descendants!

Besides this, the most successful insect species are winged. In fact, they are the *only* winged invertebrates. They can move among food sources that are too widely scattered to be exploited by other kinds of animals. This capacity for flight contributed to their success on land.

Finally, insect life cycles commonly proceed through stages that allow exploitation of different resources at different times. As an insect embryo develops, organs required for feeding and other vital activities form and become functional. Before an insect becomes an **adult** (the sexually mature form of the species), it proceeds through immature, post-embryonic stages. **Nymphs** and **pupae**, as well as larvae, are examples of the stages.

Like human infants, some insect nymphs are shaped like miniature adults; but unlike children, they undergo growth and molting (Figure 20.28*a*). Other insects go through post-embryonic stages of reactivated growth, tissue reorganization, and remodeling of body parts. Metamorphosis is the name for this resumption of growth and transformation into the adult form. Figure 20.28*b,c* shows how the transformation is more drastic in some species than in others.

The factors that contribute to insect success also make them our most aggressive competitors. Insects destroy crops, stored food, wool, paper, and timber. As they stealthily draw blood from us and from our pets, they often transmit pathogenic microorganisms. On the

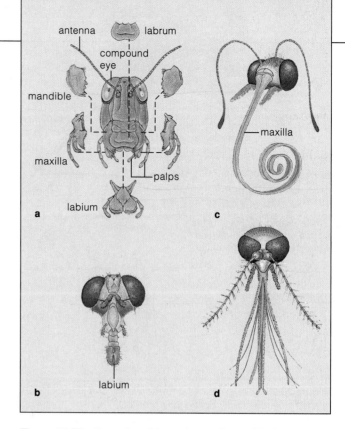

Figure 20.27 Examples of insect appendages. The headparts of (**a**) grasshoppers, a chewing insect; (**b**) flies, which sponge up nutrients with a specialized labium; (**c**) butterflies, which siphon up nectar with a specialized maxilla, and (**d**) mosquitoes, with piercing and sucking appendages.

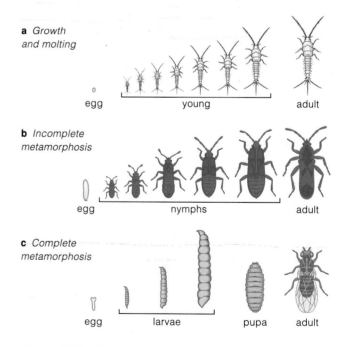

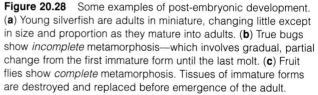

Figure 20.28 Some examples of post-embryonic development. (**a**) Young silverfish are adults in miniature, changing little except in size and proportion as they mature into adults. (**b**) True bugs show *incomplete* metamorphosis—which involves gradual, partial change from the first immature form until the last molt. (**c**) Fruit flies show *complete* metamorphosis. Tissues of immature forms are destroyed and replaced before emergence of the adult.

Figure 20.29 Insects. (**a**) Stinkbugs (order Hemiptera), newly hatched. (**b**) A honeybee (order Hymenoptera) attracting its hive mates with a dance, as described in Section 40.4. (**c**) Ladybird beetles (order Coleoptera) swarming. These beetles are raised commercially and released as biological controls of aphids and other pests. Also in this order, the scarab beetle (**d**). With more than 300,000 species, Coleoptera is the largest order of the animal kingdom. (**e**) Luna moth (order Lepidoptera) of North America. Like most other moths and butterflies, its wings and body are covered with microscopic scales. (**f**) Dragonfly (order Odonata), which swiftly captures and eats insects in midflight.

(**g**) Duck louse (order Mallophaga). It eats bits of feathers and skin. (**h**) European earwig (order Dermaptera), a common household pest. (**i**) Flea (order Siphonaptera), with strong legs for jumping onto and off animal hosts. (**j**) Mediterranean fruit fly (order Diptera). Its larvae destroy citrus fruit and other crops.

bright side, many insects pollinate flowering plants in general and crop plants in particular. And many "good" insects attack or parasitize the ones we would rather do without. We now leave our survey of insects and other arthropods with Figure 20.29, which shows representatives from some of the major orders of insects.

As a group, insects show immense variation on the basic arthropod body plan. Many have wings, and their life cycles have stages that allow exploitation of different and often widely scattered food sources. Many species produce great numbers of small individuals that pass through immature, post-embryonic stages before emergence of the adult.

20.15 THE PUZZLING ECHINODERMS

We turn now to the second lineage of coelomate animals, the deuterostomes. The major invertebrate members of this lineage include **echinoderms**. The feather star, brittle stars, sea cucumber, and sea urchin in Figure 20.30 belong to this phylum. As is true of all echinoderms, their body wall incorporates spines, spicules, or plates stiffened with calcium. (Echinodermata means spiny skinned.) These structures provide protection from enemies. Also, if you accidentally step on a sea urchin, its spines can trigger painful swelling if they break off under your skin.

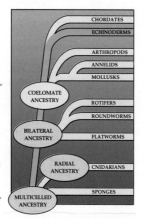

Oddly, adult echinoderms are radial with some bilateral features. Most even produce bilateral larvae during their life cycle. Did bilateral invertebrates give rise to the ancestors of echinoderms? Maybe.

Adult echinoderms have no brain. However, their decentralized nervous system allows them to respond to information about food, predators, and so forth that is coming from different directions. For instance, any arm of a sea star that senses the shell of a tasty scallop can become the leader, directing the rest of the body to move in a direction suitable for prey capture.

Figures 20.30 and 20.31 show examples of **tube feet**. These are fluid-filled, muscular structures with suckerlike adhesive disks. Sea stars use their tube feet for walking, burrowing, clinging to a rock, or gripping a clam or snail about to become a meal. Tube feet are part of a **water-vascular system** unique to echinoderms. In

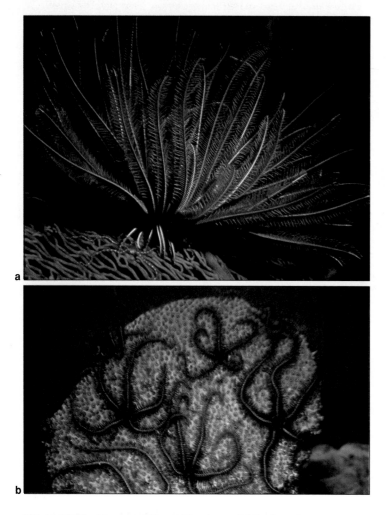

Figure 20.30 Representative echinoderms. (**a**) Feather star, with finely branched food-gathering appendages. (**b**) Brittle stars. Their arms (rays) make rapid, snakelike movements. (**c**) Sea cucumber, with rows of tube feet along its body. (**d**) Sea urchin, which moves about on spines and tube feet.

tube feet   spine

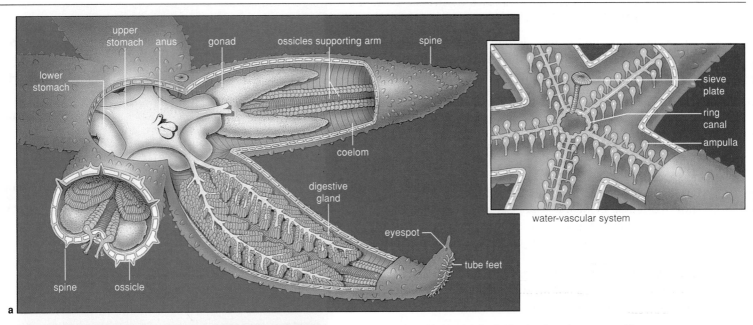

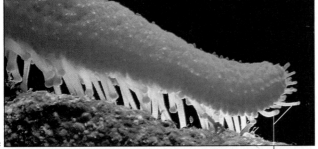

tube feet of ray

mouth, with its feeding apparatus

Figure 20.31 (**a**) Radial body plan of a sea star. The water-vascular system, in combination with tube feet, is the basis of movement. (**b**–**d**) Five-armed sea star, with closer views of some tube feet and of the feeding apparatus on the ventral surface.

sea stars, the system includes a main canal in each arm. Short side canals extend from them and deliver water to the tube feet. Each tube foot has an ampulla, a fluid-filled, muscular structure shaped rather like the rubber bulb on a medicine dropper. As an ampulla contracts, it forces fluid into the foot and causes it to lengthen.

Tube feet change shape constantly as muscle action redistributes fluid through the water-vascular system. Hundreds of tube feet may move at a time. After being released, each one swings forward, reattaches to the substrate, then swings backward and is released before swinging forward again. Their motions are splendidly coordinated, so sea stars glide rather than lurch along.

Some sea stars swallow their prey whole. But some types can push part of their stomach outside the mouth and around their prey, then start digesting the prey even before swallowing it. Sea stars get rid of coarse, undigested residues through the mouth. They do have a small anus, but this is of no help in getting rid of empty clam or snail shells.

With their curious traits, echinoderms are a suitable point of departure for this chapter. Even though we can identify broad trends in animal evolution, we should keep in mind that there are confounding exceptions to the perceived macroevolutionary patterns.

Echinoderms are coelomate animals with spines, spicules, or plates in the body wall. From the evolutionary perspective, they are a puzzling mix of bilateral and radial features.

SUMMARY

1. Animals are multicelled, aerobic heterotrophs that ingest or parasitize other organisms. Nearly all have diploid body cells, arranged as tissues, organs, and organ systems. Animals reproduce sexually and often asexually. They go through embryonic development, and most are motile during at least part of the life cycle.

2. Animals range from structurally simple placozoans and sponges to vertebrates. The species of each phylum share certain characteristics.

   a. By comparing phyla and integrating the information with fossil evidence, biologists have identified the major trends that unfolded as major groups evolved.

   b. The most revealing aspects of animal body plans are type of symmetry, gut, and cavity (if any) between the gut and body wall; whether the body has a distinct head end; and whether the body is divided into a series of segments (Figure 20.32).

3. *Trichoplax*, the only known placozoan, is the simplest animal. It consists of little more than two layers of cells, with a fluid matrix in between. *Trichoplax* shows some differentiation between cells of its two layers.

4. The sponge body is at the cellular level of construction and has no symmetry. A sponge has several kinds of cells, but these are not organized as epithelia and other tissues of more complex animals. In the body wall, collar cells trap food particles from the surrounding water. The beating of their flagella moves water through a system of pores, canals, and chambers.

5. Cnidarians (such as jellyfishes, sea anemones, and hydras) have radial symmetry. They are at the tissue level of construction. The outer and inner tissue layers are separated by mesoglea, a secreted material. Only cnidarians produce nematocysts (stinging capsules).

6. Flatworms, roundworms, rotifers, and nearly all other animals more complex than cnidarians show bilateral symmetry, and they form tissues, organs, and organ systems. The gut may be saclike, as in flatworms, but it usually is complete, with an anus as well as a mouth. Most flatworms have a coelom or false coelom (cavities that separate the gut from the body wall). Peritoneum lines a coelom.

7. Two major lineages diverged shortly after ancestral flatworms evolved. One (protostomes) gave rise to the mollusks, annelids, and arthropods. The other (deuterostomes) gave rise to echinoderms and chordates.

8. All mollusks have a fleshy, soft body, mantle, and shell (or remnant of one) but vary greatly in size, body details, and life-styles.

9. Annelids (earthworms, polychaetes, and leeches) have a segmented body, complex organs, and a series of coelomic chambers.

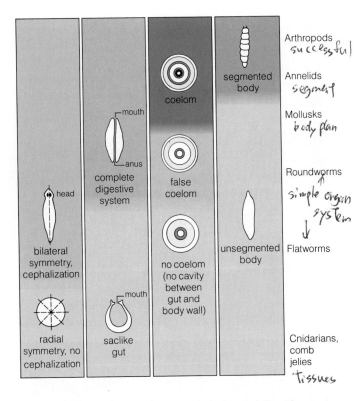

Figure 20.32 Summary of key trends in the evolution of animals, as identified by comparing body plans of major phyla. All of these features did not appear in every group.

10. Arthropods as a group are the most successful organisms in terms of diversity, numbers, distribution, defenses, and capacity to exploit food resources.

   a. Arachnids, crustaceans, insects, and other arthropods have hardened exoskeletons, modified segments, jointed appendages, highly specialized respiratory, nervous, and sensory organs, and (insects only) wings.

   b. New individuals develop by growth and molting or by proceeding through a series of immature stages, such as nymphs and larvae. Many arthropods undergo metamorphosis, with immature forms undergoing major remodeling and tissue reorganization before the adult emerges.

11. Echinoderms have spines, spicules, or plates in the body wall. Evolutionarily, the phylum is puzzling. The larvae of most species form bilateral features, but they go on to develop into basically radial adults.

Review Questions

1. Describe the features of one group of invertebrates. *330–349*

2. What is a coelom? Why was it important in the evolution of certain animal lineages? *329*

3. Name some animals with a saclike gut. Evolutionarily, what advantages does a complete gut afford? *332–333, 329*

4. Choose an insect that lives in your neighborhood and describe some of the adaptations that underlie its success. *342–347*

1. Which is *not* characteristic of the animal kingdom?
   a. multicellularity; cells form tissues, organs
   b. exclusive reliance on sexual reproduction
   c. motility at some stage of the life cycle
   d. embryonic development during the life cycle

2. Jellyfishes, sea anemones, and their relatives have _____ symmetry, and their cells form _____ .
   a. radial; mesoderm   c. radial; tissues
   b. bilateral; tissues   d. bilateral; mesoderm

3. In sheer numbers and distribution, _____ are the most successful animals.
   a. arthropods   d. sea stars
   b. sponges   e. vertebrates
   c. snails and clams

4. Bilateral, segmented bodies and hardened exoskeletons occur among the _____ .
   a. arthropods   d. sea stars
   b. sponges   e. vertebrates
   c. snails and clams

5. Which phylum contains members that are notorious for causing serious diseases in humans?
   a. cnidarians   c. segmented worms
   b. flatworms   d. chordates

6. More complex animals have a _____ between the gut and body wall.
   a. pharynx   c. coelom
   b. peritoneum   d. archenteron

7. Most animals that are more complex than cnidarians have _____ symmetry, and _____ forms in their embryos.
   a. radial; mesoderm   c. bilateral; mesoderm
   b. bilateral; endoderm   d. radial; endoderm

8. Match the terms with the appropriate groups.
   ____ sponges   a. spiny-skinned
   ____ cnidarians   b. vertebrates and kin
   ____ flatworms   c. flukes and tapeworms
   ____ roundworms   d. no tissue organization
   ____ rotifers   e. no males for some
   ____ mollusks   f. nematocysts, radial symmetry
   ____ annelids   g. hookworms, elephantiasis
   ____ arthropods   h. jointed appendages
   ____ echinoderms   i. "belly-foots" and kin
   ____ chordates   j. segmented worms

Critical Thinking

1. Recently a carnivorous sponge was discovered in an underwater cave in the Mediterranean Sea. Unlike other sponges, it has no pores and canals. Rather, branching outgrowths of its body surface have hooklike spicules projecting from them. These act rather like Velcro to trap shrimp and other animals. The outgrowths envelop the prey, which sponge cells then digest. What modifications must have evolved in the basic sponge body plan to allow such a feeding strategy? What selective advantages might such a sponge enjoy?

2. Tapeworms are hermaphroditic. What selective advantages may be associated with this characteristic?

3. People who eat raw oysters or clams that were harvested from sewage-polluted waters develop mild to severe gastrointestinal problems. Consider the feeding modes of these mollusks and develop a hypothesis to explain this outcome.

4. You are diving in the calm, warm waters behind a tropical reef. You see something that looks rather like a bright-red plant, but it has a profusion of tiny tentacles (Figure 20.33). It actually is a soft, branching coral (*Telesto*), with small individual polyps. You will never see it growing on exposed surfaces of the reef that face the open ocean. Propose two possible explanations for this.

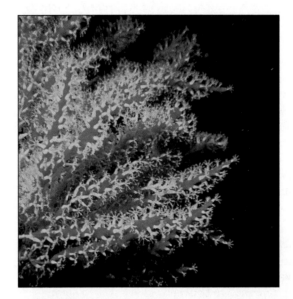

Figure 20.33 *Telesto*, a soft branching coral.

Selected Key Terms

adult *346*
animal *328*
annelid *340*
arthropod *342*
bilateral
  symmetry *328*
centipede *345*
cephalization *329*
cnidarian *332*
coelom *329*
collar cell *330*
contractile cell *333*
cuticle *335*
deuterostome *337*
echinoderm *348*
ectoderm *328*
endoderm *328*
epithelium *332*
exoskeleton *342*
flatworm *334*
ganglion *341*

gonad *333*
gut *329*
hermaphrodite *334*
hydrostatic
  skeleton *333*
invertebrate
  326, 328
larva *331*
Malpighian
  tubule *346*
medusa *332*
mesoderm *328*
metamorphosis
  342, 346
millipede *345*
mollusk *338*
molting *342*
nematocyst *332*
nephridia *341*
nerve cell *333*
nerve cord *341*

nymph *346*
organ *334*
organ system *334*
placozoan *330*
planula *333*
polyp *332*
protostome *337*
pupae *346*
radial cleavage *337*
radial
  symmetry *328*
rotifer *335*
roundworm *335*
sensory cell *333*
spiral cleavage *337*
sponge *330*
trachea *342*
tube feet *348*
vertebrate *328*
water-vascular
  system *348*

Readings

Kozloff, E. 1990. *Invertebrates*. Philadelphia: Saunders.

Pearse, V., et al. 1987. *Living Invertebrates*. Palo Alto, California: Blackwell.

Pechenik, J. 1995. *Biology of Invertebrates*. Third edition. Dubuque: Wm. C. Brown.

Ruppert, E., and R. Barnes. 1994. *Invertebrate Zoology*. Sixth edition. New York: Saunders.

21 ANIMALS: THE VERTEBRATES

Making Do (Rather Well) With What You've Got

In 1798, naturalists skeptically probed a specimen delivered to the British Museum in London, looking for signs that a prankster had slyly stitched the bill of an oversized duck onto the pelt of a small furry mammal. They were examining the remains of a platypus, a web-footed mammal about half the size of a housecat. Like other mammals, the platypus has mammary glands and hair. Like birds and reptiles, it has a cloaca (a single opening at the body surface that functions in reproduction and in excreting wastes). Like birds and most reptiles, it lays shelled eggs. Its young hatch early in development, pink and helpless (Figure 21.1*a*). Its fleshy bill looks ducklike; and its broad, flat, furry tail seems borrowed from a beaver (Figure 21.1*b*).

Anyone who assumes that the platypus doesn't quite measure up to what "an animal" is supposed to be needs a new yardstick. Its combination of traits, although unusual, happens to be exquisitely adapted for survival and reproduction under a specific set of environmental conditions.

The platypus lives in streams and lagoons in remote parts of Australia and Tasmania. It dives at night into the water and preys on small mollusks and other prey, then hides during the day in underground burrows. By holding in heat, its dense, dark fur helps maintain body temperature when the platypus stays in cold water all night long and when it rests all day in a cool burrow. The broad tail serves as a rudder and helps the animal maneuver through water; it also stores energy-rich fats. The webbed feet flare and function as paddles. Strong claws on the hind feet are great for digging. The broad, flattened bill is adapted to scooping up freshwater snails, mussels, insect larvae, worms, and shrimps. The horny pads of its jaws easily grind up the shelled and hard-cuticled meals.

a

b

Figure 21.1 One of evolution's success stories—the platypus, (**a**) raising its incompletely formed offspring in a burrow, and (**b**) underwater, eyes and ears shut, yet homing in on prey.

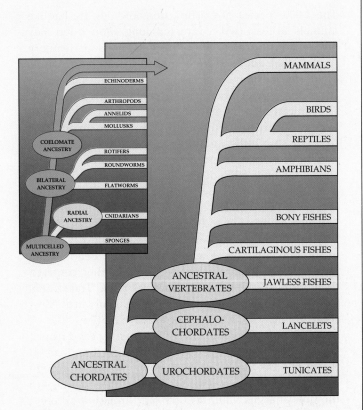

Figure 21.2 Family tree for vertebrates and other chordates.

When a platypus dives, its nostrils close, and a fleshy groove around the eyes and ears snaps shut. It doesn't use its eyes and ears to zero in on prey. Rather, it relies on about 800,000 sensory receptors in the bill. Many of these can detect tiny oscillations in water pressure when prey swim past. Other sensory receptors can detect changes in electric fields as weak as those induced by the flick of a shrimp tail.

The platypus began to evolve 100 million years ago, when the supercontinent Pangea was breaking up (Section 17.7). Its ancestors happened to be on a land mass that drifted south and eventually became what we call Australia. Geographic isolation promoted genetic divergence, and the much-ridiculed platypus body plan evolved. Yet that body plan is scarcely a failure; it has endured far longer than we humans have.

The platypus is only one of 4,500 kinds of mammals that now occupy the vertebrate branch of the animal family tree (Figure 21.2). Its other vertebrate relatives include the fishes, amphibians, reptiles, and birds described in this chapter. As you consider each group, keep a key concept in mind. *The unique traits of each kind of animal are modifications of traits that evolved earlier, in ancestral forms.* As you will see at this chapter's end, the human species speaks eloquently of this concept.

KEY CONCEPTS

1. The chordate branch of the animal family tree includes invertebrate and vertebrate species. All are bilateral animals with a supporting rod for the body (notochord), dorsal nerve cord, pharynx, and gill slits in the pharynx wall. These features typically appear in embryos or larvae; some or all may persist in the adults.

2. Existing invertebrate chordates include tunicates and lancelets. Existing vertebrates include jawless fishes, cartilaginous fishes, bony fishes, amphibians, reptiles, birds, and mammals. Another group, the placoderms, became extinct early in vertebrate history.

3. We can identify five key trends in vertebrate evolution. Structural support and movement came to depend less on the notochord and more on a backbone. Jaws evolved. A nerve cord evolved into a spinal cord and brain. Among gilled, aquatic ancestors of land vertebrates, gas exchange came to depend more on lungs and an efficient circulatory system. Also, fleshy fins with skeletal supports evolved into legs, which were further modified in amphibians, reptiles, birds, and mammals.

4. The primate branch of the mammalian lineage includes prosimians, tarsioids, and anthropoids (monkeys, apes, and humans). Apes and humans are hominoids. Only humans and some extinct, humanlike species are further classified as hominids.

5. Unlike other primates, some early hominids did not have single, narrowly specialized life-styles. They were behaviorally flexible and were able to adapt to changing environments, and to new ones.

6. The remarkable adaptability of hominids in general and humans in particular was an outcome of evolutionary modifications in bones, muscles, teeth, sensory systems, the brain, and behavior.

chordate : Animal having a notochord, dorsal nerve cord, pharynx, and gill slits in the pharynx wall.)

muscular tube by which food enters the gut

rod of stiffened tissue that serve as a supporting structure for the body

cord like communication line consisting of axons and neurons

353

Characteristics of Chordates

The preceding chapter left off with echinoderms, one of the most ancient lineages of the deuterostome branch of the animal family tree. Dominating this branch are their more recently evolved relatives, the **chordates** (phylum Chordata). Only about 2,100 species of these bilateral animals are invertebrates, like echinoderms. The vast majority—about 47,000 species—are vertebrates.

All of the "vertebrate chordates" have a backbone, either of cartilage or bone, and their brain is positioned inside a protective chamber of skull bones. Although the "invertebrate chordates" share certain features with the dominant members of this phylum, a backbone isn't one of them.

Four features are evident in chordate embryos, and in many species these features persist into adulthood. *First,* a **notochord**, a long rod of stiffened tissue (not cartilage or bone), helps support the body. *Second,* the nervous system is based on a tubular, dorsal **nerve cord** that is parallel to the notochord and gut. The anterior end of this cord expands as the embryo develops, and it becomes the brain. *Third,* a **pharynx**, which is a type of muscular tube, functions in feeding, respiration, or both. The chordate pharynx has slits in the wall. *Fourth,* a tail forms in embryos and extends past the anus.

Chordate Classification

Biologists group nearly all of the chordates into three subphyla. These are named the Urochordata (such as tunicates), Cephalochordata (lancelets), and Vertebrata (vertebrates). There are eight classes of vertebrates:

Agnatha	*Jawless fishes*
Placodermi	*Jawed, armored fishes (extinct)*
Chondrichthyes	*Cartilaginous fishes*
Osteichthyes	*Bony fishes*
Amphibia	*Amphibians*
Reptilia	*Reptiles*
Aves	*Birds*
Mammalia	*Mammals*

Appendix I includes an expanded classification scheme for vertebrates. Unit VI provides details of their body plans and functions. Here we become acquainted with major trends in their evolution. We find clues to those trends among the invertebrate chordates.

The embryos of chordates alone have this combination of features: a notochord, a tubular dorsal nerve cord, a pharynx with slits in the wall, and a tail extending past the anus.

Tunicates

The 2,000 species of existing urochordates are baglike animals, at least two centimeters long. More often they are called **tunicates**, after the gelatinous or leathery "tunic" that adults secrete around themselves. Adults of the most common species, the "sea squirts," squirt water through a siphon when something irritates them.

Tunicates live in marine habitats from the intertidal zone to surprising depths. Most adults remain attached to rocks, ship hulls, and other suitably hard substrates. Some live solitary lives; others are colonial. Figure 21.3 shows the body plan of an adult sea squirt. It develops from a bilateral, free-swimming larva that looks rather like a tadpole. (A larva, recall, is an immature stage between embryonic and adult stages of the life cycle.) Its firm, flexible notochord (a series of fluid-filled cells in the tail) functions like a torsion bar. It bends when muscles on one side of the tail or the other contract, then springs back when the muscles relax. The strong, side-to-side motion propels the animal forward. Most fishes use this same kind of propulsive motion.

Like some other animals, tunicates filter food from a current of water that is directed through some part of the body; they are **filter feeders**. Water flows through one siphon and passes through **gill slits**—openings in the thin pharynx wall. It flows out through a different siphon. The pharynx of tunicates also functions as a respiratory organ. Some of the dissolved oxygen in the water flowing through the pharynx diffuses down its concentration gradient, into adjoining blood vessels; and carbon dioxide (from aerobic respiration) diffuses down its gradient, from blood to the outgoing water.

Sea squirts provide an example of **metamorphosis**. This developmental process transforms an immature stage into the adult through major tissue remodeling and reorganization. In a sea squirt larva undergoing this process, the tail and notochord disappear. A tunic is secreted. The pharynx enlarges, and perforations in its wall are subdivided into many slits. The nerve cord regresses, leaving a greatly simplified nervous system.

Lancelets

About twenty-five species of fish-shaped, translucent animals called cephalochordates live just offshore, on seafloors around the world. Most are shorter than your little finger. Most of the time they are buried, almost up to their mouth, in sand and sediments. Their common name, **lancelets**, refers to the sharp tapering of their body at both ends. The lancelet body plan, shown in Figure 21.4*a*, clearly shows the four chordate features.

Notice the segmented pattern of muscles on both sides of the notochord. Like tunicate larvae, lancelets

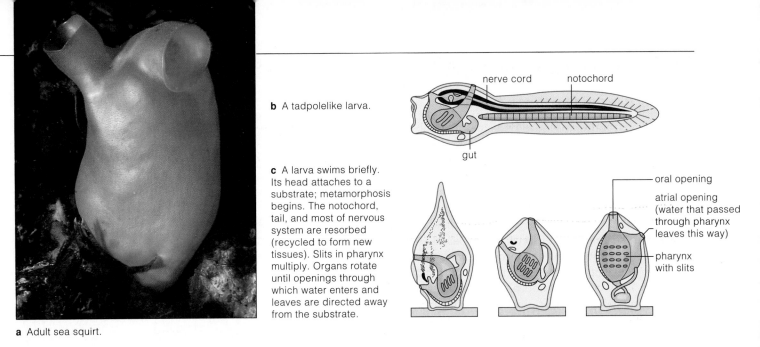

b A tadpolelike larva.

c A larva swims briefly. Its head attaches to a substrate; metamorphosis begins. The notochord, tail, and most of nervous system are resorbed (recycled to form new tissues). Slits in pharynx multiply. Organs rotate until openings through which water enters and leaves are directed away from the substrate.

a Adult sea squirt.

Figure 21.3 (**a**) Adult and (**b**) larval sea squirt, one of the tunicates. (**c**) Metamorphosis into the adult.

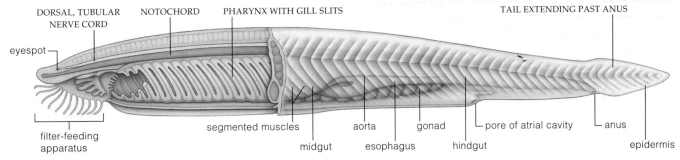

a

b

Figure 21.4 Cutaway view (**a**) and photograph (**b**) of a lancelet.

use the notochord and muscles to produce swimming motions. Unlike tunicates, they have a closed circulatory system (but no red blood cells). A complex brain is nowhere in sight, but the head end of the dorsal nerve cord is expanded and pairs of nerves extend into each muscle segment. The mode of respiration is simple yet effective for such a small body. Dissolved oxygen and carbon dioxide diffuse across its relatively thin skin.

Like tunicates, lancelets are filter-feeding animals. Cilia line their mouth cavity and create a current that draws water through it. The water then moves through the pharynx, where food becomes trapped in mucus. The trapped food is delivered to the rest of the gut. By itself, the collective beating of tiny cilia cannot deliver sufficient food to a filter-feeding animal; delivery also depends on having a large food-trapping surface area. In lancelets, the pharynx provides this. It is quite large, relative to the overall body length (Figure 21.4). Besides this, as many as 200 ciliated, food-trapping gill slits perforate its wall. As you will read in sections to follow, the pharynx turned out to be an organ with interesting evolutionary possibilities for the vertebrates.

Tunicate larvae and lancelets use their notochord and muscles for fishlike swimming. Their pharynx, a muscular tube with gill slits, has finely divided openings across its thin wall.

Like other filter-feeding animals, tunicates and lancelets filter microscopic food from a current of water that they draw through a portion of their body. Tunicate larvae also use the pharynx in respiration.

Puzzling Origins, Portentous Trends

How did vertebrates arise? We find clues in an obscure phylum, the hemichordates (*hemi-* meaning half, as in "halfway to chordates"). Evolutionarily, these marine invertebrates seem to be midway between echinoderms and chordates. They don't have a notochord, but they do have a gill-slitted pharynx and a dorsal, tubular nerve cord. Intriguingly, their larvae resemble echinoderm and tunicate larvae. Suppose, in some ancient echinoderms, regulatory gene mutations enhanced the rates at which sex organs developed. If sex organs became functional early, in a *larval* body, metamorphosis no longer would offer advantages—so the original adult form could be dispensed with. This scenario is not far-fetched. A few existing tunicates look like larvae but have sex organs; some amphibians become sexually mature even though they are larvae in other ways. These sexually precocious larvae can reproduce, generation after generation!

Even if there were tadpole-shaped chordates, how did they evolve into vertebrates? Let's start with the broad trends. One involved a shift from the notochord to a reliance on a column of separate, hard segments called **vertebrae** (singular, vertebra). The column was the start of a strong internal skeleton (endoskeleton) that muscles could work against. *The vertebral column was the foundation for fast-moving predators—some of which were ancestral to all other vertebrate animals.*

In a related trend, part of the nerve cord expanded and evolved into a brain. Expansion began after **jaws** evolved in some early fishes. They arose through modification of the first in a series of structural elements that helped support the gill slits (Figure 21.5). Jaws meant new feeding possibilities—and intensified competition among predators. Fishes able to smell or see food or predators at greater distances were favored. The brain became better at processing the new information. *A trend toward complex sensory organs and nervous systems began in fishes, and it continued among land vertebrates.*

Another trend began when paired fins evolved. **Fins** are appendages that help propel, stabilize, and guide the body in water (compare Figure 21.9). In some lineages, ventral fins became fleshy and equipped with skeletal supports—the forerunners of limbs. *Paired, fleshy fins were the starting point for the evolution of legs, arms, and wings seen among amphibians, reptiles, birds, and mammals.*

Another trend involved modifications to respiratory structures. Think of a lancelet. Except when buried in sand, oxygen and carbon dioxide simply diffuse across its body surface. In most vertebrate lineages, however, **gills** of one sort or another evolved. These respiratory structures have a moist, thin, intricately folded surface, they are richly endowed with blood vessels, and they offer a large surface area for gas exchange. For example,

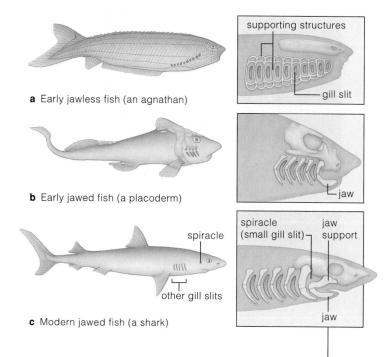

a Early jawless fish (an agnathan)

supporting structures

gill slit

b Early jawed fish (a placoderm)

jaw

spiracle

other gill slits

spiracle (small gill slit)   jaw support

c Modern jawed fish (a shark)

jaw

Figure 21.5 Comparison of gill-supporting structures in jawless fishes and jawed fishes. In the placoderms and other early jawed vertebrates, cartilage supported the rim of the mouth. In modern jawed fishes, the gill slit between the jaws and an adjacent supporting element serves as a spiracle, an opening through which water is drawn. In (**a**), the gill supports are just under the skin. In (**b**) and (**c**), they are internal to the gill surface.

five to seven pairs of a shark's gill slits are continuous with gills that extend from the pharynx to the surface of the body. When a shark opens its mouth and closes its external gill openings, the pharynx expands. Inside the mouth, oxygen diffuses *from* the water into the gills, and carbon dioxide diffuses *into* the water. Muscles make the pharynx constrict—which forces the oxygen-depleted, carbon dioxide–enriched water out through the gill slits.

As some fishes became larger and more active, gills became more efficient. But gills can't work out of water. *They stick together unless water flows through them and keeps them moist.* In fishes that were ancestral to the land vertebrates, pouches developed on the gut wall. The pouches evolved into **lungs**—internally moistened sacs for gas exchange. In a related trend, modifications to

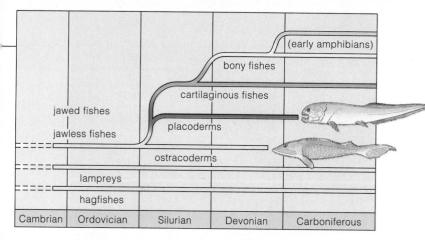

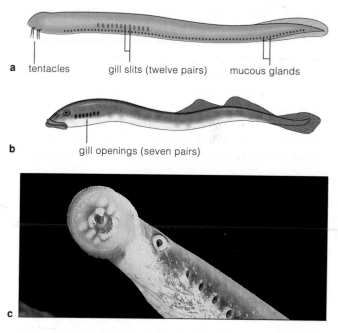

Figure 21.6 Evolutionary history of the fishes.

the heart enhanced the pumping of oxygen and carbon dioxide through the body. *Ancestors of land vertebrates relied less on gills and more on lungs. And more efficient circulatory systems accompanied the evolution of lungs.*

The First Vertebrates

Figure 21.6 gives a time frame for vertebrate evolution. Free-swimming species originated in the Cambrian and gave rise to two kinds of fishes—those without and those with jaws. One or the other kind probably gave rise to all vertebrate lineages that followed. The earliest jawless fishes (Agnatha) included **ostracoderms**. Figure 21.5*a* shows one of these bottom-dwelling filter feeders. Their skeleton probably consisted of a notochord and a protective cover for the newly enlarged brain. Armor-like plates covering the body consisted of bony tissue and dentin (a hardened tissue that vertebrate teeth still have). The armor was useful against the giant pincers of sea scorpions, but it apparently wasn't much good against jaws. Ostracoderms disappeared when jawed fishes began their adaptive radiations.

Placoderms were among the first fishes with jaws and paired fins (Figure 21.5*b*). These bottom-dwelling scavengers or predators had a notochord reinforced with bony elements. The first gill-supporting structures were enlarged, with bony projections something like teeth, and they functioned as jaws. Before placoderms, feeding strategies were limited to filtering, sucking, or rasping bits of food material. When placoderms began to bite and tear up large chunks of prey, they started an evolutionary race of offensive and defensive adaptations that has continued to the present.

Placoderms diversified, then became extinct during the Carboniferous period. New kinds of predators—the cartilaginous and bony fishes—replaced them in the seas. We will consider the living descendants of the new predators, after a look at existing jawless fishes.

The emergence of a vertebral column, jaws, paired fins, and lungs were pivotal in the evolution of certain vertebrates.

21.4 EXISTING JAWLESS FISHES

The **hagfishes** and **lampreys** are living descendants of some early jawless fishes. All seventy-five or so species have a cylindrical body with a cartilaginous skeleton and no paired fins (Figure 21.7). Most are less than 1 meter (3.3 feet) long.

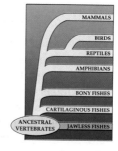

Hagfishes live in groups on the sediments of continental shelves. They prey on polychaete worms or scavenge for weakened or dead organisms. Although lacking jaws, hagfishes have sensory tentacles around the mouth and a tongue that rasps soft tissues from their prey. Hagfishes defend themselves by secreting copious amounts of sticky, smelly, slimy mucus from a series of glands along their body.

Figure 21.7 Body plan of (**a**) a hagfish and (**b**) a lamprey. (**c**) The toothed oral disk of a lamprey, pressed against aquarium glass.

Lampreys are specialized predators that are almost parasites. Their suckerlike oral disk has horny, tooth-like parts that rasp flesh from prey. Some types latch onto salmon, trout, and other commercially valuable fishes, then suck out juices and tissues. Just before the turn of the century, lampreys began invading the Great Lakes of North America. Here and elsewhere, their introduction has led to the collapse of populations of lake trout and other large fishes.

Hagfishes and lampreys get along without jaws; they latch onto prey with their efficient, specialized mouthparts.

21.5 EXISTING JAWED FISHES

Few of us pay much attention to life beneath the surface of seas and other bodies of water, so fishes are not widely recognized as being the world's dominant vertebrates. Yet their numbers exceed those of all other vertebrate groups combined. And they show far more diversity; there are more than 21,000 species of bony fishes alone.

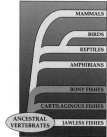

The form and behavior of a fish tell us something about the challenges it faces in water. For example, being about 800 times more dense than air, water resists rapid movements. As an adaptation to this constraint, predatory marine fishes are streamlined for pursuit. Their long, trim body reduces friction, and their tail muscles are organized for propulsive force and forward motion. Or consider how some bottom-dwelling fishes have an amazingly flattened body. We can deduce that this body plan is adaptive in hiding such fishes from predators or concealing them from prey.

A motionless trout, suspended in shallow water, is another example of adaptation to water's density. Like many fishes, it maintains neutral buoyancy with a **swim bladder**—an adjustable flotation device that exchanges gases with blood. When a trout gulps air at the water's surface, it is adjusting the volume of its swim bladder.

Cartilaginous Fishes

Cartilaginous fishes (Chondrichthyes) include about 850 species of skates, sharks, and chimaeras (Figure 21.8). Most are marine predators with a streamlined body. These fishes have pronounced fins, a skeleton of cartilage, and five to seven gill slits on both sides of the pharynx. Most species have a few scales or many rows of them. <u>Scales</u> (small, bony plates at the body surface) often can protect a fish without weighing it down.

Skates and rays are mostly bottom dwellers with flattened teeth, suitable for crushing hard-shelled prey. Both have enlarged fins that extend onto the side of the head. The largest species, the manta ray, measures up to six meters from fin tip to fin tip. A venom gland in the tail of sting rays probably helps deter predators. Other rays have electric organs in the tail or fins that can stun prey with up to 200 volts of electricity.

At fifteen meters from head to tail, some sharks are among the largest living vertebrates. As Figure 21.5c shows, sharks have formidable jaws. They continually shed and replace their sharp, triangular teeth (modified scales), which they use to grab prey and rip off chunks of flesh. Their relatively few attacks on humans have given the whole group a bad reputation. Yet, for many millions of years sharks have been eating invertebrates,

Figure 21.8 Cartilaginous fishes: (**a**) shark, (**b**) blue-spotted reef ray, and (**c**) chimaera (ratfish).

fishes, and marine mammals (including seals) but not surfboards with legs dangling over the side.

The thirty or so species of chimaeras feed mostly on mollusks. With their bulky body and slender tail, they resemble a rat (hence their common name, ratfishes). They have a venom gland in front of the dorsal fin.

Bony Fishes

Bony fishes (Osteichthyes), the most numerous and diverse vertebrates, make up all but 4 percent of the existing species of modern fishes. Their ancestors arose during the Silurian period and soon gave rise to three lineages: the ray-finned fishes, lobe-finned fishes, and lungfishes. Descendants of these early forms radiated into nearly all aquatic habitats.

Body plans vary greatly. Marine predators typically have a torpedo shape, a flexible body, and strong tail

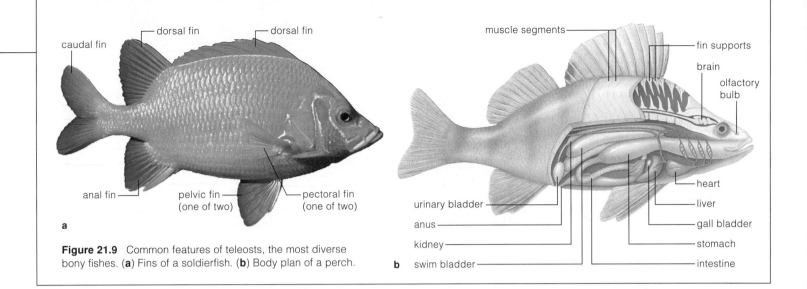

Figure 21.9 Common features of teleosts, the most diverse bony fishes. (**a**) Fins of a soldierfish. (**b**) Body plan of a perch.

Labels for (a): caudal fin, dorsal fin, dorsal fin, anal fin, pelvic fin (one of two), pectoral fin (one of two)

Labels for (b): muscle segments, fin supports, brain, olfactory bulb, heart, liver, gall bladder, stomach, intestine, urinary bladder, anus, kidney, swim bladder

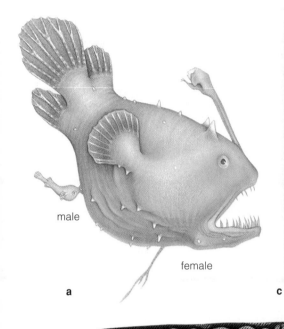

male, female, a, b, c, d, e

Figure 21.10 A few variations on the body plan of bony fishes. (**a**) A female deep-sea angler fish, with a tiny male attached to her. (**b**) Moray eel. (**c**) Sea horse. (**d**) Long-nose gar. (**e**) Coelocanth (*Latimeria*), a "living fossil" that resembles early lobe-finned fishes.

fins that aid in swift pursuit. Many reef dwellers are box shaped, with small fins; they move easily through narrow spaces. The moray eel, with its elongated, flexible body, can wriggle through mud and slip into concealing crevices. Sea horses and many bottom-dwelling species have cryptic forms that help conceal them from prey or predators. Figures 21.9 and 21.10 show examples of these body plans.

Ray-finned fishes have rays that support paired fins. (The rays originate from the dermis, one of the skin's layers.) Most have highly maneuverable fins and light, flexible scales, both of which contribute to a capacity for complex movement. An efficient respiratory system rapidly delivers oxygen to metabolically active tissues.

One group of ray-finned fishes still resembles their ancestors. It includes sturgeons and paddlefishes of the Mississippi River basin. Teleosts, the most abundant group, include salmon, tuna, rockfish, catfish, perch, minnows, moray eels, flying fish, sculpins, blennies, scorpionfish, and pikes. All are thin scaled or scaleless.

By contrast, there is only one existing species of lobe-finned fishes and three genera of lungfishes. As their name suggests, a **lobe-finned fish** is unique in having paired fins that incorporate fleshy extensions from the body. As you will see next, neither it nor the lungfishes have changed much from ancient forms.

Of all existing vertebrates, the bony fishes are the most spectacularly diverse and the most abundant.

21.6 AMPHIBIANS

Origin of Amphibians

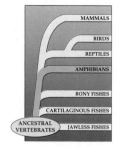

Lobe-finned fishes (Figure 21.10*e*) are living fossils—relics of a time when the first vertebrates moved onto land. Remember what living conditions were like in the Devonian period? Sea levels rose and fell repeatedly, and swamps fringing the coasts flooded and drained— many times. The ancestors of lobe-finned fishes evolved in those trying times. They must have used their lobed fins to pull themselves from dried-up ponds to ones that were still habitable. What else made their pond-to-pond lurchings possible? Those animals had sac-shaped outpouchings from the wall of the esophagus (a tube leading to the stomach)—and the sac began to be used as a way to supplement gas exchange. In other words, the ancestors of lobe-finned fishes had simple lungs.

Existing lungfishes provide more clues to how the ancestral forms might have made it through stressful times. They live in stagnant water but surface to gulp air. In dry seasons, when streams shrink to mud, a lungfish encases itself in a mixture of mud and slime that protects it from drying out until the next rainy season.

Devonian lobe-finned fishes lurched over land only as a way to reach more hospitable ponds. Yet the very act of traveling out of water favored the evolution of stronger fins and more efficient lungs. Among the evolving forms were the ancestors of amphibians. An **amphibian** is a vertebrate that is somewhere between fishes and reptiles in body plan and reproductive mode. Most amphibians have a largely bony endoskeleton and four legs (or four-legged ancestors).

Early amphibians were spending time on land by the close of the Devonian (Figure 21.11*a*). For them, life on land was dangerous—and promising. Temperatures shifted more on land than in water, air didn't support the body as water did, and water wasn't always plentiful. But air has far more oxygen, and lungs continued to evolve in ways that enhanced oxygen uptake. Besides this, circulatory systems became better at distributing oxygen to cells throughout the body. Both modifications increased the energy base for more active life-styles.

New sensory information also challenged the early amphibians. Swamp forests supported vast numbers of edible insects and other invertebrate prey. Animals with good vision, hearing, and balance—the senses that are most advantageous on land—were favored. And brain regions concerned with those senses expanded.

The salamanders, frogs and toads, and caecilians alive today are descended from those first amphibians. None has escaped water entirely. Even when gills or lungs are present, amphibians can use their thin skin as a respiratory surface. But respiratory surfaces must be kept moist—and skin dries easily. Some species live their entire lives in water; others lay their eggs in water

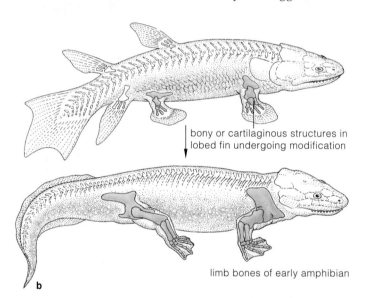

bony or cartilaginous structures in lobed fin undergoing modification

limb bones of early amphibian

b

Figure 21.11 (**a**) *Ichthyostega*, one of the first of the Devonian amphibians. Fossils of this species have been recovered in Greenland. The skull, deep tail, and fins were decidedly fishlike. Unlike fish, this species had four limbs adapted for moving on land, and a short neck intervened between its head and the rest of the body. Its vertebral column and rib cage were adapted to support the body's weight out of water. (**b**) Proposed evolution of skeletal elements inside the lobed fins of certain fishes into limb bones of early amphibians.

a

or produce aquatic larvae. Even the species that have adapted completely to dry habitats on land must lay their eggs in moist places.

Figure 21.12 Amphibians. (**a**) Terrestrial stage in the life cycle of a red-spotted salamander. (**b**) A frog, splendidly jumping. (**c**) American toad. (**d**) A caecilian.

Salamanders

Like fishes and the first amphibians, salamanders bend from side to side when they walk:

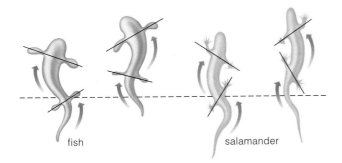

fish   salamander

Probably the first four-legged vertebrates also walked this way. Adults of some species retain several larval features. And some larvae are sexually precocious; they can breed. For example, the Mexican axolotl retains the external gills of the larval form, and the development of its teeth and bones is arrested at an early stage.

Frogs and Toads

With more than 3,000 species, frogs and toads are the most successful amphibians. Their long hindlimbs and powerful muscles allow them to catapult into the air or move forcefully through water. Most often, their sticky-tipped, prey-capturing tongue flips out from the front of the mouth. An adult eats just about any animal it can catch; only its head size dictates the upper limit of prey size. One frog has such a large head, it is called a walking mouth. Skin glands of some species produce toxins, and poisonous types often have bright coloration that advertises this. The South African clawed frog (*Xenopus laevis*) has antibiotics in its skin that protect it against the stew in its microbe-rich, swampy habitat.

Caecilians

The ancestors of caecilians lost their limbs and most of their scales, and they gave rise to decidedly worm-shaped amphibians (Figure 21.12*d*). Nearly all of the 150 or so species burrow through soft, moist soil in pursuit of insects and earthworms. A few live in shallow freshwater habitats.

Amphibians are halfway between fishes and reptiles in their body form and behavior. Regardless of how far they venture onto land, they have not fully escaped dependency on water.

21.7 REPLTILES

The Rise of Reptiles

Reptiles (Reptilia) evolved from certain amphibians during the late Carboniferous. Compared to amphibians, early reptiles chased prey with far greater cunning and speed. The muscles and bones of their jaws could apply more sustained, crushing force. Their teeth could latch onto other vertebrates as well as insects. In general, their limbs were better at supporting the body on land. Many species had more efficient circulatory, respiratory, and nervous systems.

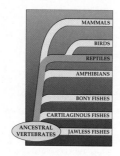

Reptiles were the first vertebrates to escape dependency on standing water. They did so through four adaptations that distinguish them from fishes and amphibians. *First*, they have tough, scaly skin that limits moisture loss. *Second*, they have a copulatory organ that permits internal fertilization. (Sperm are deposited into a female's body; they do not have to swim through water to reach eggs.) *Third*, their kidneys are good at conserving water. *Fourth*, most produce **amniote eggs**, in which an embryo develops to an advanced stage before being hatched or born into dry habitats. The eggs have membranes that retain water and protect or give metabolic support to the embryo. Most of these eggs also have a leathery or calcified shell (Figure 21.13*a*).

Reptiles underwent adaptive radiations that led to fabulously diverse forms. The group called dinosaurs evolved in the Triassic and for the next 125 million years were the dominant land vertebrates. When the Cretaceous ended abruptly, so did they (Section 17.8). The reptiles we call crocodilians, turtles, tuataras, and lizards and snakes made it to the present.

Crocodilians

Crocodiles and alligators have a slender snout, powerful jaws, and sharp teeth (Figure 21.13*b*). All live in or near water. The feared "man-eaters" of southern Asia and the Nile crocodiles weigh up to 1,000 kilograms. They drag a mammal or bird into the water, tear it apart by violently spinning about, then gulp the torn chunks. Like other reptiles, crocodilians adjust their body temperature by behavioral and physiological mechanisms. They also show complex social behavior, as when the parents guard nests and assist hatchlings into the water.

Turtles

The 250 existing species of turtles live inside a shell that is attached to the skeleton (Figure 21.13*d*). When they are threatened, most turtles pull the head and limbs

Figure 21.13 Reptiles. (**a**) Eastern hognose snakes, emerging from leathery-shelled amniote eggs. This type of egg contributed to the colonization of land by reptiles, then birds and mammals. (**b**) Spectacled caiman. The peglike upper teeth of crocodiles don't match up with the peglike lower ones. (**c**) An evolutionary history of the reptiles. (**d**) A heavy-shelled Galápagos tortoise. (**e**) Tuatara. (**f**) A frilled lizard, flaring a ruff of neck skin in a defensive display. (**g**) Rattlesnake of the American Southwest.

inside it. Turtles don't have teeth; they have tough, horny plates suitable for gripping and chewing food. They have strong jaws and often a fierce disposition that helps keep predators at bay. All turtles lay eggs on land, then leave them. Predators eat most of the eggs, so few new turtles hatch. Sea turtles have been hunted to the brink of extinction; they are endangered species.

Tuataras

The two existing species of tuataras (*Sphenodon*) live on small, windswept islands near New Zealand. Their body plan (Figure 21.13*e*) hasn't changed much since the Mesozoic. They resemble lizards, but their lineage is more ancient. Tuataras don't engage in sex until they are twenty years old. This works well because tuataras, like turtles, may live for sixty years or more.

Lizards and Snakes

About 95 percent of the existing reptiles are lizards and snakes—distant relatives of dinosaurs. Most are small, but the Komodo monitor lizard is big enough to hunt

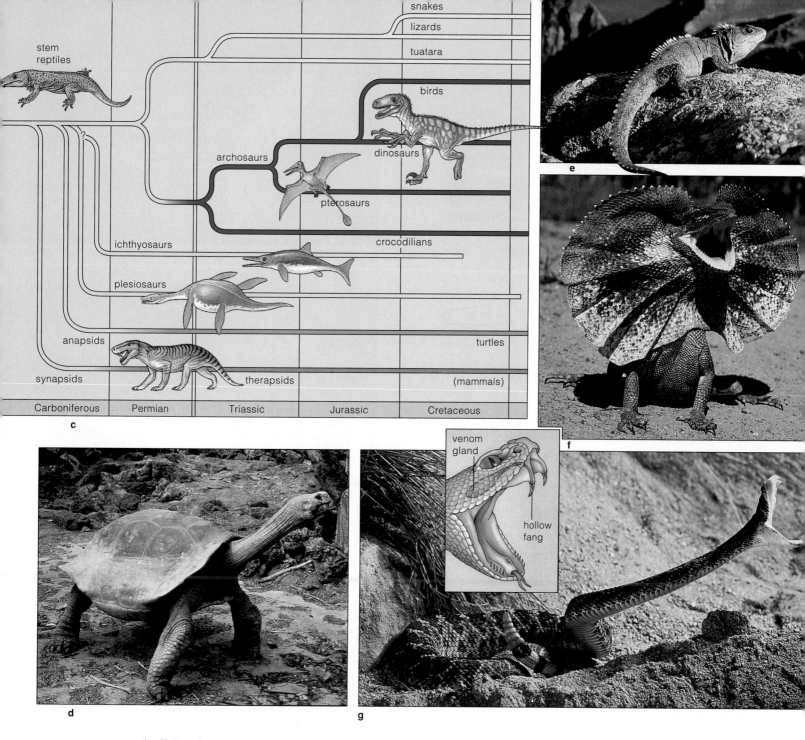

young water buffalo. The longest modern snake would stretch across 10 yards of a football field.

Most of the 3,750 kinds of lizards, such as iguanas, chameleons, and geckos, are insect eaters of deserts and tropical forests. Some species try to startle predators and intimidate rivals by flaring their throat fan (Figure 21.13*f*). Many give up their tail when a predator grabs them. The detached tail wriggles for a bit and may be distracting enough to permit a getaway.

Short-legged, long-bodied lizards of the Cretaceous gave rise to the elongated, limbless snakes. Most of the 2,300 existing species move in S-shaped waves, much like salamanders. The "sidewinders" make J-shaped movements across loose sand and sediments. Some snakes have bony remnants of the ancestral hindlimbs.

All snakes have highly movable jaws; some swallow animals wider than they are. Some species subdue prey with venom (Figure 21.13*g*). Pythons or boas coil tightly around prey and suffocate it. Snakes usually are not aggressive toward humans, but each year, as many as 40,000 people around the world die of snakebites.

With their scaly skin, reliance on internal fertilization, water-conserving kidneys, and amniote eggs, the reptiles were the first vertebrates to escape dependency on standing water.

21.8 BIRDS

By definition, **birds** are the only animals that produce feathers (Figure 21.14). **Feathers** are lightweight structures, derived from skin, that are used in flight, body insulation, or both. Judging from fossils, birds are descended from tiny reptiles that ran around on two legs 160 million years ago. Figure 14.7 shows a fossil of one of the earliest forms, *Archaeopteryx*. It conserved many reptilian traits but had feathers and other avian traits.

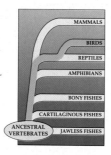

Birds still resemble reptiles in many respects. For example, they have horny beaks, scaly legs, and many of the same internal structures. They, too, lay eggs and commonly exhibit parental behavior. One of Darwin's champions, Thomas Huxley, argued that birds surely are glorified reptiles. Today, many biologists do indeed classify birds as a branch of the reptilian lineage.

Figure 21.14 Characteristics of birds. (**a**) Flight. Of all living vertebrates, only birds and bats fly by flapping their wings.

(**b**) Feathers—the defining characteristic of birds. This male pheasant has flamboyant plumage, an outcome of sexual selection. It is a native of the Himalaya Mountains of India, and is an endangered species. As is the case for many other kinds of birds, its jewel-colored feathers end up adorning humans—in this case, on the caps of native tribespeople.

(**c**) Many birds, including these Canada geese, show migratory behavior. *Animal migration* is a recurring pattern of movement between two or more locations in response to environmental rhythms. For example, seasonal change in daylength is a cue that acts on internal timing mechanisms (biological clocks), which in turn trigger physiological and behavioral changes. Such changes induce migratory birds to make round trips between distant regions that differ in climate. Canada geese spend the summer nesting in marshes and lakes in the northern United States and Canada. Their wintering grounds are in New Mexico and other parts of the southern United States.

(**d**) Speckled eggs of a magpie. All birds lay hard-shelled eggs of the sort shown in the generalized diagram.

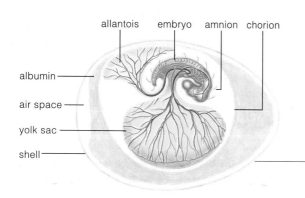

allantois · embryo · amnion · chorion

albumin

air space

yolk sac

shell

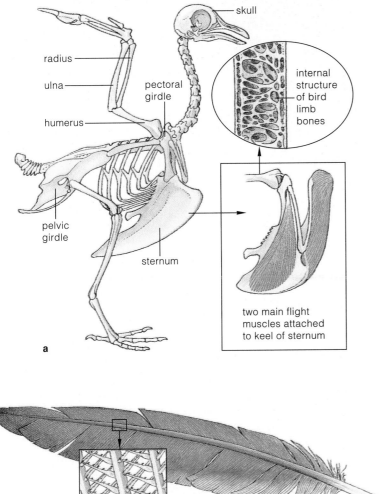

We have named almost 9,000 species of birds. They show stunning variation in their body size, proportions, coloration, and capacity for flight. One of the smallest hummingbirds barely tips the scales at 2.25 grams (0.08 ounce). The largest existing bird, the ostrich, weighs about 150 kilograms (330 pounds). Ostriches cannot fly, but they are impressively long-legged sprinters (Figure 14.2). Many birds, such as warblers and other perching types, differ markedly in feather coloration and in their territorial songs. Bird songs and other complex social behaviors are the topics of later chapters.

Flight demands high metabolic rates, which require a good deal of oxygen. Like you, all birds have a large, durable, four-chambered heart that pumps oxygen-enriched blood to the lungs. Besides this, birds have a unique respiratory system that greatly enhances oxygen uptake. This system is described in Section 29.3.

Flight also demands an airstream, low weight, and a powerful downstroke that will provide lift (a force at right angles to an airstream). The bird wing, a modified forelimb, consists of feathers, strong muscles, and light-weight bones. Bird bones are strong even though they weigh very little (because of profuse air cavities in the bone tissue). For example, the skeleton of a frigate bird, which has a 7-foot wingspan, weighs only 4 ounces. That's less than the feathers weigh!

A bird's flight muscles attach to an enlarged breast-bone and to upper limb bones adjacent to it (Figure 21.15). When they contract, they produce the powerful downstroke for flight. With its long flight feathers, the wing serves as an airfoil. Usually, a bird spreads out these feathers on a downstroke and increases the size of the surface pushing against air (Figure 21.14*a*). On the upstroke, it folds the feathers somewhat, so that each wing presents the least possible profile against the air.

Figure 21.15 (**a**) Body plan of birds, showing the large, keeled breastbone (sternum) to which flight muscles attach. (**b**) The bird wing is a system of lightweight bones and feathers. Feathers gain strength from a hollow central shaft and from tiny barbules interlocked in a latticelike array.

Of all animals, birds alone have feathers, which they use in flight, in heat conservation, and in socially significant visual displays.

21.9 MAMMALS

Mammals arose from therapsids, mammal-like reptiles that existed with dinosaurs in Carboniferous times (Figure 21.13c). When the last of the dinosaurs vanished, a great mammalian radiation began that continued into the modern era (Figure 15.9). We know of more than 4,500 existing species. They range in size from Kitti's hognosed

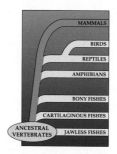

bat (1.5 grams) to whales that exceed 100 tons. Of all vertebrates, they have the most well-developed cerebral cortex—the most complex brain region for processing and integrating sensory information about the world. And of all vertebrates, only female mammals feed the young with milk, a nutritious fluid formed in mammary glands (Figure 21.16a). Females and males care for the young for an extended period and serve as models for behavior. Young mammals have an inborn capacity to learn and to repeat behaviors with survival value. Most mammals also show **behavioral flexibility**. That is, they can expand on the basics with novel forms of behavior.

Mammals have hair or thick skin that conserves heat (most whales lost the hair trait). Unlike reptiles, which generally swallow prey whole, most mammals secure, cut, and maybe chew food before swallowing. They differ from reptiles in **dentition** (the type, number, and size of teeth). Mammals have four types of upper and lower teeth that match up and work together to crush, grind, or cut food (Figure 21.16b). Their incisors (flat chisels or cones) nip or cut food. Horses and other grazing mammals have large incisors. Canines, with piercing points, are longest in meat-eating mammals. Premolars and molars (cheek teeth) are a platform with surface bumps (cusps); these crush, grind, and shear food. If a mammal has large, flat-surfaced cheek teeth, you can safely assume its ancestors evolved in places where fibrous plants were abundant foods. As you will see, fossilized jaws and teeth from human ancestors provide revealing clues to their life-styles.

We turn now to the three major groups: egg-laying, pouched, and placental mammals.

Egg-Laying Mammals

Besides the platypus, the only other type of egg-laying mammal is the spiny anteater—a burrowing animal of Australia and New Guinea that is covered with spines derived from hair (Figure 21.17a). Both types lost most or all of their teeth as they evolved. The loss correlates with specialized diets. A platypus mostly eats tiny invertebrates; spiny anteaters eat termites and ants that they capture with their long, sticky tongue. Females lay eggs but suckle their young.

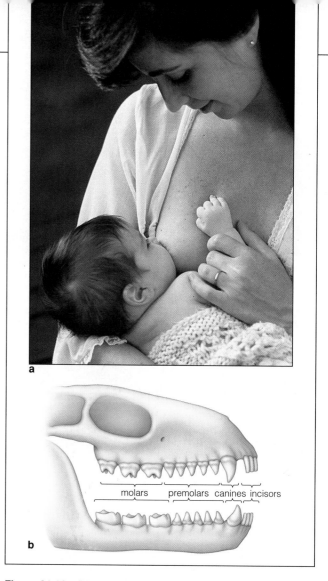

Figure 21.16 Distinctly mammalian traits. (**a**) A human baby, busily demonstrating the key defining feature. It derives nourishment from mammary glands. (**b**) Unlike reptilian teeth, the upper and lower rows of mammalian teeth match up. (Compare Figure 21.13b.)

Pouched Mammals

Nearly all of the 260 species of pouched mammals, or marsupials, are native to Australia and nearby islands; a few live in the Americas. At birth, the young are tiny, blind, and hairless. They suckle and finish their development inside a pouch on the mother's ventral surface.

For at least 50 million years, Australian marsupials evolved in relative isolation from placental mammals. Their ancestors had crossed a narrow sea that opened up between two land masses, but placental mammals stayed behind. In the absence of competition, marsupials radiated freely into adaptive zones. Now, wallabies, kangaroos, and other plant-eating mammals occupy adaptive zones comparable to the ones filled by deer, antelope, and other plant eaters on other continents. Siberian and North American forests have wolves; Australia has (or had) the Tasmanian "wolf," a marsupial that may now be extinct. Other marsupials glide like

Figure 21.17 Representative mammals. (**a**) Short-nosed anteater, (*Tachyglossus aculeatus*), an egg-laying mammal of Australia. (**b**) A pouched mammal—a female opossum with young. Placental mammals: (**c**) Manatee, which lives in the sea and eats submerged seaweed. (**d**) Camels, traversing an extremely hot desert with ease. (**e**) Bats, the only flying mammals, dominate the night sky vacated by birds. (**f**) An arctic fox. Its thick fur insulates the animal and helps it "hide in the open" (camouflage itself) from prey. In summer, the light-brown fur blends with golden-brown grasses. In winter, it turns white and blends with the snow-covered habitat.

flying squirrels; some climb like monkeys. In Australia, many native species are suffering from competition with placental mammals. Cattle, sheep, horses, rabbits, and other groups that humans have imported from other regions threaten to displace them.

Placental Mammals

In nearly all habitats on land and in water, you will find placental mammals. The **placenta**, a spongy tissue made of maternal and fetal membranes, develops inside a pregnant female's uterus. This tissue is the means by which the new individual receives nutrients and oxygen and gets rid of metabolic wastes. Placental mammals grow faster in the uterus than marsupials do in a pouch, and many are fully developed at birth.

Appendix I lists the major groups of placental mammals. Rats, mice, prairie dogs, squirrels, and other rodents are the most diverse group, followed by bats. Other familiar types are dogs, bears, cats, seals, and dolphins (carnivores), as well as horses, camels, deer, and elephants (herbivores). Manatees and anteaters are among the exotic types. The body form and function, behavior, and ecology of many of these mammals will occupy our attention in later chapters.

Mammals alone feed their young with milk from mammary glands. They have distinctive dentition and, typically, hair as well as an internal skeleton, a nerve cord, a three-part brain, and sensory organs inside the skull. Their young typically require an extended period of dependency and learning.

EVOLUTIONARY TRENDS AMONG THE PRIMATES

Having arrived at the mammalian branch of the animal family tree, we are now ready to follow it out along the finer branchings leading to primates, then to humans.

Primate Classification

The order **Primates** includes prosimians, tarsioids, and anthropoids (Table 21.1 and Appendix I). Figure 21.18 shows a few of these distinctive mammals. Prosimians are the oldest primate lineage (*pro*, before; *simian*, ape). They dominated the trees in North America, Europe, and Asia for millions of years before monkeys and apes evolved and nearly displaced them. Tarsiers of southeastern Asia are the only living tarsioids. The features of these small primates place them between prosimians and anthropoids. All monkeys, apes, and humans are **anthropoids**. In biochemistry and structure, the apes are closer to humans than to monkeys. That is why apes, humans, and extinct species of the human lineage are classified together, as **hominoids**.

By 5 million years ago, a divergence from the last shared ancestor of apes and humans was under way. All species that evolved on that separate road leading to humans are further classified as **hominids**.

Key Evolutionary Trends

Most primates live in tropical or subtropical forests, woodlands, or **savannas** (open grasslands with a few stands of trees). Like their ancient ancestors, the vast majority are tree dwellers. Yet no one feature sets "the primates" apart from other mammals. Each lineage evolved in a distinct way and has its own defining traits. Five trends define the human lineage, which is our focus here. They were set in motion when primates started adapting to a new way of life in the trees, and they contributed to the emergence of modern humans. *First*, there was less reliance on the sense of smell and more on daytime vision. *Second*, skeletal changes led to upright walking, which freed the hands for novel tasks. *Third*, changes in bones and muscles led to refined hand movements. *Fourth*, teeth became less specialized. *Fifth*, elaboration of the brain and changes in the skull led to speech. These developments became interlocked with each other and with cultural evolution.

Enhanced Daytime Vision   Early primates had an eye on each side of the head. Later ones had forward-directed eyes, an arrangement that is much better for sampling shapes and movements in three dimensions. Other modifications allowed the eyes to respond to variations in color and light intensity (dim to bright). These visual stimuli are complex, and responsiveness to them is advantageous for life in the trees.

Figure 21.18   Representative primates. (**a**) Gibbon, with body and limbs adapted for swinging arm over arm in the trees. (**b**) Spider monkey, a four-legged climber, leaper, and runner. (**c**) Tarsiers, which are vertical climbers and leapers.

Table 21.1	Primate Classification
Grouping	Representatives
Prosimians	
Lemuroids	Lemurs, lorises
Tarsioids	
Tarsioids	Tarsiers
Anthropoids	
Ceboids (New World monkeys)	Spider monkeys
Cercopithecoids (Old World monkeys)	Baboons
Hominoids:	
Hylobatids	Gibbons, siamang
Pongids	Orangutan, gorilla, chimpanzee
Hominids	Humans and earlier species of lineage

Figure 21.19 Skeletal organization and stance of a monkey, ape, and human. Monkeys climb and leap. Gorillas and chimps can use forelimbs to climb and help support their weight but are mainly ground-dwelling, knuckle-walking quadrupeds. Humans are bipedal. These modes of locomotion arose through modifications of the basic mammalian plan.

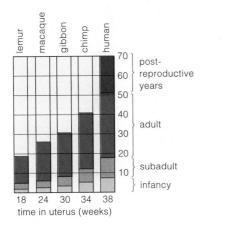

Figure 21.20 Trend toward longer life spans and longer dependency among primates.

Precision Grip and Power Grip The first mammals spread their toes apart to help support the body as they walked or ran on four legs. (Primates still spread their toes or fingers. Many make cupping motions, as when monkeys lift food to the mouth.) Among tree-dwelling primates, hand bones became modified. Now they could wrap fingers around objects (*prehensile* movements), and touch the thumb to the tip of each finger (*opposable* movements). In time, their hands became freed from load-bearing functions. Later, when hominids evolved, refinements led to the precision grip and power grip:

The new hand positions gave early humans a capacity to manufacture and use tools. They were a foundation for unique technologies and cultural development.

Teeth for All Occasions Before hominids evolved, modifications in primate jaws and teeth accompanied a shift from eating insects, to fruits and leaves, and on to a mixed diet. Later, rectangular jaws and long canines came to be further defining features of monkeys and apes. Along the road leading to humans, a bow-shaped jaw and smaller teeth of about the same length evolved.

Upright Walking A monkey's skeleton lets it climb, leap, and run on four legs, but not two. For instance, its arm and leg bones are about the same length (Figure 21.19). With its long arms, a gorilla can knuckle walk. Of all primates, though, only humans are **bipedal**; they stride freely, on two legs. The shapes and positioning of shoulder blades, pelvic girdle, and backbone allow this. (For example, compared to monkeys and apes, the backbone is shorter, S-shaped, and flexible.) By current thinking, the evolution of bipedalism was the pivotal modification in the origin of hominids.

Better Brains, Bodacious Behavior When primates moved into the trees, shifts in reproductive and social behavior followed. In many lineages, parents invested more effort in fewer offspring. They formed strong bonds with the young, maternal care grew more intense, and the learning period became longer (Figure 21.20). In time, brain modifications and behavioral complexity became highly interlocked. (Brain regions dealing with encoding and processing information expanded greatly. Novel behaviors stimulated their development—which stimulated more novel behavior, and so on.) We find evidence of such interlocking in the parallel evolution of the human brain and culture. **Culture** is the sum total of behavior patterns of a social group, passed between the generations by learning and by symbolic behavior—especially language. A capacity for language arose among ancestral humans, through changes in the skullbones and brain.

These features emerged along the evolutionary road leading to humans: upright walking, refined hand movements, generalized dentition, refined vision, and the interlocked elaboration of brain regions and cultural behavior.

FROM PRIMATES TO HOMINIDS

Origins and Early Divergences

Primates evolved from mammals more than 60 million years ago, in the Paleocene. The first ones resembled small rodents or tree shrews (Figures 21.21 and 21.22). Like tree shrews, they probably had huge appetites and foraged at night for insects, seeds, buds, and eggs beneath the trees of tropical forests. They had a long snout and a good sense of smell, suitable for snuffling food or predators. They clawed their way up through the shrubbery, although not with much speed or grace.

During the Eocene (between 54 and 38 million years ago), some primates were staying in the trees. This was a time when enhanced daytime vision, increased brain size, a shorter snout, and refined grasping movements evolved. But how did this happen?

Consider the trees. Trees offered food and safety from ground-dwelling predators. They also were a habitat of uncompromising selection. Imagine dappled sunlight, boughs swaying in the breezes, colorful fruit hidden among the leaves, and perhaps predatory birds. A long, odor-sensitive snout would not have been of much use up in the trees, where air currents disperse odors. But a brain that could assess movement, depth, shape, and color would have been a definite plus. So would a brain that could work fast when its owner was running and leaping (especially!) from branch to branch. The body's weight, the wind speed, and the distance and suitability of a destination had to be estimated all at once—and any adjustments for miscalculations had to be quick.

By at least 36 million years ago (before the dawn of the Oligocene), tree-dwelling anthropoids had evolved in the forests. One form, the squirrel-sized *Catopithecus*, was on or very close to the evolutionary road that led to monkeys, apes, and humans. It had forward-directed eyes. Given its snoutless, flattened face and upper jaw with shovel-shaped front teeth, it must have used its hands to grab fruit and insects, like modern monkeys do. Some early anthropoids lived in swamps infested

Figure 21.22 A night-foraging tree shrew of Indonesia.

with fierce, predatory reptiles. Maybe that's why they rarely ventured to the ground—and why it was imperative to think fast and to grip strongly. Slip-ups were always possible; a surprising number of primates still fall out of trees.

During the Miocene (25 to 5 million years ago), land masses were drifting into their current positions, and colossal uplifting produced high mountain ranges. The global climate became cooler and drier, and an adaptive radiation of apelike forms—*the first hominoids*—took place. By 13 million years ago, hominoids had spread into Africa, Europe, and southern Asia. Most became extinct around that time. But fossils and genetic studies point to three divergences between 10 million and 5 million years ago. Two branchings of the Miocene apes led to gorillas and chimpanzees. The third gave rise to *hominids*—including the ancestors of humans.

The First Hominids

Most of the early hominids we know about lived in the East African Rift Valley, a 3,200-kilometer fracture in the Earth's crust. By the Miocene-Pliocene boundary, the long-term cooling trend was well under way. Once-vast tropical forests—and their bounty of fruits and insects —were giving way to scattered woodlands and grassy plains adapted to pronounced seasonal shifts in water and food supplies. The African savanna had emerged.

Survivors of these changes were able to locate and exploit new foods and hide from new predators. It must have been a "bushy" time of evolution, with branchings and radiations into new adaptive zones. Why? Fossil

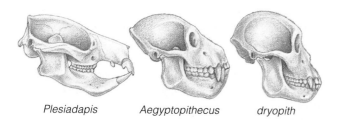

Plesiadapis   *Aegyptopithecus*   *dryopith*

Figure 21.21 Comparison of skull shape and teeth of some early primates. *Plesiadapis* (Paleocene), had rodentlike teeth. *Aegyptopithecus* (Oligocene anthropoid) probably predates the divergence leading to Old World monkeys and apes. Apelike dryopiths lived in the Miocene. The sketches are not to the same scale. *Plesiadapis* was as tiny as a tree shrew. *Aegyptopithecus* was monkey-sized and some dryopiths, chimpanzee-sized.

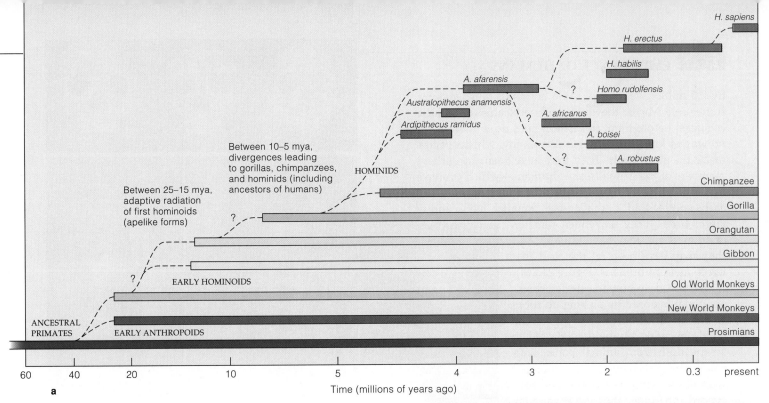

a

Time (millions of years ago)

Figure 21.23 (a) One proposed evolutionary tree for primates. (b) Remains of Lucy (*Australopithecus afarensis*). (c) Footprints made in soft damp volcanic ash 3.7 million years ago at Laetoli, Tanzania, as discovered by Mary Leakey. The arch, big toe, and heel marks are signs of bipedal hominids.

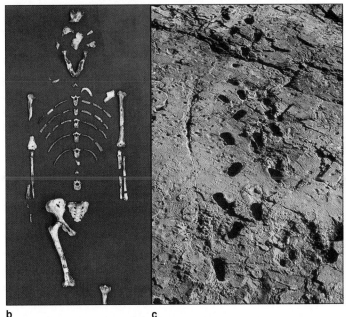

b   c

hunters have accurately dated the fragmented remains of many forms that have apelike and humanlike traits. At this writing, the earliest known hominids lived 4 to 4.5 million years ago. We call them the **australopiths** (meaning southern apes). *Australopithecus anamensis, A. afarensis,* and *A. africanus* were gracile (slightly built). *A. boisei* and *A. robustus* were robust (muscular, heavily built). Currently, *A. anamensis* is the most ancient form.

We don't know how the australopiths were related or which ones may have been ancestral to us (Figure 21.23a). But many shared three features. *First,* they walked upright, which meant their hands were free for new functions. *Second,* their jaws and large, thickly enameled teeth accommodated a variety of foods. *Third,* with an average volume of 400 cubic centimeters, their brain was not Einsteinian, but it probably was more complex than that of their forerunners. At the very least, these hominids had to be able to think ahead, to plan when and where to get different foods when seasonal supplies ran out. All three features resulted from modifications of traits that we see among other primates. *They were based on the primate heritage.*

What sort of evidence tells us this? As one example, consider the *A. afarensis* female dubbed Lucy, who lived 3.2 million years ago (Figure 21.23b). In some respects she was chimplike, and only about 3-1/2 feet tall. Yet her foot bones and leg bones had muscle attachment sites similar to yours. Her thighbones angled inward,

so her weight had to be centered directly beneath her pelvis—a sure sign of bipedalism. And the most telling sign—australopiths left footprints (Figure 21.23c).

Primates evolved from small, rodentlike mammals that moved into arboreal habitats about 60 million years ago. By 36 million years ago, their descendants included ancestors of monkeys, apes, and the hominids.

Between 10 million and 5 million years ago, a divergence led to the australopiths, the first hominids. They were apelike in some ways and humanlike in others. Unlike earlier forms, they were bipedal, had smaller canines and a more flattened face, and their brain probably was more complex.

EMERGENCE OF HUMANS

By 2.5 million years ago, some hominids were making stone tools. Maybe they had used perishable tools (such as sticks) before then, as modern apes do, but we have no way of knowing. We also don't know which species made the stone tools. They may have been the earliest members of the genus *Homo*—**humans**—which evolved by 2 million years ago. Compared to australopiths, the early species had a smaller face, smaller teeth, and a larger brain. They scavenged remains of animals and may have hunted. And they may have been on the road that led to modern humans (Figure 21.24*a*).

Maybe early humans started down a toolmaking road by picking up rocks to crack animal bones and expose the edible marrow. Maybe they started to scrape flesh from animal bones with small, sharp flakes that had fractured naturally from rocks. However they started, in time they were *shaping* stone implements. Mary Leakey was the first to discover manufactured tools at Africa's Olduvai Gorge, which cuts through a great sequence of sedimentary rock layers. The most ancient tools—crudely chipped pebbles—were buried in the deepest layers (Figure 21.24*b*). They may have been used to dig for roots, smash marrow bones, and poke insects from tree bark. More recent layers have more complex tools. Hominids were getting serious about exploiting a major food source—the huge game herds of open grasslands.

Another early human species, *H. erectus*, endured from about 1.8 million to 300,000 years ago. *H. erectus* had a long, chinless face, thick-walled skull, heavy browridge, and other archaic traits, but a larger brain (Figure 21.24*a*). *And it was the first hominid to venture out of Africa.* Fossils discovered in Southeast Asia and in the former Soviet republic of Georgia, recently dated at 1.8 million and 1.6 million years old, support this. In cold regions, individuals became much stockier, with long bones adapted to muscular stress. And life must have been strenuous. More than once, glaciers—some two miles high—advanced and later retreated in North America, northern Europe, and southern Asia.

This was the time of cultural lift-off for the human lineage. From southern Africa to England, *H. erectus* populations were using the same kinds of hand axes and other tools designed to pound, scrape, shred, cut, and whittle. *H. erectus* also learned to control fire.

Where modern humans (*H. sapiens*) arose is hotly debated. By one model, perhaps more than a million years ago *H. erectus* migrated from Africa, then distinct subpopulations of modern humans arose later through genetic divergence in different regions. By another model, modern subpopulations originated *in* Africa,

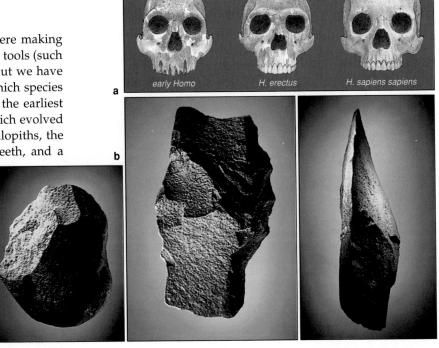

Figure 21.24 (**a**) Comparison of skulls of early and modern humans. (**b**) Three of the 37,000+ stone tools from Olduvai Gorge. A chopper, a hand ax, and a cleaver.

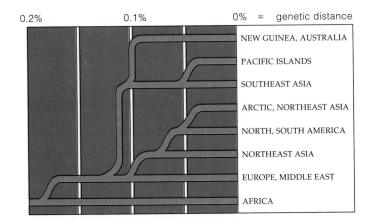

Figure 21.25 One proposed family tree for populations of modern humans (*Homo sapiens*) that are native to different regions. Branch points imply relative times of divergences. The data are based on genetic, biochemical, and immunological comparisons. They support the model of a late common origin for modern subpopulations of humans in Africa and their subsequent migrations into Europe and into Southeast Asia.

before major migrations occurred (Figure 21.25). New finds may help resolve the debate.

Whatever the outcome, we do know that *H. sapiens* arose between 300,000 and 200,000 years ago. Compared to *H. erectus*, early *H. sapiens* had smaller teeth, jaws, and often a chin. It had smaller facial bones, a larger brain, and a rounder, higher skull. And it may have had the capacity for complex language.

Figure 21.26 Cave painting from Lascaux, France.

One group of early humans, the Neandertals, lived in southern France, central Europe, and the Near East about 130,000 years ago. They had a large brain, heavy facial bones, and often a large browridge. Neandertals lived at the edge of forests, in caves, rock shelters, and open-air camps. Apparently they simply hunted or gathered food at their doorstep; they never learned to exploit the game herds. They disappeared 35,000 or 40,000 years ago, when anatomically modern humans arose. These more recent groups learned to store and share food, and they developed rich cultures, as evidenced by refined tools and artistic treasures (Figure 21.26).

From 40,000 years ago to today, human evolution has been almost entirely cultural, not biological—and so we leave the story with these conclusions: Humans spread through the world by rapidly devising *cultural* means to deal with a broad range of environments. Compared with their predecessors, they developed rich and varied cultures. They moved from "stone-age" technology to the age of "high tech." Yet hunters and gatherers persist in parts of the world, attesting to the great plasticity and depth of human adaptations.

Cultural evolution has outpaced the biological evolution of the only remaining human species (*H. sapiens*). Humans everywhere rely on cultural innovation to adapt rapidly to a broad range of environmental challenges.

SUMMARY

1. Nearly all chordate embryos have a notochord, a dorsal hollow nerve cord, a pharynx with gill slits (or hints of these), and a tail that extends past the anus. Some or all of these traits persist in the adult forms. Chordates with a backbone are known as vertebrates. Tunicates (including the sea squirts) and lancelets are invertebrate chordates.

2. The eight classes of vertebrates are jawless fishes, jawed armored fishes (extinct), cartilaginous fishes, bony fishes, amphibians, reptiles, birds, and mammals.

3. These trends occurred during vertebrate evolution: A vertebral column supplanted the notochord as a structural element against which muscles act. (This led to swift predatory animals.) Jaws evolved from gill-supporting elements. This led to increased predator-prey competition; it favored far more efficient nervous systems and sensory organs. In certain bony fishes, paired fins evolved into fleshy lobes with structural elements (forerunners of paired limbs). Lungs evolved in some fishes, enhanced respiration by gills, and proved adaptive in the invasion of land. The circulatory system became better at distributing oxygen.

4. Amphibians, the first land vertebrates, never fully escaped the water. Reptiles escaped dependency on standing water, owing to their tough, scaly skin that conserves moisture; internal fertilization; efficient water-conserving kidneys; and amniote eggs, often leathery or shelled, that protect and metabolically support the embryos.

5. Reptiles, birds, and mammals have highly efficient circulatory, respiratory, and nervous systems. Birds alone have feathers, used in flight, heat conservation, and social displays.

6. Like other mammals, primates have a complex brain and sensory organs inside a skull, mammary glands (in females), and distinctive teeth. Adults nourish, protect, and serve as behavioral models for the young. Primates include prosimians, tarsioids, and anthropoids (all monkeys, apes, and humans). Only apes and humans are hominoids. Only modern humans (*H. sapiens*) and others of their lineage (from *A. afarensis* to *H. erectus*) are further classified as hominids.

7. Humans can adapt to a wide range of environments. This capacity resulted from evolutionary modifications in certain primate lineages: less reliance on the sense of smell and more on enhanced daytime vision; a shift from a four-legged gait to bipedalism; increases in manipulative skills owing to changes in the hands, which began to be freed from load-bearing functions among tree-dwelling primates; a shift from specialized to omnivorous eating habits; and increases in brain complexity and behavior.

Review Questions

1. List and describe the features that distinguish chordates from other animals. *354*

2. List four evolutionary trends among vertebrates, then briefly describe the eight classes of vertebrates. *354, 356*

3. What is the difference between "hominoid" and "hominid"? Are we hominoids, hominids, or both? *368*

4. State the macroevolutionary trends that proved to be important in the evolution of humans. *368–369*

5. What environmental changes correlated with the great adaptive radiation of apelike forms during the Miocene? *370*

Self-Quiz *(Answers in Appendix IV)*

1. The first vertebrates were _____ .
   a. bony fishes
   b. jawless fishes
   c. jawed fishes
   d. both a and b

2. Of all existing vertebrates, _____ are the most diverse.
   a. cartilaginous fishes
   b. bony fishes
   c. amphibians
   d. reptiles
   e. birds
   f. mammals

3. Reptiles moved fully onto land owing to _____ .
   a. tough skin
   b. internal fertilization
   c. good kidneys
   d. amniote eggs
   e. both b and d
   f. all are correct

4. Various mammals _____ .
   a. hatch
   b. complete embryonic development in pouches
   c. complete embryonic development in the uterus
   d. all are correct

5. Hominids _____ .
   a. adapted to a wide range of environments
   b. adapted to a narrow range of environments
   c. had flexible bones that cracked easily
   d. were limber enough to swing through the trees

6. The first known hominids were the _____ .
   a. *Homos*
   b. dryopiths
   c. cercopiths
   d. australopiths

7. Match the organisms with the appropriate features.
   _____ jawless fishes
   _____ cartilaginous fishes
   _____ bony fishes
   _____ amphibians
   _____ reptiles
   _____ birds
   _____ mammals
   a. mammary glands, thick skin or hair
   b. respiration by skin and lungs
   c. include coelocanths
   d. include hagfishes
   e. include sharks and rays
   f. complex social behavior, feathers
   g. first with amniote eggs

Critical Thinking

1. Describe the factors that might contribute to the collapse of native fish populations in a lake, following the introduction of a novel predator, such as the lamprey.

2. Think about the flight muscles of birds and their demands for oxygen and ATP energy. What type of organelle would you expect to be profuse in these muscles? Explain your reasoning.

3. Kathie and Gary, two amateur fossil hunters, have unearthed the complete fossilized remains of a mammal. How can they determine whether their find is a herbivore, carnivore, or omnivore?

4. In Australia, many species of marsupials are competing with recently introduced placental mammals (such as rabbits) for resources—but they are not winning. Explain how it is that placental mammals that did not even evolve in the Australian habitats show greater fitness than the native mammals.

5. Fossil evidence suggests that Neandertals and archaic human populations coexisted in the same parts of the Middle East for 25,000 to 50,000 years. By one hypothesis, the Neandertals evolved in Europe, then migrated to the Middle East, whereas the archaic human populations migrated out of Africa. It appears that both used the same kinds of tools and neither developed the techniques required for cave painting or for constructing decorative artifacts. The fact of their extended coexistence suggests they did not compete with each other. However, the Neandertals became extinct, and the other group gave rise to populations of fully modern humans. Were the two groups different species?

Selected Key Terms

amniote egg *362*
amphibian *360*
anthropoid *368*
australopith *371*
behavioral flexibility *366*
bipedalism *369*
bird *364*
bony fish *358*
cartilaginous fish *358*
chordate *354*
culture *369*
dentition *366*
feather *364*
filter feeder *354*
fin *356*
gill *356*
gill slit *354*
hagfish *357*
hominid *368*
hominoid *368*
human (*Homo*) *372*
jaw *356*
lamprey *357*
lancelet *354*
lobe-finned fish *359*
lung *356*
mammal *366*
metamorphosis *354*
nerve cord *354*
notochord *354*
ostracoderm *357*
pharynx *354*
placenta *367*
placoderm *357*
primate *368*
reptile *362*
savanna *368*
scale *358*
swim bladder *358*
tunicate *354*
vertebra *356*

Readings

Gould, S. J. (general editor). 1993. *The Book of Life*. New York: Norton. Splendid, easy-to-read essays, gorgeous illustrations.

Romer, A. S., and T. S. Parsons. 1986. *The Vertebrate Body*. Sixth edition. Philadelphia: Saunders.

Weiss, M., and A. Mann. 1990. *Human Biology and Behavior*. Fifth edition. New York: Harper Collins.

FACING PAGE: *A flowering plant* (Prunus) *busily doing what it does best: producing flowers for the fine art of reproduction.*

22 PLANT TISSUES

The Greening of the Volcano

In the spring of 1980, Mount Saint Helens in the southwestern Cascades of Washington exploded and 540 million tons of ash blew skyward. Within minutes, shock waves from the blast blew down or incinerated hundreds of thousands of mature trees near the volcano's northern flank. Rivers of ash and fire surged down the slopes faster than a hundred miles an hour. Those nightmarish rivers turned into torrents of cementlike mud when the intense heat melted and released twenty billion gallons of water from the mountain's snowfields and glacial ice.

In one brief moment, nearly 100,000 acres of magnificent forests had been transformed into a scarred, barren sweep of land (Figure 22.1*a* and *b*). The aftermath of the volcano's eruption gave us a mind-numbing picture of what our world would be like without plants.

And yet, in less than a year's time, the seeds of fireweed, blackberry, and other flowering plants were sprouting near the grayed trunks of fallen trees around Mount Saint Helens. Within a decade, willows and alders were flourishing along the rivers, and low shrubs were growing across much of the land. Their presence began to provide the shade necessary for seedlings of slower growing but ultimately dominant species—the hemlocks and Douglas fir.

Figure 22.1*b* gives you a sense of the view from a ridge overlooking Spirit Lake and Mount Saint Helens not long after the eruption. Figure 22.1*c* shows the same view nine years later. Within fifty years, new forest trees will be well established. Within a century, the forest will be as it once was.

With this example, we open a unit dedicated to the most successful plants on Earth—the flowering plants. No other kind of plant surpasses them in diversity or distribution. They have distinctive patterns of structural organization and splendid modes of reproduction, growth, and development. Their structure and physiology (functioning) help them survive hostile conditions on land—even momentary takeovers by volcanoes.

This first chapter of the unit focuses mainly on the tissues and body plans of these plants. Chapter 23 explains how they take up water and nutrients, restrict water loss, and efficiently distribute organic substances throughout the plant body. Chapter 24 describes key aspects of their growth, development, and reproduction.

a

b

c

Figure 22.1 (**a,b**) A stunning reminder of what our world would be like without plants—the devastation following the eruption of Mount Saint Helens in 1980. Nothing remained of the forest that had surrounded the volcano's flanks. (**c**) Less than a decade later, plants were making a comeback.

KEY CONCEPTS

1. Angiosperms—the flowering plants—dominate the plant kingdom. They have aboveground shoots (which include stems, leaves, and flowers) and descending roots that typically grow downward and outward through soil.

2. Flowering plant tissues are organized into three main kinds of tissue systems. Ground tissue systems make up most of the plant body. Vascular tissue systems distribute water, minerals, and products of photosynthesis through roots, stems, and leaves. Dermal tissues cover and protect plant surfaces.

3. A simple plant tissue (parenchyma, sclerenchyma, or collenchyma) consists of one type of cell. Complex tissues have two or more types of cells. They include xylem and phloem (vascular tissues) and epidermis (a dermal tissue).

4. Plants grow by cell divisions and cell enlargements at meristems, which are localized regions of self-perpetuating embryonic cells. The plant's primary growth (lengthening of stems and roots) originates at apical meristems of stem and root tips. Its secondary growth (increases in diameter) originates at lateral meristems inside stems and roots.

5. Many plants show primary and secondary growth, year after year. Wood is the outcome of extensive secondary growth.

OVERVIEW OF THE PLANT BODY

Earlier, in Chapter 19, we surveyed representatives of the 295,000 known species of plants. Even that fleeting run through plant diversity reveals why no one species can be used as a typical example of plant body plans. Even so, when we hear the word "plant," we mostly think of the well-known gymnosperms (such as pine trees) and angiosperms (which include roses, cactuses, corn plants, and apple trees). **Angiosperms**, recall, are the only flower-producing plants. With 260,000 species, they dominate the plant kingdom.

Shoots and Roots

Many flowering plants have a body plan similar to that shown in Figure 22.2. Its aboveground parts, or **shoots**, consist of stems and their branchings, leaves, flowers, and other structures. Components of its stems function in support for upright growth and in the conduction of substances throughout the plant body. (Upright growth gives photosynthetic cells in young stems and leaves favorable exposure to light.) Flowers are reproductive structures that attract pollinators. **Roots** are specialized descending structures that typically penetrate soil and spread downward and outward through it. A plant's root system absorbs soil water and dissolved nutrients, and it commonly anchors aboveground parts. It also stores food, then releases it as required for root cells or for transport to aboveground parts.

Three Plant Tissue Systems

The stems, branches, leaves, and roots of a flowering plant are similar in a key respect. Their main tissues are grouped into three systems. The **ground tissue system** is the most extensive; its tissues make up the bulk of the plant body. The **vascular tissue system** contains two kinds of conducting tissues that distribute water and solutes throughout the plant body. The **dermal tissue system** covers and protects the plant's surfaces. Figure 22.2 shows the general locations of these tissue systems.

Some tissues in each system are simple, in that they contain one type of cell only. Parenchyma, collenchyma, and sclerenchyma fall in this category. Other tissues are complex, with organized arrays of two or more types of cells. Xylem, phloem, and epidermis are like this.

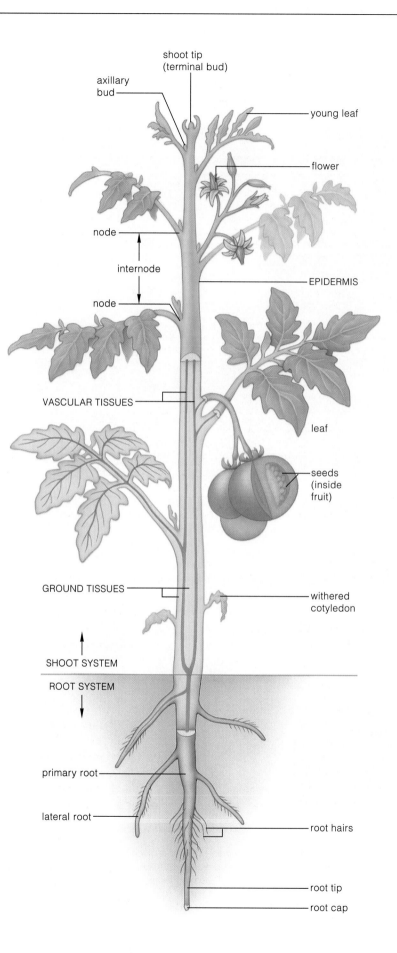

Figure 22.2 Body plan for one type of flowering plant, showing its shoot and root systems. Vascular tissues (*purple*) conduct water, nutrients, and organic substances throughout the plant body. They thread through ground tissues, which make up most of the plant body. Dermal tissues (in this case, epidermis) cover the surfaces of both the root system and shoot system.

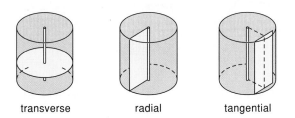

transverse   radial   tangential

Figure 22.3 Terms that identify the manner in which tissue specimens are cut from plants.

Cuts made perpendicular to the long axis of a stem or root provide *transverse* sections (also called cross-sections). Longitudinal cuts made along the radius of a root or stem provide *radial* sections. Longitudinal cuts made at right angles to the root or stem give *tangential* sections.

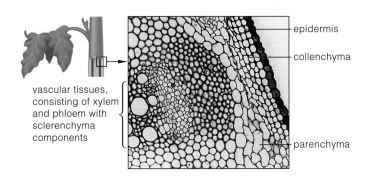

epidermis

collenchyma

vascular tissues, consisting of xylem and phloem with sclerenchyma components

parenchyma

Figure 22.4 Transverse section showing the locations of simple tissues and complex tissues in one kind of plant stem.

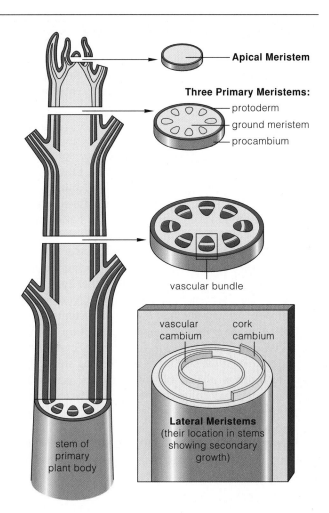

Apical Meristem

Three Primary Meristems:
protoderm
ground meristem
procambium

vascular bundle

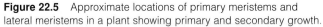

vascular cambium   cork cambium

Lateral Meristems (their location in stems showing secondary growth)

stem of primary plant body

Figure 22.5 Approximate locations of primary meristems and lateral meristems in a plant showing primary and secondary growth.

The next sections in this chapter describe the tissue organization of shoots and roots. You may find it easier to interpret photographs of this organization if you first study Figures 22.3 and 22.4. They will help you identify the way tissue specimens are cut from plants.

Meristems

As the stems and roots of flowering plants grow, they lengthen and thicken through activity at **meristems**: localized regions of self-perpetuating, embryonic cells.

Increases in length originate at **apical meristems** located inside the dome-shaped tips of stems and roots. There, cell divisions and cell enlargement give rise to three primary meristems—which go on to produce all of the specialized tissue systems (Figure 22.5). Growth at apical meristems and at tissues derived from them produces the *primary* tissues of the plant body.

Increases in diameter originate at lateral meristems, which form inside a stem or root (Figure 22.5). The two kinds of lateral meristems are called **vascular cambium**

and **cork cambium**. Growth originating here produces the plant's *secondary* tissues.

Each spring, for example, *primary* growth lengthens a maple tree's stems, branches, and roots; and *secondary* growth thickens them. Some of the new cells that form at this time perpetuate the meristems. Others differentiate and become part of the tree's specialized tissues.

Flowering plant shoots have stems that support upright growth and conduct substances, leaves that function in photosynthesis, and flowers that function in pollination. The plant's roots absorb water and solutes, anchor aboveground parts, and often store food.

A ground tissue system makes up the bulk of the plant. A vascular tissue system distributes water and solutes through it; a dermal tissue system covers and protects its surfaces.

Primary growth (stem and root lengthening) originates at apical meristems and meristematic tissues derived from them. Secondary growth (thickening of stems and roots) originates at lateral meristems (vascular cambium and cork cambium).

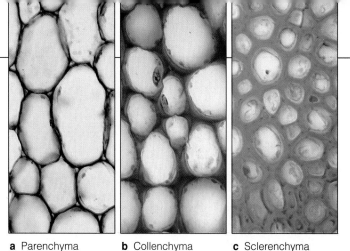

a Parenchyma   **b** Collenchyma   **c** Sclerenchyma

Figure 22.6 From the stem of a sunflower plant (*Helianthus*), examples of ground tissues, in transverse section.

22.2 TYPES OF PLANT TISSUES

We turn now to the organization and function of simple and complex tissues in plants. Figures 22.6 through 22.8 show a few examples from the main tissue categories.

Simple Tissues

Parenchyma tissues make up most of the soft, moist, primary tissues of roots, stems, leaves, flowers, and fruits. Its cells typically are thin-walled, pliable, and many-sided (Figure 22.6a). The cells are alive at maturity, and they retain the capacity to divide. Parenchymal cell divisions often heal wounded plant parts. One type of parenchyma, the mesophyll of leaves, specializes in photosynthesis. Carbon dioxide and oxygen diffuse through air spaces between its cells. Other types have roles in storage, secretion, and other tasks. Parenchyma also threads through the vascular tissue systems.

Collenchyma provides flexible support for primary tissues (Figure 22.6b). You will find patches or cylinders of this tissue near the surface of lengthening stems. This tissue also forms pliable ribs in many leaf stalks. The cells that make up collenchyma are mostly elongated, with unevenly thickened walls. Pectin inside the cell walls imparts pliability to this tissue.

Sclerenchyma supports mature plant parts and also protects many seeds. Generally, sclerenchyma cells have thick, lignin-impregnated walls (Figure 22.6c). **Lignin** strengthens and waterproofs cell walls. Without lignin, land plants would not have evolved (Section 19.1).

Sclerenchyma tissues consist of fibers or sclereids. *Fibers* are in vascular tissue systems of some stems and leaves. These long, tapered cells flex and twist without stretching. We use fibers from flax plants to make rope, paper, cloth, thread, and other commercially valuable products. *Sclereids* are shorter cells. Numerous sclereids give pears and some other fruits a gritty texture. They also form coconut shells, peach pits, and seed coats.

Complex Tissues

Unlike a ground tissue, which has one cell type only, vascular and ‌ermal tissues of flowering plants are complex, with a variety of cell types.

Vascular Tissues Two vascular tissues, called *xylem* and *phloem*, function in the distribution of substances throughout the plant. Often their conducting cells are associated with a sheath of fibers and parenchyma cells.

Xylem conducts water as well as mineral ions dissolved in it. It also helps mechanically support the plant. Figure 22.7 shows examples of its conducting cells. The cells, called vessel

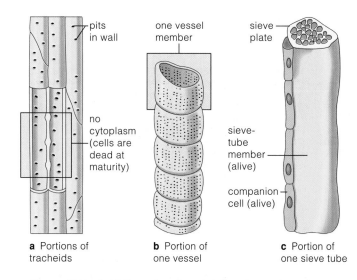

a Portions of tracheids   **b** Portion of one vessel   **c** Portion of one sieve tube

Figure 22.7 (**a**,**b**) Examples of tracheids and vessel members, the main types of cells in xylem that conduct water and dissolved mineral salts. (**c**) One type of phloem cell that conducts sugars and other organic compounds.

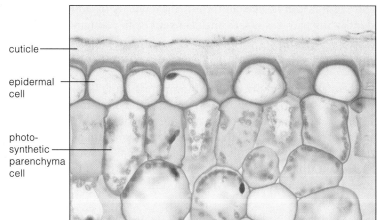

Figure 22.8 Light micrograph of a section through the upper surface of a kaffir lily leaf. The thick cuticle consists of secretions from epidermal cells. Inside the leaf are parenchyma cells that specialize in photosynthesis.

members and tracheids, are not alive at maturity. Lignified walls of these dead cells interconnect. Collectively, they form the water-conducting pipelines and they also strengthen plant parts. Water flows between the adjoining cells through pits in the walls.

Phloem conducts sugars and other solutes through the plant. Its main conducting cells, called sieve-tube members, are alive at maturity (Figure 22.7c). Adjacent cells interconnect at openings in their side walls and at their open or perforated end walls.

Sieve-tube members of leaves are loaded with sugars from the photosynthetic cells. Specialized, living parenchyma cells known as "companion" cells typically help with the loading process. Sugars travel through the phloem pipelines and are unloaded from it in any region where cells are growing or storing food. The next chapter says more about the functions of xylem and phloem.

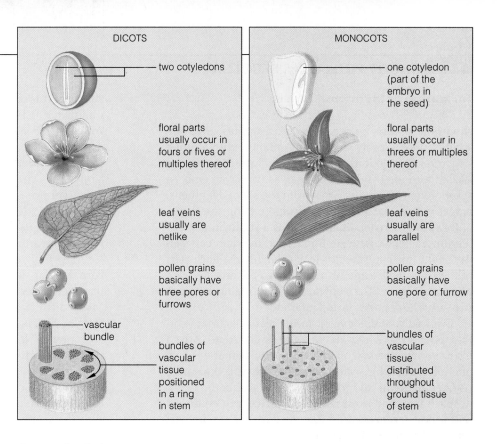

Figure 22.9 Comparison of dicots and monocots. The same simple and complex tissues occur in both classes of flowering plants, but they show some differences in structural organization.

Dermal Tissues All surfaces of primary plant parts are covered and protected by a dermal tissue system called **epidermis**. In most plants, epidermis is mainly a single layer of unspecialized cells. Waxes and cutin (a fatty substance) coat the outermost cell walls. We call the surface coating a **cuticle**. A plant cuticle functions in restricting water loss and often in resisting attacks by some microorganisms (Figure 22.8).

Stem and leaf epidermis also contains specialized cells, including pairs of guard cells. Guard cells change shape in response to changing conditions. When they do so, a gap called a **stoma** (plural, stomata) opens up between them or closes. The next chapter looks at how stomata are control points for the movement of carbon dioxide, oxygen, and water vapor across the epidermis.

As you will see later, **periderm** replaces epidermis in stems and roots showing secondary growth. Most of this protective cover consists of walls of cork cells that are no longer alive. The walls are heavily impregnated with suberin, a fatty substance.

Dicots and Monocots—Same Tissues, Different Features

The **dicots** and **monocots**, recall, are the two classes of flowering plants (Section 19.6). Most trees and shrubs other than conifers—such as maples, elms, roses, cacti,

peas, beans, lettuces, cotton, and carrots—are dicots. Lilies, orchids, palms, grasses, bamboos, wheat, corn, sugarcane, and pineapples are familiar monocots

Dicots and monocots are similar in structure and function, but they differ in some distinctive ways. For example, dicot seeds have two cotyledons and monocot seeds have one. "Cotyledons" are leaflike structures, commonly known as seed leaves. They form in seeds as part of a plant embryo, and they store or absorb food for it. After the seed germinates, the cotyledons wither, and leaves start functioning. Figure 22.9 shows other differences between dicots and monocots.

Most of the plant body (the ground tissue system) consists of simple tissues called parenchyma, collenchyma, and sclerenchyma. Each tissue is composed of one cell type only.

Xylem and phloem are vascular tissues. Xylem's pipelines of tracheids and vessel members conduct water and dissolved minerals. Phloem's sieve-tube members and companion cells interact to transport organic compounds, such as sugars.

Of two dermal tissues, epidermis covers the surfaces of the primary plant body. Periderm replaces it on plant parts showing secondary growth.

The same tissues make up the plant body in monocots and dicots, but some of these are organized in distinctive ways in each class of flowering plant.

22.3 SHOOT PRIMARY STRUCTURE

Formation of Stems and Leaves

Next time you or a friend eats bean sprouts or alfalfa sprouts, look closely at the tiny strands with leaflike structures (cotyledons) at the tip. Each of these shoots started forming inside a seed coat, as part of an embryo sporophyte. Normally, as a primary shoot lengthens, bulges called **leaf primordia** develop along the flanks of its apical meristem. Each bulge is a rudimentary leaf (Figure 22.10). Growth continues, and the stem between the newly forming tiers of leaves lengthens. Each part of the stem where one or more leaves are attached is a **node**. As Figure 22.2 shows, the stem region between two successive nodes is an "internode."

Gradually, buds develop in the leaf axils (that is, in the upper angle where leaves attach to the stem). A **bud** is an undeveloped shoot of mostly meristematic tissue, often protected by scales (modified leaves). Buds give rise to new stems and to leaves, flowers, or both.

Each **leaf** that forms is a metabolic factory, equipped with photosynthetic cells. Leaves vary greatly in size, shape, and texture. A duckweed leaf is no more than a millimeter (0.04 inch) across; some water lily leaves are 2 meters (6.5 feet) wide. Leaves resemble blades, spikes, needles, cups, feathers, tubes, and other structures. Many are smooth surfaced; others are sticky, hairy, or slimy. They may have hairs, scales, spikes, hooks, and other surface specializations; the introduction to the next chapter describes a particularly splendid example. Leaves also vary enormously in coloration, odor, and edibility (many are poisonous).

Most monocot leaves, such as those of ryegrass and corn, are flat surfaced like a knife blade. The base of the blade encircles and sheathes the stem (Figure 22.11b). Many dicot leaves have a broad blade, attached to the stem by a stalk (petiole). Often the blade is lobed. Some "compound" leaves have stalked leaflets.

As you probably know, the leaves of birches and other **deciduous plants** drop away from the stem as winter approaches. Leaves of **evergreen plants** such as camellias also drop, but they don't all drop at the same time (hence the name).

Figure 22.10 (**a**) Light micrograph of a shoot tip of *Coleus*. (**b**) Formation of leaves on a primary shoot of a dogwood tree.

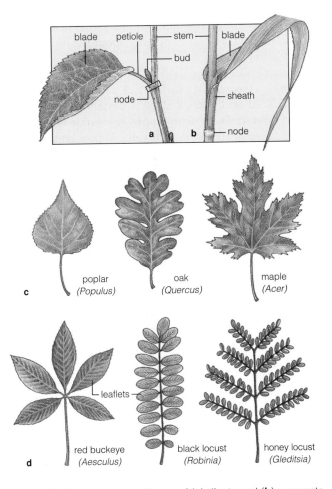

Figure 22.11 Common leaf forms of (**a**) dicots and (**b**) monocots. Examples of (**c**) simple leaves and (**d**) compound leaves.

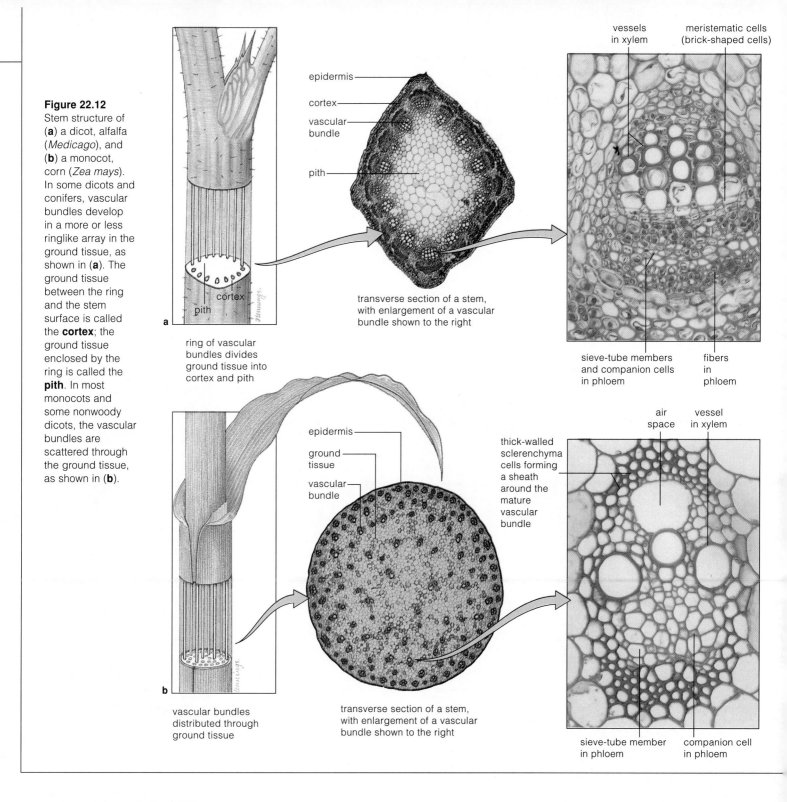

Figure 22.12
Stem structure of (**a**) a dicot, alfalfa (*Medicago*), and (**b**) a monocot, corn (*Zea mays*). In some dicots and conifers, vascular bundles develop in a more or less ringlike array in the ground tissue, as shown in (**a**). The ground tissue between the ring and the stem surface is called the **cortex**; the ground tissue enclosed by the ring is called the **pith**. In most monocots and some nonwoody dicots, the vascular bundles are scattered through the ground tissue, as shown in (**b**).

epidermis
cortex
vascular bundle
pith

a

ring of vascular bundles divides ground tissue into cortex and pith

cortex
pith

transverse section of a stem, with enlargement of a vascular bundle shown to the right

vessels in xylem
meristematic cells (brick-shaped cells)

sieve-tube members and companion cells in phloem
fibers in phloem

epidermis
ground tissue
vascular bundle

b

vascular bundles distributed through ground tissue

transverse section of a stem, with enlargement of a vascular bundle shown to the right

air space
vessel in xylem

thick-walled sclerenchyma cells forming a sheath around the mature vascular bundle

sieve-tube member in phloem
companion cell in phloem

Internal Structure of Stems

As a stem lengthens, primary meristems give rise to its ground, vascular, and dermal tissues. Its primary xylem and phloem commonly develop as **vascular bundles**. As you can see from Figure 22.12, these multistranded, sheathed cords thread lengthwise through the ground tissue system. Inside each bundle, the phloem is often positioned near the side of the sheath facing the stem surface, and the xylem is positioned near the side facing

the stem center. Figure 22.12 shows two of the common patterns of stem internal structure. The next chapter takes a look at the functioning of these vascular tissues.

Stems, leaves, and flowers of primary shoots arise at parts of apical meristem (leaf and bud primordia). Leaves of a given species have a distinctive size and shape. The stem's internal structure is distinctive also, particularly with respect to the pattern in which its vascular bundles are arranged.

A CLOSER LOOK AT LEAVES

Think about the leaves shown in Figure 22.11. Each one has a large surface area exposed to the environment and is not what you would call thickened. Its morphology affords maximum interception of sunlight (the plant's energy source) and carbon dioxide from the surrounding air (its source of carbon). If a leaf *is* thick, you can safely bet it belongs to a succulent or some other plant that lives in an arid habitat—and that it serves in water storage as well as in photosynthesis.

Similarly, the internal structure of a leaf is adapted to the plant's requirements for sunlight interception, absorption of carbon dioxide, and removal of oxygen—a by-product of photosynthesis.

Leaf Epidermis

Epidermis covers all leaf surfaces that are exposed to the surroundings (Figures 22.8 and 22.13). The cuticle covering the compact, sheetlike array of epidermal cells restricts loss of precious water. In most plants, stomata are more numerous on the lower leaf surface. Think this through, and you can deduce that stomata in the leaf of a water lily are on the upper surface, and that a submerged plant generally has no stomata at all. In arid habitats, stomata are often located in depressions in the leaf surface, together with thick-coated epidermal hairs. Their location (and the hairs) helps restrict water loss.

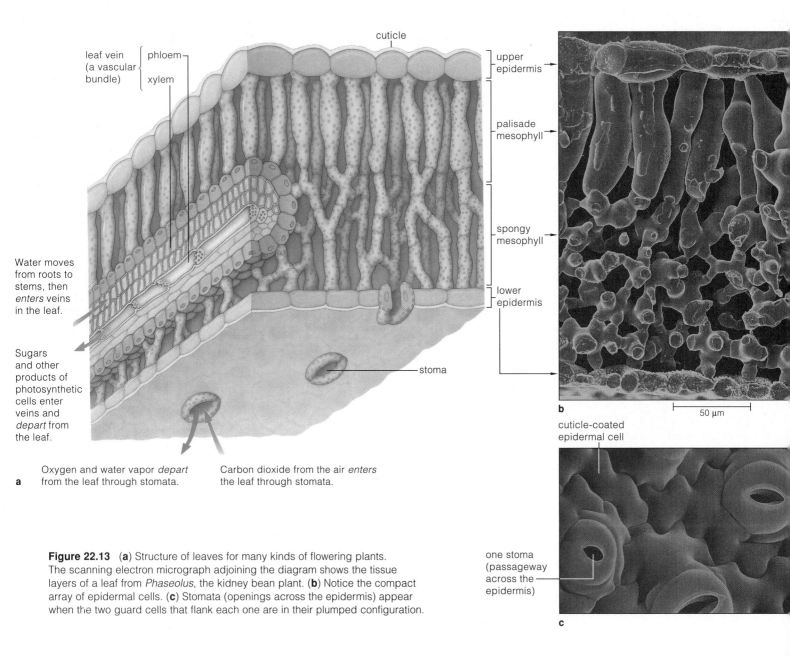

Figure 22.13 (**a**) Structure of leaves for many kinds of flowering plants. The scanning electron micrograph adjoining the diagram shows the tissue layers of a leaf from *Phaseolus*, the kidney bean plant. (**b**) Notice the compact array of epidermal cells. (**c**) Stomata (openings across the epidermis) appear when the two guard cells that flank each one are in their plumped configuration.

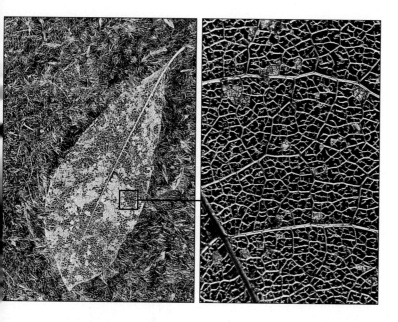

Figure 22.14 A decaying leaf, showing how veins extend throughout a leaf's photosynthetic tissue.

Mesophyll—Photosynthetic Ground Tissue

Mesophyll, a type of parenchyma with photosynthetic cells and a large volume of air spaces, extends through a leaf's interior (Figure 22.13). The air spaces, which are continuous with stomata, promote the rapid diffusion of carbon dioxide to cells and oxygen away from them. In many plants, columnar parenchymal cells attached to the upper epidermis seem densely packed, but most of their walls are exposed to air. These palisade cells have far more chloroplasts than spongy mesophyll cells below them and carry out most of the photosynthesis.

Veins—The Leaf's Vascular Bundles

A leaf's **veins** are vascular bundles, often strengthened with fibers. Their continuous strands of xylem rapidly move water and dissolved nutrients to the mesophyll cells; and continuous strands of phloem carry sugars and other substances away from them. The veins of most dicots branch lacily into minor veins that are embedded in mesophyll. In most monocots, veins are more or less similar in length and run parallel with the leaf's long axis (Figures 22.9 and 22.14).

The next section concludes our survey of leaves and invites reflection on how we use and abuse them.

A leaf's structure reflects its functions—gas exchange, nutrient delivery, and water conservation for photosynthesis inside it and the export of photosynthetic products from it.

USES AND ABUSES OF LEAVES

NICOTIANA TABACUM

Imagine a leaf-deprived grocery cart. Out go the lettuces, parsley, and spinach. No thyme, basil, and oregano for spaghetti sauce. No more mint, Darjeeling, and other teas. Beyond the table, oils from flowers scent perfumes. Eucalyptus and camphor oils find their way into medicine chests. Leaves of nightshade (including *Atropa belladonna*) provide extracts to treat Parkinson's disease. Digitalis from foxglove (*Digitalis purpurea*) leaves helps stabilize heartbeat and circulation. Juices from *Aloe vera* leaves soothe sun-damaged skin. Alkaloids from rosy periwinkle leaves are used to treat Hodgkin's disease.

Landscapers and homeowners plant shrubs next to a building to help keep it from heating up in summer and losing heat in winter. We get twine and rope from century plant leaves (*Agave*), cords and textiles from leaf fibers of Manila hemp, hats from Panamanian palm fronds, and thatched roofs from leaves of palms and grasses. Insecticides derived from Mexican cockroach plants kill fleas, lice, flies, and cockroaches. Extracts of neem tree leaves kill a variety of insects, mites, and nematodes without killing off natural predators of these common pests.

DIGITALIS PURPUREA

Through much of recorded history, people have used leaves in nonbeneficial ways. Mayans introduced early European explorers to tobacco smoking. They cultivated tobacco plants, including *Nicotiana tabacum*. Mayan priests thought that the smoke rising from pipes carried their priestly thoughts to the gods. People continue to smoke, chew, or tuck into their mouth the leaves of tobacco plants—and so become candidates for the hundreds of thousands of annual deaths from lung, mouth, and throat cancers. Heavy smoking of *Cannabis sativa*, the source of marijuana and other mind-altering products, is linked to abnormally low sperm counts. Cocaine, derived from coca leaves, is used medicinally and abused by millions of people, with devastating social and economic effects.

And what about that henbane (*Hyosyamus niger*) and belladonna? Their alkaloids have been tapped for murder as well as for medicinal purposes. Remember Hamlet's father? As he slept, he was sneakily dispatched by someone who poured a henbane solution into his ear. Remember Romeo's heartbroken, suicidal Juliet? She sipped a potion of nightshade, then dropped dead. Those self-proclaimed witches of the Middle Ages used henbane and nightshade in some of their suspect rituals. They used sticks to apply atropine-impregnated solutions to their body and so induce a sensation of weightlessness. During atropine-induced hallucinatory sprees, they assumed they were flying off to meet with demons. Hence those Halloween cartoons of witches flying on broomsticks.

ROOT PRIMARY STRUCTURE

Taproot and Fibrous Root Systems

When you think you can't bear one more gratuitously violent action-hero movie, give your mind a rest and watch a seed germinate. The first part to emerge from the seed coat is a **primary root**. In most dicot seedlings, it increases in diameter and grows downward. Later on, **lateral roots** start forming in internal tissues at an angle perpendicular to the primary root's axis, then erupt through the epidermis. The youngest lateral roots are closest to the root tip.

We call a primary root and its lateral branchings a **taproot system**. Dandelions, carrots, and oak trees are a few plants with a taproot system (Figure 22.15*a*).

What about monocots, such as rye plants and other grasses? Such plants generally have short-lived primary roots. In their place, adventitious roots arise from the young plant's stem, and then lateral roots branch from them. (*Adventitious* refers to a shoot-borne root.) All of these lateral roots are more or less similar in diameter and length. Collectively, they represent a **fibrous root system** (Figure 22.15*b*).

Internal Structure of Roots

Figure 22.16 shows the location of apical meristem in a root tip. Behind this region, cellular descendants of the apical meristem divide, enlarge, and become cells of specialized tissue systems. Notice the root cap, a dome-shaped mass of cells at the root tip. Root apical meristem produces the cap and in turn is protected by it.

A root's epidermis is the absorptive interface with soil. Some of the epidermal cells send out extensions called **root hairs** (Figure 22.16*a*). Collectively, the root

Figure 22.16 **(a)** Generalized diagram of a young primary root. **(b)** The root tip of a corn plant (*Zea mays*), longitudinal section. Individual cells of the root tip and root cap are clearly visible.

As the primary root continues to grow, it pushes the root cap forward against soil particles, and some cells tear loose. The slippery remnants lubricate the cap as the growing root advances through the soil.

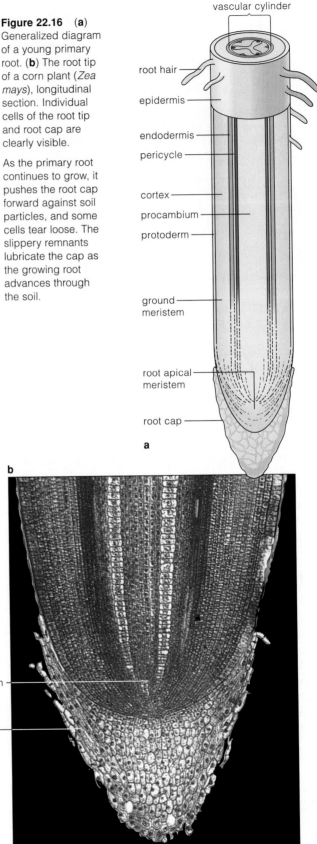

a

b

root apical system

root cap

100 µm

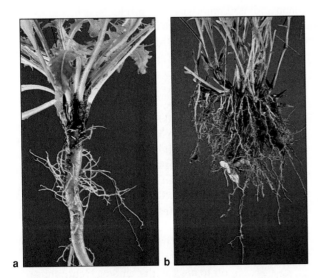

Figure 22.15 **(a)** Taproot system of a dandelion plant. **(b)** Fibrous root system of a grass plant.

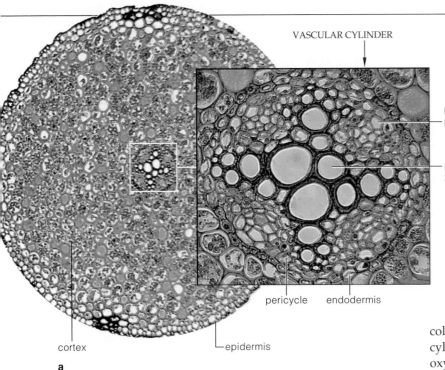

VASCULAR CYLINDER

primary
phloem

primary
xylem

pericycle   endodermis

cortex   epidermis

a

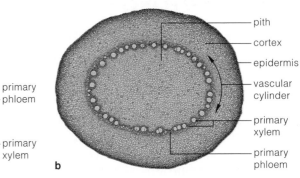

pith

cortex

epidermis

vascular
cylinder

primary
xylem

primary
phloem

b

Figure 22.17 (**a**) Young root from a buttercup (*Ranunculus*), transverse section. The close-up shows details of its vascular cylinder. (**b**) Root of a corn plant (*Z. mays*), transverse section. Notice how the vascular cylinder divides the ground tissue into cortex and pith.

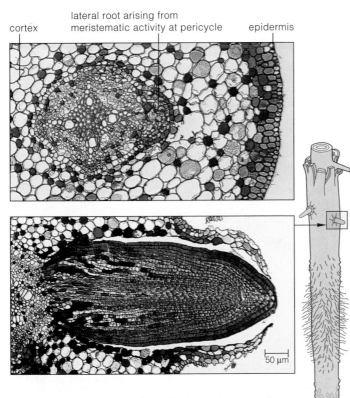

cortex

lateral root arising from
meristematic activity at pericycle

epidermis

50 µm

Figure 22.18 Lateral root formation in a primary root from a willow tree (*Salix*), transverse section.

hairs greatly increase the surface area available for taking up water and nutrients. Only a foolish or first-time gardener would yank a plant from the ground when transplanting it. He or she would tear off too much of the fragile absorptive surface.

Vascular tissues form a **vascular cylinder**, a central column inside the root. Ground tissues surrounding the cylinder are the root cortex (Figure 22.17). Dissolved oxygen easily diffuses through numerous spaces in the cortex. Like other cells in the plant, the living root cells take up oxygen and use it in aerobic respiration.

Water entering the root moves from cell to cell until it reaches the innermost part of the root cortex. This is the endodermis, a sheetlike layer, one cell thick, around the vascular cylinder. Endodermal cells have abutting, waterproof walls, so incoming water is forced to pass through their cytoplasm. As described in Chapter 23, this arrangement helps control the movement of water and dissolved substances into the vascular cylinder.

Just inside the endodermis is the pericycle. This part of the vascular cylinder has one or more layers of cells that give rise to lateral roots, which erupt through the cortex and epidermis (Figure 22.18).

Regarding the Sidewalk-Buckling, Record-Breaking Root Systems

Unless tree roots start to buckle a sidewalk or choke off a sewer line, most of us don't pay much attention to the root systems of flowering plants. Yet most manage to mine the soil to a depth of 2 to 5 meters. In hot deserts, where water is scarce, one hardy mesquite shrub is known to have sent roots down 53.4 meters (175 feet) near a streambed. Some "simple" cacti have shallow roots radiating outward for 15 meters. Someone once measured the roots of a young rye plant that had been growing for four months in only 6 liters of soil. If the surface area of that root system were laid out as a single sheet, it would occupy more than 600 square meters!

Primary growth regions of a root system provide a plant with a tremendous surface area for absorbing water and solutes.

22.7 WOODY PLANTS

Nonwoody and Woody Plants Compared

Flowering plant life cycles extend from germination to seed formation, then death. In that time, most monocots and some dicots add little or no secondary growth; they are *nonwoody* (herbaceous) plants. But many dicots (and all gymnosperms) show secondary growth during two or more growing seasons; they are *woody* plants. The ones we call **annuals** complete the life cycle in one growing season with little or no secondary growth. Corn and marigolds are like this. It takes **biennials**, such as carrots, two growing seasons to complete the life cycle. Roots, stems, and leaves form the first season; seeds and flowers form and the plant dies in the second. **Perennials** continue vegetative growth and seed formation year after year, and secondary tissues form in some. Examples are nonwoody cacti, ivy, and elms.

Secondary Growth of Stems and Roots

How do stems or roots get massive and woody? Let's focus first on an older stem. Each season, its buds give rise to primary tissues (Figure 22.19*a*). And secondary tissues originate at its **lateral meristems**, especially the vascular cambium. Fully developed vascular cambium is like a cylinder, one cell (or a few cells) thick, and its cells give rise to the secondary xylem and phloem. Xylem forms on the inner face of the vascular cambium, and phloem forms on the outer face (Figure 22.20).

Figure 22.19*b* shows the structure of a tree trunk—an older, woody stem with extensive secondary growth. Living phloem cells are restricted to a thin zone beneath the corky surface. Layers of xylem increase season after season. They usually crush the thin-walled phloem cells

Figure 22.19 (**a**) Woody twig of a walnut tree in winter, after its leaves have dropped. In spring, primary growth resumes at its buds. Secondary growth occurs in existing stem regions.

(**b**) Structure of a stem with extensive secondary growth. *Heartwood*, the mature tree's core, has no living cells. *Sapwood*, a cylindrical zone of xylem between heartwood and vascular cambium, contains some living parenchyma cells among nonliving conducting cells of xylem. Everything outside the vascular cambium is often called *bark*. Everything inside it is called *wood*. *Girdling* (stripping off the band of phloem around a tree's circumference) cuts through all vertical phloem. Photosynthetically derived food can't reach roots, which die—and so, in time, does the tree.

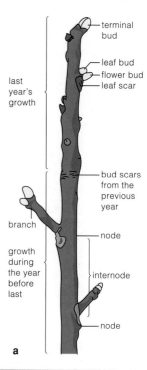

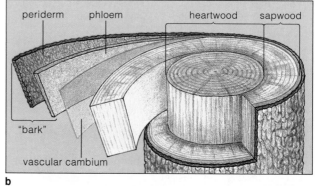

b

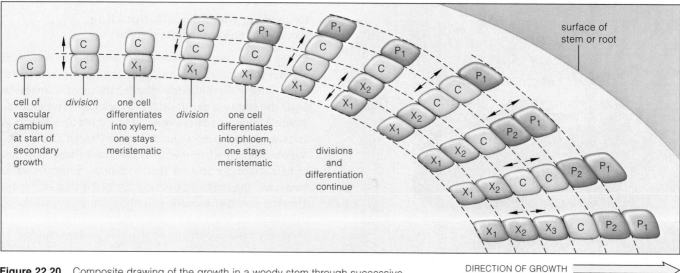

Figure 22.20 Composite drawing of the growth in a woody stem through successive seasons. Ongoing divisions displace the vascular cambium, moving its cells steadily outward as the enlarging core of xylem makes the stem or root thicker.

DIRECTION OF GROWTH ⟹

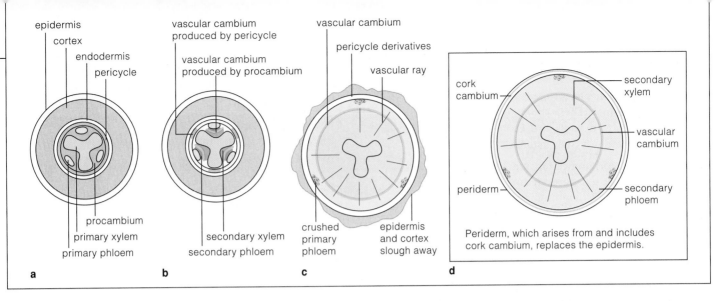

a
- epidermis
- cortex
- endodermis
- pericycle
- procambium
- primary xylem
- primary phloem

b
- vascular cambium produced by pericycle
- vascular cambium produced by procambium
- secondary xylem
- secondary phloem

c
- vascular cambium
- pericycle derivatives
- vascular ray
- crushed primary phloem
- epidermis and cortex slough away

d
- cork cambium
- secondary xylem
- vascular cambium
- periderm
- secondary phloem

Periderm, which arises from and includes cork cambium, replaces the epidermis.

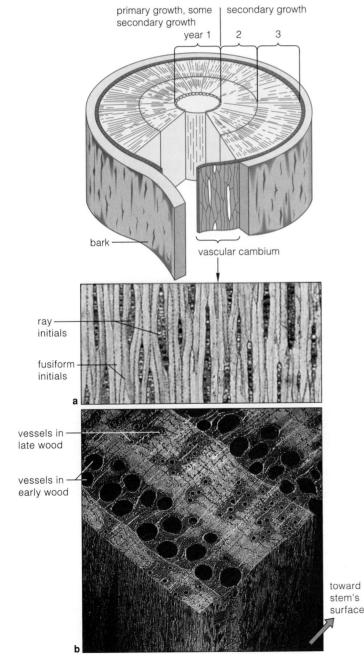

primary growth, some secondary growth | secondary growth
year 1 | 2 | 3

- bark
- vascular cambium
- ray initials
- fusiform initials
a
- vessels in late wood
- vessels in early wood
- toward stem's surface
b

Figure 22.21 Secondary growth of a root. (**a**) Tissues as primary growth ends. (**b,c**) A ring of vascular cambium forms and gives rise to secondary xylem and phloem. Cell divisions proceed parallel to the vascular cambium. The cortex ruptures as the root thickens.

Figure 22.22 *Left:* (**a**) An older stem with three annual rings. Some cells of the vascular cambium (*fusiform* initials) produce the secondary xylem and phloem that conduct water and food vertically in the stem. Other cells (*ray* initials) produce parenchyma and other cells that channel water and food laterally, and that serve as food storage centers. (**b**) Scanning electron micrograph of early and late wood in a block cut from red oak (*Quercus rubra*). Notice the different diameters of xylem's water-conducting vessels.

each year, outside the growing core of xylem. In time, pressure exerted by tissues accumulating in the stem ruptures the cortex and outer phloem. Part of the cortex and epidermis split away. But cork cambium forms and produces periderm, a corky replacement for the lost epidermis. Cork isn't the same as **bark**, which refers to all living and nonliving tissues external to the vascular cambium of woody plants. Similar events occur in roots (Figure 22.21).

Tree Rings—Annual Growth Layers

In regions having prolonged dry spells or cool winters, vascular cambium becomes inactive during parts of the year. The first xylem cells produced at the start of the growing season tend to have large diameters and thin walls. They form early wood. As the season progresses, cell diameters become smaller and their walls thicker. These cells form late wood. Those "tree rings" in a full-diameter slice from a tree trunk are alternating bands of early and late wood, which reflect light differently. The bands define the **annual growth layers** (Figure 22.22).

Woody plants add secondary growth during two or more growing seasons (nonwoody plants add little or none). Their phloem cells form outside a growing core of xylem. In time, cork cambium produces periderm in place of the epidermis.

SUMMARY

1. Monocots and dicots are two classes of flowering plants (angiosperms). Their shoots (stems, leaves, and other structures) and roots are composed of dermal, ground, and vascular tissue systems.

2. Plant growth originates at meristems, which are localized regions of self-perpetuating, embryonic cells.

   a. Primary growth (lengthening of stems and roots) originates at apical meristems in root and shoot tips.

   b. Secondary growth (increases in diameter) originates at lateral meristems (vascular cambium and cork cambium) inside stems and roots.

3. Parenchyma, sclerenchyma, and collenchyma are simple tissues, each with only one cell type (Table 22.1).

   a. Parenchyma cells, alive (metabolically active) at maturity, make up the bulk of ground tissue systems. They function in photosynthesis and other tasks.

   b. Collenchyma strengthens some growing plant parts. Sclerenchyma supports mature plant parts.

4. Complex tissues include vascular tissues (xylem and phloem) and dermal tissues (epidermis and periderm). Each consists of two or more cell types (Table 22.1).

   a. Vascular tissues distribute water and dissolved substances through the plant. Strands of conducting cells are commonly associated with sclerenchyma and parenchyma. Many of these vascular bundles thread through the ground tissues.

   b. Water-conducting cells of xylem are not alive at maturity. Their lignified, pitted walls interconnect as pipelines for water and dissolved minerals.

   c. Phloem's conducting cells are alive at maturity. The cytoplasm of adjoining cells interconnects across perforated end walls and side walls. In leaves, sugars and other photosynthetic products are loaded into these cells (often companion cells help in this) and unloaded in regions where cells are growing or storing food.

   d. Epidermis covers and protects the outer surfaces of primary plant parts. Periderm replaces epidermis on plants showing extensive secondary growth.

5. Stems function in support of upright growth and in conducting substances through the plant body by way of vascular bundles. Most monocot stems have vascular bundles distributed throughout the ground tissue. Most dicot stems have a ring of bundles that divides the ground tissue into cortex and pith.

6. Leaves contain veins and mesophyll (photosynthetic parenchyma) between the upper and lower epidermis. Air spaces around the photosynthetic cells enhance gas exchange. Water vapor and gases move across the epidermis through numerous tiny openings (stomata).

7. Roots absorb water and dissolved minerals for distribution to aboveground parts. They anchor the plant. Most store food, and some help support the shoot.

Table 22.1	Summary of Flowering Plant Tissues and Their Component Cells
Simple Tissues	
Parenchyma	Parenchyma cells
Collenchyma	Collenchyma cells
Sclerenchyma	Fibers or sclereids
Complex Tissues	
Xylem	Conducting cells (tracheids and vessel members); parenchyma cells; sclerenchyma cells
Phloem	Conducting cells (sieve-tube members); parenchyma cells; sclerenchyma cells
Epidermis	Mostly undifferentiated cells; also guard cells, other specialized cells
Periderm	Cork cells; cork cambium cells

Review Questions

1. Choose a flowering plant and list some functions of its roots and shoots. *378*

2. Name and define the basic functions of a flowering plant's three main tissue systems. *378*

3. Describe the differences between:
   a. apical and lateral meristems *379*
   b. parenchyma and sclerenchyma *380*
   c. xylem and phloem *380–381*
   d. epidermis and periderm *381*

4. Is the plant that produced this yellow flower (*Hypericum*, or St. John's wort) a dicot or monocot? Is the plant that produced this purple flower (*Iris*) a dicot or monocot? *381*

5. Which of the following stem sections is typical of most dicots? Which is typical of most monocots? Label the main tissue regions of both sections. *383*

6. Label the component parts of this three-year-old tree section. Correctly label the annual growth layers. *389*

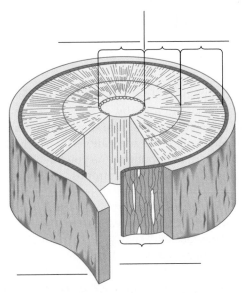

7. Identify early and late wood of this three-year-old oak stem. *389*

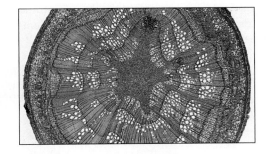

Self-Quiz *(Answers in Appendix IV)*

1. _____ and _____ are two classes of flowering plants.
   a. Angiosperms; gymnosperms   c. Shrubs; trees
   b. Monocots; dicots   d. Herbs; shrubs

2. Fleshy plant parts consist mostly of _____ cells.
   a. parenchyma   c. collenchyma
   b. sclerenchyma   d. epidermal

3. Xylem and phloem are _____ tissues.
   a. ground   c. dermal
   b. vascular   d. both b and c

4. _____ conducts water and minerals; _____ conducts food.
   a. Phloem; xylem   c. Xylem; phloem
   b. Cambium; phloem   d. Xylem; cambium

5. Roots and shoots lengthen through activity at _____ .
   a. apical meristems   c. vascular cambium
   b. lateral meristems   d. cork cambium

6. Older roots and stems thicken through activity at _____ .
   a. apical meristems   c. vascular cambium
   b. lateral meristems   d. both b and c

7. Buds give rise to _____ .
   a. leaves   c. stems
   b. flowers   d. all of the above

8. Mesophyll consists of _____ .
   a. waxes and cutin   c. photosynthetic cells
   b. lignified cell walls   d. cork but not bark

9. In early wood, cells have _____ diameters, _____ walls.
   a. small; thick   c. large; thick
   b. small; thin   d. large; thin

10. Match the terms with the appropriate plant parts.
   ____ apical meristem   a. masses of xylem
   ____ lateral meristem   b. stems, roots lengthen
   ____ xylem, phloem   c. corky covering
   ____ periderm   d. older roots, stems thicken
   ____ vascular cylinder   e. conduct water, food
   ____ wood   f. central column in roots

Critical Thinking

1. Think about the kinds of conditions that might prevail in a hot desert in New Mexico or Arizona; on the floor of a shady, moist forest in Atlanta or Oregon; and in the arctic tundra in Alaska. Also consider the kinds of animals living there. Then "design" a type of flowering plant that would be suitably adapted to each place.

2. Fitzgerald, who has an underdeveloped sense of nature, sneaks into a forest preserve at night and maliciously girdles an old-growth redwood. Then he settles down and watches the tree, waiting for it to die. But the tree does not die immediately. Explain why.

3. Sylvia lives in Santa Barbara, where droughts are common and an abundance of water is not. She replaced most of the plants in her garden with drought-tolerant species and drastically cut back the size of the lawn. The lawn doesn't get a light sprinkling every day. Sylvia only waters it twice a week in the evening, after the sun goes down. Then the lawn gets a good soak, down to a depth of several inches. Why is her strategy good for lawn grasses?

Selected Key Terms

angiosperm *378*
annual *388*
annual growth layer *389*
apical meristem *379*
bark *389*
biennial *388*
bud *382*
collenchyma *380*
cork cambium *379*
cortex *383*
cuticle *381*
deciduous plant *382*
dermal tissue system *378*
dicot *381*
epidermis *381*
evergreen plant *382*
fibrous root system *386*
ground tissue system *378*
lateral meristem *388*
lateral root *386*
leaf *382*
leaf primordium *382*
lignin *380*

meristem *379*
mesophyll *385*
monocot *381*
node *382*
parenchyma *380*
perennial *388*
periderm *381*
phloem *381*
pith *383*
primary root *386*
root *378*
root hair *386*
sclerenchyma *380*
shoot *378*
stoma (stomata) *381*
taproot system *386*
vascular bundle *383*
vascular cambium *379*
vascular cylinder *387*
vascular tissue system *378*
vein *385*
xylem *380*

Readings

Simpson, B., and M. Conner-Ogorzaly. 1995. *Economic Botany.* New York: McGraw-Hill. Fascinating book on uses and abuses of plants.

Stern, K. 1991. *Introductory Plant Biology.* Fifth edition. Dubuque, Iowa: Brown. Beautifully illustrated, accessible. Paperback.

23 PLANT NUTRITION AND TRANSPORT

Flies for Dinner

How often do we think that plants actually do anything impressive? Being mobile, intelligent, and emotional, we tend to be fascinated more with ourselves than with immobile, expressionless plants. Yet plants don't just stand around soaking up sunlight. Consider the Venus flytrap, a native of North and South Carolina. Its two-lobed, spine-fringed leaves open and close much like a steel trap (Figure 23.1a–c). This vascular plant cannot grow properly without certain mineral ions—which happen to be scarce in the soggy soils where it lives. But plenty of insects fly in from neighboring regions.

Sticky, sugary substances ooze from glands on the surface of the flytrap's two-lobed leaf. The sugary ooze entices insects to land on the leaf. As they do, they brush against tiny hairlike structures projecting from its surface—the triggers for the trap. When an insect touches two hairs at the same time, or even when it touches the same hair twice in rapid succession, the leaf snaps shut. Then digestive juices pour out from leaf cells, pool around the insect—and dissolve its minerals. In this way, the plant makes its own mineral-rich water!

The Venus flytrap is one of several kinds of plants that we call carnivorous (although it takes a flying leap of the imagination to put their extracellular digestion and absorption in the same category as the chompings of lions, dogs, and similar meat eaters). Each species of

carnivorous plant evolved in a habitat where particular minerals are hard to come by.

With this chapter we turn to some major examples of **plant physiology**—to adaptations by which plants function in their environment. As you already know, nearly all plants are photoautotrophs. Their survival depends on the uptake, distribution, and conservation of oxygen, hydrogen, and carbon, which they use in photosynthesis and aerobic respiration. Plants use water (H_2O) and gaseous oxygen (O_2) and carbon dioxide (CO_2) in the air as sources of these three elements. Their survival also depends on the uptake of at least thirteen other elements, as listed in Table 23.1. These typically are available as mineral ions dissolved in soil water. Six are **macronutrients**, which accumulate in concentrations easy to detect in tissues. Others are **micronutrients**; you can detect only traces of them in tissues.

Figure 23.1 (**a**) One of the "carnivorous" plants that turn the dinner table on animals—a Venus flytrap (*Dionaea muscipula*). (**b**,**c**) A victim of the bait-and-snare mechanism.

a

Table 23.1	**Essential Elements and Plant Function**					
Macronutrient	Some Known Functions	Some Deficiency Symptoms	**Micronutrient**	Some Known Functions	Some Deficiency Symptoms
Nitrogen	Component of proteins, nucleic acids, coenzymes, chlorophylls	Stunted growth; light-green older leaves; older leaves yellow and die (these symptoms define a condition called chlorosis)	Chlorine	Role in root and shoot growth; role in photolysis	Wilting; chlorosis; some leaves die
			Iron	Roles in chlorophyll synthesis and in electron transport	Chlorosis; yellow and green striping in grasses
Potassium	Activation of enzymes; key role in maintaining water-solute balance and so influences osmosis*	Reduced growth; curled, mottled, or spotted older leaves; burned leaf edges; weakened plant	Boron	Roles in germination, flowering, fruiting, cell division, nitrogen metabolism	Terminal buds, lateral branches die; leaves thicken, curl, and become brittle
Calcium	Regulation of many cell functions; cementing of cell walls	Leaves deformed; terminal buds die; poor root growth	Manganese	Chlorophyll synthesis; coenzyme action	Dark veins but leaves whiten and fall off
Magnesium	Component of chlorophyll; activation of enzymes	Chlorosis; drooped leaves	Zinc	Role in formation of auxin, chloroplasts, and starch; enzyme component	Chlorosis; mottled or bronzed leaves; abnormal roots
Phosphorus	Component of nucleic acids, phospholipids, ATP	Purplish veins; stunted growth; fewer seeds, fruits	Copper	Component of several enzymes	Chlorosis; dead spots in leaves; stunted growth
Sulfur	Component of most proteins, two vitamins	Light-green or yellowed leaves; reduced growth	Molybdenum	Part of enzyme used in nitrogen metabolism	Pale green, rolled or cupped leaves

*All mineral elements contribute to the water-solute balance, but potassium is notable because there is so much of it.

Yet plants, like people, don't always have unlimited supplies of required resources. For example, of every million molecules of air, only 350 are carbon dioxide. Unlike the soggy home of Venus flytraps, most soils are frequently dry. Nowhere except in an overfertilized garden does soil water contain lavish amounts of dissolved minerals. As you will see in this chapter, *many aspects of plant structure and function are responses to low environmental concentrations of essential resources.*

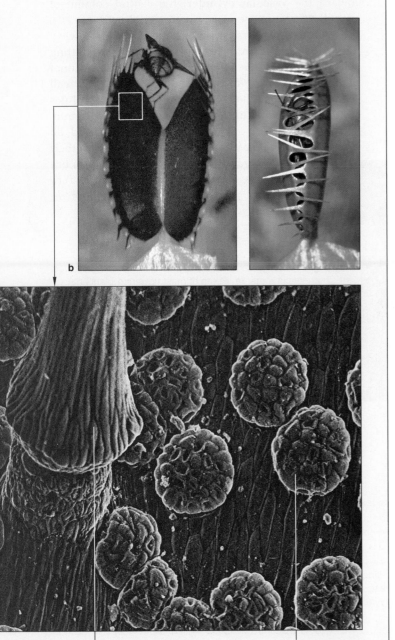

c   base of hairlike projection   secretory gland at leaf surface

KEY CONCEPTS

1. Many aspects of plant structure and function are adaptive responses to low concentrations of water, minerals, and other environmental resources.

2. A vascular system functionally connects a plant's roots, stems, and leaves. The system functions in the controlled uptake and distribution of water as well as dissolved minerals. It also functions in the distribution of sugars and other products of photosynthesis through the plant body.

3. Conservation of water, a scarce resource in most land environments, depends on the plant's cuticle and stomata. The stomata are passageways across the epidermis of leaves, mostly. They help control water loss even while allowing for gas exchange.

4. Plants lose water rapidly during the day, when their stomata remain open and so allow carbon dioxide (used in photosynthesis) to move into the leaves. Most plants conserve water by closing stomata at night.

5. Plants have mechanisms for water transport. The force of dry air causes evaporation from aboveground plant parts (transpiration). This pulls continuous columns of water molecules (which are hydrogen-bonded to one another) from roots to aboveground parts.

6. Sucrose and other organic compounds are distributed throughout the plant by translocation. In this energy-requiring process, organic compounds are loaded into conducting cells of phloem. They are unloaded at actively growing regions or storage regions of the plant body.

UPTAKE OF WATER AND NUTRIENTS AT THE ROOTS

Let's begin our tour of plant physiology with young roots that absorb water and solutes. The roots branch out through parts of the soil around them. As soil conditions change, new roots branch and replace them in different parts of the soil. It isn't that roots "explore" the soil for resources. Rather, the portions of the soil where concentrations of water and mineral ions are greater stimulate outward growth.

Absorption Routes

Recall, from the preceding chapter, that young roots absorb most water and solutes at their **root hairs**. These are slender extensions of specialized epidermal cells, and they greatly increase the surface area available for absorption (Figure 23.2). During growth, a root system may develop millions or billions of root hairs.

Water that crosses the epidermis of a root moves toward the **vascular cylinder**, the central column of vascular tissue in the root. A cylindrical layer of cells, the **endodermis**, wraps around the column. A waxy band, the **Casparian strip**, runs through the abutting cell walls in this layer (Figure 23.3). The water cannot penetrate this strip; it can only diffuse across cell wall regions that *don't* have the strip. Water and solutes reach the vascular cylinder only by moving into, through, and out of endodermal cells. And all cells, recall, have many transport proteins in their plasma membrane. These assist some solutes but not others across the membrane (Section 4.3). *They are the control points where the plant adjusts the quantity and types of solutes that are absorbed.*

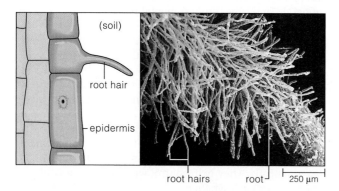

(soil)

root hair

epidermis

root hairs   root   250 μm

Figure 23.2 Sketch of a root hair (a slender extension of a specialized root epidermal cell), and a scanning electron micrograph of the root hairs of a portion of a young root.

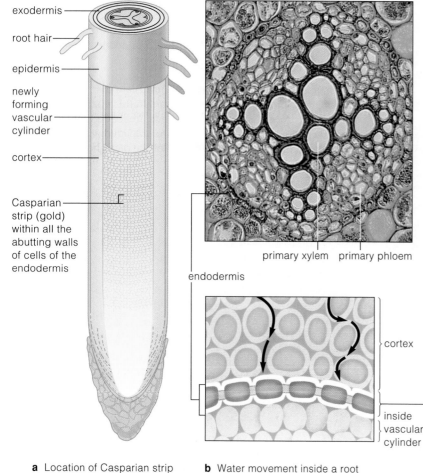

vascular cylinder (transverse section)

exodermis

root hair

epidermis

newly forming vascular cylinder

cortex

Casparian strip (gold) within all the abutting walls of cells of the endodermis

primary xylem   primary phloem

endodermis

a Location of Casparian strip   **b** Water movement inside a root

cortex

inside vascular cylinder

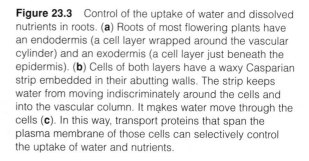

Figure 23.3 Control of the uptake of water and dissolved nutrients in roots. (**a**) Roots of most flowering plants have an endodermis (a cell layer wrapped around the vascular cylinder) and an exodermis (a cell layer just beneath the epidermis). (**b**) Cells of both layers have a waxy Casparian strip embedded in their abutting walls. The strip keeps water from moving indiscriminately around the cells and into the vascular column. It makes water move through the cells (**c**). In this way, transport proteins that span the plasma membrane of those cells can selectively control the uptake of water and nutrients.

c Water and solutes move into a root vascular cylinder only by passing through cytoplasm of endodermal cells.

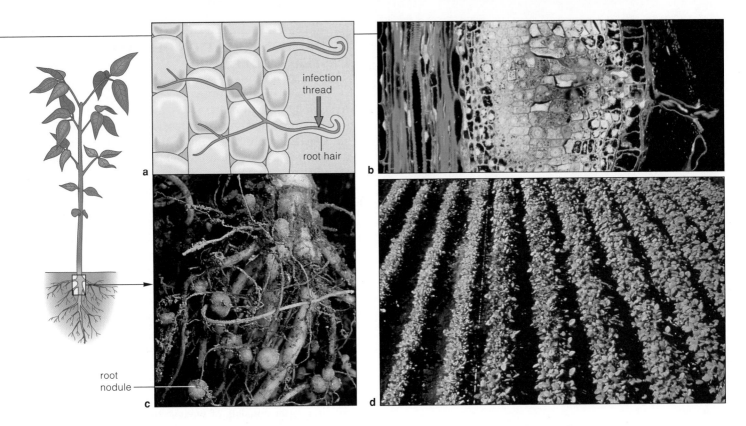

infection
thread

root hair

a

b

root
nodule

c

d

Figure 23.4 (**a**) Nutrient uptake at root nodules of legumes, which interact mutualistically with nitrogen-fixing bacteria (*Rhizobium* and *Bradyrhizobium*). When bacterial cells infect root hairs (slender extensions of root epidermal cells), the cells are induced to form an "infection thread" of cellulose deposits. The bacteria use the thread like a highway to invade cells in the root cortex. (**b,c**) Infected cells and bacteria within them divide repeatedly, forming a swollen mass that becomes a root nodule. The bacteria start fixing nitrogen when plant cell membranes surround them. The plant takes up some of the nitrogen, and the bacteria take up some photosynthetic compounds. (**d**) Soybean plants growing in nitrogen-poor soil. Plants on the right were inoculated with *Rhizobium* cells and developed root nodules.

Many flowering plants also have an **exodermis**, a layer of cells just inside their roots (Figure 23.3*a*). This layer has a Casparian strip that functions just like the one next to the root vascular cylinder.

Specialized Absorptive Structures

Certain bacteria and fungi help many plants absorb nutrients and get something in return. This two-way flow of benefits between species is a form of symbiosis called **mutualism**. Consider how nitrogen scarcity can limit the growth of crop plants. There actually is plenty of gaseous nitrogen (N≡N) in the air. But plants don't have the metabolic equipment to engage in **nitrogen fixation**. By this process, certain enzymes break apart the three covalent bonds in gaseous nitrogen and attach the nitrogen atoms to organic compounds. To get high crop yields, farmers provide crop plants with nitrogen compounds in fertilizers or they encourage the growth of nitrogen-fixing bacteria in soil. Such bacteria convert gaseous nitrogen to forms they—and plants—can use.

String beans, peas, alfalfa, clover, as well as other legumes have an advantage in this respect. Nitrogen-fixing bacteria live in their roots, in localized swellings called **root nodules** (Figure 23.4*b*). These symbiotic bacteria withdraw some organic compounds from the plant tissues, but they provide the plants with some of their fixed nitrogen in return.

Or consider **mycorrhizae** (singular, mycorrhiza). As described in Section 19.11, a mycorrhiza is a symbiotic interaction between a young root and a fungus (hence the name, which means "fungus-root"). In some interactions, fungal filaments (hyphae) form a velvety covering around the roots. Collectively, hyphae have a large surface area for absorbing minerals from a large volume of soil. The fungus gets sugars and nitrogen-containing compounds from the roots. The roots get some scarce minerals that the fungus is better able to absorb.

Plants control nutrient uptake and distribution at a cell layer near the root surface, at another layer around the root vascular cylinder, then at the plasma membranes of living cells throughout the plant body.

Root hairs, root nodules, and mycorrhizae enhance the uptake of water and scarce mineral ions at the root.

Focus on the Environment

PUTTING DOWN ROOTS

All around you, vast absorptive, anchoring networks of roots are mining the good dirt—that is, *topsoil*, a complex mixture of mineral ions and organic remains as well as sand, silt, clay, pebbles, and rocks. In deserts and some other places, good soil is only a few centimeters deep. In some grasslands, plants are lucky; the soil there may be hundreds of meters deep.

Roots do more than absorb nutrients. Collectively, they influence soil stability and nutrient storage in ecosystems. Consider the roots of trees in a young, undisturbed forest in New Hampshire. The calcium and other minerals they absorb get stored in plant tissues instead of being leached from soil during the spring rains. Researchers cleared all plants from part of the forest. Before then, the annual loss of calcium was 8 kilograms for each hectare (2.47 acres). After deforestation, the loss was six times higher.

On steep hills, even on level farmland, erosion can be severe. Soil erosion is a natural process but natural vegetation often limits its effects (Figure 23.5).

a

b

Figure 23.5 (a) Middle fork of the Salmon River in Idaho. The clear water (*right*) comes from a wilderness area. The brown-colored water (*left*) is enriched with silt from a wilderness area disturbed by cattle ranching and other human activities. The silt is affecting water quality and fish habitats. (b) In an agricultural field, gullies formed by the erosive force of water. Their very formation channels runoff from the land. As gullies grow deeper and wider, erosion becomes more and more rapid. Soil nutrients are depleted, productivity declines, and fertilizers are brought in—all for the want of plants and their roots.

23.3 CONSERVATION OF WATER IN STEMS AND LEAVES

At least 90 percent of the water that enters a leaf goes right on through it and evaporates into the surrounding air. Cells use only 2 percent of the water that stays in the leaf for photosynthesis, membrane functions, and other events. That tiny amount is vital. Plants wilt and water-dependent events are severely disrupted when water loss exceeds water uptake at roots for extended periods (Figure 4.7). Yet land plants are not entirely at the mercy of changes in the availability of soil water. They have a cuticle, and they have stomata.

The Water-Conserving Cuticle

Even mildly water-stressed plants would rapidly wilt and die without their **cuticle** (Figure 23.6). Epidermal cells secrete this translucent, water-impermeable layer, which coats their wall regions exposed to the air.

At the cuticle surface are deposits of waxes, which are water-insoluble lipids with long fatty-acid tails. The cuticle proper consists of waxes embedded in a matrix of **cutin**, an insoluble lipid polymer. Beneath the waxes, cellulose threads through the matrix. Often a layer of pectins (a type of polysaccharide) helps bind the cuticle to cell walls.

A cuticle does not bar the passage of light rays into photosynthetic parts of a plant. It does restrict water loss. It also restricts the *inward* diffusion of the carbon dioxide required for photosynthesis and the *outward* diffusion of oxygen by-products. (Recall, from Section 5.6, that a buildup of oxygen in the air spaces inside a leaf has bad effects on the rate of photosynthesis.)

Controlled Water Loss at Stomata

The cuticle-covered epidermis conserves water—but how do carbon dioxide and oxygen get past it? They do so at the tiny openings called **stomata** (singular, stoma). Water evaporates from the plant and carbon dioxide moves in mainly at these openings. A pair of highly specialized cells, the **guard cells**, define each opening (Figure 23.7). When the two cells swell with water, the force of internal fluid pressure distorts their walls. The cells bend under the pressure, and so a gap (the stoma) forms between them. When the two cells lose water and internal fluid pressure drops inside them, they collapse against each other and so close the gap.

In most plants, stomata stay open during the day, when photosynthesis takes place. They do lose water—but they gain carbon dioxide for the reactions. Stomata stay closed at night. Then, plants conserve water—and precious carbon dioxide accumulates inside the leaf as cells busily engage in aerobic respiration.

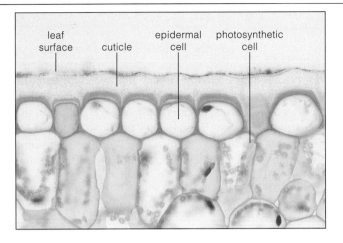

leaf surface — cuticle — epidermal cell — photosynthetic cell

Figure 23.6 Example of a waxy cuticle on the upper epidermis of a leaf. Water, carbon dioxide, and oxygen cannot cross the cuticle proper. They enter or depart from the leaf mainly at stomata.

guard cell   stoma   guard cell

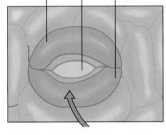

Stomatal opening: Water enters collapsed guard cells, which swell (under turgor pressure) and move apart, forming the stoma.

Stomatal closing: Water leaves swollen guard cells, which collapse against each other and thereby close the stoma.

Figure 23.7 Stomatal action. A stoma remains open or closed, depending on changes in the shape of the paired guard cells that define the opening across the cuticle-covered epidermis.

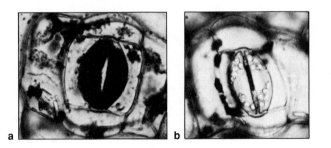

a   b

Figure 23.8 Experimental evidence for the accumulation of potassium in guard cells that are expanding with incoming water. Strips from the leaf epidermis of a dayflower (*Commelina communis*) were immersed in solutions containing dark-staining substances that bind preferentially with potassium ions. (**a**) In leaf samples having opened stomata, most of the potassium was concentrated in the guard cells. (**b**) In leaf samples having closed stomata, very little potassium was in guard cells; most of the potassium was present in normal epidermal cells.

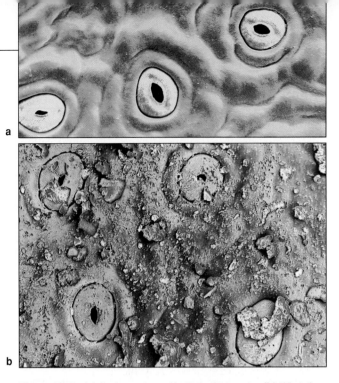

a

b

Figure 23.9 (**a**) Surface view of holly leaf stomata. (**b**) What the leaf looks like when a holly plant grows in industrialized regions. Gritty airborne pollutants clog its stomata and prevent much of the sun's rays from reaching photosynthetic cells inside the leaf.

Whether a stoma opens or closes depends on how much water and carbon dioxide are in the two guard cells. Photosynthesis starts when the sun comes up. As the morning progresses, carbon dioxide levels drop in cells—including guard cells. The decrease helps trigger active transport of potassium ions into the guard cells (Figure 23.8). So do blue wavelengths of light, which penetrate the atmosphere better as the sun arcs higher in the sky. Water follows potassium into the cells, by osmosis. The inward movement of water provides the fluid pressure to open the stoma.

Carbon dioxide levels rise when the sun goes down and photosynthesis stops. Potassium, and then water, moves out of the guard cells—and the stoma closes.

CAM plants, which include most cacti, conserve water in a different way. They open stomata at night, when they fix carbon dioxide by the special metabolic C4 pathway described in Section 5.6. The following day, when their stomata close, CAM plants use carbon dioxide in photosynthesis.

As this section makes clear, plant survival depends on stomata. Think about this when you are out and about on smog-shrouded days (Figure 23.9).

Transpiration and gas exchange take place mainly at stomata. These numerous small openings span the waxy cuticle, which covers all plant epidermal surfaces exposed to air.

Plants open and close stomata at different times to control water loss, carbon dioxide uptake, and oxygen disposal.

Transpiration Defined

By now, you have a sense of how the distribution of water and dissolved mineral ions to all living cells is central to plant growth and functioning. Let's turn now to the prevailing model of how water actually moves from a plant's roots to its stems, then into leaves.

Recall that plants use only a fraction of the water they absorb for growth and metabolism; most of the water is lost, primarily through the numerous stomata in leaves. The evaporation of water from leaves as well as from stems and other plant parts is a process called **transpiration**.

Cohesion-Tension Theory of Water Transport

This brings up an interesting question. Assuming that plants lose most of the absorbed water from leaves, how does the water actually get *to* the leaves? What gets it to the top of leafy plants, including redwoods and other trees that may grow a hundred meters tall?

In a plant's vascular tissues, water moves through the pipelines called **xylem**. Recall, from Chapter 22, that water-conducting cells of xylem are called **tracheids** and **vessel members**. Figure 23.10 shows examples. These cells are dead at maturity, and only their lignin-reinforced walls remain. Therefore, xylem's conducting cells cannot be actively pulling the water "uphill."

Some time ago, the botanist Henry Dixon came up with a good way to explain water transport in plants. By his **cohesion-tension theory**, the water in xylem is being pulled upward by the drying power of air, which creates continuous negative pressures (tensions) that extend downward from leaves to roots. Figure 23.11 illustrates this theory. Think about the following points of Dixon's explanation as you review this illustration.

1. The drying power of air causes transpiration—the evaporation of water from plant parts exposed to air. Transpiration puts the water confined in the conducting tubes of xylem in a state of tension. The tension extends from veins inside leaves, down through the stems, and on into the young roots where water is being absorbed.

2. Unbroken, fluid columns of water show *cohesion*; they resist rupturing as they are pulled upward under *tension* (compare Section 2.5). The collective strength of hydrogen bonds between water molecules, which are confined in the narrow, tubular xylem cells, imparts this cohesion.

3. As long as water molecules continue to escape from the plant, the continuous tension in the xylem permits more molecules to be pulled upward from the roots to replace them.

Hydrogen bonds are strong enough to hold water molecules together inside the xylem. But they are not strong enough to prevent the molecules from breaking

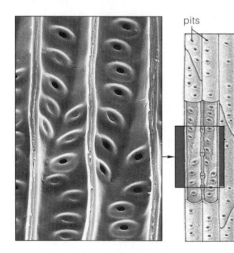

a Tracheids, with tapering, unperforated end walls. Small pits in the walls of adjoining tracheids match up. Although the pits are too small to permit rapid flow, they confine air bubbles (which obstruct water transport) to individual tracheids.

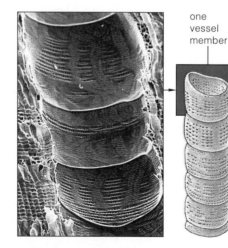

b Part of one type of vessel, with thick-walled cells (vessel members) that interconnect as a water-conducting tube.

c A vessel member with end wall perforations. These permit water (and air bubbles) to flow unimpeded through the vessel. This may be why natural selection has favored retention of tracheids and vessels in the same plant.

Figure 23.10 Examples of tracheids and vessel members. These cells are dead at maturity. Their walls remain interconnected and form the water-conducting tubes of xylem. Tracheids probably evolved before vessel members, but both are present in nearly all flowering plants.

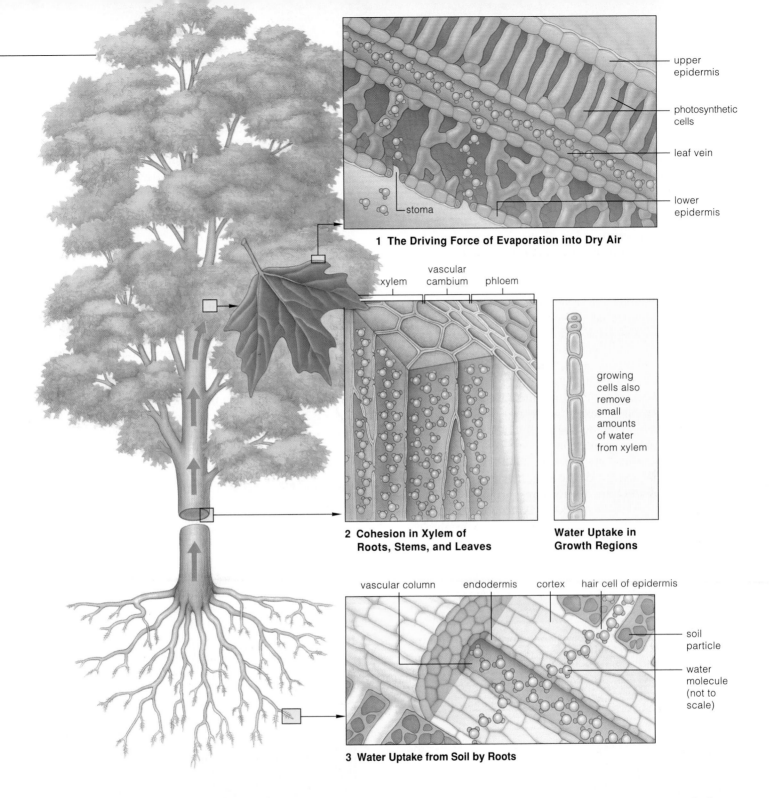

1 The Driving Force of Evaporation into Dry Air

upper epidermis
photosynthetic cells
leaf vein
lower epidermis
stoma

xylem
vascular cambium
phloem

2 Cohesion in Xylem of Roots, Stems, and Leaves

growing cells also remove small amounts of water from xylem

Water Uptake in Growth Regions

vascular column
endodermis
cortex
hair cell of epidermis
soil particle
water molecule (not to scale)

3 Water Uptake from Soil by Roots

Figure 23.11 The cohesion-tension theory of water transport.

Point 1: Transpiration (the evaporation of water from aboveground plant parts) puts the water in xylem in a state of tension that extends from leaves to roots.

Point 2: The collective strength of hydrogen bonds among water molecules, which are confined in narrow water-conducting tubes in xylem, impart cohesion to the water. Hence the narrow columns of water in xylem resist rupturing under the continuous tension.

Point 3: As long as water molecules escape by transpiration, that tension drives the uptake of replacement water molecules.

away from each other during transpiration and then escaping from leaves, through the stomata.

Transpiration is the process of evaporation from plant parts.

By the cohesion-tension theory of water transport, this process is the key source of tensions in water inside the xylem. The tensions extend from leaves to roots, and they allow columns of water molecules that are hydrogen-bonded to one another to be pulled upward through the plant body.

Whereas xylem distributes water and minerals through the plant, the vascular tissue called **phloem** distributes organic products of photosynthesis.

Like xylem, phloem consists of many conducting tubes, fibers, and strands of parenchyma cells. Unlike xylem, its conducting tubes are **sieve tubes**, which are constructed of living cells positioned side by side and end to end. The cells are sieve-tube members of the sort illustrated in Figure 23.12. Dissolved sugars and other organic compounds flow rapidly through large pores in their abutting end walls. Adjoining the sieve tubes are **companion cells**, a type of specialized parenchyma cell. As you will read shortly, companion cells help load the organic compounds into sieve tubes.

To understand what goes on in phloem, think about what happens to sucrose and other organic products of photosynthesis. Leaf cells use some of these products. The rest travel to roots, stems, buds, flowers, and fruits. In most cells in the plant, carbohydrates become stored as starch, in plastids. Proteins and fats, synthesized from the distributed amino acids and carbohydrates, become stored in many seeds. Avocados and some other fruits also accumulate fats.

Starch molecules are too large for transport across the plasma membrane of the cells that produce them. Also, they are too insoluble for transport to other plant parts. Proteins are too large and fats are too insoluble overall for transport from storage sites. *But cells can convert the storage forms of organic compounds to solutes of smaller size that are more easily transported through the phloem.* For instance, cells degrade starch to glucose units, which combine with fructose to form sucrose—a transportable sugar.

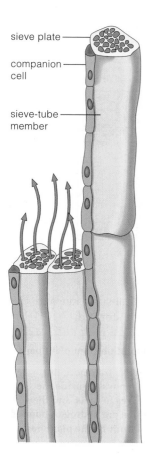

sieve plate

companion cell

sieve-tube member

Figure 23.12 Example of the cells (sieve-tube members) that interconnect to form a sieve tube—a conducting tube in phloem. The companion cells expend ATP energy to load sugars and other organic compounds into sieve-tube members adjacent to them.

Experiments with insects called aphids revealed that sucrose is the main carbohydrate transported in phloem. The experimenters exposed aphids feeding on the juices in the conducting tubes of phloem to high levels of carbon dioxide to anesthetize them. Then they severed the aphid body from the mouthparts, which were left embedded in the conducting tubes. For most of the plants studied, sucrose was the most abundant carbohydrate in the fluid being forced out of the tubes.

Translocation

The transport of sucrose and other organic compounds through the phloem of a vascular plant is known as **translocation**. Apparently, high pressure drives this process. Often the pressure is five times as high as it is in automobile tires. Aphids demonstrate this when they force their mouthpart into sieve tubes and feed on dissolved sugars. High pressure can force fluid through an aphid gut and out the other end as "honeydew" (Figure 23.13). Park a car under trees being attacked by aphids, and it may get spattered by sticky honeydew droplets—thanks to the high fluid pressure in phloem.

Pressure Flow Theory

The organic compounds in phloem are translocated along gradients of decreasing solute concentration and pressure. The flow's *source* is any region where organic compounds are being loaded into the sieve tubes. For example, photosynthetic tissues in leaves (mesophylls) are common sources. The flow ends at a *sink*, which is any plant region where organic compounds are being unloaded and used or stored. Developing flowers and fruits are common sink regions.

Why do organic compounds flow from a source to a sink? According to the **pressure flow theory**, pressure builds at the source end of a sieve-tube system and it *pushes* the solute-rich solution toward any sink, where

Figure 23.13 A honeydew droplet exuding from an aphid. The insect's tubular mouthpart penetrated a conducting tube in phloem. The sugary fluid was under such high pressure in phloem, it was forced on through the aphid gut.

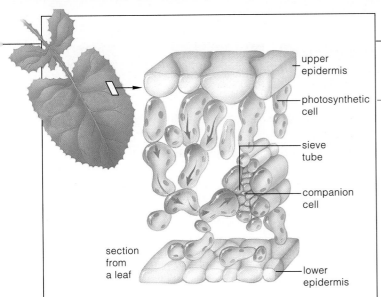

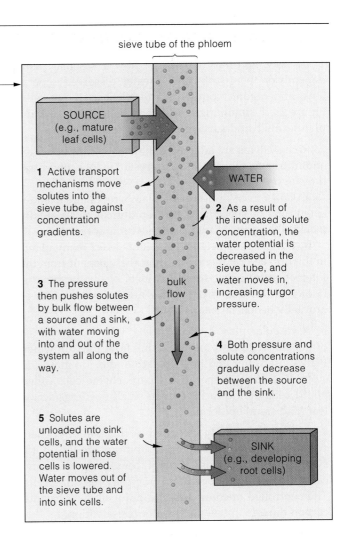

a *Loading at a source.* Photosynthetic cells in leaves are a common source of organic compounds that must be distributed through a plant. Small, soluble forms of these compounds move from the cells into phloem (a leaf vein).

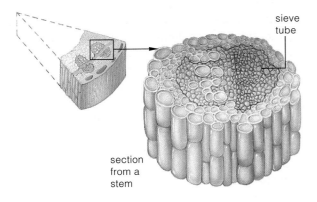

b *Translocation along a distribution path.* Fluid pressure is greatest inside sieve tubes at the source. It pushes the solute-rich fluid to a sink, which is any region where cells are growing or storing food. There, the pressure is lower because cells are withdrawing solutes from the tubes.

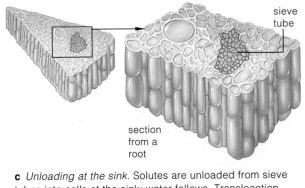

c *Unloading at the sink.* Solutes are unloaded from sieve tubes into cells at the sink; water follows. Translocation continues as long as solute concentration gradients and a pressure gradient exist between the source and the sink.

Figure 23.14 Translocation of organic compounds in sow thistle (*Sonchus*). Compare Figure 5.2 to see how translocation relates to photosynthesis in the same plant.

solutes are removed. Experimental evidence for the pressure flow theory was obtained from studies of sow thistle (*Sonchus*). You can use Figure 23.14 to track what happens after sucrose moves from photosynthetic cells into small veins in the leaves. Companion cells in the veins spend energy to load sucrose into adjacent sieve-tube members. As the sucrose concentration increases in the tubes, water also moves in, by osmosis. More internal fluid pressure is exerted against the tube walls. As the pressure increases, it pushes the sucrose-laden fluid out of the leaf, into the stem, and on to the sink.

Plants store organic compounds in the form of starch, fats, and proteins. They convert the storage forms to sucrose and other small units that are soluble and easily translocated.

Translocation (the distribution of organic compounds to different plant regions) depends on concentration and pressure gradients in the sieve-tube system of phloem.

Gradients exist as long as companion cells load compounds into the sieve tubes at sources (such as mature leaves) and as long as compounds are removed at sinks (such as roots).

SUMMARY

1. By now, you have a sense of how a vascular plant depends on the distribution of water, dissolved mineral ions, and organic compounds to all of its living cells. Figure 23.15 summarizes how that distribution sustains plant growth and survival.

2. Root systems are adapted to efficiently take up water and nutrients, which often are present in low concentrations in the surrounding soil.

   a. Collectively, a flowering plant's root hairs (slender extensions of specialized root epidermal cells) greatly increase the surface area available for absorption.

   b. Fungal and bacterial symbionts assist many plants in the uptake of mineral ions, and these benefit from the interaction by using some products of photosynthesis. Root nodules and mycorrhizae are examples of these mutually beneficial interactions.

3. A vascular plant's waxy cuticle, which covers most of its aboveground parts, is impermeable to water. Plants lose water and take up carbon dioxide mostly at stomata. A pair of specialized parenchyma cells called guard cells define each of these small openings across leaf (and stem) epidermis.

4. Stomata commonly are open during the day. Then, plants lose water but take in carbon dioxide for photosynthesis. Stomata close at night, conserving water and carbon dioxide produced by aerobically respiring cells. The controlled opening and closing of stomata balance carbon dioxide uptake with water conservation.

5. Plants distribute water and dissolved mineral ions through water-conducting tubes of the vascular tissue called xylem. The cells that form the tubes, which are called tracheids and vessel members, are dead at maturity. Only their walls remain and interconnect to form narrow, continuous pipelines.

6. Plants lose water through transpiration (evaporation of water from leaves and other parts exposed to air).

7. These are the points of the cohesion-tension theory of water transport in xylem:

   a. Inside xylem's water-conducting cells, continuous negative pressures (tensions) extend from the leaves to the roots. Transpiration causes the tension.

   b. When water molecules escape from the leaves, replacements are pulled into the leaf under tension.

   c. The collective strength of numerous hydrogen bonds between water molecules imparts cohesion that allows the water to be pulled upward as continuous fluid columns. In this case, cohesion refers to the capacity of hydrogen bonds to resist rupturing when placed under tension.

8. Plants distribute organic compounds through sieve tubes of phloem, a vascular tissue. We call the distribution of organic compounds in phloem translocation.

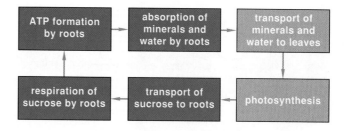

Figure 23.15 Summary of the interdependent processes that sustain the growth of vascular plants. All the living cells in plants require oxygen, carbon, hydrogen, and thirteen essential mineral ions. The cells expend ATP energy and actively take up substances at membrane proteins. In photosynthetic cells, ATP is produced during photosynthesis and aerobic respiration. In other cells, ATP forms mostly by aerobic respiration.

9. According to the pressure flow theory, translocation is driven by differences in solute concentrations and pressure between a source region and a sink region.

   a. A source is any site where organic compounds are loaded into sieve tubes. Mature leaves are examples.

   b. A sink is any site where organic compounds are unloaded from sieve tubes. Roots are examples.

   c. The solute concentration gradients and pressure gradients exist as long as companion cells expend energy to load solutes into the sieve tubes and as long as solutes are removed at the sink.

Review Questions

1. Refer to Table 23.1. What are some signs that a plant suffers a deficiency in certain mineral elements? *392*

2. What is the function of Casparian strips in the walls of root endodermal cells? *394–395*

3. Your biology instructor shows you a micrograph of the surface of a beavertail cactus (*Opuntia*). The micrograph, reproduced here as Figure 23.16, shows two cells of the epidermis that obviously are paired. Label each cell and the small gap between them. *396–397*

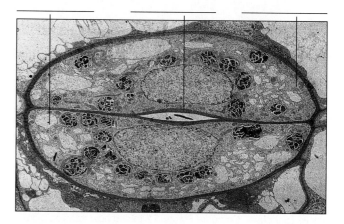

Figure 23.16 Part of the surface of a beavertail cactus stem.

4. Using botanist Henry Dixon's explanation, explain how water moves from the soil upward through leafy plants. In some cases, water moves to tree tops that are 100 meters tall. *398–399*

5. Explain the process of translocation in terms of the pressure flow theory. *400–401*

Self-Quiz *(Answers in Appendix IV)*

1. Water can be pulled up through a plant due to the cumulative strength of _____ between water molecules.

2. In plants, mineral ions _____ .
   a. have roles in metabolism
   b. help establish gradients across membranes
   c. influence water movement into cells
   d. maintain cell shape and growth
   e. all of the above

3. The nutrition of some plants depends on a root-fungus association known as _____ .
   a. root nodules   c. root hairs
   b. mycorrhizae   d. root hyphae

4. The nutrition of some plants depends on a root-bacterium association known as _____ .
   a. root nodules   c. root hairs
   b. mycorrhizae   d. root hyphae

5. Water evaporation from plant parts is called _____ .
   a. translocation   c. transpiration
   b. expiration   d. tension

6. Upward water transport to the tops of tall trees is explained by the _____ .
   a. pressure flow theory
   b. differences in source and sink solute concentrations
   c. pumping force of xylem vessels
   d. cohesion-tension theory

7. Water entering roots is forced to diffuse through the cytoplasm of endodermal cells because of _____ in their walls.
   a. cutin strips   c. Casparian strips
   b. lignin strips   d. cellulose strips

8. During the day, most plants lose _____ and take up _____ .
   a. carbon dioxide; water   c. oxygen; water
   b. water; oxygen   d. water; carbon dioxide

9. At night, plants conserve _____ and _____ accumulates.
   a. carbon dioxide; oxygen   c. oxygen; water
   b. water; oxygen   d. water; carbon dioxide

10. In phloem, organic compounds flow through _____ .
    a. collenchyma cells   c. vessels
    b. sieve tubes   d. tracheids

11. Match the concepts of plant nutrition and transport.
    _____ stomata   a. evaporation from plant parts
    _____ sink   b. balancing water loss with
    _____ root system   carbon dioxide requirements
    _____ hydrogen bonds   c. cohesion in water transport
    _____ transpiration   d. sugars unloaded from sieve tubes
    _____ translocation   e. distribution of organic
         compounds through the plant
         f. response to scarce soil nutrients

Critical Thinking

1. Home gardeners, like farmers, must make sure that their plants have access to nitrogen (from nitrogen-fixing bacteria or fertilizer). Insufficient nitrogen stunts plant growth; leaves turn yellow and die. What kinds of major biological molecules cannot be synthesized in nitrogen-deprived plants? How would this deficiency affect biosynthesis and thereby cause the observable symptoms?

2. When moving a plant from one place to another, it helps to include some native soil around the roots. Explain why, given what you know about mycorrhizae and root hairs.

3. Henry discovered a way to keep all the stomata of a plant open at all times. He also figured out how to keep all the stomata of another plant closed all the time. Both plants died. Explain why.

4. Allen is studying the rate of transpiration from leaves of tomato plants. He notices that several environmental factors, including wind and relative humidity, affect the rate. Explain why.

5. You have just returned home from a three-day vacation. By the obvious wilting, you realize you forgot to water your houseplants before you left. Being aware of the cohesion-tension theory of water transport, explain what happened to them.

6. Not having a green thumb, your friend Stephanie decides to grow zucchini plants, which are most forgiving of poor soils and amateur gardeners. She plants too many seeds and ends up with far too many plants. Among them you happen to notice a stunted plant and decide to find out what happened to it. After many experiments, you determine its leaves are producing a mutated, malfunctioning form of an enzyme that is necessary for sucrose formation. Knowing what you do about the pressure flow theory, explain why plant growth was hindered.

Selected Key Terms

CAM plant *397*	nitrogen fixation *395*
Casparian strip *394*	phloem *400*
cohesion-tension theory *398*	plant physiology *392*
companion cell *400*	pressure flow theory *400*
cuticle *396*	root hair *394*
cutin *396*	root nodule *395*
endodermis *394*	sieve tube *400*
exodermis *395*	stoma (stomata) *396*
guard cell *396*	tracheid *398*
macronutrient *392*	translocation *400*
micronutrient *392*	transpiration *398*
mutualism *395*	vascular cylinder *394*
mycorrhiza (mycorrhizae) *395*	vessel member *398*
	xylem *398*

Readings

Galston, A., P. Davies, and R. Satter. 1980. *The Life of a Green Plant.* Englewood Cliffs, New Jersey: Prentice-Hall.

Hewitt, E., and T. Smith. 1975. *Plant Mineral Nutrition.* New York: Wiley.

Salisbury, F., and C. Ross. 1992. *Plant Physiology.* Fourth edition. Belmont, California: Wadsworth.

24 PLANT REPRODUCTION AND DEVELOPMENT

A Coevolutionary Tale

Flowering plants thrive almost everywhere, from icy tundra to deserts to oceanic islands. What accounts for their distribution and diversity? *Consider the flower.* About 435 million years ago, when plants invaded the land, insects that fed on decaying plants and spores were probably right behind them. The smorgasbord of aboveground plant parts seemed to have favored natural selection of winged insects with an amazing array of sucking, piercing, and chewing mouthparts.

By 390 million years ago, the first seed-bearing plants were flourishing in humid coastal forests. Many had conelike structures where pollen grains and ovules formed. (When flowering plants reproduce, recall, sperm form inside pollen grains, then travel to the eggs, which form in ovules.) At first, pollen—which is rich in nutrients—simply may have drifted on air currents to the ovules. At some point, though, insects made the connection between "cone" and "food source." Plants lost some pollen to them but gained a reproductive advantage when pollen-dusted insects brushed against ovules. Why? Insects clambering over cones made more accurate deliveries of pollen grains than air currents alone. The tastier the pollen, the more home deliveries, and the more seeds formed. *And the greater the number of seeds, the greater the chance of reproductive success.*

What we are describing here is **coevolution**. The word refers to two or more species jointly evolving as an outcome of close ecological interactions. When one species evolves, the change affects selection pressures operating between the two, so the other also evolves. In our coevolutionary tale, plant structures that proved more attractive to pollen-delivering insects were favored. Insects quick to recognize and locate particular plants had a fine competitive edge. Hence the evolution of plants with distinctive flowers and fragrances—and of pollinators adapted to them (Figure 24.1*a* and *b*).

A **pollinator** is any agent that can transfer pollen grains to a receptive "landing platform" on a flower. Besides insects, pollinating agents include air currents, water currents, birds, bats, and various other animals.

Today you can correlate many floral features with specific pollinators. For example, flowers pollinated by beetles and flies smell like decaying meat, moist dung, or decaying litter in forests, where beetles first evolved.

Beetles don't respond to red colors, and they can drown in the nectar of flowers with large, deep tubes. Birds respond to flowers with red as well as yellow components. Red petals of the flower in Figure 24.1*b* form a tube for a tempting, sugar-rich fluid—nectar. Certain birds have a bill as long as the floral tube. This flower's reproductive parts are located where birds

Figure 24.1 (**a**) The giant saguaro of Arizona's Sonoran Desert produces large, showy white flowers at the tips of its spiny arms. Insects visit the flowers by day, and bats visit by night. The plant offers the animals nectar; and the animals transport pollen grains that stick to their body from one cactus plant to another. (**b**) Bahama woodstar sipping nectar from a hibiscus blossom. Like other hummingbirds, it forages for nectar in midflight. Its long, narrow bill coevolved with long, narrow floral tubes.

a

Figure 24.2 Shine ultraviolet light on a gold-petaled marsh marigold to reveal its bee-attracting pattern.

brush against them. Birds visiting nectar cups of many plants of this species promote cross-pollination. Bird-pollinated plants don't divert resources to producing fragrances; birds have a poor sense of smell.

Daisies and other fragrant flowers with distinctive patterns, shapes, and red or orange components attract butterflies, which forage by day. Nectar-sipping bats and most moths forage by night. They pollinate strong, sweet-smelling flowers with white or pale-colored petals, which are more visible in the dark. Moths as well as butterflies have long, narrow mouthparts, corresponding to narrow floral tubes or floral spurs. The mouthpart of a Madagascar hawkmoth uncoils 22 centimeters—the same length as the floral tube of an orchid (*Angraecum sesquipedale*)! Hawkmoths hover near the floral tubes as they draw nectar from them.

Flowers with strong, sweet odors and bright yellow, blue, and purple parts commonly attract bees. Some pigments in the flowers that absorb ultraviolet light are distributed in ways that reflect it in specific patterns. Unlike us, bees can see ultraviolet light—and they find the patterns attractive (Figure 24.2).

In short, as decorative as we find flowers to be, their structural parts are adaptations to the environment—and to other organisms that help their chance of reproductive success. This chapter focuses on their modes of reproduction, their growth, and their development.

b

KEY CONCEPTS

1. Sexual reproduction is the premier reproductive mode of flowering plant life cycles. Strawberry plants and many other species also can reproduce asexually.

2. In flowering plants, sexual reproduction involves the formation of spores and gametes, both of which develop inside the specialized reproductive shoots called flowers.

3. Microspores form in a flower's male reproductive parts. They develop into pollen grains, the sperm-bearing male gametophytes. Pollinating agents (animals, air currents, or water currents) transfer pollen grains to the female reproductive parts of flowers.

4. Megaspores form within ovules—tissue masses inside the ovary of a flower's female reproductive parts. They give rise to egg-producing female gametophytes.

5. After sperm fertilize the eggs, seeds develop. Each seed is a mature ovule. It consists of an embryo sporophyte along with tissues that function in its nutrition, protection, and dispersal. Animals often serve as dispersal agents.

6. From the time a seed germinates, hormones (signaling molecules between cells) influence plant growth and development. Each kind of hormone stimulates metabolic changes in target cells. The changes have predictable effects, as when they trigger cell division and other processes necessary for stem elongation.

7. Plant hormones help bring about genetically prescribed patterns of development, such as the extent and direction of growth of different plant parts. Hormones also adjust patterns of growth to local conditions and environmental rhythms, especially seasonal changes in daylength and temperature.

Think Sporophyte and Gametophyte

You probably don't think about this very often, but flowering plants engage in sex. Like humans, they have splendid reproductive systems that produce, nourish, and protect sperm and eggs. Like human females, they house embryos during early development. They issue floral invitations to third parties—pollinators that help sperm and egg get together. Long before we thought of it, flowering plants were using tantalizing fragrances and colors to improve the odds for sexual success.

When you hear the word "plant," you may think of something like a cherry tree (Figure 24.3a). This is an example of a **sporophyte**, a vegetative body that grows, by mitotic cell divisions, from a fertilized egg. At some point in the life cycle, the sporophyte bears **flowers**, which are shoots specialized for reproduction. In those flowers, haploid spores form and develop into haploid bodies called **gametophytes**. Sperm form inside male gametophytes; eggs form inside female gametophytes. As you know, the first cell that forms at fertilization has two sets of genetic instructions (from two gametes).

Components of Flowers

Flowers grow at floral shoots. As they do, they differentiate into nonfertile parts (sepals and petals) and fertile parts (stamens and carpels). Figures 24.3b and 24.4 show how the floral components are arranged. Directly or indirectly, all parts are attached to a receptacle, the modified base of the floral shoot. Peel open a rosebud, and you see that sepals, an outermost whorl of leaflike parts, enclose leaflike petals. Petals enclose a flower's fertile components. This is a common arrangement. The sepals are called the flower's calyx. The petals, which ring the male and female parts, are the flower's corolla.

Like leaves, sepals and petals have ground tissues, vascular tissues, and epidermis. Why are many flowers sweet smelling? Epidermal cells of the petals produce fragrant oils. What gives the petals their shimmer and color? Cells of their ground tissues contain pigments, such as carotenoids (yellow to red-orange) and anthocyanins (red to blue), and tiny, light-refracting crystals. What functions do the fragrances, colors, patterns, and arrangements of petals serve? They attract pollinators.

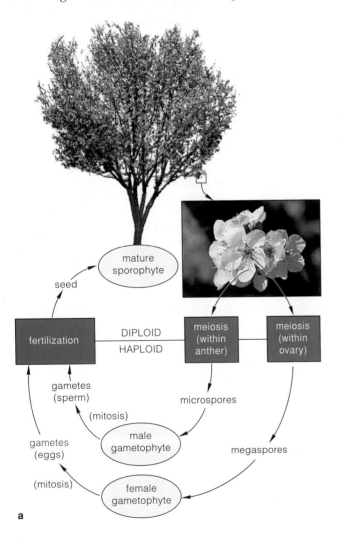

a

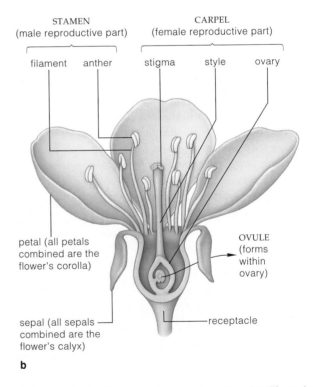

b

Figure 24.3 (**a**) Overview of typical flowering plant life cycles. (**b**) Structure of a flower—a cherry blossom (*Prunus*). As is true of the flowers of many plants, it has a single carpel (a female reproductive part) and stamens (male reproductive parts). Flowers of other plants have two or more carpels, often united as a single structure. Single or fused carpels consist of an ovary and a stigma. Often the ovary extends upward as a slender column (style).

corolla
(all petals
combined)

calyx
(all sepals
combined)

receptacle

Figure 24.4 Location of a few floral parts in one of the most prized of all cultivated plants, the rose (*Rosa*).

Where Pollen and Eggs Develop

Just inside the flower's corolla are **stamens**, the male reproductive parts. Nearly all stamens consist of an anther and a single-veined stalk (filament). An anther is internally divided into pollen sacs—chambers in which walled, haploid spores form and develop into sperm-producing bodies called **pollen grains** (Figure 24.5).

The female reproductive parts of a flower are at its center. You may have heard someone call these parts pistils, but their more recent name is **carpels**.

Some flowers have one carpel. Others have more, often fused as a compound structure. The lower portion of single or fused carpels is the **ovary**, where the eggs develop, fertilization takes place, and seeds mature. Angiosperm, a more formal name for flowering plants, refers to the carpel (from the Greek *angeion*, for vessel; and *sperma*, meaning seed). The carpel's upper portion, a stigma, is a sticky or hairy surface tissue that captures pollen grains and favors their germination. Commonly a style, a slender, upward extension of the ovary's wall, elevates the stigma.

Not every kind of flower has stamens and carpels. The species that have both male and female parts are called *perfect* flowers. Other species produce *imperfect* flowers, which contain male or female parts. In some species, such as oaks, the same individual plant bears male and female flowers. In willows and other species, they are on separate individual plants.

Sexual reproduction is the dominant reproductive mode of flowering plant life cycles. Such cycles alternate between the production of sporophytes (spore-producing bodies) and gametophytes (gamete-producing bodies).

A flowering plant sporophyte grows by mitotic division from a fertilized egg. It is all of the plant body *except* for the male and female gametophytes that form within its flowers.

Gametophytes arise from haploid spores, which form within the flower's stamens (male reproductive parts) and ovaries (female reproductive parts). Male gametophytes develop into sperm-producing pollen grains. Female gametophytes are structures in which eggs develop, fertilization takes place, and seeds mature.

POLLEN SETS ME SNEEZING

Sidney Smith in 1835 wrote, "I am suffering from my old complaint, the hay-fever . . . that sets me sneezing; and if I begin sneezing at twelve, I don't leave off till two o'clock, and am heard distinctly in Taunton [six miles distant] when the wind sets that way."

Sidney suffered what is now called *allergic rhinitis*, a common hypersensitivity to a normally harmless substance. Ragweed pollen (Figure 24.5) brought on his dreaded hay fever, but it wasn't out to get him. Every spring and summer, flowering plants release pollen grains—and in many millions of people, certain cells respond by mounting an immune response against proteins projecting from the pollen grain walls. Different types of pollen have differences in their wall proteins. The immune system of a person who is hypersensitive might zero in on some of these but not others. The misdirected response to all of them results in a runny nose, reddened and itchy eyelids, congestion, and bouts of sneezing.

Hay fever runs in families. Some individuals simply are genetically predisposed to overreact to certain kinds of pollen. Also, infections, emotional stress, even changes in air temperature may trigger immune reactions to pollen.

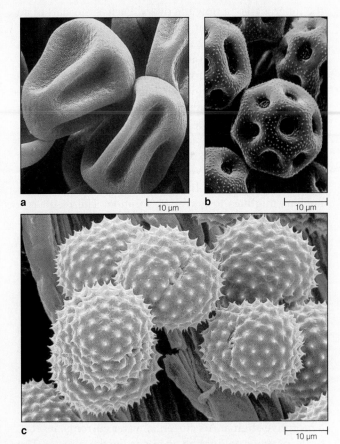

a   10 μm   b   10 μm

c   10 μm

Figure 24.5 Pollen grains from (**a**) grass, (**b**) rose, and (**c**) ragweed plants. Pollen grains of most families of plants differ in size, wall sculpturing, and number of wall pores.

24.3 A NEW GENERATION BEGINS

From Microspores to Pollen Grains

We turn now to pollen grain formation. While anthers are growing, four masses of spore-producing cells form by mitotic cell divisions. Walls develop around them. Each anther now has four chambers, called pollen sacs (Figure 24.6a). Cells in the sacs undergo meiosis and cytoplasmic division. This results in **microspores**. These haploid spores go on to develop an elaborately sculpted wall. The walled microspores divide once or twice by mitosis and become pollen grains, which now enter a period of arrested growth. In time, they will be released from the anther. Their wall components will protect them from decomposers in their surroundings.

As soon as many types of pollen grains form, they produce sperm nuclei—which are the male gametes of flowering plants. Other types don't do this until after they travel to a carpel and start growing toward its ovule. In short, a pollen grain is a mature *or* immature male gametophyte, depending on the plant species.

From Megaspores to Eggs

Meanwhile, one or more masses of cells are forming on the inner wall of a flower's ovary. Each is the start of an **ovule**—a structure that contains a female gametophyte and that may become a seed. As each cell mass grows, a tissue forms within it, and one or two protective layers (integuments) form around it. Inside the mass, a cell divides by meiosis, and four haploid spores form. The spores formed in flowering plant ovaries are generally larger than microspores and are called **megaspores**.

Commonly, all megaspores but one disintegrate. The one remaining undergoes mitosis three times without cytoplasmic division. At first it is a cell that has eight nuclei (Figure 24.6b). Its cytoplasm divides after each nucleus migrates to a specific location. The result is a seven-celled embryo sac, the female gametophyte. One cell (the "endosperm mother cell") has two nuclei and will help form **endosperm**—a nutritive tissue for the forthcoming embryo. Another cell is the egg.

From Pollination to Fertilization

Each spring, flowering plants release pollen. You are acutely aware of this reproductive event if you are one of the millions of people who suffer from hay fever, the allergic reaction to wall proteins of pollen grains released by ragweed and many other plants.

Pollination refers to the transfer of pollen grains to a receptive stigma. Air currents, water currents, insects, birds, or other pollinating agents make the transfer. The chapter introduction provided a look at the relationship between flowering plants and their pollinators.

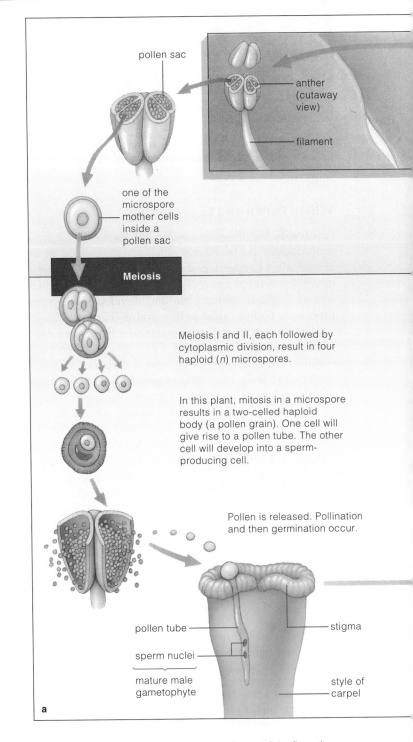

pollen sac

anther (cutaway view)

filament

one of the microspore mother cells inside a pollen sac

Meiosis

Meiosis I and II, each followed by cytoplasmic division, result in four haploid (*n*) microspores.

In this plant, mitosis in a microspore results in a two-celled haploid body (a pollen grain). One cell will give rise to a pollen tube. The other cell will develop into a sperm-producing cell.

Pollen is released. Pollination and then germination occur.

pollen tube

sperm nuclei

mature male gametophyte

stigma

style of carpel

a

Figure 24.6 Life cycle of cherry (*Prunus*), one of the flowering plants classified as a dicot. (**a**) This part of the diagram shows how pollen grains (male gametophytes) develop and germinate. (**b**) This part shows what goes on inside the ovule in this ovary.

Once a pollen grain lands on a receptive stigma, it germinates. For these plants, germination means that the pollen grain resumes growth and develops into a tubular structure. This "pollen tube" starts growing down through the tissues of the ovary, carrying sperm

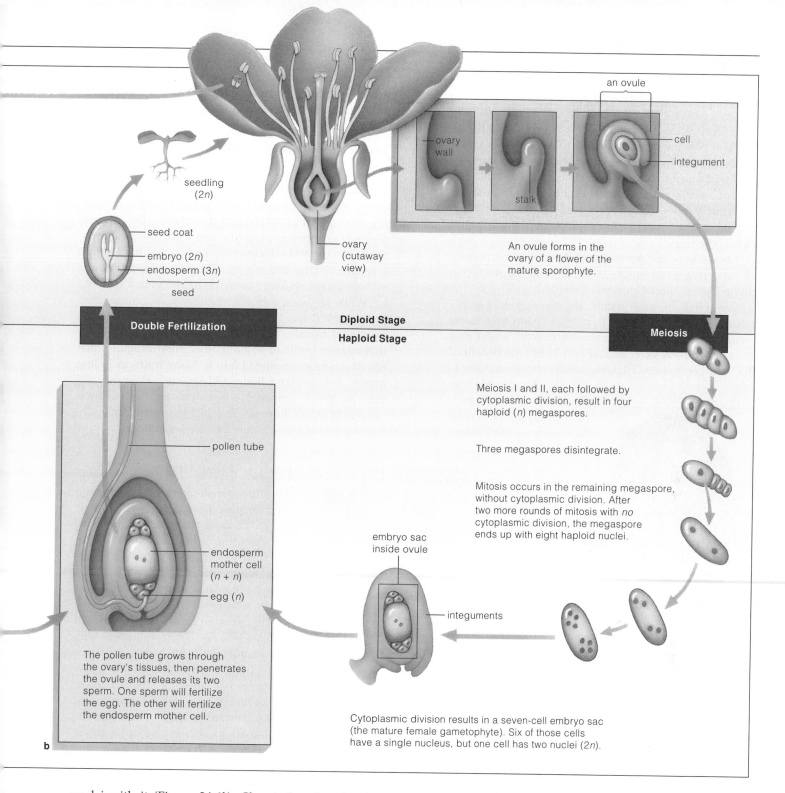

seedling
(2n)

seed coat
embryo (2n)
endosperm (3n)
seed

ovary
(cutaway
view)

an ovule

ovary
wall

stalk

cell
integument

An ovule forms in the
ovary of a flower of the
mature sporophyte.

Double Fertilization

Diploid Stage

Haploid Stage

Meiosis

Meiosis I and II, each followed by
cytoplasmic division, result in four
haploid (n) megaspores.

Three megaspores disintegrate.

Mitosis occurs in the remaining megaspore,
without cytoplasmic division. After
two more rounds of mitosis with *no*
cytoplasmic division, the megaspore
ends up with eight haploid nuclei.

pollen tube

endosperm
mother cell
(n + n)

egg (n)

embryo sac
inside ovule

integuments

The pollen tube grows through
the ovary's tissues, then penetrates
the ovule and releases its two
sperm. One sperm will fertilize
the egg. The other will fertilize
the endosperm mother cell.

b

Cytoplasmic division results in a seven-cell embryo sac
(the mature female gametophyte). Six of those cells
have a single nucleus, but one cell has two nuclei (2n).

nuclei with it (Figure 24.6b). Chemical and molecular
cues guide a pollen tube's growth through the tissues,
toward the egg chamber and sexual destiny. When the
pollen tube reaches an ovule, it penetrates the embryo
sac, its tip ruptures, and the two sperm are released.

"Fertilization" generally means fusion of a sperm
nucleus with an egg nucleus. But **double fertilization**
takes place in flowering plants. In the diploid species,
one sperm nucleus fuses with that of the egg, and so
produces a diploid (2n) zygote. Meanwhile, the other

sperm nucleus fuses with both nuclei of the endosperm
mother cell, forming a cell with a triploid (3n) nucleus.
That 3n cell gives rise to the nutritive tissue.

In flowering plants, sperm cells form within pollen grains,
the male gametophytes. Eggs develop inside ovules, which
contain the female gametophytes.

After pollination and double fertilization, an embryo and
nutritive tissue form inside the ovule, which becomes a seed.

Formation of the Embryo Sporophyte

After fertilization, the newly formed zygote embarks on a course of mitotic cell divisions that lead to a mature embryo sporophyte. Consider how an embryo forms in shepherd's purse (*Capsella*), a dicot. By the time this kind of embryo reaches the stage shown in Figure 24.7e, its primary meristems have already formed, and two cotyledons have started to develop from two lobes of embryonic tissue.

Cotyledons, or "seed leaves," develop as portions of all flowering plant embryos. Dicot embryos have two cotyledons; monocot embryos have one.

As is true of most dicots, the developing *Capsella* embryo absorbs nutrients from endosperm and stores them in its cotyledons. By contrast, in corn, wheat, and most other monocots, endosperm is not tapped until the seed germinates. Digestive enzymes become stockpiled inside the thin cotyledons of their embryos. When the enzymes do become active, nutrients stored in endosperm will be released and transferred to the growing seedling.

Formation of Seeds and Fruits

From the time a zygote forms until a mature embryo develops, the parent plant transfers nutrients to tissues of the ovule. Food reserves accumulate in the expanding endosperm or cotyledons. Eventually, the physical connection between the ovule and the ovary wall separates. The ovule's integuments thicken and harden into a seed coat. The embryo and food reserves are now a self-contained package; an ovule has become a **seed**.

While seeds are forming, changes also occur in other parts of the flower. The ovary itself expands in size, and its tissues become modified. What we call a **fruit** is a mature ovary, with or without other floral structures that have become incorporated into it. Many fruits, including apples, are juicy and fleshy. Others, such as grains and nuts, are dry. The pineapple, a multiple fruit, forms from clusters of many flowers. Table 24.1 lists some of the different kinds of fruits; Figure 24.8 shows examples.

Fruit and Seed Dispersal

All fruits function as seed-dispersing mechanisms. They coevolved with air currents, water currents, and animals. Consider the winglike extensions of the fruit of a maple (*Acer*) in Figure 24.8. When the fruit drops, its wings spin it sideways. With the help of air currents, these wings may whirl the fruit far enough away that its enclosed embryo will not have to compete with the parent plant for water, minerals, and sunlight. Other fruits taxi to new locations by adhering to feathers or fur with hooks, spines, hairs, and sticky surfaces. Embryos inside the seed coats of berries and other fleshy fruits must survive being eaten and assaulted by digestive enzymes in the animal gut. Besides digesting a fruit's

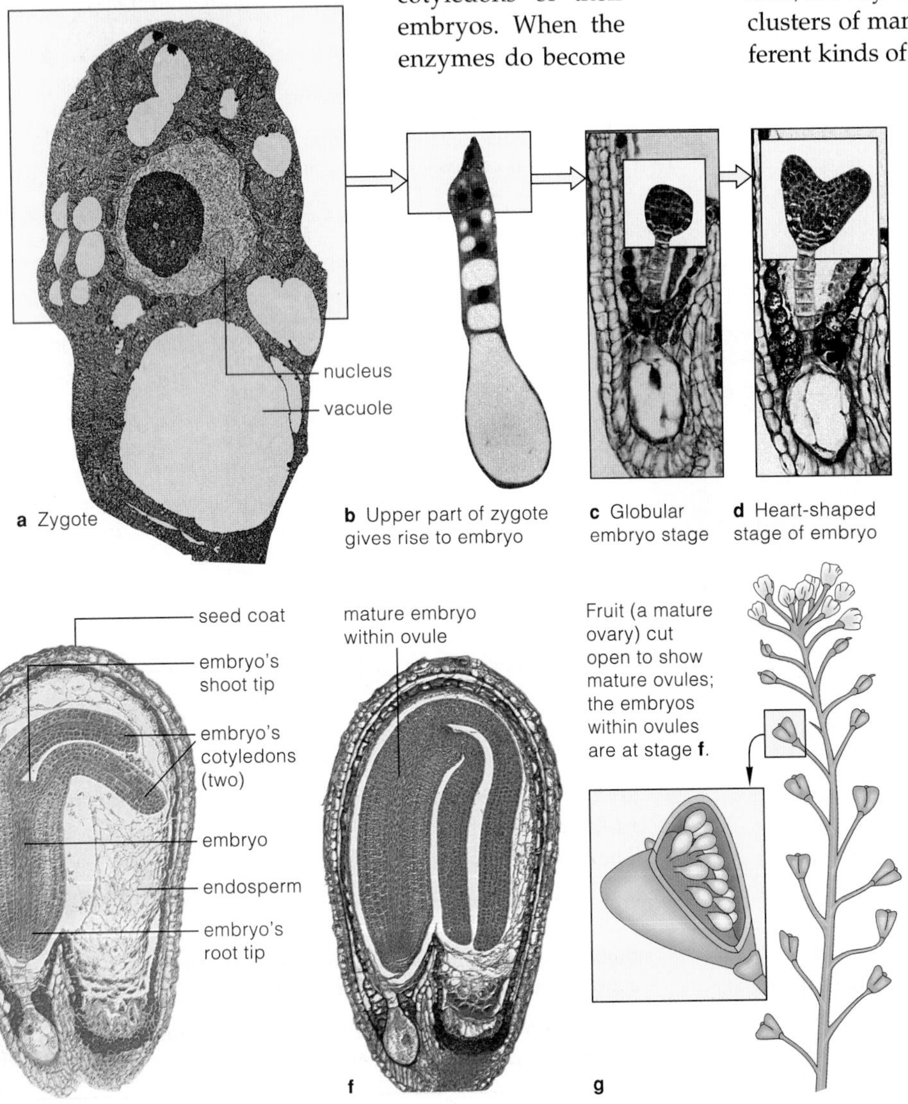

a Zygote

nucleus
vacuole

b Upper part of zygote gives rise to embryo

c Globular embryo stage

d Heart-shaped stage of embryo

seed coat
embryo's shoot tip
embryo's cotyledons (two)
embryo
endosperm
embryo's root tip

e

mature embryo within ovule

f

Fruit (a mature ovary) cut open to show mature ovules; the embryos within ovules are at stage **f**.

g

Figure 24.7 Stages in the development of shepherd's purse (*Capsella*), a dicot. The micrographs are not to the same scale. (**a**) The single-celled zygote. (**b**–**d**) The early embryo is identified by the *yellow* boxes. The row of cells below it transfers nutrients to the developing embryo from the parent plant. The embryo in (**e**) is well developed; the one in (**f**) is mature.

a Expansion of ovaries after petals drop from apple blossoms

b ⎿ individual fruit

remnants of petals and sepals

vascular bundles in outer portion of the ovary

seeds

enlarged receptacle

c seed (in carpel) ⎿ wing

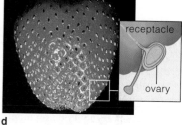

d receptacle, ovary

Figure 24.8 Fruits. (**a**) Fruit formation on an apple (*Malus*) tree. When petals drop, eggs usually are fertilized. Now the ovary and receptacle expand. Sepals and stamens remain on the immature fruit. (**b**) Pineapple (*Ananas comosus*), a multiple fruit. Besides running into the New World, Christopher Columbus also ran into cultivated pineapples, viewed by native Americans as a symbol of hospitality. To Columbus, they resembled pine cones, hence the name. At the time, native populations used pineapple juice as the base for an alcoholic beverage and for arrow tip poison. (**c**) Winged seeds of maple (*Acer*). (**d**) Strawberry (*Fragaria*), a fragrant, wildly popular, easily bruised fruit. Genetic engineering is yielding strains that retain the strong fragrance yet are firm enough to resist packing damage.

Table 24.1	Kinds of Fruits of Some Flowering Plants	
Type	Characteristics	Some Examples
Simple (formed from single carpel, or two or more united carpels of one flower)	1. Fruit wall *dry*; *split* at maturity	Pea, magnolia, tulip, mustard
	2. Fruit wall *dry*; *intact* at maturity	Sunflower, wheat, rice, maple
	3. Fruit wall *fleshy*, sometimes with leathery skin	Grape, banana, lemon, cherry, orange
Aggregate (formed from many separate carpels of single flower)	*Aggregate* (cluster) of matured ovaries (fruits), all attached to receptacle (modified stem end)	Blackberry, raspberry
Multiple (formed from carpels of several associated flowers)	*Multiple* matured ovaries massed together; may include accessory structures (such as receptacle, sepal, and base of petals)	Pineapple, fig, mulberry
Accessory (formed from one or more ovaries *plus* receptacle tissue that becomes fleshy)	1. *Simple*: a single ovary enclosed in receptacle tissue	Apple, pear
	2. *Aggregate*: swollen, fleshy receptacle with dry fruits on surface	Strawberry

flesh, the enzymes digest some of the seed coat. The digested portions make it easier for the embryo to break through its hard coat after the seeds are expelled from the animal body and start to germinate. We turn to this topic later in the chapter.

A mature ovule, which encases an embryo sporophyte and food reserves inside a hardened coat, is a seed.

A mature ovary, with or without additional floral parts that have become incorporated into it, is a fruit.

The structure of seeds and fruits enhances dispersal by wind, by water, or by particular kinds of animals.

24.5 ASEXUAL REPRODUCTION OF FLOWERING PLANTS

Asexual Reproduction in Nature

The features of sexual reproduction that we considered in the preceding sections dominate flowering plant life cycles. Bear in mind, many species also can reproduce asexually by various modes of **vegetative growth**. Table 24.2 lists some of these modes. In essence, new roots and shoots grow right out of extensions or fragments of a parent plant. As you know, such asexual reproduction proceeds by way of mitosis, so offspring are genetically identical to the parent (they are a clone).

A "forest" of quaking aspen (*Populus tremuloides*) provides us with a stunning example of vegetative reproduction. The leaves of this flowering plant species tremble even in the slightest breeze, hence the name. Figure 24.9 provides a panoramic view of the shoot systems of a single individual. Its root system gave rise to

many adventitious shoots, which became separate shoot systems. Barring rare mutations, individuals at the north end of this vast clone are genetically identical to individuals at the south end. Roots near the lake deliver water to shoot systems in much drier soil. Nutrients can travel in the opposite direction.

As long as environmental conditions favor growth and regeneration, such clones are as close as one can get to being immortal. No one knows how old aspen clones are. The oldest known clone, a ring of creosote bushes (*Larrea divaricata*) in the Mojave Desert, has been around for the past 11,700 years.

The possibilities are astounding. Raise strawberry plants and watch them send out runners (horizontal aboveground stems); then watch new roots and shoots develop at every other node. Eat an orange, and it may be descended from a tree in Southern California that reproduced by **parthenogenesis** (the development of an embryo from an unfertilized egg). Parthenogenesis may be stimulated when pollen has contacted a stigma even though a pollen tube has not grown down the style. Maybe hormones from the stigma or from pollen grains diffuse to the unfertilized egg and trigger formation of an embryo—which becomes $2n$ by fusion of products of egg mitosis. A $2n$ cell outside the gametophyte also may be stimulated to develop into an embryo.

Induced Propagation

Most houseplants, woody ornamentals, and orchard trees are clones. They often are propagated from cuttings or fragments of shoot systems. Thus, with suitable encouragement, a severed African violet leaf forms a callus from which adventitious roots develop. A callus is meristematic tissue (a mass of undifferentiated cells that retain the potential for division).

Or consider a twig or bud from one plant, grafted onto a different variety of some closely related species. Vintners in France, for example, graft prize grapevines onto disease-resistant root stock from America.

Frederick Steward and his coworkers pioneered in **tissue culture propagation**. They cultured tiny bits of phloem from the differentiated roots of carrot plants (*Daucus carota*) in rotating flasks. They used a liquid growth medium that contained sucrose, mineral ions, and vitamins. It also contained coconut milk, which Steward knew was rich in then-unidentified growth-inducing substances. As the flasks rotated, individual cells that were torn away from the tissue bits divided and formed multicelled clumps—which sometimes gave rise to roots (Figure 24.10*a*). These experiments were among the first to demonstrate that some cells of

Table 24.2	Asexual Reproductive Modes of Flowering Plants	
Mechanism	Representative	Characteristics
Vegetative reproduction on modified stems		
1. Runner	Strawberry	New plants arise at nodes on an aboveground horizontal stem
2. Rhizome	Bermuda grass	New plants arise at nodes of underground horizontal stem
3. Corm	Gladiolus	New plant arises from axillary bud on short, thick, vertical underground stem
4. Tuber	Potato	New shoots arise from axillary buds on tubers (enlarged tips of slender underground rhizomes)
5. Bulb	Onion, lily	New bulb arises from axillary bud on short underground stem
Partheno-genesis	Orange tree, rose	Embryo develops without nuclear or cellular fusion (e.g., from unfertilized haploid egg; or develops adventitiously, from tissue surrounding embryo sac)
Vegetative propagation	Jade plant, African violet	New plant develops from tissue or organ (e.g., a leaf) that drops or is separated from plant
Tissue culture propagation	Orchids, lily, tulip, wheat, rice, corn	New plant induced to arise from cell of a parent plant that is not irreversibly differentiated

Figure 24.9 A mere portion of Pando the Trembling Giant, so named by the researchers who studied its genetic makeup (Michael Grant and coworkers at the University of Colorado). This 106-acre "forest" in Utah is actually a single asexually reproducing male organism, of a type called quaking aspen (*Populus tremuloides*). A root system connects its 47,000 genetically identical shoots (the trees). The clone weighs an estimated 13,000,000+ pounds.

a

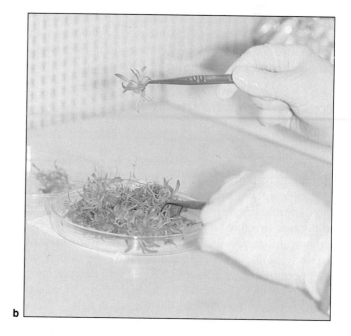

b

Figure 24.10 (a) Cultured cells from a carrot plant. Roots and shoots of embryonic plants are already forming. (b) Young orchid plants, developed from cultured meristem and young leaf primordia. Orchids are one of the most prized cultivated plants. Before meristem cloning, they were difficult to hybridize. From the time seeds form, it can take seven long years or more until a new plant flowers. In the natural habitat, seeds normally will not germinate unless they interact with a specific fungus.

tissue contain the genetic instructions required to build new individuals.

Tissue culture propagations are now done with shoot tips or other parts of individual plants. The techniques are useful when an advantageous mutant arises. One such mutant might show resistance to a disease that is particularly crippling to wild-type plants of the same species. Tissue culture propagation can lead to hundreds and even thousands of identical plants from just one mutant specimen. This technique is already being used in efforts to improve major food crops, such as corn, wheat, rice, and soybeans. It also is being used to increase production of hybrid orchids, lilies, and other valued ornamental plants (Figure 24.10*b*).

Besides reproducing sexually, flowering plants engage in asexual reproduction, as by vegetative growth.

PATTERNS OF GROWTH AND DEVELOPMENT

Seed Germination

We turn now to the patterns by which monocots and dicots grow and develop, starting with events within the seed (Figure 24.11). Before or after seed dispersal from the parent plant, an embryo sporophyte's growth idles. Later, if all goes well, it germinates. For seeds, **germination** refers to the resumption of growth after a period of arrested embryonic development. Genes *and* the environment govern this process.

Environmental factors affecting germination include soil temperature, moisture and oxygen levels, and the number of daylight hours (which varies seasonally). For example, mature seeds do not contain enough water for cell expansion or metabolism. In most places, ample water is available only on a seasonal basis, so germination must coincide with the return of spring rains. In a process called **imbibition**, water molecules move into the seed, being attracted mainly to the hydrophilic groups of stored proteins. As more water moves in, the seed swells and its coat ruptures.

Once the seed coat splits, more oxygen reaches the embryo, and aerobic respiration moves into high gear. Now the cells rapidly grow and divide. Generally, the embryonic root cells are activated first. They divide, elongate, and give rise to a **primary root**, the first root of the seedling sporophyte. Germination is over when the primary root breaks through the seed coat.

Prescribed Growth Patterns

Growth involves cell divisions and enlargements. On the average, half of the resulting daughter cells do not increase in size, but they retain the capacity to divide. The others enlarge, often by twenty times (Figure 24.12).

Water uptake drives cell enlargement. It increases internal fluid pressure and forces the cell's primary wall to expand. Imagine blowing up a balloon. If it's soft, the balloon inflates easily. Similarly, cells with a soft wall expand rapidly under pressure. However, a balloon wall gets thinner as it "grows." A cell wall doesn't. New polysaccharides are added to it, and more cytoplasm forms between the wall and the central vacuole.

Consider the growth and development of monocots and dicots. Again, the basic patterns are heritable, dictated by the plant's genes. All cells in a new plant arise from the same cell (the zygote) and so carry the same genes. But unequal cytoplasmic divisions between daughter cells lead to differences in their metabolic equipment and output. Daughter cells start interacting in different ways, through selective gene expression. Genes governing, say, synthesis of growth-stimulating hormones become activated in some cells and remain silent in others. Such events seal the developmental fate

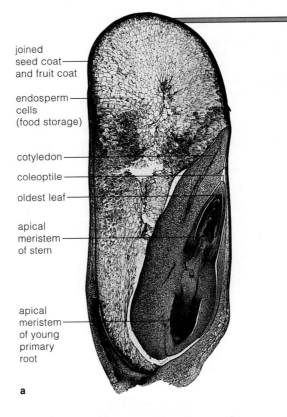

joined
seed coat
and fruit coat

endosperm
cells
(food storage)

cotyledon

coleoptile

oldest leaf

apical
meristem
of stem

apical
meristem
of young
primary
root

a

Figure 24.11 Growth and development of a monocot. Corn (*Zea mays*) is the example. (**a**) An embryo sporophyte within a corn grain (seed), longitudinal section. (**b**) Growth resumes with germination of the corn grain. Following germination, the coleoptile protects young leaves when the new seedling grows through soil. Adventitious roots develop at the coleoptile's base.

Growth and development of a dicot. The common bean plant (*Phaseolus vulgaris*) is the example. (**c**) Below the cotyledon, a hypocotyl (a hook-shaped portion of the young stem) forces a channel through soil. Food-storing cotyledons are pulled up through the channel without being torn apart. At the surface, light causes the hook to straighten. The cotyledons serve in photosynthesis for several days; then they wither and fall off. In time, foliage leaves, each divided into three leaflets, take over the task. Flowers will develop in buds at the nodes.

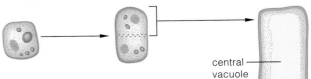

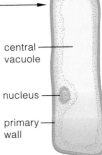

central
vacuole

nucleus

primary
wall

Meristems provide new cells for growth. At root and shoot tips, small meristematic cells double in size, then divide. One cell remains meristematic. The other grows into a mature parenchyma cell of pith or cortex, for example. Tiny vacuoles in young cells absorb water, fuse, and form the central vacuoles of these cells.

Figure 24.12 Cell divisions and cell enlargements underlying the patterns of growth and development among flowering plants.

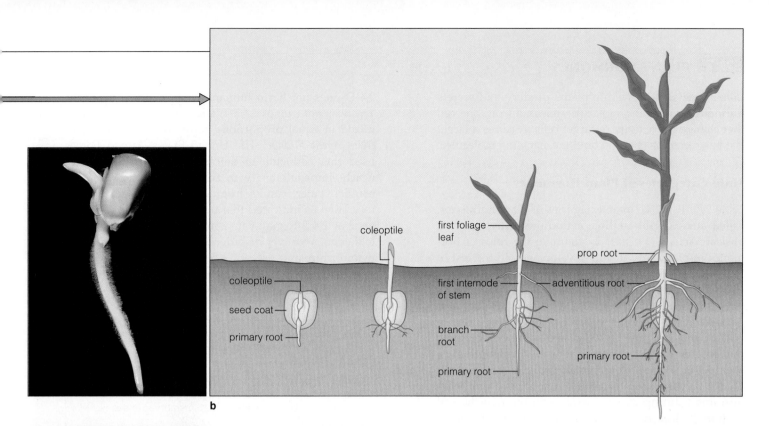

b

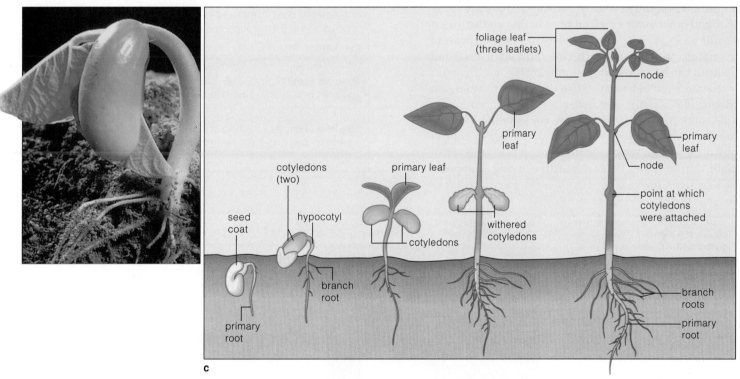

c

of various cell lineages. Their descendant cells divide in prescribed planes and expand in specific directions. The eventual outcome will be plant parts of specific shapes and functions.

Bear in mind, the prescribed patterns of growth can be adjusted in response to unusual environmental pressures. Suppose a seed germinates in a vacant lot and a heavy paper bag blows on top of it. The first shoot will bend and grow out from under the bag, toward sun-

light. Enzymes, hormones, and other gene products in shoot cells carry out this growth response.

Plant growth involves cell divisions and cell enlargements. Plant development requires cell differentiation, as brought about by selective gene expression.

Interactions among genes, hormones, and the environment govern how an individual plant grows and develops.

Like animals, all flowering plants produce and secrete **hormones**—signaling molecules released from one cell that change the activity of target cells. A cell is a target if it has receptors that can bind the signaling molecule.

Main Categories of Plant Hormones

Table 24.3 lists five main categories of plant hormones, called **auxins**, **gibberellins**, **cytokinins**, **ethylene**, and **abscisic acid**. Let's start by defining the roles of hormones in plant growth and development. In the rest of the chapter, we can look at specific examples of their effects at different stages of the life cycle.

Auxins Stem lengthening and possibly responses to gravity and light are effects of auxins. Auxins promote the lengthening of coleoptiles as well. A **coleoptile** is a sheath around the shoot of grass seedlings, such as corn plants. It helps keep the tender new shoot from being torn apart during its upward growth through the soil (Figure 24.13). Indoleacetic acid (IAA) is the most important naturally occurring auxin. Orchardists spray IAA and other auxins on fruit trees to thin overcrowded seedlings in spring. They also use auxins to prevent premature fruit drop, which cuts farm labor costs (all the fruit can be picked at the same time).

Certain synthetic auxins are used as herbicides. An **herbicide** is a compound that, at suitable concentration, kills some plants but not others. For example, farmers use 2,4-dichlorophenoxyacetic acid (2,4-D) to control broadleaf weeds, which vigorously compete with cereal plants for nutrients. After exhaustive testing, it appears 2,4-D does not harm humans when handled properly. This was not true of 2,4,5-T, a related compound. When mixed in equal proportions, the two compounds produce *Agent Orange*. The United States Armed Forces used this herbicide to defoliate thickly forested and nearly impenetrable war zones during the Vietnam conflict. Later, results from experiments with laboratory animals indicated that traces of dioxin, a contaminant of 2,4,5-T, may cause miscarriages, birth defects, leukemia, and liver disorders. The government banned manufacture and use of 2,4,5-T in the United States. In 1994, the results of comprehensive tests indicated that dioxin apparently is a carcinogenic substance.

Table 24.3 Main Plant Hormones and Some Known (or Suspected) Effects

Auxins. Promote cell elongation in coleoptiles and stems; roles in phototropism and gravitropism. Notable amounts in bud and leaf apical meristems and in embryos in seeds.

Gibberellins. Promote stem elongation; help end dormancy of seeds and buds; contribute to flowering. Notable amounts in apical meristems of buds, roots, and leaves and embryos.

Cytokinins. Promote cell division and leaf expansion; retard leaf aging. Synthesized in roots; moves elsewhere.

Abscisic acid. Promotes stomatal closure; promotes bud and seed dormancy. Notable amounts in leaves, stems, and unripened fruit.

Ethylene. Promotes ripening of fruit and abscission of leaves, flowers, and fruits. Notable amounts in fruit, stems, leaves, roots, and seeds.

Figure 24.13 Effects of auxin, a plant hormone. (**a**) Cutting from a gardenia plant (*left*), four weeks after an auxin was applied to its base. The other cutting (*right*) was untreated.

(**b**) Experiments demonstrating how IAA (an auxin) in a coleoptile tip promotes elongation of cells below it. (*1*) Cut off the tip of an oat coleoptile. The cut stump does not elongate as much as a normal oat coleoptile (*2*) used as a control. (*3*) Next, place a tiny block of agar under the cut tip and leave it for several hours. During that time, IAA diffuses into the agar. (*4*) Place the agar on another de-tipped coleoptile. Elongation proceeds about as rapidly as in an intact coleoptile growing next to it (*5*).

a treated with auxin   untreated

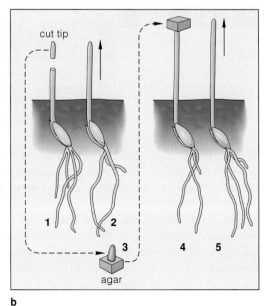

b

Gibberellins Gibberellins promote stem lengthening (Figure 24.14 and Section 24.8). They also help buds and seeds break dormancy and resume growth in the spring. Gibberellins influence the flowering process in at least some species.

Cytokinins Cytokinins stimulate cell division (hence the name, which refers to cytoplasmic division). They are most abundant in root and shoot meristems and in the tissues of maturing fruits. Natural and synthetic versions are used to prolong the shelf life of cut flowers as well as lettuces, mushrooms, and other vegetables.

Abscisic Acid Abscisic acid (ABA) helps plants adapt to seasonal changes, as by inducing bud dormancy and inhibiting cell growth or premature seed germination. ABA also contributes to stomatal closure and thereby helps conserve water when a plant is water stressed. Often ABA is applied to nursery stock that is about to be shipped, because when plants are dormant, they are more resistant to damage.

Ethylene Ethylene induces various aging responses, including fruit ripening and leaf drop. Ancient Chinese burned incense to make fruit ripen faster, although they didn't know the smoke contained ethylene. In the early 1900s, growers put citrus fruits in sheds with kerosene stoves. They thought heat did the trick; they didn't know ethylene was in the smoke. Today, to minimize losses, food distributors use ethylene to ripen green tomatoes and other fruit after they ship it. Fruit that is picked green doesn't bruise or deteriorate as fast. Oranges and other citrus fruits are exposed to ethylene to brighten the color of their rind before being displayed in the market.

Other Plant Hormones

Besides the hormones just described, root and leaf cells produce their own hormones. One or more unknown hormones may trigger flowering. Another unknown hormone associated with shoot tips blocks growth of lateral buds, an effect called apical dominance. By pinching off shoot tips, gardeners prevent the hormone from diffusing through a plant's stems and exerting its inhibitory effect. Lateral buds are free to branch out, and gardeners get bushier plants.

Like animals, flowering plants produce hormones. These signaling molecules are released from one cell type and then change the metabolic activities of target cells.

Auxins, gibberellins, cytokinins, abscisic acid, and ethylene are the main categories of plant hormones.

FOOLISH SEEDLINGS, GORGEOUS GRAPES

A few years before the American stock market grew feverishly and then collapsed catastrophically in 1929, a researcher in Japan came across a substance that caused runaway growth and collapse of rice plants. Ewiti Kurosawa was studying the "foolish seedling" effect on rice plants. Stems of rice seedlings infected with *Gibberella fujikuroi*, a fungus, elongated twice the length of stems of uninfected plants. The long, spindly, weakened stems eventually collapsed and the plants died. Kurosawa discovered that applying extracts of the fungus to plants also could trigger the disease. Years later, other researchers purified the disease-causing substance from fungal extracts. The substance was given the name gibberellin. More than eighty different kinds have now been isolated.

Expose a radish plant to a suitable concentration of a gibberellin and it may grow as large as a beach ball. Do the same to a cabbage plant and it may grow 6 feet tall. Gibberellin applications can make celery stalks longer and crispier; they can prevent the skin of navel oranges from growing old too soon in orchards. Walk past plump seedless grapes in produce bins of grocery stores and notice their market appeal (Figure 24.14). The fruits of grapevines grow in bunches along stems. Gibberellin applications cause the stems to lengthen between internodes. With more space between one another, the individual grapes grow larger. Also, air circulates more freely between them—which makes it harder for disease-causing, grape-juice-loving fungi to become established and do damage.

Figure 24.14 Seedless grapes. Gibberellin made the grape stems lengthen, which improved air circulation around the fruits and gave them more growing room. Larger grapes weigh more and delight growers and grocers, who sell grapes by the pound.

The Tropisms

Generally, the young roots of land plants grow down through soil, and the shoots grow upright through air. Both also can adjust the direction of growth in response to environmental stimuli, as when a shoot turns toward sunlight. When a root or shoot turns toward or away from an environmental stimulus, we call this a **plant tropism** (after the Greek *trope*, for turning). Hormone-mediated shifts in the rates at which different cells grow and elongate bring about this type of response, as the following examples will illustrate.

Gravitropism After a seed germinates, the first root to emerge through the seed coat always curves down. Similarly, coleoptiles or stems always curve up. Such growth responses to the Earth's gravitational force are forms of **gravitropism**.

Auxin *and* a growth-inhibiting hormone may trigger the gravitropic response in roots. Turn a young root on its side and remove its root cap, and it will *not* curve downward. Put the cap back on, and the root will do so. Elongating root cells will not stop growing if you remove the root cap; if anything, they will grow faster. Suppose a growth inhibitor that is present in root cap cells becomes *redistributed* in a root turned on its side. If gravity somehow causes the inhibitor to move out of the cap and accumulate in cells on the lower side of the root, those cells would not elongate as much as cells on the upper side—and the root would curve downward.

Generally, gravity-sensing mechanisms are based on **statoliths**, which are clusters of some type of particle. In plants, these are clusters of unbound starch grains in modified plastids. Figure 24.15*a* and *b* shows how the statoliths collect at the bottom of plastids in root cells according to which way gravity is pulling on them. These plastids settle down through the cytoplasm until

they rest upon the lowest region of the cell. Possibly the redistribution of statoliths triggers the redistribution of auxin to bring about this gravitropic response.

Similarly, Figure 24.15*c* shows how you can track what will happen if you turn a potted seedling on its side in a dark room. The stem will curve upward, even in the absence of light. Why? Elongation of cells on the upward-facing side of the horizontally positioned stem will slow down—markedly so—and elongation of cells on the lower side will rapidly increase. The different rates of elongation will be enough to make the stem turn in an upward direction. Cells on the bottom of a sideways stem must be *more* sensitive to auxin or some other hormone and those on top, *less* so.

Phototropism When stems or leaves adjust the rate and direction of growth in response to light, this is a form of **phototropism**. The adaptive advantage is obvious; photosynthesis stops—and so does plant growth—in the absence of certain wavelengths of light.

Charles Darwin was aware of phototropism. In the late 1800s, he made note of a coleoptile growing toward

Figure 24.15 Experiments demonstrating gravitropic responses by young shoots and roots. (**a**) From a corn root cap, statoliths within the plastids of cells that were elongating vertically. This light micrograph shows their orientation before the experimenter turned the root sideways. (**b**) Within seconds after becoming reoriented, the plastids had settled to the bottom of the root cap cells. The gravity-sensing mechanism based on these statoliths may be sensitive to redistribution of auxin through the root tip. A difference in auxin concentration may cause cells on the "top" of the sideways root to elongate faster than cells on the bottom. Different elongation rates could turn the root tip downward.

(**c**) As a different experiment, grow a newly emergent sunflower seedling in the dark for five days. Then turn it on its side and mark the young shoot at 0.5-centimeter intervals. A gravity-sensing mechanism will cause the stem to turn upright.

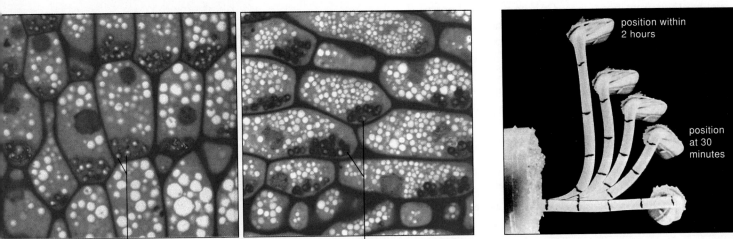

a   statoliths   b   statoliths   c

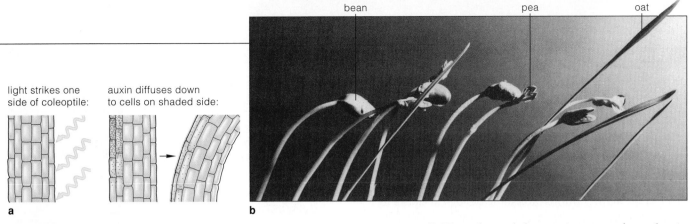

light strikes one side of coleoptile: auxin diffuses down to cells on shaded side:

a   b

bean   pea   oat

Figure 24.16 (**a**) Hormone-mediated differences in the rates of cell elongation bring about the bending of a stem toward light. (**b**) Phototropism in seedlings. The plant physiologist Frank Salisbury grew these seedlings in darkness, then allowed light to strike their right side for a few hours before photographing them.

Figure 24.17 Tendrils of a grapevine, twisting thigmotropically.

a   b   c

Figure 24.18 Effect of mechanical stress on tomato plants. (**a**) This plant (the control) grew in a greenhouse. (**b**) Each day for 28 days, this plant was mechanically shaken for 30 seconds. (**c**) This one had two shakings each day.

light striking one side of its tip. It wasn't until the 1920s that Frits Went, a graduate student in Holland, made the connection between phototropism and a growth-promoting substance. (He was the one who named the substance auxin, after the Greek *auxein*, meaning "to

increase.") Went showed that auxin moves from the tip of a coleoptile into cells less exposed to light and makes them elongate faster than cells on the illuminated side. This brings about bending toward light (Figure 24.16*a*).

Plants make the strongest phototropic response to light of blue wavelengths. Thus **flavoprotein**, a yellow pigment molecule that absorbs blue wavelengths, may be part of the phototropic bending mechanism.

You can observe a phototropic response by putting bean sprouts or seedlings of another sun-loving plant in shade, next to a window through which sunlight is streaming in. They will start curving in the direction where the most light is available (Figure 24.16*b*).

Thigmotropism Plants also shift their direction of growth when they contact solid objects. This response is called **thigmotropism** (after the Greek *thigma*, which means "touch"). Auxin and ethylene may have roles in the response. **Vines**—stems too slender or soft to grow upright without support—make this contact response. So do **tendrils**—modified leaves or stems that wrap around objects and thus help support the plant (Figure 24.17). You can observe what happens when a tendril grows against a stem of another plant. Within minutes, cells on the contact side will stop elongating and the tendril will start curling around the stem, possibly more than once. Afterward, cells on both sides will resume growth at the same rate.

Responses to Mechanical Stress

Mechanical stress, such as prevailing strong winds and grazing animals, can inhibit stem elongation and plant growth. You can see its effects on trees growing near the snowline of windswept mountains; they are stubby compared to trees of the same species that are growing at lower elevations. Similarly, plants grown outdoors commonly have shorter stems than plants grown in a greenhouse. You can observe this response to stress by shaking a plant daily for a brief period. Doing so will inhibit growth of the entire plant (Figure 24.18).

Plants adjust their direction and rate of growth in response to environmental stimuli.

BIOLOGICAL CLOCKS AND THEIR EFFECTS

Like other organisms, flowering plants have internal timing mechanisms called **biological clocks**. The clocks set the time for biochemical events that cause recurring changes in daily activities. Biological clocks also have key roles in seasonal adjustments in patterns of growth, development, and reproduction.

Phytochrome, a blue-green pigment molecule, is one of the "alarm buttons" for some biological clocks in plants. This pigment absorbs red or far-red wavelengths of light, with different results (Figure 24.19). At sunrise, when red wavelengths dominate, it converts to active molecular form (Pfr). At sunset, at night, or even in shade (where far-red wavelengths predominate), the molecule reverts to its inactive form (Pr). Phytochrome activation may induce cells to take up free calcium ions (Ca^{++}) or induces certain plant organelles to release them. Either way, when free calcium ions combine with calcium-binding proteins in cells, they initiate rhythmic leaf movements and certain other responses to light.

Rhythmic Leaf Movements

Each day, some plants position their leaves horizontally then fold them closer to stems at night (Figure 24.20). Keep such a plant in full light or darkness for a few days and it will continue to fold its leaves in the "sleep" position! These rhythmic leaf movements are one type of **circadian rhythm**—a biological activity that recurs in cycles of 24 hours or so. (*Circadian* means "about a day"). Ruth Satter, Richard Crain, and their coworkers at the University of Connecticut gathered experimental evidence that phytochrome plays a role in these leaf movements. Dr. Satter, a pioneer in the study of internal time-keeping mechanisms, was the first to correlate the sleep movements with "hands of a biological clock."

Flowering—A Case of Photoperiodism

As another example of internal time-keeping, consider that flowering plants reset biological clocks as seasons change. For example, at a certain time of year, a plant

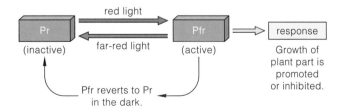

Figure 24.19 Interconversion of phytochrome from active form (Pfr) to inactive form (Pr). This blue-green pigment is part of a switching mechanism that promotes or inhibits the growth of a variety of plant parts.

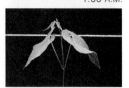

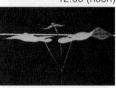
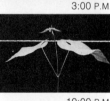

Figure 24.20 Rhythmic leaf movements. The investigator kept a bean plant in total darkness for 24 hours. Its leaves continued to move independently of sunrise (6 A.M.) and sunset (6 P.M.). Leaves folded close to a stem may keep moonlight from activating phytochrome and interrupting the dark period that triggers flowering. Or folding may reduce heat loss from leaves exposed to cold night air.

1:00 A.M.

6:00 A.M.

12:00 (noon)

3:00 P.M.

10:00 P.M.

12:00 (midnight)

starts diverting more energy to the formation of flowers. Figure 24.21 will give you an idea of how this activity and others are predictable responses to rhythmic environmental cues, such as more daylight hours and warmer temperatures in summer.

Photoperiodism refers to a biological response to change in the relative length of daylight and darkness in the 24-hour cycle. Phytochrome's active form, Pfr, may be an alarm button for the process. It may trigger enzyme synthesis in different cell types. Different enzymes certainly are needed for seed germination, stem lengthening and branching, expansion of leaves, and the formation of flowers, fruits, and seeds.

Different plants flower in response to different cues. Spinach, irises, and other "long-day" plants produce flowers in spring, when daylength exceeds a critical value (Figure 24.22*a*). Poinsettias, chrysanthemums, and cockleburs are "short-day" plants. They flower in late summer or early fall, when daylength is shorter than a critical value (Figure 24.22*b*). "Day-neutral" plants flower when mature enough to do so.

Spinach plants won't flower and produce seeds unless they are exposed to 10 hours of darkness (14 hours of daylight) for 2 weeks. That's why you would not want to start a spinach seed farm in the tropics—where this set of cues doesn't exist. By contrast, flower growers can stall the flowering process by exposing chrysanthemum plants to a flash of light nightly—thereby breaking up one long night into two "short" nights. That's why you can buy chrysanthemums in

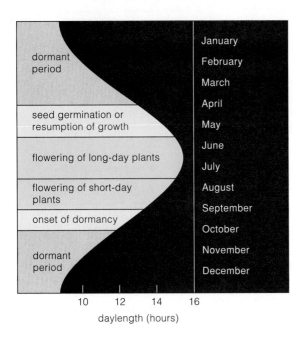

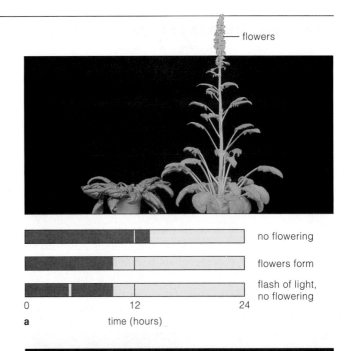

— flowers

| no flowering |
| flowers form |
| flash of light, no flowering |

0   12   24
a   time (hours)

| no flowering |
| flowers form |
| flash of light, no flowering |

0   12   24
b   time (hours)

dormant period

January
February
March
April
May
June
July
August
September
October
November
December

seed germination or resumption of growth

flowering of long-day plants

flowering of short-day plants

onset of dormancy

dormant period

10   12   14   16

daylength (hours)

Figure 24.21 Plant growth and development correlated with the number of hours of daylight, which changes through the seasons. The data are from temperate regions of North America that show seasonal shifts in rainfall and temperature.

Figure 24.22 (*Right*) Experimental results showing different flowering responses of (**a**) spinach, a long-day plant, and (**b**) chrysanthemum, a short-day plant. In each photograph, the plant at the left grew under short-day conditions; the one at the right grew under long-day conditions.

Each horizontal bar represents a 24-hour period. Hours of daylight are *yellow*; hours of darkness are *black*. As you can deduce from the diagrams, the key factor is an uninterrupted period of *darkness* of some critical value, which varies among species. Spinach will not flower unless exposed to 10 hours of darkness (14 hours of daylight) for 2 weeks. Even a brief flash of light during the critical period of darkness (the third horizontal bar) will inhibit flowering. Chrysanthemums will not flower unless exposed to an uninterrupted darkness that lasts about 10 hours each night.

spring even though they naturally bloom in autumn. Or think about cockleburs, which normally flower after a single night longer than 8.5 hours. If artificial light interrupts their dark period for even a minute, they refuse to flower. And why didn't poinsettias that were planted along California's interstate highways put out flowers? Headlights from cars and trucks zipping by during the night blocked the flowering response.

Besides phytochrome, other hormones that have not yet been identified probably influence flowering and other growth responses. Such hormones may be produced in leaves and transported to new buds. For example, if you trim all but one leaf from a cocklebur, then cover the remaining leaf with black paper for 8.5

hours, the plant will flower. If you cut off that one leaf right after the dark period is over, you won't see any cocklebur flowers.

As with other organisms, flowering plants have internal time-keeping mechanisms—biological clocks.

Phytochrome, a blue-green pigment, is part of a switching mechanism that responds to light of different wavelengths. Its active form may interact with other hormones to control which enzymes are being produced in particular cells.

Different enzymes are necessary to complete flowering and other growth responses that are influenced by sunlight and other environmental cues.

LIFE CYCLES END, AND TURN AGAIN

Senescence

At some stage in their life cycle, plants withdraw nutrients from leaves, stems, and roots, then distribute them to flowers, fruits, and seeds. Deciduous plants (which shed leaves when the growing season ends) channel nutrients to storage sites in twigs, stems, and roots before leaves die and drop. The dropping of flowers, leaves, fruits, and other plant parts is called **abscission**. Ethylene in cells near the abscission zone (where parts break off) may trigger this process (Figure 24.23), and abscisic acid may induce ethylene production.

Senescence is the sum total of processes that lead to death of a plant or some of its parts. The diversion of nutrients into reproductive parts may be a cue for leaf, stem, and root senescence. Stop the nutrient drain by removing each new flower or seed pod, and a plant's

abscission zone at base of leaf where it joins the stem

Figure 24.23 Abscission zone in maple (*Acer*). The section is through the base of a leaf petiole.

three Douglas fir plants exposed to different day-night cycles during three year-long experiments

Figure 24.24 Effect of the relative length of day and night on Douglas firs. The plant at left was exposed to 12-hour light/12-hour dark cycles for a year. Its buds became dormant; daylength was too short. The plant at right was exposed to 20-hour light/4-hour dark cycles; growth continued. The middle plant was exposed to 12-hour light/11-hour dark cycles with 1 hour of light in the middle of the dark period. The light interruption prevented bud dormancy; it caused Pfr formation at a sensitive time in the normal day-night cycle.

Figure 24.25 A profusion of new leaves harvesting sunlight—a sure sign of spring and the resurgence of plant growth.

leaves and stems stay vigorous and green much longer. Gardeners routinely remove flower buds from many kinds of plants to maintain vegetative growth.

Dormancy

In autumn, when daylight hours decrease, new growth slows or stops in many plants, even if temperatures remain mild, the sky is bright, and water is plentiful. When metabolic activities idle and growth stops under conditions that seem (to us) suitable for growth, a plant has entered **dormancy**. Its buds normally won't resume growth until early spring, for the plant has responded to a convergence of environmental cues—in this case, fewer daylight hours, longer, cold nights, and nitrogen-depleted, dry soil. Recognition of multiple dormancy cues is adaptive. If, say, temperature were the only cue, growth might continue through warm days in autumn, then winter frost would damage or kill plant parts.

Figure 24.24 shows how interrupting a long dark period with red light prevents dormancy in Douglas firs. In nature, buds may enter dormancy because less Pfr forms when daylength decreases in late summer.

A dormancy-breaking process works between fall and spring. Often temperatures become milder, rains begin, and nutrients have accumulated. With the return of favorable conditions, the cycle of life begins anew. Seeds germinate; buds resume growth and give rise to new leaves, then to flowers (Figure 24.25).

Dormancy-breaking mechanisms probably involve gibberellins and abscisic acid. But other cues, such as exposure to low winter temperatures for a certain time, vary among species. In Utah, for example, Delicious apples require 1,230 hours near 43°F (6°C) and apricots require only 720 hours. Generally, trees growing in the southern United States require less cold exposure than those growing in northern states and Canada.

Multiple environmental cues—such as changes in daylength, temperature, moisture, and nutrient availability—influence hormonal secretions that stimulate or inhibit growth processes during the life cycles of flowering plants.

SUMMARY

1. Sexual reproduction is the main reproductive mode of flowering plant life cycles. A diploid sporophyte (spore-producing plant body) alternates with haploid gametophytes (gamete-producing bodies).

    a. The sporophyte is a multicelled vegetative body with roots, stems, leaves, and, at some point, flowers. Air currents, water currents, and animals coevolved with flowers and serve as their pollinating agents. The gametophytes form in male and female floral parts.

    b. Many flowering plants also reproduce asexually. They do this naturally (as by runners, rhizomes, and bulbs) and artificially (as by cuttings and grafting).

2. Flowers typically have sepals, petals, one or more stamens (male reproductive structures), and carpels (female reproductive structures), most or all attached to a receptacle, the modified end of a floral shoot.

    a. Anthers of stamens contain pollen sacs in which cells divide by meiosis. A wall develops around each resulting haploid cell (microspore), which develops into a sperm-bearing pollen grain (male gametophyte).

    b. A carpel (or two or more fused carpels) has an ovary where eggs develop, fertilization takes place, and seeds mature. A stigma (sticky or hairy surface tissue) above the ovary captures pollen grains and promotes their germination. Ovules form on the ovary's inner wall. Each consists of a female gametophyte with an egg cell, endosperm mother cell, a surrounding tissue, and one or two protective layers called integuments.

3. At double fertilization, one sperm nucleus fuses with an egg nucleus to form a diploid zygote. The other sperm nucleus and both nuclei of another cell inside the female gametophyte fuse to form a cell that will give rise to endosperm, a nutritive tissue.

4. After fertilization, the endosperm forms, the ovule expands, the embryo sporophyte develops, and the integuments harden and thicken. A fully matured ovule is a seed (its integuments have become the seed coat). While seeds form, ovaries develop into fruits, which will help protect and disperse the seeds.

5. After dispersal, seeds germinate—the embryo inside absorbs water, resumes growth, and breaks through the seed coat. The seedling increases in volume and mass. Its tissues and organs develop. Later, fruits and new seeds form; older leaves drop. Plant hormones govern these and other patterns of growth and development. They also help adjust the patterns in response to local conditions and to environmental rhythms, including seasonal changes in daylength and temperature.

6. We know of five classes of plant hormones. Auxins and gibberellins promote stem elongation. Cytokinins promote cell division and leaf expansion, and retard leaf aging. Abscisic acid promotes bud and seed dor-mancy, and limits water loss by triggering stomatal closure. Ethylene promotes fruit ripening and abscission.

7. Plant parts make tropic responses to light, gravity, and other environmental conditions. Hormones induce a difference in the rate and direction of growth on two sides of the part, which causes it to turn or move.

8. A biological clock is an internal time-measuring mechanism that has a biochemical basis.

    a. In photoperiodism, plants respond to a change in the relative length of daylight and darkness, as occurs seasonally. The blue-green pigment phytochrome is the switching mechanism of the clock that promotes or inhibits germination, stem elongation, leaf expansion, stem branching, and flower, fruit, and seed formation.

    b. Long-day plants flower in spring or summer, when daylength is long relative to night. Short-day plants flower when daylength is relatively short. Light does not regulate the flowering of day-neutral plants.

9. Senescence is the sum total of processes leading to the death of a plant or plant structure. Dormancy is a state in which a biennial or perennial plant stops growing even when conditions appear suitable for continued growth. A decrease in Pfr levels may trigger dormancy.

Review Questions

1. Label the floral parts. Explain floral function by relating the parts to events in the life cycle of flowering plants. *406–409*

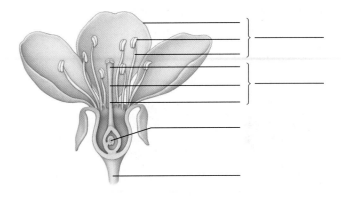

2. Distinguish between these terms:
    a. Sporophyte and gametophyte *406*
    b. Megaspore and microspore *408*
    c. Pollination and fertilization *408–409*
    d. Pollen grain and pollen tube *408–409*
    e. Ovule and female gametophyte *408*
    f. Senescence and dormancy *422*

3. Describe the steps by which a seven-cell, eight-nucleate embryo sac (a type of female gametophyte) forms. *408–409*

4. List the five known types of plant hormones and describe the function of each. *416–417*

5. What is phytochrome, and what is its role in flowering or some other process? *420–421*

6. Define plant tropism and give a specific example. *418–419*

Self-Quiz *(Answers in Appendix IV)*

1. Male gametophytes produce _____ , female gametophytes produce _____ .
   a. megaspores; eggs
   c. eggs; sperm
   b. sperm; microspores
   d. sperm; eggs

2. Seeds are mature _____ and fruits are mature _____ .
   a. ovaries; ovules
   c. ovules; ovaries
   b. ovules; stamens
   d. stamens; ovaries

3. A _____ is a closed vessel that contains an ovary in which eggs develop, fertilization occurs, and seeds mature.
   a. pollen sac
   c. receptacle
   b. carpel
   d. sepal

4. After meiosis within pollen sacs, haploid _____ form.
   a. megaspores
   c. stamens
   b. microspores
   d. sporophytes

5. Following meiosis in ovules, _____ megaspores form.
   a. two
   c. six
   b. four
   d. eight

6. The seed coat forms from which structure(s)?
   a. integuments
   c. endosperm
   b. ovary
   d. residues of sepals

7. Cotyledons develop as part of all flowering plant _____ .
   a. seeds
   c. fruits
   b. embryos
   d. ovaries

8. Plant growth depends on _____ .
   a. cell division
   c. hormones
   b. cell enlargement
   d. all of the above are correct

9. Light of _____ wavelengths causes phytochrome to switch from inactive to active form; light of _____ wavelengths has the opposite effect.
   a. red; far-red
   c. far-red; red
   b. red; blue
   d. far-red; blue

10. The flowering process is commonly a _____ response.
    a. phototropic
    c. photoperiodic
    b. gravitropic
    d. thigmotropic

11. Abscission occurs during _____ .
    a. seed germination
    c. senescence
    b. flowering
    d. dormancy

12. Match the terms with their appropriate description.
    ____ double fertilization
    a. formation of zygote and first cell of endosperm
    ____ ovule
    b. outcome of two species interacting in close ecological fashion over geologic time
    ____ mature female gametophyte
    ____ asexual reproduction
    c. contains a female gametophyte and has potential to be a seed
    ____ coevolution
    d. an embryo sac, commonly with seven cells (one has two nuclei)
    e. mitotic cell division at bud or node produces a new plant

13. Match the plant reproduction and development terms.
    ____ gibberellin
    a. promotes stem elongation
    ____ senescence
    b. unequal growth following contact with solid objects
    ____ phytochrome
    ____ phototropism
    c. switching mechanism for photoperiodism
    ____ thigmotropism
    d. response to blue light, mainly
    e. all processes leading to death of plant or plant part

Figure 24.26 Sunflowers (*Helianthus*).

Critical Thinking

1. "Solar tracking" refers to the observation that many plants are able to maintain the flat blades of their leaves at right angles to the sun throughout the day. Sunflowers are an example (Figure 24.26). This tropic response maximizes light harvest by leaves. Suggest the name of a molecule that might be involved in the response.

2. Observe several kinds of flowers growing in the area where you live. Given the coevolutionary links between flowering plants and their pollinators, describe what sorts of pollination agents your floral neighbors might depend upon.

3. Elaine, a plant physiologist, succeeded in cloning genes for insect resistance into petunia cells. How can she use tissue culture propagation to produce many petunia plants with the good genes?

4. Before they ripen and the seeds inside mature, cherries, apples, peaches, and many other fruits that are otherwise tasty to animals have bitter- or sour-tasting flesh. Develop a hypothesis of how this feature improves the odds for the plant's reproductive success.

5. Belgian scientists isolated a mutant of wall cress (*Arabidopsis thaliana*) that produces excess amounts of auxin. Predict what some of this mutant plant's phenotypic traits might be.

Selected Key Terms

abscisic acid *416*
abscission *422*
auxin *416*
biological clock *420*
carpel *407*
circadian rhythm *420*
coevolution *404*
coleoptile *416*
cotyledon *410*
cytokinin *416*
dormancy *422*
double fertilization *409*
endosperm *408*
ethylene *416*
flavoprotein *419*
flower *406*
fruit *410*
gametophyte *406*
germination *414*
gibberellin *416*
gravitropism *418*
herbicide *416*
hormone *416*
imbibition *414*
megaspore *408*
microspore *408*
ovary *407*
ovule *408*
parthenogenesis *412*
photoperiodism *420*
phototropism *418*
phytochrome *420*
plant tropism *418*
pollen grain *407*
pollination *408*
pollinator *404*
primary root *414*
seed *410*
senescence *422*
sporophyte *406*
stamen *407*
statolith *418*
tendril *419*
thigmotropism *419*
tissue culture propagation *412*
vegetative growth *412*
vine *419*

Readings

Proctor, M., and P. Yeo. 1973. *The Pollination of Flowers*. New York: Taplinger. Beautifully illustrated introduction to pollination.

Raven, P., R. Evert, and S. Eichhorn. 1992. *Biology of Plants*. Fifth edition. New York: Worth.

FACING PAGE: *How many and what kinds of body parts does it take to function as a lizard in a tropical forest? Make a list of what comes to mind as you start reading Unit VI, then see how resplendent the list can become at the unit's end.*

25 TISSUES, ORGAN SYSTEMS, AND HOMEOSTASIS

Meerkats, Humans, It's All the Same

After a cold night in Africa's Kalahari Desert, animals small enough to fit inside a coat pocket emerge stiffly from their burrows. These "meerkats" are a type of mongoose. They stand on their hind legs and face east, exposing their chilled bodies to the warm rays of the morning sun (Figure 25.1). Meerkats don't know it, but sunning behavior helps their enzymes. If their body's internal temperature were to fall below a tolerable range, enzyme activity would drop sharply. With such a change in enzyme activity, metabolism would suffer.

Once meerkats warm up, they fan out from their burrows and look for food. Into the meerkat gut go insects and an occasional lizard. These are pummeled, dissolved, and then digested into glucose and other nutritious tidbits small enough to move across the gut wall, into the bloodstream, and on to cells throughout the body. In cells, aerobic machinery cracks nutrients apart and so releases vital energy. A respiratory system works with the bloodstream to supply the machinery with oxygen and take away its carbon dioxide leftovers.

Figure 25.1 In the Kalahari Desert, gray meerkats (*Suricata suricatta*) line up and face the warming rays of the morning sun, just as they do every morning. This simple behavior helps maintain internal body temperature. How animals function in their environment is the subject of this unit.

All of this activity changes the composition and volume of the "internal environment"—blood and the fluids bathing the body's cells. Drastic changes in those fluids would kill cells, but a urinary system works to keep this from happening. Governing this system and all others are the body's command posts—a nervous system and an endocrine system. These work together and mobilize the body as a whole for everything from simple housekeeping tasks to heart-thumping flights from predators.

And so meerkats start us thinking about this unit's central topics: how an animal body is structurally put together (its *anatomy*) and how the body functions (its *physiology*). This chapter provides us with an overview of the animal tissues and organ systems that we will be considering. It also introduces the important concept of **homeostasis**—of stability in the internal environment, brought about by the intricately coordinated activities of the body's cells, tissues, organs, and organ systems.

Amazingly, the body of all complex animals consists of only four basic types of tissues. These are epithelial, connective, muscle, and nervous tissues. A **tissue** is an interacting group of cells and intercellular substances that take part in one or more specific tasks. As one example, muscle tissue takes part in contraction. An **organ** consists of different tissues that are organized in

specific proportions and patterns. Thus every vertebrate heart has predictable proportions and arrangements of epithelial, connective, muscle, and nervous tissues. An **organ system** consists of two or more organs that are interacting physically, chemically, or both in a common task, as when a heart and blood vessels circulate blood.

Cells, tissues, organs, and organ systems split up the work, so to speak, in ways that contribute to the survival of the animal as a whole. This is sometimes known as a **division of labor**. By the end of this unit, you may have an abiding appreciation of the sheer magnitude of the division of labor among the separate parts—and of the extent to which their activities are integrated. As you will see, regardless of whether you look at a flatworm or salmon, a meerkat or human, the complex animal is structurally and physiologically adapted to perform four overriding tasks:

1. *Maintain internal operating conditions within a tolerable range even as external conditions change.*

2. *Acquire nutrients and other materials, distribute them through the body, and dispose of wastes.*

3. *Protect itself against injury or attack from viruses, bacteria, and other disease-causing agents.*

4. *Reproduce, and often help nourish and protect the new individuals during their early growth and development.*

KEY CONCEPTS

1. The cells of most animals interact at three levels of organization—in tissues, many of which are combined in organs, which are components of organ systems.

2. Most animals are constructed of only four types of tissues: epithelial, connective, muscle, and nervous tissues.

3. Each cell engages in basic metabolic activities that assure its own survival. At the same time, cells of a tissue or organ perform activities that contribute to the survival of the animal as a whole.

4. The body's internal environment consists of all fluids that are not inside cells—that is, the blood and interstitial fluid.

5. The combined contributions of cells, tissues, organs, and organ systems help maintain stability in the internal environment, which is required for individual cell survival. This concept helps us understand the functions of any organ or organ system.

6. Homeostasis is the formal name for stable operating conditions in the internal environment.

25.1 EPITHELIAL TISSUE

General Characteristics

An epithelial tissue is commonly called **epithelium** (plural, epithelia). This tissue has a free surface, which faces some type of body fluid or the outside environment. *Simple* epithelium, with a single layer of cells, functions as a lining for body cavities, ducts, and tubes. *Stratified* epithelium, with two or more cell layers, typically functions in protection, as it does in skin. Figure 25.2 shows a few examples of this type of animal tissue.

All cells in epithelia are close together, with little intervening material. As is true of nearly every animal tissue, specialized junctions provide both structural and functional links between the individual cells, which absorb, synthesize, and secrete a variety of substances.

Cell-to-Cell Contacts

Figure 25.3 shows three types of cell-to-cell contacts in epithelium and other tissues. **Tight junctions** prevent substances from leaking across the tissue. **Adhering junctions** cement cells together. **Gap junctions** help cells communicate by promoting the rapid transfer of ions and small molecules among them.

Consider the epithelial lining of your stomach. If the highly acidic, gastric fluid inside the stomach were to leak through this lining, it would start digesting your body's proteins instead of those brought in with your meals. (Actually, this is an end result of a peptic ulcer, as described in Section 30.4.) Tight junctions in this

free surface of epithelium

epithelial cells
basement membrane
connective tissue

a

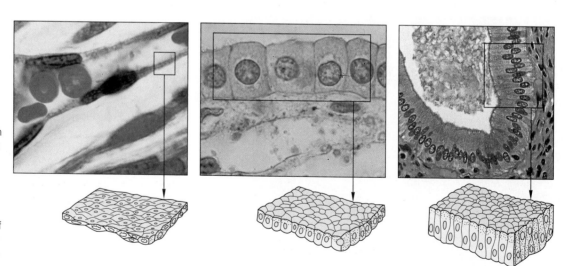

Figure 25.2 (**a**) Some of the characteristics of epithelium. All epithelia have a free surface, and a basement membrane is interposed between the opposite surface and an underlying connective tissue. The boxed diagram shows this arrangement in simple epithelium, which consists of a single layer of cells. The micrograph shows the upper portion of stratified epithelium. It has more than one layer of cells, which become more flattened near the surface.

(**b**) Examples of simple epithelium, showing the three basic cell shapes in this type of tissue.

b

TYPE: Simple squamous

DESCRIPTION: Single layer of flattened cells

COMMON LOCATIONS: Blood vessel walls; air sacs of lungs

FUNCTION: Diffusion

TYPE: Simple cuboidal

DESCRIPTION: Single layer of cubelike cells; free surface may have microvilli (absorptive structures)

COMMON LOCATIONS: Glands and nephrons (slender tubes) in kidneys

FUNCTION: Secretion, absorption

TYPE: Simple columnar

DESCRIPTION: Single layer of tall, slender cells; free surface may have microvilli

COMMON LOCATIONS: Part of lining of gut and respiratory tract

FUNCTION: Secretion, absorption

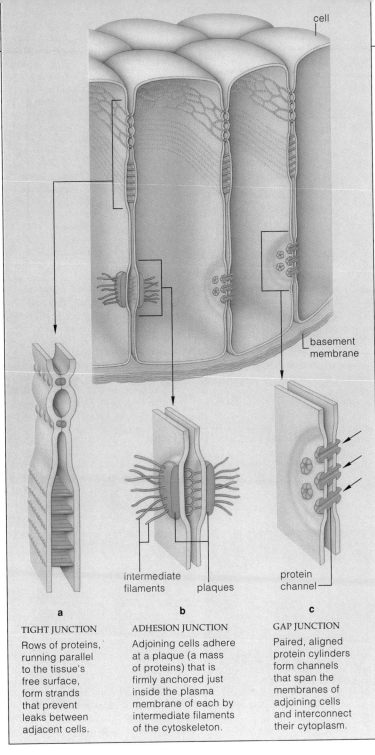

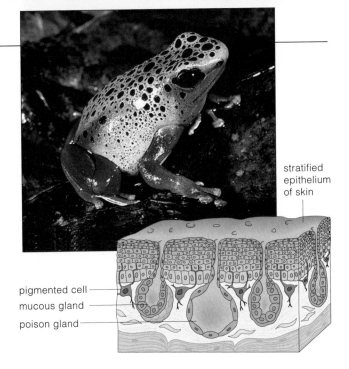

stratified
epithelium
of skin

pigmented cell
mucous gland
poison gland

Figure 25.4 Glandular epithelium of frogs. This frog, of the genus *Dendrobates*, produces one of the most lethal glandular secretions known. Some natives of a tribe in Colombia use this exocrine gland secretion to poison the tips of darts, which they shoot through blowguns. Branching into the lower epithelial layers of frog skin are pigment-rich cells that produce surface coloration. The showy, distinctive coloration of poisonous frogs, including this one, warn away potential predators.

---

a

TIGHT JUNCTION

Rows of proteins, running parallel to the tissue's free surface, form strands that prevent leaks between adjacent cells.

b

ADHESION JUNCTION

Adjoining cells adhere at a plaque (a mass of proteins) that is firmly anchored just inside the plasma membrane of each by intermediate filaments of the cytoskeleton.

c

GAP JUNCTION

Paired, aligned protein cylinders form channels that span the membranes of adjoining cells and interconnect their cytoplasm.

intermediate filaments   plaques   protein channel

Figure 25.3 Examples of cell junctions.

(**a**) In some epithelia, protein strands form tight seals that ring each cell and seal it to its neighbors. The tight junctions prevent substances from leaking across the free epithelial surface. The substances reach the tissues below only by passing through the cells, which have mechanisms that can control their passage.

(**b**) Adhesion junctions are like spot welds that "cement" cells of epithelium (and all other tissues) together so that they function as a unit. They are abundant in the skin's surface layer and other tissues subjected to abrasion.

(**c**) Gap junctions promote diffusion of ions and small molecules from cell to cell. They are abundant in the heart, liver, and other organs in which cell activities must be rapidly coordinated.

---

epithelium and others form a leakproof barrier between the cells near their free surface. As Figure 25.3 shows, certain junctions in the tissue serve as spot welds and as open channels between cells.

Glandular Epithelium

Glands are secretory cells or multicelled structures that are derived from epithelium and often remain linked to it. **Exocrine glands** secrete mucus, saliva, earwax, milk, oil, digestive enzymes, and other cell products. Usually the products are released onto a free epithelial surface through ducts or tubes. Figure 25.4 shows an example.

By contrast, **endocrine glands** have no ducts. Their products—hormones—are secreted directly into fluid bathing the gland. Typically, the bloodstream picks up hormone molecules and distributes them to target cells elsewhere in the body.

Epithelia are sheetlike tissues with one free surface. Different types line the body's surface, cavities, ducts, and tubes.

Profuse cell-to-cell contacts bind epithelial cells closely together, with little intercellular material between them.

Glands are secretory cells or multicelled structures derived from epithelium and often connected to it.

Of all tissues in complex animals, connective tissues are the most abundant and widely distributed. They range from connective tissue proper to specialized types, which include cartilage, bone, adipose tissue, and blood (Table 25.1). In all types except blood, fibroblasts and other kinds of cells secrete fibers of collagen or elastin, which are structural proteins. (This is the collagen that plastic surgeons use to plump wrinkled skin, sunken acne scars, and lips.) Fibroblasts, too, secrete modified polysaccharides. Secreted material accumulates between cells and fibers, as the tissue's "ground substance."

Connective Tissue Proper

All of these tissues have mostly the same components but in different proportions. **Loose connective tissue** has its fibers and cells loosely arranged in a semifluid ground substance (Figure 25.5a). Often it serves as a support framework for epithelium. Besides fibroblasts, it contains infection-fighting white blood cells. When small cuts or other wounds allow bacteria to breach the skin or lining of the digestive, respiratory, or urinary tracts, these cells mount an early counterattack.

Dense, irregular connective tissue contains fibers, mostly collagen-containing ones, and a few fibroblasts.

Table 25.1 Types of Connective Tissue
Connective tissue proper:
Loose connective tissue
Dense, irregular connective tissue
Dense, regular connective tissue (ligaments, tendons)
Specialized connective tissue:
Cartilage
Bone
Adipose tissue
Blood

This tissue forms protective capsules around organs that don't stretch much. It also is present in the deeper part of skin (Figure 25.5b). **Dense, regular connective tissue**, which has parallel bundles of many collagen fibers, resists being torn apart. Rows of fibroblasts often intervene between the bundles. This is true of tendons, which attach muscles to bones (Figure 25.5c) and of elastic ligaments, which attach bones to each other.

Specialized Connective Tissue

Intercellular material of **cartilage** is solid yet pliable, like solid rubber, and resists compression. The material

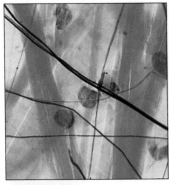
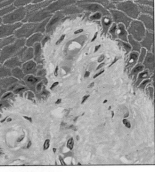
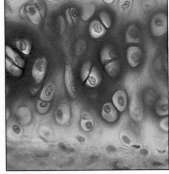

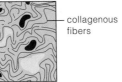

TYPE: Loose connective tissue
DESCRIPTION: Fibroblasts, other cells, plus fibers loosely arranged in semifluid ground substance
COMMON LOCATIONS: Under the skin and most epithelia
FUNCTION: Elasticity, diffusion

a

TYPE: Dense, irregular connective tissue
DESCRIPTION: Collagenous fibers, fibroblasts, less ground substance
COMMON LOCATIONS: In skin and capsules around some organs
FUNCTION: Support

b

TYPE: Dense, regular connective tissue
DESCRIPTION: Collagen fibers in parallel bundles, long rows of fibroblasts, little ground substance
COMMON LOCATIONS: Tendons, ligaments
FUNCTION: Strength, elasticity

c

TYPE: Cartilage
DESCRIPTION: Cells embedded in pliable, solid ground substance
COMMON LOCATIONS: Ends of long bones, nose, parts of airways, skeleton of vertebrate embryos
FUNCTION: Support, flexibility, low-friction surface for joint movement

d

Figure 25.5 Examples of connective tissue proper and of specialized connective tissue.

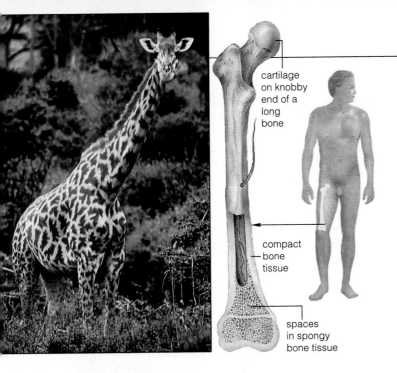

Figure 25.6 Cartilage and bone tissue. Spongy bone tissue has tiny, needlelike hard parts with spaces between. Compact bone tissue is more dense. Bone is a splendid load-bearing tissue that resists compression. It was the basis of remarkable increases in the body size of many land vertebrates, including giraffes. It gives big animals selective advantages. Among other things, they can ignore most predators with impunity, roam farther for food and water, and heat up and cool off more slowly than small animals (because of a lower surface-to-volume ratio and greater heat production).

cartilage on knobby end of a long bone

compact bone tissue

spaces in spongy bone tissue

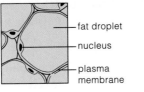

compact bone tissue

space that contained living bone cell (osteocyte)

fat droplet

nucleus

plasma membrane

TYPE: Bone tissue

DESCRIPTION: Collagen fibers, ground substance hardened with calcium

COMMON LOCATIONS: Bones of vertebrate skeleton

FUNCTION: Movement, support, protection

e

TYPE: Adipose tissue

DESCRIPTION: Large, tightly packed fat cells occupying most of ground tissue

COMMON LOCATIONS: Under skin, around heart, kidneys

FUNCTION: Energy reserves, insulation, padding

f

red blood cell   platelet   plasma   white blood cell

Figure 25.7 Some components of human blood, a vascular tissue. Its straw-colored, liquid matrix (plasma) is mostly water in which nutrients, diverse proteins, oxygen, carbon dioxide, ions, and other substances are dissolved.

is produced by cells that in time become imprisoned in small cavities in their own secretions (Figure 25.5*d*). Cartilage elements in vertebrate embryos are structural models for bones that generally replace them. Cartilage maintains the shape of the nose, outer ear, and other body parts. It cushions joints between adjacent bones of the vertebral column, limbs, hands, and elsewhere.

Bone (Figures 25.5*e* and 25.6) is the weight-bearing tissue of vertebrate skeletons, which support or protect softer tissues and organs. Minerals harden this tissue; its collagen fibers and ground substance are loaded with calcium salts. Limb bones, such as the ones in your legs, interact with muscles that are attached to them to bring about movements. Tissues in some bones also are production sites for blood cells.

Adipose tissue is so chockful of large fat cells, it no longer looks like a connective tissue (Figure 25.5*f*). The body's excess carbohydrates and proteins are converted to storage fats and tucked away in this tissue. Adipose tissue is richly supplied with blood, which serves as an immediately accessible "highway" for moving fats to and from the tissue's individual cells.

Blood, derived mainly from connective tissue, has transport functions. Circulating in its fluid medium, plasma, are a great many red blood cells, white blood cells, and platelets (Figure 25.7). Red blood cells deliver oxygen to metabolically active tissues, and carry carbon dioxide and other wastes away from them. Plasma is largely water, but it has a great number of different kinds of proteins, ions, and other substances dissolved in it. Chapter 27 describes this complex tissue.

Diverse connective tissues bind together, support, strengthen, protect, and insulate other tissues in the body.

Most connective tissues consist of protein fibers and a variety of cells in a ground substance. One type, blood, is a fluid tissue. Another type, adipose tissue, serves as a reservoir of stored energy.

In muscle tissue alone, cells contract (that is, shorten) in response to stimulation, then lengthen and so return to their uncontracted state. Many long, cylindrical cells are arranged in parallel in these tissues. Their coordinated contraction and relaxation help move the body and its individual parts. There are three types of muscle tissue, called skeletal, smooth, and cardiac muscle tissues.

Certain muscles connect with the bones of your skeleton. They are composed of **skeletal muscle tissue** (Figure 25.8a). In a typical muscle, such as the biceps, striated (striped) skeletal muscle cells are bundled together in parallel. A tough connective tissue sheath encloses several bundles of the muscle cells, as Figure 25.9 indicates. The structure and function of skeletal muscle tissue are topics of the next chapter.

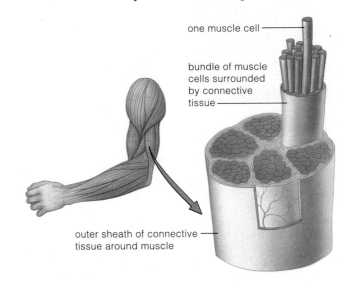

Figure 25.9 Location and general arrangement of cells in a typical skeletal muscle.

The contractile cells of **smooth muscle tissue** taper at both ends (Figure 25.8b). Cell junctions hold them together, and a connective tissue sheath encloses them. The wall of internal organs, such as blood vessels, the stomach, and the intestines contains this type of muscle tissue. Smooth muscle action is sometimes said to be "involuntary," because we usually are not able to make it contract merely by thinking about it (as we can do with skeletal muscle).

Cardiac muscle tissue is the contractile tissue of the heart (Figure 25.8c). Cell junctions fuse together the plasma membranes of cardiac muscle cells. Certain junctions at the fusion points allow the cells to contract as a unit. When one cell receives a signal to contract, its neighbors are also stimulated into contracting.

Muscle tissue, which can contract (shorten) in response to stimulation, helps move the body and specific body parts.

Skeletal muscle alone is attached to bones. Smooth muscle is a component of internal organs. Cardiac muscle makes up the contractile walls of the heart. Connective tissue sheathes all three types.

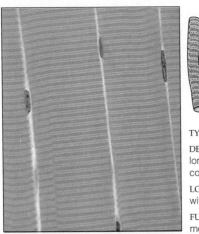
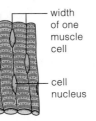

width of one muscle cell

cell nucleus

TYPE: Skeletal muscle

DESCRIPTION: Bundles of long, cylindrical, striated, contractile cells

LOCATION: Associated with skeleton

FUNCTION: Locomotion, movement of body parts

a

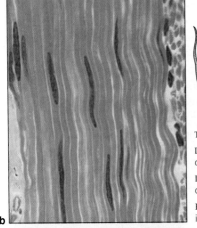
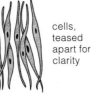

cells, teased apart for clarity

TYPE: Smooth muscle

DESCRIPTION: Contractile cells with tapered ends

LOCATION: Wall of internal organs, such as stomach

FUNCTION: Movement of internal organs

b

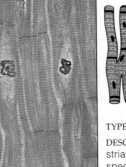

junction between adjacent cells

TYPE: Cardiac muscle

DESCRIPTION: Cylindrical, striated cells that have specialized end junctions

LOCATION: Wall of heart

FUNCTION: Pump blood within circulatory system

c

Figure 25.8 Characteristics and examples of skeletal muscle, smooth muscle, and cardiac muscle tissues.

25.4 NERVOUS TISSUE

Of all tissues, **nervous tissue** exerts greatest control over the body's responsiveness to changing conditions. This tissue consists of cells called neurons and many other cells that provide them with metabolic support. A **neuron** is an excitable cell. That is, when adequately stimulated, an electrical "message" can be propagated along the surface of its plasma membrane. This by itself wouldn't accomplish much, any more than a single ant at a picnic could accomplish much.

cell body of one motor neuron in this nervous tissue sample

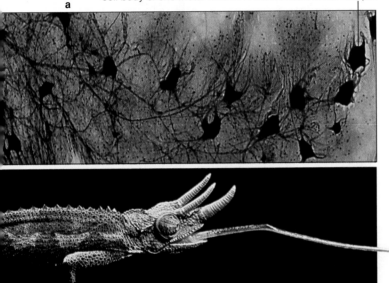

Figure 25.10 (**a**) A few motor neurons that form part of the communication lines within and between different regions of the human body. These relay signals from the brain or spinal cord to muscles and glands. Collectively, neurons sense environmental change, integrate a great number and variety of signals about changes, and initiate suitable responses. (**b**) Without them, this chameleon could not find an edible insect, calculate its distance, and command a long sticky tongue to uncoil with stunning speed.

However, complex animals have many neurons. You have more than a *hundred billion* of them, organized as communication lines that extend through your body. Some neurons detect specific changes in environmental conditions. Others coordinate immediate and long-term responses to change. Neurons of the sort shown in Figure 25.10 relay signals to muscles and glands that carry out suitable responses. How these remarkable cells function are topics of later chapters.

Neurons are the basic units of communication in nervous tissue. Different kinds detect specific stimuli, integrate information, and issue or relay commands for response.

25.5 *Focus on Science*

FRONTIERS IN TISSUE RESEARCH

As you will gather from your examination of examples throughout this unit, a tissue is more than a sum of its cells. As each new individual grows and develops, cells interact and become organized in specific ways to give rise to the body's diverse tissues. And as each *tissue* develops, its cells synthesize specific gene products that are vital for normal body functioning.

For many decades, medical researchers have attempted to find a way to construct artificial tissues in quantity in the laboratory. Today they can grow extensive sheets of epidermis from a patient and use it to regenerate skin that was destroyed through third-degree burns and other types of severe damage. They culture a small section of epidermis from the patient's skin with nutrients and growth factors. The cells proliferate and form *laboratory-grown epidermis*. When surgeons place an epidermal sheet over a wound, the cells in the sheet interact biochemically and structurally with underlying cells to regenerate the lost or damaged tissue.

On the horizon are *designer organs*—encapsulated, selected groups of living cells that can synthesize specific hormones, enzymes, and other substances. The idea is to package the cells within a bit of laboratory-grown epithelium that has been surgically snipped from a patient. Because the capsule is derived from cells of his or her own body, it will not be chemically recognized as "foreign" and attacked by the patient's immune system. As you will see in Chapter 28, such rejection of implants can have serious medical consequences.

Currently, biotechnologists are close to understanding how to synthesize the molecular cues that will allow such designer organs to stick to suitable sites in the body and become integral parts of its normal functioning.

Ultimately, the goal of this research is to put together packages of cells capable of producing specific life-saving substances that are absent in patients who suffer from genetic disorders or chronic diseases. For example, imagine the potential for people who have type I *diabetes mellitus*, a metabolic disorder that leads to an elevated blood level of glucose. Such people cannot produce insulin, the hormone that signals cells to take up glucose. They must receive insulin injections on a regular basis to prevent death. However, if a customized, insulin-secreting organ can be successfully installed inside the body, their daily injections of insulin might be a thing of the past.

25.6 ORGAN SYSTEMS

Overview of the Major Organ Systems

Figure 25.11 gives an overview of the organ systems of a typical vertebrate, the adult human. Figure 25.12 lists some terms that are used when describing the positions of the various organs, as well as the major body cavities in which they are located.

Each organ system contributes to the survival of all living cells in the body. You may think this is stretching things a bit. For example, how could muscles and bones help every tiny cell? Yet interactions between skeletal and muscular systems allow us to move about—toward sources of nutrients and water, for example. Some parts of the two systems help keep blood circulating, as when leg muscle contractions help move blood in veins back to the heart. Blood inside the circulatory system rapidly transports gases, nutrients, and other substances to cells, and transports products and wastes away from them. The respiratory system imports and exports the gases, certain skeletal muscles assist the respiratory system—and so it goes, throughout the entire body.

Figure 25.11 (*Below*) Human organ systems and their functions.

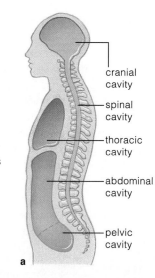

Figure 25.12 (**a**) The major cavities in the human body. (**b,c**) Directional terms and planes of symmetry for the vertebrate body. Notice how the *midsagittal* plane divides the body into right and left halves. Most vertebrates, such as fishes and rabbits, move with the main body axis parallel with Earth's surface. For them, *dorsal* pertains to their back or upper surface, and *ventral* pertains to the opposite, lower surface.

cranial cavity
spinal cavity
thoracic cavity
abdominal cavity
pelvic cavity

a

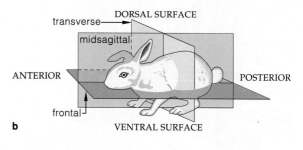

transverse
DORSAL SURFACE
midsagittal
ANTERIOR
POSTERIOR
frontal
b
VENTRAL SURFACE

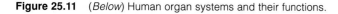

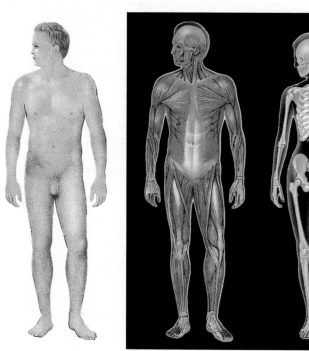

INTEGUMENTARY SYSTEM	MUSCULAR SYSTEM	SKELETAL SYSTEM	NERVOUS SYSTEM	ENDOCRINE SYSTEM	CIRCULATORY SYSTEM
Protect body from injury, dehydration, and some pathogens; control its temperature; excrete some wastes; receive some external stimuli	Move body and its internal parts; maintain posture; produce heat (by high metabolic activity)	Support and protect body parts; provide muscle attachment sites; produce red blood cells; store calcium, phosphorus	Detect both external and internal stimuli; control and coordinate responses to stimuli; integrate all organ system activities	Hormonally control body function; work with nervous system to integrate short-term and long-term activities	Rapidly transport many materials to and from cells; help stabilize internal pH and temperature

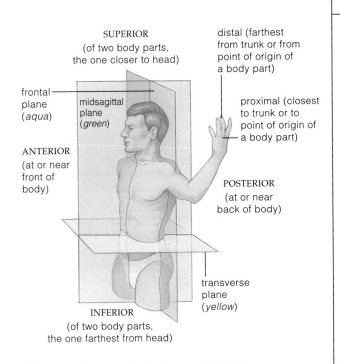

SUPERIOR
(of two body parts,
the one closer to head)

distal (farthest
from trunk or from
point of origin of
a body part)

frontal
plane
(*aqua*)

midsagittal
plane
(*green*)

proximal (closest
to trunk or to
point of origin of
a body part)

ANTERIOR
(at or near
front of
body)

POSTERIOR
(at or near
back of body)

transverse
plane
(*yellow*)

INFERIOR
(of two body parts,
the one farthest from head)

c Humans walk upright, with their main body axis
perpendicular to the ground. *Anterior* refers to the
front of the body (it corresponds to ventral, as in **b**).
Posterior refers to the back (it corresponds to dorsal).

Tissue and Organ Formation

Where do the tissues of organ systems come from? To
get a sense of how they originate, start with a sperm
and egg. Recall that sperm and eggs develop from germ
cells, which are immature reproductive cells. (All other
cells in the body are "somatic," after the Greek word for
body.) After a zygote forms at fertilization, mitotic cell
divisions form an early embryo. In vertebrates, the cells
become arranged as three primary tissues—ectoderm,
mesoderm, and endoderm. These are the embryonic
forerunners of all tissues in the adult. **Ectoderm** gives
rise to the skin's outer layer and to tissues of a nervous
system. **Mesoderm** gives rise to the tissues of muscles,
bones, and most of the circulatory, reproductive, and
urinary systems. **Endoderm** gives rise to the lining of
the digestive tract and to organs derived from it.

In general, vertebrates have the same kinds of organ systems.
Each organ system serves specialized functions, such as gas
exchange, blood circulation, and locomotion.

Vertebrate tissues, organs, and organ systems arise from three
primary tissues in the developing embryo. These primary
tissues are ectoderm, mesoderm, and endoderm.

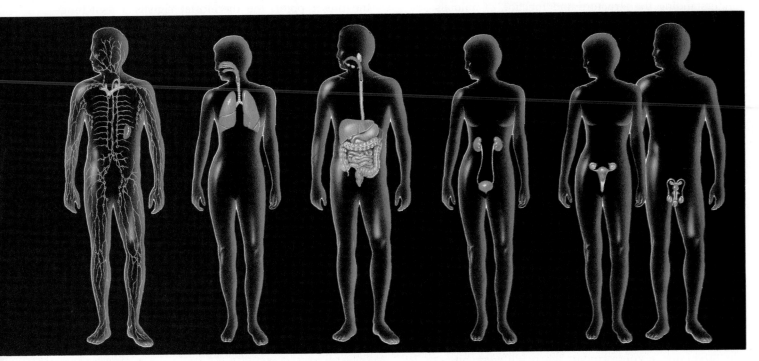

LYMPHATIC SYSTEM
Collect and return some tissue fluid to the bloodstream; defend the body against infection and tissue damage

RESPIRATORY SYSTEM
Rapidly deliver oxygen to the tissue fluid that bathes all living cells; remove carbon dioxide wastes of cells; help regulate pH

DIGESTIVE SYSTEM
Ingest food and water; mechanically, chemically break down food, and absorb small molecules into internal environment; eliminate food residues

URINARY SYSTEM
Maintain the volume and composition of internal environment; excrete excess fluid and blood-borne wastes

REPRODUCTIVE SYSTEM
Female: Produce eggs; after fertilization, afford a protected, nutritive environment for the development of new individual. *Male:* Produce and transfer sperm to the female. Hormones of both systems also influence other organ systems.

The Internal Environment

To stay alive, your cells must remain bathed in a fluid that offers nutrients and carries away metabolic wastes. In this they are no different from an amoeba or some other free-living, single-celled organism. The difference is, *trillions* of cells crowd together in your body. And they must draw nutrients from and dump wastes into the same 15 liters of fluid. That is less than 16 quarts.

The fluid *not* inside cells is **extracellular fluid**. Much of it is *interstitial*, meaning it occupies spaces between cells and tissues. The remainder is *plasma*, which is the fluid portion of blood. Interstitial fluid exchanges substances with the cells it bathes and with blood.

In functional terms, extracellular fluid is continuous with the fluid inside cells. That is why drastic changes in the fluid's composition and volume have drastic effects on cellular activities. Its ion concentrations are especially important in this regard. They must be maintained at levels that are compatible with cell survival. Otherwise, the animal itself cannot survive.

It makes no difference whether an animal is simple or complex. *The component parts of any animal work together to maintain the stable fluid environment required by all of its living cells.* This concept is absolutely central to understanding the structure and function of animals, and its key points may be summarized as follows. *First,* each cell of the animal body engages in basic metabolic activities that ensure its own survival. *Second,* the cells of a given tissue also perform one or more activities that contribute to the survival of the whole organism. *Third,* the combined contributions of individual cells, organs, and organ systems help maintain the stable **internal environment**—that is, the extracellular fluid—required for individual cell survival.

Mechanisms of Homeostasis

Homeostasis refers to stable operating conditions in the internal environment. Three components interact to maintain this state. They are called sensory receptors, integrators, and effectors. **Sensory receptors** are cells or cell parts that can detect a **stimulus**, which is a specific change in the environment. When someone kisses you, for example, there is a change in pressure on your lips. Receptors in the skin of your lips translate the stimulus into a signal that can be sent to the brain. Your brain is an **integrator**, a control point where different bits of information are pulled together in the selection of a response. It can send signals to your muscles or glands (or both). Muscles and glands are **effectors**—they carry out the response. In this particular case, the response might include flushing with pleasure and kissing the person back. Of course, you cannot engage in a kiss

indefinitely, for this would prevent you from eating and from carrying out other activities necessary to maintain operating conditions inside your body.

So how does your brain reverse the physiological changes induced by the kiss? Receptors only provide it with information about how things *are* operating. The brain also receives information about how things *should be* operating—that is, information from "set points." When physical or chemical conditions deviate sharply from a set point, the brain functions to bring them back to an effective operating range. It does this by way of signals that cause specific muscles and specific glands to increase or decrease their activity.

Feedback mechanisms are among the controls that help keep physical and chemical aspects of the body within tolerable ranges. For example, with a **negative feedback mechanism**, an activity alters a condition in the internal environment, and this triggers a response that reverses the altered condition (Figure 25.13).

Think of a furnace with a thermostat. A thermostat senses the air temperature and "compares" it against a preset point on a thermometer built into the furnace's control system. Whenever the temperature falls below the preset point, the thermostat signals a switching mechanism that turns on the heating unit. When the air becomes heated enough to match the prescribed level, the thermostat signals the switching mechanism, which shuts off the heat.

Similarly, feedback mechanisms help keep the body temperature of meerkats, humans, huskies, and many other animals near 37°C (98.6°F), even during hot or cold weather. Imagine a husky running around on a hot summer day. Its body gets hot—and receptors trigger events that slow down the whole dog *and* its cells. The husky flops down and rests in the shade of a tree. Moisture from its respiratory system evaporates from the tongue and carries some body heat with it (Figure 25.14). These and other control mechanisms counter the overheating

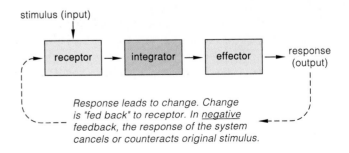

Figure 25.13 Components necessary for negative feedback at the organ level.

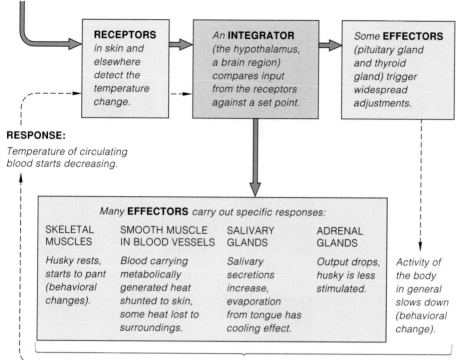

STIMULUS:

The husky is overactive on a hot, dry day and its body surface temperature rises.

RECEPTORS
in skin and elsewhere detect the temperature change.

An **INTEGRATOR**
(the hypothalamus, a brain region) compares input from the receptors against a set point.

Some **EFFECTORS**
(pituitary gland and thyroid gland) trigger widespread adjustments.

RESPONSE:

Temperature of circulating blood starts decreasing.

Many **EFFECTORS** *carry out specific responses:*

SKELETAL MUSCLES	SMOOTH MUSCLE IN BLOOD VESSELS	SALIVARY GLANDS	ADRENAL GLANDS
Husky rests, starts to pant (behavioral changes).	*Blood carrying metabolically generated heat shunted to skin, some heat lost to surroundings.*	*Salivary secretions increase, evaporation from tongue has cooling effect.*	*Output drops, husky is less stimulated.*

Activity of the body in general slows down (behavioral change).

The overall slowdown in activities results in less metabolically generated heat.

Figure 25.14 Homeostatic controls over the internal temperature of a husky's body. *Blue* arrows indicate the main control pathways. The *dashed line* shows how a feedback loop is completed.

by curbing the body's heat-generating activities and by giving up excess heat to the surroundings.

Sometimes **positive feedback mechanisms** operate. These set in motion a chain of events that *intensify* a change from an original condition—and after a limited time, the intensification reverses the change. Positive feedback is associated with instability in a system. For example, during sexual intercourse, chemical signals from a female's nervous system can induce her to make intense physiological responses to her sexual partner. These responses stimulate changes in her partner that stimulate the female even more—and so on until an explosive, climax level is reached. Normal conditions now return, and homeostasis prevails.

As another example, at the time of childbirth, a fetus exerts pressure on the wall of its mother's uterus. This stimulates the production and secretion of oxytocin, a hormone. Oxytocin causes wall muscles to contract and exert pressure on the fetus, which exerts more pressure on the wall, and so on until the fetus is expelled.

What we have been describing is a general pattern of monitoring and responding to a constant flow of information about an animal's internal and external environments. During all of this activity, organ systems operate together in astoundingly coordinated fashion.

Throughout this unit, we will be asking the following questions about their operation:

1. *What physical or chemical aspects of the internal environment are organ systems working to maintain as conditions change?*

2. *By what means are organ systems kept informed of the various changes?*

3. *By what means do they process incoming information?*

4. *What mechanisms are set in motion in response?*

As you will see in later chapters, operation of all organ systems is under neural and endocrine control.

Each living cell of the animal body engages in basic metabolic activities that ensure its own survival. Concurrently, cells of a given tissue perform one or more activities that contribute to the survival of the whole animal.

The combined contributions of cells, organs, and organ systems help maintain the stable internal environment (the extracellular fluid) required for individual cell survival.

Homeostatic control mechanisms help maintain physical and chemical aspects of the body's internal environment within ranges that are most favorable for cell activities.

SUMMARY

1. A tissue is an aggregation of cells and intercellular substances that perform a common task. An organ is a structural unit of different tissues combined in definite proportions and patterns that allow them to perform a common task. An organ system has two or more organs interacting chemically, physically, or both in ways that contribute to the survival of the body as a whole.

2. Epithelial tissues cover external body surfaces and line internal cavities and tubes. These tissues have one free surface exposed to body fluids or the environment.

3. A great variety of connective tissues bind together, support, strengthen, protect, and insulate other tissues. Most consist of fibers of structural proteins (especially collagen), fibroblasts, and other cells within a ground substance.

    a. Loose connective tissue has a semifluid ground substance. It occurs under skin and most epithelia.

    b. Dense, irregular connective tissue contains mostly collagen fibers and fibroblasts. It occurs in skin and forms protective capsules around many organs.

    c. Dense, regular connective tissue, with its parallel bundles of collagen fibers, provides structural support to tendons, skin, and other organs.

    d. Cartilage, with its solid yet pliable intercellular material, has structural and cushioning roles. Bone, the weight-bearing tissue of vertebrate skeletons, works with skeletal muscle to bring about movement.

    e. Blood, a specialized connective tissue, consists of plasma, cellular components, and dissolved substances. Adipose tissue (mostly fat cells), another specialized connective tissue, is a reservoir of stored energy.

4. Muscle tissues contract (shorten), then return to the resting position. They help move the body or parts of it. The three types are skeletal muscle, smooth muscle, and cardiac muscle tissue.

5. Nervous tissue detects and integrates information about internal and external conditions, and it governs the body's responses to change. Its neurons are the basic units of communication in nervous systems.

6. Tissues, organs, and organ systems work together to maintain the stable internal environment (that is, the extracellular fluid) required for individual cell survival. At homeostasis, conditions in the internal environment are most favorable for cell activities.

7. Feedback controls help maintain internal operating conditions for the body's cells. For example, in negative feedback, a change in some condition triggers the response that reverses the change.

8. Homeostasis depends on receptors, integrators, and effectors. Receptors detect stimuli (specific changes in the environment). Integrating centers (such as a brain) process the information and direct muscles and glands (the body's effectors) to carry out responses.

Review Questions

1. Describe the characteristics of epithelial tissue in general. Then describe the various types of epithelial tissues in terms of specific characteristics and functions. *428–429*

2. List the major types of connective tissues; add the names and characteristics of their specific types. *430–431*

3. Identify and describe the following tissues. *428–432*

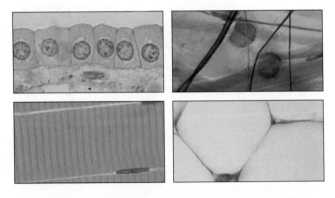

4. Identify this category of tissue and its characteristics. *433*

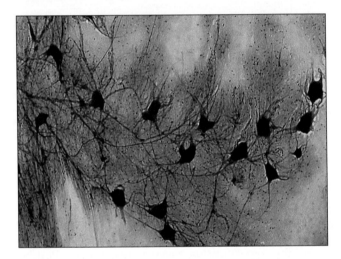

5. What type of cell serves as the basic unit of communication in nervous systems? *433*

6. Define animal tissue, organ, and organ system. List and define the functions of the major organ systems of the human body. *426–427, 434, 435*

7. Define extracellular fluid and interstitial fluid. *436*

8. Define homeostasis. *426, 436*

9. Describe two major kinds of homeostatic mechanisms of the human body. *436–437*

Self-Quiz *(Answers in Appendix IV)*

1. _____ tissues have closely linked cells and one free surface.
    a. Muscle   c. Connective
    b. Nervous   d. Epithelial

2. Most _____ tissues have cells that secrete fibers of collagen and elastin.
    a. muscle   c. connective
    b. nervous   d. epithelial

3. _____ has a semifluid ground substance and occurs under skin and most epithelia.
   a. Dense, irregular connective tissue
   b. Loose connective tissue
   c. Dense, regular connective tissue
   d. Cartilage

4. _____, a specialized connective tissue, is mostly plasma with cellular components and a variety of dissolved substances.
   a. Irregular connective tissue   c. Cartilage
   b. Blood   d. Bone

5. When you ingest excess carbohydrates and proteins, your body converts the excess amounts to storage fats, which accumulate in _____.
   a. connective tissue proper   c. adipose tissue
   b. dense connective tissue   d. both b and c

6. Components of _____ tissues detect and coordinate information about change, then control responses to those changes.
   a. Muscle   c. Connective
   b. Nervous   d. Epithelial

7. _____ tissues can shorten (contract).
   a. Muscle   c. Connective
   b. Nervous   d. Epithelial

8. Cells of complex animals _____ .
   a. survive through their own metabolism
   b. contribute to the survival of the whole animal
   c. help maintain extracellular fluid
   d. all of the above

9. At _____, physical as well as chemical aspects of the internal environment are being kept within tolerable ranges.
   a. positive feedback   c. homeostasis
   b. negative feedback   d. metastasis

10. In negative feedback mechanisms, _____ .
    a. a detected change brings about a response that tends to return internal operating conditions to the original state
    b. a detected change suppresses internal operating conditions to levels below the set point
    c. a detected change raises internal operating conditions to levels above the set point
    d. fewer solutes are fed back to the affected cells

11. Of the components that exert feedback control of organ activity, _____ detect specific changes in the environment, an _____ pulls together different bits of information and selects a suitable response, and _____ carry out the response.

12. Match the terms with the suitable description.
    ____ epithelium   a. somewhat pliable, like rubber
    ____ cartilage   b. covers or lines body surfaces
    ____ homeostasis   c. stable internal environment
    ____ muscles and glands   d. integrating center
    ____ positive feedback   e. the most common type of
    ____ negative feedback   homeostatic control mechanism
    ____ brain   f. effectors
                              g. chain of events intensifies the
                              original condition

Critical Thinking

1. *Anhidrotic ectodermal dysplasia*, a genetic disorder described in Section 12.8, is associated with a recessive allele on the mammalian X chromosome. Affected males or females have no sweat glands in tissues where the recessive allele is expressed. What type of tissue are we talking about?

2. Adipose tissue and blood are often said to be "atypical" connective tissues. Compared with other connective tissues, which of their features are *not* typical?

3. After graduating from high school, Jeff and Ryan set out on a trip through the desert roads of California and Arizona. One hot, dry morning in Joshua Tree National Monument, they saw an unusual rock formation that didn't appear to be too far from the road. They left the car and started hiking toward it. Their destination turned out to be farther away than they thought—and the sun's rays more relentless than they had anticipated. They reached the shade of the rocks in the early afternoon. Their canteen was nearly empty, and the physiological meaning of "thirst" made itself known to them in a scary way. They knew that they had to locate and drink water (or some other fluid), which is what the brain usually tells us to do when the body starts to get dehydrated. From what you read in this chapter, would you suspect that this *thirst behavior* is part of a positive or negative feedback control mechanism?

4. If you are familiar with vampire stories, which date from the Middle Ages or earlier, speculate on how they might have evolved among deeply superstitious folk who could not otherwise explain *porphyria*. This genetic disorder still affects about 1 in every 25,000 people. Those affected lack certain enzymes of a pathway leading to formation of heme, the iron-containing group of hemoglobin. A build-up of intermediates of this pathway (porphyrins) causes awful symptoms, especially after exposure to sunlight. Lesions and scars form on the skin. Hair grows thickly on the face and hands. As gums retreat from teeth, canines take on a fanglike appearance. Symptoms worsen upon exposure to many substances, including garlic and alcohol. Affected people avoid sunlight and aggravating substances, and get injections of heme from normal red blood cells.

Selected Key Terms

adhering junction 428	gap junction 428
adipose tissue 431	homeostasis 426
blood 431	integrator 436
bone 431	internal environment 436
cardiac muscle tissue 432	loose connective tissue 430
cartilage 430	mesoderm 435
dense, irregular connective tissue 430	negative feedback mechanism 436
	nervous tissue 433
dense, regular connective tissue 430	neuron 433
	organ 426
division of labor 427	organ system 427
ectoderm 435	positive feedback mechanism 437
effector 436	sensory receptor 436
endocrine gland 429	skeletal muscle tissue 432
endoderm 435	smooth muscle tissue 432
epithelium 428	stimulus 436
exocrine gland 429	tight junction 428
extracellular fluid 436	tissue 426

Readings

Bloom, W., and D. W. Fawcett. 1995. *A Textbook of Histology*. Twelfth edition. Philadelphia: Saunders.

Leeson, C. R., T. Leeson, and A. Paparo. 1988. *Textbook of Histology*. Philadelphia: Saunders.

Ross, M., L. Romrell, and G. Kaye. 1988. *Histology: A Text and Atlas*. Baltimore: Williams & Wilkins.

Vander, A., J. Sherman, and D. Luciano. 1994. "Homeostatic Mechanisms and Cellular Communication," in *Human Physiology*. Sixth edition. New York: McGraw-Hill.

26 PROTECTION, SUPPORT, AND MOVEMENT

Fancied Feathers and Muscle Mania

A male penguin has a bit of a problem when scouting for a female penguin, because superficially they both look too much alike. The only way he can find one is to drop a pebble at the feet of a likely prospect. If those feet belong to a male, the pebble may be perceived as an insult, and it may start a fight. If female, his stony overture to courtship might be ignored or fancied, depending on her receptivity at that moment.

Peacocks and many other males have no such problem. For their species, sexual dimorphism is visible and pronounced (Figure 26.1*a*). Maybe the peacock's eye-stopping feathers evolved through sexual selection, with peahens serving as the deciders of reproductive success. Charles Darwin certainly thought so. He saw extreme sexual dimorphism as the outcome of male competition for females and of female choice.

What about extreme sexual dimorphism in body size, which involves increases in muscle mass? This may be one measure of how much mammalian males invest in fighting capacity. Possibly the cost of securing and using resources for growth and maintenance of a massive body is offset by great reproductive rewards.

All of this might make you wonder: Exactly what are the "rewards" for astoundingly muscled human athletes (Figure 26.1*b*)? Is it a misdirected will to win,

or a sign of modern athletic competition? Consider this: Each year in the United States alone, a million or so athletes use anabolic steroids. The vast majority of professional football players have used them to increase "brute power." Even adolescent boys are known to use them as a way to gain a winning edge in wrestling, football, and weightlifting tournaments.

Anabolic steroids are synthetic hormones. They mimic testosterone—a sex hormone that, among other things, governs secondary sexual traits. Testosterone makes boys get a deeper voice; more hair on their face, underarms, and pubic skin; and greater muscle mass in the arms, legs, shoulders, and elsewhere. Besides this, testosterone stimulates heightened aggressive behavior, which is often associated with maleness.

Anabolic steroids stimulate the synthesis of protein molecules, including muscle proteins. Supposedly, they induce rapid gains in muscle mass and strength when taken during weight-training exercise programs. This claim is disputed; results of most studies are based on

Figure 26.1 Some phenotypic wonders of the integumentary, muscle, and skeletal systems on display. (**a**) A peacock wooing a peahen; his fanned-out tail feathers tell you she is nearby, even if out of camera range. (**b**) A human male with pumped-up biceps.

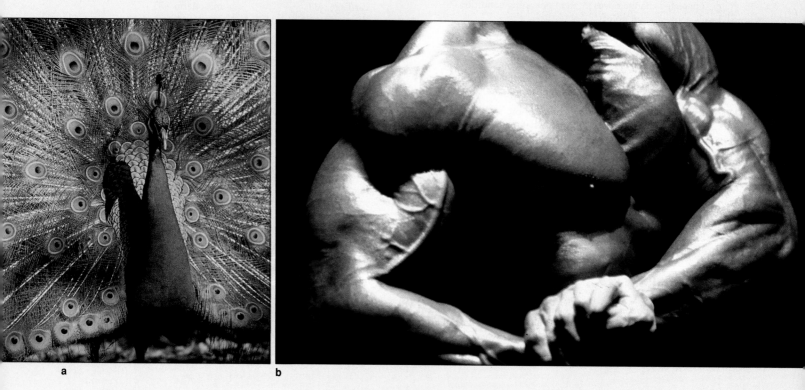

a   b

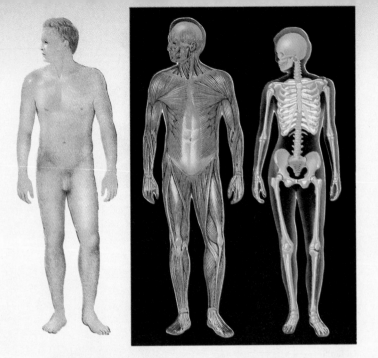

Figure 26.2 From left to right, overview of the integumentary, muscular, and skeletal system of a typical vertebrate body.

too few subjects. Even so, not everyone believes that these drugs do enough damage to outweigh the edge they presumably give in athletic competition—or the wealth and hero status they accord the "winners."

Yet steroid-using athletes suffer minor and major side effects. In men, acne, baldness, shrinking testes, and infertility are early signs of toxicity. The symptoms begin when a high blood level of anabolic steroids triggers a sharp decline in the body's production of testosterone. Anabolic steroids also may cause early heart disease. Even brief or occasional use may damage the kidneys or set the stage for cancer of the liver, testes, and prostate gland. In women, anabolic steroids deepen the voice, and they produce pronounced facial hair. Menstrual cycles become irregular. Breasts may shrink, and the clitoris may become grossly enlarged.

Not all steroid users develop severe physical side effects. Far more common are mental difficulties, called 'roid rage or *body-builder's psychosis*. In such cases, users become irritable and increasingly aggressive. Some men become wildly aggressive, uncontrollably manic, and delusional. One steroid user accelerated his car to high speed and deliberately drove it into a tree; and doesn't that make you wonder just how superior some of us are to a pebble-toting penguin.

This chapter focuses mainly on three organ systems that are responsible for the superficial features, shapes, and movements of vertebrates. Figure 26.2 is a starting point for your tour. Traveling from the outside in, this illustration gives an example of an integumentary system (skin and its derivatives), a muscular system, and a skeletal system. The story of feathers and muscles and many other splendid aspects of the vertebrate body begins here.

KEY CONCEPTS

1. Nearly all animals have an integument (an outer covering, such as skin), muscles, and a skeleton.

2. Skin protects the body from abrasion, ultraviolet radiation, bacterial attack, and other environmental insults. It also contributes to overall body functioning, as when it helps control moisture loss.

3. Many responses to changes in external and internal conditions involve movement of the animal body or parts of it. The skeletal system and muscular system interact to bring about many of these movements.

4. Three categories of skeletal systems are common in the animal kingdom. We call them hydrostatic skeletons, exoskeletons, and endoskeletons. Each has internal body fluids or structural elements, such as bones, against which the force of muscle contraction can be applied.

5. Bones—collagen-rich, mineralized organs—function in movement, protection and support of soft organs, and mineral storage. Blood cells form in some. Ligaments or cartilage bridges joints between bones, and tendons attach them to skeletal muscles.

6. Smooth muscle and cardiac muscle help move internal organs. Skeletal muscle helps move the body's limbs and other structural elements. In response to stimulation, cells of all three types of muscle tissue can contract (shorten).

7. Many threadlike structures (myofibrils) inside muscle cells are divided into sarcomeres, the basic units of contraction. Each sarcomere has parallel arrays of actin and myosin filaments. ATP-driven interactions between actin and myosin shorten sarcomeres and collectively account for contraction.

26.1 INTEGUMENTARY SYSTEM

Animals ranging from invertebrate worms to humans have an outer covering, or **integument** (after the Latin *integere*, meaning to cover). Most coverings are tough, pliable barriers against many environmental insults. As described in Section 20.10, the covering of insects, crabs, and other arthropods is a chitin-hardened **cuticle**.

The vertebrate integument, **skin**, includes structures derived from certain epidermal cells of its outer tissue layer (Figures 26.3 and 26.4). Variation exists within and between vertebrate groups. Among them we see hair, feathers, beaks, hooves, horns, claws, nails, quills, and other derived structures, and assorted epithelia and glands. Some fishes have hard scales; others have bare skin with a lathering of slime (Sections 21.4 and 21.5).

Functions of Skin

No garment ever made comes close to skin's qualities. What besides skin holds its shape after repeated stretchings and washings, blocks harmful rays from the sun, kills many bacteria on contact, holds in moisture, fixes small cuts and burns, *and* can last as long as you do? Skin also produces the vitamin D required for calcium metabolism. It plays a passive role in adjusting internal temperature; the nervous system can rapidly adjust the flow of blood (which carries metabolic heat) to and from skin's great numbers of tiny blood vessels. And signals from sensory receptor endings in skin help the brain assess what's going on in the outside world.

Structure of Skin

Skin is one of the largest organs. Your own skin weighs about 4 kilograms (9 pounds). Stretched out, its surface area would be 15 to 20 square feet. Most of your skin is thin as a paper towel. It thickens only on the foot's soles and other regions subjected to pounding or abrasion.

Skin has two regions—an outermost **epidermis** and an underlying **dermis** (Figure 26.3). The hypodermis (a tissue region below the dermis) anchors skin to underlying structures yet allows it to move a bit. Fat stored in the hypodermis insulates and cushions body parts.

Epidermis is mostly stratified epithelium, with great numbers of junctions knitting the cells together (Section 25.1). Epidermal cells originate deep in the tissue, but rapid, ongoing mitotic divisions push them toward the skin's free surface. (The rapid divisions also help skin mend quickly after cuts or burns.) Most of the cells are keratinocytes, which produce keratin (a tough, water-insoluble protein). By the time these cells reach skin's free surface, they are dead and flattened. Deeper in the epidermis, cells called melanocytes produce and donate melanin, a brownish-black pigment, to keratinocytes. Melanin screens out harmful wavelengths from the sun.

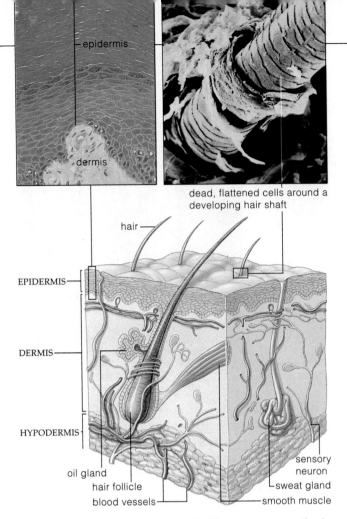

Figure 26.3 Structure of human skin. The uppermost portion is epidermis; the lower portion is dermis.

The distribution and activity of melanocytes affects skin color, as when the absence of melanin results in *albinism*. Hemoglobin (a red pigment in red blood cells) and carotene (a yellow-orange pigment) also impart color to skin. Hemoglobin shows through the epidermis and blood vessels, both of which are transparent.

The dermis is mostly dense connective tissue that resists daily stretching and other mechanical insults. Blood and lymph vessels, and receptor endings of sensory nerves, thread through the dermis. Nutrients from the bloodstream reach epidermal cells by diffusing through dermal ground tissue. Sweat glands, oil glands, and husklike cavities called hair follicles reside mostly in the dermis but are derived from epidermal cells.

Sweat gland secretions are 99 percent water, with dissolved salts, traces of ammonia and other metabolic by-products, vitamin C, and other substances. Humans have about 2.5 million sweat glands. One type abounds in the palms, soles, forehead, and armpits. These help control body temperature and even function in *cold sweats*—one of the responses you make when you are frightened, nervous, or merely embarrassed.

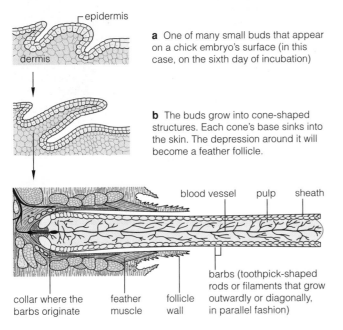

a One of many small buds that appear on a chick embryo's surface (in this case, on the sixth day of incubation)

b The buds grow into cone-shaped structures. Each cone's base sinks into the skin. The depression around it will become a feather follicle.

blood vessel   pulp   sheath

barbs (toothpick-shaped rods or filaments that grow outwardly or diagonally, in parallel fashion)

collar where the barbs originate

feather muscle

follicle wall

c A layer of thin, horny cells at a cone's surface differentiates into a sheath. An epidermal layer just beneath the sheath will give rise to the feather itself. The dermis beneath this epidermal layer is richly vascularized. It becomes the pulp, which nourishes the growing feather but does not contribute to its structure.

Figure 26.4 How feathers form.

Oil glands (sebaceous glands) lubricate and soften hair and the skin, and their secretions kill many surface bacteria. *Acne* is an inflammation of the skin that results from bacterial infection of oil gland ducts.

A **hair** is a flexible structure of mostly keratinized cells. Its root is embedded in skin; its shaft projects above the surface. Cells divide near the root's base, are pushed upward, then flatten and die. Flattened cells of the shaft's outer layer overlap like roof shingles (Figure 26.3). When mechanically abused, they tend to frizz out near the end of the hair shaft; we call these "split ends."

The average human scalp has about 100,000 hairs, although genes, nutrition, and hormones influence hair growth and density. Protein deficiency causes hair to thin (amino acids are required for keratin synthesis). So do high fever, emotional stress, and excess vitamin A intake. When the body produces abnormal amounts of testosterone, excessive hairiness (*hirsutism*) may result.

Skin consists of epidermis and dermis. The multiple layers of keratinized, melanin-shielded epidermal cells help the body conserve water, avoid damage by ultraviolet radiation, and resist mechanical stress.

Hairs, oil glands, sweat glands, and other structures derived from epidermal cells are embedded largely in the dermis. Blood vessels, lymph vessels, and receptor endings of sensory neurons also reside in the dermis.

SUNLIGHT AND SKIN

Are you someone who tans by rotating beneath the sun's rays or tanning lamps, like a chicken in an oven broiler? If so, remember what a broiler does to the chicken. The sun's ultraviolet wavelengths stimulate skin cells that produce melanin. Continued exposure increases the melanin in light skin and visibly darkens it—the sought-after "tan." Tanning does help protect against ultraviolet radiation. Even in naturally dark skin, however, prolonged exposure to sunlight causes elastin fibers in connective tissue of the dermis to clump together, so skin loses its resiliency. In time it starts to look like old shoe leather.

In this respect, tanning accelerates skin aging. As any person grows older, epidermal cells divide less often. Skin gets thinner and more susceptible to injury. Glandular secretions that once kept it soft and moistened dwindle. Collagen and elastin fibers in the dermis break down and become sparser, so skin loses elasticity and its wrinkles deepen. Besides excessive tanning, prolonged exposure to dry wind and tobacco smoke accelerates the aging process.

As if this weren't bad enough, prolonged exposure to ultraviolet radiation suppresses the immune system. For instance, infection-fighting phagocytes in the epidermis combat viruses and bacteria. Sunburns interfere with their functioning. This may be why sunburns can trigger small, painful blisters (*cold sores*) that announce the recurrence of a *Herpes simplex* infection. Nearly everyone harbors this virus. It remains hidden in the face, inside a ganglion (a cluster of neuron cell bodies). Sunburns and other stress factors can activate the virus. Virus particles move down the neurons to their endings in the skin. There they infect epithelial cells and cause painful skin eruptions.

Ultraviolet radiation also can activate proto-oncogenes that can trigger cancerous transformation of skin cells. *Epidermal skin cancers* start out as scaly, reddened bumps. They grow rapidly and can spread to adjacent lymph nodes unless they are surgically removed (Section 11.4).

Figure 26.5 How shoe-leather skin forms.

SKELETAL SYSTEMS

Skeletal Function

Many of an animal's responses to external and internal conditions involve the movement of the whole body or parts of it. The activation, contraction, and relaxation of muscle cells bring about these movements. But muscle cells alone cannot produce them. *All muscles require the presence of some medium or structural element against which the force of contraction can be applied.* A skeletal system fulfills this requirement.

Types of Skeletal Systems

Three types of skeletons are common. In a **hydrostatic skeleton**, muscles work against an internal body fluid and redistribute it within a limited space. (Like a filled waterbed, the confined fluid resists compression.) An **exoskeleton** has rigid, *external* body parts, such as a shell, that receive the applied force of contraction. An **endoskeleton** has rigid, *internal* body parts, such as bones, that receive the applied force of contraction.

Many soft-bodied invertebrates have a hydrostatic skeleton. Consider the sea anemone, with its soft, vase-shaped body and saclike gut (Figure 26.6). Its body wall incorporates longitudinal and radial muscles. Between meals, when the longitudinal muscles are contracted (shortened) and radial ones are relaxed (lengthened), a sea anemone looks short and squat. When the body lengthens into an upright feeding position, its radial muscles are contracting (and forcing some fluid out of the gut cavity), and its longitudinal ones are relaxing.

Spiders, lobsters, houseflies, and other arthropods have a hinged exoskeleton. Figure 26.7 will give you an idea of how muscles work against these hard parts.

The endoskeleton of sharks and a few other fishes is cartilage. In most vertebrates, it is mainly bone (Figures 26.8 and 26.9). For example, a human skeleton has 206 bones. Its *appendicular* portion consists of the pectoral girdles (at the shoulders), arms, hands, pelvic girdle (at the hips), legs, and feet. Its pectoral girdles have slender collarbones and flat shoulder blades, flimsily arranged. Fall on an outstretched arm and you might dislocate a shoulder or fracture a collarbone, which is the bone that is most frequently broken. The human skeleton's *axial* portion consists of skull bones, twenty-six vertebrae (bony segments of the vertebral column, or backbone), twelve pairs of ribs, and the breastbone.

The curved backbone extends from the base of the skull to the pelvic girdle, where it transmits the weight of the torso to the lower limbs. A spinal cord threads through a series of bony canals at the rear of the column. Between the vertebrae are **intervertebral disks**. These are cartilage-containing shock absorbers and flex points that permit movement. Sometimes a severe or rapid

a feeding position

b resting position

Figure 26.6 Outcomes of the force of contraction applied against the hydrostatic skeleton of sea anemones. (**a**) Radial muscles are contracted, longitudinal muscles are relaxed; the body is extended to its upright feeding position. (**b**) Radial muscles (ringing the gut cavity) are relaxed; longitudinal muscles (parallel with the body axis) are contracted. Anemones typically look like this at low tide, when currents cannot bring food morsels to them.

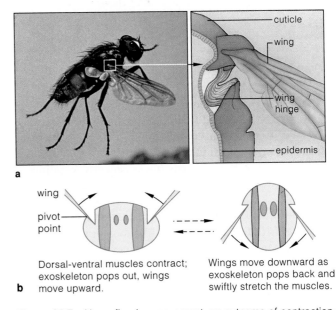

cuticle

wing

wing hinge

epidermis

a

wing

pivot point

Dorsal-ventral muscles contract; exoskeleton pops out, wings **b** move upward.

Wings move downward as exoskeleton pops back and swiftly stretch the muscles.

Figure 26.7 Housefly wing movement, an outcome of contraction of certain dorsal-ventral muscles. The contractile force works against an exoskeleton, near hinge points where wings are attached. When the muscles contract, the exoskeleton pops inward and wings move up. When the exoskeleton pops outward by elastic force, the wings move down. The quick-stretched muscles invite a fast repeat of the events.

Figure 26.8 Endoskeleton of a generalized mammal.

Bones of Skull

CRANIAL BONES
Enclose and protect brain and sensory organs

FACIAL BONES
Framework for facial area; support for teeth

Bones of Rib Cage

These bones and some vertebrae enclose and protect internal organs and assist breathing:

STERNUM (breastbone)

RIBS (twelve pairs)

Vertebral Column (backbone)

VERTEBRAE (twenty-six bones)
Enclose, protect spinal cord; support skull and upper extremities; attachment site for muscles

INTERVERTEBRAL DISKS
Fibrous, cartilaginous structures between vertebrae; they absorb movement-related stress and impart flexibility to the backbone

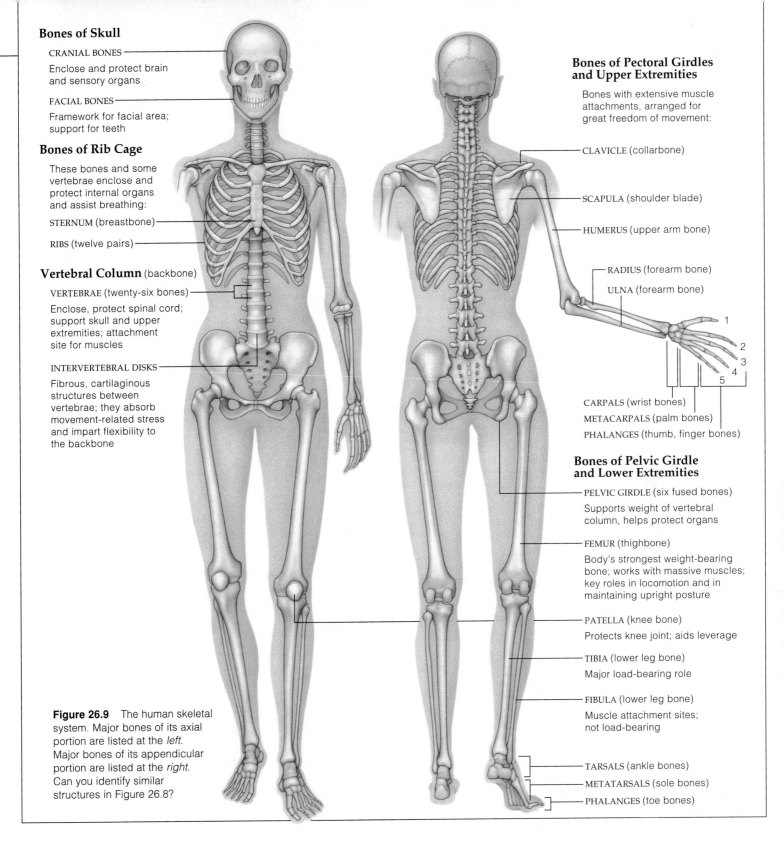

Figure 26.9 The human skeletal system. Major bones of its axial portion are listed at the *left*. Major bones of its appendicular portion are listed at the *right*. Can you identify similar structures in Figure 26.8?

Bones of Pectoral Girdles and Upper Extremities

Bones with extensive muscle attachments, arranged for great freedom of movement:

CLAVICLE (collarbone)

SCAPULA (shoulder blade)

HUMERUS (upper arm bone)

RADIUS (forearm bone)

ULNA (forearm bone)

CARPALS (wrist bones)

METACARPALS (palm bones)

PHALANGES (thumb, finger bones)

Bones of Pelvic Girdle and Lower Extremities

PELVIC GIRDLE (six fused bones)
Supports weight of vertebral column, helps protect organs

FEMUR (thighbone)
Body's strongest weight-bearing bone; works with massive muscles; key roles in locomotion and in maintaining upright posture

PATELLA (knee bone)
Protects knee joint; aids leverage

TIBIA (lower leg bone)
Major load-bearing role

FIBULA (lower leg bone)
Muscle attachment sites; not load-bearing

TARSALS (ankle bones)

METATARSALS (sole bones)

PHALANGES (toe bones)

shock forces a disk to slip out of place or rupture. Such painful, *herniated disks* are one of the less-than-favorable outcomes of bipedalism. Remember, our four-legged primate ancestors started walking upright more than 4 million years ago—and so put a pronounced S-shaped curve in their backbone. Presently, the older we get, the longer we have been fighting gravity in a compromised way—and the more back pain we suffer.

Animal skeletons have structural elements or body fluids against which the force of muscle contraction can be applied.

26.4 A CLOSER LOOK AT BONES AND THE JOINTS BETWEEN THEM

Structure and Functions of Bone

By definition, **bones** are complex organs that function in movement, protection, support, mineral storage, and blood cell formation. First, by interacting with skeletal muscles, bones maintain or change the position of body parts. Second, they are hard compartments that enclose and protect the brain, lungs, and other internal organs. Third, bones support and anchor muscles. Fourth, bone tissue is a bank for depositing and withdrawing mineral ions. In this respect it helps maintain the body's fluids and support its metabolic activities. Finally, some bones contain regions in which blood cells are produced.

Human bones range in size from tiny earbones to clublike thighbones. Bones are long, short (or cubelike), flat, and irregular. All have connective, epithelial, and bone tissues. The bone tissue is calcium hardened, with living cells and collagen fibers in a ground substance. Consider a thighbone (Figure 26.10). *Compact* bone tissue in its shaft and at its ends resists mechanical shock. The tissue has many thin, dense, cylindrical layers around small, interconnected canals. Blood vessels and nerves in these "Haversian canals" service living bone cells. *Spongy* bone tissue in the bone ends and shaft imparts strength without adding much weight. Its abundant spaces make the tissue appear spongy, but its flattened parts are firm. **Red marrow**, a major site of blood cell formation, fills the spaces in some bones, such as the breastbone. The cavities in most mature bones contain **yellow marrow**. Yellow marrow consists mostly of fat. It converts to red marrow and produces new red blood cells when blood loss from the body is severe.

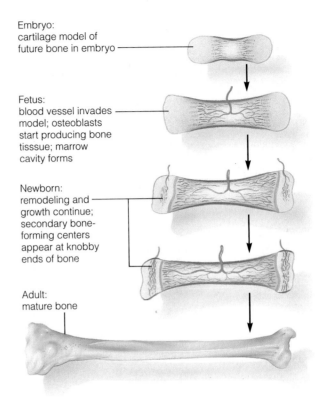

Embryo:
cartilage model of future bone in embryo

Fetus:
blood vessel invades model; osteoblasts start producing bone tisssue; marrow cavity forms

Newborn:
remodeling and growth continue; secondary bone-forming centers appear at knobby ends of bone

Adult:
mature bone

Figure 26.11 Long bone formation, starting with osteoblast activity in a cartilage model (here, already formed in the embryo). Bone-forming cells are active first in the shaft region, then at the knobby ends. In time, the only cartilage left is at the ends.

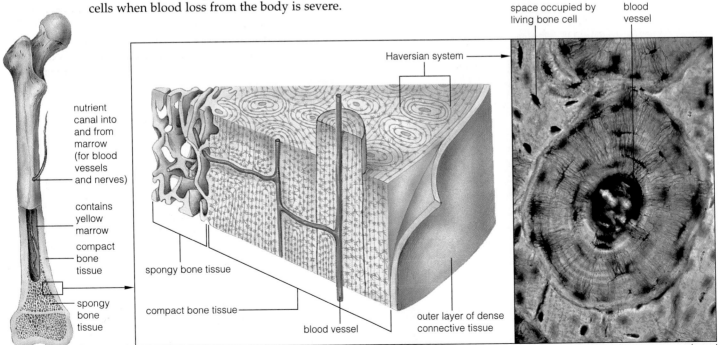

space occupied by living bone cell

blood vessel

Haversian system

nutrient canal into and from marrow (for blood vessels and nerves)

contains yellow marrow

compact bone tissue

spongy bone tissue

spongy bone tissue

compact bone tissue

blood vessel

outer layer of dense connective tissue

HAVERSIAN SYSTEM   75 µm

Figure 26.10 Bone tissue of a femur (thighbone), one of the long bones of mammals.

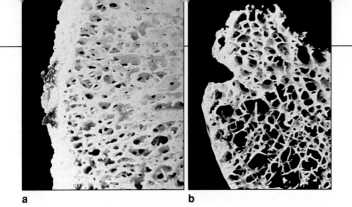

a   b

Figure 26.12 Osteoporosis. (**a**) In normal bone tissue, mineral deposits continually replace mineral withdrawals. (**b**) After the onset of osteoporosis, replacements cannot keep pace with withdrawals. In time the tissue erodes, and bones become hollow and brittle.

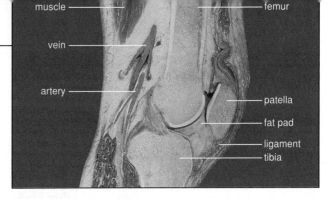

Figure 26.13 Knee joint, longitudinal section. Ligaments stabilize the joint. The patella, a small bone in one of the ligaments, protects the knee joint. A flexible, membrane-lined capsule encloses a lubricating fluid at the joint cavity.

Cartilage models for many bones form in an embryo (Figure 26.11). Bone-forming cells (osteoblasts) secrete material inside the model's shaft and on its surface. A marrow cavity opens as cartilage breaks down inside. In time, osteoblasts become surrounded by their secretions and thereafter are called osteocytes (living bone cells). Their metabolic activities maintain mature bones.

Minerals are continually deposited *and* withdrawn from bone tissue in ways that maintain the adult body's calcium levels. In this **bone tissue turnover**, bone cells secrete enzymes that digest bone tissue. The released calcium and other minerals enter interstitial fluid, then blood—which distributes them to metabolically active cells. For instance, a thighbone gets thicker and stronger as bone cells deposit minerals at the shaft's surface. At the same time, it gets less heavy as other bone cells destroy bone tissue within its shaft.

Exercising stimulates calcium deposition and tends to increase bone density. Stress or injury triggers mineral withdrawals. Also, as a person ages, the backbone, hip bones, and other bones decrease in mass, especially in women. Figure 26.12 shows the effect of *osteoporosis*. A weakened backbone may collapse, curve abnormally, and lower the rib cage, which puts stress on internal organs. Decreasing osteoblast activity, calcium loss, sex hormone deficiencies, excessive protein intake, and decreased physical activity contribute to the disorder.

Skeletal Joints

Joints, areas of contact or near-contact between bones, have distinctive connective tissue bridges. Very short connecting fibers hold bones together at *fibrous* joints. Cartilage holds them at *cartilaginous* joints. Long straps of dense connective tissue called **ligaments** bridge the gap between bones at *synovial* joints.

Fibrous joints hold teeth in their sockets. They also connect the flat skull bones of a fetus. At childbirth, the loose connections allow the bones to slide over each other and prevent skull fractures. A newborn's skull still has fibrous joints and membranous areas known as "soft spots" (fontanels). In childhood, the fibrous tissue hardens and skull bones are fused into a single unit.

Cartilaginous joints bridging vertebrae, ribs, and the breastbone permit slight movements. Synovial joints, such as knee joints, move freely. Ligaments stabilize the knee joints, as in Figure 26.13. Where one bone touches another, cartilage cushions them and absorbs shocks. A flexible capsule of dense connective tissue surrounds the region of contact. Cells of a membrane that lines the capsule's interior secrete a fluid that lubricates the joint.

Like many other joints, the knee joint is vulnerable to stress. As athletes know, this joint lets you swing, bend, and turn the long bones below it. When you run, it absorbs the force of your weight each time the foot hits the ground. Stretch or twist a knee joint suddenly and too far, and you may *strain* it. Tear its ligaments or tendons and you *sprain* it. Move the wrong way and you may dislocate the attached bones. During football and other collision sports, blows to the knee frequently sever a ligament. The severed part must be reattached surgically before ten days pass. Why? Phagocytic cells in lubricating fluid within the joint clean up after everyday wear and tear. Present them with torn ligaments, and they indiscriminately turn the tissue to mush.

Joint inflammation and degenerative disorders are collectively called "arthritis." In *osteoarthritis*, cartilage at the knees and other freely movable joints wears off as a person ages. Joints in the fingers, knees, hips, and the vertebral column are affected most often. In *rheumatoid arthritis*, synovial membranes in joints become inflamed and thickened, cartilage degenerates, and bone deposits build up. This degenerative disorder may be triggered by a bacterial or viral infection, but it also appears to have a genetic component. It can begin at any age, but symptoms usually emerge before age fifty.

Bones are collagen-rich, mineralized organs that function in movement, protection, support, storage of calcium and other minerals, and blood cell formation. Ligaments and other connective tissues hold bones together at skeletal joints.

How Muscles and Bones Interact

We turn now to **skeletal muscles**, the functional partners of bones. Each skeletal muscle contains bundles of hundreds to many thousands of muscle cells, which look like long, striped fibers. In muscle tissue, recall, these cells contract (shorten) in response to stimulation. They lengthen in response to gravity and other loads. When you dance, breathe, scribble notes, or tilt your head, contracting muscle cells are helping to move your body or to change the positions of some of its parts.

Connective tissue bundles muscle cells together and extends beyond them to form **tendons**. Tendons are cords or straps of dense connective tissue that attach muscle to bone (Figure 26.14). Most of the attachment sites are like a car's gearshift. They are a *lever system*, in which a rigid rod is attached to a fixed point but able to move about at it. Muscles connect to bones (rigid rods)

near a joint (fixed point). As the muscles contract, they transmit force to bones and make them move.

Skeletal muscles interact with one another as well as with bones. Some are arranged in pairs or groups that work together to promote the same movement. Others work in opposition, with the action of one opposing or reversing the action of another. Figure 26.15a shows how opposing muscle groups work in frog legs. Also look at Figure 26.15b, then extend your right arm. Now place your left hand over the upper arm's biceps and slowly "bend your elbow." Feel the biceps contract? Even if your biceps contracts only a bit, it can produce a large movement in the forearm bone connected to it. This is true of most leverlike arrangements.

Where tendons move against bones, sheaths help reduce the resulting friction. For example, the knees, wrists, and finger joints have sheathed tendons. Figure 26.14 shows how the sheath is a flattened membrane sac, filled with fluid, through which tendons can slide.

Bear in mind, only skeletal muscle is the functional partner of bone. Recall that smooth muscle is present mostly in the walls of internal organs, including the stomach and intestines (Section 25.3). Cardiac muscle is present only in the wall of the heart. We will consider the structure and functioning of smooth muscle and cardiac muscle in later chapters in this unit.

Figure 26.14 Tendon sheath. This is not the same as a bursa, a small sac also filled with synovial fluid. Bursae are cushions interposed *between* bone and skin or bone and tendons.

muscle
tendon (attached to bone)
tendon sheath
bone

Figure 26.15 (**a**) A frog demonstrating how a small decrease in the length of contracting muscles can produce a large movement. Its leap depends on opposing muscle groups attached to the upper limb bone of each hind leg. One muscle group pulls the limb forward and toward the body's midline. Another pulls it back and away from the body. (**b**) Two opposing muscle groups in a human arm. When a triceps contracts, the forearm extends straight out. When the triceps relaxes and its opposing partner (biceps) contracts, the elbow joint flexes and the forearm bends up.

3 Now the first muscle group in the frog's upper hindlimb contracts again, drawing the limb back toward the body.

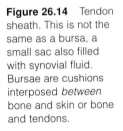
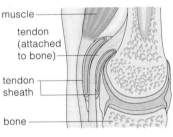

1 A muscle group attached to each upper hindlimb contracts and pulls the leg forward a bit.

2 An opposing muscle group attached to the same limb contracts forcefully and pulls the limb back. The contractile force, directed against the ground, propels the frog forward.

triceps relaxes

biceps contracts at same time, and pulls forelimb up

triceps contracts, pulls forelimb down

biceps relaxes at same time

a

b

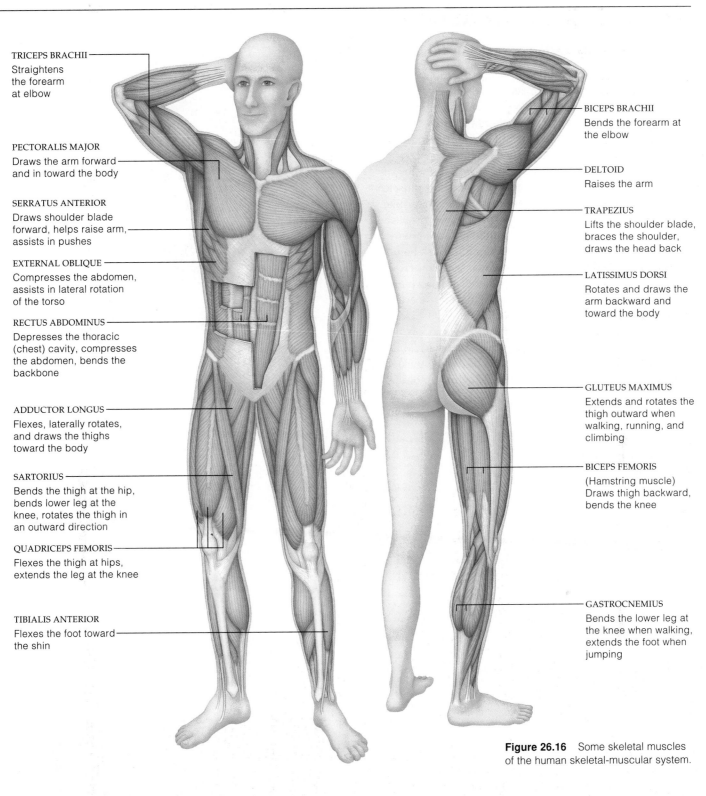

TRICEPS BRACHII
Straightens
the forearm
at elbow

PECTORALIS MAJOR
Draws the arm forward
and in toward the body

SERRATUS ANTERIOR
Draws shoulder blade
forward, helps raise arm,
assists in pushes

EXTERNAL OBLIQUE
Compresses the abdomen,
assists in lateral rotation
of the torso

RECTUS ABDOMINUS
Depresses the thoracic
(chest) cavity, compresses
the abdomen, bends the
backbone

ADDUCTOR LONGUS
Flexes, laterally rotates,
and draws the thighs
toward the body

SARTORIUS
Bends the thigh at the hip,
bends lower leg at the
knee, rotates the thigh in
an outward direction

QUADRICEPS FEMORIS
Flexes the thigh at hips,
extends the leg at the knee

TIBIALIS ANTERIOR
Flexes the foot toward
the shin

BICEPS BRACHII
Bends the forearm at
the elbow

DELTOID
Raises the arm

TRAPEZIUS
Lifts the shoulder blade,
braces the shoulder,
draws the head back

LATISSIMUS DORSI
Rotates and draws the
arm backward and
toward the body

GLUTEUS MAXIMUS
Extends and rotates the
thigh outward when
walking, running, and
climbing

BICEPS FEMORIS
(Hamstring muscle)
Draws thigh backward,
bends the knee

GASTROCNEMIUS
Bends the lower leg at
the knee when walking,
extends the foot when
jumping

Figure 26.16 Some skeletal muscles
of the human skeletal-muscular system.

Human Skeletal-Muscular System

The human body has more than 600 skeletal muscles, some superficial, others deep in the body wall. Some, such as facial muscles, attach to the skin. The trunk has muscles of the thorax, backbone, abdominal wall, and pelvic cavity. Other muscle groups attach to upper and lower limb bones. Figure 26.16 shows a few of the main skeletal muscles and lists their functions. We turn next to the mechanisms underlying their contraction.

As skeletal muscles contract, they transmit force to bones and make them move. Tendons strap skeletal muscles to bones.

26.6 MUSCLE STRUCTURE AND FUNCTION

Functional Organization of Skeletal Muscle

Bones move—they are pulled in some direction—when the skeletal muscles that are attached to them shorten. When a skeletal muscle shortens, its component muscle cells are shortening. When a muscle cell shortens, many units of contraction within that cell are shortening. The basic units of contraction are called **sarcomeres**.

Figure 26.17 gives you an idea of how bundles of cells in a skeletal muscle run parallel with the muscle itself. Inside each muscle cell are myofibrils, threadlike structures all bundled together in parallel array. Each myofibril is functionally divided into many sarcomeres, arranged one after another along its length. Dark bands called Z lines define the two ends of every sarcomere.

As Figure 26.17c shows, a sarcomere contains many filaments, oriented parallel with its long axis. Certain differences in their length and positioning give rise to the striped appearance of skeletal muscle (and cardiac muscle). Some of the filaments are thin; others are thick. Each *thin* filament is like two strands of pearls, twisted together. The "pearls" are molecules of **actin**, a globular protein with contractile functions:

— one actin molecule

} portion of one thin filament

Other proteins (coded *green*) are near actin's surface grooves. Each *thick* filament is made of molecules of **myosin**, another protein with contractile functions, in parallel array. A myosin molecule has a long tail and a double head that projects from the filament's surface:

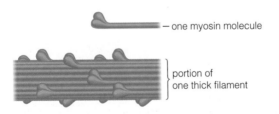

— one myosin molecule

} portion of one thick filament

Thus the myofibrils, muscle cells, and muscle bundles of a skeletal muscle all run in the same direction. What function does this consistent, parallel orientation serve? It focuses the force of muscle contraction onto a bone in a particular direction.

Sliding-Filament Model of Contraction

How do sarcomeres shorten and bring about contraction of a skeletal muscle? The answer lies with sliding and pulling interactions among the sarcomere's filaments. A

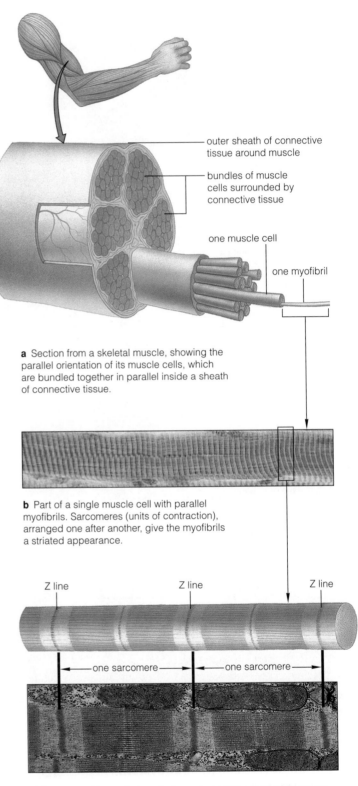

a Section from a skeletal muscle, showing the parallel orientation of its muscle cells, which are bundled together in parallel inside a sheath of connective tissue.

b Part of a single muscle cell with parallel myofibrils. Sarcomeres (units of contraction), arranged one after another, give the myofibrils a striated appearance.

c Diagram and transmission electron micrograph of two of the sarcomeres from a myofibril. This closer view shows how dark "bands," called Z lines, define the two ends of each sarcomere. Mitochondria (the oval-shaped organelles) adjacent to the sarcomeres provide ATP energy for muscle action.

Figure 26.17 Components of a skeletal muscle.

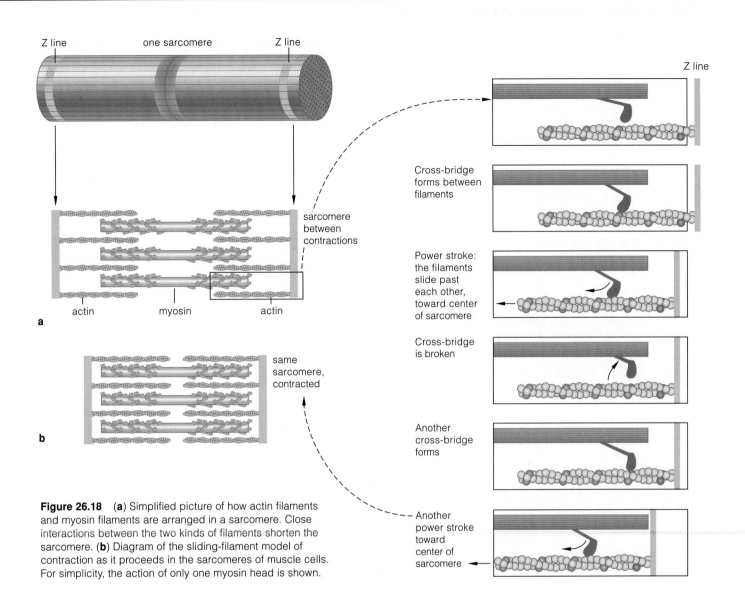

Figure 26.18 (**a**) Simplified picture of how actin filaments and myosin filaments are arranged in a sarcomere. Close interactions between the two kinds of filaments shorten the sarcomere. (**b**) Diagram of the sliding-filament model of contraction as it proceeds in the sarcomeres of muscle cells. For simplicity, the action of only one myosin head is shown.

set of actin filaments extends from each Z line partway to the center portion of each sarcomere. A set of myosin filaments partially overlaps the sets of actin but doesn't extend all the way to the Z lines (Figure 26.18). When a muscle contracts, these myosin filaments are physically sliding along and pulling the two sets of actin filaments toward the center of the sarcomere—which shortens as a result. The physical interaction is a key premise of the **sliding-filament model** of muscle contraction.

The myosin and actin filaments interact through **cross-bridge formation**. As indicated in Figure 26.18*b*, these particular cross-bridges are attachments between a myosin head and a binding site on actin. The myosin heads are activated when messages from the nervous system stimulate the muscle cell. They physically attach to an adjacent actin filament and tilt in a short power stroke, driven by ATP energy, toward the sarcomere's

center. During the power stroke, the heads pull the actin filament along with them. Another energy input makes the heads let go, attach to another region of the filament, tilt in another power stroke, and so on down the line. A single contraction of a sarcomere requires a whole series of power strokes.

A skeletal muscle shortens through the combined decreases in length of its numerous sarcomeres. Sarcomeres are the basic units of contraction.

The parallel orientation of a muscle's component parts directs the force of contraction toward a bone that must be pulled in some direction.

By energy-driven interactions between the myosin and actin filaments, the many sarcomeres of a muscle cell shorten and collectively account for its contraction.

CONTROL OF MUSCLE CONTRACTION

When skeletal muscles contract, they help move the body and its assorted parts at certain times, in certain ways—and they do so in response to commands from the nervous system. These commands, which reach the muscles by way of motor neurons, can stimulate *or* inhibit the contraction of sarcomeres in muscle cells.

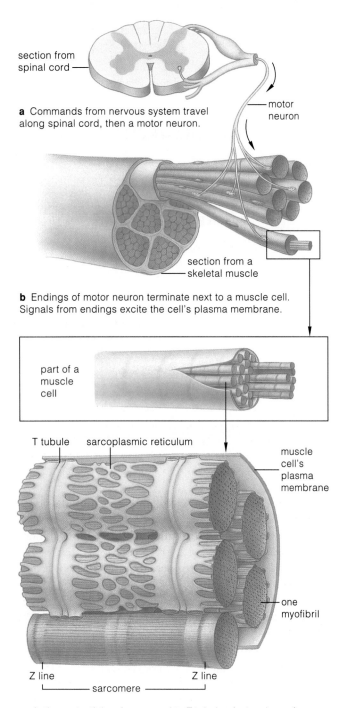

section from spinal cord

a Commands from nervous system travel along spinal cord, then a motor neuron.

motor neuron

section from a skeletal muscle

b Endings of motor neuron terminate next to a muscle cell. Signals from endings excite the cell's plasma membrane.

part of a muscle cell

T tubule   sarcoplasmic reticulum

muscle cell's plasma membrane

one myofibril

Z line   Z line
└──── sarcomere ────┘

c Action potentials arise, spread to T tubules (extensions of plasma membrane connected to calcium-storing sarcoplasmic reticulum). Arrival of action potentials induces release of calcium. Calcium diffuses into myofibrils, clears blocked sites of cross-bridge formation in sarcomeres and so brings about contraction.

Like all cells, a muscle cell shows a difference in electric charge across its plasma membrane. (That is, the cytoplasm just beneath the membrane is a bit more negative than fluid outside it.) But only in muscle cells, neurons, and other *excitable* cells does this difference in charge reverse suddenly though briefly in response to sufficient stimulation.

Figure 26.19 (*Left*) Pathway for signals from the nervous system that stimulate contraction of skeletal muscle. The plasma membrane of muscle cells surrounds their myofibrils and connects with inward-threading T tubules. These membranous tubes are close to the sarcoplasmic reticulum, a calcium-storing system that functions in the control of contraction.

The abrupt reversal in charge, an **action potential**, occurs as charged ions flow across the membrane in an accelerating way. The electrical commotion spreads without diminishing along the membrane, away from the point of stimulation.

Suppose action potentials arise in a muscle cell. They spread rapidly away from the stimulation point, then along small, tubelike extensions of the plasma membrane (Figure 26.19). The small tubes connect with a system of membrane-bound chambers that thread lacily around the muscle cell's myofibrils. That system—called the **sarcoplasmic reticulum**—takes up, stores, and releases calcium ions in controlled ways.

The arrival of action potentials causes an outward flow of calcium ions from the sarcoplasmic reticulum. The released ions diffuse into the myofibrils and reach actin filaments. Before this happened, the muscle was at rest (it was not contracting). Its actin binding sites were blocked and myosin could not form cross-bridges with them. However, the arrival of calcium ions clears the binding sites—so contraction can proceed. Afterward, the calcium ions are actively transported back into the membrane storage system.

What blocks cross-bridge binding sites in a muscle at rest? Two proteins called tropomyosin and troponin are located within or near the surface grooves of actin filaments. At a low calcium level, the proteins are joined so tightly together, the tropomyosin is forced slightly outside the groove—and in this position it blocks the cross-bridge binding site. When the calcium level rises, calcium binds to troponin and thereby alters its shape. When troponin is in its altered shape, it has a different molecular grip on tropomyosin—which is now free to move into the groove and expose the binding site.

Commands from the nervous system initiate action potentials in muscle cells. These action potentials are the signals for cross-bridge formation—hence for contraction.

PROPERTIES OF WHOLE MUSCLES

Muscle Tension and Muscle Fatigue

Whether a muscle actually shortens during cross-bridge formation depends on the external forces acting upon it. Collectively, the cross-bridges exert **muscle tension**. By definition, this is a mechanical force that a contracting muscle exerts on an object, such as a bone. Opposing it is a load—either the weight of an object or gravity's pull on the muscle. Only when muscle tension exceeds the load does a stimulated muscle shorten.

An *isometrically* contracting muscle develops tension but doesn't shorten. It supports a load in a constant position, as when you hold a glass of lemonade in front of you. An *isotonically* contracting muscle shortens and moves a load. With *lengthening* contraction, though, an external load is greater than the muscle tension, so the muscle lengthens during the period of contraction. This happens to leg muscles when you walk down stairs.

A muscle's tension depends on the cross-bridge formation in its cells and the number of cells recruited into action. Consider a **motor unit**—a motor neuron and all muscle cells that form junctions with its endings. By stimulating the motor unit with an electrical impulse, we can induce an action potential and make a recording of an isometric contraction. It takes a few milliseconds for tension to increase, then it peaks and declines. This response is a **muscle twitch** (Figure 26.20*a*). Its duration depends on the load and cell type. For example, fast-acting muscle cells rely on glycolysis (not efficient but fast) and use up ATP faster than slow-acting cells do.

Apply another stimulus before the response is over, and the muscle twitches again. "Tetanus" is a large contraction in which a motor unit is stimulated repeatedly, so twitches mechanically run together. (In a disease by the same name, toxins interfere with muscle relaxation.) Figure 26.20*d* shows a recording of tetanic contraction.

Continuous, high-frequency stimulation that keeps a muscle in a state of tetanic contraction leads to muscle fatigue—a decline in tension. After a few minutes of rest, a fatigued muscle will contract again in response to stimulation. The extent of recovery depends largely on how long and how frequently it was stimulated before. Muscles associated with brief, intense exercise (such as weightlifting) fatigue fast but recover fast. Those associated with prolonged, moderate exercise fatigue more slowly but take longer to recover, often up to 24 hours. The molecular mechanisms that cause muscle fatigue are not known, but glycogen depletion is a factor.

Effects of Exercise and Aging

A muscle's properties depend on how often, how long, and how intensely it is put to use. With regular **exercise** (that is, increased levels of contractile activity), muscle

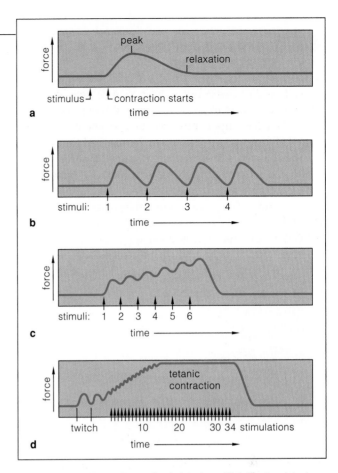

Figure 26.20 Recordings of twitches in artificially stimulated muscles. (**a**) A single twitch. (**b**) Two stimulations per second cause a series of twitches. (**c**) Six per second cause a summation of twitches. (**d**) About twenty per second cause tetanic contraction.

cells don't increase in number, but they increase in size and metabolic activity—and become more resistant to fatigue. Consider *aerobic exercise*, which is not intense but is long in duration. Such exercise increases the number of mitochondria in both fast and slow muscle cells, and it increases the blood capillaries that service them. Such physiological changes improve endurance. By contrast, *strength training* (intense, short-duration exercise such as weightlifting) affects fast-acting muscle cells. These form more myofibrils and more enzymes of glycolysis. Strong, bulging muscles of the sort shown in Figure 26.1*b* may result, although these don't have much endurance. They fatigue rapidly.

Muscle tension decreases in adult humans thirty to forty years old. These people may exercise just as long and intensely as younger ones, but their muscles can't adapt (change) in response to the same extent. Yet even some adaptation can be beneficial. Aerobic exercise improves blood circulation. And, as it turns out, even modest strength training slows the loss of muscle tissue that is an inevitable part of the aging process.

The properties of muscles vary with activity and with age.

26.9 ATP FORMATION AND LEVELS OF EXERCISE

All cells require ATP, but only in muscle cells does the demand skyrocket in so short a time. When a muscle cell at rest is called upon to contract, phosphate donations from ATP must proceed twenty to a hundred times faster. But the cell has only a small supply of ATP at the start of contractile activity. At such times, it forms ATP by a very fast reaction. An enzyme simply transfers phosphate from **creatine phosphate**, an organic compound, to ADP. The cell has about five times as much creatine phosphate as ATP, so this reaction is good for a few more muscle twitches. And it is enough to buy time for a relatively slower ATP-forming pathway to kick in (Figure 26.21).

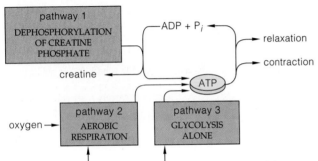

Figure 26.21 Three metabolic routes by which ATP can form in muscles in response to physical exercise.

During prolonged, moderate exercise, the oxygen-requiring reactions of aerobic respiration typically can provide most of the ATP required for contraction. For the first five to ten minutes, a muscle cell taps its store of glycogen for glucose (the starting substrate). For the next half hour or so of sustained activity, the muscle cell depends on glucose and fatty acid deliveries from the bloodstream. For contractile activity longer than this, fatty acids are the main fuel source (Section 6.6).

What happens when exercise is so intense that it exceeds the capacity of the respiratory and circulatory systems to deliver oxygen for the aerobic pathway? At such times, glycolysis alone will contribute more of the total ATP that is being formed. Remember, by this set of anaerobic reactions, a glucose molecule is only partly broken down, so the net ATP yield is small. But muscle cells can use this metabolic route as long as glycogen stores continue to provide glucose.

After intense exercise, deep, rapid breathing helps repay the body's **oxygen debt**, incurred when ATP use by muscles exceeded the aerobic pathway's deliveries.

During exercise, the availability of ATP in muscle cells affects whether contraction will proceed, and for how long.

SUMMARY

1. Most animals have an integumentary system, which covers the body's surface. An example is vertebrate skin and the structures derived from it. Skin protects against abrasion, ultraviolet radiation, dehydration, and many infectious bacteria. It helps control internal temperature (blood flow to the skin can dissipate heat). Its receptors detect stimuli in the external environment.

2. Epidermis, the outermost portion of vertebrate skin, consists mostly of multiple layers of dead, keratinized, and melanin-shielded epithelial cells. In the dermis, the inner portion, cell divisions produce replacements for cells shed on an ongoing basis. Hair, oil glands, sweat glands, and other structures derived from specialized epidermal cells are embedded mainly in the dermis.

3. Muscle contractions move body parts. The muscles require a medium or structure against which contractile force can be applied. In hydrostatic skeletons, as in sea anemones, body fluids accept the force and are redistributed inside a confined space. In exoskeletons, as in insects, rigid external body parts accept the force. In endoskeletons, rigid internal body parts (mainly bones) receive the applied force of contraction.

4. Bones have living cells in a mineralized, collagenous ground substance. Bones function in the movement, protection, and support of body parts; mineral storage; and (in some bones) blood cell formation.

5. The human skeleton's axial portion includes bones of the skull, vertebral column (backbone), ribs, and breastbone. A backbone consists of vertebrae (bony segments) cushioned by cartilaginous, intervertebral disks. The skeleton's appendicular portion includes limb bones, the pelvic girdle, and the pectoral girdle.

6. Skeletal joints are areas of contact or near-contact between bones. Short fibers, cartilage, or straplike ligaments bridge the gap between bones at different joints.

7. Cells of smooth, cardiac, and skeletal muscle contract (shorten) in response to adequate stimulation.

8. Tendons attach skeletal muscles to bones. Skeletal muscles and bones interact as a system of levers, with rigid rods (bones) moving at fixed points (joints). Many muscles work together or in opposition to bring about movement or positional changes in body parts.

9. Inside each skeletal muscle cell are many myofibrils, arranged parallel with its long axis. These threadlike structures contain actin and myosin filaments, also organized in parallel, and are functionally divided into sarcomeres, the basic units of contraction. The parallel orientation of the muscle's component parts directs the force of contraction toward a bone that must be pulled in some direction.

10. In response to stimulation, skeletal muscles shorten by decreases in the length of all of its sarcomeres.

a. The nervous system stimulates muscle cells by way of motor neuron endings that terminate on them and trigger action potentials at their plasma membrane.

b. Action potentials cause the release of calcium ions from a membrane system (sarcoplasmic reticulum) that threads around the cell's myofibrils. Calcium diffuses inside sarcomeres, binds to actin filaments, and makes the binding sites change shape. Now myosin filament heads can form cross-bridges with actin filaments.

c. Each cross-bridge is a brief attachment between a myosin head and an actin binding site. Cross-bridges form during repeated, ATP-driven power strokes. The repeated strokes make actin filaments slide past myosin filaments, and collectively they shorten the sarcomere.

11. Muscle tension refers to a mechanical force created by cross-bridge formation. The load (gravity, the weight of objects) is an opposing force. A stimulated muscle shortens only when tension exceeds the load. Levels of exercise and aging affect the properties of muscles.

12. Muscle cells obtain the ATP required for contraction by dephosphorylation of creatine phosphate (simple, fast, and good for a few seconds of contraction). Or they may obtain it by aerobic respiration (which sustains prolonged, moderate exercise) or by glycolysis (when intense exercise exceeds the body's capacity to deliver oxygen to muscle cells).

Review Questions

1. What are the functions of skin? *442–443*

2. What are the functions of bones? *446*

3. Name the three types of muscle, then state the function of each and where they are located in the body. *448*

4. How do actin and myosin interact in a sarcomere to bring about contraction? What roles do ATP and calcium play? *450–452*

Self-Quiz *(Answers in Appendix IV)*

1. Which is *not* a function of skin?
   a. resist abrasion
   b. restrict dehydration
   c. initiate movement
   d. help control temperature

2. _____ are shock pads and flex points.
   a. Vertebrae
   b. Femurs
   c. Marrow cavities
   d. Intervertebral disks

3. Blood cells form in _____ .
   a. red marrow
   b. all bones
   c. certain bones only
   d. a and c

4. In skeletal muscle, the _____ is the basic unit of contraction.
   a. myofibril
   b. sarcomere
   c. muscle fiber
   d. myosin filament

5. Muscle contraction requires _____ .
   a. calcium ions
   b. ATP
   c. action potential arrival
   d. all of the above

6. ATP for muscle contraction can be formed by _____ .
   a. aerobic respiration
   b. glycolysis
   c. creatine phosphate breakdown
   d. a and b only
   e. a and c only
   f. a, b, and c

7. Match the M words with their defining feature.
   _____ muscle
   _____ muscle twitch
   _____ muscle tension
   _____ melanin
   _____ myosin
   _____ marrow
   _____ metacarpals
   _____ myofibrils
   _____ muscle fatigue

   a. actin's partner
   b. all in the hands
   c. blood cell production
   d. decline in tension
   e. brownish-black pigment
   f. motor unit response
   g. force exerted by cross-bridges
   h. muscle cells bundled in connective tissue
   i. threadlike parts in muscle cell

Critical Thinking

1. For a young woman, the recommended daily allowance (RDA) of calcium is 800 milligrams per day. During pregnancy, the RDA is 1,200 milligrams per day. Why would a pregnant woman need the larger amount? What might happen to her bones without it?

2. Kate found a malnourished, stray cat that had just given birth to a litter of four kittens. The cat was trying to hunt, but her muscles were twitching so badly that she could scarcely walk. Kate recalled what she had learned about muscle contraction. She offered milk as well as solid food to the cat—which, after several days, regained muscle control and was able to care for her kittens. How might milk help restore muscle function?

3. Compared to most people, long-distance runners have a greater number of muscle fibers with more mitochondria. Sprinters have a greater number of muscle fibers that have more of the enzymes necessary for glycolysis but fewer mitochondria. Think about how these two forms of exercise differ and explain why the muscle fibers differ between the two kinds of runners.

Selected Key Terms

actin *450*
action potential *452*
anabolic steroid *440*
bone *446*
bone tissue turnover *447*
creatine phosphate *454*
cross-bridge formation *451*
cuticle (animal) *442*
dermis *442*
endoskeleton *444*
epidermis *442*
exercise *453*
exoskeleton *444*
hair *443*
hydrostatic skeleton *444*
integument *442*

intervertebral disk *444*
joint *447*
ligament *447*
motor unit *453*
muscle tension *453*
muscle twitch *453*
myosin *450*
oxygen debt *454*
red marrow *446*
sarcomere *450*
sarcoplasmic reticulum *452*
skeletal muscle *448*
skin *442*
sliding-filament model *451*
tendon *448*
yellow marrow *446*

Readings

Vander, A., J. Sherman, and D. Luciano. 1994. *Human Physiology.* Sixth edition. New York: McGraw-Hill.

Weeks, O. December 1989. "Vertebrate Skeletal Muscle: Power Source for Locomotion." *BioScience* 39(11): 791–798.

27 CIRCULATION

Heartworks

For Dr. Augustus Waller, Jimmie the bulldog was no ordinary pooch. Connected to wires and soaked to his ankles in buckets of salty water, Jimmie was a four-footed window to the workings of the heart (Figure 27.1a and b). Press your fingers a few inches left of the center of your chest, between the fifth and sixth ribs, to feel the repetitive thumpings of your heart. The same rhythms intrigued Waller and other nineteenth-century physiologists. They wondered: Does each beat of the heart produce a pattern of electrical currents? Could they devise a painless way to record such a current at the body surface?

That's where Jimmie and the buckets of salty water came in—for saltwater happens to be an efficient conductor of electricity. In Waller's experiment, it was so efficient that it carried faint signals from Jimmie's beating heart, through the skin of his legs, to a crude monitoring device. With that device, Waller made one of the first recordings of heart activity (Figure 27.1c). Today we call these **electrocardiograms**, or ECGs.

A graph of the normal electrical activity of your own heart would look much the same. The pattern emerged a few weeks after you started growing from a single fertilized egg inside your mother. Mitotic cell divisions produced the embryo. Early on, some embryonic cells differentiated into cardiac muscle cells—and these started to contract spontaneously. One small patch of cells took the lead, and it has functioned as your heart's pacemaker ever since.

a

JIMMIE   DR. WALLER

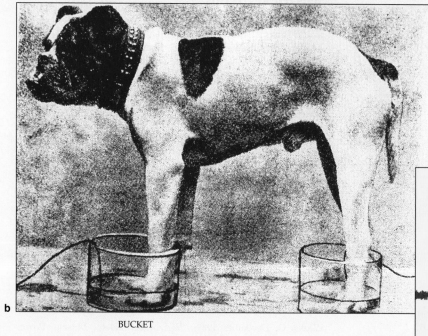

b

BUCKET

Figure 27.1 A bit of history in the making. (**a**) Augustus Waller and Jimmie, beloved pet bulldog, sharing a quiet moment in Waller's study. (**b**) Jimmie taking part in a painless experiment that yielded one of the world's first electrocardiograms (**c**).

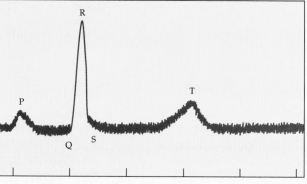

c

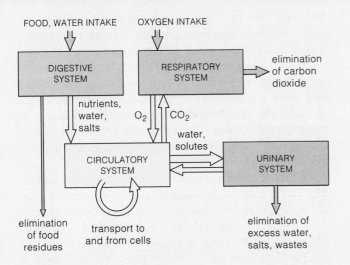

FOOD, WATER INTAKE   OXYGEN INTAKE

DIGESTIVE SYSTEM

RESPIRATORY SYSTEM

elimination of carbon dioxide

nutrients, water, salts

O_2 CO_2

water, solutes

CIRCULATORY SYSTEM

URINARY SYSTEM

elimination of food residues

transport to and from cells

elimination of excess water, salts, wastes

Figure 27.2 Diagram of the functional connections between the circulatory, respiratory, and digestive systems in transporting substances to and from all living cells in the animal body. Their integrated activities help maintain favorable operating conditions in the internal environment.

If all goes well, the patch of muscle cells will go on contracting until the day you die. They are a natural pacemaker; they set the baseline rate at which blood is pumped away from the heart, through blood vessels, then back to the heart. The rate is moderate, somewhere in the neighborhood of 70 beats a minute; but commands from the nervous system and endocrine systems adjust it all the time. When you jog, for example, your skeletal muscles demand much more blood-borne oxygen and glucose. Then, your heart may start pounding 150 times a minute to deliver sufficient blood to them.

We have come a long way from Waller and Jimmie in our monitorings of the heart. Sensors can now detect the faint signals characteristic of an impending heart attack. Internists now use computers to analyze a patient's beating heart, and they use ultrasound probes to build images of it on a video screen. Cardiologists routinely substitute battery-powered pacemakers for malfunctioning natural ones.

With this chapter, we turn to the circulatory system—the means by which substances move rapidly to and from the interstitial fluid that surrounds all living cells in the body of nearly all kinds of animals. The circulatory system continually accepts oxygen, nutrients, and many other substances that the animal secures, by way of its respiratory and digestive systems, from the outside environment. Simultaneously it picks up carbon dioxide and other wastes from cells and delivers them to the respiratory and urinary systems for disposal. Its smooth operation is absolutely central to maintaining operating conditions in the internal environment within a tolerable range—a state we call homeostasis.

KEY CONCEPTS

1. Cells survive by exchanging substances with their surroundings. In most animals, substances move rapidly to and from living cells by way of a closed circulatory system.

2. Blood, a fluid connective tissue that is commonly confined within a heart and blood vessels, is the transport medium of circulatory systems. Blood transports oxygen, carbon dioxide, plasma proteins, vitamins, hormones, lipids, and other solutes. It also transports metabolically generated heat.

3. Human blood flows to and from the heart, in two separate circuits of blood vessels. In the pulmonary circuit, the heart pumps oxygen-poor blood to the lungs, where it picks up oxygen; then blood flows back to the heart. In the systemic circuit, the heart pumps oxygen-enriched blood to all body regions. After giving up oxygen and picking up carbon dioxide in those regions, blood flows back to the heart.

4. Arteries, large-diameter blood vessels, function in the rapid transport and distribution of blood flow away from the heart to different body regions. Veins, which also have large diameters, deliver the blood back to the heart. In between are arterioles, capillaries, and venules of smaller diameters.

5. Large numbers of capillaries interconnect in capillary beds, the diffusion zones that service different tissues and organs. Collectively, capillaries in the bed slow down the flow of blood from the heart and allow time for exchanges with the surrounding interstitial fluid. The interstitial fluid exchanges substances with living cells, which it bathes.

6. Arterioles are blood vessels with adjustable diameters. They are control points for the distribution of different volumes of blood flow. In response to signals, as from the nervous system, arteriole diameters widen in some parts of the body and narrow in others. The coordinated responses divert a greater portion of blood flowing from the heart to tissues and organs that are engaging in heightened activity.

General Characteristics

Imagine that an earthquake has closed off the highways around your neighborhood. Grocery trucks can't enter and waste-disposal trucks can't leave, so food supplies dwindle and garbage piles up. Cells would face similar predicaments if your body's highways were disrupted. The highways are part of a **circulatory system**, which functions in the rapid internal transport of substances to and from cells.

The circulatory system helps maintain favorable neighborhood conditions, so to speak. This is important. Your body's differentiated cells, which perform specialized tasks, cannot fend for themselves. Different types must interact in coordinated ways to maintain the composition, volume, and temperature of a tissue fluid surrounding them, the **interstitial fluid**. A circulating connective tissue—**blood**—interacts with tissue fluid, making continual deliveries and pickups that help keep conditions tolerable for enzymes and other molecules that carry out cell activities. Together, interstitial fluid and blood are the body's "internal environment."

Blood flows within blood vessels—tubes that differ in wall thickness and diameter. A muscular pump, the **heart**, generates pressure that keeps blood flowing. Like many animals, you have a *closed* circulatory system— blood is confined within the continuously connected walls of the heart and blood vessels. This is not true of *open* circulatory systems of arthropods (such as insects) and most mollusks. In such systems, blood flows out through the vessels and into sinuses (small spaces in body tissues). There it mingles with tissue fluids, then moves back to the heart through openings in the blood vessels or the heart wall (Figure 27.3a).

Think about the overall "design" of a closed system. The heart pumps incessantly, so the volume of flow through the entire system equals the volume returned to the heart. Yet the rate and volume must be *adjusted* along the route. Large-diameter vessels must be used for rapid distribution of blood flow. But elsewhere in the system, the flow cannot be as rapid, or there would not be enough time for substances to be exchanged with interstitial fluid. The required slowdown proceeds at **capillary beds**. At such beds, the flow fans out through vast numbers of small-diameter blood vessels called **capillaries**. By dividing the blood flow, the capillaries can handle the same total volume of flow as the large-diameter vessels—but at a more leisurely pace.

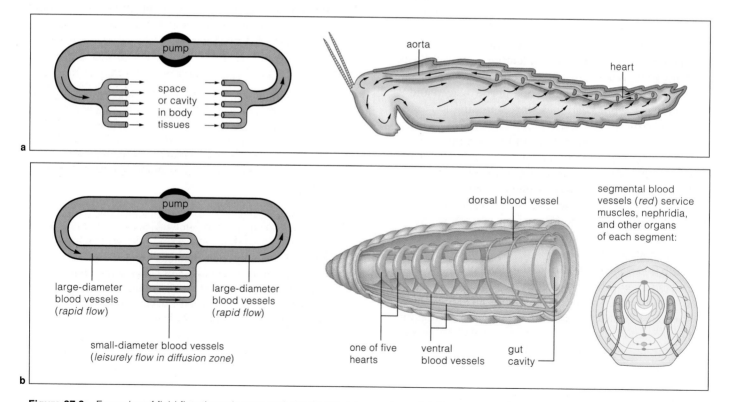

Figure 27.3 Examples of fluid flow through open and closed circulatory systems. (**a**) Open system of a grasshopper. A "heart" pumps blood through a vessel (aorta). From there, the blood moves into spaces in tissues and mingles with fluid that bathes cells. Then the blood reenters the heart through openings in its wall. (**b**) Closed system of an earthworm. Blood remains inside vessels (which are similar in function to arteries, capillaries, and veins) and inside several pairs of muscular "hearts" near the head end.

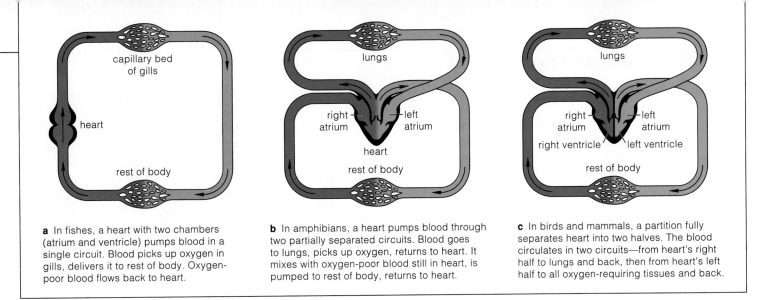

a In fishes, a heart with two chambers (atrium and ventricle) pumps blood in a single circuit. Blood picks up oxygen in gills, delivers it to rest of body. Oxygen-poor blood flows back to heart.

b In amphibians, a heart pumps blood through two partially separated circuits. Blood goes to lungs, picks up oxygen, returns to heart. It mixes with oxygen-poor blood still in heart, is pumped to rest of body, returns to heart.

c In birds and mammals, a partition fully separates heart into two halves. The blood circulates in two circuits—from heart's right half to lungs and back, then from heart's left half to all oxygen-requiring tissues and back.

Figure 27.4 Closed circulatory systems of vertebrates.

Evolution of Vertebrate Circulatory Systems

Humans and other existing vertebrates have a closed circulatory system—although the pump and plumbing of fishes, amphibians, birds, and mammals differ in their details. These differences evolved over hundreds of millions of years and corresponded to the move of some vertebrates onto land. Recall from Chapter 21 that the first vertebrates (fishes) had gills. Gills, like all respiratory structures, have a thin, moist surface that oxygen and carbon dioxide can diffuse across. Later, in fishes ancestral to land vertebrates, lungs evolved that supplemented gas exchange. Being *internally moistened* sacs, lungs had advantages for the move onto dry land. It seems selection pressure favored modifications in lungs *and* in circulatory systems, which pick up oxygen from lungs and deliver carbon dioxide wastes to them.

Consider this: In fishes, blood flows in *one* circuit (Figure 27.4*a*). Pressure generated by a two-chambered heart forces it through extensive capillary beds of gills, then the largest artery (aorta), then the capillary beds of various organs, then back to the heart. Gill capillaries present so much resistance to flow, the pressure drops considerably before the aorta. Such blood delivery is good enough for the activity level of most fishes. But it would not be sufficient for the more active life-styles of most land vertebrates.

As amphibians evolved, their heart became partitioned into right and left halves—only partially so, but enough to pump blood through *two* partially separated circuits (Figure 27.4*b*). Separation of flow continued in some reptiles (crocodilians). And it became complete in birds and mammals; their heart acts like two side-by-side pumps (Figure 27.4*c*). In birds and mammals, the heart's *right* half pumps oxygen-poor blood to the lungs, where blood picks up oxygen and gives up carbon dioxide. The freshly oxygenated blood then flows to the heart's left half. This is the **pulmonary circuit**. By contrast, in a **systemic circuit**, the *left* half of the heart pumps oxygenated blood to all tissues where oxygen is

used and carbon dioxide forms. Then the oxygen-poor blood flows to the heart's right half. The double circuit is a rapid and efficient mode of blood delivery—one that supports the high levels of activity characteristic of all vertebrates whose ancestors evolved on land.

Links With the Lymphatic System

The heart's pumping action puts pressure on blood flowing through the circulatory system. Partly because of this pressure, small amounts of water and a few of the proteins dissolved in blood are forced out of the capillaries and become part of interstitial fluid. However, an elaborate network of drainage vessels picks up excess interstitial fluid and reclaimable solutes, then returns them to the circulatory system. This network is part of the **lymphatic system**. Later in the chapter, you will see how other parts of the lymphatic system help cleanse bacteria and other disease agents from fluid being returned to the blood.

Closed circulatory systems confine blood within one or more hearts and a network of blood vessels. (In open systems, blood also intermingles with tissue fluids.)

Vertebrates have a closed circulatory system. Together with drainage vessels of the lymphatic system, this circulatory system transports substances to and from the interstitial fluid that bathes all living cells of the body.

Blood flows rapidly in large-diameter vessels between the heart and capillary beds. There, the flow rate slows, giving time for substances to be exchanged with interstitial fluid.

In fishes, blood flows in one circuit, away from and back to the heart. In birds and mammals, blood flows in two circuits, through a heart that has been partitioned as two side-by-side pumps. This double circuit supports the high levels of activity characteristic of vertebrates that evolved on land.

Functions of Blood

Blood, a connective tissue, has multiple functions. It transports oxygen, nutrients, and other solutes to cells. It carries away their metabolic wastes and secretions, including hormones. Blood helps stabilize internal pH. Blood also serves as a highway for phagocytic cells that scavenge tissue debris and fight infections. In birds and mammals, blood helps equalize body temperature. It does this by carrying excess heat from skeletal muscles and other regions of high metabolic activity to the skin, where heat can be dissipated.

Blood Volume and Composition

The volume of blood depends on body size and on the concentrations of water and solutes. Blood volume for average-size adult humans is about 6 to 8 percent of the body weight. That amounts to about four or five quarts. As for all vertebrates, the blood of humans is a sticky fluid, thicker than water and slower flowing. **Plasma**, **red blood cells**, **white blood cells**, and **platelets** are its components (Figures 27.5 and 27.6). Plasma normally accounts for 50 to 60 percent of the total blood volume.

Plasma Plasma, which is mostly water, functions as a transport medium for blood cells and platelets. It also serves as a solvent for ions and molecules, including hundreds of different kinds of plasma proteins. Some of

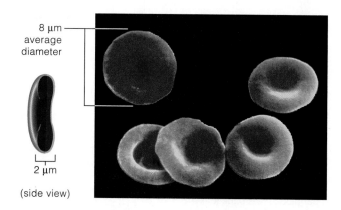

8 µm average diameter

2 µm

(side view)

Figure 27.6 Size and shape of red blood cells.

the plasma proteins transport lipids and fat-soluble vitamins through the body. Others have roles in blood clotting or in defense against pathogens. Collectively, the concentration of plasma proteins affects the blood's fluid volume, for it influences the movement of water between blood and interstitial fluid. Glucose and other simple sugars, as well as lipids, amino acids, vitamins, and hormones, are dissolved in plasma. So are oxygen, carbon dioxide, and nitrogen.

Red Blood Cells Erythrocytes, or red blood cells, are biconcave disks, like doughnuts with a squashed-in center instead of a hole (Figure 27.6). They transport the oxygen used in aerobic respiration and they carry away

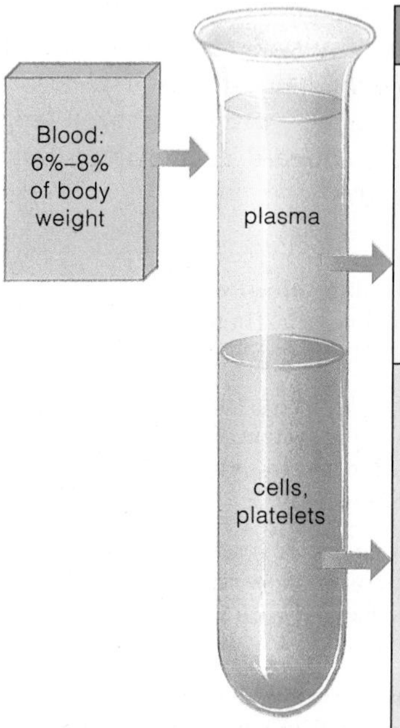

Components	Relative Amounts	Functions
Plasma Portion *(50%–60% of total volume):*		
1. Water	91%–92% of plasma volume	Solvent
2. Plasma proteins (albumin, globulins, fibrinogen, etc.)	7%–8%	Defense, clotting, lipid transport, roles in extracellular fluid volume, etc.
3. Ions, sugars, lipids, amino acids, hormones, vitamins, dissolved gases	1%–2%	Roles in extracellular fluid volume, pH, etc.
Cellular Portion *(40%–50% of total volume):*		
1. Red blood cells	4,800,000–5,400,000 per microliter	Oxygen, carbon dioxide transport
2. White blood cells:		
Neutrophils	3,000–6,750	Phagocytosis
Lymphocytes	1,000–2,700	Immunity
Monocytes (macrophages)	150–720	Phagocytosis
Eosinophils	100–360	Roles in inflammatory response, immunity
Basophils	25–90	Roles in inflammatory response, anticlotting
3. Platelets	250,000–300,000	Roles in clotting

Blood: 6%–8% of body weight

plasma

cells, platelets

Figure 27.5 Components of blood. Keep a blood sample in a test tube from clotting, and it will separate into a layer of straw-colored liquid (the plasma) that floats over the red-colored cellular portion of blood.

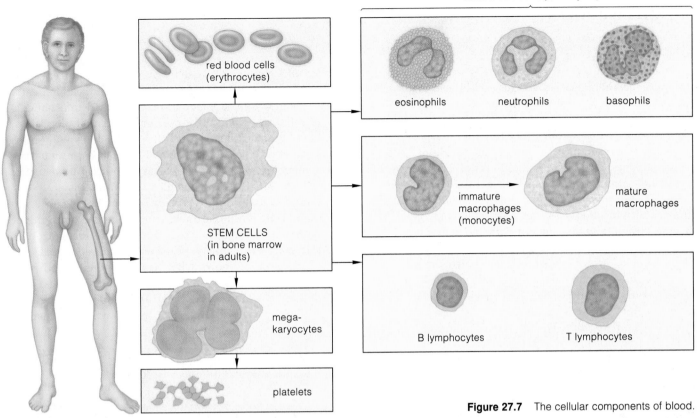

white blood cells (leukocytes)

red blood cells (erythrocytes)

eosinophils   neutrophils   basophils

STEM CELLS (in bone marrow in adults)

immature macrophages (monocytes)   mature macrophages

mega-karyocytes

B lymphocytes   T lymphocytes

platelets

Figure 27.7 The cellular components of blood.

some carbon dioxide wastes. When oxygen diffuses into blood, it binds with hemoglobin, the iron-containing pigment that gives red blood cells their color. (Here you may wish to review hemoglobin's molecular structure, as shown in Figure 28.8b.) Oxygenated blood is bright red. Poorly oxygenated blood is darker red but appears blue inside blood vessel walls near the body surface.

Red blood cells are derived from stem cells in bone marrow (Figure 27.7). Generally speaking, **stem cells** are unspecialized and retain the capacity for mitotic cell division. A portion of their daughter cells also divide, then differentiate into specialized types.

Mature red blood cells no longer have their nucleus, nor do they require it. They have enough hemoglobin, enzymes, and other proteins to function for about 120 days. At any time, phagocytes are engulfing the oldest red blood cells or the ones already dead, but ongoing replacements keep the cell count fairly stable. A **cell count** is the number of cells of a given type in a micro-liter of blood. The average number of red blood cells is 5.4 million in males and 4.8 million in females.

White Blood Cells Leukocytes, or white blood cells, arise from stem cells in bone marrow. They function in daily housekeeping and defense. Many patrol tissues, where they target or engulf damaged or dead cells and anything chemically recognized as foreign to the body.

Many others are massed together in the lymph nodes and spleen. There they divide to produce armies of cells that battle specific viruses, bacteria, and other invaders.

White blood cells differ in size, nuclear shape, and staining traits. There are five categories: neutrophils, eosinophils, basophils, monocytes, and lymphocytes (Figure 27.7). Their numbers can vary, depending on whether an individual is active, healthy, or under siege, as described in the next chapter. The neutrophils and monocytes are search-and-destroy cells. The monocytes follow chemical trails to inflamed tissues. There they develop into macrophages ("big eaters") that can engulf invaders and debris. Two classes of lymphocytes, B cells and T cells, make highly specific defense responses.

Platelets Some stem cells in bone marrow give rise to giant cells (megakaryocytes). These shed fragments of cytoplasm enclosed in a bit of plasma membrane. The membrane-bound fragments are what we call the platelets. Each platelet only lasts five to nine days, but hundreds of thousands are always circulating in blood. They can release substances that initiate blood clotting.

Blood has roles in transport, in defense, in blood clotting, and in maintaining the volume, composition, and temperature of the internal environment.

27.3 *Focus on Health*

BLOOD DISORDERS

The body continually replaces blood cells for good reason. Besides aging and dying off regularly, these cells typically encounter a variety of pathogens that use them as places to complete their life cycle. Besides this, sometimes blood cells malfunction as a result of gene mutations.

Red Blood Cell Disorders   Consider the *anemias*— disorders that result from too few red blood cells or deformed ones. Blood oxygen levels cannot be kept high enough to support normal metabolism. Fatigue, shortness of breath, and chills follow. *Hemorrhagic* anemias result from sudden blood loss, as from a severe wound; *chronic* anemias result from ongoing but slight blood loss, as from an undiagnosed bleeding ulcer, hemorrhoids, or the monthly blood loss of premenopausal women. Certain infectious bacteria and parasites replicate inside blood cells, then escape by lysis; they cause some *hemolytic* anemias. Iron deficiency causes another, because red blood cells can't produce enough normal hemoglobin. With B_{12} *deficiency* anemia, a potential hazard for strict vegetarians and alcoholics, red blood cells form but can't divide. Normally, meats, poultry, and fish in the diet provide enough vitamin B_{12}.

As described elsewhere in the book, a gene mutation that gives rise to an abnormal form of hemoglobin can result in *sickle-cell* anemia. Another gene mutation, which blocks or lowers the synthesis of the globin chains that make up hemoglobin, causes *thalassemias*. Too few red blood cells can form; those that do are thin and fragile.

Or consider the *polycythemias*—symptoms of far too many red blood cells—which make blood flow sluggish. Some bone marrow cancers can result in this condition. So can blood doping by athletes who compete in strenuous events. They withdraw their own red blood cells, store them, then reinject them a few days prior to competition. The withdrawal triggers red blood cell formation—to replace the "lost" cells—which bumps up the cell count when cells are put back in the body. The idea is to increase oxygen-carrying capacity, hence to increase endurance. This causes temporary high blood pressure and reduces blood's viscosity. Some believe the practice works. Others call it unethical; it is banned from the Olympics.

White Blood Cell Disorders   You may have heard of *infectious mononucleosis*. An Epstein-Barr virus causes this highly contagious disease, which results from too many monocytes and lymphocytes. Following a few weeks of fatigue, aches, low fever, and a chronic sore throat, the patient usually recovers. Recovery is chancy for *leukemias*, a category of cancers that suppress or impair white blood cell formation in bone marrow. Radiation therapy and chemotherapy can kill the cancer cells. The remissions (symptom-free periods) may last for months or years.

27.4 BLOOD TYPING AND BLOOD TRANSFUSIONS

Whenever blood volume or blood cell counts decline, countermeasures kick in automatically. However, if the volume were to decrease by more than 30 percent, then circulatory shock would follow and could lead to death.

Blood from donors can be transfused into patients affected by a blood disorder or substantial blood loss. Such **blood transfusions** cannot be done willy-nilly. Why not? *A potential donor and the recipient may or may not have the same kinds of recognition proteins at the surface of their red blood cells*. Some of these proteins are "self" markers; they identify the cells as belonging to one's own body. During a transfusion, if cells from a donor have the "wrong" marker, they will be recognized as foreign in the recipient, with serious consequences.

Figure 27.8 shows what happens when blood from incompatible donors and recipients intermingle. In a defensive response called **agglutination**, proteins called antibodies that are circulating in plasma act against the

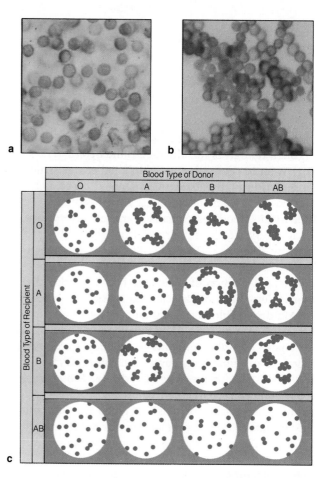

Figure 27.8   Micrographs showing (**a**) absence of agglutination in a mixture of two different but compatible blood types and (**b**) agglutination in a mixture of incompatible types. (**c**) Agglutination responses in blood types A, B, AB, and O when mixed with samples of the same and different types.

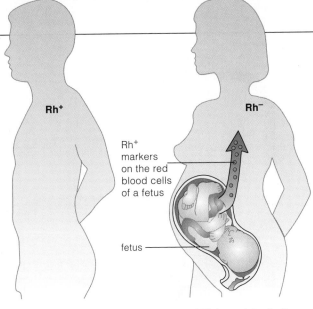
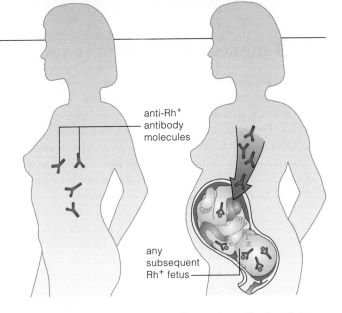

a A forthcoming child of an Rh⁻ woman and Rh⁺ man inherits the gene for the Rh⁺ marker. During pregnancy or childbirth, some of its cells bearing the marker may leak into the maternal bloodstream.

b The foreign marker stimulates her body to make antibodies. If she gets pregnant again and if this second fetus (or any other) inherits the gene for the marker, the circulating anti-Rh⁺ antibodies will act against it.

Figure 27.9 Antibody production in response to Rh⁺ markers on red blood cells of a developing fetus.

foreign cells and cause them to clump. (As described in the next chapter, antibodies bind to specific foreign markers and so target the cell or particle bearing it for destruction by the immune system.) The sheer number of "foreign" cells transfused into the recipient translates into a great many clumps, which can clog small blood vessels and damage tissues. Without treatment, death may follow. The same thing can happen during certain pregnancies, if the antibodies diffuse from the mother's circulatory system into that of her unborn child.

Based on an understanding of cell surface markers and antibodies, scientists have devised ways to analyze the forms of self markers on a person's red blood cells. Blood typing is based on these analyses.

ABO Blood Typing

Molecular variations in one kind of self marker on red blood cells are analyzed in **ABO blood typing**. The genetic basis of this variation is described in Section 9.4. People with one form of the marker are said to have type A blood; those with another form have type B blood. Some have both forms of the marker on their red blood cells; they have type AB blood. Others don't have either form of the marker; they have type O blood.

If you are type A, your antibodies ignore A markers but will act against B markers. If you are type B, your antibodies ignore B markers but will act against A markers. If you are type AB, your antibodies ignore both forms of the marker, so you can tolerate donations of type A, B, AB, or O blood. If you are type O, you have antibodies against both forms of the marker, so your options are limited to type O donations.

Rh Blood Typing

Rh blood typing is based on the presence or absence of an Rh marker (so named because it was first identified in blood samples of *Rh*esus monkeys). If you are type Rh⁺, your blood cells bear this marker at their surface. If you are type Rh⁻, they don't. Ordinarily, people do not have antibodies against Rh markers. But a recipient of transfused Rh⁺ blood produces antibodies against them, and the antibodies remain in the blood.

If an Rh⁻ woman becomes impregnated by an Rh⁺ man, there is a chance the fetus will be Rh⁺. During pregnancy or childbirth, some of its red blood cells may leak into the mother's bloodstream. If they do, her body will produce antibodies against Rh (Figure 27.9). If she becomes pregnant *again*, Rh antibodies will enter the bloodstream of this new fetus. If its blood is type Rh⁺, her antibodies will cause its red blood cells to swell, rupture, and release hemoglobin.

Erythroblastosis fetalis is a severe outcome of mixing Rh⁺ and Rh⁻ types. The fetus dies, for too many cells are destroyed. If diagnosed before birth or delivered alive, it can survive if its blood is slowly replaced with transfusions that are free of Rh antibodies. Currently, a known Rh⁻ woman can be treated right after her first pregnancy with an anti-Rh gamma globulin (RhoGam) that will protect her next fetus. The drug will inactivate any Rh⁺ fetal blood cells that are circulating in the mother's bloodstream before she can become sensitized and begin producing potentially dangerous antibodies.

To avoid the symptoms of blood incompatibilities, red blood cells should be typed before transfusions or pregnancies.

"Cardiovascular" comes from the Greek *kardia* (heart) and the Latin *vasculum* (vessel). In the human cardiovascular system, a heart pumps blood into large-diameter **arteries**. From there, blood flows into small, muscular **arterioles**, which branch into the even smaller diameter capillaries introduced earlier. Blood flows continuously from capillaries into small **venules**, then into large-diameter **veins** that return blood to the heart.

As in most vertebrates, a partition separates the heart into a double pump, which drives blood through two cardiovascular circuits (Figure 27.10). Each circuit has its own set of arteries, arterioles, capillaries, venules, and veins. The *pulmonary* circuit, a short loop, rapidly oxygenates blood. It leads from the heart's right half to capillary beds in both lungs, then returns to the heart's left half. The *systemic* circuit is a longer loop. It starts at the heart's left half and the **aorta** (the main artery carrying oxygenated blood away from the heart), branches to all organs and tissues with metabolically active cells, then converges into major veins that deliver oxygen-poor blood to the heart's right half. Figure 27.11 defines the functions of the major blood vessels of both circuits.

For most of the systemic branchings, a given volume of blood flows through one capillary bed. The branch to the intestines is one of the exceptions. First blood picks up glucose and other absorbed substances from one bed, then it moves through *another* capillary bed, in the liver—an organ with a key role in nutrition. The second bed gives the liver time to process absorbed substances.

And this brings us to an important point. All organs, not just the intestines, have different requirements for blood flow. As you know from Chapter 26, the flow to and from a skeletal muscle varies greatly, depending on what it is being called upon to do at any given time.

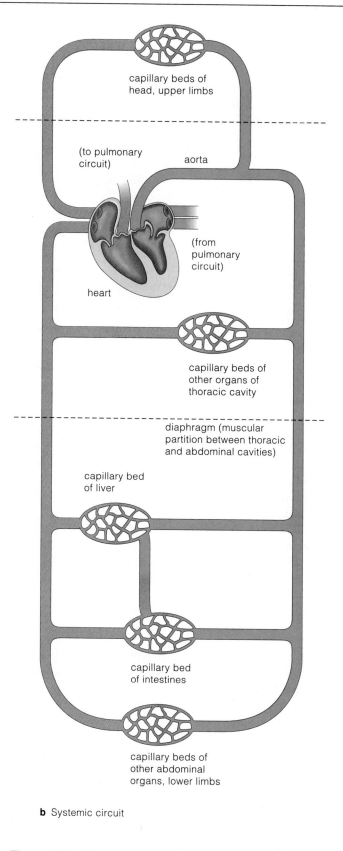

b Systemic circuit

Figure 27.10 Diagrams of the pulmonary and systemic circuits for blood flow through the human cardiovascular system. Blood vessels carrying oxygenated blood are color-coded *red*. Those carrying oxygen-poor blood are color-coded *blue*.

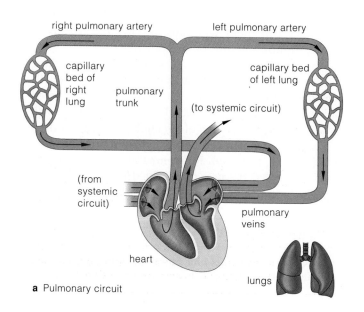

a Pulmonary circuit

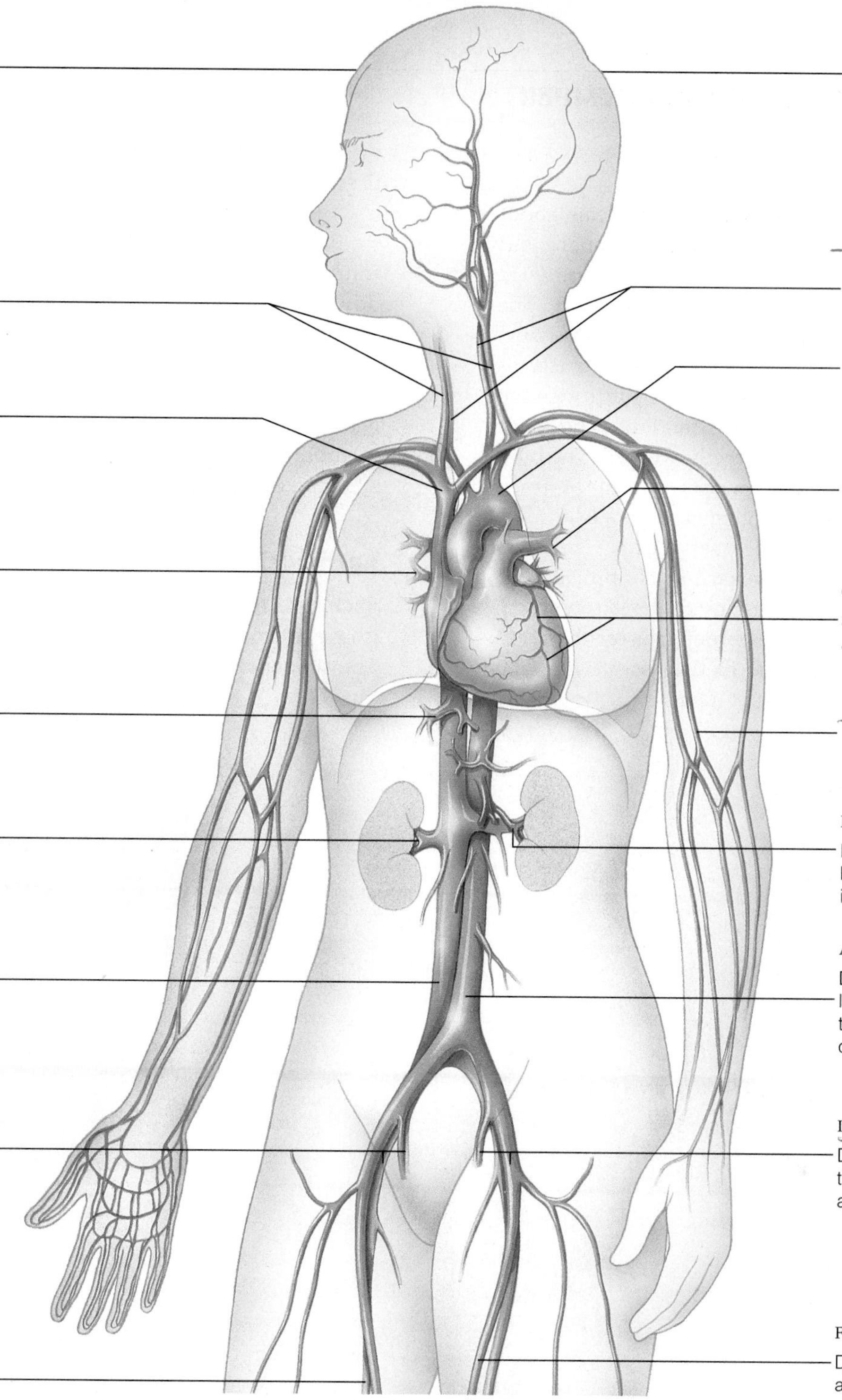

JUGULAR VEINS
Receive blood from brain and from tissues of head, neck

SUPERIOR VENA CAVA
Receives blood from veins of upper body

PULMONARY VEINS
Deliver oxygen-rich blood from the lungs to the heart

HEPATIC PORTAL VEIN
Delivers nutrient-rich blood from small intestine to liver for processing

RENAL VEIN
Carries processed blood away from the kidneys

INFERIOR VENA CAVA
Receives blood from all veins below the diaphragm

ILIAC VEINS
Carry blood away from the pelvic organs and lower abdominal wall

FEMORAL VEIN
Carries blood away from the thigh and inner knee

CAROTID ARTERIES
Deliver blood to neck, head, brain

ASCENDING AORTA
Carries oxygen-rich blood away from heart; the largest artery

PULMONARY ARTERIES
Deliver oxygen-poor blood from the heart to the lungs

CORONARY ARTERIES
Service the incessantly active cardiac muscle cells

BRACHIAL ARTERY
Delivers blood to arm, hand; blood pressure is measured here

RENAL ARTERY
Delivers blood to the kidneys for adjustments in volume, composition

ABDOMINAL AORTA
Delivers blood to arteries leading to the digestive tract, kidneys, pelvic organs, lower limbs

ILIAC ARTERIES
Deliver blood to the pelvic organs and lower abdominal wall

FEMORAL ARTERY
Delivers blood to the thigh and inner knees

Figure 27.11 Location and functions of major blood vessels of the human cardiovascular system.

The same is true of skin. When the body gets too cold, the flow of blood—and metabolically generated heat—can be diverted away from its vast capillary beds. When the body gets too hot, blood flow to the beds increases and heat radiates away from the skin's surface. Blood flow to the central command post—the brain—can't vary much; the neurons making up its communication lines cannot tolerate local decreases in blood supply.

The human cardiovascular system consists of separate pulmonary and systemic circuits for blood flow.

The pulmonary circuit is a short loop between the heart's right half, both lungs, and the heart's left half. Oxygen-poor blood flowing into this circuit rapidly picks up oxygen and gives up carbon dioxide at capillary beds in the lungs.

The much longer systemic circuit starts at the heart's left half and aorta, then branches to capillary beds of all metabolically active tissues and organs. There it gives up oxygen and takes up carbon dioxide; then it returns to the heart's right half.

Heart Structure

Think about the fact that the human heart beats about 2.5 billion times during a seventy-year life span, and you know it must be a truly durable pump. Figure 27.12 shows its structure, which attests to its durability. The pericardium, a double sac of tough connective tissue, protects the heart and anchors it to nearby structures (*peri*, around). A fluid between its layers lubricates the heart during its perpetual twisting motions. The inner layer serves as the outer part of the heart wall. The bulk of that wall, the myocardium, consists of cardiac muscle cells tethered to elastin and collagen fibers. The fibers are so densely crisscrossed, they serve as a "skeleton" against which the force of contraction is applied. The oxygen-demanding cardiac muscle cells have their own, "coronary" circulation; arteries branching off the aorta deliver blood to a capillary bed that services them. The

wall's glistening, innermost layer consists of connective tissue and endothelium, a one-cell-thick epithelial sheet. Only the heart and blood vessels have endothelium.

Each half of the heart has *two* chambers—an **atrium** (plural, atria) and a **ventricle**. Blood flows into the atria, down into the ventricles, then out through great arteries (the aorta or pulmonary trunk). Between each atrium and ventricle is an AV valve (short for atrioventricular). Between the ventricle and the artery leading out from it is a semilunar valve. These are all *one-way* valves, with membranes that rhythmically flap open, flap shut, and so help keep blood moving in one direction.

Cardiac Cycle

Each time the heart beats, its four chambers go through phases of contraction (systole) and relaxation (diastole). The sequence of contraction and relaxation is a **cardiac**

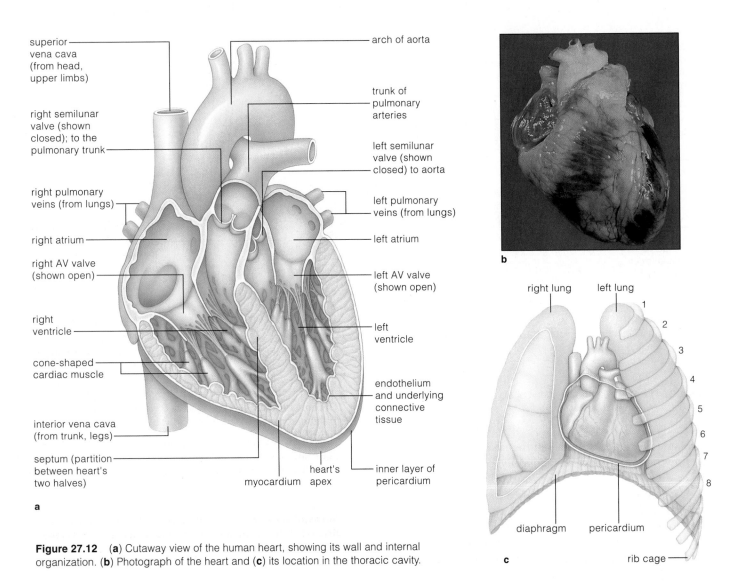

Figure 27.12 (**a**) Cutaway view of the human heart, showing its wall and internal organization. (**b**) Photograph of the heart and (**c**) its location in the thoracic cavity.

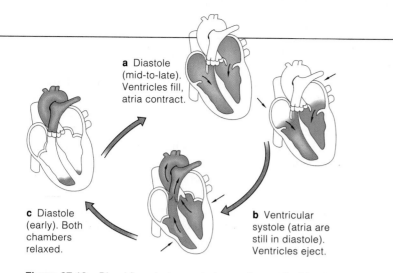

a Diastole (mid-to-late). Ventricles fill, atria contract.

b Ventricular systole (atria are still in diastole). Ventricles eject.

c Diastole (early). Both chambers relaxed.

Figure 27.13 Blood flow during part of a cardiac cycle. Blood and heart movements generate a "lub-dup" sound at the chest wall. At each "lub," AV valves are closing as ventricles contract. At each "dup," semilunar valves are closing as ventricles relax.

cycle. When relaxed, the atria fill with blood. Increasing fluid pressure forces the AV valves open. Blood flows into the ventricles, which completely fill when the atria contract (Figure 27.13a). As the filled ventricles start to contract, the rising fluid pressure forces the AV valves shut. It rises so sharply above the pressure in the great arteries, it forces the semilunar valves open—and blood leaves the heart (Figure 27.13b). The ventricles relax, the semilunar valves close, and the already filling atria are ready to repeat the cycle. Thus, in a cardiac cycle, atrial contraction simply helps fill the ventricles. *Contraction of the ventricles is the driving force for blood circulation.*

Mechanisms of Contraction

Cardiac muscle cells are striated (striped). Like skeletal muscle cells, their sarcomeres contract in response to action potentials, as the sliding-filament model predicts (Section 26.6). The required ATP energy comes from mitochondria that pepper the myocardium. But cardiac muscle cells are structurally unique. They are branching, short, and joined at their end regions. Communication junctions in the abutting regions allow action potentials to spread swiftly among cells, in waves of excitation that wash over the entire heart (Figure 27.14a).

Besides this, about 1 percent of the cardiac muscle cells *don't* contract. Instead, they function as a **cardiac conduction system**. These specialized cells initiate and propagate waves of excitation about seventy-five times a minute. They do so in a rhythmic, orderly sequence, from atria to ventricles. By their synchronized activity, the heart pumps efficiently. Each wave starts at the SA node, a cluster of cell bodies in the right atrium's wall (Figure 27.14b). The wave passes through the wall to another cell body cluster, the AV node. This is the *only* electrical bridge between atria and ventricles (which connective tissue insulates everywhere else). After the

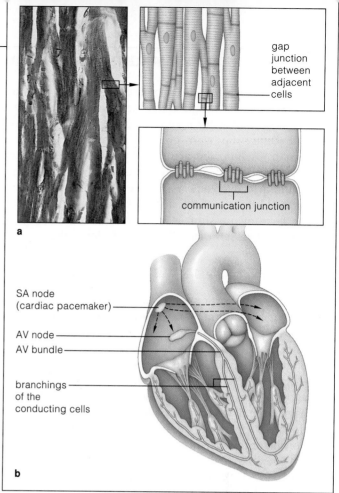

gap junction between adjacent cells

communication junction

a

SA node (cardiac pacemaker)

AV node
AV bundle

branchings of the conducting cells

b

Figure 27.14 (**a**) Communication junctions between abutting cardiac muscle cells. (**b**) Location of specialized cardiac muscle cells of the cardiac conduction system.

AV node, conducting cells are arranged as a bundle in the partition between the heart's two halves. The cells then branch, and the branches deliver the excitatory wave up the ventricle walls. The ventricles contract in response with a twisting movement, upward from the heart's apex, that ejects blood into the great arteries.

The SA node fires action potentials faster than the rest of the system and serves as the **cardiac pacemaker**. Its spontaneous, rhythmic signals are the foundation for the normal rate of heartbeat. The nervous system can only *adjust* the rate and strength of contractions dictated by the pacemaker. Even if all nerves leading to a heart are severed, the heart will keep on beating!

The heart's construction gives evidence of its role as a durable pump. Under spontaneous, rhythmic signals from the cardiac pacemaker, the heart's branching, abutting cardiac muscle cells contract in synchrony, almost as a single unit.

Although the heart has four chambers (each half has one atrium and one ventricle), contraction of the ventricles is the driving force for blood circulation away from the heart.

Blood pressure, the fluid pressure generated by heart contractions, is highest in contracting ventricles. During the time it takes for a given volume of blood to leave and reenter the heart, pressure is still high in arteries, then drops along the circuit, and is lowest in the relaxed atria (Figure 27.15a). What causes the drop? Energy is required to overcome resistance to the flow of blood; and as blood passes through different vessels, energy (in the form of pressure) is lost. Structural differences in the blood vessels themselves are a factor in this loss.

Pressure at Arteries and Arterioles

Besides transporting oxygenated blood, arteries of the systemic and pulmonary circuits are pressure reservoirs that smooth out pulsations in pressure caused by each cardiac cycle. Arteries have a thick, muscular, elastic wall (Figure 27.15b). The wall bulges a bit with each pressure surge resulting from ventricular contraction, then recoils and so forces blood on through the circuit. Arteries also have large diameters that offer little resistance to blood flow. Thus, pressure does not drop much in the arterial portion of the systemic and pulmonary circuits.

The arteries branch into smaller diameter arterioles (Figure 27.15c). Track the flow along the systemic route and you find the greatest pressure drop at arterioles, for these offer the greatest resistance to blood flow. With the slowdown, the flow can be controlled, in ways that divert greater or lesser portions of the total volume to different regions. How? Read on.

Control of Blood Pressure and Distribution

Suppose you decide to measure your blood pressure daily when you are resting, in the manner described in Figure 27.16. Assuming you are an adult in good health,

the average resting value stays fairly constant over a few weeks, even months—about 120/80 mm Hg. The nervous and endocrine systems direct the adjustments that maintain this value and so ensure good flow to all regions. Besides this, shifts in local chemical conditions can bring about immediate adjustments.

Suppose sensory receptors notify the brain of a shift in blood pressure. (Such receptors are located in cardiac muscle and in arteries, including the aorta and carotid arteries in the neck.) If the resting value *rises*, the brain commands the heart to beat more slowly and contract less forcefully. It also commands rings of smooth muscle in arteriole walls to relax. The result is **vasodilation**, an increase in blood vessel diameter. If the resting value *falls*, the brain makes the heart beat faster and contract more forcefully. And it makes rings of arteriole muscles contract. The result is **vasoconstriction**, a decrease in blood vessel diameter. Also, changes in local chemical conditions can make smooth muscle cells relax or contract. So can hormones. For instance, epinephrine can cause vasoconstriction or vasodilation. Angiotensin can bring about widespread vasoconstriction.

The nervous and endocrine systems also control the allocation of blood flow to different regions at different times. For example, after you have eaten a large meal, more blood is diverted to your digestive system. When you stay out in cold wind or snow for an extended time, blood is diverted away from the skin to deeper tissue regions, so metabolically generated heat that warms the blood can be conserved.

Local chemical controls, too, divert blood flow. For instance, when you run, skeletal muscle cells use up oxygen and release wastes, such as carbon dioxide and hydrogen ions. In response, arterioles vasodilate. Now more blood flows past the working muscles, delivering materials and carrying away products and wastes.

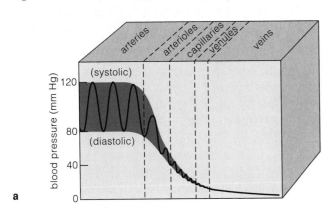

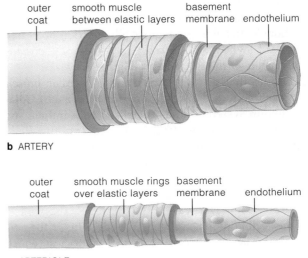

b ARTERY

c ARTERIOLE

Figure 27.15 (**a**) Blood pressure. This is a plot of measurements made of the drop in fluid pressure for a given volume of blood moving through the systemic circuit. (**b**–**e**) Structure of blood vessels. The basement membrane around the endothelium is a noncellular layer, rich with proteins and polysaccharides.

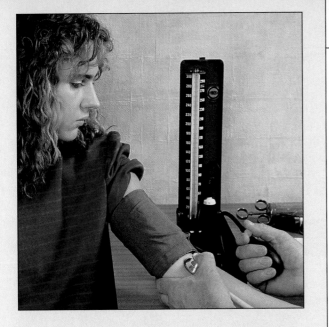

Figure 27.16 Measuring blood pressure. A hollow cuff attached to a pressure gauge is wrapped around the upper arm. Then it is inflated with air to a pressure above the highest pressure of the cardiac cycle (at systole, when the ventricles contract). Above this pressure, no sounds are heard through a stethoscope positioned below the cuff and over the artery (because no blood is flowing through the vessel).

Air in the cuff is slowly released, so some blood flows into the artery. The turbulent flow causes soft tapping sounds, and when this first occurs, the value on the gauge is the systolic pressure—about 120 mm mercury (Hg) in young adults at rest. (This means the measured pressure would force mercury to move upward 120 millimeters in a narrow glass column.) More air is released from the cuff. Just after the sounds become dull and muffled, blood flows continuously. So the turbulence and tapping sounds stop. The silence corresponds to the diastolic pressure (at the end of a cardiac cycle, just before the heart pumps out blood). Generally the reading is about 80 mm Hg. In this example, the *pulse* pressure (the difference between the highest and lowest pressure readings) is 120 – 80, or 40 mm Hg.

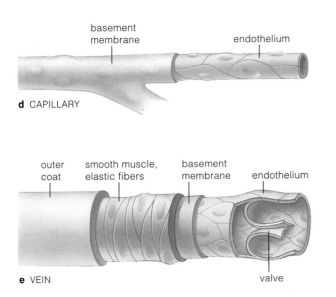

d CAPILLARY

basement membrane
endothelium

outer coat
smooth muscle, elastic fibers
basement membrane
endothelium

e VEIN
valve

Capillary Function

Capillary beds are *diffusion zones* for exchanges between blood and interstitial fluid. A lone capillary presents high resistance to flow. Its diameter is so tiny, blood cells squeeze through it single file. Yet so many capillaries weave through a bed, their *combined* diameters are greater than the diameters of arterioles leading to them. Thus they present less total resistance to flow than the arterioles, so the total drop in pressure is not great.

Capillaries thread through nearly every tissue, and at least one of them is as close as 0.001 centimeter to every living cell. Each capillary is little more than a tube of endothelial cells, a single layer thick (Figure 27.15*d*). Oxygen, carbon dioxide, and most other small solutes diffuse across the capillary wall. Certain proteins enter and leave by endocytosis or exocytosis. Ions probably cross at clefts between endothelial cells. So do white blood cells, as you will see in the chapter to follow.

The concentrations of water and solutes in blood and interstitial fluid at any given time influence the direction in which fluid flows across the capillary wall. Normally there is only a small *net* outward movement of fluid, which the lymphatic system returns to the blood. *Edema* results from an accumulation of fluid in interstitial spaces. Edema occurs to some extent during exercise, as arterioles dilate in local tissues. It is most extreme during *elephantiasis*, which is brought on by a roundworm infection (Section 20.6).

Venous Pressure

Capillaries merge into venules ("little veins"), which merge into veins, the large-diameter transport tubes to the heart (Figure 27.15*e*). Veins have valves that prevent backflow. (When gravity beckons, venous flow reverses direction and pushes the valves shut.) Their wall can bulge considerably under pressure. Collectively, veins contain 50–65 percent of the total blood volume. When blood must flow faster, smooth muscle in vein walls contracts. The walls stiffen; veins bulge less. Pressure rises and drives more blood to the heart. Pressure also rises when skeletal muscles bulge against veins during limb movements and when breathing is rapid.

The nervous system, endocrine system, and changes in local chemical conditions control the level of blood pressure and the flow distribution to different regions. The main controls operate on cardiac muscle and on smooth muscle in arterioles.

Capillary beds are diffusion zones for exchanges between blood and interstitial fluid. Venules overlap with capillaries in function.

Veins are highly distensible blood volume reservoirs and help adjust flow volume back to the heart.

CARDIOVASCULAR DISORDERS

Cardiovascular disorders are the leading cause of death in the United States. They affect at least 40 million people and kill about a million each year. The most common disorders are called *hypertension* (sustained high blood pressure) and *atherosclerosis* (progressive thickening of the arterial wall and narrowing of the arterial lumen). Both affect blood circulation and so cause most *heart attacks* (damage or death of heart muscle) and *strokes* (brain damage).

In most heart attacks, a "crushing" pain behind the breastbone lasts a half hour or more. Mild to severe pain may radiate into the left arm, shoulder, or neck. Sweating, shortness of breath, erratic heartbeat, nausea, vomiting, and dizziness or fainting may accompany an attack.

Risk Factors   Extensive research indicates the following risk factors have roles in cardiovascular disorders:

1. Smoking (Section 29.6)
2. Genetic predisposition to heart failure
3. High level of cholesterol in the blood
4. High blood pressure
5. Obesity (Section 30.11)
6. Lack of regular exercise
7. Diabetes mellitus (Section 33.7)
8. Age (the older you get, the greater the risk)
9. Gender (until age fifty, males are at much greater risk than females)

Yet people can *control* some of these factors if they exercise, eat properly, and don't smoke. For example, the fatter you get, the more blood capillaries you get (they service the increased adipose tissue masses). So your heart has to work harder to pump blood through the increasingly divided vascular circuit. As another example, nicotine in tobacco makes the adrenal glands secrete epinephrine, a hormone that constricts blood vessels and boosts the heart rate, so blood pressure rises. Carbon monoxide in cigarette smoke inhibits the binding of oxygen to hemoglobin, so the heart has to pump harder to get oxygen to cells. Smoking also predisposes individuals to atherosclerosis even when their cholesterol levels are normal.

The next sections describe some of the tissue damage that results from cardiovascular disorders.

Hypertension   Hypertension results from gradual increases in the resistance to blood flow through small arteries. In time, blood pressure remains above 140/90, even when a person is resting. Heredity may be a factor; the disorder tends to run in families. Diet is also a factor. For instance, high salt intake can raise blood pressure in susceptible people and increase the heart's workload. The heart may eventually enlarge and fail to pump blood effectively. High blood pressure also may contribute to the "hardening" of arterial walls that hampers delivery of oxygen to the brain, heart, and other vital organs.

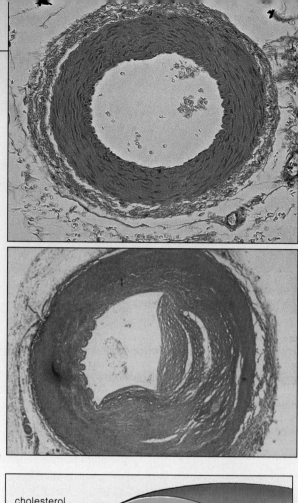

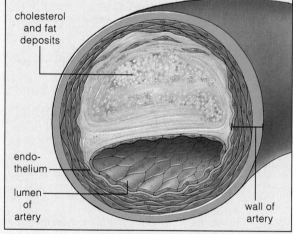

Figure 27.17   (**a**) Sections from a normal artery and one with a narrowed lumen. (**b**) Diagram of an atherosclerotic plaque.

Hypertension is called the "silent killer" because affected persons may have no outward symptoms. Even when they know their blood pressure is high, some tend to resist helpful medication, changes in diet, and regular exercise. Of 23 million hypertensive Americans, most don't undergo treatment. About 180,000 die each year.

Atherosclerosis   With arteriosclerosis, arteries thicken and lose elasticity. With atherosclerosis, this condition worsens as cholesterol and other lipids build up in the wall of arteries and cause the lumen to narrow.

Recall, from the story at the start of Chapter 13, that the liver produces enough cholesterol to satisfy the body's needs. Food intake increases the cholesterol level. When circulating in blood, cholesterol is bound to proteins as *low-density lipoproteins* (LDLs). These bind to receptors on cells throughout the body. Cells take up LDLs and their cholesterol cargo for use in cell activities. The excess is attached to proteins as *high-density lipoproteins* (HDLs) and transported back to the liver, where it can be metabolized.

For a variety of reasons, not enough LDL is removed from the blood in some people. As the blood level of LDL increases, so does the risk of atherosclerosis. With their bound cholesterol, LDLs *infiltrate* arterial walls. Abnormal smooth muscle cells multiply; connective tissue components increase in the walls. Cholesterol accumulates in cells and extracellular spaces of the wall's endothelial lining. On top of the lipids, calcium deposits actually form microscopic slivers of bone. A fibrous net forms over the entire mass. This *atherosclerotic plaque* sticks out into the arterial lumen (Figure 27.17a,b).

The plaque's bony slivers rip the endothelium. Platelets gather at the damaged site, secreting chemicals that initiate clot formation. The condition worsens as fatty globules in the plaque become oxidized. Many take on a form that resembles surface components of common bacteria— including a type that instigates formation of bonelike calcium deposits in the lungs. As an awful consequence, the call goes out to bacteria-fighting monocytes. Soon an inflammatory response gets under way—and certain chemicals released during the fray activate the genes for bone formation. Normally, this is a good thing; it helps wall off invaders and prevents infection from spreading. It is bad news for arteries.

As plaques and clots grow, they can narrow or block an artery. Blood flow to the tissues serviced by the artery may decrease to a trickle or stop. A clot that stays in place is a *thrombus*. If it becomes dislodged and travels the bloodstream, it is an *embolus*. Coronary arteries have narrow diameters. They are extremely vulnerable to clogging by a plaque or clot. When they narrow to one-quarter of their former diameter, the outcome ranges from *angina pectoris* (mild chest pains) to a heart attack.

Atherosclerosis involving coronary arteries can be diagnosed through *stress electrocardiograms*. These are recordings of the electrical activity of the cardiac cycle while a person is exercising on a treadmill. It also can be diagnosed by *angiography*. This procedure involves injections of a dye that is opaque to x-rays. Severe blockage may require surgery. In *coronary bypass surgery*, a section of an artery from the chest is stitched to the aorta and to the coronary artery below the narrowed or blocked region (Figure 27.18). In *laser angioplasty*, laser beams vaporize the plaques. In *balloon angioplasty*, a small balloon is inflated within a blocked artery to break up plaques.

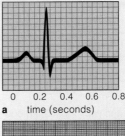

aorta

blocked portion of coronary artery

grafted artery

Figure 27.18 Two coronary bypasses (color-coded *green*).

Arrhythmias ECGs can be used to detect *arrhythmias*, irregular heart rhythms (Figure 27.19). Arrhythmias are not always abnormal. Endurance athletes may have a below-average resting cardiac rate (*bradycardia*). As an adaptation to ongoing strenuous exercise, their nervous system has adjusted the cardiac pacemaker's rate of contraction downward. Exercise or stress often causes 100+ heartbeats a minute (*tachycardia*). *Atrial fibrillation*, an irregular heartbeat, affects more than 10 percent of the elderly and young people with various heart diseases. A coronary occlusion or some other disorder may cause irregular rhythms that can rapidly lead to a dangerous *ventricular fibrillation*. In portions of the ventricles, cardiac muscle contracts haphazardly, and blood pumping suffers. Within seconds, the person loses consciousness, which may signify impending death. A strong electric shock to the chest may restore normal cardiac function.

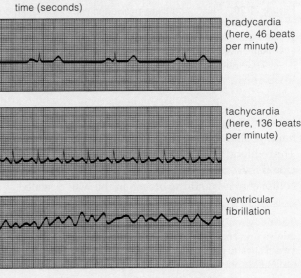

0   0.2   0.4   0.6   0.8

a   time (seconds)

Figure 27.19 (**a**) An ECG of a single, normal human heartbeat and (**b–d**) three examples of arrhythmias.

b   bradycardia (here, 46 beats per minute)

c   tachycardia (here, 136 beats per minute)

d   ventricular fibrillation

27.9 HEMOSTASIS

Small blood vessels are vulnerable to ruptures, cuts, and other damage. **Hemostasis** repairs the damage and prevents blood loss. This process involves blood vessel spasm, platelet plug formation, and blood coagulation. First, smooth muscle in a damaged wall contracts in an automatic response called a spasm. The blood vessel constricts, so blood flow through it temporarily stops. Second, platelets clump together as a temporary plug

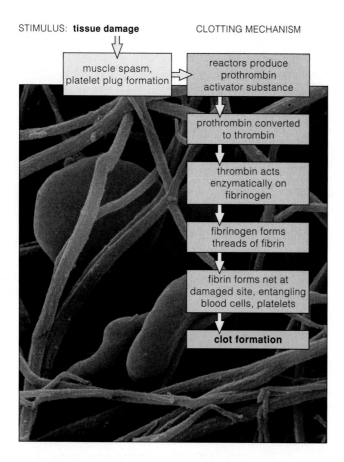

STIMULUS: **tissue damage**   CLOTTING MECHANISM

muscle spasm, platelet plug formation → reactors produce prothrombin activator substance

prothrombin converted to thrombin

thrombin acts enzymatically on fibrinogen

fibrinogen forms threads of fibrin

fibrin forms net at damaged site, entangling blood cells, platelets

clot formation

Figure 27.20 Plasma proteins involved in clot formation. Blood coagulates when damage exposes collagen fibers in blood vessel walls. This initiates a series of reactions that cause rod-shaped plasma proteins (fibrinogens) to stick together as long, insoluble threads. These stick to the exposed collagen, forming a net that traps blood cells and platelets. The entire mass is a clot.

in the damaged wall. They also release substances that help prolong the spasm and attract more platelets. Third, blood coagulates (converts to a gel) and forms a clot (Figure 27.20). Finally, the clot retracts, forming a compact mass, and the breach in the wall is sealed.

The body routinely repairs damage to small blood vessels and so prevents blood loss. The repair process includes blood vessel spasm, platelet plug formation, and coagulation.

27.10 LYMPHATIC SYSTEM

We conclude this chapter with a brief section on the manner in which the lymphatic system supplements blood circulation. But think of this section as a bridge to the next chapter, on immunity, for the lymphatic system also helps defend the body against injury and attack. This system consists of drainage vessels, lymphoid organs, and lymphoid tissues. Tissue fluid that has moved into the vessels is called lymph.

Lymph Vascular System

A portion of the lymphatic system consists of tubes that collect and transport water and solutes from interstitial fluid to ducts of the circulatory system. This portion is the **lymph vascular system**. Its main components are **lymph capillaries** and **lymph vessels** (Figure 27.21).

The lymph vascular system serves three functions. First, its vessels are drainage channels for water and plasma proteins that have leaked away from blood at capillary beds and that must be delivered back to the blood circulation. Second, it also takes up fats that the body has absorbed from the small intestine and delivers them to the blood circulation, in the manner described in Section 30.5. Third, it delivers pathogens, foreign

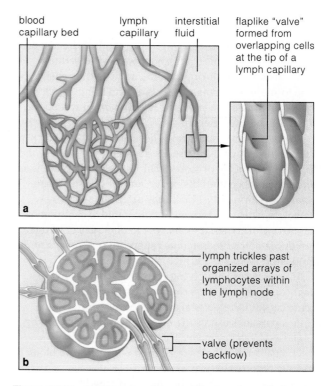

blood capillary bed   lymph capillary   interstitial fluid   flaplike "valve" formed from overlapping cells at the tip of a lymph capillary

a

lymph trickles past organized arrays of lymphocytes within the lymph node

valve (prevents backflow)

b

Figure 27.21 (a) Diagram of some of the lymph capillaries at the start of the drainage network called the lymph vascular system. (b) Cutaway diagram of a single lymph node. Its inner compartments are packed with organized arrays of infection-fighting white blood cells.

TONSILS
Defense against bacteria and other foreign agents

RIGHT LYMPHATIC DUCT
Drains right upper portion of the body

THYMUS GLAND
Site where certain white blood cells acquire means to chemically recognize specific foreign invaders

THORACIC DUCT
Drains most of the body

SPLEEN
Major site of antibody production; disposal site for old red blood cells and foreign debris; site of red blood cell formation in the embryo

SOME OF THE LYMPH VESSELS
Return excess interstitial fluid and reclaimable solutes to the blood

SOME OF THE LYMPH NODES
Filter bacteria and many other agents of disease from lymph

BONE MARROW
Marrow in some bones is production site for infection-fighting blood cells (as well as red blood cells and platelets)

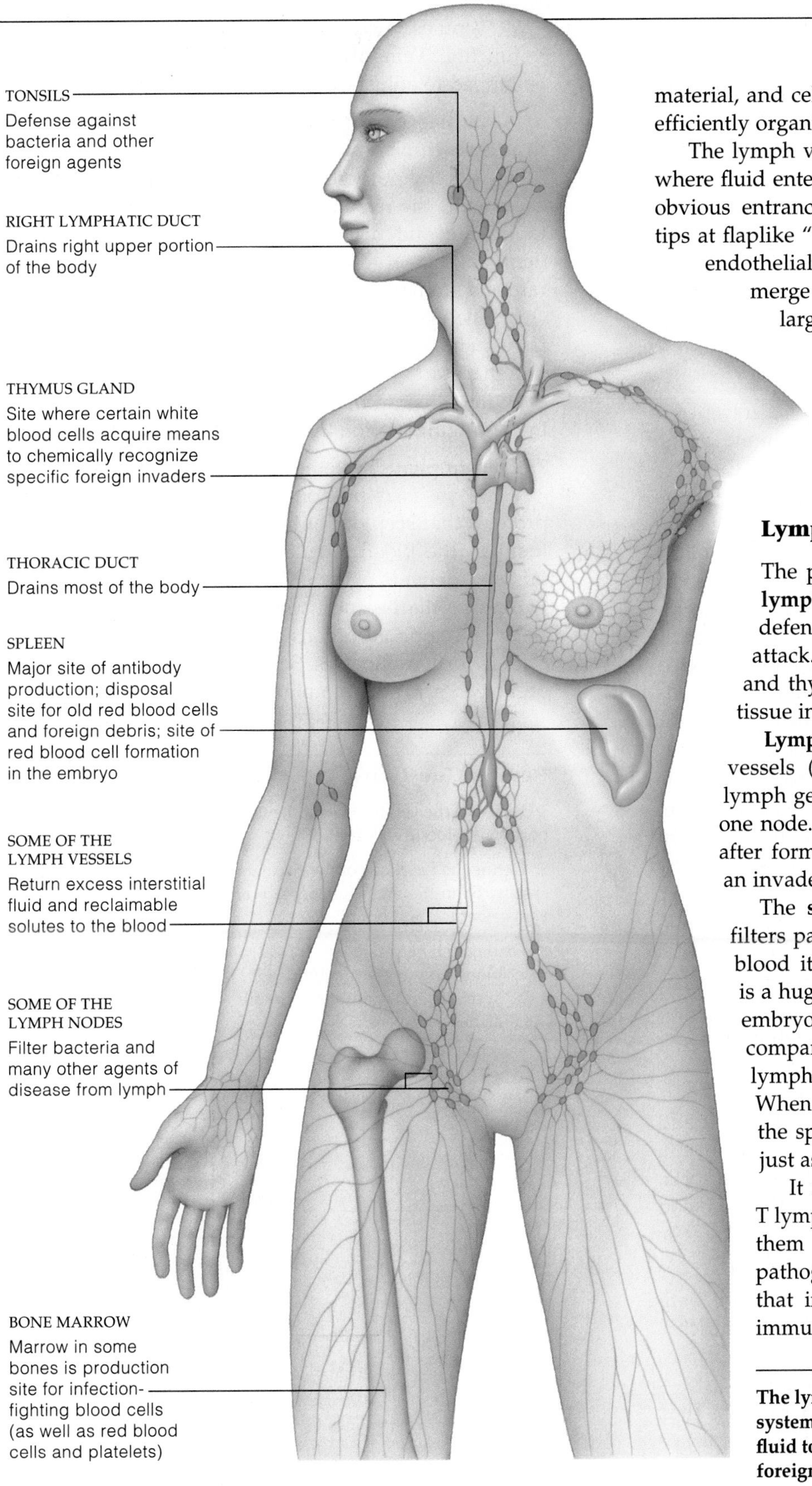

Figure 27.22 Components of the human lymphatic system and their functions. The *green* dots show some of the major lymph nodes. Patches of lymphoid tissue in the small intestine and in the appendix also are part of the lymphatic system.

material, and cellular debris from the body's tissues to efficiently organized disposal centers (lymph nodes).

The lymph vascular system starts at capillary beds, where fluid enters the lymph capillaries. These have no obvious entrance; water and solutes move into their tips at flaplike "valves" that are regions of overlapping endothelial cells (Figure 27.21*a*). Lymph capillaries merge into lymph vessels. These are vessels of larger diameter. They contain some smooth muscle in their wall, as well as valves in the lumen that prevent backflow. They converge into collecting ducts that drain into veins in the lower neck (Figure 27.22).

Lymphoid Organs and Tissues

The portion of the lymphatic system called **lymphoid organs and tissues** has roles in defending the body against damage and attack. It includes the lymph nodes, spleen, and thymus, as well as tonsils and patches of tissue in the small intestine and appendix.

Lymph nodes occur at intervals along lymph vessels (Figure 27.22). Before entering blood, lymph gets filtered as it trickles through at least one node. Many lymphocytes reside in the nodes after forming in bone marrow. If they recognize an invader, they multiply rapidly and destroy it.

The **spleen**—the largest lymphoid organ— filters pathogens and aged blood cells from the blood itself. One compartment (the red pulp) is a huge reservoir of red blood cells. In human embryos, it also produces these cells. The other compartment (the white pulp) has masses of lymphocytes associated with blood vessels. When specific invaders become trapped in the spleen, these lymphocytes destroy them, just as they do in lymph nodes.

It is in the **thymus gland** that immature T lymphocytes differentiate in ways that allow them to recognize and respond to specific pathogens. The thymus produces hormones that influence these events. It is central to immunity, the focus of the chapter to follow.

The lymph vascular portion of the lymphatic system returns water and solutes from tissue fluid to the blood. It also delivers fats and foreign material and debris to lymph nodes.

The system's lymph nodes and other lymphoid organs function in defending the body against tissue damage and infectious diseases.

SUMMARY

1. The closed circulatory system of humans and other vertebrates consists of a heart (a muscular pump), blood vessels (arteries, arterioles, capillaries, venules, and veins), and blood. The system functions in rapid internal transport of substances to and from cells.

2. Blood helps maintain favorable conditions for cells. This fluid connective tissue consists of plasma, red and white blood cells, and platelets. It delivers oxygen and other substances to the interstitial fluid around cells. It also picks up cell products and wastes from that fluid.

   a. Plasma, the liquid portion of blood, is a transport medium for blood cells and platelets. It is a solvent for plasma proteins, simple sugars, lipids, amino acids, mineral ions, vitamins, hormones, and several gases.

   b. Red blood cells transport oxygen from the lungs to all body regions. They are packed with hemoglobin, an iron-containing pigment molecule that binds reversibly with oxygen. Red blood cells also transport some carbon dioxide from interstitial fluid to the lungs.

   c. Some phagocytic white blood cells cleanse tissues of dead cells, cellular debris, and anything else detected as not belonging to the body. Other white blood cells (lymphocytes) form great armies that destroy specific bacteria, viruses, and other disease agents.

3. The human heart is an incessantly beating double pump. Each half of the heart has two chambers (an atrium and a ventricle). Blood flows into the atria, then the ventricles, then on into the great arteries. One-way valves enforce the one-way flow.

4. The heart's partition separates blood flow into two circuits, one pulmonary and the other systemic.

   a. The pulmonary circuit loops between the heart and lungs. Oxygen-poor blood from the systemic veins enters the *right* atrium, is pumped through pulmonary arteries to both lungs, picks up oxygen, then flows through pulmonary veins to the heart's left atrium.

   b. The systemic circuit loops between the heart and all body tissues. Oxygenated blood in the *left* atrium flows into the left ventricle, is pumped into the aorta, then is distributed to capillary beds. There, the blood gives up oxygen and picks up carbon dioxide. Systemic veins return it to the heart's right atrium.

5. Ventricular contraction drives blood through both circuits. Blood pressure is high in contracting ventricles. It drops successively in arteries, arterioles, capillaries, venules, and veins. It is lowest in relaxed atria.

   a. Arteries are an elastic pressure reservoir. They smooth out pressure changes resulting from heartbeats and so smooth out blood flow through capillaries.

   b. Arterioles are control points for distributing different volumes of blood to different regions.

   c. Capillary beds are diffusion zones where blood and interstitial fluid exchange substances.

   d. Venules overlap capillaries and veins somewhat in function. Veins are a blood volume reservoir that can be tapped to adjust the flow volume back to the heart.

6. A cardiac conduction system is the basis for the heart's rhythmic, spontaneous contractions.

   a. Although 99 percent of the cardiac muscle cells are contractile, the remainder are not. They are specialized to initiate and distribute action potentials, independently of the nervous system, which can only adjust the rate and strength of the basic contractions.

   b. The system's SA node fires action potentials the fastest and is the cardiac pacemaker. Waves of excitation that start here wash over the atria, down the heart's partition, then up the ventricles. The ventricles contract in a wringing motion that ejects blood from the heart.

7. The lymphatic system has these functions:

   a. Its vascular portion takes up water and plasma proteins that seep out of blood capillaries, then returns them to the blood circulation. It transports absorbed fats and delivers agents of disease to disposal centers. Lymph capillaries and lymph vessels are components.

   b. Its lymphoid organs and tissues have production centers for lymphocytes. Some are battlegrounds where organized arrays of lymphocytes fight disease agents.

Review Questions

1. Describe the cellular components of blood. Describe the plasma portion of blood. *460–461*

2. Define the functions of the circulatory system and the lymphatic system. State the main functions of arteries, arterioles, capillaries, veins, venules, and lymph vessels. *458–459, 468–469, 472–473*

3. Distinguish between:
   a. blood and interstitial fluid *458*
   b. systemic and pulmonary circuits *459*
   c. atrium and ventricle *466*

4. Label the heart's components. *466*

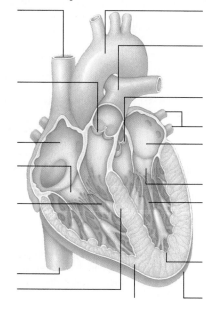

Self-Quiz *(Answers in Appendix IV)*

1. Cells directly exchange substances with _____ .
   a. blood vessels   c. interstitial fluid
   b. lymph vessels   d. both a and b

2. Which are *not* components of blood?
   a. plasma
   b. blood cells and platelets
   c. gases and other dissolved substances
   d. all of the above are components of blood

3. The _____ produces red blood cells, which transport _____ and some _____ .
   a. liver; oxygen; mineral ions
   b. liver; oxygen; carbon dioxide
   c. bone marrow; oxygen; hormones
   d. bone marrow; oxygen; carbon dioxide

4. The _____ produces white blood cells, which function in _____ and _____ .
   a. liver; oxygen transport; defense
   b. lymph nodes; oxygen transport; pH stabilization
   c. bone marrow; housekeeping; defense
   d. bone marrow; pH stabilization; defense

5. In the pulmonary circuit, the heart's _____ half pumps blood to lungs, then _____ blood flows to the heart.
   a. right; oxygen-poor   c. right; oxygen-rich
   b. left; oxygen-poor   d. left; oxygen-rich

6. In the systemic circuit, the heart's _____ half pumps _____ blood to all body regions.
   a. right; oxygen-poor   c. right; oxygen-rich
   b. left; oxygen-poor   d. left; oxygen-rich

7. Blood pressure is high in _____ and lowest in _____ .
   a. arteries; veins
   b. arteries; relaxed atria
   c. arteries; ventricles
   d. arterioles; veins

8. _____ contraction drives blood through the pulmonary circuit and the systemic circuit; blood pressure is highest in contracting _____ .
   a. Atrial; ventricles   c. Ventricular; arteries
   b. Atrial; atria   d. Ventricular; ventricles

9. Which of the following is *not* a function of the lymphatic system?
   a. delivers disease agents to disposal centers
   b. produces lymphocytes
   c. delivers oxygen to cells
   d. returns water and plasma proteins to blood

10. Match the type of blood vessel with its major function.
    ____ arteries   a. diffusion
    ____ arterioles   b. control of blood volume distribution
    ____ capillaries
    ____ venules   c. transport, blood volume reservoirs
    ____ veins
                          d. overlap capillary function
                          e. transport and pressure reservoirs

11. Match the components with their most suitable description.
    ____ capillary beds   a. two atria, two ventricles
    ____ lymph vascular   b. bathes body's living cells
         system   c. driving force for blood
    ____ heart chambers   d. zones of diffusion
    ____ blood   e. starts at capillary beds
    ____ heart contractions   f. fluid connective tissue
    ____ interstitial fluid

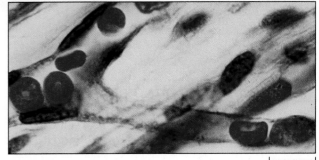

Figure 27.23 Light micrograph of some human blood vessels.

Critical Thinking

1. Shirley, who is utilizing a light microscope to examine a human tissue specimen, sees red blood cells moving single file through thin-walled tubes. She makes a photomicrograph of this (Figure 27.23). What type of blood vessel has she captured on film?

2. In individuals who have weak or leaky valves in their veins, fluid pressure associated with the backflow of blood causes venous walls below the valves to bulge outward. In time, the walls become stretched and flabbily distorted, a condition called *varicose veins*. Some people are genetically predisposed to develop the condition, but the cumulative mechanical stress associated with prolonged standing, pregnancy, and aging can contribute to it. With chronic varicosity, the legs themselves become swollen. Explain why this might happen. Also explain why veins close to the leg surface are more susceptible to varicosity than those deeper in the leg tissues.

3. Infection by a hemolytic bacterium (*Streptococcus pyogenes*) may trigger an inflammation that ultimately damages valves in the heart. The disease symptoms of *rheumatic fever* follow. Explain how this disease must affect the heart's functioning. Also describe what kinds of symptoms would arise as a consequence of this.

Selected Key Terms

ABO blood typing *463*
agglutination *462*
aorta *464*
arteriole *464*
artery *464*
atrium *466*
blood *458*
blood pressure *468*
blood transfusion *462*
capillary *458*
capillary bed *458*
cardiac conduction system *467*
cardiac cycle *466*
cardiac pacemaker *467*
cell count *461*
circulatory system *458*
electrocardiogram *456*
heart *458*
hemostasis *472*
interstitial fluid *458*
lymph capillary *472*

lymph node *473*
lymph vascular system *472*
lymph vessel *472*
lymphatic system *459*
lymphoid organ *473*
lymphoid tissue *473*
plasma *460*
platelet *460*
pulmonary circuit *459*
red blood cell *460*
Rh blood typing *463*
spleen *473*
stem cell *461*
systemic circuit *459*
thymus gland *473*
vasoconstriction *468*
vasodilation *468*
vein *464*
ventricle *466*
venule *464*
white blood cell *460*

Readings

Eisenberg, M. S., et al. May 1986. "Sudden Cardiac Death." *Scientific American* 254(5): 37–43.

Robinson, T. F., et al. June 1986. "The Heart as a Suction Pump." *Scientific American* 254(6): 84–91.

28 IMMUNITY

Russian Roulette, Immunological Style

Until about a century ago, smallpox epidemics swept repeatedly through the world's cities. Some outbreaks were so severe that only half of those who were stricken survived. Survivors were left with permanent scars on the face, neck, shoulders, and arms—but they seldom contracted the disease again. They were "immune" to smallpox.

No one knew what caused smallpox. But the idea of acquiring immunity was dreadfully appealing. In twelfth-century China, healthy people gambled with deliberate infections. They sought out individuals with mild cases of smallpox (who were only mildly scarred), then removed crusts from the scars, ground them up, and inhaled the powder.

By the seventeenth century, Mary Montagu, wife of the English ambassador to Turkey, was championing inoculation. She went so far as to poke bits of smallpox scabs into her children's skin. Others soaked threads in fluid from the sores, then poked the threads into the skin.

Individuals who survived these practices acquired immunity to smallpox—but many developed raging infections. As if the odds were not dangerous enough, the crude inoculation procedures also invited the acquisition of several other infectious diseases.

While this immunological version of Russian roulette was going on, Edward Jenner was growing up in the English countryside. At the time, it was common knowledge that people who contracted cowpox never got smallpox. (Cowpox is a mild disease that can be transmitted from cattle to humans.) No one thought much about this until 1796, when Jenner, by now a physician, injected material from a cowpox sore into the arm of a healthy boy. Six weeks later, after the reaction subsided, Jenner injected some material from *smallpox* sores into the boy (Figure 28.1a). He had hypothesized that the earlier injection might provoke immunity to smallpox—and fortunately he was right. The boy remained free of smallpox.

The French mocked Jenner's procedure, calling it "vaccination" (which translates as "encowment"). Much later a French chemist, Louis Pasteur, devised similar procedures for other diseases. Pasteur also called his procedures vaccinations, and only then did the term become respectable.

By Pasteur's time, improved microscopes were revealing diverse bacteria, fungal spores, and other previously invisible forms of life. As Pasteur himself discovered, microorganisms abound in ordinary air.

a

Figure 28.1 (**a**) Statue honoring Edward Jenner's development of an immunization procedure against smallpox, one of the most dreaded diseases in human history. (**b**) Micrograph of a white blood cell being attacked by the virus (*blue particles*) that causes AIDS. Immunologists are working to develop weapons against this modern-day scourge.

Did some cause diseases? Probably. Could they settle into food or drink and cause it to spoil? He proved that they did.

Pasteur also found a way to kill most of the suspect disease agents in food or beverages. As he and others knew, boiling killed these agents. He also knew that you can't boil wine—or beer or milk, for that matter—and end up with the same beverage. He devised a way to heat these beverages at a temperature low enough not to ruin them but high enough to kill most of the microorganisms. We still depend on this antimicrobial method, which was named pasteurization in his honor.

In the late 1870s Robert Koch, a German physician, linked a specific microorganism to a specific disease—namely, anthrax. In one experiment, Koch had injected blood from animals with anthrax into healthy ones. The recipients of the injection ended up with blood that teemed with cells of a bacterium (*Bacillus anthracis*)—and they developed anthrax. Even more convincing, injections of bacterial cells that were cultured outside the body also caused the disease!

Thus, by the beginning of the twentieth century, the promise of understanding the basis of infectious disease and immunity loomed large—and the battles

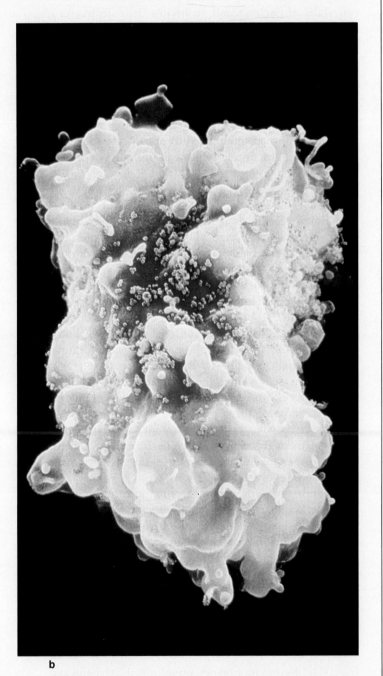

b

against those diseases were about to begin in earnest. Since that time, spectacular advances in microscopy, biochemistry, and molecular biology have increased our understanding of the body's defenses. We now have greater insight into its responses to tissue damage in general—and into its immune responses to specific pathogens or tumor cells. These responses are the focus of this chapter.

1. The vertebrate body has physical, chemical, and cellular defenses against pathogenic microorganisms, malignant tumor cells, and other agents that can destroy tissues, even the individual itself.

2. During early stages of tissue invasion and damage, white blood cells and plasma proteins take part in a rapid, nonspecific counterattack. We call this an inflammatory response. Phagocytic white blood cells ingest the invaders. The plasma proteins promote phagocytosis, and some also destroy invaders directly.

3. If the invasion persists, certain white blood cells make immune responses. These cells can chemically recognize distinct configurations on molecules that are abnormal or foreign to the body, such as those on bacteria and viruses. If the foreign or abnormal molecule triggers an immune response, it is called an antigen.

4. In one type of immune response, some of the white blood cells produce great quantities of antibodies. Antibodies are molecules that bind to a specific antigen and tag it for destruction. In another type of immune response, executioner cells directly destroy body cells that have become abnormal, as by infection or by a tumor-producing process.

You continually cross paths with amazingly diverse **pathogens**—the viruses, bacteria, fungi, protozoans, parasitic worms, and other agents that cause diseases. You and other vertebrates coevolved with most of them, so you need not lose sleep over this. Your body surface bars most pathogens. Chemical weapons and certain white blood cells (leukocytes) that are not picky about their targets destroy many pathogens that do breach the surface barriers. Other white blood cells zero in on the rest. Table 28.1 lists these three lines of defense.

Surface Barriers to Invasion

Most often, pathogens cannot get past skin or the other linings of body surfaces. Think of skin as a habitat of low moisture, low pH, and thick layers of dead cells. Normally harmless bacteria tolerate these conditions. Few pathogens can compete with dense populations of established types unless conditions change. Repeatedly subject your toes to warm, damp shoes, for example, and you may be inviting certain fungi to penetrate the sodden, weakened tissues and cause *athlete's foot.*

Similarly, resident bacteria on the mucous lining of the gut and vagina help protect you—as when lactate, a fermentation product of *Lactobacillus* populations in the vagina, helps maintain a low pH that most bacteria and fungi cannot tolerate. Barriers in branching, tubular airways leading to your lungs stop airborne pathogens. Air rushing down the tubes flings bacteria against their sticky, mucus-coated walls. In that mucus are protective substances such as **lysozyme**, an enzyme that digests cell walls of bacteria and so invites their death. As a final touch, broomlike cilia in the airways sweep out the trapped and enzymatically whapped pathogens.

Lysozyme and other substances in tears, saliva, and gastric fluid offer more protection, as when tears give the eyes a sterile washing. Urine's low pH and flushing action help bar pathogens from moving into the urinary tract. Diarrhea can flush out irritating pathogens. It must be controlled in children to prevent dehydration, but blocking its action in adults can prolong infection.

Nonspecific and Specific Responses

All animals react to tissue damage. Even simple aquatic invertebrates have phagocytic cells and antimicrobial substances, including lysozymes. But the more complex the animal body, the more complex are the systems that defend it. Think back on the evolution of the circulatory and lymphatic systems in vertebrates that invaded the land (Section 27.1). As circulation of body fluids became more efficient, so did the means for defending the body. Phagocytic cells as well as plasma proteins could move rapidly to tissues under attack and they could intercept pathogens trickling along the vascular highways.

Vertebrates came to be equipped with sets of plasma proteins. Some of these promote rapid clot formation after tissue damage; others destroy invading pathogens or target them for phagocytosis. Besides these general responses, exquisitely focused responses to specific dangers also evolved.

In short, *internal defenses against a great variety of pathogens are in place even before damage occurs.* Ready and waiting are specialized white blood cells as well as plasma proteins. They take part in a *nonspecific* response to tissue damage in general, not to one pathogen or another. Other white blood cells may recognize unique molecular configurations on a *specific* pathogen. If they do, the resulting "immune" response will run its course whether tissues are damaged or not.

Table 28.1 The Vertebrate Body's Three Lines of Defense Against Pathogens
Barriers at Body Surfaces (*nonspecific* targets):
Intact skin; mucous membranes at other body surfaces
Infection-fighting chemicals in tears, saliva, etc.
Normally harmless bacterial inhabitants of body surfaces that outcompete pathogenic visitors
Flushing effect of tears, saliva, urination, and diarrhea
Nonspecific Responses (*nonspecific* targets):
Inflammation
1. Fast-acting white blood cells (neutrophils, eosinophils, and basophils)
2. Macrophages (also take part in immune responses)
3. Complement proteins, blood-clotting proteins, and other infection-fighting substances
Organs with pathogen-killing functions (e.g., lymph nodes)
Immune Responses (*specific* targets):
White blood cells (macrophages, T cells, B cells)
Communication signals (e.g., interleukins) and chemical weapons (e.g., antibodies, perforins)

Intact skin, mucous membranes, antimicrobial secretions, and other barriers at the body surface constitute the first line of defense against invasion and tissue damage.

Inflammation and other internal, nonspecific responses to invasion are the second line of defense.

Immune responses against specific invaders, as executed by armies of white blood cells and their chemical weapons, are the third line of defense.

28.2 COMPLEMENT PROTEINS

A set of plasma proteins has roles in both nonspecific and specific defenses. Collectively, these proteins are called the **complement system**. About twenty kinds of complement proteins circulate in the blood in inactive form. If even a few molecules of one kind are activated, they trigger a huge cascade of reactions. They activate many molecules of another complement protein. Each

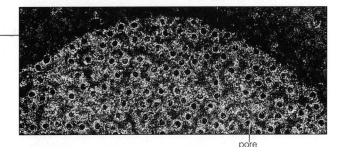

pore

Figure 28.3 Micrograph of a cell surface, showing pores formed by membrane attack complexes.

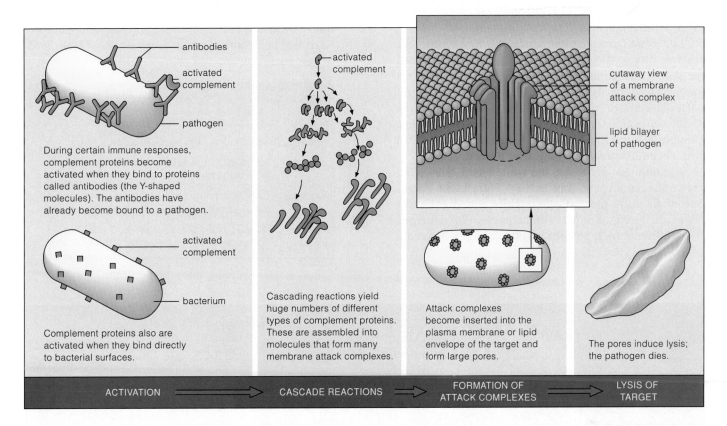

antibodies
activated complement
pathogen

During certain immune responses, complement proteins become activated when they bind to proteins called antibodies (the Y-shaped molecules). The antibodies have already become bound to a pathogen.

activated complement
bacterium

Complement proteins also are activated when they bind directly to bacterial surfaces.

activated complement

Cascading reactions yield huge numbers of different types of complement proteins. These are assembled into molecules that form many membrane attack complexes.

Attack complexes become inserted into the plasma membrane or lipid envelope of the target and form large pores.

cutaway view of a membrane attack complex
lipid bilayer of pathogen

The pores induce lysis; the pathogen dies.

ACTIVATION → CASCADE REACTIONS → FORMATION OF ATTACK COMPLEXES → LYSIS OF TARGET

Figure 28.2 Formation of membrane attack complexes. One reaction pathway starts as complement proteins bind to bacterial surfaces; another operates during immune responses to specific invaders. Both pathways produce membrane pore complexes that induce lysis in pathogens.

of these activates many molecules of another kind of protein at the next reaction step, and so on. Deployment of all these molecules has the following effects.

Some of the complement proteins join together to form pore complexes. These molecular structures have an interior channel (Figures 28.2 and 28.3). They get inserted into the plasma membrane of many pathogens, forming pores that induce lysis. **Lysis**, recall, is a gross structural disruption of a plasma membrane or cell wall that leads to cell death. Pore complexes also become inserted into the wall of Gram-negative bacteria. Such cell walls consist of a lipid-rich, outer surface above a peptidoglycan layer. Lysozyme molecules can diffuse through the pores and digest the peptidoglycan.

Some activated proteins promote inflammation, a nonspecific defense response described next. Through their cascades of reactions, they create concentration gradients that attract phagocytic white blood cells to an irritated or damaged tissue. The complement proteins also encourage phagocytes to dine. They can bind to the surface of many invaders. When they do this, an invader ends up with a complement "coat." The coat can adhere to phagocytes, and this is rather like putting a basted turkey on a dinner table.

In such ways, complement proteins target many bacteria, parasitic protistans, and enveloped viruses.

The complement system, a set of twenty or so kinds of plasma proteins circulating in blood, takes part in cascades of reactions that help defend against many bacteria, some parasitic protistans, and enveloped viruses.

The complement system takes part in both nonspecific and specific defenses.

28.3 INFLAMMATION

The Roles of Phagocytes and Their Kin

Certain white blood cells take part in an initial response to tissue damage. White blood cells, recall, arise from stem cells in bone marrow. Many of these cells circulate in blood and lymph. A great many take up stations in the lymph nodes as well as the spleen, liver, kidneys, lungs, and brain. (Here you may wish to refer to Figure 27.7, which shows the components of blood, and to 27.22, which shows the lymphatic system.)

Like SWAT teams, three kinds of white blood cells react swiftly to danger in general but are not adapted for sustained battles. **Neutrophils**, the most abundant kind, phagocytize bacteria. They ingest, kill, and digest bacterial cells to simple molecular bits. **Eosinophils** secrete enzymes that make holes in parasitic worms. **Basophils** secrete histamine and other substances that help keep inflammation going after it starts.

Although slower to act, the white blood cells called **macrophages** are the "big eaters." Figure 28.4 shows one of them. A macrophage engulfs and digests just about any foreign agent. It also helps clean up damaged tissues. Immature macrophages circulating in blood are called monocytes.

The Inflammatory Response

An inflammatory response develops when something damages or kills cells of any given tissue. Infections, punctures, burns, and other insults are the triggers. By a mechanism known as **acute inflammation**, fast-acting phagocytes, complement proteins, and other plasma proteins escape from the bloodstream at capillary beds in the damaged tissue, and there they enter interstitial fluid. Localized redness, swelling, heat, and pain are signs that acute inflammation is under way (Table 28.2).

Table 28.2	Localized Signs of Inflammation and Their Causes
Redness	Vasodilation, increased blood flow to site
Warmth	Vasodilation, greater flow of blood carrying more metabolic heat to site
Swelling	Capillaries made more permeable, so plasma and leukocytes leak out; also vasodilation
Pain	Increased fluid pressure and local chemical signals stimulate nocireceptors (pain receptors)

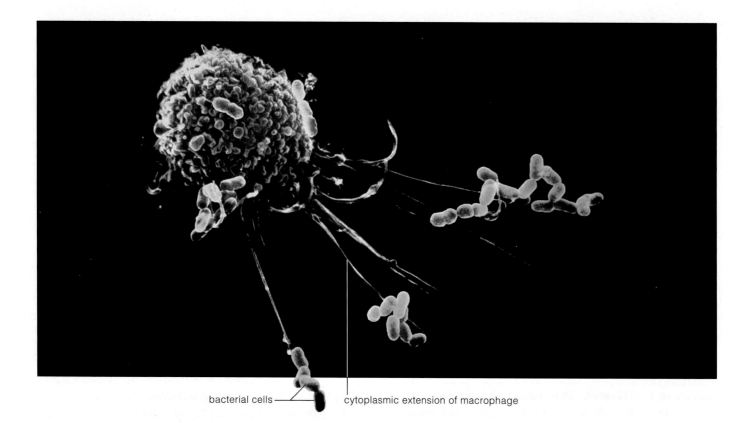

bacterial cells —— cytoplasmic extension of macrophage

Figure 28.4 Scanning electron micrograph of a macrophage, probing its surroundings with cytoplasmic extensions. This phagocytic cell engulfs bacteria that make contact with its surface.

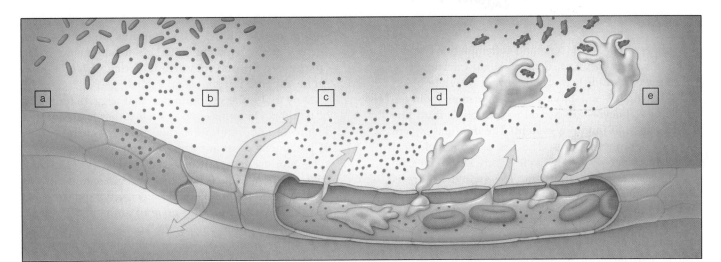

a Bacteria invade a tissue. They kill cells or release harmful metabolic by-products.

b The substances released by bacteria and by damaged or killed body cells accumulate in the tissue.

c The substances make the tissue's small blood vessels more permeable. Plasma fluid and various plasma proteins escape into the tissue.

d Some plasma proteins attack bacteria. Others create chemical gradients that facilitate migration of phagocytes to the tissue. Still others repair tissue damage (as by clotting mechanisms).

e Phagocytic white blood cells engulf bacteria.

Mast cells, which dwell in tissues and function like basophils, act during an inflammatory response. They release **histamine** and other chemicals into interstitial fluid. Their secretions promote vasodilation of small blood vessels threading through the tissue that is under attack. (Vasodilation means a blood vessel enlarges in diameter when smooth muscle in its wall relaxes.) As a result, these vessels become engorged with blood—which carries infection-fighting weapons. The tissues redden, swell, and become warmer. Blood, remember, carries metabolic heat.

In the walls of the engorged vessels, endothelial cells pull apart slightly as the junctions between them weaken. The vessels become "leaky" to plasma, some of which enters the tissue. Swelling and pain follow. Voluntary movements that might aggravate the pain are avoided, and this behavior promotes tissue repair.

Within hours of the first physiological responses to tissue damage, neutrophils are squeezing out through gaps in the blood vessel walls. They swiftly go to work. Monocytes arrive later, differentiate into macrophages, and engage in more sustained action (Figure 28.5).

While macrophages are busily engulfing pathogens, they secrete several substances. Among the secretions are **interleukins**, which are important communication signals between white blood cells. One kind, called interleukin-1, also signals a brain region that controls body temperature. Interleukin-1 induces an increase in the "set point" on the body's thermostat. A *fever* is a body temperature that has reached the higher set point.

Figure 28.5 Acute inflammation. In this example, a bacterial invasion induces chemical changes in a local tissue that trigger increased blood flow to the region. Small blood vessels become more permeable. Plasma fluid, certain plasma proteins, and phagocytic white blood cells leave the blood and enter the tissue. The invaders are killed and the tissue can be repaired.

A fever of about 39°C (100°F) is not a bad thing. It increases body temperature to a level that is "too hot" for the functioning of most pathogens. It also promotes an increase in a host's defense activities. Interleukin-1 induces drowsiness. Drowsiness reduces the body's demands for energy, so more energy can be diverted to the tasks of defense and tissue repair. Macrophages take part in the cleanup and repair operations. So do blood-clotting proteins that can help patch up blood vessels, in the manner described in Section 27.9.

An inflammatory response develops in a local tissue when cells are damaged or killed, as by infection.

Small blood vessels in the tissue dilate. The cells making up their wall pull apart slightly, so these blood vessels become leaky to plasma and white blood cells.

Neutrophils, eosinophils, basophils, and then macrophages leave the bloodstream. They defend the damaged tissue and help repair it.

Histamine, complement proteins, interleukins, blood-clotting proteins, and other substances take part in the inflammatory response or in the mopping-up and repair operations.

28.4 THE IMMUNE SYSTEM

Defining Features

Sometimes physical barriers and inflammation are not enough to overwhelm an invader, so an infection may become well established. Then, armies of white blood cells known as **B** and **T lymphocytes** form and join the battle. Lymphocytes are central to the body's third line of defense—the **immune system**. The immune system has two defining features. The first is immunological *specificity*, whereby lymphocytes zero in on specific pathogens and eliminate them. The second feature is immunological *memory*, whereby some lymphocytes that form during a first-time confrontation are set aside for a future battle with the same pathogen.

The operating principle for the system is this: *Each kind of cell, virus, or substance bears unique molecular configurations that give it a unique identity.* The unique configurations on an individual's own cells serve as *self* markers. Lymphocytes recognize self markers and will normally ignore them. They also can recognize *nonself* molecular configurations, which are unique to specific foreign agents. When they do, B and T lymphocytes are stimulated to divide repeatedly, by way of mitosis. And huge populations form.

As the divisions proceed, subpopulations of the new cells become specialized to respond to the foreign agent in different ways. Some consist of *effector* cells—fully differentiated cells that engage and destroy the enemy.

MHC marker that designates self (only on body's own cells)

T and B cells ignore this

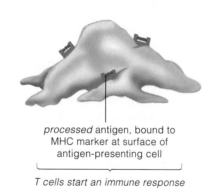

processed antigen, bound to MHC marker at surface of antigen-presenting cell

T cells start an immune response

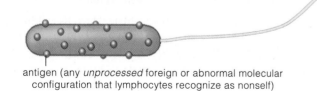

antigen (any *unprocessed* foreign or abnormal molecular configuration that lymphocytes recognize as nonself)

B cells start an immune response

Figure 28.6 Molecular cues that T and B cells either ignore or recognize as a signal to initiate immune responses.

Other subpopulations consist of *memory* cells, which enter a resting phase. Instead of attacking the specific agent that triggered the initial response, memory cells "remember" it. And they will undertake a larger, more rapid response if it shows up again.

Any molecular configuration that triggers formation of lymphocyte armies and is their target is an **antigen**. The most important are certain proteins at the surface of pathogens or tumor cells. As you will see, lymphocytes produce receptor molecules that can bind to these configurations. This is how they recognize nonself.

In short, immunological specificity and memory involve three events: *recognition* of antigen, *repeated cell divisions* that form huge populations of lymphocytes, and *differentiation* into subpopulations of effector and memory cells with receptors for one kind of antigen.

Antigen-Presenting Cells—The Triggers for Immune Responses

The plasma membrane of every nucleated cell in an individual incorporates a variety of proteins. Among these proteins are **MHC markers**, named after the genes that encode instructions for making them. Certain MHC markers are common at the surface of all the nucleated cells. Others are unique to the body's macrophages and lymphocytes.

Suppose bacterial cells enter a cut into your finger. Some end up in local lymph nodes where macrophages join the fray. They engulf the foreign cells and process their antigen. Digestive enzymes cleave the antigen molecules into small fragments, which become attached to MHC molecules. These **antigen-MHC complexes** move in vesicles to the macrophage surface, where they are displayed.

Any cell that processes and displays antigen with a suitable MHC molecule is an **antigen-presenting cell**. And when a cell bears processed antigen complexed with a certain MHC molecule, lymphocytes take notice (Figure 28.6). *This is the antigen recognition that promotes the cell divisions by which great armies of lymphocytes form.*

Key Players in Immune Responses

The same kinds of white blood cells are called into action during each immune response. Figure 28.7 is an overview of how the cells interact. In brief, recognition of antigen-MHC complexes activates **helper T cells**. These produce and secrete substances that induce *any* responsive T or B lymphocyte to divide and give rise to large populations of effector cells and memory cells. Recognition also activates **cytotoxic T cells**. These can eliminate infected body cells or tumor cells by "touch killing." When they touch a target, they deliver cell-

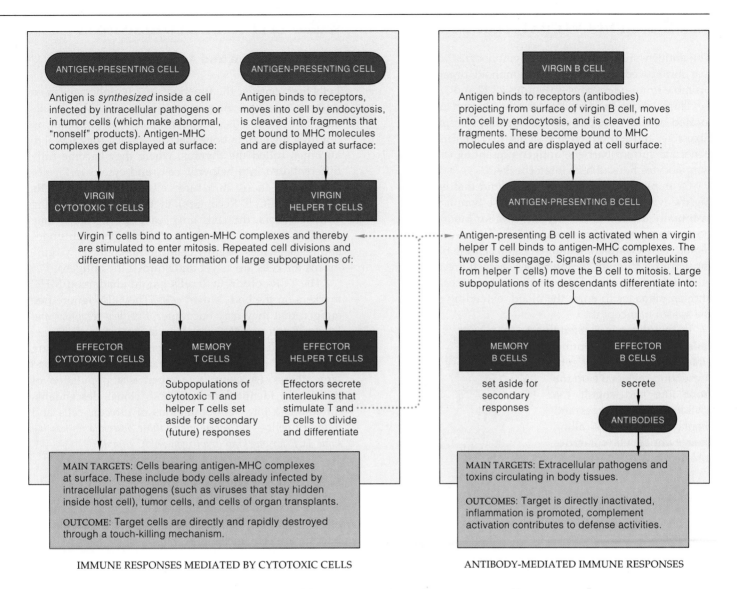

ANTIGEN-PRESENTING CELL

Antigen is *synthesized* inside a cell infected by intracellular pathogens or in tumor cells (which make abnormal, "nonself" products). Antigen-MHC complexes get displayed at surface:

VIRGIN CYTOTOXIC T CELLS

ANTIGEN-PRESENTING CELL

Antigen binds to receptors, moves into cell by endocytosis, is cleaved into fragments that get bound to MHC molecules and are displayed at surface:

VIRGIN HELPER T CELLS

Virgin T cells bind to antigen-MHC complexes and thereby are stimulated to enter mitosis. Repeated cell divisions and differentiations lead to formation of large subpopulations of:

EFFECTOR CYTOTOXIC T CELLS

MEMORY T CELLS

EFFECTOR HELPER T CELLS

Subpopulations of cytotoxic T and helper T cells set aside for secondary (future) responses

Effectors secrete interleukins that stimulate T and B cells to divide and differentiate

MAIN TARGETS: Cells bearing antigen-MHC complexes at surface. These include body cells already infected by intracellular pathogens (such as viruses that stay hidden inside host cell), tumor cells, and cells of organ transplants.

OUTCOME: Target cells are directly and rapidly destroyed through a touch-killing mechanism.

IMMUNE RESPONSES MEDIATED BY CYTOTOXIC CELLS

VIRGIN B CELL

Antigen binds to receptors (antibodies) projecting from surface of virgin B cell, moves into cell by endocytosis, and is cleaved into fragments. These become bound to MHC molecules and are displayed at cell surface:

ANTIGEN-PRESENTING B CELL

Antigen-presenting B cell is activated when a virgin helper T cell binds to antigen-MHC complexes. The two cells disengage. Signals (such as interleukins from helper T cells) move the B cell to mitosis. Large subpopulations of its descendants differentiate into:

MEMORY B CELLS

EFFECTOR B CELLS

set aside for secondary responses

secrete

ANTIBODIES

MAIN TARGETS: Extracellular pathogens and toxins circulating in body tissues.

OUTCOMES: Target is directly inactivated, inflammation is promoted, complement activation contributes to defense activities.

ANTIBODY-MEDIATED IMMUNE RESPONSES

Figure 28.7 Overview of key interactions among B and T lymphocytes during an immune response. Most often, both types of white blood cells are activated when an antigen has been detected. An antigen is any large molecule that lymphocytes recognize as not being "self" (normal body molecules). A first-time encounter with antigen elicits a *primary* response. A subsequent encounter with the same type of antigen elicits a *secondary* immune response. This response is larger and more rapid. Memory cells that formed but that were not used during the first battle can immediately engage in the second one.

killing chemicals into it. By contrast, **B cells** produce antigen-binding receptor molecules called **antibodies**. When a response is under way, effector B cells secrete staggering numbers of antibody molecules. Only B cells are the basis of *antibody-mediated* responses.

Control of Immune Responses

Antigen provokes an immune response—and removal of antigen stops it. For example, by the time the tide of battle turns, the effector cells and their secretions have already destroyed most antigen-bearing agents in the body. With fewer antigen molecules around to stimulate

the cells, the response declines, then it stops. As a final example, inhibitory signals from cells with suppressor functions also help shut down the immune response.

Antigens are nonself molecular configurations which, when recognized by certain lymphocytes, trigger immune responses. Helper T cells, cytotoxic T cells, B cells, and their secretions execute these responses.

Following antigen recognition, large T and B cell armies form through repeated mitotic cell divisions. These differentiate into subpopulations of effector cells and memory cells, all of which are sensitized to that one kind of antigen.

28.5 LYMPHOCYTE BATTLEGROUNDS

The antigen-presenting cells and lymphocytes we have just introduced interact within lymphoid organs that promote immune responses (Figures 27.22 and 28.8).

Think about the tonsils and other lymph nodules located beneath mucous membranes of the respiratory, digestive, and reproductive systems. Just after invaders penetrate surface barriers, antigen-presenting cells and lymphocytes housed here intercept them.

Or think about antigen in tissue fluid that is entering the lymph vascular system. Because lymph vessels eventually drain into the expressways for blood transport, antigen could become distributed to every body region. However, before antigen can reach the blood, it must trickle through lymph nodes—which are packed with defending cells. Even in those few cases where antigen manages to enter the blood, defending cells in the spleen intercept it.

In lymph nodes, defending cells are organized for utmost effectiveness. The antigen-presenting cells make up the front line and engulf the invaders. They process and display antigen, thus calling their lymphocyte comrades into action.

Figure 28.8 Organized arrays of antigen-presenting cells and lymphocytes in lymph nodes.

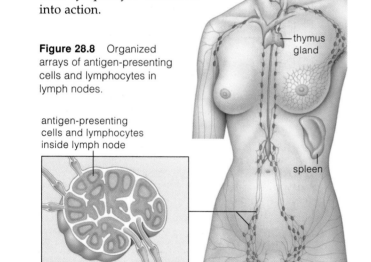

antigen-presenting cells and lymphocytes inside lymph node

tonsils

thymus gland

spleen

The cell divisions that produce subpopulations of effector and memory cells proceed in lymph nodes. As lymph drains through, it moves effector activities to the back of the organ and beyond. And all the while, virgin and memory cells circulate through the lymph node, reconnoitering at the front line.

Antigen-presenting cells and lymphocytes intercept and battle pathogens in organized ways within lymphoid organs and tissues, especially the lymph nodes.

28.6 CELL-MEDIATED RESPONSES

T Cell Formation and Activation

Let's first consider the functions of T lymphocytes in an immune response. As you read in Section 27.2, T cells arise from stem cells in bone marrow. However, they do not fully develop in bone marrow. Rather, they travel to an organ called the thymus, where they become fully differentiated into helper T cells and cytotoxic T cells. Specifically, these immature cells acquire their **TCRs** (short for *T-Cell Receptors*) in the thymus. Bristling with receptors, the cells now leave the thymus. They circulate in blood or take up stations in lymph nodes and the spleen as virgin T cells. In this context, "virgin" means the cells are as yet undisturbed (by antigen).

The TCRs of virgin T cells ignore unadorned MHC markers on the body's own cells. They also ignore free antigen that they may encounter. *But they recognize and bind with antigen-MHC complexes at the surface of antigen-presenting cells.* As you can see from Figure 28.9, binding stimulates T cells to divide repeatedly and give rise to large clones. (A clone, remember, is a population of genetically identical cells.) These clonal descendants differentiate into subpopulations of effector cells and memory cells—*and every one of those descendants has the same TCR for one kind of antigen-MHC complex.*

Functions of Effector T Cells

What actions do subpopulations of effector T cells take? Effector helper T cells secrete communication signals—interleukins. Their secretions fan repeated cell divisions and differentiation of responsive T and B cells, as described shortly. The effector cytotoxic T cells respond to antigen-MHC complexes on body cells that already have been infected by intracellular pathogens (such as viruses) and on tumor cells. The complex serves as a "double signal" to destroy the cells that bear it.

Effector cytotoxic T cells destroy infected cells with a touch-kill mechanism. They secrete **perforins**, protein molecules that form doughnut-shaped pores in a target cell's plasma membrane. (The pores look similar to the ones shown in Figure 28.3.) These effectors also secrete chemicals that induce cell death. By a process called apoptosis, a target cell commits suicide, so to speak. Its cytoplasm is expelled, its organelles are disrupted, and its DNA becomes fragmented. Having made its lethal hit, the cytotoxic T cell disengages quickly and moves on to new targets (Figure 28.9e).

Cytotoxic T cells also contribute to the rejection of tissue and organ transplants. Parts of MHC markers on donor cells are different enough from the recipient's to be recognized as antigens—but other parts are similar enough to complete the double signal. MHC typing and matching donors to recipients minimize the risk.

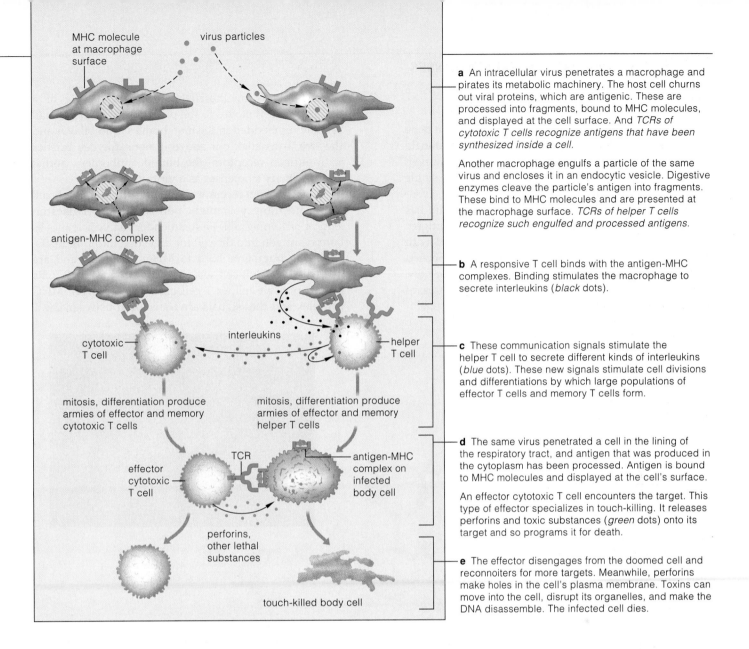

a An intracellular virus penetrates a macrophage and pirates its metabolic machinery. The host cell churns out viral proteins, which are antigenic. These are processed into fragments, bound to MHC molecules, and displayed at the cell surface. And *TCRs of cytotoxic T cells recognize antigens that have been synthesized inside a cell.*

Another macrophage engulfs a particle of the same virus and encloses it in an endocytic vesicle. Digestive enzymes cleave the particle's antigen into fragments. These bind to MHC molecules and are presented at the macrophage surface. *TCRs of helper T cells recognize such engulfed and processed antigens.*

b A responsive T cell binds with the antigen-MHC complexes. Binding stimulates the macrophage to secrete interleukins (*black* dots).

c These communication signals stimulate the helper T cell to secrete different kinds of interleukins (*blue* dots). These new signals stimulate cell divisions and differentiations by which large populations of effector T cells and memory T cells form.

d The same virus penetrated a cell in the lining of the respiratory tract, and antigen that was produced in the cytoplasm has been processed. Antigen is bound to MHC molecules and displayed at the cell's surface.

An effector cytotoxic T cell encounters the target. This type of effector specializes in touch-killing. It releases perforins and toxic substances (*green* dots) onto its target and so programs it for death.

e The effector disengages from the doomed cell and reconnoiters for more targets. Meanwhile, perforins make holes in the cell's plasma membrane. Toxins can move into the cell, disrupt its organelles, and make the DNA disassemble. The infected cell dies.

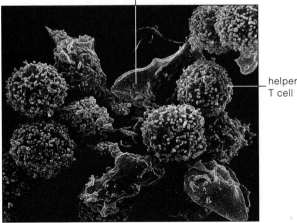

Figure 28.9 Example of a T cell-mediated immune response. In this case, the response involves an antigen-MHC complex that activates T cells. Scanning electron micrograph shows helper T cells associating with an antigen-presenting macrophage.

Regarding the Natural Killer Cells

Other cytotoxic cells, including **natural killer cells (NK cells)**, also arise from stem cells in bone marrow. They appear to be lymphocytes, but not T or B cells. Their arousal does not depend on a double signal (that is, an antigen-MHC complex). NK cells reconnoiter for tumor cells and virus-infected cells, then touch-kill them. They possibly recognize odd molecular configurations at the surface of their targets.

T cells arise in bone marrow. Later on, in the thymus, they acquire TCRs (receptors for self markers and bound antigen).

Effector helper T cells secrete interleukins that trigger the cell divisions and differentiation into armies against specific antigens. Effector cytotoxic cells touch-kill infected cells or tumor cells, even foreign cells of transplants.

ANTIBODY-MEDIATED RESPONSES

B Cells and the Targets of Antibodies

Like T cells, the B cells also arise from stem cells in bone marrow and start down a pathway that will culminate in full differentiation. Along *their* pathway, however, B cells start synthesizing numerous copies of a single kind of antibody molecule.

Although all antibodies are proteins, each kind has binding sites that only match up to a particular antigen. They are more or less Y-shaped, with a tail and with two arms that bear identical antigen receptors. Section 28.9 provides a closer look at these molecules. For now, we can simply think of them as Y-shaped structures, of the sort shown in Figure 28.10.

Each newly synthesized antibody molecule of a maturing B cell moves to the plasma membrane. Its tail becomes embedded in the membrane's lipid bilayer and the two arms stick out above it. Soon the cell bristles with antigen receptors (the bound antibodies), and it joins the body's defenses as a virgin B cell.

When antigen receptors lock onto a target, the B cell does something you might not expect. It becomes an *antigen-presenting* cell. First, an endocytic vesicle moves bound antigen into the cell for digestion into fragments. Next, the fragments bind to MHC molecules and are presented at the B cell surface. Now suppose that TCRs of a responsive helper T cell bind to the antigen-MHC complex and that signals are transferred between the T

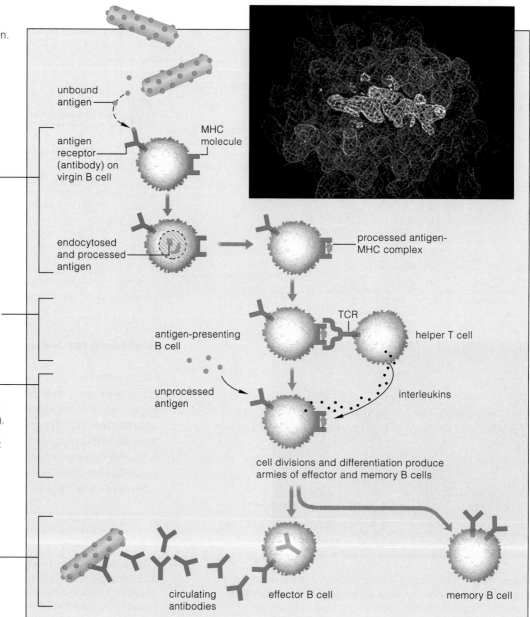

Figure 28.10 Example of an antibody-mediated immune response to a bacterial invasion. The inset is a computer model of an antigen fragment (*pink*) that is bound to the cleft of an MHC protein.

a A virgin B cell encounters unbound antigen in tissue fluid. Antigen receptors (membrane-bound antibody molecules) bind the antigen, which moves into the cell by endocytosis and is processed. The processed antigen is displayed with MHC molecules at the cell surface. The B cell has become an antigen-presenting cell.

b The TCRs of a helper T cell bind to antigen-MHC complexes on the B cell. Binding activates the T cell *and* stimulates the B cell to prepare for mitosis. Then the cells disengage.

c More antigen, which has not been processed, binds to the B cell. Meanwhile, the helper T cell secretes interleukins (*black* dots). Both events trigger repeated cell divisions and differentiations that produce large armies of memory B cells *and* antibody-secreting effector B cells.

d Antibody molecules released from the effector B cells enter body fluids. When they contact a bacterial cell that is a target, they will bind to antigen on its surface. This tags the cell for destruction (*compare* Figures 28.2 and 28.3).

Labels in figure:
unbound antigen
antigen receptor (antibody) on virgin B cell
MHC molecule
endocytosed and processed antigen
processed antigen-MHC complex
antigen-presenting B cell
TCR
helper T cell
unprocessed antigen
interleukins
cell divisions and differentiation produce armies of effector and memory B cells
circulating antibodies
effector B cell
memory B cell

and B cell. The cells soon disengage. When the B cell encounters unprocessed antigen, its surface antibodies bind it. The binding, in combination with interleukins secreted from nearby helper T cells, drives the B cell to mitosis. Its clonal descendants differentiate into effector and memory B cells. The effectors (also called plasma cells) produce and secrete huge numbers of antibody molecules. When freely circulating antibody molecules bind antigen, they tag an invader for destruction, as by phagocytes and complement activation.

The main targets of antibody-mediated responses are extracellular pathogens and toxins, which are freely circulating in tissues or body fluids. Antibodies can't bind to pathogens or toxins hidden in a host cell.

The Immunoglobulins

During immune responses, B cells produce four classes of antibodies in abundance and lesser quantities of another. Collectively, the five classes of antibodies are called **immunoglobulins**, or **Igs**. They are the protein products of gene shufflings that proceed while B cells mature and while an immune response is under way. The molecules in each class have antigen-binding sites *and* other sites with specialized functions.

IgM antibodies are the first to be secreted during immune responses. They trigger complement cascades, and they bind targets together in clumps—which are more handily eliminated by phagocytes. (Remember the agglutination responses described in Section 27.4?) *IgD* antibodies associate with IgM on virgin B cells, but their function is not yet understood.

IgG antibodies activate complement proteins and neutralize many toxins. These long-lasting antibodies are the only ones to cross the placenta. They can protect the fetus and newborn with the mother's acquired immunities. IgGs also are secreted into the early milk produced by mammary glands, then are absorbed into the suckling newborn's bloodstream.

IgA antibodies enter mucus-coated surfaces of the respiratory, digestive, and reproductive tracts, where they can neutralize infectious agents. Mother's milk delivers them to the mucous lining of a newborn's gut.

IgE triggers inflammation after attacks by parasitic worms and other pathogens. As described later, it also figures in allergies. The tails of IgE antibodies bind to basophils and mast cells, and the antigen receptors face outward. Antigen binding induces basophils and mast cells to release substances that promote inflammation.

Antibodies that are secreted by B cells bind to antigens of extracellular pathogens or toxins and tag them for disposal, as by phagocytes and complement activation.

CANCER AND IMMUNOTHERAPY

Carcinomas, sarcomas, leukemia—these chilling words refer to malignant tumors in skin, bone, and other tissues. Such tumors arise when viral attack, irradiation, or chemicals alter genes and cells turn cancerous (Sections 7.5 and 12.9). The transformed cells divide repeatedly. Unless they are destroyed or surgically removed, they kill the individual.

Often the surface of a transformed cell bears abnormal proteins and protein fragments bound to MHC. Although defense responses to the transformed cells can make the tumor regress, they can be inadequate. Also, some tumors release many copies of the abnormal proteins. If these saturate antigen receptors on the defenders, the tumor may escape detection. If the tumor hides long enough and reaches a certain critical mass, it may overwhelm the immune system's capacity for an effective response.

Researchers hope to develop procedures to enhance immunological defenses against tumors as well as against certain pathogens. This prospect is called *immunotherapy*.

Monoclonal Antibodies   Injections of mass-produced antibodies against tumor-specific antigens might help a cancer patient. But mature B cells don't live long enough in culture to mass-produce antibody. Also, being end cells, they can't reproduce. Cesar Milstein and Georges Kohler showed how to make antibody "factories." They injected an antigen into a mouse, which made antibodies against it. They fused antibody-producing B cells from the mouse with cells extracted from B cell tumors. Some descendants of the hybrid cells divided nonstop, and they, too, produced the antibody. Clones of such hybrid cells are now being maintained indefinitely—and they make identical copies of antibodies in useful amounts. Their products are known as *monoclonal antibodies*.

A few cancer patients were inoculated with monoclonal antibodies that are expected to home in on the malignant tumors. In some cases, cell-killing chemicals have been artificially attached to such antibodies. At this writing, the success of clinical trials is limited.

Multiplying the Tumor Killers   Lymphocytes often infiltrate tumors. Researchers remove them from a tumor and expose them to an interleukin (lymphokine). The result is a population of tumor-infiltrating lymphocytes with enhanced killing abilities. These *LAK* cells (short for lymphokine-activated killers) appear to be somewhat effective when injected back into the patient.

Therapeutic Vaccines   On the horizon are *therapeutic vaccines* that can serve as wake-up calls against elusive tumor cells. One idea is to genetically engineer antigen to be more obvious to killer lymphocytes. For example, Lynn Spitler is already tinkering with a prostate-specific antigen that is present on all prostate cancers.

Formation of Antigen-Specific Receptors

The variety of antigens in your surroundings is mind-boggling. Collectively, however, antigen receptors of all T and B cell populations in your body show staggering diversity—enough to recognize about a *billion* different antigens by one estimate. How does the diversity arise?

For the answer, start with the knowledge that all antigen receptors of a single T or B cell are identical—and all are proteins. For example, a Y-shaped antibody molecule consists of four polypeptide chains bonded together (Figure 28.11). Certain parts of each chain, the *variable* regions, fold in ways that produce grooves and bumps with a certain charge distribution. Only antigen that has complementary grooves, bumps, and charge distribution will be able to bind with them.

Receptor diversity begins with rearrangements in the DNA that codes for such variable regions. For example, take a look at Figure 28.12. As a B cell matures, one of many DNA sequences called V segments becomes joined at random to one of distant DNA

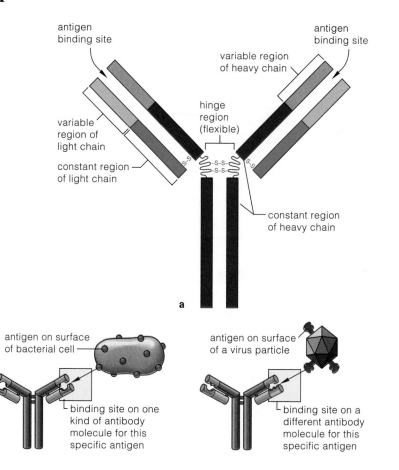

Figure 28.11 Antibody structure. (**a**) Each antibody molecule consists of four polypeptide chains, often bonded together in a Y shape. (**b**) At the antigen binding sites, antigen fits into grooves and onto protrusions.

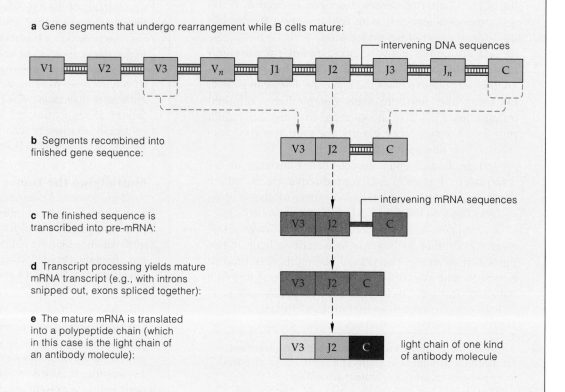

Figure 28.12 The generation of antibody diversity. Antibodies are proteins, and protein-building instructions are encoded in genes. In the chromosomes with antibody genes, extensive DNA regions contain different versions of segments that code for the variable regions of an antibody molecule. The different V and J segments shown are examples; they code for the variable region of a light chain (compare Figure 28.11*a*). As each B cell matures, a recombination event occurs in this region. In this case, any one of the V segments may be joined to any one of the J segments. Afterward, the intervening DNA is excised. The new sequence is joined to a C (*Constant*) segment, and this completes a rearranged antibody gene—which also will be found in all of the descendants of that cell.

a Gene segments that undergo rearrangement while B cells mature:

V1 V2 V3 V_n J1 J2 J3 J_n C — intervening DNA sequences

b Segments recombined into finished gene sequence:

V3 J2 C

c The finished sequence is transcribed into pre-mRNA:

V3 J2 C — intervening mRNA sequences

d Transcript processing yields mature mRNA transcript (e.g., with introns snipped out, exons spliced together):

V3 J2 C

e The mature mRNA is translated into a polypeptide chain (which in this case is the light chain of an antibody molecule):

V3 J2 C — light chain of one kind of antibody molecule

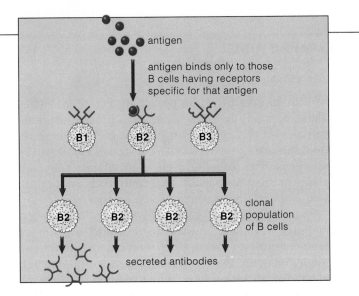

Figure 28.13 Clonal selection of a B cell that produced the specific antibody that can combine with a specific antigen. Only antigen-selected B cells (and T cells) are activated and give rise to a clonal population of immunologically identical cells.

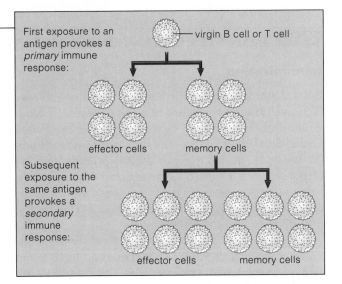

Figure 28.14 Immunological memory. Not all B and T cells are used in a primary immune response to an antigen. A large number continue to circulate as memory cells, which become activated during a secondary immune response.

sequences called J segments. The DNA that intervenes between the joined V and J segments loops out and is excised. In this way, the B cell ends up with a unique DNA sequence.

The same kinds of DNA rearrangements also help produce the variable regions of the TCR molecules of maturing T cells.

Some time ago, Mafarlane Burnet developed a *clonal selection* hypothesis that helped point the way to our current view of receptor diversity. He proposed that antigen "chooses" (binds to) one lymphocyte from all the various types in the body, because that lymphocyte has the receptor specific for it. Repeated mitotic cell divisions then give rise to a clone of cells that carry out the response (Figure 28.13).

Immunological Memory

The clonal selection theory explains how an individual can have "immunological memory" of a first encounter with antigen. The term refers to the body's capacity to make a *secondary* immune response to any subsequent encounter with the same type of antigen that provoked the primary response (Figure 28.14).

Memory cells that form during a primary immune response do not engage in battle. They circulate for years or for decades. Compared to the virgin cells that initiate a primary response, these patrolling battalions have far more cells, so they intercept antigen far sooner. Effector cells form sooner, in greater numbers, so the infection is terminated before the host gets sick. Even greater numbers of memory T and B cells form during a

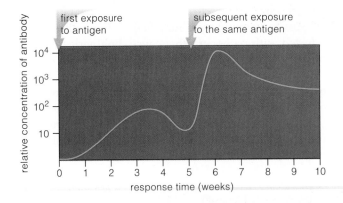

Figure 28.15 Differences in magnitude and duration between a primary and a secondary immune response to the same antigen. In this example, the primary response peaked 24 days after it started. The secondary response peaked after only 7 days (the span between weeks 5 and 6). Antibody concentration during the secondary response was 100 times greater (10^2 to 10^4).

secondary response. Figure 28.15 shows an example of this. In evolutionary terms, these advance preparations against subsequent encounters with a pathogen bestow a splendid survival advantage on the individual.

Recombination of segments drawn at random from receptor-encoding regions of DNA helps give each T or B cell a gene sequence for one of a billion possible antigen receptors.

Specificity means an antigen-selected cell will give rise to a clone of cells that will react only with the selecting antigen.

Memory means the individual has the capacity to make a secondary immune response to the pathogen that caused the primary response. A secondary response is greater and faster.

Immunization

Immunization refers to various processes that promote increased immunity against specific diseases. In *active* immunization, an antigen-containing preparation called a **vaccine** is either taken orally or injected into the body (Figure 28.16). A first injection elicits a primary immune response. Later, a subsequent injection (booster) elicits a secondary response, with the formation of more effector cells and memory cells that can provide long-lasting protection against the disease.

Many vaccines are manufactured from weakened or killed pathogens, as when Sabine polio vaccine is made from weakened polio virus particles. Other vaccines are based on inactivated forms of natural toxins, such as a bacterial toxin that causes tetanus. Still others are made with harmless genetically engineered viruses that have genes from three or more different viruses inserted in their DNA or RNA. After a person is vaccinated with an engineered virus, the incorporated genes are expressed, antigens are produced—and immunity is established.

Passive immunization may help individuals already infected with pathogens that cause diphtheria, tetanus, measles, hepatitis B, and some other diseases. A person receives injections of purified antibody. The best source of this is another individual who already has produced a large amount of the required antibody. The effects are not lasting, because the person's own B cells are not producing antibodies. But injected antibody molecules may help counter the immediate attack.

Allergies

In 8 to 10 percent of the people in the United States alone, normally harmless substances provoke immune responses that cause inflammation, excess mucus secretion, and other problems. Such substances are known as **allergens**. The response itself is an **allergy**. Common allergens are pollen, many drugs and foods, dust mites, fungal spores, insect venom, and cosmetics.

Some people are genetically inclined to develop allergies. Infections, emotional stress, or changes in air temperature also may trigger reactions that otherwise might not occur. Upon exposure to certain antigens, IgE antibodies are secreted and bind to mast cells. When the IgE binds antigen, mast cells secrete prostaglandins, histamine, and other substances that fan inflammation. They also stimulate mucus secretion and cause airways to constrict. In *asthma* and *hay fever*, stuffed sinuses, a drippy nose, labored breathing, and sneezing are the main symptoms of the allergic response (Figure 28.17).

In some cases, inflammatory reactions wash through the body and trigger a life-threatening condition called *anaphylactic shock*. For example, a person who is allergic to wasp or bee venom can die within minutes of a single sting. Airways to the lungs constrict massively. Fluid escapes swiftly from dilated, grossly permeable blood capillaries. Blood pressure plummets and may lead to circulatory collapse.

Antihistamines (anti-inflammatory drugs) often may relieve the mild, short-term symptoms of allergies. Over time, a patient may try a desensitization program. First, skin tests identify the offending allergens. Inflammatory responses

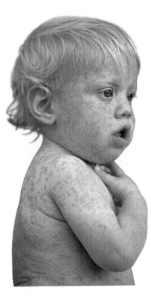

Age	Recommended Vaccines
At birth	Hepatitis B
2 months	Hepatitis B, DPT (diphtheria, whooping cough, tetanus), Hib (*Hemophilus influenzae*)
2–4 months	Hepatitis B
4 months	Polio, DPT, Hib
6–18 months	Hepatitis B, polio
6 months	DPT, Hib
12–15 months	Hib, MMR (measles, mumps, rubella)
12–18 months	DPT
4–6 years	Polio, DPT, MMR (final dose of MMR can be at 11–12 years)
11–12 years	DT (diptheria, tetanus)

Figure 28.16 From the Centers for Disease Control and Prevention, the 1995 immunization guidelines for children living in the United States. Pediatricians routinely immunize infants and children during office visits. Low-cost or free vaccinations are available at many community clinics and health departments.

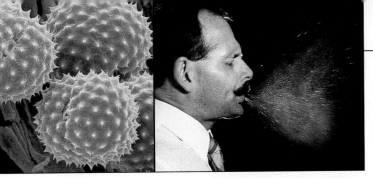

RAGWEED POLLEN . . .   . . . AND SOMETHING IT CAN PROVOKE

Figure 28.17 One of the effects of pollen and other allergens in sensitive people. Allergy sufferers who moved to deserts to escape pollen brought it with them. Half the human population in Tucson, Arizona, is now sensitized to pollen from olive and mulberry trees, planted far and wide in cities and the suburbs.

to some can be blocked if the patient can be stimulated to make IgG instead of IgE. Larger doses of specific allergens are administered gradually. Each time, the body makes more circulating IgG and memory cells. The IgG binds with allergen that it encounters and blocks its attachment to IgE—and thereby blocks inflammation.

Autoimmune Disorders

In an **autoimmune response**, the immune system is unleashed against self antigens. Consider *Grave's disorder*, in which the body produces thyroid hormones in excess. Feedback mechanisms control the production of these hormones, which influence metabolic rates and the development of many tissues. Affected individuals produce antibodies that bind to receptors on cells that produce the hormones. The antibodies don't respond to feedback controls, and they trigger overproduction of thyroid hormones. Disease symptoms typically include elevated metabolic rates, heart fibrillations, excessive sweating, nervousness, and weight loss.

Or consider *myasthenia gravis*, a progressive weakening of muscles. Antibodies that bind to acetylcholine receptors on skeletal muscle cells cause the disorder.

Finally, consider the chronic inflammation of skeletal joints associated with *rheumatoid arthritis*. Patients are genetically predisposed to this disorder. Macrophages, T cells, and B cells get activated by antigens associated with the joints. Immune responses are made against the body's own collagen molecules and apparently against antibody that has bound to an as yet unknown antigen. Complement activation and inflammation cause more damage in tissues of the joints. So do skewed repair mechanisms. Eventually, the joints fill with synovial membrane cells and become immobilized.

Deficient Immune Responses

When the body has inadequate numbers of functioning lymphocytes, immune responses are not effective. Such *severe combined immunodeficiencies* (SCIDs) result from

Figure 28.18 Example of a SCID (severe combined immunodeficiency). Ashanthi DeSilva was born without an immune system. She has a mutated gene for ADA (adenosine deaminase), an enzyme. Without ADA, her cells cannot break down adenosine. As a result, a reaction product accumulates—and it is toxic to lymphocytes. Symptoms of the disorder are caused by infections that cannot be controlled. They include high fever, severe ear and lung infections, diarrhea, and an inability to gain weight.

Ashanthi's parents consented to the first federally approved gene therapy for humans. Researchers used genetic engineering methods to splice the ADA gene into the genetic material of a harmless virus. They used the modified virus rather like a hypodermic needle. They allowed it to deliver copies of the "good" gene into Ashanthi's bone marrow cells. Some of the cells incorporated the gene in their DNA and started to synthesize the missing enzyme. At this writing, Ashanthi is now nine years old. Like other ADA-deficient patients who have started treatment, she is doing well.

Researchers also enlist bone marrow stem cells from blood in the umbilical cord of affected newborns. (The cord, which connects the fetus to the placenta during pregnancy, is discarded after childbirth.) They expose the cells to viruses that deliver copies of the ADA gene into them and to factors that stimulate growth and division. The cells are reinserted into the newborns.

heritable disorders as well as from various assaults on the body by outside agents. Either way, deficient or nonexistent immune responses make the person highly vulnerable to infections that are not life threatening to the general population. Figure 28.18 describes one of the heritable disorders. A viral infection causes *AIDS* (acquired immunodeficiency syndrome). Section 28.11 describes how the virus (HIV) replicates inside certain lymphocytes and destroys the body's capacity to fight infections. We return to this topic in Section 34.19.

Immunization programs boost immunity to specific diseases.

Certain heritable disorders or attacks by certain pathogens and other agents can result in misdirected, compromised, or nonexistent immunity.

AIDS—THE IMMUNE SYSTEM COMPROMISED

Characteristics of AIDS   AIDS is a constellation of disorders that follow an infection by a pathogen called the human immunodeficiency virus (HIV). This virus cripples the immune system, and the body becomes highly susceptible to usually harmless infections and some otherwise rare forms of cancer.

At this writing, researchers have not developed any vaccine that will work against the known forms of this virus (HIV-1 and HIV-2). Also at this writing, *there is no cure for those already infected.*

By current estimates, more than a million Americans are infected. Worldwide, an estimated 22.2 million people are infected; 6 million are already dead. By the turn of the century, the number of infected people may be as high as 110 million.

At first an infected person might appear to be in good health, suffering no more than a bout of "the flu." Then he or she starts displaying symptoms that foreshadow AIDS. Typically, symptoms include persistent weight loss, fever, fatigue, bed-drenching night sweats, and many enlarged lymph nodes. In time, diseases resulting from certain opportunistic infections are signs of AIDS. The diseases are rare in the population at large. They include yeast infections of the mouth, esophagus, vagina, and elsewhere, as well as a form of pneumonia caused by *Pneumocystis carinii*. Spots resembling bruises may appear, especially on legs and feet. These are signs of Kaposi's sarcoma, a form of cancer that develops from endothelial cells of blood vessels. The immune system cannot control the infections or cancers, which end up killing the person.

How HIV Replicates   HIV infects antigen-presenting macrophages and helper T cells (these are also called CD4 lymphocytes). HIV is one of the retroviruses. Each virus particle has an outermost lipid envelope—a bit of plasma membrane that surrounded the particle when it departed from an infected cell. Spiking outward from the envelope are HIV proteins that had become inserted into it. Beneath the lipid envelope, two protein coats—the core proteins—surround two strands of RNA and several copies of reverse transcriptase, an enzyme (compare Figure 18.13).

Once inside a host cell, the viral enzyme uses the RNA as a template for making DNA, which then is inserted into a host chromosome (Figure 28.19). Transcription yields copies of viral RNA. Some transcripts are translated into viral proteins. Others get enclosed (as hereditary material) in the proteins when new virus particles are put together. The particles bud from the host cell's plasma membrane and are released (Figure 28.20) to start a new round of infection. With each round, more macrophages, antigen-presenting cells, and helper T cells are impaired or killed.

The viral genes inserted into the DNA of some host cells remain inactive. But they may be activated during a later round of infection.

A Titanic Struggle Begins   Infection marks the onset of a titanic battle between the enemy and the immune system. The host immune system produces antibodies in response to HIV antigenic proteins. (The antibodies are the basis of diagnostic tests to identify HIV infection.) Helper T and cytotoxic T cells also are produced. By some estimates, however, HIV infects about 2 billion helper T cells and produces 100 million to 1 billion virus particles per day during certain phases of the infection. Every two

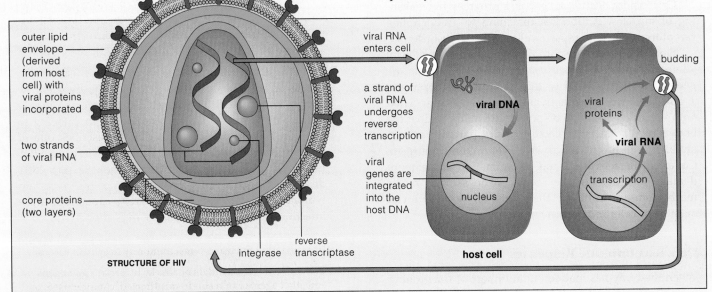

outer lipid envelope (derived from host cell) with viral proteins incorporated

two strands of viral RNA

core proteins (two layers)

integrase

reverse transcriptase

STRUCTURE OF HIV

viral RNA enters cell

a strand of viral RNA undergoes reverse transcription

viral genes are integrated into the host DNA

viral DNA

nucleus

budding

viral proteins

viral RNA

transcription

host cell

Figure 28.19   Life cycle of HIV, one of the retroviruses.

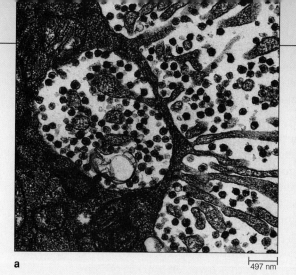

a

497 nm

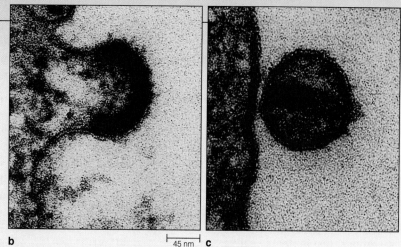

b

45 nm

c

days, the immune system destroys about half of these viruses and replaces half of the helper T cells lost in the battle. Huge reservoirs of HIV and masses of infected T cells accumulate in lymph nodes. As the battle proceeds, the number of virus particles in the general circulation rises. Gradually, the numbers tilt. The body produces fewer and fewer helper T cells to replace the ones it lost. Although it may take a decade or more, the erosion of the helper T cell count inevitably causes the body to lose its capacity to mount effective immune responses.

Some viruses, including the measles virus, produce far more virus particles in a given day, but the immune system usually wins out. Other viruses, including herpes viruses, can lurk in the body for a lifetime, but the immune system keeps them in check. With HIV, the immune system loses the struggle, then infections and tumors kill the person.

How HIV Is Transmitted Like any human virus, HIV requires a medium that allows it to leave one host, survive in the environment into which it is released, then enter another host.

HIV is transmitted when body fluids of an infected person enter another person's tissues. Initially in the United States, transmission occurred most often between males who engaged in homosexual activities, especially anal intercourse. It has spread among intravenous drug abusers who share blood-contaminated syringes and needles. It has spread in the heterosexual population, increasingly by vaginal intercourse.

HIV has traveled from infected mothers to offspring during pregnancy, birth, and breast-feeding. Before 1985, contaminated blood supplies accounted for some AIDS cases before health care providers implemented screening. Tissue transplants have caused four infections in 1991. In several developing countries, health care providers have spread HIV by way of contaminated transfusions and reuse of unsterile needles and syringes.

The molecular structure of HIV is not stable outside the human body, so transmission must be direct. Currently, we have no evidence that HIV is effectively transmitted by way of food, air, water, casual contact, or insect bites. The

Figure 28.20 (a) Transmission electron micrograph of HIV particles (black specks) escaping from an infected cell. (b,c) A virus particle budding from the host cell's plasma membrane.

virus *has* been isolated from human blood, semen, vaginal secretions, saliva, tears, breast milk, amniotic fluid, urine, and cerebrospinal fluid. It is probably present in other fluids, secretions, and excretions. However, only infected blood, semen, vaginal secretions, and breast milk contain the virus in concentrations that seem to be high enough for successful transmission.

Regarding Prevention and Treatment Developing effective drugs or vaccines against HIV is a formidable challenge. Its replication mechanisms, combined with a staggering number of replications in an infected person, promote high mutation rates in the viral genome. Drug therapy, remember, favors drug resistance—and fans the evolution of drug-resistant HIV populations. Some drugs used alone or in combination can slow replication. AZT (azidothymidine), ddI (dideoxyinosine), and protease inhibitors are examples. But they can't *cure* an infected person, because they can't eliminate the HIV genes that have become incorporated into his or her DNA. At this writing, we have no basis for assuming a cure is possible.

High mutation rates have another worrisome outcome: they give rise to variations in HIV antigens. This makes it difficult for researchers to select effective antigens for vaccines. Another problem is the scarceness of HIV strains with poor cell-killing abilities. Such strains are central to producing antigen that can stimulate the formation of cytotoxic T cell armies—the best protection against HIV. However, over a decade ago, many people in Australia received blood transfusions from a donor whose infection had not been diagnosed. At this writing, neither they nor the donor show any immunodeficiency. And they all carry HIV with a similar defect in the same gene (the nef gene).

In short, *until researchers develop effective vaccines and treatments, checking the spread of HIV depends absolutely on persuading people to avoid or modify social behaviors that put them at risk.* We return to this topic in Section 34.19.

SUMMARY

1. Vertebrates fend off many pathogens with physical and chemical barriers at body surfaces. They also are protected by the nonspecific and specific responses of the white blood cells listed in Table 28.3.

a. Nonspecific responses to irritation or damage of tissues include inflammation and involve organs with phagocytic functions, such as the spleen and liver.

b. Immune responses are made against specific pathogens, foreign cells, or abnormal body cells.

2. Intact skin and mucous membranes that line body surfaces are physical barriers to infection. Glandular secretions (as in tears, saliva, and gastric fluid) are examples of chemical barriers. So are the metabolic products of resident bacteria on body surfaces.

3. Tissue redness, warmth, swelling, and pain are signs of inflammation. The inflammatory response begins with changes in blood flow to a damaged tissue.

a. Pathogens as well as dead or damaged body cells release substances that make blood capillaries leaky. White blood cells enter the tissue and destroy invaders.

b. Plasma proteins enter the tissue. Complement proteins bind and induce lysis of pathogens, and they attract phagocytes. Blood-clotting proteins help repair damaged blood vessels.

4. Immune responses have these characteristics:

a. Each response is triggered by antigen, a unique molecular configuration that lymphocytes recognize as foreign (nonself). And it shows specificity, meaning it is directed against one antigen alone.

b. Each response also shows memory. A subsequent encounter with the same antigen will trigger a more rapid, secondary response, of greater magnitude.

c. An immune response normally is not made against the body's own self-marker proteins.

5. Macrophages and other antigen-presenting cells process and display fragments of antigen with their own MHC markers. Lymphocytes have receptors that can bind to antigen-MHC complexes. Binding is the start signal for an immune response.

6. An immune response starts with recognition of antigen. It proceeds through repeated cell divisions that form clones of B and T lymphocytes, which differentiate into subpopulations of effector and memory cells. Signals (e.g., interleukins) among white blood cells drive the responses. Effector helper T and cytotoxic T cells, as well as effector B cells and their antibodies, act at once. Memory cells are set aside for secondary responses.

7. T cells arise in bone marrow but continue to develop in the thymus, where they acquire TCRs. These T-cell receptors will be able to recognize and bind antigen-MHC complexes at the surface of antigen-presenting cells. B cells arise in bone marrow. While they mature, they start to synthesize antigen receptors (antibodies) that become positioned at their surface.

8. Effector cytotoxic T cells directly destroy virus-infected cells, tumor cells, and cells of tissue or organ transplants. Effector B cells (plasma cells) produce and secrete great numbers of antibodies that freely circulate.

9. Antibodies are protein molecules, often Y-shaped, and each has binding sites for one kind of antigen. Only B cells produce them. When antibody binds to antigen, toxins may be neutralized, pathogens may be tagged for destruction, or the attachment of pathogens to body cells may be prevented.

10. In active immunization, vaccines provoke immune responses, with production of effector and memory cells. In passive immunization, injections of purified antibodies help the individual through an infection.

11. An allergic reaction is an immune response to a generally harmless substance. Autoimmune responses are misguided attacks by lymphocytes on the body's own antigens. An immunodeficiency is a weakened or nonexistent capacity to mount an immune response.

Table 28.3	Summary of Major White Blood Cells and Their Roles in Defense
Cell Type	Main Characteristics
MACROPHAGE	Phagocyte; has roles in nonspecific defense responses; presents antigen to T cells; cleans up and helps repair tissue damage
NEUTROPHIL	Fast-acting phagocyte; takes part in inflammation, not in sustained responses; most effective against bacteria
EOSINOPHIL	Secretes enzymes that attack certain parasitic worms
BASOPHIL AND MAST CELL	Secrete histamines and other substances that act on small blood vessels, thereby producing inflammation; also contribute to allergic reactions
LYMPHOCYTES:	(All take part in most immune responses; following antigen recognition, all form clonal populations of effector cells and memory cells.)
B cell	Effectors secrete four classes of antibodies (IgA, IgE, IgG, and IgM) that protect the host in specialized ways
Helper T cell	Effectors secrete interleukins that stimulate rapid divisions and differentiation of both B cells and T cells
Cytotoxic T cell	Effectors kill infected cells, tumor cells, and foreign cells by a touch-kill mechanism
NATURAL KILLER (NK) CELLS	Cytotoxic cell of undetermined affiliation (may be a variety of lymphocyte); kills virus-infected cells and tumor cells

Review Questions

1. While jogging barefoot along a seashore, your toes accidentally land on a jellyfish. Soon the bottoms of your toes are swollen, red, and very warm to the touch. Describe the events that result in these signs of inflammation. *480–481*

2. Distinguish between:
   a. neutrophil and macrophage *480*
   b. cytotoxic T cell and natural killer cell *482, 485*
   c. effector cell and memory cell *482*
   d. antigen and antibody *482, 483*

3. Describe how a macrophage becomes an antigen-presenting cell. *482*

4. Why is a vaccine to control AIDS so elusive? *492–493*

Self-Quiz *(Answers in Appendix IV)*

1. _____ are barriers to pathogens at body surfaces.
   a. Intact skin and mucous membranes
   b. Tears, saliva, and gastric fluid
   c. Resident bacteria
   d. Urine flow
   e. All of the above

2. Macrophages are derived from _____ .
   a. lymphocytes   d. monocytes
   b. basophils   e. eosinophils
   c. neutrophils

3. Activated complement functions in defense by _____ .
   a. neutralizing toxins   c. promoting inflammation
   b. enhancing resident   d. forming holes in memory
   bacteria   lymphocyte membranes

4. _____ are certain molecules that lymphocytes recognize as foreign and that elicit an immune response.
   a. Interleukins   d. Antigens
   b. Antibodies   e. Histamines
   c. Immunoglobulins

5. Immunoglobulins designated _____ increase antimicrobial activity in mucus.
   a. IgA   d. IgM
   b. IgE   e. IgD
   c. IgG

6. Antibody-mediated responses work best against _____ .
   a. intracellular pathogens   d. both b and c
   b. extracellular pathogens   e. all of the above
   c. extracellular toxins

7. The most important antigens are _____ .
   a. nucleotides   c. steroids
   b. triglycerides   d. proteins

8. _____ would be a target of an effector cytotoxic T cell.
   a. Extracellular virus particles in blood
   b. A virus-infected cell
   c. Parasitic flukes in the liver
   d. Bacterial cells in pus
   e. Pollen grains in nasal mucus

9. Development of a secondary immune response is based on populations of _____ .
   a. memory cells
   b. circulating antibodies
   c. effector B cells
   d. effector cytotoxic T cells
   e. mast cells

10. Match the immunity concepts.
    ____ inflammation   a. neutrophil
    ____ antibody secretion   b. effector B cell
    ____ a phagocyte   c. nonspecific response
    ____ immune memory   d. deliberately provoking
    ____ vaccination   an immune response
    ____ allergy   e. basis of secondary response
                               f. nonprotective immune
                               response

Critical Thinking

1. Edward Jenner lucked out. He performed a potentially harmful experiment on a young boy who managed to survive it. What would happen if a would-be Jenner tried to do the same thing today?

2. Before each flu season starts, you get an influenza vaccination. This year you come down with "the flu" anyway. What do you suppose happened? There are at least three explanations.

3. As described in Section 18.7, infection by the *Ebola* virus results in a hemorrhagic fever with a 90 percent mortality rate. A patient received a transfusion of blood serum from another who survived the disease. Explain why this might increase chances of survival.

4. Rob bought a bumper sticker that read, "Have you thanked your resident bacteria today?" Explain why he appreciates the bacteria that normally reside on the body's skin and mucous membranes.

5. Ellen developed *chicken pox* when she was in kindergarten. Later in life, when her children developed chicken pox, she remained healthy. She did so even though she was exposed to countless virus particles daily. Explain why.

6. Quickly review Section 25.7 on homeostasis. Then write a short essay on how the immune response contributes to homeostasis.

Selected Key Terms

acute inflammation *480*	immune system *482*
allergen *490*	immunization *490*
allergy *490*	immunoglobulin (Ig) *487*
antibody *483*	interleukin *481*
antigen *482*	lysis *479*
antigen-MHC complex *482*	lysozyme *478*
antigen-presenting cell *482*	macrophage *480*
autoimmune response *491*	MHC marker *482*
B cell *483*	natural killer cell *485*
B lymphocyte *482*	neutrophil *480*
basophil *480*	pathogen *478*
complement system *479*	perforin *484*
cytotoxic T cell *482*	T lymphocyte *482*
eosinophil *480*	TCR *484*
helper T cell *482*	vaccine *490*
histamine *481*	

Readings

Edelson, R., and J. Fink. June 1985. "The Immunologic Function of Skin." *Scientific American* 252(6): 46–53.

Nowak, M., and A. McMichael. August 1995. "How HIV Defeats the Immune System." *Scientific American.* 273(2): 58–65.

Tizard, I. 1995. *Immunology: An Introduction.* Fourth edition. Philadelphia: Saunders.

29 RESPIRATION

Conquering Chomolungma

To experienced climbers, Chomolungma may be the ultimate challenge (Figure 29.1). The summit of this Himalayan mountain, also known as Everest, is 9,700 meters (29,108 feet) above sea level. It is the highest place on earth. Iced-over vertical rock, driving winds, blinding blizzards, and heart-stopping avalanches await the challengers. So does the extreme danger that oxygen-poor air poses to the brain.

Most of us live at low elevations. Of the air we breathe, one molecule in five is oxygen. When we travel to places 2,400 meters (about 8,000 feet) or more above sea level, the breathing game changes. The Earth's gravitational pull is weaker and gas molecules spread out more. This can result in *hypoxia*—cellular oxygen deficiency. Sensing the deficiency, the brain makes us hyperventilate, or breathe much faster and more deeply than normal. Hyperventilating can be worrisome above 3,300 meters (10,000 feet). It may cause significant ion imbalances in cerebrospinal fluid. This can trigger heart palpitations, shortness of breath, headaches, nausea, and even vomiting—strong clues that our cells are, in a manner of speaking, screaming for oxygen.

By living for months at high elevations, climbers can adapt to the thinner air. For example, their red blood cell count increases. Even so, climbers never get as much oxygen as, say, a llama does (Figure 29.2). While llamas are growing up, a great profusion of air sacs and blood vessels forms within their lungs. Llama heart ventricles enlarge more than ours and can pump larger volumes of blood. And llama hemoglobin picks up oxygen much more efficiently at lower air pressure than ours does.

Chomolungma's base camp is 6,300 meters (19,000 feet) above sea level. At 7,000 meters, oxygen and other gaseous molecules are extremely diffuse. The oxygen scarcity and low air pressure combine to make small blood vessels leaky. More plasma escapes through gaps between endothelial cells that make up the blood vessel walls. In the brain and lungs, tissues swell with excess fluid. If the edema is not reversed, climbers will become comatose and die. When stricken, they inhale bottled oxygen and get zipped inside airtight bags. Rescuers use a device that pumps in oxygen and removes carbon dioxide until the "air" inside the bag more closely approximates the air at 2,400 meters.

Figure 29.1 A climber inching toward Chomolungma's summit, where oxygen is brutally scarce.

Figure 29.2 Llamas, with long-lasting adaptations to the thinner air high in the Peruvian Andes.

Few of us will ever find ourselves near the peak of Chomolungma, pushing our reliance on oxygen to the limits. Here in the lowlands, disease, smoking, and other environmental insults push it in more ordinary ways, although the risks can be just as great.

The point is this: Regardless of the species, animal body plans are adapted to the oxygen levels in their habitat. One way or another, by a physiological process called **respiration**, the plans allow oxygen to move into the animal and carbon dioxide to move out. Why is this important? Animals, remember, have great energy demands. Their cells use a lot of oxygen—especially for aerobic respiration, the metabolic pathway that produces a lot of energy and carbon dioxide wastes.

This chapter samples a few **respiratory systems**, which function in the exchange of gases between the body and the environment. Together with other organ systems, they also contribute to maintaining internal operating conditions for the body's cells (Figure 29.3).

1. Of all organisms, multicelled animals require the most energy. At the cellular level, the energy comes mainly from aerobic respiration, a metabolic pathway that requires oxygen and produces carbon dioxide wastes.

2. In a physiological process called respiration, animals move oxygen into their internal environment and give up carbon dioxide to the external environment.

3. Oxygen diffuses into the body as a result of a pressure gradient. The pressure of this gas is higher in air than it is in metabolically active tissues, where cells rapidly use oxygen. Carbon dioxide follows a gradient in the other direction. Its pressure is higher in tissues, where it is a by-product of metabolism, than it is in the air.

4. In most respiratory systems, oxygen and carbon dioxide diffuse across a respiratory surface, such as the thin, moist epithelium in the human lungs. Blood flowing through the body's circulatory system picks up oxygen and gives up carbon dioxide at this respiratory surface.

5. Respiratory systems differ in their adaptations for increasing gas exchange efficiency. They differ also in how they match air flow to blood flow.

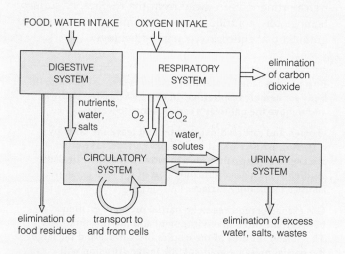

Figure 29.3 Interactions between the respiratory system and other organ systems in complex animals.

The Basis of Gas Exchange

In all animals, the process of respiration is based on the tendency of oxygen and carbon dioxide to diffuse down their concentration gradients—or, as we say for gases, their *pressure* gradients. When molecules of either gas are more concentrated in the surrounding air or water, more pressure is available to drive them into the body. When the molecules are more concentrated inside, more pressure is available to drive them outside.

This is not to say both of these gases exert the *same* pressure. Pump air into a flat tire near a beach in San Diego or Miami, and you will be filling it with about 78 percent nitrogen, 21 percent oxygen, 0.04 percent carbon dioxide, and 0.96 percent of other gases. This is true of dry air anywhere at sea level. The numbers tell you that oxygen exerts only part of the total pressure on the tire wall, and its "partial pressure" is greater than that of carbon dioxide. Said another way, atmospheric pressure at sea level is about 760 mm Hg, as measured by a mercury barometer. (Figure 29.4 shows a diagram of this device.) Therefore, the partial pressure of oxygen is (760 × 21/100), or about 160 mm Hg. The partial pressure of carbon dioxide is about 0.3 mm Hg.

Gases enter and leave the animal body by crossing a **respiratory surface**. A respiratory surface is a thin layer of epithelium or some other tissue. It must be kept moist at all times, for gaseous molecules cannot diffuse across it unless they are dissolved in fluid. What dictates the number of gas molecules moving across a respiratory surface in a given time? According to Fick's law, the more extensive the surface area and the larger the partial pressure gradient, the faster will be the diffusion rate.

Factors That Influence Gas Exchange

Surface-to-Volume Ratio All animal body plans promote favorable rates of inward diffusion of oxygen and outward diffusion of carbon dioxide. For example, animals with no respiratory organs are tiny, tubelike, or flattened, and gases diffuse directly across the body surface. These body plans meet a constraint imposed by the surface-to-volume ratio, as described in Section 3.2. To recap the key point about this, imagine a flatworm starts to grow in all directions, like an inflating balloon. The worm's surface area does not increase at the same rate as its volume. And once its girth exceeds a single millimeter, the diffusion distance between the body's surface and internal cells is so great, the flatworm dies.

Ventilation A large-bodied, highly active animal has great demand for gas exchange, more than diffusion alone can satisfy. A variety of wonderful adaptations make the exchange rate more efficient. For example,

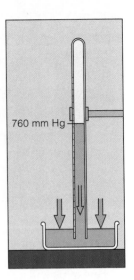

Figure 29.4 Atmospheric pressure as measured with a device called a mercury barometer. Part of the device is a glass tube in which the height of a column of mercury (Hg) can either increase or decrease, depending on air pressure outside the device.

At sea level, the mercury rises to about 760 millimeters (29.91 inches) from the base of the tube. At this level, the pressure that the column of mercury exerts inside the tube is equal to the atmospheric pressure on the outside.

760 mm Hg

above the respiratory organs (gills) of a trout, tissue flaps move back and forth, stirring the surrounding water. Stirred water puts more dissolved oxygen closer to the gills and carries more carbon dioxide away from them. Vertebrate circulatory systems rapidly deliver oxygen to cells and transport carbon dioxide to gills or lungs for disposal. You breathe to ventilate your lungs.

Transport Pigments Rates of gas exchange get a boost with respiratory pigments, mainly **hemoglobin**, that help maintain the steep pressure gradients across a respiratory surface. For example, at the respiratory surfaces in human lungs, where the concentration of oxygen is high, each hemoglobin molecule binds loosely with up to four oxygen molecules. As the circulatory system carries that same hemoglobin molecule past an oxygen-poor tissue, the oxygen is released. Thus, by transporting oxygen away from the respiratory surface, hemoglobin contributes to maintaining the pressure gradient that entices oxygen into the lungs.

By a process called respiration, animals take oxygen into the body for aerobic respiration, an energy-releasing pathway, and remove the pathway's carbon dioxide wastes.

Oxygen and carbon dioxide enter and leave the body by diffusing across a moist respiratory surface. Like other atmospheric gases, they tend to move down their pressure gradients. Each gas exerts only part of the total pressure across a respiratory surface.

Gas exchange requires steep partial pressure gradients between the internal environment and the surroundings. The greater the area of the respiratory surface and the larger the partial pressure gradient, the faster diffusion will proceed.

29.2 INVERTEBRATE RESPIRATION

Flatworms, earthworms, and many other invertebrates are not massive, and their life-styles don't require high metabolic rates (Figure 29.5*a*). Their demands for gas exchange are simply met by **integumentary exchange**, in which gases diffuse directly across the body's surface covering (integument). This mode of respiration works as long as the surface stays moist; invertebrates that rely solely on it are restricted to aquatic or damp habitats. Integumentary exchange supplements other modes of respiration in amphibians and some other large animals.

Many marine and freshwater invertebrates have moist, thin-walled respiratory organs called **gills**. When gill walls are highly folded, the increased respiratory surface area enhances exchange rates between blood (or some other body fluid) and the surroundings. Figure 29.5*b* shows the folded gill of a sea hare (*Aplysia*). By supplementing integumentary exchange, the gill helps provide adequate oxygen for this rather large mollusk; some sea hares are 40 centimeters (nearly 16 inches) long.

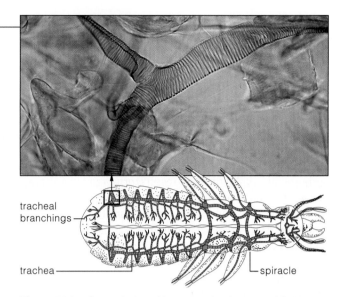

Figure 29.6 General plan of insect tracheal systems. Chitin rings reinforce many of the branching tubes.

tracheal branchings

trachea

spiracle

Invertebrates of dry habitats have small, thick, or hardened integuments that are not well endowed with blood vessels. Although their integuments do conserve precious water, these aren't good respiratory surfaces. Such land-dwelling animals have *internal* respiratory surfaces. For example, most spiders have book lungs—respiratory organs with thin, folded walls that resemble book pages (Figure 20.23*d*). Most insects, millipedes, centipedes, and some spiders have a system of internal tubes that function in **tracheal respiration**.

Consider the tracheal system of an insect (Figure 29.6). Small openings perforate the insect integument. Each opening, a spiracle, is the start of a tube that branches inside the body. The last branchings dead-end at a fluid-filled tip, where gases diffuse directly into tissues. The tips of the tubes are especially profuse in muscle and other tissues with high oxygen demands.

We find hemoglobin or other respiratory pigments in many invertebrates, although these are rare in insects. As is true of vertebrates, respiratory pigments increase the capacity of body fluids to transport oxygen. Finally, in invertebrates with a well-developed head, oxygenated blood tends to circulate first through the head end, then through the rest of the body.

Flatworms and some other invertebrates that are not massive and that live in aquatic or moist habitats use integumentary exchange, in which oxygen and carbon dioxide diffuse directly across the body surface.

Most marine invertebrates and many freshwater types have gills of one sort or another. These respiratory organs have moist, thin, and often highly folded walls.

Most insects, millipedes, centipedes, and some spiders use tracheal respiration. Gases flow through open-ended tubes that start at the body surface and end directly in tissues.

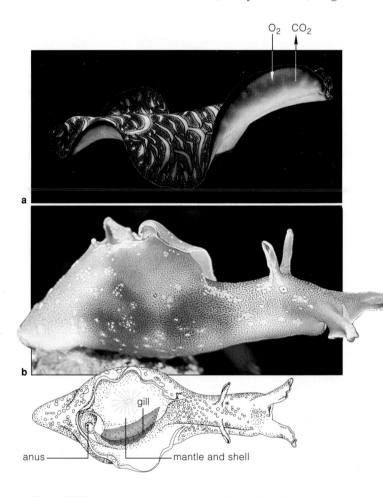

O₂   CO₂

a

b

gill

anus

mantle and shell

Figure 29.5 Invertebrates of aquatic habitats. (**a**) A flatworm, small enough to get along well without an oxygen-transporting circulatory system. Dissolved oxygen in their aquatic habitats reaches individual cells simply by diffusing across the body surface. (**b**) A sea hare (*Aplysia*), one of the gastropods.

Gills of Fishes and Amphibians

Gills persist in many vertebrate lineages. A few kinds of fish larvae and a few amphibians have *external* gills that project into the surrounding water. Adult fishes have a pair of *internal* gills. These are rows of slits or pockets at the back of the mouth that extend to the body surface (Figure 29.7*a*). All of these respiratory organs have walls of moist, thin, vascularized epithelium.

In fishes, water flows into the mouth and pharynx, then over arrays of filaments in the gills (Figure 29.7*b*). Blood vessels thread through respiratory surfaces in each filament. First the water flows past a vessel that is carrying blood back to the rest of the body. This blood has less oxygen than the water does, so oxygen diffuses into the blood. Then the same volume of water flows over a vessel carrying blood into the gills. The water has already given up oxygen, but it still has more than blood inside this vessel does—so more oxygen diffuses into the filament. Movement of two fluids in opposing directions is known as **countercurrent flow**. Through this mechanism, a fish extracts about 80 to 90 percent of the dissolved oxygen flowing past. That is more than the fish would get from a one-way flow mechanism, at far less energy cost.

Lungs

Some fishes and all amphibians, birds, and mammals have a pair of **lungs**, which are internal respiratory surfaces in the shape of a cavity or sac. Lungs originated in some fish lineages more than 450 million years ago, as pouches off the anterior part of the gut wall.

They probably evolved rapidly by natural selection, for lungs afforded major advantages. They increased the surface area for gas exchange in oxygen-poor habitats, and they worked better than gills could in the move onto land. Gills stick together and can't function unless water flows through them and keeps them moist.

Lungfishes of oxygen-poor habitats still have gills; they also use tiny lungs as backups. Amphibians never completed the transition to land. Their skin serves as a respiratory surface (integumentary exchange), in salamanders especially. Frogs and toads rely more on small lungs for oxygen uptake; most of the carbon dioxide still diffuses across the skin. Frogs also are heavy-duty breathers; they *force* air into the lungs, then empty them by contracting muscles of the body wall (Figure 29.8).

Frogs use their lungs for another function—sound production. So do all mammals except whales. Sound originates near the entrance to the larynx, an airway leading to the lungs. Here, part of a mucous membrane

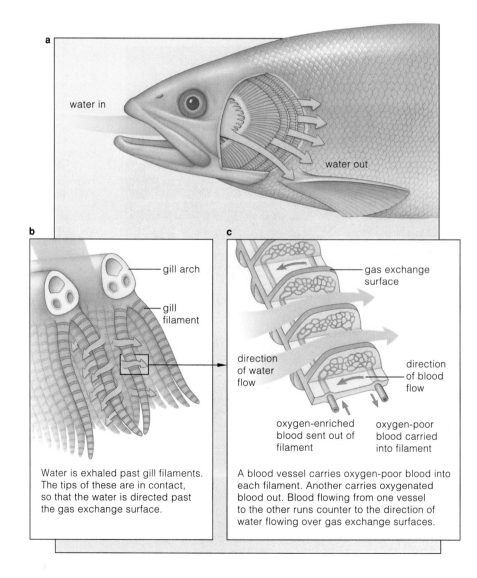

Figure 29.7 One kind of fish gill. (**a**) One of a pair of gills, each located under a bony lid (removed for this sketch).

(**b**) Filaments in a fish gill have vascularized respiratory surfaces. The tips of neighboring filaments touch, so water flowing over them is directed past gas exchange surfaces before being exhaled. (**c**) A blood vessel carries oxygen-poor blood into a filament; another carries oxygenated blood away from it. Blood flowing from one vessel to the other runs counter to the direction of water flowing over the gas exchange surfaces. This arrangement favors the movement of oxygen (down its partial pressure gradient) into the blood.

water in

water out

gill arch

gill filament

gas exchange surface

direction of water flow

direction of blood flow

oxygen-enriched blood sent out of filament

oxygen-poor blood carried into filament

Water is exhaled past gill filaments. The tips of these are in contact, so that the water is directed past the gas exchange surface.

A blood vessel carries oxygen-poor blood into each filament. Another carries oxygenated blood out. Blood flowing from one vessel to the other runs counter to the direction of water flowing over gas exchange surfaces.

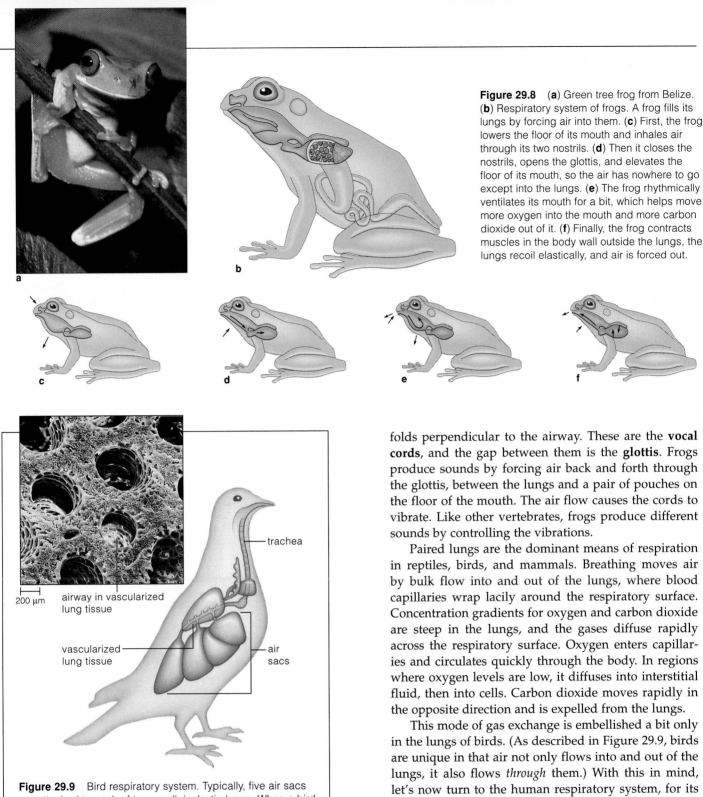

Figure 29.8 (**a**) Green tree frog from Belize. (**b**) Respiratory system of frogs. A frog fills its lungs by forcing air into them. (**c**) First, the frog lowers the floor of its mouth and inhales air through its two nostrils. (**d**) Then it closes the nostrils, opens the glottis, and elevates the floor of its mouth, so the air has nowhere to go except into the lungs. (**e**) The frog rhythmically ventilates its mouth for a bit, which helps move more oxygen into the mouth and more carbon dioxide out of it. (**f**) Finally, the frog contracts muscles in the body wall outside the lungs, the lungs recoil elastically, and air is forced out.

200 µm

airway in vascularized lung tissue

trachea

vascularized lung tissue

air sacs

Figure 29.9 Bird respiratory system. Typically, five air sacs are attached to each of two small, inelastic lungs. When a bird inhales, air is drawn into air sacs through tubes, open at both ends, that thread through the vascularized lung tissue. This tissue is the respiratory surface, where gases are exchanged.

When the bird exhales, air is blown out of the sacs, through the small tubes, and out of the trachea. Thus, air is not merely drawn into the bird lungs. Air is continuously drawn through them and across the respiratory surface. This ventilating system is unique, and it supports the high metabolic rates that birds require for flight and other energy-intensive activities.

folds perpendicular to the airway. These are the **vocal cords**, and the gap between them is the **glottis**. Frogs produce sounds by forcing air back and forth through the glottis, between the lungs and a pair of pouches on the floor of the mouth. The air flow causes the cords to vibrate. Like other vertebrates, frogs produce different sounds by controlling the vibrations.

Paired lungs are the dominant means of respiration in reptiles, birds, and mammals. Breathing moves air by bulk flow into and out of the lungs, where blood capillaries wrap lacily around the respiratory surface. Concentration gradients for oxygen and carbon dioxide are steep in the lungs, and the gases diffuse rapidly across the respiratory surface. Oxygen enters capillaries and circulates quickly through the body. In regions where oxygen levels are low, it diffuses into interstitial fluid, then into cells. Carbon dioxide moves rapidly in the opposite direction and is expelled from the lungs.

This mode of gas exchange is embellished a bit only in the lungs of birds. (As described in Figure 29.9, birds are unique in that air not only flows into and out of the lungs, it also flows *through* them.) With this in mind, let's now turn to the human respiratory system, for its operating principles apply to most vertebrates.

A countercurrent flow mechanism in fish gills compensates for low oxygen levels in aquatic habitats. Internal air sacs—lungs—are more efficient in dry land habitats. Amphibians rely on integumental exchange; they also force air into and out of small lungs. Paired lungs are the dominant means of respiration in reptiles, birds, and mammals.

From Airways To the Lungs

It will take at least 300 million breaths to get you to age seventy-five. You may find yourself going without food for a few hours or days. But stop breathing even for five minutes and normal brain function is over.

Take a deep breath, then look at Figure 29.10 to get an idea of where the air will travel in your respiratory system. Unless you are out of breath and panting, the air has just entered two nasal cavities, not your mouth. There it is warmed and picks up moisture from mucus. Ciliated epithelium and hairs in the nasal cavities filter dust and particles from it. Now the air is poised at the **pharynx**, or throat. This is the entrance to the **larynx**, the airway with the vocal cords (Figure 29.11). When muscles of the larynx contract or relax, elastic ligaments in the cords

tighten or slacken, and change how much the cords are stretched. The nervous system coordinates the muscle action that narrows or widens the glottis and thereby controls which sounds are produced. Right now the **epiglottis**, a tissue flap at the start of the larynx, is pointing up, so air moves into the **trachea** (windpipe). When you swallow, the epiglottis points down and so closes off the entrance to the trachea. At such times, food or fluid being swallowed enters the esophagus, a tube connecting the pharynx with the stomach.

The trachea branches into two airways, one leading into the tissue of each lung. Each airway is a **bronchus** (plural, bronchi). Its epithelial lining has a profusion of cilia and mucus-secreting cells. This lining is a barrier to infection. Bacteria and airborne particles stick in the mucus, then the cilia sweep debris-laden mucus toward the mouth. Where it goes from there really is up to you, but possibly the sidewalk is not a suitable destination.

ORAL CAVITY
Supplemental airway when breathing is labored

EPIGLOTTIS
Closes off larynx during swallowing

PLEURAL MEMBRANES
Membranes that separate lungs from other organs; also form a thin, fluid-filled cavity that facilitates breathing

LUNG (ONE OF A PAIR)
Lobed, elastic organ of breathing that enhances gas exchange between the body and the outside air

INTERCOSTAL MUSCLES
Rib cage muscles with roles in breathing

DIAPHRAGM
Muscle sheet between the chest cavity and abdominal cavity with roles in breathing

NASAL CAVITY
Chamber in which air is warmed, moistened, and initially filtered; and in which sounds resonate

PHARYNX (THROAT)
Airway connecting the nasal cavity and mouth with the larynx; enhances speech sounds; connects also with the esophagus that leads to the stomach

LARYNX (VOICE BOX)
Airway where sound is produced and where breathing is blocked during swallowing movements

TRACHEA (WINDPIPE)
Airway connecting the larynx with two bronchi that lead into the lungs

BRONCHIAL TREE
Increasingly branched airways starting with the bronchi and ending at air sacs (alveoli)

ALVEOLI
Thin-walled air sacs where oxygen diffuses into the internal environment and carbon dioxide diffuses out

a

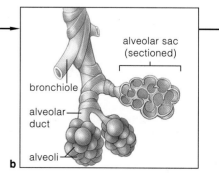

alveolar sac (sectioned)

bronchiole

alveolar duct

alveoli

b

Figure 29.10 (**a**) Components of the human respiratory system and their functions. Also shown are the diaphragm and other structures with secondary roles in respiration. (**b,c**) Location of alveoli relative to the lung capillaries.

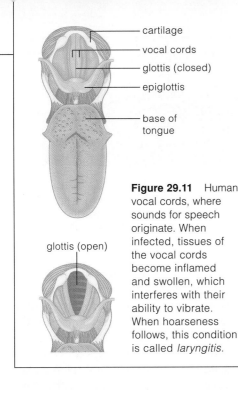

cartilage
vocal cords
glottis (closed)
epiglottis
base of tongue

glottis (open)

Figure 29.11 Human vocal cords, where sounds for speech originate. When infected, tissues of the vocal cords become inflamed and swollen, which interferes with their ability to vibrate. When hoarseness follows, this condition is called *laryngitis*.

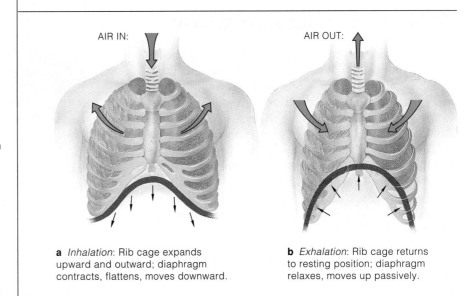

AIR IN:

AIR OUT:

a *Inhalation*: Rib cage expands upward and outward; diaphragm contracts, flattens, moves downward.

b *Exhalation*: Rib cage returns to resting position; diaphragm relaxes, moves up passively.

Figure 29.12 Changes in chest cavity size during (**a**) inhalation and (**b**) exhalation. The *blue* line marks the change in the diaphragm's position.

Human lungs are elastic, cone-shaped organs of gas exchange. They are located within the rib cage, to the left and right of the heart and above the **diaphragm**, a muscular partition between the chest cavity and the abdominal cavity. A thin, pleural membrane lines the outer surface of the lungs and the inner surface of the chest cavity wall. Visualize the lungs as two baseballs pushed into a partly inflated balloon. The balls take up so much space, they press the balloon's opposing sides together. Similarly, the pleural membrane is saclike, with the chest wall and the lungs pressing its opposing surfaces together. A thin film of lubricating fluid separates the two membrane surfaces and decreases friction between them. During *pleurisy*, a respiratory ailment, the pleural membrane becomes inflamed and swollen, friction follows, and breathing can be painful.

Inside each lung, air moves through finer and finer branchings of a "bronchial tree" (Figure 29.10*a*). These airways are **bronchioles**. Their endings, the *respiratory bronchioles*, have cup-shaped outpouchings from their walls. Each outpouching is an **alveolus** (plural, alveoli). The lungs have about 300 million of them. Most often, alveoli are clustered as a larger pouch, an alveolar sac.

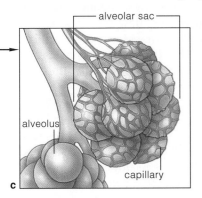

alveolar sac

alveolus

capillary

c

The sacs are the major sites of gas exchange with lung capillaries (Figure 29.10*b* and *c*). Collectively, they offer a tremendous surface area for gas exchanges with blood. If all your alveoli were stretched out in one layer, they would cover the floor of a racquetball court!

Ventilation

Breathing is a way to ventilate the lungs. Air is *inhaled* (drawn into the airways) and then *exhaled* (expelled from them). The chest cavity's volume increases and decreases in a rhythmic way. Every time you change the cavity's volume, you reverse the pressure gradients between the lungs and air outside your body. Gases in the respiratory tract follow those gradients.

As you start to inhale, the dome-shaped diaphragm contracts and flattens, and skeletal muscles lift the ribs upward and outward (Figure 29.12). Then, as the chest cavity expands, the rib cage moves away slightly from the lung surface. Pressure is lowered in the potential space between each lung and the pleural membrane. This pressure is transmitted to air spaces inside the alveoli, and it is lower than the atmospheric pressure at the mouth. Fresh air follows the pressure gradient. It flows down the airways, almost into the respiratory bronchioles, and causes the lungs to expand.

When you start to exhale, elastic tissue in the lungs and chest wall recoils passively, so the volume of the chest cavity decreases. The decrease compresses the air in the alveolar sacs. At that time, the air pressure in the sacs is greater than the atmospheric pressure, so now air follows the gradient—out of the lungs.

When oxygen demands increase, as during exercise, skeletal muscles of the rib cage and abdomen are stimulated to contract more often and more forcefully. As they do, they hasten the air flow.

Breathing reverses pressure gradients between the lungs and air outside the body. Oxygen and carbon dioxide follow their gradients, to and from exchange sites at alveoli in the lungs.

29.5 GAS EXCHANGE AND TRANSPORT

Each cup-shaped alveolus is a single layer of epithelial cells, surrounded by a thin basement membrane. Lung capillaries also have thin walls, and only a thin film of interstitial fluid separates them from alveoli. Thus gases diffuse easily in both directions (Figure 29.13).

Passive diffusion alone is enough to carry oxygen and carbon dioxide across the respiratory surface, to and from the bloodstream. You cannot depend on the blood plasma alone to transport adequate amounts of oxygen and carbon dioxide. You also must depend on hemoglobin molecules in your red blood cells to bind and transport both gases. Hemoglobin boosts oxygen transport from the lungs by 70 times and boosts carbon dioxide transport away from tissues by 17 times.

Oxygen Transport

Inhaled air that reaches the alveoli has plenty of oxygen and not much carbon dioxide. The opposite is true of blood in the lung capillaries. Thus, in the lungs, oxygen diffuses into the plasma portion of blood and then into red blood cells, where it rapidly binds with hemoglobin. A hemoglobin molecule with oxygen bound to it is called **oxyhemoglobin**, or HbO_2. The amount of HbO_2 that forms depends on the partial pressure of oxygen. The higher the pressure, the more oxygen will be picked up. This will continue until all of the hemoglobin binding sites are saturated.

HbO_2 molecules hold onto oxygen rather weakly. They give it up in tissues where the partial pressure of oxygen is lower than in the lungs. They give it up even faster in tissues where the blood is warmer, the pH lower, and the partial pressure of carbon dioxide high. Such conditions exist in contracting muscle and other metabolically whipped-up tissues.

Carbon Dioxide Transport

Now think of a tissue where carbon dioxide's partial pressure is higher than in the blood flowing through the adjacent capillaries. Carbon dioxide diffuses into these capillaries, which transport it toward the lungs.

About 7 percent of the dissolved carbon dioxide remains in plasma. Another 23 percent or so binds with hemoglobin, forming carbaminohemoglobin ($HbCO_2$). But most of it—about 70 percent—is transported in the form of bicarbonate (HCO_3^-). This bicarbonate forms after carbon dioxide combines with water. The resulting carbonic acid dissociates (separates) into bicarbonate and hydrogen ions (H^+):

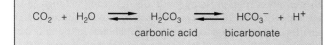

$$CO_2 + H_2O \rightleftharpoons \underset{\text{carbonic acid}}{H_2CO_3} \rightleftharpoons \underset{\text{bicarbonate}}{HCO_3^-} + H^+$$

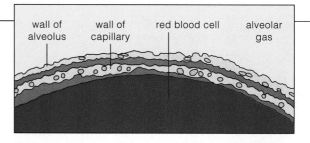

Figure 29.13 What a section through an alveolus and an adjacent lung capillary would look like. Compared to the red blood cell's diameter, the diffusion distance across the capillary wall, the interstitial fluid, and the alveolar wall is small.

wall of alveolus · wall of capillary · red blood cell · alveolar gas

In blood plasma, this reaction converts only 1 of every 1,000 carbon dioxide molecules, which doesn't amount to much. It's a different story in red blood cells. In these cells, the enzyme **carbonic anhydrase** enhances the reaction rate by 250 times! Most of the carbon dioxide that is not bound to hemoglobin becomes converted to carbonic acid. The enzyme-mediated conversion makes the blood level of carbon dioxide drop swiftly. This helps maintain the gradient that keeps carbon dioxide diffusing from interstitial fluid into the bloodstream.

What happens to the bicarbonate ions that form in the reactions? They tend to move out of the red blood cells and into blood plasma. What about the hydrogen ions? Hemoglobin acts as a buffer for them and keeps the blood from becoming too acidic. A buffer, recall, is a molecule that combines with or releases hydrogen ions in response to changes in cellular pH.

The reactions are reversed in the alveoli, where the partial pressure of carbon dioxide is lower than it is in neighboring blood capillaries. Now carbon dioxide and water form and diffuse into alveolar sacs. From there, carbon dioxide is exhaled from the body.

Matching Air Flow With Blood Flow

Gas exchange is most efficient when the rate of air flow matches the rate of blood flow. The nervous system and local chemical control mechanisms interact to bring the rates into balance by adjustments in individual tissues as well as in tissues through the body as a whole.

Some local controls work on alveoli. Suppose something really frightens you. Your heart pounds and your breaths are shallow. When blood flow is too fast and air flow is too sluggish, carbon dioxide disposal suffers. But the increase in the blood level of carbon dioxide affects smooth muscle in bronchiole walls. The wall diameter widens, enhancing the air flow.

And local controls work at lung capillaries. When air flow is too great relative to blood flow, oxygen levels rise in parts of the lungs. This affects smooth muscle in blood vessel walls. The wall diameter widens, so blood flow increases. If the volume of air flow is too small, the diameter shrinks, and blood flow decreases.

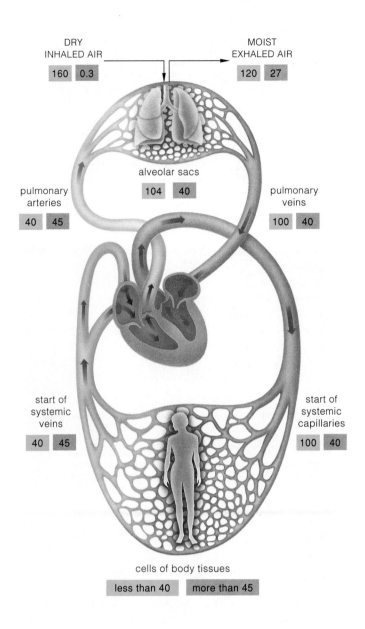

DRY INHALED AIR
| 160 | 0.3 |

MOIST EXHALED AIR
| 120 | 27 |

alveolar sacs
| 104 | 40 |

pulmonary arteries
| 40 | 45 |

pulmonary veins
| 100 | 40 |

start of systemic veins
| 40 | 45 |

start of systemic capillaries
| 100 | 40 |

cells of body tissues
| less than 40 | more than 45 |

Figure 29.14 Partial pressure gradients for oxygen (*blue* boxes) and carbon dioxide (*pink* boxes) in the respiratory tract.

The nervous system monitors arterial blood for the entire body. Some sensory receptors notify the brain if carbon dioxide levels rise in the blood; others detect a decrease in the partial pressure of oxygen. Receptors at carotid arteries to the brain and in the wall of an artery near the heart are examples. The brain responds by commanding muscles in the diaphragm and chest wall to alter their activity. In this way, it adjusts the rate and depth of breathing.

The partial pressure gradients for oxygen and for carbon dioxide through the human respiratory system are summarized in Figure 29.14.

Carbon Monoxide Poisoning

As this discussion makes clear, *gases move from regions of higher to lower partial pressure.* That's why you get light-headed when you first travel to high altitudes. Oxygen's partial pressure decreases with altitude, and you don't function as well when the pressure gradient between the air and your lungs is lower than what you normally encounter. Cellular oxygen deficiency also arises when its partial pressure in arterial blood falls as a result of *carbon monoxide poisoning.* Carbon monoxide, a colorless, odorless gas, is in vehicle exhaust fumes and smoke from burning tobacco, coal, or wood. It binds at least 200 times more strongly to hemoglobin than oxygen does. Even very small amounts can tie up half the body's hemoglobin—and dangerously impair the delivery of oxygen to tissues.

Regarding Gaseous Nitrogen and the Bends

When outside pressure is great enough, mismatched gas flow and air flow also can play havoc with deep-sea divers. Water pressure increases greatly with depth. Professional divers take along tanks of compressed air (air under pressure). They also make their ascents from the depths with utmost caution.

Why the worry? The increased pressure in deep waters causes more gaseous nitrogen (N_2) than usual to become dissolved in body tissues. During an ascent, the pressure decreases, so N_2 moves out of the tissues and into the bloodstream. If it is too rapid, N_2 enters the blood faster than the lungs can dispose of it. Then, bubbles of nitrogen may form in the blood and tissues. Too many bubbles cause pain, especially at joints. Hence the common name, "the bends," for what is otherwise known as *decompression sickness.* If bubbles obstruct blood flow to the brain, deafness, impaired vision, and paralysis may result. At depths of about 150 meters, high partial pressures of N_2 produce euphoria, even drunkenness. Divers have been known to offer their air-tank's mouthpiece to a fish.

Driven by its partial pressure gradient, oxygen diffuses from alveolar air spaces, through interstitial fluid, and into lung capillaries. Carbon dioxide, driven by its partial pressure gradient, diffuses in the opposite direction.

Hemoglobin in red blood cells greatly enhances the oxygen-carrying capacity of blood.

Most carbon dioxide is transported in the blood in the form of bicarbonate. Nearly all of the bicarbonate forms through an enzyme-mediated reaction in red blood cells.

Through neural controls and local controls (as in the lungs), the rates of air flow and blood flow can be adjusted in ways that enhance gas exchange.

WHEN THE LUNGS BREAK DOWN

In large cities, in certain workplaces, even near a cigarette smoker, airborne particles and certain gases are present in abnormal amounts, and they put extra workloads on the respiratory system.

Bronchitis   Ciliated, mucous epithelium lines your bronchioles (Figure 29.15). It is one of the built-in defenses that protect you from respiratory infections. Toxins in cigarette smoke and other airborne pollutants irritate this lining and may lead to *bronchitis*. With this respiratory ailment, excess mucus is secreted when epithelial cells that line the airways become irritated. As mucus accumulates, so do bacteria and other particles stuck in it. Coughing brings up some of the gunk. If the irritation persists, so does coughing. Initial attacks of bronchitis are treatable. When the aggravation is allowed to continue, bronchioles become inflamed. Bacteria, chemical agents, or both attack cells of the bronchiole walls. Ciliated cells are destroyed, and mucus-secreting cells multiply. Fibrous scar tissue forms; in time it may narrow or obstruct the airways.

Emphysema   When bronchitis persists, thick mucus clogs the airways. Tissue-destroying bacterial enzymes attack the thin, stretchable walls of alveoli. As the walls break down, inelastic fibrous tissue surrounds them. Gas exchange now proceeds at fewer alveoli, which become enlarged. In time, the balance between air flow and blood flow is permanently compromised because the lungs remain distended and inelastic.

By comparing Figure 29.16*a* and *b*, you may get a sense of what goes on. Running, walking, even exhaling become difficult. These are symptoms of *emphysema*, a respiratory ailment that affects about 1.3 million people in the United States alone.

A few people are genetically predisposed to develop emphysema. They don't have a workable gene for an enzyme, antitrypsin, that can inhibit bacterial attacks on alveoli. Poor diet and chronic (persistent or recurring) colds and other respiratory infections also make people susceptible to emphysema later in life. *But smoking is the major cause of the disease.* Emphysema can develop slowly, over twenty or thirty years. When detected too late, the damage to lung tissue cannot be repaired.

Effects of Cigarette Smoke   Worldwide, smoking now kills 3 million people every year. If current patterns do not change, then by the time today's young smokers reach middle age, 10 million may be dying annually because of the habit. That is one person every three seconds. Even now in the United States, one of 50 million cigarette-puffing people dies of emphysema, chronic bronchitis, or heart disease every thirteen seconds. For every eight of them, one nonsmoker dies of ailments brought on by *secondhand smoke*—of prolonged exposure to tobacco smoke in the

Figure 29.15   Scanning electron micrograph of cilia (*gold*) and mucus-secreting cells (*rust-colored*) in the respiratory tract.

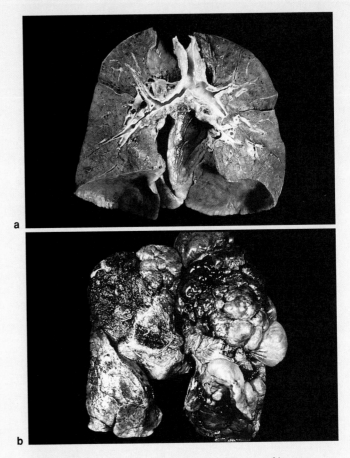

Figure 29.16   (**a**) Normal appearance of tissues of human lungs. (**b**) Lungs from someone affected by emphysema.

surroundings. And how much thought is given to children who breathe secondhand smoke? Parents and others who otherwise care about children are indirectly causing them to suffer more allergies and lung ailments. Smoking is the major cause of lung cancer, which is now the leading cause of death among women. Yet every day, 3,000 to 5,000 Americans light a cigarette for the first time. Children spend a billion dollars a year on cigarettes. The economy shrinks by 22 billion a year because of direct medical costs of treating smoke-induced respiratory disorders.

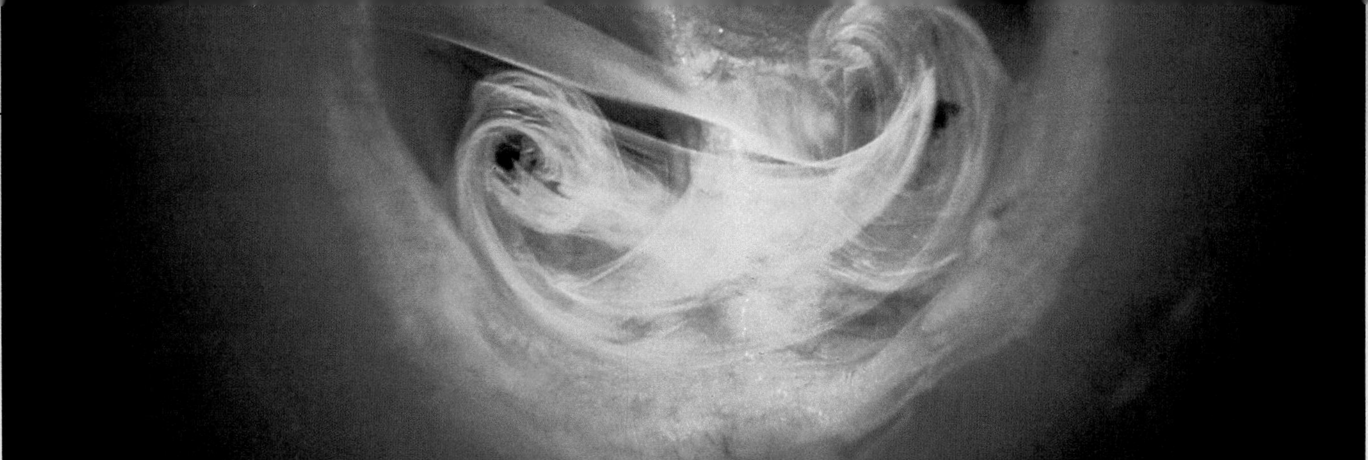

Risks Associated with Smoking:	Reducing the Risks by Quitting:
SHORTENED LIFE EXPECTANCY: Nonsmokers live 8.3 years longer on average than those who smoke two packs daily from the midtwenties on.	Cumulative risk reduction; after 10 to 15 years, life expectancy of ex-smokers approaches that of nonsmokers.
CHRONIC BRONCHITIS, EMPHYSEMA: Smokers have 4–25 times more risk of dying from these diseases than do nonsmokers.	Greater chance of improving lung function and slowing down rate of deterioration.
LUNG CANCER: Cigarette smoking is the major cause of lung cancer.	After 10 to 15 years, risk approaches that of nonsmokers.
CANCER OF MOUTH: 3–10 times greater risk among smokers.	After 10 to 15 years, risk is reduced to that of nonsmokers.
CANCER OF LARYNX: 2.9–17.7 times more frequent among smokers.	After 10 years, risk is reduced to that of nonsmokers.
CANCER OF ESOPHAGUS: 2–9 times greater risk of dying from this.	Risk proportional to amount smoked; quitting should reduce it.
CANCER OF PANCREAS: 2–5 times greater risk of dying from this.	Risk proportional to amount smoked; quitting should reduce it.
CANCER OF BLADDER: 7–10 times greater risk for smokers.	Risk decreases gradually over 7 years to that of nonsmokers.
CORONARY HEART DISEASE: Cigarette smoking is a major contributing factor.	Risk drops sharply after a year; after 10 years, risk reduced to that of nonsmokers.
EFFECTS ON OFFSPRING: Women who smoke during pregnancy have more stillbirths, and weight of liveborns averages less (hence, babies are more vulnerable to disease, death).	When smoking stops before fourth month of pregnancy, risk of stillbirth and lower birthweight eliminated.
IMPAIRED IMMUNE SYSTEM FUNCTION: Increase in allergic responses, destruction of defensive cells (macrophages) in respiratory tract.	Avoidable by not smoking.
BONE HEALING: Evidence suggests that surgically cut or broken bones require up to 30 percent longer to heal in smokers, possibly because smoking depletes the body of vitamin C and reduces the amount of oxygen reaching body tissues. Reduced vitamin C and reduced oxygen interfere with production of collagen fibers, a key component of bone. Research in this area is continuing.	Avoidable by not smoking.

Figure 29.17 From the American Cancer Society, a list of the risks incurred by smoking and the benefits of quitting. The photograph shows swirls of cigarette smoke poised at the entrance to the two bronchi that lead into the lungs.

Consider how cigarette smoke leads to lung damage. Noxious particles in the smoke from just one cigarette immobilize cilia in the bronchioles for several hours. The particles also stimulate mucus secretions that in time clog the airways. They can kill infection-fighting macrophages in the respiratory tract. What starts as "smoker's cough" can end in bronchitis and emphysema.

Or consider how cigarette smoke contributes to lung cancer. Within the body, certain compounds in coal tar and cigarette smoke become converted to highly reactive intermediates. These are the carcinogens; they provoke uncontrolled cell divisions in lung tissues. On the average, 90 of every 100 smokers who develop lung cancer will die

from it. If you now smoke, or if you are thinking about starting or quitting, you may wish to give serious thought to the information presented in Figure 29.17.

Effects of Marijuana Smoke In 1994, 17 million Americans smoked *pot*—marijuana (*Cannabis*)—at least once to induce light-headed euphoria. The number is rising, especially in junior high and high schools. About 1.5 million are chronic users; euphoria fades to apathy, depression, and fatigue, so they keep smoking just to keep from feeling "wasted." Besides psychological dependency, chronic use leads to throat irritation, persistent coughing, bronchitis, and emphysema.

SUMMARY

1. Aerobic respiration is the main metabolic pathway that provides enough energy for active life-styles. It uses oxygen and produces carbon dioxide. The process by which the animal body as a whole acquires oxygen and disposes of carbon dioxide is called respiration.

2. Air is a mixture of oxygen, carbon dioxide, and other gases, each exerting a partial pressure. Each gas tends to move from areas of higher to lower partial pressure. Respiratory systems make use of this tendency.

3. In respiratory systems, oxygen and carbon dioxide diffuse across a respiratory surface, a moist, thin layer of epithelium that is interposed between the external and internal environments.

4. Modes of respiration differ among animal groups.

a. Small invertebrates or those with a small body mass use integumentary exchange (oxygen and carbon dioxide diffuse across the body surface). This mode also persists in some large animals, including amphibians.

b. Many marine and some freshwater invertebrates have gills, respiratory organs with moist, thin, often highly folded walls. Most insects and certain spiders use tracheal respiration. Gases flow through open-ended tubes from the body surface directly to tissues. Most spiders have book lungs, with leaflike folds.

c. Fish have greater energy requirements than small invertebrates. They have gills in which a countercurrent flow mechanism compensates for low oxygen levels in aquatic habitats. Paired lungs are the dominant means of respiration in reptiles, birds, and mammals.

5. The airways of the human respiratory system are the nasal cavities, pharynx, larynx, trachea, bronchi, and bronchioles. Many millions of alveoli at the end of the terminal bronchioles are the main sites of gas exchange.

6. Breathing ventilates the lungs. During inhalation, the chest cavity expands, lung pressure decreases below atmospheric pressure, and air flows into the lungs. The events are reversed during normal exhalation.

7. Driven by its partial pressure gradient, oxygen in the lungs diffuses from alveolar air spaces into the lung capillaries. Then it diffuses into red blood cells and binds weakly with hemoglobin. In tissues where cells are metabolically active, hemoglobin gives up oxygen, which diffuses out of the capillaries, across interstitial fluid, and into the cells.

8. Driven by its partial pressure gradient, the carbon dioxide in tissues diffuses from cells, across interstitial fluid, and into the bloodstream. Most reacts with water molecules to form bicarbonate. Then the reactions are reversed in the lungs. There, carbon dioxide diffuses from lung capillaries into the air spaces of the alveoli, then is exhaled.

Review Questions

1. A few of your friends who have not taken a biology course ask you what insect lungs look like. What is your answer (assuming your instructor is listening)? *499*

2. Describe the features of the respiratory surface that are common to all respiratory systems. *498*

3. Distinguish between:
   a. aerobic respiration and respiration *497*
   b. pharynx and larynx *502*
   c. bronchiole and bronchus *502, 503*
   d. pleural sac and alveolar sac *503*

4. Explain why humans (Figure 29.18) can't survive on their own, for very long, under water. *502*

Figure 29.18 Human female at play in the domain of aquatic animals.

5. Label the components of the human respiratory system and the structures that enclose it. *502*

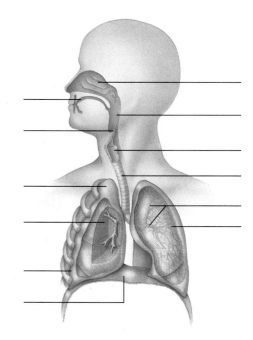

1. A partial pressure gradient of oxygen exists between _____ .
   a. air and lungs
   b. lungs and metabolically active tissues
   c. air at sea level and air at high altitudes
   d. all of the above

2. The _____ is an airway that connects the nose and mouth with the _____ .
   a. oral cavity; larynx   c. trachea; pharynx
   b. pharynx; trachea   d. pharynx; larynx

3. In humans, oxygen in the air must diffuse across _____ as it follows its partial pressure gradient into the internal environment.
   a. pleural sacs   c. a moist respiratory surface
   b. alveolar sacs   d. both b and c

4. Each human lung encloses a _____ .
   a. diaphragm   c. pleural sac
   b. bronchial tree   d. both b and c

5. Gas exchange occurs at the _____ .
   a. two bronchi   c. alveolar sacs
   b. pleural sacs   d. both b and c

6. Breathing _____ .
   a. ventilates the lungs
   b. draws air into airways
   c. expels air from airways
   d. causes reversals in pressure gradients
   e. all of the above

7. When you inhale, the diaphragm _____ .
   a. curves upward   c. relaxes
   b. flattens   d. remains stationary

8. After oxygen diffuses into lung capillaries, it diffuses into _____ and binds with _____ .
   a. interstitial fluid; red blood cells
   b. interstitial fluid; carbon dioxide
   c. red blood cells; hemoglobin
   d. red blood cells; carbon dioxide

9. Most carbon dioxide in blood is in the form of _____ .
   a. carbon dioxide   c. carbonic acid
   b. carbon monoxide   d. bicarbonate

10. Match the components with their descriptions.
   _____ trachea   a. airway leading into a lung
   _____ pharynx   b. throat
   _____ alveolus   c. fine bronchial tree branching
   _____ hemoglobin   d. windpipe
   _____ bronchus   e. respiratory pigment
   _____ bronchiole   f. site of gas exchange

Critical Thinking

1. Some cigarette manufacturers have conducted public relations campaigns urging their customers to "smoke responsibly." What are some social and biological issues in this controversy? How do these issues apply to the nonsmoking spouse or children of a smoker? To nonsmoking patrons in a restaurant? To the unborn child of a pregnant smoker? In your opinion, what behavior would constitute "responsible smoking"?

2. People occasionally poison themselves with carbon monoxide by building a charcoal fire in an enclosed area. Assuming that help arrives in time, what would be the best treatment: placing the victim outdoors in fresh air or rapidly administering pure oxygen? Explain how you arrived at your answer.

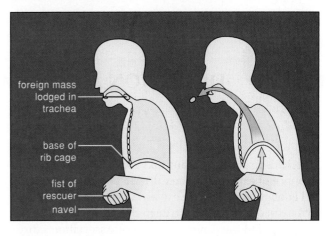

Figure 29.19 Heimlich maneuver to dislodge food stuck in the trachea. Stand behind the victim, make a fist with one hand, then position the fist, thumb-side in, against the victim's abdomen. The fist must be slightly above the navel and well below the rib cage. Next, press the fist into the abdomen with a sudden upward thrust. Repeat the thrust several times if needed. The maneuver can be performed on someone who is standing, sitting, or lying down.

3. When you swallow food, muscle contractions force the epiglottis down, to its closed position. This prevents food from going down the trachea and blocking gas flow. Yet each year, several thousand choke to death after food enters the trachea and blocks air flow for as little as four or five minutes. The Heimlich maneuver—*which is an emergency procedure only*—often can dislodge food from the esophagus (Figure 29.19). When correctly performed, it elevates the diaphragm forcibly, causing a sharp decrease in the chest cavity volume and a sudden increase in alveolar pressure. Air forced up the trachea by the increased pressure may be enough to dislodge the obstruction. Once the obstacle is dislodged, a physician must see the person at once. An inexperienced rescuer can inadvertently cause internal injuries or crack a rib.

Think about the current social climate in which even frivolous lawsuits abound. Would you risk performing the Heimlich maneuver to save the life of a relative? Of a stranger? Why or why not?

Selected Key Terms

alveolus *503*	larynx *502*
bronchiole *503*	lung *500*
bronchus *502*	oxyhemoglobin *504*
carbonic anhydrase *504*	pharynx *502*
countercurrent flow *500*	respiration *497*
diaphragm *503*	respiratory surface *498*
epiglottis *502*	respiratory system *497*
gill *499*	trachea *502*
glottis *501*	tracheal respiration *499*
hemoglobin *498*	vocal cord *501*
integumentary exchange *499*	

Readings

American Cancer Society. *Dangers of Smoking, Benefits of Quitting, and Relative Risks of Reduced Exposure.* Revised edition. New York: American Cancer Society.

Vander, A., J. Sherman, and D. Luciano. 1990. *Human Physiology: The Mechanisms of Body Function.* Fifth edition. New York: McGraw-Hill. Clear introduction to the respiratory system.

West, J. 1994. *Respiratory Physiology: The Essentials.* Fifth edition. Baltimore: Williams & Wilkins. Paperback.

30 DIGESTION AND HUMAN NUTRITION

Lose It—And It Finds Its Way Back

America's fixation on Beautiful People who have nary an ounce of extra fat on their utterly perfect selves is bad enough. After all, we can always rationalize our own extra ounce with the truism that beauty is only skin deep. But the American Medical Association's dire warnings of the Fat Connection to atherosclerosis, heart attacks, strokes, colon cancer, and other bad ailments is beyond rationalization. By its current standards, the proportion of body fat relative to total tissue mass should be 18 to 24 percent for a human female who is less than thirty years old. For a male, it should be no more than 12 to 18 percent. Yet an estimated 34 million Americans don't even come close to the standards.

What's a person to do? No matter how much we might lose by dieting, the lost weight just seems to find its way back. This physiological dilemma is an outcome of our evolutionary heritage. Like most other mammals, we have a goodly amount of adipose tissue with great numbers of fat-storing cells. Collectively, the cells are an adaptation for survival. They serve as an energy warehouse that may help the individual through times when food simply is not available. Once a fat-storing cell forms, fluctuations in food intake affect how empty or full it gets—but the cell itself is in the body to stay.

Dieting starts to empty the fat warehouse. The brain interprets this as "starvation" and triggers a slowdown in metabolism. Now the body uses energy far more efficiently, even for the basics—breathing, circulating blood, digesting food, keeping warm or cool enough, and so on. Said another way, it requires less food to do the same things!

As you've probably heard, dieting should go hand in hand with exercise. But skeletal muscles also adapt to "starvation" and burn less energy than before, so the going is slow. Jog for four hours or play tennis for eight hours straight, and you may only lose a pound of fat.

Meanwhile, the appetite surges. If and when the diet ends, "starved" fat cells quickly refill. In 1995, after ten years of careful study, researchers determined that lost weight *will* find its way back unless individuals eat less and exercise moderately throughout their lifetime.

Intriguingly, the researchers also found that a weight gain triggers an *increase* in metabolism. The body uses 15 to 20 percent more energy than before—until weight drops back to what it was before. Apparently, body fat tends to be maintained at a constant level that is, to a large extent, genetically prescribed for each individual. Researchers are now pinpointing *which* genes make this happen, as you will read at the chapter's end.

Figure 30.1 Mirror, mirror, on the wall, who is fairest of them all? (And by whose standards?)

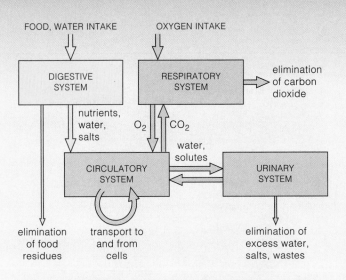

FOOD, WATER INTAKE   OXYGEN INTAKE

DIGESTIVE SYSTEM

RESPIRATORY SYSTEM → elimination of carbon dioxide

nutrients, water, salts

O_2 CO_2

water, solutes

CIRCULATORY SYSTEM ⇄ URINARY SYSTEM

elimination of food residues   transport to and from cells   elimination of excess water, salts, wastes

Figure 30.2 Links between the digestive, respiratory, circulatory, and urinary systems. These organ systems work together to supply cells with raw materials and eliminate wastes.

We still don't fully understand the emotional aspects of weight gains and losses. You have only to consider your reaction to the two gentlemen in Figure 30.1 to know this is so. You have only to consider *anorexia nervosa*, a potentially fatal eating disorder that is based on abnormal perception of body weight. Overwhelming fear of being fat *and* hungry leads some people to starve themselves and often to overexercise. Some anorexics fear growing up or maturing sexually. Others impose irrational expectations on themselves. Is this a rare disorder? No. Anorexia nervosa is especially common among women in their teens and early twenties.

Finally, consider that at least 20 percent of college-age women suffer to varying degrees from *bulimia*, an out-of-control, "oxlike appetite." Some start the routine because it seems like an easy way to lose weight. Others suffer severe emotional stress. In an hour-long eating binge, a bulimic may take in 50,000 kilocalories or more —then vomit or use laxatives to purge the body of the ingested food. They may not even like to eat, but the purging relieves them of anger and frustration. Some bulimics binge-purge once a month. Others do this several times a day. Do they know that the repeated purgings can damage the gut? Or that chronic vomiting brings gastric fluid into the mouth and so erodes teeth to stubs? Or that, at its extreme, bulimia causes the stomach to rupture, and the heart or kidneys to fail?

And with these sobering thoughts in mind, we start a tour of **nutrition**. The word encompasses processes by which an animal ingests and digests food, absorbs the released nutrients for later conversion to the body's own carbohydrates, lipids, proteins, and nucleic acids. The processes proceed at the **digestive system**. This body tube or cavity reduces food to particles, then to molecules small enough to be absorbed into the internal environment, and it eliminates the unabsorbed residues. Other organ systems, especially those shown in Figure 30.2, contribute to the nutritional processes.

KEY CONCEPTS

1. Interactions among the digestive, circulatory, respiratory, and urinary systems supply the body's cells with raw materials, dispose of wastes, and maintain the volume and composition of extracellular fluid.

2. Most digestive systems have specialized parts for food transport, processing, and storage. Different parts mechanically break apart and chemically break down food, absorb breakdown products, and eliminate the unabsorbed residues.

3. To maintain an acceptable body weight and overall health, energy intake must balance energy output (by way of metabolic activity, physical exertion, and so on). Complex carbohydrates provide the most glucose, which typically is the body's main source of immediately usable energy.

4. Nutrition involves the intake of vitamins, minerals, and certain amino acids and fatty acids that the body itself cannot produce.

Incomplete and Complete Systems

Recall, from Chapter 20, that we don't see many organ systems until we get to the flatworms. These are among the invertebrates that have an **incomplete digestive system**—a saclike, branching gut cavity with a single opening at the start of a pharynx (Figure 30.3a). After entering the sac, food is partly digested and circulated to cells even as residues are being sent back out.

Two-way traffic through a single opening satisfies the nutritional requirements of flatworms, sponges, cnidarians, and a few other obscure invertebrates that are not what you would call structurally complicated. It would not work well enough for other animals, which have a **complete digestive system**. This is a tube with an opening at one end (a mouth) for food intake and an opening at the other end (an anus) for eliminating unabsorbed residues. In between the openings, the tube itself is subdivided into regions that are specialized for the one-way transport, processing, and storage of raw materials.

As an example, think about the complete digestive system of a frog (Figure 30.3b). In between the frog's mouth and anus are a pharynx, stomach, and small and large intestines. Functionally connected with that small intestine are a liver, gallbladder, and pancreas, which are organs with accessory roles in digestion. Similarly, the bird shown in Figure 30.3c also has a complete digestive system with regional specializations.

Regardless of its complexity, a complete digestive system carries out five overall tasks:

1. **Mechanical processing and motility.** Movements that break up, mix, and propel food material.

2. **Secretion.** Release of digestive enzymes and other substances into the space inside the tube.

3. **Digestion.** Breakdown of food into particles, then into nutrient molecules small enough to be absorbed.

4. **Absorption.** Passage of digested nutrients and fluid across the tube wall and into body fluids.

5. **Elimination.** Expulsion of the undigested and unabsorbed residues from the end of the gut.

Remember, a digestive system does not act alone. In most animals, a circulatory system distributes absorbed nutrients to cells throughout the body. A respiratory system helps cells use the nutrients by supplying them with oxygen (for aerobic respiration) and taking away carbon dioxide wastes. A urinary system counters the variations in the kinds and amounts of absorbed nutrients. It helps maintain the composition and volume of the internal environment (Figure 30.2).

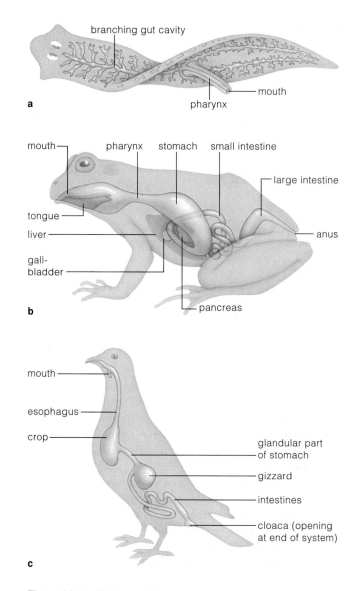

Figure 30.3 (a) Incomplete digestive system of a flatworm, with its two-way traffic of food and undigested material through one opening. (b,c) Two examples of a complete digestive system— a tube with regional specializations and an opening at each end.

Correlations With Feeding Behavior

The specializations of any digestive system correlate with feeding behavior. Consider the pigeon (Figure 30.3c). A long tube (esophagus) leads from its bill, a food-*gathering* region, to a compact food-*processing* region centered in the body mass. This adaptation is favorable for flight. As for other seed-eating birds, a large *crop*, a stretchable storage organ, balloons out from the esophagus. It allows the bird to zoom down on food, eat a lot, and maybe make a quick getaway, before predators become aware of it. Like other birds active during the day, it fills the crop before the sun goes down, then releases food during the first few

a

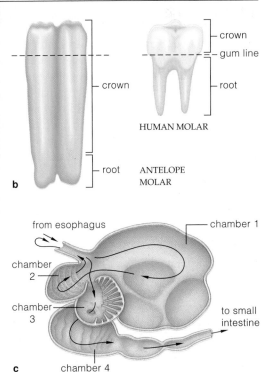

b

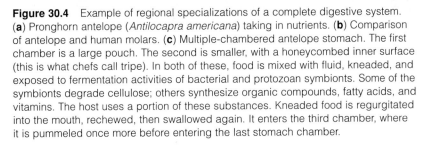

c

Figure 30.4 Example of regional specializations of a complete digestive system. (a) Pronghorn antelope (*Antilocapra americana*) taking in nutrients. (b) Comparison of antelope and human molars. (c) Multiple-chambered antelope stomach. The first chamber is a large pouch. The second is smaller, with a honeycombed inner surface (this is what chefs call tripe). In both of these, food is mixed with fluid, kneaded, and exposed to fermentation activities of bacterial and protozoan symbionts. Some of the symbionts degrade cellulose; others synthesize organic compounds, fatty acids, and vitamins. The host uses a portion of these substances. Kneaded food is regurgitated into the mouth, rechewed, then swallowed again. It enters the third chamber, where it is pummeled once more before entering the last stomach chamber.

hours after dark. This shrinks the length of overnight fasting. The first part of its stomach has a glandular lining that secretes enzymes and other substances that aid in digestion. The second part, a *gizzard*, is a muscular organ that grinds food much as teeth and jaws do. The intestines are proportionally shorter than they are in ducks, ostriches, and other birds that feed on plant parts rich in tough, fibrous cellulose—which requires a much longer processing time.

We find a similar correlation among pronghorn antelope (*Antilocapra americana*). Fall through winter, along the mountain ridges from central Canada down into northern Mexico, these animals browse on wild sage. Come spring, they move to open grasslands and deserts, there to browse on new growth (Figure 30.4). Coyotes, bobcats, and golden eagles prey on the young, so while antelope are eating, they keep an eye out for danger. And can they eat and run! With bursts of speed that can reach 95 kilometers per hour, they can leave predators in the dust.

Now think of your own cheek teeth—molars—each with a flattened crown that acts as a grinding platform. The crown of an antelope's molars dwarfs yours (Figure 30.4b). Why the difference? You probably do not rub your mouth against the ground while eating. But an antelope does, and abrasive bits of soil enter its mouth along with tough plant material. Its teeth wear down more rapidly than yours do, and natural selection has favored more crown to wear down.

Antelopes are **ruminants**, a type of hoofed mammal with multiple stomach chambers in which cellulose is

slowly broken down. The breakdown gets under way in the first two of four stomach sacs (Figure 30.4c). There, bacterial symbionts produce cellulose-digesting enzymes, to the antelope's benefit as well as their own. As enzyme action proceeds, the antelope regurgitates, then rechews the contents of the first two stomach sacs, then swallows again. (This is what "chewing cud" means.) Pummeling the plant material more than once exposes more surface area to the enzymes and gives them more time to act.

Thus the elaborate stomach of ruminants accepts a steady flow of plant material during lengthy feeding times, then slowly liberates nutrients during times of rest. Predatory or scavenging mammals have different specializations. They gorge on food when they can get it, then may not eat again for some time. Part of their digestive system stores food being gulped down too fast to be digested and absorbed. Other, accessory parts of the digestive system help assure that there will be an adequate distribution of nutrients between meals.

An incomplete digestive system is a saclike body cavity. Food and residues enter and leave through a single opening. A complete digestive system is a tube with two openings (mouth and anus) and regional specializations in between.

Complete digestive systems carry out the following tasks in controlled ways: Movements that break up, mix, and propel food material; secretion of digestive enzymes and other substances; breakdown of food; absorption of nutrients; and expulsion of undigested and unabsorbed residues.

30.2 OVERVIEW OF THE HUMAN DIGESTIVE SYSTEM

Let's look at your own body as an example of the kinds of organs in a complete digestive system. The human digestive system is a tube with two openings and many specialized organs (Figure 30.5). Stretch out the tube of an adult, and it would extend 6.5 to 9 meters (21 to 30 feet). From beginning to end, mucus-coated epithelium lines all surfaces facing the lumen (the space within the tube). The thick, moist mucus protects the tube's wall and enhances diffusion across its inner lining. Substances advance in one direction, from the mouth through the pharynx, esophagus, and gastrointestinal tract (gut). The **gut** starts at the stomach and extends through the small intestine, large intestine (colon), and rectum, to the anus. The salivary glands, gallbladder, liver, and pancreas are accessory organs that secrete substances into different parts of the tube.

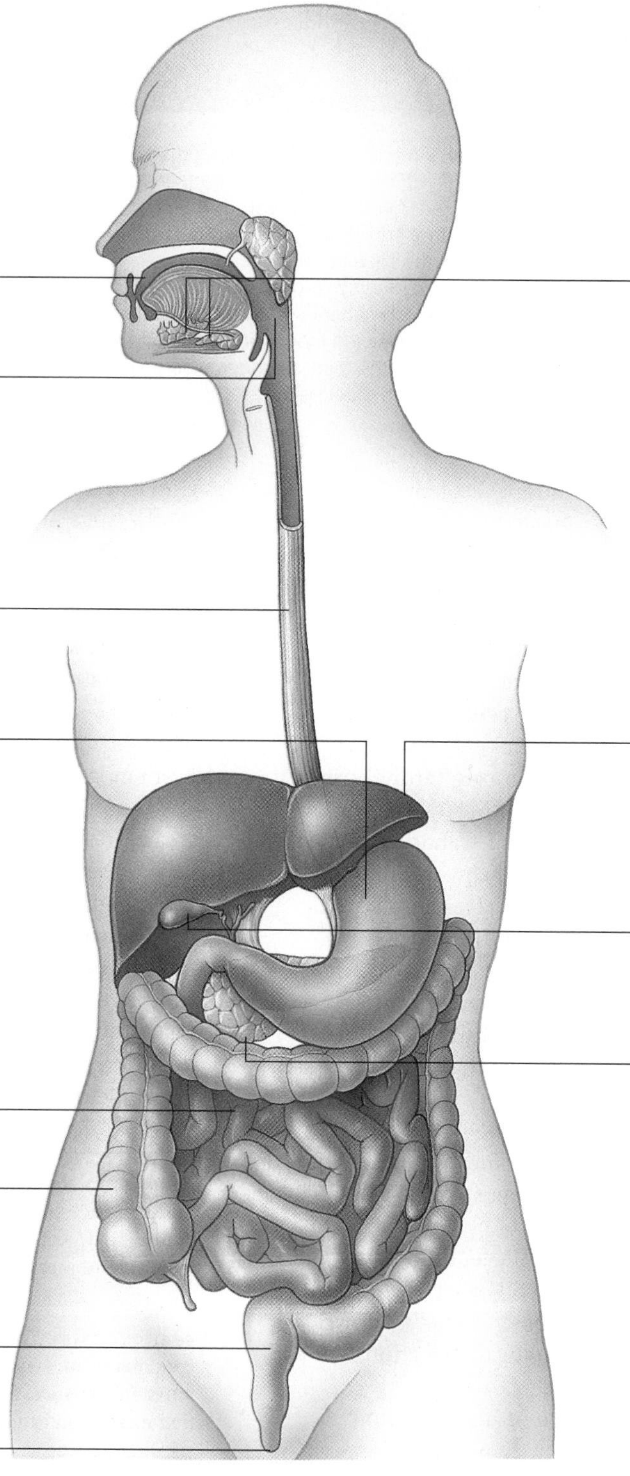

Major Components:

MOUTH (ORAL CAVITY)
Entrance to system; food is moistened and chewed; polysaccharide digestion starts.

PHARYNX
Entrance to tubular part of system (and to respiratory system); moves food forward by contracting sequentially.

ESOPHAGUS
Muscular, saliva-moistened tube that moves food from pharynx to stomach.

STOMACH
Muscular sac; stretches to store food taken in faster than can be processed; gastric fluid mixes with food and kills many pathogens; protein digestion starts.

SMALL INTESTINE
First part (duodenum, C-shaped, about 10 inches long) receives secretions from liver, gallbladder, and pancreas.
In second part (jejunum, about 3 feet long), most nutrients are digested and absorbed.
Third part (ileum, 6–7 feet long) absorbs some nutrients; delivers unabsorbed material to large intestine.

LARGE INTESTINE (COLON)
Concentrates and stores undigested matter by absorbing mineral ions, water; about 5 feet long: divided into ascending, transverse, and descending portions.

RECTUM
Distension stimulates expulsion of feces.

ANUS
End of system; terminal opening through which feces are expelled.

Accessory Organs:

SALIVARY GLANDS
Glands (three main pairs, many minor ones) that secrete saliva, a fluid with polysaccharide-digesting enzymes, buffers, and mucus (which moistens and lubricates food).

LIVER
Secretes bile (for emulsifying fat); roles in carbohydrate, fat, and protein metabolism.

GALLBLADDER
Stores and concentrates bile that the liver secretes.

PANCREAS
Secretes enzymes that break down all major food molecules; secretes buffers against HCl from the stomach.

Figure 30.5 Overview of the component parts of the human digestive system and their functions. The organs with accessory roles in digestion are also listed.

30.3 INTO THE MOUTH, DOWN THE TUBE

Food is chewed and polysaccharide breakdown begins in the **oral cavity**, or **mouth**. Thirty-two teeth typically project into an adult's mouth (Figure 30.6). Each **tooth** has an enamel coat (hardened calcium deposits), dentin (a thick, bonelike layer), and an inner pulp (with nerves and blood vessels). It is an engineering marvel, able to withstand years of chemical insults and mechanical stress. Recall, from Section 21.9, that the chisel-shaped incisors shear off chunks of food. Cone-shaped canines tear it. Premolars and molars, with broad crowns and rounded cusps, are good at grinding and crushing food.

Also in the mouth is a **tongue**, an organ consisting of membrane-covered skeletal muscles that function in positioning food in the mouth, swallowing, and speech. On the tongue's surface are circular structures with taste buds embedded in their tissues (Figure 30.7). Each bud contains sensory receptors that can detect chemical

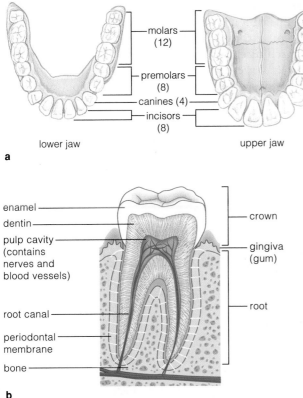

a

b

Figure 30.6 (**a**) Number and arrangement of human teeth. (**b**) Closer look at a molar's two main regions (the crown and root). Enamel caps the crown. It consists of calcium deposits and is the hardest substance in the body.

Normally harmless bacteria live on and between teeth. Daily flossing, gentle brushing, and avoidance of too many sweets help keep their populations in check. In the absence of such preventive measures, conditions favor bacterial infections. These may result in tooth decay (*caries*), inflamed gums (*gingivitis*), or both. Infections also can spread to the periodontal membrane, which anchors teeth to the jawbone. In *periodontal disease*, the infection slowly destroys the bone tissue around a tooth.

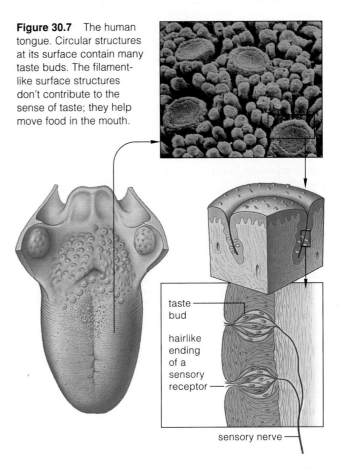

Figure 30.7 The human tongue. Circular structures at its surface contain many taste buds. The filament-like surface structures don't contribute to the sense of taste; they help move food in the mouth.

differences in dissolved substances. The brain uses this information to give rise to our sense of taste.

Chewing mixes food with **saliva**. This fluid contains an enzyme (salivary amylase), a buffer (bicarbonate, or HCO_3^-), mucins, and water. Salivary glands, beneath and in back of the tongue, produce and secrete saliva through ducts to the free surface of the mouth's lining. Salivary amylase breaks down starch. The HCO_3^- helps maintain the mouth's pH when you eat acidic foods. Mucins (modified proteins) help form the mucus that binds food into a softened, lubricated ball, or bolus.

When you swallow, contractions of tongue muscles force boluses into the **pharynx**, the tubular entrance to the esophagus *and* the trachea, an airway to the lungs. As food leaves the pharynx, the epiglottis (a flaplike valve) closes off the trachea and prevents breathing. That is why you normally don't choke on food (Figure 29.19). The **esophagus** connects the pharynx with the stomach. Contractions of its muscular wall propel food past a sphincter (a ring of muscles) into the stomach.

Through the action of the mouth's teeth and tongue, food gets chewed, mixed with saliva, and bound into soft, lubricated balls that will be propelled through tubes to the stomach. Enzymes in saliva start the digestion of polysaccharides.

We arrive now at the premier food-processing organs, the stomach and small intestine. Both have layers of smooth muscles, the contractions of which break apart, mix, and directionally move food (Figure 30.8). The lumen of both organs receives digestive enzymes and other secretions that play central roles in the chemical breakdown of food into fragments, then into nutrient molecules small enough to be absorbed. Table 30.1 lists the names and sources of the major digestive enzymes. Carbohydrate breakdown, recall, *starts* in the mouth. Protein breakdown *starts* in the stomach. But digestion of nearly all carbohydrates, lipids, proteins, and nucleic acids proceeds to completion in the small intestine.

The Stomach

The **stomach**, a muscular, stretchable sac (Figure 30.8*a*), serves three major functions. First, it mixes and stores ingested food. Second, its secretions help dissolve and degrade the food, proteins especially. Third, it helps control passage of food into the small intestine.

Glandular cells of the stomach's lining secrete two liters or so of hydrochloric acid (HCl), pepsinogens, mucus, and other substances every day. The substances form the **gastric fluid** (stomach fluid), H⁺ and Cl⁻, and this helps dissolve bits of food to form a liquid mixture called **chyme**. The acidity kills many pathogens (and causes *heartburn* when it backs up into the esophagus).

How does protein digestion start in the stomach? The high acidity alters the structure of proteins and exposes the peptide bonds. It also converts pepsinogens to active enzymes (pepsins) that cleave the bonds. The protein fragments accumulate in the stomach's lumen. Meanwhile, glandular cells of the stomach's lining are secreting gastrin. And *this* hormone stimulates the cells that secrete HCl and pepsinogen.

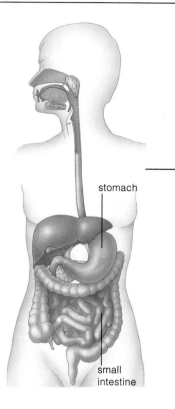

stomach

small intestine

A *peptic ulcer*, an eroded wall region in the stomach or small intestine, results from insufficient secretion of mucus and buffers (or too much pepsin). Heredity, chronic emotional stress, smoking, and excess intake of alcohol and aspirin may contribute to this disorder. Also, nearly everyone who has intestinal ulcers and 70 percent of those who have stomach ulcers are infected by *Helicobacter pylori*. A full course of antibiotic therapy can cure peptic ulcers in patients who test positive for this bacterium.

The stomach empties by waves of contraction and relaxation, of a type called peristalsis. The waves mix chyme and gain force as they approach a sphincter at the stomach's end (Figure 30.8*b*). Arrival of a strong contraction closes the sphincter, so most of the chyme is squeezed back. A small amount moves into the small intestine. In time, most moves out of the stomach.

Table 30.1 Major Digestive Enzymes and Their Breakdown Products				
Enzyme	Source	Where Active	Substrate	Main Breakdown Products*
Carbohydrate Digestion:				
Salivary amylase	Salivary glands	Mouth	Polysaccharides	Disaccharides
Pancreatic amylase	Pancreas	Small intestine	Polysaccharides	Disaccharides
Disaccharidases	Intestinal lining	Small intestine	Disaccharides	Monosaccharides (e.g., glucose)
Protein Digestion:				
Pepsins	Stomach lining	Stomach	Proteins	Protein fragments
Trypsin and chymotrypsin	Pancreas	Small intestine	Proteins	Protein fragments
Carboxypeptidase	Pancreas	Small intestine	Protein fragments	Amino acids
Aminopeptidase	Intestinal lining	Small intestine	Protein fragments	Amino acids
Fat Digestion:				
Lipase	Pancreas	Small intestine	Triglycerides	Free fatty acids, monoglycerides
Nucleic Acid Digestion:				
Pancreatic nucleases	Pancreas	Small intestine	DNA, RNA	Nucleotides
Intestinal nucleases	Intestinal lining	Small intestine	Nucleotides	Nucleotide bases, monosaccharides

*Yellow parts of table identify breakdown products that can be absorbed into the internal environment.

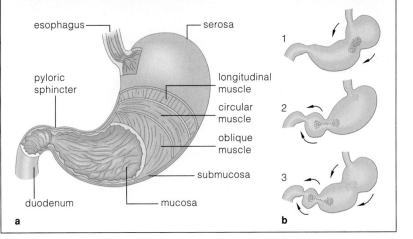

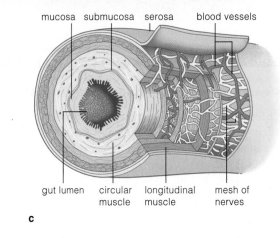

Figure 30.8 (**a**) Structure of the stomach. (**b**) Peristaltic wave down the stomach. (**c**) Structure of the human small intestine. Generally, the gut wall starts with the mucosa (an epithelium and underlying layer of connective tissue) that faces the lumen. The submucosa is a connective tissue layer with blood and lymph vessels and a mesh of nerves that function in local control over digestion. The wall has layers of smooth muscle that differ in orientation. The serosa is an outer layer of connective tissue.

The Small Intestine

The small intestine has three regions: the duodenum, jejunum, and ileum (Figure 30.5). Each day, the stomach as well as ducts from three organs, the **pancreas**, **liver**, and **gallbladder**, deliver about 9 liters of fluid to the duodenum. At least 95 percent of the fluid is absorbed across the epithelial lining of the small intestine.

Enzymes secreted from cells of the intestinal lining and from the pancreas digest food to monosaccharides (including glucose), free fatty acids, monoglycerides, amino acids, nucleotides, and nucleotide bases. For example, like pepsin secreted from cells in the stomach lining, two pancreatic enzymes (trypsin, chymotrypsin) digest protein molecules into peptide fragments, then another cleaves the fragments to free amino acids. The pancreas also secretes bicarbonate, a buffer that helps neutralize HCl arriving from the stomach.

The Role of Bile in Fat Digestion

Fat digestion depends on more than enzyme action. It depends also on **bile**, a fluid that the liver secretes on a continual basis. Bile contains bile salts, bile pigments, cholesterol, and lecithin (one of the phospholipids). When food is not traveling through the gut, a sphincter closes off the main bile duct from the liver. At such times, bile backs up into the saclike gallbladder, where it is stored and concentrated.

By a process called **emulsification**, bile salts speed up fat digestion. Most fats in our diet are triglycerides. Triglyceride molecules are insoluble in water, and they tend to aggregate into large fat globules in the chyme. As movements of the intestinal wall agitate chyme, fat globules break up into small droplets, which become coated with bile salts. Bile salts carry negative charges, so the coated droplets repel each other and so remain separated. This suspension of fat droplets, formed by mechanical and chemical action, is the "emulsion."

Compared to fat globules, emulsion droplets give fat-digesting enzymes a much greater surface area to act upon. Thus triglycerides can be broken down much more rapidly to fatty acids and monoglycerides.

Controls Over Digestion

Homeostatic controls, recall, work to counter changes in the internal environment. By contrast, controls over digestion act *before* food is absorbed into the internal environment. The nervous and endocrine systems, and a mesh of nerves in the gut wall, exert the control.

For example, incoming food distends the stomach wall. Signals from sensory receptors spread through the mesh of nerves and may reach the brain. Either way, they lead to muscle contractions in the gut wall or secretion of enzymes, hormones, and other substances into the gut lumen. Stomach emptying depends on the chyme's volume and composition. For example, a large meal activates more receptors in the stomach wall, so contractions get more forceful and emptying proceeds faster. High acidity or a high fat content in the small intestine calls for hormone secretions that trigger a slowdown in stomach emptying, so food is not moved faster than it can be processed. Fear, depression, and other emotional upsets also trigger such slowdowns.

Gastrointestinal hormones are part of the controls. If chyme contains amino acids and peptides, cells of the stomach lining will secrete the gastrin that stimulates secretion of acid into the stomach. Secretin prods the pancreas to secrete bicarbonate. CCK (cholecystokinin) enhances secretin's action and causes the gallbladder to contract. When the small intestine contains glucose and fat, GIP (glucose insulinotropic peptide) calls for insulin secretion, which stimulates cells to absorb glucose.

Carbohydrate breakdown starts in the mouth, and protein breakdown starts in the stomach. It is in the small intestine that most large organic compounds are digested to molecules small enough to be absorbed into the internal environment.

30.5 ABSORPTION IN THE SMALL INTESTINE

Structure Speaks Volumes About Function

Unlike the stomach, which can only absorb alcohol and a few other substances, the small intestine is the main site of absorption of the vast majority of nutrients. What makes it so?

Consider Figure 30.9, which shows the structure of the intestinal wall. First take a look at the exuberant folds of the mucosa, which project into the lumen. Then look closer and you see amazing numbers of even tinier projections from each one of the large folds. Look closer still and you see that the epithelial cells at the surface of these tiny projections have a brushlike crown of even tinier projections, all exposed to the intestinal lumen.

What's the significance of all this convolution? For the answer, think back on the description of the surface-to-volume ratio in Section 3.2. Collectively, all of the projections from the intestinal mucosa greatly, hugely, ENORMOUSLY increase the surface area that is available for interacting with chyme and for absorbing nutrients from it. Without this vast surface area, the absorption would proceed hundreds of times more slowly, which of course would not be enough to sustain human life.

As indicated in Figure 30.9a, the tiny absorptive structures on each fold of the intestinal mucosa are **villi** (singular, villus). Although each villus is only about a millimeter long, they number in the millions, and their very density gives the mucosa a velvety appearance. Small as they are, each villus houses an arteriole, a venule, and a lymph vessel (Figure 30.9d). These vessels function in the movement of substances to and from the general circulation.

At the free surface of the cells that make up the epithelial lining of each villus are **microvilli** (singular, microvillus). Each cell has about 1,700 of these ultrafine, threadlike projections of the plasma membrane. Hence its name, "brush border" cell. The ones near the tip of the villus produce some digestive enzymes, as listed in Table 30.1.

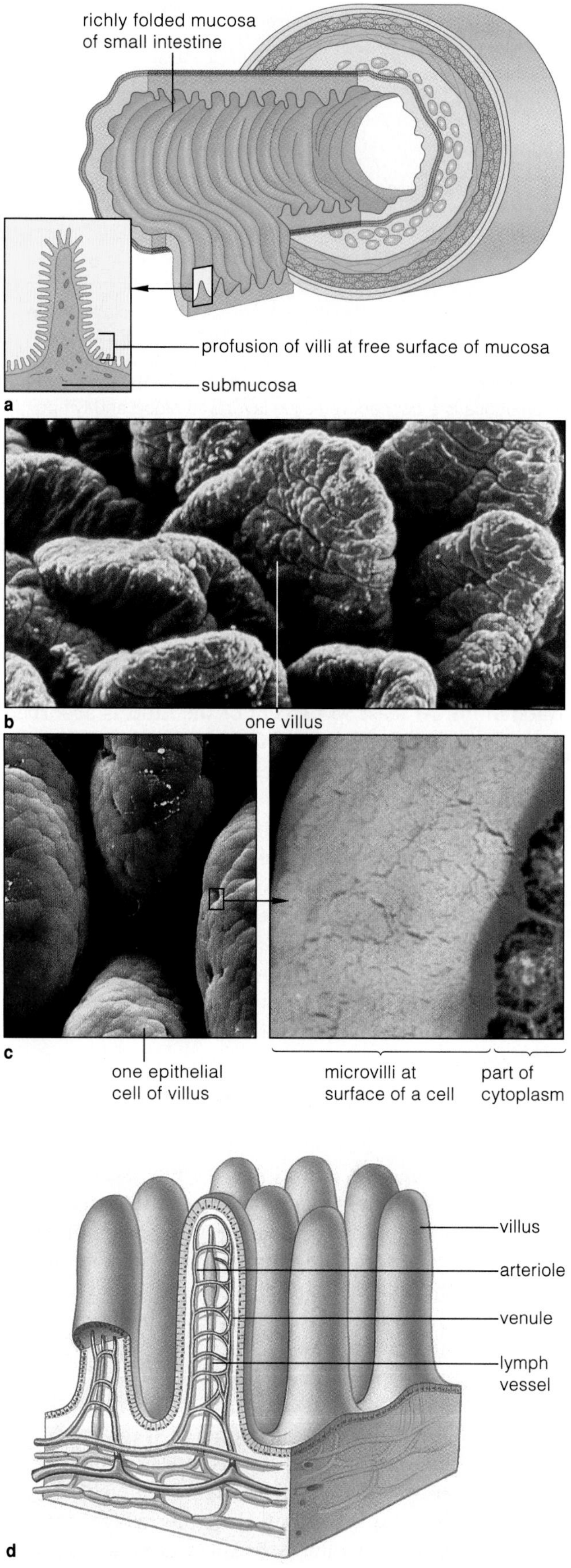

richly folded mucosa
of small intestine

profusion of villi at free surface of mucosa

submucosa

a

b   one villus

c   one epithelial cell of villus   microvilli at surface of a cell   part of cytoplasm

Figure 30.9 Surface structures of the mammalian small intestine at increasing magnifications.

(**a,b**) Surface view of the deep, permanent folds of the intestinal mucosa. Each fold bears a profusion of fingerlike absorptive structures called villi. (**c**) Individual epithelial cells at the free surface of a villus. Each has a dense crown of microvilli facing the intestinal lumen.

(**d**) Blood and lymph vessels in intestinal villi. Monosaccharides and most amino acids that cross the intestinal lining enter blood vessels. Fats enter lymph vessels.

villus

arteriole

venule

lymph vessel

d

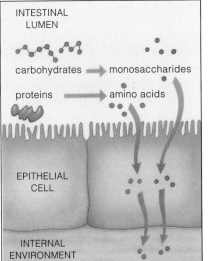

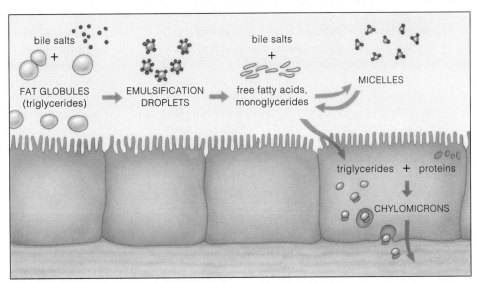

a Digestion of carbohydrates to monosaccharides, and proteins to amino acids; completed by enzymes from the pancreas and from brush border cells.

b Active transport of monosaccharides and amino acids across the plasma membrane of cells of intestinal lining, then out of the cells and into the internal environment.

c Emulsification. Intestinal wall movements break up fat globules into small, emulsification droplets. Bile salts prevent the globules from re-forming. Pancreatic enzymes digest the droplets to fatty acids and monoglycerides.

d Formation of micelles (as bile salts combine with digestion products and phospholipids). Products readily slip into and out of micelles.

e Concentration of monoglycerides and fatty acids in micelles enhances gradients that lead to their diffusion across lipid bilayer of plasma membrane of the lining's cells.

f Reassembly of fat digestion products into triglycerides in cells of intestinal lining. These become coated with proteins, then are expelled (by exocytosis) into the internal environment.

Figure 30.10 Digestion and absorption in the small intestine.

Mechanisms of Absorption

Absorption, recall, is the passage of nutrients, water, salts, and vitamins into the internal environment. The vast absorptive surface of the intestinal wall facilitates the process, but so does the action of smooth muscle in the intestinal wall. During *segmentation*, rings of circular muscle contract and relax repeatedly. This creates an oscillating (back and forth) movement that constantly mixes and forces the contents of the lumen against the wall's absorptive surface:

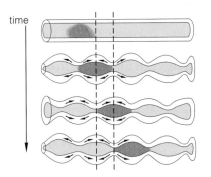

By the time food is halfway through the small intestine, most has been broken down and digested. Some breakdown products, including monosaccharides and amino acids, are absorbed in straightforward fashion. So are water and mineral ions, which cross by way of osmosis. Transport proteins in the plasma membrane of

brush border cells actively shunt these digestion products across the intestinal lining. By contrast, bile salts assist the absorption of fatty acids and monoglycerides. These products of fat digestion diffuse across the lipid bilayer of the plasma membrane (Figure 30.10).

By a process called **micelle formation**, bile salts combine with products of fat digestion, forming tiny droplets (micelles). Product molecules in the micelles continuously exchange places with product molecules that are dissolved in the chyme. When concentration gradients favor it, the molecules diffuse out of micelles, and into epithelial cells. There, fatty acids and monoglycerides recombine, thus forming triglycerides. Then the triglycerides combine with proteins into particles (chylomicrons) that leave the cells, by exocytosis, and enter the internal environment.

Once they have been absorbed, glucose and amino acids enter blood vessels. The triglycerides enter lymph vessels, which eventually drain into blood vessels.

With its richly folded mucosa, millions of villi, and hundreds of millions of microvilli, the small intestine offers a vast surface area for absorbing nutrients.

Substances pass through the brush border cells that line the free surface of each villus by active transport, osmosis, and diffusion across the lipid bilayer of plasma membranes.

DISPOSITION OF ABSORBED ORGANIC COMPOUNDS

Earlier in the book, in Section 6.6, you considered a few mechanisms that govern metabolism—specifically, the disposition of glucose and other organic compounds in the body as a whole. Figure 6.11 provided an example of the conversion pathways by which carbohydrates, fats, and proteins can be broken apart to molecules that can serve as intermediates in the ATP-producing pathway of aerobic respiration. Here, Figure 30.11 rounds out the picture by showing all of the major routes by

Between meals, the body taps into the fat stores. In adipose tissue, cells dismantle fats to glycerol and fatty acids and release these into blood. In the liver, cells break apart glycogen and release the glucose units, which also enter the blood. Body cells take up the fatty acids and glucose and use them for ATP production.

Bear in mind, the liver does more than store and interconvert organic compounds. For example, it helps maintain their concentrations in blood. It inactivates

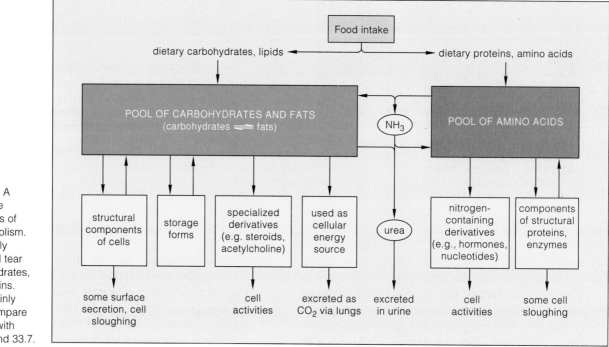

Figure 30.11 A summary of the major pathways of organic metabolism. Cells continually synthesize and tear down carbohydrates, fats, and proteins. Urea forms mainly in the liver. Compare this summary with Sections 6.6 and 33.7.

which organic compounds obtained from the diet are shuffled and reshuffled in the body as a whole.

At any time, all living cells in the body are breaking apart most of their own supplies of carbohydrates, lipids, and proteins, then using the breakdown products as energy sources or as building blocks. The nervous system and endocrine system work together to integrate this massive molecular turnover, although a treatment of how they do this would be beyond the scope of this book.

For now, simply consider a few key points about organic metabolism. When you eat, your body builds up pools of materials. Excess amounts of carbohydrates and other organic compounds absorbed from the gut are transformed mostly into fats, which become stored in adipose tissue. Some are converted to glycogen in the liver and in muscle tissue. *While* the compounds are being absorbed and stored, most cells use glucose as the main energy source. There is no net breakdown of protein in muscle or other tissues during this period.

most hormone molecules, which are sent to the kidneys for excretion, in urine. The liver also removes worn-out blood cells from the circulation and inactivates many potentially toxic compounds. For example, ammonia (NH_3), produced during amino acid breakdown, can be toxic in high concentrations. The liver converts NH_3 to urea. This is a less toxic waste product, and it leaves the body by way of the kidneys, in urine.

Right after a meal, the body's cells take up the glucose being absorbed from the gut and use it as an energy source. Excess amounts of glucose and other organic compounds become converted mainly to fats, which get stored in adipose tissue. Some of the excess is converted to glycogen, which gets stored mainly in the liver and in muscle tissue.

***Between* meals, the body taps the fat reservoirs as the main energy source. Fats get converted to glycerol and fatty acids, both of which can enter ATP-producing pathways.**

30.7 THE LARGE INTESTINE

What happens to the material *not* absorbed in the small intestine? It moves into the large intestine, or **colon**. The colon concentrates and stores feces: a mixture of water, undigested and unabsorbed matter, and bacteria. The mixture gets concentrated as water moves across the lining of the colon. The lining's cells actively transport sodium ions out of the lumen. As ion concentrations in the lumen drop, the water concentration increases. Thus water also moves out of the lumen, by osmosis.

The colon starts at a cup-shaped pouch known as the cecum. The **appendix** is a slender projection from that pouch (Figure 30.12). The appendix has no known digestive functions. It may have roles in defense against infections. The colon ascends on the right side of the abdominal cavity, continues across to the other side, then descends and connects with a short tube called the rectum (Figure 30.5). When feces distend the rectal wall, this triggers a reflex pathway that leads to expulsion of feces from the body. The nervous system controls the expulsion by stimulating or inhibiting the contraction of a muscle **sphincter** at the anus, the terminal opening of the gut.

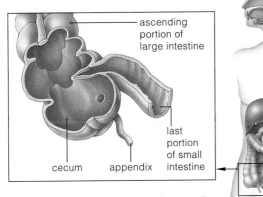

ascending portion of large intestine

last portion of small intestine

cecum   appendix

Figure 30.12 The cecum and appendix, the start of the large intestine (colon).

The volume of fiber and other undigested material that can't be decreased by absorption in the colon is called **bulk**. As bulk increases, it puts more pressure on the colon wall. Without adequate bulk, it takes longer for feces to move through the colon, with irritating effects. People who have a fiber-poor diet are more likely to suffer *appendicitis*, an inflammation of the appendix. If an infected appendix ruptures, normally harmless bacterial residents of the colon can enter the abdominal cavity and cause truly severe problems.

The large intestine (colon) functions in the absorption of water and mineral ions from the gut lumen. It also functions in the compaction of undigested residues into feces, for expulsion at the terminal opening of the gut.

30.8 Focus on Health

CANCER IN THE SYSTEM

We of the United States are one of the world's wealthiest and best-fed populations. Yet along with affluence, we have picked up some bad eating habits. We skip meals, eat too much and too fast when we do sit down at the table, and generally give our gut erratic workouts. Worse, our diet tends to be rich in sugar, cholesterol, and salt, and low in bulk. The problem with too little bulk comes from the longer transit time of feces through the colon. The longer this irritating, potentially carcinogenic material is in contact with the colon wall, the more damage it can do.

Telltale symptoms of *colon cancer* are a change in bowel functioning, rectal bleeding, and blood in the feces. Some individuals appear to have a hereditary predisposition to develop colon cancer, but a low-fiber diet puts any person at risk. Colon cancer is almost nonexistent in rural India and Africa, where people can't afford to eat much more than fiber-rich whole grains. When these people move to cities in more affluent nations, the incidence of colon cancer increases among them. Is diet a factor? Probably. Does a change in diet affect the diversity and distribution of the gut's bacteria? Possibly. Even emotional stress of life in complex societies compounds the problem.

Drink a lot of alcohol and eat a lot of smoked, pickled, and salted foods and you raise the risk of *stomach cancer*. Smoke and you do more than blacken your lungs. You set up conditions that favor *pancreatic cancer* and *esophageal cancer*.

Think about it. In 1993, cancer of the colon and rectum hit nearly 150,000 of us. Of those, 27,000 are already dead. And what about the 15,000 deaths from stomach cancers? The 7,600 from esophageal cancer? The 13,000 from liver cancer or 25,000 from pancreatic cancer?

The American Cancer Society tells us to eat plenty of vegetables, fruits, and fiber and to cut back on fatty foods. Eat more cabbage, kale, broccoli, cauliflower, and other cruciferous vegetables because they appear to contain cancer-fighting compounds. As described in Section 30.10, beta carotene and vitamins E and C may do the same.

Finally, put the questions to yourself. Have you looked upon a bowl of bran cereal with the same passion as you look upon, say, french fries and ice cream, hamburgers and chocolate mousse? Have you thought twice before lighting up or repeatedly getting drunk? Now ask again—knowing your gut depends on the answers.

You grow and maintain yourself by eating certain foods that are suitable sources of energy and raw materials. Nutritionists measure the energy in **kilocalories**. Each kilocalorie is 1,000 calories of heat energy (the amount needed to raise the temperature of 1 kilogram of water by 1°C). The raw materials will now be described.

New, Improved Food Pyramids

A few million years ago, our hominid ancestors dined mostly on fresh fruits and other fibrous plant material. From this nutritional beginning, humans in many parts of the world have become habituated to low-fiber, high-fat diets. And yet, from elementary school onward, you probably have encountered "food pyramids," or charts of presumably well-balanced diets. Nutritionists revise the charts on an ongoing basis to reflect the findings of additional research. Figure 30.13 shows the chart most recently favored. For an adult male of average size, the daily portions of different foods should provide the body with these proportions of organic compounds:

Complex carbohydrates:	58–60 percent
Proteins (less for females):	12–15 percent
Fats and other lipids:	20–25 percent

Carbohydrates

Starch and, to a lesser extent, glycogen should be the main carbohydrates in the human diet. These complex carbohydrates are readily broken apart into glucose units, which are the body's main energy source. Starch is abundant in fleshy fruits, cereal grains, and legumes, including beans and peas.

The foods that are rich in complex carbohydrates have an additional advantage—they typically are high in fiber. Section 30.8 highlighted the rather sobering consequences of a low-fiber diet. Epidemiologist Denis Burkitt may have been only half-joking when he put it this way: If you pass a small volume of feces, you have large hospitals.

The carbohydrates called simple sugars don't have the fiber of complex carbohydrates. Neither do they have the vitamins and minerals found in whole foods. Each week, the average American eats as much as 2 pounds of refined sugar (sucrose). You may think this a far-fetched statement, but start taking a closer look at the ingredients listed on the packages of cereal, frozen dinners, soft drinks, and other prepared foods. Many of the sugars are "hidden" as corn syrup, corn sweeteners, dextrose, and so on.

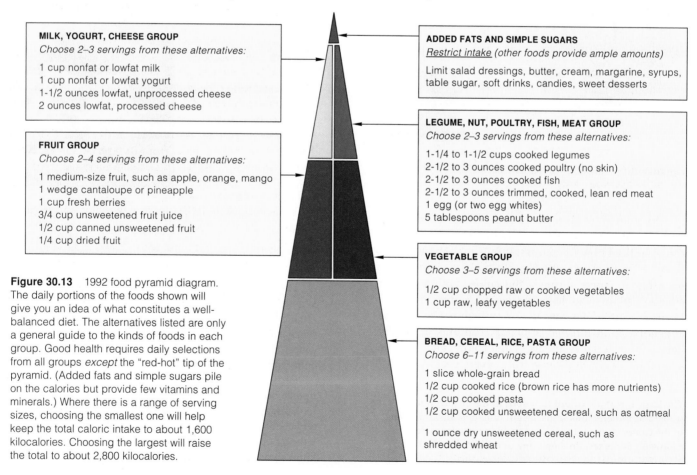

Figure 30.13 1992 food pyramid diagram. The daily portions of the foods shown will give you an idea of what constitutes a well-balanced diet. The alternatives listed are only a general guide to the kinds of foods in each group. Good health requires daily selections from all groups *except* the "red-hot" tip of the pyramid. (Added fats and simple sugars pile on the calories but provide few vitamins and minerals.) Where there is a range of serving sizes, choosing the smallest one will help keep the total caloric intake to about 1,600 kilocalories. Choosing the largest will raise the total to about 2,800 kilocalories.

No Limiting Amino Acid	Low in Lysine	Low in Methionine, Other Sulfur-Containing Amino Acids	Low in Tryptophan
legumes: soybean tofu soy milk cereal grains: wheat germ milk cheeses (except cream cheese) yogurt eggs meats	legumes: peanuts cereal grains: barley buckwheat corn meal oats rice rye wheat nuts, seeds: almonds cashews coconut English walnuts hazelnuts pecans pumpkin seeds sunflower seeds	legumes: beans (dried) black-eyed peas garbanzos lentils lima beans mung beans peanuts nuts: hazelnuts fresh vegetables: asparagus broccoli green peas mushrooms parsley potatoes soybeans Swiss chard	legumes: beans (dried) garbanzos lima beans mung beans peanuts cereal grains: corn meal nuts: almonds English walnuts fresh vegetables: corn green peas mushrooms Swiss chard

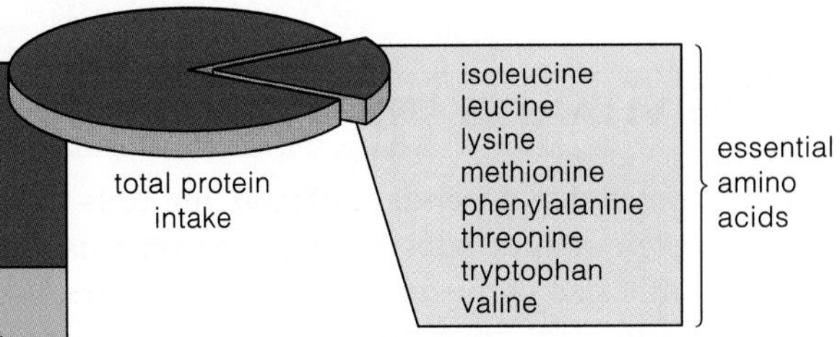

total protein intake

isoleucine
leucine
lysine
methionine
phenylalanine
threonine
tryptophan
valine

} essential amino acids

Figure 30.14 Essential amino acids—a small portion of the total protein intake. All eight must be available at the same time, in certain amounts, if cells are to build their own proteins. Milk and eggs have high amounts of all eight in proportions that humans require; they are among the complete proteins.

Nearly all plant proteins are incomplete, so vegetarians should construct their diet carefully to avoid protein deficiency. For example, they can combine different foods from the three columns of incomplete proteins shown to the left. Also, vegetarians who avoid dairy products and eggs should take vitamin B_{12} and vitamin B_2 (riboflavin) supplements. Animal protein is a luxury in most traditional societies. Yet their cuisines include good combinations of plant proteins, including rice/beans, chili/cornbread, tofu/rice, and lentils/wheat bread.

Lipids

The body can't function without fats and other lipids. For instance, the phospholipid lecithin is a required component of cell membranes. Fats serve as energy reserves, they cushion many organs, and they provide insulation beneath the skin. Dietary fats also help the body store fat-soluble vitamins. Yet the body can synthesize most of its own fats, including cholesterol, from protein and carbohydrates. The ones it cannot produce are called **essential fatty acids**, but whole foods can provide plenty of them. Linoleic acid is an example. One teaspoon a day of corn oil, olive oil, or some other polyunsaturated fat in food provides enough of it.

Currently, butter and other fats make up 40 percent of the average diet in the United States. Most of the medical community agrees that the proportion should be less than 30 percent.

Like other animal fats, butter is a saturated fat that tends to raise the level of cholesterol in the blood. As described in earlier chapters, cholesterol is used in the synthesis of bile acids and steroid hormones, and it is a required component of our cell membranes. However, too much cholesterol may interfere seriously with the circulation of blood through the body. Work is under way to develop edible but nondigestible oils, such as *sucrose polyester*, for those individuals who have great difficulty cutting back on the fats.

Proteins

The amino acid components of dietary proteins are required for the body's own protein-building programs.

Of the twenty common types, eight are known as the **essential amino acids**. Our body cells can't synthesize them, so we must obtain them from food. These eight amino acids are methionine (or its metabolic equivalent, cysteine), isoleucine, leucine, lysine, phenylalanine (or tyrosine), threonine, tryptophan, and valine.

Most animal proteins are "complete," meaning that their ratios of amino acids match human nutritional needs. Plant proteins are "incomplete," meaning they lack one or more of the essential amino acids. To get suitable amounts of amino acids, vegetarians should eat certain combinations of different plants (Figure 30.14).

A measure called **net protein utilization** (NPU) is used to compare proteins from different sources. NPU values range from 100 (all essential amino acids present in ideal proportions) to 0 (one or more absent; when eaten alone, the protein is not complete).

Because enzymes and other proteins are vital for the body's structure and function, protein-deficient diets may have severe consequences. Protein deficiency is most damaging to fetuses and infants, for the brain grows and develops rapidly early in life. Unless the mother-to-be eats adequate amounts of protein just before and just after birth, her newborn will be affected by irreversible mental retardation. Even mild protein starvation can retard growth, and it can affect mental and physical performance.

Individuals grow and maintain themselves in good health by eating certain amounts of complex carbohydrates, lipids, and proteins that provide them with adequate raw materials and energy, as measured in kilocalories.

Metabolic activity depends on small amounts of more than a dozen organic substances called **vitamins**. Most plants synthesize all of these substances. Animals generally have lost the ability to do so, and they must obtain vitamins from food.

At the minimum, human cells require the thirteen different vitamins listed in Table 30.2. Each vitamin has specific metabolic roles. Many reactions require several vitamins, so the absence of one affects the functions of others. Metabolism also depends on certain inorganic substances called **minerals**. For example, most of your cells use calcium and magnesium in many reactions. All cells require iron in electron transport chains. Red blood cells can't function without iron in hemoglobin, the oxygen-carrying pigment in blood. Neurons simply cannot function without sodium and potassium.

Table 30.2   Vitamins: Sources, Functions, and Effects of Deficiencies or Excesses*

Vitamin	Common Sources	Main Functions	Signs of Severe Long-Term Deficiency	Signs of Extreme Excess
Fat-Soluble Vitamins:				
A	Its precursor comes from beta-carotene in yellow fruits, yellow or green leafy vegetables; also in fortified milk, egg yolk, fish liver	Used in synthesis of visual pigments, bone, teeth; maintains epithelia	Dry, scaly skin; lowered resistance to infections; night blindness; permanent blindness	Malformed fetuses; hair loss; changes in skin; liver and bone damage; bone pain
D	D_3 formed in skin and in fish liver oils, egg yolk, fortified milk; converted to active form elsewhere	Promotes bone growth and mineralization; enhances calcium absorption	Bone deformities (rickets) in children; bone softening in adults	Retarded growth; kidney damage; calcium deposits in soft tissues
E	Whole grains, dark green vegetables, vegetable oils	Possibly inhibits effects of free radicals; helps maintain cell membranes; blocks breakdown of vitamins A and C in gut	Lysis of red blood cells; nerve damage	Muscle weakness, fatigue, headaches, nausea
K	Colon bacteria form most of it; also in green leafy vegetables, cabbage	Blood clotting; ATP formation via electron transport	Abnormal blood clotting; severe bleeding (hemorrhaging)	Anemia; liver damage and jaundice
Water-Soluble Vitamins:				
B_1 (thiamin)	Whole grains, green leafy vegetables, legumes, lean meats, eggs	Connective tissue formation; folate utilization; coenzyme action	Water retention in tissues; tingling sensations; heart changes; poor coordination	None reported from food; possible shock reaction from repeated injections
B_2 (riboflavin)	Whole grains, poultry, fish, egg white, milk	Coenzyme action	Skin lesions	None reported
Niacin	Green leafy vegetables, potatoes, peanuts, poultry, fish, pork, beef	Coenzyme action	Contributes to pellagra (damage to skin, gut, nervous system, etc.)	Skin flushing; possible liver damage
B_6	Spinach, tomatoes, potatoes, meats	Coenzyme in amino acid metabolism	Skin, muscle, and nerve damage; anemia	Impaired coordination; numbness in feet
Pantothenic acid	In many foods (meats, yeast, egg yolk especially)	Coenzyme in glucose metabolism, fatty acid and steroid synthesis	Fatigue, tingling in hands, headaches, nausea	None reported; may cause diarrhea occasionally
Folate (folic acid)	Dark green vegetables, whole grains, yeast, lean meats; colon bacteria produce some folate	Coenzyme in nucleic acid and amino acid metabolism	A type of anemia; inflamed tongue; diarrhea; impaired growth; mental disorders	Masks vitamin B_{12} deficiency
B_{12}	Poultry, fish, red meat, dairy foods (not butter)	Coenzyme in nucleic acid metabolism	A type of anemia; impaired nerve function	None reported
Biotin	Legumes, egg yolk; colon bacteria produce some	Coenzyme in fat, glycogen formation and in amino acid metabolism	Scaly skin (dermatitis), sore tongue, depression, anemia	None reported
C (ascorbic acid)	Fruits and vegetables, especially citrus, berries, cantaloupe, cabbage, broccoli, green pepper	Collagen synthesis; possibly inhibits effects of free radicals; structural role in bone, cartilage, and teeth; role in carbohydrate metabolism	Scurvy, poor wound healing, impaired immunity	Diarrhea, other digestive upsets; may alter results of some diagnostic tests

*The guidelines for appropriate daily intakes are being worked out by the Food and Drug Administration.

Table 30.3　Major Minerals: Sources, Functions, and Effects of Deficiencies or Excesses*

Mineral	Common Sources	Main Functions	Signs of Severe Long-Term Deficiency	Signs of Extreme Excess
Calcium	Dairy products, dark green vegetables, dried legumes	Bone, tooth formation; blood clotting; neural and muscle action	Stunted growth; possibly diminished bone mass (osteoporosis)	Impaired absorption of other minerals; kidney stones in susceptible people
Chloride	Table salt (usually too much in diet)	HCl formation in stomach; contributes to body's acid-base balance; neural action	Muscle cramps; impaired growth; poor appetite	Contributes to high blood pressure in susceptible people
Copper	Nuts, legumes, seafood, drinking water	Used in synthesis of melanin, hemoglobin, and some transport chain components	Anemia, changes in bone and blood vessels	Nausea, liver damage
Fluorine	Fluoridated water, tea, seafood	Bone, tooth maintenance	Tooth decay	Digestive upsets; mottled teeth and deformed skeleton in chronic cases
Iodine	Marine fish, shellfish, iodized salt, dairy products	Thyroid hormone formation	Enlarged thyroid (goiter), with metabolic disorders	Goiter
Iron	Whole grains, green leafy vegetables, legumes, nuts, eggs, lean meat, molasses, dried fruit, shellfish	Formation of hemoglobin and cytochrome (transport chain component)	Iron-deficiency anemia, impaired immune function	Liver damage, shock, heart failure
Magnesium	Whole grains, legumes, nuts, dairy products	Coenzyme role in ATP-ADP cycle; roles in muscle, nerve function	Weak, sore muscles; impaired neural function	Impaired neural function
Phosphorus	Whole grains, poultry, red meat	Component of bone, teeth, nucleic acids, ATP, phospholipids	Muscular weakness; loss of minerals from bone	Impaired absorption of minerals into bone
Potassium	Diet provides ample amounts	Muscle and neural function; roles in protein synthesis and body's acid-base balance	Muscular weakness	Muscular weakness, paralysis, heart failure
Sodium	Table salt; diet provides ample to excessive amounts	Key role in body's acid-base balance; roles in muscle and neural function	Muscle cramps	High blood pressure in susceptible people
Sulfur	Proteins in diet	Component of body proteins	None reported	None likely
Zinc	Whole grains, legumes, nuts, meats, seafood	Component of digestive enzymes; roles in normal growth, wound healing, sperm formation, and taste and smell	Impaired growth, scaly skin, impaired immune function	Nausea, vomiting, diarrhea; impaired immune function and anemia

*The guidelines for appropriate daily intakes are being worked out by the Food and Drug Administration.

People who are in good health get all the vitamins and minerals they need from a balanced diet of whole foods. Generally speaking, specific vitamin and mineral supplements are necessary only for strict vegetarians, the elderly, and people suffering from a chronic illness or taking medication that affects the use of specific nutrients. For example, in one comprehensive study, vitamin K supplements helped older women retain calcium and so diminish bone loss (osteoporosis) by 30 percent. Preliminary research suggests two vitamins (C and E) and beta-carotene (the precursor for vitamin A) may inactivate free radicals—those rogue metabolic fragments that may contribute to aging and may impair immune function.

No one should take massive doses of any vitamin or mineral supplement except under medical supervision.

As Tables 30.2 and 30.3 indicate, excess amounts of many vitamins and minerals are harmful. For example, large doses of at least two vitamins (A and D) actually can damage the body. Like all fat-soluble vitamins, excess amounts can accumulate in tissues and interfere with normal metabolic function. Similarly, sodium is present in plant and animal tissues and is a component of table salt. Sodium has roles in the body's salt–water balance, muscle activity, and nerve function. However, prolonged, excessive intake of sodium may contribute to high blood pressure.

Severe shortages or self-prescribed, massive excesses of vitamins and minerals can disturb the delicate balances in body function that promote health.

TANTALIZING ANSWERS TO WEIGHTY QUESTIONS

In Pursuit of the "Ideal" Weight   To keep your body functioning normally over long periods of time—while simultaneously maintaining an acceptable weight—*caloric intake must be balanced with energy output.*

How many kilocalories should you take in each day to maintain a particular weight? First, multiply that weight (in pounds) by 10 if you are not active physically, by 15 if you are moderately active, and by 20 if you are highly active. Then, depending on your age, subtract the following amount from the value obtained from the first step:

Age:	25–34	*Subtract:*	0
	35–44		100
	45–54		200
	55–64		300
	Over 65		400

For example, if you wish to weigh 120 pounds and are very active, 120 × 20 = 2,400 kilocalories. If you are thirty-five years old, you would take in (2,400 − 100) or 2,300 kilocalories a day. Such calculations provide a rough estimate of caloric intake. Other factors, such as height, must be considered also. An active person who is 5 feet, 2 inches tall doesn't need as much energy as an active person who weighs the same but is 6 feet tall.

The question becomes this: *Why* do you wish to weigh that amount? Are you merely fearful of looking fat—that is, obese? By definition, **obesity** is an excess of fat in adipose tissues, most often caused by imbalances between caloric intake and energy output. As we mentioned in passing at the start of this chapter, one standard of what constitutes obesity is simply cultural. And it varies from one culture to the next. For example, one female student who despaired over her plumpness took part in a graduate studies program in Africa. There, great numbers of males considered her one of the most desirable females on the planet.

Traditionally, insurance companies used a different standard. They developed charts mostly to identify overweight individuals who might be insurance risks. The medical profession also has used these same charts.

By 1995, researchers were questioning the insurance charts. For example, after a fourteen-year study of nearly 116,000 women, researchers at Harvard University found a strong link between the risk of heart attack and gaining even 11 to 18 pounds in adult life. Women who gained weight after age 18 were by far the group at greatest risk.

In making their correlation, the Harvard researchers took into account age, smoking habits, family history of heart disorders, the use of postmenopausal hormones, and other factors that could influence the risk. They also found conclusive evidence that exercise is *the* most crucial factor in controlling weight gain and thereby avoiding heart attacks.

Generally, thinner people simply live longer. You can use the chart in Figure 30.15 to get an idea of how a 1995 definition of "thin" applies to you.

Figure 30.15   How to estimate "ideal" weights for adults. The values given are consistent with the long-term Harvard study into the link between excessive weight and increased risk of heart disorders. Depending on certain factors (such as having a small, medium, or large frame), the "ideal" may vary by plus or minus 10 percent.

Weight Guideline for Women:

Starting with an ideal weight of 100 pounds for a woman who is 5 feet tall, add five additional pounds for each additional inch of height. Examples:

Height (feet)	Weight (pounds)
5' 2"	110
5' 3"	115
5' 4"	120
5' 5"	125
5' 6"	130
5' 7"	135
5' 8"	140
5' 9"	145
5' 10"	150
5' 11"	155
6'	160

Weight Guideline for Men:

Starting with an ideal weight of 106 pounds for a man who is 5 feet tall, add six additional pounds for each additional inch of height. Examples:

Height (feet)	Weight (pounds)
5' 2"	118
5' 3"	124
5' 4"	130
5' 5"	136
5' 6"	142
5' 7"	148
5' 8"	154
5' 9"	160
5' 10"	166
5' 11"	172
6'	178

My Genes Made Me Do It As you've probably noticed, some people have far more trouble keeping off excess weight than others do. Actually, about one of every three Americans is like this. To some extent, age and emotional state have something to do with it. More importantly, *genes* have a lot to do with it.

Researchers have suspected as much for a long time. Consider their studies of certain identical twins (who are born with identical genes). As a result of family problems, the twins had been separated at birth and were raised apart, in different households. Even so, at adulthood, the body weights of those separated twins were similar! People, it seemed, were born with a "set point" for body fat. Could it be that whatever set point an individual inherits is the one he or she will be stuck with for life?

Through many studies and experiments, researchers gathered strong evidence that ultimately confirmed this suspicion. It all started in 1950 with an exceptionally obese mouse (Figure 30.16). But it wasn't until 1995 that molecular geneticists identified one of its genes that has profound influence over the set point for body fat. They named it the *ob gene* (guess why). What were the clues that led to it? The nucleotide sequence in the gene region of obese mice differed from that in normal mice. The gene is active in adipose tissue, nowhere else. Its biochemical information translates into a type of protein that cells can release into the bloodstream—that is, the gene might specify a hormonal signal.

Jeffrey Friedman discovered the hormone in 1994 and named it **leptin**. A nearly identical hormone was isolated in humans. Leptin is only one of a number of factors that influence the brain's commands to suppress or whip up appetite, but it's a crucial one. By assessing the blood concentrations of incoming hormonal signals, an appetite control center in a brain region (the hypothalamus) "decides" whether the body has taken in enough food to provide enough fat for the day. If so, commands go out to increase metabolic rates—and to stop eating.

If the *ob* gene is mutated, then its expression alone might be enough to disrupt the control center, so that a mouse's appetite skyrockets and metabolic furnaces burn less. By extension, the same might happen in a human with a similar gene mutation even if he or she fully understands the inherent health risks. (In one experiment that supports the hypothesis, researchers injected obese mice with leptin. The mice quickly shed their excess weight.) Will the researchers go on to develop therapies for obesity in humans? Maybe, but not for five to ten years. Hormonal control in humans is tricky business.

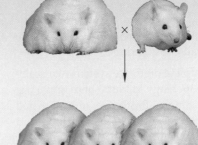

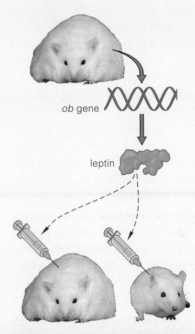

a *1950.* Researchers at the Jackson Laboratories in Maine notice that one of their laboratory mice is extremely obese, with an uncontrollable appetite. Through crossbreeding of this apparent mutant individual with a normal mouse, they produce a strain of obese mice.

b *Late 1960s.* Douglas Coleman of the Jackson Laboratories surgically joins the bloodstreams of an obese mouse and a normal one. The obese mouse now loses weight. Coleman hypothesizes that a factor circulating in blood may be influencing its appetite, but he is not able to isolate it.

c *1994.* Late in the year, Jeffrey Friedman of Rockefeller University discovers a mutated form of what is now called the *ob* gene in obese mice. Through DNA cloning and gene sequencing, he defines the protein that the mutated gene encodes. The protein, now called leptin, is a hormone that influences the brain's commands to suppress appetite and increase metabolic rates.

d *1995.* Three different research teams develop and use genetically engineered bacteria to produce leptin which, when injected in obese and normal mice, triggers significant weight loss, apparently without harmful side effects.

ob gene

leptin

Figure 30.16 Chronology of key developments that revealed the identity of a key factor in the genetic basis of body weight.

SUMMARY

1. *Nutrition* involves all the processes by which the body takes in, digests, absorbs, and uses food.

2. Mammals have a complete digestive system—basically a tube that has two openings (mouth and anus), as well as regional specializations. Protective, mucus-coated epithelium lines all exposed surfaces of the tube and facilitates diffusion across the tube wall.

3. As summarized in Table 30.4, the human digestive system includes the mouth, pharynx, esophagus, stomach, small intestine, large intestine (colon), rectum, and anus. Salivary glands and the liver, gallbladder, and pancreas have accessory roles in the system's functions.

4. These activities proceed in the digestive system:

   a. Mechanical processing and motility (movements that break up, mix, and propel food material).

   b. Secretion (release of digestive enzymes and other substances from the pancreas, liver, and glandular epithelium into the gut lumen).

   c. Digestion (breakdown of food into particles, then into nutrient molecules small enough to be absorbed).

   d. Absorption (diffusion or transport of digested organic compounds, fluid, and ions from the gut lumen into the internal environment).

   e. Elimination (the expulsion of undigested as well as unabsorbed residues at the end of the system).

5. Starch digestion starts in the mouth, and protein digestion starts in the stomach. Digestion is completed and most nutrients are absorbed in the small intestine. The pancreas secretes the main digestive enzymes. Bile from the liver assists in fat digestion.

6. In absorption, cells of the intestinal lining actively transport glucose and most amino acids out of the gut lumen. Fatty acids and monoglycerides diffuse across the lipid bilayer of these cells. In the cells' cytoplasm they are recombined as triglycerides, then are released, by exocytosis, into interstitial fluid.

7. The nervous system, endocrine system, and nerve plexuses in the gut wall interact to govern activities of the digestive system. Many controls operate in response to the volume and composition of food passing through the stomach and intestines. The controls trigger changes in muscle activity and in the rate at which hormones and enzymes are secreted.

8. Many nutritionists recommend a daily food intake in certain proportions. For example, for an adult male of average body weight: 58–60 percent complex carbohydrates, 12–15 percent protein, and 20–25 percent fats and other lipids. A well-balanced diet of whole foods normally provides all required vitamins and minerals.

9. To maintain a suitable body weight and overall health, caloric intake must balance energy output.

Table 30.4	Summary of the Digestive System
MOUTH (oral cavity)	Start of digestive system, where food is chewed and moistened. Polysaccharide digestion begins here
PHARYNX	Entrance to tubular part of digestive and respiratory systems
ESOPHAGUS	Muscular tube, moistened by saliva, that moves food from pharynx to stomach
STOMACH	Sac where food mixes with gastric fluid and protein digestion begins; stretches to store food taken in faster than can be processed; gastric fluid destroys many microbes
SMALL INTESTINE	The first part (duodenum) receives secretions from the liver, gallbladder, and pancreas. Most nutrients are digested and absorbed in the second part (jejunum). Some nutrients are absorbed in the last part (ileum), which delivers unabsorbed material to the colon
LARGE INTESTINE (colon)	Concentrates and stores undigested matter (by absorbing mineral ions and water)
RECTUM	Distension triggers expulsion of feces
ANUS	Terminal opening of digestive system

Accessory Organs:	
SALIVARY GLANDS	Glands (three main pairs, many minor ones) that secrete saliva, a fluid with polysaccharide-digesting enzymes, buffers, and mucus (which moistens and lubricates ingested food)
PANCREAS	Secretes enzymes that digest all major food molecules; secretes buffers against HCl from the stomach
LIVER	Secretes bile (used in fat emulsification); roles in carbohydrate, fat, and protein metabolism
GALLBLADDER	Stores and concentrates bile from the liver

Review Questions

1. Choose an animal with a complete digestive system and correlate some organs of the system with feeding behavior. *512–513*

2. What are the main functions of the stomach? the small intestine? the large intestine (colon)? *516, 517–519, 521*

3. Name the kinds of breakdown products that are small enough to be absorbed across the intestinal lining and into the internal environment. *(see Table 30.1)*

4. Name and define the key functions of the organs of the human digestive system, shown in the diagram on the next page. *514*

Self-Quiz *(Answers in Appendix IV)*

1. The _____ maintains the internal environment, supplies cells with raw materials, and disposes of metabolic wastes.
   a. digestive system
   b. circulatory system
   c. respiratory system
   d. urinary system
   e. interaction of all of the systems listed does this

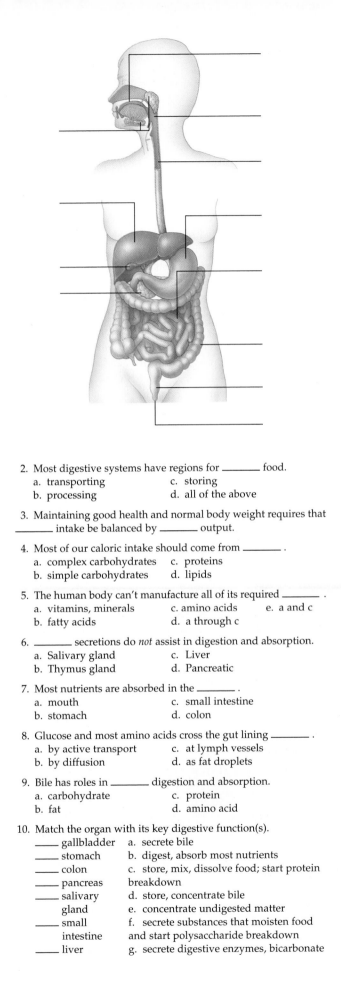

2. Most digestive systems have regions for _____ food.
   a. transporting   c. storing
   b. processing   d. all of the above

3. Maintaining good health and normal body weight requires that _____ intake be balanced by _____ output.

4. Most of our caloric intake should come from _____ .
   a. complex carbohydrates   c. proteins
   b. simple carbohydrates   d. lipids

5. The human body can't manufacture all of its required _____ .
   a. vitamins, minerals   c. amino acids   e. a and c
   b. fatty acids   d. a through c

6. _____ secretions do *not* assist in digestion and absorption.
   a. Salivary gland   c. Liver
   b. Thymus gland   d. Pancreatic

7. Most nutrients are absorbed in the _____ .
   a. mouth   c. small intestine
   b. stomach   d. colon

8. Glucose and most amino acids cross the gut lining _____ .
   a. by active transport   c. at lymph vessels
   b. by diffusion   d. as fat droplets

9. Bile has roles in _____ digestion and absorption.
   a. carbohydrate   c. protein
   b. fat   d. amino acid

10. Match the organ with its key digestive function(s).
    ____ gallbladder   a. secrete bile
    ____ stomach   b. digest, absorb most nutrients
    ____ colon   c. store, mix, dissolve food; start protein
    ____ pancreas   breakdown
    ____ salivary   d. store, concentrate bile
         gland   e. concentrate undigested matter
    ____ small   f. secrete substances that moisten food
         intestine   and start polysaccharide breakdown
    ____ liver   g. secrete digestive enzymes, bicarbonate

Critical Thinking

1. A glassful of whole milk contains lactose, proteins, butterfat (mostly triglycerides), vitamins, and minerals. Explain what will happen to each component in your digestive tract.

2. As a person ages, the number of body cells steadily decreases, and energy needs decline. If you were planning an older person's diet, what foods would you emphasize, and why? Which ones would you deemphasize?

3. Using Section 30.11 as a reference, determine your ideal weight and design a well-balanced program of diet and exercise that will help you achieve or maintain that weight.

4. Often, holiday meals are larger than everyday ones and have a high fat content. After stuffing themselves at an early dinner on Thanksgiving Day, Richard and other members of his family feel uncomfortably full for the rest of the afternoon. Based on what you have learned about controls over digestion, propose a biochemical explanation for their discomfort.

5. Your digestive system affords some effective protection against many pathogenic bacteria that can contaminate the kinds of food you eat. Explain some of the ways it can destroy or wash away these microorganisms. (You may wish to refer to Section 28.1, also.)

6. In 1996 in Great Britain, 160,000 cattle were staggering, drooling, and showing other symptoms of a fatal disease called bovine spongiform encephalopathy, or *mad cow disease*. The culprit? Rogue proteins—prions (pronounced PREE-ons, for proteinaceous infectious particles). The prions triggered conversion of normal proteins in nerve cells—and spongy holes and amyloid deposits formed in cattle brains. Worldwide, 1 in 1 million people have developed the related *Creutzfeldt-Jacob disease*. Symptoms include loss of vision and speech, rapid mental deterioration, and spastic paralysis, but it may be decades before this happens. Thus, in Great Britain, ground-up tissues of sheep that had died of scrapie (another prion-caused disease) were used as a supplement in cattle feed. The practice was banned in 1988—but prions were already in the food chain. Ten young Britons (and, coincidentally, a woman in Stockton, California) are diagnosed with the disease at this writing. In the United States, the FDA is moving to prohibit *all* protein supplements for sheep, cattle, and other ruminants. Investigate how widespread this practice has been in the United States and how closely it has been monitored.

Selected Key Terms

appendix *521*	gallbladder *517*	net protein
bile *517*	gastric fluid *516*	utilization *523*
bulk *521*	gut *514*	nutrition *511*
chyme *516*	incomplete digestive	obesity *526*
colon *521*	system *512*	pancreas *517*
complete digestive	kilocalorie *522*	pharynx *515*
system *512*	leptin *527*	ruminant *513*
digestive system *511*	liver *517*	saliva *515*
emulsification *517*	micelle	sphincter *521*
esophagus *515*	formation *519*	stomach *516*
essential amino	microvillus *518*	tongue *515*
acid *523*	mineral *524*	tooth *515*
essential fatty	mouth (oral	villus *518*
acid *523*	cavity) *515*	vitamin *524*

Reading

Wardlaw, G., P. Insel, and M. Seyler. 1992. *Contemporary Nutrition: Issues and Insights.* St. Louis: Mosby.

31 THE INTERNAL ENVIRONMENT

Tale of the Desert Rat

Look closely at a fish or some other marine animal, and you find that its cells are exquisitely adapted to life in a salty fluid. Yet just imagine. About 375 million years ago, some animals left the seas for life on land. Actually, they brought salty fluid along with them, as an *internal environment* for their cells. Even so, it wasn't easy. On land they found intense sunlight, more pronounced swings in temperature, dry winds, water of dubious salt content, and sometimes no water at all.

How did those pioneers conserve or replace the water and specific salts they lost through everyday activities? How did they manage to stay warm when their surroundings got too cold or too hot? They surely did so, otherwise they could not have maintained the volume, composition, and temperature of their internal environment. In other words, *how did the land-dwelling descendants of marine animals maintain internal operating conditions and so prevent cellular anarchy?*

Any of their existing descendants provides you with some answers. Think of a kangaroo rat in an isolated desert of New Mexico (Figure 31.1*a*). After a brief rainy season, the sun bakes the sand for months. The only obvious water is imported, sloshing in canteens of an occasional researcher or tourist. Yet with nary a sip of free water, this tiny mammal counters the threat to its internal environment.

The kangaroo rat waits out the daytime heat in a burrow, then forages in the cool of night for dry seeds and maybe a succulent. It is not sluggish about this. It hops rapidly and far, searching for seeds and fleeing from coyotes and snakes. All that hopping requires ATP energy and water. Seeds, chockful of energy-rich carbohydrates, supply both. Metabolic reactions that release energy from carbohydrates and other organic compounds also yield water. Each day, this "metabolic water" represents about 12 percent of your own total

Figure 31.1 **(a)** Kangaroo rat, master of water conservation in the desert. **(b)** Notice the differences in the ways it gains and loses water, compared to a human. And notice, for both the rodent and the human, *the losses balance out the gains*—just as they do in all animals. How this happens, and why it absolutely must happen, is the first focus of this chapter.

	Kangaroo Rat	Human
Water Gain (milliliters):		
by ingesting solids	6.0	850
by ingesting liquids	0	1,400
by way of metabolism	54.0	350
	60.0	2,600
Water Loss (milliliters):		
in urine	13.5	1,500
in feces	2.6	200
by evaporation	43.9	900
	60.0	2,600

a

b

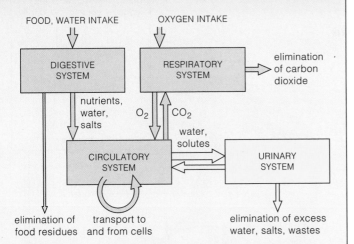

FOOD, WATER INTAKE   OXYGEN INTAKE

DIGESTIVE SYSTEM

RESPIRATORY SYSTEM → elimination of carbon dioxide

nutrients, water, salts O_2 CO_2

water, solutes

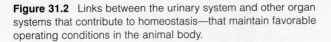

CIRCULATORY SYSTEM ⇄ URINARY SYSTEM

elimination of food residues   transport to and from cells   elimination of excess water, salts, wastes

Figure 31.2 Links between the urinary system and other organ systems that contribute to homeostasis—that maintain favorable operating conditions in the animal body.

water intake. It represents a whopping 90 percent of the total for a kangaroo rat.

Inside its cool burrow, the kangaroo rat conserves and recycles water. As the animal breathes cool air into its warm lungs, water vapor condenses on the epithelial lining inside its nose—so some water can diffuse back into the body. Also, after a night of foraging, the animal empties its cheek pouches of seeds—which soak up water vapor that does escape by dripping out from the nose. Eating dripped-upon seeds recycles the water.

The kangaroo rat can't lose water by perspiring; it has no sweat glands. It can lose water by urinating— but specialized kidneys don't let it piddle away much. Kidneys filter the blood's water and dissolved salts (solutes). They adjust *how much water* and *which solutes* return to the blood or leave the body as urine.

Overall, kangaroo rats and all other animals take in enough water and solutes to replace the daily losses (Figure 31.1*b*). How they accomplish these balancing acts will be our initial focus in the chapter. Later, we will take a look at some of the means by which mammals withstand hot, cold, and sometimes unpredictable changes in environmental temperatures on land.

As a starting point, remind yourself of the kinds of fluids inside most animals. **Interstitial fluid** fills the spaces between living cells and other components of tissues. Another fluid, **blood**, transports substances to and from all tissue regions by way of a circulatory system. Taken together, the interstitial fluid and blood are the **extracellular fluid**. In most animals, a well-developed urinary system helps keep the volume and composition of extracellular fluid within tolerable ranges. As you will see, other organ systems, especially those indicated in Figure 31.2, interact with the urinary system in the performance of this homeostatic task.

KEY CONCEPTS

1. Animals continually gain and lose water and dissolved substances (solutes). They continually produce metabolic wastes. Even with all the inputs and outputs, the overall volume and composition of the body's extracellular fluid remain relatively constant.

2. In humans, as in other vertebrates, a urinary system is crucial to balancing the intake and output of water and solutes. The urinary system eliminates excess water and solutes, and it conserves water when necessary.

3. Kidneys are blood-filtering organs, and the urinary system has a pair of them. Packed inside each kidney are a great number of tubelike structures called nephrons.

4. At its cup-shaped beginning, each nephron receives water and solutes from an adjacent set of blood capillaries. The nephron returns most of the filtrate to a second set of capillaries intertwined around its tubular parts.

5. The water and solutes not returned to the blood leave the body as a fluid called urine. At any given time, control mechanisms influence whether the urine is concentrated or dilute. Two hormones, ADH and aldosterone, have key roles in these adjustments.

6. The internal body temperature of animals depends on the balance between heat produced through metabolism, heat absorbed from the environment, and heat lost to the environment.

7. The internal body temperature is maintained within a favorable range through controls over metabolic activity and adaptations in structure, physiology, and behavior.

The Challenge—Unwanted Shifts in Extracellular Fluid

Different solid foods and fluids intermittently hit your gut. Afterward, differing amounts of absorbed water, nutrients, and other substances move into your blood, then into interstitial fluid. Such events could shift the volume and composition of extracellular fluid beyond tolerable limits. Thus the body must make adjustments to balance its gains and losses. *Within a given time frame, it must take in as much water and solutes as it gives up.*

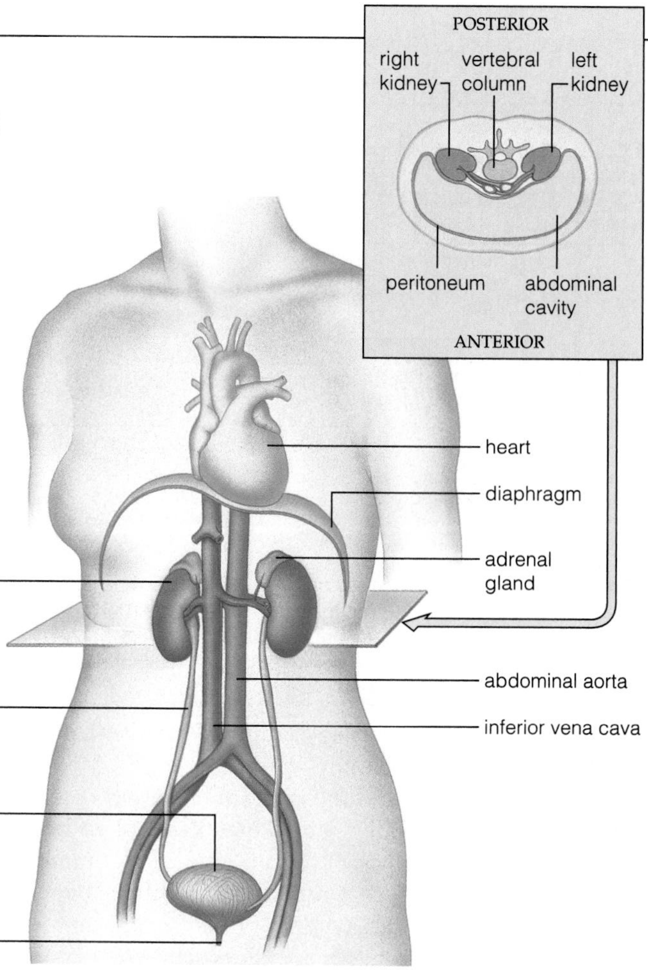

POSTERIOR
right kidney — vertebral column — left kidney
peritoneum — abdominal cavity
ANTERIOR

KIDNEY (one of a pair)
Constantly filters water and all solutes except proteins from blood; reclaims water and solutes as the body requires and excretes the remainder, as urine

URETER (one of a pair)
Channel for urine flow from a kidney to the urinary bladder

URINARY BLADDER
Stretchable container for temporarily storing urine

URETHRA
Channel for urine flow between the urinary bladder and body surface

heart
diaphragm
adrenal gland
abdominal aorta
inferior vena cava

Figure 31.3 Organs of the human urinary system and their functions. The two kidneys, two ureters, and urinary bladder are located *outside* the membranous lining (peritoneum) of the abdominal cavity. Compare Figure 20.3.

Water Gains and Losses Think of a human or some other mammal. It *gains* water mainly by two processes:

Absorption from gut
Metabolism

The body absorbs a lot of water from solids and liquids in the gut. In land mammals, a thirst mechanism affects how much water enters the gut in the first place. When the body loses too much water, such animals seek out streams, water holes, and so on. We will look at the thirst mechanism later in the chapter. Water also forms as a by-product of many metabolic reactions.

A mammal normally *loses* water by these processes:

Urinary excretion
Evaporation from lungs and skin
Sweating, in mammals that sweat
Elimination (in feces)

Urinary excretion affords the most control over water loss. This process eliminates excess water and solutes as urine, a fluid that forms in a urinary system such as that shown in Figure 31.3. Some water also evaporates from the respiratory surface in lungs and, in some species, it departs in sweat. A healthy body loses very little water from the gut (most is absorbed, not eliminated in feces).

Solute Gains and Losses How do mammals *gain* solutes? They do so mainly by four processes:

Absorption from gut
Secretion from cells
Respiration
Metabolism

Nutrients and mineral ions are absorbed from the gut. Cellular secretions and wastes, such as carbon dioxide, enter interstitial fluid, then blood. Oxygen enters blood by way of the respiratory system.

The body *loses* solutes in these ways:

Urinary excretion
Respiration
Sweating

Some mammals lose minerals in sweat. All exhale carbon dioxide, the most abundant waste. Their urine contains wastes formed by the breakdown of organic compounds. For example, ammonia forms as amino groups are split from amino acids, then **urea**—a major waste—forms in the liver when two ammonia molecules join with carbon dioxide. Uric acid from nucleic acid breakdown, wastes from hemoglobin breakdown (which give urine much of its color), drugs, and food additives also leave in urine.

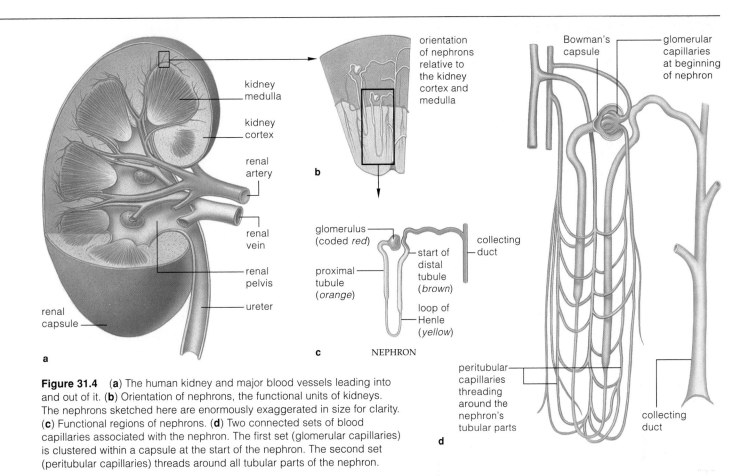

kidney
medulla

kidney
cortex

renal
artery

renal
vein

renal
pelvis

ureter

renal
capsule

a

b orientation
of nephrons
relative to
the kidney
cortex and
medulla

glomerulus
(coded *red*)

proximal
tubule
(*orange*)

start of
distal
tubule
(*brown*)

collecting
duct

loop of
Henle
(*yellow*)

c NEPHRON

Bowman's
capsule

glomerular
capillaries
at beginning
of nephron

peritubular
capillaries
threading
around the
nephron's
tubular parts

collecting
duct

d

Figure 31.4 (**a**) The human kidney and major blood vessels leading into and out of it. (**b**) Orientation of nephrons, the functional units of kidneys. The nephrons sketched here are enormously exaggerated in size for clarity. (**c**) Functional regions of nephrons. (**d**) Two connected sets of blood capillaries associated with the nephron. The first set (glomerular capillaries) is clustered within a capsule at the start of the nephron. The second set (peritubular capillaries) threads around all tubular parts of the nephron.

Components of the Urinary System

Mammals mainly counter shifts in the composition and volume of extracellular fluid by a **urinary system**. This consists of two kidneys, two ureters, a urinary bladder, and a urethra. The **kidneys** are a pair of bean-shaped organs, about as big as an average fist (Figures 31.3 and 31.4). Each has an outer capsule of connective tissue. Blood capillaries (small-diameter blood vessels) thread through its two inner regions—the cortex and medulla.

Kidneys filter water, mineral ions, organic wastes, and other substances out of blood, then they adjust the filtrate's composition and return all but about 1 percent to the blood. This tiny portion of unreclaimed water and solutes is urine. By definition, **urine** is a fluid that rids the body of water and solutes that are in excess of the amounts required to maintain extracellular fluid.

Urine flows from a kidney into a **ureter**, one of the two tubular channels to the **urinary bladder**. Here, urine is briefly stored before moving into the **urethra**, a muscular tube that opens at the body surface. Flow from the bladder (urination) is a reflex action. When the bladder is filled, smooth muscle of its balloonlike wall contracts and a sphincter around its neck opens, so urine is forced out through the urethra. Contraction of skeletal muscle surrounding the urethra is under voluntary control, and it can prevent urination.

Nephrons—Functional Units of Kidneys

A human kidney has more than a million **nephrons**, slender tubules packed inside lobes that extend from the cortex down through the medulla. At the nephrons, water and solutes are filtered from blood—and the amount sent back to blood is adjusted.

Each nephron starts as a **Bowman's capsule**. Here its wall cups around *glomerular* capillaries. The cupped wall region and blood vessels are a blood-filtering unit called a **glomerulus** (Figure 31.4c). Next, the nephron has a **proximal tubule** (closest to the capsule), then a hairpin-shaped **loop of Henle** and **distal tubule** (most distant from the capsule). It ends as a **collecting duct**, which is part of a system that leads into the kidney's central cavity (renal pelvis) and the entrance to a ureter.

Blood does not give up all of its water and solutes. The unfiltered part flows into a second set of capillaries around the nephron's tubular parts. In these *peritubular* capillaries, the blood reclaims water and solutes, then flows into veins and back to the general circulation.

A urinary system counters unwanted shifts in the volume and composition of extracellular fluid. In its paired kidneys, water and solutes are filtered from blood. The body reclaims most of this, but any excess leaves the kidneys as urine.

Urine forms in nephrons by three processes: filtration, tubular reabsorption, and tubular secretion. All depend on properties of cells that make up the nephron wall. The cells differ in membrane transport mechanisms and permeability from one part of the nephron to the next.

Blood pressure generated by the heart's contractions drives **filtration**, which proceeds at the glomerulus (Figure 31.5). Pressure "filters" blood by forcing water and all solutes except proteins out of the glomerular capillaries. This protein-free filtrate moves from the cupped part of the nephron into the proximal tubule.

Tubular reabsorption proceeds along the nephron's tubular regions. Here, most of the filtrate's water and solutes move out of the nephron's lumen (the space enclosed by its wall). Then they move into the nearby peritubular capillaries (Table 31.1 and Figure 31.6).

Tubular secretion proceeds at the tubule wall but in the opposite direction of reabsorption. Solutes move out of peritubular capillaries, then into cells of the wall—which secrete them into the nephron's lumen. The main solutes are ions of hydrogen (H^+) and potassium (K^+). The process also helps prevent some metabolic wastes (such as uric acid) and foreign substances (such as drugs) from accumulating in blood.

Factors Influencing Filtration

About 1.5 liters (1-1/2 quarts) of blood flow through an adult's kidneys every minute. Of this, 120 milliliters of water and small solutes are filtered into the nephrons. That's 180 liters of filtrate per day! The high filtration rate is possible mainly because glomerular capillaries are 10–100 times more permeable than other capillaries to water and small solutes. Also, blood pressure is high in these capillaries. Arterioles that deliver blood to the glomerulus have a wider diameter and less resistance to flow than most of the other arterioles in the body. So hydrostatic pressure generated by heart contractions does not drop as much as it does elsewhere.

At any time, the blood flow to the kidneys affects the filtration rate. Neural, endocrine, and local controls maintain the flow even when blood pressure changes. For example, if you run a race or dance until dawn, the nervous system will divert an above-normal volume of blood away from kidneys, toward the heart and skeletal muscles. It directs coordinated vasoconstriction and vasodilation in different parts of the body, so that less of the blood flow reaches the kidneys. As one more example, cells in the walls of arterioles

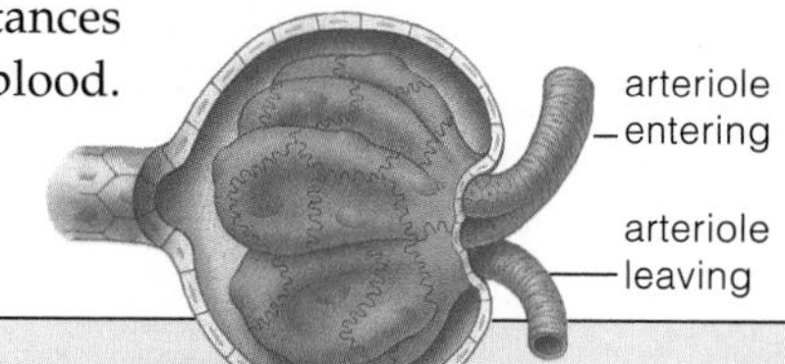

arteriole entering

arteriole leaving

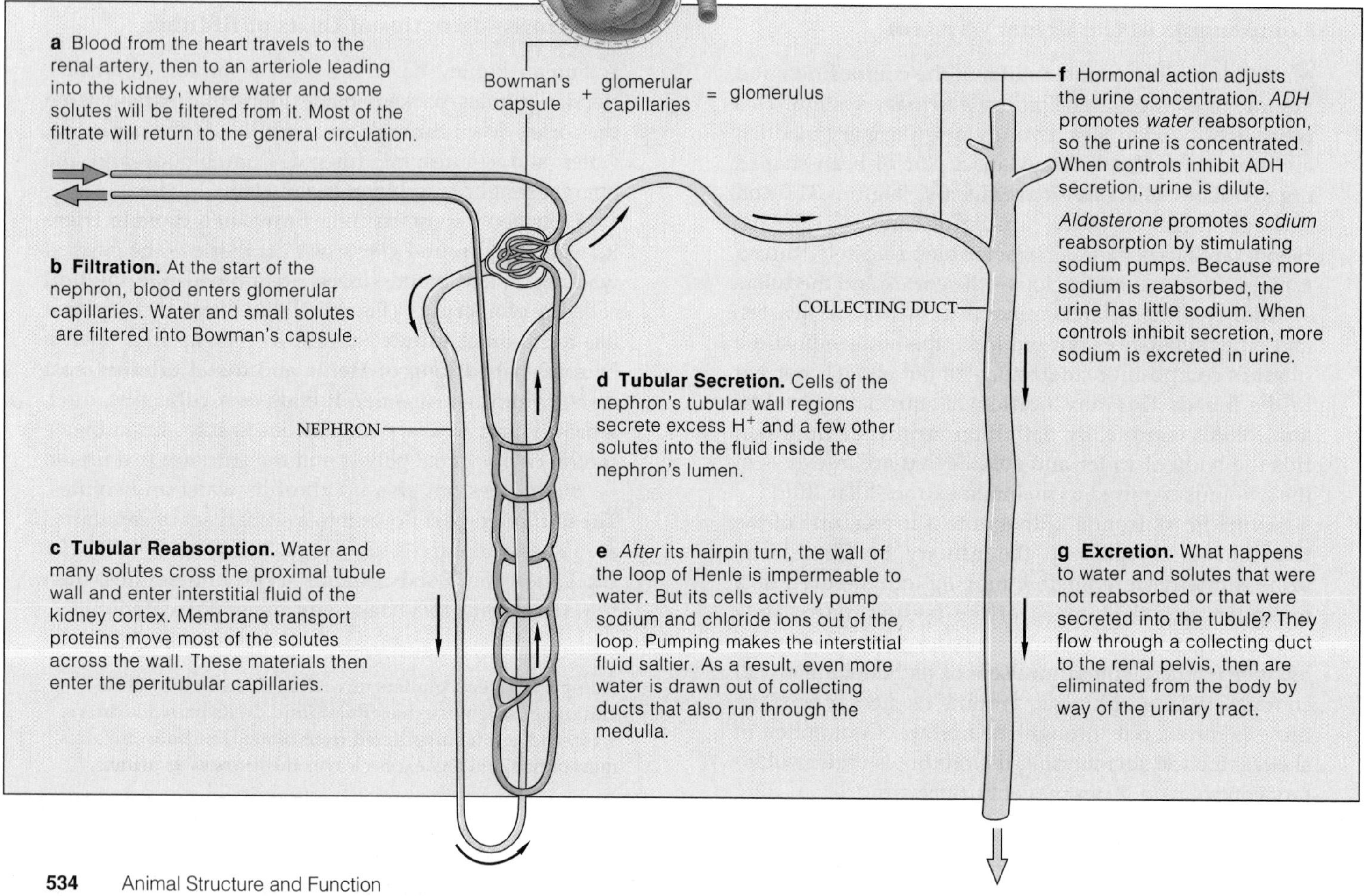

Figure 31.5 Processes of urine formation.

a Blood from the heart travels to the renal artery, then to an arteriole leading into the kidney, where water and some solutes will be filtered from it. Most of the filtrate will return to the general circulation.

Bowman's capsule + glomerular capillaries = glomerulus

f Hormonal action adjusts the urine concentration. *ADH* promotes *water* reabsorption, so the urine is concentrated. When controls inhibit ADH secretion, urine is dilute.

b Filtration. At the start of the nephron, blood enters glomerular capillaries. Water and small solutes are filtered into Bowman's capsule.

Aldosterone promotes *sodium* reabsorption by stimulating sodium pumps. Because more sodium is reabsorbed, the urine has little sodium. When controls inhibit secretion, more sodium is excreted in urine.

COLLECTING DUCT

d Tubular Secretion. Cells of the nephron's tubular wall regions secrete excess H^+ and a few other solutes into the fluid inside the nephron's lumen.

NEPHRON

c Tubular Reabsorption. Water and many solutes cross the proximal tubule wall and enter interstitial fluid of the kidney cortex. Membrane transport proteins move most of the solutes across the wall. These materials then enter the peritubular capillaries.

e *After* its hairpin turn, the wall of the loop of Henle is impermeable to water. But its cells actively pump sodium and chloride ions out of the loop. Pumping makes the interstitial fluid saltier. As a result, even more water is drawn out of collecting ducts that also run through the medulla.

g Excretion. What happens to water and solutes that were not reabsorbed or that were secreted into the tubule? They flow through a collecting duct to the renal pelvis, then are eliminated from the body by way of the urinary tract.

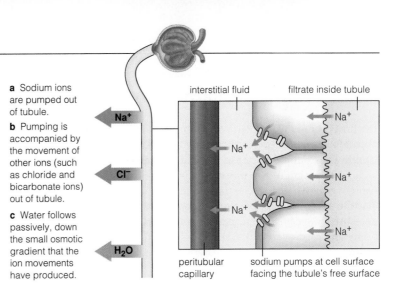

Table 31.1	Average Daily Reabsorption Values for a Few Substances			
	Water (liters)	Glucose (grams)	Sodium (grams)	Urea (grams)
Filtered:	180	180	630	54
Excreted:	1.8	none	3.2	30
Reabsorbed:	99%	100%	99.5%	44%

a Sodium ions are pumped out of tubule.

b Pumping is accompanied by the movement of other ions (such as chloride and bicarbonate ions) out of tubule.

c Water follows passively, down the small osmotic gradient that the ion movements have produced.

Figure 31.6 (*Right*) Sodium pumping associated with reabsorption of sodium and water in the kidney.

leading to glomeruli respond to pressure changes. With lower blood pressure, they vasodilate, so more blood flows into kidneys. With elevated blood pressure, they vasoconstrict, so less blood flows in.

Reabsorption of Water and Sodium

Reabsorption Mechanism Kidneys precisely adjust how much water and sodium ions the body excretes or conserves. Suppose you drink too much or too little water or wolf down salty potato chips or lose too much sodium in sweat. Responses start promptly, as filtrate enters proximal tubules of nephrons. Cells in the tubule wall actively transport some sodium out of the filtrate into interstitial fluid. Other ions follow—and then water follows the ions, by osmosis. This wall region is highly permeable to water. In fact, about two-thirds of the filtrate's water is reabsorbed here (Figure 31.5c,d).

In the medulla, interstitial fluid is saltiest around the hairpin turn of the loop of Henle. Water moves out of the filtrate by osmosis *before* the turn. The fluid left behind gets saltier until it matches the interstitial fluid. Water cannot cross the loop's wall *after* the turn. But sodium is pumped out, by active transport mechanisms (Figure 31.5e). The interstitial fluid gets saltier—so more water is drawn out of filtrate just entering the loop.

Fluid arriving at the distal tubule is dilute. The stage is now set for adjustments that can lead to highly dilute or concentrated urine—or anywhere in between.

Hormone-Induced Adjustments Cells of the distal tubules and collecting ducts have hormone receptors, for ADH and aldosterone. When the extracellular fluid volume falls, the hypothalamus (a brain region) triggers the secretion of **ADH** (antidiuretic hormone). In effect, ADH makes the tubule walls more permeable to water (Figure 31.5h). More water is reabsorbed, so urine gets more concentrated. When the body holds excess water, ADH secretion decreases. The walls get less permeable, less water can be reabsorbed, and urine remains dilute.

Aldosterone promotes sodium reabsorption. The extracellular fluid volume decreases when too much sodium is lost. Sensory receptors in the heart and blood vessels detect the decrease and signal glandular cells in the walls of the arteriole entering the glomerulus. These cells secrete renin, an enzyme that splits off part of a plasma protein. A second reaction converts the protein fragment to a hormone (angiotensin II), which acts on aldosterone-secreting cells of the adrenal cortex. This is the outer portion of a gland perched on each kidney (Figure 31.4). Aldosterone stimulates cells of the distal tubules and collecting ducts to reabsorb sodium faster, so less sodium is excreted. Conversely, when the body holds excess sodium, aldosterone secretion is inhibited. Less sodium is reabsorbed, and more is excreted.

Thirst Behavior A decline in blood's volume and a rise in its solute level also trigger **thirst behavior**, by which water is sought and ingested. Besides signaling brain regions concerned with ADH production, sensory receptors also signal brain centers that mediate thirst. So does angiotensin II, which acts directly on the brain. And so does a cottony-dry mouth, an early sign of dehydration that triggers profound thirst.

A concentrated or dilute urine forms in kidneys by filtration, tubular reabsorption, and tubular secretion.

Filtration rates depend mainly on heart contractions, which generate high hydrostatic pressure at the glomerulus of the nephron. They also depend on neural, endocrine, and local control of blood flow directed to them at a given time.

Reabsorption, which can be adjusted by hormonal controls, helps maintain extracellular fluid. The adjustments rid the body of suitable amounts of water and solutes, in the form of dilute or concentrated urine.

ADH promotes water conservation and concentrated urine. When ADH secretion is inhibited, urine is dilute.

Aldosterone promotes sodium conservation. When its secretion is inhibited, more sodium is excreted in urine.

WHEN THE KIDNEYS BREAK DOWN

By this point in the chapter, you probably have sensed that normal kidney function is absolutely central to good health. For example, suppose something interferes with sodium excretion. When that happens, the body's sodium content rises. So does the extracellular fluid volume. This leads to a rise in blood pressure. Abnormally high blood pressure, or *hypertension*, can adversely affect the kidneys, brain, and cardiovascular system. Restricted intake of sodium chloride—table salt—lessens the problem.

As another example, uric acid, calcium salts, and other wastes can settle out of urine and collect in the renal pelvis as *kidney stones*. These hard deposits may become lodged in the ureter (or urethra), then disrupt urine flow. They usually are passed naturally in urine. If not, they must be removed by medical or surgical procedures.

Whether by illness or as a result of accidents, about 13 million people in the United States alone suffer from kidney malfunctions that seriously disrupt controls over the volume and composition of extracellular fluid. Among other things, the by-products of protein breakdown may accumulate to toxic levels in the bloodstream of these people. Nausea, fatigue, and loss of memory are some of the typical outcomes. In advanced cases, death may follow.

A *kidney dialysis machine* often can restore the proper solute balances. Like the kidney itself, this machine helps maintain the extracellular fluid by selectively removing solutes from blood and adding solutes to it. The term "dialysis" refers to an exchange of substances across an artificial membrane that is interposed between solutions that differ in composition.

In *hemodialysis*, a clinician connects the machine to an artery or to a vein. Then the patient's blood is pumped through tubes made of a material that is similar to sausage casing or cellophane. The tubes are submerged in a warm saline bath. The particular mix of salts, glucose, and other substances of the bath sets up the correct concentration gradients with blood. The blood then returns to the body. In *peritoneal dialysis*, fluid of the proper composition is introduced into the patient's abdominal cavity, left in place for a specific length of time, then drained out. In this case, the peritoneum itself—that is, the cavity's lining—serves as the membrane for dialysis. For kidney dialysis to have optimum effect, it must be performed three times a week. Each time, the procedure takes about four hours, for the patient's blood must circulate repeatedly in order to improve the solute concentrations in her or his body.

Bear in mind, kidney dialysis is used as a temporary measure in reversible kidney disorders. In chronic cases, the procedure must be used for the rest of the patient's life or until a transplant operation provides her or him with a functional kidney. With treatment and controlled diets, many patients can resume fairly normal activity.

31.4 THE BODY'S ACID-BASE BALANCE

Besides maintaining the volume and composition of extracellular fluid, kidneys help keep it from becoming too acidic or too basic (alkaline). The overall **acid-base balance** is maintained by controlling the concentrations of hydrogen ions (H^+) and other dissolved ions. Buffer systems, respiration, and urinary excretion provide the control. A **buffer system** consists of weak acids or bases that can minimize changes in pH by reversibly latching onto and releasing ions.

Normally, the extracellular pH of the human body must be maintained between 7.37 and 7.43. As you know, acids lower the pH and bases raise it. A variety of acidic and basic substances enter the blood. They do so by absorption from the gut and as a result of normal metabolism. Typically, cell activities produce an excess of acids, these dissociate into H^+ and other fragments, and pH decreases. The effect is minimized when excess hydrogen ions react with buffer molecules. An example is the *bicarbonate–carbon dioxide* buffer system:

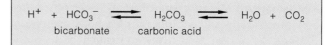

In this case, the buffering system neutralizes the excess hydrogen ions, and the carbon dioxide that forms in the reactions is eventually exhaled from the lungs. Like other buffer systems in the body, however, this one has a temporary effect. It does not eliminate the excess hydrogen ions. Only the urinary system can do this and so restore the buffers.

The same reactions proceed in reverse in cells of the nephron's tubular walls. The HCO_3^- resulting from the reverse reactions moves into interstitial fluid, then into the capillaries around the nephron. After this, it moves into the general circulation, where it buffers excess acid. The H^+ that formed in the cells is secreted into the nephron. There, it may combine with bicarbonate ions to form CO_2—which can be returned to the blood and exhaled by the lungs. The H^+ also may combine with phosphate ions or ammonia (NH_3), then leave the body in urine.

The kidneys, along with buffering systems (which neutralize acids) and the respiratory system, help keep the extracellular fluid from becoming too acidic or too basic (alkaline).

The bicarbonate–carbon dioxide buffer system temporarily neutralizes excess hydrogen ions. The urinary system alone eliminates excess hydrogen ions and restores these buffers.

The bicarbonate–carbon dioxide buffer system is one of the key mechanisms that help maintain the acid-base balance.

31.5 ON FISH, FROGS, AND KANGAROO RATS

Now that you have an idea of how your own body maintains water and solute levels, consider what goes on in some other vertebrates, including that kangaroo rat hopping about at the start of the chapter.

In freshwater, the bony fishes and amphibians gain water and lose solutes (Figure 31.7*a*). They don't gain water by drinking it, however. Water moves into the internal environment by osmosis—in fishes, across thin gill membranes, and in adult amphibians, across the skin. Excess water leaves as dilute urine, formed in a pair of kidneys. In both groups of vertebrates, solute losses are balanced by solutes gained from food and by the inward pumping of sodium by certain cells.

Body fluids of herring, snapper, and other marine fishes are about three times less salty than seawater. Such fishes lose water by osmosis, and they replace it by drinking more. They excrete ingested solutes against concentration gradients (Figure 31.7*b*). Fish kidneys do not have loops of Henle, so urine cannot ever become saltier than body fluids. Cells in the fish gills actively pump out most of the excess solutes in blood.

Figure 31.7*c* describes the wonderful water-solute balancing act in a salmon—a type of fish that makes its home in both freshwater *and* seawater.

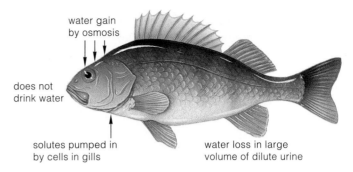

a Freshwater bony fish (body fluids far saltier than surroundings)

water gain by osmosis

does not drink water

solutes pumped in by cells in gills

water loss in large volume of dilute urine

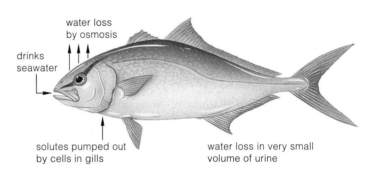

b Marine bony fish (body fluids less salty than surroundings)

water loss by osmosis

drinks seawater

solutes pumped out by cells in gills

water loss in very small volume of urine

Figure 31.7 (**a**,**b**) Water-solute balance in fishes.

(**c**) Water-solute balance by salmon, a type of fish that lives in saltwater and in freshwater. Salmon hatch in streams and later move downstream to the seas, where they feed and mature. They return to home streams to spawn.

For most salmon, salt tolerance is one outcome of changes in the concentrations of certain hormones. The changes seem to be triggered by increasing daylength in spring. Prolactin, a pituitary hormone, plays a role in sodium retention in freshwater habitats. When a freshwater fish has its pituitary gland removed, it dies from sodium loss— but that fish will live if prolactin is administered to it.

Cortisol, a steroid hormone secreted by the adrenal cortex, is crucial to the development of salt tolerance in salmon. Cortisol secretions correlate with an increase in sodium excretion, in sodium-potassium pumping by cells in the salmon's gills, and in absorption of ions and water in the gut. In young salmon, cortisol secretion increases prior to the seaward movement—and so does salt tolerance.

salmon avoiding grizzly while maintaining solute-water balance

c

And about that kangaroo rat! Proportionally, the loops of Henle of its nephrons are amazingly long, compared to yours. This means a great deal of sodium gets pumped out of the nephron. And so the solute concentration in the interstitial fluid around the loops becomes high. The osmotic gradient between the fluid in the loops and the urine is *so* steep, nearly all of the water that does reach the equally long collecting ducts gets reabsorbed. In fact, kangaroo rats give up only a tiny volume of urine, and it's three to five times more concentrated than the concentrated urine of humans.

The urinary systems of vertebrates differ in their details, but each kind functions to balance gains in water and solutes with losses of water and solutes.

Enzyme-mediated reactions proceed simultaneously in the millions or billions of cells of a large-bodied animal. The enzymes of most animals typically stay functional within the range of 0°–40°C (32°–104°F). Above 41°C or so, chemical bonds holding an enzyme molecule in its required shape are disrupted, and so is its function. When the temperature drops by 10 degrees, the rate of enzyme activity commonly plummets by 50 percent or more. Thus metabolism—and life itself—depends on maintaining the **core temperature** within the range of tolerance of the body's enzymes. "Core" refers to the internal temperature of the animal body, as opposed to temperatures of the tissues near its surface.

Heat Gains and Heat Losses

Each metabolic reaction generates heat. If heat were to accumulate internally, an animal's core temperature would steadily rise. But a warm body tends to lose heat to a cooler environment. And core temperature holds steady when the rate of heat loss balances the rate of metabolic heat production. Generally, the heat content of a complex animal depends on the balance between heat gains and losses, as summarized here:

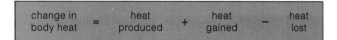

In this equation, the gains and losses are exchanges at the body surfaces, such as skin and respiratory surfaces. Four processes—radiation, conduction, convection, and evaporation—drive these exchanges.

With **radiation**, an animal gains heat after exposure to radiant energy (as from sunlight) or to any surface that is warmer than its own surface temperature.

With **conduction**, an animal directly gains or gives up heat from a solid object in contact with it. The outcome depends on the thermal gradient between them.

With **convection**, moving air or water transfers heat. This process involves conduction (heat moves down the thermal gradient between the body and air or water next to it). It also involves mass transfer, with currents carrying heat away from or toward the body.

With **evaporation**, a liquid converts to gaseous form and heat is lost in the process. The liquid's heat content provides energy for the conversion.

Ectotherms, Endotherms, and In-Betweens

Animals can adjust the amount of heat lost or gained through changes in behavior and physiology, although some are better equipped than others to do so. Like most animals, lizards, snakes, and reptiles have low

metabolic rates and poor insulation (Figure 31.8a). They rapidly absorb and gain heat, especially the small ones. Protecting the core temperature depends mainly on heat gains from the environment, not from metabolism. Hence we classify these animals as **ectotherms**, which means "heat from outside."

When outside temperatures change, an ectotherm must alter its behavior. This is **behavioral temperature regulation**. For example, the iguana shown at the start of this unit basks on warm rocks, thus gaining heat by conduction. It keeps reorienting its body to expose the most surface area to the sun's infrared radiation. It loses heat after sunset. Before its metabolic rates slow, it crawls inside crevices or under rocks, where heat loss is not as great and it is not as vulnerable to predators.

Most birds and mammals are **endotherms** ("heat from within"). With high metabolic rates, they can stay active under a wide temperature range. (Compared to a foraging lizard of the same weight, a foraging mouse uses up to thirty times *more* energy.) Core temperatures of these animals depend on a balancing of metabolism, controlled heat loss and conservation, and complex behavior. Adaptations in morphology help conserve or dissipate heat associated with the high metabolic rates. Feathers, fur, fat layers, and clothing reduce heat loss (Figure 31.8b). Some mammals in cold habitats have more massive bodies than close relatives in warmer places. They have a greater volume of cells for generating metabolic heat and less surface area for losing it.

Certain birds and mammals are **heterotherms**. At some times, their core temperature shifts (as it does in ectotherms). At other times, heat exchange is controlled (as in endotherms). For example, given their small size, hummingbirds have high metabolic rates. They locate and sip nectar only in the day. At night, they may shut down almost entirely. Their metabolic rates plummet, and they may get almost as cool as their surroundings. This way, they conserve precious energy.

In general, ectotherms are at an advantage in warm, humid tropics. They need not spend much energy to maintain core temperatures, and more energy can be devoted to reproduction and other tasks. In numbers and diversity, reptiles far exceed mammals in tropical regions. Endotherms have the advantage in moderate to cold environments. With their high metabolic rates, arctic hares and some other endotherms occupy even polar regions, where you would never find a lizard.

Responses to Cold Stress

Suppose receptors at the surface of a mammal's body detect a drop in outside temperature. The first response might be **peripheral vasoconstriction**—the diameters of blood vessels in skin would constrict and reduce blood's convective delivery of heat to body surfaces. (When your fingers or toes get chilled, all but 1 percent of blood that otherwise would flow to their skin is diverted.) Also, muscle contractions can make hairs "stand up" and create a layer of still air next to skin. This **pilomotor response** can reduce convective and radiative heat loss. Behavioral changes might minimize exposed surface areas and reduce heat loss, as when cats curl up or when you hold both arms tightly against the body. Prolonged cold can trigger **shivering**. Rhythmic tremors begin as skeletal muscles contract about ten to twenty times a second; this increases heat production by several times. Shivering costs energy and is not effective for long.

Failure to defend against cold results in *hypothermia*, a condition in which the core temperature falls below normal. In humans, a drop of only a few degrees affects brain function and leads to confusion; further cooling leads to coma and death. Many mammals recover from profound hypothermia. But frozen cells may die unless tissues thaw under close medical supervision. Tissue destruction through localized freezing is called *frostbite*.

Responses to Heat Stress

Peripheral vasodilation and evaporative heat loss are the main responses to heat stress. With **peripheral vasodilation**, blood vessels in skin dilate. More blood flows from deeper body regions to skin, where excess heat is dissipated. With **evaporative heat loss**, water molecules escape from body surfaces and carry energy away with them (Section 2.5). For animals on land, humidity and air movements govern evaporation rates. If the air is saturated (when the local relative humidity is 100 percent), water will not evaporate. If it's hot and dry, evaporative water loss occurs.

In humans and some other mammals, **sweat glands** move water and specific solutes through pores to the skin surface. An average-size human can produce 1 to 2 liters of sweat in an hour. For every liter of sweat that evaporates, 600 kilocalories of heat energy are lost. During intense exercise, this mechanism balances the high rates of heat production in skeletal muscle. With extreme sweating, as might occur in a marathon race, the body loses sodium chloride as well as water. Such losses disrupt extracellular fluid and cause the runner to collapse and faint. In itself, sweat dripping from skin doesn't dissipate heat; water in sweat must evaporate.

Sometimes peripheral blood flow and evaporative heat loss cannot counter heat stress, and *hyperthermia* results—the core temperature increases above normal. For humans and other endotherms, an increase of only a few degrees above normal can be dangerous. During

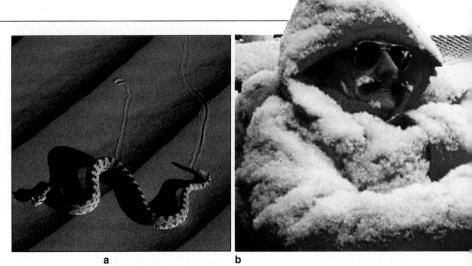

a   **b**

Figure 31.8 (**a**) A sidewinder making its signature J-shaped track across hot desert sand at dusk. As near as you can tell, how might this rattlesnake be gaining and losing heat? (**b**) How might the intrepid tourist sitting on a deck chair of a ship off the coast of Antarctica be gaining and losing heat?

a *fever*, the hypothalamus has reset a "thermostat" that dictates what the core temperature is supposed to be. Normal responses to heat stress operate, but they are carried out to maintain a *higher* temperature. At the onset of fever, heat loss decreases, heat production increases, and a person feels chilled. When the fever "breaks," peripheral vasodilation and sweating increase as the hypothalamus attempts to restore the normal core temperature. At such times, the person feels warm.

Fever may be an important defense mechanism against infections, and perhaps against cancer. During infection, macrophages and other cells secrete signaling molecules, including interleukin-1 and interferons. The secretions somehow influence the hypothalamus. We know that aspirin and other drugs that interfere with synthesis of prostaglandins can block their influence. (Prostaglandins injected directly into the hypothalamus can induce fever.) The controlled elevation in the core temperature during a fever seems to enhance the body's immune response. Therefore, the widespread practice of administering aspirin and other drugs may interfere with beneficial effects of minor fevers. But without question, medication is called for if a fever approaches dangerously high levels.

The internal, core temperature of an animal's body is being maintained when heat gains and heat losses are in balance.

Metabolic reactions generate heat inside the body. Radiation, conduction, and convection can move heat down thermal gradients that exist between the body and its surroundings. Evaporative heat loss carries heat away from the body.

Besides being morphologically adapted to their habitats, animals can make behavioral and physiological adjustments to environmental temperatures.

SUMMARY

Control of Extracellular Fluid

1. For cells inside the animal body, the environment consists of certain types and amounts of substances that are dissolved in water. The extracellular fluid fills tissue spaces and blood vessels. Its volume and composition are maintained only when the animal's daily intake and output of water and solutes are in balance. In mammals, the following processes maintain the balance:

   a. Water is gained by absorption from the gut and by metabolism. It is lost by urinary excretion, evaporation from lungs and skin, sweating, and elimination of feces.

   b. Solutes are gained by absorption from the gut, secretion, respiration, and metabolism. They are lost by excretion, respiration, and sweating.

   c. Losses of water and solutes are controlled mainly by adjusting the volume and composition of urine.

2. The urinary system of vertebrates consists of two kidneys, two ureters, a urinary bladder, and a urethra.

3. Kidneys have many nephrons that filter blood and form urine. Each nephron interacts intimately with two sets of blood capillaries (glomerular and peritubular).

   a. The start of a nephron is cup-shaped (Bowman's capsule). It continues as three tubular regions (proximal tubule, loop of Henle, and distal tubule, which empties into a collecting duct).

   b. Together, the Bowman's capsule and the set of highly permeable, glomerular capillaries within it are a blood-filtering unit (glomerulus).

   c. Blood pressure forces water and small solutes out of the capillaries, into the cup. Most of the filtrate is reabsorbed by the tubules and returned to the blood. A portion is excreted as urine.

4. Urine forms in the nephron by three processes:

   a. Filtration of blood at the glomerulus, which puts water and small solutes into the nephron.

   b. Reabsorption. Water and solutes to be retained leave the nephron's tubular parts and enter capillaries that thread around them. A small volume of water and solutes remains in the nephron.

   c. Secretion. A few substances can leave peritubular capillaries and enter the nephron, for disposal in urine.

5. Urine is made more concentrated or less so by the action of two hormones that act on cells making up the wall of distal tubules and collecting ducts. One, ADH, conserves water by enhancing reabsorption across the nephron wall. Inhibition of ADH allows more water to be excreted. The other hormone, aldosterone, conserves sodium by enhancing reabsorption. The inhibition of aldosterone allows more sodium to be excreted.

6. The urinary system, together with the respiratory system, has a central role in maintaining the body's overall acid-base balance.

Control of Body Temperature

1. Maintaining an animal's core (internal) temperature depends on balancing metabolically produced heat and the heat absorbed from and lost to the environment.

2. Animals exchange heat with their environment by these processes:

   a. Radiation. Emission of infrared and other wavelengths. Radiant energy can be absorbed at the body surface, then converted to heat energy.

   b. Conduction: Direct transfer of heat energy from one object to another object in contact with it.

   c. Convection. Heat transfer by air or water currents; involves conduction and mass transfer of heat-bearing currents away from or toward the animal body.

   d. Evaporation. Conversion of liquid to a gas, driven by energy inherent in the heat content of the liquid. Some animals lose heat by evaporative water loss.

3. Core temperatures depend on metabolic rates and on anatomical, behavioral, and physiological adaptations.

   a. For ectotherms, core temperature depends more on heat exchange with the environment than on metabolic heat.

   b. For endotherms, core temperature depends largely on high metabolic activity and precise controls over heat produced and heat lost.

   c. For heterotherms, core temperature fluctuates some of the time, and controls over heat balance come into play at other times.

Review Questions

1. State the function of the urinary system in terms of gains and losses for the internal environment. Then name the components of the mammalian urinary system and state their overall functions. *532–533*

2. Define filtration, tubular reabsorption, and secretion. How does urine formation help maintain the internal environment? *534*

3. Label the component parts of the kidney and the nephron. *533*

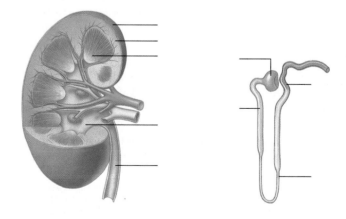

4. Which hormone promotes water conservation? Which hormone promotes sodium conservation? *535*

Self-Quiz (Answers in Appendix IV)

1. In mammals, water intake depends on _____ .
   a. absorption from gut   c. a thirst mechanism
   b. metabolism   d. all of the above

2. In mammals, water is lost by way of the _____ .
   a. skin   d. urinary system
   b. respiratory system   e. c and d
   c. digestive system   f. a through d

3. Water and small solutes return to blood during _____ .
   a. filtration   c. tubular secretion
   b. tubular reabsorption   d. both a and b

4. Water as well as small solutes leave blood during _____ .
   a. filtration   c. tubular secretion
   b. tubular reabsorption   d. both a and c

5. A few substances move out of the capillaries threading around tubular parts of the nephron. These substances are moved into the nephron during _____ .
   a. filtration   c. tubular secretion
   b. tubular reabsorption   d. both a and c

6. A nephron's reabsorption mechanism depends on _____ .
   a. osmosis across nephron wall
   b. active transport of sodium across nephron wall
   c. a steep solute concentration gradient
   d. all of the above

7. _____ promotes water conservation.
   a. ADH   c. Low extracellular fluid volume
   b. Aldosterone   d. Both a and c

8. _____ enhances sodium reabsorption.
   a. ADH   c. Low extracellular fluid volume
   b. Aldosterone   d. Both b and c

9. Match the term with the most suitable description.
   ____ glomerulus   a. surrounded by saltiest fluid
   ____ distal tubule   b. extra-long loops of Henle
   ____ loop of Henle   c. involves buffer systems
   ____ acid-base balance d. blood-filtering unit
   ____ kangaroo rat   e. ADH, aldosterone act here

10. Match the term with the most suitable description.
    ____ ectotherm   a. heat transfer by air or water currents
    ____ endotherm   b. body temperature fluctuates some of
    ____ evaporation   the time, is controlled other times
    ____ heterotherm   c. emission of wavelength energy
    ____ radiation   d. direct heat transfer between one
    ____ conduction   object and another in contact with it
    ____ convection   e. metabolism dictates core temperature
                        f. environment dictates core temperature
                        g. conversion of liquid to gas

Critical Thinking

1. Fatty tissue holds the kidneys in place. Extremely rapid weight loss may cause the tissue to shrink and the kidneys to slip from their normal position. If slippage puts a kink in one or both ureters and blocks urine flow, what might happen to the kidneys?

2. Drink a quart of water in an hour. What changes can you expect in your kidney function and in urine composition?

3. In 1912, the ocean liner *Titanic* left Europe on her maiden voyage across the Atlantic to America. In that same year, a chunk of the leading edge of a Greenland glacier broke off and floated out to sea. Late at night, off the Newfoundland coast, the iceberg and the *Titanic* made an ill-fated rendezvous (Figure 31.9). The *Titanic* was

Figure 31.9 Artist's rendition of the sinking of the *Titanic*, based on eyewitness accounts.

said to be unsinkable. Survival drills had been neglected. There were not enough lifeboats to hold even half the 2,200 passengers. The *Titanic* sank in about 2-1/2 hours. Within two hours, rescue ships were on the scene, but 1,517 bodies were recovered from a calm sea. All the dead had on life jackets. None had drowned. Probably they died from _____ . If so, how did their blood flow, metabolism, and skeletal muscle action change prior to death?

4. When iguanas have an infection, they rest for a very long time in the sun. Propose a hypothesis to explain why.

5. Out on a first date, Jon takes Geraldine's hand in a darkened theater. "Aha!" he thinks. "Cold hands, warm heart!" What does this tell us about the regulation of core temperature, let alone Jon?

Selected Key Terms

Water-Solute Balance:
acid-base balance 536
ADH 535
aldosterone 535
blood 531
Bowman's capsule 533
buffer system 536
collecting duct 533
distal tubule 533
extracellular fluid 531
filtration 534
glomerulus 533
interstitial fluid 531
kidney 533
loop of Henle 533
nephron 533
proximal tubule 533
tubular reabsorption 534
tubular secretion 534
urea 532
ureter 533
urethra 533
urinary bladder 533

urinary excretion 532
urinary system 533
urine 533

Temperature Control:
behavioral temperature
   regulation 538
conduction 538
convection 538
core temperature 538
ectotherm 538
endotherm 538
evaporation 538
evaporative heat loss 539
heterotherm 538
peripheral vasoconstriction 538
peripheral vasodilation 539
pilomotor response 539
radiation 538
shivering 539
sweat gland 539
thirst behavior 535

Readings

Flieger, K. March 1990. "Kidney Disease: When Those Fabulous Filters Are Foiled." *FDA Consumer* 24: 26–29.

Smith, H. 1961. *From Fish to Philosopher*. New York: Doubleday.

Vander, A., J. Sherman, and D. Luciano. 1994. *Human Physiology*. Sixth edition. New York: McGraw-Hill, Chapter 15.

32 NEURAL CONTROL AND THE SENSES

Why Crack The System?

Suppose your biology instructor asks you to volunteer for an experiment. You will get a microchip implanted in your brain. It will make you feel really good. But it may mess up your health, lop ten years off your life, and destroy a good part of your brain. Your behavior will change for the worse, so you might have trouble completing school, getting or keeping a job, or even having a normal family life.

The longer the chip is implanted, the less you will want to give it up. You won't get paid. You will pay the experimenter—first at bargain rates, then a little more each week. The chip is illegal. If you get caught using it, you and the experimenter will go to jail.

Sometimes Jim Kalat, a professor at North Carolina State University, proposes this experiment—which of course is hypothetical. Hardly any students volunteer. Then he substitutes *drug* for microchip and *dealer* for experimenter—and an amazing number of students come forward! Like 30 million other Americans, the "volunteers" seem ready to engage in self-destructive uses of drugs that alter emotional and behavioral states.

The destruction shows up in unexpected places. Each year, for instance, about 300,000 newborns are

already addicted to crack—thanks to their addicted mothers. *Crack* is a cheap form of cocaine. It causes relentless stimulation of brain regions that govern our sense of pleasure. It dampens normal urges to eat and sleep, and blood pressure rises. Elation and sexual desire intensify. In time, though, the brain cells that produce the stimulatory chemicals can't keep up with the abnormal demand. The chemical vacuum makes crack users frantic, then profoundly depressed. Only crack makes them feel good again.

Addicted babies cannot know all of this. They can only quiver with "the shakes" and respond to the world with chronic irritation. And they are abnormally small. While they were developing inside their mother, their body tissues simply were not provided with enough oxygen and nutrients. As one of its side effects, crack causes blood vessels to constrict—and maternal blood vessels are the only supply lines that a developing individual has.

Paradoxically, crack babies are exceptionally fussy, yet they cannot respond to rocking and other forms of stimulation that normally have soothing effects. It may be a year or more before they recognize even their own

Figure 32.1 Owners of an evolutionary treasure—a complex brain that is the foundation for our memory and reasoning, and our future.

mother. Without treatment, they are likely to remain emotionally unstable, prone to aggressive outbursts and stony silences. Why? The mother's drug habit crippled their nervous system.

Think about it. The **nervous system** evolved as a way to sense and respond, with exquisite precision, to changing conditions inside and outside of the body. Awareness of sounds and sights, of odors, of hunger and passion, fear and rage—all such things begin with the flow of information along the communication lines of the nervous system.

Do those communication lines remain silent until they receive outside signals, much as telephone lines wait to carry calls from all over the country? Absolutely not. Even before you were born, those excitable cells called neurons became organized in vast gridworks— and started chattering among themselves. All through your life, in moments of danger or reflection, supreme excitement or sleep, their chattering has never ceased.

In most animals, three classes of neurons *collectively* monitor conditions and issue commands for responsive actions that benefit the body as a whole. The **sensory neurons** respond to specific stimuli, such as pressure, light, and other forms of energy. They relay information about the stimulus to integration centers. Your own spinal cord and brain are examples of such centers, where **interneurons** receive and process sensory input, integrate it with stored information about previous experiences, then influence the activity of other neurons. Finally, **motor neurons** relay commands away from the brain and spinal cord to muscles or glands. Muscles and glands, the body's effectors, carry out the responses:

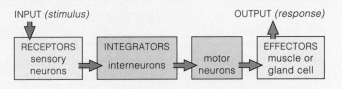

Altogether, the communication lines make up less than half the volume of your nervous system. The balance consists mostly of **neuroglia**, amazingly specialized cells that variously protect neurons and provide them with structural and functional support.

This chapter can give you insight into the structure and function of neurons, then of the nervous system. Use these insights to think about what can happen when their operation falters, whether by disease or mutation, or by deliberate and abusive use of drugs.

KEY CONCEPTS

1. Neurons are the basic units of communication of most nervous systems. Collectively, they detect and integrate information about external and internal conditions, then select or control muscles and glands in ways that produce suitable responses.

2. The inside of a neuron is negatively charged relative to the outside. When a neuron is stimulated, this polarity of charge across the membrane may briefly and abruptly reverse. Such reversals are called action potentials.

3. Action potentials are the basis of information flow through the nervous system. They are self-propagating along an individual neuron, but they can't cross the small gaps between most neurons. Chemical signals bridge the gaps and stimulate or inhibit the adjoining neuron, muscle cell, or gland cell.

4. The simplest nervous systems are the nerve nets of radial animals. The vertebrate nervous system shows pronounced cephalization and bilateral symmetry. It includes a brain, spinal cord, and many paired nerves. The brain, the most complex integrative center, receives, processes, and responds to information.

5. Sensory systems are part of the nervous system. Each consists of specific types of sensory receptors, nerve pathways from those receptors to the brain, and brain regions that deal with specific sensory information.

6. A stimulus is a form of energy that activates a specific sensory receptor. The somatic sensations include touch, pressure, temperature, pain, and muscle sense. The special senses include taste, smell, hearing, and vision.

32.1 INVERTEBRATE BEGINNINGS

Life-Styles and Nervous Systems

Even if you stare at it for hours on end, a sea urchin will never dazzle you with spectacular bursts of speed or precision acrobatics. Because it seems only a bit livelier than the pincushion it resembles, you might suspect that this invertebrate is brainless, and in fact it is. But sea urchins do have a nervous system, even if it is a decentralized one.

At this point in the book, you already know that nearly all animals have some type of nervous system. Depending on the animal, these systems incorporate dozens to more than a hundred billion communication cells—neurons or simpler types of excitable "nerve cells." These cells are oriented relative to one another, as part of signal conducting and information-processing pathways that extend throughout the body. *Together, they detect changing conditions outside and inside the body, then elicit suitable responses from muscles and glands.*

To appreciate any animal's nervous system, you have to ask, *What does the animal do?* Many sea urchins spend most of their lives moving about slowly on little tube feet, scraping bits of algae off submerged rocks with a marvelously toothy feeding apparatus (Figure 32.2). When neurons located in body tissues respond to information, signals travel along three sets of nerves. A **nerve** is a cablelike bundle of long, slender extensions from many neurons. The nerves in sea urchins deal mainly with sensory functions, motor functions, and (you guessed it) getting food into the gut. Information flow through the brainless body is not highly focused. It travels in all directions from a point of stimulation. You might not find this type of nervous system very impressive. But if you did little more than inch about eating algae, it would be quite sufficient.

Regarding the Nerve Net

Animals first evolved in the seas, and it is in the seas that we still find the animals with the simplest nervous systems. They are sea anemones, jellyfishes, and other cnidarians (Figure 32.3). All of these invertebrates show radial symmetry, with their body parts arranged about a central axis, like bike wheel spokes. Cnidarians do not have nerves. Information travels in a rather diffuse way along a **nerve net**, which is a loose mesh of nerve cells associated with epithelial tissue. Cells of this type of nervous system interact with sensory and contractile cells (also located in epithelial tissue) and thereby form **reflex pathways**. In such pathways, sensory stimulation directly causes simple, stereotyped movements.

For example, in jellyfishes, reflex pathways permit slow swimming and keep the body right-side up. A reflex pathway dealing with feeding behavior extends

Figure 32.2 A spiny sea urchin, right-side up and flipped over to show its toothy, all-important mouth.

from sensory receptors present in the tentacles, along nerve cells, to contractile cells around the mouth.

On the Importance of Having a Head

Flatworms, recall, are the simplest animals with a bilateral nervous system (Figure 32.4). Bilateral symmetry means having equivalent parts on the left and right sides of the body's midsagittal plane. Both sides have the same array of muscles that move the body, the same array of nerves to control the muscles, and so on. The ladderlike, bilateral system of a flatworm includes two nerve cords and **ganglia** (singular, ganglion)—clusters of nerve cell bodies. Some ganglia form an integration center—a rudimentary **brain**—in the flatworm head. They coordinate signals from paired sensory organs, including two eyespots, and provide some control over the nerve cords.

Did bilateral nervous systems evolve from nerve nets? Maybe. Consider planulas, a type of self-feeding larval stage that develops in the life cycles of certain cnidarians. Like flatworms, most planulas use small cilia to swim or crawl about:

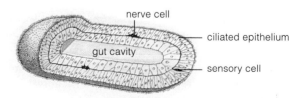

nerve cell
ciliated epithelium
gut cavity
sensory cell

Imagine a few ancient planulas crawling about on the seafloor in Cambrian times. By chance, mutations of

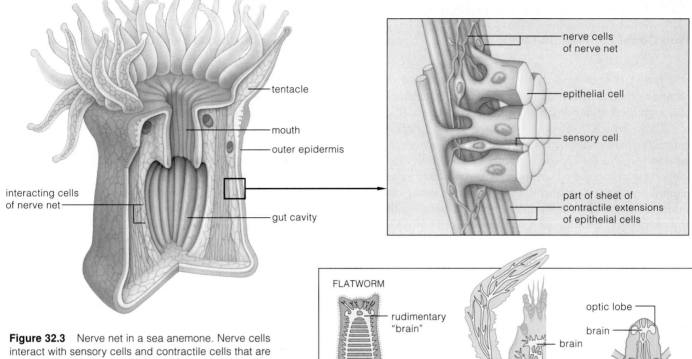

Figure 32.3 Nerve net in a sea anemone. Nerve cells interact with sensory cells and contractile cells that are arranged loosely in an epithelium between the outer layer and jellylike middle layer (mesoglea) of the body wall.

regulatory genes blocked the transformation into adults, yet the reproductive organs still matured. (This is not a far-fetched scenario; it occurs among certain existing animals, such as salamanders.) The planulas kept on crawling, they reproduced —and so passed on the mutated genes responsible for their permanently forward mobility.

Planulalike animals that crawled far enough could put themselves in new, potentially splendid or dangerous situations. Having a lot of sensory cells concentrated at the trailing end of the body wouldn't do the animal much good. But if they were to become concentrated at the leading end, this would increase the capacity for fast, effective responses to stimuli. Natural selection would have favored such a concentration of sensory cells at the leading end. And this might have been the beginning of cephalization (formation of a head region) and bilateral symmetry.

Now jump a bit forward in time, to the emergence of the first vertebrates—filter-feeders or scavengers that moved around very little or not at all (Section 21.3). Some evolved into swift, jawed predators—and their nervous system, with simple reflex pathways, became more intricately wired. Later, when some vertebrates invaded the land, their senses of smell, hearing, and vision became keener. Muscles evolved in specialized ways. The brain itself grew thicker, with many more neurons that could better integrate the richer sensory information and issue commands for more intricate responses. As you will see later in the chapter, patterns

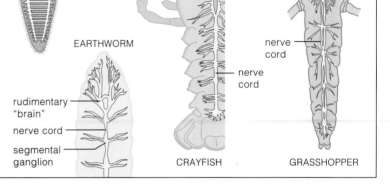

Figure 32.4 Bilateral nervous system of a few invertebrates. The sketches are not to the same scale.

of bilateral symmetry and cephalization have echoes in the pairs of nerves, pairs of muscles, paired sensory structures, and paired brain centers of you and all other vertebrates.

For any animal, the organization of its nervous system and the distribution of sensory organs are correlated with body symmetry.

Radial nervous systems are decentralized; they can respond to stimuli coming from any direction.

Bilateral nervous systems include centralized masses of nervous tissue, ranging from simple ganglia to a complex brain, and communication lines (nerves) that are similarly arranged on both sides of the body.

Functional Zones of a Neuron

To understand how your own nervous system works, you can start with how its neurons function. Neurons have a nucleated cell body and cytoplasmic extensions, although these differ greatly in number and length (Figure 32.5). Typically, the cell body and certain slender extensions called **dendrites** are *input* zones, where the neuron receives information. Another slender but often longer extension called an **axon** is a *conducting* zone. Signals arising at a neuron's *trigger zone* are propagated rapidly along an axon. Except in sensory neurons, the trigger zone is at the junction between the cell body and the axon. An axon's branched endings are *output* zones, where messages are sent to other cells.

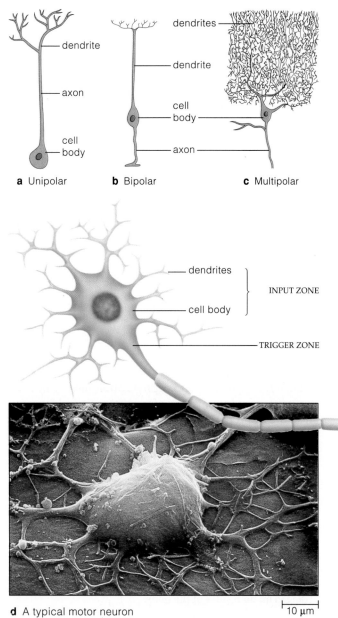

a Unipolar   **b** Bipolar   **c** Multipolar

d A typical motor neuron   10 µm

A Neuron At Rest, Then Moved To Action

Different forms of signals arise in the nervous system. Let's start with one that begins and ends on the same neuron, and is not transferred to another cell.

When a neuron is not being bothered, a difference in electric charge is being maintained across its plasma membrane. The cytoplasmic fluid next to the membrane remains negatively charged, compared to the interstitial fluid right outside. We measure these charges in units called millivolts. The amount of energy inherent in the steady voltage difference across the plasma membrane is the **resting membrane potential**. For many neurons, that amount is about −70 millivolts.

Suppose a weak signal reaches a patch of membrane in a neuron's input zone. The voltage difference across the patch changes only slightly, if at all. By contrast, a strong signal might trigger an **action potential**—an abrupt, short-lived reversal in the voltage difference across the plasma membrane. For a fraction of a second, the inside becomes positive with respect to the outside. The reversal invites an action potential at the adjoining patch of membrane, which invites another at the next patch, and so on away from the initiation point. In short, *stimulation of a neuron disturbs the distribution of electric charge across its plasma membrane.*

Restoring and Maintaining Readiness

How does a neuron restore and maintain the voltage difference across each patch of plasma membrane in between action potentials? It relies on two membrane properties. First, a membrane has a lipid bilayer, which bars the passage of potassium ions (K^+), sodium ions (Na^+), and other charged substances. Thus, the neuron can build up differences in ion concentrations across the membrane. Second, ions flow from one side to the other through the interior of transport proteins that span the bilayer—and that flow can be controlled (Figure 32.6).

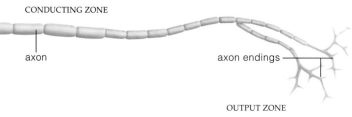

Figure 32.5 (**a–c**) Classification of neurons based on the number of cytoplasmic extensions of the cell body. *Unipolar* cells have one extension (axon) with dendritic branchings. *Bipolar* cells have one dendrite and one axon. Many sensory neurons are like this. *Multipolar* cells, with one axon and many dendrites, predominate in vertebrate nervous systems. (**d**) Functional zones of a motor neuron, a multipolar cell.

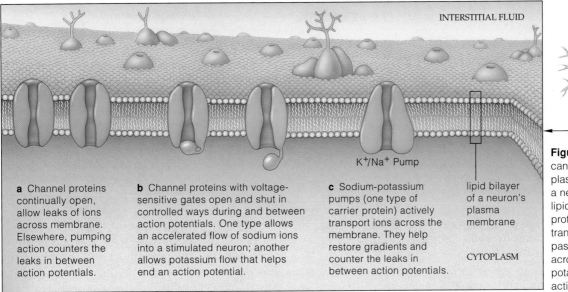

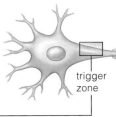

trigger zone

a Channel proteins continually open, allow leaks of ions across membrane. Elsewhere, pumping action counters the leaks in between action potentials.

b Channel proteins with voltage-sensitive gates open and shut in controlled ways during and between action potentials. One type allows an accelerated flow of sodium ions into a stimulated neuron; another allows potassium flow that helps end an action potential.

c Sodium-potassium pumps (one type of carrier protein) actively transport ions across the membrane. They help restore gradients and counter the leaks in between action potentials.

lipid bilayer of a neuron's plasma membrane

CYTOPLASM

K⁺/Na⁺ Pump

Figure 32.6 How ions can move across the plasma membrane of a neuron. Spanning the lipid bilayer are channel proteins (a type of transport protein) that passively move ions across and sodium-potassium pumps that actively move ions across.

Suppose a motor neuron has 15 sodium ions inside the membrane for every 150 outside. Suppose it has 150 potassium ions inside for every 5 on the outside. We can depict each ion's concentration gradient in this way (from the large to the small letters):

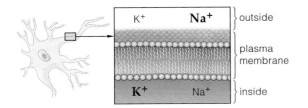

Gradients such as these determine the net direction in which sodium and potassium ions diffuse. The ions can diffuse through the interior of channel proteins, a type of transport protein in the membrane. Some of the channels never shut, so ions leak (diffuse) through them all of the time. Others have molecular gates, which can open only after the neuron is adequately stimulated.

Suppose that motor neuron is not being stimulated. Its sodium channels are shut, so sodium ions can't rush inside. Some potassium is leaking out through a few open channels and making the cytoplasm a bit more negative, so some potassium is attracted back in. When the inward pull of electric charge balances the outward force of diffusion, there is no more net movement of potassium. The concentration and electric gradients now existing across the plasma membrane will permit the neuron to respond to stimulation.

After the gradients have reversed during an action potential, **sodium-potassium pumps** restore them. As you read in Section 4.3, these transport proteins span the plasma membrane. When they get an energy boost from ATP, they actively transport potassium into the

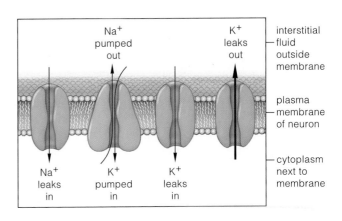

Figure 32.7 Pumping and leaking processes that influence the distribution of sodium and potassium ions across the plasma membrane of a neuron at rest. Arrow widths indicate the magnitude of the movements. Notice how the total inward and outward movements for each kind of ion are balanced.

neuron and sodium out at the same time. These pumps must maintain as well as restore the gradients. Why? Even in a resting neuron, a tiny fraction of the outward-leaking potassium isn't attracted back in. Besides this, a tiny fraction of sodium leaks in, through a few open channels (Figure 32.7). If the leaks went unattended, the crucial gradients would gradually disappear.

A neuron not being disturbed maintains a resting membrane potential—a voltage difference across its plasma membrane. An action potential is an abrupt, short-lived reversal in that voltage difference in response to adequate stimulation.

After an action potential, sodium-potassium pumps restore and maintain the resting membrane potential.

Approaching Threshold

Stimulate a neuron at its input zone and you disturb the ion balance across the membrane, but not much. For instance, suppose your toes tap a cat and put a bit of pressure on its skin. In tissues beneath the skin surface are receptor endings—input zones of sensory neurons. Patches of plasma membrane at the receptor endings deform under the pressure. Some ions now flow across, so the voltage difference across the membrane changes slightly. The pressure produced a graded, local signal.

Graded means that signals arising at an input zone can vary in magnitude. Such signals can be large or small, depending on the intensity of the stimulus.

Local means the signals don't spread far from the point of stimulation. It takes specialized types of ion channels to propagate a signal along the membrane, and input zones simply don't have them.

However, when a stimulus is intense or long lasting, graded signals can spread out from the input zone and into an adjacent trigger zone. At this zone, a certain minimum amount of change in the voltage difference across the plasma membrane can trigger an action potential. That amount is known as the *threshold* level. Threshold can be reached at any membrane patch that has voltage-sensitive, gated channels for sodium ions.

As Figure 32.8 indicates, a stimulus causes sodium ions to flow across the membrane, into the neuron. With this influx of positively charged ions, the cytoplasmic

side of the membrane becomes less negative. This causes more gates to open, more sodium to enter, and so on. The ever increasing inward flow of sodium is a good example of positive feedback, whereby an event intensifies as a result of its own occurrence:

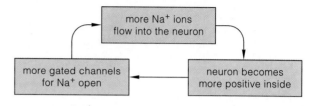

At threshold, the opening of more sodium gates no longer depends on the strength of the stimulus. The positive-feedback cycle is now under way, so that the inward-rushing sodium itself is enough to cause more sodium gates to open.

An All-or-Nothing Spike

Figure 32.9 shows a recording of the voltage difference across the plasma membrane before, during, and after an action potential. Notice how the membrane potential spikes once threshold is reached. Every single action potential in the neuron spikes to the same level above threshold as an *all-or-nothing* event. That is, once the positive-feedback cycle starts, nothing will stop the full spiking. If the threshold is not reached, however, the disturbance to the plasma membrane will subside when the stimulus is removed.

Each spike lasts only for a millisecond or so. Why? At the membrane site of a charge reversal, the gated sodium channels closed and shut off the sodium inflow. Also, about halfway through the reversal, potassium channels opened, so many more potassium ions flowed

Figure 32.8 Propagation of an action potential along the axon of a motor neuron.

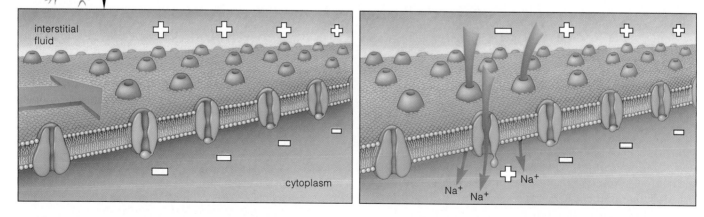

a Membrane at rest (inside negative with respect to the outside). An electrical disturbance (*red arrow*) spreads from an input zone to an adjacent trigger region of the membrane, which has a great number of gated sodium channels.

b A strong disturbance initiates an action potential. Sodium gates open. The sodium inflow decreases the negativity inside the neuron. This causes more gates to open, and so on, until threshold is reached and the voltage difference across the membrane reverses.

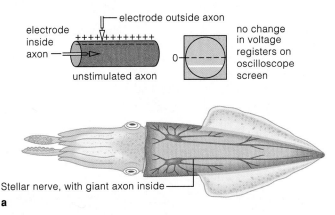

electrode inside axon

electrode outside axon

no change in voltage registers on oscilloscope screen

unstimulated axon

a

b stimulated axon

recording of voltage change

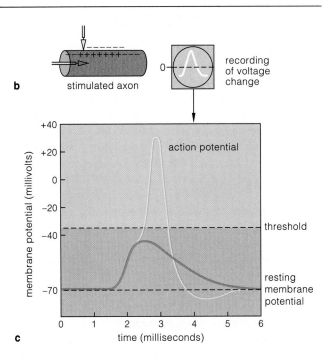

c

Figure 32.9 Action potentials. (**a**) When early researchers studied neural function, the squid *Loligo* provided them with evidence of action potential spiking. The squid's "giant" axons were large enough to slip electrodes inside. (**b**) When such an axon was stimulated, electrodes positioned on the inside and outside detected voltage changes, which showed up as deflections in a beam of light across the screen of an oscilloscope. (**c**) This is a typical waveform (*yellow* line) for an action potential. The *red* line represents a recording of a local signal that did not reach the threshold of an action potential; spiking did not occur.

Stellar nerve, with giant axon inside

out and restored the original voltage difference across the membrane. And sodium-potassium pumps restored the ion gradients. Later on, after the resting membrane potential has been restored, most potassium gates close and sodium gates are in their initial state, ready to be opened with the arrival of a suitable disturbance.

Propagation of Action Potentials

The membrane disturbances leading up to an action potential are self-propagating, and they don't diminish in magnitude. As they spread to an adjacent membrane patch, an equivalent number of gated channels open. With this new disturbance, gated channels open in the *next* adjacent patch, then the next, and so on.

For a brief period after a membrane patch has been excited, it is insensitive to stimulation. Its sodium gates are inactivated and ions cannot move past them. This is the reason why action potentials do not spread back to the trigger zone (the site where they were initiated) but rather are self-propagating away from it.

The cytoplasm just inside the plasma membrane of a neuron at rest is more negative than the interstitial fluid just outside the membrane.

During an action potential, the inside is more positive than the outside at a disturbed patch of membrane.

Following an action potential, resting conditions are restored at the membrane patch.

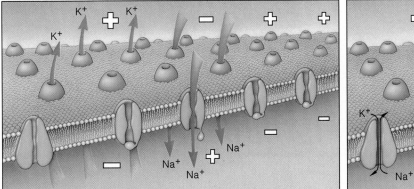

c With the reversal, sodium gates shut and potassium gates open (*at purple arrows*). Potassium follows its gradient out of the neuron. Voltage is restored. The disturbance triggers an action potential at the adjacent site, and so on, away from the point of stimulation.

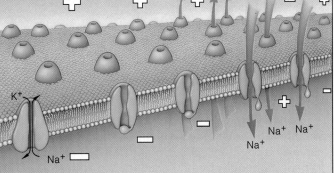

d Following each action potential, the inside of the plasma membrane becomes negative once again. But the sodium and potassium concentration gradients are not yet fully restored. Active transport at sodium-potassium pumps restores them.

CHEMICAL SYNAPSES

When action potentials reach a neuron's output zone, they usually don't proceed farther than this. But their arrival may induce the neuron to release one or more **neurotransmitters**. These are signaling molecules that diffuse across junctions called **chemical synapses**. The junctions are narrow clefts between one neuron's output zone and an input zone of an adjacent cell (Figure 32.10). Some clefts occur between two neurons, others between a neuron and a muscle cell or gland cell.

At each chemical synapse, one of the two cells stores neurotransmitter molecules inside synaptic vesicles in its cytoplasm. Think of it as the *pre*synaptic cell. Here, gated channels for calcium ions span the membrane, and they open when an action potential arrives. There are more calcium ions outside the cell. When they flow into the cell (down their gradient), the synaptic vesicles are induced to fuse with the plasma membrane, so that neurotransmitter is released into the synaptic cleft.

The neurotransmitter molecules diffuse through the cleft. On the *post*synaptic cell membrane are protein receptors that bind specific neurotransmitters. Binding changes the shape of these receptors and opens up a channel through their interior. Ions cross the membrane by diffusing through the channels (Figure 32.10*d*).

A postsynaptic cell's response depends on the type and concentration of neurotransmitter in the cleft, what kinds of receptors the cell bears, and the number and responsiveness of gated channels in its membrane. Such factors influence whether a neurotransmitter will have an *excitatory* effect and help drive the postsynaptic cell's membrane toward the threshold of an action potential. They also influence whether it will have an *inhibitory* effect and drive the membrane away from threshold.

Consider acetylcholine (ACh), a neurotransmitter. It has excitatory and inhibitory effects on the brain, spinal cord, glands, and muscles. For example, it acts at the chemical synapse between a motor neuron and muscle cell (Figure 32.10). ACh released from the motor neuron diffuses across the cleft and binds to receptors on the muscle cell membrane. It has excitatory effects on this kind of cell. It triggers action potentials, which in turn initiate muscle contraction (Section 26.7).

A Smorgasbord of Signals

Acetylcholine is only one of a veritable smorgasbord of signals that neurons deliver to target cells. For example, serotonin acts on neurons in brain regions that govern sleeping, sensory perception, temperature control, and emotions. Norepinephrine works in brain regions that control emotions, dreaming, and waking up. Dopamine also works in brain regions dealing with emotions. GABA (gamma aminobutyric acid) is the most common inhibitory signal in the brain. Antianxiety drugs, such

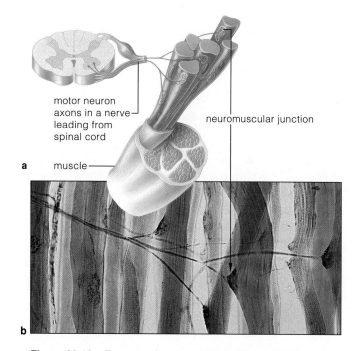

Figure 32.10 Example of a chemical synapse. (**a**,**b**) Drawing and light micrograph of neuromuscular junctions—regions of chemical synapsing between axon endings of a motor neuron and a muscle cell. *Facing page:* (**c**) The plasma membrane of this motor neuron and muscle cell face each other across the junction between them. Arrival of an action potential at an axon ending triggers the release of neurotransmitter molecules from the neuron. (**d**) The molecules diffuse through the junction and bind to receptors on gated channel proteins of the muscle cell membrane. The gates open, and ions flow inward. The inflow triggers a graded potential at the membrane site.

as Valium, may exert their effects by enhancing GABA's effects. Except for ACh, these neurotransmitters and others are amino acids or are derived from them.

Signaling molecules known as neuromodulators can magnify or reduce the effects of a neurotransmitter on neighboring or distant neurons. They include substance P, which induces pain perception, and endorphins—the natural pain killers that inhibit the release of substance P from sensory nerves. Neuromodulators might also influence memory and learning, sexual activity, control of body temperature, and emotional states.

Synaptic Integration

Anywhere from 1,000 to 10,000 communication lines form synapses with a typical neuron in your brain. And your brain contains at least 100 billion neurons. As long as you are alive, those neurons continually hum with messages about doing what it takes to be a human.

At any moment, a great number of excitatory and inhibitory signals are washing over the input zones of a postsynaptic cell. Some signals drive its membrane closer to threshold; others maintain the resting level or

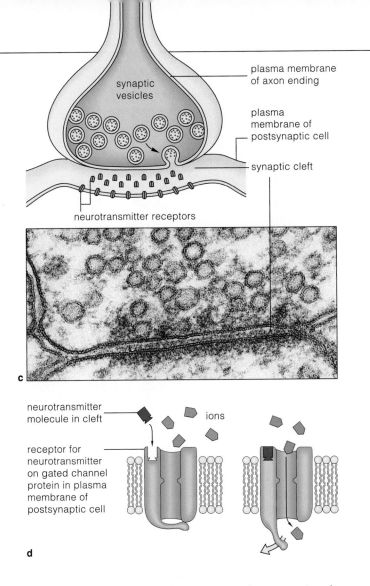

synaptic vesicles

plasma membrane of axon ending

plasma membrane of postsynaptic cell

synaptic cleft

neurotransmitter receptors

c

neurotransmitter molecule in cleft

ions

receptor for neurotransmitter on gated channel protein in plasma membrane of postsynaptic cell

d

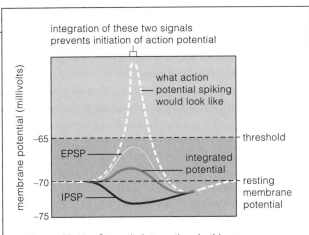

integration of these two signals prevents initiation of action potential

what action potential spiking would look like

threshold

EPSP

integrated potential

resting membrane potential

IPSP

membrane potential (millivolts)

−65

−70

−75

Figure 32.11 Synaptic integration. In this case, an excitatory synapse and an inhibitory synapse nearby are activated at the same time. The *yellow* line shows how an EPSP of a certain magnitude would register on an oscilloscope screen *if it were acting alone*. The *purple* line shows what the effect would be for a single IPSP. The *red* line shows their effect on a postsynaptic cell membrane when they arrive at the same time. When these two signals are integrated (the *red* line), threshold is not reached, so an action potential cannot be initiated in the target cell.

drive it away from threshold. Said another way, signals compete for control of the neuron's membrane.

All synaptic signals are graded potentials. The ones we call EPSPs (for excitatory postsynaptic potentials) have a *depolarizing* effect. This simply means they bring the membrane closer to threshold. IPSPs (inhibitory postsynaptic potentials) may have a *hyperpolarizing* effect (driving the membrane away from threshold) or may help maintain the membrane at its resting level.

With **synaptic integration**, competing signals that reach an input zone of a neuron at the same time are summed. This summation process is the means by which two or more signals arriving at a neuron become dampened or reinforced, suppressed or sent onward, to other cells in the body. Figure 32.11 shows what two separate recordings of an EPSP and an IPSP might look like—and includes a recording of their summation.

Integration occurs if neurotransmitter molecules from several presynaptic cells reach a neuron's input zone at the same time. It also occurs if neurotransmitter has been released swiftly and repeatedly from a single presynaptic cell that has been whipped into a frenzy of excitability by a rapid series of action potentials.

Removing Neurotransmitter From the Synaptic Cleft

The flow of information through the nervous system depends on the prompt, highly controlled removal of neurotransmitter molecules from synaptic clefts. Some amount of these molecules simply diffuses out of the cleft. Enzymes cleave others right in the cleft, as when acetylcholinesterase breaks apart ACh. Also, transport proteins actively pump the molecules back into the presynaptic cells or into neighboring neuroglial cells.

What happens if neurotransmitter accumulates in the cleft? As one example, cocaine blocks the uptake of dopamine. The dopamine that lingers in synaptic clefts just keeps on stimulating target cells. The stimulation produces euphoria (intense pleasure) at first, but then has disastrous effects, as you read in the introduction. You'll come across other examples later in the chapter.

Neurotransmitters are signaling molecules that bridge the gap (synaptic cleft) between two neurons or between a neuron and a muscle cell or gland cell.

Neurotransmitters have excitatory or inhibitory effects on different kinds of receiving cells. Synaptic integration is the moment-by-moment combining of excitatory and inhibitory signals acting on a postsynaptic cell.

With this summation process, messages traveling through the nervous system can be reinforced or downplayed, sent onward or suppressed. The process is essential for normal body functioning.

Blocks and Cables of Neurons

Through synaptic integration, signals arriving at any neuron in the body can be reinforced or dampened, sent on or suppressed. What determines the direction in which a given signal will travel? That depends on the organization of neurons in different body regions.

For example, your brain deals with its staggering numbers of neurons in a manner analogous to block parties. Regional blocks of hundreds or thousands of neurons receive excitatory and inhibitory signals. They integrate signals entering the block, then send out new ones in response. Some regions have neurons organized as divergent circuits, as when their processes fan out from one block and form connections with many others. Different regions have neurons arranged in convergent circuits, with signals from many sent on to just a few. In still other regions, neurons synapse back on themselves and repeat signals like gossip that just won't go away. Such neurons form reverberating circuits. They include the ones that make your eye muscles rhythmically twitch while you sleep.

In the cablelike nerves, long axons of many sensory neurons, motor neurons, or both permit long-distance communication between the brain, spinal cord, and the rest of the body. Connective tissue bundles most of the axons in parallel array (Figure 32.12). Each axon has a **myelin sheath** that enhances the rate of action potential propagation. The sheath is a series of Schwann cells, a type of neuroglial cell, wrapped like jellyrolls around the long axon. An exposed node, or gap, separates each cell from adjacent ones. Here, voltage-sensitive, gated sodium channels pepper the plasma membrane (Figure 32.13). The sheathed regions in between nodes hamper ion movements across the membrane. Ion disturbances tend to flow along the membrane until the next node in line. At each node, ion flow can produce a new action potential. In large sheathed axons, action potentials are propagated at a remarkable 120 meters per second.

With *multiple sclerosis*, myelin sheaths in the spinal cord's nerve tracts degenerate slowly but inevitably. Gene mutation may predispose a person to the disease, but viral infection may be the actual trigger. Symptoms include weakening of muscles, fatigue, and numbness.

Reflex Arcs

Figure 32.14 provides a specific example of the direction of information flow through nervous systems. It shows how sensory and motor neurons of particular nerves take part in the stretch reflex. Reflexes, remember, are simple movements made in response to specific sensory information. In the simplest reflex arcs, sensory neurons synapse directly on motor neurons.

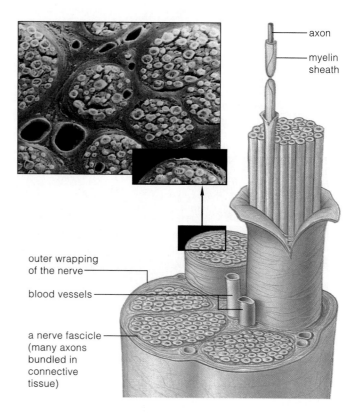

Figure 32.12 Structure of a nerve. Axons in the nerve are bundled together inside wrappings of connective tissue.

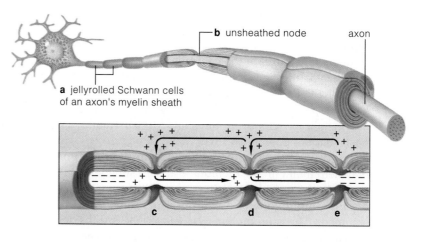

Figure 32.13 High-speed travel in the nervous system—how an action potential is propagated along sheathed neurons. (**a**) A myelin sheath is a series of Schwann cells, each wrapped like a jellyroll around an axon. (**b**) Each jellyroll blocks ion movements across the membrane, but ions can cross at nodes in between. The unsheathed nodes have dense arrays of gated sodium channels. (**c,d**) A disturbance caused by an action potential spreads down the axon. When it reaches a node, sodium gates open, sodium ions rush inward, and another action potential results. (**e**) The new disturbance spreads rapidly to the next node and triggers another action potential, and so on down the line.

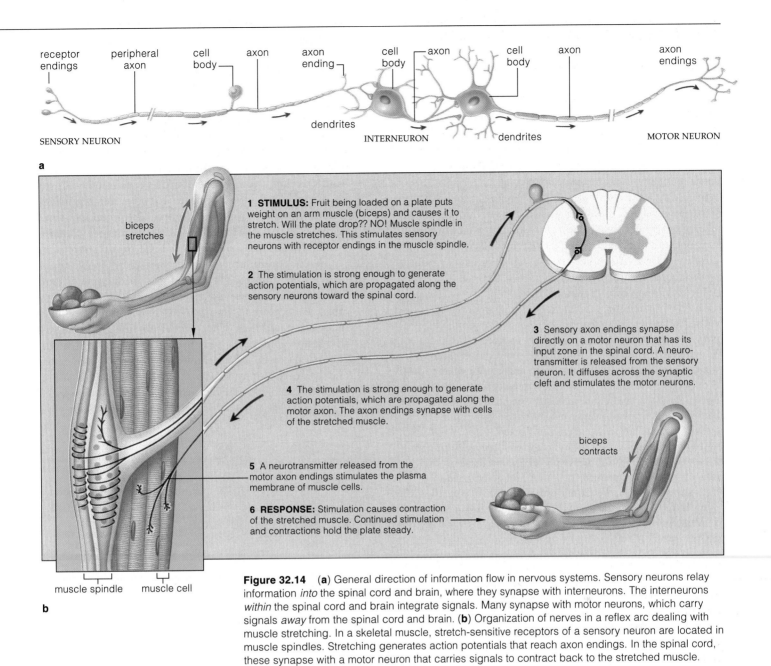

receptor endings — peripheral axon — cell body — axon — axon ending — cell body — axon — cell body — axon — axon endings

dendrites

SENSORY NEURON   INTERNEURON   dendrites   MOTOR NEURON

a

biceps stretches

1 STIMULUS: Fruit being loaded on a plate puts weight on an arm muscle (biceps) and causes it to stretch. Will the plate drop?? NO! Muscle spindle in the muscle stretches. This stimulates sensory neurons with receptor endings in the muscle spindle.

2 The stimulation is strong enough to generate action potentials, which are propagated along the sensory neurons toward the spinal cord.

3 Sensory axon endings synapse directly on a motor neuron that has its input zone in the spinal cord. A neuro-transmitter is released from the sensory neuron. It diffuses across the synaptic cleft and stimulates the motor neurons.

4 The stimulation is strong enough to generate action potentials, which are propagated along the motor axon. The axon endings synapse with cells of the stretched muscle.

biceps contracts

5 A neurotransmitter released from the motor axon endings stimulates the plasma membrane of muscle cells.

6 RESPONSE: Stimulation causes contraction of the stretched muscle. Continued stimulation and contractions hold the plate steady.

muscle spindle   muscle cell

b

Figure 32.14 (**a**) General direction of information flow in nervous systems. Sensory neurons relay information *into* the spinal cord and brain, where they synapse with interneurons. The interneurons *within* the spinal cord and brain integrate signals. Many synapse with motor neurons, which carry signals *away* from the spinal cord and brain. (**b**) Organization of nerves in a reflex arc dealing with muscle stretching. In a skeletal muscle, stretch-sensitive receptors of a sensory neuron are located in muscle spindles. Stretching generates action potentials that reach axon endings. In the spinal cord, these synapse with a motor neuron that carries signals to contract back to the stretched muscle.

The stretch reflex works to contract a muscle after gravity or some other load has caused the muscle to stretch. Suppose you hold out a large bowl and keep it stationary as someone puts peaches into it. The peaches add weight to the bowl, and when your hand starts to drop, a muscle in your arm (the biceps) is stretched.

In the muscle, stretching activates receptor endings that are a part of muscle spindles—sensory organs in which specialized cells are enclosed in a sheath that runs parallel with the muscle. The receptor endings are the input zones of sensory neurons, the axons of which synapse with motor neurons within the spinal cord (Figure 32.14). Axons of the motor neurons lead back to the stretched muscle. Action potentials that reach the axon endings trigger the release of ACh, which initiates

contraction. As long as receptor activity continues, the motor neurons are excited further, and this allows them to maintain your hand's position.

In the vast majority of reflexes, sensory neurons also make connections with a number of interneurons, which then activate or suppress all the motor neurons necessary for a coordinated response.

Vertebrates have interneurons organized in information-processing blocks. Their brain and spinal cord communicate with the rest of the body through cablelike nerves that have long axons of sensory neurons, motor neurons, or both.

Reflex arcs, in which sensory neurons synapse directly on motor neurons, are the simplest paths of information flow.

32.6 *Focus on Health*

CASES OF SKEWED INFORMATION FLOW

What happens if something disrupts information flow in the nervous system? Consider *Clostridium botulinum*, an anaerobic, endospore-forming bacterium that lives in soil. It can cause the disease *botulism*. When its endospores contaminate improperly stored, preserved, or canned food, they germinate and produce a dangerous toxin, which people can ingest and absorb. The toxin can bind to neurons that synapse with muscle cells and block their release of acetylcholine (ACh). Once this happens, muscles cannot contract. Progressively they become flaccid and paralyzed. Death may follow within ten days, usually from respiratory and cardiac failure. Recovery is possible if antitoxins are quickly administered.

A related bacterium, *C. tetani*, lives in the gut of horses, cattle, and other grazing animals, even many people. Its endospores survive in soil, especially manure-rich ones. They persist for years if sunlight and oxygen don't reach them. They resist disinfectants, heat, and boiling water. If they enter the human body through a deep puncture or cut, they germinate in anaerobic, dead tissues. These bacteria don't spread from the dead tissue. They produce a toxin that the blood or nerves deliver to the spinal cord and brain. The toxin acts on interneurons that help control motor neurons in the spinal cord. It blocks release of inhibitory neurotransmitters (GABA and glycine) and frees the motor neurons from normal inhibitory control. Symptoms of the disease *tetanus* are about to begin.

Muscles overstimulated for four to ten days stiffen and go into spasm. They cannot be released from contraction, so prolonged, spastic paralysis follows. Fists and jaws may stay clenched (the disease is also called lockjaw). The back may arch—permanently. Spasms may be strong enough to break bones. When respiratory and cardiac muscles also become paralyzed, death nearly always follows.

Vaccines were not available for soldiers of early wars, when dead cavalry horses and manure littered battlefields (Figure 32.15). Today, vaccines have all but eradicated tetanus in the United States.

Figure 32.15 Painting of a young victim of a contaminated battle wound, as he lay dying in a military hospital.

32.7 FUNCTIONAL DIVISIONS OF VERTEBRATE NERVOUS SYSTEMS

The vertebrate nervous system, recall, evolved from a nerve cord, which persists in embryos as a neural tube. This fluid-filled tube gives rise to the spinal cord and brain. Cerebrospinal fluid in its interconnected canals and cavities bathes the brain and spinal cord; it also cushions them from sudden, jarring movements. As the brain develops, it expands to become the hindbrain, midbrain, and forebrain (Figure 32.16). In all three regions, the most ancient nervous tissue still contains basic reflex centers, and we call it the "brain stem."

In the hindbrain, we find the medulla oblongata, cerebellum, and pons. The **medulla oblongata** contains reflex centers for respiration, circulation, and other vital tasks. It coordinates movements and complex reflexes (such as coughing). The **cerebellum** integrates signals from eyes, inner ears, and muscle spindles with motor commands from the forebrain to coordinate movement and balance. It may be crucial in human language and other forms of mental dexterity. Bands of axons extend from it into the **pons**. Although lodged in the brain stem,

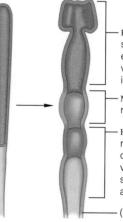

FOREBRAIN. Receives, integrates sensory information from nose, eyes, and ears; in land-dwelling vertebrates, contains the highest integrating centers

MIDBRAIN. Coordinates reflex responses to sight, sounds

HINDBRAIN. Reflex control of respiration, blood circulation, other basic tasks; in complex vertebrates, coordination of sensory input, motor dexterity, and possibly mental dexterity

(start of spinal cord)

dorsal, hollow nerve cord expanded regionally and increased in complexity

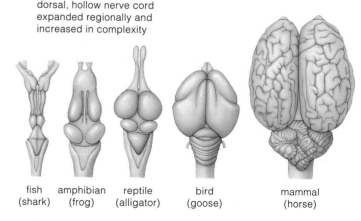

fish (shark) amphibian (frog) reptile (alligator) bird (goose) mammal (horse)

Figure 32.16 Evolutionary trend toward an expanded, more complex brain, as suggested by comparing existing vertebrates. These dorsal views are not to the same scale.

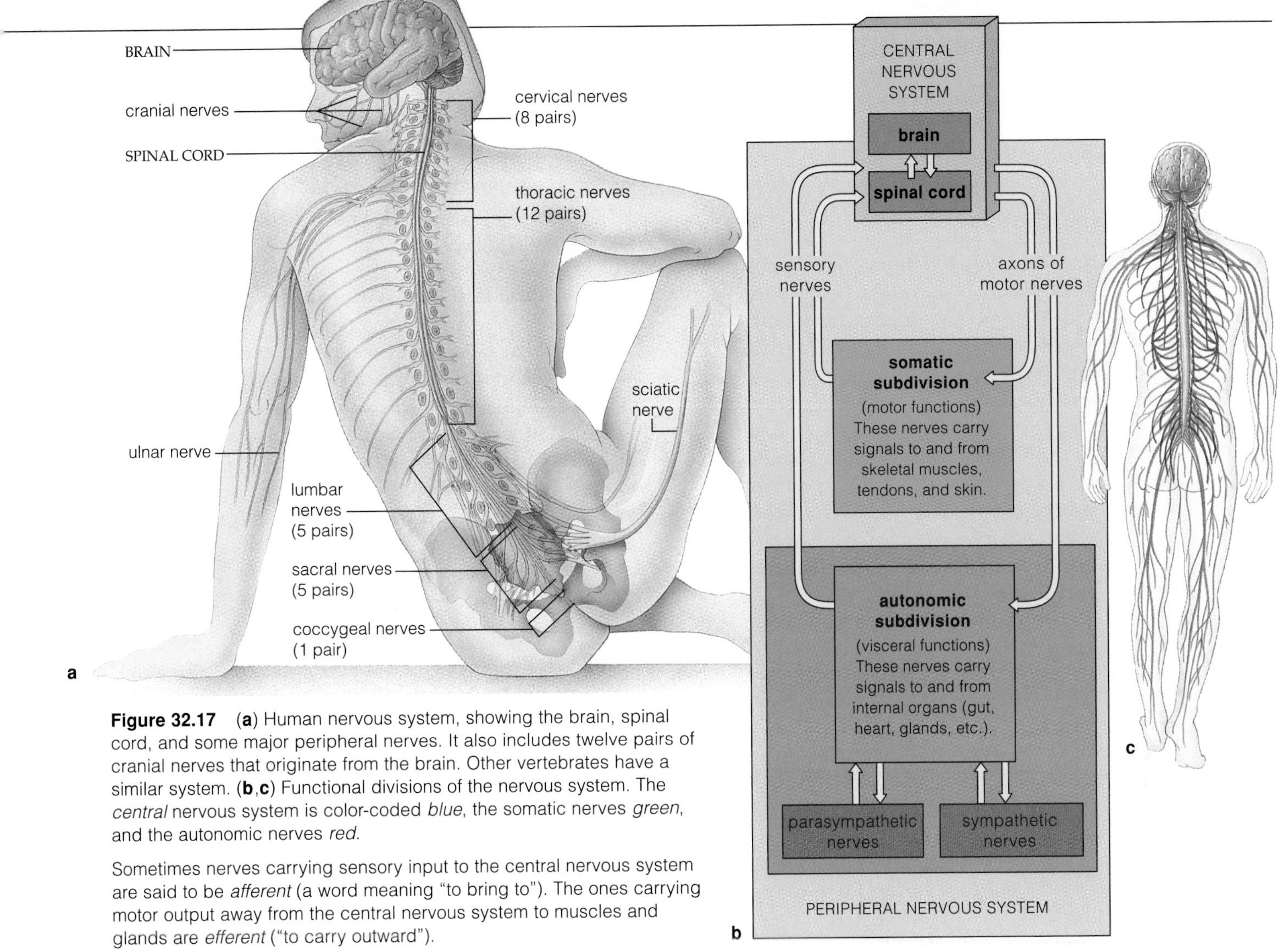
BRAIN
cranial nerves
SPINAL CORD
cervical nerves (8 pairs)
thoracic nerves (12 pairs)
sciatic nerve
ulnar nerve
lumbar nerves (5 pairs)
sacral nerves (5 pairs)
coccygeal nerves (1 pair)
a
CENTRAL NERVOUS SYSTEM
brain
spinal cord
sensory nerves
axons of motor nerves
somatic subdivision
(motor functions)
These nerves carry signals to and from skeletal muscles, tendons, and skin.
autonomic subdivision
(visceral functions)
These nerves carry signals to and from internal organs (gut, heart, glands, etc.).
parasympathetic nerves
sympathetic nerves
PERIPHERAL NERVOUS SYSTEM
b
c

Figure 32.17 (**a**) Human nervous system, showing the brain, spinal cord, and some major peripheral nerves. It also includes twelve pairs of cranial nerves that originate from the brain. Other vertebrates have a similar system. (**b**,**c**) Functional divisions of the nervous system. The *central* nervous system is color-coded *blue*, the somatic nerves *green*, and the autonomic nerves *red*.

Sometimes nerves carrying sensory input to the central nervous system are said to be *afferent* (a word meaning "to bring to"). The ones carrying motor output away from the central nervous system to muscles and glands are *efferent* ("to carry outward").

the pons directs the signal traffic between the cerebellum and the forebrain's higher integrating centers.

The midbrain coordinates reflex responses to sights and sounds. In fish and amphibians, its roof (tectum) deals with most sensory inputs and motor responses. A frog with its tectum intact but surgically deprived of higher brain centers still functions normally. In most vertebrates, but not mammals, two optic lobes in the midbrain deal with input from the eyes. In mammals, the eyes form direct connections with the forebrain.

For much of vertebrate history, the forebrain was a pair of small bumps on the brain stem that integrated crucial olfactory input—smells of predators, prey, and potential mates—and coordinated responses to it. These outgrowths, called the **cerebrum**, became the highest integrating center in mammals. Another region, the **thalamus**, coordinates sensory input and acts as a relay station for signals entering the cerebrum. A subregion of the thalamus, the **hypothalamus**, monitors internal organs and the behavior related to organ activities, such as hunger, thirst, temperature regulation, and sex.

Figure 32.17 shows the human nervous system, which we functionally divide into central or peripheral regions. All interneurons are confined to the **central nervous system**—the spinal cord and brain. The brain receives, integrates, stores, and retrieves input and coordinates responses by adjusting activities through the body. Axons of sensory and motor neurons belong to the **peripheral nervous system**, the communication lines through the rest of the body that carry signals into and out of the central nervous system.

The vertebrate brain is subdivided into three regions: the hindbrain, forebrain, and midbrain. In all three, the most ancient tissue (the brain stem) affords reflex control over the most basic functions necessary for survival.

We functionally subdivide the nervous system into central and peripheral regions. The central nervous system consists of the brain and spinal cord. The peripheral nervous system consists of nerves that thread through the rest of the body and carry signals into and out of the central region.

THE MAJOR EXPRESSWAYS

Let's now take a look at the peripheral nervous system and spinal cord—which are the major expressways for information flow through the body.

Peripheral Nervous System

Somatic and Autonomic Subdivisions   In humans, the peripheral nervous system includes thirty-one pairs of *spinal* nerves, which connect with the spinal cord. The system also includes twelve pairs of *cranial* nerves, which connect directly with the brain.

Cranial and spinal nerves can be further classified according to function. The ones that carry signals about moving your head, trunk, and limbs are the **somatic nerves**. The sensory axons inside these nerves deliver information from receptors in the skin, skeletal muscles, and tendons to the central nervous system. Their motor axons deliver commands from the brain and spinal cord to the body's skeletal muscles.

By contrast, spinal and cranial nerves dealing with smooth muscle, cardiac (heart) muscle, and glands are the **autonomic nerves**. They deal with the visceral parts of the body—in other words, its internal organs and structures. Autonomic nerves carry signals to and from these organs.

Sympathetic and Parasympathetic Nerves   Figure 32.18 shows the two categories of autonomic nerves. We call them parasympathetic and sympathetic. They normally work antagonistically, with signals from one opposing signals from the other. Both carry excitatory and inhibitory signals to internal organs. Often their signals arrive at the same time at muscle or gland cells and so compete for control over them. In such cases, synaptic integration at the cellular level leads to minor adjustments in the organ's level of activity.

Parasympathetic nerves dominate when the body is not receiving much outside stimulation. They tend to

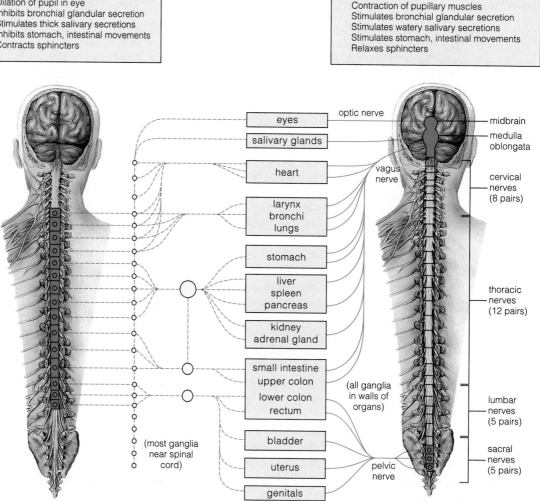

SYMPATHETIC OUTFLOW:

Examples of effects:
Dilation of pupil in eye
Inhibits bronchial glandular secretion
Stimulates thick salivary secretions
Inhibits stomach, intestinal movements
Contracts sphincters

PARASYMPATHETIC OUTFLOW:

Examples of effects:
Contraction of pupillary muscles
Stimulates bronchial glandular secretion
Stimulates watery salivary secretions
Stimulates stomach, intestinal movements
Relaxes sphincters

Figure 32.18 Autonomic nervous system. This diagram shows the major sympathetic nerves and parasympathetic nerves leading out from the central nervous system to some major organs. Remember, there are *pairs* of both kinds of nerves, servicing the right and left halves of the body. The ganglia are simply clusters of cell bodies of the neurons that are bundled together in nerves.

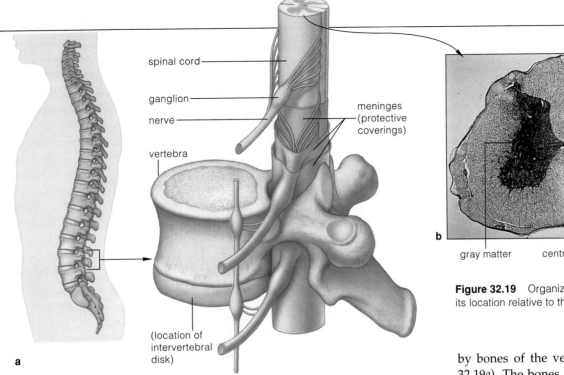

spinal cord

ganglion

nerve

vertebra

meninges (protective coverings)

(location of intervertebral disk)

a

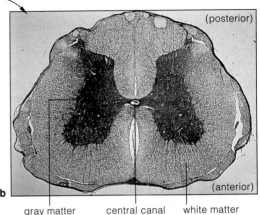

(posterior)

b

(anterior)

gray matter   central canal   white matter

Figure 32.19 Organization of the spinal cord and its location relative to the vertebral column.

slow down the body overall and divert energy to basic "housekeeping" tasks, such as digestion.

Sympathetic nerves dominate in times of sharpened awareness, excitement, or danger. They tend to shelve housekeeping tasks. At the same time, they prepare the animal to fight or escape (when threatened) or to frolic (as in play or sexual behavior). Right now, sympathetic nerves are commanding your heart to beat a bit faster, and parasympathetic nerves are commanding it to beat a bit slower. Integration of the opposing signals adjusts the heart rate. If something scares or excites you, the parasympathetic input to the heart drops. Sympathetic signals cause the release of epinephrine, which makes your heart beat faster. It also makes you breathe faster and start to sweat. In this state of intense arousal, you are primed to fight (or play) hard or to get away fast. Hence the name *fight-flight response.*

Suppose the stimulus for the fight-flight response ends. Sympathetic activity may decrease abruptly, and parasympathetic activity may rise suddenly. You might observe this "rebound effect" after someone has been instantly mobilized to rush onto a highway to save a child from an oncoming car. The person may well faint as soon as the child has been swept out of danger.

The Spinal Cord

By definition, the **spinal cord** is a vital expressway for signals between the peripheral nervous system and the brain. Also, sensory and motor neurons make direct reflex connections in the spinal cord. (The stretch reflex, shown in Figure 32.14, arcs through the spinal cord this way.) The spinal cord threads through a canal formed by bones of the vertebral column (Figure 32.19a). The bones and attached ligaments protect the cord. So do the meninges, three tough, tubelike coverings around the spinal cord and brain. The coverings are not impervious to all attacks. *Meningitis,* an often-fatal disease, results from viral or bacterial infection. Disease symptoms include severe headaches, fever, a stiff neck, nausea, and vomiting.

Signals travel rapidly up and down the spinal cord in bundles of axons with glistening myelin sheaths. The sheaths distinguish the cord's "white matter" from its "gray matter" (Figure 32.19b). The gray matter consists of dendrites, cell bodies of neurons, and neuroglial cells. It plays an important role in controlling reflexes for limb movements (as when you walk or wave your arms) and organ activity (such as bladder emptying).

Experiments with frogs provide evidence for these pathways. Between the frog's spinal cord and brain are neural circuits that deal with straightening legs after they have been bent. If these circuits are severed at the base of the brain, the legs become paralyzed—but only for about a minute. So-called extensor reflex pathways in the spinal cord recover quickly and have the frog hopping about in no time. Recovery is minimal after similar damage in humans and other primates—the vertebrates with the greatest cephalization.

The nerves of the peripheral nervous system connect the brain and spinal cord with the rest of the body.

Its somatic nerves deal with skeletal muscle movements. Its autonomic nerves deal with the functions of internal organs, such as the heart and glands.

The spinal cord is a vital expressway for signals between the peripheral nerves and the brain. Some of its interneurons also exert direct control over certain reflex pathways.

32.9 THE HUMAN BRAIN

Brain Components and Their Functions

Study any nervous system or even the best computers, and you realize the human brain is the most complex integration center of all. It only weighs 1,300 grams (3 pounds) in a person of average size, and more than half of that mass consists of neuroglial cells. But remember, the human brain also has at least *100 billion* interneurons!

Figure 32.20 shows the brain's components and lists their functions. The cerebrum, housed in a chamber of hard skull bones, resembles the much-folded nut in a hard walnut shell. Interneurons here contribute most to your humanness. A long fissure divides the cerebrum into left and right "cerebral hemispheres." Most of the brain centers dealing with analytical skills, speech, and mathematics reside in the left half. Those dealing with spatial abilities, music, and other nonverbal (abstract) skills reside mainly in the right half. Each hemisphere receives, processes, and coordinates responses to sensory input mostly from the opposite side of the body. For example, signals about pressure on the right arm reach the left half. Signals flowing in both directions through the corpus callosum (a transverse band of nerve tracts) coordinate the functions of both hemispheres.

At or near the surface of the cerebrum is the **cerebral cortex**, a thin layer of gray matter (cell bodies of interneurons) above the brain's axons. Different parts of it receive and process different signals. EEGs as well as PET scans have localized the activities in each hemisphere to four tissue lobes—the occipital, temporal, parietal, and frontal lobes. Figure 32.21 gives a few examples. (EEG, short for electroencephalogram, is simply a recording of summed electrical activity in a given brain region. Section 2.2 describes PET scans.) The occipital lobe at the back of each hemisphere has centers for vision. Severe blows to it may destroy part of the visual field (the outside world that the individual can see). The temporal lobe near each temple deals with hearing and associations related to vision. A blow here won't leave you blind, but it can impair your ability to recognize complex visual patterns, such as human faces. Centers in this lobe also influence emotional behavior. Compared to the olfactory centers of early vertebrates, the lobe's centers for vision and hearing have become far more important among humans and other primates.

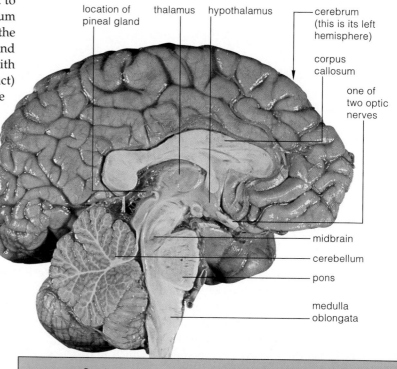

location of pineal gland · thalamus · hypothalamus · cerebrum (this is its left hemisphere) · corpus callosum · one of two optic nerves · midbrain · cerebellum · pons · medulla oblongata

left cerebral hemisphere · right cerebral hemisphere

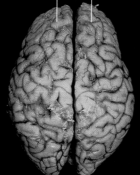

Figure 32.20 Human brain. (**a**) Its cerebral hemispheres, separated by a longitudinal fissure. This is a dorsal view. (**b**) The left cerebral hemisphere, sagittal view. The reticular formation extends through all three brain regions listed at the right; it carries signals between the upper spinal cord into the cerebrum and back.

FOREBRAIN	Cerebrum	Highest coordination of sensory and motor functions, association, memory, abstract thought
	Thalamus	Major coordination of sensory signals; relay station for sensory signals to cerebrum
	Hypothalamus	Neural-endocrine coordination of organ activities (e.g., digestion), related behavior; role in emotions
	Limbic system	Scattered brain centers; with hypothalamus, governs emotions and related activities
	Pituitary gland	Master endocrine gland governing growth, metabolism, etc.; controlled by hypothalamus
	Pineal gland	Control of some recurring, rhythmic behaviors
MIDBRAIN	Tectum	Reflex coordination of sensory input; nerve tracts between thalamus and cerebrum
HINDBRAIN	Cerebellum	Smooths, coordinates movements and balance
	Pons	Nerve tracts connect cerebrum with cerebellum and connect forebrain with spinal cord
	Medulla oblongata	Reflex centers for respiration, blood circulation, etc.; nerve tracts between pons and spinal cord

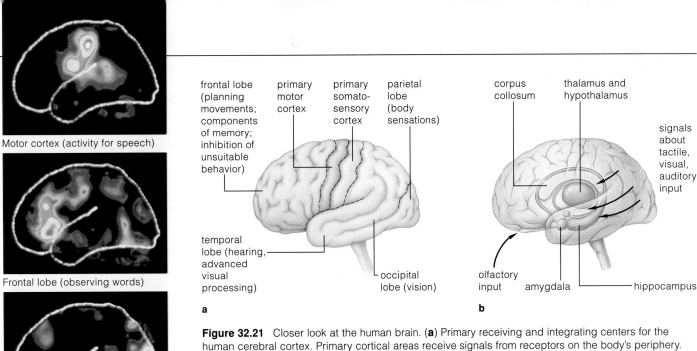

Motor cortex (activity for speech)

Frontal lobe (observing words)

Visual cortex (activity for vision)

a

frontal lobe (planning movements; components of memory; inhibition of unsuitable behavior)

primary motor cortex

primary somato-sensory cortex

parietal lobe (body sensations)

temporal lobe (hearing, advanced visual processing)

occipital lobe (vision)

b

corpus collosum

thalamus and hypothalamus

signals about tactile, visual, auditory input

olfactory input

amygdala

hippocampus

Figure 32.21 Closer look at the human brain. (**a**) Primary receiving and integrating centers for the human cerebral cortex. Primary cortical areas receive signals from receptors on the body's periphery. Association areas coordinate and process sensory input from different receptors. The PET scans show which brain regions are active during the performance of specific tasks—speaking, hearing, generating, and observing words.

(**b**) The limbic system, our "emotional brain," which includes the thalamus, hypothalamus, amygdala, and hippocampus. The limbic system also has roles in memory.

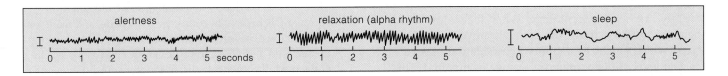

alertness

0 1 2 3 4 5 seconds

relaxation (alpha rhythm)

0 1 2 3 4 5

sleep

0 1 2 3 4 5

In the parietal lobe is the somatosensory cortex—the main receiving area for signals from skin and joints. In the frontal lobe, the motor cortex generates signals for motor responses. Thumb, finger, and tongue muscles get much of the brain's attention. This gives you an idea of how much control is required for hand movements and verbal expression. The prefrontal cortex is crucial in planning movements and inhibiting unsuitable behavior. It may deal with some aspects of memory. It also may interact with the cerebellum to control motor abilities associated with language and some thought processes.

The **limbic system** consists of several brain regions that deal with emotions and memory (Figure 32.21*b*). It is distantly related to olfactory lobes and still deals with the sense of smell. (That's one reason why you may feel warm and fuzzy when your brain recalls the cologne of a delightful person who wore it.)

Connections from the cerebral cortex as well as from ancient brain centers pass through the hypothalamus, the gatekeeper of the limbic system. These connections correlate organ activities with self-gratifying behavior, such as eating and sex. They might make you feel that, say, your heart and stomach are on fire—with passion or indigestion. And by these connections, neural signals based on reasoning in the cerebral cortex can override or dampen rage, hatred, and other "gut reactions."

Figure 32.22 EEG patterns correlated with a few states of consciousness. Vertical bars indicate a range of 50 microvolts of electrical activity. Irregular graph lines indicate electrical responses with time.

Concerning the Reticular Formation

You may be thinking that the brain is tidily subdivided into three regions. This is not the case. An evolutionary ancient mesh of interneurons, the **reticular formation**, extends from the uppermost part of the spinal cord to the cerebral cortex. It is a low-level pathway to motor centers of the medulla oblongata and the spinal cord. By activating centers in the cerebral cortex, it also helps govern the nervous system as a whole. Part of the mesh affects levels of consciousness by releasing serotonin. High levels of this neurotransmitter promote sleep and drowsiness (Figure 32.22). Substances from another center inhibit its effects and make you wake up. We see such changes as altered wavelike patterns in EEGs.

The human cerebrum contains the highest integrative centers and still makes evolutionarily ancient connections with the brain stem and spinal cord.

32.10 MEMORY

One of the most astonishing aspects of the human brain is its capacity to accumulate **memory**—that is, to store and retrieve information about the individual's unique experiences. Sensory input triggers each memory. When information encoded in its signals arrives at association centers in the cerebral cortex, it gets linked into larger informational packages, much as chemical bonds link atoms into compounds. These packages move to other brain regions for storage, but in response to retrieval cues, they can be recovered later. This memory process starts before birth, and it continues throughout life.

According to one prevailing theory, the brain stores information in stages. Short-term storage is a fleeting stage of neural excitation. It lasts a few seconds to a few hours and is limited to a few bits of information—numbers, words of a sentence, and so on. In long-term storage, seemingly unlimited amounts of information get tucked away on a more or less permanent basis. The theory helps explain *retrograde amnesia*, which follows a severe head blow and loss of consciousness. Memory of what happened during a period preceding the blow is lost. Yet memories of events before then often survive.

Information of most interest to a person moves most rapidly into long-term storage. When that information heightens brain activity, the epinephrine level in blood rises, which enhances memory consolidation. Above a certain level, though, memory storage is impaired. This is one reason why a person has trouble remembering details about a situation that brought on extreme panic.

Information unused for decades still can be recalled, so memory must be encoded in a form that resists degradation. As many experiments indicate, memory resides in neurons that have changed, chemically and physically, as a result of altered synaptic connections. For example, after laboratory mice were raised without any visual stimulation whatsoever, synapses withered in their visual cortex. Where withering had occurred, the functional link between neurons was weakened or broken. As another example, intensively stimulated synapses form stronger connections, grow in size, or sprout buds or spines to form more connections.

The capacity to recover memories diminishes with *Alzheimer's disease*, which usually has its onset in later life. Affected people often can remember long-standing information, such as their Social Security number, but they have trouble remembering what has just happened to them. In time they grow confused, depressed, and unable to complete a train of thought.

The brain accumulates and stores information in retrievable form. Its short-term memory is rapid and limited to a few bits of information. Its long-term memory requires chemical and physical alterations in synaptic connections.

32.11 *Focus on Health*

DRUGS, THE BRAIN, AND BEHAVIOR

Broadly speaking, a drug is any substance introduced into the body to provoke a specific physiological response. Some drugs help a person cope with illness or emotional stress. Others act on the brain, artificially fanning pleasure associated with sex and other self-gratifying behaviors.

Many drugs are habit-forming. That is, even if the body is able to function well without them, a person continues to use drugs for the real or imagined relief they afford. Often the body develops tolerance of such drugs; it takes larger or more frequent doses to produce the same effect. Habituation and tolerance are signs of *drug addiction*—the chemical dependence on a drug (Table 32.1). *With addiction, the drug has assumed an "essential" biochemical role in the body.* Abruptly deprive addicts of the drug, and they go through physical pain and mental anguish. The entire body goes through a period of biochemical upheavals. Stimulants, depressants, hypnotics, narcotic analgesics, hallucinogens, and psychedelics have such effects.

Stimulants Caffeine, nicotine, amphetamines, cocaine, and other stimulants increase alertness and body activity. Then they cause depression. Caffeine, a truly common stimulant, is present in coffee, tea, chocolate, and many soft drinks. Low doses arouse the cerebral cortex first, leading to increased alertness and restlessness. Higher doses acting at the medulla oblongata disrupt motor coordination and mental coherence. Nicotine, found in tobacco, has powerful effects on the nervous system. It mimics acetylcholine and directly stimulates a variety of sensory receptors. Its short-term effects include water retention, irritability, increased heart rate and blood pressure, and gastric upsets.

Amphetamines, including the one called "speed," mimic two neurotransmitters, dopamine and norepinephrine. In time, the brain produces less and less of these signaling molecules and depends more on artificial stimulation.

Table 32.1   Warning Signs of Drug Addiction*
1. Tolerance—it takes greater amounts of the drug to produce the same effect.
2. Habituation—continued drug use over time to maintain self-perception of functioning normally.
3. Inability to stop or curtail use of the drug, even if there is persistent desire to do so.
4. Concealment—not wanting others to know of the drug use.
5. Extreme or dangerous behavior to get and use the drug—as by stealing, asking more than one doctor to write drug prescriptions, or jeopardizing employment by drug use at work.
6. Deterioration of professional and personal relationships.
7. Anger and defensive behavior when someone suggests there may be a problem.
8. Preference of drug use over previously customary activities.

* Three or more of these signs may be cause for concern.

Possibly 2 million Americans are cocaine abusers. This drug gives a rush of pleasure by *blocking* reabsorption of dopamine, norepinephrine, and other neurotransmitters. Because these molecules accumulate in synaptic clefts, they incessantly stimulate receptor cells for an extended period. Heart rate and blood pressure rise; sexual appetite increases. In time the molecules diffuse away. However, neurons can't synthesize replacements fast enough to counter the loss. The sense of pleasure evaporates as the now-hypersensitive receptor cells demand stimulation. After prolonged, heavy use of cocaine, "pleasure" cannot be experienced. Addicts become anxious and depressed. They lose weight and cannot sleep properly. Their immune system weakens, and heart abnormalities set in.

Granular cocaine, which is inhaled (snorted), has been around for some time. Abusers burn crack cocaine and inhale the smoke (Figure 32.23). As suggested at the start of this chapter, crack is extremely addictive. Its highs are higher, but the crashes are more devastating.

Depressants, Hypnotics   These drugs lower activity in nerves and parts of the brain. Some act at synapses in the reticular formation and in the thalamus. Depending on the dosage, the responses range from emotional relief, drowsiness, sleep, anesthesia, and coma to death. Low doses act more on inhibitory synapses, so the person feels excited or euphoric at first. But increased doses suppress excitatory synapses as well, and this leads to depression. Depressants and hypnotics have additive effects. One amplifies another, as when alcohol plus barbiturates increases the depression.

Alcohol (ethyl alcohol) acts directly on the plasma membranes to alter cell function. Some people mistakenly think of it as a harmless stimulant because it produces an initial "high." But alcohol is one of the most powerful psychoactive drugs and is a major factor in many deaths. Small amounts of alcohol even over the short term can produce disorientation, uncoordinated motor functions, and diminish a person's judgment. Long-term addiction can permanently damage the liver (cirrhosis).

Analgesics   When stress leads to physical or emotional pain, the brain can produce natural analgesics—pain relievers. Endorphins and enkephalins are examples. They act on many parts of the nervous system, including brain centers that deal with emotions and pain. The narcotic analgesics, including codeine and heroin, sedate the body and relieve pain. They are among the most addictive substances known. Deprivation following massive doses of heroin leads to fever, chills, hyperactivity and anxiety, violent vomiting, cramping, and diarrhea.

Psychedelics, Hallucinogens   These drugs skew sensory perception by interfering with the activity of

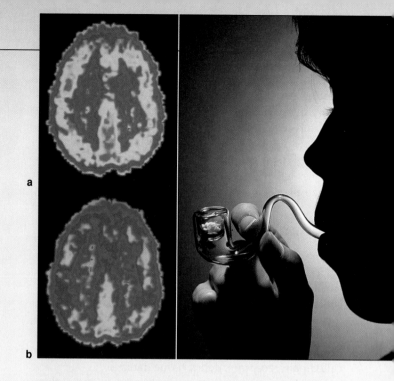

Figure 32.23   Smoking crack puts cocaine in the brain in less than eight seconds. (**a**) PET scan of a horizontal section of the brain, showing normal activity. (**b**) PET scan of a comparable section showing cocaine's effect. *Red* marks greatest activity; *yellow, green,* and *blue* mark successively inhibited activity.

acetylcholine, norepinephrine, or serotonin. LSD (lysergic acid diethylamide) alters serotonin's influence on inducing sleep, on temperature regulation, and on sensory perception. Even in small doses, LSD warps perceptions. For example, some users "perceived" they could fly and "flew" off buildings.

Marijuana, another hallucinogen, is made from crushed leaves, flowers, and stems of the plant *Cannabis*. In low doses it is like a depressant. It slows down but does not impair motor activity; it relaxes the body and elicits mild euphoria. It also causes disorientation, increased anxiety bordering on panic, delusions (including paranoia), and hallucinations.

Like alcohol, marijuana affects the performance of complex tasks, such as driving a car. In one study, pilots showed a marked deterioration in instrument-flying ability for more than two hours after smoking marijuana. Over time, marijuana smoking can suppress the immune system and impair mental functions.

Each of us possesses a body of great complexity. Its architecture, its functioning are legacies of millions of years of evolution. Its nervous system is unparalleled in the living world. One of its most astonishing products is language—the encoding of shared experiences of groups of individuals in time and space. Through the evolution of our nervous system, the sense of history was born, and the sense of destiny. Through this system we can ask how we have come to be what we are, and where we are headed from here. Perhaps the sorriest consequence of drug abuse is its implicit denial of this legacy—the denial of self when we cease to ask, and cease to care.

32.12 SENSORY SYSTEMS

We turn now to the **sensory systems**, which detect and notify the spinal cord and brain of specific changes outside and inside the body. Such a change is a **stimulus** (plural, stimuli). The systems all have sensory receptors, nerve pathways from receptors to the brain, and brain regions that process and translate sensory information into a **sensation**—a conscious awareness of a stimulus.

Types of Sensory Receptors

Different kinds of receptors respond to different stimuli. **Chemoreceptors** detect the chemical energy of specific substances that become dissolved in fluid next to them. **Mechanoreceptors** detect forms of mechanical energy, including pressure. **Photoreceptors** detect the energy of visible light or ultraviolet light. **Thermoreceptors** detect infrared energy—that is, heat. **Nociceptors**, sometimes called pain receptors, detect tissue damage.

Many sensory receptors are located in more than one body region and contribute to somatic sensations. Others are restricted to special locations, such as eyes and ears. We say they contribute to the special senses.

Depending on their kinds and numbers of receptors, animals sample the environment in different ways and differ in their awareness of it. Unlike bees, you have no receptors for ultraviolet light, so you cannot see many flowers the way they do (Figure 24.2). Unlike pythons, you, bees, and bats do not have receptors for detecting warm-blooded prey in the dark (Figure 32.24).

Sensory Pathways

Before sensory axons can carry signals from receptors, a stimulus energy must be converted to action potentials. When a stimulus disturbs the plasma membrane of a receptor ending, it triggers a local, graded potential. Remember, such signals don't spread far from the point of stimulation, and they vary in magnitude. Suppose a stimulus is intense or repeated fast enough so that the local signals give rise to action potentials. The action potentials self-propagate from receptor endings to the axon endings of sensory neurons (Figure 32.14a). These synapse with interneurons (as well as motor neurons) that may be part of nerve tracts leading to the brain.

Action potentials are not like an ambulance siren; they don't vary in amplitude. How, then, does the brain assess the nature of a stimulus? That depends on *which* sensory area receives signals from nerve pathways, the *frequency* of signals traveling on each of the pathway's axons, and the *number* of axons recruited into action.

First, specific sensory areas of the brain interpret action potentials only in certain ways. That's why you "see stars" when an eye is poked, even in the dark. The mechanical pressure on its photoreceptors generated

Figure 32.24 Thermoreceptors in rows of pits above and below a python's mouth. Body heat (infrared energy) of prey—small, night-foraging mammals—activates the receptors, which signal the brain. The brain assesses the signals and directs the aim of a strike with stunning accuracy. An edible, motionless frog does not have the same stimulatory effect; the same snake will slither past it. Frog skin is cool and blends with background colors. The snake does not have receptors to detect the frog or a neural program to respond to it.

signals, which an optic nerve delivered to a sensory area that interprets incoming signals as "light." If those signals were somehow rerouted to the sensory area for sound, then you would "hear" the poke in the eye.

Second, when a stimulus is stronger, receptors fire more frequently. The exact same receptor detects the sounds of a throaty whisper and a wild screech. But the brain senses the difference through variations in the frequency of signals that the receptor is sending to it.

Third, stronger stimulation recruits more receptors. Tap a small spot on your hand and you activate a few receptors. Press hard and you activate many more in a larger area. Consider these action potential recordings from a single mechanoreceptor in a human hand:

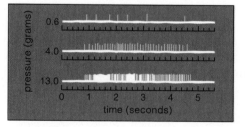

Each vertical white line represents an action potential. Their increases in frequency correspond to increases in stimulus strength.

A sensory system has sensory receptors for specific stimuli, nerve pathways that conduct signals from receptors to the brain, and brain regions where the signals are processed.

The brain senses a stimulus based on which nerve pathways carry the signals, the frequency of signals from each axon of that pathway, and the number of axons recruited.

32.13 SOMATIC SENSATIONS

When you become aware of touch, pressure, heat, cold, and pain, these are *somatic* sensations. So is awareness of movements and positions in space. These sensations start with receptor endings in surface tissues, in skeletal muscles, and in walls of internal organs. Information travels into the spinal cord and to the somatosensory cortex of the cerebral hemispheres, where neurons are laid out like maps corresponding to the body surface. Map regions with the largest areas correspond to body parts having the most sensory acuity and requiring the most intricate control. Such body parts include fingers, thumbs, and lips (Figure 32.25).

Muscle Sense

Sensing limb motions and the body's position in space requires mechanoreceptors in skeletal muscle, joints, tendons, ligaments, and skin. The stretch receptors of muscle spindles, shown in Figure 32.14b, are examples. These sensory organs run parallel with skeletal muscle cells. Their responses to stimulation depend upon how much and how fast the muscle stretches.

Pressure, Temperature, and Pain

Near the body's surface, free nerve endings and other mechanoreceptors abound. The tips of the tongue and fingers have the greatest number and show the greatest sensory acuity. Some mechanoreceptors in skin, such as Pacinian corpuscles, are encapsulated in epithelial or connective tissue (Figure 32.26). Mechanoreceptors near the body surface respond easily to applied pressure. Many fire action potentials when the stimulus is first encountered and when it ends. They make you aware of touch, even light tickles and vibrations. Many others will fire action potentials for as long as the stimulus is present. They keep you aware of ongoing pressure.

When the skin's temperature doesn't change much, thermoreceptors fire signals steadily. If it increases, so will the firing frequency. Some heat receptors are free nerve endings. Cold receptors have yet to be identified.

Pain is the perception of injury to some body region. Some free nerve endings and other nociceptors detect any stimulus strong enough to damage tissues around them. They respond to strong mechanical stimulation, intense heat or cold, and chemical irritation. Responses to pain often depend on the brain's ability to identify the affected tissue. Get hit in the face with a snowball and you "feel" the contact on facial skin. For reasons not fully understood, the brain sometimes associates pain with a tissue located some distance away from an internal injury. This is referred pain. For example, a heart attack can be felt as pain in skin above the heart and along the left shoulder and arm.

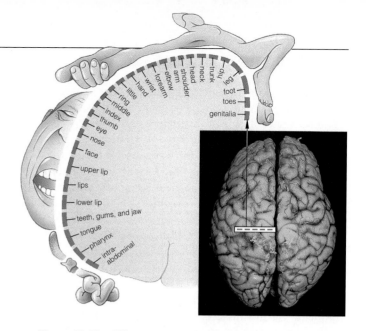

Figure 32.25 Differences in the representation of different body parts in the primary somatosensory cortex. This region is a strip of cerebral cortex, a little wider than 2.5 centimeters (about an inch), from the top of the head to above the ear.

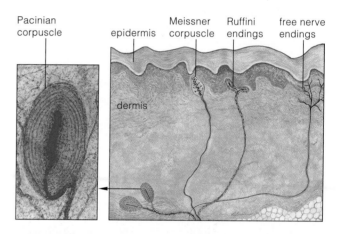

Figure 32.26 Tactile receptors in human skin. Free nerve endings detect changes in temperature, light pressure, and pain. Pacinian corpuscles detect vibration. Meissner corpuscles detect the onset and the end of sustained pressure. Ruffini endings react continually to ongoing stimulation.

Bear in mind, "perception" is not the equivalent of sensation; it is understanding what a sensation means. The jerkings of a live lobster cooking in boiling water suggest its brain senses a stimulus, but a lobster reacts the same way to any strong stimulus, as when you grip it firmly. Probably the lobster doesn't "understand" its predicament or "feel" pain, as humans do.

Diverse mechanoreceptors detect touch, pressure, heat and cold, pain, limb motions, and changes in the body's position in space. Responses to signals from these receptors give rise to somatic sensations.

32.14 HEARING AND BALANCE

Many arthropods and most vertebrates hear sounds by means of acoustical organs. For example, land-dwelling mammals have a pair of distinctive **ears**. Small organs, flush with the body surface, serve comparable functions in insects and amphibians. Such organs are on the front legs of crickets and near the hind legs of grasshoppers.

Hearing starts with acoustical receptors, which are vibration-sensitive mechanoreceptors. A vibration is a wavelike form of mechanical energy, as when hand clapping produces waves of compressed air. Each time hands clap together, molecules are forced outward, so a low-pressure state is created in the region they vacated. The pressure variations can be depicted as a wave form, with amplitude corresponding to loudness. The sound's frequency is the number of wave cycles per second. Each cycle extends from the start of one wave peak to the start of the next. The more cycles per second, the higher the sound's frequency and perceived pitch:

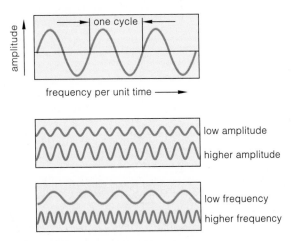

Sound waves arriving at an acoustical organ reach a membrane that has acoustical receptors attached to it. Hair cells are such receptors in vertebrates. Directly or indirectly, vibrations make them bend. Deformation may start a flow of information along an auditory nerve leading from the receptors to the brain.

Echolocation

Figure 32.27 describes how the brain of a bat responds to signals that are generated by way of **echolocation**—the use of echoes from ultrasounds that the animal itself produces. Ultrasounds are extremely high-frequency waves, in the 100-decibel range, which humans cannot even hear.

In the seas, dolphins and whales emit ultrasounds that travel through water. By assessing the frequency variations in the echoes, these aquatic mammals can pinpoint the distance and direction of movement of one another and of predators or prey.

Figure 32.27 Echolocating bat. Nearly all bats sleep by day and spread their webbed wings at dusk. Different kinds take to the air in search of nectar, fruit, frogs, or insects. Many sensory receptors in their eyes, nose, ears, mouth, and skin are not that different from yours. Others are the basis of echolocation. By this hearing mechanism, bats emit calls, as shown here. When sound waves of the calls bounce off insects, trees, and other objects, sensory receptors detect the echoes.

An echolocating bat emits a steady stream of about ten clicking sounds per second. These intense "ultrasounds" are above the range of sound waves that receptors in human ears detect. If the bat hears a pattern of distant echoes from, say, an airborne mosquito or moth, it increases the rate of ultrasonic clicks to as many as 200 per second, faster than a machine gun can fire bullets. In the few milliseconds of silence between these clicks, receptors detect the echoes and send signals about them to the brain. The brain constructs a "map" of sounds that the bat uses to maneuver itself as it flies through the night world.

The Human Ear

The human ear has regions that receive, amplify, and sort out sound waves. Figure 32.28 shows the location of their specialized hair cells. Prolonged exposure to some intense sounds permanently damages these acoustical receptors, which are not adapted to amplified music, jet planes, and other recent developments.

Organs of Balance

Most animals right themselves after they tilt or turn upside-down. They do this by assessing displacement from their natural, *equilibrium* position, when the body is balanced in relation to gravity, velocity, acceleration, and other forces that influence position and movement. The sense of balance depends on organs of equilibrium that have hair cells. Such organs occur in a system of canals and fluid-filled sacs in the internal ear of animals ranging from amphibians to mammals (Figure 32.28b). Some detect rotational, accelerated motions of the head. Others detect linear movements in the head's position.

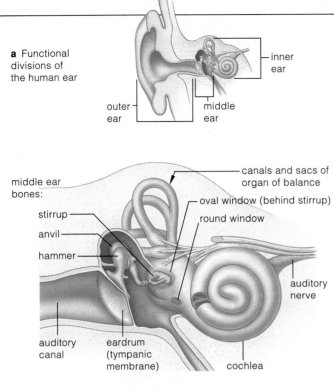

a Functional divisions of the human ear

inner ear

outer ear

middle ear

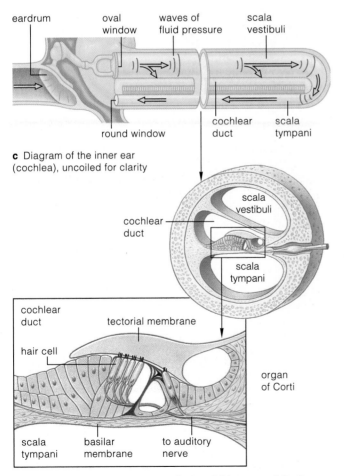

middle ear bones:

canals and sacs of organ of balance

oval window (behind stirrup)

round window

stirrup

anvil

hammer

auditory nerve

auditory canal

eardrum (tympanic membrane)

cochlea

b Structure of the middle ear and the inner ear

eardrum

oval window

waves of fluid pressure

scala vestibuli

cochlear duct

scala tympani

round window

c Diagram of the inner ear (cochlea), uncoiled for clarity

scala vestibuli

cochlear duct

scala tympani

cochlear duct

tectorial membrane

hair cell

organ of Corti

scala tympani

basilar membrane

to auditory nerve

d Location of acoustical receptors (hair cells) in organ of Corti

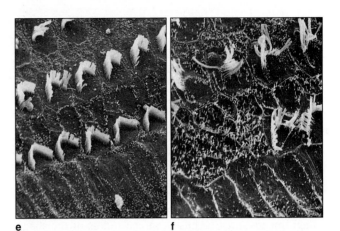

e

f

Figure 32.28 Case study: Sensory reception in the human ear. (**a**) Each ear has three regions (outer, middle, and inner ear), which receive, amplify, and sort out acoustical stimuli. An ear's components are shown in (**b**). External flaps of the *outer* ear collect sound waves, which move into an auditory canal and then arrive at the eardrum (the tympanic membrane).

(**c**) Sound waves cause the eardrum to vibrate. In the *middle* ear, small bones pick up the vibrations and amplify the stimulus by transmitting the force of pressure waves to a smaller surface, the oval window. The "window" is an elastic membrane over the entrance to the coiled *inner* ear (cochlea). When the oval window bows in and out, it produces fluid pressure waves in two ducts (scala vestibuli and scala tympani). The waves reach another membrane, the round window. When this membrane bulges under pressure, fluid moves back and forth in the inner ear.

(**d**) The cochlear duct (third duct of the coiled inner ear) sorts out pressure waves. Its basilar membrane starts out narrow and stiff. It is broader and flexible deep in the coil. At different points, it vibrates more strongly to sounds of different frequencies.

Many hair cells are positioned on the basilar membrane, below a jellylike flap (tectorial membrane). If the flap vibrates, hair cells move and release neurotransmitter, which may generate action potentials that travel through an auditory nerve to the brain.

Results of an experiment on the effect of intense sound on the inner ear. (**e**) From a guinea pig's organ of Corti, two rows of the hair cells that project into the tectorial membrane. (**f**) The same organ after 24 hours of exposure to noise levels comparable to extremely loud music (2,000 cycles per second at 120 decibels).

Motion sickness may result when monotonous linear, angular, or vertical motion overstimulates hair cells in the canals of the organs of balance. It also may result from conflicting signals from the ears and eyes about motion or the head's position. People who are carsick, airsick, or seasick suffer nausea and may vomit; action potentials triggered by the sensory input reach a brain center that governs the vomiting reflex.

In vertebrates, the senses of hearing and balance begin with mechanoreceptors called hair cells, which are attached to certain membranes of organs within a pair of ears.

Requirements for Vision

All organisms are sensitive to light. Even a single-celled amoeba abruptly stops moving when you shine a light on it. Many invertebrates have photoreceptor cells, but not all of them see as you do. They may detect a change in light intensity, as when a fish swims above them, but they can't discern the size or shape of objects. What we call *vision* requires **eyes**—visual organs that incorporate a tissue of densely packed photoreceptors known as a **retina**. And it requires image formation in brain centers that can not only receive but also interpret patterns of stimulation coming in from different parts of the eye. Such images must be based on awareness of the shape, brightness, position, and movement of visual stimuli.

Mollusks are the simplest animals with a visual sense (Figure 32.29*a–c*). Some have a transparent cover (cornea) over a lens: a transparent body that focuses light onto the retina. Octopuses and other cephalopods have a ring of contractile tissue (iris) in each eye. An iris can be adjusted to admit more or less light through the pupil, an opening at its center. Like you, cephalopods have what's called camera eyes. Light entering the eye chamber through a small opening is focused at a lens, then strikes a light-sensitive surface. Crustaceans and insects have **compound eyes**, with densely packed units of photoreceptors called ommatidia (Figure 32.29*d*).

Structure and Function of Vertebrate Eyes

Figure 32.30 shows the components of the human eye, and Table 32.2 lists their functions. As is true of most vertebrate eyes, it has three layers. The outermost layer consists of a sclera and cornea. The middle layer consists of a choroid and lens, as well as semifluid and jellylike substances. A retina is the key component of the inner layer. It is at the back of your eyeball and more toward the eyeball's roof in certain birds (Figure 32.29*e*).

The sclera is the dense, fibrous "white" of the eye; it protects most of the eyeball. The cornea covers the rest of it. In the eye's middle layer, a clear fluid called the aqueous humor bathes the lens. A jellylike substance, the vitreous body, fills the chamber behind the lens. The choroid, a dark-pigmented tissue, absorbs light that has not been absorbed by photoreceptors. Thus it prevents light from scattering inside the eyeball.

Suspended behind the cornea is a doughnut-shaped, pigmented iris (after the Latin *irid*, referring to the rings of a rainbow). The dark "hole" in its center is the pupil, the entrance for light. When bright light hits the eye, circular muscles in the iris contract and so shrink the size of the pupil. In dim light, radial muscles contract and so enlarge the pupil. Behind the iris is a lens with onionlike layers of transparent proteins.

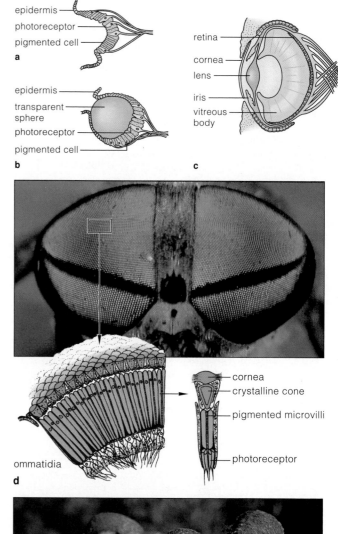

Figure 32.29 A sampling of invertebrate and vertebrate photoreceptors. Longitudinal sections of (**a**) a limpet eyespot, an epidermal depression with light-sensitive receptors; (**b**) an abalone eye; and (**c**) an octopus eye. (**d**) Compound eyes of a deerfly. Its crystal-like cones of photoreceptor units (ommatidia, singular, ommatidium) act like a lens to focus light on pigmented photoreceptor cells behind them. (**e**) In owls and other birds of prey, photoreceptors are concentrated more on the roof of the eyeball. These birds look down more than they look up when they are in flight, scanning the ground for a meal. When they are on the ground, they cannot see something overhead very easily unless they turn their head almost upside-down.

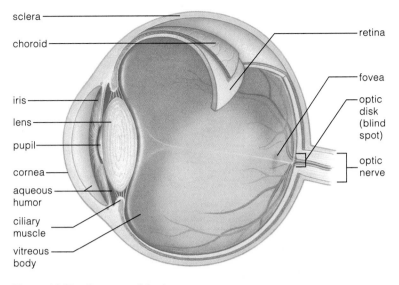

sclera
choroid
retina
iris
lens
pupil
cornea
aqueous humor
ciliary muscle
vitreous body
fovea
optic disk (blind spot)
optic nerve

Figure 32.30 Structure of the human eye.

Table 32.2	Components of the Vertebrate Eye
Outer Layer	
Sclera	Protect eyeball
Cornea	Focus light
Middle Layer	
Choroid:	
Pigmented tissue	Prevent light scattering
Iris	Control incoming light
Pupil	Let light enter
Lens	Finely focus light on photoreceptors
Aqueous humor	Transmit light, maintain pressure
Vitreous body	Transmit light, support lens and eyeball
Inner Layer	
Retina	Absorb, convert light
Fovea	Increase visual acuity
Start of optic nerve	Transmit signals toward brain

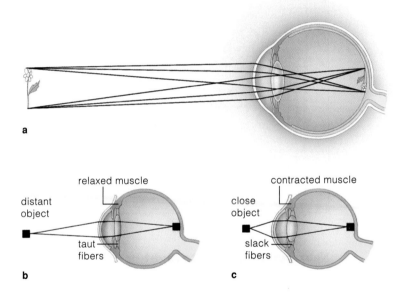

a

relaxed muscle
distant object
taut fibers
b

contracted muscle
close object
slack fibers
c

Figure 32.31 (a) Pattern of retinal stimulation. Being curved, the cornea changes the trajectories of light rays as they enter the eye (compare Section 3.3). The pattern is upside-down and reversed left to right, compared to the stimulus in the visual field.

Focusing mechanisms. A ciliary muscle encircles and attaches to the lens. (b) When this muscle relaxes, the lens flattens; the focal point moves farther back and brings distant objects into focus. (c) When it contracts, the lens bulges; the focal point moves closer and brings close objects into focus.

In *astigmatism*, the cornea is unevenly curved and cannot bend light rays to the same focal point. *Nearsightedness* often occurs when the eyeball's horizontal axis is longer than the vertical axis. It also occurs when the ciliary muscle contracts too strongly and focuses images of distant objects in front of the retina. In *farsightedness*, the eyeball's vertical axis is longer than the horizontal axis (or the lens is "lazy"), so close images are focused behind the retina.

Vision is impossible *unless* light rays converge at the back of the eye, there to stimulate the retina in a distinct pattern (Figure 32.31*a*). Because the rays from sources at varying distances from the eye strike the cornea at different angles, they could become focused at different distances behind it. But adjustments in the position or shape of the lens focus all incoming stimuli onto the retina. We call these adjustments visual accommodation. Without them, light rays from distant objects would be improperly focused in front of the retina, and rays from close objects would be focused behind it.

In fishes and reptiles, eye muscles move the lens forward or back, like a camera's focusing apparatus. Extending the distance between the lens and the retina moves the focal point forward; shrinking the distance moves it back. By contrast, in birds and mammals, the shape of the lens is adjusted. A ciliary muscle encircles the lens and attaches to it by fiberlike ligaments (Figure 32.30). When the muscle relaxes, the lens flattens and thus moves the focal point farther back (Figure 32.31*b*). When the muscle contracts, the lens bulges and moves the focal point forward (Figure 32.31*c*).

Injuries, diseases, and aging can disrupt vision. For example, *histoplasmosis*, a fungal infection of the lungs, damages the retina. *Herpes* infections can cause corneal ulcers. Aging can bring on *cataracts*, a gradual clouding of the lens. In *glaucoma*, fluid pressure builds up inside the eyeball and vision deteriorates. Early detection and therapy or surgery often can alleviate these problems.

Vision starts at an adjustable lens that can focus rays of light onto a retina—a light-sensitive tissue of photoreceptors in the eyes of many invertebrates and all vertebrates.

32.16 CASE STUDY—FROM SIGNALING TO VISUAL PERCEPTION

We conclude this chapter with one of the best examples of neuronal architecture—the sensory pathway from the retina to the brain. This is how raw visual information is received, transmitted, and combined; and it leads to awareness of light and shadows, of colors, of near and distant objects in the outside world.

Organization of the Retina

Information flow begins as light strikes the retina. The retina's basement layer, a pigmented epithelium, covers the choroid. Resting on this epithelium are the densely packed photoreceptors called **rod cells** and **cone cells** (Figure 32.32). Rod cells detect very dim light. At night, they contribute to coarse perception of movement by responding to changes in light intensity across the field of vision. Cone cells detect bright light. They contribute to sharp daytime vision and color perception.

Distinct layers of neurons are located above the rods and cones. The first to accept the visual information from the photoreceptors are bipolar sensory neurons, then ganglion cells—the axons of which form the two optic nerves to the brain (Figure 32.33).

Before signals depart from the retina, they converge dramatically. The input from 125 million photoreceptors converges on a mere 1 million ganglion cells. Signals also flow laterally among horizontal cells and amacrine cells. Both kinds of neurons act in concert to dampen or enhance the signals before ganglion cells get them. Thus, *a great deal of synaptic integration and processing goes on even before visual information is sent to the brain.*

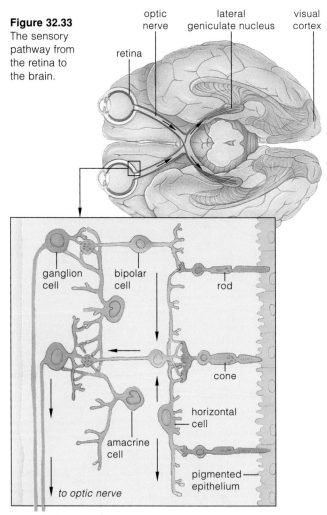

Figure 32.33 The sensory pathway from the retina to the brain.

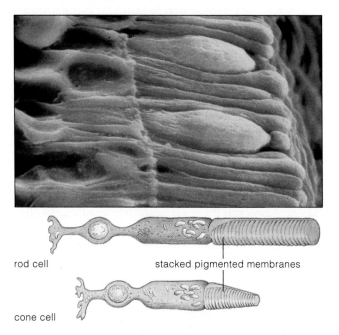

rod cell

stacked pigmented membranes

cone cell

Figure 32.32 Mammalian photoreceptors: rods and cones.

Neuronal Responses to Light

Rod Cells A rod cell's outer segment contains several hundred membrane disks, each peppered with about 10^8 molecules of a visual pigment called rhodopsin. The membrane stacking and the extremely high density of pigments greatly increase the chances of intercepting packets of light energy—that is, photons of particular wavelengths. The action potentials that result from the absorption of even a few photons can lead to conscious awareness of objects in dimly lit surroundings, as in dark rooms or late at night.

Rhodopsin effectively absorbs wavelengths in the blue-to-green portion of the visible spectrum (Section 5.2). Absorption changes the shape of the pigment. This triggers a cascade of reactions that alter activity at ion channels and ion pumps across the rod cell's plasma membrane. Gated sodium channels close, the voltage difference changes across the membrane, and a graded potential results. This potential reduces the ongoing release of a neurotransmitter with inhibitory effects on

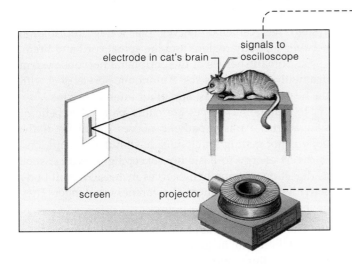

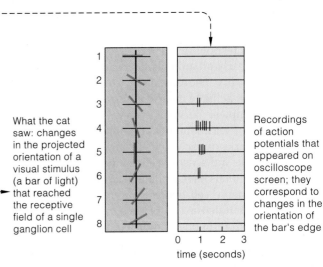

What the cat saw: changes in the projected orientation of a visual stimulus (a bar of light) that reached the receptive field of a single ganglion cell

Recordings of action potentials that appeared on oscilloscope screen; they correspond to changes in the orientation of the bar's edge

time (seconds)

Figure 32.34 Example of experiments into the nature of the receptive fields for visual stimuli. David Hubel and Torsten Wiesel implanted an electrode in an anesthetized cat's brain. They placed the cat in front of a small screen upon which different patterns of light were projected—in this case, a hard-edged bar. Light or shadow falling on part of the screen excited or inhibited signals sent to a single neuron in the visual cortex.

Tilting the bar at different angles produced changes in the neuron's activity. A vertical bar image produced the strongest signal (*numbered 5 in the sketch*). When the bar image tilted slightly, signals were less frequent. When it tilted past a certain angle, signals stopped.

adjacent sensory neurons. Released from inhibition, these start sending signals about the visual stimulus on toward the brain.

Cone Cells The sense of color and of daytime vision starts with photon absorption by red, green, and blue cone cell, each with a different kind of visual pigment. Here again, photon absorption reduces the release of a neurotransmitter that otherwise inhibits the neurons that are adjacent to the photoreceptors.

Near the center of the retina is a funnel-shaped depression called the fovea. This is the retinal area with the greatest density of photoreceptors and the greatest visual acuity. Its cone cells are the ones that discriminate most precisely between adjacent points in space.

In *red-green color blindness,* some or all cone cells that normally respond to light of red or green wavelengths are missing from the retina. It is a common, X-linked recessive trait that primarily affects males. Color-blind persons in general have trouble distinguishing red from green in dim light. Some cannot distinguish between the two even in bright light. The rare few who are fully color blind have only one of three kinds of pigments that selectively respond to light of red, green, or blue wavelengths. They see the world only in shades of gray.

Responses at Receptive Fields Collectively, neurons respond to light in organized ways. The retinal surface is organized into receptive fields, restricted areas that influence the activity of individual sensory neurons. For example, for each ganglion cell the field is a tiny circle. Some cells respond best to a tiny spot of light, ringed by dark, in the field's center. Others respond to motion, a rapid change in light intensity, or a spot of one color. In one experiment, cells generated signals about a suitably oriented bar (Figure 32.34). Such cells do not respond to diffuse, uniform illumination, which is just as well; doing so would produce a confusing array of signals to the brain.

On To the Visual Cortex The part of the outside world that an individual actually sees is its visual field. The right side of both retinas intercepts light from the left half of the visual field; the left side intercepts light from the *right* half. The optic nerve leading out of each eye delivers the signals about a stimulus from the *left* visual field to the right cerebral hemisphere, and signals from the *right* to the left hemisphere (Figure 32.33).

Axons of the optic nerves end in a layered brain region, the lateral geniculate nucleus. Each layer has a map corresponding to receptive fields of the retina. Its interneurons deal with one aspect of the stimulus—form, movement, depth, color, texture, and so on. After initial processing, all the visual signals travel rapidly, at the same time, to different parts of the visual cortex. There, final integration produces organized electrical activity—and the sensation of sight.

Each eye is an outpost of the brain, collecting and analyzing information about the distance, shape, brightness, position, and movement of visual stimuli. Its sensory pathway starts at the retina and proceeds along an optic nerve to the brain.

Organization of visual signals begins at receptive fields in the retina, is further processed in the lateral geniculate nucleus, and is finally integrated in the visual cortex.

SUMMARY

1. The nervous system senses, interprets, and issues commands for responses to specific aspects of the environment. In nearly all animals, its communication lines consist of sensory neurons, interneurons, and motor neurons, which activate muscle and gland cells.

   a. The simplest nervous systems are nerve nets, such as those of sea anemones and other radial animals. These meshworks of nerve cells form reflex connections with contractile and sensory cells of the epithelium.

   b. Most animals have a bilateral, cephalized nervous system, with nerve cell bodies concentrated as ganglia and a brain. Bundled axons of sensory neurons, motor neurons, or both form one or more cordlike nerves.

2. A stimulus is a form of energy that the body detects by specific receptor regions of sensory neurons.

3. A neuron's dendrites and cell body are input zones. Arriving signals may spread to a trigger zone (the start of an axon) and trigger an action potential that is propagated to the neuron's output zone (its axon endings).

4. A difference in electric charge (a voltage difference) is maintained across the membrane of an undisturbed neuron; this is the resting membrane potential. The cytoplasmic fluid next to the membrane is negatively charged with respect to interstitial fluid just outside. An action potential is an abrupt, short-lived reversal in this voltage difference in response to adequate stimulation.

   a. Stimulation of a patch of membrane at an input zone gives rise to local, graded potentials. These vary in magnitude and do not spread far. A disturbance caused by a number of graded potentials may spread to the trigger zone and drive the membrane to the threshold of an action potential.

   b. The disturbance makes many gated channels for sodium ions open in an accelerating, self-propagating way and abruptly reverses the voltage difference across the membrane. Accelerated openings proceed in patch after patch of membrane, to the neuron's output zone.

   c. In between action potentials, sodium-potassium pumps counter ion leaks across the membrane and so maintain the gradients. After an action potential, they restore the gradients.

5. Action potentials arriving at an output zone trigger release of neurotransmitters. These diffuse across chemical synapses, where the neuron forms a junction with another neuron, a muscle cell, or a gland cell. They may excite that cell's membrane (drive it closer to threshold) or inhibit it (drive it away from threshold).

6. Integration is the moment-by-moment summation of excitatory and inhibitory signals acting at all the synapses on a neuron. It is the means by which information is played down or reinforced, suppressed or sent on to other neurons of the nervous system.

7. In vertebrates, the central nervous system consists of the brain and spinal cord. The peripheral nervous system has cablelike nerves that carry information between all body regions and the spinal cord and brain.

   a. Somatic nerves of the peripheral nervous system deal with skeletal muscles. Autonomic nerves deal with the heart, lungs, glands, and other internal organs.

   b. Autonomic nerves are either parasympathetic or sympathetic. Parasympathetic nerves dominate if the body is not getting much outside stimulation, and tend to divert energy to basic housekeeping activities, such as digestion. Sympathetic nerves increase overall body activities during heightened awareness or danger. They govern the fight-flight response.

8. The vertebrate brain has three main divisions:

   a. Hindbrain: includes the medulla oblongata, pons, and cerebellum and contains reflex centers for vital functions and muscle coordination.

   b. Midbrain: functions in coordinating and relaying visual and auditory information.

   c. Forebrain: includes the thalamus, hypothalamus, limbic system, and cerebrum. The thalamus relays sensory signals and helps coordinate motor responses. The hypothalamus monitors internal organs; governs thirst, sexual activity, and other behaviors related to organ functions; and is gatekeeper to the limbic system, which has roles in learning, memory, and emotions. The cerebral cortex interacts with other brain centers to receive, process, and integrate new stimuli with memories of past events, and coordinate motor responses.

9. Receptor endings of sensory neurons or specialized cells adjacent to them detect specific stimuli, such as light. Brain centers interpret variations in stimulus intensity through the frequency and number of action potentials arriving from a particular nerve pathway.

10. Somatic sensations include pressure, temperature, pain, and muscle sense. Receptors associated with these are not localized in one organ or tissue. For example, muscles throughout the body have stretch receptors.

11. The special senses include taste, smell, hearing and balance, and vision. Receptors associated with them are restricted to particular organs, such as eyes.

Review Questions

1. Define sensory neuron, interneuron, and motor neuron. *543*

2. Label the functional zones of this motor neuron. *546*

3. Distinguish between the following:
   a. central and peripheral nervous system *555*
   b. cranial and spinal nerves *556*
   c. somatic and autonomic nerves *556*
   d. parasympathetic and sympathetic nerves *556–557*

Self-Quiz *(Answers in Appendix IV)*

1. Action potentials occur when _____ .
   a. a neuron is adequately stimulated
   b. sodium gates open in an accelerating way
   c. sodium-potassium pumps kick into action
   d. both a and b

2. Compared to its surroundings, an undisturbed neuron carries a slight _____ charge inside the plasma membrane.
   a. positive
   b. negative
   c. graded, local
   d. both b and c

3. The resting membrane potential is maintained by _____ .
   a. ion leaks
   b. ion pumps
   c. neurotransmitters
   d. both a and b

4. Extending from the uppermost part of the spinal cord and on through the brain stem to the cerebral cortex is the _____ .
   a. reticular formation
   b. blood-brain barrier
   c. olfactory lobe
   d. tectum

5. Each sensory system consists of _____ .
   a. nerve pathways from specific receptors to the brain
   b. sensory receptors
   c. brain regions that deal with sensory information
   d. all of the above are components

6. _____ detect mechanical energy associated with changes in pressure, position, or acceleration.
   a. Chemoreceptors
   b. Mechanoreceptors
   c. Photoreceptors
   d. Thermoreceptors

7. Match the terms with the suitable descriptions.
   ____ synaptic integration
   ____ muscle spindle
   ____ graded, local potential
   ____ action potential

   a. occurs at neuron's input zone
   b. spatial, temporal summation of all signals arriving at neuron
   c. arises at trigger zone
   d. stretch-sensitive receptor

8. Match the component with its main functions.
   ____ spinal cord
   ____ medulla oblongata
   ____ hypothalamus
   ____ limbic system
   ____ cerebral cortex

   a. receives, integrates sensory input with stored information
   b. monitors internal organs, related behavior, influences emotions
   c. our emotional brain
   d. reflex control of respiration, blood circulation, other basic activities
   e. connects brain with trunk, limbs, neck; makes direct reflex connections

Critical Thinking

1. To Kim, a graduate student studying in Tanzania, the flamingos closest to him along the shore of a lake are in sharp focus, but the ones flying some distance away look fuzzy (Figure 32.35). Is Tim nearsighted or farsighted, and is the focal point of light rays from the flamingos falling in front of or behind his retinas?

2. When Jennifer was six years old, a man lost control of his car and hit a tree in front of her house. She ran to the car and screamed when she saw blood from a deep head wound dripping onto a huge bouquet of red roses. Thirty-five years years later, someone gave her a bottle of "Tea Rose" perfume for a present. When she

Figure 32.35 Flamingos in Tanzania, East Africa.

sniffed the perfume, she became frightened and anxious. A few minutes later she had a vivid recollection of the accident. Explain this incident in terms of what you have learned about memory.

3. Eric typically drinks a cup of coffee nearly every hour, all day long. Today, by midafternoon, he is having trouble concentrating on his studies, and he feels tired and rather clumsy. Drinking another cup of coffee didn't make him more alert. Explain how the caffeine in coffee might produce such symptoms.

4. Wayne, on standby for the last flight from San Francisco to New York, was assigned the last available seat on the plane. It was in the last row, where vibrations and noise from the engines are most pronounced. When he got off the plane in New York, Wayne was speaking very loudly and having trouble hearing what others were saying. The next day, things were back to normal. Explain what happened to his sense of hearing during the flight.

Selected Key Terms

Neural Function:
action potential *546*
axon *546*
chemical synapse *550*
dendrite *546*
interneuron *543*
motor neuron *543*
myelin sheath *552*
nerve *544*
neuroglia *543*
neurotransmitter *550*
reflex pathway *544*
resting membrane potential *546*
sodium-potassium pump *547*
synaptic integration *551*

Nervous Systems:
autonomic nerve *556*
brain *544*
central nervous system *555*
cerebellum *554*
cerebral cortex *558*
cerebrum *555*
ganglion *544*
hypothalamus *555*
limbic system *559*
medulla oblongata *554*
memory *560*

nerve net *544*
nervous system *543*
parasympathetic nerve *556*
peripheral nervous system *555*
pons *554*
reticular formation *559*
somatic nerve *556*
spinal cord *557*
sympathetic nerve *557*
thalamus *555*

Sensory Systems:
chemoreceptor *562*
compound eye *566*
cone cell *568*
ear *564*
echolocation *564*
eye *566*
mechanoreceptor *562*
nociceptor (pain) *562*
photoreceptor *562*
retina *566*
rod cell *568*
sensation *562*
sensory neuron *543*
sensory system *562*
stimulus *562*
thermoreceptor *562*

Reading

Entire September 1992 issue. "The Mind and Brain." *Scientific American* 267(3).

33 ENDOCRINE CONTROL

Hormone Jamboree

In the 1960s, at her forest camp by the shore of Lake Tanganyika in Tanzania, the primatologist Jane Goodall let it be known that bananas were available. One of the first chimpanzees attracted to the delicious food was a female—Flo, as she came to be called (Figure 33.1). Flo brought along her offspring, an infant female and a juvenile male, and exhibited commendable parental behavior toward them.

Three years passed, and Goodall observed that Flo's preoccupation with motherhood had now given way to a preoccupation with sex. She also observed that male chimpanzees followed Flo to the camp and stayed for more than the bananas.

Sex, Goodall discovered, is *the* premier force in the social life of chimpanzees. These primates do not mate for life as eagles do, or wolves. Before the rainy season begins, mature females that are entering their fertile cycle become sexually active. In response to changing concentrations of hormones in their bloodstream, their external sexual organs become swollen and vividly pink, and this is a visual signal to males. The swellings are the flags of sexual jamborees, of great gatherings of highly stimulated chimps in which any males present may copulate in sequence with the same females.

The gathering of many flag-waving females draws together individuals that forage alone or in small family groups for most of the year. It reestablishes the bonds that unite their rather fluid community. As infants and juveniles play with one another and with the adults, their aggressive and submissive jostlings help map out future dominance hierarchies. Consider Flo, a high-ranking member of the social hierarchy. Through her sexual attractiveness and direct solicitations, she built alliances with many male chimps. Through her status and aggressive behavior, she helped her male offspring win confrontations with other young male chimps.

The hormone-induced swelling during a female's fertile period lasts somewhere between ten and sixteen days—yet she is fertile for only one to five days. Sex hormones induce the swelling even after a female has

a

become pregnant. Almost certainly, sexual selection has favored the prolonged swelling. Males groom a sexually attractive female more often, protect her, give her more food, and let her tag along to new foraging sites. The more that males accept a female, the higher she rises in the social hierarchy—and the more her individual offspring benefit.

Through their effects, hormones help orchestrate the growth, development, and reproductive cycles of nearly all animals, from the invertebrate worms to chimpanzees to humans. They influence minute-by-minute and day-to-day metabolic functions. Through interplays with one another and with the nervous system, hormones have profound influence over the physical appearance, the well-being, and the behavior of individuals. And even the behavior of individuals affects whether they will survive, either on their own or in social groups.

This chapter focuses mainly on hormones—on their sources, targets, and interactions as well as the mechanisms involved in their secretion. If the details start to seem remote, remember that this is the stuff of life. Hormones underwrote Flo's appearance, behavior, and rise through the chimpanzee social hierarchy—and just imagine what they have been doing for you.

Figure 33.1 (**a**) Jane Goodall in Gombe National Park, near the shores of Lake Tanganyika, scouting for chimpanzees. (**b**) Flo and three of her offspring.

b

KEY CONCEPTS

1. Hormones and other signaling molecules have roles in integrating the activities of many individual cells in ways that benefit the whole body. Only cells with molecular receptors for a specific hormone are its targets.

2. Hormones operate by inducing changes in the activities of target cells. Many types influence gene transcription and protein synthesis in their targets. Other types cause alterations in existing molecules and structures in the cell. Some may even exert effects by binding with and altering membrane characteristics.

3. Some hormones help the body adjust to short-term shifts in diet and in levels of activity. Others help induce long-term adjustments in cell activities that bring about bodily growth, development, and reproduction.

4. Among vertebrates, the hypothalamus and pituitary gland interact in ways that coordinate the activities of many endocrine glands. Together, they exert control over many of the body's functions.

5. Neural signals, hormonal signals, chemical changes, and environmental cues are the triggers for hormonal secretion.

Hormones and Other Signaling Molecules

Throughout their lives, cells must respond to changing conditions by taking up and releasing various chemical substances. In vertebrates, the responses of millions to many billions of cells must be integrated in ways that benefit the whole body.

Integration of cell activities depends on signaling molecules. These include hormones, neurotransmitters, local signaling molecules, and pheromones. Each type of signaling molecule acts on target cells, which are any cells that have receptors for it and that may alter their activities in response to it. A target may or may not be adjacent to the cell that sends the signal.

By definition, **hormones** are secretory products of endocrine glands, endocrine cells, and some neurons that the bloodstream distributes to nonadjacent target cells. They are this chapter's focus.

By contrast, *neurotransmitters* are released from the axon endings of neurons. They act swiftly on target cells by diffusing across a tiny gap that separates them. Chapter 32 describes their sources and their action. *Local signaling molecules*, released by many types of body cells, alter conditions within localized regions of tissues. Prostaglandins, for instance, target smooth muscle cells in bronchiole walls, which then constrict or dilate and so alter air flow in lungs.

Pheromones, which are the secretions of particular exocrine glands, diffuse through water or air to targets outside the animal body. Pheromones act on cells of other individuals of the same species and thereby help integrate social behavior. Chapter 40 provides some fine examples of these hormonelike substances. For now, simply consider the effect of bombykol, a sex-attracting pheromone, on olfactory receptors of a male silk moth. Contact between a receptor and merely one bombykol molecule per second sends action potentials to the moth brain. This pheromone can help a male moth locate a female of the same species in the dark, even if she is more than a kilometer upwind from him.

Discovery of Hormones and Their Sources

The word "hormone" dates back to the early 1900s. Then, W. Bayliss and E. Starling were trying to find out what triggers the secretion of pancreatic juices when food travels through the canine gut. As they knew already, acids mix with food in the stomach, and the pancreas secretes an alkaline solution after the acidic mixture moves into the small intestine. Was the nervous system or something else stimulating the pancreatic response?

To find the answer, Bayliss and Starling blocked nerves—but not blood vessels—leading to a laboratory

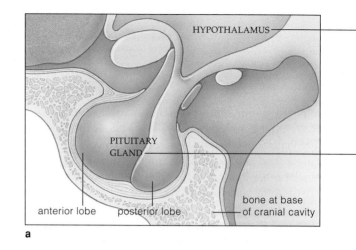

Figure 33.2 (**a**) An important neural-endocrine control center. The pituitary gland interacts intimately with the hypothalamus, a brain region that secretes some hormones. (**b**) *Facing page:* Overview of the major components of the human endocrine system and the primary effects of their main hormonal secretions. The system also includes endocrine cells of many organs, including the liver, kidneys, heart, small intestine, and skin.

animal's upper small intestine. Later, when acidic food entered the small intestine, the pancreas still secreted the alkaline solution. Even more telling, extracts of cells from the intestinal lining—a glandular epithelium—also induced the response. Glandular cells had to be the source of the pancreas-stimulating substance.

The substance came to be called secretin. Proof of its existence and mode of action confirmed a centuries-old idea: *Internal secretions that are picked up by the bloodstream influence the activities of the body's organs.* Starling coined the word "hormone" for such internal glandular secretions (after *hormon*, meaning to set in motion).

Later, researchers identified many other kinds of hormones and their sources. Figure 33.2 is a simplified picture of the locations of the following major sources of hormones in the human body. Bear in mind, these sources are typical of most vertebrates:

Pituitary gland

Adrenal glands (*two*)

Pancreatic islets (*numerous cell clusters*)

Thyroid gland

Parathyroid glands (*in humans, four*)

Pineal gland

Thymus gland

Gonads (*two*)

Endocrine cells of the stomach, small intestine, liver, kidneys, heart, placenta, skin, adipose tissue, and other organs

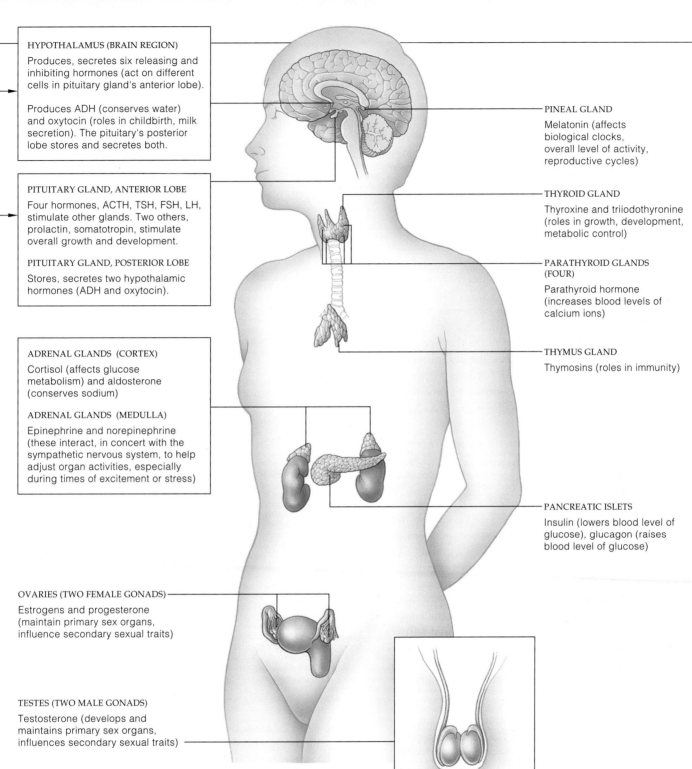

HYPOTHALAMUS (BRAIN REGION)

Produces, secretes six releasing and inhibiting hormones (act on different cells in pituitary gland's anterior lobe).

Produces ADH (conserves water) and oxytocin (roles in childbirth, milk secretion). The pituitary's posterior lobe stores and secretes both.

PITUITARY GLAND, ANTERIOR LOBE

Four hormones, ACTH, TSH, FSH, LH, stimulate other glands. Two others, prolactin, somatotropin, stimulate overall growth and development.

PITUITARY GLAND, POSTERIOR LOBE

Stores, secretes two hypothalamic hormones (ADH and oxytocin).

ADRENAL GLANDS (CORTEX)

Cortisol (affects glucose metabolism) and aldosterone (conserves sodium)

ADRENAL GLANDS (MEDULLA)

Epinephrine and norepinephrine (these interact, in concert with the sympathetic nervous system, to help adjust organ activities, especially during times of excitement or stress)

OVARIES (TWO FEMALE GONADS)

Estrogens and progesterone (maintain primary sex organs, influence secondary sexual traits)

TESTES (TWO MALE GONADS)

Testosterone (develops and maintains primary sex organs, influences secondary sexual traits)

PINEAL GLAND

Melatonin (affects biological clocks, overall level of activity, reproductive cycles)

THYROID GLAND

Thyroxine and triiodothyronine (roles in growth, development, metabolic control)

PARATHYROID GLANDS (FOUR)

Parathyroid hormone (increases blood levels of calcium ions)

THYMUS GLAND

Thymosins (roles in immunity)

PANCREATIC ISLETS

Insulin (lowers blood level of glucose), glucagon (raises blood level of glucose)

b

Collectively, the body's sources of hormones came to be called the **endocrine system**. The name implies that there is a separate control system for the body, apart from the nervous system. (*Endon* means within; *krinein* means separate.) Later, however, biochemical research and electron microscopy studies revealed that endocrine sources and the nervous system function in intricately connected ways, as you will see shortly.

Integration of cell activities depends on hormones and other signaling molecules. Each type of signaling molecule acts on target cells, which are any cells that have receptors for it and that may alter their activities in response to it.

Components of the endocrine system and certain neurons produce and secrete hormones, which the bloodstream takes up and distributes to nonadjacent target cells.

33.2 SIGNALING MECHANISMS

The Nature of Hormone Action

Hormones and other signaling molecules interact with protein receptors of target cells, and their interactions have diverse effects on physiological processes. Some kinds of hormones induce target cells to increase their uptake of glucose, calcium, or some other substance from the surroundings. Other kinds stimulate or inhibit target cells into altering their rates of protein synthesis, modifying the structure of proteins or other elements in the cytoplasm, or even changing cell shape.

Two factors exert considerable influence over the responses to hormonal signals. *First,* different hormones activate different cellular mechanisms. *Second,* not all cell types are equipped to respond to a given signal. For example, many cell types have receptors for cortisol, so this hormone has widespread effects on the body. By contrast, only a few cell types have the receptors for hormones that stimulate a highly directed response.

Let's now briefly consider the effects of two main categories of these signaling molecules—the steroid and peptide hormones (Table 33.1).

Characteristics of Steroid Hormones

Steroid hormones, recall, are lipid-soluble molecules derived from cholesterol (Section 2.9). The cholesterol remodeling proceeds at multiple-enzyme systems in the endoplasmic reticulum and mitochondria of steroid-secreting cells. We find such cells in adrenal glands and primary reproductive organs. Testosterone, one of the sex hormones, is an example of the final products.

Consider testosterone's effects on the development of the secondary sexual traits associated with maleness. The developmental steps will proceed as normal only if target cells have working receptors for testosterone. In a genetic disorder called *testicular feminization syndrome,* these receptors are defective. Genetically, an affected

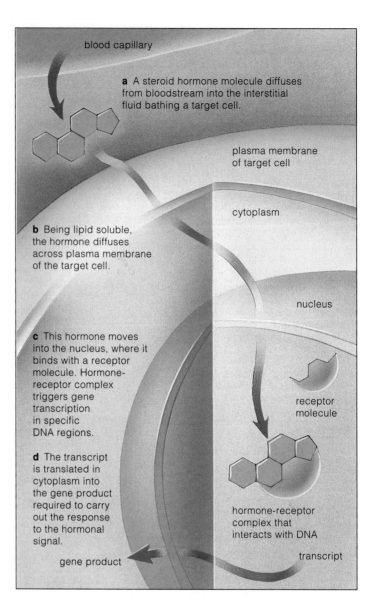

a A steroid hormone molecule diffuses from bloodstream into the interstitial fluid bathing a target cell.

blood capillary

plasma membrane of target cell

cytoplasm

b Being lipid soluble, the hormone diffuses across plasma membrane of the target cell.

nucleus

c This hormone moves into the nucleus, where it binds with a receptor molecule. Hormone-receptor complex triggers gene transcription in specific DNA regions.

receptor molecule

d The transcript is translated in cytoplasm into the gene product required to carry out the response to the hormonal signal.

hormone-receptor complex that interacts with DNA

gene product

transcript

Figure 33.3 Example of a mechanism by which a steroid hormone initiates changes in a target cell's activities.

Table 33.1	Two Main Categories of Hormones
Type	Examples
Steroid and steroidlike hormones	Estrogens (feminizing effects), progestins (related to pregnancy), androgens (such as testosterone; masculinizing effects), cortisol, aldosterone. In addition, thyroid hormones and vitamin D act like steroids
Peptide hormones:	
Peptides	Glucagon, ADH, oxytocin, TRH
Proteins	Insulin, somatotropin, prolactin
Glycoproteins	FSH, LH, TSH

person is male (XY); he has functional testes that are able to secrete testosterone. However, target cells can't respond to the hormone, so the secondary sexual traits that do develop are like those of females.

How does a steroid hormone such as testosterone exert effects on cells? Being lipid soluble, it may diffuse directly across the lipid bilayer of a target cell's plasma membrane (Figure 33.3). Once inside the cytoplasm, the hormone molecule usually moves into the nucleus and binds to some type of protein receptor. In some cases, it binds to a receptor molecule in the cytoplasm, then the hormone-receptor complex moves into the nucleus. The configuration of the complex allows it to interact with a specific region of the cell's DNA. Different complexes

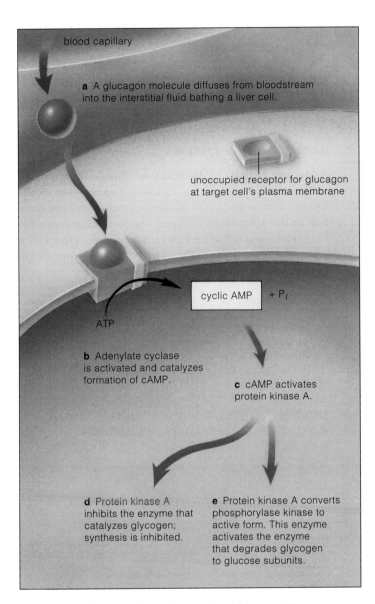

blood capillary

a A glucagon molecule diffuses from bloodstream into the interstitial fluid bathing a liver cell.

unoccupied receptor for glucagon at target cell's plasma membrane

cyclic AMP $+ P_i$

ATP

b Adenylate cyclase is activated and catalyzes formation of cAMP.

c cAMP activates protein kinase A.

d Protein kinase A inhibits the enzyme that catalyzes glycogen; synthesis is inhibited.

e Protein kinase A converts phosphorylase kinase to active form. This enzyme activates the enzyme that degrades glycogen to glucose subunits.

Figure 33.4 Example of a mechanism by which a peptide hormone initiates changes in a target cell's activities. Binding of the hormone glucagon at a receptor initiates reactions inside the cell. In this case, a second messenger (cyclic AMP) relays the signal into the cell interior.

inhibit or stimulate transcription of certain gene regions into mRNA. The translation of such mRNA transcripts results in enzymes and other proteins that can carry out a response to the hormonal signal.

Steroid hormones may also exert effects in another way. It now appears they can bind to cell membranes and alter membrane properties, in ways that alter the target cell's functional responses.

Finally, thyroid hormones and vitamin D are not steroid hormones, but they act like them. In fact, the genes coding for their receptors are part of the same group that codes for steroid hormones.

Characteristics of Peptide Hormones

Traditionally, **peptide hormones** have been defined as water-soluble signaling molecules that may incorporate anywhere from 3 to 180 amino acids. Various peptides, polypeptides, and glycoproteins fall in this category. All peptide hormones issue signals right at a receptor of a target cell's plasma membrane. The signals activate specific membrane-bound enzyme systems that initiate reactions leading to the cellular response.

Consider a liver cell with receptors for glucagon, a peptide hormone. Each receptor spans the plasma membrane, and part of it extends into the cytoplasm. After a receptor binds glucagon, the response requires assistance from a **second messenger**. Such messengers are small molecules in the target cell's cytoplasm, and they relay a signal from a hormone-receptor complex at the plasma membrane into the cell interior. In this case, cAMP (cyclic adenosine monophosphate) is the second messenger. As Figure 33.4 indicates, glucagon binding triggers activity at a membrane-bound enzyme system. An activated form of an enzyme (adenylate cyclase) converts ATP to cyclic AMP, which starts a cascade of reactions. Many molecules of cyclic AMP form and act as signals for the conversion of many molecules of an enzyme (a protein kinase) to active form. These act on other enzymes, and so on until a final reaction converts stored glycogen in the cell to glucose. In short order, the number of molecules involved in the final response to the glucagon-receptor complex is enormous.

Or consider a muscle cell with receptors for insulin, a protein hormone. Among other things, a signal from a hormone-receptor complex triggers the movement of molecules of glucose transporter proteins through the cytoplasm and insertion into the plasma membrane, so that the cell can take up glucose faster. In addition, it switches on enzymes of glucose metabolism.

Bear in mind, there are other hormone categories, including catecholamines such as epinephrine. Like glucagon, epinephrine combines with specific surface receptors and triggers the release of cyclic AMP as a second messenger that assists in the cellular response.

Hormones interact with receptors located either at the plasma membrane or within the cytoplasm of target cells.

Steroid hormone-receptor complexes typically enter the cell and interact with the target cell's DNA. Some also interact with membranes and alter membrane functions.

Peptide hormones do not exert their effects by entering cells. When they bind to a membrane receptor, the binding itself is a signal for enzyme-mediated, intracellular events. Often a second messenger in the cytoplasm relays the signal into the cell interior.

Deep in the brain is the **hypothalamus**. This brain region monitors internal organs and activities related to their functioning, such as eating and sexual behavior. It also secretes some hormones. Suspended from its base by a slender stalk of tissue is a pea-size lobed gland. The hypothalamus and this **pituitary gland** interact as a major neural-endocrine control center.

The pituitary's *posterior* lobe secretes two hormones that are synthesized by the hypothalamus. Its *anterior* lobe produces and secretes its own hormones, most of which regulate the release of hormones from other endocrine glands (Table 33.2). The pituitary of many vertebrates—but not humans—also has an intermediate lobe. In many cases, this lobe secretes a hormone that governs reversible changes in skin or fur color.

Posterior Lobe Secretions

Figure 33.5*a* shows the cell bodies of certain neurons in the hypothalamus. Their axons extend down into the posterior lobe, then terminate next to a capillary bed. The neurons produce antidiuretic hormone (ADH) and oxytocin, then store them in the axon endings. After either hormone is released into the interstitial fluid, it diffuses into capillaries, then travels the bloodstream to its targets. ADH acts on cells of nephrons and collecting ducts in the kidneys. Recall that kidneys filter blood and rid the body of excess water and salts (as urine). ADH promotes water reabsorption when the body must conserve water. Oxytocin has roles in reproduction. It triggers contractions of the uterus during labor. It also causes milk release when offspring are being nursed.

Anterior Lobe Secretions

Anterior Pituitary Hormones Inside the pituitary stalk, a capillary bed picks up hormones secreted by the hypothalamus and delivers them into a *second* capillary bed in the anterior lobe. There, the hormones leave the bloodstream and then act on target cells. As Figure 33.6 shows, different cells of the anterior pituitary secrete six hormones that they themselves produce:

Corticotropin	ACTH
Thyrotropin	TSH
Follicle-stimulating hormone	FSH
Luteinizing hormone	LH
Prolactin	PRL
Somatotropin (or growth hormone)	STH (or GH)

ACTH acts on adrenal glands, and TSH on the thyroid gland, as described shortly. FSH and LH have roles in reproduction, the central topic of Chapter 34.

Table 33.2	Hormones Released From the Mammalian Pituitary Gland			
Pituitary Lobe	Secretions	Designation	Main Targets	Primary Actions
Posterior Nervous tissue (extension of hypothalamus)	Antidiuretic hormone	ADH	Kidneys	Induces water conservation as required during control of extracellular fluid volume and solute concentrations
	Oxytocin	OCT	Mammary glands	Induces milk movement into secretory ducts
			Uterus	Induces uterine contractions
Anterior Mostly glandular tissue	Corticotropin	ACTH	Adrenal cortex	Stimulates release of adrenal steroid hormones
	Thyrotropin	TSH	Thyroid gland	Stimulates release of thyroid hormones
	Gonadotropins:			
	Follicle-stimulating hormone	FSH	Ovaries, testes	In females, stimulates estrogen secretion, egg maturation; in males, helps stimulate sperm formation
	Luteinizing hormone	LH	Ovaries, testes	In females, stimulates progesterone secretion, ovulation, corpus luteum formation; in males, stimulates testosterone secretion, sperm release
	Prolactin	PRL	Mammary glands	Stimulates and sustains milk production
	Somatotropin (also called growth hormone)	STH (GH)	Most cells	Promotes growth in young; induces protein synthesis, cell division; roles in glucose, protein metabolism in adults
Intermediate* Glandular tissue, mostly	Melanocyte-stimulating hormone	MSH	Pigmented cells in skin, other surface coverings	Induces color changes in response to external stimuli; affects behavior

* Present in most vertebrates (not humans). MSH is associated with the anterior lobe in humans.

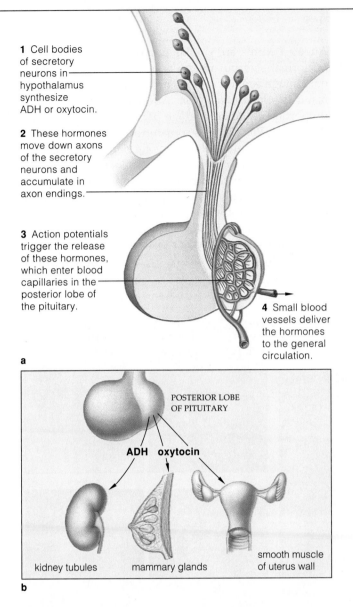

1 Cell bodies of secretory neurons in hypothalamus synthesize ADH or oxytocin.

2 These hormones move down axons of the secretory neurons and accumulate in axon endings.

3 Action potentials trigger the release of these hormones, which enter blood capillaries in the posterior lobe of the pituitary.

4 Small blood vessels deliver the hormones to the general circulation.

a

POSTERIOR LOBE OF PITUITARY

ADH **oxytocin**

kidney tubules   mammary glands   smooth muscle of uterus wall

b

Figure 33.5 (**a**) Functional links between the hypothalamus and the posterior lobe of the pituitary gland. (**b**) Main targets of the posterior lobe secretions.

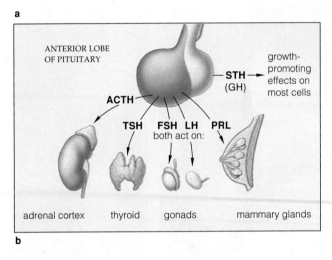

1 Cell bodies of secretory neurons in hypothalamus secrete releasing and inhibiting hormones.

2 First capillary bed, in base of hypothalamus, picks up hormones.

3 Hormones are delivered into second capillary bed, in anterior lobe of pituitary.

4 Releasing or inhibiting hormones diffuse out of the capillaries, act on endocrine cells in the anterior lobe.

5 Hormones secreted from anterior lobe cells enter venules that lead to the general circulation.

a

ANTERIOR LOBE OF PITUITARY

ACTH

TSH **FSH LH** **PRL**
both act on:

STH (GH) → growth-promoting effects on most cells

adrenal cortex   thyroid   gonads   mammary glands

b

Figure 33.6 (**a**) Functional links between the hypothalamus and the anterior lobe of the pituitary gland. (**b**) Main targets of the anterior lobe secretions.

Somatotropin affects body tissues in general (Table 33.2 and Figure 33.6*b*). It works by stimulating protein synthesis and cell division, and therefore has profound influence over growth, especially of cartilage and bone.

Prolactin also has general effects. But it is better known for its role in stimulating and sustaining milk production in mammary glands, after other hormones have primed the tissues. In some species, prolactin also affects hormone production in the ovaries.

About the Hypothalamic Triggers Most of the hypothalamic hormones acting in the anterior lobe of the pituitary are *releasers*; they stimulate the secretion of hormones from their target cells. For example, the one called GnRH (gonadotropin-releasing hormone) brings

about secretion of FSH and LH, which are classified as gonadotropins. Similarly, the releasing hormone called TRH stimulates the secretion of thyrotropin.

But some hypothalamic hormones are *inhibitors* of secretion from their targets in the anterior pituitary. For instance, the one known as somatostatin brings about a decrease in somatotropin and thyrotropin secretion.

The hypothalamus and pituitary gland interact to secrete eight different hormones. Five of these stimulate or inhibit the release of hormones from other endocrine glands. Of the other three, ADH has targets in the kidneys, oxytocin has targets in the uterus and mammary glands, and somatotropin (growth hormone) targets body cells in general.

EXAMPLES OF ABNORMAL PITUITARY OUTPUT

The body does not churn out enormous quantities of hormone molecules. Roger Guilleman and Andrew Schally realized this when they isolated the first known releasing hormone. After four years of dissecting 500 tons of sheep brains, then 7 tons of hypothalamic tissue, they extracted a single milligram of TSH. Yet normal body function depends on those tiny amounts.

Endocrine glands release their tiny but significant secretions mainly in short bursts. Elegant controls over the frequency of those bursts prevent underproduction or overproduction of a given hormone. If something interferes with the controls, disorders may result.

For example, when anterior pituitary cells produce too much somatotropin in childhood, *gigantism* results.

Affected adults are similar proportionally to a person of normal size but much larger (Figure 33.7*a*). When not enough somatotropin is produced in childhood, the result is *pituitary dwarfism*. Affected adults are similar proportionally to an average person but much smaller (Figure 33.7*b*). What if somatotropin output becomes excessive in adulthood, when long bones no longer can lengthen? Then, *acromegaly* results. Bone, cartilage, and other connective tissues in the hands, feet, and jaws thicken abnormally. So do epithelia of the skin, nose, eyelids, lips, and tongue (Figure 33.7*c*).

As another example, ADH secretion can diminish or stop when the pituitary's posterior lobe is damaged, as by a blow to the head or by a brain tumor. This is

age nine   sixteen

thirty-three   fifty-two

a   b   c

Figure 33.7 Effects of overproduction and underproduction of hormones.

(a) Manute Bol, an NBA center, is 7 feet 6-3/4 inches tall owing to excessive STH production during his childhood. **(b)** Effect of somatotropin (STH) on body growth. The male at the center of this photograph is affected by gigantism, which resulted from excessive STH production in childhood. The person at the right displays pituitary dwarfism, which resulted from underproduction of STH in childhood. The person at the left is average in size.

(c) Acromegaly, which resulted from excessive production of STH during adulthood. Before this female reached maturity, she was symptom-free.

one cause of *diabetes insipidus*. The symptoms include excretion of large volumes of dilute urine, which may cause life-threatening dehydration. Diabetes insipidus responds to hormone replacement therapy based on injections or nasal spray applications of synthetic ADH.

Generally, endocrine glands release very small amounts of hormones in short bursts, the frequency of which depends on control mechanisms. When controls fail, the resulting oversecretion or undersecretion may cause disorders.

33.5 SOURCES AND EFFECTS OF OTHER HORMONES

Table 33.3 lists hormones from endocrine sources other than the pituitary. The remainder of this chapter will provide you with a few examples of their effects and of the controls over their output. The examples will make more sense if you keep the following points in mind.

First, hormones often interact with one another. In other words, one or more hormones may oppose, add to, or prime target cells for another hormone's effects. *Second*, negative feedback mechanisms often control the secretions. When a hormone's concentration increases or decreases in some body region, the change triggers

events that respectively dampen or stimulate further secretion. *Third*, a target cell may react differently to a hormone at different times. Its response depends on the hormone's concentration *and* on the functional state of the cell's receptors. *Fourth*, environmental cues may be important mediators of hormonal secretion.

The secretion of a hormone and its effects are influenced by hormone interactions, feedback mechanisms, variations in the state of target cells, and in some cases environmental cues.

Table 33.3 Hormone Sources Other Than the Mammalian Hypothalamus and Pituitary

Source	Secretion(s)	Main Targets	Primary Actions
Adrenal cortex	Glucocorticoids (including cortisol)	Most cells	Promote protein breakdown and conversion to glucose
	Mineralocorticoids (including aldosterone)	Kidney	Promote sodium reabsorption (sodium conservation); help control the body's salt-water balance
Adrenal medulla	Epinephrine (adrenaline)	Liver, muscle, adipose tissue	Raises blood level of sugar, fatty acids; increases heart rate and force of contraction
	Norepinephrine	Smooth muscle of blood vessels	Promotes constriction or dilation of blood vessel diameter
Thyroid	Triiodothyronine, thyroxine	Most cells	Regulate metabolism; have roles in growth, development
	Calcitonin	Bone	Lowers calcium level in blood
Parathyroids	Parathyroid hormone	Bone, kidney	Elevates calcium level in blood
Gonads:			
Testes (in males)	Androgens (including testosterone)	General	Required in sperm formation, development of genitals, maintenance of sexual traits; growth, development
Ovaries (in females)	Estrogens	General	Required for egg maturation and release; preparation of uterine lining for pregnancy and its maintenance in pregnancy; genital development; maintenance of sexual traits; growth, development
	Progesterone	Uterus, breasts	Prepares, maintains uterine lining for pregnancy; stimulates breast development
Pancreatic islets	Insulin	Liver, muscle, adipose tissue	Lowers sugar level in blood
	Glucagon	Liver	Raises sugar level in blood
	Somatostatin	Insulin-secreting cells	Influences carbohydrate metabolism
Thymus	Thymosins, etc.	Lymphocytes	Have roles in immune responses
Pineal	Melatonin	Gonads (indirectly)	Influences daily biorhythms, seasonal sexual activity
Endocrine cells of stomach, small intestine	Gastrin, secretin, etc.	Stomach, pancreas, gallbladder	Stimulate activities of stomach, pancreas, liver, gallbladder required for food digestion, absorption
Liver	Somatomedins	Most cells	Stimulate cell growth and development
Kidneys	Erythropoietin	Bone marrow	Stimulates red blood cell production
	Angiotensin*	Adrenal cortex, arterioles	Helps control blood pressure, secretion of aldosterone (hence sodium reabsorption)
	1,25-hydroxyvitamin D_3* (calcitriol)	Bone, gut	Enhances calcium resorption from bone and calcium absorption from gut
Heart	Atrial natriuretic hormone	Kidney, blood vessels	Increases sodium excretion; lowers blood pressure

* Kidneys produce *enzymes* that modify precursors of this substance, which then enters the general circulation as an activated hormone.

By considering just a few of the endocrine glands listed in Table 33.3, you can sense how feedback mechanisms control hormonal secretions. Briefly, the hypothalamus, pituitary, or both signal these glands to alter secretory activity. The outcome is a change in the concentration of the secreted hormone in blood or elsewhere. With the shift in chemical information, a feedback mechanism trips into action and blocks or promotes further change.

With **negative feedback**, an increase or decrease in the concentration of a secreted hormone triggers events that *inhibit* further secretion. With **positive feedback**, an increase in the concentration of a secreted hormone triggers events that *stimulate* further secretion.

Negative Feedback From the Adrenal Cortex

Humans have a pair of adrenal glands, one above each kidney (Figure 33.2b). Some cells of the **adrenal cortex**, the outer portion of each gland, secrete hormones such as glucocorticoids. The glucocorticoids help maintain the concentration of glucose in blood and help suppress inflammatory responses. Cortisol, for instance, blocks the uptake and use of glucose by muscle cells. It also stimulates liver cells to form glucose from amino acids.

A negative feedback mechanism operates when the glucose level in blood declines below a set point. This chemical condition is known as *hypoglycemia*. Take a look at Figure 33.8. When the hypothalamus detects the

decrease, it secretes CRH in response. This releasing hormone stimulates the anterior pituitary to secrete corticotropin (ACTH). In turn, ACTH stimulates cells of the adrenal cortex to secrete cortisol—which helps raise the glucose level in blood. How? Cortisol stops muscle cells from taking up glucose that blood is delivering through the body. These cells are major glucose users.

During severe stress, painful injury, or a prolonged illness, the nervous system overrides feedback control of cortisol secretion. It initiates a *stress response* in which cortisol helps to suppress inflammation. If unchecked, prolonged inflammation damages tissues. That is why cortisol-like drugs are often prescribed for asthma and other chronic inflammatory disorders.

Local Feedback in the Adrenal Medulla

The **adrenal medulla** is the inner portion of the adrenal gland (Figure 33.8). It has hormone-secreting neurons that release epinephrine and norepinephrine. (These substances are neurotransmitters in some contexts and hormones in others.) Suppose the axons of sympathetic nerves carry hypothalamic signals that call for secretion of norepinephrine. Molecules of norepinephrine collect in the synaptic cleft between the axon endings and the target cells. In this case, a localized, negative feedback mechanism operates at receptors on the axon endings. The excess norepinephrine binds to the receptors—and causes a shutdown of its further release.

Epinephrine and norepinephrine help adjust blood circulation as well as fat and carbohydrate metabolism in times of excitement or stress. They increase heart rate, dilate and constrict arterioles in different regions, and dilate airways to the lungs. The regulated activity shunts more of the blood volume to heart and muscle cells from other regions, and more oxygen flows to energy-demanding cells through the body. These are features of the *fight-flight response* (Section 32.8).

Skewed Feedback From the Thyroid

The human **thyroid gland** is located at the base of the neck in front of the trachea, or windpipe (Figures 33.2b and 33.9a,b). Thyroxine and triiodothyronine, its main hormones, have widespread effects. In their absence, many tissues cannot develop normally. Also, the overall metabolic rates of warm-blooded animals, including humans, depend on them. The importance of feedback control over the secretion of these hormones is brought into sharp focus by cases of abnormal thyroid output.

Consider how thyroid hormone synthesis requires iodine, which we get from food. Iodine is converted to an iodized form, iodide, when absorbed from the gut. Without iodide, the blood levels of thyroid hormones

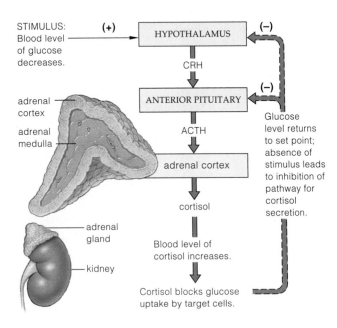

Figure 33.8 Location of the adrenal glands—one above each kidney. The diagram shows a negative feedback loop that governs secretion of cortisol from the adrenal cortex.

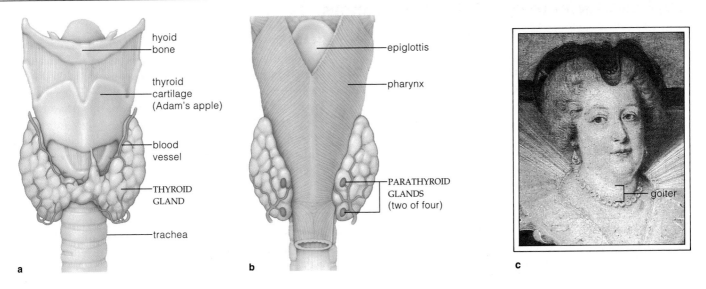

Figure 33.9 Human thyroid gland: (**a**) anterior view, and (**b**) posterior view showing the location of four parathyroid glands. (**c**) A mild case of goiter, displayed by Maria de Medici in the year 1625. A rounded neck was considered to be a sign of great beauty during the late Renaissance. It occurred regularly in parts of the world where iodine supplies were insufficient for normal thyroid function.

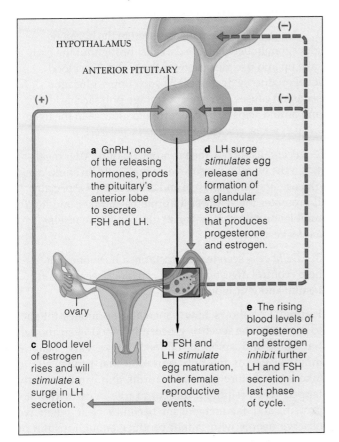

a GnRH, one of the releasing hormones, prods the pituitary's anterior lobe to secrete FSH and LH.

d LH surge *stimulates* egg release and formation of a glandular structure that produces progesterone and estrogen.

c Blood level of estrogen rises and will *stimulate* a surge in LH secretion.

b FSH and LH *stimulate* egg maturation, other female reproductive events.

e The rising blood levels of progesterone and estrogen *inhibit* further LH and FSH secretion in last phase of cycle.

Figure 33.10 Feedback loops to the hypothalamus and pituitary gland from the ovaries during the menstrual cycle—a recurring reproductive event. Positive feedback triggers egg release from an ovary. Negative feedback after its release prevents release of another egg until the cycle is completed.

decline. The anterior pituitary responds by secreting thyroid-stimulating hormone (TSH). But without iodine, thyroid hormones can't be made. The feedback signal continues—and so does TSH secretion. Excess TSH in blood overstimulates the thyroid gland, which enlarges in response. The enlargement is a form of *goiter*. Goiter resulting from iodine deficiency is no longer common in countries where people use iodized salt (Figure 33.9c).

When blood levels of thyroid hormones are too low, *hypothyroidism* results. Hypothyroid adults often are overweight, sluggish, dry-skinned, intolerant of cold, confused, and depressed. Affected women often have menstrual disturbances. When blood levels of thyroid hormones are too high, *hyperthyroidism* results. Affected adults show an increased heart rate, elevated blood pressure, intolerance of heat, profuse sweating, and weight loss even if caloric intake increases. Typically they are nervous, agitated, and have trouble sleeping.

Feedback Control of the Gonads

Gonads are *primary* reproductive organs, which produce gametes and also synthesize and secrete sex hormones. Testes (singular, testis) in males and ovaries in females are examples. Testes secrete testosterone; ovaries secrete estrogens and progesterone. All of these sex hormones influence secondary sexual traits (as they did for the chimps described earlier), and feedback controls govern their secretion. Figure 33.10 is a preview of the feedback loops from ovaries to the hypothalamus and pituitary during the menstrual cycle—a key topic of Chapter 34.

Feedback mechanisms control the secretions from endocrine glands. In many cases, feedback loops to the hypothalamus, pituitary, or both govern the secretory activity.

With negative feedback, further secretion of a hormone slows down. With positive feedback, further secretion is enhanced.

Some endocrine glands or cells don't respond primarily to signals from other hormones or nerves. They respond homeostatically to chemical change in the immediate surroundings, as the following examples illustrate.

Secretions From Parathyroid Glands

Humans have four **parathyroid glands** located on the posterior surface of the thyroid (Figure 33.9b). These glands secrete parathyroid hormone (PTH), the main regulator of the calcium level in blood. Calcium ions, recall, have roles in muscle contraction, enzyme action, blood clot formation, and other tasks. The parathyroids secrete PTH in response to a low calcium level in blood. Their secretory activity slows when the calcium level rises. PTH acts on cells of the skeleton and kidneys.

PTH induces living bone cells (osteocytes and osteoclasts) to secrete enzymes that digest bone tissue and so release calcium and other minerals to interstitial fluid, then to blood. It enhances calcium reabsorption from the filtrate flowing through kidney nephrons. PTH also prods some kidney cells to secrete enzymes that act on blood-borne precursors of the active form of vitamin D_3, a hormone (Table 33.3). The vitamin D_3 stimulates intestinal cells to increase calcium absorption from the gut lumen. In a child with vitamin D deficiency, insufficient calcium and phosphorus are absorbed, so rapidly growing bones develop improperly. The resulting bone disorder, *rickets*, is characterized by bowed legs, a malformed pelvis, and in many cases a malformed skull and rib cage (Figure 33.11).

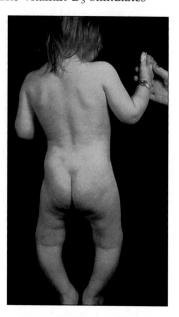

Figure 33.11 A child who is affected by rickets.

Effects of Local Signaling Molecules

Many cells detect changes in the surrounding chemical environment and alter their activity, often in ways that counteract or amplify the change. The cells secrete **local signaling molecules**, the effects of which are confined to the immediate vicinity of change. Target cells take up most signaling molecules so rapidly that few enter the general circulation.

Prostaglandins are examples of signaling molecules. Cells in many tissues continually produce and release a variety of prostaglandins. But the rate of synthesis often increases in response to local chemical changes. The next chapter includes a fine example of this response.

Growth factors are other examples. They influence growth by regulating the rate at which cells divide. Thus an epidermal growth factor (EGF), discovered by Stanley Cohen, influences growth of many cell types. Nerve growth factor (NGF), discovered by Rita Levi-Montalcini, helps assure the survival of neurons and influences the direction of their growth in an embryo. One experiment demonstrated that certain immature neurons will survive indefinitely in tissue culture when NGF is present but will die within a few days if it is not.

Secretions From Pancreatic Islets

The pancreas is a gland with exocrine and endocrine functions. Its *exocrine* cells secrete digestive enzymes. Its *endocrine* cells are located in about 2 million clusters within the pancreas. Each small cluster, a **pancreatic islet**, contains three types of hormone-secreting cells:

1. *Alpha* cells in the pancreas secrete the hormone glucagon. In between meals, cells throughout the body take up and use glucose from the blood. The blood level of glucose decreases. At such times, glucagon secretion causes glycogen (a storage polysaccharide) and amino acids to be converted to glucose in the liver. In such ways, *glucagon raises the glucose level.*

2. *Beta* cells secrete the hormone insulin. After meals, when the blood glucose level is high, insulin stimulates glucose uptake by muscle and adipose cells especially. It promotes the synthesis of proteins and fats, and it inhibits protein conversion to glucose. Thus, *insulin lowers the glucose level.*

3. *Delta* cells secrete somatostatin, a hormone that helps control digestion. It also can block secretion of insulin and glucagon.

Figure 33.12 shows how pancreatic hormones interact to maintain the level of glucose in blood even though the times and amounts of food intake vary. Bear in mind, insulin is the only hormone that prods cells to take up and store glucose in forms that can be rapidly tapped when required. Its central role in carbohydrate, protein, and fat metabolism becomes clear when we observe people who cannot produce enough insulin or who lack body cells that can respond to it.

For example, insulin deficiency may lead to *diabetes mellitus*, a disorder in which excess glucose accumulates in blood, then in urine. Urination becomes excessive, so the body's water-solute balance is disrupted. Affected

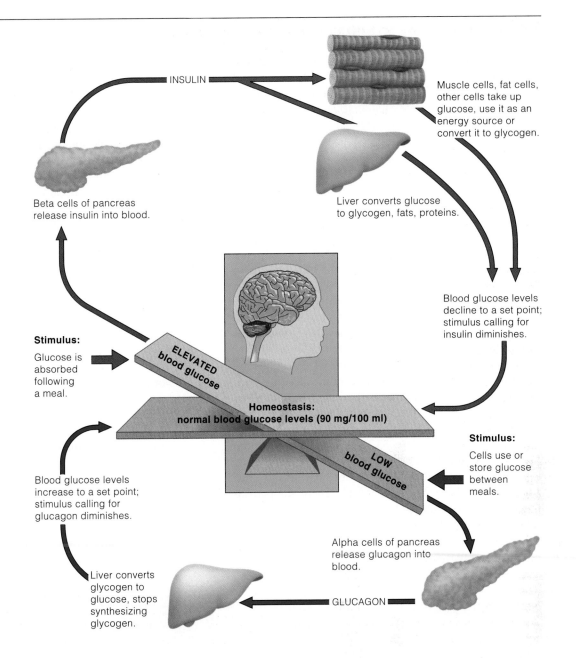

Figure 33.12 Some of the homeostatic controls over glucose metabolism.

Following a meal, glucose enters the bloodstream faster than cells can use it. The blood glucose level rises, and pancreatic beta cells are stimulated to secrete insulin. Insulin's targets (mainly liver, fat, and muscle cells) use glucose or store excess amounts as glycogen.

Between meals, the blood glucose level decreases. Pancreatic alpha cells are stimulated to secrete glucagon. Target cells that respond to this hormone convert glycogen back to glucose, which enters the blood.

Glucose metabolism also is subject to indirect controls. For example, the hypothalamus commands the adrenal medulla to secrete hormones that speed the conversion of glycogen to glucose in the liver and slow the reverse process, especially in cells of the liver, adipose tissue, and muscle tissue.

INSULIN

Muscle cells, fat cells, other cells take up glucose, use it as an energy source or convert it to glycogen.

Beta cells of pancreas release insulin into blood.

Liver converts glucose to glycogen, fats, proteins.

Blood glucose levels decline to a set point; stimulus calling for insulin diminishes.

Stimulus: Glucose is absorbed following a meal.

ELEVATED blood glucose

Homeostasis: normal blood glucose levels (90 mg/100 ml)

LOW blood glucose

Stimulus: Cells use or store glucose between meals.

Blood glucose levels increase to a set point; stimulus calling for glucagon diminishes.

Alpha cells of pancreas release glucagon into blood.

Liver converts glycogen to glucose, stops synthesizing glycogen.

GLUCAGON

people become dehydrated and thirsty—abnormally so. Without a steady supply of glucose, their body cells start depleting their own fats and proteins as sources of energy. Weight loss is one outcome. Ketones, which are normal acidic products of fat breakdown, accumulate in the blood and in urine. The accumulation of ketones contributes to water loss and profoundly alters the body's acid-base balance. The imbalance disrupts brain function. In extreme cases, death may follow.

In "type 1 diabetes," the body mistakenly mounts an autoimmune response against its insulin-secreting beta cells. Certain lymphocytes identify the beta cells as "foreign" and destroy them. A combination of genetic susceptibility and environmental triggers produces the disorder, which is less common but more immediately dangerous than other forms of diabetes. Usually, the symptoms first appear in childhood and adolescence (the disorder is also known as juvenile-onset diabetes). Type 1 diabetic patients survive with insulin injections.

In "type 2 diabetes," insulin levels are close to or above normal—but the target cells cannot respond to insulin. As affected persons grow older, their beta cells produce less and less insulin. Type 2 diabetes usually is manifested during middle age. Affected persons lead normal lives by controlling their diet and weight, and sometimes by taking drugs to enhance insulin action or secretion.

The secretions from some endocrine glands and cells are direct homeostatic responses to a change in the local chemical environment.

HORMONAL RESPONSES TO ENVIRONMENTAL CUES

This last section of the chapter invites you to reflect on a key point. An individual's growth, development, and reproduction begin with genes and hormones, and so does behavior. *But certain environmental factors commonly influence gene expression and hormonal secretion—and they do so in predictable ways.* Chapter 40, on animal behavior, analyzes this influence in detail. For now, it is enough to consider the following examples.

Daylength and the Pineal Gland

Embedded in the brain is a photosensitive organ, the **pineal gland** (Figure 32.21). In the absence of light, the gland secretes the hormone melatonin. Thus the level of melatonin in the blood varies from day to night, and with the seasons. The variations influence the growth and development of gonads—the primary reproductive organs. In a variety of species, they have important roles in reproductive cycles and reproductive behavior.

Consider the hamster. In winter, when night-time darkness is longest, the blood level of melatonin is high and sexual activity is suppressed. In the summer, when daylength is longest, the melatonin level is low—and hamster sex reaches its peak. Or consider a male white-throated sparrow (Figure 33.13a). In fall and winter, melatonin indirectly suppresses growth of its gonads by inhibiting gonadotropin secretion. It does so until spring, when days start to lengthen. Then, stepped-up gonadal activity leads to production of hormones that influence singing behavior, as described in Section 40.1. With his distinctive song, the male sparrow defines his territory and may hold the interest of a mate.

Does melatonin also influence human behavior? Perhaps. Clinical observations and studies suggest that decreased melatonin secretion may trigger puberty, the age during which human reproductive organs and structures start to mature. For example, in cases where disease caused the destruction of an individual's pineal gland, puberty began prematurely.

Melatonin is known to act on certain neurons that lower your body's core temperature and that make you drowsy after sunset, when light is waning. At sunrise, when melatonin secretion slows, your core temperature increases and you wake up and become active.

An internal, biological clock governs the cycle of sleep and arousal. It seems to tick in synchrony with daylength. Think of night workers who try to sleep in the morning but end up staring groggily at sunbeams on the ceiling. Think of travelers from the United States to Paris who go through four days of "jet lag." Two or three hours past midnight they are sitting up in bed, wondering where the coffee and croissants are. Two hours past noon they are ready for bed. They will shift to a new routine when the melatonin signals arrive at their target neurons on Paris time.

In winter, some individuals experience *winter blues*. They get abnormally depressed, go on carbohydrate binges, and have an almost overwhelming desire to sleep (Figure 33.13b). Winter blues might arise when a biological clock gets out of sync with the seasonally shorter daylengths. Intriguingly, clinically administered doses of melatonin make the seasonal symptoms worse. And exposure to intense light—which can shut down pineal activity—may lead to dramatic improvement.

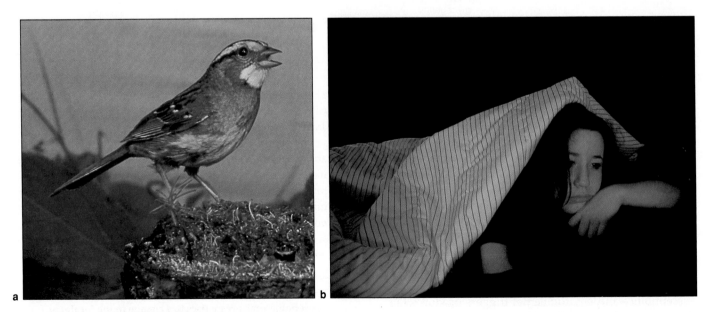

Figure 33.13 (**a**) A male white-throated sparrow, belting out a song that began, indirectly, with an environmentally induced decline in melatonin secretion. (**b**) Annie blanketing her winter blues.

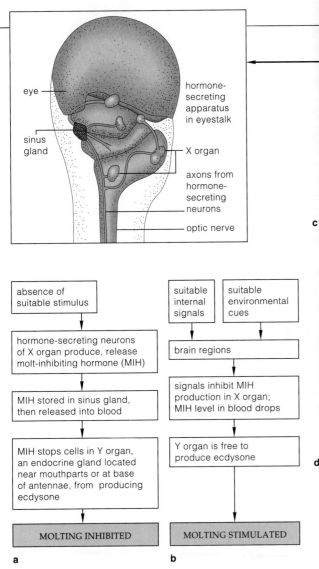

absence of suitable stimulus	**suitable internal signals** / **suitable environmental cues**
↓	↓ ↓
hormone-secreting neurons of X organ produce, release molt-inhibiting hormone (MIH)	brain regions
↓	↓
MIH stored in sinus gland, then released into blood	signals inhibit MIH production in X organ; MIH level in blood drops
↓	↓
MIH stops cells in Y organ, an endocrine gland located near mouthparts or at base of antennae, from producing ecdysone	Y organ is free to produce ecdysone
↓	↓
MOLTING INHIBITED	**MOLTING STIMULATED**
a	**b**

Figure 33.14 (**a,b**) Steps in the hormonal control of molting in crustaceans, such as crabs (**c**). The steps differ a bit in insects, which do not use a molt-inhibiting hormone. Rather, stimulation of the insect brain causes certain neurons to secrete ecdysiotropin. This hormone induces different neurons to produce still another hormone that targets ecdysone-producing cells in prothoracic glands. (**d**) This insect, a cicada, is emerging from its old cuticle.

Comparative Look at a Few Invertebrates

Although this chapter's focus has been on vertebrates, don't lose sight of the fact that all organisms produce signaling molecules of one sort or another. Consider the hormonal control of **molting**, a periodic discarding and replacement of a hardened cuticle that otherwise would limit increases in body mass. As described in Section 20.10, molting occurs during the life cycle of all insects, crustaceans, and other invertebrates with thick cuticles.

Although details vary from group to group, molting is largely under the control of ecdysone. This steroid hormone is derived from cholesterol and is chemically related to many important vertebrate hormones. In insects and crustaceans, molting glands produce and store ecdysone, then release it for distribution through the body at molting time. Hormone-secreting neurons in the brain seem to regulate its release. The hormone-secreting neurons seem to respond to a combination of environmental cues, such as light and temperature, and internal signals.

Figure 33.14 gives examples of the control steps, which differ in crustaceans and insects.

During premolt and molting periods, coordinated interactions among ecdysone and other hormones bring about major structural and physiological changes. The interactions cause the old cuticle to detach from the epidermis and muscles. They induce the dissolving and recycling of the cuticle's inner layers. The interactions also trigger shifts in metabolism and in the composition and volume of the internal environment. They promote cell divisions, secretions, and pigment formation, all of which go into producing a new cuticle. Simultaneously, hormonal interactions control heart rate, muscle action, color changes, and other physiological processes.

Environmental cues, such as changes in light intensity from day to night and seasonal changes in daylength, influence certain hormonal secretions.

SUMMARY

1. The cells of complex animals continually exchange substances with the body's internal environment. Their myriad withdrawals and secretions are integrated in ways that ensure cell survival through the whole body.

2. Integration of cell activities requires the stimulatory or inhibitory effects of signaling molecules.

a. Signaling molecules are chemical secretions from a cell that adjust the behavior of other, target cells.

b. Any cell with molecular receptors for a signaling molecule is a target. The target cells may or may not be next to the cell sending the signal.

c. There are different kinds of signaling molecules. Hormones as well as neurotransmitters, local signaling molecules, and pheromones are the main kinds.

d. Certain steroids, steroidlike molecules, amines, peptides, proteins, and glycoproteins are hormones.

3. In target cells, hormones influence gene activation, protein synthesis, and alterations in existing enzymes, membranes, and other cellular components. Hormones exert their physiological effects through interactions with specific protein receptors at the plasma membrane or in the cytoplasm of target cells.

a. Steroid hormones are lipid soluble. The hormone-receptor complexes interact with DNA and possibly with membranes.

b. Protein hormones are water soluble. Some enter the cytoplasm complexed with receptors. Others are assisted by membrane transport proteins and by second messengers in the cytoplasm, some of which trigger the response.

4. The hypothalamus and pituitary gland interact to integrate many body activities.

a. ADH and oxytocin, two hypothalamic hormones, are stored in the posterior lobe of the pituitary and are secreted from it. ADH influences the extracellular fluid volume. Oxytocin has roles in reproduction and other events.

b. The hypothalamic hormones called releasing and inhibiting hormones control secretions from different cells of the anterior lobe of the pituitary gland.

c. Of six hormones produced in the anterior lobe, two (prolactin and somatotropin, or growth hormone) have general effects on tissues. Four other anterior lobe hormones (ACTH, TSH, FSH, and LH) act on specific endocrine glands and structures, such as the adrenal cortex, thyroid, and gonads.

5. Responses to hormones are influenced by hormone interactions and feedback loops to the hypothalamus and pituitary. They may be influenced by variations in hormone concentrations. They also may be influenced by the number, kind, and state of hormone receptors on the target cell.

6. Many cellular responses to hormones help the body adjust to short-term shifts in diet and levels of activity. Other kinds help bring about long-term adjustments for growth, development, and reproduction.

a. In general, the secretion of hormones such as insulin and parathyroid hormone can change rapidly when the extracellular concentration of some substance must be controlled homeostatically.

b. Other hormones, including somatotropin, have prolonged, gradual, and often irreversible effects, such as those on development.

Review Questions

1. Name the main endocrine glands and state where each is located in the human body. *574–575*

2. Distinguish among hormones, neurotransmitters, local signaling molecules, and pheromones. *574*

3. Which secretions of the posterior lobe of the pituitary gland have the targets indicated? (Fill in the blanks; see pages *578–579*.)

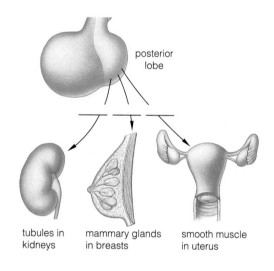

tubules in kidneys　　mammary glands in breasts　　smooth muscle in uterus

4. Which secretions of the anterior lobe of the pituitary gland have the targets indicated? (Fill in the blanks; see pages *578–579*.)

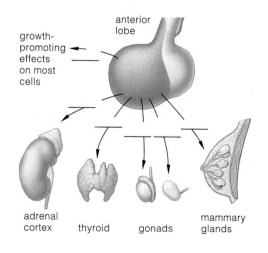

adrenal cortex　　thyroid　　gonads　　mammary glands

5. A hormone molecule binds to a receptor on a cell membrane. It does not enter the cell, however. Rather, binding activates a second messenger inside the cell that triggers an amplified response to the hormonal signal. State whether the signaling molecule is a steroid hormone or a protein hormone. *576–577*

Self-Quiz *(Answers in Appendix IV)*

1. _____ are molecules released from a signaling cell that have effects on target cells.
   a. Hormones
   b. Neurotransmitters
   c. Local signaling molecules
   d. Pheromones
   e. both a and b
   f. a through d

2. Hormones are products of _____ .
   a. endocrine glands and cells
   b. some neurons
   c. exocrine cells
   d. a and b
   e. a and c
   f. a, b, and c

3. ADH and oxytocin are hypothalamic hormones secreted from the pituitary's _____ lobe.
   a. anterior
   b. posterior
   c. intermediate
   d. secondary

4. _____ has effects on body tissues in general.
   a. ACTH
   b. TSH
   c. Prolactin
   d. Somatotropin

5. Which do *not* stimulate hormone secretions?
   a. neural signals
   b. local chemical changes
   c. hormonal signals
   d. environmental cues
   e. all of the above can stimulate hormone secretion

6. _____ lowers blood sugar levels; _____ raises it.
   a. Glucagon; insulin
   b. Insulin; glucagon
   c. Gastrin; insulin
   d. Gastrin; glucagon

7. The pituitary detects a rising hormone concentration in blood and inhibits the gland secreting the hormone. This is a _____ feedback loop.
   a. positive
   b. negative
   c. long-term
   d. b and c

8. Second messengers include _____ .
   a. steroid hormones
   b. protein hormones
   c. cyclic AMP
   d. both a and b

9. Match the hormone source with the closest description.
   ____ adrenal gland
   ____ thyroid gland
   ____ parathyroids
   ____ pancreatic islets
   ____ pineal gland
   ____ thymus gland
   a. affected by daylength
   b. key roles in immunity
   c. raise blood calcium level
   d. epinephrine source
   e. insulin, glucagon
   f. hormones require iodine

Critical Thinking

1. The zebra being nursed in Figure 33.15 is too young to nourish itself by eating grasses. Its source of nutrients is its mother's milk. Explain how secretions from the hypothalamus and both lobes of the pituitary gland influence the production and secretion of milk.

2. In winter, with its far fewer daylight hours compared to the summer, Maxine became very depressed, craved carbohydrate-rich foods, and stopped exercising regularly. And she put on a great deal of weight. Her doctor diagnosed her condition as *seasonal affective disorder* (SAD)—the winter blues. Maxine was advised to purchase a cluster of intense, broad-spectrum lights and to sit near them at least an hour every day. The cloud of depression started to lift quickly. Use your understanding of the secretory activity of the

Figure 33.15 Female zebra nursing her offspring.

pineal gland to explain why Maxine's symptoms appeared and why the prescribed therapy worked.

3. Mary is affected by *type I insulin-dependent diabetes*. One day, after injecting herself with too much insulin, she starts to shake and feels confused. Her doctor recommends a glucagon injection. What caused her symptoms? How would the injection help?

4. Through recombinant DNA technology, somatotropin (growth hormone) is now available commercially for treating pituitary dwarfism. Although it is illegal to do so, some athletes are starting to use somatotropin instead of anabolic steroids. (Here you may wish to compare the introduction to Chapter 26.) Why are they doing so? Somatotropin can't be detected by the test procedures employed in sports medicine. Explain how the athletes might believe somatotropin can improve their performance.

5. *Osteoporosis* is a condition in which loss of calcium results in thin, brittle bones. In combination with other treatments, vitamin D$_3$ injections are sometimes recommended. Explain why.

Selected Key Terms

adrenal cortex *582*
adrenal medulla *582*
endocrine system *575*
gonad *583*
hormone *574*
hypothalamus *578*
local signaling molecule *584*
molting *587*
negative feedback *582*
pancreatic islet *584*
parathyroid gland *584*
peptide hormone *577*
pineal gland *586*
pituitary gland *578*
positive feedback *582*
second messenger *577*
steroid hormone *576*
thymus gland *574*
thyroid gland *582*

Readings

Goodall, J. 1986. *The Chimpanzees of Gombe*. Cambridge, Massachusetts: Belknap Press of Harvard University Press.

Goodman, H. 1994. *Basic Medical Endocrinology*. Second edition. New York: Raven Press.

Hadley, M. 1995. *Endocrinology*. Fourth edition. Englewood Cliffs, New Jersey: Prentice-Hall.

Snyder, S. October 1985. "The Molecular Basis of Communication Between Cells." *Scientific American* 253(4): 132–141.

34 REPRODUCTION AND DEVELOPMENT

From Frog to Frog and Other Mysteries

With a quavering, low-pitched call that only a female of its kind could find seductive, a male frog proclaims the onset of warm spring rains, of ponds, of sex in the night. By August the summer sun will have parched the earth, and his pond dominion will be gone. But tonight is the hour of the frog!

Through the dark, a hormone-primed female moves toward the vocal male. They meet; they dally in the behaviorally prescribed ways of their species. Then he clamps his forelegs above her swollen abdomen and gives her a prolonged squeeze (Figure 34.1). Out into the water streams a ribbon of hundreds of eggs; the male blankets them with a milky cloud of sperm. Soon afterward, tiny fertilized eggs —zygotes—are suspended in the water.

For the leopard frog, *Rana pipiens*, a drama begins to unfold that has been reenacted each spring, with only minor variations, for many millions of years. Within a few hours after fertilization, each zygote divides into two cells, then into four, then many more. In less than twenty hours after fertilization, mitotic cell divisions produce a ball of cells, no larger than the zygote. This is the early embryo.

Soon the cells are signaling one another with chemical substances. They continue to divide, but many of their descendants now change shape and migrate to prescribed positions in the embryo. Amazingly, those cells—so recently descended from a single zygote—are becoming different from one another in their appearance and function.

Through their associations, the cells form distinctive layers of embryonic tissues, then embryonic organs. Certain cells at the surface of the embryo interact with the interior cells beneath them—and together they give rise to a pair of eyes. Within the embryo a tiny heart is forming. Very soon now, it will start an incessant, rhythmic beating. In less than one week's time, the embryo has become a tadpole—a swimming, algae-eating larva. Several months pass. Legs start to grow. The tail shortens, then disappears. The mouth develops jaws that snap shut on insects and worms. Eventually the transformations result in an adult frog. With luck the frog will avoid predators, disease, and other threats in the months ahead. And in time it may even call out quaveringly across a moonlit pond, and the life cycle will begin again.

Some time ago, you also started a developmental journey when a tiny zygote carved itself up into a ball of cells. In your case, some cells became arranged as a thin outer layer and others, as a bunch huddled inside. In a stunning feat of self-organization, the huddled bunch transformed itself into a three-layered, oval disk. It was smaller than the head of a pin. Yet the disk transformed itself into a pale, crescent-shaped embryo. Three weeks into the journey, your embryonic body had the stamp of "vertebrate" on it. A mere five weeks after that, it was a recognizable human in the making!

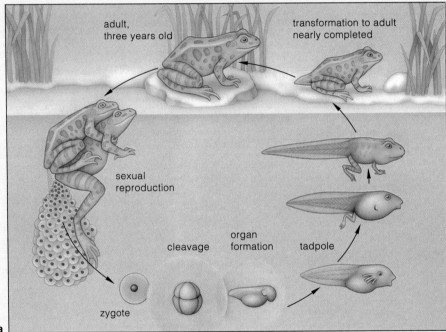

Figure 34.1 Stages in the development of *Rana pipiens*, the leopard frog. (**a**) Frog life cycle. (**b**) A male clasps a female in a behavior called amplexus. When the female releases her eggs into the water, the male releases sperm over the eggs. (**c**) Frog embryos suspended in water. (**d**) A larval form called a tadpole. (**e**) A transitional form between the tadpole and the young adult (**f**).

With this chapter we turn to one of life's greatest dramas—the development of offspring in the image of their parents. The guiding question is this: *How does the single-celled zygote of a frog, human, or any other complex animal become transformed into all the specialized cells and structures of the adult form?* The answers will start to emerge through a brief survey of basic developmental principles. Quite probably, you will find the answers coming into sharp focus when we turn, mid-chapter, to the story of human reproduction and development.

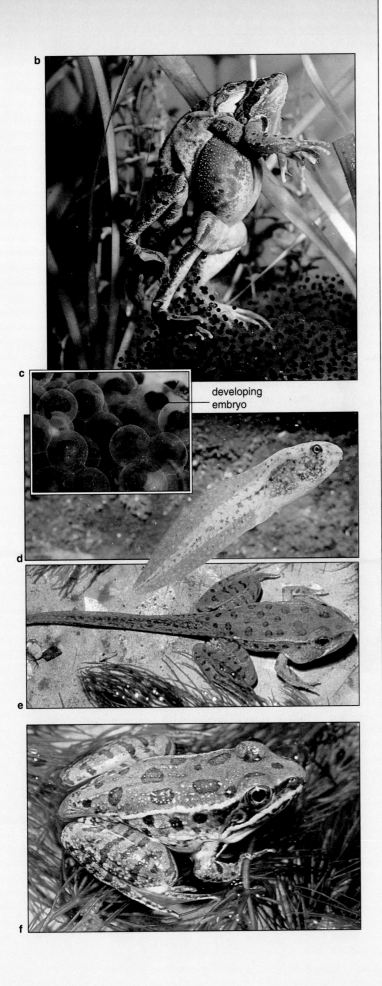

developing
embryo

1. Sexual reproduction dominates the life cycles of nearly all animals. The separation into sexes requires specialized reproductive structures, hormonal control mechanisms, and forms of behavior. Having two distinct sexes affords a great selective advantage—variation in traits among the offspring. This advantage offsets the biological cost of the separation.

2. The life cycles of humans and many other animals proceed through six stages of embryonic development— gamete formation, fertilization, cleavage, gastrulation, organ formation, and growth and tissue specialization. Each stage of embryonic development builds on the tissues and structures that formed during the stage that preceded it.

3. In a developing embryo, the fate of each type of cell depends first on cleavage (when daughter cells inherit different regions of the fertilized egg's cytoplasm), and then on cell interactions. These two activities are the basis of cell differentiation and morphogenesis.

4. With cell differentiation, each cell type selectively uses certain genes and synthesizes proteins not found in other types, and so becomes unique in structure and function. With morphogenesis, tissues and organs change in size, shape, and proportion. They also become organized relative to one another in prescribed patterns.

5. The human reproductive system consists of a pair of primary reproductive organs (testes in males, ovaries in females), accessory glands, and ducts. Testes produce sperm on a continual basis; ovaries produce eggs on a cyclic basis. Both also produce sex hormones.

6. The sex hormones testosterone, LH, and FSH control male reproductive functions. Estrogens, progesterone, FSH, and LH control female reproductive functions.

In earlier chapters, you read about the cellular basis of **sexual reproduction**, in which offspring form by way of meiosis, gamete formation, and fertilization. You also read about **asexual reproduction**, in which offspring form by means other than gamete formation. Consider now a few structural, behavioral, and ecological aspects of these two different reproductive modes.

Think of a new sponge growing asexually from a fragment of its parent's body. Or think of how some flatworms undergo fission. Their body can constrict below the midsection. The part behind this constriction grips a substrate and starts a tug-of-war with the part in front of it. After a few hours, it splits off. Both parts go their separate ways, regenerate the missing part, and become a whole worm. In such cases of asexual reproduction, all of the offspring are genetically the same as parents, or nearly so. The absence of variation is most useful when the individual's gene-encoded traits are highly adaptive to a limited, more or less consistent set of environmental conditions.

But most animals live where opportunities, resources, and danger vary in complex ways. Such animals mainly use sexual reproduction, with offspring getting different mixes of alleles from male and female parents. The resulting variation in traits improves the likelihood that at least some offspring will be able to survive and reproduce if the environment changes.

Separation into sexes is not without cost. Sex means constructing special reproductive structures. It means engaging in behavior, such as courtship, that promotes fertilization. It means having built-in controls to synchronize the timing of sperm and egg production, sexual readiness, even parental behavior.

Consider the question of *reproductive timing*. How do the mature sperm in one individual become available *exactly* when eggs mature in a different individual? Timing requires energy outlays for sensory structures, and it involves hormonal control mechanisms in each parent. The parents must produce mature gametes in response to the same cues, such as the seasonal change in daylength, that mark the onset of the most suitable time for reproduction. For example, male and female moose become sexually active in late summer and early fall. The timing assures that their offspring will be born the following spring—when the weather improves and food will be plentiful for many months.

Or consider the challenge of finding and recognizing a potential mate of the same species. Different species invest energy when they synthesize chemical signaling

Figure 34.2 Like most mammals, human females retain fertilized eggs, and their tissues nourish the embryo until it is liveborn.

molecules, structural signals such as feathers of certain colors and patterns, and complex sensory receptors that will be able to pick up the signals being sent. Besides this, males often expend astonishing amounts of energy on courtship routines, as you will read in Chapter 40.

Assuring survival of offspring is also costly. Many invertebrates and bony fishes release eggs and motile sperm into water. The odds for fertilization would not be good if adults produced only *one* sperm or *one* egg each season. Such species invest energy to make hundreds of thousands of gametes. As another example, nearly all land animals depend upon internal fertilization—the union of sperm and egg *inside* the female's body. They invest energy to construct elaborate reproductive organs, such as a penis (by which the male deposits its sperm inside the female) and a uterus (a type of chamber in the female where the embryo grows and develops).

Finally, animals set aside energy in forms that can *nourish some number of offspring*. For example, nearly all animal eggs contain **yolk**, a protein-rich, lipid-rich substance that nourishes the embryo. The eggs of certain species have more yolk than others. Sea urchins make small eggs with little yolk, release large numbers of them, and limit the biochemical investment in each one. Later on, each fertilized egg develops into a self-feeding, free-moving larva in less than one day. Predators consume most of the eggs, so bestowing as little as possible on as many gametes as possible pays off, in terms of reproductive success.

By contrast, mother birds lay truly yolky eggs. Yolk nourishes the bird embryo through a long development period, inside an eggshell that forms after fertilization. Your mother put tremendous demands on her body to protect and nourish you through the months of early development inside her body, starting from the time you were an egg with almost no yolk. You implanted yourself in her uterus, then physical exchanges with her tissues supported you all through an extended pregnancy (Figure 34.2).

As these few examples suggest, animals show great diversity in reproduction and development. However, some patterns are widespread in the animal kingdom, and they will serve as a framework for our reading.

Separation into male and female sexes requires special reproductive structures, control mechanisms, and behaviors. A selective advantage—variation in traits among offspring—offsets the biological costs associated with the separation.

34.2 STAGES OF DEVELOPMENT—AN OVERVIEW

You don't look like a frog. You didn't look like one when you and the frog were embryos, either. **Embryos** are transitional forms on the road from a fertilized egg to the adult. Yet despite the differences in appearance, it is possible to identify certain patterns in the way that the embryos of nearly all species of animals develop.

Figure 34.3 is an overview of the stages of animal development. With **gamete formation**, the first stage, eggs or sperm develop in reproductive organs in a parent's body. **Fertilization** starts when the plasma membranes of a sperm and an egg fuse. It ends when the egg nucleus and sperm nucleus fuse. With this fusion we have a zygote, the first cell of a new animal. The next stage, **cleavage**, is a program of mitotic cell divisions that will convert a zygote into a multicelled embryo. In many species, cleavage produces a blastula, a tiny ball of cells that is no wider in diameter than the zygote was.

As cleavage draws to a close, the pace of cell division slackens. The embryo enters **gastrulation**, which is a stage of major cellular reorganization. The newly formed cells become arranged into two or three primary tissues (often called germ layers), which will give rise to all tissues of the adult:

1. **Ectoderm.** This is the *outermost* primary tissue layer, the one that forms first in the embryos of every animal. Ectoderm is the embryonic forerunner of the cell lineages that give rise to tissues of the nervous system and to the outer layer of the integument.

2. **Endoderm.** Endoderm is the *innermost* primary tissue layer. It is the embryonic forerunner of the gut's inner lining and of the organs derived from it.

3. **Mesoderm.** This *intermediate* primary tissue layer is the forerunner of muscle; of circulatory, reproductive, and excretory organs; of most of the skeleton; and of connective tissue layers of the gut and integument. Hundreds of millions of years ago, the origin of mesoderm was a pivotal step in the evolution of nearly all large, complex animals.

The primary tissue layers split into subpopulations of cells. This marks the onset of **organ formation**. By now,

different sets of cells have become unique in structure and function, and their descendants are giving rise to different kinds of tissues and organs. During the final stage, called **growth and tissue specialization**, organs increase in size and gradually assume their specialized functions. This stage continues into adulthood.

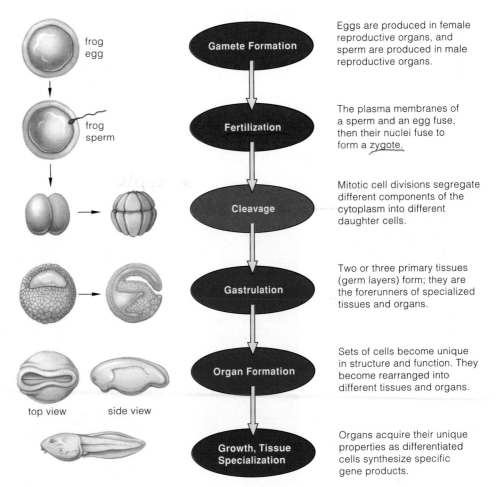

Figure 34.3 Overview of the stages of animal development. We use a few forms that appear in the frog life cycle as examples.

Section 34.3 includes photographs of representative stages of embryonic development. Take a moment to study them, for they reinforce a key concept. Structures that form during one stage serve as the foundation for the stage that comes after it. Successful development depends absolutely on the formation of these structures according to normal patterns, in a prescribed sequence.

Each stage of embryonic development builds on structures that were formed during the stage preceding it. Development cannot proceed properly unless each stage is successfully completed before the next begins.

A VISUAL TOUR OF FROG AND CHICK DEVELOPMENT

In the sections that follow, you will be reading about the key mechanisms underlying the different stages of animal development. You also will be considering some of the experiments that provided crucial insights into these mechanisms.

When studying the details, it is sometimes easy to lose sight of the magnitude of the transformations and of how rapidly certain stages can proceed. The selection of photographs and sketches in Figures 34.4 and 34.5 may help you get a better sense of what goes on.

Figure 34.5 Onset of organ formation in a chick embryo during the first seven days of development.

The heart begins to beat between thirty and thirty-six hours. You may have observed such embryos at the yolk surface of raw, fertilized eggs.

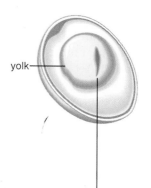

yolk

In chick eggs, cleavage produces a layer of cells (the chick blastula) at the surface of the yolk. The cuts do not penetrate the dense yolk.

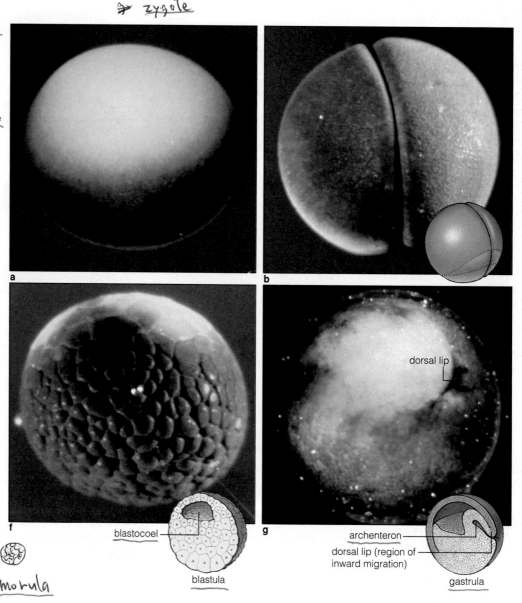

zygote

animal pole

gray crescent

vegetal pole

Figure 34.4 Micrographs and diagrams of the early embryonic development of a frog. For these micrographs, the jellylike layer surrounding the egg has been removed, with the exception of the micrograph in (**i**).

(**a**) Within about an hour after fertilization, the gray crescent establishes the body axis for the embryo, and gastrulation will begin here.

(**b–f**) Cleavage leads to a blastula, a ball of cells in which a cavity (blastocoel) has appeared.

(**g,h**) Cells move about and become rearranged during gastrulation. Tissue layers form; a primitive gut cavity (archenteron) develops.

(**i,j**) Neural developments now take place, and the fluid-filled body cavity in which vital organs will be suspended appears. Cell differentiation proceeds, moving the embryo on its way to becoming a functional larval form.

morula

a

b

blastocoel

blastula

f

g

dorsal lip

archenteron

dorsal lip (region of inward migration)

gastrula

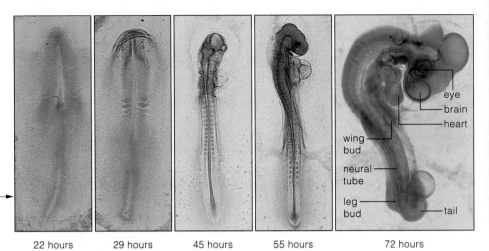

22 hours 29 hours 45 hours 55 hours 72 hours 168 hours (seven days old)

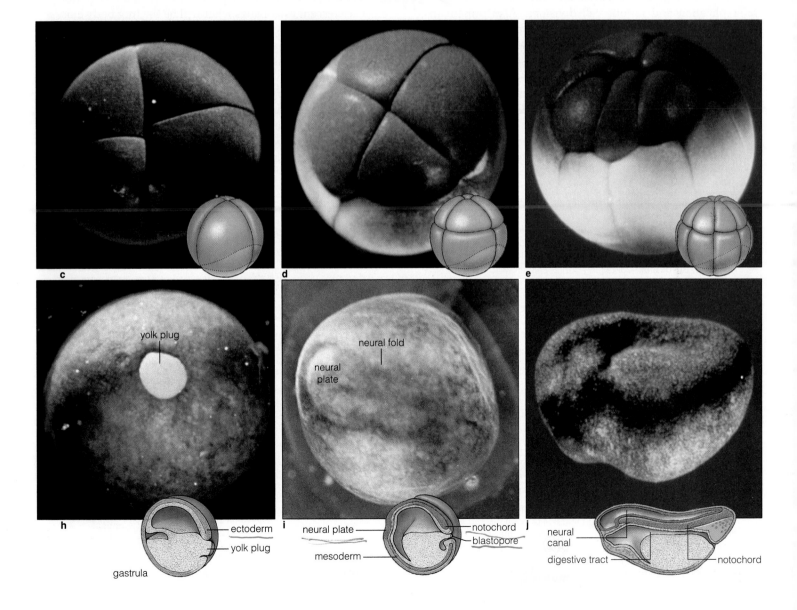

Information in the Egg Cytoplasm

You probably don't have an arm attached to your nose or toenails growing from your navel. To a large extent, the patterning of body parts for any complex animal, including yourself, is mapped out in the cytoplasm of an immature egg, or **oocyte**, even before a sperm enters the picture. A **sperm**, recall, consists of paternal DNA and a bit of equipment that helps the sperm reach and penetrate an egg. Compared to a sperm, an oocyte is much larger and more complex (compare Figure 8.9).

As an oocyte matures, its volume increases. And it stockpiles maternal instructions that will influence the embryo's shape and arrangement of its parts. Enzymes, mRNA transcripts, and other components embody the instructions—and they become distributed in different regions of the cytoplasm. After fertilization, maternal instructions will be used for the initial rounds of DNA replication and mitotic cell divisions. Also present in the cytoplasm are microtubules and other structures, oriented in directions that will affect the early divisions. How? When each cell divides, it does so at a prescribed angle relative to any adjacent cells—based partly on the orientation of microtubules of the mitotic spindle.

The amount and distribution of yolk in the egg also will influence cleavage. For example, the position of both the yolk and the nucleus in a frog oocyte imparts polarity to it. The *animal* pole is simply the one closest to the nucleus. Opposite is the *vegetal* pole, where yolk and other substances accumulate. All animal eggs show some polarity—that is, two identifiable poles.

Reorganization of the Egg Cytoplasm at Fertilization

When a sperm fertilizes an egg, it triggers structural reorganization in the egg cytoplasm. You can observe indirect signs of this reorganization in frog eggs. They have pigment granules in their cortex, which includes the plasma membrane and the cytoplasm just beneath it. The granules are concentrated near one pole of the egg; yolk is concentrated near the other. At fertilization, part of the granule-containing cortex shifts away from the yolk to expose lighter colored cytoplasm in a gray, crescent-shaped area. The fertilized egg now has a **gray crescent**, of intermediate pigmentation, near its equator. The crescent establishes what will be the frog body axis.

In itself, the gray crescent is not evidence of regional differences in maternal messages. Such evidence comes from experiments of the sort shown in Figure 34.6b. It also comes from observing embryos in which localized cytoplasmic differences are pronounced enough to be tracked easily during development. For example, if you were to continue tracking the development of fertilized frog eggs, you would see that a gray crescent is *always* the site where gastrulation normally begins.

Figure 34.6 Examples of experiments that illustrate how the cytoplasm of a fertilized egg must have localized differences that help determine the fate of cells in a developing embryo. The cortex of frog eggs contains granules of dark pigment, concentrated near one pole of the egg. At fertilization, a portion of the granule-containing cortex shifts toward the point of sperm entry. The shift exposes lighter colored, yolky cytoplasm in a crescent-shaped gray area:

Normally, the first mitotic cell division after fertilization divides the gray crescent between two daughter cells.

(**a**) In one experiment, the first two daughter cells of the embryo were separated from each other. Each still gave rise to a complete tadpole.

(**b**) In another experiment, a fertilized egg was manipulated so that the cleavage plane of the first cell division missed the gray crescent entirely. Only one of the two daughter cells received the gray crescent. It alone developed into a normal tadpole. The daughter cell deprived of substances in the gray crescent gave rise to a ball of undifferentiated cells.

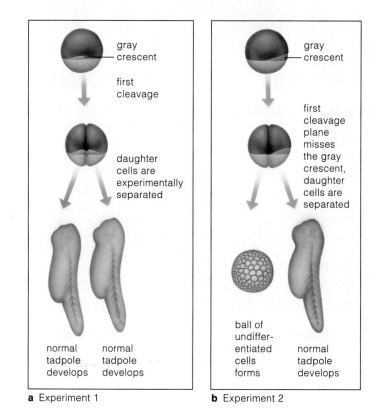

a Experiment 1

b Experiment 2

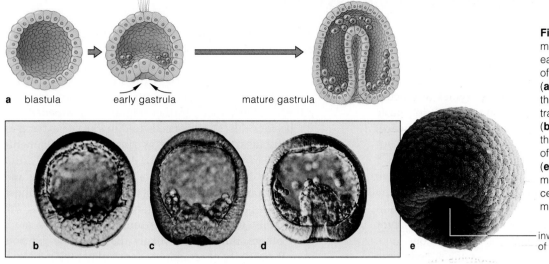

a blastula   early gastrula   mature gastrula

b   c   d   e

Figure 34.7 Diagrams and micrographs of steps in the early embryonic development of a sea urchin (*Lytechinus*). (**a**) The ball of cells (blastula) that results from cleavage is transformed into a gastrula. (**b–d**) Micrographs showing the inward migration of cells of a gastrula in cross-section. (**e**) This scanning electron micrograph shows the surface cells and the region of inward migration.

inward migration of surface cells

Cleavage

Once fertilization is over, mitotic cell divisions carve up the zygote. A cleavage furrow, recall, forms during each mitotic cell division (Section 7.4). The furrow defines the actual plane where the cytoplasm will pinch in two. Usually the embryo does not increase in size early in cleavage. Taken together, the daughter cells collectively occupy the same volume as did the zygote, although those cells differ in size, shape, and activity.

Simply by virtue of their location, the cells that form during cleavage receive different maternal instructions. This outcome, **cytoplasmic localization**, helps seal the developmental fate of each cell's descendants. Only the portion of cytoplasm allocated to one cell may contain molecules of, say, a protein that can activate the gene coding for a certain hormone. Descendants of that cell alone will end up producing the hormone.

In many animal groups, successive cleavages result in a **blastula**, a ball of cells with a fluid-filled cavity (Figure 34.7). But cleavage patterns differ, often because eggs are not alike in the density and distribution of yolk. A sea urchin egg has little yolk, and a symmetrical blastula develops from it. The concentrated yolk of a frog egg impedes cleavage near the vegetal pole, so that a fluid-filled cavity forms only near the animal pole (Figure 34.4*f*). Eggs of reptiles, birds, and most fishes have a large volume of yolk that restricts cleavage to a tiny, caplike region at the animal pole. Cleavage of such eggs produces two flattened layers, perched on the yolk surface, with a thin cavity in between (Figure 34.5).

The blastula stage of mammals, called a **blastocyst**, has *two* distinct regions. Some of its cells form a hollow sphere that is not part of the embryo. A small cluster of different cells attaches to the inner blastocyst wall. As you will read shortly, the embryo develops from this "inner cell mass."

Gastrulation

As cleavage draws to a close, the pace of cell division slackens and cells start to migrate into new positions. Gastrulation, a stage of major cell rearrangements, is under way. The embryo's size increases little, if at all. In sea urchins, surface cells migrate inward and form a lining for a cavity that will develop into a gut. In other species, the migrations produce a crucial long axis. For example, in vertebrate embryos, this axis defines where the neural tube—the forerunner of a brain and spinal cord—will form (Section 32.7).

This last example underscores the significance of gastrulation. Nearly all animals have an internal region of cells, tissues, and organs that function in digestion and absorption of nutrients. They have a surface region with tissues that afford protection of internal parts and with sensory receptors that detect outside changes. In between these regions, most animals have numerous organs, such as those dealing with structural support, movement, and blood circulation. All of these regions arise from the three primary tissue layers—endoderm, ectoderm, and mesoderm. *This three-layered organization is typical of most animals, and it arises through gastrulation.*

Even before an immature egg undergoes fertilization, localized differences are established in its cytoplasm. These will help seal the fate of cells in the forthcoming embryo.

After fertilization, cleavage produces a number of daughter cells at prescribed locations in the embryo.

The particular cleavage planes that form dictate which maternal instructions each cell will inherit. They also influence its size and spatial position.

Cytoplasmic localization influences how each cell will interact with others during the gastrula stage—and so on through subsequent stages of development.

Cell Differentiation

All cells of an embryo descend from the same zygote, so they have the same number and kinds of genes. They all activate the genes for histones, enzymes of glucose metabolism, and other proteins that are absolutely basic to cell survival. However, from gastrulation onward, certain groups of genes are activated in some cells but not others. When a cell selectively activates genes and synthesizes proteins not found in other cell types, we call this process **cell differentiation**. You read about the molecular basis of selective gene expression in Section 12.8. In brief, it results in proteins that are the basis for distinctive cell structures, products, and functions.

For example, when your eye lenses developed, some cells started synthesizing crystallin, a family of proteins that become incorporated in transparent fibers of the lens. Only those cells could activate the required genes. Long crystallin fibers formed, and they forced the cells to lengthen and flatten. Collectively, the differentiated cells impart unique optical properties to the lenses. And these crystallin-producing cells are only one of 150 or so differentiated cell types now present in your body.

As many experiments tell us, nearly all cells become fully differentiated without loss of genetic information. For example, John Gurdon removed the nucleus from unfertilized eggs of the African clawed frog (*Xenopus laevis*). Then he isolated intestinal cells from tadpoles of the same species. He ruptured their plasma membrane, left the nucleus and most of the cytoplasm intact, and inserted the nucleus into the enucleated egg. In some cases, that nucleus directed the developmental steps leading to a complete frog! The intestinal cell had the same genes as the zygote, hence all the genes required to form all of the required cell types. Its genes had not been lost when it differentiated into an intestinal cell.

Morphogenesis

As an embryo develops, tissues and organs change in size, shape, and proportion, and so become distinctively organized tissues and organs. This process involves cell divisions, tissue growth, cell migrations, and changes in cell size and shape. It involves the folding of sheetlike tissues as well as the controlled death of cells at certain locations in the embryo. We call it **morphogenesis**.

In active cell migration, cells send out pseudopods and move along prescribed routes to their destination, then they establish contact with cells already there. For example, this is how the forerunners of neurons make interconnections when a nervous system is developing.

How do cells "know" where to move? They respond to adhesive cues, as when migrating Schwann cells stick to adhesion proteins on the surface of axons but not on

blood vessels, and they respond to chemical gradients. Their migrations may be coordinated by the synthesis, release, deposition, and removal of specific substances in the extracellular matrix. Adhesive cues also tell the cells when to stop. Cells migrate to regions of strongest adhesion. Once there, further migration is impeded.

Besides this, sheets of cells fold inward or expand as microtubules lengthen and as rings of microfilaments constrict in cells (Figure 34.8). Agents that can disrupt these tiny tubes and filaments control the folding and expansion. Such localized growth contributes to changes in the size, shape, and proportion of body parts. We don't fully understand why some tissues grow more than others. But it seems likely that selective controls over the activity of regulatory genes play major roles.

Finally, *programmed* cell death eliminates tissues and cells that are used for only short periods in the embryo. Such events are genetically programmed. For example, initially a human embryo's hands and feet look like paddles (Figure 7.8). Then certain cells die on cue and so produce separate toes and fingers. Between the time a death signal is sent and the time of cell death, protein synthesis declines rapidly in the doomed cells. In some humans and some mice, a gene mutation blocks cell death in the paddles, and the digits remain webbed.

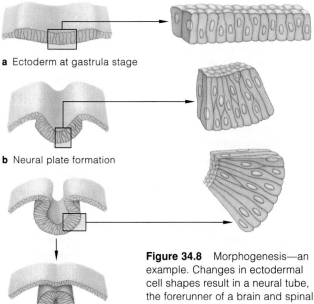

a Ectoderm at gastrula stage

b Neural plate formation

neural tube

c Neural tube formation

Figure 34.8 Morphogenesis—an example. Changes in ectodermal cell shapes result in a neural tube, the forerunner of a brain and spinal cord. When gastrulation ends, ectoderm is a uniform sheet of cells. In some cells, microtubules lengthen, and the elongating cells form a neural plate. In other cells, microfilament rings at one end constrict and the cells become wedge shaped. *Their* part of the ectodermal sheet folds over the neural plate to form the tube.

Pattern Formation

The sculpting of nondescript clumps of embryonic cells into specialized tissues and organs follows an ordered, spatial pattern. We call this **pattern formation**. It begins with local differences in the egg cytoplasm, then in the local environments of embryonic cells. The differences give rise to asymmetries in gene expression. Different gene products are the basis of signals among cells that are sent and received at prescribed times, in sequential fashion. The increasingly diverse cellular responses to these signals fill in details of the body plan.

Consider how the positioning of bones in a chick wing depends on a cue from the local environment. A wing forms as a primordial bud, through cell divisions in mesoderm and ectoderm. If you surgically remove certain groups of ectodermal cells before the bud grows fully, further development here ceases:

normal wing

surgically altered wing

Rapid cell divisions in mesoderm at the tip of the early bud generate tissues that elongate the developing limb. As cells exit from this zone, they give rise to the tissues of the limb's outermost parts. Cells pushed out early on give rise to bones closest to the trunk of the body. Cells that depart much later are using their genes differently. They are biochemically different, and they respond to different signals. These cells give rise to the outermost bones and other structures of the limb.

The term **embryonic induction** refers to a change in the developmental fate of embryonic tissues as brought about by exposure to a chemical signal released from an adjacent tissue. For example, **morphogens** are slowly degradable proteins and other chemical substances that form a concentration gradient as they diffuse from an inducing tissue into adjoining tissues. The signal they represent is strongest at the start of the gradient and weakens with distance. Thus, cells at different distances receive different information about their position in the embryo, and this guides differences in gene activity.

Morphogens and other inducers switch on blocks of genes in sequence. Among their targets are **homeobox genes**, which occur in animals as evolutionarily distant as *Caenorhabditis elegans* (a roundworm), fruit flies, and vertebrates. The gene products interact with regulatory elements in DNA to activate or inhibit blocks of genes in similar ways in groups of cells along the embryo's anterior-posterior axis.

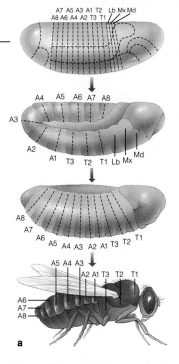

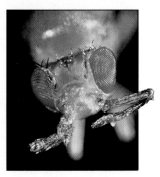

b During a larval stage, a cluster of cells that normally gives rise to antennae was surgically removed and exposed to abnormal tissue environments. Then the cluster was reinserted into a different place in a larva undergoing metamorphosis. Instead of antennae, legs appeared on the head.

Figure 34.9 (**a**) Fate map of a *Drosophila* zygote. Dashed lines indicate regions that will develop into different body segments, many with specialized appendages. Each segment's fate is sealed at the time of cytoplasmic localization. (**b**) Experimental evidence of embryonic induction in *Drosophila*.

For example, look at the *Drosophila* zygote shown in Figure 34.9*a*. A "fate map" of its surface shows where each kind of differentiated cell in the forthcoming body segments will originate. Before embryonic cells show any sign of their fate, different homeobox genes are activated in successive stripes along the body's long axis. In one stripe, protein products of the genes assign the cells to develop into parts of the head. In another stripe, they assign cells to develop into a thorax, and so on. Mutated homeobox genes can transform one body segment into the likeness of another. Thus mutations in the *antennapedia* gene result in abnormal responses to inducer signals. In one case, a mutated gene product activated the wrong set of homeobox genes in cells that should produce two antennae on the head. They gave rise to a pair of legs instead (Figure 34.9*b*)!

In cell differentiation, a cell selectively uses certain genes and makes proteins not found in other cell types. It uses the proteins for distinct cell structures, products, and functions.

Morphogenesis refers to programmed changes in the size, shape, and proportion of an embryo, and to the specialization and positioning of its tissues in space.

Embryonic tissues and organs form in a sequential, ordered pattern. Pattern formation arises from local differences in the egg cytoplasm and in the environment of embryonic cells. At prescribed times, different cell lineages respond in different ways to positional cues and inducer signals. Through their responses, they fill in the details of the body plan.

So far, we have looked at the basic principles of animal reproduction and development. The remainder of the chapter shows how you can apply these principles to humans, starting with the male reproductive system.

As Figure 34.10 and Table 34.1 show, an adult male has a pair of gonads, of a type called **testes** (singular, testis). These are his primary reproductive organs, the equivalent of ovaries in females. Testes produce sperm and sex hormones. The hormones are central to male reproductive function. They also are necessary for the development of *secondary* sexual traits. Such traits do not play a direct role in reproduction but are distinctly associated with maleness (or femaleness). Examples are the amount of body fat, hair, and skeletal muscle.

Where Sperm Form

Until human embryos are seven weeks old, gonads all look the same. After that, activation of genes on the sex chromosomes and hormonal signals bring about the formation of testes *or* ovaries, in the manner described in Section 10.3. In an embryo destined to become male, a pair of testes form on the abdominal cavity wall. Before birth, his testes descend into the scrotum, which is an outpouching of skin below the pelvic region. Gonads are fully formed at birth. They reach full size and become functional twelve to sixteen years later.

Figure 34.10 shows the position of the scrotum in an adult male. For sperm cells to develop properly, the temperature in the scrotum's interior must remain a few degrees cooler than the rest of the body's normal core temperature. A control mechanism, which operates by stimulating or inhibiting contractions of smooth muscles in the scrotum, helps assure that the internal temperature does not stray far from 95°F. When the air just outside the body is cold, contractions draw the pouch closer to the (warmer) body. When it is warm outside, the muscles relax and thereby lower the pouch.

Packed inside each testis are a great number of small, highly coiled tubes. These are the seminiferous tubules. Sperm formation begins in the tubules, in the manner described in Section 34.7.

Where Semen Forms

Mammalian sperm travel from the testes, through a series of ducts that lead to the urethra. They are not quite mature when they begin the trip in a pair of long, coiled ducts. These initial ducts are the epididymides (singular, epididymis). Secretions from glandular cells in the duct wall trigger actions that put the finishing touches on sperm. Until the time that sperm depart from the body, they will be stored in the last stretch of each epididymis.

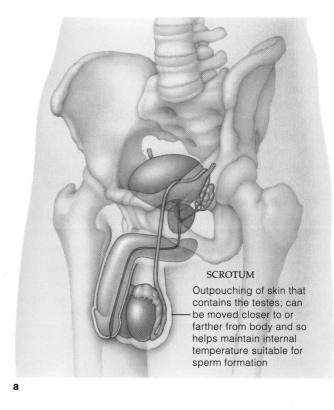

SCROTUM
Outpouching of skin that contains the testes; can be moved closer to or farther from body and so helps maintain internal temperature suitable for sperm formation

a

Figure 34.10 Components of the reproductive system of the human male and their functions.

Table 34.1	Organs and Accessory Glands of the Male Reproductive Tract
Organs:	
Testis (2)	Sperm, sex hormone production
Epididymis (2)	Sperm maturation site and storage
Vas deferens (2)	Rapid transport of sperm
Ejaculatory duct (2)	Conduction of sperm
Penis	Organ of sexual intercourse
Accessory Glands:	
Seminal vesicle (2)	Secretion of large part of semen
Prostate gland	Secretion of part of semen
Bulbourethral gland (2)	Production of lubricating mucus

When a male becomes sexually aroused, muscle contractions in the walls of reproductive organs propel sperm into and through a pair of thick-walled tubes, the vas deferentia (singular, vas deferens). From there, contractions propel sperm through a pair of ejaculatory ducts, then through the urethra. This last tube extends through the penis, the male sex organ, and opens at its tip. The urethra, recall, is a duct that also functions in urine excretion.

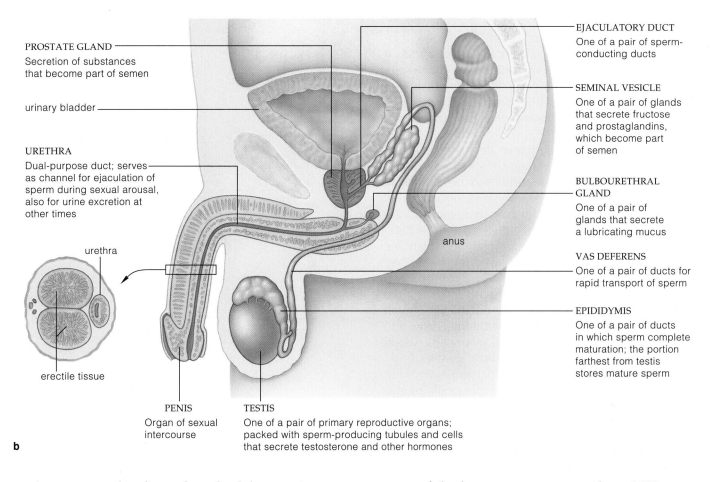

PROSTATE GLAND
Secretion of substances
that become part of semen

urinary bladder

URETHRA
Dual-purpose duct; serves
as channel for ejaculation of
sperm during sexual arousal,
also for urine excretion at
other times

urethra

erectile tissue

EJACULATORY DUCT
One of a pair of sperm-
conducting ducts

SEMINAL VESICLE
One of a pair of glands
that secrete fructose
and prostaglandins,
which become part
of semen

BULBOURETHRAL
GLAND
One of a pair of
glands that secrete
a lubricating mucus

anus

VAS DEFERENS
One of a pair of ducts for
rapid transport of sperm

EPIDIDYMIS
One of a pair of ducts
in which sperm complete
maturation; the portion
farthest from testis
stores mature sperm

PENIS
Organ of sexual
intercourse

TESTIS
One of a pair of primary reproductive organs;
packed with sperm-producing tubules and cells
that secrete testosterone and other hormones

b

As sperm travel to the urethra, glandular secretions become mixed with them. The result is semen, a thick fluid that the penis eventually expels during sexual engagement. Early in the formation of semen, a pair of seminal vesicles secrete fructose, a sugar that nourishes sperm. Seminal vesicles also secrete prostaglandins that can induce muscle contractions. Possibly these signaling molecules take effect during sexual activity, when they trigger contractions in the female's reproductive tract and thereby assist sperm movement through it. Secretions from a prostate gland probably help buffer the acidic conditions that sperm encounter within the female tract. The vaginal pH is about 3.5–4.0, but sperm motility improves at pH 6. Two bulbourethral glands secrete some mucus-rich fluid into the urethra when the male is sexually aroused.

Cancers of the Prostate and Testis

Until recently, cancers of the male reproductive tract did not get much media coverage in the United States, even though *prostate cancer* alone kills 40,000 older men annually. Yet the mortality rate for breast cancer, which is well publicized, is not much higher. You also may be surprised to learn that *testicular cancer* is a frequent

cause of death among young men. About 5,000 cases are diagnosed each year in the United States alone. During early stages, these cancers are painless. If not detected in time, they can spread silently into lymph nodes in the abdomen, chest, neck, and, eventually, the lungs. Once testicular cancer has metastasized, it kills as many as 50 percent of those stricken.

Doctors can detect prostate cancer in older men through physical examination and blood tests for a rise in prostate-specific antigen. Also, once a month from high school onward, men should examine each testis after a warm bath or shower, when scrotal muscles are relaxed. The testis should be rolled gently between the thumb and forefinger to check for any enlargement, hardening, or lump. Such changes may or may not cause discomfort. But they must be reported so that a physician can make a complete examination. Treatment of testicular cancer has one of the highest success rates when the cancer is caught before it has spread.

Human males have a pair of testes—primary reproductive organs that produce sperm and sex hormones—as well as accessory glands and ducts. The hormones influence sperm formation and the development of secondary sexual traits.

MALE REPRODUCTIVE FUNCTION

Sperm Formation

Each testis is about 5 centimeters (less than 2 inches) long. Even so, 125 meters of seminiferous tubules are packed into it! As many as 300 wedge-shaped lobes, of the sort shown in Figures 34.11 and 34.12, partition its interior. Two or three tubules are coiled against one another inside each lobe.

Just inside each tubule's wall are undifferentiated cells called spermatogonia (singular, spermatogonium). Ongoing cell divisions force them away from the wall, toward the interior. During their forced departure, the cells become transformed into primary spermatocytes. As Figure 34.13*a* indicates, they undergo meiosis I, the result being secondary spermatocytes. Although the immature cells are now haploid, each chromosome they contain is still in the duplicated state, consisting of two sister chromatids. (Here you may wish to review the diagram in Section 8.3.) The sister chromatids of each chromosome separate from each other during meiosis II. The resulting cells become haploid spermatids, which gradually develop into sperm—the male gametes.

The entire process of sperm formation takes nine to ten weeks. All the while, the developing cells receive nourishment and chemical signals from adjacent *Sertoli* cells. These are the only other type of cell located inside the seminiferous tubules.

From puberty onward, human males continuously produce sperm. On any given day, many millions of these gametes are in different stages of development. Each mature sperm is a flagellated cell that has a core of microtubules in its long tail (Figure 34.13*b*). An enzyme-containing cap called an acrosome covers most of the head region, which is mainly a DNA-packed nucleus. Enzymes of the cap will help the sperm penetrate the extracellular material around an egg at fertilization. In a midpiece just behind the head, arrays of mitochondria supply energy for the tail's whiplike movements.

Hormonal Controls

Coordinated secretions of LH, FSH, and testosterone govern reproductive function in males. Take a look at Figure 34.12, which shows the location of *Leydig* cells in tissue between the lobes in testes. Leydig cells secrete **testosterone**. This steroid hormone is absolutely central to the growth, form, and functions of the reproductive tract in males. Besides having the main role in sperm formation, it stimulates sexual and aggressive behavior. It promotes development of secondary sexual traits in males, including stepped-up growth of facial hair and a deepening of the voice at puberty.

LH and **FSH**, recall, are secreted by the anterior lobe of the pituitary gland (Section 33.3). They were initially

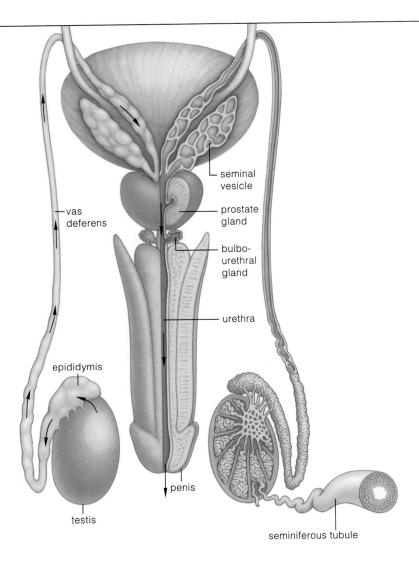

Figure 34.11 Male reproductive tract, posterior view. The arrows show the route that sperm take before ejaculation from a sexually aroused male.

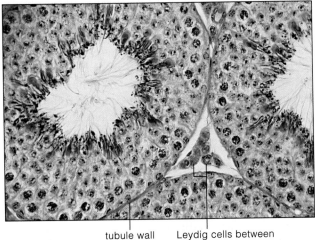

Figure 34.12 Micrograph of cells inside seminiferous tubules. Portions of three neighboring tubules are shown in cross-section.

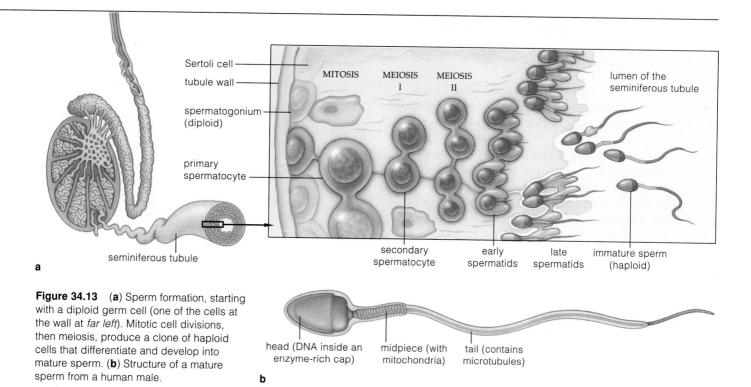

Sertoli cell
tubule wall
spermatogonium (diploid)
primary spermatocyte
seminiferous tubule

a

MITOSIS   MEIOSIS I   MEIOSIS II

lumen of the seminiferous tubule

secondary spermatocyte   early spermatids   late spermatids   immature sperm (haploid)

Figure 34.13 (**a**) Sperm formation, starting with a diploid germ cell (one of the cells at the wall at *far left*). Mitotic cell divisions, then meiosis, produce a clone of haploid cells that differentiate and develop into mature sperm. (**b**) Structure of a mature sperm from a human male.

head (DNA inside an enzyme-rich cap)   midpiece (with mitochondria)   tail (contains microtubules)

b

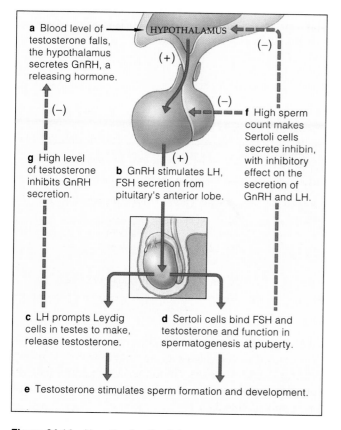

a Blood level of testosterone falls, the hypothalamus secretes GnRH, a releasing hormone.

HYPOTHALAMUS

(+)   (−)

(−)

(−)

f High sperm count makes Sertoli cells secrete inhibin, with inhibitory effect on the secretion of GnRH and LH.

g High level of testosterone inhibits GnRH secretion.

(+)

b GnRH stimulates LH, FSH secretion from pituitary's anterior lobe.

c LH prompts Leydig cells in testes to make, release testosterone.

d Sertoli cells bind FSH and testosterone and function in spermatogenesis at puberty.

e Testosterone stimulates sperm formation and development.

Figure 34.14 Negative feedback loops to the hypothalamus and pituitary gland from the testes. Through these loops, excess testosterone production shuts off the mechanisms leading to its production. This helps maintain the testosterone level in amounts required for sperm formation.

named for their effects in females. (One abbreviation stands for *Luteinizing Hormone*, the other for *Follicle-Stimulating Hormone*.) Later on, researchers discovered that the molecular structure of LH and FSH is identical in both males and females.

The hypothalamus controls secretions of LH, FSH, and testosterone—and thus controls sperm formation. In response to a low level of testosterone in the blood and other factors, the hypothalamus secretes GnRH. As you can see from Figure 34.14, this releasing hormone prompts the pituitary's anterior lobe to release LH and FSH, which have targets in the testes. LH stimulates Leydig cells in the testes to secrete testosterone, which stimulates diploid germ cells to become sperm. Sertoli cells have FSH receptors. FSH is crucial to establishing spermatogenesis at the time of puberty, but researchers don't know whether this hormone is essential for the normal functioning of mature testes in humans.

A high testosterone level in blood has an inhibitory effect on GnRH release. Also, when the sperm count is high, Sertoli cells release inhibin, a protein hormone that acts on the hypothalamus and pituitary to inhibit the release of GnRH and FSH. And so, feedback loops to the hypothalamus kick in—with a resulting decrease in testosterone secretion and sperm formation.

Sperm formation depends on the hormones LH, FSH, and testosterone. Negative feedback loops from the testes to the hypothalamus and pituitary gland control their secretion.

The Reproductive Organs

We turn now to the reproductive system of a human female. Figure 34.15 shows its components and lists their functions. Table 34.2 summarizes the major organs of this system. The primary reproductive organs, a pair of **ovaries**, produce the eggs and secrete sex hormones. After an immature egg, or oocyte, is released from an ovary, it enters the adjacent, billowing entrance of an oviduct. A female has a pair of oviducts. They serve as channels to the uterus, which is a hollow, pear-shaped organ where embryos can grow and develop.

The uterus has a thick layer of smooth muscle (the myometrium) in its wall. As you will see, the wall's inner lining, the **endometrium**, is central to embryonic development. It consists of connective tissue, glands, and blood vessels. The narrowed portion of the uterus is called the cervix. A muscular tube, the vagina, extends from the cervix to the body's surface. This tube receives sperm and functions as part of the birth canal.

At the body surface are external genitals (vulva) that include organs for sexual stimulation. Outermost are a pair of fat-padded skin folds, the labia majora. They enclose a smaller pair of skin folds (labia minora) that are highly vascularized but have no fatty tissue. The smaller folds partly enclose the clitoris, a sex organ that is sensitive to stimulation. At the body's surface, the urethra's opening is about midway between the clitoris and the vaginal opening.

Overview of the Menstrual Cycle

Most mammalian females follow an *estrous* cycle. They are fertile and in heat (sexually receptive to males) only at certain times of year. By contrast, female primates, including humans, follow a **menstrual cycle**. They are fertile intermittently, on a cyclic basis, and heat is not synchronized with the fertile periods. Said another way, female primates of reproductive age can get pregnant only at certain times, but they may be receptive to sex at any time.

During each menstrual cycle, an oocyte matures and escapes from an ovary. Also, the endometrium becomes primed to receive and nourish a forthcoming embryo, *if* fertilization takes place. However, if the oocyte does not get fertilized, blood-rich fluid (four to six tablespoons total) starts flowing out through the vaginal canal. This menstruation means "there is no embryo at this time," and it marks the first day of a new cycle. The uterine nest is being sloughed off and is about to be constructed once again.

The events just sketched out proceed through three phases. First is the *follicular* phase. This is the time of menstruation, endometrial breakdown and rebuilding,

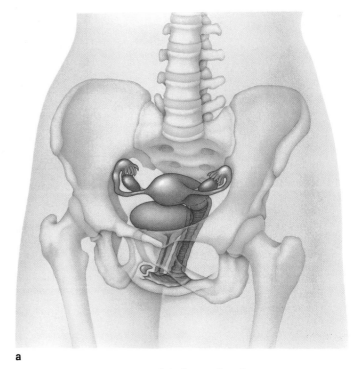

a

Figure 34.15 Components of the human female reproductive system and their functions.

Table 34.2	Female Reproductive Organs
Ovaries	Oocyte production, sex hormone production
Oviducts	Conduction of oocyte from ovary to uterus
Uterus	Chamber in which new individual develops
Cervix	Secretion of mucus that enhances sperm movement into uterus and (after fertilization) reduces the embryo's risk of bacterial infection
Vagina	Organ of sexual intercourse; birth canal

and maturation of an oocyte. The next phase, *ovulation*, is restricted to the release of an oocyte from the ovary. During the *luteal* phase, an endocrine structure (corpus luteum) forms, and the endometrium is primed for pregnancy (Table 34.3).

All three phases are governed by feedback loops to the hypothalamus and pituitary gland from the ovaries. As you will see, FSH and LH promote cyclic changes in ovaries. These hormones also stimulate the ovaries to secrete sex hormones—**estrogens** and **progesterone**.

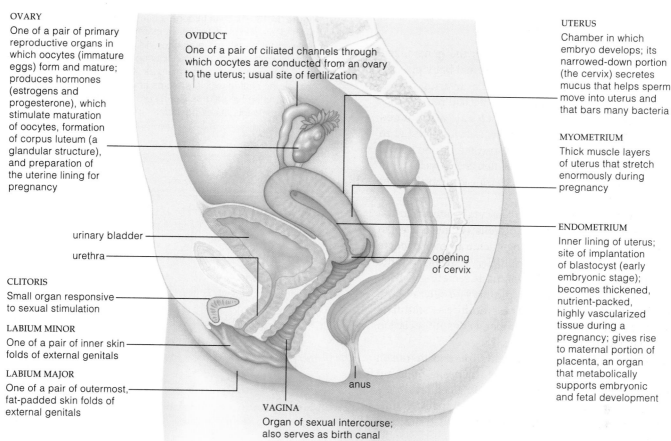

OVARY
One of a pair of primary reproductive organs in which oocytes (immature eggs) form and mature; produces hormones (estrogens and progesterone), which stimulate maturation of oocytes, formation of corpus luteum (a glandular structure), and preparation of the uterine lining for pregnancy

OVIDUCT
One of a pair of ciliated channels through which oocytes are conducted from an ovary to the uterus; usual site of fertilization

UTERUS
Chamber in which embryo develops; its narrowed-down portion (the cervix) secretes mucus that helps sperm move into uterus and that bars many bacteria

MYOMETRIUM
Thick muscle layers of uterus that stretch enormously during pregnancy

urinary bladder

urethra

opening of cervix

ENDOMETRIUM
Inner lining of uterus; site of implantation of blastocyst (early embryonic stage); becomes thickened, nutrient-packed, highly vascularized tissue during a pregnancy; gives rise to maternal portion of placenta, an organ that metabolically supports embryonic and fetal development

CLITORIS
Small organ responsive to sexual stimulation

LABIUM MINOR
One of a pair of inner skin folds of external genitals

LABIUM MAJOR
One of a pair of outermost, fat-padded skin folds of external genitals

anus

VAGINA
Organ of sexual intercourse; also serves as birth canal

b

Table 34.3 Events of the Menstrual Cycle

Phase	Events	Days of Cycle*
Follicular phase	Menstruation; endometrium breaks down	1–5
	Follicle matures in ovary; endometrium rebuilds	6–13
Ovulation	Oocyte released from ovary	14
Luteal phase	Corpus luteum forms, secretes progesterone; the endometrium thickens and develops	15–28

*Assuming a 28-day cycle.

FSH and LH secretions also promote the cyclic changes in the endometrium.

A human female's menstrual cycles start between ages ten and sixteen. Each cycle lasts about twenty-eight days, but this is simply the average. It runs longer for some women and shorter for others. Menstrual cycles continue until a woman is in her late forties or early fifties, when her supply of eggs is dwindling and hormonal secretions slow down. This is the onset of *menopause*, the twilight of reproductive capacity.

You may have heard about *endometriosis*, a condition that arises when endometrial tissue abnormally spreads and grows outside the uterus. Endometrial scar tissue may form on ovaries or oviducts and lead to infertility. In the United States, 10 million women may be affected annually. Possibly the condition arises when menstrual flow backs up through the oviducts and spills into the pelvic cavity. Or perhaps some embryonic cells became positioned in the wrong place before birth and were stimulated to grow during puberty, when sex hormones became active. Whatever the case, estrogen still acts on cells in the mislocated tissue. The resulting symptoms include pain during menstruation, sex, or urination.

Ovaries, the primary reproductive organs in females, produce oocytes (immature eggs) and sex hormones. Endometrium lines the uterus, a chamber in which embryos develop.

Secretion of the sex hormones, estrogens and progesterone, is coordinated on a cyclic basis through the reproductive years.

A cycle starts with menstruation, breakdown and rebuilding of the endometrium, and maturation of an oocyte. After release of a mature egg (ovulation), it ends when a corpus luteum forms and the endometrium is primed for pregnancy.

Cyclic Changes in the Ovary

Take a moment to review Figure 8.9, the generalized picture of meiosis in an oocyte. A normal baby girl has about 2 million primary oocytes in her ovaries. By the time she is seven years old, about 300,000 remain (her body resorbed the rest). These oocytes already entered meiosis I, but then the division process was arrested in a programmed way. Meiosis will resume in one oocyte at a time, starting with the first menstrual cycle. Only about 400 to 500 oocytes will be released during her reproductive years.

Figure 34.16 shows a *primary* oocyte located near the surface of an ovary. A layer of cells (granulosa cells) surrounds and nourishes it. Together, a primary oocyte and the cell layer are called a follicle. At the start of the menstrual cycle, the hypothalamus is secreting GnRH, which stimulates the anterior pituitary to release both FSH and LH (Figure 34.17). These hormones stimulate the follicle to grow. FSH, remember, is the abbreviation for follicle-stimulating hormone.

The oocyte starts to increase in size, and more cell layers form around it. Glycoprotein deposits build up between the oocyte and the surrounding cell layers. As they do so, they widen the space between them. In time the deposits form a noncellular coating, called the zona pellucida, around the oocyte.

FSH and LH cause cells outside the zona pellucida to secrete estrogens. An estrogen-containing fluid accumulates in the follicle, and estrogen levels in the blood start to rise. About eight to ten hours before its release from the ovary, the oocyte completes meiosis I, then its cytoplasm divides. There are now two cells in the follicle. One, the *secondary* oocyte, ends up with nearly all of the cytoplasm. The other cell is the first polar body. Chromosomes are distributed between the two cells in such a way that the secondary oocyte is haploid—which is the chromosome number required of gametes.

About midway through the cycle, the pituitary gland detects the rising level of estrogen and responds with a brief outpouring of LH. This triggers cellular contractions, which make the fluid-engorged follicle balloon outward and then rupture. The fluid escapes and carries the secondary oocyte with it (Figure 34.17). *Thus the midcycle surge of LH triggered ovulation—the release of a secondary oocyte from the ovary.*

Figure 34.16 Section through a human ovary. A follicle at a given location in the ovary stays there throughout the menstrual cycle and does not move around; this sketch merely shows the order in which events proceed. During the first phase of the cycle, the follicle grows and matures. At ovulation (the second phase), the mature follicle ruptures and releases a secondary oocyte. During the third phase, a corpus luteum forms from the follicle's remnants. It self-destructs if pregnancy does not occur.

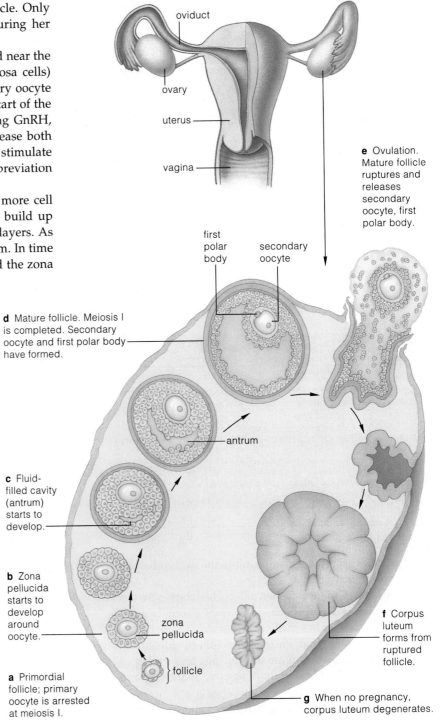

e Ovulation. Mature follicle ruptures and releases secondary oocyte, first polar body.

d Mature follicle. Meiosis I is completed. Secondary oocyte and first polar body have formed.

c Fluid-filled cavity (antrum) starts to develop.

b Zona pellucida starts to develop around oocyte.

a Primordial follicle; primary oocyte is arrested at meiosis I.

f Corpus luteum forms from ruptured follicle.

g When no pregnancy, corpus luteum degenerates.

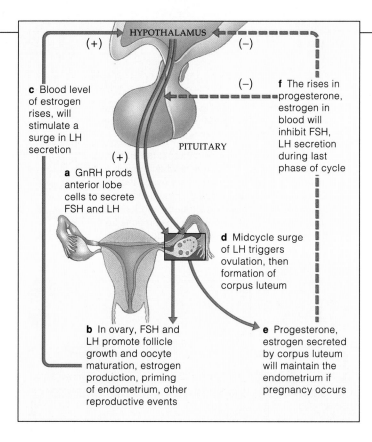

a GnRH prods anterior lobe cells to secrete FSH and LH

c Blood level of estrogen rises, will stimulate a surge in LH secretion

HYPOTHALAMUS

PITUITARY

f The rises in progesterone, estrogen in blood will inhibit FSH, LH secretion during last phase of cycle

d Midcycle surge of LH triggers ovulation, then formation of corpus luteum

b In ovary, FSH and LH promote follicle growth and oocyte maturation, estrogen production, priming of endometrium, other reproductive events

e Progesterone, estrogen secreted by corpus luteum will maintain the endometrium if pregnancy occurs

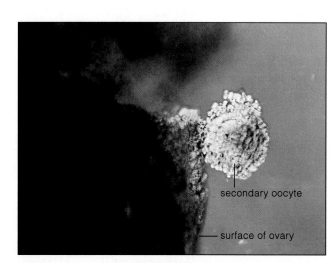

secondary oocyte

surface of ovary

Figure 34.17 Feedback control of hormonal secretion during a menstrual cycle. A positive feedback loop from an ovary to the hypothalamus causes a surge in LH secretion. This surge triggers ovulation. The micrograph above shows a secondary oocyte being released from an ovary at this time. Afterward, negative feedback loops to the hypothalamus and pituitary inhibit FSH secretion. They prevent another follicle from maturing until the cycle is completed.

Cyclic Changes in the Uterus

The estrogens released during the early phase of the menstrual cycle also help pave the way for pregnancy. They stimulate growth of the endometrium and its glands. Just before the midcycle surge of LH, cells of the follicle start secreting some progesterone as well as estrogens. Blood vessels grow rapidly in the thickened endometrium. Then, at the time of ovulation, estrogens act on tissue around the cervical canal, the narrowed-down portion of the uterus that leads to the vagina. The cervix now starts to secrete large quantities of a thin, clear mucus—an ideal medium for sperm travel.

Following ovulation, another structure dominates the cycle. The granulosa cells left behind in the follicle differentiate and form a yellowish glandular structure called the **corpus luteum** (meaning "yellow body"). The corpus luteum forms as an outcome of the midcycle surge of LH. Hence the name, luteinizing hormone.

The corpus luteum secretes progesterone and some estrogen. Progesterone prepares the reproductive tract for the arrival of a blastocyst—the tiny ball of cells that develops from a fertilized egg. For example, it causes cervical mucus to become thick and sticky. The mucus may prevent normal bacterial inhabitants of the vagina from entering the uterus. Progesterone also maintains the endometrium during a pregnancy.

A corpus luteum persists for about twelve days. All the while, the hypothalamus signals for minimal FSH

secretion and so stops other follicles from developing. If a blastocyst does not burrow into the endometrium, the corpus luteum self-destructs during the last days of the cycle. It does so by secreting prostaglandins, which apparently disrupt its own functioning.

After this, progesterone and estrogen levels in blood decline rapidly, and so the endometrium starts to break down. Deprived of oxygen and nutrients, the lining's blood vessels constrict, and tissues die. Blood escapes from the ruptured walls of weakened capillaries. The blood and sloughed endometrial tissues make up the menstrual flow, which continues for three to six days. Then the cycle begins anew, and rising estrogen levels stimulate the repair and growth of the endometrium.

By menopause, the supply of oocytes is dwindling, hormone secretions slow down, and in time menstrual cycles (and fertility) are over. The eggs that are still to be released late in a woman's life are at some risk of alterations in chromosome number or structure when meiosis resumes. A newborn with Down syndrome is one of the possible outcomes (Section 10.7).

During a menstrual cycle, FSH and LH stimulate growth of a follicle—a primary oocyte and its surrounding layer of cells—in an ovary. A midcycle surge of LH triggers ovulation.

Early on, estrogens call for endometrial repair and growth. Then estrogens and progesterone prepare the endometrium and other parts of the reproductive tract for pregnancy.

By now, you probably have come to the conclusion that the menstrual cycle is not a simple tune on a biological banjo. It is a full-blown hormonal symphony!

Before continuing with your reading, take a moment to review Figure 34.18. It may leave you with a better understanding of the integrated changes in hormone levels during each menstrual cycle, as correlated with the cyclic changes that they bring about in the ovary and uterus.

Figure 34.18 Changes in the ovary and uterus, correlated with changing hormone levels during each turn of the menstrual cycle.

Green arrows indicate which hormones dominate the cycle's first phase (the time when the follicle matures), then the second phase (when the corpus luteum forms). (**a,b**) FSH and LH secretions bring about changes in ovarian structure and function. (**c,d**) Estrogen and progesterone secretions from the ovary cause changes in the endometrium.

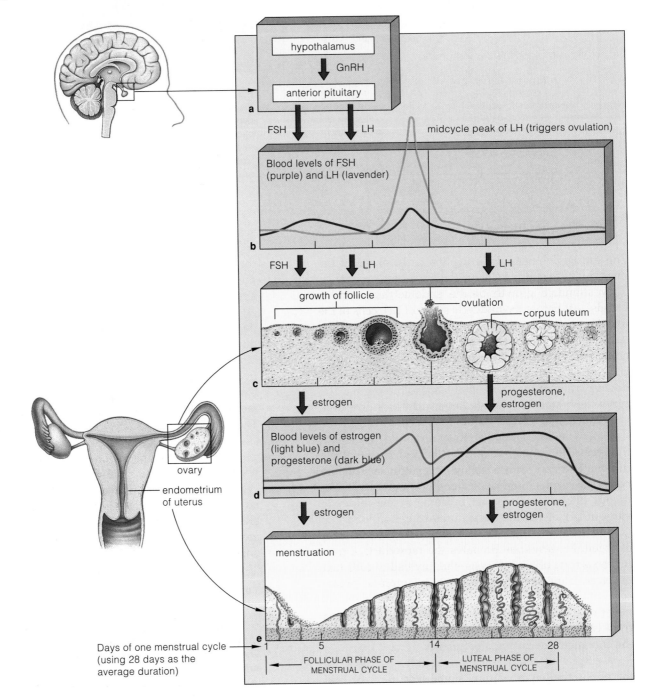

PREGNANCY HAPPENS

Sexual Intercourse

Suppose a secondary oocyte is on its way down an oviduct when a female and male are engaged in sexual intercourse (coitus). Within seconds of sexual arousal, the male's penis undergoes changes that will assist its penetration into the vaginal canal. The penis contains cylinders of spongy tissue (Figure 34.10). One cylinder has a mushroom-shaped tip (glans penis) loaded with sensory receptors, which friction activates. In sexually unaroused males, the penis is limp because large blood vessels leading into the cylinders are constricted. In aroused males, blood flows into the cylinders faster than it flows out. As blood collects in the spongy tissue, the penis lengthens and stiffens.

At coitus, pelvic thrusts stimulate the penis as well as the female's vaginal wall and clitoris. Mechanical stimulation causes rhythmic, involuntary contractions in the male reproductive tract. They force sperm to move rapidly out of each epididymis, and they force the contents of seminal vesicles and the prostate gland into the urethra. And now the semen is ejaculated into the vagina. During an ejaculation, a sphincter closes and prevents urination.

The muscular contractions, ejaculation, and sensations of release, warmth, and relaxation define a male's orgasm. A female's orgasm involves involuntary contractions of the uterus and vagina, intense awareness of the vagina, and sensations of relaxation and warmth. Even if the female does not get that excited, she can still get pregnant.

Fertilization

Now sperm are in the vagina. A single ejaculation can put 150 million to 350 million there. If they arrive a few days before or after ovulation, or anytime in between, fertilization may be the outcome. Within thirty minutes after ejaculation, muscle contractions move the sperm deeper into the female reproductive tract. Only a few hundred sperm actually reach the upper portion of the oviduct, where fertilization most often takes place.

The stunning micrograph on page 113 shows living sperm at the surface of a secondary oocyte. When they contact an oocyte, sperm release acrosomal enzymes that clear a path through its zona pellucida (Figure 34.19). Although many sperm might get this far, usually

Figure 34.19 Fertilization. (**a**) A number of sperm surround a secondary oocyte. Acrosomal enzymes clear a path through the zona pellucida. (**b**) When a sperm manages to penetrate the secondary oocyte, cortical granules inside the egg cytoplasm release substances that make the zona pellucida impenetrable to other sperm. Penetration also stimulates the second meiotic division of the oocyte's nucleus. (**c**) The sperm tail degenerates. The sperm nucleus enlarges and fuses with the oocyte nucleus. (**d**) With that fusion, fertilization is over. The zygote has formed.

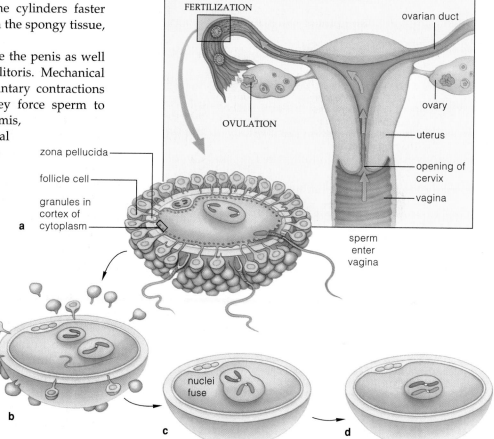

only one fuses with the oocyte. (Only its nucleus and centrioles do not degenerate in the oocyte's cytoplasm.) This event stimulates the secondary oocyte and the first polar body to complete meiosis II. There are now three polar bodies and a mature egg, or **ovum** (plural, ova). As the sperm and egg nuclei fuse, their chromosomes restore the diploid number for a brand new zygote.

The intense physiological events that accompany coitus have one function: to put sperm on a collision course with an egg. Fertilization—the fusion of a sperm nucleus and egg nucleus—results in a diploid zygote.

34.12 FORMATION OF THE EARLY EMBRYO

Pregnancy lasts an average of thirty-eight weeks from the time of fertilization. It takes about two weeks for the blastocyst to form. The time span from the third to the end of the eighth week is the **embryonic period**, when the major organ systems form. When it ends, the new individual has distinctly human features and is called a fetus. The **fetal period** extends from the start of the ninth week until birth. During this period, organs grow in size and assume specialized functions.

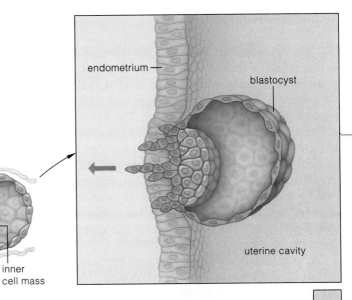

endometrium

blastocyst

uterine cavity

e DAY 5. A blastocyst has now developed from the morula. It has a surface layer of cells, a fluid-filled cavity, and an inner cell mass.

inner cell mass

d DAY 4. By 96 hours, divisions produce a ball of sixteen to thirty-two cells. It looks like a tiny mulberry and is called a morula (after the Latin *morum*, meaning mulberry). The morula gives rise to the embryo and attached membranes.

f DAYS 6–7. The blastocyst attaches to the endometrium, and it starts to burrow into it. With these activities, implantation is under way.

actual size

c DAY 3. By 72 hours, divisions have produced a ball of six to twelve cells.

b DAY 2. The second division, which is completed after about 40 hours, produces the four-cell stage.

a DAY 1. Within 24 hours after fertilization, cleavage (a series of controlled mitotic cell divisions) begins. The first cut is along a plane at right angles to the zygote's equator, in line with the polar bodies.

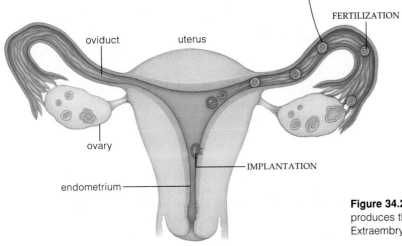

oviduct   uterus

ovary

FERTILIZATION

IMPLANTATION

endometrium

We often call the first three months of pregnancy the first trimester. The second trimester extends from the start of the fourth month to the end of the sixth month. The third trimester extends from the seventh month until birth. Starting with Figure 34.20, the next series of illustrations shows characteristic features of the new individual at progressive stages of development.

Early Cleavage and Implantation

By three to four days after fertilization, the zygote is moving through the oviduct, and cleavage has begun. By the time the cluster of dividing cells reaches the uterus, it is a solid ball, the morula (Figure 34.20d). A fluid-filled cavity forms inside the ball, which thereby becomes a blastocyst. A human blastocyst consists of a surface layer of cells and an interior cluster called the inner cell mass.

Six or seven days after fertilization, **implantation** is under way. By this process, the blastocyst adheres to the uterine lining, some of its cells send out projections that invade the mother's tissues, and connections start forming that will metabolically support the developing embryo through the many months ahead. While the invasion is proceeding, the inner cell mass becomes transformed into two cell layers having a flattened,

Figure 34.20 From fertilization through implantation. Early on, cleavage produces the blastocyst, a stage that develops into a disklike early embryo. Extraembryonic membranes (the amnion, chorion, and yolk sac) start forming.

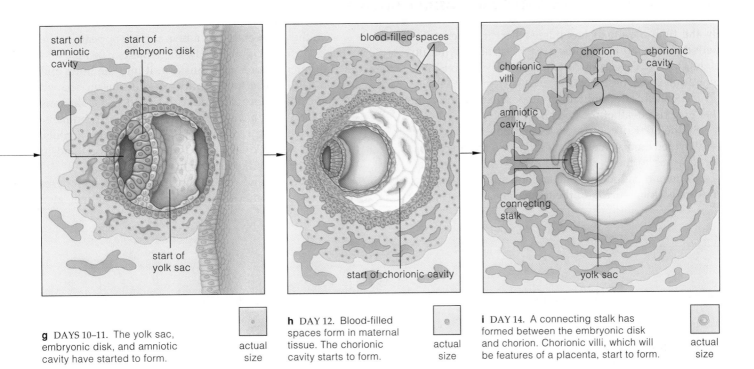

g DAYS 10–11. The yolk sac, embryonic disk, and amniotic cavity have started to form.

actual size

h DAY 12. Blood-filled spaces form in maternal tissue. The chorionic cavity starts to form.

actual size

i DAY 14. A connecting stalk has formed between the embryonic disk and chorion. Chorionic villi, which will be features of a placenta, start to form.

actual size

somewhat circular shape. Together, they represent the embryonic disk—and in short order they will give rise to the embryo proper.

Extraembryonic Membranes

As implantation progresses, membranes start to form outside the embryo. First a fluid-filled, *amniotic* cavity opens up between the embryonic disk and part of the blastocyst's surface (Figure 34.20*g*). Then cells migrate around the cavity wall and form the **amnion**. This is a membrane that will enclose the embryo. Fluid in the amniotic cavity will serve as a buoyant cradle where the embryo can grow, move freely, and be protected from sudden temperature shifts and mechanical impacts.

While the amnion forms, other cells migrate around the inner wall of the blastocyst's first cavity. They form a lining that becomes the **yolk sac**. This extraembryonic membrane speaks of the evolutionary heritage of land vertebrates (Section 21.7). For most animal species that produce shelled eggs, the sac holds nutritive yolk. In humans, some of the yolk sac becomes a site of blood cell formation, and some will give rise to germ cells, the forerunners of gametes.

Before the blastocyst is fully implanted, spaces open in maternal tissues and fill with blood seeping in from ruptured capillaries. Inside the blastocyst, another cavity opens up around the amnion and yolk sac. Fingerlike projections start to form on the cavity's lining, which is the **chorion**. This new membrane will become part of a spongy, blood-engorged tissue called the placenta.

Later, after the blastocyst becomes fully implanted, another extraembryonic membrane will form as an outpouching of the yolk sac. This membrane will become the **allantois**. It serves different functions in different groups. Among reptiles, birds, and some mammals, the allantois serves in respiration and storage of metabolic wastes. In humans, it serves in early blood formation and formation of the urinary bladder.

One more point should be made here. Cells of the blastocyst secrete a hormone **HCG** (*Human Chorionic Gonadotropin*), which stimulates the corpus luteum to keep on secreting progesterone and estrogen. Thus the blastocyst itself prevents menstrual flow and works to avoid being sloughed off until the placenta takes over the task, some eleven weeks later. By the start of the third week, HCG can be detected in the mother's blood or urine. At-home *pregnancy tests* are based on a "dipstick" that changes color when HCG is present in urine.

Six or seven days after fertilization, the blastocyst implants itself in the endometrium. Projections from its surface invade maternal tissues, and connections start to form that in time will metabolically support the developing embryo.

Some parts of the blastocyst give rise to the amnion, yolk sac, chorion, and allantois. These extraembryonic membranes have different roles. But together they are vital for the structural and functional development of the embryo.

After the *embryonic* period (from the third to the end of the eighth week), an embryo has distinctively human features. The *fetal* period extends from the ninth week to birth.

By the time a woman has missed her first menstrual period, gastrulation is under way. Gastrulation, recall, is the stage of development when mitotic cell divisions, migrations, and tissue folding result in primary tissue layers—which later split into subpopulations of cells that give rise to all of the body's organs.

By the start of the third week of development, the two-layered embryonic disk, described in the preceding section, has formed (Figure 34.21). It consists only of ectoderm and endoderm. The amnion and chorion, two fluid-filled sacs, surround the embryonic disk except at

On the eighteenth day, a neural plate appears at the disk's surface (Figure 34.21). Regional expansions and cell differentiations will transform it into the brain and spinal cord. Early in the third week, a sausage-shaped outpouching appears on the yolk sac. It is the allantois (after the Greek *allas*, meaning "sausage"). In humans, the outpouching remains small. The human allantois, recall, has early roles in the formation of blood and in the emergence of the urinary bladder.

Toward the end of the third week, some mesoderm gives rise to somites—paired segments that will give

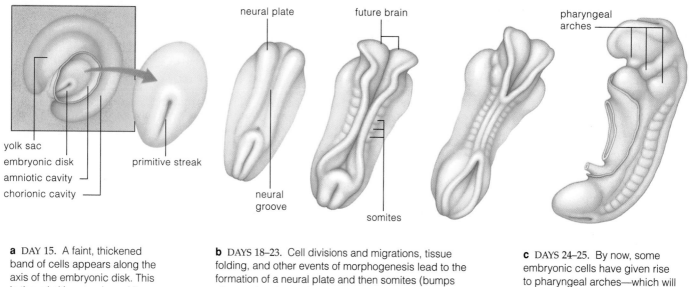

a DAY 15. A faint, thickened band of cells appears along the axis of the embryonic disk. This is the primitive streak, and it marks the onset of gastrulation in vertebrate embryos.

b DAYS 18–23. Cell divisions and migrations, tissue folding, and other events of morphogenesis lead to the formation of a neural plate and then somites (bumps of mesoderm) near the embryo's dorsal surface. The neural plate will give rise to the brain and spinal cord. Somites will give rise to most of the skeleton's axial portion, skeletal muscles, and much of the dermis.

c DAYS 24–25. By now, some embryonic cells have given rise to pharyngeal arches—which will contribute to the formation of the face, neck, mouth, nasal cavities, larynx, and pharynx.

Figure 34.21 Hallmarks of the embryonic period of humans and other vertebrates. A primitive streak appears, then a notochord. A neural plate forms, then somites and pharyngeal arches. These are dorsal views (of the embryo's back). Compare the diagrams with the photographs in Figure 34.23.

the point where a stalk connects the disk to the inner wall of the chorionic cavity.

Now, rapidly dividing and migrating cells cause the disk's midline to thicken slightly. This faint, "primitive streak," which will lengthen and thicken more the next day, marks the onset of gastrulation (Figure 34.21). It establishes the embryo's long axis and its forthcoming bilateral symmetry. In other words, *the embryonic disk is undergoing morphogenetic events that will provide the body with the basic form that is characteristic of all vertebrates.*

Soon mesoderm starts to form inside the disk. Some of it folds into a tube that becomes the notochord. A human notochord is a structural framework; the bony segments of the vertebral column will form around it.

rise to most bones and skeletal muscles of the head and trunk, as well as to the dermis overlying these regions. Pharyngeal arches start to form. Small spaces open up in parts of the mesoderm. In time, all of these spaces will interconnect and form the coelomic cavity.

During the third week after fertilization (a time when the woman has missed her first menstrual period), the basic vertebrate body plan emerges in the new individual.

A primitive streak, neural plate, somites, and pharyngeal arches form during the embryonic period of all vertebrates. Formation of the primitive streak establishes the body's long axis and its forthcoming bilateral symmetry.

ON THE IMPORTANCE OF THE PLACENTA

Even before the onset of the embryonic period, the extraembryonic membranes have been collaborating with the uterus to sustain the embryo's rapid growth. By the third week, tiny fingerlike projections from the chorion have grown profusely into the maternal blood that has pooled in endometrial spaces. The projections, the chorionic villi, enhance the exchange of substances between the mother and the new individual. They are functional components of the **placenta**. This blood-engorged organ consists of both endometrial tissue and extraembryonic membranes.

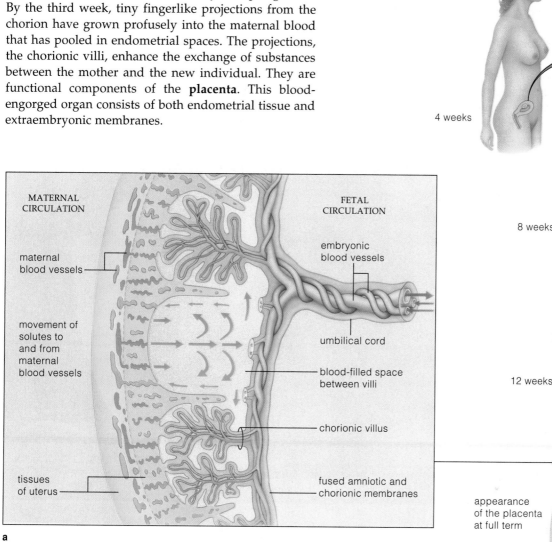

4 weeks

8 weeks

12 weeks

appearance of the placenta at full term

b

Figure 34.22 (**a**) Relationship between fetal and maternal blood circulation in a full-term placenta (**b**). Blood vessels extend from the fetus, through the umbilical cord, and into chorionic villi. Maternal blood spurts into spaces between the villi. Oxygen, carbon dioxide, and other small solutes diffuse across the placental membrane surface. There is no gross intermingling of the two bloodstreams.

At full term, the placenta will make up a fourth of the inner surface of the woman's uterus (Figure 34.22).

The placenta is the body's way of sustaining the new individual while allowing its blood vessels to develop apart from the mother's. Oxygen and nutrients diffuse out of the maternal blood vessels, across the placenta's blood-filled spaces, then into embryonic blood vessels. (These vessels converge in the umbilical cord, the lifeline between the placenta and the new individual.) Carbon dioxide and other wastes diffuse in the other direction.

Maternal lungs and kidneys quickly dispose of them. After the third month, the placenta secretes progesterone and estrogens—and so maintains the uterine lining.

The placenta is a blood-engorged organ of endometrial and extraembryonic membranes. It permits exchanges between the mother and the new individual without intermingling their bloodstreams. Thus it sustains the individual *and* allows its blood vessels to develop apart from the mother's.

EMERGENCE OF DISTINCTLY HUMAN FEATURES

By the end of the fourth week of the embryonic period, the embryo has grown to 500 times its original size. The placenta has been sustaining the growth spurt, but now the pace slows as the embryo embarks on a course of cell differentiation, morphogenesis, and pattern formation. Limbs form, fingers and toes are sculpted from embryonic paddles, and the umbilical cord forms. The circulatory system becomes more intricate. Much attention is devoted to the all-important head; its growth now surpasses that of any other body region (Figure 34.23). The embryonic period ends as the eighth week closes. No longer is the embryo merely "a vertebrate." By now its features clearly define it as a human fetus.

Figure 34.23 (**a**) Human embryo at four weeks. As for all vertebrates, it has a tail and pharyngeal arches. (**b**) Embryo at five to six weeks after fertilization. (**c**) Embryo at the boundary between embryonic and fetal periods. It now has human features. It floats in amniotic fluid. The chorion normally covers the amniotic sac but has been pulled aside here. (**d**) Fetus at sixteen weeks. In the fetal period, movements begin as soon as nerves make functional connections with developing muscles. Legs kick, arms wave, fingers grasp, the mouth puckers. These reflex actions will be vital skills in the world outside the uterus.

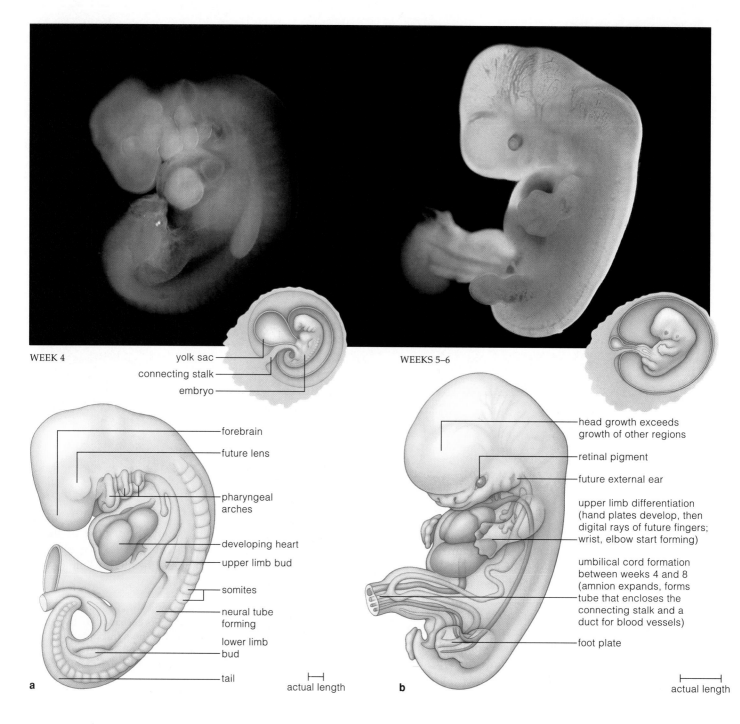

WEEK 4

yolk sac
connecting stalk
embryo

WEEKS 5–6

forebrain

future lens

pharyngeal arches

developing heart

upper limb bud

somites

neural tube forming

lower limb bud

tail

actual length

a

head growth exceeds growth of other regions

retinal pigment

future external ear

upper limb differentiation (hand plates develop, then digital rays of future fingers; wrist, elbow start forming)

umbilical cord formation between weeks 4 and 8 (amnion expands, forms tube that encloses the connecting stalk and a duct for blood vessels)

foot plate

actual length

b

During the second trimester, the fetus is moving facial muscles. It frowns; it squints. It busily practices the sucking reflex, as shown in the next section. Now the mother can easily sense movements of the fetal arms and legs. When the fetus is five months old, she can hear its heart through a stethoscope positioned on her abdomen. Soft, fuzzy hair (the lanugo) covers the fetal body. The skin is wrinkled, reddish, and protected from abrasion by a thick, cheesy coating. In the sixth month, delicate eyelids and eyelashes form. During the seventh month, the eyes open.

A fetus born prematurely before twenty-two weeks have passed cannot survive. The situation also is grave for births before twenty-eight weeks, mainly because lungs have not developed sufficiently. Even with the best medical care, premature infants have difficulty breathing and maintaining a normal core temperature. By the ninth month, the survival rate is 95 percent.

During the fetal period, primordial tissues that formed in the early embryo become sculpted into distinctly human features.

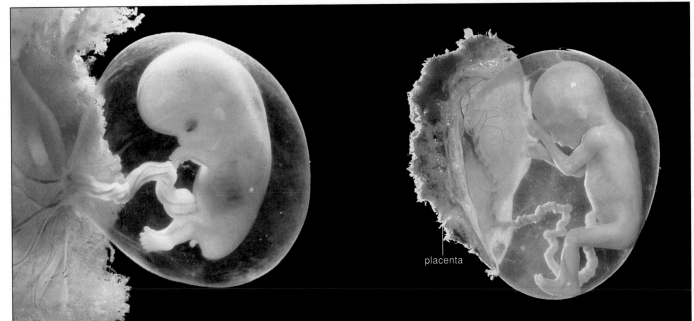

placenta

WEEK 8

final week of embryonic period; embryo looks distinctly human compared to other vertebrate embryos

upper and lower limbs well formed; fingers and then toes have separated

primordial tissues of all internal, external structures now developed

tail has become stubby

c |— actual length —|

WEEK 16
Length: 16 centimeters
         (6.4 inches)
Weight: 200 grams
         (7 ounces)

WEEK 29
Length: 27.5 centimeters
         (11 inches)
Weight: 1,300 grams
         (46 ounces)

WEEK 38 (full term)
Length: 50 centimeters
         (20 inches)
Weight: 3,400 grams
         (7.5 pounds)

During fetal period, length measurement extends from crown to heel (for embryos, it is the longest measurable dimension, as from crown to rump).

d

MOTHER AS PROTECTOR, PROVIDER, POTENTIAL THREAT

A woman who decides to become pregnant is committing a large part of her body's resources and functions to the development of a new individual. From fertilization until birth, her future child is absolutely at the mercy of her diet, health habits, and life-style (Figure 34.24).

Some Nutritional Considerations How does a pregnant woman best provide nutrients for the embryo, and then the fetus? Normally, the same balanced diet that is good for her provides her future child with all required carbohydrates, lipids, and proteins. (Chapter 30 is a good starting point for your understanding of nutritional requirements.) The woman's own need for vitamins and minerals increases during pregnancy. The placenta absorbs enough for her embryo from the bloodstream, except for folic acid. She can reduce the risk that her embryo will develop severe neural tube defects by taking more B-complex vitamins (under medical supervision) before conception and early in pregnancy. Besides this, one study showed

that women who smoked while pregnant *and* their fetuses had depressed blood levels of vitamin C even when their vitamin C intake was identical to that of a control group. Smoking may affect utilization of other nutrients as well.

A pregnant woman also must eat enough so that her body weight increases by between 20 and 25 pounds, on average. If her weight gain is a great deal lower than that, she is stacking the deck against the fetus. Compared to newborns of normal weight, the significantly underweight ones have more postdelivery complications. They are also at greater risk of showing mental impairment later in life.

As birth approaches, a fetus makes greater nutritional demands of the mother. Clearly her diet will profoundly influence the remaining developmental events. The brain, like most of the other fetal organs, is especially vulnerable in the weeks just before and after birth, when it undergoes its greatest expansion. By now, its neurons have formed. That's why poor nutrition now will have repercussions on intelligence and other brain functions later in life.

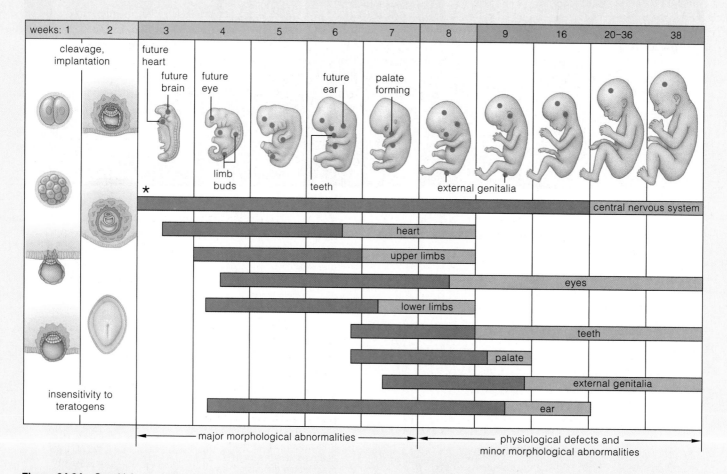

Figure 34.24 Sensitivity to teratogens during pregnancy. *Teratogens* are drugs and any other environmental factors that may induce deformities in the embryo or fetus. They usually have no effect before the onset of organ formation. After that, they can block or abnormally stimulate growth as well as the programmed remodeling and resorption of tissues.

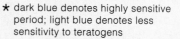 ★ dark blue denotes highly sensitive period; light blue denotes less sensitivity to teratogens

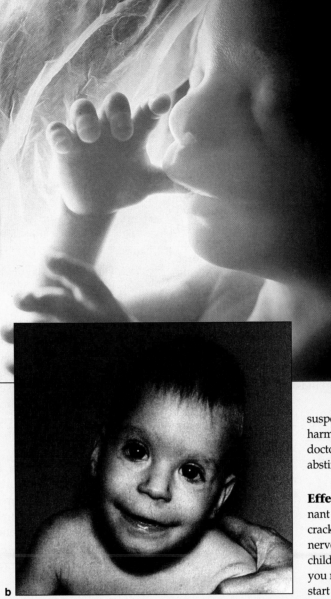

Risk of Infections The antibodies circulating in a pregnant woman's blood continually move across the placenta. They protect the new individual from all but the most serious bacterial infections. Certain viral diseases can be dangerous during the first six weeks after fertilization—a critical time of organ formation. Suppose a pregnant woman contracts *rubella* (German measles) during this critical period. There is a 50 percent chance some organs won't form properly. For example, if she gets infected when embryonic ears are forming, her newborn may be deaf. If she contracts this disease during or after the fourth month, it will have no notable effect. But she can avoid the risk entirely; vaccination *before* pregnancy can prevent rubella.

Effects of Prescription Drugs
A pregnant woman should not take *any* drugs without close medical supervision. Think about what happened when *thalidomide* was prescribed in Europe. Women who had used this tranquilizer during the first trimester gave birth to infants who had missing or badly deformed arms and legs. The drug was withdrawn from the market. However, other tranquilizers, sedatives, and barbiturates are still being prescribed. There is a risk that they may cause similar, although less severe, damage. Even certain anti-acne drugs increase the risk of facial and cranial deformities. Tetracycline, a commonly prescribed antibiotic, yellows teeth. Streptomycin causes hearing problems and may affect the nervous system.

Effects of Alcohol As a fetus grows, its physiology becomes increasingly like the mother's. Alcohol passes freely across the placenta and has the same effect on the fetus as it has on her. *Fetal alcohol syndrome* (FAS) is a set of deformities that result from excessive alcohol intake during pregnancy. Symptoms include facial deformities, poor coordination, and, sometimes, heart defects (Figure 34.25*b*). In 1992, nearly 4 of every 10,000 newborns were diagnosed with FAS, and the rate is rising. About 60–70 percent of newborns of alcoholic women have FAS. Some researchers

Figure 34.25
(**a**) The fetus at eighteen weeks. (**b**) An infant with FAS. Symptoms include a small head size, low and prominent ears, improperly developed cheekbones, and a wide, smooth upper lip. The child can expect growth problems and abnormalities of the nervous system. This disorder affects about 1 in 750 newborns in the United States.

suspect any alcohol at all may be harmful to the fetus. Increasingly, doctors are urging near- or total abstinence during pregnancy.

Effects of Cocaine A pregnant woman who uses cocaine, crack especially, disrupts the nervous system of her future child as well as her own. Here you may wish to read again the start of Chapter 32.

Effects of Cigarette Smoke Cigarette smoke impairs fetal growth and development. Besides this, a long-term study at Toronto's Hospital for Sick Children showed that toxic elements in tobacco also accumulate in the fetuses of pregnant nonsmokers who are exposed to *second-hand smoke* at home or work.

Daily smoking during pregnancy leads to underweight newborns—even when the woman's weight, nutritional status, and all other relevant variables match those of pregnant nonsmokers. Smoking has other effects. In Great Britain, all infants born the same week were tracked for seven years. Those of smokers were smaller, ended in more postdelivery deaths, and had twice as many heart abnormalities. At age seven, their average "reading age" was nearly half a year behind children of nonsmokers.

The mechanisms by which smoking affects the fetus are not known. But its demonstrated effects are further evidence that the placenta, marvelous structure that it is, cannot prevent all assaults on the fetus that the human mind can dream up.

Giving Birth and Nourishing the Newborn

Pregnancy normally ends thirty-eight weeks after the time of fertilization, give or take a few weeks. The birth process is called labor (for good reason, as any mother will tell you). It starts with contractions of the uterus. For the next two to eighteen hours, the contractions increase in frequency and intensity. Inside the cervix, the canal that the fetus will pass through dilates fully. The amnion usually ruptures just before birth, and "water" (amniotic fluid) gushes from the vagina. Most of the time, the fetus is expelled within an hour after full dilation (Figure 34.26). Then, contractions force fluid, blood, and the placenta from the body, typically within fifteen minutes after birth. The umbilical cord—the lifeline to the mother—is severed. The newborn embarks on a nurtured existence in the outside world.

During the pregnancy, estrogen and progesterone had been stimulating the growth of mammary glands and ducts in the mother's breasts (Figure 34.27). After birth, the hormone-primed glands produce milk, by the

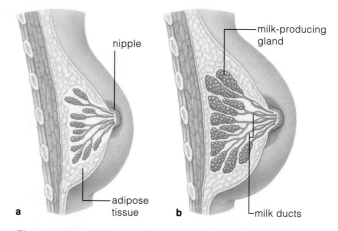

Figure 34.27 (**a**) Breast of a woman who is not pregnant. (**b**) Breast of a lactating woman.

process of lactation. For the first few days, the glands produce a fluid rich in proteins and lactose. Then the anterior pituitary secretes prolactin, which stimulates milk production (Section 33.3). As a newborn suckles, the pituitary releases oxytocin. This hormone triggers contractions that force milk into breast tissue ducts and that shrink the uterus back to normal size.

Each year in the United States, well over 100,000 women develop *breast cancer*. Obesity, high cholesterol, and high estrogen levels contribute to the cancerous transformation. With early detection and treatment, the chances for cure are excellent. Once a month, about a week after menstruating, a woman should examine her breasts. She can contact her physician or the American Cancer Society (listed in the telephone directory) for a pamphlet on recommended examination procedures.

Postnatal Development, Aging, and Death

After birth, a new individual follows a course of further growth and development that leads to the adult, the mature form of the species. Table 34.4 summarizes the *prenatal* (before birth) stages and *postnatal* (after birth) stages. Figure 34.28 is an example of the proportional changes that occur in the body as the life cycle unfolds. Postnatal growth is most rapid between ages thirteen and nineteen. Then, secretions of sex hormones step up and bring about the emergence of secondary sexual traits as well as sexual maturity. Not until adulthood are the bones fully mature. Body tissues normally are maintained in top condition during early adulthood. As the years pass, it becomes more and more difficult for the body to maintain and repair existing

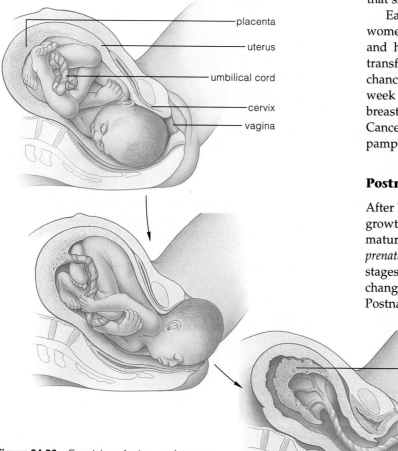

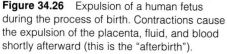

Figure 34.26 Expulsion of a human fetus during the process of birth. Contractions cause the expulsion of the placenta, fluid, and blood shortly afterward (this is the "afterbirth").

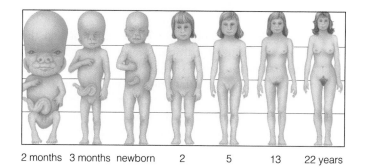

2 months 3 months newborn 2 5 13 22 years

Figure 34.28 Observable, proportional changes in the human body during prenatal and postnatal growth. Changes in overall physical appearance are gradual but quite noticeable until the teenage years. For example, the head becomes proportionally smaller, compared to what it was during the embryonic period. The legs become longer and the trunk becomes shorter.

Table 34.4	Stages of Human Development
Prenatal Period	
Zygote	Single cell resulting from fusion of sperm nucleus and egg nucleus at fertilization
Morula	Solid ball of cells produced by cleavages
Blastocyst	Ball of cells with surface layer, fluid-filled cavity, and inner cell mass
Embryo	All developmental stages from two weeks after fertilization until end of eighth week
Fetus	All developmental stages from the ninth week to birth (about thirty-eight weeks after fertilization)
Postnatal Period	
Newborn	Individual during the first two weeks after birth
Infant	Individual from two weeks to about fifteen months after birth
Child	Individual from infancy to about ten or twelve years
Pubescent	Individual at puberty, when secondary sexual traits develop; girls between ten and fifteen years, boys between twelve and sixteen years
Adolescent	Individual from puberty until about three or four years later; physical, mental, emotional maturation
Adult	Early adulthood (between eighteen and twenty-five years); bone formation and growth finished. Changes proceed very slowly afterward
Old age	Aging follows late in life

tissues, so the body gradually deteriorates. By processes collectively known as **aging**, the body's cells gradually start to break down. This leads to structural changes and gradual loss of bodily functions. All animals that show extensive cell differentiation undergo aging.

No one knows what causes aging, although research gives us interesting things to think about. For example, many years ago, Paul Moorhead and Leonard Hayflick cultured normal human embryonic cells. All of the cell lines proceeded to divide about fifty times—then the entire population died off. Hayflick took some cultured cells that were partway through the series of divisions and froze them for several years. Afterward, he thawed the cells and placed them in a culture medium. The cells proceeded to complete the cycle of fifty doublings— whereupon they all died on schedule.

Such experiments imply that normal types of cells have a limited division potential. That is, mitosis may be genetically programmed to decline at a certain time in the life cycle. This is certainly true of neurons. They stop undergoing mitotic cell division after birth, then deteriorate over time. But does the shutdown of mitosis *cause* aging or is it a *result* of aging?

DNA's capacity for self-repair may be gradually lost through an accumulation of gene mutations. As you read in the Chapter 4 introduction, free radicals—rogue molecular fragments that bombard DNA and the other biological molecules—are associated with aging. Also, researchers have just now correlated *Werner's syndrome,* an aging disorder, with a mutation in the gene coding for helicase. This enzyme is vital for DNA replication *and* repair operations. Whether by free radical attacks or by accumulated replication errors, *structural changes in DNA could endanger the production of functional enzymes and other proteins required for normal life processes.*

Think about how cells depend on smooth exchanges of materials between the cytoplasm and extracellular fluid. Also think about collagen, a structural component

of bone and other connective tissues throughout the body. If something shuts down or mutates the collagen-encoding genes, the absent or skewed gene product would affect the flow of oxygen, nutrients, hormones, and so forth to and from cells. The repercussions would ripple through the entire body. Finally, think about a deterioration in genes for membrane proteins that serve as self markers. If markers change, do T lymphocytes then perceive the body's own cells as foreign and attack them? Perhaps such autoimmune responses increase over time. The increase would invite the greater vulnerability to disease and stress we associate with old age.

A person who sticks to a low-fat diet and a program of low-impact, aerobic exercise throughout adulthood is more likely to reduce some of the common effects of aging. Regular exercise improves cardiovascular and respiratory functions. It also minimizes bone loss in older adults by helping to maintain bone density.

Yet, as Lewis Thomas wrote so eloquently, perhaps we should learn to accept the cycling of life and give up the notion that aging and death are catastrophes, or avoidable, or even strange. We might even find comfort in recognizing our connection to the greater picture— that we all go down together, in the best of company.

The human life cycle flows naturally from the time of birth, growth, and development, to production of the individual's own offspring, and on through aging to the time of death.

34.18 CONTROL OF HUMAN FERTILITY

Some Ethical Considerations

The transformation of a zygote into an adult of intricate detail raises profound questions. *When does development begin?* As you have seen, key developmental events unfold even before fertilization. *When does life begin?* During her lifetime, a woman can produce as many as 500 eggs, all of which are alive. During one ejaculation, a man can release a quarter of a billion sperm, which are alive. Even before sperm and egg merge by chance and establish the genetic makeup of a new individual, they are as much alive as any other form of life. It is scarcely tenable, then, to say life begins at fertilization. *Life began billions of years ago; and each gamete, each zygote, each sexually mature individual is but a fleeting stage in the continuation of that beginning.*

This greater perspective on life cannot diminish the meaning of conception. It is no small thing to entrust a new individual with the gift of life, wrapped in the unique evolutionary threads of our species and handed down through an immense sweep of time.

Yet how can we reconcile the marvel of individual birth with growing awareness of the astounding birth rate for our whole species? While this book is being written, an average of 10,700 newborns enter the world every single hour. By the time you go to bed tonight, there will be 257,000 more people on Earth than there were last night at that hour. Within a week, the number will reach 1,800,000—about as many people as there are now in the entire state of Massachusetts. *Within one week.* Worldwide, human population growth is rapidly outstripping resources, and each year millions face the horrors of starvation. Living as we do on one of the most productive continents on Earth, few of us know what it means to give birth to a child, to give it the gift of life, and have no food to keep it alive.

And how can we reconcile the marvel of birth with the confusion that surrounds unwanted pregnancies? Even highly developed countries do not have adequate educational programs concerning fertility. And a great number of people are not inclined to exercise control. Each year in the United States alone, there are 1 million teenage pregnancies. (Many parents actually encourage early boy-girl relationships without thinking through the great risks of premarital intercourse and unplanned pregnancy. Advice is often condensed to a terse "Don't do it. But if you do it, be careful!") And each year, there are 1.6 million abortions among all age groups.

The motivation to engage in sex has been evolving for more than 500 million years. A few centuries of moral and ecological arguments for its suppression have not stopped all that many unwanted pregnancies. Besides this, complex social factors have contributed to a population growth rate that is out of control.

How will we reconcile our biological past and the need for a stabilized cultural present? Whether and how human fertility is to be controlled is one of the most volatile issues of our time. We will return to this issue in the next chapter, in the context of principles that govern the growth and stability of populations. Here, we briefly consider some control options.

Birth Control Options

The most effective method of birth control is complete *abstinence*, no sexual intercourse whatsoever. Data show that it is unrealistic to expect many people to practice it.

A less reliable form of abstinence is the *rhythm method*. The idea is to avoid intercourse in the woman's fertile period, starting a few days before ovulation and ending a few days after. The woman identifies and tracks her fertile period. She keeps records of the length of her menstrual cycles, takes her temperature each morning when she wakes up, or both. (The body's core temperature rises by one-half to one degree just before a fertile period.) But ovulation may not be regular, and miscalculations are frequent. Also, sperm deposited in the vagina a few days before ovulation may survive until ovulation. The rhythm method *is* inexpensive; it costs nothing after you buy a thermometer. It doesn't require fittings and periodic medical checkups. But its practitioners run a risk of pregnancy (Figure 34.29).

Withdrawal, or removing the penis from the vagina before ejaculation, dates back at least to biblical times. But withdrawal requires very strong willpower, and the practice may fail anyway. Fluid released from the penis just before ejaculation may contain some sperm.

Douching (rinsing the vagina with a chemical right after intercourse) is next to useless. Sperm move past the cervix and out of reach of the douche within ninety seconds after ejaculation.

Controlling fertility by surgical intervention is less chancy. In *vasectomy*, a physician makes a tiny incision in a man's scrotum, then severs and ties off each vas deferens. The operation takes only twenty minutes and requires only a local anesthetic. After the operation, sperm cannot leave the testes and cannot be present in semen. So far, there is no firm evidence that vasectomy disrupts hormonal functions or adversely affects sexual activity. Vasectomies can be reversed. But half of those who submit to surgery later develop antibodies against sperm and may not be able to regain fertility.

Females may have a *tubal ligation*. By this surgical intervention, oviducts are cauterized or cut and tied off, usually in a hospital. When the operation is performed correctly, tubal ligation is the most effective means of birth control. A few women have recurring pain in the pelvic area. The operation sometimes can be reversed.

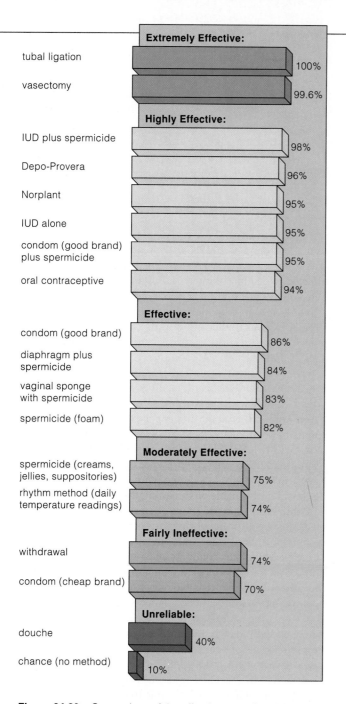

Extremely Effective:

tubal ligation — 100%
vasectomy — 99.6%

Highly Effective:

IUD plus spermicide — 98%
Depo-Provera — 96%
Norplant — 95%
IUD alone — 95%
condom (good brand) plus spermicide — 95%
oral contraceptive — 94%

Effective:

condom (good brand) — 86%
diaphragm plus spermicide — 84%
vaginal sponge with spermicide — 83%
spermicide (foam) — 82%

Moderately Effective:

spermicide (creams, jellies, suppositories) — 75%
rhythm method (daily temperature readings) — 74%

Fairly Ineffective:

withdrawal — 74%
condom (cheap brand) — 70%

Unreliable:

douche — 40%
chance (no method) — 10%

Figure 34.29 Comparison of the effectiveness of methods of contraception in the United States. Percentages are based on the number of unplanned pregnancies per 100 couples who used only that method of birth control for a year. For example, "94% effectiveness" for oral contraceptives (the Pill) means that 6 of every 100 women will still become pregnant, on the average.

Other, less drastic methods of controlling fertility involve physical or chemical barriers to prevent sperm from entering the uterus and moving to the ovarian ducts. *Spermicidal foam* and *spermicidal jelly* are toxic to sperm. They are transferred from an applicator into the vagina just before intercourse. These products are not always reliable unless used with another device, such as a diaphragm or condom.

A *diaphragm* is a flexible, dome-shaped device. It is inserted into the vagina and positioned over the cervix before intercourse. It is relatively effective when fitted initially by a doctor, used with foam or jelly before each sexual contact, inserted correctly each time, and left in place for a prescribed length of time.

Condoms—thin, tight-fitting sheaths worn over the penis during intercourse—are about 85 to 93 percent reliable. Only those made of latex provide protection against sexually transmitted diseases (Section 34.19). However, condoms often tear and leak, at which time they become absolutely useless.

The *birth control pill* is an oral contraceptive made of synthetic estrogens and progestins (progesterone-like hormones). Taken daily except for the last five days of a menstrual cycle, it suppresses oocyte maturation and ovulation. Often it corrects erratic menstrual cycles and reduces cramps, but in some women it causes nausea, weight gain, tissue swelling, and headaches. With more than 50 million women taking it, "the Pill" is the most-used fertility control method. It is 94 percent effective. Earlier formulations were linked to blood clots, high blood pressure, and maybe breast cancer. Lower doses of newer formulations might lessen the risk of breast and endometrial cancer. The possibility of a correlation between the Pill and breast cancer is still under study.

Progestin injections or implants inhibit ovulation. A *Depo-Provera* injection works for three months and is 96 percent effective. *Norplant* (six rods implanted under the skin) works for five years and is 95 percent effective. Both may cause sporadic, heavy bleeding, and doctors have some trouble surgically removing Norplant rods.

A pregnancy test doesn't register positive until after implantation. According to one view, a woman is not pregnant until that time. From this viewpoint, RU-486, the *morning-after pill*, intercepts pregnancy. It interferes with hormonal signals that control the events between ovulation and implantation. Three pills, taken within seventy-two hours after intercourse, block fertilization or prevent a blastocyst from burrowing into the uterine lining. RU-486 has side effects (nausea, vomiting, and breast tenderness, for the most part) but not all women experience them. However, RU-486 disrupts complex hormonal interactions and should only be taken under medical supervision. As is true of oral contraceptives, it may trigger elevated blood pressure, blood clots, or breast cancer in some women. At this writing, RU-486 is available in Europe. In the United States, its use is still a subject of controversy.

Whether and how human fertility is to be controlled is a volatile issue. It centers on reconciling our biological past with seriously divergent views about our cultural present.

SEXUALLY TRANSMITTED DISEASES

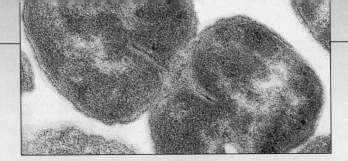

Figure 34.30 Bacterial agents of gonorrhea.

During his or her lifetime, at least one of every four people in the United States who engage in sexual intercourse is likely to become infected by pathogens that cause **sexually transmitted diseases** (STDs). Even now, STDs have reached epidemic proportions; over 56 million are already infected. Two-thirds of those infected are under age twenty-five. One-fourth are teenagers. Women and children are hardest hit. Ominously, antibiotic-resistant strains of bacteria are on the rise, and some viral diseases simply cannot be cured.

Urban poverty, prostitution, intravenous drug abuse, and sex-for-drugs are fanning the epidemic. Yet STDs also are rampant in high schools and colleges, where students too often think, "It can't happen to me." They reject the idea that no sex—abstinence—is the only safe sex. In one poll of high school students, two-thirds of the respondents said they don't use condoms. More than 40 percent of them have two or more sex partners.

The economics of this health problem are staggering. In 1993, the Centers for Disease Control related to us the annual cost of treating the most prevalent STDs: *Herpes*, $759 million; gonorrhea, $1 billion; chlamydial infection, $2.4 billion; and pelvic inflammatory disease, $4.2 billion. This does not include the accelerating cost of treating AIDS. In many developing countries, AIDS alone threatens to overwhelm health-care delivery systems and to unravel decades of economic progress.

The social consequences are sobering. Mothers bestow a chlamydial infection on one of every twenty newborns in the United States. They bestow type II *Herpes* virus on 1 in 10,000 newborns; one-half of the infected babies die, and one-fourth have severe neural defects. Each year 1 million women develop pelvic inflammatory disease, and 100,000 to 150,000 become infertile.

AIDS   Someone can get infected by HIV, the human immunodeficiency virus, and not even know it. But the infection marks the start of a titanic battle that the immune system almost certainly will lose (Section 28.11). At first there may be no outward symptoms. Five to ten years later, a set of chronic disorders develops. Collectively, these are evidence of *AIDS* (acquired immune deficiency syndrome). When the immune system finally gives up, the stage is set for opportunistic infections. Normally harmless, resident bacteria are the first to take advantage of the lowered resistance. Dangerous pathogens also take their toll. In time, the immune-compromised person simply is overwhelmed.

Most commonly, HIV spreads by vaginal, anal, and oral intercourse and by IV drug users. Most infections occur by the transfer of blood, semen, urine, or vaginal secretions between people. HIV enters the internal environment of a new host through cuts or abrasions on the penis, vagina, rectum, and maybe in membranes in the mouth.

Today there is no vaccine against HIV and no effective treatment for AIDS. If you get it, you die. *There is no cure.*

HIV wasn't identified until 1981, but it was present in some parts of Central Africa for at least several decades. In the 1970s and early 1980s, it spread to the United States and other developed countries. Most of those initially infected were male homosexuals. Now a large part of the heterosexual population is infected or at risk. Worldwide, as many as 13 million may now be HIV infected. By early 1992, AIDS was the second leading cause of death among men and the fifth leading cause among women between ages 25 and 44.

Free or low-cost, confidential testing for HIV exposure is available at public health facilities and at many doctor's offices. People who are worried should know there may be a time lag between their first exposure and the first test to come out positive. It takes a few weeks to six months or more before detectable amounts of antibodies will form in response to infection. (The presence of antibodies indicates exposure to the virus.) Anyone who tests positive should be considered capable of spreading the virus.

Public education programs attempt to stop the spread of HIV. Most health-care workers advocate safe sex, yet there is confusion about what "safe" means. Many advocate use of high-quality latex condoms *and* a spermicide containing nonoxynol-9 to help block transmission of the virus. Even then, there is a slight risk. No one should participate in open-mouthed kissing with someone who tests positive for HIV. Caressing isn't risky *if* there are no lesions or cuts where body fluids that harbor the virus can enter the body. Pronounced lesions caused by other sexually transmitted diseases may increase susceptibility to HIV infection.

In sum, AIDS reached epidemic proportions mainly for three reasons. First, it took a while to discover that the virus travels in semen, blood, and vaginal fluid and that *behavioral* controls can limit its spread. Second, it took time to develop ways to test symptom-free carriers, who infect others. Third, many still don't realize that the medical, economic, and social consequences affect everyone.

Genital Warts   We know of more than sixty types of the human papillomaviruses (HPV). A few cause benign, bumplike growths called *genital warts*. HPV infection of the genitals and anus has become the most prevalent STD in the United States. Type 16 HPV does not usually cause obvious warts but may be linked to precancerous sores and cancers of the cervix, vagina, vulva, penis, and anus. In one Seattle study, 22 percent of female college students who were examined tested positive for the virus.

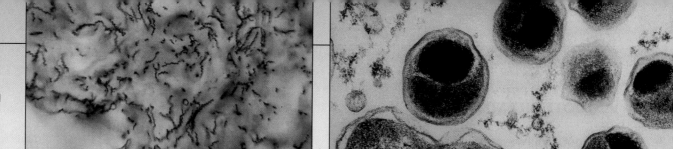

Figure 34.31 Bacterial agents of (**a**) syphilis and (**b**) chlamydia. **a** **b**

Gonorrhea During intercourse, *Neisseria gonorrhoeae* (Figure 34.30) can enter the body at mucous membranes of the urethra, cervix, and anal canal. This bacterium causes *gonorrhea*. An infected female might only notice a slight vaginal discharge or burning sensation while urinating. If the infection spreads to her oviducts, it may induce severe cramps, fever, vomiting, and scar tissue formation, which may cause sterility. Males have more obvious symptoms. Within a week after infection, the penis discharges yellow pus. Urinating is more frequent and may be painful.

This STD is rampant even though prompt treatment quickly cures it. Why? Women experience no troubling symptoms during early stages. Also, infection doesn't confer immunity to the bacterium, perhaps because there are sixteen or more different strains of it. Contrary to common belief, someone can get gonorrhea over and over again. Also, oral contraceptives, used so widely, encourage infection by altering vaginal pH. Populations of resident bacteria decline—and *N. gonorrhoeae* is free to move in.

Syphilis *Syphilis*, a dangerous STD, is caused by a spirochete, *Treponema pallidum* (Figure 34.31a). Sex with an infected partner puts the motile, spiral bacterium on the surface of genitals or into the cervix, vagina, or oral cavity. It can enter the body through tiny epidermal cuts. One to eight weeks later, new treponemes are twisting about in a flattened, painless chancre (local ulcer). This first chancre is a symptom of the primary stage of syphilis. By then, treponemes are in the blood. The treponemes can cross the placenta of an infected, pregnant woman. The result will be a miscarriage, stillbirth, or syphilitic newborn.

Usually a chancre heals, but in mucous membranes, joints, bones, eyes, spinal cord, and brain, treponemes are multiplying. More chancres and a skin rash develop in this highly infectious, secondary stage of syphilis. Symptoms subside; the immune system successfully fights the disease in about 25 percent of the cases. Another 25 percent remain infected but symptom-free. In the rest, minor to significant lesions and scars appear in the skin, liver, bones, aorta, and other internal organs. Few treponemes are present in this tertiary stage. However, the host's immune system is hypersensitive to them. Chronic immune reactions may severely damage the brain and spinal cord and lead to general paralysis.

Probably because the symptoms are so alarming, more people seek early treatment for syphilis than for gonorrhea. Later stages require prolonged treatment.

Pelvic Inflammatory Disease *Pelvic inflammatory disease* (PID) is one of the most serious complications of gonorrhea, chlamydial infections, and other STDs. It also can arise when normal bacterial residents of the vagina ascend to the pelvic region. Most often, the uterus, oviducts, and ovaries are affected. There is bleeding and vaginal discharge. Pain in the lower abdomen may be so severe, infected women often think they are having an attack of acute appendicitis. The oviducts may become scarred, and so invite abnormal pregnancies as well as sterility.

Genital Herpes About 25 million Americans have *genital herpes*, caused by the Type II *Herpes simplex* virus. Infection requires direct contact with active *Herpes* viruses or sores that contain them. Mucous membranes of the mouth or genitals are highly susceptible to invasion. Symptoms often are mild or absent. Among infected women, small, painful blisters may appear on the vulva, cervix, urethra, or anal tissues. Among men, blisters form on the penis and anal tissues. Within three weeks, the virus enters latency, and the sores crust over and heal.

The virus is reactivated sporadically. Each time, it causes new, painful sores at or near the original site of infection. Sexual intercourse, menstruation, emotional stress, or other infections can trigger *Herpes* infections. Acyclovir, an antiviral drug, decreases the healing time and often decreases the pain and viral shedding.

Chlamydial Infection *Chlamydia trachomatis* (Figure 34.31b), a parasitic bacterium, spends part of its life cycle in cells of the genital and urinary tracts. It causes several diseases, including *NGU* (chlamydial nongonococcal urethritis). NGU is far more common than syphilis or gonorrhea. *N. gonorrhoeae* and *C. trachomatis* often are transmitted at the same time. Prompt doses of penicillin cure the gonorrhea—but not NGU, which requires both tetracycline and sulfonamide.

NGU leads to inflammation of the cervix and, in both sexes, the urethra. Infected people may notice a burning sensation while urinating. But when they are symptom-free, they don't seek treatment and complications develop. In males, the prostate becomes swollen and inflamed. In females, the infection may spread into the uterus and oviducts to cause pelvic inflammatory disease.

Symptom-free infected individuals are unwittingly spreading destructive chlamydial infections through all ethnic groups, among poor and affluent alike.

MEDICAL INTERVENTIONS THAT PROMOTE OR END PREGNANCY

Because of sterility or infertility, about 15 percent of all couples in the United States cannot conceive. For example, hormonal imbalances in the female may prevent ovulation, or the male's sperm count may be so low that fertilization is next to impossible without medical intervention.

In vitro fertilization—external conception—may be possible if the couple can produce normal sperm and oocytes. First the woman receives injections of a hormone that prepares her ovaries for ovulation. Afterward, the preovulatory oocyte is removed from her with a suction device. Meanwhile, sperm from the male are placed in a solution that simulates fluid in oviducts. A few hours after the sperm and suctioned oocyte meet in a petri dish, fertilization may result. Twelve hours later, the zygote is transferred to a solution that will support its development. Two to four days after that, the blastocyst is transferred to the female's uterus.

Each attempt at in vitro fertilization costs about 8,000 dollars, and most attempts aren't successful. In 1994, each "test-tube" baby cost the nation's health-care system about 60,000 to 100,000 dollars, on average. The childless couple may believe no cost is too great. But in an era of increased population growth and shrinking medical coverage, is the cost too great for society to bear? Court battles are being waged over this issue.

At the other extreme is **abortion**, the dislodging and removal of the blastocyst, embryo, or fetus from the uterus. At one time in the United States, abortions were forbidden by law unless the pregnancy endangered the mother's life. Later the Supreme Court ruled that the government does not have the right to forbid abortions during the early stages of pregnancy (typically up to five months). Before this ruling, there were dangerous, traumatic, and often fatal attempts to abort embryos, either by pregnant women themselves or by quacks.

From a clinical standpoint, vacuum suctioning and other methods make abortion painless for the woman, rapid, and free of complications when performed during the first trimester. Abortions performed in the second and third trimesters are extremely controversial unless the mother's life is threatened. Even so, for both medical and humanitarian reasons, the majority of people in this country generally agree that the preferred route to birth control is not through abortion. Rather, it is through sexually responsible behavior that prevents unwanted pregnancy from happening in the first place.

This biology textbook can't offer you the "right" answer to a moral question, for all the reasons given in Section 1.8. It *can* provide you with a serious, detailed description of how a new human individual develops. Your choice of the "right" answer to the question of the morality of abortion will be just that—your choice—and one that can be based on objective insights into the nature of life.

SUMMARY

General Principles of Reproduction and Development

1. Sexual reproduction is the dominant reproductive mode in the animal kingdom. Specialized reproductive structures and forms of behavior assist fertilization and support the offspring.

2. Among animals, embryonic development commonly proceeds through six stages:

   a. Gamete formation, when oocytes (immature eggs) and sperm form and develop in reproductive organs. Molecular and structural components are stockpiled and become localized in different parts of the oocyte.

   b. Fertilization, from the time a sperm penetrates an egg to the fusion of sperm and egg nuclei that results in a zygote (fertilized egg).

   c. Cleavage, when a zygote undergoes mitotic cell divisions that form the early multicelled embryo. Most genes are inactive at this time. The fate of cell lineages is established in part by their position in the developing embryo. It also is established in part by cytoplasmic localization (the allocation of different components of the egg cytoplasm to different daughter cells).

   d. Gastrulation, when the organizational framework of the whole animal is laid out. Endoderm, ectoderm, and (in most species) mesoderm form. All the tissues of the adult body will develop from these primary tissue layers (also called germ layers).

   e. The onset of organ formation. The different organs start developing by a tightly orchestrated program of cell differentiation and morphogenesis.

   f. Growth and tissue specialization, when organs enlarge overall and acquire their specialized chemical and physical properties. The maturation of tissues and organs continues into post-embryonic stages.

3. In cell differentiation, a cell selectively uses certain genes and synthesizes proteins not found in other cell types. Subpopulations of specialized cells with structural, biochemical, and functional differences result.

4. In morphogenesis, the size, shape, and proportion of tissues and organs change and become organized in prescribed, orderly patterns. Pattern formation involves cell divisions, migrations, and changes in shape; it also involves localized growth and controlled cell death.

5. Cytoplasmic localization and then intricate chemical and physical interactions among embryonic cells are the foundations for cell differentiation and morphogenesis.

   a. In embryonic induction, the developmental fate of embryonic tissues changes following exposure to chemical signals released from an adjacent tissue.

   b. Signals activate or suppress blocks of genes that are the foundation for formation of specific structures and organs. Morphogens are examples of signals. They act on homeobox gene complexes that govern formation of a sequence of body parts along the body's long axis.

Human Reproduction and Development

1. Humans have a pair of primary reproductive organs (sperm-producing testes in males and egg-producing ovaries in females), accessory ducts, and glands. Testes and ovaries also produce sex hormones that influence reproductive functions and secondary traits.

   a. Sperm form in response to LH, FSH, and testosterone secretions that are part of feedback loops from the testes to the hypothalamus and anterior pituitary.

   b. Oocytes mature and are released, and the lining of the uterus (endometrium) undergoes cyclic changes in response to secretions of FSH, LH, estrogens, and progesterone. The secretions are part of feedback loops from the ovaries to the hypothalamus and anterior pituitary.

2. These events occur during a menstrual cycle:

   a. A follicle (an oocyte surrounded by a cell layer) matures in an ovary. The endometrium starts to rebuild. (It breaks down each cycle if pregnancy does not occur.)

   b. A midcycle surge of LH triggers ovulation (the release of a secondary oocyte from the ovary).

   c. A corpus luteum forms from the remnants of the follicle. It secretes progesterone and some estrogen that prime the endometrium for fertilization. If fertilization occurs, the corpus luteum is maintained.

3. After fertilization, cleavage produces a blastocyst that becomes implanted in the endometrium. Three primary tissue layers form that will give rise to all organs. They are ectoderm, endoderm, and mesoderm.

4. Four extraembryonic membranes form as embryos of humans (and other vertebrates) develop:

   a. The amnion becomes a fluid-filled sac around the embryo, which it protects from mechanical shocks, abrupt temperature changes, and drying out.

   b. The yolk sac stores nutritive yolk in most shelled eggs, but in humans, part becomes a major site of blood formation and some of its cells give rise to germ cells. (Eventually germ cells give rise to sperm and eggs.)

   c. The chorion is protective membrane around the embryo and the other membranes. It also becomes a major component of the placenta.

   d. In humans, the allantois functions in early blood formation and in formation of the urinary bladder.

5. The placenta, a blood-engorged organ, is composed of endometrium and extraembryonic membranes. It allows embryonic blood vessels to develop independently of the mother's, even while allowing oxygen, nutrients, and wastes to diffuse between them.

6. Whether and how to control human sexuality and fertility are major ethical issues of our time.

Review Questions

1. Define and describe the key events that unfold during gamete formation, fertilization, cleavage, gastrulation, and the onset of organ formation. *593, 596–599*

2. Define cell differentiation, morphogenesis, and pattern formation. Describe processes on which they are based. *598–599*

3. Label the component parts of the human male and the human female reproductive systems and state their functions: *602–606*

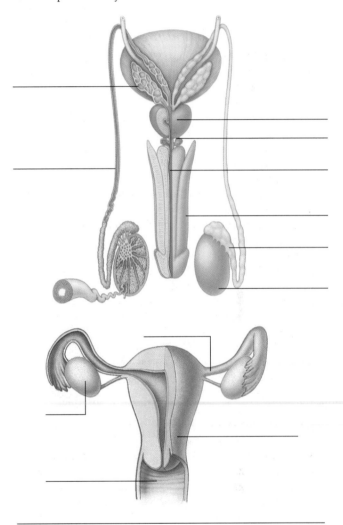

Self-Quiz *(Answers in Appendix IV)*

1. Sexual reproduction among animals _____ .
   a. is biologically costly
   b. is diverse in its details
   c. has evolutionary advantages
   d. all of the above

2. At cleavage, daughter cells are allocated different components of the fertilized egg. This is called _____ .
   a. cytoplasmic localization
   b. embryonic induction
   c. cell differentiation
   d. morphogenesis

3. The primary tissue layers first appear in the _____ .
   a. egg cortex
   b. blastula
   c. gastrula
   d. primary organs

4. During development, the formation of subpopulations of different cell types is the outcome of _____ .
   a. selective gene expression
   b. cell differentiation
   c. pattern formation
   d. morphogenesis
   e. a and b are correct
   f. metamorphosis

5. The specialized tissues and body parts that form in an embryo become positioned relative to one another in orderly sequence. This feat of self-organization is called _____ .
   a. gastrulation   c. morphogenesis
   b. pattern formation   d. metamorphosis

6. The _____ and the pituitary gland control secretion of sex hormones from the testes and the ovaries.

7. _____ secretions govern reproductive function in males.
   a. FSH   c. Testosterone   e. both b and c
   b. LH   d. both a and b   f. all are correct

8. During a menstrual cycle, a midcycle surge of _____ triggers ovulation.
   a. estrogen   c. LH
   b. progesterone   d. FHS

9. During a menstrual cycle, _____ and _____ prime the uterus for pregnancy.
   a. FSH; LH   c. estrogens; progesterone
   b. FSH; testosterone   d. estrogens only

10. In implantation, a _____ burrows into the endometrium.
    a. zygote   c. blastocyst
    b. gastrula   d. morula

11. The _____ , a fluid-filled sac, surrounds and protects the embryo from mechanical shocks and keeps it from drying out.
    a. yolk sac   c. amnion
    b. allantois   d. chorion

12. At full term, a placenta _____ .
    a. is composed of extraembryonic membranes
    b. directly connects maternal and fetal blood vessels
    c. keeps maternal and fetal blood vessels separated

13. (A) _____ form(s) in all vertebrate embryos.
    a. neural plate   c. pharyngeal arches   e. a through c
    b. somites   d. primitive streak   f. a through d

14. Distinctly human features emerge in the embryo by the end of the _____ week following fertilization.
    a. second   c. fourth   e. eighth
    b. third   d. fifth   f. sixteenth

15. Match the term with the most suitable description.
    ____ seminiferous tubule   a. glandular structure formed from follicle remnants
    ____ allantois   b. most bones, skeletal muscles of head and trunk form from them
    ____ corpus luteum   c. helps form blood, urinary bladder in humans
    ____ somites
    ____ yolk sac   d. blood cell formation site, source of germ cells
         e. sperm form here

Critical Thinking

1. Steve and Alina experimentally divide an amphibian egg so the gray crescent is wholly within one of the two cells formed. They separate the two cells, and only the one with the gray crescent forms an embryo with a long axis, notochord, nerve cord, and back musculature. The other cells form a shapeless mass of immature gut and blood cells. Explain these outcomes.

2. At the first cleavage, a human zygote spontaneously splits into two separate cells. Both continue on the prescribed developmental program. The result is *identical twins*—two normal, genetically

identical individuals. *Nonidentical twins* form when two different eggs are fertilized at the same time by two different sperm. On the basis of this information, explain why nonidentical twins show considerable genetic variability and identical twins do not.

3. Infection by the rubella virus appears to inhibit mitosis. Serious birth defects result when a pregnant woman is infected during the first trimester, but not later. Review the developmental events that unfold during pregnancy and explain why this is so.

4. Imagine you are an obstetrician advising a woman who has just learned she is pregnant. What instructions would you provide concerning her diet and behavior during the pregnancy?

5. In the United States, teenage pregnancies and STD infections are rampant. Suppose the office of the Surgeons General asks you to take part in a task force that will recommend practices that might reduce the incidence of both. What practices might meet with the greatest success? What kind of enthusiasm or resistance might they provoke among the teenagers in your community? Among adults? Explain why.

Selected Key Terms

abortion 624	growth and tissue specialization 593
aging 619	HCG 611
allantois 611	homeobox genes 599
amnion 611	implantation 610
asexual reproduction 592	in vitro fertilization 624
blastocyst 597	LH 602
blastula 597	menstrual cycle 604
cell differentiation 598	mesoderm 593
chorion 611	morphogen 599
cleavage 593	morphogenesis 598
corpus luteum 607	oocyte 596
cytoplasmic localization 597	organ formation 593
ectoderm 593	ovary 604
embryo 592	ovum 609
embryonic induction 599	pattern formation 599
embryonic period 610	placenta 613
endoderm 593	progesterone 604
endometrium 604	sexual reproduction 592
estrogen 604	sexually transmitted disease 622
fertilization 593	sperm 596
fetal period 610	testis (testes) 600
FSH 602	testosterone 602
gamete formation 593	yolk 592
gastrulation 593	yolk sac 611
gray crescent 596	

Readings

Browder, L., C. Erickson, and W. Jeffrey. 1991. *Developmental Biology*. Third edition. Philadelphia: Saunders.

Gilbert, S. 1994. *Developmental Biology*. Fourth edition. Sunderland, Massachusetts: Sinauer.

Larsen, W. 1993. *Human Embryology*. New York: Churchill Livingston. Up-to-date and informative, but its clinical descriptions are not for the faint of heart. Paperback.

McGinnis, W., and M. Kuziora. February 1994. "The Molecular Architects of Body Design." *Scientific American* 270(2): 58–66.

FACING PAGE: *Two organisms—a fox in the shadows cast by a snow-dusted spruce tree. What are the nature and consequences of their interactions with each other, with other organisms, and with their environment? By the end of this last unit, you possibly will see worlds within worlds in such photographs.*

35 POPULATION ECOLOGY

Tales of Nightmare Numbers

Across from Sausalito, California, the steep flanks of Angel Island rise from the waters of San Francisco Bay (Figure 35.1*a*). The island, set aside as a game reserve, escaped urban development. But it did not escape the descendants of a few deer that well-meaning nature lovers shipped over in the early 1900s. With no natural predators to keep them in check, the few deer became many—far too many for the limited food supply of their isolated habitat. Yet the island attracted a steady supply of picnickers from the mainland. They felt sorry for the malnourished animals and made sure to load the picnic baskets with extra food for them.

In fact, they imported so much food that scrawny deer kept on living and reproducing. In time, the herd nibbled away the native grasses, the roots of which had helped slow soil erosion on the steep hillsides. Deer chewed off all the new leaves of seedlings; they killed small trees by stripping away the bark and its phloem. The herd was destroying the island habitat. Out of desperation, game managers proposed using a few skilled hunters to thin the herd, but they were strongly

denounced as being cruel. They proposed importing a few coyotes to thin the herd naturally. Animal rights advocates opposed this solution, also.

As a compromise, about 200 of the 300+ deer were captured, loaded onto a boat, and shipped to suitable mainland habitats. A number of them received collars with radio transmitters so that they could be tracked after their release. In less than sixty days, dogs, coyotes, bobcats, hunters, and speeding cars and trucks had killed off most of them. In the end, relocating each surviving deer had cost taxpayers almost 3,000 dollars. The State of California refused to do it again. And no one else, anywhere, volunteered to pick up future tabs.

It wasn't difficult to define the boundaries of Angel Island or track its inhabitants, so it's easy to draw a lesson from this tale: *A population's growth depends on the resources of its environment. And attempts to "beat nature" by altering the sometimes cruel outcome of limited resources can only postpone the inevitable.* Does this same lesson apply to other populations, in other places? Yes, but other considerations may obscure the essence of it.

Figure 35.1 (**a**) Angel Island, which turned out to be a laboratory for studying population growth. (**b**) Bathers crowding the banks of the Ganges River in India—a tiny sampling of a human population that has now surpassed 5.7 billion. In this chapter we turn to the principles that govern the growth and sustainability of all populations.

a

b

Consider this next tale. When 1995 drew to a close, there were more than 5.7 billion people on Earth. About 2 billion already live in poverty. Each year, 40 million more join their ranks. Next to China, India is the most populous country, with 931 million inhabitants (Figure 35.1b). By 2025, it may reach 1.4 billion. Forty percent of the people live in rat-infested shantytowns, without adequate food or water. They are forced to wash clothes and dishes in open sewers. Croplands available to feed them shrink by 365 acres a day, on the average. Why? Irrigated soils become too salty when they drain poorly and there isn't enough water to flush away the salts.

Can wealthier, less densely populated nations help? After all, they use most of the world's resources. Maybe they should learn to get by more efficiently, on less. For example, people could restrict their diet to cereal grains and water, give up their private cars, living quarters, air conditioning, televisions, and dishwashers, stop taking vacations, stop laundering so much, close the malls, restaurants, and theaters at night, and so on.

Maybe wealthier nations also should donate more surplus food than they already do to less fortunate ones. Then again, would huge donations help, or would they encourage dependency and spur more increases in population size? And what if surpluses run out?

It is a monumental dilemma. At one extreme, the redistribution of resources on a global scale would allow the greatest number of people to survive, but at the lowest comfort level. At the other extreme, foreign aid rationed only to nations that restrict population growth would allow fewer individuals to be born, but their lives would have greater quality.

Currently, the United States foreign aid program is based on two premises: (1) that individuals of every nation have an irrevocable right to bear children, even if unrestricted reproduction ruins the environment that must sustain them; and (2) that because human life is precious above all else, the wealthiest nations have an absolute moral obligation to save lives everywhere.

Regardless of the positions that nations take on this issue, ultimately they must come to terms with the fact that *certain principles govern the growth and sustainability of populations over time*. These principles are the bedrock of **ecology**—the systematic study of how organisms interact with one another and with their physical and chemical environment. Ecological interactions start within and between populations; and they extend on up through communities, ecosystems, and the biosphere. They are the focus of this last unit of the book.

KEY CONCEPTS

1. Certain ecological principles govern the growth and sustainability of all populations, including our own.

2. A population may show a pattern of an exponential growth, by which it increases in size by ever larger amounts over specified intervals of time.

3. A population may show a pattern of logistic growth. By this pattern, the density of a population is low at a given point in time. Then the population rapidly increases in numbers, and finally it levels off in size as resource scarcity limits its further increase or triggers a decline in numbers.

4. Some populations show large fluctuations in numbers that cannot be explained by a single growth model.

5. All populations face limits to growth. The reason is no environment can indefinitely sustain a continuously increasing number of individuals. Competition for resources, disease, predation, and other factors act as controls over population growth. The controls vary in their relative effects on populations of different species, and they vary over time.

35.1 CHARACTERISTICS OF POPULATIONS

Let's start with the relationships that influence the size, structure, and distribution of populations. Later, we will apply the basic principles of population growth to the past, present, and future of the human species.

A **population**, recall, is a group of individuals of the same species occupying a given area. The type of place where the species normally lives is called its **habitat**. We define **population size** as the number of individuals in the gene pool, and **population density** as the number of individuals in a specified area or volume, such as the number of guppies in a liter of water in a stream.

Population distribution is the general pattern in which individuals are dispersed in the habitat. Uniform or random dispersion is rare. Most commonly, you will find individuals clumped together at specific sites:

nearly uniform   random   clumped

Why is clumping the most common pattern? A species is adapted to a limited set of conditions, which usually are patchy through a habitat. (Some patches offer more shade, better hideouts, and so on.) Also, social groups are common. And offspring of many species are unable to disperse over large distances. For example, sponge larvae swim weakly—and not far from a parent sponge.

A population also has an **age structure**—the number of individuals in each of several to many age categories. For example, we may divide all of the individuals into *pre-reproductive*, *reproductive*, and *post-reproductive* ages. Those in the first category have a capacity to produce offspring when they mature. Together with actually and potentially reproducing members in the next category, we count them as part of the **reproductive base**.

Bear in mind, your perception of a population may be influenced by how wide a view you take. Look at a few square meters, and acorns under an oak tree seem randomly spaced. Look at a forest of mixed hardwood trees, and you see they are clustered near parent trees. Also, population characteristics can vary over time. For example, few places yield abundant resources all year long, and many animals move from one habitat into a neighboring habitat with the changing seasons.

Each population has a characteristic size, density, distribution pattern, and age structure. The scale of an investigation and the time of year it is undertaken may influence perceptions of population distribution.

35.2 POPULATION SIZE AND EXPONENTIAL GROWTH

Over a specified time interval, any change in the size of a population depends on how many individuals enter and leave. *Births* and *immigration* (the arrival of new residents) increase population size. By contrast, *deaths* and *emigration* (individuals move out) decrease the size.

For our purposes, assume immigration is balancing emigration over time, so that we can ignore the effects of both on population size. By doing so, we can define **zero population growth** as a circumstance in which the number of births and number of deaths are in balance. Population size would be stabilized for that interval of time, with no overall increase or decrease.

Births, deaths, and other variables that may affect population size can be measured in terms of rates per individual—that is, as "per capita" rates. Think of 2,000 mice living in a cornfield. The females produce a litter twenty or so days after fertilization, nurse offspring for a month or so, then get pregnant again. If they bore 1,000 mice in a month's time, the birth rate would be $1,000/2,000 = 0.5$ per mouse per month. If 200 of the 2,000 died during the same interval, the death rate would be $200/2,000 = 0.1$ per mouse per month.

If we assume the birth rate and death rate remain constant, we can combine both into a single variable—the **net reproduction per individual per unit time**, or *r* for short. For the mouse population in the cornfield, *r* is $0.5 - 0.1 = 0.4$ per mouse per month. This example gives us a way to represent population growth:

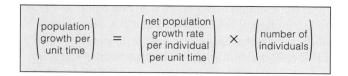

$$\begin{pmatrix} \text{population} \\ \text{growth per} \\ \text{unit time} \end{pmatrix} = \begin{pmatrix} \text{net population} \\ \text{growth rate} \\ \text{per individual} \\ \text{per unit time} \end{pmatrix} \times \begin{pmatrix} \text{number of} \\ \text{individuals} \end{pmatrix}$$

or, more simply, $G = rN$.

As the next month begins, 2,800 mice are scurrying about. With the net increase of 800 fertile critters, the reproductive base has expanded. If *r* does not change, population size will expand this month also, by a net increase of $0.4 \times 2,800 = 1,120$ mice. The population is now 3,920. In the months ahead, it just so happens that *r* remains constant, and a growth pattern emerges. As Figure 35.2a shows, *in less than two years from the time we first started counting, the number of mice running around in the cornfield increased from 2,000 to more than a million!*

Plot these monthly increases against time, as in Figure 35.2b, and you end up with a graph line in the shape of a "J." When population growth over increments of time plots out as a J-shaped curve, you know you are tracking **exponential growth**.

"Exponential" refers to a relationship in which one variable increases much faster than another in a specific

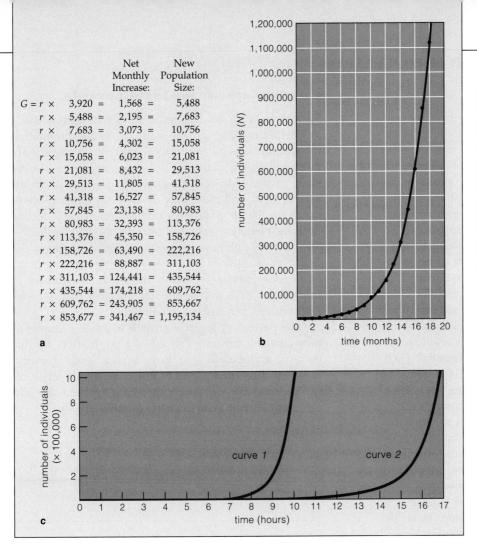

		Net Monthly Increase:		New Population Size:
$G = r \times$	3,920	=	1,568	=	5,488
$r \times$	5,488	=	2,195	=	7,683
$r \times$	7,683	=	3,073	=	10,756
$r \times$	10,756	=	4,302	=	15,058
$r \times$	15,058	=	6,023	=	21,081
$r \times$	21,081	=	8,432	=	29,513
$r \times$	29,513	=	11,805	=	41,318
$r \times$	41,318	=	16,527	=	57,845
$r \times$	57,845	=	23,138	=	80,983
$r \times$	80,983	=	32,393	=	113,376
$r \times$	113,376	=	45,350	=	158,726
$r \times$	158,726	=	63,490	=	222,216
$r \times$	222,216	=	88,887	=	311,103
$r \times$	311,103	=	124,441	=	435,544
$r \times$	435,544	=	174,218	=	609,762
$r \times$	609,762	=	243,905	=	853,667
$r \times$	853,677	=	341,467	=	1,195,134

a

b

Figure 35.2 (**a**) Data showing the net monthly increases in a population of field mice in a cornfield. The numbers from start to finish show a pattern that is typical of exponential growth. (**b**) A graph of the data yields a J-shaped growth curve. (**c**) This graph shows that deaths can slow the rate of increase but cannot in themselves stop exponential growth. Here, growth curve *1* represents a population of bacterial cells that reproduced every half hour. Growth curve *2* represents a different population in which cells divided every half hour, but 25 percent died between divisions.

c

mathematical way. In this case, a population's size is a variable that depends on how many individuals make up the reproductive base in successive increments of time. *The larger the reproductive base, the greater will be the expansion in population size during each specified interval.*

Let's consider other aspects of exponential growth by supplying a lone bacterium in a culture flask with all the nutrients required for growth. Thirty minutes later, the one cell divides in two. Thirty minutes pass, the two cells divide, and so on every thirty minutes. Assuming no cells die between divisions, the population size will double in each interval—from 1 to 2, then 4, 8, 16, 32, and so on. The length of time it takes for a population to double in size is its **doubling time**.

The larger this population gets, the more cells there are to divide. After 9-1/2 hours (nineteen doublings), it has more than 500,000 bacterial cells. After 10 hours (twenty doublings), it has more than 1 million. Plot the doublings in size against time, and you get the J-shaped curve that is typical of unrestricted, exponential growth (curve 1 in Figure 35.2c).

To examine the effect of deaths on the growth rate, let's start over with one bacterium. Assume 25 percent of the cells die in each thirty-minute interval. At this rate, it takes about 17 hours (instead of 10 hours) for the population size to reach a million. *But the deaths changed only the time scale for population growth.* As you can see from Figure 35.2c, you still end up with a J-shaped curve (curve 2).

Finally, imagine that a population lives in a place where conditions are ideal. Every one of its individuals has adequate shelter, food, and other vital resources. No predators, pathogens, or pollutants lurk in the habitat. The population may well display its **biotic potential**, which is the maximum rate of increase per individual under ideal conditions.

Each species has a characteristic maximum rate of increase. For many bacteria, it is 100 percent every 30 minutes or so. For humans and other large mammals, it is between 2 and 5 percent per year. The *actual* rate depends on the age at which each generation starts to reproduce, how often each individual reproduces, and how many offspring are produced. Now think about this. A human female is biologically capable of bearing twenty or more children, yet in each generation, many females do not reproduce at all. The human population has not been displaying its biotic potential. Nevertheless, since the mid-eighteenth century, its growth has been exponential, for reasons that will soon be apparent.

With exponential growth, the size of a population expands by ever increasing increments during successive time intervals as the reproductive base gets larger.

A plot of population size against time has a characteristic J-shaped curve if the population is growing exponentially.

For as long as the per capita birth rate remains even slightly above the per capita death rate, a population will grow exponentially.

Limiting Factors

Most of the time, environmental circumstances prevent any population from fulfilling its biotic potential. That is why sea stars, the females of which can produce 2,500,000 eggs each year, do not fill up the oceans with sea stars. That also is why humans will never fill up the planet, even with our capacity for exponential growth.

In natural environments, complex interactions exist within and between populations of different species, so it is not easy to identify the factors working to limit population growth. To get a sense of what some of the factors might be, start again with a lone bacterium in a culture flask, where you can control the variables. First you enrich the culture medium with glucose and other nutrients required for bacterial growth. Then you allow bacterial cells to reproduce for many generations. At first the growth pattern appears to be exponential. Then growth slows, and after that, population size remains rather stable. But after the stable period, the population size starts to decline rapidly—and all the bacteria die.

What happened? As the population expanded by ever increasing amounts, the growing number of cells used more and more nutrients. When the nutrients dwindled, cell divisions decreased. When the supply was exhausted, the bacterial cells starved to death.

Any essential resource that is in short supply is a **limiting factor** on population growth. Food, minerals of certain types, refuge from predators, living space, and a pollution-free environment are examples. The number of such factors can be huge, and their effects can vary. Even so, one factor alone is often enough to put the brakes on population growth at any given time.

Suppose you kept on freshening the supply of all required nutrients for the growing bacterial culture. After an episode of exponential growth, the population would still crash. Like all organisms, bacteria produce metabolic wastes. The wastes produced by such a huge population of bacterial cells would be so great, they would drastically alter conditions in the culture. By its own activities, the population would end up polluting the surrounding environment—and put a stop to its exponential growth.

Carrying Capacity and Logistic Growth

Now visualize a small population in which individuals are dispersed through the habitat. As the population increases in size, more and more individuals must share nutrients, living quarters, and other resources. As the share available to each one shrinks, fewer individuals may be born, and more may die by starvation or lack of nutrients. The population's rate of growth will decline until the births are balanced or outnumbered by deaths. The final population size will largely depend on the *sustainable* supply of resources. **Carrying capacity** is the name ecologists have given to the maximum number of individuals of a population (or species) that a given environment can sustain indefinitely.

The pattern of **logistic growth** is a fine example of how carrying capacity can affect a population. By this pattern, a population at low density starts growing slowly in size, then grows rapidly, and finally levels off in size once the carrying capacity is reached. We can represent logistic growth in this simplified way:

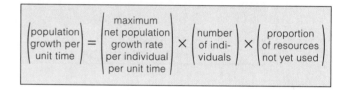

or $G = r_{max} N [(K - N)/K]$. The K designates carrying capacity. The term inside the parentheses is close to 1 when a population is small. It approaches zero when population size is close to carrying capacity.

A plot of logistic growth gives an S-shaped curve (Figure 35.3). Such curves are only an approximation of what goes on in nature. For example, a population that grows too rapidly can overshoot the carrying capacity. The death rate skyrockets and the birth rate plummets. These two responses drive the number of individuals down to the carrying capacity—or lower (Figure 35.4).

Figure 35.3 Idealized S-shaped curve characteristic of logistic growth. After a rapid growth phase (time *b* to time *c*), growth slows and the curve flattens out as the carrying capacity is reached (to time *d*). Variations can pop up in S-shaped growth curves, as when changes in the environment bring about a decreased carrying capacity (time *d* to *f*). This happened to the human population of Ireland before 1900, when late blight, a disease caused by a water mold, destroyed potato crops that were the mainstay of the diet (Section 18.9).

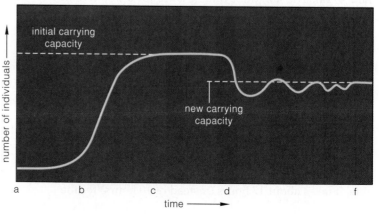

Figure 35.4 Carrying capacity and a reindeer herd. In 1910, four male and twenty-two female reindeer were introduced on St. Matthew Island in the Bering Sea. In less than thirty years, the herd increased to 2,000. Its members had to compete for dwindling vegetation, and overgrazing destroyed most of it. In 1950, the herd plummeted to eight members. The growth pattern reflects how population size overshot the carrying capacity, then crashed.

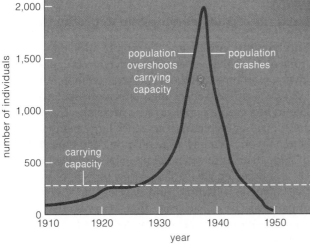

European cities where humans were crowded together, sanitary conditions were poor, and rats were abundant. By the time the epidemic subsided, urban populations in Europe had declined by 25 million.

Both diseases are still threats. For example, in 1994 in India, they raced through rat-infested cities where garbage and animal carcasses had piled up for months in the streets. Terrified residents fled by the thousands; some carried the pathogen with them as far away as London. Global panic ensued before concerted efforts to burn the garbage, poison the rats and fleas, and rapidly dispense antibiotics averted a pandemic.

Density-Dependent Controls

The logistic growth equation describes *density-dependent* population growth. When a population is not large, it may grow rapidly. When it finally bumps up against its carrying capacity, so to speak, it stops growing, grows very little, or even declines in size. However, carrying capacity is not the only factor that influences the size of a population. Other natural controls come into play that can put the number of individuals *below* the maximum sustainable level.

Consider how high density and overcrowding put individuals at greater risk of being killed. Under such conditions, predators, parasites, and pathogens interact much more intensely with the population and therefore bring about a decline in its density. Once population size declines, density-dependent interactions relax and the population may grow once again.

Bubonic plague and *pneumonic plague* are examples of density-dependent controls. A bacterium, *Yersinia pestis,* invades the blood and causes both of these dangerous diseases. *Y. pestis* persists in a large reservoir of rats, rabbits, and some other small mammals. Hungry fleas transmit it to new hosts. The bacterium multiplies in the flea gut and blocks digestion, the fleas attempt to feed more and more often, and so the disease spreads.

A devastating episode of bubonic plague occurred in the fourteenth century. The plague swept through

Density-Independent Factors

Sometimes events result in more deaths or fewer births regardless of population density. For example, every year, monarch butterflies migrate from Canada to spend the winter in forested mountains of Mexico—but logging has opened up stands of trees, which normally buffer temperatures. In late December 1995 the deforestation, in combination with a sudden freeze, killed millions of butterflies. They were *density-independent* factors.

Similarly, heavy applications of pesticides in your backyard may kill most insects, mice, cats, birds, and other animals—and they will do so regardless of how dense their populations are.

Resources in short supply put limits on population growth. Together, all of the limiting factors acting on a population dictate how many individuals can be sustained.

Carrying capacity **is the maximum number of individuals of a population that can be sustained indefinitely by the resources in a given environment. The number may rise or fall with changes in resource availability.**

A low-density population may increase slowly in size, go through a rapid growth phase, then level off in size once the carrying capacity is reached. This is a logistic growth pattern.

Density-dependent controls and density-independent factors bring about decreases in population size.

35.4 LIFE HISTORY PATTERNS

So far, we have been considering populations as if all of their individuals are identical during a given time span. For most species, however, different individuals are at different stages of development, so they are interacting in different ways with other organisms and with the environment. Let's look at a few examples of the kinds of environmental variables that influence age-specific, life history patterns.

Life Tables

Each species has a characteristic life span, although few individuals ever reach the maximum age possible. Why? Death looms more at some ages than at others. Also, individuals tend to reproduce or leave their population at certain ages, which vary from one species to the next.

Age-specific patterns in populations first intrigued life insurance and health insurance companies. They also interest ecologists, who typically track a **cohort**—a group of individuals from the time of birth until the last one dies. They also track the number of offspring born to individuals during each age interval. A **life table** lists the completed data on a population's age-specific death schedule. Such a schedule is often converted to a more cheery "survivorship" schedule, which lists the number of individuals that reach some specified age (x). The life table in Table 35.1 was constructed for the 1989 human population of the United States.

To ecologists, dividing a population into age classes and assigning birth rates and mortality risks to each class have practical applications. Unlike a crude census (head count), the data can serve as a basis for informed policy decisions on a variety of issues. Such concerns include pest management, conservation of endangered species, and social planning for human populations. For example, birth and death schedules for the spotted owl figured in the court decisions that halted logging in old-growth forests, the owl's habitat.

Patterns of Survivorship and Reproduction

Survivorship curves are the graph lines that emerge from plots of the age-specific survival of a cohort in a given environment. Three types are common in nature.

Type I curves reflect high survivorship until fairly late in life, then a large increase in deaths. Such curves are typical of elephants and other large mammals that bear only one or a few large offspring at a time and then provide them with extended parental care (Figure 35.5a). For example, female elephants give birth to only four or five calves in a lifetime, and they devote several years of parental care to each one. Type I curves are typical of human populations also, provided people have access to good health care services. Historically, and wherever health care is poor today, infant deaths cause a sharp drop at the start of the curve. Following the drop, the curve then levels off from childhood to early adulthood.

Type II curves reflect a fairly constant death rate at all ages. They are typical of organisms that are just as likely to be killed or to die of disease at any age. This is true of lizards, small mammals, and some songbirds (Figure 35.5b).

Type III curves typify populations where the death rate is high early in life. They are characteristic of species that produce many small offspring and then invest little, if any, parental care. Figure 35.5c shows how the curve plummets for sea stars. Although sea stars produce mind-boggling numbers of tiny larvae, these must rapidly eat, grow, and finish developing on their own without support, protection, or guidance from the parents. Corals and other animals quickly eat most of them. Plummeting survivorship curves are characteristic of many other kinds of marine invertebrates, most insects, and many fishes, plants, and fungi.

At one time, ecologists thought that selection processes favored *either* the early, rapid production of many small offspring *or* the late production of only a few large ones. These two patterns are now known to be extremes, at opposite ends of a range of possible life histories. Also, both patterns—

Age Interval (category for individuals between the two ages listed)	Survivorship (number alive at start of age interval, per 100,000 individuals)	Mortality (number dying during the age interval)	Life Expectancy (average lifetime remaining at start of age interval)	Reported Live Births for Total Population
0–1	100,000	896	75.3	
1–5	99,104	192	75.0	
5–10	98,912	117	71.1	
10–15	98,795	132	66.2	11,486
15–20	98,663	429	61.3	506,503
20–25	98,234	551	56.6	1,077,598
25–30	97,683	606	51.9	1,263,098
30–35	97,077	737	47.2	842,395
35–40	96,340	936	42.5	293,878
40–45	95,404	1,220	37.9	44,401
45–50	94,184	1,766	33.4	1,599
50–55	92,418	2,727	28.9	
55–60	89,691	4,334	24.7	
60–65	85,357	6,211	20.8	
65–70	79,146	8,477	17.2	
70–75	70,669	11,470	13.9	
75–80	59,199	14,598	10.9	
80–85	44,601	17,448	8.3	
85 +	27,153	27,153	6.2	Total: 4,040,958

Table 35.1 Life Table for the United States Human Population, 1989*

*Compiled by Marion Hansen, based on data from U.S. Bureau of the Census, *Statistical Abstract of the United States*, 1992 (edition 112).

Figure 35.5 Three generalized types of survivorship curves. Type I populations show high survivorship until some age, then high mortality. Type II populations show a fairly constant death rate; Type III populations show low survivorship early in life.

and intermediate ones—may be evident in different populations of the same species, as the next section makes clear.

Tracking a cohort (a group of individuals from birth until the last one dies) reveals patterns of reproduction, death, and migration that are typical of the populations of a species.

Survivorship curves can reveal differences in age-specific survival among species. In some cases, such differences exist even between different populations of the same species.

THE GUPPIES OF TRINIDAD

Several years ago on Trinidad, a Caribbean island, John Endler and David Reznick were netting guppies—small, live-bearing fishes that darted about in shallow mountain streams. In time the researchers identified environmental variables that affect the life history patterns of the guppies.

Male guppies are smaller than the females, with bright-colored scales, and they stop growing at maturity. The females are drab colored and continue to grow larger as they reproduce. In some streams, a small fish (killifish) preys successfully on immature guppies but not on larger adults. Elsewhere a larger fish (a pike-cichlid) preys on mature, larger guppies and tends not to hunt small ones.

As Reznick and Endler predicted from their knowledge of natural selection, predation has profoundly influenced the life history patterns of guppies. In streams where the pike-cichlids reign supreme, individual guppies mature faster and are smaller at maturity, compared to those in killifish streams. They also reproduce at a younger age, produce far more offspring, and do so more often.

Could other, unknown differences between the streams be influencing the life history patterns? To check this out, the researchers shipped groups of guppies from both kinds of streams back to the laboratory. They raised them for two generations under identical conditions, in the absence of predation. The populations showed the same differences as natural populations. Thus the differences appear to have a genetic basis and are subject to natural selection.

The researchers also raised guppies alone, with killifish, or with pike-cichlids through many generations. As predicted, guppies of populations subjected to killifish were larger at maturity. Those of populations subjected to pike-cichlids were smaller and matured earlier (Figure 35.6).

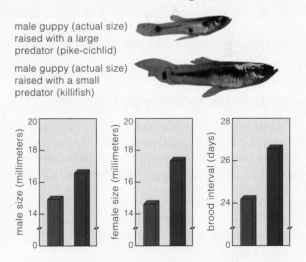

male guppy (actual size) raised with a large predator (pike-cichlid)

male guppy (actual size) raised with a small predator (killifish)

Figure 35.6 Experimental evidence of natural selection among populations of guppies (*Poecilia reticulata*) subjected to different predation pressures. Differences arise in body size and intervals between broods for guppies raised with pike-cichlids (*green* bars) or killifish (*red* bars). Pike-cichlids prey on larger guppies; killifish prey on smaller ones, and so select for the differences.

35.6 HUMAN POPULATION GROWTH

In 1995, the human population reached 5.7 billion. That year, rates of increase in different nations ranged from less than 1 percent to over 3 percent—which averages out to 1.55 percent annually. If the current rate holds, another 88.4 million will be added in 1996 (5.7 billion × 1.55% = 88,400,000). Visualize enough people to make another Los Angeles every two weeks or another New York City every four weeks, and you get the picture. Now think about this: The annual additions to the base population mean the *same* percent increase will result in a *larger* absolute increase each year into the future.

Our staggering population growth continues even though as many as 2 billion are already malnourished or starving, without clean drinking water and adequate shelter. It continues when 1.5 billion are already going without the benefits of health care delivery and sewage treatment facilities. And it continues mainly in regions that are already overcrowded. About 4.5 billion are now clumped together on 10 percent of the land, and about 3 billion crowd together within 300 miles of the seas.

Suppose it were possible, by monumental efforts, to double the food supply to keep pace with growth. We would do little more than maintain marginal living conditions for most. Annual deaths from starvation could still be 20 million to 40 million. Even this would come at great cost, for we are drastically modifying the environment that must sustain us. Salted-out cropland, desertification, deforestation, pollution—these are some of the consequences you will read about in Chapter 39, and they do not bode well for our future.

For a while, it would be like the Red Queen's garden in Lewis Carroll's *Through the Looking Glass*, where one is forced to run as fast as one can to remain in the same place. But what happens when our population doubles again? Can you brush this picture aside as being too far in the future to warrant concern? *It is no further removed from you than the sons and daughters of the next generation.*

How We Began Sidestepping Controls

How did we get into this predicament? For most of its history, the human population grew slowly. Over the past two centuries, increases in growth rates became astounding. There are three possible reasons for this:

1. Humans steadily developed the capacity to expand into new habitats and new climate zones.

2. Humans increased the carrying capacity in their existing habitats.

3. Human populations sidestepped limiting factors.

The human population did all three of these things. Consider the first point. Early humans lived mostly in the open grasslands called savannas. They were vegetarians who added scavenged bits of meat to their diet. By 200,000 years ago, small bands of hunters and gatherers had emerged. By 40,000 years ago, hunter-gatherers had spread through much of the world.

Most other species could not have expanded into such a broad range of habitats. Humans, with their truly complex brains, could use learning and memory to figure out how to build fires, assemble shelters, make clothing and tools, and plan community hunts. Learned experiences did not die with individuals. They spread quickly from one band to another through language—the capacity for extraordinary cultural communication. Thus, *the human population expanded into diverse new environments in an extremely short time, compared with the long-term geographic dispersal of other kinds of organisms.*

What about the second possibility? About 11,000 years ago, humans began to shift away from hunting and gathering to agriculture. They stopped following game herds and moving about to harvest seasonal fruits and grains. They settled down and developed a more dependable basis for life in more favorable settings. A pivotal factor was the domestication of wild grasses, including species ancestral to modern wheats and rice. People now harvested, stored, and planted seeds in one place. They domesticated animals and kept them close to home for food and pulling plows. They dug ditches and diverted water to irrigate crops.

Productivity increased through the new agricultural practices. With larger, more dependable food supplies, population growth rates increased. As towns and cities developed, a social hierarchy emerged that provided a labor base for more intensive agriculture. Much later, food supplies increased further by the use of fertilizers, herbicides, and pesticides. Transportation improved, and so did food distribution. Thus, *even at its simplest, the management of food supplies, through agriculture, increased the carrying capacity for the human population.*

What about the third possibility—did we sidestep limiting factors? Consider what happened as medical practices and sanitary conditions improved. Until about 300 years ago, contagious diseases, malnutrition, and poor hygiene kept death rates high enough to more or less balance birth rates. Contagious diseases, which are density-dependent factors, swept unchecked through crowded settlements and cities that were infested with fleas and rodents. Then came plumbing and methods of sewage treatment. Over time, vaccines, antitoxins, and antibiotics were developed as weapons against many pathogens. The death rates dropped sharply. Births now began to exceed deaths—and rapid population growth was under way.

In the mid-eighteenth century, people discovered how to harness the energy stored in fossil fuels, starting with coal. Within a few decades, large industrialized

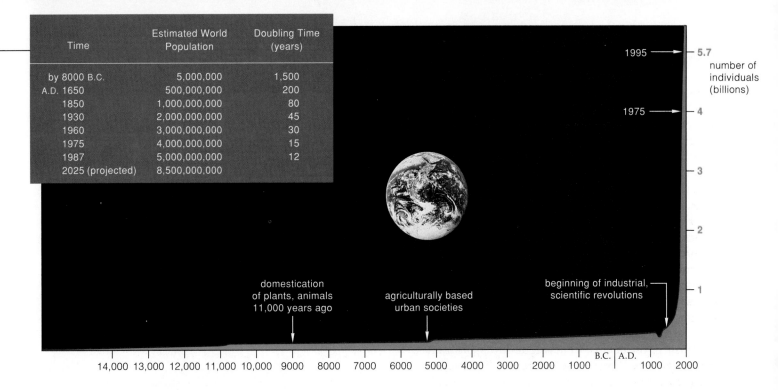

Time	Estimated World Population	Doubling Time (years)
by 8000 B.C.	5,000,000	1,500
A.D. 1650	500,000,000	200
1850	1,000,000,000	80
1930	2,000,000,000	45
1960	3,000,000,000	30
1975	4,000,000,000	15
1987	5,000,000,000	12
2025 (projected)	8,500,000,000	

Figure 35.7 Growth curve for the human population. The graph's vertical axis represents world population, in billions. (The slight dip between the years 1347 and 1351 is the time when 25 million people died in Europe as a result of bubonic plague.) The accelerated growth pattern over the past two centuries has been sustained by agricultural revolutions, industrialization, and improvements in health care. The list in the *blue* box tells us how long it took for the human population to double in size at different times in its history.

societies started forming in western Europe and North America. Then efficient technologies developed after World War I. Factories mass-produced cars, tractors, and other affordable goods. Machines replaced many of the farmers needed to produce food, and fewer farmers were able to support a larger population.

Thus, *by bringing many disease agents under control and by tapping into concentrated, existing forms of energy, humans sidestepped factors that had previously limited their population growth.*

Present and Future Growth

Where have the farflung migrations and the spectacular advances in agriculture, industrialization, and health care taken us? Starting with early *Homo,* it took about 2 million years for the first populations of humans to reach 1 billion. As you can see from Figure 35.7, it took only 80 years to reach the second billion, 30 years to reach the third, 15 years to reach the fourth—and only 12 years to reach the fifth billion!

From what the principles governing population growth tell us—and unless technological breakthroughs once again increase the carrying capacity—we should expect a dramatic increase in death rates. *Although the*

stupendously accelerated growth of the human population continues, it cannot be sustained indefinitely.

Besides having adverse effects on resource supplies, these skyrocketing numbers invite density-dependent controls. For example, the largest cholera epidemic of this century has been sweeping through South Asia. Like the six previous epidemics, it began in India. And it may spread through Africa, the Middle East, and then into Mediterranean countries before finally peaking. This epidemic alone may claim 5 million lives. Existing vaccines do not work against a current strain of *Vibrio cholerae,* the bacterial pathogen. People get infected by drinking water or eating food that is contaminated by raw sewage. *V. cholerae* multiplies in the gut. There it produces a toxin that triggers diarrhea and massive fluid loss. Two to seven days later, untreated people can die from extreme dehydration.

For centuries, *V. cholerae* has thrived and mutated in sewage-rich rivers in India (Figure 35.1). In Calcutta's slums, millions are forced to bathe in stagnant ponds and polluted waterways. In 1992, cholera struck tens of thousands in that city alone.

At this writing, a new era of human migration has begun. By some estimates, economic hardship and civil strife have put 50 million people on the move within and between countries. Will their relocation prove peaceable? Where will they find sustainable supplies of food, clean water, and other basic resources?

Through expansion into new habitats, cultural intervention, and technological innovation, the human population has temporarily skirted environmental resistance to growth. Its accelerated growth cannot be sustained indefinitely.

35.7 CONTROL THROUGH FAMILY PLANNING

Figure 35.8 shows the 1995 annual rates of increase for populations in different parts of the world. If the annual current average rate of 1.55 percent is maintained, the human population may approach *8.5 billion* just thirty years from now. It is mind-numbing to think about the resources required to sustain that many people. We will have to increase food production, supplies of drinkable water, energy reserves, and supplies of wood and other materials to meet everyone's basic needs—something we are not even doing now. The gross manipulation of resources is likely to intensify pollution, which almost certainly will adversely affect our water supplies, the atmosphere, and productivity on land and in the seas.

Today there is growing realization that population growth, resource depletion, pollution, and the quality of life are interconnected. As evidence of this, consider that most governments are trying to lower birth rates, as through **family planning programs**. The programs educate individuals about choosing how many children they will have, and when. They vary in details from country to country, but all offer information on the available methods of fertility control, as described in Section 34.18. Carefully developed and administered programs may bring about a decline in birth rates.

To achieve zero population growth, the average "replacement rate" would be slightly higher than two children per couple, inasmuch as some female children die before reaching reproductive age. The replacement rate is about 2.5 children per woman in less developed countries, and 2.1 in more developed countries. Yet even if each couple on the planet decided to have only two children, the human population would keep on growing for another sixty years! Why? An immense number of existing children will soon be reproducing!

A more useful measure of global trends is the **total fertility rate**—the average number of children born to women during their reproductive years, as estimated on the basis of current age-specific birth rates. In 1995,

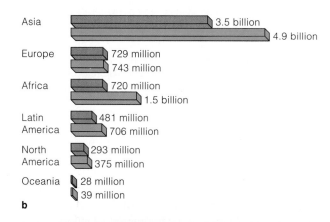

b

Figure 35.8 (**a**) The 1995 average annual population growth rate in different parts of the world. (**b**) Population sizes in 1995 (*blue* bars) and projected for the year 2025 (*red* bars).

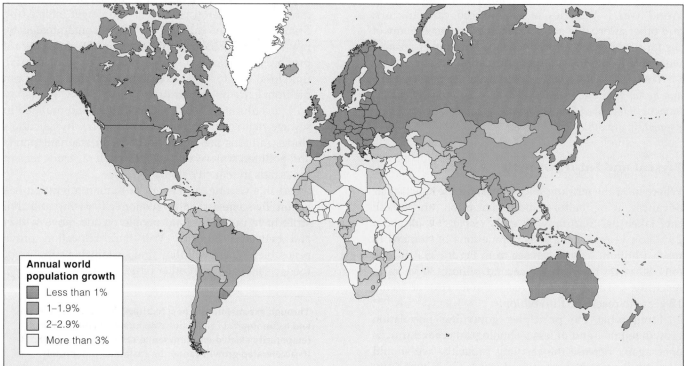

Annual world population growth

- Less than 1%
- 1–1.9%
- 2–2.9%
- More than 3%

a

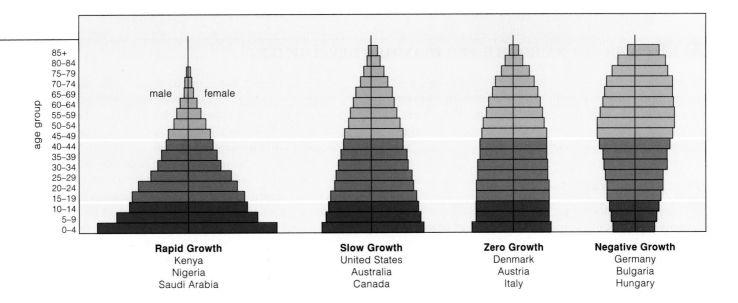

the average rate was 3.1 children per woman. This is an impressive decline from 1950, when the rate was 6.5, but it is still far above the replacement level.

Take a look at Figure 35.9, which shows the age structure diagrams for populations that are growing at different rates. In these diagrams, 15–44 is the average range for childbearing years. The diagram for Kenya and other rapidly growing populations has a broad base. The central part of each diagram includes women and men of reproductive age. Besides this, the lower portion includes an even larger number of children who will be moving into the reproductive category during the next fifteen years. Said another way, *more than one-third of the world population is in the broad pre-reproductive base.* This gives you an idea of the magnitude of the effort it will take to control population growth.

One way to slow the birth rate is to bear children in the early thirties, rather than in the mid-teens or early twenties. Delayed reproduction slows the rate of growth and lowers the average number of children in families. In China, for example, the government has established the world's most extensive family planning program. It strongly discourages premarital sex, vigorously urges postponing marriage until later in life, and encourages limiting family size to one child only (Figure 35.10). Married couples have access to free contraceptives, abortion, and sterilization. Paramedics and mobile units ensure access in remote rural areas. Couples who pledge to have only one child receive more food, free medical care, better housing, and salary bonuses. Their child will be granted free tuition and preferential treatment when he or she enters the job market. Those who break the pledge forego benefits and pay higher taxes. Are these measures inhumane? Think about this: Between 1958 and 1962 alone, an estimated 30 million Chinese died of starvation as a result of widespread famine.

Since 1972, China's total fertility rate declined from 5.7 to 1.9. Even so, the population time bomb has not stopped ticking. Its population is now 1.22 billion—and

Figure 35.9 Age structure diagrams for countries with rapid, slow, zero, and negative population growth rates. *Dark green* signifies pre-reproductive years; *purple*, reproductive years; and *light blue*, post-reproductive years. The portion of the population to the left of the vertical axis in each diagram represents males; the portion to the right represents females. Bar widths correspond to the proportion of males and females within each age group.

Figure 35.10 Alongside a city street in China, a billboard reminding couples to produce no more than one child. Couples who comply are rewarded economically and socially; those who do not suffer economic penalties.

340 million of its young women are moving into the reproductive age category. By the year 2025, China is projected to have a population of 1.5 billion.

Family planning programs on a global scale can help stabilize the size of the human population.

Even if it reaches a level of zero population growth, the human population will continue to grow for sixty years, for its reproductive base already consists of a staggering number of individuals.

Today, there is greater awareness of the connection between population growth rates and economic development. When individuals are economically secure, they seem to be under less pressure to produce large numbers of children to help them survive.

Demographic Transition Model

Changes in population growth can be correlated with changes that typically unfold in four stages of economic development. This is the premise of the **demographic transition model** (Figure 35.11).

According to this model, living conditions are very difficult, even harsh, during a *preindustrial* stage. Birth rates are high, but so are death rates. Therefore, the rate of population growth is low.

Next, during the *transitional* stage, industrialization begins, food production rises, and health care improves. Although death rates drop, the birth rates remain fairly high. As a result, the population grows rapidly. Growth continues at high rates over a long interval. Annual growth rates tend to be 2.5 to 3 percent, on the average. As living conditions improve and birth rates begin to decline, growth starts to level off.

Population growth slows dramatically during the *industrial* stage, when industrialization is in full swing. The slowdown emerges mostly because people move from the countryside to cities—and urban couples tend to control family size. Many couples decide that raising more than a few children is costly and puts them at a disadvantage in the accumulation of goods. (This goal is a sticky, related issue, as you will read shortly.)

In the *postindustrial* stage, zero population growth is reached. The birth rate falls below the death rate, and population size slowly decreases.

Today, the United States, Canada, Australia, Japan, nations of the former Soviet Union, and most countries of western Europe are in the industrial stage. Their growth rate is slowly decreasing. In Germany, Bulgaria, Hungary, and some other countries, the deaths exceed births, and the populations are getting smaller.

Mexico and other less developed countries are in the transitional stage. And they don't have enough skilled workers to complete the transition to a fully industrial economy. Fossil fuels and other resources that drive industrialization are being used up there as well as in the industrialized countries. Fuel costs might become prohibitive for countries at the bottom of the economic ladder even before they can enter the industrial stage.

If population growth keeps outpacing economic growth, death rates will increase. Thus many countries may now be stuck in the transitional stage. Some may return to the harsh conditions of the preceding stage.

Figure 35.11
Demographic transition model of changes in growth characteristics and size of populations, as correlated with changing economic development. The model explains changes that have occurred in western Europe and other industrialized regions.

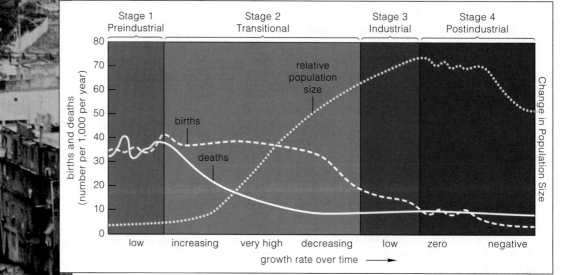

Enormous disparities in economic development are driving forces for immigration and emigration. Think of how in 1995 the United States acquired nearly 900,000 legal immigrants and political refugees and at least 300,000 illegal immigrants, mainly from Latin America and Asia. Immigrants accounted for 40 percent of the nation's increase in size that year alone.

Or think of how California's population increased by 12 *million* between 1970 and 1996. With its 2.5 percent growth rate, mainly among legal and illegal newcomers, it may jump by another 16 million by the year 2020. For some time, California has been a symbol of the good life, drawing people from the world over. Yet its schools are overcrowded, social service funds are shrinking, sewage treatment plants are nearing capacity, and water shortages are severe and frequent. Salinization of land and air pollution are eroding the agricultural base that is a key factor in California's economic health.

Claiming that population growth affects economic health, many governments restrict immigration. Only the United States, Canada, Australia, and a few other countries do allow large annual increases. Elsewhere, appalling living conditions, hard government policies, and civil strife promote emigration of far more people than the better-off nations can handle.

A Question of Resource Consumption

This chapter opened with a brief look at the conditions that confront most people in India, with its whopping 16 percent of the human population. By comparison, the United States has merely 4.7 percent. Yet which country is the most "overpopulated"—not in terms of numbers, but rather in terms of resource consumption and instigation of environmental damage?

The highly industrialized United States produces 21 percent of all goods and services. Its people consume fifty times as much as the average person in India. It uses 25 percent of the available processed minerals and nonrenewable energy resources. It also generates at least 25 percent of the global pollution and trash. By contrast, India produces about 1 percent of all goods and services. It uses 3 percent of the available minerals and nonrenewable energy resources; and it generates only about 3 percent of the pollution and trash.

Extrapolating from these numbers, Tyler Miller, Jr., estimated that it would take 12.9 billion impoverished individuals in India to have as much impact on the environment as 258 million Americans.

Differences in population growth among countries correlate with levels of economic development, hence with economic security (or lack of it) of individuals.

For us, as for all species, the biological implications of extremely rapid growth are truly staggering. Yet so are the social implications of what will happen when (and if) the human population declines to the point of zero population growth—and stays there.

For instance, as you read earlier, most individuals of an actively growing population fall in the younger age brackets. Over time, if living conditions assure constant growth, the age distribution for the population as a whole will guarantee availability of a future work force. This has social implications. Why? *It takes a large work force to support individuals in the older age brackets*. In the United States, older, nonproductive people expect their government to provide them with subsidized medical care, low-cost housing, and other social programs. Through advances in medicine and hygiene, they now live far longer than they did when the nation's social security program was set up—so their cash benefits far exceed the required contributions that they made to the program when *they* were younger.

If the population ever does reach and maintain zero population growth over time, a larger proportion of the population will end up in the older age brackets. Will all those nonproductive members continue to get goods and services if productive ones are asked to carry a greater and greater share of the economic burden? This is not an abstract question. Put it to yourself. How much economic hardship are you willing to bear for the sake of your parents? Your grandparents? How much will your children be willing or able to bear for you?

We have arrived at a major turning point, not only in our biological evolution but in our cultural evolution as well. The decisions awaiting us are among the most difficult we will ever have to make, yet it is clear that they must be made, and soon.

All species face limits to growth. We might think we are different from the rest, for our special ability to undergo rapid cultural evolution has allowed us to postpone the action of most of the factors that limit growth. But the key word here is *postpone*. No amount of cultural intervention can hold back the ultimate check of limited resources and a damaged environment.

We have sidestepped a number of the smaller laws of nature. In doing so, we have become more vulnerable to those laws which cannot be repealed. Today there may be only two options available. Either we make a global effort to limit population growth in accordance with environmental carrying capacity, or we wait until the environment does it for us.

In the final analysis, no amount of cultural intervention can repeal the ultimate laws governing population growth, as imposed by the carrying capacity of the environment.

SUMMARY

1. A population is a group of individuals of the same species occupying a given area. It has a characteristic size, density, distribution, and age structure.

2. The growth rate for a population during a specified interval can be determined by calculating the rates of birth, death, immigration, and emigration. To simplify the calculations, we can put aside effects of immigration and emigration, and combine the birth and death rates into a variable r (net reproduction per individual per unit time). Then we can represent population growth as $G = rN$, where N is the number of individuals during the specified interval.

   a. In cases of exponential growth, the population's reproductive base increases and its size expands by ever increasing increments during successive intervals. This trend plots out as a J-shaped growth curve.

   b. As long as the per capita birth rate remains even slightly above the per capita death rate, a population will grow exponentially.

   c. In logistic growth, a low-density population slowly increases in size, goes through a rapid growth phase, then levels off in size once carrying capacity is reached.

3. Carrying capacity is the name ecologists give to the maximum number of individuals in a population that can be sustained indefinitely by the resources available in their environment.

4. The availability of sustainable resources, predation, competition, and other factors that limit growth dictate population size during a specified interval. Limiting factors vary in their relative effects and they vary over time, so population size also changes over time.

5. Limiting factors such as competition for resources, disease, and predation are density dependent. Density-independent factors, such as weather on a rampage, tend to increase the death rate or decrease the death rate more or less independently of population density.

6. Patterns of reproduction, death, and migration vary over the life span characteristic of a species. Besides this, environmental variables help shape the life history (age-specific) patterns.

7. The human population now exceeds 5.7 billion. Its rate of growth varies from below zero in a few developed countries to more than 3 percent per year in some less developed countries. In 1995 the annual growth rate for the entire human population was 1.55 percent.

8. Rapid growth of the human population during the past two centuries was possible largely because of a capacity to expand into new environments, and because of agricultural and technological developments that increased the carrying capacity. Ultimately, we must confront the reality of the carrying capacity and limits to our population growth.

Review Questions

1. Define population size, population density, population distribution, and age structure. *630*

2. Why do populations that are not restricted in some way tend to grow exponentially? *630–631*

3. Define carrying capacity, then describe its effect as evidenced by a logistic growth pattern. *632*

4. Give examples of the limiting factors that come into play when a population of mammals (for example, rabbits or humans) reaches very high density. *633*

5. At present growth rates, how long will it take before the human population has another billion individuals added to it? *636*

6. How did earlier human populations expand steadily into new environments? How did they increase the carrying capacity of their habitats? How have they avoided some of the limiting factors on population growth? Is the avoidance an illusion? *636–637*

Self-Quiz *(Answers in Appendix IV)*

1. _____ is the study of how organisms interact with one another as well as with their physical and chemical environment.

2. A _____ is a group of individuals of the same species that occupy a certain area.

3. The rate at which a population grows or declines depends upon the rate of _____ .
   a. births   c. immigration   e. all of the above
   b. deaths   d. emigration

4. Populations grow exponentially when _____ .
   a. birth rate exceeds death rate and neither changes
   b. death rate remains above birth rate
   c. immigration and emigration rates are equal
   d. emigration rates exceed immigration rates
   e. both a and c

5. For a given species, the maximum rate of increase per individual under ideal conditions is the _____ .
   a. biotic potential   c. environmental resistance
   b. carrying capacity   d. density control

6. Resource competition, disease, and predation are _____ controls on population growth rates.
   a. density-independent   c. age-specific
   b. population-sustaining   d. density-dependent

7. Which of the following factors does *not* affect sustainable population size?
   a. predation   c. resources   e. all of the above can
   b. competition   d. pollution   affect population size

8. In 1995, the average annual growth rate for the human population was _____ percent.
   a. 0   c. 1.55   e. 2.7
   b. 1.05   d. 1.6   f. 4.0

9. Match each term with its most suitable description.
   ____ carrying capacity   a. disease, predation
   ____ exponential growth   b. depends on rates of birth and
   ____ population   death, as well as on emigration
        growth rate   and immigration
   ____ density-dependent   c. maximum number of
        controls   individuals sustainable by an
                                     environment's resources
                                  d. population growth plots out
                                     as J-shaped curve

Critical Thinking

1. If house cats that have not been neutered or spayed live up to their biotic potential, two can be the start of many kittens—12 the first year, 72 the second year, 429 the third, 2,574 the fourth, 15,416 the fifth, 92,332 the sixth, 553,019 the seventh, 3,312,280 the eighth, and 19,838,741 kittens in the ninth year. Is this a case of logistic growth? Exponential growth? Irresponsible cat owners?

2. If a third of the world population is now below age fifteen, what effect will this age distribution have on the future growth rate of the human population? What sorts of humane recommendations would you make that would encourage individuals of this age group to limit the number of children they plan to have? What are some of the social, economic, and environmental factors that might keep them from following those recommendations?

3. Write a short essay about a population having one of the two following age structures. Describe what might happen to younger and older age groups when individuals move into new categories.

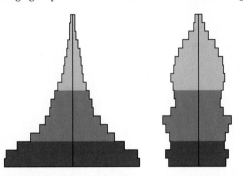

4. Figure 35.12 charts the legal immigration to the United States between 1820 and 1995. (The Immigration Reform and Control Act of 1986 accounted for the most recent dramatic increase; it granted legal status to illegal immigrants who could prove they had lived in the country for years.) An economic downturn during the 1980s and 1990s fanned resentment against newcomers. Many now say legal immigration should be restricted to 300,000–450,000 annually and we should crack down on the illegal immigrants. Others argue such a policy would diminish our reputation as a land of opportunity and discriminate against legal immigrants during crackdowns on others of the same ethnic background. Do some research on this topic, then write an essay on the pros and cons of both positions.

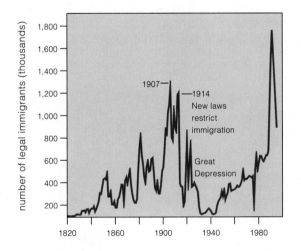

Figure 35.12 Chart of legal immigration to the United States between 1820 and 1995.

5. In his book *Environmental Science*, Miller points out that the unprecedented projected increase in human population from 5 billion to 10 billion by the year 2050 raises serious questions. Will there be enough food, energy, water, and other resources to sustain twice as many people? Will governments around the world be able to assure adequate education, housing, medical care, and other social services for all of them? Certain computer models suggest the answers are no (Figure 35.13). Yet some claim that we can adapt socially and politically to an even more crowded world, assuming that harvests improve through more technological innovation, that every inch of arable land be put under cultivation, and that everyone eats only grain. There are no easy answers to the questions. If you have not yet been doing so, start following the arguments in your local newspapers, in magazines, and on television. This will allow you to become an informed participant in a global debate that surely will have impact on your future.

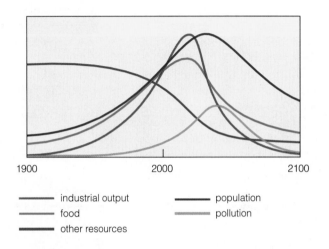

industrial output   population
food   pollution
other resources

Figure 35.13 Computer-based projection of what might happen if the size of the human population continues to skyrocket without dramatic policy changes and technological innovations. The assumptions are that the population has already overshot the carrying capacity and that current trends continue unchanged.

Selected Key Terms

Readings

Cohen, J. E. 1995. *How Many People Can the Earth Support?* New York: Norton. Analysis of past human population growth, attempts to predict future growth, and efforts to determine the biosphere's sustainable human population. No pat answer to title's question.

Miller, G. T. 1996. *Environmental Science.* Sixth edition. Belmont, California: Wadsworth.

36 COMMUNITY INTERACTIONS

No Pigeon Is An Island

Flying through the rain forests of New Guinea is an extraordinary pigeon with cobalt blue feathers and lacy plumes on its head (Figure 36.1). It is about as big as a turkey, and it flaps so slowly and noisily that its flight sounds like an idling truck. Like eight species of smaller pigeons living in the same forest, it perches on branches to eat fruit. How is it possible that *nine* species of large and small fruit-eating pigeons live in the space of the same forest? Wouldn't you think that competition for food would leave one the winner? In

Figure 36.1 The turkey-size Victoria crowned pigeon, one of nine pigeon species in the same tropical rain forest of New Guinea. Within this habitat, each species has its own niche.

fact, in that rain forest, every species lives, grows, and reproduces in a characteristic way, as defined by its relationships with other individuals and with the surroundings.

Big pigeons perch on the sturdiest branches when they feed, and they eat big fruit. Smaller pigeons, with their smaller bills, cannot open big fruit. They eat small fruit hanging from slender branches—which are not sturdy enough to support the weight of a turkey-size pigeon. The species of trees in the forest differ with respect to the diameter of fruit-bearing branches and the size of fruit, so they attract pigeons with different characteristics. In such ways, the fruit supply in the forest is partitioned among the nine pigeon species.

And how do individual trees benefit from enticing pigeons to dine? Think about it. The seeds inside their fruits have tough coats, which can resist the action of digestive enzymes in the pigeon gut. During the time it takes for ingested seeds to travel through the gut, the pigeons fly about, so they dispense seed-containing droppings in more than one place. In this way, pigeons tend to disperse seeds away from the parent plants. Later, when seedlings grow, the odds are better that at least some of them won't have to compete with their parents for sunlight, water, and nutrients. Seeds that drop close to home can't compete in any significant way with the resource-gathering capacity of mature trees, which already have extensive roots and leafy crowns.

In the same forest, leaf-eating, fruit-munching, and bud-nipping insects interact with other organisms and the surroundings in certain ways. So do nectar-drinking, flower-pollinating bats, birds, and insects. So do great numbers of beetles, worms, and other invertebrates that busily extract energy from the remains and wastes of other organisms on the forest floor. By their activities, they cycle nutrients back to the trees.

Like humans, then, no pigeon is an island, isolated from the rest of the living world. The nine species of New Guinea pigeons eat fruit of different sizes. They disperse seeds from different sorts of trees. Dispersal influences where new trees will grow and where the decomposers will flourish. Ultimately, tree distribution and the decomposition activities influence how the entire forest community is organized.

Directly or indirectly, interactions among coexisting populations organize the community to which they belong. With this chapter, we turn to community interactions that influence all populations over time and in the space of their environment.

KEY CONCEPTS

1. A habitat is the type of place where individuals of a species normally live. A community is an association of all the populations of species that occupy the same habitat.

2. Each species in a community has its own niche, defined as the sum of activities and relationships in which its individuals engage as they secure and use the resources required for their survival and reproduction.

3. Community structure starts with the adaptive traits that give each species the capacity to respond to the habitat's physical and chemical features, and to levels and patterns of resource availability over time.

4. Interactions among species influence the structure of a community. The interactions include mutually beneficial activities, competition, predation, and parasitism.

5. Community structure also depends on the geographic location and size of the habitat, the rates at which various species arrive and disappear, and the history of physical disturbances to the habitat.

6. The first species to occupy a particular type of habitat are replaced by others, which are replaced by others, and so on in sequence. This process, called primary succession, produces a climax community—a stable, self-perpetuating array of species that are in balance with one another and with the environment.

7. Different stages of succession often exist in the same habitat, owing to local differences in the soil and other environmental factors, recurring disturbances such as seasonal fires, and chance events.

Think of a clownfish darting above a coral reef, a maple tree on a Vermont hillside, or a mole burrowing in your lawn. The type of place where you will normally find a clownfish, maple, or mole is its **habitat**. The habitat of an organism is characterized by physical and chemical features, such as temperature and salinity, and by the array of other species living in it. Directly or indirectly, the populations of all species in a habitat associate with one another as a **community**.

A community's structure arises from five factors. *First*, interactions between climate and topography help dictate the habitat's temperatures, rainfall, soil types, and other conditions. *Second*, the kinds and amounts of food and other resources that become available through the year influence which species can live there. *Third*, individuals of each species have adaptive traits that allow them to survive and exploit specific resources in the habitat. *Fourth*, species in the habitat interact, as by competition, predation, and mutually helpful activities. *Fifth*, community structure is influenced by the overall pattern and actual history of changing population sizes, the arrival and disappearances of species, and physical disturbances to the habitat.

Together, these factors help dictate the number of species at different "feeding levels," starting with the producers and continuing through levels of consumers. They influence population sizes. They also help dictate the overall number of species. For example, high solar radiation, warm temperatures, and high humidity in tropical habitats favor growth of many kinds of plants, which support many kinds of animals. Conditions in arctic habitats do not favor great numbers of species.

Chapters that follow deal with energy flow through feeding levels and the geographic factors influencing community structure. Here we begin with interactions among species, using the niche concept as our guide.

The Niche

If the organisms within a community all share the same habitat—that is, if they all have the same "address"—in what respects do they differ? Each kind is distinct in terms of its "profession" within the community—that is, in the sum of activities and relationships in which it engages to secure and use the resources necessary for its survival and reproduction. This is its **niche**.

For each species, the *potential* niche is the one that could prevail in the absence of competition and other factors that could constrain its acquisition and use of resources. However, as you will see, such constraining factors do come into play in all communities. They tend to bring about a more constrained, *realized* niche that shifts in large and small ways over time, as individuals of the species respond to a mosaic of changes.

Table 36.1 Types of Two-Species Interactions*		
Type of Interaction	Direct Effect on Species 1	Direct Effect on Species 2
Neutral relationship	0	0
Commensalism	+	0
Mutualism	+	+
Interspecific Competition	−	−
Predation	+	−
Parasitism	+	−

*0 means no direct effect on population growth; + means positive effect; − means negative effect.

Categories of Species Interactions

Dozens to hundreds of species interact in diverse ways, even in simple communities. In spite of this diversity, we can identify six categories of interactions that have different effects on population growth (Table 36.1).

Two species that don't interact directly are said to have a neutral relationship. But even though, say, eagles and meadow grasses have no direct effect on each other, relations with other species indirectly link them. Eagles preying on grass-eating rabbits help grasses—which feed rabbits and help provide eagles with fattened prey.

Commensalism directly helps one species but does not affect the other much, if at all. For example, by roosting in trees, some birds gain home bases. The trees get nothing but are not harmed. In **mutualism**, benefits flow both ways between the interacting species. (Don't think of this as cozy cooperation; the benefits flow from a two-way exploitation.) In **interspecific competition**, disadvantages flow both ways between species. Finally, **predation** and **parasitism** are interactions that directly benefit one species (either the predator or the parasite) and directly hurt the other (the prey or host).

Commensalism, mutualism, and parasitism are all forms of **symbiosis**, which means "living together." For at least part of the life cycle, individuals of one species live near, in, or on individuals of another species.

A habitat is the type of place where individuals of a species normally live. The association of all populations in a habitat represents a community.

Community structure arises from the habitat's physical and chemical features, resource availability over time, adaptive traits of its members, how those members interact, and the history of the habitat and its occupants.

A niche is the sum of all activities and relationships in which individuals of a species engage as they secure and use the resources necessary for their survival and reproduction.

When two species interact, each may have neutral, positive, or negative effects on the other.

MUTUALISM

Mutualistic interactions, in which positive benefits flow both ways, abound in nature. The chapter introduction, for example, described how trees provide New Guinea pigeons with food and how the pigeons help disperse seeds from the trees to new germination sites. Such interactions exist between many other kinds of plants and animals. As another example, most of the flowering plants and their pollinators are mutualists.

young roots (Sections 19.11 and 23.1). The interaction arises when hyphae of a fungal species penetrate the roots of certain plants and become virtual extensions of them. The plant comes to depend on the fungus to take up enough vital nutrients to maintain growth. In turn, the fungus depends on some of the sugar molecules that the photosynthetic plant produces. In fact, when something interferes with the plant's functioning and

Figure 36.2 One mutualistic interaction in the high desert of Colorado.

(**a**) Different kinds of flowering plants of the genus *Yucca* are each pollinated exclusively by one species of yucca moth (**b**). This insect cannot complete its embryonic development in any other plant.

The adult stage of the moth life cycle coincides with blossoming of yucca flowers. By using her specialized mouthparts, the female moth gathers sticky pollen and rolls it into a ball. Then she flies to another flower. She pierces the wall of the flower's ovary, where seeds will develop, and lays eggs inside. As she crawls out of the flower, she pushes a ball of pollen onto a pollen-receiving surface.

Pollen grains germinate and grow down through the ovary tissues, carrying sperm to the flower's eggs. After fertilization, seeds develop. Meanwhile, the moth eggs develop to the larval stage. (**c**) When larvae emerge, they eat a few seeds, then gnaw their way out of the ovary. The seeds that the larvae didn't eat give rise to new yucca plants.

Some forms of mutualism are *obligatory*. That is, the individuals of one species cannot grow and reproduce unless they spend their entire life with individuals of the other species, in intimate dependency. This is true of the interaction between yucca plants and yucca moths. Each species of yucca plant is pollinated exclusively by one species of yucca moth. Moth larvae can grow only in yucca plants; they eat only yucca seeds (Figure 36.2).

Mycorrhizae are another example of obligatory mutualism. A mycorrhiza, recall, is an intermeshing of two kinds of absorptive structures: fungal hyphae and

photosynthesis stops, the fungus will stop producing spores, and this affects its reproductive success. Finally, reflect once more on the apparent endosymbiotic origins of eukaryotes (Section 17.5). If cells of different species of bacteria had not evolved in such intimate, mutually beneficial ways, you and all other eukaryotic organisms would not even be around today.

In forms of mutualism, each of the participating species reaps benefits from the interaction.

Categories of Competition

As you probably concluded from the preceding chapter, *intraspecific* competition can be fierce. The term refers to competition among individuals of the same population or species. By contrast, *interspecific* competition—that is, between populations of different species—is usually not as intense. Why? The requirements of two species might be similar, but they never will be as close as they are for individuals of the same species.

Competitive interactions take two forms. In certain cases, all individuals have equal access to a required resource but some are better than others at exploiting it. This interaction tends to reduce the supply of a shared, limited resource. (If you and a friend both use straws to share a small milkshake, you might not get nearly as much if your friend has a jumbo straw.) In other cases, some of the individuals control access to a resource. They partially or wholly prevent others from using the resource regardless of its scarcity or abundance. (Even if you shared a 10-gallon milkshake, you would object if your friend pinched your straw.)

Competition abounds in nature. Competition forces different chipmunk species to use different mountain habitats (Figure 36.3). Reef corals poison and grow over other species of reef corals. A strangler fig tree wraps around other trees as a framework for its own growth and eventually kills them. From early spring until late summer in the Rocky Mountains, a male broadtailed hummingbird will chase other males and females of its own species out of his blossom-rich territory. Come August, however, *rufous* hummingbirds from the Pacific Northwest travel through the Rockies as they migrate to wintering grounds in Mexico. At such times, males of this more aggressive species evict the male broadtails from broadtail territories along the migratory route.

alpine chipmunk   lodgepole chipmunk   yellow pine chipmunk   least chipmunk

Figure 36.3 Competition in nature. On the eastern slopes of the Sierra Nevada, different chipmunk species occupy different habitats. The alpine habitat is at the highest elevation. Below this are the lodgepole pine, piñon pine, and then the sagebrush habitats. The least chipmunk lives in sagebrush at the base of the mountains. Its adaptations would allow it to move up into the piñon pine habitat, but the aggressively competitive behavior of yellow pine chipmunks in that habitat won't let it. Dietary habits keep the yellow pine chipmunk out of the sagebrush habitat.

Competitive Exclusion

To a greater or lesser extent, any two species differ in their adaptations for getting food or avoiding enemies, so one usually competes more effectively for scarce resources. The two are less likely to coexist in the same habitat when they use resources in very similar ways. G. Gause demonstrated this by growing two species of *Paramecium* separately, then together (Figure 36.4). Both species exploited the same food (bacterial cells) and competed intensely for it. The test results suggested

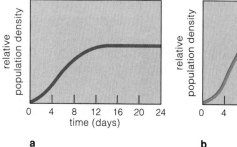

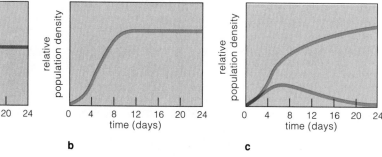

Figure 36.4 Competitive exclusion, as demonstrated by two protistan species that compete for the same food resource.

Paramecium caudatum (**a**) and *P. aurelia* (**b**) were grown apart in separate culture tubes and established stable populations, as the S-shaped growth curves show. Then the populations were grown together, and *P. aurelia* (*blue* curve in **c**) drove the other species toward extinction (*red* curve in **c**).

This experiment and others suggest that two species cannot coexist indefinitely in the same habitat if they require identical resources. If their requirements do not overlap much, one might influence the population growth rate of the other, but both may still coexist.

Figure 36.5 Two coexisting salamander species: individuals of (**a**) *Plethodon glutinosus* complex and (**b**) *P. jordani*.

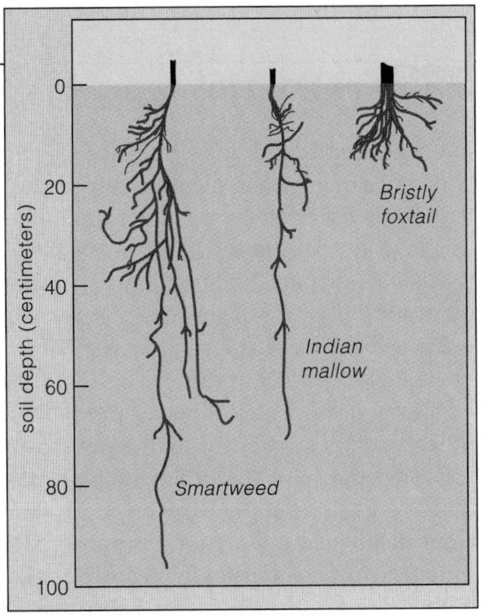

Figure 36.6 Resource partitioning by three annual plants in a plowed, abandoned field.

that two species that require identical resources cannot coexist indefinitely. Many other studies also support the concept, which is now called **competitive exclusion**.

In other experiments, Gause used two other species of *Paramecium* with less overlap in their requirements. When grown together, one species tended to feed on bacteria suspended in the liquid in the culture tube. The other tended to feed on yeast cells at the bottom of the tube. The rate of population growth was slower for both species—but not enough for one to exclude the other. The two species continued to coexist.

Field experiments also reveal effects of competition. For example, N. Hairston studied salamanders in the Great Smoky Mountains and Balsam Mountains (Figure 36.5). *Plethodon glutinosus* lives at lower elevations than its relative, *P. jordani*, but the ranges overlap in some places. Hairston removed one or the other species from different test plots in overlap areas. He also left some plots untouched, as controls. After five years, nothing had changed in the control plots; the two species were coexisting. In plots that had been cleared of one species (*P. jordani*), the other (*P. glutinosus*) *had* increased in numbers. Plots cleared of *P. glutinosus* had a greater proportion of young *P. jordani* salamanders—evidence of a growing population. Where the species coexist in nature, then, competitive interactions seem to suppress the growth rate of both populations.

Resource Partitioning

Think back on the nine species of fruit-eating pigeons in the same forest in New Guinea. They require the same resource—fruit. Yet they overlap only slightly in their use of this resource, because each specializes in fruits of a particular size. All together, they are a good example of **resource partitioning**, whereby competing species coexist by *subdividing* a category of similar resources.

A similar competitive situation arose among three species of annual plants in a plowed, abandoned field. As for other plants, all three required sunlight, water, and dissolved mineral salts. Yet each species was able to exploit a different portion of the habitat (Figure 36.6). Drought-tolerant foxtail grasses have a shallow, fibrous root system that quickly absorbs rainwater. They grew where soil moisture varied from day to day. Mallow plants, with a taproot system, grew where deeper soil was moist early in the growing season but drier later. Smartweed has a taproot system that branches in topsoil and in soil below the roots of the other species. It grew where soil was continuously moist.

In short, species may coexist in the same habitat even if their niches do overlap. Many species compete when experimentally grown together but not in their natural habitats, where they tend to use different foods or other resources. Hairston's salamanders compete yet coexist, although at suppressed population sizes. New Guinea pigeons coexist through resource partitioning. Birds and people overlap in their need for oxygen, but atmospheric oxygen is so abundant there is no need to compete for it.

In some competitive situations, all individuals have equal access to a required resource, but some are better than others at exploiting it. In other situations, some individuals control access to a resource.

The more two species in the same habitat differ in their use of resources, the more likely they can coexist.

Two competing species may coexist by sharing a resource in different ways or at different times.

36.4 PREDATION AND PARASITISM

Let's use two broad definitions for interactions between consumers and their victims. A **predator** is an animal that feeds on other living organisms (its prey) but does not take up residence on or in them. Its prey may or may not die from the interaction. A **parasite** lives in or on other living organisms (its hosts) and typically feeds on specific tissues for a good part of its life cycle. Its hosts may or may not die as a result of the interaction.

Many of the adaptations of predators, parasites, and their victims arose through **coevolution**. This term refers to the joint evolution of two (or more) species that exert selection pressure on each other through their close ecological interaction. Suppose, as an outcome of mutation, a new, heritable means of defense appears in a prey organism and spreads through the population. Some predator organisms respond better than others to the new defense. They tend to eat more, and have better luck surviving and reproducing. In time the forms of traits that are most efficient at overcoming the new prey defense increase in frequency in the population. Now the better predators exert selective pressure that favors better defenses among prey—and so on through time.

Figure 36.7 Idealized cycling of abundances of predator and prey. (For clarity, this diagram exaggerates the predator density; predators usually are less common than their prey throughout the cycle.) The pattern arises through time lags in predator responses to changes in prey abundance. At time *a*, prey density is low; predators have more difficulty securing food and their population is declining. In response to the predator decline, prey start increasing. But predators do not start increasing until they start reproducing (time *b*). Both populations grow until the greater number of predators causes the prey population to decline (time *c*). Predators continue to increase and take out more prey, but the lower prey density leads to starvation among them and their growth rate slows (time *d*). At time *e*, a new cycle begins.

Dynamics of Predator-Prey Interactions

In any specified interval, the outcome of predator-prey interactions depends in part on the carrying capacity of the prey population. (*Carrying capacity*, remember, is the maximum number of individuals that the resources in a given environment can maintain indefinitely.) It also depends on the reproductive rates of predator and prey populations. And the outcome depends on responses of individual predators to increases in prey density.

At times when predation keeps a prey population from exceeding the carrying capacity, both populations tend to coexist at fairly steady levels. Then, more prey are around, and predators reproduce promptly and eat more. Population densities fluctuate when predators do not reproduce as fast as prey, when they can eat only so many prey organisms in a given interval, and when the carrying capacity for the prey population is high.

Figure 36.7 shows idealized cyclic changes brought about by time lags in a predator's response to changes in prey abundance. In nature, we sometimes find such a correspondence between changes in predator and prey populations, but other factors also are involved.

For example, Charles Krebs led a long-term study in the Yukon to identify the basis of the ten-year cycle in populations of the snowshoe hare and its predators (Figure 36.8). For eight years, the researchers tracked hare densities in 1-square-kilometer experimental plots

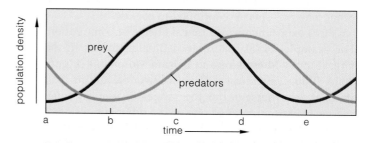

Figure 36.8 Predator-prey interactions between Canadian lynx and snowshoe hares. An apparent correspondence between the abundances of both populations is based on counts of pelts sold by trappers to Hudson's Bay Company over ninety years. The *dashed* line tracks lynx abundances, and the *solid* line, the hare abundances. This figure is a good test of whether you readily accept someone else's conclusions without questioning their scientific basis. (Remember the discussion of scientific methods in Chapter 1?) What other factors might have influenced the cycle? Did weather vary, with more severe winters imposing greater demand for food (required for animals to keep warmer) and higher death rates? Did the lynx have to compete with other predators for prey? Did predators turn to alternate prey during low points of the hare cycle? Did trapping increase with rising fur prices in Europe, and did it decrease as the pelt supply outstripped the demand?

and control plots. In some plots, electrified fences kept out mammalian predators but not the hares. In others, hares were supplied with extra food. In still other plots, fertilizers enhanced plant growth for the herbivorous hares. During the cyclic peak and decline, hare densities increased in predator-free plots and in plots with extra food. In plots where both conditions occurred, densities were thirty-six times greater than in control plots. By itself, increased vegetation (in the fertilized plots) had negligible effect on hare densities. A simple predator-prey model or plant-herbivore model is not enough to explain the results. This cycle seems to turn on plants, herbivores, and carnivores—a *three-level* interaction.

Parasite-Host Interactions

Flukes and tapeworms are fine examples of parasites—consumers that live on or in a host organism and draw nutrients from its tissues. As Section 20.6 describes, sometimes these parasites and others indirectly cause death when secondary infections kill a weakened host. In terms of reproductive success, this is not good for the parasite; hosts that live longer can sustain and disperse more of its offspring. Thus parasites tend to coevolve with hosts in ways that produce less-than-fatal effects. Death usually occurs only when a parasite attacks a novel host (with no coevolved defenses against it) or when many individual parasites infect the same host.

Parasites as Biological Control Agents

Many parasites exert control over increases in insect populations. Many types are commercially raised and released as an alternative to chemical pesticides. Less

than 20 percent of these qualify as *effective* defenses against insect pests. As C. Huffaker and C. Kennett point out, effective control agents have five attributes. They are well adapted to host species and the habitat. They are good at intercepting hosts. Their population growth rate is higher than that of the host species, and the offspring are mobile enough for adequate dispersal. Finally, the lag time between their responses to changes in the numbers of the host population is minimal.

Releasing more than one kind of biological control agent in an area may trigger competition among them and lessen their overall level of effectiveness. Besides, a shotgun approach to biological control is risky. What if the parasites attack other, nontargeted species? In 1983, F. Howarth reported that native populations of moths and butterflies in the Hawaiian Islands were declining, partly because of parasitic wasps that were released as controls over something else. We return to the effects of such species introductions later in the chapter.

Some predator and prey populations coexist at more or less steady population levels. Others undergo recurring cycles of abundance and population crashes, erratic population cycles, or prey extinction.

When a predator population keeps a prey population from overshooting the carrying capacity, both may coexist at stable levels. Delays in a predator's response to changes in prey abundance contribute to the cyclic or irregular fluctuations.

Predator-prey or plant-herbivore models alone cannot explain the cyclic changes. Other factors, such as changes in the prey's food supply, also appear to contribute to the cycles.

Like predators and their prey victims, parasites and their hosts are locked in long-term, coevolutionary contests.

THE COEVOLUTIONARY ARMS RACE

Populations of other species are part of the environment of any organism, and the ones that interact as predators and prey exert continual selection pressure on each other. One must defend itself, the other must overcome the defenses. This is the basis of a coevolutionary arms race that has resulted in some truly amazing adaptations.

Camouflage   Consider prey species that **camouflage** themselves—they can hide in the open. Such organisms have adaptations in form, patterning, color, and behavior that permit them to blend with their surroundings and escape detection. Figure 36.9 shows classic examples, including a desert plant (*Lithops*) that looks like a small rock. *Lithops* only flowers during a brief rainy season. That is when other plants grow profusely (and possibly divert plant eaters from *Lithops*) and free water instead of juicy plant tissues is available to quench an animal's thirst.

Warning Coloration   Now consider prey species that are bad-tasting, toxic, or able to inflict pain on attackers. Many of the toxic types have conspicuous patterns and colors that serve as warning signals to predators. Maybe a young, inexperienced bird will spear a yellow-banded wasp or orange-patterned monarch butterfly—once. But it will quickly learn to associate the distinctive colors and patterning with pain or digestive upsets.

Species that are truly dangerous or repugnant make little or no effort to conceal themselves. Skunks are like this. So are frogs of the genus *Dendrobates*; they are among the most vivid and most poisonous organisms (Figure 25.4).

Mimicry   Many tasty but weaponless prey organisms closely resemble unpalatable or dangerous species. Such resemblances are a type of **mimicry**. Figure 36.10 shows how closely some mimics physically resemble repugnant species that serve as their models. But mimicry has many forms. For example, in "speed" mimicry, a sluggish, easy-to-catch prey organism closely resembles swift species that predators long ago gave up trying to catch.

Moment-of-Truth Defenses   When their luck runs out, cornered prey organisms may resort to a last-ditch trick. Suppose a leopard outruns a baboon. By turning abruptly and displaying formidable canines, the baboon may startle and confuse the predator long enough for a chance at a getaway (Figure 36.7). Other cornered animals release chemicals as warning odors, repellants, or poisons. Earwigs, skunks, and stink beetles produce awful odors. Several beetles take aim and let loose with noxious sprays.

Similarly, many plants emit predator repellants. Tannins in the foliage and seeds of certain plants taste bitter and make the plant tissues hard to digest. Make the mistake of nibbling on the yellow petals of a buttercup (*Ranunculus*) and you will badly irritate the lining of your mouth.

Figure 36.9   A few prey organisms demonstrating the fine art of camouflage.

(**a**) What bird??? When a predator approaches its nest, the least bittern stretches its neck (which is colored much like the surrounding withered reeds), thrusts its beak upward, and sways gently like reeds in the wind. (**b**) An unappetizing bird dropping? No. The body coloration of this caterpillar and the stiff positions it takes help it hide in the open from birds that prey on it. (**c**) Find the plants (*Lithops*) hidden from herbivores owing to their stonelike form, pattern, and coloring.

The model . . .

a

. . . and
two of its
mimics:

b

c

d   e

Figure 36.10 Mimicry. Many predators avoid prey that taste awful, secrete toxins, or inflict painful bites or stings. Such prey often display bright colors, bold markings, or both; many don't even bother to hide. Many prey species unrelated to the dangerous or unpalatable ones have evolved striking behavioral and morphological resemblance to them. (**a**) This aggressive yellow jacket is the likely model for similar-looking but edible wasps (**b**) and beetles (**c**). The inedible butterfly in (**d**) is a model for the edible mimic *Dismorphia* (**e**).

a

b

c

Figure 36.11 Adaptive responses of predators to prey defenses. (**a**) Which parts are flowers and which parts are praying mantid? (**b**) Certain beetles spray noxious chemicals at their attackers, which works some of the time. But grasshopper mice plunge the chemical-spraying tail end of their prey into the ground and feast on the head end. (**c**) Find a scorpionfish, a venomous predator with camouflaging fleshy flaps, multiple colors, and profuse spines.

Adaptive Responses to Prey   In the coevolutionary arms race, predators counter prey defenses with their own marvelous adaptations, including clever avoidance of repellants, stealth, and camouflage.

Remember those beetles that direct sprays of noxious chemicals at their attackers? Grasshopper mice grab such beetles, plunge the "sprayer" end into the ground, and feast on the unprotected head (Figure 36.11*b*). Chameleons

are able to remain motionless for extended intervals. Prey may not even "see" them until zapped by the amazing chameleon tongue (Figure 25.10). And it is no accident that stealthy predators blend with their background. Think of polar bears camouflaged against snow, golden tigers against tall-stalked, golden grasses, and pastel insects against pastel flowers. And hope you never step on a dangerous scorpionfish, concealed on the seafloor.

FORCES CONTRIBUTING TO COMMUNITY STABILITY

The Successional Model

How do communities come into being? By the classical model, known as **ecological succession**, a community develops in sequence, from pioneers to an end array of species that remain in equilibrium over some region. A **pioneer species** is an opportunistic colonizer of vacant or vacated habitats, notable for its high dispersal rate and rapid growth. Over time, more competitive species replace the pioneers, then are themselves replaced until the array of species stabilizes under the conditions that prevail in the habitat. This persistent array of species is the **climax community**.

A process of *primary* succession begins as pioneer species colonize a barren habitat. A new volcanic island is such a habitat. So is land, once covered with ice for thousands of years, that is exposed as glaciers retreat (Figure 36.12). The typical pioneer species are small plants with brief life cycles, adapted to grow in exposed areas with intense sunlight, swings in temperature, and nutrient-deficient soil. Each year they produce great numbers of small seeds, which are quickly dispersed.

Once established, the pioneers improve conditions for other species and often set the stage for their own replacement. Many types are mutualists with nitrogen-fixing bacteria and initially outcompete other plants in nitrogen-poor habitats. Then their accumulated wastes and remains add volume to the soil and enrich it with nutrients—which allows other species to take hold. The pioneers also form low-growing mats that shelter seeds of later species—yet cannot shade out seedlings. In time, later successional species crowd out the pioneers, whose seeds travel as fugitives on the wind or water—destined, perhaps, for a new but temporary habitat.

In *secondary* succession, a disturbed area within a community recovers and moves again toward the climax state. The pattern is typical of abandoned fields, burned forests, and storm-battered intertidal zones. It emerges after falling trees open part of an established forest's canopy. Sunlight reaches seeds and seedlings that are already on the forest floor and spurs their growth.

By one hypothesis, the colonizers *facilitate* their own replacement. By another, the earliest colonizers compete against species that could replace them, so the sequence of succession depends on who gets there first.

The Climax-Pattern Model

At one time, some ecologists thought the same general type of community always develops in a given region because of constraints imposed by climate. However, stable communities *other* than "the climax community" often persist in a region, as when tallgrass prairie extends into the deciduous forests of Indiana.

Figure 36.12 (**a**) Primary succession in Alaska's Glacier Bay area. The glacier shown here has been retreating since 1794. (**b**) As a glacier retreats, meltwater tends to leach the newly exposed soil of minerals, including nitrogen. Less than ten years ago, this nutrient-poor soil was still buried under ice. (**c,d**) The first invaders are seeds of mountain avens (*Dryas*) that drift in on the winds. Mountain avens is a pioneer species that benefits from the nitrogen-fixing activities of mutualistic microbes. It grows and spreads rapidly over glacial till.

(**e**) Within twenty years, shrubby, deciduous alders take hold. They are symbionts with nitrogen-fixing microbes. Cottonwood and willows also become established (**f**). In time, the alders form dense thickets (**g**). As the thickets mature, cottonwood as well as hemlock trees grow rapidly. So do a few evergreen spruce trees. (**h**) By eighty years, spruce crowd out the mature alders. (**i**) In areas deglaciated for more than a century, dense forests of Sitka spruce and western hemlock dominate. Nitrogen reserves are depleted. Much of the biomass is tied up in peat—excessively moist, compressed organic matter that resists decomposition and forms a thick mat on the forest floor.

By a **climax-pattern model**, a community is adapted to many environmental factors—topography, climate, soil, wind, species interactions, recurring disturbances, chance events, and so on—that vary in their influence over a region. As a result, even in the same region you may see one climax stage extending into another along gradients of influential environmental conditions.

Cyclic, Nondirectional Changes

Many changes recur in patches of a given habitat, over and over again. Such small-scale changes contribute to the internal dynamics of the community as a whole. Observe a disturbed patch of habitat, and you might conclude that great shifts in species composition are afoot. Observe the community on a larger scale, and you might find that the overall composition includes all of the pioneer species that are colonizing the patch as well as the dominant climax species.

Consider a tropical forest, which develops through phases of colonization by pioneer species, increases in

b

e

h

c

f

d

g

i

species diversity, and then reaches maturity. Whipping sporadically through the successional pattern, however, are heavy winds that cause treefalls. Where trees fall, gaps open in the forest canopy so more light can reach the forest floor. In that local patch, conditions favor the growth of previously suppressed small trees and the germination of pioneers or shade-intolerant species.

Or consider the groves of sequoia trees in the Sierra Nevada in California. Some trees of this type of climax community are giants, more than 4,000 years old. Their persistence depends partly on recurring brush fires that sweep through parts of the forests. Sequoia seeds only germinate in the absence of smaller, shade-tolerant plant species. Too much litter on the forest floor inhibits their germination. Modest fires eliminate trees and shrubs that compete with young sequoias but don't damage the sequoias themselves. (Mature sequoias have very thick bark that burns poorly and insulates the living phloem cells of the giant trees against modest heat damage.) At one time, fires were prevented in many sequoia groves

in national and state parks—not just accidental fires from campsites and discarded cigarettes, but also natural fires touched off by lightning. Litter tends to build up when small fires are prevented. It allows fire-susceptible species to take hold, and underbrush thickens. The dense underbrush prevents the germination of sequoia seeds and fuels hotter fires that can damage the giant trees. Currently, rangers set controlled fires to eliminate underbrush and promote the conditions in the habitat that are favorable for cyclic replacements.

Community structure is an outcome of a balance of forces, including predation and competition, operating over time.

A climax community is a stable, self-perpetuating array of species in equilibrium with one another and their habitat.

Similar climax stages can persist along gradients dictated by environmental factors and by species interactions. Also, recurring, small-scale changes are built into the overall workings of many communities.

COMMUNITY INSTABILITY

The preceding sections might lead you to believe that all communities become stabilized in predictable ways. This is not always the case. Community stability is an outcome of forces that have come into uneasy balance. Resources are sustained, as long as populations do not flirt dangerously with the carrying capacity. Predators and prey coexist, as long as neither wins. Competitors have no sense of fair play. Mutualists are stingy, as when plants produce as little nectar as required to attract pollinators, and pollinators take as much nectar as they can for the least effort.

Over the short term, disturbances can hamper the growth of some populations. And long-term changes in climate or some other environmental variable also can have destabilizing effects. If instability becomes great enough, the community may change in ways that will persist even when the disturbance ends or is reversed. If some of its member species happen to be rare or don't compete well with others, they may become extinct.

How Keystone Species Tip the Balance

The uneasy balancing of forces in a community comes into sharp focus with studies of a **keystone species**—a dominant species that can dictate community structure. Roger Paine identified a keystone species by removal experiments in intertidal zones along North America's west coast. Pounding surf, tides, and storms continually disturb these zones, and living space is scarce. In his control plots, Paine included a sea star (*Pisaster*) and its invertebrate prey. The sea star proved to be a keystone species; it controls the abundances of resident mussels (*Mytilus*), limpets, chitons, and assorted barnacles.

After Paine removed sea stars from experimental plots, mussels took over and crowded out seven other invertebrate species. Mussels are the main prey of sea stars but are the strongest competitors when sea stars are absent. Predation (by sea stars) normally maintains the diversity of prey species by preventing competitive exclusion (by mussels). Remove the sea stars, and the community shrinks from fifteen species to eight.

Or consider periwinkles (*Littorina littorea*), an alga-eating relative of land snails. Jane Lubchenko found that periwinkles can increase *or* decrease the diversity of algal species in different settings. In tidepools, they eat a dominant algal species (*Enteromorpha*). By doing so, they help many other, less competitive algal species survive. By contrast, on rocks exposed only at high tide, *Chondrus* and other red algae dominate. Periwinkles leave these tough, unpalatable species alone—and eat the competitively weak algal species that they pass up in tidepools. Thus, periwinkles increase algal diversity in tidepools and reduce it on recurringly emergent rocks (Figure 36.13).

Figure 36.13 Effect of competition and predation on community structure. (**a**) Periwinkles (*Littorina littorea*) influence the number of algal species in certain marine habitats, depending on where they graze. (**b**) *Chondrus* and (**c**) *Enteromorpha*, two algae in their natural habitat. (**d**) In tidepools, periwinkles graze on the dominant alga (*Enteromorpha*) and so favor less competitive algae that might otherwise be overwhelmed. (**e**) But on rocks exposed only at low tide, they do not eat the dominant species (*Chondrus* and other red algae), so they decrease algal diversity.

How Species Introductions Tip the Balance

Finally, consider what can happen when some residents of established communities move from their home range and successfully take up residence elsewhere. We call this **geographic dispersal**, and it can occur in three ways. First, over the generations, a population may expand its home range by gradually diffusing into outlying regions that prove hospitable. Second, individuals may be rapidly transported across great distances. This *jump* dispersal

Table 36.2 Detrimental Effects of Some Species Introduced Into the United States

Species Introduced	Origin	Mode of Introduction	Outcome
Water hyacinth	South America	Intentionally introduced (1884)	Clogged waterways; shading out of other vegetation
Dutch elm disease: *Ophiostoma ulmi* (the fungal pathogen)	Europe	Accidentally imported on infected elm timber (1930)	Destruction of millions of elms; great disruption of forest ecology
Bark beetle (carrier of the pathogen)		Accidentally imported on unbarked elm timber (1909)	
Chestnut blight fungus	Asia	Accidentally imported on nursery plants (1900)	Destruction of nearly all eastern American chestnuts; disruption of forest ecology
Argentine fire ant	Argentina	In coffee shipments from Brazil? (1891)	Crop damage; destruction of native ant communities; death of ground-nesting birds
Japanese beetle	Japan	Accidentally imported on irises or azaleas (1911)	Defoliation of more than 250 plant species, including commercially important species such as citrus
Sea lamprey	North Atlantic Ocean	Through Erie Canal (1860s), then through Welland Canal (1921)	Destruction of lake trout and lake whitefish in Great Lakes
European starling	Europe	Released intentionally in New York City (1890)	Competition with native songbirds; crop damage; transmission of swine diseases; airport runway interference; extremely noisy and messy in large flocks
House sparrow	England	Released intentionally (1853)	Crop damage; displacement of native songbirds; transmission of some diseases

commonly takes the individual across a region where it could not survive, as when a ship transports an insect from the mainland to Maui. Third, with imperceptible slowness, a population may move out from its home range over geologic time, as by continental drift.

The dispersal and colonization of vacant places can be amazingly rapid as well as successful. Consider one of Amy Schoener's experiments in the Bahamas. She set out plastic sponges on the barren sandy floor of Bimini Lagoon. How fast did aquatic species take up residence in the chambers of these artificial hotels? Schoener recorded occupancy by 220 species in less than 30 days.

Think about the 4,500 or so nonindigenous species that have become successfully established in the United States following jump dispersal. These are just the ones we know about. Some, including soybeans, rice, wheat, corn, and potatoes, have been put to good use as food resources. Most of the rest have disrupted community organization and agriculture (Table 36.2).

One arrival was a blue-flowered aquatic plant, the water hyacinth. In the 1880s, someone displayed this South American plant at an exposition in New Orleans. Flower fanciers took home clippings and set them out in ponds and streams. Unchecked by natural predators, the fast-growing hyacinths spread through nutrient-rich waters, displaced many native species, then choked off rivers and canals. They have spread as far west as San Francisco. Another arrival, kudzu, is a well-behaved legume in Asia. It now blankets trees, hills, and houses in the southern United States.

In another wholesale disaster, in 1859 a landowner in northern Australia released two dozen wild English rabbits. Good food and good sport hunting, that was the idea. Perfect habitat with no natural predators—that

was the reality. Six years later, the landowner had killed 20,000 rabbits and was besieged by 20,000 more. The rabbits displaced livestock, even kangaroos. Did the construction of a 2,000-mile-long fence protect western Australia? No. Rabbits made it to the other side before workers completed the fence. In 1951, someone brought in the myxoma virus by way of mildly infected South American rabbits, its normal hosts. Lacking coevolved defenses, the English rabbits died in droves. As you might expect, natural selection has since favored the rapid growth of populations of virus-resistant rabbits. This coevolutionary tale is still unfolding.

If you live in the northeastern United States, you know about imported gypsy moths. They escaped from a research facility near Boston in 1869, and today their descendants defoliate entire forests. You might know about zebra mussels, which probably entered the Great Lakes on a cargo ship's hull. They attach to and block the water intake pipes for cities. The estimated damage is set at 5 billion dollars. And *Cryphonectria parasitica*, a fungus introduced to North America almost a century ago, killed nearly all American chestnut trees. Finally, remember the African bees released in South America? As you read in Chapter 6, their Africanized, "killer bee" descendants traveled north, through Mexico and on into the United States. Swarms of them have already killed several hundred people in Latin America. At this writing, these aggressive bees also have attacked 140 Americans, one of whom died from multiple stings.

Long-term shifts in climate, the rapid introduction and successful establishment of a new species, and other disturbances can permanently alter community structure.

As you might well conclude from your consideration of the preceding discussions, biodiversity differs greatly from one habitat to another. Besides this, as many field investigations by ecologists indicate, biodiversity differs greatly between geographic regions.

Mainland and Marine Patterns

The most striking pattern of biodiversity is related to distance from the equator. For most groups of plants and animals, the number of coexisting species on land and in the seas is highest in the tropics. The number systematically declines toward the poles. Figure 36.14 shows two examples. What factors underlie this pattern of biodiversity? Three are paramount.

First, tropical latitudes intercept more sunlight of consistently greater intensity, rainfall is higher, and the growing season is longer. Thus, in the tropics, *resource availability tends to be higher and more reliable.* All year long, different trees in humid tropical forests put out new leaves, flowers, and fruit. Year in and year out, they support many diverse herbivores, nectar foragers, and fruit consumers. In temperate and arctic regions, such specializations would never evolve.

Second, *species diversity might be self-reinforcing.* The diversity of tree species in tropical forests is far greater than in comparable forests at higher latitudes. When more plant species compete and coexist, more species of herbivores evolve, partly because no one herbivore can overcome the chemical defenses of all kinds of plants. Typically, then, more predators and parasites evolve in response to the diversity of prey and hosts. The same is true of diversity on tropical reefs.

Third, evolutionary history tells us this: *The rates of speciation in the tropics have exceeded those of background*

extinction. Decreases in biodiversity have occurred, but mainly during mass extinctions. As you read in Section 15.3, global temperatures drop at such times. Species adapted to cool climates survive; those in the tropics have nowhere to go. Bear in mind, however, millions of species in tropical forests may disappear in the next decade, for reasons described in chapters to follow.

Island Patterns

Islands often serve as terrific laboratories for studying biodiversity. For example, in 1965, a volcanic eruption formed a new island southwest of Iceland. Within six months, bacteria, fungi, seeds, flies, and some seabirds were established on it. One vascular plant appeared after two years, and a moss two years after that (Figure 36.15). As soils improved, the number of plant species increased. All species were colonists from Iceland. None originated on the island, which was named Surtsey.

Yet the number of new species will not increase indefinitely. Why not? Studies of community patterns on islands around the world provide us with two ideas.

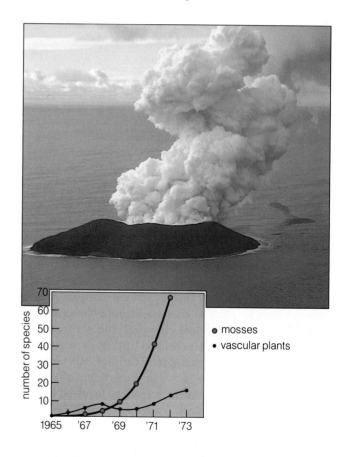

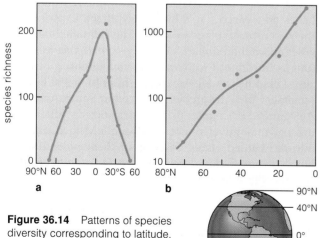

Figure 36.14 Patterns of species diversity corresponding to latitude, as represented by (**a**) ants and (**b**) breeding birds of North and Central America.

Figure 36.15 Surtsey, a volcanic island, at the time of its formation. Such islands are natural laboratories for ecologists. The chart shows the number of species of mosses and vascular plants recorded on the new island from 1965 to 1973.

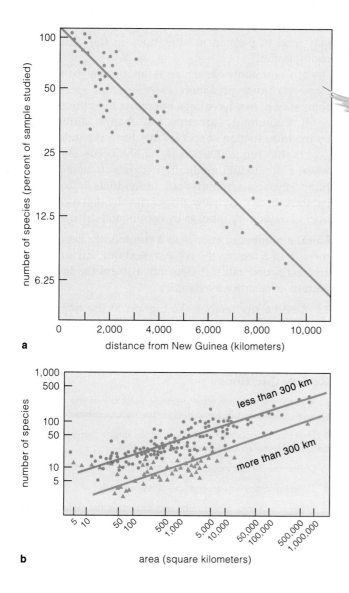

a

number of species (percent of sample studied)

distance from New Guinea (kilometers)

b

number of species

area (square kilometers)

less than 300 km

more than 300 km

c

Figure 36.16 (**a**) The distance effect. Biodiversity on islands of a given size declines with increasing distance from the source of colonizing species. Each graph point denotes the number of species of land birds of lowland areas of South Pacific islands. The huge island of New Guinea is the source of colonists for these islands, each of which is at least 500 kilometers away. To correct for differences in island size (area), the actual number of bird species on each of these distant islands is expressed as a percentage of the number of bird species on an island of equivalent size close to New Guinea.

(**b**) The area effect. Among islands similarly distant from the source of colonizing species, the larger islands support larger numbers of species. The graph points denote the number of bird species on tropical and subtropical islands. The *solid green* circles and the upper line are for islands less than 300 kilometers from their source of colonists. The *orange* triangles and the lower line denote islands more than 300 kilometers from source areas. Thus this graph incorporates data on the distance effect as well as on the area effect.

(**c**) One of the travel agents for jump dispersals. Seabirds that island-hop over long distances might have a few seeds stuck in their feathers, as shown here. If those seeds successfully germinate in a new island community, they will give rise to a population of new immigrants.

First, islands far from a source of potential colonists receive few colonizing species, and the few that arrive are adapted for long-distance dispersal (Figure 36.16). This is called the **distance effect**. Second, larger islands tend to support more species than smaller islands at equivalent distances from source areas. This is the **area effect**. Larger islands tend to offer more habitats; most have more complex topography and are higher above sea level. Thus they favor species diversity. Also, being bigger targets, they may intercept more colonists.

Most importantly, extinctions suppress biodiversity on small islands. Their smaller populations are far more vulnerable to storms, volcanic eruptions, diseases, and random shifts in birth and death rates. As for any island, the number of species reflects a balance between immigration rates for new species and extinction rates for established ones. Small islands that are distant from

a source of colonists have low immigration rates and high extinction rates, so they support few species once this balance has been struck for their populations.

For most groups of organisms, the number of coexisting species is highest in the tropics and systematically declines toward the poles.

A region's biodiversity depends on many factors, such as its climate, topographical variation, possibilities for dispersal, and evolutionary history—including extinctions and the time available for speciation.

Disturbances to the habitat tend to work against competitive exclusion and therefore favor increases in biodiversity.

Biodiversity on islands represents a balance between the immigration rates for new species and extinction rates for established species.

SUMMARY

1. A habitat is the type of place where individuals of a given species normally live—that is, their "address." A community consists of all populations of all species that occupy a habitat and that directly or indirectly associate with one another.

2. Each species has its own niche, or "profession," in the community. This is defined as the sum of activities and relationships in which its members engage as they secure and use the resources they require to survive and reproduce.

3. Mutualism, commensalism, competition, predation, and parasitism are species interactions that directly or indirectly link the populations in a community.

4. Two species that require the same limited resource tend to compete, as by using the resource as rapidly or efficiently as possible or by interfering with use of it.

   a. According to the competitive exclusion concept, if two (or more) species require identical resources, they cannot coexist indefinitely.

   b. Species are more likely to coexist if they differ in their use of resources. Also, they may coexist by using a shared resource in different ways or at different times.

5. Some predator and prey populations may coexist at stable levels if predators keep prey from overshooting the carrying capacity. Other predator and prey populations show recurring or erratic cycles of abundance and crashes, owing to such factors as delays in predator responses to changes in prey density, and changes in the prey's food supply.

6. Predators and their prey coevolve. After some novel, heritable trait appears and gives prey an edge in their contest, selection pressure operates on the population of predators, which also may evolve in response. The same happens when a novel, heritable trait appears in the predator population.

   a. Evolved prey defenses include threat displays, chemical weapons, mimicry, and camouflaging.

   b. Predators overcome prey defenses by adaptive behavior (including stealth), camouflaging, and so on.

7. Parasites and their hosts coevolve in ways that favor resistant hosts and only moderately harmful parasites.

8. By the classical model of ecological succession, a community develops in predictable sequence, from its pioneer species to an end array of species that persists over an entire region.

9. A stable, self-perpetuating array of species that are in equilibrium with one another and with a particular environment is called a climax community.

10. Similar climax stages may persist within the same region, yet show variation as a result of environmental gradients and species interactions.

11. Recurring, small-scale changes are a part of the internal dynamics of communities. Other changes, such as long-term shifts in climate or species introductions, may lead to permanent alterations in the community configuration.

   a. Community structure is an uneasy balance of forces, including predation (as by keystone species) and competition, that have been operating over time.

   b. Community structure can change permanently by the introduction of species that have expanded their geographic range. Dispersals might occur slowly, as when a population gradually expands its home range. Jump dispersals rapidly put individuals into distant habitats. Extremely slow dispersals also have occurred over evolutionary time, as by continental drift.

12. The number of species in a community depends on the size of a region, the colonization rate, disturbances, and extinction rates. It depends also on the level and pattern of resource availability.

13. Biodiversity tends to be highest in the tropics and to systematically decline toward polar regions, except during times of mass extinction.

Review Questions

1. What is the difference between the habitat and the niche of a species? Why is it difficult to define "the human habitat"? *646*

2. Describe competitive exclusion. How might two species that compete for the same resource coexist? *648–649*

3. Define the difference between a predator and a parasite. *650*

4. Define primary and secondary succession. *654*

5. What is a climax community, and how does the climax-pattern model help explain its structure? *654–655*

Self-Quiz (Answers in Appendix IV)

1. A habitat _____ .
   a. has distinguishing physical and chemical features
   b. is where individuals of a species normally live
   c. is occupied by various species
   d. both a and b
   e. a through c are correct

2. A niche _____ .
   a. is the sum of activities and relationships in a community by which individuals of a species secure and use resources
   b. is unvarying for a given species
   c. shifts in large and small ways
   d. both a and b
   e. both a and c

3. A two-way flow of benefits in mutualistic interactions between species is an outcome of _____ .
   a. close cooperativeness   c. resource partitioning
   b. two-way exploitation   d. competitive coexistence

4. Two species in the same habitat can coexist when they_____ .
   a. differ in their use of resources
   b. share the same resource in different ways
   c. use the same resource at different times
   d. all are correct

5. A predator population and prey population _____ .
   a. always coexist at relatively stable levels
   b. may undergo cyclic or irregular changes in density
   c. cannot coexist indefinitely in the same habitat
   d. both b and c

6. All parasites _____ .
   a. tend to kill their hosts   c. consume host tissues
   b. can kill novel hosts   d. both b and c

7. In _____ , a disturbed site in a community recovers and moves again toward the climax state.
   a. the area effect   c. primary succession
   b. the distance effect   d. secondary succession

8. The number of coexisting species is _____ in the tropics and _____ toward the poles.
   a. highest; declines systematically
   b. lowest; increases systematically
   c. highest; declines sporadically
   d. lowest; increases sporadically

9. Match the terms with the most suitable descriptions.
   _____ jump dispersal
   _____ area effect
   _____ pioneer species
   _____ climax community
   _____ keystone species

   a. opportunistic colonizer of vacant or newly vacated places
   b. dominates community structure
   c. rapid transport over great distance
   d. more biodiversity on large islands than small ones at same distance from source areas
   e. stable, self-perpetuating array of species

Critical Thinking

1. Think of possible examples of competitive exclusion besides the ones used in the chapter, as by considering some of the animals and plants living in your own neighborhood.

2. Several years ago, someone thought it would be a great idea to introduce Nile perch into Lake Victoria in East Africa. People had been using simple, traditional methods of fishing for thousands of years, but now they were taking too many fish:

Soon there would be too few fish to feed the local populations and no excess catches to sell. But Lake Victoria is a big lake, and the Nile perch is a big fish. It seemed an ideal combination to attract commercial fishermen, with big nets, from the outside world.

Native fishermen favored cichlids, fishes that eat mostly detritus and aquatic plants. The Nile perch eats cichlids as well as other fishes. Having had no prior evolutionary experience with Nile perch, the 200 coexisting, native species of cichlids had no defenses against it. The Nile perch ate its way through cichlid populations and destroyed the lake's biodiversity. Levels of dissolved oxygen plummeted near the lake bottom and contributed to frequent fish kills. By 1990, fishermen were catching mostly Nile perch, and now there are signs the Nile perch population is about to crash.

As if that weren't enough, the Nile perch is an oily fish. Unlike cichlids, which can be sun-dried, it must be preserved by smoking, and smoking requires firewood. The people started cutting down more trees in local forests, and trees are not rapidly renewable resources. People living near Lake Victoria never liked to eat Nile perch anyway; they preferred the flavor and texture of cichlids.

Explain how an ecologist might have helped avoid this mess.

3. Somewhere between predators and parasites are the *parasitoids*, insect larvae that always kill what they eat. Peter Price studied the evolutionary effects of a parasitoid wasp that lays eggs on sawfly cocoons. The sawflies lay their eggs in trees. In time, fertilized eggs give rise to fly larvae, which drop to the forest floor. The larvae spin cocoons after they burrow into leaf litter. Some burrow deeper than others. The first adult sawflies emerge from the cocoons that were the least deeply buried. Later in the season, more sawflies emerge from the more deeply buried ones.

These parasitoid wasps tend to lay eggs on the cocoons closest to the surface of the leaf litter. They exert strong selection pressure on the sawfly population, for the deep-burrowing sawfly individuals are more likely to escape detection. There are only so many flies near the surface, so the wasp able to locate cocoons deeper in the litter will be competitive in securing food for her larvae. Explain how the host species stays ahead in this coevolutionary contest.

Selected Key Terms

area effect *659*	interspecific competition *646*
camouflage *652*	keystone species *656*
climax community *654*	mimicry *652*
climax-pattern model *654*	mutualism *646*
coevolution *650*	niche *646*
commensalism *646*	parasite *650*
community *646*	parasitism *646*
competitive exclusion *649*	pioneer species *654*
distance effect *659*	predation *646*
ecological succession *654*	predator *650*
geographic dispersal *656*	resource partitioning *649*
habitat *646*	symbiosis *646*

Readings

Begon, M., J. Harper, and C. Townsend. 1990. *Ecology: Individuals, Populations, and Communities.* Second edition. Sunderland, Massachusetts: Sinauer.

Krebs, C. 1994. *Ecology.* Fourth edition. New York: HarperCollins.

Krebs, C., et al. 25 August 1995. "Impact of Food and Predation on the Snowshoe Hare Cycle." *Science* 269: 1112–1115.

Moore, P. 1987. "What Makes a Forest Rich?" *Nature* 329: 292.

Smith, R. 1992. *Elements of Ecology.* Third edition. New York: HarperCollins.

Worthington, E. B. 1994. "African Lakes Reviewed: Creation and Destruction of Biodiversity." *Environmental Conservation* 21(3): 201–204.

37 ECOSYSTEMS

Crêpes for Breakfast, Pancake Ice for Dessert

Think of Antarctica, and you think of ice. Mile-thick slabs of the stuff hide all but a small fraction of a vast continent whipped by fierce winds and kept frozen by murderously low temperatures, on the order of −100°F. Yet, on patches of exposed rocky soil and on nearby islands, mosses and lichens grow. There, during the breeding season, various penguins as well as seals form great noisy congregations, then reproduce and raise offspring. They cruise offshore or venture out in the open ocean in pursuit of food—krill, fishes, and squids.

The first explorers of Antarctica called it "The Last Place on Earth." For all of its harshness, though, the eerie, unspoiled beauty fired the imagination of those first travelers and others who followed them.

In 1961, thirty-eight nations signed a treaty to set aside Antarctica as a reserve for scientific research. This was the start of scientific outposts—and of the trashing of Antarctica. At first, the researchers discarded a few oil drums and old tires. Then prefabricated villages went up. Onto the ice and into the water went garbage and used equipment. Just offshore, marine life became acquainted with sewage and chemical wastes.

Antarctica also became a destination of cruise ships (Figure 37.1a). Every summer thousands of tourists,

a

Figure 37.1 (a) A boatload of tourists crossing a channel of "pancake ice" off the coast of Antarctica.
(b) On the shore of an island near Antarctica, tourists get close to nature—maybe too close.

fortified by three sumptuous meals a day plus snacks, are ferried from ship to land across channels glistening with pancake ice. They trample the sparse vegetation and bob around the penguins, cameras clicking. We don't know what the long-term impact of tourism will be, but it pales in comparison to other vulnerabilities. Nations looking for new sources of food started licking their chops over Antarctica's krill. Word got out about potentially rich deposits of uranium, oil, and gold—and by 1988, treaty nations were poised to authorize digs and drillings. Of course, oil spills or commercial krill harvesting could easily destroy the fragile ecosystem.

b

For example, the tiny, shrimplike krill are all that Adélie penguins eat. They are a key food resource for baleen whales and other marine animals that are, in turn, food for still others.

In 1991, the treaty nations thought about all of this and imposed a fifty-year ban on mineral exploration. Research stations started to bury or incinerate wastes, treat raw sewage, and take other steps to curb pollution. Tour operators promised to supervise tourists. Krill are not being harvested on a massive scale. Yet.

Because Antarctica seems so remote from the rest of the world, it is easier to see how life is interconnected there. We can ask whether harm to one species or one habitat will lead to collapse of the whole, and be fairly sure of the answer. What about places not as sharply defined? Does interconnectedness prevail there, also? Are other places as vulnerable to disturbance or more resilient? These topics will now occupy our attention.

KEY CONCEPTS

1. An ecosystem is an association of organisms and their physical environment, interconnected by an ongoing flow of energy and a cycling of materials through it.

2. Every ecosystem is an *open* system, with inputs and outputs of both energy and nutrients.

3. Energy flows in one direction through an ecosystem. Most commonly, the flow begins when photosynthetic autotrophs harness sunlight energy and convert it to forms that they and other organisms of the ecosystem are able to use. These autotrophs are the primary producer organisms for the ecosystem.

4. Energy-rich organic compounds become stored in the tissues of primary producers. They are the foundation for the ecosystem's food webs, which consist of consumers, decomposers, and detritivores at different feeding levels.

5. Water and major nutrients, such as carbon and nitrogen, move from the physical environment, through organisms, then back to the environment. Their movements, known as biogeochemical cycles, are global in scale.

6. Human activities are disrupting the natural cycles of nature and so are endangering ecosystems.

Overview of the Participants

Diverse natural systems abound on the Earth's surface. In climate, landforms, soil, vegetation, animal life, and other features, deserts differ from hardwood forests, which differ from tundra and prairies. In biodiversity and physical properties, seas differ from reefs, which differ from lakes. *Yet despite the differences, such systems are alike in many aspects of their structure and function.*

With few exceptions, each system runs on energy that plants and other photosynthesizers capture from the sun. Photosynthesizers, recall, are the most common autotrophs (self-feeders). They convert sunlight energy to chemical energy, which they use to construct organic compounds from inorganic raw materials. By securing energy from the environment, autotrophs are **primary producers** for the entire system (Figure 37.2).

All other organisms in the system are heterotrophs, not self-feeders. They extract energy from compounds that primary producers put together. **Consumers** feed on the tissues of other organisms. The consumers called *herbivores* eat plants, *carnivores* eat animals, *omnivores* eat both, and *parasites* extract energy from living hosts that they live in or on. Still other heterotrophs, called **decomposers**, include fungi and bacteria that obtain energy by breaking down the remains or products of organisms. Others, the **detritivores**, obtain nutrients from decomposing particles of organic matter. Crabs and earthworms are examples of detritivores.

Autotrophs secure nutrients as well as energy for the entire system. During growth, they take up water and carbon dioxide (as sources of oxygen, carbon, and hydrogen) and dissolved minerals (including nitrogen and phosphorus). Such materials are building blocks for carbohydrates, lipids, proteins, and nucleic acids. When decomposers and detritivores consume organic matter, they break it down to small inorganic molecules. Unless something removes the molecules from the system, as by runoff from a meadow, the autotrophs can use the molecules again as nutrients.

What we have just described in broad outline is an ecosystem. An **ecosystem** is an array of organisms and their physical environment, all interacting through a one-way flow of energy and a cycling of materials. It is an open system, unable to sustain itself. Each ecosystem requires an *energy input* (as from the sun) and often *nutrient inputs* (as from a creek that delivers dissolved minerals to a lake). It has energy and nutrient outputs. Energy cannot be recycled; in time, most of the energy that autotrophs fixed is lost to the environment, mainly as metabolic heat. Nutrients do get cycled, but some are still lost. Most of this chapter deals with the inputs, internal transfers, and outputs of ecosystems.

Structure of Ecosystems

Trophic Levels   The organisms of an ecosystem can be classified in terms of their functions in a hierarchy of feeding relationships, called **trophic levels** (from *troph*, meaning nourishment). "Who eats whom?" we can ask. As organism *B* eats organism *A*, energy gets transferred to *B* from *A*. All organisms at a given trophic level are the same number of transfer steps away from an energy input into the ecosystem. For example, Table 37.1 lists

Figure 37.2 Model of an ecosystem. Energy flows in only one direction—*through* the ecosystem. Nutrients are cycled *within* the ecosystem, among its varied autotrophs and heterotrophs.

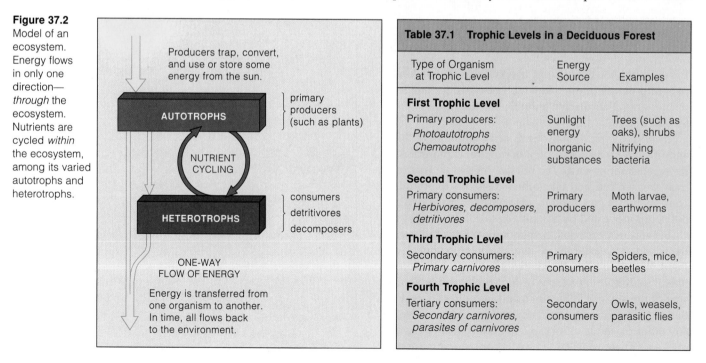

Table 37.1 Trophic Levels in a Deciduous Forest		
Type of Organism at Trophic Level	Energy Source	Examples
First Trophic Level		
Primary producers:		
Photoautotrophs	Sunlight energy	Trees (such as oaks), shrubs
Chemoautotrophs	Inorganic substances	Nitrifying bacteria
Second Trophic Level		
Primary consumers:		
Herbivores, decomposers, detritivores	Primary producers	Moth larvae, earthworms
Third Trophic Level		
Secondary consumers:		
Primary carnivores	Primary consumers	Spiders, mice, beetles
Fourth Trophic Level		
Tertiary consumers:		
Secondary carnivores, parasites of carnivores	Secondary consumers	Owls, weasels, parasitic flies

Producers trap, convert, and use or store some energy from the sun.

AUTOTROPHS — primary producers (such as plants)

NUTRIENT CYCLING

HETEROTROPHS — consumers, detritivores, decomposers

ONE-WAY FLOW OF ENERGY

Energy is transferred from one organism to another. In time, all flows back to the environment.

Figure 37.3 Food web in the waters of the Antarctic. Many more participants, including decomposers, are not shown.

organisms of a forest ecosystem. Being closest to the energy input into this system (sunlight), trees and other primary producers are at the *first* trophic level. Primary consumers—which feed directly on primary producers—include herbivores (such as leaf-eating larvae of moths) and detritivores (such as earthworms that feed on fallen leaves). Primary consumers are at the *second* trophic level. At the *third* level, we find primary carnivores (a variety of beetles, spiders, and birds) that prey on primary consumers. At the *fourth* level are secondary carnivores, such as owls, that feed on species of the third trophic level.

By this classification, then, all organisms at a given trophic level have the same sets of predators, prey, or both. But remember this qualification: many decomposers, people, and other omnivores feed at several trophic levels, so they must be partitioned among levels or assigned to one of their own.

Food Webs Often a straight-line sequence of who eats whom in an ecosystem is called a **food chain**. You will have a hard time finding such simple, isolated cases. Why? Most often, the same food source is a part of more than one chain,

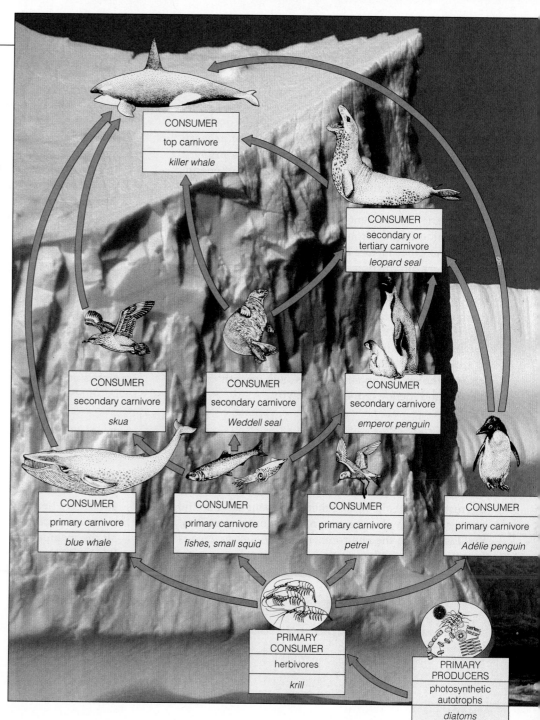

especially when it is at low trophic levels. It is much more accurate to think of food chains as *cross-connecting* with one another—that is, as **food webs**. Figure 37.3 shows part of a food web in the seas around Antarctica.

A bad-luck story about a fisherman can clarify the difference between a food chain and food web. Suppose a fisherman nets fish that were feeding on algae near the ocean's surface. At lunchtime, he cooks some fish but later loses his footing and falls into the water, where sharks lurk. You might think this is a simple food chain: algae ⟶ fish ⟶ fisherman ⟶ shark. But alternate feeding relationships cut into it. Crustaceans also were grazing on algae, small squids and midsized fishes were feeding on the crustaceans, and larger fishes were

feeding on smaller ones. Sharks may have been about to feed on those larger and midsized fishes. The fisherman ate cooked fish and onions, and therefore shifted between trophic levels (as carnivore and herbivore). He was more omnivorous than this, for he sipped wine—derived from the fermenting activities of the decomposer organisms called yeasts.

An ecosystem consists of producers, consumers, decomposers, and detritivores and the physical environment, connected by one-way energy flow and a cycling of materials.

A food web is a network of crossing, interlinked food chains involving primary producers, consumers, and decomposers.

Primary Productivity

To get an idea of how energy flow is studied, consider a land ecosystem with multicelled plants as the primary producers. The rate at which the ecosystem's primary producers capture and store a given amount of energy in a specified time interval is the **primary productivity**. How much energy actually gets stored depends on how many plants are present and on the balance between photosynthesis and aerobic respiration in the plants.

Several other factors help determine the amount of net primary production, its seasonal patterns, and its distribution through the habitat. And they do so in ecosystems on land and in the seas (Figure 37.4). For example, the size and form of the primary producers affect how much they accomplish. So does availability of minerals, the range of temperatures, and the amount of sunlight and rainfall during a growing season. The harsher the environment, the less new growth on the plants—and the lower the productivity.

Major Pathways of Energy Flow

In what direction does energy flow through ecosystems on land? Plants fix only a small part of the energy from the sun. They store half of that in new tissues but lose the rest as metabolic heat. Other organisms tap into the energy stored in plant tissues, remains, or wastes. They, too, lose heat to the environment. *All of these heat losses represent a one-way flow of energy out of the ecosystem.*

Energy from a primary source flows in one direction through two kinds of food webs. In **grazing food webs**, the energy flows from plants to herbivores, and then through an assortment of carnivores. In **detrital food webs**, it flows mainly from plants through detritivores and decomposers. Usually the two kinds of food webs cross-connect, as when a herring gull of a grazing food web eats a crab of a detrital food web (Figure 37.5).

The amount of energy moving through food webs differs from one ecosystem to the next and often varies with the seasons. In most cases, however, most of the net primary production passes through detrital food webs. You may doubt this; after all, when cattle graze heavily on pasture plants, about half the net primary production enters a grazing food web. But cattle don't use all the stored energy. Quantities of undigested plant parts and feces become available for decomposers and detritivores. Also consider marshes. There, most of the stored energy is not used until parts of plants die and become available for detrital food webs.

Ecological Pyramids

Often ecologists will represent the trophic structure of an ecosystem in the form of an **ecological pyramid**. In such pyramids, the primary producers form a base for successive tiers of consumers above them.

Some pyramids are based on biomass (the weight of all the members at each trophic level). For example, for Silver Springs, Florida, a small aquatic ecosystem, the

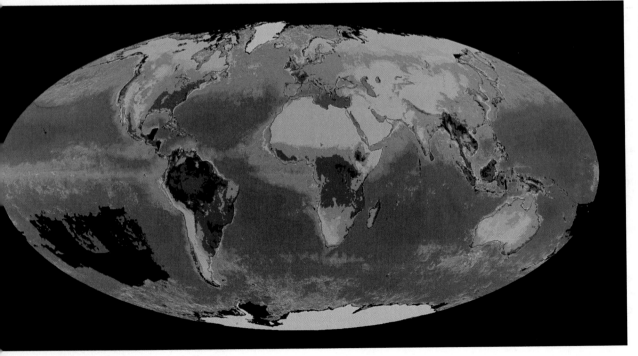

Figure 37.4 A summary of three years of satellite data on the Earth's primary productivity. *Dark green* denotes rain forests and other highly productive regions. *Yellow* denotes deserts (low productivity). The productivity in oceans, ranging from high to low, is coded *red* down through *orange, yellow, green,* and *blue*.

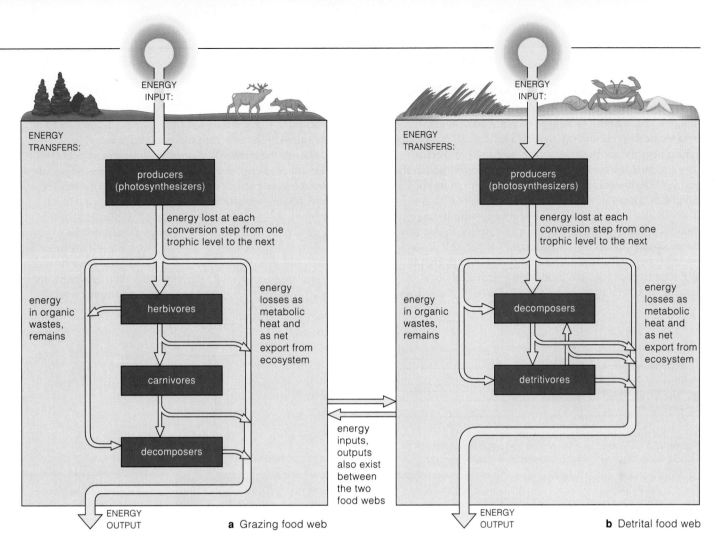

Figure 37.5 One-way flow of energy through two kinds of cross-connected food webs in ecosystems.

pyramid of biomass (measured as grams/square meter during one specified interval) would look like this:

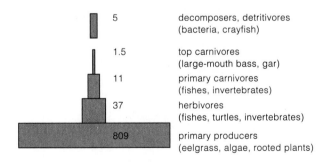

5	decomposers, detritivores (bacteria, crayfish)
1.5	top carnivores (large-mouth bass, gar)
11	primary carnivores (fishes, invertebrates)
37	herbivores (fishes, turtles, invertebrates)
809	primary producers (eelgrass, algae, rooted plants)

Some pyramids of biomass are "upside-down," with the smallest tier on the bottom. Think of a small pond. Its biomass of fast-growing and rapidly reproducing phytoplankton may support a greater biomass of zoo-plankton in which individuals are bigger, grow more slowly, and consume less energy per unit of weight.

An **energy pyramid** is a more useful way to depict an ecosystem's trophic structure. Such pyramids show the energy losses at each transfer to a different trophic level in the ecosystem. They have a large energy base at the bottom and are always "right-side up." As you will see from the next section, they provide a better picture of how energy flows in ever diminishing amounts through successive trophic levels of the ecosystem.

Energy flows into food webs of ecosystems from an outside source (mainly the sun). Energy leaves ecosystems mainly by losses of metabolic heat, which each organism generates.

Gross primary productivity is the total rate of photosynthesis in an ecosystem during a specified interval. *Net* primary productivity is the rate of energy storage in plant tissues in excess of the rate of aerobic respiration by primary producers. Heterotrophic consumption affects the rate of energy storage.

Living tissues of photosynthesizers are the basis of grazing food webs. The remains of photosynthesizers and consumers are the basis of detrital food webs.

The loss of metabolic heat and the shunting of food energy into organic wastes mean that usable energy flowing through consumer trophic levels declines at each energy transfer.

ENERGY FLOW AT SILVER SPRINGS, FLORIDA

Imagine you are with ecologists who are bent on gathering data to construct an energy pyramid for a small freshwater spring over the course of one year. You observe them as they measure the energy that each type of individual in the spring takes in, loses as metabolic heat, stores in its body tissues, and loses in waste products. You see that they multiply the energy per individual by population size, then they calculate energy inputs and outputs. In this way, the ecologists are able to express the flow per unit of water (or land) per unit of time. The energy pyramid in Figure 37.6 summarizes the data from a long-term study of a grazing food web in an aquatic ecosystem—Silver Springs,

Florida. The larger diagram in Figure 37.7 shows some of the calculations on which the pyramid is based.

Given the metabolic demands of organisms and the amount of energy shunted into their organic wastes, only about 6 to 16 percent of the energy entering one trophic level becomes available for organisms at the next level. Because the efficiency of the energy transfers is so low, ecosystems in general have no more than four consumer trophic levels.

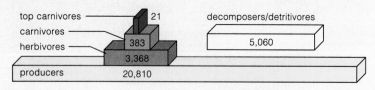

Figure 37.6 Pyramid of energy flow through Silver Springs, Florida, as measured in kilocalories/square meter/year.

Figure 37.7 Breakdown of the annual energy flow through Silver Springs, Florida, as measured in kilocalories/square meter/year.

The primary producers in this small spring are mostly aquatic plants. The carnivores are insects and small fishes; top carnivores are larger fishes. The energy source (sunlight) is available throughout the year. The spring's detritivores and decomposers cycle organic compounds from the other trophic levels.

The producers trapped 1.2 percent of the incoming solar energy, and only a little more than a third of this amount became fixed in new plant biomass (4,245 + 3,368). The producers used more than 63 percent of the fixed energy for their own metabolism.

About 16 percent of the fixed energy was transferred to herbivores, and most of it was used for metabolism or transferred to detritivores and decomposers.

Of the energy that did get transferred to herbivores, only 11.4 percent reached the next trophic level (carnivores). These carnivores used all but about 5.5 percent, which was transferred to top carnivores.

By the end of the specified time interval, all 5,060 kilocalories that were transferred through the system appeared as metabolically generated heat.

Bear in mind, this diagram is oversimplified, for no community is isolated from others. New individual organisms and substances continually drop from overhead leaves and branches into the springs. Also, organisms and substances are gradually lost by way of a stream that leaves the springs.

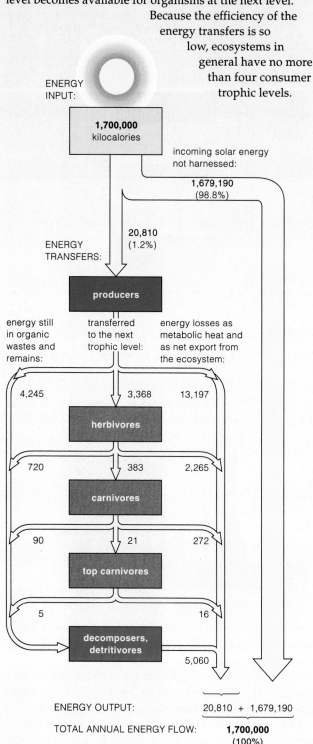

BIOGEOCHEMICAL CYCLES—AN OVERVIEW

Availability of *nutrients* as well as energy profoundly influences the structure of an ecosystem. Its photosynthetic producers require carbon, hydrogen, and oxygen, which they get from water and air. They require nitrogen, phosphorus, and other mineral ions. A scarcity of even one of these minerals has widespread adverse effects, for it lowers the ecosystem's primary productivity.

In a **biogeochemical cycle**, ions or molecules of a nutrient are transferred from the environment to organisms, then back to the environment—part of which functions as a vast reservoir for them. The transfers of nutrients into and out of that reservoir are usually less than the exchanges with organisms and among organisms.

Figure 37.8 is a simple model of the relationship between the geochemical part of the cycles and most ecosystems. This model is based on four factors. *First*, the elements that producer organisms use as nutrients usually are available to them in the form of mineral ions, such as ammonium (NH_4^+). *Second*, the nutrient reserves of an ecosystem are maintained by inputs from the physical environment and by cycling activities of decomposers and detritivores. *Third*, the amount of a nutrient being cycled through most of the major ecosystems is greater than the amount that enters and departs in a given year. *Fourth*, rainfall or snowfall, metabolism (such as nitrogen fixation), and the gradual weathering of rocks are common sources of inputs to an ecosystem's nutrient reserves. For ecosystems on land, the losses of mineral ions through runoff are typical outputs from the nutrient reserves.

There are three categories of biogeochemical cycles, based on the part of the environment that contains the greatest portion of the specified ion or molecule. As you will see, in the *hydrologic* cycle, oxygen and hydrogen move in the form of water molecules. This movement is also called the global water cycle. In *atmospheric* cycles, a large percentage of the nutrient is in the form of an atmospheric gas. For example, this is true of gaseous

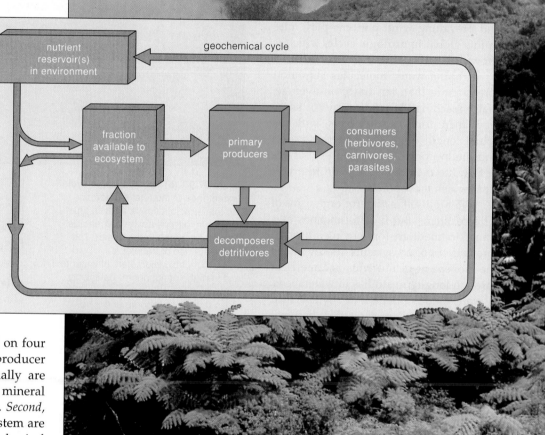

Figure 37.8 A generalized model of nutrient flow through a land ecosystem. The overall movement of nutrients from the physical environment, through organisms, and back to the environment constitutes a biogeochemical cycle.

forms of nitrogen and carbon (mainly carbon dioxide). *Sedimentary* cycles deal with phosphorus and other solid nutrients that do not have gaseous forms. Such nutrients move from land to the seafloor and "return" to land only through geological uplifting—which may take millions of years. Their largest storehouse is the Earth's crust.

Primary productivity—hence the structure of an ecosystem—depends largely on the availability of nutrients.

In a biogeochemical cycle, the ions or molecules of a nutrient move slowly through the environment, then rapidly among organisms, then back to the environmental reservoir for them.

Driven by solar energy, the waters of the Earth move slowly and on a vast scale from the ocean, into the atmosphere, to the land, and back to the ocean. Water evaporating into the atmosphere remains aloft as vapor, clouds, and ice crystals. It falls to Earth as precipitation—rain and snow, for the most part. Ocean currents and prevailing wind patterns play roles in this global **hydrologic cycle**, which is shown in Figure 37.9.

Airborne water molecules stay aloft for no more than ten days, on average. Water reaching the land as rain or snow stays there for 10 to 120 days, with the actual length depending on the season and the location. Then water evaporates or flows to the ocean, the main reservoir from which it evaporates.

Water in itself is vital for organisms of all ecosystems. But it also functions as a transport medium for moving nutrients into and out of ecosystems. Its key role in the movement of nutrients became clear through long-term studies in watersheds. A **watershed** is a region of any specified size in which all precipitation becomes funneled into a single stream or river. Figure 37.10 gives a simple model of the pathways that water can follow through a watershed.

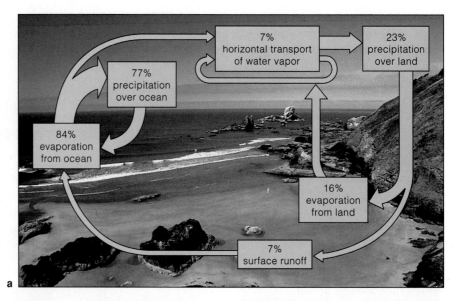

a

Figure 37.9 (**a**) The hydrologic cycle. Percentages of the total evaporation and total precipitation at a given time are shown, as well as the percentage of runoff from land to the ocean. The annual net rate of transfer is 37.3×10^3 cubic kilometers of water to land from the atmosphere. Balancing this is a comparable net loss from the ocean to the atmosphere and a net gain of that amount by the ocean (through runoff).

(**b**) Global water budget. Values reflect the annual movement of water into and out of the atmosphere.

	Annual Volume (10^{18} grams)
Ocean	1,380,000
Sedimentary layers	210,000
Evaporation from ocean	319
Precipitation over ocean	283
Precipitation over land	95
Evaporation from land	59
Runoff and groundwater	36
Atmospheric water vapor	13

b

Figure 37.10 Model of the movement of water through a watershed.

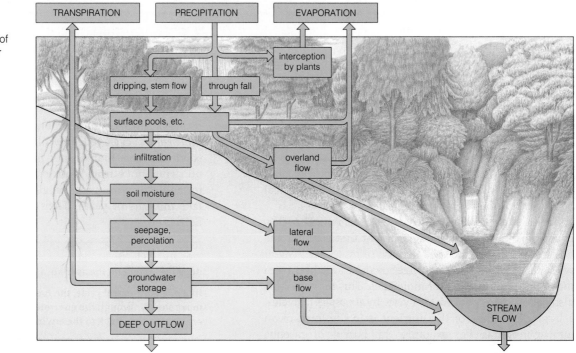

a

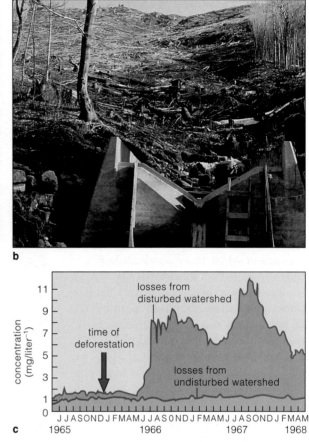

b

Figure 37.11 Studies of disturbances to a forest ecosystem. (**a**) In certain experimental watersheds in the Hubbard Brook Valley, New Hampshire, researchers studied the effects of deforestation. (**b**) All water that was to drain from an area being studied had to flow over the V-notched concrete structure, where measurements were taken. This area was deforested, then herbicides were applied to prevent regrowth for three years.

(**c**) The *green* arrow marks the time of deforestation. Concentrations of calcium ions and other mineral ions in water moving through the concrete catchment were compared against those in water passing through a control catchment in an undisturbed part of the same region. As you can see, calcium losses were six times greater from the deforested region.

c

(graph:) concentration (mg/liter^{-1}) versus time — losses from disturbed watershed; time of deforestation; losses from undisturbed watershed. J J A S O N D J F M A M J J A S O N D J F M A M J J A S O N D J F M A M — 1965 1966 1967 1968

A watershed may be any size that is of interest to an investigator. To give but two examples, the Mississippi River watershed extends across roughly one-third of the continental United States, whereas the watersheds at Hubbard Brook Valley in the White Mountains of New Hampshire are 14.6 hectares (36 acres), on the average. Most of the water entering a watershed seeps into soil or becomes surface runoff that moves into streams (Figure 37.11). Plants take up water and dissolved minerals from the soil, then they lose water by transpiration.

Measurements of inputs and outputs at watersheds have practical application. For example, cities that draw on the surface supplies in watersheds can adjust their water usage on the basis of seasonal variations in the volume of water. Watershed studies also revealed the importance of vegetation in the movement of nutrients through the ecosystem phase of biogeochemical cycles.

For example, you might think that water draining a watershed would rapidly leach calcium ions and other minerals. Yet in studies of young, undisturbed forests in the Hubbard Brook watersheds, each hectare lost only about 8 kilograms of calcium—and rainfall and the weathering of rocks brought in calcium replacements.

Tree roots were also "mining" the soil, so calcium was being stored in a growing biomass of tree tissues.

In some experimental watersheds, nutrient outputs shifted. Researchers stripped the vegetation cover but did not disturb the soil. Then they applied herbicides to the soil for three years to prevent regrowth. After that time, stream outflow from the experimental watersheds carried more than *six times* as much calcium as the stream outflow from undisturbed ones (Figure 37.11*c*).

Because calcium and other nutrients move so slowly through geochemical cycles, deforestation may disrupt nutrient availability for an entire ecosystem. This is true of forests that cannot regenerate themselves over the short term. Coniferous forests of the Pacific Northwest and elsewhere are like this.

In the hydrologic cycle, water slowly moves on a global scale from the ocean reservoir, through the atmosphere, onto land, then back to the ocean.

In ecosystems on land, plants stabilize the soil and absorb dissolved minerals. By doing so, they minimize the loss of soil nutrients in runoff from land.

37.6 CARBON CYCLE

In one of the most vital of all atmospheric cycles, carbon moves from vast reservoirs —the ocean and atmosphere—on through organisms in ecosystems, then back to the reservoirs. Figure 37.12*a* shows the global movement, which is known as the **carbon cycle**.

Carbon enters the atmosphere each second. It does so as countless living cells engage in aerobic respiration, as fossil fuels burn, and as volcanoes erupt and release carbon from rocks in the Earth's crust. The world ocean holds most of the carbon, in dissolved form (Figure 37.12*b*). The atmosphere, soils, and plant biomass represent the next largest holding stations for carbon. Most atmospheric carbon is in the form of carbon dioxide (CO_2).

By engaging in carbon dioxide fixation, photosynthetic autotrophs lock up billions of metric tons of carbon atoms in organic compounds every year (Section 5.7). The average time that an ecosystem holds any one carbon atom varies greatly. In tropical rain forests, organic remains decompose rapidly, so not much carbon is tied up in litter on the soil surface. In marshes, bogs, and other anaerobic habitats, decomposers cannot break down organic compounds to smaller bits, so carbon slowly accumulates in compressed organic matter, such as peat.

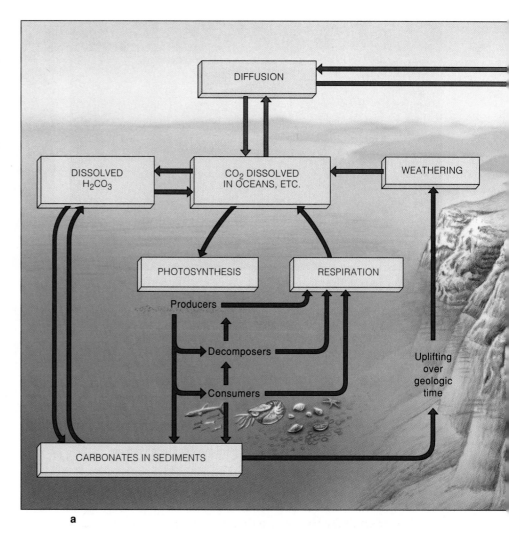

a

Figure 37.12 (**a**) Global carbon cycle. The portion of the diagram on this page illustrates the movement of carbon through typical marine ecosystems. The portion of the diagram on the facing page illustrates the movement of carbon through ecosystems on land.

(**b**) The present-day global carbon budget. The photograph to the far right shows a Los Angeles freeway system under its self-generated blanket of smog at twilight. The exhaust from vehicles as well as fossil fuel burning by industries and homes adds carbon and other substances to the atmosphere.

The values in (**b**) are reported in W. Schlesinger, 1991, *Biogeochemistry: An Analysis of Global Change*, New York: Academic Press.

	Amount per Year (10^{15} grams)
Carbon reservoirs and holding stations:	
Dissolved in ocean	38,000
Present in soil	1,500
Present in atmosphere	720
Plant biomass	560
Annual fluxes:	
From atmosphere to plants (carbon fixation)	120
From atmosphere to ocean	107
To atmosphere from ocean	105
To atmosphere from plants	60
To atmosphere from soil	60
To atmosphere from fossil fuel burning	5
To atmosphere from net destruction of plants	2
To ocean from runoff	0.4
Burial in ocean sediments	0.1

b

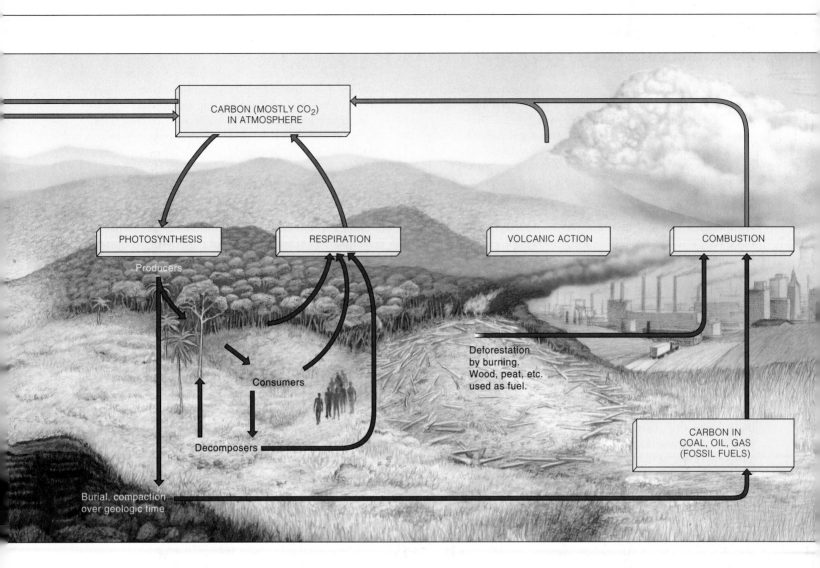

CARBON (MOSTLY CO₂) IN ATMOSPHERE

PHOTOSYNTHESIS

Producers

RESPIRATION

VOLCANIC ACTION

COMBUSTION

Consumers

Decomposers

Deforestation by burning. Wood, peat, etc. used as fuel.

CARBON IN COAL, OIL, GAS (FOSSIL FUELS)

Burial, compaction over geologic time

In food webs of ancient aquatic ecosystems, carbon became incorporated into shells and other hard parts. Later, when shelled organisms died, they sank through the water and became buried in sediments. In deeper sediments, carbon remained buried for many millions of years, until geologic forces raised part of the seafloor above the ocean surface (Section 17.3). Also over great time spans, the carbon-containing compounds of many ancient forests were converted to reserves of petroleum, coal, and gas—which we presently tap as fossil fuels.

Now fossil fuel burning and other human activities are putting more carbon into the atmosphere than can be cycled naturally to the ocean reservoir and to other holding stations (Figure 37.12b). Because this increase amplifies the greenhouse effect, it may be contributing to global warming. The section to follow describes this effect and some possible outcomes of its modification.

Extensive fossil fuel burning and other human activities may be contributing to imbalances in the global carbon budget.

FROM GREENHOUSE GASES TO A WARMER PLANET?

The Greenhouse Effect   Atmospheric concentrations of gaseous molecules play a profound role in shaping the average temperature near the Earth's surface. That temperature has enormous effect on the global climate.

Countless molecules of carbon dioxide, water, ozone, methane, nitrous oxide, and chlorofluorocarbons are key players in the interactions that dictate global temperature. Collectively, they act somewhat like a pane of glass in a greenhouse—hence their name, the "greenhouse gases." Wavelengths of visible light can get around them and reach the Earth's surface. However, greenhouse gases impede the escape of longer, infrared wavelengths—that is, heat—from the Earth into space. How? The gaseous molecules absorb these wavelengths, then reradiate much of the absorbed energy back toward Earth (Figure 37.13). As an outcome of the constant bombardment, absorption, and reradiation of heat energy, the greenhouse gases cause heat to build up in the lower atmosphere. The **greenhouse effect** is the name for this warming action.

Global Warming Defined   Without the action of greenhouse gases, the Earth's surface would be cold and lifeless. However, there can be too much of a good thing. Largely as an outcome of human activities, greenhouse gases are building to atmospheric levels that are higher than they were in the past (Figure 37.14). They may be contributing to long-term higher temperatures at the Earth's surface—an effect known as **global warming**.

What is so alarming about a warmer planet? Suppose the temperature of the lower atmosphere were to rise by only 4°C (7°F). The increase might cause sea levels to rise by about 2 feet, or 0.6 meter. Why? Temperatures near the ocean surface would increase—and water expands when heated. Also, global warming could make glaciers and the polar ice sheets melt faster. The volume of water released this way alone would flood low coastal regions.

Imagine a long-term rise in sea level, combined with high tides and storm waves. The waterfronts of Vancouver, Seattle, San Diego, New York, Boston, Galveston, Hilo, and all other cities perched near the rim of the world ocean would be submerged. So would the agricultural lowlands and deltas in India, China, and Bangladesh—where much of the world's rice is grown. Huge tracts of Florida and Louisiana would face saltwater intrusions. And what if global warming disturbs regional patterns of precipitation and temperature? We might expect deserts to expand and the interiors of the great continents to become drier than they already are. Will nations controlling the great grain belts be affected? Will other nations be better or worse off? Imagine the economic and political consequences.

Some predicted effects of climate change are emerging. For example, in 1996, we found that the water 200 meters below the Arctic icecap is 1°C warmer than it was just five years ago. If rapid warming continues, the icecap will disappear in the next century. It may already be disrupting a vast, deep current in the North Atlantic that affects circulation through the world ocean. Also in the 1990s, record-breaking hurricanes, flooding, and droughts have adversely affected crop production around the world. And beaches have been shrinking by two to three feet a year.

Evidence of an Intensified Greenhouse Effect   In the late 1950s, some researchers on a mountaintop in the Hawaiian Islands started measuring the concentrations of different greenhouse gases. Their monitoring activities are still going on. They chose the remote site because it was free of local contamination, and it was representative of average conditions for the Northern Hemisphere.

a Rays of sunlight penetrate the lower atmosphere and warm the Earth's surface.

b The Earth's surface radiates heat (infrared wavelengths) to the lower atmosphere. Some heat escapes into space. But greenhouse gases and water vapor absorb some infrared wavelengths and reradiate a portion back toward the Earth.

c As concentrations of greenhouse gases increase in the atmosphere, more heat is trapped near the Earth's surface. The surface temperature of the world ocean rises, more water evaporates into the atmosphere, and the Earth's surface temperature rises.

Figure 37.13   The greenhouse effect, as executed mainly by carbon dioxide, water, ozone, methane, nitrous oxide, and chlorofluorocarbons in the atmosphere.

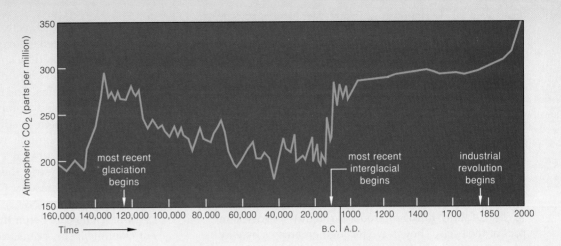

Figure 37.14 Chart of the shifts in the atmospheric concentrations of carbon dioxide, as correlated with the most recent glaciation and interglacial periods during the past 160,000 years.

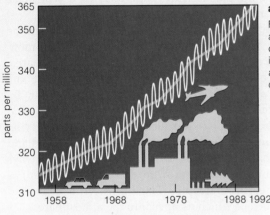

a CARBON DIOXIDE (CO$_2$)
Fossil fuel burning and deforestation are contributing most to increases in the atmospheric levels of this gas.

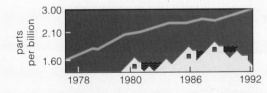

b CHLOROFLUOROCARBONS (CFCs)
Until recent restrictions, these were prevalent in plastic foams, refrigerators, air conditioners, and industrial solvents.

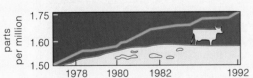

c METHANE (CH$_4$)
Termite activity and anaerobic bacteria in swamps, landfills, and the stomachs of cattle and other ruminants produce large quantities of methane.

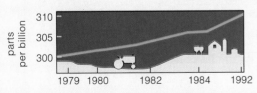

d NITROUS OXIDE (N$_2$O)
Denitrifying bacteria produce nitrous oxide as a metabolic by-product. The gas also is released in great amounts from fertilizers and animal wastes, as in livestock feedlots.

Figure 37.15 Recently documented increases in atmospheric concentrations of four greenhouse gases.

Consider what they found out about carbon dioxide alone. In the Northern Hemisphere, atmospheric levels of carbon dioxide follow the annual cycle of plant growth. They are lower in summer, when photosynthesis rates are highest. They are higher in winter, when photosynthesis slows but aerobic respiration continues. In Figure 37.15a, the troughs and peaks around the graph line show the annual lows and highs. For the first time, scientists had a picture of the integrated effects of the carbon balances for land and water ecosystems of an entire hemisphere. Notice the midline of the troughs and peaks in the cycle. *It steadily increased.* Many scientists take this as evidence of a carbon dioxide buildup that may intensify the greenhouse effect over the next century.

Probably the global burning of fossil fuels is contributing most to the rise in carbon dioxide levels. Deforestation is adding to it; carbon is released when wood burns. Especially during the past four decades, vast tracts of the world's great forests have been cleared and burned at astounding rates (Section 39.4). Also, as plant biomass plummets, the global absorption of carbon dioxide in photosynthesis may decline.

Will atmospheric levels of greenhouse gases continue to increase until the middle of the twenty-first century? Will global temperature rise by several degrees? If a trend is already in motion, we will not be able to reverse it simply by waiting until the last minute to stop fossil fuel burning and deforestation.

Among many scientists, there is widespread agreement that nations must begin preparing for the consequences. For example, we might step up genetic engineering studies to develop drought-resistant and salt-resistant plants. Such plants may prove crucial in regions of saltwater intrusions and climatic change.

Nitrogen, a component of all proteins and nucleic acids, moves in a great atmospheric cycle called the **nitrogen cycle**. The largest reservoir, the atmosphere, is about 80 percent gaseous nitrogen (N_2, or $N\equiv N$). Only certain bacteria, volcanic action, and lightning can break the triple covalent bonds of this molecule and convert it to forms that enter food webs. Today, nearly all nitrogen in soils has been put there by nitrogen-fixing organisms. Soils lose it through activities of bacteria that "unfix" the fixed nitrogen. They lose more through leaching of soils, although this puts nitrogen into streams, lakes, and other aquatic ecosystems (Figure 37.16).

Let's follow nitrogen atoms through the ecosystem part of the cycle. They move through organisms by way of nitrogen fixation, assimilation and biosynthesis, decomposition, ammonification, and nitrification.

In **nitrogen fixation**, a few kinds of bacteria convert N_2 to ammonia (NH_3), which dissolves in the cytoplasm to form ammonium (NH_4^+). *Anabaena, Nostoc,* and other cyanobacteria are nitrogen fixers of aquatic ecosystems. *Rhizobium* and *Azotobacter* fix nitrogen in many land ecosystems. Collectively they fix 200 million metric tons of nitrogen each year! Plants assimilate and use fixed nitrogen in the biosynthesis of amino acids, proteins,

and nucleic acids. Directly or indirectly, plant tissues are the only nitrogen source for animals.

With **decomposition** and **ammonification**, bacteria and fungi break down nitrogen-containing wastes and remains of organisms. The decomposers use part of the released proteins and amino acids for their metabolism. Most of the nitrogen is still in the decay products, as ammonia or ammonium, which plants take up. Also, nitrifying bacteria act on ammonia or ammonium. In **nitrification**, they strip these compounds of electrons, and nitrite (NO_2^-) results. Other bacteria use the nitrite and produce nitrate (NO_3^-), which plants take up.

Some plants are better than others at fixing nitrogen. For example, peas, beans, clover, and other legumes are mutualists with nitrogen-fixing bacteria. And most land plants are mutualists with fungi, forming mycorrhizae that enhance nitrogen uptake (Section 23.1).

The ammonium, nitrite, and nitrate that form in the nitrogen cycle are vulnerable to leaching and runoff. With leaching, recall, soilwater moves out of an area, which thereby loses the nutrients dissolved in it. Some nitrogen also is lost to the air during **denitrification**. By this process, bacteria convert nitrate or nitrite to N_2 and some nitrous oxide (N_2O). Ordinarily, most denitrifying bacteria rely on aerobic respiration. In waterlogged and poorly aerated soil, they switch to anaerobic pathways and use nitrate, nitrite, or nitrous oxide (not oxygen) as the final electron acceptor. Fixed nitrogen is converted to N_2, much of which escapes into the atmosphere.

Humans also affect the cycle in natural ecosystems. For example, nitrogen-containing fertilizers and fossil fuel burning release pollutants that increase soil acidity,

Figure 37.16 The nitrogen cycle in a land ecosystem. The atmosphere is the largest reservoir of nitrogen, which all organisms require for biosynthesis.

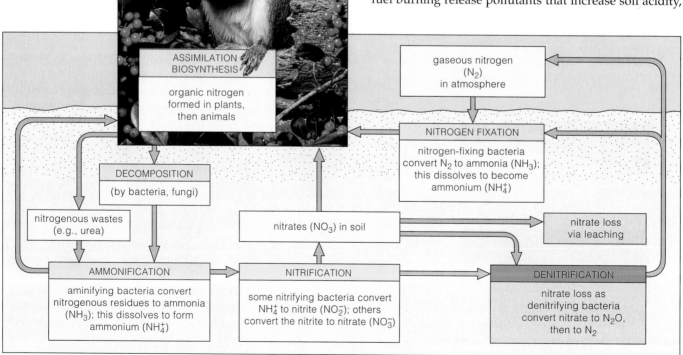

ASSIMILATION BIOSYNTHESIS
organic nitrogen formed in plants, then animals

gaseous nitrogen (N_2) in atmosphere

NITROGEN FIXATION
nitrogen-fixing bacteria convert N_2 to ammonia (NH_3); this dissolves to become ammonium (NH_4^+)

DECOMPOSITION
(by bacteria, fungi)

nitrogenous wastes (e.g., urea)

nitrates (NO_3^-) in soil

nitrate loss via leaching

AMMONIFICATION
aminifying bacteria convert nitrogenous residues to ammonia (NH_3); this dissolves to form ammonium (NH_4^+)

NITRIFICATION
some nitrifying bacteria convert NH_4^+ to nitrite (NO_2^-); others convert the nitrite to nitrate (NO_3^-)

DENITRIFICATION
nitrate loss as denitrifying bacteria convert nitrate to N_2O, then to N_2

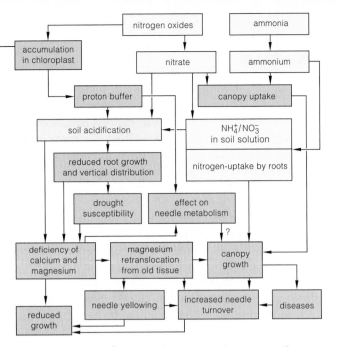

Figure 37.17 How nitrogen oxides, ammonia, ozone, sulfur oxides, and other air pollutants are accelerating the decline of spruce forests. Seedlings growing in acidified soils develop magnesium deficiencies if ammonium levels are high. Needles yellow and drop; photosynthesis suffers. Trees weakened by nutrient imbalances become susceptible to diseases.

which affects root absorption of magnesium, calcium, and potassium ions (Figure 37.17). As another example, nitrogen is lost by poor agricultural practices that amplify soil erosion and leaching. In Europe and North America, farmers rotate their crops, as when they alternate wheat with legumes. Crop rotation and other conservation practices have kept the soils stable and productive, often for thousands of years. But farmers also use heavy applications of nitrogen-rich fertilizers to compensate for nitrogen losses. (With each harvest, nitrogen departs, in tissues of harvested plants.) New strains of crop plants are bred for an ability to take up fertilizers, and crop yields per hectare have doubled, even quadrupled, over the past forty years. Will pest control and soil management technologies sustain high yields indefinitely? We're not certain (Chapter 39).

Finally, we can't get something for nothing. It takes energy to make fertilizer—not from unending sunlight but from fossil fuels. Few once believed that fossil fuel supplies might run out, so fertilizer costs weren't a big concern. It still is common to put more energy into soil (as in fertilizers) than we get out of it (as food). As long as human population growth continues to skyrocket, farmers will be racing to grow as much food as they can for as many people as possible. Soil enrichment with fertilizers is part of the race, as it is now being run.

Nitrogen enters ecosystems on land by way of precipitation and nitrogen-fixing bacteria. Erosion, leaching, and activities of denitrifying bacteria contribute to nitrogen losses.

37.9 PHOSPHORUS CYCLE

We conclude our look at biogeochemical cycling with one of the sedimentary cycles. In the **phosphorus cycle**, phosphorus moves from land to sediments in the seas then back to the land (Figure 37.18). The Earth's crust is the main storehouse for this mineral and for others.

In rock formations on land, phosphorus is typically in the form of phosphates. By the natural processes of weathering and erosion, phosphates enter rivers and streams that transport them to the ocean. There, mainly on continental shelves, phosphorus accumulates with other minerals as insoluble deposits. Millions of years pass. Where crustal plates uplift part of the seafloor, the phosphates become exposed on drained land surfaces. Over time, weathering releases them from the exposed rocks—and the cycle's geochemical phase begins again.

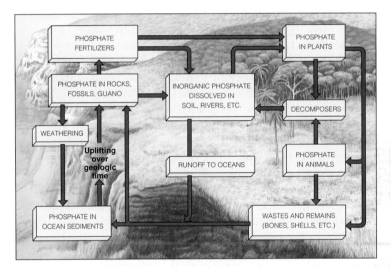

Figure 37.18 The phosphorus cycle.

The ecosystem phase of the cycle is more rapid than the long-term geochemical phase. All organisms require phosphorus for synthesizing phospholipids, NADPH, ATP, nucleic acids, and other compounds. Plants take up dissolved, ionized forms of phosphorus so rapidly and efficiently, they often reduce its soil concentrations to extremely low levels. Herbivores obtain phosphorus by eating plants; carnivores get it by eating herbivores. Both excrete phosphorus as a waste product in urine and feces. Decomposition also releases phosphorus to the soil. Plants then take up phosphorus and so recycle it rapidly within the ecosystem.

The Earth's crust is the primary reservoir for phosphorus and for other minerals that move through ecosystems as part of sedimentary cycles.

37.10 PREDICTING THE IMPACT OF CHANGE IN ECOSYSTEMS

Ecosystem Modeling

We now come full circle to the premise of the story that opened this chapter—that disturbances to one part of an ecosystem can have unexpected effects on other, seemingly unrelated parts.

One approach to predicting unforeseen effects of disturbances is through **ecosystem modeling**. By this method, researchers identify crucial bits of information about different ecosystem components. Then they rely on computer programs and models in order to combine the information. They use the resulting data to predict the outcome of the next disturbance.

For example, an analysis of which species feed on others in the Antarctic food web shown in Figure 37.3 can be turned into a series of equations that describe how many individuals of each population are being consumed. Such equations can be used to predict, say, the impact on the ecosystem if humans overharvest whales or greatly expand the harvesting of krill.

As ecologists attempt to deal with larger and more complex ecosystems, it becomes far more difficult and expensive to run experiments in the field. A modern temptation is to run them instead on the computer. This is a valid exercise *if* the computer model adequately represents the system. The danger is that investigators may not have identified all of the key relationships in the ecosystem and incorporated them accurately into the model. The most crucial fact may be one that we do not yet know, as the following study makes clear.

Biological Magnification—A Case Study

DDT, the first of the synthetic organic pesticides, was first used during World War II. In mosquito-infested, tropical regions of the Pacific, people were vulnerable to a dangerous disease, malaria. DDT helped control the mosquitoes that were transmitting the sporozoan disease agent (*Plasmodium japonicum*). In war-ravaged cities of Europe, other people were suffering from the crushing headaches, fevers, and rashes associated with typhus. DDT helped control the body lice that were transmitting *Rickettsia rickettsii*, the bacterial agent of this terrible disease. After the war, it seemed like a good idea to use DDT against many insects that were agricultural or forest pests, transmitters of pathogens, or merely nuisances in homes and gardens.

DDT is a relatively stable hydrocarbon compound. It is nearly insoluble in water, so you might think that it would stay put and act only where applied. But winds can carry DDT in vapor form; water can transport fine particles of it. DDT also is highly soluble in fats, so it accumulates in the tissues of organisms. Thus, as we now know, DDT can show **biological magnification**.

The term refers to an increase in the concentration of a nondegradable (or slowly degradable) substance in the organisms at higher trophic levels in a food web. Most of the DDT from all organisms that a consumer feeds on during its lifetime becomes concentrated in its tissues. Besides this, many organisms partially metabolize DDT to modified compounds, such as DDE, with different but still disruptive effects. Both DDT and the modified compounds are metabolically disruptive or toxic to *many* aquatic and terrestrial animals.

After World War II, DDT began to move through the global environment, infiltrate food webs, and affect organisms in ways that no one had predicted. In cities where DDT was sprayed to control Dutch elm disease, songbirds started dying. In streams flowing through forests where DDT was sprayed to control pests called spruce budworms, salmon started dying. In croplands sprayed to control one kind of pest, new kinds of pests moved in. *DDT was indiscriminately killing off the natural predators that had been keeping pest populations in check!* It took no great flying leap of the imagination to make the connection. All of those organisms were dying at the same time—and in the same places where DDT had been sprayed.

Then side effects of biological magnification started showing up in places that were far removed from the areas of DDT application—*and much later in time*. Most devastated were species at the top of food webs, such as bald eagles, peregrine falcons, ospreys, and brown pelicans. It happens that a product of DDT breakdown interferes with physiological processes. As one of the consequences, birds produced eggs with brittle shells—and many chick embryos didn't make it to hatching time. Some species were at the brink of extinction.

Since the 1970s, DDT has been banned in the United States, except for restricted applications where public health is endangered. Many of the hardest hit species have partially recovered in numbers. Yet even today, some kinds of birds are still laying thin-shelled eggs. They are picking up DDT at their winter ranges in Latin America. And as recently as 1990, the California State Department of Health recommended that a fishery off the coast of Los Angeles be closed. DDT from industrial waste discharges that ended twenty years before is still contaminating that ecosystem.

Disturbances to one aspect of an ecosystem typically have unexpected effects on other, seemingly unrelated parts.

Predictions of the consequences, as by ecosystem modeling, must identify all of the key relationships in the ecosystem and incorporate them accurately into the model.

SUMMARY

1. An ecosystem is an array of producers, consumers, detritivores, and decomposers and their environment, all interacting through a flow of energy and a cycling of materials. It is an open system, with inputs and outputs of energy and nutrients.

    a. With few exceptions, sunlight is the initial energy source and photoautotrophs are the primary producers. They convert the energy of sunlight to ATP and other forms that are used to synthesize organic compounds from simple inorganic substances.

    b. Primary producers also assimilate many of the nutrients required by all other members of the system.

2. We classify organisms of an ecosystem in terms of their functions in a hierarchy of feeding relationships called trophic levels. Producers are the first level, then come tiers of heterotrophic consumers (the herbivores, carnivores, and omnivores). Decomposers (many fungi and bacteria), detritivores (which eat particles of dead or decomposing matter), and many other organisms that get energy from more than one source cannot be assigned to a single trophic level.

3. Isolated food chains (straight-line sequences of who eats whom in an ecosystem) are rare in nature. They cross-connect with one another, as food webs.

4. The rate at which primary producers capture and store a given amount of energy in a given time interval is the primary productivity. (The total rate is the *gross* primary productivity. With respect to producers themselves, the rate of energy storage in excess of the rate of aerobic metabolism is the *net* primary productivity.)

5. Energy fixed by autotrophs passes through grazing food webs and detrital food webs. Typically, both food webs are interconnected in the same ecosystem. They lose energy (as heat) through metabolism. The amount of useful energy flowing through their consumer levels declines at each energy transfer. It declines through the loss of metabolically generated heat and as food energy is shunted into organic wastes.

6. In biogeochemical cycles, substances move from the physical environment, to organisms, then back to the environment. Water moves in a hydrologic cycle. Many substances that occur mainly in gaseous phases move in atmospheric cycles. Phosphorus and other minerals move in sedimentary cycles. Nutrients tend to move slowly through the geochemical phase of a cycle but rapidly between organisms and the environment.

7. Land ecosystems have predictable rates of nutrient losses that generally increase when the land is cleared or otherwise disturbed. Fossil fuel burning and other human activities are having adverse effects not only on ecosystems but on the biosphere, as when they have been amplifying the greenhouse effect.

Review Questions

1. Define an ecosystem and its trophic levels. *664*

2. Characterize grazing and detrital food webs. *666*

3. Describe one of the biogeochemical cycles. *669*

4. Define and describe the connections among nitrogen fixation, nitrification, ammonification, and denitrification. *676–677*

Self-Quiz *(Answers in Appendix IV)*

1. Ecosystems have _____ .
    a. energy inputs and outputs   c. one trophic level
    b. nutrient cycling but not outputs   d. a and b

2. Trophic levels can be described as _____ .
    a. structured feeding relationships
    b. who eats whom in an ecosystem
    c. a hierarchy of energy transfers
    d. all of the above

3. Primary productivity is affected by _____ .
    a. photosynthesis and respiration by plants
    b. how many plants are neither eaten nor decomposed
    c. rainfall and temperature
    d. all of the above

4. Match the ecosystem terms with the suitable description.
    _____ producers   a. herbivores, carnivores, omnivores
    _____ consumers   b. feed on partly decomposed matter
    _____ decomposers   c. break down organic remains, products
    _____ detritivores   d. photoautotrophs

Critical Thinking

1. Imagine and describe an extreme situation in which you would be a participant in a food chain rather than a food web.

2. Marguerite is growing a vegetable garden in Maine. What are the variables that can affect its net primary production?

3. List agricultural products and manufactured goods you depend on. Are any implicated in the amplified greenhouse effect?

4. Of all crops in the United States, less than 10 percent are grown without pesticide or fungicide applications. The *organically grown produce* costs more, spoils faster, and is notably nonuniform in size and appearance. Do you buy such produce? Why or why not?

Selected Key Terms

ammonification *676*	food chain *665*
biogeochemical cycle *669*	food web *665*
biological magnification *678*	global warming *674*
carbon cycle *672*	grazing food web *666*
consumer *664*	greenhouse effect *674*
decomposer *664*	hydrologic cycle *670*
decomposition *676*	nitrification *676*
denitrification *676*	nitrogen cycle *676*
detrital food web *666*	nitrogen fixation *676*
detritivore *664*	phosphorus cycle *677*
ecological pyramid *666*	primary producer *664*
ecosystem *664*	primary productivity *666*
ecosystem modeling *678*	trophic level *664*
energy pyramid *667*	watershed *670*

Reading

Krebs, C. 1994. *Ecology.* Fourth edition. New York: Harper Collins.

38 THE BIOSPHERE

Does a Cactus Grow in Brooklyn?

Suppose you live in the American Southwest but find yourself touring one of the deserts of Africa. There you come across a flowering plant with spines, tiny leaves, and columnlike, fleshy stems—just like some cactus plants back home (Figure 38.1). Or suppose you live in the coastal hills of California and decide to tour the Mediterranean coast, the southern tip of Africa, or even central Chile. There you come across many-branched woody plants—very much like the many-branched chaparral plants back home.

In both cases, the plants are separated by enormous geographic and evolutionary distances. Why, then, are they so much alike? The question intrigues you, so you decide to compare their locations on a global map. As you quickly discover, the American and African desert plants grow about the same distance from the equator. Chaparral plants and their distant look-alikes grow along the western or southern coasts of continents between latitudes 30° and 40°. As Charles Darwin and other naturalists did long ago, you have just stumbled onto one of many predictable patterns in the world distribution of species.

In part, "accidents of history" put many species in particular places. Go back to Pangea's colossal breakup, more than 100 million years ago. When chunks of that supercontinent began to drift apart, many species had no choice but to go along for the ride. Over evolutionary time, the travelers were dispersed to different isolated locations. And there they proceeded to undergo genetic divergence

a

b

Figure 38.1 Morphological convergence—clue to interactions between life and the environment. (**a**) *Echinocerus*, of the cactus family, grows in the deserts of the American Southwest. (**b**) In deserts of southwestern Africa we find *Euphorbia*, of the spurge family (Euphorbiaceae). These plants might appear to be related, but their lineages are geographically and evolutionarily distant. Long ago, body parts of plants in both lineages were put to similar uses in similar environments—and they ended up resembling each other.

from the parent populations. Among these changing populations were the ancestors of eucalyptus trees, wombats, and kangaroos on the chunk that became Australia.

But species also owe their distribution to topography, climate, and species interactions. With diligence, you might grow a cactus under artificial lights in a heated room in Brooklyn or some other New York City borough. Plant that cactus outside and it won't last one winter.

This last example reminds us that we humans tinker with the distribution of species. Not all of our tinkering is as harmless as growing a cactus in Brooklyn. Think of our predatory effects on the world's fisheries or the effects of our pesticide battles with insect competitors for food. Earlier chapters provided you with a general picture of predation, competition, and other species interactions. Consider now the physical forces shaping the biosphere itself. This will serve as a foundation for addressing the impact of the human species on the biosphere—the topic of the chapter to follow.

Start with the definition of the **biosphere**—the sum total of all places in which organisms live. It includes the *hydrosphere*—the waters of the Earth, including the ocean, polar ice caps, and other forms of liquid and frozen water. It includes all the soils and sediments of the *lithosphere*—the outer, rocky layer of the Earth. It extends into the lower *atmosphere*, which is made up of gases and airborne particles that envelop the Earth. The air in the upper atmosphere, about 17 kilometers above the Earth's surface, is too thin to support life.

In this vast biosphere you will find ecosystems that range from continent-straddling forests to tiny pools in cup-shaped leaves of plants. As you will see, climate profoundly influences all ecosystems except for a few at hydrothermal vents on the ocean floor.

Climate refers to the average weather conditions—such as temperature, humidity, wind speed, cloud cover, and rainfall—over time. Many factors contribute to climate. The main ones are variations in the amount of incoming solar radiation, the Earth's daily rotation and its path around the sun, the world distribution of continents and oceans, and the elevations of land masses. These factors interact to produce prevailing winds and ocean currents that influence the global patterns of climate. Climate affects the physical and chemical development of sediments and soils.

Together, climate and the composition of sediments and soils affect the growth of primary producers—and, through them, the distribution of entire ecosystems.

KEY CONCEPTS

1. Solar energy is the main energy source for ecosystems. It also influences the global distribution of ecosystems. It does so by continually providing heat energy that warms the atmosphere and drives the Earth's weather systems.

2. Global air circulation patterns, ocean currents, and diverse topographic features interact to produce regional variations in patterns of temperature and rainfall. These patterns influence the composition of soils and sediments, the growth and distribution of primary producers, and, through them, the distribution of ecosystems.

3. A biome is a large, regional unit of land that can be characterized by the climax vegetation of the ecosystems within its boundaries. Deserts and broadleaf forests are examples. Their distribution corresponds roughly with regional variations in climate, topography, and soil type.

4. The water provinces cover more than 70 percent of the Earth's surface. They include the world ocean, inland seas, and bodies of standing freshwater and moving freshwater. Each freshwater and marine ecosystem has gradients in light availability, temperature, and dissolved gases. The gradients, which vary daily and seasonally, influence primary productivity and the composition of species.

Each winter, Pacific Grove, California, is host to great gatherings of tourists and monarch butterflies (Figure 38.2). Similarly, caribou, Canadian geese, sea turtles, and whales undertake vast migrations, moving to and from overwintering grounds. Other kinds of animals stay put, having adaptations in form, physiology, and behavior that allow them to endure seasonal change. Throughout Canada and the United States, flowering plants form leaves, flower, bear fruit, and drop leaves. In the ocean, uncountable numbers of photosynthetic microorganisms undergo seasonal bursts of primary productivity. Clearly, organisms are exquisitely attuned to climatic conditions, which differ from one region to the next and with the changing seasons. The conditions begin with the incoming rays from the sun.

Of the total amount of solar radiation arriving at the outer atmosphere, only about half gets through to the Earth's surface. Atmospheric molecules of ozone (O_3) and oxygen (O_2) absorb most of the wavelengths of ultraviolet radiation, which are lethal for most forms of life. Absorption is greatest in the **ozone layer**, a region that ranges between 17 and 25 kilometers above sea level where ozone is most concentrated (Figure 38.3). Clouds, bits of dust, and water vapor in the atmosphere absorb some of the other wavelengths or reflect them back into space.

Radiation that does penetrate the atmosphere warms the surface of the Earth, which gives up heat by way of radiation or evaporation. Molecules of the lower atmosphere absorb some heat, then reradiate part of it toward the Earth. The effect is somewhat like heat retention in a greenhouse, which lets in the sun's rays while retaining heat that is being lost from the plants and soil inside (Section 37.7). Why is the greenhouse effect important? *Heat energy derived from the sun warms the atmosphere—and it ultimately drives the Earth's weather systems.*

The effects of incoming rays from the sun differ from one latitude to the next. The rays are more concentrated at the equator than at the poles—so air gets heated more at the equator (Figure 38.4). The global pattern of air circulation starts as warm equatorial air rises and spreads northward and southward. Then the Earth's rotation modifies the circulation into worldwide belts of prevailing east and west winds. Taken together, the differences in solar heating at different latitudes and modified air circulation patterns define the world's major **temperature zones** (Figure 38.5a).

Global air circulation patterns also give rise to differences in rainfall at different latitudes. Think about this: Warm air can hold more moisture than cool air. When air heats up at the equator, it picks up moisture and rises to cooler altitudes. And then it gives up moisture as rain. That rainfall can support luxuriant tropical forests. The

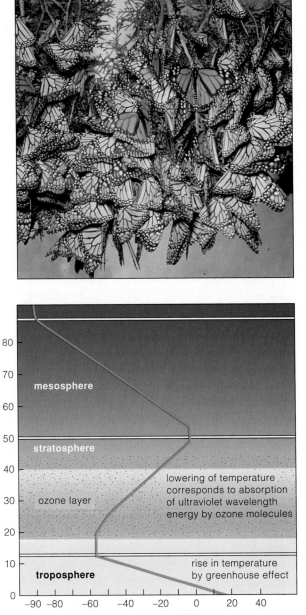

Figure 38.2 Monarch butterflies, migratory insects that gather in winter in the trees of California coastal regions and Central Mexico. The monarchs typically travel hundreds of kilometers south to these regions, which are cool and humid in winter; if they stayed in their breeding grounds, they would risk being killed by far more severe conditions.

Figure 38.3 Earth's atmosphere. Most of the global air circulation proceeds in the troposphere, where temperature decreases rapidly with altitude. Most of the ultraviolet wavelengths in the sun's rays are absorbed in the upper atmosphere, mainly at the ozone layer (which ranges between 17 and 25 kilometers above sea level).

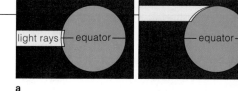

a

Figure 38.4 Global air circulation patterns, as brought about by three interrelated factors. *First*, the sun's rays are less spread out in equatorial regions than in polar regions (**a**). Warm equatorial air rises and spreads northward and southward to produce the initial pattern of air circulation (**b**).

Second, the nonuniform distribution of land masses creates variations in air pressure. Land absorbs and gives up heat faster than the ocean, so some parcels of air rise (or sink) faster than others. Air pressure decreases where air rises (and it increases where air sinks). These differences give rise to winds that disrupt the overall movement from the equator to the poles.

Third, the Earth's rotation and overall shape introduce easterly and westerly deflections in wind directions. Each time the ball-shaped Earth makes a full rotation, its surface turns faster beneath air masses at the equator and slower beneath those at the poles. Therefore, a rising air mass can't really move "straight north" or "straight south." The air mass is deflected to the east or west. This is the source of prevailing east and west winds (**c**).

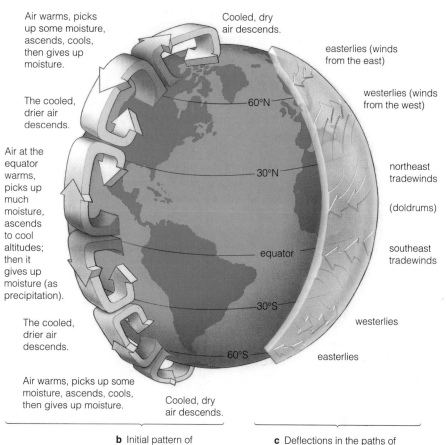

b Initial pattern of air circulation

c Deflections in the paths of air flow near the Earth's surface

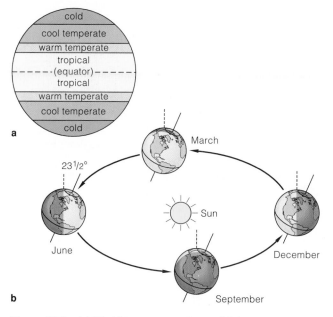

a

b

Figure 38.5 (**a**) World temperature zones. (**b**) Annual variation in the amount of incoming solar radiation. The northern end of the Earth's fixed axis tilts toward the sun in June and away from it in December. Thus the position of the equator relative to the boundary of illumination between day and night varies annually. Such variations in the intensity and duration of daylight lead to seasonal variations in temperature in the two hemispheres. The seasonal change becomes more pronounced with distance from the equator. It is greatest in the central regions of continents, where the moderating effects of oceans are minimal.

now-drier air moves away from the equator, then it descends at latitudes of about 30°. As it descends, the air gets warmer and drier. Deserts tend to form at these latitudes. Air even farther from the equator picks up moisture, ascends to high altitudes, and creates another moist belt at latitudes of about 60°. Then air descends in polar regions, where the low temperatures and almost nonexistent precipitation give rise to the cold, dry, polar deserts.

Finally, the amount of solar radiation reaching the surface varies seasonally, owing to the Earth's rotation around the sun (Figure 38.5*b*). This leads to seasonal changes in daylength, prevailing wind directions, and temperature. And so primary productivity alternately rises and falls on land and in the seas—and migrations shift many kinds of animals to new locations.

Latitudinal differences in the amount of solar radiation reaching the Earth produce global air circulation patterns, which produce latitudinal belts of temperature and rainfall.

The Earth's annual rotation around the sun introduces seasonal changes in winds, temperatures, and rainfall.

These factors influence the locations of different ecosystems.

Ocean Currents and Their Effects

The **ocean** is one continuous body of water that covers 71 percent of the Earth's surface. As the uppermost 10 percent circulates in currents, it distributes nutrients in marine ecosystems and affects regional climates. Solar heating and wind friction drive the surface currents.

Latitudinal and seasonal variations in solar heating cause the ocean water to warm and cool on a vast scale. Water increases in volume when heated, and it decreases in volume when cooled. At the equator, the sea level is about 8 centimeters (3 inches) higher than it is at the poles. The sheer volume of water associated with this "slope" is enough to get the surface waters moving in response to gravity.

circulation is clockwise in the Northern Hemisphere. In the Southern Hemisphere, it is counterclockwise. Swift, deep, and narrow currents of nutrient-poor waters run parallel with the east coasts of continents. For example, every second, the Gulf Stream moves 55 million cubic meters of warm water northward, along the eastern coast of North America. Slower, shallow, and broad currents paralleling the west coasts of continents move cold water toward the equator. As you will see, these currents have major roles in moving deep, nutrient-rich waters to the surface.

Why do cool, mild, foggy summers prevail along the Pacific Northwest coast? Track the currents shown in Figure 38.6. The California Current moves cold water near the coast as it flows toward the equator. Winds

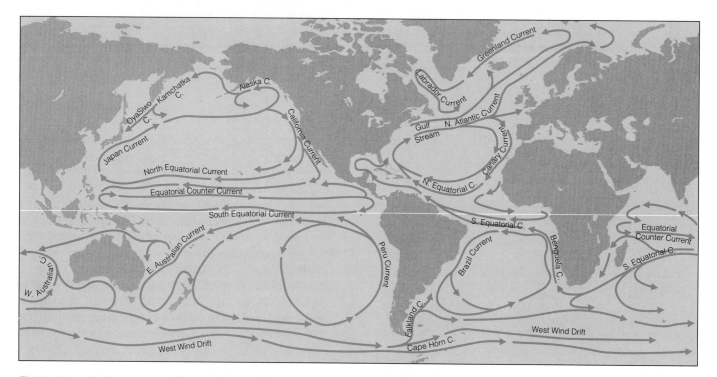

Figure 38.6 Surface currents of the world ocean. Warm water moves from the equator toward the poles. The direction of flow depends on winds, the Earth's rotation, gravity, the shape of ocean basins, and the positions of land masses.

Surface waters tend to move from the equator to the poles—and they warm air parcels above them during the journey. At midlatitudes, about *10 million billion* calories of heat energy get transferred from warm water to air every second!

The "tug" of mainly the trade winds and westerlies causes a rapid, mass flow of water. The Earth's rotation, the positions of land masses, even the shapes of ocean basins influence the direction and properties of these currents. As Figure 38.6 indicates, the global pattern of

approaching the coast give up heat to the cold waters and so become cooler. Why is Washington, D.C., so muggy in summer? The air above the Gulf Stream gains heat and moisture—which winds from the south and east carry to the city. Compared to Ontario in central Canada, why are the winters milder in London and Edinburgh even though they are at the same latitude? The North Atlantic Current picks up warm water from the Gulf Stream, flows past northwestern Europe—and gives up heat energy to prevailing winds.

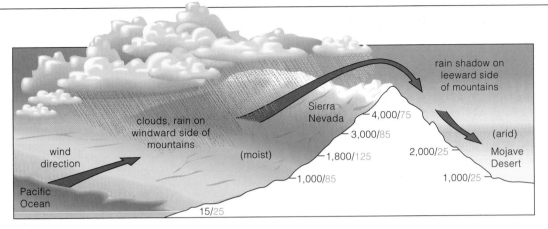

Figure 38.7 Rain shadow effect, a reduction of rainfall on the side of high mountains facing away from prevailing wind. Only plants adapted to arid or semiarid conditions grow in such places. *Blue* numbers are the average yearly precipitation (in centimeters), measured at different locations on both sides of the mountain range. *Black* numbers signify elevation (in meters).

Regarding Rain Shadows and Monsoons

Mountains, valleys, and other aspects of topography influence regional climates. "Topography" means the physical features of a region, such as its elevation.

For example, imagine a warm air mass picking up moisture off California's western coast. It moves inland and reaches the Sierra Nevada, a mountain range that parallels the coast. The air cools as it ascends to higher altitudes, and loses moisture as rain (Figure 38.7). The vegetation belts at different elevations correspond to changes in air temperature and moisture. The western base of the range supports plants that can survive the semiarid conditions. At higher elevations of the range, moisture levels and cool temperatures support forests of deciduous and evergreen trees. The subalpine belt higher up supports only evergreen trees adapted to the rigors of a cold climate. Only low, nonwoody plants and dwarfed shrubs can grow above the subalpine belt.

After air flows over the mountain crests and starts to descend, it gets warmer and can hold more moisture. Now air retains moisture; it even draws *more* moisture out of plants and soil. The result is a **rain shadow**—an arid or semiarid region of sparse rainfall on the leeward side of tall mountains. (Leeward means the direction not facing a wind; windward means the direction from which the wind is blowing.) The mountain ranges of Europe, the Himalayas of Asia, and the Andes of South America also create rain shadows. On Hawaii, tropical forests thrive on the windward side of high volcanoes and conditions are extremely arid on the leeward side.

Now consider **monsoons**, patterns of air circulation that affect continents lying poleward of warm oceans. Intense heating of the land creates low pressures that draw in moisture-laden air from the oceans, as from the Bay of Bengal into India and Bangladesh, or from the Gulf of Mexico into the American Southwest. Tucson's fantastic summer thunderstorms are one outcome. Or consider the doldrums, an equatorial zone where the direct overhead sun creates intense heating and heavy rain, and where trade winds converge. The zone moves north and south seasonally. It creates the patterns of wet and dry seasons we see, for example, in East Africa.

Finally, the recurring sea breezes along coastlines are "mini-monsoons." Here again, the heat capacities of land and water differ. In the morning, the temperature of water does not rise as fast as it does on land. As warmed air above the land rises, cooler marine air moves in. After sunset, the land loses heat faster, and land breezes flow in the reverse direction (Figure 38.8).

Surface ocean currents, in combination with global air circulation patterns, influence regional climates and help distribute nutrients in marine ecosystems.

Atmospheric circulation patterns, ocean currents, and landforms interact in ways that influence regional temperatures and moisture levels—which affect the distribution and dominant features of ecosystems.

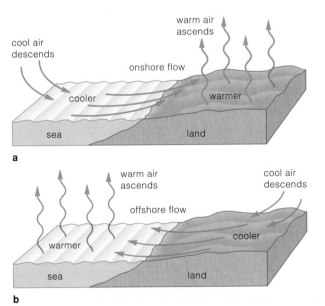

Figure 38.8 Coastal breeze in the afternoon (**a**) and at night (**b**).

THE WORLD'S BIOMES

Biogeographic Distribution

The circulation patterns in the atmosphere and at the ocean surface, in concert with topography, give rise to regional differences in temperature and moisture. This information helps explain why it is that tundra, forests, grasslands, and deserts form in some places but not others. It also is the focus of **biogeography**—the study of the global distribution of species, each of which is adapted to regional conditions.

For example, the information provides clues to why many evolutionarily distant species look alike. Such species are often results of convergent evolution, which is described in Section 16.3. Briefly, their ancestors were not closely related but lived in similar environments and faced similar pressures. By natural selection, they underwent similar kinds of modifications and ended up looking alike. The ancestors of those cactus and spurge plants shown in Figure 38.1 both evolved in hot, dry deserts where water is scarce. Fleshy plant stems that have thick cuticles conserve precious water—and rows of sharp spines help prevent herbivores from chewing on juicy plant parts.

Biogeographic Realms and Biomes

Long ago W. Sclater, then Alfred Wallace, attached the name **biogeographic realms** to six vast land areas, each with distinguishing plants and animals (Figure 38.9a).

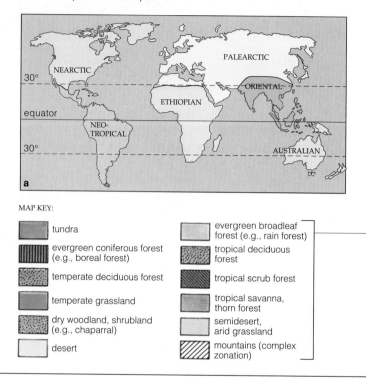

Figure 38.9 (**a**) Major biogeographic realms. (**b**) Distribution of the world's major biomes. The overall pattern corresponds roughly with the distribution patterns for climate and soil type. Compare this map with the two-page photograph of the Earth's surface that precedes Chapter 1.

MAP KEY:

tundra

evergreen coniferous forest (e.g., boreal forest)

temperate deciduous forest

temperate grassland

dry woodland, shrubland (e.g., chaparral)

desert

evergreen broadleaf forest (e.g., rain forest)

tropical deciduous forest

tropical scrub forest

tropical savanna, thorn forest

semidesert, arid grassland

mountains (complex zonation)

High ←   Water Availability

alpine tundra

montane coniferous forest

deciduous forest

tropical forest

tropical forest   temperate deciduous forest   boreal coniferous forest

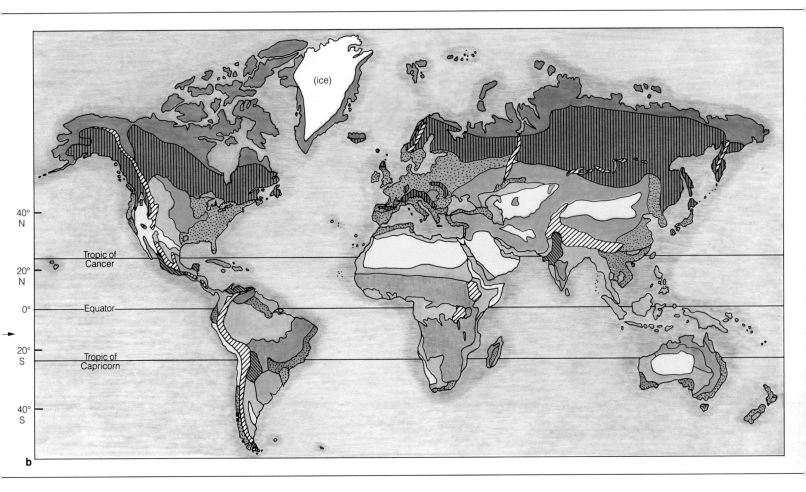

b

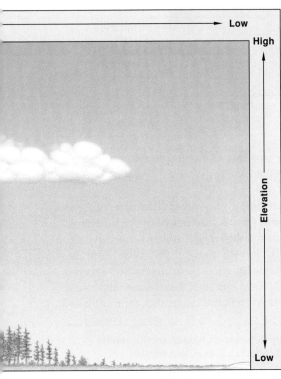

arctic tundra

Figure 38.10 The changes in plant form along some environmental gradients for North America. Shown here, gradients in water availability and elevation that influence primary productivity. The mean annual temperature, soil drainage, and other factors also have effects on plant form.

The realms retain their distinct identity partly because of climate. They tend to maintain that identity if oceans, mountain ranges, and other major barriers restrict gene flow and so keep their component species isolated.

Each realm may be further divided into biomes. A **biome** is a large region of land that is characterized mainly by the climax vegetation of ecosystems within its boundaries. Take a moment to look over the map of biomes in Figure 38.9b, then see how it correlates with the information in Figure 38.10. As you might deduce, distinctive types of biomes prevail at certain latitudes and elevations. Thus biomes dominated by species of short plants prevail in dry regions, at high elevations, and at high latitudes. Biomes dominated by tall, leafy plant species prevail at tropical and temperate latitudes, and at low elevations with warm temperatures and high rainfall. As you will see next, soils also affect the distribution of biomes.

The distribution and key features of biomes are an outcome of temperatures, soils, and moisture levels (which vary with latitude and elevation), and of evolutionary history.

38.4 SOILS OF MAJOR BIOMES

Soil is a complex mixture of rock, mineral ions, and organic matter in some state of physical and chemical breakdown. The rocks range from coarse-grained gravel to sand, silt, and finely grained clay. Water, air, and diverse organisms occupy spaces among the particles and grains. Collectively the organic matter, which is in various stages of decomposition, is called humus.

Figure 38.11 shows some vertical structures of soils, or soil profiles. Topsoil, the upper region with the most humus, is most vulnerable to weathering. It may be less than a centimeter deep on steep slopes to more than a meter deep in grasslands. Water that percolates down from topsoil deposits clay, minerals, and some organic matter in the next layer. Beneath this are weathered, broken rocks from which true soil forms.

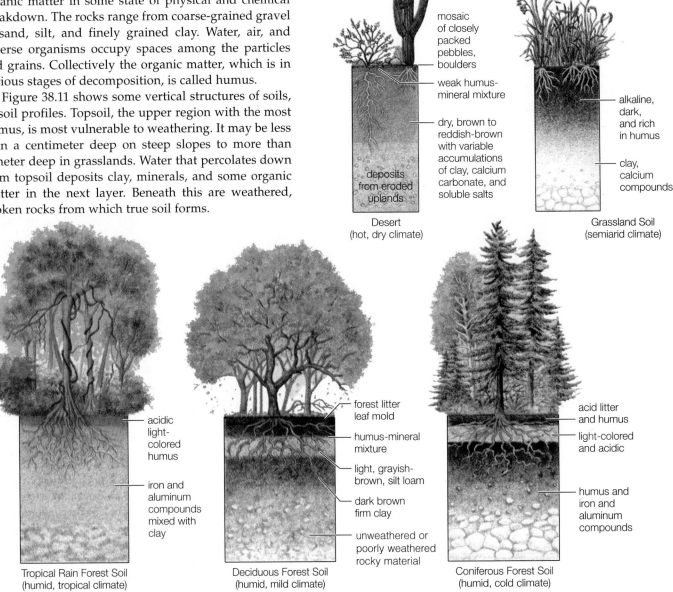

Figure 38.11 Soil characteristics that influence the distribution of ecosystems on land.

The growth of most plants suffers in poorly aerated, poorly draining soils. Loam topsoils have the best mix of sand, silt, and clay for agriculture. They have enough coarse particles to promote good drainage and finer particles to retain water-soluble mineral ions that serve as nutrients for plant growth. Gravelly or sandy soils allow rapid leaching that can deplete them of mineral ions. Clay soils with fine, closely packed particles are poorly aerated and do not drain well. Few plants can grow in waterlogged clay soils.

As farmers know, soil affects primary productivity. They grow most crops in cleared, former grasslands. Many burn or clear-cut tropical forests for agriculture (Section 39.4). But these biomes have very little topsoil above a poorly draining clay layer. Clearing exposes the topsoil, and heavy rains leach most of the nutrients.

We turn now to the deserts, shrublands, woodlands, grasslands, various forests, and tundra. Bear in mind, none of these major biomes is uniform throughout. Within its borders, local climates, landforms, and other physical features favor patches of distinct communities.

Primary productivity depends on the soil profile and its proportions of sand, silt, clay, gravel, and humus.

38.5 DESERTS

Deserts form in lands with less than 10 centimeters or so of annual rainfall and high potential for evaporation. Such conditions prevail at latitudes of about 30° north and south. There we find the great deserts of the American Southwest, northern Chile, Australia, northern and southern Africa, and Arabia. Farther north are the high deserts of eastern Oregon, and Asia's vast Gobi and the Kyzyl-Kum east of the Caspian Sea. Rain shadows are the main reason why these northern deserts are so arid.

the rains. Deep-rooted species, including mesquite and cottonwood, may grow near the very few streambeds that have a permanent underground water supply.

Of all land surfaces, more than one-third is arid or semiarid, without enough rainfall to support crops. The crops growing in California's Imperial Valley and some other deserts require unrelenting soil management and extensive irrigation. Without drainage programs, such croplands become unproductive from waterlogging and

Deserts do not have lush vegetation. Rain falls in heavy, brief, infrequent pulses that swiftly erode the exposed topsoil. Humidity is so low that the sun's rays easily penetrate the air and quickly heat the ground surface—which at night radiates heat and cools quickly.

Although arid or semiarid conditions do not favor large, leafy plants, deserts show plenty of diversity. In a patch of Arizona's desert, you may come across various deep-rooted evergreen shrubs, such as creosote, and fleshy-stemmed, shallow-rooted cacti (Figure 38.12). You may see tall saguaro, short prickly pear, and ocotillo, which drops leaves more than once a year, then grows new ones after a rain. Desert perennials and annuals flower profusely, spectacularly, but briefly after

Figure 38.12 Warm desert near Tucson, Arizona. The plants include creosote bushes, multistemmed ocotillo, columnlike saguaro cacti, and prickly pear cacti with rounded pads.

salt buildup. Alarmingly, many parts of the world are undergoing **desertification**—the wholesale conversion of grasslands and similarly productive biomes to dry wastelands. This trend is described in Section 39.6.

Where the potential for evaporation greatly exceeds sparse rainfall, deserts form. This condition prevails at latitudes 30° north and south and in rain shadows.

DRY SHRUBLANDS, DRY WOODLANDS, AND GRASSLANDS

Dry shrublands and **dry woodlands** prevail in western or southern coastal regions of the continents between latitudes 30° and 40°. These semiarid regions get more rain than deserts, but not much more. Most rain falls in mild winters; summers are long, hot, and dry. Often, dominant plants have hard, tough, evergreen leaves.

We find examples of dry shrublands, which get less than 25 to 60 centimeters of rain per year, in California, South Africa, and Mediterranean regions. Their exotic local names include fynbos and chaparral. California has 2.4 million hectares (6 million acres) of chaparral. In summer, lightning-sparked, wind-driven firestorms can sweep through these biomes (Figure 38.13). Shrubs with highly flammable leaves burn swiftly to the ground. Yet they are highly adapted to episodes of fire and quickly resprout from their root crowns. Trees do not fare as well during the firestorms. The shrubs, which "feed" the fires, have the competitive edge.

Dry woodlands dominate where annual rainfall is about 40 to 100 centimeters. The dominant trees can be tall, but they do not form a dense, continuous canopy. Eucalyptus woodlands of southwestern Australia and oak woodlands of California and Oregon are like this.

Grasslands sweep across much of the interior of continents, in the zones between deserts and temperate forests. Warm temperatures prevail in the summer, and winters are extremely cold. The 25 to 100 centimeters of annual rainfall prevent deserts from forming but are not enough to support forests. Drought-tolerant plants can survive strong winds, light and infrequent rainfall, and rapid evaporation. Dominant animals are grazing and burrowing types. The grazing activities, together with periodic fires, keep shrublands and forests from encroaching on the fringes of many grasslands.

In North America, the main grasslands are *shortgrass* prairie and *tallgrass* prairie. Usually they form where the land is flat or rolling, as in Figure 38.14a. Roots of perennial plants extend profusely through the topsoil. During the 1930s, shortgrass prairie of the American Great Plains was overgrazed and plowed under to grow wheat, which requires more water than this region sometimes receives. Strong winds, prolonged droughts, and unsuitable farming practices turned much of the prairie into a Dust Bowl (Section 39.6). John Steinbeck's *The Grapes of Wrath* and James Michener's *Centennial*, which are two historical novels, speak eloquently of the consequences.

Also in the interior of North America, tallgrass prairie once extended westward from zones of temperate deciduous forests. Diverse legumes and composites, including daisies, thrived in the interior, with its richer topsoil and slightly more frequent rainfall. Farmers converted most of the original tallgrass prairie for agriculture, but large areas are now being restored (Figure 38.14b).

Between the tropical forests and deserts of Africa, South America, and Australia are broad belts of grasslands with a smattering of shrubs or trees. These are the **savannas** (Figure 38.14c). Rainfall averages 90 to 150 centimeters a year, and prolonged seasonal droughts are common. Where rainfall is low, fast-growing grasses dominate. Acacia and other shrubs grow in regions that get slightly more moisture. Where the rainfall is higher,

Figure 38.13 California chaparral. The dominant plants are multibranched, woody, and no more than a few meters tall. Without periodic fires, they form a nearly impenetrable vegetation cover. The boxed inset shows a firestorm racing through a chaparral-choked canyon above Malibu.

Figure 38.14 (**a**) Rolling shortgrass prairie to the east of the Rocky Mountains. Once, bison were the dominant large herbivore of North American grasslands. At one time, the astoundingly abundant grasses supported 60 million of these hefty ungulates. (Ungulates are hooved, plant-eating mammals.)

(**b**) A rare patch of natural tallgrass prairie in eastern Kansas.

(**c**) The African savanna—warm grasslands with scattered stands of shrubs and trees. More varieties and greater numbers of large ungulates live here than anywhere else. They include migratory wildebeests (shown in this photograph), giraffes, Cape buffalo, zebras, and impalas.

savannas grade into tropical woodlands with tall, coarse grasses, shrubs, and low trees. *Monsoon* grasslands form in southern Asia where heavy rains alternate with a dry season. Dense stands of tall, coarse grasses form, then die back and often burn in the dry season.

The plants of dry shrublands, dry woodlands, and grasslands are supremely adapted to surviving episodes of drought and fire, strong winds, and grazing animals.

TROPICAL RAIN FORESTS AND OTHER BROADLEAF FORESTS

In forest biomes, tall trees grow close together and form a fairly continuous canopy over a broad stretch of land. In general, there are three types of trees. Which type prevails in a region depends partly on distance from the equator. The evergreen broadleafs dominate between latitudes 20° north and south. Deciduous broadleafs are common at moist, temperate latitudes where winters are not severe. Evergreen conifers are common at high, colder latitudes and in mountains of temperate zones.

Evergreen broadleaf forests sweep across tropical zones of Africa, the East Indies and Malay Archipelago, southeast Asia, South America, and Central America. Annual rainfall can exceed 200 centimeters and is never less than 130 centimeters. One forest biome, the **tropical rain forest**, requires regular and heavy rainfall, an annual mean temperature of 25°C, and humidity of 80 percent or more (Figure 38.15). In this highly productive forest, evergreen trees produce new leaves and shed old

bromeliad

Figure 38.15 Tropical forests, with their spectacular biodiversity. Among the millions of known species are jaguars and *Rafflesia*, a leafless, foul-smelling plant with a fly-pollinated flower that can be three meters across. Bromeliads, orchids, and other plants grow on tree branches, obtaining minerals from organic remains of leaves, insects, and other litter that have dissolved in tiny pockets of water.

Spring

Summer

Winter

Autumn

ones through the year, so these forests produce far more litter than others. Decomposition and mineral cycling are rapid in the hot, humid climate. Soils are weathered, humus-poor, and poor reservoirs of nutrients. We will look more closely at these forests in the next chapter.

Leaving tropical rain forests, we enter regions where temperatures stay mild but rainfall dwindles during part of the year. This is the start of **deciduous broadleaf forests**. Trees of *tropical* deciduous forests drop some or all of their leaves during a pronounced dry season. The *monsoon* forests of India and southeastern Asia also have such trees. Farther north, in the temperate zone, rainfall is even lower, and temperatures become cold during the winter. *Temperate* deciduous forests prevail here (Figure 38.16). Decomposition is not as rapid as in the humid tropics, and many nutrients are conserved in accumulated litter on the forest floor.

Figure 38.16 A temperate deciduous forest south of Nashville, Tennessee, in spring, summer, autumn, and winter.

Complex forests of ash, beech, birch, chestnut, elm, and deciduous oaks once stretched across northeastern North America, Europe, and east Asia. They declined drastically when farmers cleared the land. In North America, introduced diseases destroyed nearly all of the chestnuts and many elms (Table 36.2). Now, maple and beech predominate in the Northeast. Farther west, oak-hickory forests prevail, then oak woodlands that eventually grade into tallgrass prairie.

In forest biomes, conditions favor dense stands of tall trees that form a continuous canopy over a broad region.

38.8 CONIFEROUS FORESTS

Conifers (cone-bearing trees) are the primary producers of **coniferous forests**. Most have needle-shaped leaves with a thick cuticle and recessed stomata—adaptations that help the trees conserve water through dry times. They dominate the boreal forests, montane coniferous forests, temperate rain forests, and pine barrens.

Boreal forests stretch across northern Europe, Asia, and North America. (Such a forest also is called *taiga*, meaning "swamp forest.") Most are in glaciated regions with cold lakes and streams (Figure 38.17a). It rains mostly in summer, and evaporation is low in the cool summer air. The cold, dry winters are more severe in

We also find coniferous forests in some temperate lowlands. One such temperate rain forest, paralleling the coast from Alaska into northern California, includes some of the world's tallest trees—Sitka spruce to the north and redwoods to the south (Figure 36.12i). Heavy logging destroyed much of this biome. One bright note: In 1994, a major timber company surrendered its rights to log one of the largest intact temperate rain forests outside the tropics. This forest, with trees 800 years old, cloaks the Kitlope Valley of British Columbia.

Sandy, nutrient-poor soil of New Jersey's coastal plain supports **pine barrens**—scrub forests in which

Figure 38.17 (**a**) Spruce-dominated boreal forest. (**b**) Montane coniferous forest of Yosemite Valley in the Sierra Nevada, California.

eastern parts of these biomes than in the west, where oceanic winds moderate the climate. Spruce and balsam fir dominate the boreal forests of North America. Pines, birches, and aspens take hold in burned or logged areas. Acidic bogs, which are dominated by peat mosses, shrubs, and stunted trees, prevail in the poorly drained areas. Boreal forests become much less dense to the north, where they grade into arctic tundra.

In the Northern Hemisphere, montane coniferous forests extend southward through the great mountain ranges. Spruce and fir dominate in the north and at higher elevations. They give way to fir and pines in the south and at lower elevations (Figure 38.17b).

grasses and low shrubs grow beneath the open stands of pitch pine and oak trees. Pine barrens are adapted to recover quickly from frequent lightning-triggered fires. Managed pine plantations replaced most of the natural pine forests that once dominated the coastal plains of North and South Carolina, Georgia, and Florida.

Coniferous forests prevail in regions in which a cold, dry season alternates with a cool, rainy season.

38.9 TUNDRA

Tundra is derived from *tuntura*, a Finnish word for the great treeless plain between the polar ice cap and belts of boreal forests in Europe, Asia, and North America. Much of this *arctic* tundra is flat, windswept, and wet (Figure 38.18*a*). Temperatures are cool in summer and below freezing in winter, so not much water evaporates. Little rain or snow falls. Sunlight is nearly continuous during the summer months. Then, short plants grow and flower profusely, and seeds ripen fast.

Snow does not cloak arctic tundra all year long, but summers are too short to thaw much more than surface soil. Just beneath the surface is a perpetually frozen

b

a

Figure 38.18 (**a**) Extensive ponding in the arctic tundra. The rain and snowmelt cannot percolate downward because of the permafrost. (**b**) Short, hardy plants typical of alpine tundra.

layer, the **permafrost**. It is more than 500 meters thick in some places. Because permafrost prevents drainage, the soil above remains waterlogged. Anaerobic conditions and low temperatures limit nutrient cycling. Organic matter decomposes so slowly, it accumulates in soggy masses of organic material. All but about 5 percent of the carbon in the arctic tundra is locked up in peat.

A similar type of biome prevails at high elevations in mountains throughout the world, although there is no permafrost beneath the soil. Figure 38.18*b* shows an example of this *alpine* tundra. Dominant plants often form low cushions and mats that are able to withstand the buffeting of strong winds. Winter temperatures fall below freezing. Even in the summer, shaded patches of snow persist. The thin, fast-draining soil of an alpine tundra is nutrient-poor, so primary productivity is low.

At high latitudes with short, cool summers and long, very cold winters, we find arctic tundra. In high mountains where seasonal changes vary with latitude but where the climate is too cold to support forests, we find alpine tundra.

38.10 FRESHWATER PROVINCES

The freshwater and saltwater provinces are far more extensive than the Earth's biomes. They include lakes, rivers, ponds, estuaries, and wetlands. They include rocky and sandy shores, coral reefs, parts of the open ocean, even hydrothermal vents on the ocean floor. "Typical" examples are difficult to find. Some ponds can be waded across; Lake Baikal in Siberia is more than 1.7 kilometers deep. All of the aquatic ecosystems have gradients in light penetration, temperature, and dissolved gases, but these differ greatly from one to the next. All we can do here is sample the diversity.

Lake Ecosystems

A **lake** is a body of freshwater with littoral, limnetic, and profundal zones (Figure 38.19). Its shallow, usually well-lit littoral extends around the shore to the depth at which rooted aquatic plants stop growing. Diversity is greatest here. The limnetic, the open, sunlit water past the littoral, extends to a depth where photosynthesis is insignificant. Here, **plankton** (aquatic communities of tiny organisms) abound. The lake's *phyto*plankton include green algae, diatoms, and cyanobacteria. Its *zoo*plankton include rotifers, copepods, and many other heterotrophs. All open water below the depth at which wavelengths suitable for photosynthesis can penetrate is the profundal. Detritus sinks through this zone to bottom sediments, with their communities of bacterial decomposers that release nutrients into the water.

Seasonal Changes in Lakes In temperate regions, where summers are warm and winters cold, lakes show seasonal changes in density and temperature from the surface to the bottom. A layer of ice forms over many of them in midwinter. Water near the freezing point is the least dense and accumulates beneath the ice. Water at 4°C, is the most dense. It collects in deeper layers that are a bit warmer than the surface layer in midwinter.

In spring, daylength increases and the air is not as cold. As lake ice melts, the surface water slowly warms to 4°C, and temperatures become uniform throughout. Winds blowing over the surface now cause a **spring overturn**. In such overturns, strong vertical movements carry dissolved oxygen from a lake's surface layer to its depths, and nutrients released by decomposition are brought from the bottom sediments to the surface layer.

By midsummer, the surface layer is above 4°C and the lake has a thermocline. This middle layer changes abruptly in temperature and prevents vertical mixing. Being warmer and less dense, surface water floats on the thermocline (Figure 38.20). In the water beneath it, decomposers may deplete the dissolved oxygen. Come autumn, the upper layer cools, becomes denser, then sinks, and so the thermocline vanishes. During this **fall**

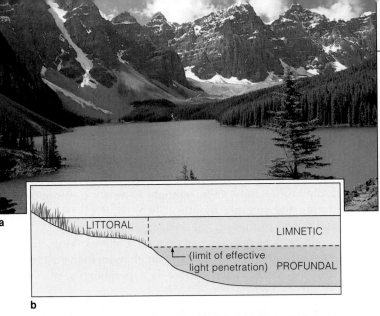

a

b

Figure 38.19 (**a**) In the Canadian Rockies, a lake basin formed by glacial action during the last ice age. (**b**) Lake zonation. The littoral includes all areas around the edge of the lake, from the shore to the depth where aquatic plants stop growing. The profundal includes areas below the depth of light penetration. Above the profundal are open, sunlit waters of the limnetic.

overturn, the water mixes vertically, so that dissolved oxygen moves down and nutrients move up.

Primary productivity corresponds with the seasons. After a spring overturn, the longer daylengths and cycled nutrients support higher rates of photosynthesis. Phytoplankton and rooted aquatic plants quickly take up phosphorus, nitrogen, and other nutrients. During the growing season, the thermocline cuts off vertical mixing. Nutrients tied up in the remains of organisms sink to deeper waters, so photosynthesis declines. By late summer, the shortages are limiting photosynthesis. After the fall overturn, nutrient cycling drives another burst of primary productivity. But with shorter daylight hours, the burst doesn't last long. Not until spring will primary productivity increase again.

Trophic Nature of Lakes The topography, climate, and geologic history of a lake dictate the kinds and numbers of its residents, how they are dispersed, and how nutrients are cycled among them. Soils of the lake basin and surrounding regions contribute to the type and amount of nutrients available to support them.

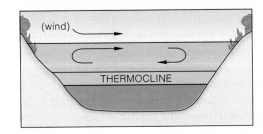

Figure 38.20 Thermal layering in a temperate lake in summer.

Geologic processes produce a lake, as when advancing glaciers carve basins in the Earth. When a glacier retreats, water collects in an exposed basin. Over geologic time, erosion, sedimentation, and other events alter the dimensions of the lake, which usually ends up filled or drained completely.

Climate, soil, and basin shape interact and produce conditions that range from oligotrophy to eutrophy. *Oligotrophic* lakes often are deep, poor in nutrients, and low in primary productivity. By contrast, *eutrophic* lakes are often shallow, nutrient enriched, and high in primary productivity.

An accumulation of sediments may cause conditions within the lake to progress from oligotrophy to eutrophy, and then on to a final successional stage with a completely filled-in basin—hence no more lake.

Human activities can disrupt the geologic, climatic, and biological interactions that determine the trophic condition of lakes. For example, dumping sewage into a lake or logging over the surrounding land can lead to **eutrophication**. The term refers to nutrient enrichment of a lake (or some other body of water) that typically reduces transparency and favors a community that is dominated by phytoplankton. Many field experiments nicely demonstrate this outcome (Figure 38.21).

As Val Smith once documented, people caused eutrophication of Lake Washington in Seattle—and they also reversed it. From 1941 to 1963, the lake received too much phosphorus-rich sewage. The sewage discharge promoted large blooms of cyanobacteria, which formed thick mats over the lake each summer and so made it useless for recreation. With phosphorus enrichment, nitrogen became the limiting resource in the lake. Then cyanobacteria—which are superior competitors for nitrogen by their ability to fix N_2—became dominant. From 1963 to 1968, sewage discharges were cut back, and finally stopped. By 1975, the lake had almost fully recovered, and lower density populations of diatoms and green algae became dominant.

Stream Ecosystems

The flowing-water ecosystems called **streams** start out as freshwater springs or seeps. They grow and merge as they flow downslope, then often combine to form a river. Between the headwaters and the river's end, we find three kinds of habitats—riffles, pools, and runs.

Figure 38.21 Experiment demonstrating lake eutrophication in Ontario, Canada. Researchers stretched a plastic curtain across a channel between two basins of the same lake. They added phosphorus, carbon, and nitrogen to one basin (*in background*) and carbon and nitrogen to the other (*in foreground*). Within two months, the phosphorus-enriched basin showed accelerated eutrophication; a dense algal bloom turned the water green.

Figure 38.22 A fast-flowing stream near the Sierra Nevada.

Riffles are shallow, turbulent stretches where water flows swiftly over a rough bottom of sand and rock. Figures 7.1 and 38.22 show examples. Pools have deep water flowing slowly over a smooth, sandy, or muddy bottom. The runs are smooth-surfaced but fast-flowing stretches over bedrock or rock and sand.

A stream's average flow volume and temperature depend on rainfall, snowmelt, geography, altitude, and even the shade cast by plants. Its solute concentrations are influenced by the streambed's composition as well as by agricultural, industrial, and urban wastes.

Especially within forests, streams import most of the organic matter that supports food webs. Where trees cast shade, photosynthesis is diminished. Most of the organic matter is litter that enters detrital food webs. Aquatic organisms continually take up and release nutrients as water flows downstream. (Nutrients move upstream only as components of migrating fishes and other animals.) Think of nutrients as spiraling between water and aquatic organisms as the stream flows on its one-way course, usually to the sea.

Ever since cities formed, streams have been sewers for industrial and municipal wastes. The wastes, as well as other pollutants from poorly managed farmlands, choked many streams with sediments and chemically poisoned them. Streams are resilient, however, and they show impressive recovery when pollution is controlled.

The freshwater and saltwater provinces are far more extensive than the Earth's biomes. All the aquatic ecosystems in these provinces have characteristic gradients in light penetration, temperature, and dissolved gases.

Beyond land's end are two vast provinces of the open ocean. Together, they cover 71 percent of the Earth's surface. The *benthic* province includes all sediments and rocks of the ocean bottom (Figure 38.23). It starts with continental shelves and extends to deep-sea trenches. The *pelagic* province is the entire volume of ocean water. Its neritic zone is all the water above the continental shelves. Its oceanic zone is the water of the ocean basin.

Walk along the ocean and you may be humbled by its vastness (Figure 38.24). You see an immense expanse stretching out to the distant horizon, with no mountains and plains and valleys to break it up visually into something less overwhelming. What you do not see are the *submerged* mountains and valleys and plains—which are as varied as the ones configuring the dry continents.

Primary Productivity in the Ocean

In the ocean's upper surface waters, photosynthesis proceeds on a stupendous scale, just as it does on land. The primary productivity varies seasonally, just as it does on land (Section 5.7 and Figure 38.25). Its vast "pastures" of phytoplankton are the start of food webs that include zooplankton, copepods, shrimplike krill, whales, squids, and fishes. Organic remains and wastes from these marine communities sink through the water. They become the basis of detrital food webs for most communities in the benthic province.

In the 1970s, biologists reevaluated their notions about the ocean's primary productivity. They realized that as much as 70 percent of it may be the contribution of **ultraplankton**. Most of these tiny photosynthetic bacterial cells are only about 2 micrometers wide (that's 40 millionths of an inch), and some are smaller than that. In tropical and subtropical seas once thought to be nearly devoid of producers, 0.035 gram (one *ounce*) of water holds as many as 3 million of these cells.

Hydrothermal Vents

In 1977, researchers studying the ocean floor near the Galápagos Islands discovered a distinct ecosystem. In the Galápagos Rift, a volcanically active boundary between two of the earth's crustal plates, they found

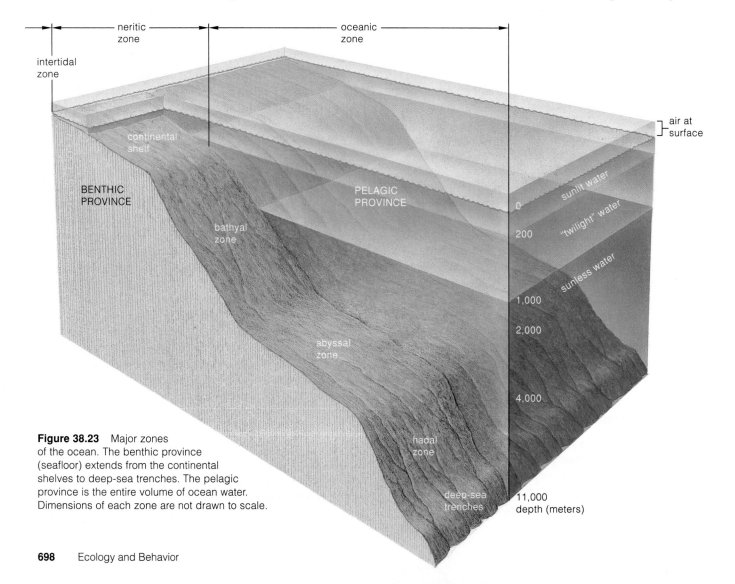

Figure 38.23 Major zones of the ocean. The benthic province (seafloor) extends from the continental shelves to deep-sea trenches. The pelagic province is the entire volume of ocean water. Dimensions of each zone are not drawn to scale.

a

b

c

Figure 38.24 (**a**) Looking out from a rocky shore toward the open ocean. (**b**) Whale breaching (leaping out of the water) near a British Columbia coastline. Views at a hydrothermal vent ecosystem on the ocean floor, crustaceans (**c**) and tube worms (**d**).

d

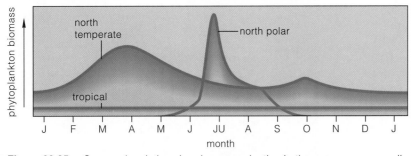

Figure 38.25 Seasonal variations in primary production in the ocean corresponding to latitude. The strongest peak in the graph line for north polar and temperate seas corresponds to phytoplankton blooms, brought about by a seasonal increase in daylength. The smaller peak corresponds to an increase in nutrient availability, something like the fall overturn in lakes. In most tropical seas, daylength and nutrient availability do not vary much; neither does primary production.

communities thriving near **hydrothermal vents**. At this boundary, near-freezing water seeps into fissures and becomes heated to very high temperatures. As pressure forces heated water upward, minerals are leached from rocks before the water spews out through vents in the seafloor. The hydrothermal outpouring is enriched with zinc, iron, copper sulfides, and calcium and magnesium sulfates. These settle out, forming rich mineral deposits. Hydrogen sulfide in the deposits are energy sources for chemoautotrophic bacteria that are the starting point for hydrothermal vent communities. The bacteria are primary producers in a food web that includes tube worms, crustaceans, clams, and fishes (Figure 38.24c,d).

Other hydrothermal vent ecosystems have been found in the South Pacific near Easter Island; the Gulf of California, about 150 miles south of the tip of Baja California, Mexico; and the Atlantic. In 1990, a team of United States and Russian scientists located one in Lake Baikal, the world's deepest lake. This lake basin seems

to be splitting apart (hence the vents) and may mark the beginning of a new world ocean.

Did life originate near such hospitable places on the seafloor? Certainly conditions at the surface of the early Earth were about as inhospitable as you might imagine. At the least, the first cells would have been protected from destructive forms of radiation that bombarded the Earth before the oxygen-rich atmosphere formed. Were they slowly carted closer to the surface by the uplifting of the seafloor during episodes of crustal crunchings? These are some of the unanswered questions that some evolutionary detectives are now asking.

Although we are most familiar with features of the land, a single great ocean dominates the Earth's surface. Its primary productivity is staggering, and seasonal. Communities thrive near hydrothermal vents on the ocean floor, which suggests to some that life itself may have originated deep in the ocean.

38.12 CORAL REEFS AND BANKS

The Master Builders

Each wave-resistant formation called a **coral reef** began as the remains of countless organisms accumulated. The hard parts of corals became the structural foundation. Coralline algae added to it; deposits of calcium and magnesium carbonates harden the cell walls of these red algae, including *Corallina* (Figure 38.26). Secretions of other organisms helped cement things together.

The massive, pocketed spine of an existing reef is home to hundreds of species of corals, and a staggering variety of red algae and other organisms. Figure 38.26 and Figure 20.8 only hint at the wealth of warning colors, spines, tentacles, and stealthy behavior—clues to fierce competition for resources among species that are packed together in limited space.

Fringing reefs, barrier reefs, and atolls are the main reef formations (Figure 38.26). Most of the substantial reefs form in clear, warm waters between latitudes 25° north and south. Farther north or south, solitary corals or small colonies of them have formed coral banks in temperate waters, even in cold waters of continental shelves. Among these are the smooth, vertical banks in cold, deep waters near Japan, California, England, and New Zealand. *Lophelia*, a colonial branched coral, has built great banks in the cold waters of Norway's fjords.

The Once and Future Reefs

Colorful dinoflagellates often live as symbionts in the tissues of reef-building corals. In return for protection, they provide a coral polyp with oxygen and recycle its mineral wastes. When stressed, the polyps expel their protistan symbionts. When stressed for more than a few months, they die, and only bleached hard parts remain. Abnormal, widespread bleaching in the Caribbean and tropical Pacific began in the 1980s. So did an increase in sea surface temperature, and this may be a key stress factor. If, as marine biologists Lucy Bunkley-Williams and Ernest Williams suggest, the damage is an outcome of global warming, the future looks grim for the reefs.

Human activities are destroying reefs in more direct ways than this. Where raw sewage and other pollutants are being discharged into the nearshore waters around populated islands, including some in the Hawaiian chain, reefs are dead or dying. Where fishermen from Japan, Indonesia, and Kenya move in for profit, reef life is being decimated. No simple nets for these fellows. They drop dynamite or cyanide in the water, so fish hiding among the corals float dead to the surface. They squirt sodium cyanide into the water, and stunned but not dead fish float to the surface. These are the tropical

coralline alga

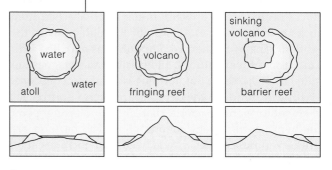

coral reef   island   lagoon   open ocean

Figure 38.26 A sampling of the astounding diversity of tropical reefs. The diagrams show three types of coral reefs. *Atolls* are ring-shaped coral reefs that enclose or almost enclose a shallow lagoon. *Fringing reefs* form next to the land's edge in regions of limited rainfall, as on the leeward side of tropical islands. *Barrier reefs*, as shown in Figure 20.8, form around islands or parallel with the shore of a continent. A calm lagoon forms behind them.

lionfish

crab

moray eel

chambered nautilus

sea anemone

daisy coral

pillar coral

crown-of-thorns sea star

fish in pet stores as well as the rare fish brought live to Japanese restaurants, where they are dispatched and served up as exorbitantly expensive status symbols. On small, native-owned islands, fishing rights are traded for small sums, then the destroyed reefs cannot sustain the small human populations that once depended on them for their own sustenance.

Coral reefs and banks show great diversity, and vulnerability.

Remarkable ecosystems exist in estuaries and intertidal zones along the coasts. Like freshwater ecosystems, they differ in physical and chemical properties, such as light penetration and water temperature, depth, and salinity.

Estuaries

An **estuary** is a partly enclosed coastal region where seawater swirls and mixes with nutrient-rich freshwater from rivers, streams, and runoff from the surrounding land. The confined conditions, slow mixing of water, and tidal action combine to trap the dissolved nutrients. The continually freshened water replenishes nutrients and allows estuaries to support productive ecosystems.

Primary producers are phytoplankton, salt-tolerant plants that can withstand submergence at high tide, and algae living in mud and on plants. Much of the primary production enters detrital food webs in which bacterial and fungal decomposers are the first to feed. Detritus (and bacteria on it) is food for nematodes, snails, crabs, and fish. Filter feeders such as clams eat food particles suspended in the slowly moving water. So many larval and juvenile stages of invertebrates and some fishes are present that estuaries are often called marine nurseries. Also, many migratory birds use estuaries as rest stops.

Chesapeake Bay, Mobile Bay, and San Francisco Bay are broad, shallow estuaries. Estuaries in Alaska and British Columbia are narrow and deep; so are Norway's fjords. In Texas and Florida, estuaries lie behind long spits of sand and mud. The New England coast has many estuarine salt marshes (Figure 38.27). In tropical regions, we find mangrove swamps that function much like salt marshes in estuarine ecology.

Today, many estuarine ecosystems are under serious assault from sewage, agricultural runoff, industrial, urban, and suburban wastes, and upstream diversion of freshwater for human use. Normal conditions, such as suitable salinity levels, cannot be maintained without inflows of unpolluted freshwater.

The Intertidal Zone

Along rocky and sandy coastlines we find ecosystems of the **intertidal zone**, which is not renowned for its creature comforts. Waves batter its resident organisms. Tides alternately submerge and expose them. The higher they are, the more they may dry out, freeze in winter, or bake in summer, and the less food comes their way. The lower they are, the more they must compete in limited spaces. At low tides, birds, rats, and raccoons move in to feed on them. High tides bring the predatory fishes.

Generalizing about coastlines isn't easy, for waves and tides constantly resculpt them. One feature that both rocky and sandy shores do share is vertical zonation.

Rocky shores often have three zones (Figure 38.28a). The highest zone, the upper littoral, is submerged only during the highest tide of the lunar cycle; it is sparsely populated. The midlittoral (middle zone) is submerged during the highest regular tide and exposed during the lowest. In its tidepools we typically find red, brown, and green algae, hermit crabs, nudibranchs, sea stars, and small fishes (Figure 38.28b). Diversity is greatest in the lowest zone, the lower littoral, which is exposed only during the lowest tide of the lunar cycle. In all three shore zones, rapid erosion prevents detritus from accumulating, so grazing food webs prevail.

Waves and currents continually rearrange stretches of loose sediments, called *sandy* and *muddy* shores. Few large plants grow in these unstable places, so you won't find many grazing food webs. Detrital food webs start with the organic debris imported from offshore or nearby landforms. Vertical zonation is less obvious than along rocky shores. Below the low tide mark in temperate areas are blue crabs and sea cucumbers. Marine worms, crabs, and other invertebrates live between the high and low tide marks. At night, at the high tide mark, beach hoppers and ghost crabs pop out of their burrows, seeking food.

Figure 38.27 Salt marsh of a New England estuary where *Spartina*, a marsh grass, is the main producer. Its microbe-enriched litter feeds consumers in the creeks and sound.

SALT MARSH (estuary)

open ocean   sound   shallow bay   creek   tidal river

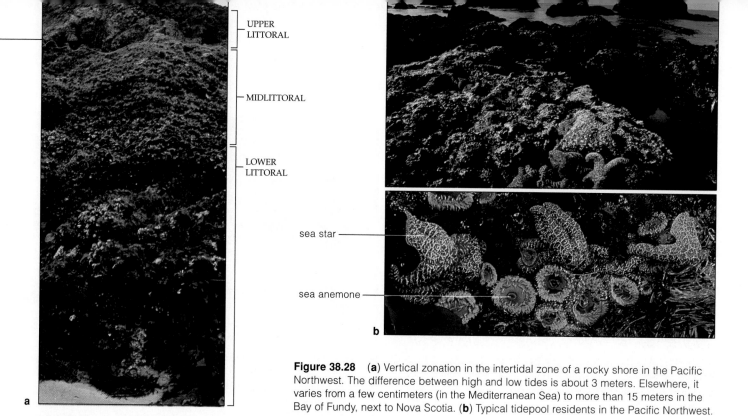

Figure 38.28 (a) Vertical zonation in the intertidal zone of a rocky shore in the Pacific Northwest. The difference between high and low tides is about 3 meters. Elsewhere, it varies from a few centimeters (in the Mediterranean Sea) to more than 15 meters in the Bay of Fundy, next to Nova Scotia. (b) Typical tidepool residents in the Pacific Northwest.

Upwelling Along Coasts

In the Northern Hemisphere, prevailing winds from the north parallel the west coasts of continents and tug on the ocean surface. Wind friction causes surface waters to begin moving. Under the force of the Earth's rotation, the moving water is deflected westward, away from a coast, and cold, deep, often nutrient-rich water moves in vertically to replace it (Figure 38.29). When cold, deep water moves up this way, we call it **upwelling**. It happens in equatorial currents as well as along the coasts of continents in both hemispheres, and it cools the air above it. For example, those fogbanks that form along the California coast are one outcome of upwelling.

a Wind from the north starts surface ocean water moving. **b** Force of Earth's rotation deflects the moving water westward.

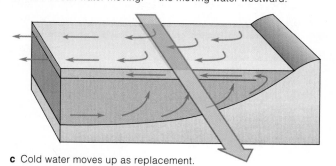

c Cold water moves up as replacement.

Figure 38.29 Upwelling along the west coasts in the Northern Hemisphere. Prevailing winds from the north start surface water moving. The force of Earth's rotation deflects the moving water westward. Cold, deeper water moves up to replace it.

In the Southern Hemisphere, the commercial fishing industries of Peru and Chile depend on wind-induced upwelling. When prevailing coastal winds blow in from the south and southeast, they tug surface water away from shore; and cold, deeper water that the Humboldt Current delivers to the continental shelf moves to the surface. Tremendous amounts of nitrate and phosphate are pulled up, then carried northward by the cold Peru Current. Phytoplankton that depend on these nutrients are the basis of one of the world's richest fisheries.

Every three to seven years, however, warm surface waters of the western equatorial Pacific move eastward. This massive displacement of warm water affects the prevailing wind direction. The eastward flow speeds up so much that it influences the movement of water along the coastlines of Central and South America.

Waters driven toward any coastline will be forced downward and will flow seaward along the continental shelf. Near the coast of Peru, prolonged "downwelling" displaces the cooler waters of the Humboldt Current—and this prevents upwelling. The phenomenon, which local fishermen named **El Niño**, has catastrophic effects on productivity, on birds that feed on anchovetas and other fishes, and on fishing industries. The next section provides a closer look at the El Niño phenomenon.

Shifts in the circulation patterns of the ocean and atmosphere can have repercussions on the functioning of ecosystems that are global in scope.

EL NIÑO AND SEESAWS IN THE WORLD'S CLIMATES

We turn, finally, to a story that will reinforce the unifying concept of this chapter—that conditions in the atmosphere, in the ocean, and on land are interconnected in ways that profoundly influence the world of life.

Our story begins in the winter of 1982–1983. Record rainfall caused massive mudslides along the California coast. Month after month, storms caused severe flooding along the normally arid and semiarid coasts of Ecuador and Peru. In India, the life-sustaining monsoon rains hardly materialized. Devastating droughts hit natural ecosystems and farms of Australia and the Hawaiian Islands. An ongoing drought in Africa intensified, with a resulting amplification of human starvation and death.

The massive dislocations in the world's patterns of rainfall started with changes in sea surface temperatures

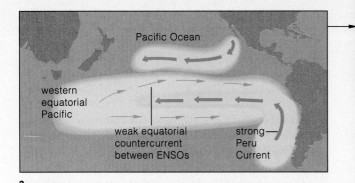

a

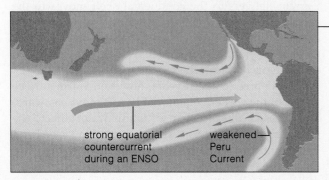

b

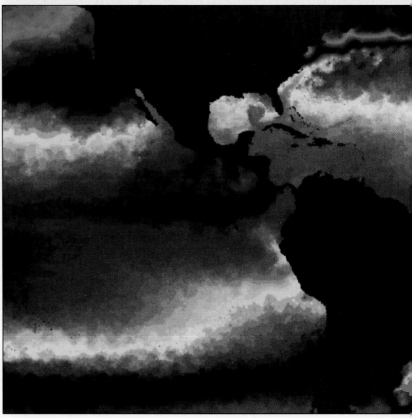

Figure 38.30 (**a**) Distribution of sea surface temperatures for the Pacific between ENSOs. (In the satellite image above right, recorded on May 31, 1988, the warmest water is *dark red*, and progressively cooler water is *yellow*, then *green*.) The western equatorial Pacific had the warmest water. A tongue of relatively cold surface water, which originated with coastal upwelling, flowed westward along the equator from South America.

(**b**) Distribution of sea surface temperatures during an ENSO, as recorded on May 13, 1992. Trade winds weakened and then reversed, leading to a massive, eastward movement of warm water along the equator. Surface waters of the central and eastern Pacific became warmer. Notice the tongue of recently upwelled water that extends eastward along the equator.

and air circulation patterns, which induced drought-related conditions on land. The changes produce a climatic event called the *El Niño Southern Oscillation* (ENSO).

"Southern Oscillation" refers to a recurring seesaw in atmospheric pressure in the western equatorial Pacific. The region is the world's largest reservoir of warm water. More warm air rises here than anywhere else, and heavy rainfall releases much of the heat energy that drives the world's air circulation system. Pulses of heat from the Earth—possibly at clusters of a thousand or more active volcanoes that were recently discovered on the deep ocean floor—may trigger the Southern Oscillation. Whatever the cause, heat moves upward and warms the surface waters.

Normally, the warm reservoir and the heavy rainfall associated with it move westward (Figure 38.30*a*). But now they move east (Figure 38.30*b*). Prevailing surface winds in the western equatorial Pacific pick up speed. The stronger winds have a pronounced effect on "dragging" the ocean surface waters to the east. The upper ocean currents are affected to the extent that the westward transport of water slows down and the eastward transport increases. *More warm water in the vast reservoir moves east—and so on in a feedback loop between the ocean and the atmosphere.*

As you read earlier, the reversal in the usual westward flow of air and water displaces the cold, deep Humboldt Current—and so stops the upwelling of nutrients along the western coast of South America. Peruvian fishermen call the warm, nutrient-poor current from the east "El Niño." Usually it reaches their coast around Christmas (hence the name, "the little one," a reference to the baby Jesus).

An ENSO is a recurring event. About 2,000 kilometers west of Peru, the sea level rises. Sea surface temperatures increase, often by 7°C (11°F). Evaporation increases and air pressure drops. And the humid updrafts trigger violent storms along coasts and often more rain in inland regions.

The wave of warm water from the 1982 El Niño was 20 centimeters (8 inches) high, it traveled about 5 miles an hour, and some of its effects lasted twelve years. It reached South America in two months, spread along the coasts, and raised the sea level as far away as Alaska. Then it slowly moved back across the Pacific. Nine years later, it was still displacing part of a major warm current north of its eastward path from Japan. Two more El Niños followed this one—which only now is subsiding in the Bering Sea. Each time, winds and rainfall patterns around the world change. And so does productivity. Thus, for 1970 to 1993, a plot of corn yields in Zimbabwe exactly matches a plot of the rise and fall of sea surface temperatures in the Pacific.

Mathematical models are used to study these and other episodes of climatic change a few seasons in advance. More reliable forecasting from this ecosystem modeling should follow when more and better observations are made of the interrelated systems of the ocean, land, and atmosphere.

SUMMARY

1. The biosphere encompasses only the Earth's waters, lower atmosphere, and uppermost portions of its crust in which organisms live. Energy flows one-way through the biosphere, and materials move through it on a grand scale to influence ecosystems everywhere.

2. The distribution of species through the biosphere is an outcome of the Earth's history, its climate, and its topography, as well as interactions among species.

3. Climate refers to the average weather conditions, including temperature, humidity, wind velocity, cloud cover, and rainfall over time. It results from differences in the amount of solar radiation reaching equatorial and polar regions, the Earth's daily rotation and its annual path around the sun, the distribution of continents and oceans, and the elevation of land masses.

4. Interacting climatic factors produce prevailing winds and ocean currents, which shape the global weather patterns. The weather affects the composition of soil, sedimentation, and water availability in ecosystems—which in turn affect the growth and distribution of their primary producers.

5. The world's land masses are classified as six major biogeographic realms. Each is more or less isolated by oceans, mountain ranges, or desert barriers, which tend to restrict gene flow between other realms. As a result, each tends to maintain its characteristic array of species.

6. A biome (a category of major ecosystems on land) is shaped by regional variations in climate, landforms, and soil composition. Plant species that are adapted to a certain set of conditions dominate each biome. Deserts, dry shrublands, dry woodlands, grasslands, broadleaf forests (such as tropical rain forests), coniferous forests, and tundra are major types.

7. Water provinces cover more than 71 percent of the Earth's surface. They include standing freshwater (such as lakes), running freshwater (such as streams), as well as the world ocean and seas. All aquatic ecosystems show gradients in light penetration, water temperature, salinity, and dissolved gases. These factors vary daily and seasonally, and they influence primary productivity.

8. Estuaries, intertidal zones, rocky and sandy shores, tropical reefs, and regions of the open ocean are major marine ecosystems. Photosynthetic activity is greatest in shallow coastal waters and in regions of upwelling. Upwelling is an upward movement of deep, cool ocean water that often carries nutrients to the surface.

9. The interrelatedness of ocean surface temperatures, the atmosphere, and the land is evident in studies of the ENSO (the El Niño Southern Oscillation). Abnormally severe drought conditions in many parts of the world accompany this recurring event.

Review Questions

1. Define atmosphere, lithosphere, and hydrosphere. *681*

2. List the major interacting factors that influence climate. *681*

3. List some of the ways in which air currents, ocean currents, or both may influence the ecosystem in which you live. *683–685*

4. Define soils, then explain how the composition of regional soils affects ecosystem distribution. *688–689*

5. Indicate where the following biomes tend to be located around the world, then describe some of their defining features. *689–695*
    a. deserts   e. deciduous broadleaf forests
    b. dry shrublands   f. coniferous forests
    c. grasslands   g. alpine tundra
    d. evergreen broadleaf forests h. arctic tundra

6. Characterize the littoral, limnetic, and profundal zones of a large temperate lake in terms of seasonal primary productivity. *696*

7. Define the two major provinces of the world ocean. *698*

8. What points favor the hypothesis that life may have originated near hydrothermal vents on the ocean floor? *699*

9. Define and characterize the following ecosystems. *700–702*
    a. estuary   c. sandy shore
    b. rocky shore   d. coral reef

10. Describe an ENSO event and some of its consequences in both hemispheres. *704–705*

Self-Quiz *(Answers in Appendix IV)*

1. Solar radiation drives the distribution of weather systems and so influences the distribution of _____ .
    a. every single ecosystem
    b. ecosystems on land only
    c. all ecosystems except those at hydrothermal vents

2. The _____ is a shield against ultraviolet wavelengths from the sun.
    a. upper atmosphere   c. ozone layer
    b. lower atmosphere   d. greenhouse effect

3. Regional variations in patterns of rainfall and temperature depend on _____ .
    a. global air circulation   c. topography
    b. ocean currents   d. all of the above

4. A rain shadow is a reduction in rainfall on the _____ of a mountain range.
    a. windward side   c. highest elevation
    b. leeward side   d. lowest elevation

5. Biogeographic realms are _____ .
    a. land and water provinces c. divided into biomes
    b. six major land provinces d. b and c

6. Biome distribution corresponds roughly with regional variations in _____ .
    a. climate   c. topography
    b. soils   d. all of the above

7. _____ are highly adapted to episodes of fire.
    a. Dry shrublands   c. Pine barrens
    b. Grasslands   d. All of the above

8. During _____ , deeper, often nutrient-rich water moves to the surface.
    a. spring overturns   c. upwellings
    b. fall overturns   d. all of the above

9. Match the terms with the suitable description.
    ____ boreal forest   a. high productivity, poor cycling
    ____ permafrost   of mineral nutrients
    ____ chaparral   b. "swamp forest"
    ____ desert   c. common at moist, temperate
    ____ deciduous   latitudes having mild winters
         broadleaf forest   d. evaporation greatly exceeds
    ____ tropical   sparse, infrequent rainfall
         rain forest   e. a type of dry shrubland
                        f. feature of arctic tundra

10. Match the terms with the suitable description.
    ____ plankton   a. deep, cool, often nutrient-rich
    ____ upwelling   ocean water moves upward
    ____ eutrophication   b. sediments, rocky formations
    ____ estuary   of the ocean bottom
    ____ benthic   c. seawater, freshwater slowly mix
         province   d. nutrient enrichment of body
                        of water; reduced transparency,
                        phytoplankton blooms
                        e. community of mostly microscopic
                        floating or weakly swimming species

Critical Thinking

1. Kangaroos are furry, pouched mammals and raccoons are furry placental mammals. Explain why Australia but not North America is the original home of kangaroos (Figure 38.31*a*) and why the reverse is true of raccoons (Figure 38.31*b*). Use your understanding of geologic history when putting together your explanation.

2. Think about the world distribution of land masses and ocean water, then develop an explanation of why grassland biomes tend to form in the interior of continents and not along their coasts.

Figure 38.31 (**a**) A female kangaroo with offspring in her ventral pouch. (**b**) Two young raccoons.

a

b

a

Oligotrophic Lake	Eutrophic Lake
Deep, steeply banked	Shallow with broad littoral
Large deep-water volume relative to surface-water volume	Small deep-water volume relative to surface-water volume
Highly transparent	Limited transparency
Water blue or green	Water green to yellow or brownish-green
Low nutrient content	High nutrient content
Oxygen abundant through all levels throughout year	Oxygen depleted in deep water during summer
Not much phytoplankton; green algae and diatoms dominant	Abundant masses of phytoplankton; cyanobacteria dominant
Abundant aerobic decomposers favored in profundal zone	Anaerobic decomposers
Low biomass in profundal	High biomass in profundal

b

Figure 38.32 (**a**) Crater Lake, Oregon, a collapsed volcanic cone now filled with water from rains and melted snow. Like the other volcanoes of the Cascade range, it started forming at the dawn of the Cenozoic, when crustal plates underwent major reorganization. (**b**) Characteristics of oligotrophic and eutrophic lakes.

3. The United States Forest Service and the Geophysical Dynamics Laboratory ran supercomputer programs in order to predict some possible outcomes of a *global warming* trend (here you may wish to review Section 37.7). According to the results, by the year 2030, the United States will experience recurring, devastating forest fires. Severe storms and more frequent rainfall in the western states will accelerate erosion, especially along the coasts and in steep foothills and mountain ranges. Increases in the deposition of sediments will have bad effects on rivers, streams, and estuaries. Dry shrublands and woodlands might replace hardwood forests in Minnesota, Iowa, Wisconsin, and in parts of Missouri, Illinois, Indiana, and Michigan. The breadbasket of the American Midwest will shift north into Canada. Think about where you live now and where you plan to live in the future. How would such changes affect you? Should nations get serious about finding ways to counter global warming? Or do you believe that the scientists who are concerned about this are merely alarmist and that nothing bad will happen? Explain why you have reached your conclusion.

4. Think about the ocean, which covers the majority of the Earth's surface. Its deepest regions are remote from human populations, to say the least. Now think about a modern-day problem—where to dispose of nuclear wastes and other hazardous materials. You will be reading about this serious problem in the next chapter; for now, simply be aware that the United States government has stockpiles of very dangerous wastes and has nowhere to put them. Would it be feasible, let alone ethical, to use the deep ocean as a "burying ground" for them? Why or why not?

5. Observe or do some research on conditions in a lake near your home or a place where you vacation. If lakes aren't part of your life, think of Oregon's Crater Lake instead (Figure 38.32*a*). Using the data in Figure 38.32*b* as a guide, would you conclude Crater Lake is oligotrophic or eutrophic? Will it remain so over time?

6. *Wetlands* are transitional zones between aquatic and terrestrial ecosystems in which plants are adapted to grow in periodically or permanently saturated soil. They include *marshes*, in which 15 to 300 centimeters of water cover the land. They include *riparian zones* along the fringes of most rivers, such as the shrubby bottom lands of the southern United States, and 14 million hectares of *mangrove swamps*, with their diverse species of mangrove trees and shrubs. Many species require wetlands to complete at least part of their life cycle. Besides this, different wetlands serve as traps for sediments and waterborne pollutants, buffer zones for controlling erosion and flooding, and often as carbon dioxide sinks. Wetlands everywhere are being converted for agriculture, home building, and other human activities. Today only 50 percent of the spectacular wetlands called the Florida Everglades remains. California has only 9 percent of its wetlands intact; Indiana and Missouri have only 10 percent. Does the great ecological value of wetlands for the nation as a whole outweigh rights of private citizens—who hold title to many of the remaining wetlands in the United States? Should the private owners be forced to transfer title to the government? If so, who should decide the land's value *and* pay for it? Or would such seizure of private property violate the United States Constitution?

Selected Key Terms

biogeographic realm *686*
biogeography *686*
biome *687*
biosphere *681*
boreal forest *694*
climate *681*
coniferous forest *694*
coral reef *700*
deciduous broadleaf forest *693*
desert *689*
desertification *689*
dry shrubland *690*
dry woodland *690*
El Niño *703*
estuary *702*
eutrophication *697*
evergreen broadleaf forest *692*
fall overturn *696*
grassland *690*

hydrothermal vent *699*
intertidal zone *702*
lake *696*
monsoon *685*
ocean *684*
ozone layer *682*
permafrost *695*
pine barren *694*
plankton *696*
rain shadow *685*
savanna *690*
soil *688*
spring overturn *696*
stream *697*
temperature zone *682*
tropical rain forest *692*
tundra *695*
ultraplankton *698*
upwelling *703*

Readings

Brown, J., and A. Gibson. 1983. *Biogeography*. St. Louis: Mosby.

Garrison, T. 1996. *Oceanography*. Second edition. Belmont, California: Wadsworth.

Smith, R. 1996. *Ecology and Field Biology*. Fifth edition. New York: Harper Collins.

39 HUMAN IMPACT ON THE BIOSPHERE

An Indifference of Mythic Proportions

Of all the concepts introduced in the preceding chapter, the one that should be foremost in your mind is this: The atmosphere, ocean, and land interact in ways that help dictate living conditions throughout the biosphere. Driven by energy streaming in from the sun, these stupendous interactions give rise to globe-spanning temperatures and circulation patterns upon which life ultimately depends. With this chapter, we turn to a related concept of equal importance. Simply put, we have become major players in these interactions even before we fully comprehend how they work.

To gain perspective on what is happening, think of something we take for granted—the air around us. The composition of the present-day atmosphere is a result of geologic and metabolic events—most notably, photosynthesis—that began billions of years ago. The first humans, recall, evolved about 2 million years ago. Like us, they breathed oxygen from an atmosphere of ancient origins. Like us, they were protected from the sun's ultraviolet radiation by an ozone shield in the stratosphere. The size of their population was not much

to speak of, and their effect on the biosphere was trivial. About 10,000 years ago, however, agriculture began in earnest, and it laid the foundation for huge increases in population size. A few centuries ago, medical and industrial revolutions expanded that foundation—and human population growth skyrocketed in a mere blip of evolutionary time.

Today we are extracting huge amounts of energy and resources from the environment and giving back monumental amounts of wastes. As we do so, we are destabilizing ecosystems everywhere, even though the magnitude of change may not even be recognized when measured on the scale of a lifetime (Figure 39.1).

In a few developed countries, population growth has more or less stabilized, and the resource use per individual has dropped a bit. But resource utilization levels in those places are already high. In developing countries in Central America, Africa, and elsewhere, population sizes and resource consumption are rapidly growing, even though hundreds of millions of people are already malnourished or starving to death.

1900   1940   1954   1962

Many of the problems sketched out in this chapter are not going to disappear tomorrow. It will take many decades, even centuries, to reverse some trends that are already in motion—and not everyone is ready to make the effort. A few enlightened individuals in Michigan or Alberta or New South Wales can commit themselves to resource conservation and to minimizing pollution. But scattered attempts will not be enough. Individuals of all nations will unite to reverse global trends only when they perceive that the dangers of *not* doing so outweigh the personal benefits of ignoring them.

Does this seem pessimistic? Think of the exhaust fumes released into the air each time you drive a car or truck. Think of oil refineries, food-processing plants, and paper mills that supply you with goods—and also release chemical wastes into the nation's waterways. Think of Mexico and other developing countries that produce cheap food by using an unskilled labor force and toxic pesticides. Unregulated pesticide applications poison the people who work the land, the land itself, and sometimes people who buy the exported produce.

Who changes behavior first? We have no answer to the question. We can suggest, however, that a strained biosphere can rapidly impose an answer upon us. As an individual, you might choose to cherish or brood about or ignore any aspect of the world of life. Whatever your choice may be, the bottom line is that you and all other organisms are in this together. Our lives interconnect, to degrees that we are only now starting to comprehend.

Figure 39.1 A track of human population growth in one small part of the world only, based on historical data and satellite imaging. *Red* denotes areas of dense human settlement in and around the San Francisco Bay Area and Sacramento since 1900.

KEY CONCEPTS

1. Human population growth has been skyrocketing ever since the mid-eighteenth century. At present, we have the population size, technology, and cultural inclination to use energy and alter the environment at astonishing rates.

2. Pollutants are substances with which ecosystems have had no prior evolutionary experience, in terms of kinds or amounts, so adaptive mechanisms are not in place that can deal with them. In a more restricted sense, pollutants are substances that have accumulated in amounts that have adverse effects on human health, activities, or survival.

3. Many pollutants, including those involved in forming smog or acid rain, exert regionally harmful effects. The effects of other pollutants, including chlorofluorocarbons that attack the ozone layer in the stratosphere, are global in scale.

4. Conversion of marginal lands for agriculture, rampant deforestation, and other practices required to meet the demands of the growing human population are leading to loss of soil fertility and desertification. They are also resulting in a decline in the quality and quantity of one of the most crucial of all resources—fresh water.

5. Ultimately, the world of life depends on energy inputs from the sun. That energy drives the complex interactions among the atmosphere, ocean, and land. Our activities are disrupting the interactions in ways that may have serious consequences in the near future.

6. We as a species must come to terms with the principles of energy flow and the principles of resource utilization that govern all systems of life on Earth.

39.1 AIR POLLUTION—PRIME EXAMPLES

Let's start this survey of the impact of human activities by defining pollution, which is central to the problem. **Pollutants** are substances with which ecosystems have had no prior evolutionary experience, in terms of kinds or amounts, so adaptive mechanisms that can deal with them are not in place. From the human perspective, they are substances that accumulate to levels that have adverse effects on our health, activities, or survival.

Air pollutants are prime examples. As listed in Table 39.1, they include carbon dioxide, oxides of nitrogen and sulfur, and chlorofluorocarbons. Also among them are photochemical oxidants, formed as the sun's rays interact with certain chemicals. The United States alone releases 700,000 metric tons of air pollutants *each day*. Whether these remain concentrated at the source or are dispersed during a given interval depends on the local climate and topography, as you will now see.

Smog

During a **thermal inversion**, weather conditions trap a layer of cool, dense air under a layer of warm air. If the trapped air contains pollutants, winds cannot disperse them, and they may accumulate to dangerous levels. Thermal inversions have been key factors in some of the worst air pollution disasters, because they intensify an atmospheric condition called smog (Figure 39.2).

Two types of smog—gray air or brown air—form in major cities. Where winters are cold and wet, **industrial smog** develops as a gray haze over industrialized cities that burn coal and other fossil fuels for manufacturing, heating, and generating electric power. Burning releases airborne pollutants, such as dust, smoke, soot, ashes, asbestos, oil, bits of lead and other heavy metals, and sulfur oxides. If not dispersed by winds and rain, these emissions may reach lethal concentrations. Industrial smog caused 4,000 deaths during the 1952 air pollution disaster in London. Before coal burning was restricted, New York, Pittsburgh, and Chicago also were gray-air cities. Today, most industrial smog forms in the cities of China, India, and other developing countries, as well as in coal-dependent countries of eastern Europe.

In warm climates, **photochemical smog** develops as a brown haze over large cities. It becomes concentrated when the surrounding land forms a natural basin, as it does near Los Angeles and Mexico City (Figure 39.2). The key culprit is nitric oxide. After it is released from vehicles, it reacts with oxygen in the air and so forms nitrogen dioxide. When exposed to sunlight, nitrogen dioxide reacts with hydrocarbons, and photochemical oxidants result. (Most hydrocarbons come from spilled or partially burned gasoline.) The main oxidants are ozone and **PANs** (peroxyacyl nitrates). Even traces of PANs can sting eyes, irritate lungs, and damage crops.

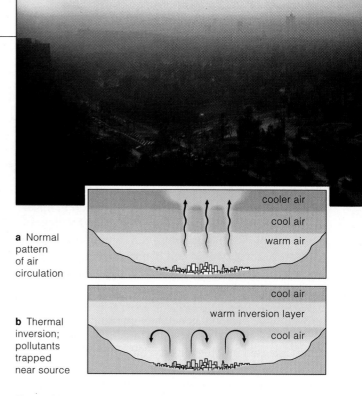

a Normal pattern of air circulation

cooler air
cool air
warm air

b Thermal inversion; pollutants trapped near source

cool air
warm inversion layer
cool air

Figure 39.2 (**a**) Normal pattern of air circulation in smog-forming regions. (**b**) Air pollutants trapped under a thermal inversion layer. The photograph shows Mexico City on an otherwise bright, sunny morning. Topography, staggering numbers of people and vehicles, and industry combine to make its air among the world's smoggiest. Breathing this air is like smoking two packs of cigarettes a day.

Table 39.1	Major Classes of Air Pollutants
Carbon oxides	Carbon monoxide (CO), carbon dioxide (CO_2)
Sulfur oxides	Sulfur dioxide (SO_2), sulfur trioxide (SO_3)
Nitrogen oxides	Nitric oxide (NO), nitrogen dioxide (NO_2), nitrous oxide (N_2O)
Volatile organic compounds	Methane (CH_4), benzene (C_6H_6), chlorofluorocarbons (CFCs)
Photochemical oxidants	Ozone (O_3), peroxyacyl nitrates (PANs), hydrogen peroxide (H_2O_2)
Suspended particles	Solids (dust, soot, asbestos, lead, etc.), liquid droplets (sulfuric acid, oils, pesticides, etc.)

Acid Deposition

Oxides of sulfur and nitrogen are among the worst air pollutants. Coal-burning power plants, metal smelters, and factories emit most sulfur dioxides. Motor vehicles, power plants that burn gas and oil, and nitrogen-rich fertilizers produce nitrogen oxides. In dry weather, fine particles of oxides may briefly stay airborne, then fall to Earth as **dry acid deposition**. When they dissolve in atmospheric water, they form weak solutions of sulfuric and nitric acids. Strong winds may distribute them over great distances. If they fall to Earth in rain and snow, we call this wet acid deposition, or **acid rain**. The pH of normal rainwater is about 5. Acid rain can be 10 to 100

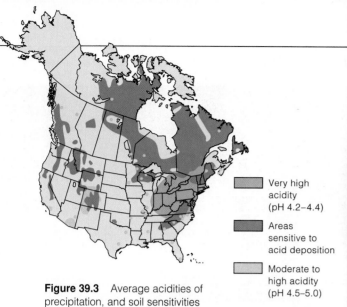

Figure 39.3 Average acidities of precipitation, and soil sensitivities to acid deposition, for regions of North America in 1984. Prolonged exposure to air pollutants is contributing to the rapid destruction of trees in New York's Whiteface Mountains (**a**), Germany (**b**), and elsewhere. The weakened trees are more vulnerable to drought, disease, and insect attack.

Very high acidity (pH 4.2–4.4)

Areas sensitive to acid deposition

Moderate to high acidity (pH 4.5–5.0)

a   b

times more acidic than this—even as potent as lemon juice! The deposited acids eat away at marble buildings, metals, rubber, plastic, nylon stockings, and many other materials. They can significantly disrupt the physiology of organisms and the chemistry of ecosystems.

Depending on their soil type and vegetation cover, some regions are far more sensitive than others to acid deposition (Figure 39.3). Highly alkaline soil neutralizes acids before they enter streams and lakes of watersheds. Water having a high carbonate content also neutralizes acids. However, in many watersheds throughout much of northern Europe and southeastern Canada, as well as in scattered regions of the United States, very thin soils overlie solid granite and do not buffer the acids much.

Rain in much of eastern North America is thirty to forty times more acidic than it was several decades ago. Crop yields are diminishing. Fish populations already have vanished from more than 200 lakes in New York's Adirondack Mountains. By some predictions, fish will disappear from 48,000 lakes in Ontario, Canada, within the next two decades. Pollution from industrial regions is contributing to the destruction of forest trees and of mycorrhizae that support new growth (Section 19.11).

Researchers confirmed long ago that emissions from power plants, factories, and vehicles are the key sources of air pollutants. In 1995, researchers from the Harvard School of Public Health and Brigham Young University reported this finding from a comprehensive air quality study: Whack a year or so off your life span if you live in cities with the dirtiest air—especially air with fine particles of dust, soot, smoke, or acid droplets. Smaller particles are more easily inhaled and can damage lung tissue. Figure 39.4 summarizes air-quality data gathered in 1992 for the continental United States.

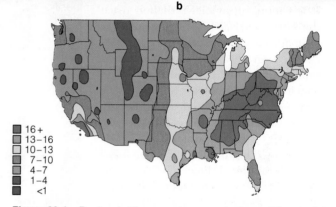

16 +
13–16
10–13
7–10
4–7
1–4
<1

Figure 39.4 Regional differences in concentrations of fine particles (2.5 μm or less) in micrograms per cubic meter of air, measured in 1992. Data are averages; they understate peaks in individual cities.

At one time, the world's tallest smokestack, in the Canadian province of Ontario, accounted for 1 percent by weight of the annual worldwide emissions of sulfur dioxide. But Canada receives more acid deposition from industrialized regions of the midwestern United States than it sends across its southern border. Most of the air pollutants in Scandinavian countries, the Netherlands, Austria, and Switzerland arise in industrialized regions of western and eastern Europe. *Prevailing winds—hence air pollutants—do not stop at national boundaries.*

Pollutants are substances with which ecosystems have had no prior evolutionary experience, in terms of kinds or amounts, so adaptive mechanisms are not in place to deal with them.

An accumulation of pollutants can harm organisms, as when they reach levels that adversely affect human health.

Smog formation and acid deposition in specific regions are examples of air pollution, although prevailing winds often distribute such pollutants beyond regional boundaries.

Smog forms mainly as a result of fossil fuel burning in urban and industrialized regions. Airborne acidic pollutants drift down to Earth as dry particles or as components of acid rain.

OZONE THINNING—GLOBAL LEGACY OF AIR POLLUTION

Now think about the ozone layer, almost twice as high above sea level as the top of Mount Everest, the highest place on Earth. Each September through mid-October, the layer gets thinner at high latitudes. Actually, this seasonal **ozone thinning** is so pronounced, it was once known as an "ozone hole." In 1995, the thinning above Antarctica spanned an area twice as big as Europe. The thinning at high northern latitudes exceeded 10 percent. Sixty years from today, the protective layer may be reduced by 30 percent or more over heavily populated regions of North America, Europe, and Asia.

Why is ozone reduction alarming? It lets more ultraviolet radiation reach the Earth's surface. This may lead to dramatic increases in cancers of the skin, cataracts of the eyes, and weakened immune systems. Ultraviolet radiation also impacts on photosynthesis. Drastic drops in the oxygen-releasing activity of phytoplankton alone could alter the composition of the atmosphere, given their staggering numbers (Section 5.7).

Chlorofluorocarbons (CFCs) are the major factors in the ozone reduction. These compounds of chlorine, fluorine, and carbon are odorless and invisible. You find them in refrigerators and air conditioners (as the coolants), solvents, and plastic foams. CFCs slowly escape into air and resist breakdown. When a free CFC molecule absorbs ultraviolet light, it gives up a chlorine atom. If chlorine reacts with ozone, this yields oxygen and chlorine monoxide—which can react with free oxygen to release another chlorine atom. Each chlorine atom thus released can convert 10,000 ozone molecules or more to oxygen!

The chlorine monoxide levels above Antarctica are 100 to 500 times higher than at mid-latitudes. Why? High-altitude ice clouds form there in winter. Winds rotate around the South Pole for most of the winter and isolate the ice clouds from other latitudes (Figure 39.5). This also happens, on a lesser scale, in the Arctic. Ice provides a surface that promotes breakdown of chlorine compounds. Chlorine is free to destroy ozone when the air warms during the Antarctic spring. Hence the ozone thinning.

CFCs aren't the only ozone eaters. Methyl bromide, a potent fungicide widely used in agriculture, is even better at it. Although it persists only a short time in the

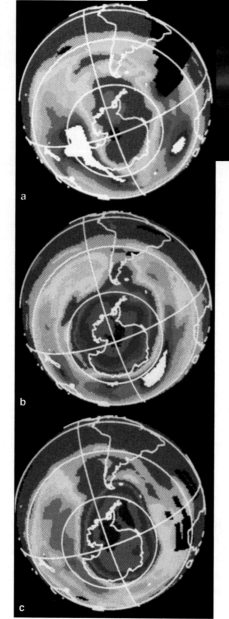

Figure 39.5 Pronounced seasonal ozone thinning over Antarctica in (**a**) 1979, (**b**) 1991, and (**c**) 1992. The *magenta* and *purple* color coding signify the lowest ozone values. The photograph above shows the ice clouds that have a role in the ozone thinning each spring.

atmosphere, it will account for about 15 percent of the ozone thinning in future years unless its production stops.

A few scientists dismiss the threat of the chemicals that contain chlorine and bromine. But the vast majority of those who have studied ecosystem modeling programs conclude that these chemicals do pose long-term threats—not only to human health and certain crops but also to animal life in general. Substitutes are now available for most applications of CFCs, and others are being developed.

Under international agreement, the production of CFCs has been phased out in developed countries. Developing countries will phase them out by 2010. Under a recent accord, methyl bromide production will be phased out by 2010. Such international accords are steps in the right direction. Of course, assuming the goals are met, it still will be about 50 years before the ozone layer is restored to 1985 levels and another 100 to 200 years to full recovery, to levels that prevailed before 1950. In the meantime, you and the children and grandchildren of future generations will be living with the destructive effects of these air pollutants. At least corrective action is under way.

Air pollution can have global repercussions, as when CFCs and other compounds contribute to a thinning of the ozone layer that shields life from the sun's ultraviolet radiation.

WHERE TO PUT SOLID WASTES, WHERE TO PRODUCE FOOD

Oh Bury Me Not, In Our Own Garbage

In natural ecosystems, one organism's wastes serve as resources for others, so the by-products of existence are cycled through the system. In developing countries, many resources are scarce. People conserve what they can and discard little. In the United States and some other developed countries, most people use something once, discard it, then buy another. Each year, millions of metric tons of solid wastes are dumped, burned, and buried. Paper products make up half the total volume, which also includes 50 billion nonreturnable cans and bottles. Each week, paper manufactured from 500,000 trees ends up in a Sunday newspaper to Americans. If every reader recycled merely one of ten newspapers, 25 million trees a year could be left standing. Recycling the paper would cut the airborne pollutants released during paper manufacturing by 95 percent and require 30 to 50 percent less energy than making new paper.

A throwaway mentality is unique in the world of life. We are condoning the monumental accumulation of solid wastes in human ecosystems and natural ones. Bury the garbage in landfills? Then what happens when the space around cities runs out? Besides this, landfills "leak" and in time threaten groundwater supplies. Burn the wastes in inefficient incinerators? These spew great volumes of pollutants and ashes into the atmosphere.

Today, recycling is affordable and feasible. On an individual basis, we can help bring about change by refusing to buy goods that are excessively wrapped, packaged in nondegradable containers, and designed for one-time use. Individuals can participate in curbside recycling, by which they presort their recyclable wastes. Such action apparently is more efficient than reliance on huge resource recovery centers. These require so much trash to turn a profit that owners may actually end up encouraging the throwaway mentality. The huge centers also produce a toxic ash that must be disposed of in landfills—which eventually leak.

Converting Marginal Lands for Agriculture

The human population already uses nearly 21 percent of the Earth's land surfaces for cropland or for grazing. Another 28 percent is said to be potentially suitable for agriculture, but productivity would be so low that the conversion may not be worth the cost (Figure 39.6).

Asia and several other heavily populated regions now experience recurring, severe food shortages. Yet more than 80 percent of their productive land is already under cultivation. Scientists have made valiant efforts to improve crop production on existing land. Under the banner of the **green revolution**, their research has been directed toward (1) improving the genetic character of

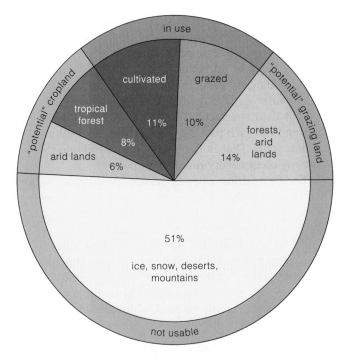

Figure 39.6 Classification of land with respect to its usability for agriculture. Theoretically, clearing tropical forests and irrigating marginal land could more than double the world's cropland. Doing so would destroy valuable forest resources, severely damage the environment, and possibly cost more than it is worth.

crop plants for higher yields and (2) exporting modern agricultural practices and equipment to the developing countries. Many of these countries rely on *subsistence* agriculture, which runs on energy inputs from sunlight and human labor. They also rely very heavily on *animal-assisted* agriculture, with energy inputs from oxen and other draft animals. By contrast, *mechanized* agriculture requires massive inputs of fertilizers, pesticides, and ample irrigation to sustain high-yield crops. It requires fossil fuel energy to drive farm machines. Crop yields are four times as high. But the modern practices use up 100 times more energy and minerals. Also, there are signs that limiting factors are coming into play to slow down further increases in crop yields.

Pressures for increased food production are greatest in parts of Central and South America, Asia, the Middle East, and Africa where human populations are rapidly expanding into marginal lands. Repercussions extend beyond national boundaries, as you will see next.

Our astounding population growth has impact on the Earth's land. We generate huge amounts of solid wastes but reuse or recycle very little. We also are making the energetically and environmentally costly move of expanding into marginal lands for food production.

DEFORESTATION—CONCERTED ASSAULT ON FINITE RESOURCES

At one time, tropical forests cloaked regions that were, collectively, twice the size of Europe. For ten thousand years or more, these forests endured in rich complexity, as the homes of an estimated 50 to 90 percent of all land-dwelling species. In less than four decades, we destroyed more than half of these ancient forests, and most of their spectacular arrays of species may be lost forever. With each passing year, we log an additional 38 million acres. That is the equivalent of leveling thirty-four city blocks every minute.

The destruction extends beyond the tropics. Today, highly mechanized logging operations are proceeding in temperate forests throughout the United States, Canada, Europe, Siberia, and elsewhere.

We have a name for the removal of all trees from large tracts of land for logging, agriculture, and grazing operations. It is **deforestation**. Why are we doing this? Paralleling the rapid increases in the size of the human population are rapidly increasing demands for lumber, fuel, and other forest products, as well as for cropland and grazing land. More and more people are competing for diminishing resources, often for economic profit—but also because alternative ways of life simply are not available to individuals and families.

The world's great forests have profound influences on ecosystems. Like giant sponges, the watersheds of forested regions absorb, hold, and then release water gradually. By intervening in the downstream flow of water, they help control soil erosion, flooding, and the accumulation of sediments that can clog rivers, lakes, and reservoirs. When the vegetation cover gets stripped away, the exposed soil becomes vulnerable to leaching of nutrients and erosion, especially on steep slopes.

Today, deforestation is greatest in Brazil, Indonesia, Colombia, and Mexico. If the clearing and destruction continue at present rates, only Brazil and Zaire will have large tropical forests in the year 2010. By 2035, most of their forests will be gone, also.

Figures 39.7 and 39.8 give close-up and panoramic views of what is now happening in South America's Amazon basin. Figure 19.1c is one example of the effects of deforestation in North America.

In tropical regions, clearing forests for agriculture leads to a long-term loss in productivity. The irony is that tropical forests are one of the worst places to grow crops or to raise pasture animals. In intact forests, litter does not build up, for the high temperatures and the frequent, heavy rains favor rapid decomposition of organic remains and wastes. As fast as nutrients are released, trees and other plants take them up, so deep, nutrient-rich soils simply cannot form.

Long before the advent of extensive mechanized logging, people were practicing **shifting**

Figure 39.7 Countries permitting the largest destruction of tropical forests. *Red* shading denotes where 2,000 to 14,800 square kilometers are deforested annually. *Orange* denotes "moderate" deforestation (100 to 1,900 square kilometers).

Figure 39.8 The vast Amazon River basin of South America in September 1988. Its features were completely obscured by smoke from fires that had been set to clear tropical forests, pasturelands, and croplands during the dry season.

Smoke extends to the Andes Mountains on the western horizon, about 650 miles (1,046 kilometers) away. The smoke cover was the largest that astronauts had ever observed. It extended almost 175 million square kilometers (1,044,000 square miles).

The smoke plume near the center of the photograph alone covered an area comparable to the extensive forest fire in Yellowstone National Park in that same year.

Massive deforestation is not confined to equatorial regions. For example, during the past century, 2 million acres of redwood forests along the coast of California have been logged over. Most of the destruction of such temperate forests has been occurring since 1950, owing to the widespread use of chainsaws and tractors and the practice of exporting many of the logs to lumbermills overseas, where wages are low.

cultivation (once called slash-and-burn agriculture). They cut and burn trees, then till ashes into the soil. The nutrient-rich ashes can sustain crops for one to several seasons. Afterward, cleared plots are abandoned when leaching renders the soil infertile. Shifting cultivation on widely scattered, small plots does not necessarily damage forest ecosystems much. Soil fertility plummets with increases in population size. Then, larger areas are cleared, and plots are cleared again at shorter intervals.

Now think about the larger picture. Deforestation alters the rates of evaporation, transpiration, runoff, and possibly the regional patterns of rainfall. For example, trees release between 50 and 80 percent of the water vapor above tropical forests. In logged-over regions, annual precipitation declines, and rain rapidly runs off the bare, nutrient-poor soil. As a deforested region gets hotter and drier, soil fertility and moisture decline. In time, sparse grassland or desertlike conditions might prevail instead of a rich forest biome.

Finally, consider that tropical forests absorb much of the sunlight reaching equatorial regions of the Earth's surface. Deforested land is shinier, so to speak, and it reflects more incoming energy back into space. Also, by their photosynthetic activity, the great numbers of trees in these vast biomes help sustain the global cycling of carbon and oxygen. During extensive tree harvesting or burning, carbon stored in the tree biomass is released to the atmosphere as carbon dioxide. Thus deforestation factors into amplification of the greenhouse effect.

Once-vast forests helped sustain rapid increases in the size of the human population. Recent and highly mechanized modes of deforestation are rapidly depleting these finite resources.

YOU AND THE TROPICAL RAIN FOREST

By far, tropical rain forests contain the greatest variety and numbers of hungry insects and the world's biggest ones. They are home to the most species of birds and to plants with the largest flowers. Slinking or leaping or bounding through the forest canopy and understory are splendidly varied monkeys, tapirs, and jaguars in South America and apes, okapi, and leopards in Africa. Specimens of shrubs and trees from these forests have made their way into homes and offices—*Ficus benjamina*, rubber plants, orchids, and tree ferns, to name a few. In the forests, vines twist around tree trunks and grow toward sunlight. Orchids, mosses, lichens, and other plants grow on tree branches, absorbing the minerals that rainwater delivered to them. Entire communities of microorganisms, insects, spiders, and amphibians live, breed, and die in the small pools of water that collect in furled leaves.

Developing countries in Latin America, Africa, and Southeast Asia have the fastest-growing populations but limited food, fuel, and lumber. Thanks to the invention of chainsaws, tractors, and logging trucks, most of their forests will probably disappear within your lifetime.

Why does it matter? For purely ethical reasons, many individuals condemn the obliteration of so much of the Earth's biodiversity. For practical reasons, the destruction is impacting on your own life. Just think about what you eat. Nearly all human populations rely on a very limited number of species of crop plants and livestock—which are vulnerable to continually evolving pathogens. Developing new or hybrid strains makes our food base less vulnerable. And genetic engineers and tissue-culture specialists can tap tropical rain forests for new genetic "resources."

Genetic researchers also can tap the rain forests for the development of more effective antibiotics and vaccines. Alkaloids from tropical plants already are used in treating cardiovascular disorders and cancer. Aspirin, the most widely used drug in the world, is based on a chemical "blueprint" of a compound extracted from tropical willow leaves. Coffee, bananas, cinnamon, cocoa, sweeteners, Brazil nuts, and many other spices and foods we take for granted originated in the tropics. So did many ornamental plants. So did latex, gums, resins, dyes, waxes, and oils used in tires, shoes, ice cream, toothpaste, shampoo, condoms, cosmetics, perfumes, and compact discs.

And if aspirin or condoms don't catch your attention, think about this: Staggering numbers of burning trees release tons of air pollutants—including carbon dioxide—that are changing the air you breathe. The carbon dioxide also may help warm up the planet during your lifetime.

Figure 39.9 El Yunque tropical rain forest, Puerto Rico.

39.6 TRADING GRASSLANDS FOR DESERTS

Long-term shifts in climate can convert grasslands to deserts. So can human populations. **Desertification** is the name for the conversion of large tracts of natural grasslands to a more desertlike state. It applies also when conversions of rain-fed or irrigated croplands result in a 10 percent or greater decline in agricultural productivity. In the past fifty years, 9 million square kilometers worldwide have become desertified. At least

of their water from plants. They also are better water conservers; they lose little in feces, compared to cattle.

In 1978 a biologist, David Holpcraft, began ranching antelopes, zebras, giraffes, ostriches, and other native herbivores. He raised cattle as control groups in order to compare costs and meat yields on the same land. The native herds increased and yielded tasty meat. Range conditions did not deteriorate; they improved. Vexing

Figure 39.10 An awe-inspiring dust storm approaching Prowers County, Colorado, in 1934.

The Great Plains of the American Midwest are dry, windy grasslands that are subjected to severe, recurring droughts. Extensive conversion of these grasslands to agriculture began in the 1870s. Overgrazing left the ground bare across vast tracts. In May 1934, a cloud of topsoil that blew off the land blanketed the entire eastern portion of the United States, giving the Great Plains a dubious new name—the Dust Bowl. About 3.6 million hectares (9 million acres) of cropland were destroyed. Today, without large-scale irrigation and intensive conservation farming, desertlike conditions could prevail.

Figure 39.11 Desertification in the Sahel, a region of West Africa that forms a belt between the hot, dry Sahara Desert and tropical forests. This savanna country is undergoing rapid desertification as a result of overgrazing, overfarming, and prolonged drought.

200,000 square kilometers are still being converted annually. Prolonged droughts accelerate the process, as one did in the Great Plains years ago (Figure 39.10). At present, overgrazing of livestock on marginal lands is the main cause of large-scale desertification.

In Africa, for example, there are too many cattle in the wrong places. Cattle require more water than the region's native wild herbivores, so they move back and forth between grazing areas and watering holes more frequently. As they do this, they trample grasses and compact the soil surface (Figure 39.11). By contrast, gazelles and other native herbivores get most (if not all)

problems remained. African tribes have their own idea of what constitutes "good" meat, and some view cattle as the symbol of wealth in their society.

Without irrigation and conservation practices, grasslands that were converted for agriculture often end up as deserts.

39.7 A GLOBAL WATER CRISIS

The Earth has a tremendous supply of water, but most is too salty for human consumption or for agriculture. Imagine all that water in a bathtub. Withdraw all of the fresh, renewable portion (from lakes, rivers, reservoirs, groundwater, and other sources of surface water) and it would barely fill a teaspoon.

Why not consider **desalination**—the removal of salt from seawater? The supply of seawater is essentially unlimited. Desalination processes are available. They either distill seawater or force it through membranes (a method known as reverse osmosis). Costly fuel energy drives these processes, so they may be feasible only in Saudi Arabia and a few other countries with limited population sizes, large energy reserves, and lots of cash. In some situations they may be the only alternative to running out of water, as Santa Barbara and some other California cities nearly did during a prolonged drought. Yet desalination cannot solve the core problem. It may never be cost effective for large-scale agriculture, and it produces mountains of salts.

Figure 39.12 Irrigated crops in the Sahara Desert, in Algeria.

Consequences of Heavy Irrigation

Large-scale agriculture accounts for nearly two-thirds of the human population's use of freshwater. In many cases, irrigation water from surface sources is piped into vast fields where water-demanding crops could not grow on their own. At its most extreme, irrigation turns some hot desert regions into lush gardens, although believing these can be maintained over the long term is rather delusional (Figure 39.12).

Irrigation itself can change the land's suitability for agriculture. Concentrations of mineral salts typically are high in piped-in water. In regions where soil drains poorly, evaporation may cause **salination**—that is, salt buildup in the soil. Salination can stunt the growth and eventually kill crop plants, and so decrease yields.

Land that drains poorly also becomes waterlogged. As water accumulates underground, it slowly raises the **water table**, the upper limit at which the ground is fully saturated. When the water table is close to the ground's surface, soil gets saturated with saline water, which can damage plant roots. Properly managing the water-soil system can correct the salination and waterlogging. The economic cost of doing so is high.

Worse yet, water tables are subsiding. The Ogallala aquifer provides irrigation water for 20 percent of all croplands in the United States (Figure 39.13). Today, overdrafts (the amount nature does not replenish) have depleted half of it. Similarly, in the San Joaquin Valley of California, so much groundwater has been removed for irrigation that the water table's surface has subsided by as much as twenty feet in some areas.

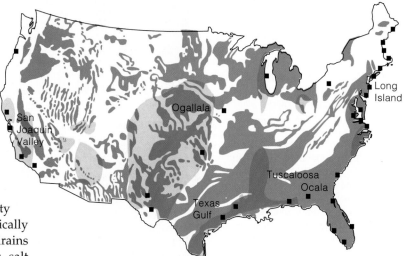

Figure 39.13 Underground aquifers (*blue*) containing 95 percent of all freshwater in the United States. *Gold* indicates regions where aquifers are being depleted, mainly for agriculture. *Black* boxes indicate aquifers being contaminated by saltwater intrusion.

Water Pollution

Water pollution amplifies the problem of water scarcity. Human sewage, animal wastes, and toxic chemicals make water unfit to drink, even to swim in. Pollutants encourage contamination by pathogens. Agricultural runoff pollutes water with sediments, pesticides, and plant nutrients. Power plants and factories pollute water with chemicals, radioactive materials, and excess heat (thermal pollution).

Pollutants collect in lakes, rivers, and bays before reaching the oceans. Many cities throughout the world dump untreated sewage into their coastal waters. Cities

Figure 39.14 An experimental wastewater treatment facility in Rhode Island. Treatment begins when sewage flows into rows of large water tanks in which water hyacinths, cattails, and other aquatic plants are growing. Decomposers in the tank degrade wastes—which contain nutrients that promote plant growth. Heat from incoming sunlight speeds the decomposition. From these tanks, water flows through an artificial marsh of sand, gravel, and bulrushes that filter out algae and organic wastes. Then it flows into aquarium tanks where zooplankton and snails consume microorganisms suspended in the water—and where zooplankton become food for crayfishes, tilapia, and other fishes that can be sold as bait. After ten days, the now-clear water flows into a second artificial marsh for final filtering and cleansing.

along rivers and harbors maintain shipping channels by dredging the polluted muck and barging it out to sea. They also barge sewage sludge: coarse, settled solids that contain bacteria, viruses, and toxic metals.

In the United States, about 15,000 facilities partially treat liquid wastes from 70 percent of the population and 87,000 industries. The remaining wastes are mostly from suburban and rural populations. These are treated in lagoons or septic tanks or are directly discharged—untreated—into waterways.

There are three levels of **wastewater treatment**. In *primary* treatment, screens and settling tanks remove sludge, which is dried, burned, dumped in landfills, or treated further. Chlorine often is used to kill pathogens. It doesn't kill them all, and it reacts with some industrial chemicals to produce cancer-causing substances.

In *secondary* treatment, microbial populations break down organic matter after primary treatment but before chlorination. The wastewater trickles through microbe-containing gravel beds or is aerated in tanks and seeded with microbes. Toxic solutes can poison the microbial helpers. At such times, the facilities are shut down until the microbial populations are reestablished. Primary and secondary treatments remove most suspended solids and oxygen-demanding wastes, but not all of the nitrogen, phosphorus, and toxic substances, such as heavy metals and pesticides. Usually the water gets chlorinated before being released into the waterways.

Tertiary treatment adequately reduces pollution but is largely experimental and expensive. It is applied to only 5 percent of the nation's wastewater.

In short, most wastewater is not treated adequately. A pattern gets repeated thousands of times along our waterways. Water for drinking is drawn upstream from a city, and wastes from industry and sewage treatment are discharged downstream. It takes no great leap of the imagination to see that water pollution intensifies as rivers flow to the oceans. In Louisiana, waters drained from the central states flow toward the Gulf of Mexico. Its pollution levels are high enough to threaten public health. Water destined for drinking does get treated to remove pathogens, but treatment does not remove toxic wastes dumped by numerous factories upstream.

This rather bleak picture might be numbing to most of us, but not to biologist John Todd. He constructed experimental wastewater treatment facilities in several greenhouses and artificial lagoons (Figure 39.14). When it works properly, the solar-aquatic treatment system produces water fit to drink. Such natural alternatives cannot work for large urban areas. But they are an attractive alternative for small towns and rural areas.

The Coming Water Wars

If the current rates of population growth and water depletion hold, the amount of freshwater available for each person on the planet will be 55 to 66 percent less than what it was in 1976. Already in the past decade, thirty-three nations have been engaged in conflicts over reductions in water flow, pollution, and silt buildup in major aquifers, rivers, and lakes. The United States and Mexico, Pakistan and India, and Israel and the occupied territories are among the squabblers.

Remember the Persian Gulf War, mainly about oil? Unless we pull off a blue revolution equivalent to the green one, we may be in for upheavals and wars over water rights. Does this sound far-fetched? By building dams and irrigation systems at the headwaters of the Tigris and Euphrates rivers, Turkey can, in the view of one of its dam-site managers, stop the water flow into Syria and Iraq for as long as eight months "to regulate their political behavior." Regional, national, and global planning for the future is long overdue.

Water, not oil, may become the most important fluid of the twenty-first century. National, regional, or global policies for water usage and water rights have yet to be developed.

Paralleling the J-shaped curve of human population growth is a dramatic rise in total and per capita energy consumption. It is due to increased numbers of energy users and to extravagant consumption and waste. For example, in one of the most pleasant of all climates, a major university constructed seven- and eight-story buildings with narrow, sealed windows. The windows can't be opened to catch prevailing ocean breezes. The buildings and windows were not designed or aligned to use sunlight for passive solar heating and breezes for passive cooling. Massive energy-demanding cooling and heating systems were installed.

When you hear talk of abundant energy supplies, bear in mind that there is a huge difference between the *total* and the *net* amounts available. Net energy is the amount left over after subtracting the energy used to locate, extract, transport, store, and deliver the stuff to consumers. Some energy sources, such as direct solar energy, are renewable. Others, such as petroleum and coal, are not. Currently, 83 percent of the energy stores being tapped are in the second category (Figure 39.15).

Fossil Fuels

Fossil fuels are nonrenewable resources, the legacy of forests that disappeared hundreds of millions of years ago (Section 19.3). Carbon-containing remains of those forests were buried and compressed in sediments, then transformed into coal, petroleum (oil), and natural gas.

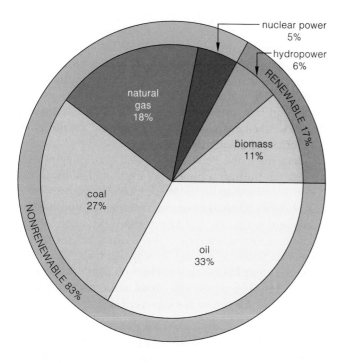

Figure 39.15 World consumption of nonrenewable and renewable energy sources in 1991.

Even with strict conservation, we may use up the known petroleum and natural gas reserves in the next century. As known reserves run out in accessible areas, we explore wilderness areas in Alaska and other fragile environments, such as continental shelves. Net energy declines as cost of extraction and transportation to and from remote areas increases. Environmental costs of extraction and transportation escalate. The damage and clean-up costs following the 11-million-gallon spill from the tanker *Valdez* off Alaska's coast is a classic example.

What about coal? In theory, world reserves can meet the energy needs of the human population for at least several centuries. But coal burning has been the main source of air pollution. Most of the known coal reserves contain low-quality, high-sulfur material. Unless sulfur is removed before or after burning, sulfur dioxides are released into the air. They add to the global problem of acid deposition. Fossil fuel burning also releases carbon dioxide and adds to the greenhouse effect.

Extensive strip mining of coal reserves close to the Earth's surface carries its own problems. It reduces the land available for agriculture, grazing, and wildlife. Most strip mines are located in arid and semiarid lands, where the absence of sufficient water supplies and poor soils make restoration efforts difficult.

Nuclear Energy

Nuclear Reactors As Hiroshima burned in 1945, the world recoiled in horror from the destructive force of nuclear energy. By the 1950s, however, people were touting nuclear energy as an instrument of progress. Many of the energy-poor industrialized nations, France included, now depend heavily on nuclear power. Even so, construction of new nuclear plants has been delayed or cancelled in most countries. Since 1970 in the United States alone, plans to build 117 nuclear power plants were cancelled. Other plants were abandoned before they could be completed. A few are being converted at great cost to fossil fuel burning. What happened? The cost, efficiency, safety record, and environmental impact of reliance on nuclear power came into question.

By 1990, nuclear power generated electricity at a cost slightly above that of coal-burning plants. Putting aside other factors, its electricity-generating costs are now lower. However, by the year 2000, solar energy with natural gas backup also should cost less.

What about safety? Less radioactivity and carbon dioxide normally escape from nuclear plants than from coal-burning plants of the same capacity, and they emit no sulfur dioxide. The danger lies in their potential for **meltdown**. When nuclear fuel breaks down, it releases considerable heat. Typically, water circulating over the nuclear fuel absorbs heat and so produces steam that

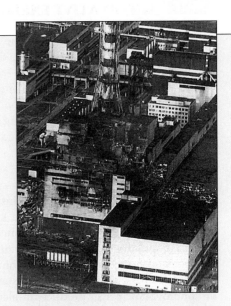

Figure 39.16 Incident at Chernobyl. On April 26, 1986, errors in judgment during a routine test procedure resulted in runaway reactions, explosions, and a complete core meltdown at the Chernobyl power plant in the Ukraine. Helicopter pilots who were supposed to drop 5,000 tons of lead, sand, clay, and other materials on the blazing core to suppress further release of radioactivity missed the target. Unimpeded and uncovered, nuclear fuel burned for nearly ten days, right on through a six-foot-thick steel and gravel barrier beneath it.

Between 185 and 250 million curies of radioactive matter may have escaped in those ten days alone. Inhaling as little as ten-millionths of a curie of plutonium can cause cancer. Thirty-one people died at once; others died of radiation sickness in the following weeks. Inhabitants of entire villages were relocated; their former homes were bulldozed under. In time, concrete entombed the 180 tons of partially burned nuclear fuel. In 1994, 11,000 square meters of holes in the concrete were still allowing rainwater, air, and other materials to enter or escape.

Afterward, the number of people opposed to nuclear plants rose sharply, even in France. You get a sense of why opposition increased by studying these maps of the global distribution of radioactive fallout within two weeks of the meltdown. That fallout put 300–400 million people at risk for leukemia and other radiation-induced disorders. Throughout Europe, hundreds of millions of dollars were lost when the fallout made crops and livestock unfit for consumption.

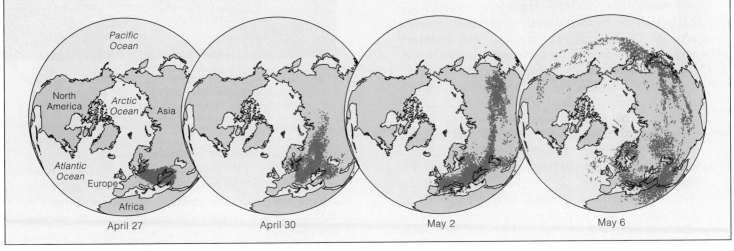

April 27   April 30   May 2   May 6

drives electricity-generating turbines. If a leak were to develop in the circulating water system, water levels might plummet around the fuel, which might then heat past its melting point. On the generator floor, melting fuel would instantly convert the remaining water to steam. Combined with other reactions, steam formation could blow the system apart and release radioactive material. And an overheated core could melt through its thick concrete containment slab. Figure 39.16 describes an incident that yielded evidence of the consequences.

Nuclear Waste Disposal Unlike coal, nuclear fuel cannot be burned to harmless ashes. After three years or so, the fuel elements are spent, but they still contain uranium fuel as well as hundreds of new radioisotopes that formed in the reactions. The wastes are extremely radioactive and dangerous. They get extremely hot as they undergo radioactive decay, so they are plunged at once into water-filled pools. The water cools them and keeps radioactive material from escaping. Even after

being stored for several months, the isotopes remaining are lethal. Some must be kept isolated for at least 10,000 years. If a certain isotope of plutonium (^{239}Pu) is not removed, the wastes must be kept isolated for a quarter of a million years! After nearly fifty years of research, scientists still cannot agree on what is the best way to store high-level radioactive wastes. Even if they could, there is no politically acceptable solution. No one wants radioactive wastes anywhere near where they live.

Finally, as if we don't have enough to worry about, following the Soviet Union's breakup, some underpaid workers of a Russian nuclear power plant have been selling fuel elements on the black market. The buyers? Some developing nations that want to produce nuclear weapons—and possibly deliver them into the hands of terrorist organizations.

The nuclear genie is out of the bottle, exploitable by the best and worst elements of the human population.

ALTERNATIVE ENERGY SOURCES

Less than thirty years from now, the projected size of the human population will be such that the demands for fuel will increase by 30 percent—and for electricity by 265 percent. More efficient use and conservation of our existing energy sources alone won't do the trick. We must make the transition to reliance on alternative sources of energy, of the sort described next.

Solar-Hydrogen Energy

Each year, incoming sunlight contains about ten times more energy than that in all of the known fossil fuel reserves. That's 15,000 times as much energy as the human population uses now. Isn't it about time we start to collect it in earnest, in something besides crops?

For example, when exposed to sunlight, electrodes in "photovoltaic cells" produce an electric current that splits water molecules into oxygen and hydrogen gas (H_2)—which can be used directly as fuel or to generate electricity. The technology to tap such **solar-hydrogen energy** has been around since the 1940s (Figure 39.17a). Such energy is stored efficiently for as long as required. It costs less to distribute H_2 than electricity, and water is the only by-product of using it. Space satellites run on it. Here on Earth, fossil fuels are still "cheaper."

Unlike fossil fuels, however, sunlight and seawater are virtually unlimited resources. And the technology's potential to protect the environment is staggering. The environmental scientist G. Tyler Miller, Jr., puts it this way: "If we make the transition to an energy-efficient solar-hydrogen age, we can say goodbye to smog, oil spills, acid rain, and nuclear energy, and perhaps to global warming. The reason is simple. As hydrogen burns in air, it reacts with oxygen gas to produce water vapor—not a bad thing to have coming out of tailpipes, chimneys, and smokestacks." Also, if this technology becomes cost-effective for developing countries, the great forests now being destroyed for timber and fuel might still be around for future generations.

Recently in the United States, the largest supplier of natural gas joined forces with a manufacturer of photovoltaic cells. They intend to build a solar facility in the Nevada desert, one that will generate enough energy to supply a city of 100,000 at less cost than for electricity generated by fossil fuels. If their plan succeeds, they may revolutionize the energy industry.

Wind Energy

As you know, solar energy also is converted into the mechanical energy of winds. Where winds that travel faster than 7.5 meters per second prevail, we find cost-effective wind turbines (Figure 39.17b). California gets 1 percent of its electricity from "wind farms." Possibly

Figure 39.17 Harnessing solar energy. (**a**) Electricity-producing photovoltaic cells in panels that collect sunlight energy. (**b**) Wind turbines, which exploit the air circulation patterns that arise from latitudinal variations in the intensity of incoming sunlight.

the winds of North and South Dakota alone can meet all but 20 percent of the current energy needs of the United States. Wind energy has potential for islands and other remote areas far from utility grids. One drawback is that winds don't blow on a regular schedule, so they can't be used as an exclusive or major energy source.

Fusion Power

The sun's gravitational force is enough to compress atomic nuclei to high densities, and its temperatures are high enough to force atomic nuclei to fuse. We call this **fusion power**. Similar conditions do not exist on Earth, but maybe we can mimic them. Researchers confine a certain fuel—a heated gas of two isotopes of hydrogen—in magnetic fields, then hit it with lasers. The fuel implodes, it is compressed to extremely high densities—and energy is released. The more energetic the lasers, the greater the compression, and the more the fuel will burn. The bad news is that, although the amount of energy released has been steadily increasing, it will be at least fifty years before fusion reactors are operating and costs will probably be high. The good news is, that's about the time fossil fuels start running out.

Sunlight may end up sustaining the energy needs of the human population in more ways than one.

BIOLOGICAL PRINCIPLES AND THE HUMAN IMPERATIVE

Molecules, single cells, tissues, organs, organ systems, multicelled organisms, populations, communities, and then ecosystems and the biosphere. These are architectural systems of life, assembled in increasingly complex ways over the past 3.8 billion years. We are latecomers to this immense biological building program. Yet within the relatively short span of 10,000 years, our activities have been changing the very character of the land, ocean, and atmosphere, even the genetic character of species.

It would be presumptuous to think we alone have had profound effects on the world of life. As long ago as the Proterozoic era, photosynthetic organisms irrevocably changed the course of biological evolution by gradually enriching the atmosphere with oxygen. In the past as well as the present, competitive adaptations assured the rise of some groups, whose dominance assured the decline of others. Change is nothing new to this biological building program. What *is* new is the capacity of a species—our own—to comprehend what is going on.

We now have the population size, the technology, and the cultural inclination to use up energy and modify the environment at frightening rates. And where will rampant, accelerated change lead us? Will feedback controls begin to operate as they do, for example, when population growth exceeds the carrying capacity of the environment? In other words, will negative feedback controls come into play and keep things from getting too far out of hand?

Feedback control will not be enough, for it operates only when deviation already exists. The human population growth rates and patterns of resource consumption are founded on an illusion of unlimited resources and a forgiving environment. A prolonged, global shortage of food or the passing of a critical threshold for the global climate can come too fast to be corrected. At some point, deviations may have too great an impact to be reversed.

What about feedforward mechanisms that might serve as early warning systems? For example, when sensory receptors in skin sense a drop in outside air temperature, each sends messages to the nervous system. That system responds by triggering mechanisms that raise the core temperature before the body itself becomes dangerously chilled. If we develop feedforward control mechanisms, maybe we could begin corrective measures before we alter the environment too significantly.

By themselves, feedforward controls won't work, for they start operating when change is already under way. Think of the DEW line—the Distant Early Warning system. It is like a vast sensory receptor, one that can detect the launching of intercontinental ballistic missiles against North America. By the time the system actually detects what it is designed to detect, it may be too late to stop widespread destruction.

It would be naive to assume we can ever reverse who we are at this point in evolutionary time, to de-evolve ourselves culturally and biologically into becoming less complex in the hope of averting disaster. However, there is reason to believe we can avert disaster by using a third kind of control mechanism—a capacity to anticipate events before they happen. We are not locked into responding only after irreversible change has begun. We have the capacity to anticipate the future—it is the essence of our visions of utopia or of nightmarish hell. *We all have the capacity to adapt to a future that we can partly shape.*

We can, for example, stop trying to "beat nature" and learn to work with it. Individually and collectively, we can work to develop long-term policies at the local, regional, and global levels—policies that take into account the biotic and abiotic limits on population growth. Far from being a surrender, this would be one of the most complex and intelligent behaviors of which we are capable.

Having a capacity to adapt and actually using it are not the same thing. We have already put the world of life on dangerous ground because we have not yet mobilized ourselves as a species to work toward self-control.

Our survival depends on predicting possible futures. It depends on designing and constructing ecosystems that are in harmony with our definition of basic human values and with the biological models available to us. Human values can change; our expectations can and must be adapted to biological reality. *For the principles of energy flow and resource utilization, which govern the survival of all systems of life, do not change.* It is our biological and cultural imperative that we come to terms with these principles, and ask ourselves what our long-term contribution will be to the world of life.

SUMMARY

1. Accompanying the extremely rapid growth of the human population are increases in energy demands and in environmental pollution.

2. Pollutants are substances with which ecosystems have had no prior evolutionary experience (in terms of kinds and amounts) and so have no mechanisms for absorbing or cycling them. Many pollutants result from human activities, and they adversely affect the health, activities, or survival of human populations.

3. Smog, a form of air pollution, arises in industrialized and urban regions that rely on fossil fuels. It becomes especially concentrated in land basins having thermal inversions (a trapping of a layer of cool, dense air under a warm air layer). Industrial coal-burning regions with cold, wet winters produce industrial smog. Large cities with many vehicles in warm climates produce photochemical smog, mainly when sunlight makes nitric oxide (emitted from vehicles) react with hydrocarbons to form photochemical oxidants such as PANs.

4. During dry weather, acidic air pollutants, especially oxides of nitrogen and sulfur, fall to Earth as dry acid deposition. They also dissolve in atmospheric water, then fall to Earth as wet acid deposition, or acid rain.

5. Seasonal thinning of the ozone layer at high latitudes has become pronounced as CFCs (chlorofluorocarbons) and other air pollutants rise to the stratosphere, where they deplete ozone and allow more harmful ultraviolet radiation from the sun to reach the Earth's surface.

6. Human population growth presently depends on the expansion of agriculture, made possible by large-scale irrigation and extensive applications of fertilizers and pesticides. Global freshwater supplies are limited, yet they are being polluted by agricultural runoff (which includes sediments as well as pesticides and fertilizers), industrial wastes, and human sewage.

7. Human populations are damaging land surfaces by:

   a. Accepting the dumping, burning, or burial of solid wastes rather than making concerted efforts to recycle or reuse materials and to reduce waste.

   b. Engaging in rampant deforestation (destruction of vast tracts of tropical and temperate forest biomes).

   c. Contributing to desertification (the large-scale conversion of natural grasslands, croplands, or grazing lands to desertlike conditions).

8. Energy supplies in the form of fossil fuels are nonrenewable, dwindling, and environmentally costly to extract and use. Nuclear energy in itself is less polluting, but the costs and risks associated with fuel containment and with storing radioactive wastes are enormous. The challenge is to develop affordable alternatives based on renewable resources, such as solar energy.

Review Questions

1. Define pollution and give specific examples of pollutants. *710*

2. Distinguish among the following conditions. *710*
   a. industrial smog   c. dry acid deposition
   b. photochemical smog   d. wet acid deposition

3. Define CFCs and describe how they have contributed to seasonal thinning of the ozone layer in the stratosphere. *712*

4. Which human activity uses the most freshwater? *718*

5. What percent of the Earth's land masses is under cultivation? What percent is available for new cultivation? *713*

6. Define and describe possible consequences of deforestation and of desertification. *714–717*

Self-Quiz *(Answers in Appendix IV)*

1. Since the mid-eighteenth century, human population growth has been _____ .
   a. leveling off   c. accelerating
   b. growing slowly   d. not much to speak of

2. Pollutants disrupt ecosystems because _____ .
   a. their components differ from those of natural substances
   b. only humans have uses for them
   c. there are no evolved mechanisms to deal with them
   d. their only effect is on ecosystems, not humans

3. During a thermal inversion, weather conditions trap a layer of _____ air under a layer of _____ air.
   a. warm; cool   c. warm; sooty
   b. cool; warm   d. cool; sooty

4. _____ is (are) a case of regional air pollution.
   a. Smog   c. Ozone layer thinning
   b. Acid rain   d. a and b

5. _____ is (are) a case of air pollution with global effects.
   a. Smog   c. Ozone layer thinning
   b. Acid rain   d. b and c

6. Two-thirds of the freshwater used annually goes to _____ .
   a. urban centers   c. treatment facilities
   b. agriculture   d. a and c

7. The upper limit at which the ground is fully saturated with water is called _____ .
   a. groundwater   c. the water table
   b. an aquifer   d. the salination limit

8. Energy from fossil fuels is _____ ; their extraction and use come at _____ cost to the environment.
   a. renewable; low   c. renewable; high
   b. nonrenewable; low   d. nonrenewable; high

9. Nuclear energy normally pollutes _____ than fossil fuels; it poses _____ dangers than fossil fuels.
   a. less; lesser   c. more; lesser
   b. more; greater   d. less; greater

10. Match each term with the most suitable description.
    ____ desertification   a. probably one of our best options
    ____ deforestation   b. soil loss, watershed damage,
    ____ green   altered rainfall patterns follow
         revolution   c. attempt to improve crop production
    ____ solar-hydrogen   on existing land
         power   d. conversion of large tracts
                                of natural grasslands to
                                a more desertlike state

Critical Thinking

1. Investigate where the water supply for your own city comes from and where it has been. You may find the answer illuminating.

2. Make a list of advantages you personally enjoy as a member of an affluent, industrialized society. List some of the drawbacks. Do you believe the benefits outweigh the costs? This is not a trick question.

3. List activities you pursue each day that might be contributing to regional or global pollution. Also list ways in which you might help reduce that contribution.

4. Kristen, a recent college graduate, is finding her idealism on a collision course with reality. She strongly believes people who live in the United States are obliged to make the world a level playing field for all human beings, with equality in resources, health, education, economic security, and a pristine environment for all. Yet she also understands that the sheer size of the human population makes this impossible. Kristen recently said she cannot be party to hard choices and actions that go against her ideals and just wants nature to solve the problem for us. Comment on this true story.

5. It has been said that economic wars, more than military battles, will determine the winners and losers among nations in the near future. The economic growth of certain nations, including some in the former Soviet Union and in the Far East, has had devastating impact on the environment. Elsewhere, maintaining standards of environmental protection adds to the cost of goods produced and puts the practicing nations at serious disadvantage in this global competition. Should the United States loosen some of its existing environmental laws to help assure its economic survival? Why or why not? Can you think of pressures that might be imposed on environmentally indifferent nations to encourage them to change their harmful practices?

6. Populations of every species use resources and produce wastes. Use Figure 39.18 as a starting point for a brief essay about the acquisition and uses of energy and materials, including wastes, in a human population that is concentrated in a large city. Contrast your description with the energy flow and materials cycling that proceed in the natural population of some other organism.

Figure 39.18 City as ecosystem.

Selected Key Terms

acid rain *710*
chlorofluorocarbon (CFC) *712*
deforestation *714*
desalination *718*
desertification *717*
dry acid deposition *710*
fossil fuel *720*
fusion power *722*
green revolution *713*
industrial smog *710*
meltdown *720*

ozone thinning *712*
PAN (peroxyacyl nitrate) *710*
photochemical smog *710*
pollutant *710*
salination *718*
shifting cultivation *714*
solar-hydrogen energy *722*
thermal inversion *710*
wastewater treatment *719*
water table *718*

Readings

Collins, M. 1990. *The Last Rain Forests.* New York: Oxford University Press.

Frosh, R. September 1995. "The Industrial Ecology of the Twenty-First Century." *Scientific American* 273(3): 178–181.

Gruber, D. 1989. "Biological Monitoring and Water Resources." *Endeavour* 13(3): 135–140.

Hoagland, W. September 1995. "Solar Energy." *Scientific American* 273(3): 170–173.

Miller, G. T., Jr. 1996. *Living in the Environment.* Ninth edition. Belmont, California: Wadsworth. Chapter 4 is especially valuable reading. This author consistently puts information from numerous sources into a current, accessible survey of the present and future state of the environment.

Mohnen, V. August 1988. "The Challenge of Acid Rain." *Scientific American* 259(2): 30–38.

Plucknett, D., and D. Winkelmann. September 1995. "Technology for a Sustainable Agriculture." *Scientific American* 273(3): 182–186.

Western, D., and M. Pearl. 1989. *Conservation for the Twenty-First Century.* New York: Oxford University Press.

Wilson, E. 1988. *Biodiversity.* Washington, D.C.: National Academy of Sciences.

40 ANIMAL BEHAVIOR

Deck the Nest With Sprigs of Green Stuff

In 1890, bird fanciers imported more than a hundred starlings (*Sturnus vulgaris*) from Europe and released them in New York City's Central Park. In less than a century, the descendants of the introduced species had expanded their geographic range from coast to coast. They are so good at gleaning food from the agricultural fields of North America that they have outmultiplied native birds. They also have evicted great numbers of them from scarce nesting sites in the cavities of trees.

Once a male and female starling commandeer a tree cavity, they build a nest of dry grass and twigs (Figure

Figure 40.1 (**a**) A most excellent fumigator in nature—the European starling (*Sturnus vulgaris*). It combats infestations of mites (**b**) by decorating previously owned nests with fresh sprigs of wild carrot (**c**).

(**d**) Results of one experiment. In this case, nests designated *A* were kept free of fresh green sprigs of wild carrot and other plants that contain certain chemicals. The chemicals can prevent baby mites from reaching adulthood. Other experimental nests, designated *B*, had fresh sprigs added every seven days.

Head counts of mites (*Ornithonyssus sylviarum*) infesting the nests were made at the time the starling chicks left the nest, twenty-one days after this particular experiment was under way.

a

b

c

40.1*a*). Theirs is no ordinary nest. They *decorate* the nest bowl with green sprigs, freshly plucked.

Why do starlings do this? Do the sprigs of greenery camouflage the nest from predators? Not likely. The nests are already concealed—inside the tree cavities. Do the sprigs function as insulative material that might help keep the forthcoming eggs warm? Actually, the still-moist, green plant parts would promote heat *loss*, not heat conservation. Well, then, do the green sprigs combat parasites, such as mites (Figure 40.1*b*)? Mites parasitize birds and infest their nest cavities. In short

order, even a few tiny mites can produce thousands of descendants. When present in large numbers, mites can suck enough blood from a nestling to weaken it—and interfere with its growth, development, and survival.

Larry Clark and Russell Mason decided to test the third hypothesis. As these biologists knew, starlings do not weave just any green plant material into the nests. They choosily favor the leaves of certain plants, such as wild carrot (Figure 40.1*c*). So Clark and Mason built a set of experimental nests, some with freshly cut wild carrot leaves and some without. Then they removed the natural nests that pairs of starlings had constructed and had already started using. Fifty percent of the nesting pairs received replacement nests decorated with sprigs of wild carrot. No sprigs decorated the replacement nests for the other pairs of starlings.

Figure 40.1*d* shows the results. The number of mites in the greenery-free nests was consistently greater than in nests with greenery. At the end of one experiment, sprig-free nests teemed with an average of 750,000 mites. The ones with sprigs contained a mere 8,000.

Shoots of wild carrot happen to contain a highly aromatic steroid compound. Almost certainly, the compound repels herbivores and thereby helps the plant survive. By coincidence, the compound also prevents mites from becoming sexually mature— and so it prevents mite population explosions in the nests festooned with wild carrot.

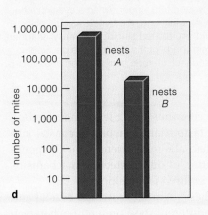

Far from being a trivial behavior, then, "decorating" a nest with a bit of aromatic greenery functions to fumigate the nest—and increase a starling's chance of producing healthy, surviving offspring.

And so starlings lead us into the fascinating world of behavioral research. As you will see, some of the behavioral studies focus on internal mechanisms that enable individuals to behave as they do. Other studies focus on the adaptive value of some behavior to an individual's reproductive success.

As you know, information encoded in genes directs the formation of the tissues and organs that make up the animal body—including its nervous system. That system detects, processes, and integrates information about specific stimuli in both the internal and external environments, then commands muscles and glands to make suitable responses. Other gene products called hormones contribute to the responses. Thus, by giving rise to the nervous and endocrine systems, *genes are the foundation for the observable, coordinated responses that animals make to stimuli.* We call these diverse responses **animal behavior**.

KEY CONCEPTS

1. "Behavior" refers to the coordinated responses that an animal makes to stimuli, which are specific changes inside and outside of its body. The responses an animal makes are instinctive, learned, or a combination of both.

2. Forms of behavior have a heritable (genetic) basis, as when instructions encoded in certain genes govern the development of the nervous and endocrine systems. These organ systems allow an animal to detect, process, and issue commands for behavioral responses to stimuli.

3. Through mechanisms underlying instinctive behavior, an animal can make complete responses to certain key stimuli the first time it encounters them. Through mechanisms that underlie learned behavior, an animal can modify its behavior after acquiring and processing information from certain experiences.

4. Like other traits having a genetic basis, forms of behavior have evolved by way of natural selection—the measure of which is reproductive success. Thus sexual selection and other evolutionary processes have favored behavioral mechanisms that enhance the ability of individuals to pass on their genes to offspring.

5. Evolved modes of communication underlie social behavior. A communication signal is an action that has net benefits for both the sender and the receiver of the signal.

6. Social life has costs and benefits, as measured by an individual's ability to pass on its genes to offspring. Not every environment favors the evolution of social behavior. Under some circumstances, solitary individuals can leave more descendants than social ones would be able to do.

7. Altruism means helping others in ways that sacrifice personal reproductive success. The evolution of altruism requires circumstances in which individuals can propagate their genes indirectly, by helping their relatives reproduce successfully.

THE HERITABLE BASIS OF BEHAVIOR

Genes and Behavior

An animal's nervous system, recall, is wired to detect, interpret, and issue commands for response to stimuli— that is, to specific changes in the external and internal environment. All of the steps required to assemble and to operate the system's receptors, nerve pathways, and brain are specified, one way or another, by the animal's genes. If we define an animal's **behavior** as coordinated responses to stimuli, then its behavior starts with genes.

Stevan Arnold found evidence of the genetic basis of behavior by studying the feeding preferences of coastal and inland populations of a snake species in California. Garter snakes near the coast prefer banana slugs (Figure 40.2*a*). Snakes living inland prefer tadpoles and small fishes. Offer them a banana slug and they ignore it.

In one set of experiments, Arnold offered captive newborn garter snakes a chunk of slug as the first meal. Offspring of coastal snakes usually ate the chunk. They even flicked their tongue at cotton swabs drenched in essence of slug. (A snake "smells" by tongue-flicking, which draws chemical odors into the mouth.) But the offspring of snakes living inland ignored the swabs, and only rarely did they eat slug meat. Here was a clear difference between captive baby snakes with no prior, direct experience with slugs. Arnold concluded that the snakes had been programmed before birth to accept or

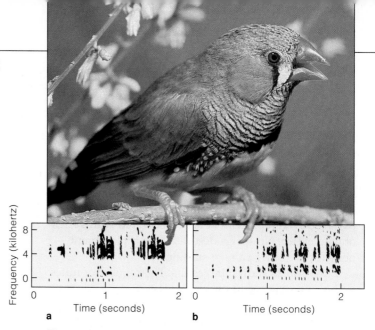

Figure 40.3 Hormones and the territorial song of zebra finches. (**a**) Sound spectrogram of the male's song. The spectrogram is a visual record of each note's pitch (its frequency, as measured in kilohertz). (**b**) Sound spectrogram of a female zebra finch that had been experimentally converted to a singer. When a nestling, she received extra estrogen. At adulthood, the female also received a testosterone implant—and she started singing. Normally, estrogen organizes the song system (parts of the brain) in the developing embryos of songbirds, and testosterone activates it in adult males.

reject the slugs; they certainly didn't learn their feeding preferences through taste trials. Maybe the coastal and inland populations had allelic differences for the genes that affect the formation of odor-detecting mechanisms when a garter snake embryo is developing.

To test his hypothesis, Arnold crossbred coastal and inland snakes. If the different food preferences between snake populations has a genetic basis, then the hybrid offspring might make an *intermediate* response to slug chunks and odors. The results of the crosses matched this prediction. Compared with typical newborn inland snakes, many baby snakes of mixed parentage tongue-flicked more often at the slug-scented cotton swabs— but less often than typical newborn coastal snakes did.

Hormones and Behavior

Gene products called hormones also contribute to bird song and other kinds of behavior. Think of a male zebra finch, such as the one in Figure 40.3*a*. Each spring he repeats the same song over and over again with such clarity and such consistency that you might wonder how he does it. His singing is an outcome of seasonal differences in the secretion of melatonin—a hormonal product of the pineal gland. A high level of melatonin in blood suppresses the growth and functions of song-bird gonads. Its effect is reversed when photoreceptors of cells in the pineal gland absorb sunlight energy and issue signals that call for decreased melatonin secretion.

Figure 40.2 (**a**) Banana slug, food for (**b**) a grown-up garter snake of coastal California. (**c**) A newborn garter snake from a coastal population, tongue-flicking at a cotton swab drenched with banana slug fluids.

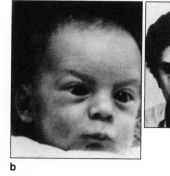
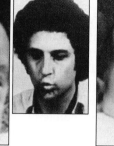

a
b

Figure 40.4 Instinctive responses in humans. (**a**) Simple stimuli trigger smiling in very young infants. (**b**) Somewhat older infants instinctively imitate facial expressions of adults.

Winter has fewer hours of daylight than spring, and there isn't enough light to inhibit melatonin secretion. When daylength does increase in spring, the gonads are released from hormonal suppression. They grow in size and step up *their* secretion of estrogen and testosterone—which trigger gender-related differences in bird singing behavior. How? While the embryos of most songbirds are developing, estrogen controls formation of a **song system**—a set of brain regions that will govern muscles of a vocal organ. The brain regions differ notably in size and structure between male and female birds.

Female songbirds carry sex chromosomes XY, and the males carry XX. Certain genes on the Y chromosome prevent the females from making estrogen. By contrast, even before a male bird hatches, a high estrogen level in his blood stimulates the development of a masculinized

a
b

Figure 40.5 (**a**) A newly hatched cuckoo making a complex, innate behavioral response to an environmental cue—spherical objects in the nest. The European cuckoo lays eggs in the nests of other species. Even before a cuckoo hatchling opens its eyes, it responds to the shape of the host's eggs by shoving them out of the nest. (**b**) The foster parents keep feeding the usurper.

brain. Later, his gonads enlarge. As the breeding season begins, his gonads step up their testosterone secretion. When this hormone binds to receptors on cells in the song system, it induces metabolic changes that will prepare the male to sing when he is suitably stimulated.

Instinctive Behavior Defined

With their tongue-flicking, body orientation, and strikes at prey, newborn garter snakes offer a fine example of instinctive responses. With **instinctive behavior**, the nervous system allows an animal to carry out complex *but stereotyped* responses to specific and often simple environmental cues. Similarly, when human infants are two or three weeks old, they smile instinctively when an adult's face comes close to their own (Figure 40.4*a*). Infants even make the same response to a simplified stimulus—a flat, face-size mask with two dark spots where eyes would be on a human face. A mask with one "eye" won't do the trick. As Figure 40.4*b* suggests, older infants, too, show instinctive behavior.

Finally, consider the cuckoo, a social parasite. Adult females lay their eggs in nests of other species. If newly hatched cuckoos eliminate the natural-born offspring, they will end up getting the undivided attention of their unsuspecting foster parents. Although blind when hatched, they will maneuver any egg they contact onto their back—and push it out of the nest (Figure 40.5).

Such instinctive responses are **fixed action patterns**. A well-defined, simple stimulus triggers them, and once set in motion, they are performed in their entirety.

Clearly, then, animals have genetically based means to respond automatically to certain environmental cues. As you will see next, however, they also have the means to process information about specific experiences—then use the information to vary or change their responses. The outcome is what we call learned behavior.

Genes underlie animal behavior—coordinated responses to stimuli. Certain gene products are essential in constructing and operating the nervous system, which governs behavior. Other gene products—hormones—also influence mechanisms required for particular forms of behavior.

40.2 LEARNED BEHAVIOR

Let's now consider **learned behavior**, whereby animals process and integrate information gained from specific experiences, then use it to vary or change responses to stimuli. Think of a young toad, with a nervous system that commands it to flip its sticky tongue instinctively at any dark object moving across its field of vision. In the toad world, such objects are usually edible insects. Suppose one dark object is a bumblebee that stings the toad's tongue. The toad's experience leads it to avoid "bumblebee-size black-and-yellow-banded objects that sting." Or think of **imprinting**, a time-dependent form of learning. Imprinting requires exposure to key stimuli in the environment, usually during a sensitive period when the animals are young. Figure 40.6 gives a classic example. Table 40.1 lists other examples.

Don't fall into the trap of thinking that genes alone specify instinctive behavior and that the environment alone governs learned behavior. Behavior is an outcome of gene expression *and* individual experiences with the environment. Thus, all male white-crown sparrows have a genetic capacity to sing—but birds in different habitats use variations ("dialects") of the species song. As Peter Marler discovered, the males acquire the full song by *hearing it* ten to fifty days after hatching. For this sensitive period, a learning mechanism is primed to receive a bit of information from the environment.

Besides this, the male birds learn parts of a song by picking up acoustical cues from other males. Marler raised nestlings to maturity in soundproof chambers so they would not hear adult males singing. At maturity, the captive males produced a song with none of the detailed structure of a typical adult's song. Marler also exposed isolated captives to recordings of white-crown sparrows *and* song sparrows. At maturity, the captives sang only the white-crown song. They even mimicked the dialects. Evidently, the songs depend partly on a genetically based capacity to learn from acoustical cues.

This isn't the whole story, however. In still another experiment, Marler allowed young, hand-reared white-

Table 40.1   A Few Categories of Learned Behavior

Imprinting. For many animals, this is a time-dependent form of learning that requires exposure to key stimuli, usually early in development. Thus baby geese formed an attachment to the ethologist Konrad Lorenz or to any moving object if they were separated from the mother shortly after hatching (Figure 40.6). But they did this only after exposure to a moving object at a short, sensitive period early in life. At that time only, they were primed to learn a crucial bit of information—the identity of the individual they would follow in the months ahead. In nature, that individual is normally the mother or father.

Imprinting affects the behavior of many birds—as when they direct sexual attention to members of the species they had been sexually imprinted on in their youth (normally their own species, for they encounter the mother soon after hatching).

Classical Conditioning. Ivan Pavlov's classic experiments with dogs are an example. Dogs salivate just before eating. Pavlov's dogs were conditioned to salivate—even in the absence of food. They did so in response to the sound of a bell or a flash of light that was initially associated with the presentation of food. In this case, the animals learned to associate an automatic, unconditioned response with a novel stimulus that does not normally trigger the response.

Operant Conditioning. An animal learns to associate a voluntary activity with its consequences, as when a toad learns to avoid stinging insects after attempts to eat them.

Habituation. An animal learns by experience *not* to respond to a situation if the response has neither positive nor negative consequences. Thus many species of birds living in cities learn not to flee from people who pose no threat to them.

Spatial or Latent Learning. By inspecting its environment, an animal acquires a mental map of some region, often by learning the position of local landmarks. Thus blue jays can store information about the position of dozens, if not hundreds, of places where they have stashed food.

Insight Learning. An animal abruptly solves some problem without trial-and-error attempts at the solution. Chimpanzees often exhibit insight learning in captivity when they suddenly solve a novel problem their captors devise for them. Some chimps abruptly stacked and stood on several boxes, *and* used a stick to reach bananas suspended out of their reach.

crowns to interact with a "social tutor" of a different species, as opposed to listening to the taped songs. The males tended to learn the tutor's song. Their learning mechanisms must be primed to be influenced by social experience as well as by acoustical cues.

In instinctive behavior, animals make complex, stereotyped responses to specific, often simple environmental cues. In learned behavior, responses may vary or change as a result of individual experiences in the environment.

Whether instinctive or learned, behavior develops through interactions between genes *and* environmental experiences.

Figure 40.6   No one can tell these imprinted baby geese that Konrad Lorenz is not Mother Goose, as Table 40.1 explains.

THE ADAPTIVE VALUE OF BEHAVIOR

If you accept that genes directly or indirectly govern forms of behavior, then it follows that forms of behavior are subject to evolution by natural selection. **Natural selection**, recall, is the result of differences in survival and reproduction among individuals of a population that differ from one another in heritable traits. Some versions of a trait are better than others at helping the individual survive and reproduce. Thus alleles for those versions increase in a population, and other alleles do not. (Alleles are different molecular forms of the same gene.) In time, such genetic changes lead to increased fitness—an increase in adaptation to the environment.

By using the theory of evolution by natural selection as our starting point, then, we should be able to identify adaptive forms of behavior—and to discern how they bestow reproductive benefits that offset reproductive costs or disadvantages that might be associated with them. It makes no difference whether we focus on some behavior of a solitary animal or of animals in a social group. Either way, if the behavior is adaptive, it must promote the survival and production of offspring of the *individual*. Keep this in mind as you read through the following terms, which you will come across repeatedly in the remainder of this chapter:

1. **Reproductive success**. Individual survival and production of offspring.

2. **Adaptive behavior**. Any behavior that promotes the propagation of an individual's genes and tends to occur at increased frequency in successive generations.

3. **Social behavior**. Cooperative, interdependent relationships among individuals of the same species.

4. **Selfish behavior**. Within a population, any behavior that increases an individual's chances to produce or protect offspring of its own, regardless of the consequences for the population.

5. **Altruistic behavior**. Within a population, a self-sacrificing behavior. The individual behaves in a way that helps others but that decreases its own chances to produce offspring.

When biologists speak of selfish or altruistic behavior, they don't mean an individual is consciously aware of what it is doing or of the behavior's reproductive goal. (A lion doesn't have to know that eating zebras is good for its reproductive success. Its nervous system simply calls for hunting behavior when the lion is hungry and sees a zebra.) The behavior persists in the population because the genes responsible for it are persisting also.

For example, Norwegian lemmings disperse from their population when density skyrockets and food is scarce. Many die during the exodus. Are they helping the species by committing suicide to get rid of "excess" individuals? Or do they die by starvation, predation, or accidental drowning as they disperse to places where they may reproduce? A wonderfully instructive cartoon by Gary Larson shows lemmings plunging over a cliff above water, presumably in the act of suicide—but one has an inflated inner tube about its waist! If many of the lemmings are altruistic, then only "selfish" lemmings would reproduce—and over time the genes underlying altruism would disappear from the population. More likely, individuals disperse to less crowded places, where they have a better chance to survive and reproduce.

As a more detailed example, in northern forests, ravens scavenge carcasses of deer, elk, or moose, which are few and far between. When one of these large birds comes across a carcass—even in winter, when food is scarce—it often calls out loudly and attracts a crowd of similarly hungry ravens. The calling behavior puzzled Bernd Heinrich, for it seems to go against the caller's interests. Wouldn't a quiet raven eat more, increase its odds of surviving, and help it leave more descendants than an "unselfish" vocalizer? If the behavior is an outcome of natural selection, then the cost of lost calories and nutrients must be offset by a reproductive benefit for the individual caller. But what benefit?

Maybe a lone bird picking at a carcass is vulnerable to predators that lie in wait for it. If that is so, then other ravens attracted by the calls could help keep an eye out for danger. But ravens are large, agile birds with very few known enemies. Heinrich stealthily watched ravens feed singly and in pairs at a carcass. He never saw a predator attack any of them.

Then he realized that territory may have something to do with it. A **territory** is an area that one or more individuals defend against competitors. After hauling a cow carcass into a Maine forest, Heinrich observed that single or paired ravens don't always vocally advertise food. Maybe those silent ravens were adults that had already staked out a large territory—which happened to include the spot where he put the carcass.

A pair of ravens would gain nothing by attracting others to their territory. But what if that Maine forest had been *subdivided* into territories and was defended by powerful adults? A wandering young bird would be lucky to eat at all in an aggressive pair's territory. But recruiting a gang of other, nonterritorial ravens might overwhelm the resident pair's defensive behavior.

As it turns out, only the wandering ravens advertise carcasses. Basically, their calling behavior is selfish, and adaptive. It gives them a shot at otherwise off-limits food—and so promotes reproductive success.

Behavioral biologists generally find it more profitable to look for evidence of natural selection of the *individual's* traits than for something that benefits the species as a whole.

40.4 COMMUNICATION SIGNALS

Social behavior can hold together two to many billions of animals of the same species, as in the case of humans. Such behavior may be lifelong or as brief as a one-time sexual encounter. Regardless of the form it takes, such behavior requires **communication signals**. It must have **signalers**—individuals whose actions or cues can induce behavioral changes in other individuals of the same species, which are designated **signal receivers**. Figures 40.7 through 40.9 show a few representatives.

Communication signals evolve or persist when they tend to increase the reproductive success of the sender and the receiver. If a signal proves disadvantageous to either party, then natural selection will tend to favor the individuals that either do not send this particular signal or do not respond to it.

Consider how communication signals function in a termite colony in Queensland, Australia. Narrow, brittle tunnels radiate out from the colony along the trunk of a dead eucalyptus tree. Chip a hole in a tunnel, and pale worker termites inside start banging their head against its wall. The vibrations alert soldier termites, which run to the breach and make a defensive stand (Figure 40.7). Each soldier has a swollen, eyeless head that tapers to a pointed "nose." Disturb a soldier, and it will shoot thin jets of silvery goo out of its nose! Such silvery strands release volatile odors that attract more soldiers to battle the danger, which most often is an invasion by ants.

The termite head-banging is an **acoustical signal**—a sound with precise, species-specific information. For another example, think of a male zebra finch singing to attract a female and to secure his territory. Or think of a

Figure 40.7 Soldier termites, responding to an acoustical signal of a dangerous breach in one of their foraging tunnels.

tungara frog, calling at night to females and rival males. Its call is a distinctive "whine," followed by an equally distinctive "chuck."

What about the odors from soldier termites? These are **pheromones**—chemical signals between individuals of the same species. As you know, chemical odors from potential food and danger were the most important stimuli that early animals had to deal with. As animals

a

Figure 40.8 (**a**) Exposing the canines, a threat display by a male baboon. This visual signal of aggressive intent can resolve conflict without a fight. (**b**) A male albatross in a courtship display. He spreads his wings and points his head to the sky in a visual signal to the female. (See also Figure 15.4.) Such displays precede copulation. They have visual, acoustical, and tactile components.

b

Figure 40.9 Dances of honeybees, which convey information through tactile signals. (**a**) Bees trained to visit feeding stations close to a hive perform a *round* dance on the honeycomb. Worker bees that maintain contact with the forager through the dance will search for food close to the hive.

(**b**) Bees trained to visit feeding stations more than 100 meters from the hive perform a *waggle* dance. During the dance, the bee makes a straight run and waggles its abdomen. The slower the waggles during this part of the dance, the more distant the food. (**c**) As discovered by Karl von Frisch, the orientation of the straight run varies, depending on the direction in which food is located. When he put a dish of honey on a direct line between the hive and the sun, foragers that located it returned to the hive and oriented their straight runs right up the honeycomb. When he put the honey at right angles to a line drawn between the hive and the sun, the foragers made their straight runs at 90 degrees to vertical. Thus, a honeybee "recruited" into searching for food could orient its flight *with respect to the sun and the hive*—and so waste less time and energy during the search.

Waggle dancers also vary the speed of their dance in relation to the distance of the food source from the hive. When a site is 150 meters from a hive, the dance is much faster, with more waggles per straight run, compared with a dance concerning a food source that is 500 meters away.

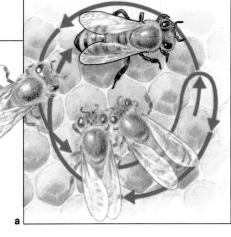

a

b

When bee moves straight down comb, recruits fly to source directly away from the sun.

When bee moves to right of vertical, recruits fly at 90° angle to right of the sun.

When bee moves straight up comb, recruits fly straight toward the sun.

c

evolved, most of them also began to rely on *signaling* pheromones that bring about an immediate behavioral response from a receiver. Some, such as termite alarm signals, stimulate or suppress aggressive or defensive behavior. Others, such as the bombykol molecules that female silk moths release, are sex attractants (Section 33.1). By contrast, *priming* pheromones, as discovered in rodents, elicit a generalized physiological response. For example, a chemical odor from the urine of some male mice will trigger and enhance estrus in female mice.

Animals active during daylight hours tend to use **visual signals**—observable actions or cues. A dominant male baboon will "yawn" and expose his formidable canines if a rival for a receptive female confronts him (Figure 40.8a). His yawn, a *threat* display, may precede a physical attack on the rival. But suppose the rival backs down. This signaler benefits by retaining access to the female without a fight. The signal receiver also benefits by avoiding a serious beating, infection, or even death.

Visual signals also are key components of **courtship displays**, which allow animals to assess the sexual overtures of potential partners. Albatross and other birds are renowned for such displays (Figure 40.8b). Fireflies and some other nocturnal animals use bioluminescent flashes (Section 4.9). The male firefly's light-generating organ emits a flashing signal. A few seconds later, a receptive female of the same species may answer with a flash. The two flash back and forth until they meet.

With **tactile signals**, a signaler touches the receiver in socially significant ways. For example, after finding a source of pollen or nectar, a foraging honeybee returns to its colony (hive) and performs a complex dance. It

moves in circles, jostling in the dark through a crowd of workers. Other bees may follow and maintain physical contact with the dancer. In this way, honeybees acquire information about the general location, distance, and direction of pollen or nectar (Figure 40.9).

One last point. Head-banging and odorous goo are communication signals *between* termites. Is a scent from an ant also a signal for termites, which respond to it? The scent may indeed have some evolved function, but announcing an invading ant is not one of them. When a soldier termite detects the scent and kills the ant, the termite is an **illegitimate receiver** of a communication signal *meant for individuals of a different species*. The scent serves to identify the ant as a member of an ant colony and induces cooperative behavior from other ants. Or consider **illegitimate signalers**. Certain assassin bugs hook dead and drained bodies of termite prey on their dorsal surface and so acquire the odor of victims. By deceptively signaling that they "belong" to a termite colony, they hunt termite victims more easily. Another illegitimate signaler is the female of some predatory fireflies. If one observes a flash from a male firefly, she flashes back. If she can lure him into attack range, he becomes her meal—an evolutionary cost of having an otherwise useful response to a come-hither signal.

A communication signal between individuals of the same species is an action that has a net beneficial effect on both the signaler and the signal receiver.

Natural selection tends to favor communication signals that promote reproductive success.

For reasons that we need not explore here, most people find the mating and parenting behaviors of different animals fascinating. How useful is selection theory in helping us interpret such behavior? Let's take a look.

Selection Theory and Mating Behavior

Competition among members of one sex for access to mates is common. So is choosiness in selecting a mate. Recall, from Section 14.9, that such activities are forms of a microevolutionary process called **sexual selection**. Sexual selection favors traits that give the individual a competitive edge in reproductive success.

Typically, male animals produce great numbers of very tiny sperm, and the females produce considerably fewer but larger eggs. Reproductive success for a male generally depends on how many eggs he fertilizes. For a female, success depends largely on how many eggs she produces or how many offspring she can care for. In most cases, the prime factor that influences her sexual preference is the quality of a mate, not the quantity of partners. Hangingflies, sage grouse, and bison (Figures 40.10 through 40.12) are splendid examples of how such females dictate the rules of male competition. They also illustrate how males employ tactics that may help them fertilize as many eggs as possible.

Female hangingflies (*Harpobittacus apicalis*) select the males that offer them superior material goods. And so you might observe males capturing and killing a moth

Figure 40.10 A male hangingfly dangling a moth as a nuptial present for a future mate. Females of certain hangingfly species choose sexual partners on the basis of the size of prey that males offer to them.

Figure 40.11 Dancing display by a male sage grouse, performed at his own vigorously defended spot inside a compact mating area called a lek. Females (the smaller brown birds) at the lek observe the prancing males before choosing the one they will mate with.

or some other insect. After males do this, they release a sex pheromone—a chemical signal—to attract females to the "nuptial gift" (Figure 40.10). A female chooses males that offer the larger, calorie-rich offerings. She permits a male to mate—but only *after* she's been eating a gift for five minutes or so. Then she will accept sperm and store it in a reproductive organ—but only as long as the food holds out. Before twenty minutes are up, she can break off the mating at any point. If the female does so, she might well mate again and accept another male's sperm—and dilute the reproductive success of her first partner.

In western regions of North and South Dakota, female sage grouse (*Centrocercus urophasianus*) are dispersed among stands of sagebrush. You never find them far from the protective cover of sage during spring courtship and summer nesting. Male sage grouse make no attempt to maintain a large territory. In the breeding season they congregate in a **lek**, a communal display ground. There, each male stakes out a few square meters as his territory. Females are attracted to the lek—not to feed or nest, but rather to observe the males. With tail feathers erect and neck pouches puffed, each male emits booming calls and stamps about like a big wind-up toy on his display ground (Figure 40.11). Female sage grouse tend to select and mate with one male only. Then they go off to nest by themselves in sagebrush, unassisted by a sexual partner. Many females often choose the same male, so most of the males never mate.

As a final example, when the females of a species cluster in defendable groups at a time when they are sexually receptive, you probably will observe male competition for access to the clusters. Competition for ready-made harems has favored combative male lions, sheep, elk, and bison, to name a few such animals (Figures 1.6*f* and 40.12).

Costs and Benefits of Parenting

What happens after mating, when the offspring arrive? Until the offspring have developed enough to survive on their own, the parents of some species will care for them. For example, adult Caspian terns (Figure 40.13) incubate the eggs, shelter the nestlings and feed them, then accompany and protect them after they start to fly. Such parental behavior comes at a reproductive cost; it

Figure 40.12 Sexual competition between male bison, which are fighting for access to a cluster of females.

Figure 40.13 Male and female Caspian terns, protecting their chicks. Parental care has costs as well as benefits.

drains time and energy that might otherwise be given to improving their own chances of living to reproduce another time. Yet for many species, parenting improves the likelihood that the current generation of offspring will survive. The benefit of devoting time and energy to immediate reproductive success outweighs the cost of reduced reproductive success at some later time.

Selection theory applies to some aspects of mating behavior, as when sexual selection favors traits that give the individual a competitive edge in reproductive success. It applies also to parental behavior that contributes to reproductive success.

40.6 COSTS AND BENEFITS OF BELONGING TO SOCIAL GROUPS

Survey the animal kingdom and you find a range of social behaviors. For some species, individuals spend most of their lives alone or in small family groups. For some other species, individuals live in huge groups of thousands of related individuals. Termite and honeybee societies are like this. Individuals of still other species live in social units composed of unrelated individuals. Populations of the human species are like this.

Given such differences, how might we gain insight into the basis for any one social group? To find answers, evolutionary biologists commonly take a **cost-benefit approach**. That is, they explore the costs and benefits of social life in terms of *reproductive success of the individual*, as measured by that individual's contribution to the gene pool of the next generation.

Disadvantages to Sociality

In large populations, and in certain environments, the benefits of social life come at great reproductive cost to each individual. For example, herring gulls nest in huge colonies, where breeding pairs will cannibalize the eggs or young chicks of their neighbors in an instant if given the opportunity to do so. Besides this, herring gulls deplete available food resources faster than they would if their colonies were less crowded. Also, overcrowding promotes the spread of contagious diseases.

Competition for resources and greater vulnerability to diseases are typical of most large colonies, including those of cliff swallows, prairie dogs, and penguins (Figure 40.14). Under such living conditions, the individual and its offspring are more likely to be weakened by the pathogens and parasites that are so readily transmitted from host to host in large groups. Similarly, plagues can spread like wildfire through densely crowded human populations. This is especially true in settlements and cities where infestations of rats and fleas are chronic, and where plumbing, sewage treatment, and modern medical care are absent or inadequate (Section 35.6).

Advantages of Sociality

Given the substantial costs to individual reproductive success, why would the individuals of any species live close to others in a particular environment? When they do, it might be that the benefits of social life are great enough to overcome the costs.

Cooperative Predator Avoidance First, consider that a group of animals acting cooperatively against a predator can reduce the net risk to any one individual. For example, for a flock or herd that is vulnerable to predation, simply having more pairs of eyes to scan the surrounding area helps individuals detect the predators

Figure 40.14 A colony of royal penguins on Macquarie Island, between New Zealand and Antarctica.

Figure 40.15 Social defensive behavior of musk oxen (*Ovibos moschatus*). In the presence of a perceived threat—usually wolf predators—the adults form a circle around their young. They face outward, and the "ring of horns" successfully deters the wolves.

Figure 40.16 Social defensive behavior of Australian sawfly caterpillars, which live together in clumps. These caterpillars will regurgitate and hold fluid in their mouth (the yellow blobs), where it serves to repel predators that try to grab them. After the danger passes, they swallow the fluid, which is toxic to most animals.

sooner. The individuals of a social group also may join together in a counterattack or in defensive behavior, as the musk oxen in Figure 40.15 are doing.

The biologist Birgitta Sillen-Tullberg found tangible evidence of the benefits of sociality. She was studying Australian sawfly caterpillars, which live together in clumps (Figure 40.16). When something disturbs the caterpillars, they collectively rear up and writhe about, all the while regurgitating partially digested food. Their food of choice is eucalyptus leaves. These leaves are impregnated with chemical compounds that are toxic to most animals—including insect-eating songbirds.

Sillen-Tullberg hypothesized that individual sawfly caterpillars benefit from the coordinated act of repelling bird predators. She used her hypothesis to predict that the birds are more likely to consume a solitary caterpillar, not a cluster of them. To test her prediction, she offered young, hand-reared Great Tits (*Parus major*) a chance to feed on caterpillars, which she offered either one by one or in a group of twenty per offering. She did this for a standard number of presentations. Ten birds that were offered one individual at a time consumed an average of 5.6 caterpillars. But ten birds that were each offered a clump of caterpillars only ate an average of 4.1. As was expected for this experiment, individuals were somewhat safer in a group than on their own.

The Selfish Herd Simply by their physical location within a group, some individuals form a living shield against predation on others in the group. They are part of a **selfish herd**—a simple society held together by reproductive self-interest, although not consciously so.

Investigators tested the selfish-herd hypothesis for male bluegill sunfishes, which build adjacent nests on the bottom of lakes. Males use their fins to hollow out a depression in lake mud, where females deposit eggs.

If a colony of bluegill males is a selfish herd, then we can predict competition for the "safe" sites—at the center of the colony. Compared to the periphery, eggs laid in nests at the center are less likely to be attacked by snails and largemouth bass. The competition does indeed exist. The largest, most powerful males tend to claim central locations. Other, smaller males assemble around them and bear the brunt of predatory attacks. Even so, they are better off in the group than on their own, fending off a bass singlehandedly, so to speak.

We may evaluate the costs and benefits of social life in terms of reproductive success, as measured by the genes that each individual contributes to the next generation.

Dominance Hierarchies

As you have seen, individuals of a selfish herd make no personal sacrifice for the others; the personal benefits of living with others simply seem to outweigh the costs. By contrast, some individuals may *help* others survive and reproduce. Suppose their helpful behavior is a cost of belonging to a social group—and maybe they don't give up their chance at reproductive success entirely. Let's look at two groups to see if this hypothesis holds.

In a baboon troop, individuals help one another, but reproductive opportunity is unequal. Some individuals give up safe sleeping places, choice bits of food, and even receptive females to others upon receiving a threat signal from another troop member. In fact, we find a **dominance hierarchy**, in which some individuals have adopted a subordinate status to others (Figure 40.17). Jane Goodall also documented such hierarchies during her studies of chimpanzee society (page 572).

Similarly, the members of a wolf pack show helpful behavior, as when they fend off predators and share food. Only *one* male and *one* female wolf produce pups. Others of the pack don't breed, but they bring back food to members that stay inside the den and guard the pups.

Why do subordinate adults that don't breed remain in a social group and make sacrifices for dominant peers? As is true of small bluegill sunfish, the benefits from group living probably offset their sacrifices. Besides, challenging a strong member may result in injuries that could shorten a life.

Also, it may not be possible to survive alone, outside the group. A solitary baboon surely quickens the pulse of the first leopard that sees it. Self-sacrificing behavior may even give subordinates a chance to reproduce if they live long enough and if predation or weakness in old age removes dominant peers. Some subordinate wolves and baboons do move up the social ladder if dominant members slip down a rung or fall off. Thus, acceptance of subordinate status might pay off in the long term for the patient individual.

The Evolution of Altruism

A subordinate animal that gives way to a dominant one is acting in its own self-interest. What about altruistic animals? We find them among many vertebrate groups. They reach extremes in certain insect societies, as when a worker bee plunges its stinger into an invader of the hive, and thereby commits suicide to defend others.

If altruistic individuals do not contribute their genes to the next generation, how does the genetic basis for altruistic behavior persist over evolutionary time? By one theory, individuals can pass on genes indirectly if they help *relatives* survive and reproduce. Remember, when a sexually reproducing parent helps offspring, it isn't helping *exact* genetic copies of itself. Each offspring has only half its genes. If other individuals of the group have the same ancestors, they also share genes with the parents. Genetically, two siblings are as similar as a parent and one of its offspring. Nephews and nieces are like their uncle in about one-fourth of their genes.

According to William Hamilton's **theory of indirect selection**, genes associated with caring for relatives that are not direct descendants—not one's own offspring— are favored in certain situations. Think of this form of altruistic behavior as an extension of parenting. For example, suppose an uncle helps his niece survive long

Figure 40.17
Appeasement behavior between baboons. Notice the assured position of the dominant animal (*left*) and the abject stare and groveling posture of the subordinate one, who is making little conciliatory smacking noises with its lips.

Figure 40.18 Life in a honeybee colony.
(**a**) A court of sterile worker daughters surrounds this queen bee, the only fertile female in the hive. They feed her and relay her pheromones throughout the hive, and these influence the activities of all members. The queen is much larger than the workers, in part because her ovaries are fully developed (unlike those of her sterile daughters).

(**b**) At certain times of year, stingless drones develop and mature. They do not work for the colony. Instead, drones attempt to mate with the queens of other hives.

(**c**) The hive contains between 30,000 and 50,000 worker bees. They feed bee larvae, clean and maintain the hive, and construct new honeycomb from wax secretions.
(**d**) The workers store honey or pollen in the honeycomb, which also houses new bee generations from the egg stage, through a series of larval stages, to the emergence of pupae and then the adults.

Young workers feed the larvae. Each adult worker lives for about six weeks in the spring and summer. It can survive about four months in an overwintering colony.

Workers also engage in scent-fanning, another cooperative action. Fanned air passes over the bee's exposed scent gland. Pheromones released from the gland help other bees orient to the hive's entrance, which bees pass through on their way to or from foraging expeditions. When foraging bees return to the hive after locating a rich source of nectar or pollen, they perform a dance that recruits more workers into taking off for the source (Figure 40.9).

In other helpful actions, (**e**) worker bees transfer food to one another, and (**f**) worker females at the hive entrance serve as guards that sacrifice themselves to repel intruders.

a

b

c

egg   nearly mature pupa   f
d

e

enough to reproduce. He has made an indirect genetic contribution to the next generation, as measured in the genes he and his relatives share. Altruism costs him; he may lose his own opportunities to reproduce. But if the cost is less than the benefit, the action will propagate the uncle's genes and favor the spread of his kind of altruism in the species. If an uncle saves two nieces, this is equivalent to saving his own daughter.

Thus, workers of insect societies indirectly promote their "self-sacrifice" genes through altruistic behavior directed toward relatives. Colonies of honeybees, ants, and termites are actually large families (Figure 40.18). The family's worker force labors on behalf of siblings, some of which are the future kings and queens. When a guard bee drives her stinger into a raccoon, she dies—

but siblings perpetuate some of her genes. What about vertebrate societies? Sterility and extreme self-sacrifice are rare among these groups. The known exceptions include naked mole-rats, the type of mammal described in the concluding section of this chapter.

In dominance hierarchies, subordinate individuals might be better off with the group than without it—and they may be presented with an opportunity to reproduce later on.

Altruistic behavior may persist when individuals pass on genes indirectly, by helping relatives survive and reproduce.

By one theory of indirect selection, genes associated with altruistic behavior that is directed toward relatives may be favored.

ABOUT THOSE SELF-SACRIFICING NAKED MOLE-RATS

We don't mean to be at all judgmental about this, but the naked mole-rats (*Heterocephalus glaber*) look rather like bucktoothed sausages with wrinkled, pink skin and a few sprouts of hairs (Figure 40.19). Their behavior is even more fascinating than their looks. Why? Unlike any other known vertebrate, many naked mole-rat individuals spend their entire lives as nonbreeding helpers in their social group.

These highly social mammals live in arid regions of eastern Africa, in cooperatively excavated burrows. They always live in clans—small social units of anywhere from 25 to 300 individuals. In each clan, you will find only a single reproducing female, and she mates with one to three males. All other members of the clan care for the "queen" and "king" (or kings) and their offspring.

The nonreproducing "diggers" busily excavate extensive subterranean tunnels and special chambers, which serve as living rooms or waste-disposal centers. They also locate large tubers that grow underground. These they chop up into edible bits, which they deliver to the queen, her retinue of males, and her offspring.

Besides this, digger mole-rats deliver food to certain other helpers that spend time loafing about, shoulder to shoulder and belly to back, with the reproductive royals. Usually the "loafer" mole-rats are larger than the diggers. And they aren't really loafers. They will spring to action when a snake or some other enemy threatens the colony. At great personal risk, they will collectively chase away or attack and kill the predator.

If helpful members fail to have offspring that would perpetuate the genetic basis for helping, then maybe their behavior persists because they are related in some way to offspring of the mole-rats that *do* reproduce. By applying the theory of indirect selection to naked mole-rats, we can formulate a hypothesis with a testable prediction:

1. If helpful behavior is genetically advantageous to some individuals in a naked mole-rat colony (*the hypothesis*), then it follows that the helpers will be related to the reproductive members of the colony that benefit from their altruism (*the prediction*).

2. Determine the genetic relationship of the subordinate, sterile helpers to the dominant, reproducing queen and king or kings. (*This would be an appropriate test of the prediction.*)

Could the stated hypothesis be tested by affixing a marker that would allow us to establish the identity of each individual in a naked mole-rat clan and, in time, to establish a family tree for the group? Probably not. It isn't likely that we could identify and track all the individuals, given the intricate, hidden tunnels of their burrows.

H. Kern Reeve and his colleagues decided on a more practical approach. They relied on *DNA fingerprinting*, a method of establishing degrees of genetic relatedness among individuals. As described in Section 13.3, this laboratory method starts with the formation of restriction fragments of DNA molecules. Then technicians construct a visual record of different sets of fragments of the DNA from different individuals. Essentially, a set of identical twins would have identical DNA fingerprints (their DNA would have the same base sequence). Individuals having the same father and mother would have similar DNA— thus similar but not identical DNA fingerprints. On the average, DNA fingerprints of genetically unrelated individuals should differ much more than those of siblings or other relatives.

When Reeve constructed DNA fingerprints for naked mole-rats, he discovered that all of the individuals from a given clan are *very* close relatives, and they are very different genetically from members of other clans. These findings suggest that each naked mole-rat clan is highly inbred, a result of generations of brother-sister, mother-son, or father-daughter matings. As you know from Section 14.9, inbreeding among the individuals of a population results in extremely reduced genetic variability.

Therefore, a self-sacrificing naked mole-rat is helping to perpetuate a very high proportion of the forms of genes (alleles) that it carries. It turns out that the genotypes of helpers and the helped might be as much as 90 percent identical!

Figure 40.19 A peek at a few naked mole-rats (*Heterocephalus glaber*).

If we can analyze the evolutionary basis of the behavior of termites, naked mole-rats, and other animals, would it not be rewarding to analyze such a basis of human behavior also? Many people resist the idea. It seems they believe that attempts to identify the adaptive value of a particular human trait is an attempt to define its moral or social advantage. Clearly, however, there is a difference between trying to explain something in terms of its evolutionary history and attempting to justify it. "Adaptive" does not mean "morally right." It means valuable with respect to the transmission of an individual's genes.

Consider the case of altruistic behavior that we call **adoption**—the acceptance of offspring of other individuals as one's own. If we wish to gain understanding of this behavior, we might use evolutionary theory to formulate a testable hypothesis about it. But this does not mean we are passing judgment on whether adoption is moral or even socially desirable. These are two entirely separate issues —about which biologists have no more to say than anybody else.

Figure 40.20 Two adult emperor penguins competing to adopt an orphan, which penguins accept as substitute offspring.

Start with a premise that all adaptations have costs and benefits. One cost is that certain adaptive behaviors may be *redirected* under rare or unusual circumstances. In many species, adults that have lost their offspring will, if presented with a substitute, adopt it. Cardinals have been known to feed goldfish. A whale tried to lift a log out of water, as if it were a distressed infant that needed help in reaching the water's surface. Emperor penguins fight to adopt orphans (Figure 40.20).

As John Alcock suggests, such examples of redirected parental behavior can be a starting point for studying *adoptive* behavior among humans. Suppose such behavior arises through a frustrated desire to bear children (the hypothesis). If this is so, then we can make the following prediction: Couples who have lost an only child or who are sterile should be especially prone to adopt strangers.

Now consider this. For most of their evolutionary history, humans have lived in small groups and, later, small villages. They probably had few opportunities or showed little inclination to adopt strangers. They were likely to accept the child of a deceased relative. But how do adoptive parents gain genetic representation in the next generation? According to theories of natural selection and indirect selection, individuals act in ways that promote genetic self-interest. So we might predict that parents with dependent, care-requiring children of their own are much less likely to adopt substitutes, compared with adults who have no children or who already raised children and are living by themselves. We could test the prediction by piecing together a large enough sampling of information, ideally from diverse existing human cultures, on possible connections between adoption and a childless condition.

Indirect selection also favors adults who focus their parenting behavior on relatives and thereby perpetuate their shared genes in an indirect way. We might even predict that such adults are more likely to be related to the adopted child than we would expect by chance alone. A researcher, Joan Silk, did test this prediction in some traditional societies. As she found out, people adopt related children far more often than nonrelated ones. It is especially in the large, industrialized societies with welfare agencies and other means of adoption assistance that individuals become parents of nonrelated children. Such societies are recent evolutionary novelties. Parenting mechanisms evolved in the past, and it may be that their redirection toward nonrelatives says more about our evolutionary history than it does about the transmission of one's genes.

The point is this: Evolutionary hypotheses about the adaptive value of behavior lend themselves to testing. And through such testing, we can gain understanding about the evolution of human behavior.

It is possible to test evolutionary hypotheses regarding the adaptive value of human behaviors.

There may be greater acceptance of such testing when more people come to understand that adaptive behavior and socially desirable behavior are separate issues.

In biology, "adaptive" means only that some specified trait has proved beneficial in the transmission of the individual's genes that are responsible for that trait.

SUMMARY

1. Behavior refers to observable, coordinated responses that animals make to external and internal stimuli. It originates with genes that directly or indirectly specify the products required for the development and operation of the nervous and endocrine systems. The systems interact to govern responses to stimuli.

2. Complex but stereotyped responses to specific, often simple environmental stimuli are instinctive behaviors. Responses that vary or change as a result of individual experiences are learned behaviors. Both are developmental outcomes of genetic and environmental inputs.

3. Behavior having a genetic basis is subject to evolution by natural selection. It evolved as a result of individual differences in reproductive success in past generations, when its reproductive benefits exceeded its reproductive costs (or disadvantages).

4. Discriminating among competing explanations for a particular behavior requires formulating hypotheses to generate predictions, which can be tested by collecting new information and by checking whether results match the expected observations.

5. Members of the same species often create obstacles to one another's reproductive success, as by selective mate choice and by competition for access to mates.

6. Social behavior can involve cooperative interdependency among individuals of a species. It is promoted by communication signals—actions or cues sent by one individual (a signaler) that can change the behavior of another of the same species (a signal receiver). Such a signal benefits both the signaler and the receiver.

7. Different animals use chemical, visual, acoustical, and tactile signals.

   a. Pheromones function in chemical communication. Signaling pheromones, including sex attractants and termite alarm signals, can induce immediate change in the behavior of a signal receiver. Priming pheromones elicit a generalized physiological response by the receiver.

   b. Visual signals (observable actions or cues) are key components of courtship displays and threat displays.

   c. Acoustical signals (sounds with precise, species-specific information) include the mating calls of frogs.

   d. Tactile signals are specific kinds of physical contact between a signaler and receiver. Interaction among honeybees during a round dance is an example.

8. Costs and benefits of social life are reflected in the individual's reproductive success, as measured by its genetic contribution to the next generation.

9. Sociality carries high risks of contagious disease and competition for limited resources. Benefits outweigh the costs of social life under some circumstances, as when predation pressure is severe.

10. In self-sacrificing (altruistic) behavior, individuals give up chances to reproduce while helping others of their social group. But in most social groups, individuals do not sacrifice reproductive chances to help others.

   a. Helpful acts often may be the result of dominant individuals forcing subordinates to relinquish food or some other resource.

   b. When subordinates give in to dominant members of their group, they may receive compensatory benefits of group living, such as safety from predators.

   c. In some species, subordinates may eventually reproduce if they live long enough and if dominant ones slip down in the hierarchy or die off.

11. By one theory of indirect selection, genes associated with self-sacrificing behavior that helps nondescendant relatives are favored.

12. Indirect selection can lead to extreme altruism. This is evident among workers of some species of social insects and naked mole-rats. Such workers normally do not reproduce. They help their reproducing siblings survive and therefore pass on "by proxy" the genes underlying the development of altruism.

Review Questions

1. What role does the environment play in the development of an instinct, such as the egg-ejecting behavior of baby cuckoos? *729*

2. Contrast altruistic behavior with selfish behavior. Why does either form of behavior persist in a population? *731*

3. Give an example of feeding or mating behavior that can be explained in terms of natural (individual) selection. *731, 734–735*

4. How are threat displays examples of cooperation? *733*

5. Is parental behavior always adaptive? *735*

6. Why don't members of a selfish herd live apart if each one is "trying to take advantage" of the others? *737*

7. How can a self-sacrificing individual of a social group still pass on more of its genes than a noncooperative individual? *738–739*

8. How do some sterile individuals propagate their genes? *738–740*

Self-Quiz *(Answers in Appendix IV)*

1. "Starlings festoon their nests with sprigs of wild carrot and so minimize nest mites." This statement is _____ .
   a. an untested hypothesis   c. a test of a hypothesis
   b. a prediction   d. a proximate conclusion

2. Genes affect the behavior of individuals by _____ .
   a. influencing the development of nervous systems
   b. affecting the kinds of hormones in individuals
   c. governing the development of muscles and skeletons
   d. all of the above

3. Many kinds of female mammals live in herds. Which of these mating systems might you see in such herds?
   a. a lek mating system
   b. males competing for clusters of receptive females
   c. territorial defense of feeding sites by males
   d. males capturing food to lure females

4. The statement that lemmings dispersing from an overcrowded population bring it in line with available resources _____ .
   a. has been proven to be true
   b. is consistent with Darwinian evolutionary theory
   c. is based on the theory of evolution by group selection
   d. is supported by the finding that most animals behave altruistically during their lives

5. Social behavior evolves because _____ .
   a. social animals are more advanced than solitary ones
   b. under some ecological conditions, the costs of social life are less than its benefits to the group
   c. under some ecological conditions, the costs of social life are less than its benefits to the individual
   d. predator pressures always favor social living

6. The theory of evolution by indirect selection _____ .
   a. focuses on evolution for group benefit
   b. explains that altruism is always adaptive
   c. examines the consequences of the effects of differences among individuals on the reproductive success of relatives
   d. shows how families with many descendants are superior to self-sacrificing families

7. Altruism is helpful behavior that _____ .
   a. cannot evolve
   b. can spread through a species only if it raises the altruist's reproductive success
   c. reduces an individual's reproductive success
   d. always helps spread the altruist's genes

8. The genetic similarity between an uncle and his nephew is _____ .
   a. the same as between a parent and his offspring
   b. greater than between two full siblings
   c. dependent on how many other nephews the uncle has
   d. less than that between a mother and her daughter

9. Match the terms with their most suitable description.
   _____ fixed action pattern
   _____ altruism
   _____ basis of instinctive *and* learned behavior
   _____ imprinting
   _____ behavior

   a. time-dependent form of learning requiring exposure to key stimulus
   b. genes and environmental experience
   c. coordinated responses to stimuli
   d. instinctive response triggered by a well-defined, often simple stimulus
   e. assisting another individual at one's own expense

Critical Thinking

1. You see a rooster wading out to mallard ducks with apparently amorous intent. What probably happened to him early in life?

2. A large caterpillar in the tropics responds to being poked by partly letting go of its substrate and puffing up part of its body:

Propose a possible mechanism that underlies this behavior. Also propose how the behavior has adaptive value. How would you test both hypotheses?

3. A hyena scent-marks plants in its territory by releasing specific chemicals from certain glands. What evidence would you need to demonstrate that this action is an evolved communication signal?

Selected Key Terms

acoustical signal *732*
adoption *741*
animal behavior *727*
behavior *728*
communication signals *732*
cost-benefit approach *736*
courtship display *733*
dominance hierarchy *738*
fixed action pattern *729*
illegitimate receiver *733*
illegitimate signaler *733*
imprinting *730*
instinctive behavior *729*

learned behavior *730*
lek *735*
natural selection *731*
pheromone *732*
selfish herd *737*
sexual selection *734*
signal receiver *732*
signaler *732*
song system *729*
tactile signal *733*
territory *731*
theory of indirect selection *738*
visual signal *733*

Readings

Alcock, J. 1993. *Animal Behavior: An Evolutionary Approach*. Fifth edition. Sunderland, Massachusetts: Sinauer.

Dawkins, R. 1989. *The Selfish Gene*. New York: Oxford University Press.

Frisch, K. von. 1961. *The Dancing Bees*. New York: Harcourt Brace Jovanovich. A classic on the behavior of honeybees.

Le Meho, Y. 1977. "Emperor Penguin: A Strategy to Live and Breed in the Cold." *American Scientist* 65(6): 680–693.

Wilson, E. O. 1975. *Sociobiology: The New Synthesis*. Cambridge, Massachusetts: Harvard University Press.

APPENDIX I
A BRIEF CLASSIFICATION SCHEME

The classification scheme that follows is a composite of several that are used in microbiology, botany, and zoology. The major groupings are more or less agreed upon. There is not always agreement, however, on what to call a given grouping or where it might fit within the overall hierarchy. There are several reasons for this.

First, the fossil record varies in its quality and in its completeness. Therefore, the relationship of one group to others is sometimes open to interpretation. Comparative studies at the molecular level are firming up the picture, but this work is still under way.

Second, ever since the time of Linnaeus, classification schemes have been based on the perceived morphological similarities and differences among organisms. Although some original interpretations are now open to question, we are so used to thinking about organisms in certain ways that reclassification often proceeds slowly. Traditionally, for example, birds and reptiles have been considered to be separate classes (Reptilia and Aves). And yet there are now compelling arguments for grouping the lizards and snakes together in one class—and the crocodilians, dinosaurs, *and* birds in a different class.

Finally, researchers in microbiology, mycology, botany, zoology, and the other fields of biological inquiry have inherited a wealth of literature, based on classification schemes that were developed over time in each of those fields. Many of these individuals see no good reason to give up established terminology and so disrupt access to the past. Until recently, for example, botanists were using *division*, and zoologists *phylum*, for taxa that are equivalent in the hierarchy. Until recently, opinions were polarized with respect to an entire kingdom—the Protista, certain members of which could just as easily be grouped in the kingdoms of plants, fungi, or animals. Indeed, the term protozoan is a holdover from an earlier scheme in which the amoebas and some other single-celled organisms were ranked as simple animals.

Given all the problems, why do we bother imposing artificial frameworks on the history of life? We do this for the same reason that a writer might decide to break up the history of civilization into several volumes, a number of chapters, and many paragraphs. Both efforts are attempts to impart obvious structure to what might otherwise be an overwhelming body of information.

Finally, bear in mind that we include this classification scheme primarily for your reference purposes. It is by no means complete. Numerous existing and extinct organisms of the so-called lesser phyla are not represented here. Our strategy is to focus mainly on organisms mentioned in the text. A few examples of organisms also are listed under the entries.

SUPERKINGDOM PROKARYOTA. Prokaryotes. Single-celled organisms with DNA concentrated in a cytoplasmic region, not in a membrane-bound nucleus.

KINGDOM MONERA. Bacteria, either single cells or simple associations of cells. Both autotrophs and heterotrophs (refer to Table 18.2). *Bergey's Manual of Systematic Bacteriology*, the authoritative reference in the field, calls this "a time of taxonomic transition." It groups bacteria mainly on the basis of form, physiology, and behavior, not on phylogeny. The scheme presented here does reflect the growing evidence of evolutionary relationships for at least some bacterial groups.

SUBKINGDOM ARCHAEBACTERIA. Methanogens, halophiles, thermophiles. Strict anaerobes, distinct from other bacteria in cell wall, membrane lipids, ribosomes, and RNA sequences. *Methanobacterium, Halobacterium, Sulfolobus.*

SUBKINGDOM EUBACTERIA. Gram-negative and gram-positive forms. Peptidoglycan in cell wall. Photosynthetic autotrophs, chemosynthetic autotrophs, and heterotrophs.

PHYLUM GRACILICUTES. Typical Gram-negative, thin wall. Autotrophs (photosynthetic and chemosynthetic) and heterotrophs. *Anabaena* and other cyanobacteria. *Escherichia, Pseudomonas, Neisseria, Myxococcus.*
PHYLUM FIRMICUTES. Typical Gram-positive, thick wall. Heterotrophs. *Bacillus, Staphylococcus, Streptococcus, Clostridium, Actinomycetes.*
PHYLUM TENERICUTES. Gram-negative, wall absent. Heterotrophs (saprobes, pathogens). *Mycoplasma.*

SUPERKINGDOM EUKARYOTA. Eukaryotes (single-celled and multicelled organisms. Cells typically have a nucleus (enclosing the DNA) and other membrane-bound organelles.

KINGDOM PROTISTA. Mostly single-celled eukaryotes, some colonial forms. Diverse autotrophs and heterotrophs (Table 18.4). Many lineages apparently related evolutionarily to certain plants, fungi, and possibly animals.

PHYLUM CHYTRIDIOMYCOTA. Chytrids. Heterotrophs; saprobic decomposers, parasites. *Chytridium.*
PHYLUM OOMYCOTA. Water molds. Heterotrophs. Decomposers, some parasites. *Saprolegnia, Phytophthora, Plasmopara.*
PHYLUM ACRASIOMYCOTA. Cellular slime molds. Heterotrophs with amoeboid and spore-bearing stages. *Dictyostelium.*
PHYLUM SARCODINA. Amoeboid protozoans. Heterotrophs. Soft-bodied; some shelled. Amoebas, foraminiferans, radiolarians, heliozoans. *Amoeba, Entomoeba.*
PHYLUM MYXOMYCOTA. Plasmodial slime molds. Heterotrophs with amoeboid and spore-bearing stages. *Physarum.*
PHYLUM CILIOPHORA. Ciliated protozoans. Heterotrophs. Distinctive arrays of cilia, used as motile structures. *Paramecium,* hypotrichs.
PHYLUM MASTIGOPHORA. Flagellated protozoans. Some free-living, many internal parasites; all with one to several flagella. *Trypanosoma, Trichomonas, Giardia.*
SPOROZOANS. Parasitic protozoans, many intracellular. "Sporozoans" is the common name with no formal taxonomic status. *Plasmodium, Toxoplasma.*
PHYLUM EUGLENOPHYTA. Euglenoids. Mostly heterotrophs, some photosynthetic types. Flagellated. *Euglena.*

PHYLUM PYRRHOPHYTA.   Dinoflagellates. Photosynthetic, mostly, but some heterotrophs. *Gymnodinium breve.*

PHYLUM CHRYSOPHYTA.   Golden algae, yellow-green algae, diatoms. Photosynthetic. Some flagellated, others not. *Mischococcus, Synura, Vaucheria.*

PHYLUM RHODOPHYTA.   Red algae. Photosynthetic, nearly all marine, some freshwater. *Porphyra. Bonnemaisonia, Euchema.*

PHYLUM PHAEOPHYTA.   Brown algae. Photosynthetic, nearly all temperate or marine waters. *Macrocystis, Fucus, Sargassum, Ectocarpus, Postelsia.*

PHYLUM CHLOROPHYTA.   Green algae. Photosynthetic. Most freshwater, some marine or terrestrial. *Chlamydomonas, Spirogyra, Ulva, Volvox, Codium, Halimeda.*

KINGDOM FUNGI.   Mostly multicelled eukaryotes. Heterotrophs (mostly saprobes, some parasites). Major decomposers of nearly all communities. Reliance on extracellular digestion of organic matter and absorption of nutrients by individual cells.

PHYLUM ZYGOMYCOTA.   Zygomycetes. All produce nonmotile spores. Bread molds, related forms. *Rhizopus, Philobolus.*

PHYLUM ASCOMYCOTA.   Ascomycetes. Sac fungi. Most yeasts and molds; morels, truffles. *Saccharomycetes, Morchella Neurospora, Sarcoscypha. Claviceps, Ophiostoma.*

PHYLUM BASIDIOMYCOTA.   Basidiomycetes. Club fungi. Mushrooms, shelf fungi, stinkhorns. *Agaricus, Amanita, Puccinia, Ustilago.*

IMPERFECT FUNGI.   Sexual spores absent or undetected. The group has no formal taxonomic status. If better understood, a given species might be grouped with sac fungi or club fungi. *Verticillium, Candida, Microsporum, Histoplasma.*

LICHENS.   Mutualistic interactions between a fungus and a cyanobacterium, green alga, or both. *Usnea, Cladonia.*

KINGDOM PLANTAE.   Nearly all multicelled eukaryotes. Nearly all photosynthetic autotrophs.

PHYLUM RHYNIOPHYTA.   Earliest known vascular plants; extinct. *Cooksonia, Rhynia.*

PHYLUM TRIMEROPHYTA.   Trimerophytes. *Psilophyton.*

PHYLUM PROGYMNOSPERMOPHYTA. Progymnosperms. Ancestral to early seed-bearing plants; extinct. *Archaeopteris.*

PHYLUM CHAROPHYTA.   Stoneworts.

PHYLUM BRYOPHYTA.   Liverworts, hornworts, mosses. *Marchantia, Polytrichum, Sphagnum.*

PHYLUM PSILOPHYTA.   Whisk ferns. *Psilotum.*

PHYLUM LYCOPHYTA.   Lycophytes, club mosses. *Lycopodium, Selaginella.*

PHYLUM SPHENOPHYTA.   Horsetails. *Equisetum.*

PHYLUM PTEROPHYTA.   Ferns.

PHYLUM PTERIDOSPERMOPHYTA.   Seed ferns. Fernlike gymnosperms; extinct.

PHYLUM CYCADOPHYTA.   Cycads. *Zamia.*

PHYLUM GINKGOPHYTA.   Ginkgo. *Ginkgo.*

PHYLUM GNETOPHYTA.   Gnetophytes. *Ephedra, Welwitchia.*

PHYLUM CONIFEROPHYTA.   Conifers.
  Family Pinaceae. Pines, firs, spruces, hemlock, larches, Douglas firs, true cedars. *Pinus.*
  Family Cupressaceae. Junipers, cypresses. *Juniperus.*
  Family Taxodiaceae. Bald cypress, redwoods, Sierra bigtree, dawn redwood. *Sequoia.*
  Family Taxaceae. Yews.

PHYLUM ANTHOPHYTA.   Flowering plants.
  Class Dicotyledonae. Dicotyledons (dicots). Some families of several different orders are listed:
    Family Nymphaeaceae. Water lilies.
    Family Papaveraceae. Poppies.
    Family Brassicaceae. Mustards, cabbages, radishes.
    Family Malvaceae. Mallows, cotton, okra, hibiscus.
    Family Solanaceae. Potatoes, eggplant, petunias.
    Family Salicaceae. Willows, poplars.
    Family Rosaceae. Roses, apples, almonds, strawberries.
    Family Fabaceae. Peas, beans, lupines, mesquite.
    Family Cactaceae. Cacti.
    Family Euphorbiaceae. Spurges, poinsettia.
    Family Cucurbitaceae. Gourds, melons, cucumbers, squashes.
    Family Apiaceae. Parsleys, carrots, poison hemlock.
    Family Aceraceae. Maples.
    Family Asteraceae. Composites. Chrysanthemums, sunflowers, lettuces, dandelions.
  Class Monocotyledonae. Monocotyledons (monocots). Some families of several different orders are listed:
    Family Liliaceae. Lilies, hyacinths, tulips, onions, garlic.
    Family Iridaceae. Irises, gladioli, crocuses.
    Family Orchidaceae. Orchids.
    Family Arecaceae. Date palms, coconut palms.
    Family Cyperaceae. Sedges.
    Family Poaceae. Grasses, bamboos, corn, wheat, sugarcane.
    Family Bromeliaceae. Bromeliads, pineapples, Spanish moss.

KINGDOM ANIMALIA.   Multicelled eukaryotes. Heterotrophs (herbivores, carnivores, omnivores, parasites, detritivores).

PHYLUM PLACOZOA.   Small, organless marine animal. *Trichoplax.*

PHYLUM MESOZOA.   Ciliated, wormlike parasites, about the same level of complexity as *Trichoplax.*

PHYLUM PORIFERA.   Sponges.

PHYLUM CNIDARIA.
  Class Hydrozoa. Hydrozoans. *Hydra, Obelia, Physalia.*
  Class Scyphozoa. Jellyfishes. *Aurelia.*
  Class Anthozoa. Sea anemones, corals. *Telesto.*

PHYLUM CTENOPHORA.   Comb jellies. *Pleurobrachia.*

PHYLUM PLATYHELMINTHES.   Flatworms.
  Class Turbellaria. Triclads (planarians), polyclads. *Dugesia.*
  Class Trematoda. Flukes. *Schistosoma.*
  Class Cestoda. Tapeworms. *Taenia.*

PHYLUM NEMERTEA.   Ribbon worms.

PHYLUM NEMATODA.   Roundworms. *Ascaris, Trichinella.*

PHYLUM ROTIFERA.   Rotifers.

PHYLUM MOLLUSCA.   Mollusks.
  Class Polyplacophora. Chitons.
  Class Gastropoda. Snails (periwinkles, whelks, limpets, abalones, cowries, conches, nudibranchs, tree snails, garden snails), sea slugs, land slugs.
  Class Bivalvia. Clams, mussels, scallops, cockles, oysters, shipworms.
  Class Cephalopoda. Squids, octopuses, cuttlefish, nautiluses. *Loligo.*

PHYLUM BRYOZOA.   Bryozoans (moss animals).

PHYLUM BRACHIOPODA.   Lampshells.

PHYLUM ANNELIDA.   Segmented worms.
  Class Polychaeta. Mostly marine worms.
  Class Oligochaeta. Mostly freshwater and terrestrial worms, but many marine. *Lumbricus* (earthworms).
  Class Hirudinea. Leeches.

PHYLUM TARDIGRADA.   Water bears.

PHYLUM ONYCHOPHORA.   Onychophorans. *Peripatus.*

PHYLUM ARTHROPODA.
  Subphylum Trilobita. Trilobites; extinct.
  Subphylum Chelicerata. Chelicerates. Horseshoe crabs, spiders, scorpions, ticks, mites.
  Subphylum Crustacea. Shrimps, crayfishes, lobsters, crabs, barnacles, copepods, isopods (sowbugs).
  Subphylum Uniramia.
    Superclass Myriapoda. Centipedes, millipedes.
    Superclass Insecta.
      Order Ephemeroptera. Mayflies.
      Order Odonata. Dragonflies, damselflies.
      Order Orthoptera. Grasshoppers, crickets, katydids.
      Order Dermaptera. Earwigs.
      Order Blattodea. Cockroaches.

Order Mantodea. Mantids.
Order Isoptera. Termites.
Order Mallophaga. Biting lice.
Order Anoplura. Sucking lice.
Order Homoptera. Cicadas, aphids, leafhoppers, spittlebugs.
Order Hemiptera. Bugs.
Order Coleoptera. Beetles.
Order Diptera. Flies.
Order Mecoptera. Scorpion flies. *Harpobittacus.*
Order Siphonaptera. Fleas.
Order Lepidoptera. Butterflies, moths.
Order Hymenoptera. Wasps, bees, ants.

PHYLUM ECHINODERMATA. Echinoderms.
  Class Asteroidea. Sea stars.
  Class Ophiuroidea. Brittle stars.
  Class Echinoidea. Sea urchins, heart urchins, sand dollars.
  Class Holothuroidea. Sea cucumbers.
  Class Crinoidea. Feather stars, sea lilies.
  Class Concentricycloidea. Sea daisies.

PHYLUM HEMICHORDATA. Acorn worms.

PHYLUM CHORDATA. Chordates.
  Subphylum Urochordata. Tunicates, related forms.
  Subphylum Cephalochordata. Lancelets.
  Subphylum Vertebrata. Vertebrates.
    Class Agnatha. Jawless vertebrates (lampreys, hagfishes).
    Class Placodermi. Jawed, heavily armored fishes; extinct.
    Class Chondrichthyes. Cartilaginous fishes (sharks, rays, skates, chimaeras).
    Class Osteichthyes. Bony fishes.
      Subclass Dipnoi. Lungfishes.
      Subclass Crossopterygii. Coelacanths, related forms.
      Subclass Actinopterygii. Ray-finned fishes.
        Order Acipenseriformes. Sturgeons, paddlefishes.
        Order Salmoniformes. Salmon, trout.
        Order Atheriniformes. Killifishes, guppies.
        Order Gasterosteiformes. Seahorses.
        Order Perciformes. Perches, wrasses, barracudas, tunas, freshwater bass, mackerels.
        Order Lophiiformes. Angler fishes.
    Class Amphibia. Mostly tetrapods; embryo enclosed in amnion.
      Order Caudata. Salamanders.
      Order Anura. Frogs, toads.
      Order Apoda. Apodans (caecilians).
    Class Reptilia. Skin with scales, embryo enclosed in amnion.
      Subclass Anapsida. Turtles, tortoises.
      Subclass Lepidosaura. *Sphenodon,* lizards, snakes.
      Subclass Archosaura. Dinosaurs (extinct), crocodiles, alligators.
    Class Aves. Birds. (In more recent schemes, dinosaurs, crocodilians, and birds are grouped in the same category.)
      Order Struthioniformes. Ostriches.
      Order Sphenisciformes. Penguins.
      Order Procellariiformes. Albatrosses, petrels.
      Order Ciconiiformes. Herons, bitterns, storks, flamingoes.
      Order Anseriformes. Swans, geese, ducks.
      Order Falconiformes. Eagles, hawks, vultures, falcons.
      Order Galliformes. Ptarmigan, turkeys, domestic fowl.
      Order Columbiformes. Pigeons, doves.
      Order Strigiformes. Owls.
      Order Apodiformes. Swifts, hummingbirds.
      Order Passeriformes. Sparrows, jays, finches, crows, robins, starlings, wrens.
    Class Mammalia. Skin with hair; young nourished by milk-secreting glands of adult.
      Subclass Prototheria. Egg-laying mammals (duckbilled platypus, spiny anteaters).
      Subclass Metatheria. Pouched mammals or marsupials (opossums, kangaroos, wombats).
      Subclass Eutheria. Placental mammals.
        Order Insectivora. Tree shrews, moles, hedgehogs.
        Order Scandentia. Insectivorous tree shrews.
        Order Chiroptera. Bats.
        Order Primates.
          Suborder Strepsirhini (prosimians). Lemurs, lorises.
          Suborder Haplorhini (tarsioids and anthropoids).
            Infraorder Tarsiiformes. Tarsiers.
            Infraorder Platyrrhini (New World monkeys).
              Family Cebidae. Spider monkeys, howler monkeys, capuchin.
            Infraorder Catarrhini (Old World monkeys and hominoids).
              Superfamily Cercopithecoidea. Baboons, macaques, langurs.
              Superfamily Hominoidea. Apes and humans.
                Family Hylobatidae. Gibbons.
                Family Pongidae. Chimpanzees, gorillas, orangutans.
                Family Hominidae. Humans and most recent ancestors of humans.
        Order Carnivora. Carnivores.
          Suborder Feloidea. Cats, civets, mongooses, hyenas.
          Suborder Canoidea. Dogs, weasels, skunks, otters, raccoons, pandas, bears.
        Order Proboscidea. Elephants; mammoths (extinct).
        Order Sirenia. Sea cows (manatees, dugongs).
        Order Perissodactyla. Odd-toed ungulates (horses, tapirs, rhinos).
        Order Artiodactyla. Even-toed ungulates (camels, deer, bison, sheep, goats, antelopes, giraffes).
        Order Edentata. Anteaters, tree sloths, armadillos.
        Order Tubulidentata. African aardvarks.
        Order Cetacea. Whales, porpoises.
        Order Rodentia. Most gnawing animals (squirrels, rats, mice, guinea pigs, porcupines).

APPENDIX II
UNITS OF MEASURE

Metric-English Conversions

Length

English		Metric
inch	=	2.54 centimeters
foot	=	0.30 meter
yard	=	0.91 meter
mile (5,280 feet)	=	1.61 kilometer

To convert	multiply by	to obtain
inches	2.54	centimeters
feet	30.00	centimeters
centimeters	0.39	inches
millimeters	0.039	inches

Weight

English		Metric
grain	=	64.80 milligrams
ounce	=	28.35 grams
pound	=	453.60 grams
ton (short) (2,000 pounds)	=	0.91 metric ton

To convert	multiply by	to obtain
ounces	28.3	grams
pounds	453.6	grams
pounds	0.45	kilograms
grams	0.035	ounces
kilograms	2.2	pounds

Volume

English		Metric
cubic inch	=	16.39 cubic centimeters
cubic foot	=	0.03 cubic meter
cubic yard	=	0.765 cubic meters
ounce	=	0.03 liter
pint	=	0.47 liter
quart	=	0.95 liter
gallon	=	3.79 liters

To convert	multiply by	to obtain
fluid ounces	30.00	milliliters
quart	0.95	liters
milliliters	0.03	fluid ounces
liters	1.06	quarts

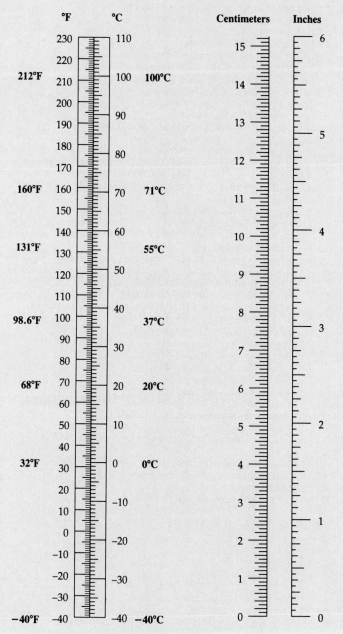

To convert temperature scales:

Fahrenheit to Celsius: $°C = 5/9\,(°F - 32)$

Celsius to Fahrenheit: $°F = 9/5\,(°C) + 32$

APPENDIX III
ANSWERS TO GENETIC PROBLEMS

CHAPTER 9

1. a. *AB*

   b. *AB* and *aB*

   c. *Ab* and *ab*

   d. *AB, aB, Ab,* and *ab*

2. a. All the offspring will be *AaBB*.

   b. 25% *AABB*; 25% *AaBB*; 25% *AABb*; 25% *AaBb*

   c. 25% *AaBb*; 25% *Aabb*; 25% *aaBb*; 25% *aabb*

   d.
1/16	*AABB*	(6.25%)
1/8	*AaBB*	(12.5%)
1/16	*aaBB*	(6.25%)
1/8	*AABb*	(12.5%)
1/4	*AaBb*	(25%)
1/8	*aaBb*	(12.5%)
1/16	*AAbb*	(6.25%)
1/8	*Aabb*	(12.5%)
1/16	*aabb*	(6.25%)

3. Yellow is recessive. Because the first-generation plants have a green phenotype and must be heterozygous, green must be dominant over the recessive yellow.

4. a. *ABC*

   b. *ABc* and *aBc*

   c. *ABC, aBC, ABc,* and *aBc*

   d. *ABC, aBC, AbC, abC, ABc, aBc, Abc,* and *abc*

5. The F_1 plants must all be heterozygous for both genes. When these plants self-pollinate, 1/4 (or 25%) of the F_2 plants will be heterozygous for both genes.

6. a. The mother must be heterozygous for both genes; the father is homozygous recessive for both of these genes. The first child is homozygous recessive for both genes.

   b. The probability that a second child will not be a tongue roller and will have detached earlobes is 1/4 (25%).

7. a. Bill: *Aa Ee Ss*
   Marie: *AA EE SS*

   b. The probability that each child will have Bill's phenotype is 100 percent. Because Marie's phenotype is the same as Bill's, the probability that each child will have Marie's phenotype is also 100 percent.

   c. Zero.

8. a. The mother must be heterozygous ($I^A i$). The male with type B blood could have fathered the child if he, too, were heterozygous ($I^B i$).

   b. If the male were heterozygous, then he *could* be the father. However, *any* male who carried the *i* allele also could be the father, so one can't say that this particular individual absolutely must be. This would include the males with type O blood (*ii*), the heterozygotes with type A blood ($I^A i$), and the heterozygotes with type B blood ($I^B i$).

9. The most direct way to do this would be to allow mice from a true-breeding white-furred strain to mate with mice from a true-breeding brown-furred strain. (True-breeding strains typically are homozygous for the alleles influencing the trait being studied.) All the F_1 offspring from this cross will be heterozygous for the alleles that influence brown or white fur color. The phenotype of the F_1 offspring should be recorded, then the offspring should be allowed to mate with one another to produce an F_2 population. Assuming that only a single gene locus is involved, several outcomes are possible, and they include the following:

   All F_1 offspring are brown. The F_2 offspring segregate 3 brown: 1 white. *Conclusion*: Segregation is occurring at a single locus where the allele specifying brown is dominant to the allele specifying white.

   All F_1 offspring are white. The F_2 offspring segregate 3 white: 1 brown. *Conclusion*: Segregation is occurring at a single locus where the allele specifying white is dominant to the allele specifying brown.

   All F_1 offspring are tan. The F_2 offspring segregate 1 brown: 2 tan: 1 white. *Conclusion*: Segregation is occurring at a single gene locus where the alleles show incomplete dominance.

10. If the allele that causes the hip disorder is recessive, you *cannot* guarantee that Dandelion's puppies will not carry the allele, inasmuch as Dandelion may be a heterozygous carrier of that allele.

11. Fred should allow his black guinea pig to mate with a white one. This would be a testcross.

    The white one has the genotype *ww*, and the black one might be homozygous dominant (*WW*) or heterozygous (*Ww*). If any offspring from the cross are white, then his black guinea pig is *Ww*. If all offspring are black after Fred allows enough matings to produce a significant number of offspring, then there is a good probability that his guinea pig is *WW*.

    (If ten of the offspring are all black and none is white, then there is about a 99.9 percent chance that his guinea pig is *WW*. The greater the number of offspring, the greater the degree of confidence in the conclusion.)

12. a. $R^1R^1 \times R^1R^2$ yields 1/2 red flowers, and 1/2 pink.

b. $R^1R^1 \times R^2R^2$ yields all pink flowers.

c. $R^1R^2 \times R^1R^2$ yields 1/4 red, 1/2 pink, and 1/4 white.

d. $R^1R^2 \times R^2R^2$ yields 1/2 pink and 1/2 white.

13. 9/16 walnut (pea-rose); 3/16 rose; 3/16 pea; 1/16 single.

14. Both parents are heterozygotes (Hb^AHb^S).

a. 1/4

b. 1/4

c. 1/2

15. The mating is M^LM. Genotypes of the offspring will be 1/4 MM, 1/2 M^LM, and 1/4 M^LM^L. Phenotypes will be:

1/4 homozygous dominant (MM) survivors

1/2 homozygous (M^LM) survivors

1/4 lethal (M^LM^L) nonsurvivors

The probability that any one kitten among the surviving progeny will be heterozygous is 2/3.

16. a. Both parents must be heterozygous (Aa). They may produce albino (aa) and unaffected (AA or Aa) children.

b. Both parents and all their children must be aa.

c. The albino male must be aa. Because there is an albino child, the female must be heterozygous Aa. If she were AA, they would have no albino children. The one albino child must be aa; the three unaffected children must be Aa. There is a 50 percent chance that any child will be albino. The 3 : 1 ratio is not surprising, given the small number of progeny.

17. a. All offspring of the mating will have an intermediate amount of pigmentation corresponding to genotype $A^1A^2B^1B^2$.

b. All possible genotypes could occur among offspring of this mating, in these expected proportions:

1/16	$A^1A^1B^1B^1$	(dark red)
1/8	$A^1A^1B^1B^2$	(medium-dark red)
1/16	$A^1A^1B^2B^2$	(medium red)
1/8	$A^1A^2B^1B^1$	(medium-dark red)
1/4	$A^1A^2B^1B^2$	(medium red)
1/8	$A^1A^2B^2B^2$	(light red)
1/16	$A^2A^2B^1B^1$	(medium red)
1/8	$A^2A^2B^1B^2$	(light red)
1/16	$A^2A^2B^2B^2$	(white)

CHAPTER 10

1. a. Males inherit their X chromosome from their mother.

b. With respect to an X-linked gene, a male can produce two kinds of gametes. One kind will have a Y chromosome without the gene. The other kind of gamete will contain an X chromosome—hence it will contain the X-linked gene.

c. All gametes of a female who is homozygous for an X-linked gene will have an X chromosome, hence that gene.

d. A female who is heterozygous for an X-linked gene will produce two kinds of gametes. One kind will have an X chromosome with one type of allele. The other kind will have an X chromosome with the other allele.

2. a. Because this gene is only carried on Y chromosomes, we would not expect females to have hairy pinnae, because females normally do not have Y chromosomes.

b. Because sons always inherit a Y chromosome from their father and daughters never do, a male having hairy pinnae will always transmit the gene for the trait to his sons and never to his daughters.

3. A 0 percent crossover frequency means that 50 percent of the gametes will be AB and 50 percent will be ab.

4. The rare vestigial-winged fruit flies may have arisen as a result of the x-rays, which might have induced a deletion of the dominant allele from one of their chromosomes. An alternative explanation is that the irradiation might have induced a mutation in the dominant allele.

5. The likelihood that crossing over will occur between the genes is low if they are very close together, and it is higher if the distance between them is greater.

6. a. Either anaphase I or anaphase II of meiosis.

b. If a translocation resulted in the attachment of most of chromosome 21 to the end of a normal chromosome 14, then the new individual would develop Down syndrome. The cells of that individual would contain the attached chromosome 21 in addition to two normal chromosomes 21. The chromosome number would still be 46.

7. By using c as the symbol for color blindness and C for normal color vision, the cross between these animals can be diagrammed as follows:

$$C(Y) \text{ female} \times Cc \text{ male}$$

In mugwumps, a son receives one sex-linked allele from each of his parents, but a daughter inherits her unpaired sex-linked allele from her father. In this cross, half of the sons will be CC and half Cc, but none will be color blind. Of the daughters, half will be $C(Y)$ and half $c(Y)$. There is a 50 percent chance that a daughter will be color blind. Note: this sex chromosome designation is not the same as it is in humans. But this is the way it is not only for mugwumps, but also for all birds, moths, butterflies, and some other organisms; the Y chromosome is associated with females, not males.

8. To produce a daughter who will be affected by childhood muscular dystrophy, the prospective father and the mother must carry the allele. Male with the allele have muscular dystrophy and rarely father children.

APPENDIX IV
ANSWERS TO SELF-QUIZZES

Chapter 1
1. cell
2. Metabolism
3. Homeostatsis
4. adaptive
5. mutation
6. d
7. d
8. d
9. c, e, d, b, a

Chapter 2
1. d
2. b
3. d
4. a
5. d
6. c, e, b, d, a

Chapter 3
1. c
2. d
3. d
4. d
5. e, d, a, b, c

Chapter 4
1. c
2. a
3. d
4. d
5. b
6. a
7. d
8. a
9. c
10. d
11. c, f, a, d, e, b

Chapter 5
1. Carbon dioxide; sunlight
2. d
3. c
4. b
5. d
6. c
7. c
8. c, d, e, b, a

Chapter 6
1. d
2. c
3. b
4. c
5. b
6. d
7. c
8. b
9. d
10. b, c, a, d

Chapter 7
1. b
2. a
3. b
4. b
5. c
6. a
7. d
8. b
9. d, b, c, a

Chapter 8
1. d
2. d
3. a
4. b
5. d
6. c
7. c
8. d
9. d, a, c, b

Chapter 9
1. a
2. b
3. a
4. c
5. b
6. a
7. d
8. b, d, a, c

Chapter 10
1. c
2. c
3. e
4. c
5. d
6. d
7. d
8. c, e, d, b, a

Chapter 11
1. c
2. d
3. d
4. c
5. a
6. d
7. b, e, d, c, a

Chapter 12
1. d
2. b
3. c
4. c
5. a
6. d
7. d
8. d
9. d
10. b
11. e, d, c, b, a

Chapter 13
1. Plasmids
2. c
3. a
4. b
5. a
6. b
7. b
8. d
9. b
10. d, c, f, e, b, a

Chapter 14
1. populations
2. c
3. b
4. c
5. c, a, d, b

Chapter 15
1. e
2. d
3. d
4. d
5. c
6. a
7. c
8. c
9. b
10. d, a, b, e, c

Chapter 16
1. b
2. a
3. b
4. c
5. c
6. e, b, f, c, a, d

Chapter 17
1. e
2. c
3. b
4. d
5. c
6. b
7. d
8. b, d, e, a, c

Chapter 18
1. c
2. d
3. d
4. d, a, b, e, c

Chapter 19
1. b
2. b
3. b
4. c
5. b
6. a
7. a
8. c, e, g, h, f, a, b, d
9. d, f, c, e, a, b

Chapter 20
1. b
2. c
3. a
4. a
5. b
6. c
7. c
8. d, f, c, g, e, i, j, h, a, b

Chapter 21
1. b
2. b
3. f
4. d
5. a
6. d
7. d, e, c, b, g, f, a

Chapter 22
1. b
2. a
3. b
4. c
5. a
6. d
7. d
8. c
9. d
10. b, d, e, c, f, a

Chapter 23
1. hydrogen bonds
2. e
3. b
4. a
5. c
6. d
7. c
8. d
9. d
10. b
11. b, d, f, c, a, e

Chapter 24
1. d
2. c
3. b
4. b
5. b
6. a
7. b
8. d
9. a
10. c
11. c
12. a, c, d, e, b
13. a, e, c, d, b

Chapter 25
1. d
2. c
3. b
4. b
5. c
6. b
7. a
8. d
9. c
10. a
11. receptors, integrator
    effectors
12. b, a, c, f, g, e, d

Chapter 26
1. c
2. d
3. d
4. b
5. d
6. f
7. h, f, g, e, a, c, b, i, d

Chapter 27
1. c
2. d
3. d
4. c
5. c
6. d
7. b
8. d
9. c
10. e, b, a, d, c
11. d, e, a, f, c, b

Chapter 28
1. e
2. d
3. c
4. d
5. a
6. d
7. d
8. b
9. a
10. c, b, a, e, d, f

Chapter 29
1. d
2. d
3. d
4. b
5. c
6. e
7. b
8. c
9. d
10. d, b, f, e, a, c

Chapter 30
1. e
2. d
3. caloric intake;
    energy output
4. a
5. d
6. b
7. c
8. a
9. b
10. d, c, e, g, f, b, a

Chapter 31
1. d
2. f
3. b
4. a
5. c
6. d
7. d
8. d
9. d, e, a, c, b
10. f, e, g, b, c, d, a

Chapter 32
1. d
2. b
3. d
4. a
5. d
6. b
7. b, d, a, c
8. e, d, b, c, a

Chapter 33
1. f
2. d
3. b
4. d
5. e
6. b
7. b
8. c
9. d, f, c, e, a, b

Chapter 34
1. d
2. a
3. c
4. e
5. b
6. hypothalamns
7. f
8. c
9. c
10. c
11. c
12. c
13. f
14. e
15. e, c, a, b, d

Chapter 35
1. Ecology
2. population
3. e
4. a
5. a
6. d
7. e
8. c
9. c, d, b, a

Chapter 36
1. e
2. e
3. b
4. d
5. b
6. d
7. d
8. a
9. c, d, a, e, b

Chapter 37
1. a
2. d
3. d
4. d, a, c, b

Chapter 38
1. c
2. c
3. d
4. b
5. d
6. d
7. d
8. d
9. b, f, e, d, c, a
10. e, a, d, c, b

Chapter 39
1. c
2. c
3. b
4. d
5. c
6. b
7. c
8. d
9. d
10. d, b, c, a

Chapter 40
1. a
2. d
3. b
4. b
5. c
6. c
7. c
8. d
9. d, e, b, a, c

APPENDIX V
A CLOSER LOOK AT SOME MAJOR METABOLIC PATHWAYS

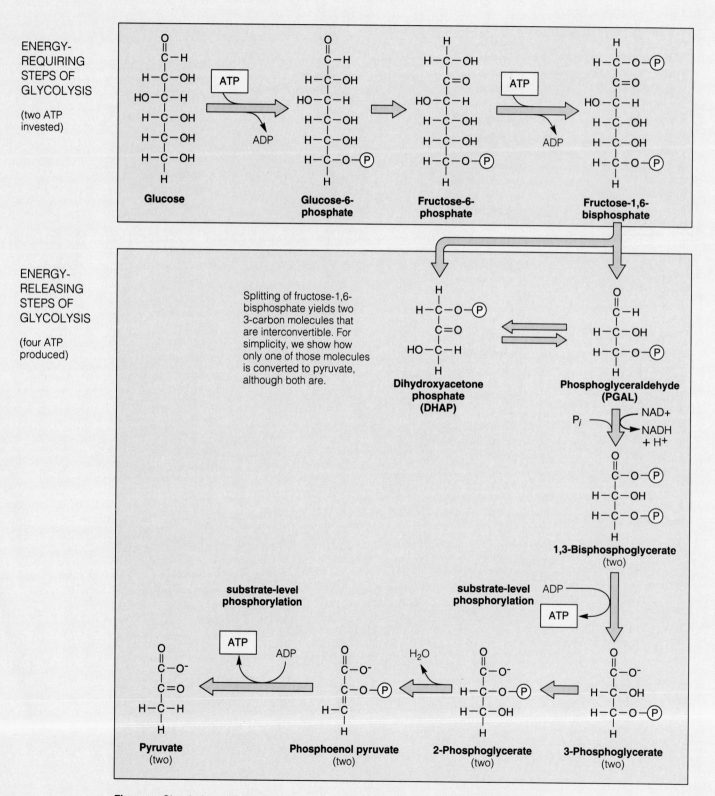

ENERGY-REQUIRING STEPS OF GLYCOLYSIS

(two ATP invested)

ENERGY-RELEASING STEPS OF GLYCOLYSIS

(four ATP produced)

Glucose

Glucose-6-phosphate

Fructose-6-phosphate

Fructose-1,6-bisphosphate

Splitting of fructose-1,6-bisphosphate yields two 3-carbon molecules that are interconvertible. For simplicity, we show how only one of those molecules is converted to pyruvate, although both are.

Dihydroxyacetone phosphate (DHAP)

Phosphoglyceraldehyde (PGAL)

1,3-Bisphosphoglycerate (two)

substrate-level phosphorylation

3-Phosphoglycerate (two)

2-Phosphoglycerate (two)

Phosphoenol pyruvate (two)

substrate-level phosphorylation

Pyruvate (two)

Figure a Glycolysis, ending with two 3-carbon pyruvate molecules for each 6-carbon glucose entering the reactions. The *net* energy yield is two ATP molecules (two invested, four produced).

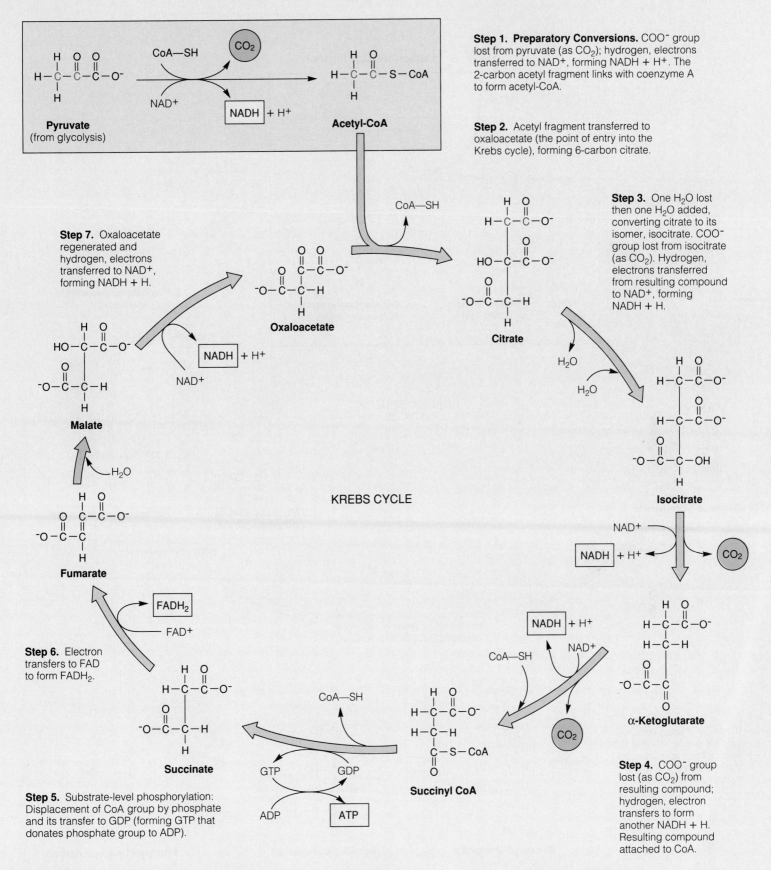

Step 1. Preparatory Conversions. COO^- group lost from pyruvate (as CO_2); hydrogen, electrons transferred to NAD^+, forming $NADH + H^+$. The 2-carbon acetyl fragment links with coenzyme A to form acetyl-CoA.

Step 2. Acetyl fragment transferred to oxaloacetate (the point of entry into the Krebs cycle), forming 6-carbon citrate.

Step 3. One H_2O lost then one H_2O added, converting citrate to its isomer, isocitrate. COO^- group lost from isocitrate (as CO_2). Hydrogen, electrons transferred from resulting compound to NAD^+, forming $NADH + H$.

Step 7. Oxaloacetate regenerated and hydrogen, electrons transferred to NAD^+, forming $NADH + H$.

Step 6. Electron transfers to FAD to form $FADH_2$.

Step 5. Substrate-level phosphorylation: Displacement of CoA group by phosphate and its transfer to GDP (forming GTP that donates phosphate group to ADP).

Step 4. COO^- group lost (as CO_2) from resulting compound; hydrogen, electron transfers to form another $NADH + H$. Resulting compound attached to CoA.

KREBS CYCLE

Pyruvate (from glycolysis)

Acetyl-CoA

Oxaloacetate

Citrate

Isocitrate

α-Ketoglutarate

Succinyl CoA

Succinate

Fumarate

Malate

Figure b Krebs cycle (citric acid cycle). Red identifies carbon entering the cycle by way of acetyl-CoA.

A11

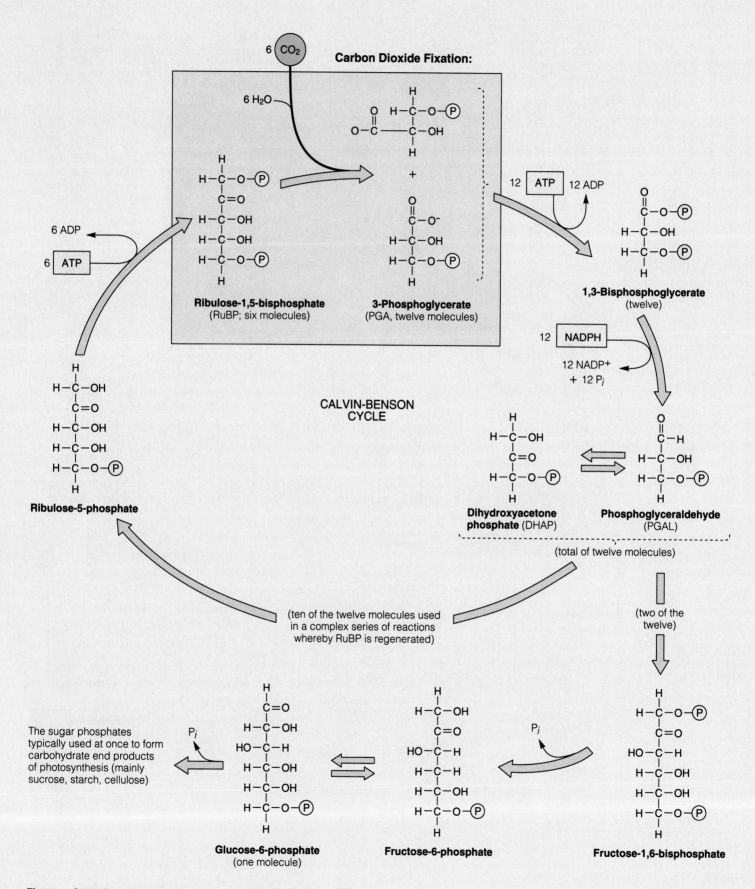

Figure c Calvin-Benson cycle of the light-independent reactions of photosynthesis.

APPENDIX VI
PERIODIC TABLE OF THE ELEMENTS

Group

Noble Gases

Atomic number → 11
Symbol → Na
Atomic mass → 22.99

Atomic masses are based on carbon-12. Numbers in parentheses are mass numbers of most stable or best known isotopes of radioactive elements.

Transition Elements

	IA(1)	IIA(2)	IIIB(3)	IVB(4)	VB(5)	VIB(6)	VIIB(7)	VIII (8)	VIII (9)	VIII (10)	IB(11)	IIB(12)	IIIA(13)	IVA(14)	VA(15)	VIA(16)	VIIA(17)	(18)
1	1 H 1.008																	2 He 4.003
2	3 Li 6.941	4 Be 9.012											5 B 10.81	6 C 12.01	7 N 14.01	8 O 16.00	9 F 19.00	10 Ne 20.18
3	11 Na 22.99	12 Mg 24.31											13 Al 26.98	14 Si 28.09	15 P 30.97	16 S 32.06	17 Cl 35.45	18 Ar 39.95
4	19 K 39.10	20 Ca 40.08	21 Sc 44.96	22 Ti 47.90	23 V 50.94	24 Cr 52.00	25 Mn 54.94	26 Fe 55.85	27 Co 58.93	28 Ni 58.7	29 Cu 63.55	30 Zn 65.38	31 Ga 69.72	32 Ge 72.59	33 As 74.92	34 Se 78.96	35 Br 79.90	36 Kr 83.80
5	37 Rb 85.47	38 Sr 87.62	39 Y 88.91	40 Zr 91.22	41 Nb 92.91	42 Mo 95.94	43 Tc 98.91	44 Ru 101.1	45 Rh 102.9	46 Pd 106.4	47 Ag 107.9	48 Cd 112.4	49 In 114.8	50 Sn 118.7	51 Sb 121.8	52 Te 127.6	53 I 126.9	54 Xe 131.3
6	55 Cs 132.9	56 Ba 137.3	57* La 138.9	72 Hf 178.5	73 Ta 180.9	74 W 183.9	75 Re 186.2	76 Os 190.2	77 Ir 192.2	78 Pt 195.1	79 Au 197.0	80 Hg 200.6	81 Tl 204.4	82 Pb 207.2	83 Bi 209.0	84 Po (210)	85 At (210)	86 Rn (222)
7	87 Fr (223)	88 Ra 226.0	89** Ac (227)	104 Unq (261)	105 Unp (262)	106 Unh (263)	107 Uns (262)	108 Uno (265)	109 Une (266)									

Period

Inner Transition Elements

Lanthanide Series 6 *	58 Ce 140.1	59 Pr 140.9	60 Nd 144.2	61 Pm (145)	62 Sm 150.4	63 Eu 152.0	64 Gd 157.3	65 Tb 158.9	66 Dy 162.5	67 Ho 164.9	68 Er 167.3	69 Tm 168.9	70 Yb 173.0	71 Lu 175.0
Actinide Series 7 **	90 Th 232.0	91 Pa 231.0	92 U 238.0	93 Np 237.0	94 Pu (244)	95 Am (243)	96 Cm (247)	97 Bk (247)	98 Cf (251)	99 Es (252)	100 Fm (257)	101 Md (258)	102 No (259)	103 Lr (260)

A GLOSSARY OF BIOLOGICAL TERMS

ABO blood typing Method of characterizing an individual's blood according to whether one or both of two protein markers, A and B, are present at the surface of red blood cells. The O signifies that neither marker is present.

abortion Spontaneous or induced expulsion of the embryo or fetus from the uterus.

abscisic acid (ab-SISS-ik) Plant hormone that promotes stomatal closure, bud dormancy, and seed dormancy.

abscission (ab-SIH-zhun) [L. *abscindere*, to cut off] The dropping of leaves, flowers, fruits, or other plant parts due to hormonal action.

absorption Of complex animals, the movement of nutrients, fluid, and ions across the gut lining and into the internal environment.

accessory pigment A light-trapping pigment that contributes to photosynthesis by extending the range of usable wavelengths beyond those absorbed by the chlorophylls.

acid [L. *acidus*, sour] A substance that releases hydrogen ions (H^+) in water.

acid rain The falling to earth of snow or rain that contains sulfur and nitrogen oxides. Also called wet acid deposition (as opposed to dry acid deposition, or the falling to earth of airborne particles of sulfur and nitrogen oxides).

acoelomate (ay-SEE-luh-mate) Type of animal that has no fluid-filled cavity between the gut and body wall.

acoustical signal Sounds that are used as a communication signal.

actin (AK-tin) A globular contractile protein. In muscle cells, actin interacts with another protein, myosin, to bring about contraction.

action potential An abrupt, brief reversal in the steady voltage difference across the plasma membrane (that is, the resting membrane potential) of a neuron and some other cells.

activation energy The minimum amount of collision energy required to bring reactant molecules to an activated condition (the transition state) at which a reaction will proceed spontaneously. Enzymes enhance reaction rates by lowering the activation energy (they put substrates on a precise collision course).

active site A crevice on the surface of an enzyme molecule where a specific reaction is catalyzed.

active transport The pumping of one or more specific solutes through a transport protein that spans the lipid bilayer of a cell membrane. Most often, the solute is transported against its concentration gradient. The protein is activated by an energy boost, as from ATP.

adaptation [L. *adaptare*, to fit] In evolutionary biology, the process of becoming adapted (or more adapted) to a given set of environmental conditions. Of sensory neurons, a decrease in the frequency of action potentials (or their cessation) even when a stimulus is maintained at constant strength.

adaptive behavior A behavior that promotes the propagation of an individual's genes and that tends to increase in frequency in a population over time.

adaptive radiation A burst of speciation events, with lineages branching away from one another as they partition the existing environment or invade new ones.

adaptive trait Any aspect of form, function, or behavior that helps an organism survive and reproduce under a given set of environmental conditions.

adaptive zone A way of life, such as "catching insects in the air at night." A lineage must have physical, ecological, and evolutionary access to an adaptive zone to become a successful occupant of it.

adenine (AH-de-neen) A purine; a nitrogen-containing base found in nucleotides.

adenosine diphosphate (ah-DEN-uh-seen die-FOSS-fate) ADP, a molecule involved in cellular energy transfers; typically formed by hydrolysis of ATP.

adenosine phosphates Any of several relatively small molecules, some of which function as chemical messengers within and between cells, and others that function as energy carriers.

adenosine triphosphate *See* ATP.

ADH Antidiuretic hormone. Produced by the hypothalamus and released by the posterior pituitary, it stimulates reabsorption in the kidneys and so reduces urine volume.

adipose tissue A type of connective tissue having an abundance of fat-storing cells and blood vessels for transporting fats.

ADP Adenosine diphosphate. A nucleotide coenzyme that accepts unbound phosphate or a phosphate group to become ATP.

ADP/ATP cycle In cells, a mechanism of ATP renewal. When ATP donates a phosphate group to other molecules (and so energizes them), it reverts to ADP, then forms again by phosphorylation of ADP.

adrenal cortex (ah-DREE-nul) Outer portion of the adrenal gland; its hormones have roles in metabolism, inflammation, maintaining extracellular fluid volume, and other functions.

adrenal medulla Inner region of the adrenal gland; its hormones help control blood circulation and carbohydrate metabolism.

aerobic respiration (air-OH-bik) [Gk. *aer*, air, + *bios*, life] The main energy-releasing metabolic pathway of ATP formation, in which oxygen is the final acceptor of electrons stripped from glucose or some other organic compound. The pathway proceeds from glycolysis through the Krebs cycle and electron transport phosphorylation. A typical net yield is 36 ATP for each glucose molecule.

age structure Of a population, the number of individuals in each of several or many age categories.

agglutination (ah-glue-tin-AY-shun) Clumping together of foreign cells that have invaded the body (as pathogens or in tissue grafts or transplants). Clumping is induced by cross-linking between antibody molecules that have already latched onto antigen at the surface of the foreign cells.

aging A range of processes, including the breakdown of cell structure and function, by which the body gradually deteriorates. All organisms showing extensive cell differentiation undergo aging.

AIDS Acquired immunodeficiency syndrome. A set of chronic disorders following infection by the human immunodeficiency virus (HIV), which destroys key cells of the immune system.

alcoholic fermentation Anaerobic pathway of ATP formation in which pyruvate from glycolysis is broken down to acetaldehyde, which accepts electrons from NADH to become ethanol, and NAD^+ is regenerated. Its net yield is two ATP.

aldosterone (al-DOSS-tuh-rohn) Hormone secreted by the adrenal cortex that helps regulate sodium reabsorption.

allantois (ah-LAN-twahz) [Gk. *allas*, sausage] Of vertebrates, one of four extraembryonic membranes that form during embryonic development. It functions in respiration and storage of metabolic wastes in reptiles, birds, and some mammals. In humans, it functions in early blood formation and development of the urinary bladder.

allele (uh-LEEL) For a given location on a chromosome, one of two or more slightly different molecular forms of a gene that code for different versions of the same trait.

allele frequency Of a given gene locus, the relative abundances of each kind of allele carried by the individuals of a population.

allergy An immune response made against a normally harmless substance.

allopatric speciation [Gk. *allos*, different, + *patria*, native land] Speciation that follows geographic isolation of populations of the same species.

allosteric control (AL-oh-STARE-ik) Control over a metabolic reaction or pathway that operates through the binding of a specific substance at a control site on a specific enzyme.

alpine tundra A type of biome that exists at high elevations in mountains throughout the world.

altruistic behavior (al-true-ISS-tik) Self-sacrificing behavior; the individual behaves in a way that helps others but decreases its own chances of reproductive success.

alveolar sac (al-VEE-uh-lar) Any of the pouch-like clusters of alveoli in the lungs; the major sites of gas exchange.

alveolus (ahl-VEE-uh-lus), plural **alveoli** [L. *alveus*, small cavity] Any of the many cup-shaped, thin-walled outpouchings of respiratory bronchioles. A site where oxygen diffuses from air in the lungs to the blood, and carbon dioxide diffuses from blood to the lungs.

amino acid (uh-MEE-no) A small organic molecule having a hydrogen atom, an amino group, an acid group, and an R group covalently bonded to a central carbon atom. The subunit of polypeptide chains, which represent the primary structure of proteins.

ammonification (uh-moan-ih-fih-KAY-shun) Together with decomposition, a process by which certain bacteria and fungi break down nitrogen-containing wastes and remains of other organisms.

amnion (AM-nee-on) Of land vertebrates, one of four extraembryonic membranes. It becomes a fluid-filled sac in which the embryo (and fetus) can grow, move freely, and be protected from sudden temperature shifts and impacts.

amniote egg A type of egg, often with a leathery or calcified shell, that contains extraembryonic membranes, including the amnion. An adaptation that figured in the vertebrate invasion of land.

amphibian A type of vertebrate somewhere between fishes and reptiles in body plan and reproductive mode; salamanders, frogs and toads, and caecilians are the existing groups.

anaerobic pathway (an-uh-ROW-bik) [Gk. *an*, without, + *aer*, air] Metabolic pathway in which a substance other than oxygen serves as the final acceptor of electrons that have been stripped from substrates.

analogous structures Body parts, once different in separate lineages, that were put to comparable uses in similar environments and that came to resemble one another in form and function. They are evidence of morphological convergence.

anaphase (AN-uh-faze) The stage at which microtubules of a spindle apparatus separate sister chromatids of each chromosome and move them to opposite spindle poles. During anaphase I of *meiosis*, the two members of each pair of homologous chromosomes separate. During anaphase II, sister chromatids of each chromosome separate.

aneuploidy (AN-yoo-ploy-dee) A change in the chromosome number following inheritance of one extra or one less chromosome.

angiosperm (AN-gee-oh-spurm) [Gk. *angeion*, vessel, and *spermia*, seed] A flowering plant.

animal A heterotroph that eats or absorbs nutrients from other organisms; is multicelled, usually with tissues arranged in organs and organ systems; is usually motile during at least part of the life cycle; and goes through a period of embryonic development.

annelid The type of invertebrate classified as a segmented worm; an oligochaete (such as an earthworm), leech, or polychaete.

annual plant A flowering plant that completes its life cycle in one growing season.

anther [Gk. *anthos*, flower] In flowering plants, the pollen-bearing part of the male reproductive structure (stamen).

antibiotic A normal metabolic product of certain microorganisms that kills or inhibits the growth of other microorganisms.

antibody [Gk. *anti*, against] Any of a variety of Y-shaped receptor molecules with binding sites for specific antigens. Only B cells produce antibodies, then position them at their surface or secrete them.

anticodon In a tRNA molecule, a sequence of three nucleotide bases that can pair with an mRNA codon.

antigen (AN-tih-jen) [Gk. *anti*, against, + *genos*, race, kind] Any molecular configuration that is recognized as foreign to the body and that triggers an immune response. Most antibodies are protein molecules at the surface of infectious agents or tumor cells.

antigen-presenting cell A macrophage or some other white blood cell that engulfs and digests antigen, then displays it with certain MHC molecules at its surface. Recognition of antigen-MHC complexes by T and B cells triggers an immune response.

aorta (ay-OR-tah) [Gk. *airein*, to lift, heave] Main artery of systemic circulation; carries oxygenated blood away from the heart to all body regions except the lungs.

apical dominance The inhibitory influence of a terminal bud on the growth of lateral buds.

apical meristem (AY-pih-kul MARE-ih-stem) [L. *apex*, top, + Gk. *meristos*, divisible] Of most plants, a mass of self-perpetuating cells responsible for primary growth (elongation) at root and shoot tips.

appendicular skeleton (ap-en-DIK-yoo-lahr) In vertebrates, bones of the limbs, hips, and shoulders.

appendix A slender projection from the cup-shaped pouch at the start of the colon.

archaebacteria One of three great prokaryotic lineages that arose early in the history of life; now represented by methanogens, halophiles, and thermophiles.

arctic tundra A type of biome that lies between the polar ice cap and boreal forests of North America, Europe, and Asia.

arteriole (ar-TEER-ee-ole) Any of the blood vessels between arteries and capillaries. They are control points where the volume of blood delivered to different body regions can be adjusted.

artery Any of the large-diameter blood vessels that conduct oxygen-poor blood to the lungs and oxygen-enriched blood to all body tissues. Their thick, muscular wall allows them to smooth out pulsations in blood pressure caused by heart contractions.

arthropod An invertebrate having a hardened exoskeleton, specialized segments, and jointed appendages. Spiders, crabs, and insects are examples.

asexual reproduction Mode of reproduction by which offspring arise from a single parent, and inherit the genes of that parent only.

atmosphere A region of gases, airborne particles, and water vapor enveloping the earth; 80 percent of its mass is distributed within seventeen miles of the earth's surface.

atmospheric cycle A biogeochemical cycle in which the atmosphere is the largest reservoir of an element. The carbon and nitrogen cycles are examples.

atom The smallest unit of matter that is unique to a particular element.

atomic number The number of protons in the nucleus of each atom of an element; it differs for each element.

ATP Adenosine triphosphate (ah-DEN-uh-seen try-FOSS-fate). A nucleotide composed of adenine, ribose, and three phosphate groups. As the main energy carrier in cells, it directly or indirectly delivers energy to or picks up energy from nearly all metabolic pathways.

atrium (AYE-tree-um) Of the human heart, one of two chambers that receive blood. *Compare* ventricle.

australopith (OHSS-trah-low-pith) [L. *australis*, southern, + Gk. *pithekos*, ape] Any of the earliest known species of hominids, that is, the first species of the evolutionary branch leading to humans.

autoimmune response Misdirected immune response in which lymphocytes mount an attack against normal body cells.

autonomic nervous system (auto-NOM-ik) Those nerves leading from the central nervous system to the smooth muscle, cardiac muscle, and glands of internal organs and structures, that is, to the visceral portion of the body.

autosomal dominant inheritance Condition arising from the presence of a dominant allele on an autosome (not a sex chromosome). The allele is always expressed to some extent, even in heterozygotes.

autosomal recessive inheritance Condition arising from a recessive allele on an autosome (not a sex chromosome). Only recessive homozygotes show the resulting phenotype.

autosome Any of the chromosomes that are of the same number and kind in both males and females of the species.

autotroph (AH-toe-trofe) [Gk. *autos*, self, + *trophos*, feeder] An organism able to build its own large organic molecules by using carbon dioxide and energy from the physical environment. Photosynthetic autotrophs use sunlight energy; chemosynthetic autotrophs extract energy from chemical reactions involving inorganic substances. *Compare* heterotroph.

auxin (AWK-sin) Any of a class of growth-regulating hormones in plants; auxins promote stem elongation as one effect.

axial skeleton (AX-ee-uhl) In vertebrates, the skull, backbone, ribs, and breastbone (sternum).

axon Of a neuron, a long, cylindrical extension from the cell body, with finely branched endings. Action potentials move rapidly, without alteration, along an axon; their arrival at axon endings may trigger the release of neurotransmitter molecules that influence an adjacent cell.

B lymphocyte, or **B cell** The only white blood cell that produces antibodies, then positions them at the cell surface or secretes them as weapons in immune responses.

bacterial conjugation The transfer of plasmid DNA from one bacterial cell to another.

bacterial flagellum Of many bacterial cells, a whiplike motile structure that does not contain a core of microtubules.

bacteriophage (bak-TEER-ee-oh-fahj) [Gk. *baktērion*, small staff, rod, + *phagein*, to eat] Category of viruses that infect bacterial cells.

balanced polymorphism The maintenance of two or more forms of a trait in fairly stable proportions over generations.

Barr body In the cells of female mammals, a condensed X chromosome that was inactivated during early embryonic development.

basal body A centriole which, after having given rise to the microtubules of a flagellum or cilium, remains attached to its base in the cytoplasm.

base A substance that releases OH⁻ in water.

base pair A pair of hydrogen-bonded nucleotide bases in two strands of nucleic acids. In a DNA double helix, adenine pairs with thymine, and guanine with cytosine. When an mRNA strand forms on a DNA strand during transcription, uracil (U) pairs with the DNA's adenine.

base sequence The particular order in which one nucleotide base follows the next in a strand of DNA or RNA. The order differs to some extent for each kind of organism.

basophil Fast-acting white blood cells that secrete histamine and other substances during inflammation.

behavior, animal A response to external and internal stimuli, following integration of sensory, neural, endocrine, and effector components. Because behavior has a genetic basis, it is subject to natural selection and commonly can be modified through experience.

benthic province All of the sediments and rocky formations of the ocean bottom; begins with the continental shelf and extends down through deep-sea trenches.

biennial A flowering plant that lives through two growing seasons.

bilateral symmetry Body plan in which the left and right halves of the animal are mirror-images of each other.

binary fission Of bacteria, a mode of asexual reproduction in which the parent cell replicates its single chromosome, then divides into two genetically identical daughter cells.

biogeochemical cycle The movement of an element such as carbon or nitrogen from the environment to organisms, then back to the environment.

biogeographic realm [Gk. *bios*, life, + *geographein*, to describe the surface of the earth] Any of six major land regions, each having distinguishing types of plants and animals and generally retaining its identity because of climate and geographic barriers to gene flow.

biological clocks Internal time-measuring mechanisms that have roles in adjusting an organism's daily activities, seasonal activities, or both in response to environmental cues.

biological magnification The increasing concentration of a nondegradable or slowly degradable substance in body tissues as it is passed along food chains.

biological systematics Branch of biology that assesses patterns of diversity based on information from taxonomy, phylogenetic reconstruction, and classification.

bioluminescence A flashing of light that emanates from an organism when excited electrons of luciferins, or highly fluorescent substances, return to a lower energy level.

biomass The combined weight of all the organisms at a particular trophic (feeding) level in an ecosystem.

biome A broad, vegetational subdivision of a biogeographic realm shaped by climate, topography, and composition of regional soils.

biosphere [Gk. *bios*, life, + *sphaira*, globe] All regions of the earth's waters, crust, and atmosphere in which organisms live.

biosynthetic pathway A metabolic pathway in which small molecules are assembled into lipids, proteins, and other large organic molecules.

biotic potential Of a population, the maximum rate of increase per individual under ideal conditions.

bipedalism A habitual standing and walking on two feet, as by ostriches and humans.

bird A type of vertebrate, the only one having feathers, with strong resemblances and evolutionary connections to reptiles.

blastocyst (BLASS-tuh-sist) [Gk. *blastos*, sprout, + *kystis*, pouch] In mammalian development, a blastula stage consisting of a hollow ball of surface cells and an inner cell mass.

blastula (BLASS-chew-lah) An embryonic stage consisting of a ball of cells produced by cleavage.

blood A fluid connective tissue composed of water, solutes, and formed elements (blood cells and platelets); it carries substances to and from cells and helps maintain an internal environment that is favorable for cell activities.

blood pressure Fluid pressure, generated by heart contractions, that keeps blood circulating.

blood-brain barrier Set of mechanisms that helps control which blood-borne substances reach neurons in the brain.

bone The mineral-hardened connective tissue of bones.

bones In vertebrate skeletons, organs that function in movement and locomotion, pro-tection of other organs, mineral storage, and (in some bones) blood cell production.

bottleneck An extreme case of genetic drift. A catastrophic decline in population size leads to a random shift in the allele frequencies among survivors. Because the population must rebuild from so few individuals, severely limited genetic variation may be an outcome.

Bowman's capsule The cup-shaped portion of a nephron that receives water and solutes filtered from blood.

brain Of most nervous systems, the most complex integrating center; it receives, processes, and integrates sensory input and issues coordinated commands for response.

brainstem The vertebrate midbrain, pons, and medulla oblongata, the core of which contains the reticular formation that helps govern activity of the nervous system as a whole.

bronchiole Of most vertebrates, a component of the finely branched bronchial tree inside each lung.

bronchus, plural **bronchi** (BRONG-CUSS, BRONG-kee) [Gk. *bronchos*, windpipe] Tubelike branchings of the trachea that lead into the lungs of most vertebrates.

brown alga A type of aquatic plant, found in nearly all marine habitats, that has an abundance of xanthophyll pigments.

bryophyte A nonvascular land plant that requires free water to complete its life cycle.

bud An undeveloped shoot of mostly meristematic tissue; often covered and protected by scales (modified leaves).

buffer A substance that can combine with hydrogen ions, release them, or both. Buffers help stabilize the pH of blood and other fluids.

bulk Of human digestion, a volume of fiber and other undigested material that absorption processes in the colon cannot decrease.

bulk flow In response to a pressure gradient, a movement of more than one kind of molecule in the same direction in the same medium (as in blood, sap, or air).

C4 pathway Of many plants, a pathway of photosynthesis in which carbon dioxide is fixed twice, in two different cell types. Carbon dioxide accumulates in the leaf and helps counter photorespiration. The first compound formed is the 4-carbon oxaloacetate.

Calvin-Benson cycle Cyclic reactions that are the "synthesis" part of the light-independent reactions of photosynthesis. In land plants, RuBP, or some other compound to which carbon has been affixed, undergoes rearrangements that lead to formation of a sugar phosphate and to regeneration of the RuBP. The cycle runs on ATP and NADPH from light-dependent reactions.

CAM plant A plant that conserves water by opening stomata only at night, when it fixes carbon dioxide by way of a C4 pathway.

cambium (KAM-bee-um), plural **cambia** In vascular plants, one of two types of meristems that are responsible for secondary growth (increases in stem and root diameter). Vascular cambium gives rise to secondary xylem and phloem; cork cambium gives rise to periderm.

camouflage An outcome of an organism's form, patterning, color, or behavior that helps it blend with its surroundings and escape detection.

cancer A type of malignant tumor, the cells of which show profound abnormalities in the plasma membrane and cytoplasm, abnormal growth and division, and weakened capacity for adhesion within the parent tissue (leading to metastasis). Unless eradicated, cancer is lethal.

canine A pointed tooth that functions in piercing food.

capillary [L. *capillus*, hair] A thin-walled blood vessel that functions in the exchange of gases and other substances between blood and interstitial fluid.

capillary bed A diffusion zone, consisting of numerous capillaries, where substances are exchanged between blood and interstitial fluid.

carbohydrate [L. *carbo*, charcoal, + *hydro*, water] A simple sugar or large molecule composed of sugar units. All cells use carbohydrates as structural materials, energy stores, and transportable forms of energy. The three classes of carbohydrates include: monosaccharides, oligosaccharides, or polysaccharides.

carbon cycle A biogeochemical cycle in which carbon moves from its largest reservoir in the atmosphere, through oceans and organisms, then back to the atmosphere.

carbon dioxide fixation First step of the light-independent reactions of photosynthesis. Carbon (from carbon dioxide) becomes affixed to a carbon compound (such as RuBP) that can enter the Calvin-Benson cycle.

carcinogen (kar-SIN-uh-jen) An environmental agent or substance, such as ultraviolet radiation, that can trigger cancer.

cardiac cycle (KAR-dee-ak) [Gk. *kardia*, heart, + *kyklos*, circle] The sequence of muscle contraction and relaxation constituting one heartbeat.

cardiac pacemaker Sinoatrial (SA) node; the basis of the normal rate of heartbeat. The self-excitatory cardiac muscle cells that spontaneously generate rhythmic waves of excitation over the heart chambers.

cardiovascular system Of most animals, an organ system that is composed of blood, one or more hearts, and blood vessels and that functions in the rapid transport of substances to and from cells.

carnivore [L. *caro*, *carnis*, flesh, + *vovare*, to devour] An animal that eats other animals; a type of heterotroph.

carotenoids (kare-OTT-en-oyds) Light-sensitive, accessory pigments that transfer absorbed energy to chlorophylls. They absorb violet and blue wavelengths but transmit red, orange, and yellow.

carpel (KAR-pul) The female reproductive part of a flower; sometimes called a pistil. The lower portion of a single carpel (or of a structure composed of two or more carpels) is an ovary, where eggs develop and are fertilized and where seeds mature. The upper portion has a stigma (a pollen-capturing surface tissue) and often a style (a slender extension of the ovary wall).

carrier protein Type of transport protein that binds specific substances and changes shape in ways that shunt the substances across a plasma membrane. Some carrier proteins function passively, others require an energy input.

carrying capacity The maximum number of individuals in a population (or species) that can be sustained indefinitely by a given environment.

cartilage A type of connective tissue with solid yet pliable intercellular material that resists compression.

Casparian strip In the exodermis and endodermis of roots, a waxy band that acts as an impermeable barrier between the walls of abutting cells.

cDNA Any DNA molecule copied from a mature mRNA transcript by way of reverse transcription.

cell [L. *cella*, small room] The smallest living unit; an organized unit that can survive and reproduce on its own, given DNA instructions and suitable environmental conditions, including appropriate sources of energy and raw materials.

cell count The number of cells of a given type in a microliter of blood.

cell cycle Events during which a cell increases in mass, roughly doubles its number of cytoplasmic components, duplicates its DNA, then undergoes nuclear and cytoplasmic division. It extends from the time a new cell is produced until it completes its own division.

cell differentiation The developmental process in which different cell types activate and suppress a fraction of their genes in different ways and so become specialized in composition, structure, and function. Regulatory proteins, enzymes, hormonal signals, and control sites built into DNA interact to bring about this selective gene expression.

cell junction Of multicelled organisms, a point of contact that physically links two cells or that provides functional links between their cytoplasm.

cell plate Of a plant cell undergoing cytoplasmic division, a disklike structure that forms from remnants of the spindle; it becomes a crosswall that partitions the cytoplasm.

cell theory A theory in biology, the key points of which are that (1) all organisms are composed of one or more cells, (2) the cell is the smallest unit that still retains a capacity for independent life, and (3) all cells arise from preexisting cells.

cell wall A rigid or semirigid wall outside the plasma membrane that supports a cell and imparts shape to it; a cellular feature of plants, fungi, protistans, and most bacteria.

central nervous system The brain and spinal cord of vertebrates.

central vacuole Of mature, living plant cells, a fluid-filled organelle that stores amino acids, sugars, ions, and toxic wastes. Its enlargement during growth causes the increases in surface area that improve nutrient uptake.

centriole (SEN-tree-ohl) A cylinder of triplet microtubules that gives rise to the microtubules of cilia and flagella.

centromere (SEN-troh-meer) [Gk. *kentron*, center, + *meros*, a part] A small, constricted region of a chromosome having attachment sites for microtubules that help move the chromosome during nuclear division.

cephalization (sef-ah-lah-ZAY-shun) [Gk. *kephalikos*, head] During the evolution of bilateral animals, the concentration of sensory structures and nerve cells in the head.

cerebellum (ser-ah-BELL-um) [L. diminutive of *cerebrum*, brain] Hindbrain region with reflex centers for maintaining posture and refining limb movements.

cerebral cortex Thin surface layer of the cerebral hemispheres. Some regions of the cortex receive sensory input, others integrate information and coordinate appropriate motor responses.

cerebrospinal fluid Clear extracellular fluid that surrounds and cushions the brain and spinal cord.

cerebrum (suh-REE-bruhm) Part of the vertebrate forebrain that originally integrated olfactory input and selected motor responses to it. In mammals, it evolved into the most complex integrating center.

channel protein Type of transport protein that serves as a pore through which ions or other water-soluble substances move across the plasma membrane. Some channels remain open, while others are gated and open and close in controlled ways.

chemical bond A union between the electron structures of two or more atoms or ions.

chemical synapse (SIN-aps) [Gk. *synapsis*, union] A small gap, the synaptic cleft, that separates two neurons (or a neuron and a muscle cell or gland cell) and that is bridged by neurotransmitter molecules released from the presynaptic neuron.

chemiosmotic theory (kim-ee-OZ-MOT-ik) Theory that an electrochemical gradient across a cell membrane drives ATP formation. Metabolic reactions cause hydrogen ions (H^+) to accumulate in a compartment formed by the membrane. The combined force of the resulting concentration and electric gradients propels hydrogen ions down the gradient, through channel proteins. Through enzyme action at these proteins, ADP and inorganic phosphate combine to form ATP.

chemoreceptor (KEE-moe-ree-sep-tur) Sensory receptor that detects chemical energy (ions or molecules) dissolved in the surrounding fluid.

chemosynthetic autotroph (KEE-moe-sin-THET-ik) One of a few kinds of bacteria able to synthesize its own organic molecules using carbon dioxide as the carbon source and certain inorganic substances (such as sulfur) as the energy source.

chlorofluorocarbon (KLORE-oh-FLOOR-oh-car-bun), or **CFC** One of a variety of odorless, invisible compounds of chlorine, fluorine, and carbon, widely used in commercial products, that are contributing to the destruction of the ozone layer above the earth's surface.

chlorophylls (KLOR-uh-fills) [Gk. *chloros*, green, + *phyllon*, leaf] Light-sensitive pigment molecules that absorb violet-to-blue and red wavelengths but that transmit green. Certain chlorophylls donate the electrons required for photosynthesis.

chloroplast (KLOR-uh-plast) An organelle that specializes in photosynthesis in plants and certain protistans.

chordate An animal having a notochord, a dorsal hollow nerve cord, a pharynx, and gill slits in the pharynx wall for at least part of its life cycle.

chorion (CORE-ee-on) Of placental mammals, one of four extraembryonic membranes; it becomes a major component of the placenta. Absorptive structures (villi) that develop at its surface are crucial for the transfer of substances between the embryo and mother.

chromatid Of a duplicated eukaryotic chromosome, one of two DNA molecules and its associated proteins. One chromatid remains attached to its "sister" chromatid at the centromere until they are separated from each other during a nuclear division; then each is a separate chromosome.

chromosome (CROW-moe-some) [Gk. *chroma*, color, + *soma*, body] Of eukaryotes, a DNA molecule with many associated proteins. A bacterial chromosome does not have a comparable profusion of proteins associated with the DNA.

chromosome number Of eukaryotic species, the number of each type of chromosome in all cells except dividing germ cells or gametes.

chytrid A type of single-celled fungus of muddy and aquatic habitats.

ciliated protozoan One of four major groups of protozoans.

cilium (SILL-ee-um), plural **cilia** [L. *cilium*, eyelid] Of eukaryotic cells, a short, hairlike projection that contains a regular array of microtubules. Cilia serve as motile structures, help create currents of fluids, or are part of sensory structures. They typically are more profuse than flagella.

circadian rhythm (ser-KAYD-ee-un) [L. *circa*, about, + *dies*, day] Of many organisms, a cycle of physiological events that is completed every 24 hours or so, even when environmental conditions remain constant.

circulatory system Of multicelled animals, an organ system consisting of a muscular pump (heart, most often), blood vessels, and blood; the system transports materials to and from cells and often helps stabilize body temperature and pH.

cladistics An approach to biological systematics in which organisms are grouped according to similarities that are derived from a common ancestor.

cladogram Branching diagram that represents patterns of relative relationships between organisms based on discrete morphological, physiological, and behavioral traits that vary among taxa being studied.

classification system A way of organizing and retrieving information about species.

cleavage Stage of animal development when mitotic cell divisions convert a zygote to a ball of cells, the blastula.

cleavage furrow Of an animal cell undergoing cytoplasmic division, a shallow, ringlike depression that forms at the cell surface as contractile microfilaments pull the plasma membrane inward. It defines where the cytoplasm will be cut in two.

climate Prevailing weather conditions for an ecosystem, including temperature, humidity, wind speed, cloud cover, and rainfall.

climax community Following primary or secondary succession, the array of species that remains more or less steady under prevailing conditions.

clonal selection theory Theory that lymphocytes activated by a specific antigen rapidly multiply and differentiate into huge subpopulations of cells, all having the parent cell's specificity against that antigen.

cloned DNA Multiple, identical copies of DNA fragments that have been inserted into plasmids or some other cloning vector.

club fungus One of a highly diverse group of multicelled fungi, the reproductive structures of which have microscopic, club-shaped cells that produce and bear spores.

cnidarian A radially symmetrical invertebrate, usually marine, that has tissues (not organs), and nematocysts. Two body forms (medusae and polyps) are common. Jellyfishes, corals, and sea anemones are examples.

coal A nonrenewable source of energy that formed more than 280 million years ago from submerged, undecayed plant remains that were buried in sediments, compressed, then compacted further by heat and pressure.

codominance Condition in which a pair of nonidentical alleles are both expressed even though they specify two different phenotypes.

codon One of a series of base triplets in an mRNA molecule, most of which code for a sequence of amino acids of a specific polypeptide chain. (Of sixty-four codons, sixty-one specify different amino acids and three of these also serve as start signals for translation; one other serves only as a stop signal for translation.)

coelum (SEE-lum) [Gk. *koilos*, hollow] Of many animals, a type of body cavity located between the gut and body wall and having a distinctive lining (peritoneum).

coenzyme A type of nucleotide that transfers hydrogen atoms and electrons from one reaction site to another. NAD$^+$ is an example.

coevolution The joint evolution of two or more closely interacting species; when one species evolves, the change affects selection pressures operating between the two species, so the other also evolves.

cofactor A metal ion or coenzyme; it helps catalyze a reaction or serves briefly as an agent that transfers electrons, atoms, or functional groups from one substrate to another.

cohesion Condition in which molecular bonds resist rupturing when under tension.

cohesion theory of water transport Theory that water moves up through vascular plants due to hydrogen bonding among water molecules confined inside the xylem pipelines. The collective cohesive strength of those bonds allows water to be pulled up as columns in response to transpiration (evaporation from leaves).

collenchyma One of the simple tissues of flowering plants; lends flexible support to primary tissues, such as those of lengthening stems.

colon (CO-lun) The large intestine.

commensalism [L. *com*, together, + *mensa*, table] Two-species interaction in which one species benefits significantly while the other is neither helped nor harmed to any notable extent.

communication signal Of social animals, an action or cue sent by one member of a species (the signaler) that can change the behavior of another member (the signal receiver).

community The populations of all species occupying a habitat; also applied to groups of organisms with similar life-styles in a habitat (such as the bird community).

companion cell A specialized parenchyma cell that helps load dissolved organic compounds into the conducting cells of the phloem.

comparative morphology [Gk. *morph*, form] Anatomical comparisons of major lineages.

competitive exclusion Theory that populations of two species competing for a limited resource cannot coexist indefinitely in the same habitat; the population better adapted to exploit the resource will enjoy a competitive (hence reproductive) edge and will eventually exclude the other population from the habitat.

complement system A set of about twenty proteins circulating in blood plasma with roles in nonspecific defenses and in immune responses. Some induce lysis of pathogens, others promote inflammation, and others stimulate phagocytes to engulf pathogens.

compound A substance in which the relative proportions of two or more elements never vary. Organic compounds have a backbone of carbon atoms arranged as a chain or ring structure. The simpler, inorganic compounds do not have comparable backbones.

concentration gradient A difference in the number of molecules (or ions) of a substance between two adjacent regions, as in a volume of fluid.

condensation reaction Enzyme-mediated reaction leading to the covalent linkage of small molecules and, often, the formation of water as a by-product.

cone cell In the vertebrate eye, a type of photoreceptor that responds to intense light and contributes to sharp daytime vision and color perception.

conifer A type of plant belonging to the dominant group of gymnosperms; mostly evergreen, woody trees and shrubs with pollen- and seed-bearing cones.

connective tissue proper A category of animal tissues, all having mostly the same components but in different proportions. These tissues contain fibroblasts and other cells, the secretions of which form fibers (of collagen and elastin) and a ground substance (of modified polysaccharides).

consumers [L. *consumere*, to take completely] Of ecosystems, heterotrophic organisms that obtain energy and raw materials by feeding on the tissues of other organisms. Herbivores, carnivores, omnivores, and parasites are examples.

continuous variation A more or less continuous range of small differences in a given trait among all the individuals of a population.

contractile vacuole (kun-TRAK-till VAK-you-ohl) [L. *contractus*, to draw together] In some protistans, a membranous chamber that takes up excess water in the cell body, then contracts, expelling the water outside the cell through a pore.

control group In a scientific experiment, a group used to evaluate possible side effects of a test involving an experimental group. Ideally, the control group should differ from the experimental group only with respect to the variable being studied.

convergence, morphological *See* morphological convergence.

cork cambium A type of lateral meristem that produces a tough, corky replacement for epidermis on parts of woody plants showing extensive secondary growth.

corpus callosum (CORE-pus ka-LOW-sum) A band of axons (200 million in humans) that functionally link two cerebral hemispheres.

corpus luteum (CORE-pus LOO-tee-um) A glandular structure; it develops from cells of a ruptured ovarian follicle and secretes progesterone and some estrogen, both of which maintain the lining of the uterus (endometrium).

cortex [L. *cortex*, bark] In general, a rindlike layer; the kidney cortex is an example. In vascular plants, ground tissue that makes up most of the primary plant body, supports plant parts, and stores food.

cotyledon A seed leaf, which develops as part of a plant embryo; cotyledons provide nourishment for the germinating seedling.

courtship display Social behavior by which individuals assess and respond to sexual overtures of potential partners.

covalent bond (koe-VAY-lunt) [L. *con*, together, + *valere*, to be strong] A sharing of one or more electrons between atoms or groups of atoms. When electrons are shared equally, the bond is nonpolar. When electrons are shared unequally, the bond is polar—slightly positive at one end and slightly negative at the other.

cross-bridge formation Of muscle cells, the interaction between actin and myosin filaments that is the basis of contraction.

crossing over During prophase I of meiosis, an interaction between a pair of homologous chromosomes. Their nonsister chromatids break at the same place along their length and exchange corresponding segments at the break points. Crossing over breaks up old combinations of alleles and puts new ones together in chromosomes.

culture The sum total of behavior patterns of a social group, passed between generations by learning and by symbolic behavior, especially language.

cuticle (KEW-tih-kull) A body covering. Of land plants, a covering of waxes and lipid-rich cutin deposited on the outer surface of epidermal cell walls. Of annelids, a thin, flexible surface coat. Of arthropods, a hardened yet lightweight covering with protein and chitin components that functions as an external skeleton.

cycad A type of gymnosperm of the tropics and subtropics; slow growing, with massive, cone-shaped structures that bear ovules or pollen.

cyclic AMP (SIK-lik) Cyclic adenosine monophosphate. A nucleotide that has roles in intercellular communication, as when it serves as a second messenger (a cytoplasmic mediator of a cell's response to signaling molecules).

cyclic pathway of ATP formation Photosynthetic pathway in which excited electrons move from a photosystem to an electron transport system, and back to the photosystem. The electron flow contributes to the formation of ATP from ADP and inorganic phosphate.

cyst Of some microorganisms, a walled, resting structure that forms during the life cycle.

cytochrome (SIGH-toe-krome) [Gk. *kytos*, hollow vessel, + *chrōma*, color] Iron-containing protein molecule; a component of electron transport systems used in photosynthesis and aerobic respiration.

cytokinesis (SIGH-toe-kih-NEE-sis) [Gk. *kinesis*, motion] Cytoplasmic division; the splitting of a parental cell into daughter cells.

cytokinin (SIGH-tow-KY-nin) Any of the class of plant hormones that stimulate cell division, promote leaf expansion, and retard leaf aging.

cytomembrane system [Gk. *kytos*, hollow vessel] Organelles, functioning as a system to modify, package, and distribute newly formed proteins and lipids. Endoplasmic reticulum, Golgi bodies, lysosomes, and a variety of vesicles are its components.

cytoplasm (SIGH-toe-plaz-um) [Gk. *plassein*, to mold] All cellular parts, particles, and semifluid substances enclosed by the plasma membrane except for the region of DNA (which in eukaryotes, is the nucleus).

cytosine (SIGH-toe-seen) A pyrimidine; one of the nitrogen-containing bases in nucleotides.

cytoskeleton Of eukaryotic cells, an internal "skeleton." Its microtubules and other components structurally support the cell, organize and move its internal components. The cytoskeleton also helps free-living cells move through their environment.

cytotoxic T cell A T lymphocyte that eliminates infected body cells or tumor cells with a single hit of toxins and perforins.

decomposers [L. *de-*, down, away, + *companere*, to put together] Of ecosystems, heterotrophs that obtain energy by chemically breaking down the remains, products, or wastes of other organisms. Their activities help cycle nutrients back to producers. Certain fungi and bacteria are examples.

deforestation The removal of all trees from a large tract of land, such as the Amazon Basin or the Pacific Northwest.

degradative pathway A metabolic pathway by which molecules are broken down in stepwise reactions that lead to products of lower energy.

deletion A change in a chromosome's structure after one of its regions is lost as a result of irradiation, viral attack, chemical action, or some other factor.

demographic transition model Model of human population growth in which changes in the growth pattern correspond to different stages of economic development. These are a preindustrial stage, when birth and death rates are both high, a transitional stage, an industrial stage, and a postindustrial stage, when the death rate exceeds the birth rate.

denaturation (deh-NAY-chur-AY-shun) Of any molecule, the loss of three-dimensional shape following disruption of hydrogen bonds and other weak bonds.

dendrite (DEN-drite) [Gk. *dendron*, tree] A short, slender extension from the cell body of a neuron.

denitrification (DEE-nite-rih-fih-KAY-shun) The conversion of nitrate or nitrite by certain bacteria to gaseous nitrogen (N_2) and a small amount of nitrous oxide (N_2O).

density-dependent controls Factors such as predation, parasitism, disease, and competition for resources, which limit population growth by reducing the birth rate, increasing the rates of death and dispersal, or all of these.

density-independent controls Factors such as storms or floods that increase a population's death rate more or less independently of its density.

dentition (den-TIH-shun) The type, size, and number of an animal's teeth.

dermal tissue system Of vascular plants, the tissues that cover and protect the plant surfaces.

dermis The layer of skin underlying the epidermis, consisting mostly of dense connective tissue.

desert A type of biome that exists where the potential for evaporation greatly exceeds rainfall and vegetation cover is limited.

desertification (dez-urt-ih-fih-KAY-shun) The conversion of grasslands, rain-fed cropland, or irrigated cropland to desertlike conditions, with a drop in agricultural productivity of 10 percent or more.

detrital food web Of most ecosystems, the flow of energy mainly from plants through detritivores and decomposers.

detritivores (dih-TRY-tih-vorez) [L. *detritus*; after *deterere*, to wear down] Of ecosystems, heterotrophs that consume dead or decomposing particles of organic matter. Earthworms, crabs, and nematodes are examples.

deuterostome (DUE-ter-oh-stome) [Gk. *deuteros*, second, + *stoma*, mouth] Any of the bilateral animals, including echinoderms and chordates, in which the first indentation in the early embryo develops into the anus.

diaphragm (DIE-uh-fram) [Gk. *diaphragma*, to partition] Muscular partition between the thoracic and abdominal cavities, the contraction and relaxation of which contribute to breathing. Also, a contraceptive device used temporarily to prevent sperm from entering the uterus during sexual intercourse.

dicot (DIE-kot) [Gk. *di*, two, + *kotylēdōn*, cup-shaped vessel] Short for dicotyledon; class of flowering plants characterized generally by seeds having embryos with two cotyledons (seed leaves), net-veined leaves, and floral parts arranged in fours, fives, or multiples of these.

differentiation *See* cell differentiation.

diffusion Net movement of like molecules (or ions) down their concentration gradient. In the absence of other forces, molecular motion and random collisions cause their net outward movement from one region into a neighboring region where they are less concentrated (because collisions are more frequent where the molecules are most crowded together).

digestive system An internal tube or cavity from which ingested food is absorbed into the internal environment; often divided into regions specialized for food transport, processing, and storage.

dihybrid cross An experimental cross in which offspring inherit two gene pairs, each consisting of two nonidentical alleles.

dinoflagellate A photosynthetic or heterotrophic protistan, often flagellated, that is a component of plankton.

diploid number (DIP-loyd) For many sexually reproducing species, the chromosome number of somatic cells and of germ cells prior to meiosis. Such cells have two chromosomes of each type (that is, pairs of homologous chromosomes). *Compare* haploid number.

directional selection Of a population, a shift in allele frequencies in a steady, consistent direction in response to a new environment or to a directional change in the old one. The outcome is that forms of traits at one end of the range of phenotypic variation become more common than the intermediate forms.

disaccharide (die-SAK-uh-ride) [Gk. *di*, two, + *sakcharon*, sugar] A type of simple carbohydrate, of the class called oligosaccharides; two monosaccharides covalently bonded.

disruptive selection Of a population, a shift in allele frequencies to forms of traits at both ends of a range of phenotypic variation and away from intermediate forms.

distal tubule The tubular section of a nephron most distant from the glomerulus; a major site of water and sodium reabsorption.

divergence Accumulation of differences in allele frequencies between populations that have become reproductively isolated from one another.

divergence, morphological *See* morphological divergence.

diversity, organismic Sum total of variations in form, function, and behavior that have accumulated in different lineages. Those variations generally are adaptive to prevailing conditions or were adaptive to conditions that existed in the past.

DNA Deoxyribonucleic acid (dee-OX-ee-RYE-bow-new-CLAY-ik). For all cells (and many viruses), the molecule of inheritance. A category of nucleic acids, each usually consisting of two nucleotide strands twisted together helically and held together by hydrogen bonds. The nucleotide sequence encodes the instructions for assembling proteins, and, ultimately, new individuals of a particular species.

DNA-DNA hybridization *See* nucleic acid hybridization.

DNA fingerprint Of each individual, a unique array of RFLPs, resulting from the DNA sequences inherited (in a Mendelian pattern) from each parent.

DNA library A collection of DNA fragments produced by restriction enzymes and incorporated into plasmids.

DNA ligase (LYE-gaze) Enzyme that seals together the new base-pairings during DNA replication; also used by recombinant DNA technologists to seal base-pairings between DNA fragments and cut plasmid DNA.

DNA polymerase (poe-LIM-uh-raze) Enzyme that assembles a new strand on a parent DNA strand during replication; also takes part in DNA repair.

DNA probe A short DNA sequence that has been assembled from radioactively labeled nucleotides and that can base-pair with part of a gene under investigation.

DNA repair Following an alteration in the base sequence of a DNA strand, a process that restores the original sequence, as carried out by DNA polymerases, DNA ligases, and other enzymes.

DNA replication Of cells, the process by which the hereditary material is duplicated for distribution to daughter nuclei. An example is the duplication of eukaryotic chromosomes during interphase, prior to mitosis.

dominance hierarchy Form of social organization in which some members of the group have adopted a subordinate status to others.

dominant allele In a diploid cell, an allele that masks the expression of its partner on the homologous chromosome.

dormancy [L. *dormire*, to sleep] Of plants, the temporary, hormone-mediated cessation of growth under conditions that might appear to be quite suitable for growth.

double fertilization Of flowering plants only, the fusion of one sperm nucleus with the egg nucleus (to produce a zygote), *and* fusion of a second sperm nucleus with the two nuclei of the endosperm mother cell, which gives rise to triploid (3*n*) nutritive tissue.

doubling time The length of time it takes for a population to double in size.

drug addiction Chemical dependence on a drug, following habituation and tolerance of

it; the drug takes on an "essential" biochemical role in the body.

dry shrubland A type of biome that exists where annual rainfall is less than 25 to 60 centimeters and where short, woody, multi-branched shrubs predominate; chaparral is an example.

dry woodland A type of biome that exists when annual rainfall is about 40 to 100 centimeters; there may be tall trees, but these do not form a dense canopy.

dryopith A type of hominoid, one of the first to appear during the Miocene about the time of the divergences that led to gorillas, chimpanzees, and humans.

duplication A change in a chromosome's structure resulting in the repeated appearance of the same gene sequence.

early *Homo* A type of early hominid that may have been the maker of stone tools that date from about 2.5 million years ago.

echinoderm A type of invertebrate that has calcified spines, needles, or plates on the body wall. It is radially symmetrical but with some bilateral features. Sea stars and sea urchins are examples.

ecology [Gk. *oikos*, home, + *logos*, reason] Study of the interactions of organisms with one another and with their physical and chemical environment.

ecosystem [Gk. *oikos*, home] An array of organisms and their physical environment, all of which interact through a flow of energy and a cycling of materials.

ecosystem modeling Analytical method of predicting unforeseen effects of disturbances to an ecosystem, based on computer programs and models.

ectoderm [Gk. *ecto*, outside, + *derma*, skin] Of animal embryos, the outermost primary tissue layer (germ layer) that gives rise to the outer layer of the integument and to tissues of the nervous system.

effector Of homeostatic systems, a muscle (or gland) that responds to signals from an integrator (such as the brain) by producing movement (or chemical change) that helps adjust the body to changing conditions.

effector cell Of the differentiated subpopulations of lymphocytes that form during an immune response, the type of cell that engages and destroys the antigen-bearing agent that triggered the response.

egg A type of mature female gamete; also called an ovum.

El Niño A recurring, massive displacement to the east of warm surface waters of the western equatorial Pacific, which in turn displaces the cooler waters of the Humboldt Current off the coast of Peru.

electron Negatively charged unit of matter, with both particulate and wavelike properties, that occupies one of the orbitals around the atomic nucleus. Atoms can gain, lose, or share electrons with other atoms.

electron transport phosphorylation (FOSS-for-ih-LAY-shun) Final stage of aerobic respiration, in which ATP forms after hydrogen ions and electrons (from the Krebs cycle) are sent through a transport system that gives up the electrons to oxygen.

electron transport system An organized array of enzymes and cofactors, bound in a cell membrane, that accept and donate electrons in sequence. When such systems operate, hydrogen ions (H^+) flow across the membrane, and the flow drives ATP formation and other reactions.

element Any substance that cannot be decomposed into substances with different properties.

embryo (EM-bree-oh) [Gk. *en*, in, + probably *bryein*, to swell] Of animals generally, the stage formed by way of cleavage, gastrulation, and other early developmental events. Of seed plants, the young sporophyte, from the first cell divisions after fertilization until germination.

embryo sac The female gametophyte of flowering plants.

emulsification Of chyme in the small intestine, a suspension of droplets of fat coated with bile salts.

end product A substance present at the end of a metabolic pathway.

endangered species A species poised at the brink of extinction, owing to the extremely small size and severely limited genetic diversity of its remaining populations.

endergonic reaction (en-dur-GONE-ik) Chemical reaction showing a net gain in energy.

endocrine gland Ductless gland that secretes hormones into interstitial fluid, after which they are distributed by way of the bloodstream.

endocrine system System of cells, tissues, and organs that is functionally linked to the nervous system and that exerts control by way of its hormones and other chemical secretions.

endocytosis (EN-doe-sigh-TOE-sis) Movement of a substance into cells; the substance becomes enclosed by a patch of plasma membrane that sinks into the cytoplasm, then forms a vesicle around it. Phagocytic cells also engulf pathogens or prey in this manner.

endoderm [Gk. *endon*, within, + *derma*, skin] Of animal embryos, the inner primary tissue layer, or germ layer, that gives rise to the inner lining of the gut and organs derived from it.

endodermis A sheetlike wrapping of single cells around the vascular cylinder of a root; it functions in controlling the uptake of water and dissolved nutrients. An impermeable barrier (Casparian strip) prevents water from passing between the walls of abutting endodermal cells.

endometrium (EN-doh-MEET-ree-um) [Gk. *metrios*, of the womb] Inner lining of the uterus, consisting of connective tissues, glands, and blood vessels.

endoplasmic reticulum or **ER** (EN-doe-PLAZ-mik reh-TIK-yoo-lum) An organelle that begins at the nucleus and curves through the cytoplasm. In rough ER (which has many ribosomes on its cytoplasmic side), many new polypeptide chains acquire specialized side chains. In many cells, smooth ER (with no attached ribosomes) is the main site of lipid synthesis.

endoskeleton [Gk. *endon*, within, + *skleros*, hard, stiff] In chordates, the internal framework of bone, cartilage, or both. Together with skeletal muscle, supports and protects other body parts, helps maintain posture, and moves the body.

endosperm (EN-doe-sperm) Nutritive tissue that surrounds and serves as food for a flowering plant embryo and, later, for the germinating seedling.

endospore Of certain bacteria, a resistant body that forms around DNA and some cytoplasm; it germinates and gives rise to new bacterial cells when conditions become favorable.

endosymbiosis A permanent, mutually beneficial interdependency between two species, one of which resides permanently inside the other's body.

energy The capacity to do work.

energy carrier A molecule that delivers energy from one metabolic reaction site to another. ATP is the most widely travelled of these; it readily donates energy to nearly all metabolic reactions.

energy flow pyramid A pyramid-shaped representation of an ecosystem's trophic structure, illustrating the energy losses at each transfer to a different trophic level.

entropy (EN-trow-pee) A measure of the degree of disorder in a system (how much energy has become so disorganized and dispersed, usually as heat, that it is no longer readily available to do work).

enzyme (EN-zime) One of a class of proteins that greatly speed up (catalyze) reactions between specific substances, usually at their functional groups. The substances that each type of enzyme acts upon are called its substrates.

eosinophil Fast-acting, phagocytic white blood cell that takes part in inflammation but not in immune responses.

epidermis The outermost tissue layer of a multicelled plant or animal.

epiglottis A flaplike structure at the start of the larynx, the position of which directs the movement of air into the trachea or food into the esophagus.

epistasis (eh-PISS-tih-sis) A type of gene interaction, whereby two alleles of a gene influence the expression of alleles of a different gene.

epithelium (EP-ih-THEE-lee-um) An animal tissue consisting of one or more layers of adhering cells that covers the body's external surfaces and lines its internal cavities and tubes. Epithelium has one free surface; the opposite surface rests on a basement membrane between it and an underlying connective tissue. Epidermis or skin is an example.

equilibrium, dynamic [Gk. *aequus*, equal, + *libra*, balance] The point at which a chemical reaction runs forward as fast as it runs in reverse; thus the concentrations of reactant molecules and product molecules show no net change.

erythrocyte (eh-RITH-row-site) [Gk. *erythros*, red, + *kytos*, vessel] Red blood cell.

esophagus (ee-SOF-uh-gus) Tubular portion of a digestive system that receives swallowed food and leads to the stomach.

essential amino acid Any of eight amino acids that certain animals cannot synthesize for themselves and must obtain from food.

essential fatty acid Any of the fatty acids that certain animals cannot synthesize for themselves and must obtain from food.

estrogen (ESS-trow-jen) A sex hormone that helps oocytes mature, induces changes in the uterine lining during the menstrual cycle and pregnancy, and maintains secondary sexual traits; also influences bodily growth and development.

estrus (ESS-truss) [Gk. *oistrus*, frenzy] For mammals generally, the cyclic period of a female's sexual receptivity to the male.

estuary (EST-you-ehr-ee) A partly enclosed coastal region where seawater mixes with freshwater from rivers, streams, and runoff from the surrounding land.

ethylene (ETH-il-een) Plant hormone that stimulates fruit ripening and triggers abscission.

eubacteria The subkingdom of all bacterial species except the archaebacteria; one of the three great prokaryotic lineages that arose early in the history of life.

euglenoid A type of flagellated protistan, most of which are photosynthesizers in stagnant or freshwater ponds.

eukaryotic cell (yoo-CARRY-oh-tic) [Gk. *eu*, good, + *karyon*, kernel] A type of cell that has a "true nucleus" and other distinguishing membrane-bound organelles. *Compare* prokaryotic cell.

eutrophication Nutrient enrichment of a body of water, such as a lake, that typically results in reduced transparency and a phytoplankton-dominated community.

evaporation [L. *e-*, out, + *vapor*, steam] Conversion of a substance from the liquid to the gaseous state; some or all of its molecules leave in the form of vapor.

evolution, biological [L. *evolutio*, act of unrolling] Change within a line of descent over time. A population is evolving when some forms of a trait are becoming more or less common, relative to the other kinds of traits. The shifts are evidence of changes in the relative abundances of alleles for that trait, as brought about by mutation, natural selection, genetic drift, and gene flow.

evolutionary tree A treelike diagram in which branches represent separate lines of descent from a common ancestor.

excitatory postsynaptic potential or **EPSP** One of two competing signals at an input zone of a neuron; a graded potential that brings the neuron's plasma membrane closer to threshold.

excretion Any of several processes by which excess water, excess or harmful solutes, or waste materials leave the body by way of the urinary system or certain glands.

exergonic reaction (EX-ur-GONE-ik) A chemical reaction that shows a net loss in energy.

exocrine gland (EK-suh-krin) [Gk. *es*, out of, + *krinein*, to separate] Glandular structure that secretes products, usually through ducts or tubes, to a free epithelial surface.

exocytosis (EK-so-sigh-TOE-sis) Movement of a substance out of a cell by means of a transport vesicle, the membrane of which fuses with the plasma membrane, so that the vesicle's contents are released outside.

exodermis Layer of cells just inside the root epidermis of most flowering plants; helps control the uptake of water and solutes.

exon Of eukaryotic cells, any of the nucleotide sequences of a pre-mRNA molecule that are spliced together to form the mature mRNA transcript and are ultimately translated into protein.

exoskeleton [Gk. *exo*, out, + *skléros*, hard, stiff] An external skeleton, as in arthropods.

experiment A test in which some phenomenon in the natural world is manipulated in controlled ways to gain insight into its function, structure, operation, or behavior.

exploitation competition Interaction in which both species have equal access to a required resource but differ in how fast or efficiently they exploit it.

exponential growth (EX-po-NEN-shul) Of populations, a pattern of growth based on a constant rate of multiplication over increments of time. One variable (population size) increases much faster than another variable (the number of individuals in the reproductive base) in a specific mathematical way. The larger the reproductive base, the greater the expansion in population size.

extinction, background A steady rate of species turnover that characterizes lineages through most of their histories.

extinction, mass An abrupt increase in the rate at which major taxa disappear, with several taxa being affected simultaneously.

extracellular fluid In animals generally, all the fluid not inside cells; includes plasma (the liquid portion of blood) and interstitial fluid (which occupies the spaces between cells and tissues).

extracellular matrix A material, largely secreted, that helps hold many animal tissues together in certain shapes; it consists of fibrous proteins and other components in a ground substance.

FAD Flavin adenine dinucleotide, a nucleotide coenzyme. When delivering electrons and unbound protons (H^+) from one reaction to another, it is abbreviated $FADH_2$.

fall overturn The vertical mixing of a body of water in autumn. Its upper layer cools, increases in density, and sinks; dissolved oxygen moves down and nutrients from bottom sediments are brought to the surface.

family pedigree A chart of genetic relationships of the individuals in a family through successive generations.

fat A lipid with a glycerol head and one, two, or three fatty acid tails. The tails of saturated fats have only single bonds between carbon atoms and hydrogen atoms attached to all other bonding sites. Tails of unsaturated fats additionally have one or more double bonds between certain carbon atoms.

fatty acid A long, flexible hydrocarbon chain with a —COOH group at one end.

feedback inhibition Of cells, a control mechanism by which the production (or secretion) of a substance triggers a change in some activity that in turn shuts down further production of the substance.

fermentation [L. *fermentum*, yeast] A type of anaerobic pathway of ATP formation, it starts with glycolysis, ends when electrons are transferred back to one of the breakdown products or intermediates, and regenerates the NAD^+ required for the reaction. Its net yield is two ATP per glucose molecule degraded.

fern One of the seedless vascular plants, mostly of wet, humid habitats; requires free water to complete its life cycle.

fertilization [L. *fertilis*, to carry, to bear] Fusion of a sperm nucleus with the nucleus of an egg, which thereupon becomes a zygote.

fever A body temperature higher than a set point that is preestablished in the brain region governing temperature.

fibrous root system Of most monocots, all the lateral branchings of adventitious roots, which arose earlier from the young stem.

filtration Of urine formation, the process by which blood pressure forces water and solutes out of glomerular capillaries and into the cupped portion of a nephron wall (Bowman's capsule).

fin Of fishes generally, an appendage that helps propel, stabilize, and guide the body through water.

first law of thermodynamics [Gk. *therme*, heat, + *dynamikos*, powerful] Law stating that the total amount of energy in the universe remains constant. Energy cannot be created and existing energy cannot be destroyed. It can only be converted from one form to another.

fish An aquatic animal of the most ancient vertebrate lineage; jawless fishes (such as lampreys and hagfishes), cartilaginous fishes (such as sharks), and bony fishes (such as coelacanths and salmon) are the three existing groups.

fixed action pattern An instinctive response that is triggered by a well-defined, simple stimulus and that is performed in its entirety once it has begun.

flagellated protozoan A member of one of four major groups of protozoans, many of which cause serious diseases.

flagellum (fluh-JELL-um), plural **flagella** [L. whip] Tail-like motile structure of many free-living eukaryotic cells; it has a distinctive 9 + 2 array of microtubules.

flatworm A type of invertebrate having bilateral symmetry, a flattened body, and a saclike gut; a turbellarian, fluke, or tapeworm.

flower The reproductive structure that distinguishes angiosperms from other seed plants and often attracts pollinators.

fluid mosaic model Model of membrane structure in which proteins are embedded in a lipid bilayer or attached to one of its surfaces. The lipid molecules give the membrane its basic structure, impermeability to water-soluble molecules, and (through packing variations and movements) fluidity. Proteins carry out most membrane functions, such as transport, enzyme action, and reception of signals or substances.

follicle (FOLL-ih-kul) In a mammalian ovary, a primary oocyte (immature egg) together with the surrounding layer of cells.

food chain A straight-line sequence of who eats whom in an ecosystem.

food web A network of cross-connecting, interlinked food chains, encompassing primary producers and an array of consumers, detritivores, and decomposers.

forebrain Brain region that includes the cerebrum and cerebral cortex, the olfactory lobes, and the hypothalamus.

forest A type of biome where tall trees grow together closely enough to form a fairly continuous canopy over a broad region.

fossil Recognizable evidence of an organism that lived in the distant past. Most fossils are skeletons, shells, leaves, seeds, and tracks that were buried in rock layers before they decomposed.

fossil fuel Coal, petroleum, or natural gas; a nonrenewable source of energy formed in sediments by the compression of carbon-containing plant remains over hundreds of millions of years.

founder effect An extreme case of genetic drift. By chance, a few individuals that leave a population and establish a new one carry fewer (or more) alleles for certain traits. Increased variation between the two populations is one outcome. Limited genetic variability in the new population is another.

free radical A highly reactive, unbound molecular fragment with the wrong number of electrons.

fruit [L. after *frui*, to enjoy] Of flowering plants, the expanded and ripened ovary of one or more carpels, sometimes with accessory structures incorporated.

FSH Follicle-stimulating hormone. Produced and secreted by the anterior lobe of the pituitary gland, this hormone has roles in the reproductive functions of both males and females.

functional group An atom or group of atoms that is covalently bonded to the carbon backbone of an organic compound and that influences its behavior.

Fungi The kingdom of fungi.

fungus A eukaryotic heterotroph that uses extracellular digestion and absorption; it secretes enzymes able to break down an external food source into molecules small enough to be absorbed by its cells. Saprobic types feed on nonliving organic matter; parasitic types feed on living organisms. Fungi as a group are major decomposers.

gall bladder Organ of the digestive system that stores bile secreted from the liver.

gamete (GAM-eet) [Gk. *gametes*, husband, and *gametē*, wife] A haploid cell that functions in sexual reproduction. Sperm and eggs are examples.

gamete formation Generally, the formation of gametes by way of meiosis. Of animals, the first stage of development, in which sperm or

eggs form and mature within reproductive tissues of parents.

gametophyte (gam-EET-oh-fite) [Gk. *phyton*, plant] The haploid, multicelled, gamete-producing phase in the life cycle of most plants.

ganglion (GANG-lee-un), plural **ganglia** [Gk. *ganglion*, a swelling] A distinct clustering of cell bodies of neurons in regions other than the brain or spinal cord.

gastrulation (gas-tru-LAY-shun) Of animals, the stage of embryonic development in which cells become arranged into two or three primary tissue layers (germ layers); in humans, the layers are an inner endoderm, an intermediate mesoderm, and a surface ectoderm.

gene [short for German *pangan*, after Gk. *pan*, all + *genes*, to be born] A unit of information about a heritable trait that is passed on from parents to offspring. Each gene has a specific location on a chromosome.

gene flow A microevolutionary process; a physical movement of alleles out of a population as individuals leave (emigrate) or enter (immigrate), the outcome being changes in allele frequencies.

gene frequency More precisely, allele frequency: the relative abundances of all the different alleles for a trait that are carried by the individuals of a population.

gene locus A given gene's particular location on a chromosome.

gene mutation [L. *mutatus*, a change] Change in DNA due to the deletion, addition, or substitution of one to several bases in the nucleotide sequence.

gene pair In diploid cells, the two alleles at a given locus on a pair of homologous chromosomes.

gene pool Sum total of all genotypes in a population. More accurately, allele pool.

gene therapy Generally, the transfer of one or more normal genes into the body cells of an organism in order to correct a genetic defect.

genetic code [After L. *genesis*, to be born] The correspondence between nucleotide triplets in DNA (then in mRNA) and specific sequences of amino acids in the resulting polypeptide chains; the basic language of protein synthesis.

genetic disorder An inherited condition that results in mild to severe medical problems.

genetic drift A microevolutionary process; a change in allele frequencies over the generations due to chance events alone.

genetic engineering Altering the information content of DNA through use of recombinant DNA technology.

genetic equilibrium Hypothetical state of a population in which allele frequencies for a trait remain stable through the generations; a reference point for measuring rates of evolutionary change.

genetic recombination Presence of a new combination of alleles in a DNA molecule compared to the parental genotype; the result of processes such as crossing over at meiosis, chromosome rearrangements, gene mutation, and recombinant DNA technology.

genome All the DNA in a haploid number of chromosomes of a given species.

genotype (JEEN-oh-type) Genetic constitution of an individual. Can mean a single gene pair or the sum total of the individual's genes. *Compare* phenotype.

genus, plural **genera** (JEEN-US, JEN-er-ah) [L. *genus*, race, origin] A taxon into which all species exhibiting certain phenotypic similarities and evolutionary relationship are grouped.

geologic time scale A time scale for earth history, the subdivisions of which have been refined by radioisotope dating work.

germ cell Of animals, one of a cell lineage set aside for sexual reproduction; germ cells give rise to gametes. *Compare* somatic cell.

germ layer Of animal embryos, one of two or three primary tissue layers that form during gastrulation and that gives rise to certain tissues of the adult body. *Compare* ectoderm; endoderm; mesoderm.

germination (jur-min-AY-shun) Generally, the resumption of growth following a rest stage; of seed plants, the time at which an embryo sporophyte breaks through its seed coat and resumes growth.

gibberellin (JIB-er-ELL-un) Any of a class of plant hormones that promote stem elongation.

gill A respiratory organ, typically with a moist, thin vascularized layer of epidermis that functions in gas exchange.

ginkgo A type of gymnosperm with fan-shaped leaves and fleshy coated seeds; now represented by a single species of deciduous trees.

gland A secretory cell or multicelled structure derived from epithelium and often connected to it.

glomerular capillaries The set of blood capillaries inside Bowman's capsule of the nephron.

glomerulus (glow-MARE-you-luss) [L. *glomus*, ball] The first portion of the nephron, where water and solutes are filtered from blood.

glucagon (GLUE-kuh-gone) Hormone that stimulates conversion of glycogen and amino acids to glucose; secreted by alpha cells of the pancreas when the flow of glucose decreases.

glyceride (GLISS-er-eyed) One of the molecules, commonly called fats and oils, that has one, two, or three fatty acid tails attached to a glycerol backbone. They are the body's most abundant lipids and its richest source of energy.

glycerol (GLISS-er-oh) [Gk. *glykys*, sweet, + L. *oleum*, oil] A three-carbon molecule with three hydroxyl groups attached; together with fatty acids, a component of fats and oils.

glycogen (GLY-kuh-jen) In animals, a storage polysaccharide that is a main food reserve; can be readily broken down into glucose subunits.

glycolysis (gly-CALL-ih-sis) [Gk. *glykys*, sweet, + *lysis*, loosening or breaking apart] Initial reactions of both aerobic and anaerobic pathways by which glucose (or some other organic compound) is partially broken down to pyruvate, with a net yield of two ATP. Gly-

colysis proceeds in the cytoplasm of all cells, and oxygen has no role in it.

gnetophyte A type of gymnosperm limited to deserts and tropics.

Golgi body (GOHL-gee) Organelle in which newly synthesized polypeptide chains as well as lipids are modified and packaged in vesicles for export or for transport to specific locations within the cytoplasm.

gonad (GO-nad) Primary reproductive organ in which gametes are produced.

graded potential Of neurons, a local signal that slightly changes the voltage difference across a small patch of the plasma membrane. Such signals vary in magnitude, depending on the stimulus. With prolonged or intense stimulation, they may spread to a trigger zone of the membrane and initiate an action potential.

granum, plural **grana** Within many chloroplasts, any of the stacks of flattened, membranous compartments with chlorophyll and other light-trapping pigments and reaction sites for ATP formation.

grassland A type of biome with flat or rolling land, 25 to 100 centimeters of annual rainfall, warm summers, and often grazing and periodic fires that regenerate the dominant species.

gravitropism (GRAV-ih-TROPE-izm) [L. *gravis*, heavy, + Gk. *trepein*, to turn] The tendency of a plant to grow directionally in response to the earth's gravitational force.

gray matter Of vertebrates, the dendrites, neuron cell bodies, and neuroglial cells of the spinal cord and cerebral cortex.

grazing food web Of most ecosystems, the flow of energy from plants to herbivores, then through an array of carnivores.

green alga One of a group or division of aquatic plants with an abundance of chlorophylls a and b; early members of its lineage may have given rise to the bryophytes and vascular plants.

green revolution In developing countries, the use of improved crop varieties, modern agricultural practices (including massive inputs of fertilizers and pesticides), and equipment to increase crop yields.

greenhouse effect Warming of the lower atmosphere due to the presence of greenhouse gases—carbon dioxide, methane, nitrous oxide, ozone, water vapor, and chlorofluorocarbons.

ground meristem (MARE-ih-stem) [Gk. *meristos*, divisible] Of vascular plants, a primary meristem that produces the ground tissue system, hence the bulk of the plant body.

ground substance Of certain animal tissues, the intercellular material made up of cell secretions and other noncellular components.

ground tissue system Tissues that make up the bulk of the vascular plant body; parenchyma is the most common of these.

guanine A nitrogen-containing base; present in one of the four nucleotide building blocks of DNA and RNA.

guard cell Either of two adjacent cells having roles in the movement of gases and water vapor across leaf or stem epidermis. An opening (stoma) forms when both cells swell with water and move apart; it closes when they lose water and collapse against each other.

gut A body region where food is digested and absorbed; of complete digestive systems, the gastrointestinal tract (the portions from the stomach onward).

gymnosperm (JIM-noe-sperm) [Gk. *gymnos*, naked, + *sperma*, seed] A plant that bears seeds at exposed surfaces of reproductive structures, such as cone scales. Pine trees are examples.

habitat [L. *habitare*, to live in] The type of place where an organism normally lives, characterized by physical features, chemical features, and the presence of certain other species.

hair cell Type of mechanoreceptor that may give rise to action potentials when bent or tilted.

halophile A type of archaebacterium that lives in extremely salty habitats.

haploid number (HAP-loyd) The chromosome number of a gamete which, as an outcome of meiosis, is only half that of the parent germ cell (it has only one of each pair of homologous chromosomes). *Compare* diploid number.

HCG Human chorionic gonadotropin. A hormone that helps maintain the lining of the uterus during the menstrual cycle and during the first trimester of pregnancy.

heart Muscular pump that keeps blood circulating through the animal body.

helper T cell One of the T lymphocytes; when activated, it produces and secretes interleukins that promote formation of huge populations of effector and memory cells for immune responses.

hemoglobin (HEEM-oh-glow-bin) [Gk. *haima*, blood, + L. *globus*, ball] Iron-containing, oxygen-transporting protein that gives red blood cells their color.

hemostasis (HEE-mow-STAY-sis) [Gk. *haima*, blood, + *stasis*, standing] Stopping of blood loss from a damaged blood vessel through coagulation, blood vessel spasm, platelet plug formation, and other mechanisms.

herbivore [L. *herba*, grass, + *vovare*, to devour] Plant-eating animal.

heterocyst (HET-er-oh-sist) Of some filamentous cyanobacteria, a type of thick-walled, nitrogen-fixing cell that forms when nitrogen is scarce.

heterotroph (HET-er-oh-trofe) [Gk. *heteros*, other, + *trophos*, feeder] Organism that cannot synthesize its own organic compounds and must obtain nourishment by feeding on autotrophs, each other, or organic wastes. Animals, fungi, many protistans, and most bacteria are heterotrophs. *Compare* autotroph.

heterozygous condition (HET-er-oh-ZYE-guss) [Gk. *zygoun*, join together] For a given trait, having nonidentical alleles at a particular locus on a pair of homologous chromosomes.

hindbrain One of the three divisions of the vertebrate brain; the medulla oblongata, cerebellum, and pons; includes reflex centers for respiration, blood circulation, and other basic functions; also coordinates motor responses and many complex reflexes.

histone Any of a class of proteins that are intimately associated with DNA and that are largely responsible for its structural (and possibly functional) organization in eukaryotic chromosomes.

homeostasis (HOE-me-oh-STAY-sis) [Gk. *homo*, same, + *stasis*, standing] Of multicelled organisms, a physiological state in which the physical and chemical conditions of the internal environment are being maintained within tolerable ranges.

homeostatic feedback loop An interaction in which an organ (or structure) stimulates or inhibits the output of another organ, then shuts down or increases this activity when it detects that the output has exceeded or fallen below a set point.

hominid [L. *homo*, man] All species on the evolutionary branch leading to modern humans. *Homo sapiens* is the only living representative.

hominoid Apes, humans, and their recent ancestors.

Homo erectus A hominid lineage that emerged between 1.5 million and 300,000 years ago and that may include the direct ancestors of modern humans.

Homo sapiens The hominid lineage of modern humans that emerged between 300,000 and 200,000 years ago.

homologous chromosome (huh-MOLL-uh-gus) [Gk. *homologia*, correspondence] Of sexually reproducing species, one of a pair of chromosomes that resemble each other in size, shape, and the genes they carry, and that line up with each other at meiosis I. The X and Y chromosomes differ in these respects but still function as homologues.

homologous structures The same body parts, modified in different ways, in different lines of descent from a common ancestor.

homozygous condition (HOE-moe-ZYE-guss) Having two identical alleles at a given locus (on a pair of homologous chromosomes).

homozygous dominant condition Having two dominant alleles at a given locus (on a pair of homologous chromosomes).

homozygous recessive condition Having two recessive alleles at a given gene locus (on a pair of homologous chromosomes).

hormone [Gk. *hormon*, to stir up, set in motion] Any of the signaling molecules secreted from endocrine glands, endocrine cells, and some neurons that the bloodstream distributes to nonadjacent target cells (any cell having receptors for that hormone).

horsetail One of the seedless vascular plants, which require free water to complete the life cycle; only one genus has survived to the present.

human genome project A basic research project in which researchers throughout the world are working together to sequence the estimated 3 billion nucleotides present in the DNA of human chromosomes.

hydrogen bond Type of chemical bond in which an atom of a molecule interacts weakly with a neighboring atom that is already taking part in a polar covalent bond.

hydrogen ion A free (unbound) proton; a hydrogen atom that has lost its electron and so bears a positive charge (H^+).

hydrologic cycle A biogeochemical cycle, driven by solar energy, in which water moves slowly through the atmosphere, on or through surface layers of land masses, to the ocean, and back again.

hydrolysis (high-DRAWL-ih-sis) [L. *hydro*, water, + Gk. *lysis*, loosening or breaking apart] Enzyme-mediated reaction in which covalent bonds break, splitting a molecule into two or more parts, and H^+ and OH^- (derived from a water molecule) become attached to the exposed bonding sites.

hydrophilic substance [Gk. *philos*, loving] A polar substance that is attracted to the polar water molecule and so dissolves easily in water. Sugars are examples.

hydrophobic substance [Gk. *phobos*, dreading] A nonpolar substance that is repelled by the polar water molecule and so does not readily dissolve in water. Oil is an example.

hydrosphere All liquid or frozen water on or near the earth's surface.

hydrothermal vent ecosystem A type of ecosystem that exists near fissures in the ocean floor and is based on chemosynthetic bacteria that use hydrogen sulfide as the energy source.

hypha (HIGH-fuh), plural **hyphae** [Gk. *hyphe*, web] Of fungi, a generally tube-shaped filament with chitin-reinforced walls and, often, reinforcing cross-walls; component of the mycelium.

hypodermis A subcutaneous layer having stored fat that helps insulate the body; although not part of skin, it anchors skin while allowing it some freedom of movement.

hypothalamus [Gk. *hypo*, under, + *thalamos*, inner chamber or possibly *tholos*, rotunda] Of vertebrate forebrains, a brain center that monitors visceral activities (such as salt-water balance, temperature control, and reproduction) and that influences related forms of behavior (as in hunger, thirst, and sex).

hypothesis A possible explanation of a specific phenomenon.

immune response A series of events by which B and T lymphocytes recognize a specific antigen, undergo repeated cell divisions that form huge lymphocyte populations, and differentiate into subpopulations of effector and memory cells. Effector cells engage and destroy antigen-bearing agents. Memory cells enter a resting phase and are activated during subsequent encounters with the same antigen.

immunization Various processes, including vaccination, that promote increased immunity against specific diseases.

immunoglobulins (Ig) Four classes of antibodies, each with binding sites for antigen and binding sites used in specialized tasks. Examples are IgM antibodies (first to be secreted during immune responses) and IgG antibodies (which activate complement proteins and neutralize many toxins).

implantation A process by which a blastocyst adheres to the endometrium and begins to establish connections by which the mother and embryo will exchange substances during pregnancy.

imprinting Category of learning in which an animal that has been exposed to specific key stimuli early in its behavioral development forms an association with the object.

incisor A tooth, shaped like a flat chisel or cone, used in nipping or cutting food.

incomplete dominance Of heterozygotes, the appearance of a version of a trait that is somewhere between the homozygous dominant and recessive conditions.

independent assortment Mendelian principle that each gene pair tends to assort into gametes independently of other gene pairs located on nonhomologous chromosomes.

indirect selection A theory in evolutionary biology that self-sacrificing individuals can indirectly pass on their genes by helping relatives survive and reproduce.

induced-fit model Model of enzyme action whereby a bound substrate induces changes in the shape of the enzyme's active site, resulting in a more precise molecular fit between the enzyme and its substrate.

industrial smog A type of gray-air smog that develops in industrialized regions when winters are cold and wet.

inflammation, acute In response to tissue damage or irritation, fast-acting phagocytes and plasma proteins, including complement proteins, leave the bloodstream, then defend and help repair the tissue. Proceeds during both nonspecific and specific (immune) defense responses.

inheritance The transmission, from parents to offspring, of structural and functional patterns that have a genetic basis and are characteristic of each species.

inhibiting hormone A signaling molecule produced and secreted by the hypothalamus that controls secretions by the anterior lobe of the pituitary gland.

inhibitor A substance that can bind with an enzyme and interfere with its functioning.

inhibitory postsynaptic potential, or **IPSP** Of neurons, one of two competing types of graded potentials at an input zone; tends to drive the resting membrane potential away from threshold.

instinctive behavior A complex, stereotyped response to a particular environmental cue that often is quite simple.

insulin Hormone that lowers the glucose level in blood; it is secreted from beta cells of the pancreas and stimulates cells to take up glucose; also promotes protein and fat synthesis and inhibits protein conversion to glucose.

integration, neural [L. *integrare*, to coordinate] Moment-by-moment summation of all excitatory and inhibitory synapses acting on a neuron; occurs at each level of synapsing in a nervous system.

integrator Of homeostatic systems, a control point where different bits of information are pulled together in the selection of a response. The brain is an example.

integument Of animals, a protective body covering such as skin. Of flowering plants, a protective layer around the developing ovule; when the ovule becomes a seed, its integument(s) harden and thicken into a seed coat.

integumentary exchange (in-teg-you-MEN-tuh-ree) Of some animals, a mode of respiration in which oxygen and carbon dioxide diffuse across a thin, vascularized layer of moist epidermis at the body surface.

interference competition Interaction in which one species may limit another species' access to some resource regardless of whether the resource is abundant or scarce.

interleukin One of a variety of communication signals, secreted by macrophages and by helper T cells, that drive immune responses.

intermediate compound A compound that forms between the start and the end of a metabolic pathway.

intermediate filament A cytoskeletal component that consists of different proteins in different types of animal cells.

interneuron Any of the neurons in the vertebrate brain and spinal cord that integrate information arriving from sensory neurons and that influence other neurons in turn.

internode In vascular plants, the stem region between two successive nodes.

interphase Of cell cycles, the time interval between nuclear divisions in which a cell increases its mass, roughly doubles the number of its cytoplasmic components, and finally duplicates its chromosomes (replicates its DNA). The interval is different for different species.

interspecific competition Two-species interaction in which both species can be harmed due to overlapping niches.

interstitial fluid (IN-ter-STISH-ul) [L. *interstitus*, to stand in the middle of something] In multicelled animals, that portion of the extracellular fluid occupying spaces between cells and tissues.

intertidal zone Generally, the area on a rocky or sandy shoreline that is above the low water mark and below the high water mark; organisms inhabiting it are alternately submerged, then exposed, by tides.

intervertebral disk One of a number of disk-shaped structures containing cartilage that serve as shock absorbers and flex points between bony segments of the vertebral column.

intraspecific competition Interaction among individuals of the same species that are competing for the same resources.

intron A noncoding portion of a newly formed mRNA molecule.

inversion A change in a chromosome's structure after a segment separated from it was then inserted at the same place, but in reverse. The reversal alters the position and order of the chromosome's genes.

invertebrate Animal without a backbone.

ion, negatively charged (EYE-on) An atom or a compound that has gained one or more electrons, and hence has acquired an overall negative charge.

ion, positively charged An atom or a compound that has lost one or more electrons, and hence has acquired an overall positive charge.

ionic bond An association between ions of opposite charge.

isotonic condition Equality in the relative concentrations of solutes in two fluids; for two fluids separated by a cell membrane, there is no net osmotic (water) movement across the membrane.

isotope (EYE-so-tope) For a given element, an atom with the same number of protons as the other atoms but with a different number of neutrons.

J-shaped curve A curve, obtained when population size is plotted against time, that is characteristic of unrestricted, exponential growth.

joint An area of contact or near-contact between bones.

karyotype (CARRY-oh-type) Of eukaryotic individuals (or species), the number of metaphase chromosomes in somatic cells and their defining characteristics.

keratin A tough, water-insoluble protein manufactured by most epidermal cells.

keratinization (care-AT-in-iz-AY-shun) Process by which keratin-producing epidermal cells of skin die and collect at the skin surface as keratinized "bags" that form a barrier against dehydration, bacteria, and many toxic substances.

kidney In vertebrates, one of a pair of organs that filter mineral ions, organic wastes, and other substances from the blood, and help regulate the volume and solute concentrations of extracellular fluid.

kilocalorie 1,000 calories of heat energy, or the amount of energy needed to raise the temperature of 1 kilogram of water by 1°C; the unit of measure for the caloric value of foods.

kinetochore A specialized group of proteins and DNA at the centromere of a chromosome that serves as an attachment point for several spindle microtubules during mitosis or meiosis. Each chromatid of a duplicated chromosome has its own kinetochore.

Krebs cycle Together with a few conversion steps that precede it, the stage of aerobic respiration in which pyruvate is completely broken down to carbon dioxide and water. Coenzymes accept the unbound protons (H^+) and electrons stripped from intermediates during the reactions and deliver them to the next stage.

lactate fermentation Anaerobic pathway of ATP formation in which pyruvate from glycolysis is converted to the three-carbon compound lactate, and NAD^+ (a coenzyme used in the reactions) is regenerated. Its net yield is two ATP.

lactation The production of milk by hormone-primed mammary glands.

lake A body of fresh water having littoral, limnetic, and profundal zones.

lancelet An invertebrate chordate having a body that tapers sharply at both ends, segmented muscles, and a full-length notochord.

large intestine The colon; a region of the gut that receives unabsorbed food residues from the small intestine and concentrates and stores feces until they are expelled from the body.

larva, plural **larvae** Of animals, a sexually immature, free-living stage between the embryo and the adult.

larynx (LARE-inks) A tubular airway that leads to the lungs. In humans, contains vocal cords, where sound waves used in speech are produced.

lateral meristem Of vascular plants, a type of meristem responsible for secondary growth; either vascular cambium or cork cambium.

lateral root Of taproot systems, a lateral branching from the first, primary root.

leaching The movement of soil water, with dissolved nutrients, out of a specified area.

leaf For most vascular plants, a structure having chlorophyll-containing tissue that is the major region of photosynthesis.

learned behavior The use of information gained from specific experiences to vary or change a response to stimuli.

lethal mutation A gene mutation that alters one or more traits in such a way that the individual inevitably dies.

LH Leutinizing hormone. Secreted by the anterior lobe of the pituitary gland, this hormone has roles in the reproductive functions of both males and females.

lichen (LY-kun) A symbiotic association between a fungus and a captive photosynthetic partner such as a green alga.

life cycle A recurring, genetically programmed frame of events in which individuals grow, develop, maintain themselves, and reproduce.

life table A tabulation of age-specific patterns of birth and death for a population.

ligament A strap of dense connective tissue that bridges a joint.

light-dependent reactions First stage of photosynthesis in which the energy of sunlight is trapped and converted to the chemical energy of ATP alone (by the cyclic pathway) or ATP and NADPH (by the noncyclic pathway).

light-independent reactions Second stage of photosynthesis, in which sugar phosphates form with the help of the ATP (and NADPH, in land plants) that were produced during the first stage. The sugar phosphates are used in other reactions by which starch, cellulose, and other end products of photosynthesis are assembled.

lignification Of mature land plants, a process by which lignin is deposited in secondary cell walls. The deposits impart strength and rigidity by anchoring cellulose strands in the walls, stabilize and protect other wall components, and form a waterproof barrier around the cellulose. Probably a key factor in the evolution of vascular plants.

lignin A substance that strengthens and waterproofs cell walls in certain tissues of vascular plants.

limbic system Brain regions that, along with the cerebral cortex, collectively govern emotions.

limiting factor Any essential resource that is in short supply and so limits population growth.

lineage (LIN-ee-age) A line of descent.

linkage The tendency of genes located on the same chromosome to end up in the same gamete. For any two of those genes, the probability that crossing over will disrupt the linkage is proportional to the distance separating them.

lipid A greasy or oily compound of mostly carbon and hydrogen that shows little tendency to dissolve in water, but that dissolves in nonpolar solvents (such as ether). Cells use lipids as energy stores and structural materials, especially in membranes.

lipid bilayer The structural basis of cell membranes, consisting of two layers of mostly phospholipid molecules. Hydrophilic heads force all fatty acid tails of the lipids to become sandwiched between the hydrophilic heads.

liver Glandular organ with roles in storing and interconverting carbohydrates, lipids, and proteins absorbed from the gut, maintaining blood; disposing of nitrogen-containing wastes; and other tasks.

local signaling molecules Secretions from cells in many different tissues that alter chemical conditions in the immediate vicinity where they are secreted, then are swiftly degraded.

locus (LOW-cuss) The specific location of a particular gene on a chromosome.

logistic population growth (low-JIS-tik) Pattern of population growth in which a low-density population slowly increases in size, goes through a rapid growth phase, then levels off once the carrying capacity is reached.

loop of Henle The hairpin-shaped, tubular region of a nephron that functions in reabsorption of water and solutes.

lung An internal respiratory surface in the shape of a cavity or sac.

lycophyte A type of seedless vascular plant of mostly wet or shade habitats; requires free water to complete its life cycle.

lymph (LIMF) [L. *lympha*, water] Tissue fluid that has moved into the vessels of the lymphatic system.

lymph capillary A small-diameter vessel of the lymph vascular system that has no pronounced entrance; tissue fluid moves inward by passing between overlapping endothelial cells at the vessel's tip.

lymph node A lymphoid organ that serves as a battleground of the immune system; each lymph node is packed with organized arrays of macrophages and lymphocytes that cleanse lymph of pathogens before it reaches the blood.

lymph vascular system [L. *lympha*, water, + *vasculum*, a small vessel] The vessels of the lymphatic system, which take up and transport excess tissue fluid and reclaimable solutes as well as fats absorbed from the digestive tract.

lymphatic system An organ system that supplements the circulatory system. Its vessels take up fluid and solutes from interstitial fluid and deliver them to the bloodstream; its lymphoid organs have roles in immunity.

lymphocyte Any of various white blood cells that take part in nonspecific and specific (immune) defense responses.

lymphoid organs The lymph nodes, spleen, thymus, tonsils, adenoids, and other organs with roles in immunity.

lysis [Gk. *lysis*, a loosening] Gross structural disruption of a plasma membrane that leads to cell death.

lysosome (LYE-so-sohm) The main organelle of digestion, with enzymes that can break down polysaccharides, proteins, nucleic acids, and some lipids.

lysozyme An infection-fighting enzyme that digests bacterial cell walls. Present in mucous membranes that line the body's surfaces.

lytic pathway During a viral infection, viral DNA or RNA quickly directs the host cell to produce the components necessary to produce new virus particles, which are released by lysis.

macroevolution The large-scale patterns, trends, and rates of change among groups of species.

macrophage One of the phagocytic white blood cells. It engulfs anything detected as foreign. Some also become antigen-presenting cells that serve as the trigger for immune responses by T and B lymphocytes. *Compare* antigen-presenting cell.

mammal A type of vertebrate; the only animal having offspring that are nourished by milk produced by mammary glands of females.

mass extinction An abrupt rise in extinction rates above the background level; a catastrophic, global event in which major taxa are wiped out simultaneously.

mass number The total number of protons and neutrons in an atom's nucleus. The relative masses of atoms are also called atomic weights.

maternal chromosome One of the chromosomes bearing the alleles that are inherited from a female parent.

mechanoreceptor Sensory cell or cell part that detects mechanical energy associated with changes in pressure, position, or acceleration.

medulla oblongata Part of the vertebrate brainstem with reflex centers for respiration, blood circulation, and other vital functions.

medusa (meh-DOO-sah) [Gk. *Medousa*, one of three sisters in Greek mythology having snake-entwined hair; this image probably evoked by the tentacles and oral arms extending from the medusa] Free-swimming, bell-shaped stage in cnidarian life cycles.

megaspore Of gymnosperms and flowering plants, a haploid spore that forms in the ovary; one of its cellular descendants develops into an egg.

meiosis (my-OH-sis) [Gk. *meioun*, to diminish] Two-stage nuclear division process in which the chromosome number of a germ cell is reduced by half, to the haploid number. (Each daughter nucleus ends up with one of each type of chromosome.) Meiosis is the basis of gamete formation and (in plants) of spore formation. *Compare* mitosis.

meltdown Events which, if unchecked, can blow apart a nuclear power plant, with a release of radioactive material into the environment.

membrane excitability A membrane property of any cell that can produce action potentials in response to appropriate stimulation.

memory The storage and retrieval of information about previous experiences; underlies the capacity for learning.

memory cell One of the subpopulations of cells that form during an immune response and that enters a resting phase, from which it is released during a secondary immune response.

memory lymphocyte Any of the various B or T lymphocytes of the immune system that are formed in response to invasion by a foreign agent and that circulate for some period, available to mount a rapid attack if the same type of invader reappears.

Mendel's theory of independent assortment Stated in modern terms, during meiosis, the gene pairs of homologous chromosomes tend to be stored independently of how gene pairs on other chromosomes are sorted for forthcoming gametes. The theory does not take into account the effects of gene linkage and crossing over.

Mendel's theory of segregation Stated in modern terms, diploid cells have two of each kind of gene (on pairs of homologous chromosomes), and the two segregate during meiosis so that they end up in different gametes.

menopause (MEN-uh-pozz) [L. *mensis*, month, + *pausa*, stop] End of the period of a human female's reproductive potential.

menstrual cycle The cyclic release of oocytes and priming of the endometrium (lining of the uterus) to receive a fertilized egg; the complete cycle averages about 28 days in female humans.

menstruation Periodic sloughing of the blood-enriched lining of the uterus when pregnancy does not occur.

mesoderm (MEH-so-derm) [Gk. *mesos*, middle, + *derm*, skin] In most animal embryos, a primary tissue layer (germ layer) between ectoderm and endoderm. Gives rise to muscle; organs of circulation, reproduction, and excretion; most of the internal skeleton (when present); and connective tissue layers of the gut and body covering.

mesophyll Of vascular plants, a type of parenchyma tissue with photosynthetic cells and an abundance of air spaces.

messenger RNA A linear sequence of ribonucleotides transcribed from DNA and translated into a polypeptide chain; the only type of RNA that carries protein-building instructions.

metabolic pathway One of many orderly sequences of enzyme-mediated reactions by which cells normally maintain, increase, or decrease the concentrations of substances.

Different pathways are linear or circular, and often they interconnect.

metabolism (meh-TAB-oh-lizm) [Gk. *meta*, change] All controlled, enzyme-mediated chemical reactions by which cells acquire and use energy. Through these reactions, cells synthesize, store, break apart, and eliminate substances in ways that contribute to growth, survival, and reproduction.

metamorphosis (met-uh-MOR-foe-sis) [Gk. *meta*, change, + *morphe*, form] Transformation of a larva into an adult form.

metaphase Of mitosis or meiosis II, the stage when each duplicated chromosome has become positioned at the midpoint of the microtubular spindle, with its two sister chromatids attached to microtubules from opposite spindle poles. Of meiosis I, the stage when all pairs of homologous chromosomes are positioned at the spindle's midpoint, with the two members of each pair attached to opposite spindle poles.

metazoan Any multicelled animal.

methanogen A type of archaebacterium that lives in oxygen-free habitats and that produces methane gas as a metabolic by-product.

MHC marker Any of a variety of proteins that are self-markers. Some occur on all body cells of an individual; others are unique to the macrophages and lymphocytes.

micelle formation Formation of a small droplet that consists of bile salts and products of fat digestion (fatty acids and monoglycerides) and that assists in their absorption from the small intestine.

microevolution Changes in allele frequencies brought about by mutation, genetic drift, gene flow, and natural selection.

microfilament [Gk. *mikros*, small, + L. *filum*, thread] In animal cells, one of a variety of cytoskeletal components. Actin and myosin filaments are examples.

microorganism An organism, usually single-celled, that is too small to be observed without the aid of a microscope.

microspore Of gymnosperms and flowering plants, a haploid spore, encased in a sculpted wall, that develops into a pollen grain.

microtubular spindle Of eukaryotic cells, a bipolar structure composed of organized arrays of microtubules that forms during nuclear division and that moves the chromosomes.

microtubule Hollow cylinder of mainly tubulin subunits; a cytoskeletal element with roles in cell shape, motion, and growth and in the structure of cilia and flagella.

microtubule organizing center, or **MTOC** Small mass of proteins and other substances in the cytoplasm; the number, type, and location of MTOCs determine the organization and orientation of microtubules.

microvillus (MY-crow-VILL-us) [L. *villus*, shaggy hair] A slender, cylindrical extension of the animal cell surface that functions in absorption or secretion.

midbrain Of vertebrates, a brain region that evolved as a coordination center for reflex responses to visual and auditory input; together with the pons and medulla oblongata,

part of the brainstem, which includes the reticular formation.

migration Of certain animals, a cyclic movement between two distant regions at times of year corresponding to seasonal change.

mimicry (MIM-ik-ree) Situation in which one species (the mimic) bears deceptive resemblance in color, form, and/or behavior to another species (the model) that enjoys some survival advantage.

mineral An inorganic substance required for the normal functioning of body cells.

mitochondrion (MY-toe-KON-dree-on), plural **mitochondria** Organelle that specializes in ATP formation; it is the site of the second and third stages of aerobic respiration, an oxygen-requiring pathway.

mitosis (my-TOE-sis) [Gk. *mitos*, thread] Type of nuclear division that maintains the parental chromosome number for daughter cells. It is the basis of bodily growth and, in many eukaryotic species, asexual reproduction.

molar One of the cheek teeth; a tooth with a platform having cusps (surface bumps) that help crush, grind, and shear food.

molecular clock With respect to the presumed regular accumulation of neutral mutations in highly conserved genes, a way of calculating the time of origin of one species or lineage relative to others.

molecule A unit of matter in which chemical bonding holds together two or more atoms of the same or different elements.

mollusk A type of invertebrate having a tissue fold (mantle) draped around a soft, fleshy body; snails, clams, and squids are examples.

molting The shedding of hair, feathers, horns, epidermis, or a shell (or some other exoskeleton) in a process of growth or periodic renewal.

Monera The kingdom of bacteria.

moneran A bacterium; a single-celled prokaryote.

monocot (MON-oh-kot) Short for monocotyledon; a flowering plant in which seeds have only one cotyledon, whose floral parts generally occur in threes (or multiples of three), and whose leaves typically are parallel-veined. *Compare* dicot.

monohybrid cross [Gk. *monos*, alone] An experimental cross in which offspring inherit a pair of nonidentical alleles for a single trait being studied, so that they are heterozygous.

monophyletic group A set of independently evolving lineages that share a common evolutionary heritage.

monosaccharide (MON-oh-SAK-ah-ride) [Gk. *monos*, alone, single, + *sakharon*, sugar] The simplest carbohydrate, with only one sugar unit. Glucose is an example.

monosomy Abnormal condition in which one chromosome of diploid cells has no homologue.

morphogenesis (MORE-foe-JEN-ih-sis) [Gk. *morphe*, form, + *genesis*, origin] Processes by which differentiated cells in an embryo become organized into tissues and organs, under genetic controls and environmental influences.

morphological convergence A macroevolutionary pattern of change in which separate lineages adopt similar lifestyles, put comparable body parts to similar uses, and in time resemble one another in structure and function. Analogous structures are evidence of this pattern.

morphological divergence A macroevolutionary pattern of change from a common ancestral form. Homologous structures are evidence of this pattern.

motor neuron A type of neuron; it delivers signals from the brain and spinal cord that can stimulate or inhibit the body's effectors (muscles, glands, or both).

mouth An oral cavity; in human digestion, the site where polysaccharide breakdown begins.

multicelled organism An organism that has differentiated cells arranged into tissues, organs, and often organ systems.

multiple allele system Three or more different molecular forms of the same gene (alleles) that exist in a population.

muscle fatigue A decline in tension of a muscle that has been kept in a state of tetanic contraction as a result of continuous, high-frequency stimulation.

muscle tension A mechanical force, exerted by a contracting muscle, that resists opposing forces such as gravity and the weight of objects being lifted.

muscle tissue Tissue having cells able to contract in response to stimulation, then passively lengthen and so return to their resting stage.

mutagen (MEW-tuh-jen) An environmental agent that can permanently modify the structure of a DNA molecule. Certain viruses and ultraviolet radiation are examples.

mutation [L. *mutatus*, a change, + *-ion*, result or a process or an act] A heritable change in the DNA. Generally, mutations are the source of all the different molecular versions of genes (alleles) and, ultimately, of life's diversity. *See also* lethal mutation; neutral mutation.

mutualism [L. *mutuus*, reciprocal] An interaction between two species that benefits both.

mycelium (my-SEE-lee-um), plural **mycelia** [Gk. *mykes*, fungus, mushroom, + *helos*, callus] A mesh of tiny, branching filaments (hyphae) that is the food-absorbing part of a multicelled fungus.

mycorrhiza (MY-coe-RISE-uh) "Fungus-root;" a symbiotic arrangement between fungal hyphae and the young roots of many vascular plants. The fungus obtains carbohydrates from the plant and in turn releases dissolved mineral ions to the plant roots.

myelin sheath Of many sensory and motor neurons, an axonal sheath that affects how fast action potentials travel; formed from the plasma membranes of Schwann cells that are wrapped repeatedly around the axon and are separated from each other by a small node.

myofibril (MY-oh-FY-brill) One of many threadlike structures inside a muscle cell; each is functionally divided into sarcomeres, the basic units of contraction.

myosin (MY-uh-sin) A type of protein with a head and long tail. In muscle cells, it interacts with actin, another protein, to bring about contraction.

NAD$^+$ Nicotinamide adenine dinucleotide; a nucleotide coenzyme. When carrying electrons and unbound protons (H$^+$) between reaction sites, it is abbreviated NADH.

NADP$^+$ Nicotinamide adenine dinucleotide phosphate; a phosphorylated nucleotide coenzyme. When carrying electrons and unbound protons (H$^+$) between reaction sites, it is abbreviated NADPH$_2$.

nasal cavity Of a respiratory system, the region where air is warmed, moistened, and filtered of airborne particles and dust.

natural selection A microevolutionary process; a difference in survival and reproduction among members of a population that vary in one or more traits.

negative feedback mechanism A homeostatic feedback mechanism in which an activity changes some condition in the internal environment and so triggers a response that reverses the changed condition.

nematocyst (NEM-ad-uh-sist) [Gk. *nema*, thread, + *kystis*, pouch] Of cnidarians only, a stinging capsule that assists in prey capture and possibly protection.

nephridium (neh-FRID-ee-um), plural **nephridia** Of earthworms and some other invertebrates, a system of regulating water and solute levels.

nephron (NEFF-ron) [Gk. *nephros*, kidney] Of the vertebrate kidney, a slender tubule in which water and solutes filtered from blood are selectively reabsorbed and in which urine forms.

nerve Cordlike communication line of the peripheral nervous system, composed of axons of sensory neurons, motor neurons, or both packed within connective tissue. In the brain and spinal cord, similar cord-like bundles are called nerve pathways or tracts.

nerve cord Of many animals, a cordlike communication line consisting of axons of neurons.

nerve impulse *See* action potential.

nerve net Cnidarian nervous system.

nervous system System of neurons oriented relative to one another in precise message-conducting and information-processing pathways.

nervous tissue A type of connective tissue composed of neurons.

net energy Of energy resources available to the human population, the amount of energy that is left over after subtracting the energy used to locate, extract, transport, store, and deliver energy to consumers.

net population growth rate per individual (r) Of population growth equations, a single variable in which birth and death rates, which are assumed to remain constant, are combined.

neuroglial cell (NUR-oh-GLEE-uhl) Of vertebrates, one of the cells that provide structural and metabolic support for neurons and that collectively represent about half the volume of the nervous system.

neuromodulator Type of signaling molecule that influences the effects of transmitter substances by enhancing or reducing membrane responses in target neurons.

neuromuscular junction Chemical synapses between axon terminals of a motor neuron and a muscle cell.

neuron A nerve cell; the basic unit of communication in nervous systems. Neurons collectively sense environmental change, integrate sensory inputs, then activate muscles or glands that initiate or carry out responses.

neurotransmitter Any of the class of signaling molecules that are secreted from neurons, act on immediately adjacent cells, and are then rapidly degraded or recycled.

neutral mutation A gene mutation that has neither harmful nor helpful effects on the individual's ability to survive and reproduce.

neutron Unit of matter, one or more of which occupies the atomic nucleus, that has mass but no electric charge.

neutrophil Fast-acting, phagocytic white blood cell that takes part in inflammatory responses against bacteria.

niche (NITCH) [L. *nidas*, nest] Of a species, the full range of physical and biological conditions under which its members can live and reproduce.

nitrification (nye-trih-fih-KAY-shun) A chemosynthetic process in which certain bacteria strip electrons from ammonia or ammonium present in soil. The end product, nitrite (NO$_2^-$), is broken down to nitrate (NO$_3^-$) by different bacteria.

nitrogen cycle Biogeochemical cycle in which the atmosphere is the largest reservoir of nitrogen.

nitrogen fixation Process by which a few kinds of bacteria convert gaseous nitrogen (N$_2$) to ammonia. This dissolves rapidly in their cytoplasm to form ammonium, which can be used in biosynthetic pathways.

NK cell Natural killer cell, possibly of the lymphocyte lineage, that reconnoiters and kills tumor cells and infected body cells.

nociceptor A receptor, such as a free nerve ending, that detects any stimulus causing tissue damage.

node In vascular plants, a point on a stem where one or more leaves are attached.

noncyclic pathway of ATP formation (non-SIK-lik) [L. *non*, not, + Gk. *kylos*, circle] Photosynthetic pathway in which excited electrons derived from water molecules flow through two photosystems and two transport chains, and ATP and NADPH form.

nondisjunction Failure of one or more chromosomes to separate properly during mitosis or meiosis.

nonsteroid hormone A type of water-soluble hormone, such as a protein hormone, that cannot cross the lipid bilayer of a target cell. These hormones enter the cell by receptor-mediated endocytosis, or they bind to receptors that activate membrane proteins or second messengers within the cell.

notochord (KNOW-toe-kord) Of chordates, a rod of stiffened tissue (not cartilage or bone) that serves as a supporting structure for the body.

nuclear envelope A double membrane (two lipid bilayers and associated proteins) that is the outermost portion of a cell nucleus.

nucleic acid (new-CLAY-ik) A long, single- or double-stranded chain of four different kinds of nucleotides joined one after the other at their phosphate groups. They differ in which nucleotide base follows the next in sequence. DNA and RNA are examples.

nucleic acid hybridization The base-pairing of nucleotide sequences from different sources, as used in genetics, genetic engineering, and studies of evolutionary relationship based on similarities and differences in the DNA or RNA of different species.

nucleoid Of bacteria, a region in which DNA is physically organized apart from other cytoplasmic components.

nucleolus (new-KLEE-oh-lus) [L. *nucleolus*, a little kernel] Within the nucleus of a nondividing cell, a site where the protein and RNA subunits of ribosomes are assembled.

nucleosome (NEW-klee-oh-sohm) Of eukaryotic chromosomes, one of many organizational units, each consisting of a small stretch of DNA looped twice around a "spool" of histone molecules, which another histone molecule stabilizes.

nucleotide (NEW-klee-oh-tide) A small organic compound having a five-carbon sugar (deoxyribose), nitrogen-containing base, and phosphate group. Nucleotides are the structural units of adenosine phosphates, nucleotide coenzymes, and nucleic acids.

nucleotide coenzyme A protein that transports hydrogen atoms (free protons) and electrons from one reaction site to another in cells.

nucleus (NEW-klee-us) [L. *nucleus*, a kernel] Of atoms, the central core of one or more positively charged protons and (in all but hydrogen) electrically neutral neutrons. In eukaryotic cells, a membranous organelle that physically isolates and organizes the DNA, out of the way of cytoplasmic machinery.

nutrition All those processes by which food is selectively ingested, digested, absorbed, and later converted to the body's own organic compounds.

obesity An excess of fat in the body's adipose tissues, caused by imbalances between caloric intake and energy output.

oligosaccharide A carbohydrate consisting of a short chain of two or more covalently bonded sugar units. One subclass, disaccharides, has two sugar units. *Compare* monosaccharide; polysaccharide.

omnivore [L. *omnis*, all, + *vovare*, to devour] An organism able to obtain energy from more than one source rather than being limited to one trophic level.

oncogene (ON-coe-jeen) Any gene having the potential to induce cancerous transformations in a cell.

oocyte An immature egg.

oogenesis (oo-oh-JEN-uh-sis) Formation of a female gamete, from a germ cell to a mature haploid ovum (egg).

operator A short base sequence between a promoter and the start of a gene; interacts with regulatory proteins.

operon Of transcription, a promoter-operator sequence that services more than a single gene. The lactose operon of *E. coli* is an example.

orbitals Volumes of space around the nucleus of an atom in which electrons are likely to be at any instant.

organ A structure of definite form and function that is composed of more than one tissue.

organ formation Stage of development in which primary tissue layers (germ layers) split into subpopulations of cells, and different lines of cells become unique in structure and function; foundation for growth and tissue specialization, when organs acquire specialized chemical and physical properties.

organ system Two or more organs that interact chemically, physically, or both in performing a common task.

organelle Of cells, an internal, membrane-bounded sac or compartment that has a specific, specialized metabolic function.

organic compound In biology, a compound assembled in cells and having a carbon backbone, often with carbon atoms arranged as a chain or ring structure.

osmosis (oss-MOE-sis) [Gk. *osmos*, act of pushing] Of cells, the tendency of water to move through channel proteins that span a membrane in response to a concentration gradient, fluid pressure, or both. Hydrogen bonds among water molecules prevent water *itself* from becoming more or less concentrated; but a gradient may exist when the water on either side of the membrane has more substances dissolved in it.

ovary (OH-vuh-ree) In female animals, the primary reproductive organ in which eggs form. In seed-bearing plants, the portion of the carpel where eggs develop, fertilization takes place, and seeds mature. A mature ovary (and sometimes other plant parts) is a fruit.

oviduct (OH-vih-dukt) Duct through which eggs travel from the ovary to the uterus. Formerly called Fallopian tube.

ovulation (AHV-you-LAY-shun) During each turn of the menstrual cycle, the release of a secondary oocyte (immature egg) from an ovary.

ovule (OHV-youl) [L. *ovum*, egg] Before fertilization in gymnosperms and angiosperms, a female gametophyte with egg cell, a surrounding tissue, and one or two protective layers (integuments). After fertilization, an ovule matures into a seed (an embryo sporophyte and food reserves encased in a hardened coat).

ovum (OH-vum) A mature female gamete (egg).

oxidation-reduction reaction An electron transfer from one atom or molecule to another. Often hydrogen is transferred along with the electron or electrons.

ozone hole A pronounced seasonal thinning of the ozone layer in the lower stratosphere above Antarctica.

pancreas (PAN-cree-us) Gland that secretes enzymes and bicarbonate into the small intestine during digestion, and that also secretes the hormones insulin and glucagon.

pancreatic islets Any of the two million clusters of endocrine cells in the pancreas, including alpha cells, beta cells, and delta cells.

parasite [Gk. *para*, alongside, + *sitos*, food] An organism that obtains nutrients directly from the tissues of a living host, which it lives on or in and may or may not kill.

parasitism A two-species interaction in which one species directly harms another that serves as its host.

parasitoid An insect larva that grows and develops inside a host organism (usually another insect), eventually consuming the soft tissues and killing it.

parasympathetic nerve Of the autonomic nervous system, any of the nerves carrying signals that tend to slow the body down overall and divert energy to basic tasks; also work continually in opposition with sympathetic nerves to bring about minor adjustments in internal organs.

parathyroid glands (PARE-uh-THY-royd) In vertebrates, endocrine glands embedded in the thyroid gland that secrete parathyroid hormone, which helps restore blood calcium levels.

parenchyma Most abundant of the simple tissues in flowering plant roots, stems, leaves, and other parts. Its cells function in photosynthesis, storage, secretion, and other tasks.

parthenogenesis Development of an embryo from an unfertilized egg.

passive immunity Temporary immunity conferred by deliberately introducing antibodies into the body.

passive transport Diffusion of a solute through a channel or carrier protein that spans the lipid bilayer of a cell membrane. Its passage does not require an energy input; the protein passively allows the solute to follow its concentration gradient.

paternal chromosome One of the chromosomes bearing alleles that are inherited from a male parent.

pathogen (PATH-oh-jen) [Gk. *pathos*, suffering, + *-genēs*, origin]. An infectious, disease-causing agent, such as a virus or bacterium.

pattern formation Of animals, mechanisms responsible for specialization and positioning of tissues during embryonic development.

PCR Polymerase chain reaction. A method used by recombinant DNA technologists to amplify the quantity of specific fragments of DNA.

peat An accumulation of saturated, undecayed remains of plants that have been compressed by sediments.

pedigree A chart of genetic connections among individuals, as constructed according to standardized methods.

pelagic province The entire volume of ocean water; subdivided into neritic zone (relatively shallow waters overlying continental shelves) and oceanic zone (water over ocean basins).

penis A male organ that deposits sperm into a female reproductive tract.

perennial [L. *per-*, throughout, + *annus*, year] A flowering plant that lives for three or more growing seasons.

perforin A type of protein, produced and secreted by cytotoxic cells, that destroys antigen-bearing targets.

pericycle (PARE-ih-sigh-kul) [Gk. *peri-*, around, + *kyklos*, circle] Of a root vascular cylinder, one or more layers just inside the endodermis that gives rise to lateral roots and contributes to secondary growth.

periderm Of vascular plants showing secondary growth, a protective covering that replaces epidermis.

peripheral nervous system (per-IF-ur-uhl) [Gk. *peripherein*, to carry around] Of vertebrates, the nerves leading into and out from the spinal cord and brain and the ganglia along those communication lines.

peristalsis (pare-ih-STAL-sis) A rhythmic contraction of muscles that moves food forward through the animal gut.

peritoneum A lining of the coelom that also covers and helps maintain the position of internal organs.

peritubular capillaries The set of blood capillaries that threads around the tubular parts of a nephron; they function in reabsorption of water and solutes back into the body and in secretion of hydrogen ions and some other substances in the forming urine.

permafrost A permanently frozen, water-impenetrable layer beneath the soil surface in arctic tundra.

peroxisome Enzyme-filled vesicle in which fatty acids and amino acids are digested first into hydrogen peroxide (which is toxic), then to harmless products.

PGA Phosphoglycerate (FOSS-foe-GLISS-er-ate). A key intermediate in glycolysis and in the Calvin-Benson cycle.

PGAL Phosphoglyceraldehyde. A key intermediate in glycolysis and in the Calvin-Benson cycle.

pH scale A scale used to measure the concentration of free hydrogen ions in blood, water, and other solutions; pH 0 is the most acidic, 14 the most basic, and 7, neutral.

phagocyte (FAG-uh-sight) [Gk. *phagein*, to eat, + *kytos*, hollow vessel] A macrophage or certain other white blood cells that engulf and destroy foreign agents.

phagocytosis (FAG-uh-sigh-TOE-sis) [Gk. *phagein*, to eat, + *kytos*, hollow vessel] Engulfment of foreign cells or substances by amoebas and some white blood cells by means of endocytosis.

pharynx (FARE-inks) A muscular tube by which food enters the gut; in land vertebrates, the dual entrance for the tubular part of the digestive tract and windpipe (trachea).

phenotype (FEE-no-type) [Gk. *phainein*, to show, + *typos*, image] Observable trait or traits of an individual; arises from interactions between genes, and between genes and the environment.

pheromone (FARE-oh-moan) [Gk. *phero*, to carry, + *-mone*, as in hormone] A type of signaling molecule secreted by exocrine glands that serves as a communication signal between individuals of the same species. Signaling pheromones elicit an immediate

behavioral response. Priming pheromones elicit a generalized physiological response.

phloem (FLOW-um) Of vascular plants, a tissue with living cells that interconnect and form the tubes through which sugars and other solutes are conducted.

phospholipid A type of lipid that is the main structural component of cell membranes. Each has a hydrophobic tail (of two fatty acids) and a hydrophilic head that incorporates glycerol and a phosphate group.

phosphorus cycle Movement of phosphorus from rock or soil through organisms, then back to soil.

phosphorylation (FOSS-for-ih-LAY-shun) The attachment of unbound (inorganic) phosphate to a molecule; also the transfer of a phosphate group from one molecule to another, as when ATP phosphorylates glucose.

photochemical smog A brown-air smog that develops over large cities when the surrounding land forms a natural basin.

photolysis (foe-TALL-ih-sis) [Gk. *photos*, light, + *-lysis*, breaking apart] A reaction sequence of the noncyclic pathway of photosynthesis, triggered by photon energy, in which water is split into oxygen, hydrogen, and electrons.

photoperiodism A biological response to a change in the relative length of daylight and darkness.

photoreceptor Light-sensitive sensory cell.

photosynthesis The trapping of sunlight energy and its conversion to chemical energy (ATP, NADPH, or both), followed by synthesis of sugar phosphates that become converted to sucrose, cellulose, starch, and other end products. It is the main biosynthetic pathway by which energy and carbon enter the web of life.

photosynthetic autotroph An organism able to synthesize all organic molecules it requires using carbon dioxide as the carbon source and sunlight as the energy source. All plants, some protistans, and a few bacteria are photosynthetic autotrophs.

photosystem One of the clusters of light-trapping pigments embedded in photosynthetic membranes. Photosystem I operates during the cyclic pathway; photosystem II operates during both the cyclic and noncyclic pathways.

phototropism [Gk. *photos*, light, + *trope*, turning, direction] Adjustment in the direction and rate of plant growth in response to light.

photovoltaic cell A device that converts sunlight energy into electricity.

phycobilins (FIE-koe-BY-lins) A class of light-sensitive, accessory pigments that transfer absorbed energy to chlorophylls. They are abundant in red algae and cyanobacteria.

phylogeny Evolutionary relationships among species, starting with most ancestral forms and including the branches leading to their descendants.

phytochrome Light-sensitive pigment molecule, the activation and inactivation of which triggers plant hormone activities governing leaf expansion, stem branching, stem length and often seed germination and flowering.

phytoplankton (FIE-toe-PLANK-tun) [Gk. *phyton*, plant, + *planktos*, wandering] A freshwater or marine community of floating or weakly swimming photosynthetic autotrophs, such as cyanobacteria, diatoms, and green algae.

pigment A light-absorbing molecule.

pineal gland (py-NEEL) A light-sensitive endocrine gland that secretes melatonin, a hormone that influences reproductive cycles and the development of reproductive organs.

pioneer species Typically small plants with short life cycles that are adapted to growing in exposed, often windy areas with intense sunlight, wide swings in air temperature, and soils deficient in nitrogen and other nutrients. By improving conditions in areas they colonize, pioneers invite their own replacement by other species.

pituitary gland Of endocrine systems, a gland that interacts with the hypothalamus to coordinate and control many physiological functions, including the activity of many other endocrine glands. Its posterior lobe stores and secretes hypothalamic hormones; the anterior lobe produces and secretes its own hormones.

placenta (play-SEN-tuh) Of the uterus, an organ composed of maternal tissues and extraembryonic membranes (the chorion especially); it delivers nutrients to the fetus and accepts wastes from it, yet allows the fetal circulatory system to develop separately from the mother's.

plankton [Gk. *planktos*, wandering] Any community of floating or weakly swimming organisms, mostly microscopic, living in freshwater and saltwater environments. *See* phytoplankton; zooplankton.

plant The type of eukaryotic organism, usually multicelled, that is a photosynthetic autotroph—it uses sunlight energy to drive the synthesis of all its required organic compounds from carbon dioxide, water, and mineral ions. Only a few nonphotosynthetic plants obtain nutrients by parasitism and other means.

Plantae The kingdom of plants.

plasma (PLAZ-muh) Liquid component of blood; consists of water, various proteins, ions, sugars, dissolved gases, and other substances.

plasma cell Of immune systems, any of the anitbody-secreting daughter cells of a rapidly dividing population of B cells.

plasma membrane Of cells, the outermost membrane. Its lipid bilayer structure and proteins carry out most functions, including transport across the membrane and reception of extracellular signals.

plasmid Of many bacteria, a small, circular molecule of extra DNA that carries only a few genes and replicates independently of the bacterial chromosome.

plasmodesma (PLAZ-moe-DEZ-muh) Of multicelled plants, a junction between linked walls of adjacent cells through which nutrients and other substances flow.

plasticity Of the human species, the ability to remain flexible and adapt to a wide range of environments.

plate tectonics Arrangement of the earth's outer layer (lithosphere) in slablike plates, all in motion and floating on a hot, plastic layer of the underlying mantle.

platelet (PLAYT-let) Any of the cell fragments in blood that release substances necessary for clot formation.

pleiotropy (PLEE-oh-troe-pee) [Gk. *pleon*, more, + *trope*, direction] A type of gene interaction in which a single gene exerts multiple effects on seemingly unrelated aspects of an individual's phenotype.

polar body Any of three cells that form during the meiotic cell division of an oocyte; the division also forms the mature egg, or ovum.

pollen grain [L. *pollen*, fine dust] Depending on the species, the immature or mature, sperm-bearing male gametophyte of gymnosperms and flowering plants.

pollen sac In anthers of flowers, any of the chambers in which pollen grains develop.

pollen tube A tube formed after a pollen grain germinates; grows down through carpel tissues and carries sperm to the ovule.

pollination Of flowering plants, the arrival of a pollen grain on the landing platform (stigma) of a carpel.

pollutant Any substance with which an ecosystem has had no prior evolutionary experience, in terms of kinds or amounts, and that can accumulate to disruptive or harmful levels. Can be naturally occurring or synthetic.

polymer (POH-lih-mur) [Gk. *polus*, many, + *meris*, part] A molecule composed of three to millions of small subunits that may or may not be identical.

polymerase chain reaction or **PCR** DNA amplification method; DNA containing a gene of interest is split into single strands, which enzymes (polymerases) copy; the enzymes also act on the accumulating copies, multiplying the gene sequence by the millions.

polymorphism (poly-MORE-fizz-um) [Gk. *polus*, many, + *morphe*, form] Of a population, the persistence through the generations of two or more forms of a trait, at a frequency greater than can be maintained by new mutations alone.

polyp (POH-lip) Vase-shaped, sedentary stage of cnidarian life cycles.

polypeptide chain Three or more amino acids joined by peptide bonds.

polyploidy (POL-ee-PLOYD-ee) A change in the chromosome number following inheritance of three or more of each type of chromosome.

polysaccharide [Gk. *polus*, many, + *sakcharon*, sugar] A straight or branched chain of hundreds of thousands of covalently linked sugar units, of the same or different kinds. The most common polysaccharides are cellulose, starch, and glycogen.

polysome Of protein synthesis, several ribosomes all translating the same messenger RNA molecule, one after the other.

population A group of individuals of the same species occupying a given area.

population density The number of individuals of a population that are living in a specified area or volume.

population distribution The general pattern of dispersion of individuals of a population throughout their habitat.

population size The number of individuals that make up the gene pool of a population.

positive feedback mechanism Homeostatic mechanism by which a chain of events is set in motion that intensifies a change from an original condition; after a limited time, the intensification reverses the change.

post-translational controls Of eukaryotes, controls that govern modification of newly formed polypeptide chains into functional enzymes and other proteins.

predation A two-species interaction in which one species (the predator) directly harms the other (its prey).

predator [L. *prehendere*, to grasp, seize] An organism that feeds on and may or may not kill other living organisms (its prey); unlike parasites, predators do not live on or in their prey.

prediction A claim about what you can expect to observe in nature if a theory or hypothesis is correct.

premolar One of the cheek teeth; a tooth having a platform with cusps (surface bumps) that can crush, grind, and shear food.

pressure flow theory Of vascular plants, a theory that organic compounds move through phloem because of gradients in solute concentrations and pressure between source regions (such as photosynthetically active leaves) and sink regions (such as growing plant parts).

primary growth Plant growth originating at root tips and shoot tips.

primary immune response Actions by white blood cells and their products elicited by a first-time encounter with an antigen; includes both antibody-mediated and cell-mediated responses.

primary productivity, gross Of ecosystems, the rate at which the producer organisms capture and store a given amount of energy during a specified interval.

primary productivity, net Of ecosystems, the rate of energy storage in the tissues of producers in excess of their rate of aerobic respiration.

primate The mammalian lineage that includes prosimians, tarsioids, and anthropoids (monkeys, apes, and humans).

probability With respect to any chance event, the most likely number of times it will turn out a certain way, divided by the total number of all possible outcomes.

procambium (pro-KAM-bee-um) Of vascular plants, a primary meristem that gives rise to the primary vascular tissues.

producers, primary Of ecosystems, the organisms that secure energy from the physical environment, as by photosynthesis or chemosynthesis.

progesterone (pro-JESS-tuh-rown) Female sex hormone secreted by the ovaries.

prokaryotic cell (pro-CARRY-oh-tic) [L. *pro*, before, + Gk. *karyon*, kernel] A bacterium; a single-celled organism that has no nucleus or any of the other membrane-bound organelles characteristic of eukaryotic cells.

promoter Of transcription, a base sequence that signals the start of a gene; the site where RNA polymerase initially binds.

prophase Of mitosis, the stage when each duplicated chromosome starts to condense, microtubules form a spindle apparatus, and the nuclear envelope starts to break up.

prophase I Of meiosis, the stage at which the microtubular spindle starts to form, the nuclear envelope starts to break up, and each duplicated chromosome also condenses and pairs with its homologous partner. At this time, their sister chromatids typically undergo crossing over and genetic recombination.

prophase II Of meiosis, a brief stage after interkinesis during which each chromosome still consists of two chromatids.

protein Large organic compound composed of one or more chains of amino acids held together by peptide bonds. Proteins have unique sequences of different kinds of amino acids in their polypeptide chains; such sequences are the basis of a protein's three-dimensional structure and chemical behavior.

Protista The kingdom of protistans.

protistan (pro-TISS-tun) [Gk. *prōtistos*, primal, very first] Single-celled eukaryote.

proto-oncogene A gene sequence similar to an oncogene but that codes for a protein required in normal cell function; may trigger cancer, generally when specific mutations alter its structure or function.

proton Positively charged particle, one or more of which is present in the atomic nucleus.

protostome (PRO-toe-stome) [Gk. *proto*, first, + *stoma*, mouth] A bilateral animal in which the first indentation in the early embryo develops into the mouth. Includes mollusks, annelids, and anthropods.

protozoan A type of protistan, some predatory and others parasitic; so named because they may resemble the single-celled heterotrophs that presumably gave rise to animals.

proximal tubule Of a nephron, the tubular region that receives water and solutes filtered from the blood.

pulmonary circuit Blood circulation route leading to and from the lungs.

Punnett-square method A way to predict the possible outcome of a mating or an experimental cross in simple diagrammatic form.

purine Nucleotide base having a double ring structure. Adenine and guanine are examples.

pyrimidine (phi-RIM-ih-deen) Nucleotide base having a single ring structure. Cytosine and thymine are examples.

pyruvate (PIE-roo-vate) A compound with a backbone of three carbon atoms. Two pyruvate molecules are the end products of glycolysis.

r Designates net population growth rate; the birth and death rates are assumed to remain constant and so are combined into this one variable for population growth equations.

radial symmetry Body plan having four or more roughly equivalent parts arranged around a central axis.

radioisotope An unstable atom that has dissimilar numbers of protons and neutrons and that spontaneously decays (emits electrons and energy) to a new, stable atom that is not radioactive.

rain shadow A reduction in rainfall on the leeward side of high mountains, resulting in arid or semiarid conditions.

reabsorption Of urine formation, the diffusion or active transport of water and usable solutes out of a nephron and into capillaries leading back to the general circulation; regulated by ADH and aldosterone.

receptor Of cells, a molecule at the surface of the plasma membrane or in the cytoplasm that binds molecules present in the extracellular environment. The binding triggers changes in cellular activities. Of nervous systems, a sensory cell or cell part that may be activated by a specific stimulus.

receptor protein Protein that binds a signaling molecule such as a hormone, then triggers alterations in cell behavior or metabolism.

recessive allele [L. *recedere*, to recede] In heterozygotes, an allele whose expression is fully or partially masked by expression of its partner; fully expressed only in the homozygous recessive condition.

recognition protein Protein at cell surface recognized by cells of like type; helps guide the ordering of cells into tissues during development and functions in cell-to-cell interactions.

recombinant technology Procedures by which DNA (genes) from different species may be isolated, cut, spliced together, and the new recombinant molecules multiplied in quantity in a population of rapidly dividing cells such as bacteria.

red blood cell Erythrocyte; an oxygen-transporting cell in blood.

red marrow A substance in the spongy tissue of many bones that serves as a major site of blood cell formation.

reflex [L. *reflectere*, to bend back] A simple, stereotyped movement elicited directly by sensory stimulation.

reflex pathway [L. *reflectere*, to bend back] Type of neural pathway in which signals from sensory neurons directly stimulate or inhibit motor neurons, without intervention by interneurons.

refractory period Of neurons, the period following an action potential at a given patch of membrane when sodium gates are shut and potassium gates are open, so that the patch is insensitive to stimulation.

regulatory protein A protein that enhances or suppresses the rate at which a gene is transcribed.

releasing hormone A hypothalamic signaling molecule that stimulates or slows down secretion by target cells in the anterior lobe of the pituitary gland.

repressor protein Regulatory protein that provides negative control of gene activity by

preventing RNA polymerase from binding to DNA.

reproduction In biology, processes by which a new generation of cells or multicelled individuals is produced. Sexual reproduction requires meiosis, formation of gametes, and fertilization. Asexual reproduction refers to the production of new individuals by any mode that does not involve gametes.

reproduction, sexual Mode of reproduction that begins with meiosis, proceeds through gamete formation, and ends at fertilization.

reproductive isolating mechanism Any aspect of structure, functioning, or behavior that restricts gene flow between two populations.

reproductive isolation An absence of gene flow between populations.

reproductive success The survival and production of the offspring of an individual.

reptile A type of carnivorous vertebrate; its ancestors were the first vertebrates to escape dependency of standing water, largely by means of internal fertilization and amniote eggs. They include turtles, crocodiles, lizards and snakes, and tuataras.

resource partitioning A community pattern in which similar species generally share the same kind of resource in different ways, in different areas, or at different times.

respiration [L. *respirare*, to breathe] In most animals, the overall exchange of oxygen from the environment for carbon dioxide wastes from cells by way of circulating blood. *Compare* aerobic respiration.

respiratory surface The surface, such as a thin epithelial layer, that gases diffuse across to enter and leave the animal body.

respiratory system An organ system that functions in respiration.

resting membrane potential Of neurons and other excitable cells that are not being stimulated, the steady voltage difference across the plasma membrane.

restriction enzymes Class of bacterial enzymes that cut apart foreign DNA injected into them, as by viruses; also used in recombinant DNA technology.

reticular formation Of the vertebrate brainstem, a major network of interneurons that helps govern activity of the whole nervous system.

reverse transcriptase Viral enzyme required for reverse transcription of mRNA into DNA; used in recombinant DNA technology.

reverse transcription Assembly of DNA on a single-stranded mRNA molecule by viral enzymes.

RFLPs Restriction fragment length polymorphisms. Of DNA samples from different individuals, slight but unique differences in the banding pattern of fragments of the DNA that have been cut with restriction enzymes.

Rh blood typing A method of characterizing red blood cells on the basis of a protein that serves as a self-marker at their surface; Rh$^+$ signifies its presence and Rh$^-$, its absence.

rhizoid Rootlike absorptive structure of some fungi and nonvascular plants.

ribosomal RNA (rRNA) Type of RNA molecule that combines with proteins to form ribosomes, on which the polypeptide chains of proteins are assembled.

ribosome In all cells, the structure at which amino acids are strung together in specified sequence to form the polypeptide chains of proteins. An intact ribosome consists of two subunits, each composed of ribosomal RNA and protein molecules.

RNA Ribonucleic acid. A category of single-stranded nucleic acids that function in processes by which genetic instructions are used to build proteins.

rod cell A vertebrate photoreceptor sensitive to very dim light and that contributes to coarse perception of movement.

root hair Of vascular plants, an extension of a specialized root epidermal cell; root hairs collectively enhance the surface area available for absorbing water and solutes.

root nodule A localized swelling on the roots of certain legumes and other plants that contain symbiotic, nitrogen-fixing bacteria.

roots Descending parts of a vascular plant that absorb water and nutrients, anchor aboveground parts, and usually store food.

rotifer An invertebrate common in food webs of lakes and ponds.

roundworm A type of parasitic or scavenging invertebrate with a bilateral, cylindrical, cuticle-covered body, usually tapered at both ends. Some cause diseases in humans.

RuBP Ribulose biphosphate. A compound with a backbone of five carbon atoms that is required for carbon fixation in the Calvin-Benson cycle of photosynthesis.

S-shaped curve A curve, obtained when population size is plotted against time, that is characteristic of logistic growth.

sac fungus A type of fungus, usually multicelled, with spores that develop inside the cells of reproductive structures shaped like globes, flasks, or dishes. Yeasts (single-celled) are also in this group.

salination A salt buildup in soil as a result of evaporation, poor drainage, and often the importation of mineral salts in irrigation water.

salivary gland Any of the glands that secrete saliva, a fluid that initially mixes with food in the mouth and starts the breakdown of starch.

salt An ionic compound formed when an acid reacts with a base.

saltatory conduction In myelinated neurons, rapid, node-to-node hopping of action potentials.

saprobe Heterotroph that obtains its nutrients from nonliving organic matter. Most fungi are saprobes.

sarcomere (SAR-koe-meer) Of vertebrate muscles, the basic unit of contraction; a region of myosin and actin filaments organized in parallel between two Z lines of a myofibril inside a muscle cell.

sarcoplasmic reticulum (sar-koe-PLAZ-mik reh-TIK-you-lum) In muscle cells, a membrane system that takes up, stores, and releases the calcium ions required for cross-bridge formation in sarcomeres, hence for contraction.

Schwann cells Specialized neuroglial cells that grow around neuron axons, forming a myelin sheath.

sclerenchyma One of the simple tissues of flowering plants, generally with cells having thick, lignin-impregnated walls. It supports mature plant parts and often protects seeds.

sea squirt An invertebrate chordate with a leathery or jellylike tunic and bilateral, free-swimming larvae that look like tadpoles. Ancient sea squirts may have resembled animals that gave rise to vertebrates.

second law of thermodynamics Law stating that the spontaneous direction of energy flow is from organized (high-quality) to less organized (low-quality) forms. With each conversion, some energy is randomly dispersed in a form, usually heat, that is not as readily available to do work.

second messenger A molecule inside a cell that mediates and generally triggers amplified response to a hormone.

secondary immune response Rapid, prolonged response by white blood cells, memory cells especially, to a previously encountered antigen.

secondary sexual trait A trait that is associated with maleness or femaleness but that does not play a direct role in reproduction.

secretion Generally, the release of a substance for use by the organism producing it. (Not the same as *excretion*, the expulsion of excess or waste material.) Of kidneys, a regulated stage in urine formation, in which ions and other substances move from capillaries into nephrons.

sedimentary cycle A biogeochemical cycle without a gaseous phase; the element moves from land to the seafloor, then returns only through long-term geological uplifting.

seed Of gymnosperms and flowering plants, a fully mature ovule (contains the plant embryo), with its integuments forming the seed coat.

segmentation Of earthworms and many other animals, a series of body units that may be externally similar to or quite different from one another.

segregation, Mendelian principle of [L. *se-*, apart, + *grex*, herd] The principle that diploid organisms inherit a pair of genes for each trait (on a pair of homologous chromosomes) and that the two genes segregate during meiosis and end up in separate gametes.

selective gene expression Of multicelled organisms, activation or suppression of a fraction of the genes in unique ways in different cells, leading to pronounced differences in structure and function among different cell lineages.

selfish behavior A behavior by which an individual protects or increases its own chance of producing offspring, regardless of the consequences of the group to which it belongs.

selfish herd A simple society held together by reproductive self-interest.

semen (SEE-mun) [L. *serere*, to sow] Sperm-bearing fluid expelled from a penis during male orgasm.

semiconservative replication [Gk. *hēmi*, half, + L. *conservare*, to keep] Reproduction of a DNA molecule when a complementary strand forms on each of the unzipping strands of an existing DNA double helix, the outcome being two "half-old, half-new" molecules.

senescence (sen-ESS-cents) [L. *senescere*, to grow old] Sum total of processes leading to the natural death of an organism or some of its parts.

sensation The conscious awareness of a stimulus.

sensory neuron Any of the nerve cells that act as sensory receptors, detecting specific stimuli (such as light energy) and relaying signals to the brain and spinal cord.

sensory system The "front door" of a nervous system; that portion of a nervous system that receives and sends on signals of specific changes in the external and internal environments.

sessile animal Animal that remains attached to a substrate during some stage (often the adult) of its life cycle.

sex chromosome Of most animals and some plants, a chromosome whose presence determines a new individual's gender. *Compare* autosomes.

sexual dimorphism Phenotypic differences between males and females of a species.

sexual reproduction Production of offspring from the union of gametes from two parents, by way of meiosis, gamete formation, and fertilization.

sexual selection A microevolutionary process; natural selection favoring a trait that gives the individual a competitive edge in reproductive success.

shifting cultivation The cutting and burning of trees, followed by tilling of ashes into the soil; once called slash-and-burn agriculture.

shoots The above-ground parts of vascular plants.

sieve tube member Of flowering plants, a cellular component of the interconnecting conducting tubes in phloem.

sink region In a vascular plant, any region using or stockpiling organic compounds for growth and development.

sister chromatids Of a duplicated chromosome, two DNA molecules (and associated proteins) that remain attached at their centromere only during nuclear division. Each ends up in a separate daughter nucleus.

skeletal muscle In vertebrates, an organ that contains hundreds to many thousands of muscle cells, arranged in bundles that are surrounded by connective tissue. The connective tissue extends beyond the muscle (as tendons that attach it to bone).

sliding filament model Model of muscle contraction, in which myosin filaments physically slide along and pull two sets of actin filaments toward the center of the sarcomere, which shortens. The sliding requires ATP energy and cross-bridge formation between the actin and myosin.

slime mold A type of heterotrophic protistan with a life cycle that includes free-living cells that at some point congregate and differentiate into spore-bearing structures.

small intestine Of vertebrates, the portion of the digestive system where digestion is completed and most nutrients absorbed.

smog, industrial Gray-colored air pollution that predominates in industrialized cities with cold, wet winters.

smog, photochemical Form of brown, smelly air pollution occurring in large cities with warm climates.

social behavior Cooperative, interdependent relationships among animals of the same species.

social parasite Animal that depends on the social behavior of another species to gain food, care for young, or some other factor to complete its life cycle.

sodium-potassium pump A transport protein spanning the lipid bilayer of the plasma membrane. When activated by ATP, its shape changes and it selectively transports sodium ions out of the cell and potassium ions in.

solute (SOL-yoot) [L. *solvere*, to loosen] Any substance dissolved in a solution. In water, this means spheres of hydration surround the charged parts of individual ions or molecules and keep them dispersed.

solvent Fluid in which one or more substances is dissolved.

somatic cell (so-MAT-ik) [Gk. *sōma*, body] Of animals, any cell that is not a germ cell (which gives rise to gametes).

somatic nervous system Those nerves leading from the central nervous system to skeletal muscles.

sound system Of birds, the brain regions that govern muscles of the vocal organ.

source region Of vascular plants, any of the sites of photosynthesis.

speciation (spee-cee-AY-shun) The evolutionary process by which species originate. One speciation route starts with divergence of two reproductively isolated populations of a species. They become separate species when accumulated differences in allele frequencies prevent them from interbreeding successfully under natural conditions. Speciation also may be instantaneous (by way of polyploidy, especially among self-fertilizing plants).

species (SPEE-sheez) [L. *species*, a kind] Of sexually reproducing organisms, a unit consisting of one or more populations of individuals that can interbreed under natural conditions to produce fertile offspring that are reproductively isolated from other such units.

sperm [Gk. *sperma*, seed] A type of mature male gamete.

spermatogenesis (sperm-AT-oh-JEN-ih-sis) Formation of a mature sperm from a germ cell.

sphere of hydration Through positive or negative interactions, a clustering of water molecules around the individual molecules of a substance placed in water. *Compare* solute.

sphincter (SFINK-tur) Ring of muscle between regions of a tubelike system (as between the stomach and small intestine).

spinal cord Of central nervous systems, the portion threading through a canal inside the vertebral column and providing direct reflex connections between sensory and motor neurons as well as communication lines to and from the brain.

spindle apparatus A type of bipolar structure that forms during mitosis or meiosis and that moves the chromosomes. It consists of two sets of microtubules that extend from the opposite poles and that overlap at the spindle's equator.

spleen One of the lymphoid organs; it is a filtering station for blood, a reservoir of red blood cells, and a reservoir of macrophages.

sponge An invertebrate having a body with no symmetry and no organs; a framework of glassy needles and other structures imparts shape to it. Distinctive for its food-gathering, flagellated collar cells.

sporangium (spore-AN-gee-um), plural **sporangia** [Gk. *spora*, seed] The protective tissue layer that surrounds haploid spores in a sporophyte.

spore Of land plants, a type of resistant cell, often walled, that forms between the time of meiosis and fertilization. It germinates and develops into a gametophyte, the actual gamete-producing body. Of most fungi, a walled, resistant cell or multicelled structure, produced by mitosis or meiosis, that can germinate and give rise to a new mycelium.

sporophyte [Gk. *phyton*, plant] Of plant life cycles, a vegetative body that grows (by mitosis) from a zygote and that produces the spore-bearing structures.

sporozoan One of four categories of protozoans; a parasite that produces sporelike infectious agents called sporozoites. Some cause serious diseases in humans.

spring overturn Of certain lakes, the movement of dissolved oxygen from the surface layer to the depths and movement of nutrients from bottom sediments to the surface.

stabilizing selection Of a population, a persistence over time of the alleles responsible for the most common phenotypes.

stamen (STAY-mun) Of flowering plants, a male reproductive structure; commonly consists of pollen-bearing structures (anthers) on single stalks (filaments).

start codon Of protein synthesis, a base triplet in a strand of mRNA that serves as the start signal for mRNA translation.

stem cell Of animals, one of the unspecialized cells that replace themselves by ongoing mitotic divisions; portions of their daughter cells also divide and differentiate into specialized cells.

steroid (STAIR-oid) A lipid with a backbone of four carbon rings and with no fatty acid tails. Steroids differ in their functional groups. Different types have roles in metabolism, intercellular communication, and (in animals) cell membranes.

steroid hormone A type of lipid-soluble hormone, synthesized from cholesterol, that diffuses directly across the lipid bilayer of a target cell's plasma membrane and that binds with a receptor inside that cell.

stigma Of many flowering plants, the sticky or hairy surface tissue on the upper portion of the ovary that captures pollen grains and favors their germination.

stimulus [L. *stimulus*, goad] A specific change in the environment, such as a variation in light, heat, or mechanical pressure, that the body can detect through sensory receptors.

stoma (STOW-muh), plural **stomata** [Gk. *stoma*, mouth] A controllable gap between two guard cells in stems and leaves; any of the small passageways across the epidermis through which carbon dioxide moves into the plant and water vapor moves out.

stomach A muscular, stretchable sac that receives ingested food; of vertebrates, an organ between the esophagus and intestine in which considerable protein digestion occurs.

stop codon Of protein synthesis, a base triplet in a strand of mRNA that serves as the stop signal for translation, so that no more amino acids are added to the polypeptide chain.

stream A flowing-water ecosystem that starts out as a freshwater spring or seep.

stroma [Gk. *strōma*, bed] Of chloroplasts, the semifluid interior between the thylakoid membrane system and the two outer membranes; the zone where sucrose, starch, and other end products of photosynthesis are assembled.

stromatolite Of shallow seas, layered structures formed from sediments and large mats of the slowly accumulated remains of photosynthetic populations.

substrate A reactant or precursor molecule for a metabolic reaction; a specific molecule or molecules that an enzyme can chemically recognize, bind briefly to itself, and modify in a specific way.

substrate-level phosphorylation The direct, enzyme-mediated transfer of a phosphate group from the substrate of a reaction to another molecule. An example is the transfer of phosphate from an intermediate of glycolysis to ADP, forming ATP.

succession, primary (suk-SESH-un) [L. *succedere*, to follow after] Orderly changes from the time pioneer species colonize a barren habitat through replacements by various species until the climax community, when the composition of species remains steady under prevailing conditions.

succession, secondary Orderly changes in a community or patch of habitat toward the climax state after having been disturbed, as by fire.

surface-to-volume ratio A mathematical relationship in which volume increases with the cube of the diameter, but surface area increases only with the square. Of growing cells, the volume of cytoplasm increases more rapidly than the surface area of the plasma membrane that must service the cytoplasm. Because of this constraint, cells generally remain small or elongated, or have elaborate membrane foldings.

survivorship curve A plot of the age-specific survival of a group of individuals in a given environment, from the time of their birth until the last one dies.

symbiosis (sim-by-OH-sis) [Gk. *sym*, together, + *bios*, life, mode of life] A state in which two or more species live together in close association (one lives with, in, or on the other). Commensalism, mutualism, and parasitism are examples of symbiotic interactions.

sympathetic nerve Of the autonomic nervous system, any of the nerves generally concerned with increasing overall body activities during times of heightened awareness, excitement, or danger; also work continually in opposition with parasympathetic nerves to bring about minor adjustments in internal organs.

sympatric speciation [Gk. *sym*, together, + *patria*, native land] Speciation that follows after ecological, behavioral, or genetic barriers arise within the boundaries of a single population. This can happen instantaneously, as when polyploidy arises in a type of flowering plant that can self-fertilize or reproduce asexually.

synaptic integration (sin-AP-tik) The moment-by-moment combining of excitatory and inhibitory signals arriving at a trigger zone of a neuron.

systematics Branch of biology that deals with patterns of diversity among organisms in an evolutionary context; its three approaches include taxonomy, phylogenetic reconstruction, and classification.

systemic circuit (sis-TEM-ik) Circulation route in which oxygen-enriched blood flows from the lungs to the left half of the heart, through the rest of the body (where it gives up oxygen and takes on carbon dioxide), then back to the right side of the heart.

T lymphocyte A white blood cell with roles in immune responses.

tactile signal A physical touching that carries social significance.

taproot system A primary root and its lateral branchings.

target cell Of hormones and other signaling molecules, any cell having receptors to which they can bind.

taxonomy (tax-ON-uh-mee) Approach in biological systematics that involves identifying organisms and assigning names to them.

telophase (TEE-low-faze) Of mitosis, the final stage when chromosomes decondense into threadlike structures and two daughter nuclei form. Of meiosis I, the stage when one of each pair of homologous chromosomes has arrived at one or the other end of the spindle pole. At telophase II, chromosomes decondense and four daughter nuclei form.

telophase II Of meiosis, final stage when four daughter nuclei form.

temperate pathway A viral infection that enters a latent period; the host is not killed outright.

tendon A cord or strap of dense connective tissue that attaches muscle to bones.

territory An area that one or more individuals defend against competitors.

test An attempt to produce actual observations that match predicted or expected observations.

testcross Experimental cross to reveal whether an organism is homozygous dominant or heterozygous for a trait. The organism showing dominance is crossed to an individual known to be homozygous recessive for the same trait.

testis, plural **testes** Male gonad; primary reproductive organ in which male gametes and sex hormones are produced.

testosterone (tess-TOSS-tuh-rown) In male mammals, a major sex hormone that helps control male reproductive functions.

tetanus Of muscles, a large contraction in which repeated stimulation of a motor unit causes muscle twitches to mechanically run together. In a disease by the same name, toxins prevent muscle relaxation.

theory A testable explanation of a broad range of related phenomena. In modern science, only explanations that have been extensively tested and can be relied upon with a very high degree of confidence are accorded the status of theory.

thermal inversion Situation in which a layer of dense, cool air becomes trapped beneath a layer of warm air; can cause air pollutants to accumulate to dangerous levels close to the ground.

thermophile A type of archaebacterium that lives in hot springs, highly acidic soils, and near hydrothermal vents.

thermoreceptor Sensory cell that can detect radiant energy associated with temperature.

thigmotropism (thig-MOTE-ruh-pizm) [Gk. *thigm*, touch] Of vascular plants, growth orientation in response to physical contact with a solid object, as when a vine curls around a fencepost.

threshold Of neurons and other excitable cells, a certain minimum amount by which the voltage difference across the plasma membrane must change to produce an action potential.

thylakoid membrane system Of chloroplasts, an internal membrane system commonly folded into flattened channels and disks (*grana*) and containing light-absorbing pigments and enzymes used in the formation of ATP, NADPH, or both during photosynthesis.

thymine Nitrogen-containing base in some nucleotides.

thymus gland A lymphoid organ with endocrine functions; lymphocytes of the immune system multiply, differentiate, and mature in its tissues, and its hormone secretions affect their functions.

thyroid gland Of endocrine systems, a gland that produces hormones that affect overall metabolic rates, growth, and development.

tissue Of multicelled organisms, a group of cells and intercellular substances that function together in one or more specialized tasks.

tonicity The relative concentrations of solutes in two fluids, such as inside and outside a cell. When solute concentrations are isotonic (equal in both fluids), water shows no net osmotic movement in either direction. When one fluid is hypotonic (has less solutes than the other), the other is hypertonic (has more

solutes) and is the direction in which water tends to move.

tooth Of the mouth of various animals, one of the hardened appendages used to secure or mechanically pummel food; sometimes used in defense.

tracer A radioisotope used to label a substance so that its pathway or destination in a cell, organism, ecosystem, or some other system can be tracked, as by scintillation counters that detect its emissions.

trachea (TRAY-kee-uh), plural **tracheae** An air-conducting tube that functions in respiration; of land vertebrates, the windpipe, which carries air between the larynx and bronchi.

tracheal respiration Of insects, spiders, and some other animals, a respiratory system consisting of finely branching tracheae that extend from openings in the integument and that dead-end in body tissues.

tracheid (TRAY-kid) Of flowering plants, one of two types of cells in xylem that conduct water and dissolved minerals.

transcript-processing controls Of eukaryotic cells, controls that govern modification of new mRNA molecules into mature transcripts before shipment from the nucleus.

transcription [L. *trans*, across, + *scribere*, to write] Of protein synthesis, the assembly of an RNA strand on one of the two strands of a DNA double helix; the base sequence of the resulting transcript is complementary to the DNA region on which it was assembled.

transcriptional controls Of eukaryotic cells, controls influencing when and to what degree a particular gene will be transcribed.

transfer RNA (tRNA) Of protein synthesis, any of the type of RNA molecules that bind and deliver specific amino acids to ribosomes *and* pair with mRNA code words for those amino acids.

translation Of protein synthesis, the conversion of the coded sequence of information in mRNA into a particular sequence of amino acids to form a polypeptide chain; depends on interactions of rRNA, tRNA, and mRNA.

translational controls Of eukaryotic cells, controls governing the rates at which mRNA transcripts that reach the cytoplasm will be translated into polypeptide chains at ribosomes.

translocation Of cells, a change in a chromosome's structure following the insertion of part of a nonhomologous chromosome into it. Of vascular plants, conduction of organic compounds through the plant body by way of the phloem.

transpiration Evaporative water loss from stems and leaves.

transport control Of eukaryotic cells, controls governing when mature mRNA transcripts are shipped from the nucleus into the cytoplasm.

transposable element DNA element that can spontaneously "jump" to new locations in the same DNA molecule or a different one. Such elements often inactivate the genes into which they become inserted and give rise to observable changes in phenotype.

trisomy (TRY-so-mee) Of diploid cells, the abnormal presence of three of one type of chromosome.

trophic level (TROE-fik) [Gk. *trophos*, feeder] All the organisms in an ecosystem that are the same number of transfer steps away from the energy input into the system.

tropical rain forest A type of biome where rainfall is regular and heavy, the annual mean temperature is 25°C, and humidity is 80 percent or more; characterized by great biodiversity.

tropism (TROE-prizm) Of vascular plants, a growth response to an environmental factor, such as growth toward light.

true-breeding Of sexually reproducing organisms, a lineage in which the offspring of successive generations are just like the parents in one or more traits.

tumor A tissue mass composed of cells that are dividing at an abnormally high rate.

turgor pressure (TUR-gore) [L. *turgere*, to swell] Internal pressure applied to a cell wall when water moves by osmosis into the cell.

uniformitarianism The theory that existing geologic features are an outcome of a long history of gradual changes, interrupted now and then by huge earthquakes and other catastrophic events.

upwelling An upward movement of deep, nutrient-rich water along coasts to replace surface waters that winds move away from shore.

uracil (YUR-uh-sill) Nitrogen-containing base found in RNA molecules; can base-pair with adenine.

ureter A tubular channel for urine flow between the kidney and urinary bladder.

urethra A tubular channel for urine flow between the urinary bladder and an opening at the body surface.

urinary bladder A distensible sac in which urine is temporarily stored before being excreted.

urinary excretion A mechanism by which excess water and solutes are removed by way of a urinary system.

urinary system An organ system that adjusts the volume and composition of blood, and so helps maintain extracellular fluid.

urine Fluid formed by filtration, reabsorption, and secretion in kidneys; consists of wastes, excess water, and solutes.

uterus (YOU-tur-us) [L. *uterus*, womb] Chamber in which the developing embryo is contained and nurtured during pregnancy.

vaccine An antigen-containing preparation, swallowed or injected, that increases immunity to certain diseases. It induces formation of huge armies of effector and memory B and T cell populations.

vagina Part of a female reproductive system that receives sperm, forms part of the birth canal, and channels menstrual flow to the exterior.

variable Of a scientific experiment, the only factor that is not exactly the same in the experimental group as it is in the control group.

vascular bundle Of vascular plants, the arrangement of primary xylem and phloem into multistranded, sheathed cords that thread lengthwise through the ground tissue system.

vascular cambium Of vascular plants, a lateral meristem that increases stem or root diameter.

vascular cylinder Of plant roots, the arrangement of vascular tissues as a central cylinder.

vascular plant Plant having tissues that transport water and solutes through well-developed roots, stems, and leaves.

vascular tissue system Xylem and phloem; the conducting tissues that distribute water and solutes through the body of vascular plants.

vein Of the circulatory system, any of the large-diameter vessels that lead back to the heart; of leaves, one of the vascular bundles that thread through photosynthetic tissues.

ventricle (VEN-tri-kuhl) Of the vertebrate heart, one of two chambers from which blood is pumped out. *Compare* atrium.

venule A small blood vessel that accepts blood from capillaries and delivers it to a vein; also overlaps capillaries somewhat in function.

vernalization Of flowering plants, stimulation of flowering by exposure to low temperatures.

vertebra, plural **vertebrae** Of vertebrate animals, one of a series of hard bones arranged with intervertebral disks into a backbone.

vertebrate Animal having a backbone of bony segments, the vertebrae.

vesicle (VESS-ih-kul) [L. *vesicula*, little bladder] Within the cytoplasm of cells, one of a variety of small membrane-bound sacs that function in the transport, storage, or digestion of substances or in some other activity.

vessel member One of the cells of xylem, dead at maturity, the walls of which form the water-conducting pipelines.

villus (VIL-us), plural **villi** Any of several types of absorptive structures projecting from the free surface of an epithelium.

viroid An infectious nucleic acid that has no protein coat; a tiny rod or circle of single-stranded RNA.

virus A noncellular infectious agent, consisting of DNA or RNA and a protein coat; can replicate only after its genetic material enters a host cell and subverts its metabolic machinery.

vision Precise light focusing onto a layer of photoreceptive cells that is dense enough to sample details concerning a given light stimulus, followed by image formation in the brain.

visual signal An observable action or cue that functions as a communication signal.

vitamin Any of more than a dozen organic substances that animals require in small amounts for normal cell metabolism but generally cannot synthesize for themselves.

vocal cord One of the thickened, muscular folds of the larynx that help produce sound waves for speech.

water mold A type of saprobic or parasitic fungus that lives in fresh water or moist soil.

water potential The sum of two opposing forces (osmosis and turgor pressure) that can cause the directional movement of water into or out of a walled cell.

water table The upper limit at which the ground in a specified region is fully saturated with water.

watershed Any specified region in which all precipitation drains into a single stream or river.

wax A type of lipid with long-chain fatty acid tails that help form protective, lubricating, or water-repellent coatings.

white blood cell Leukocyte; of vertebrates, any of the macrophages, eosinophils, neutrophils, and other cells which, together with their products, comprise the immune system.

white matter Of spinal cords, major nerve tracts so named because of the glistening myelin sheaths of their axons.

wild-type allele Of a population, the allele that occurs normally or with greatest frequency at a given gene locus.

wing Of birds, a forelimb of feathers, powerful muscles, and lightweight bones that functions in flight. Of insects, a structure that develops as a lateral fold of the exoskeleton and functions in flight.

X chromosome Of humans, a sex chromosome with genes that cause an embryo to develop into a female, provided that it inherits a pair of these.

X-linked gene Any gene on an X chromosome.

X-linked recessive inheritance Recessive condition in which the responsible, mutated gene occurs on the X chromosome.

xylem (ZYE-lum) [Gk. *xylon*, wood] Of vascular plants, a tissue that transports water and solutes through the plant body.

Y chromosome Of humans, a sex chromosome with genes that cause the embryo that inherited it to develop into a male.

Y-linked gene Any gene on a Y chromosome.

yellow marrow A fatty tissue in the cavities of most mature bones that produces red blood cells when blood loss from the body is severe.

yolk sac Of land vertebrates, one of four extraembryonic membranes. In most shelled eggs, it holds nutritive yolk. In humans, part becomes a site of blood cell formation and some of its cells give rise to the forerunners of gametes.

zero population growth A population for which the number of births is balanced by the number of deaths over a specified period, assuming immigration and emigration also are balanced.

zooplankton A freshwater or marine community of floating or weakly swimming heterotrophs, mostly microscopic, such as rotifers and copepods.

zygospore-forming fungus A type of fungus for which a thick spore wall forms around the zygote; this resting spore germinates and gives rise to stalked, spore-bearing structures.

zygote (ZYE-goat) The first cell of a new individual, formed by the fusion of a sperm nucleus with the nucleus of an egg (fertilization).

CREDITS AND ACKNOWLEDGMENTS

Front Matter

Pages vi–vii David Macdonald / **Pages viii–ix** James M. Bell/Photo Researchers / **Pages x–xi** S. Stammers/SPL/Photo Researchers / **Pages xii–xiii** John Alcock / **Pages xiv–xv** Gary Head / **Pages xvi–xvii** Lennart Nilsson from *Behold Man*, ©1974 Albert Bonniers Forlag and Little, Brown and Company, Boston / **Pages xviii–xix** Lennart Nilsson from *A Child Is Born*, ©1966, 1977 Dell Publishing Company, Inc. / **Pages xx–xxi** / Jim Doran

Page 1 Tom Van Sant/The GeoSphere Project, Santa Monica, CA

Chapter 1

1.1 Frank Kaczmarek / **Page 4** Art by American Composition and Graphics / **1.3** (a) Walt Anderson/Visuals Unlimited; (b) Gregory Dimijian/Photo Researchers; (c) Alan Weaving/Ardea, London / **1.4** Jack deConingh / **1.5** J. A. Bishop and L. M. Cook / **1.6** (a) Tony Brain/SPL/Photo Researchers; (b) M. Abbey/Visuals Unlimited; (c), (e) Edward S. Ross; (d) Dennis Brokaw; (f), (g) Pat & Tom Leeson/Photo Researchers / **1.7** Levi Publishing Company / **1.8** Photograph Jack deConingh; art by Raychel Ciemma

Page 15 James M. Bell/Photo Researchers

Chapter 2

2.1 Martin Rogers/FPG / **2.2** Jack Carey / **Page 19** (left) Kingsley R. Stern; (right) (a) Hank Morgan/Rainbow; (b) Dr. Harry T. Chugani, M.D., UCLA School of Medicine / **2.8** Art by Palay/ Beaubois / **2.9** Photograph Richard Riley/FPG; art by Raychel Ciemma / **2.10** H. Eisenbeiss/Frank Lane Picture Agency / **2.11** Art by Raychel Ciemma / **2.13** Michael Grecco/Picture Group / **Page 27** Art by Raychel Ciemma / **2.20** David Scharf/Peter Arnold, Inc. / **2.21** Art by Precision Graphics / **2.22** Clem Haagner/Ardea, London / **2.23** (a) Art by Precision Graphics; (c) micrograph Lewis L. Lainey; (d) Larry Lefever/Grant Heilman; (e) Kenneth Lorenzen / **2.24, 2.25** Art by Precision Graphics / **2.27** Art by Palay/Beaubois / **2.28, 2.29** Art by Precision Graphics

Chapter 3

3.1 (a) (left) National Library of Medicine; (right) Armed Forces Institute of Pathology; (b) The Bettmann Archive; (d) The Francis A. Countway Library of Medicine / **Page 42** Art by Raychel Ciemma / **3.2** Art by Raychel Ciemma / **3.3** Art by Raychel Ciemma; micrograph Driscoll, Youngquist, and Baldeschwieler/Cal Tech/Science Source/Photo Researchers / **3.4** Jeremy Pickett-Heaps, School of Botany, University of Melbourne / **3.5** Art by Raychel Ciemma and American Composition & Graphics / **3.6** Micrograph M. C. Ledbetter, Brookhaven National Laboratory; art by Raychel Ciemma and American Composition & Graphics / **3.7** Micrograph G. L. Decker; art by Raychel Ciemma and American Composition & Graphics / **3.8** Micrograph Stephen L. Wolfe / **3.9** (a) (left) Don W. Fawcett/Visuals Unlimited; (right) A. C. Faberge, *Cell and Tissue Research*, 151:403–415, 1974 / **3.10** Art by Raychel Ciemma and American Composition & Graphics / **3.11** (a), (b) Micrographs Don W. Fawcett/Visuals Unlimited; (below) art by Robert Demarest / **3.12** Micrograph Gary W. Grimes; art by Robert Demarest after a model by J. Kephart / **3.13** Art by Leonard Morgan; (c) micrographs M. M. Perry and A. B. Gilbert / **3.15** Micrograph Keith R. Porter / **3.16** Micrograph L. K. Shumway; (below) (right) art by Palay/Beaubois / **3.17** Art by Raychel Ciemma / **3.18** (a) J. Victor Small and Gottfried Rinnerthaler; art by Precision Graphics / **3.19** (a) Art by Precision Graphics after Stephen L. Wolfe, *Molecular and Cellular Biology*, Wadsworth, 1993; (b) C. J. Brokaw; (c) Sidney L. Tamm / **3.20** (a) Art by Raychel Ciemma; (b) micro-graph G. Cohen–Bazire; (c) K. G. Murti/Visuals Unlimited; (d) R. Calentine/ Visuals Unlimited; (e) Gary Gaard and Arthur Kelman

Chapter 4

4.1 Gary Head / **4.2** Evan Cerasoli / **4.3** (above) NASA; (below) Manfred Kage/Peter Arnold, Inc. / **4.4** Art by Palay/Beaubois / **4.6** Micrographs M. Sheetz, R. Painter, and S. Singer, *Journal of Cell Biology*, 70:193, by copyright permission of The Rockefeller University Press / **4.7** (a) Photographs Frank B. Salisbury / **4.8** Art by Raychel Ciemma / **4.15** (a), (b) Thomas A. Steitz; art by Palay/Beaubois / **4.17** (b) Photograph Douglas Faulkner/Sally Faulkner Collection / **4.20** Art by Raychel Ciemma and American Composition and Graphics after B. Alberts et al., *Molecular Biology of the Cell*, Garland Publishing Co., 1983 / **4.21** Art by Raychel Ciemma and American Composition and Graphics / **4.22** (a), (b) © Oxford Scientific Films/Animals Animals; (c) © Raymond Mendez/Animals Animals; (d) Keith V. Wood

Chapter 5

5.1 Richard Surman/Tony Stone Images / **5.2** (a) Hans Reinhard/Bruce Coleman Ltd.; (e), (f) micrographs David Fisher; art by Raychel Ciemma and American Composition and Graphics / **5.3** (a) Barker-Blakenship/FPG; (b) art by Precision Graphics after Govindjee / **5.4** Photograph Carolina Biological Supply Company; art by Raychel Ciemma / **5.5** Larry West/FPG / **5.6** Art by Raychel Ciemma / **5.7** Art by Illustrious, Inc. / **5.8** E. R. Degginger / **5.9** Art by Raychel Ciemma and Precision Graphics / **5.10** Art by Precision Graphics / **5.11** (a) Art by Raychel Ciemma / **5.12** NASA / **5.13** Art by Raychel Ciemma and Precision Graphics

Chapter 6

6.1 Stephen Dalton/Photo Researchers / **6.2** (above) Dennis Brokaw; (a) (below) Janeart/Image Bank; (b) Paolo Fioratti; art by Precision Graphics / **6.3** (right) Art by Palay/Beaubois / **6.4** (a) Micrograph Keith R. Porter; (b) art by L. Calver; (c) art by Raychel Ciemma / **6.7** Art by Raychel Ciemma / **6.10** (b) Adrian Warren/Ardea, London; (c) David M. Phillips/Visuals Unlimited / **6.11** Photograph Gary Head / **Page 110** R. Llewellyn/Superstock, Inc.

Page 113 © Lennart Nilsson

Chapter 7

7.1 (left and right above) Chris Huss; (left inset and right below) Tony Dawson / **7.2** C. J. Harrison et al., *Cytogenetics and Cell Genetics* 35:21–27, copyright 1983 S. Karger A. G., Basel / **7.4** Andrew S. Bajer, University of Oregon / **7.5** Micrographs Ed Reschke; art by Raychel Ciemma / **7.6** Micrograph B. A. Palevitz and E. H. Newcomb, University of Wisconsin/BPS/Tom Stack & Associates / **7.7** Micrographs H. Beams and R. G. Kessel, *American Scientist*, 64:279–290, 1976 / **7.8** (a–c),(e) Lennart Nilsson from *A Child Is Born* © 1966, 1967 Dell Publishing Company, Inc.; (d) Lennart Nilsson from *Behold Man*, © 1974 by Albert Bonniers Forlag and Little, Brown and Company, Boston

Chapter 8

8.1 (a) Jane Burton/Bruce Coleman Ltd.; (b) Dan Kline/Visuals Unlimited / **8.2** Art by Precision Graphics / **8.3** CNRI/SPL/Photo Researchers / **8.4** Art by Raychel Ciemma / **8.5** Art by Raychel Ciemma and American Composition & Graphics / **8.6** Art by Raychel Ciemma / **8.9** Micrograph David M. Phillips/Visuals Unlimited / **8.10** Art by Raychel Ciemma

Chapter 9

9.1 (left) Frank Trapper/Sygma; (center) Focus on Sports; (right above) Fabian/Sygma; (right below) Moravian Museum, Brno / **9.2** Photograph Jean M. Labat/Ardea, London; art by Jennifer Wardrip / **9.5, 9.7** Art by Hans & Cassady, Inc. / **9.8** Art by Raychel Ciemma / **9.10** Photographs William E. Ferguson / **9.12** Micrographs Stanley Flegler/Visuals Unlimited / **9.13** (a), (b) Michael Stuckey/Comstock Inc.; (c) Russ Kinne/Comstock Inc. / **9.14** David Hosking / **9.15** Tedd Somes / **9.16** (top to bottom) Frank Cezus; Frank Cezus; Michael Keller; Ted Beaudin; Stan Sholik/all FPG / **9.17** (a) Dan Fairbanks, Brigham Young University / **9.18** Photograph Jane Burton/Bruce Coleman Ltd.; art by D. & V. Hennings / **9.19** After John G. Torrey, *Development in Flowering Plants*, by permission of Macmillan Publishing Company, copyright © 1967 by John G. Torrey / **9.20** Evan Cerasoli / **9.21** Leslie Faltheisek/Clacritter Manx / **9.22** Joe McDonald/Visuals Unlimited

Chapter 10

10.1 Eddie Adams/AP Photo / **10.2** (b) Photograph Omikron/Photo Researchers / **10.4** From Lennart Nilsson, *A Child Is Born*, © 1966, 1977 Dell Publishing Company, Inc. (b) Redrawn by Robert Demarest by permission from page 126 of Michael Cummings, *Human Heredity: Principles and Issues*, Third Edition. Copyright © 1994 by West Publishing Company. All rights reserved.; (c) art by Robert Demarest after Patten, Carlson, and others / **10.5** Photograph Carolina Biological Supply Company / **10.8** (b) Photograph Dr. Victor A. McKusick; (c) Steve Uzzell / **10.12** After Victor A. McKusick, *Human Genetics*, Second edition, copyright 1969. Reprinted by permission of Prentice-Hall, Inc., Engelwood Cliffs, NJ; photograph The Bettmann Archive / **10.13** Art by Raychel Ciemma / **10.14** (a) Cytogenetics Laboratory, University of California, San Francisco; (b) after Collman and Stoller, *American Journal of Public Health*, 52, 1962; photographs (above) courtesy of Peninsula Association for Retarded Children and Adults, San Mateo Special Olympics, Burlingame, CA; (below) used by permission of Carole Iafrate / **10.15** (a) Courtesy of G. H. Valentine; (c) C. J. Harrison / **10.16** Art by Raychel Ciemma / **10.17** Bonnie Kamin/Stuart Kenter Associates / **10.18** Carolina Biological Supply Company

Chapter 11

11.1 (above) A. Lesk/SPL/Photo Researchers; (below) A. C. Barrington Brown © 1968 J. D. Watson / **11.2** Art by Raychel Ciemma / **11.3** (b) Micrograph Lee D. Simon/Science Source/Photo Researchers / **11.5** Micrograph Biophoto Associates/SPL/Photo Researchers / **11.6** Art by Precision Graphics / **11.8** (a) (left) C. J. Harrison et al., *Cytogenetics and Cell Genetics*, 35:21–27, copyright 1983 S. Karger A. G., Basel; (b) micrograph B. Hamkalo; (c) micrograph O. L. Miller, Jr. and Steve L. McKnight; (b-d) art by Nadine Sokol / **11.9** U. K. Laemmli from *Cell*, 12:817–828, copyright 1977 by MIT/Cell Press / **11.10** (a) Ken Greer/Visuals Unlimited; (b) Biophoto Associates/Science Source/ Photo Researchers; (c) James Stevenson/SPL/Photo Researchers; (d) Gary Head

Chapter 12

12.1 (above) Kevin Magee/Tom Stack & Associates; (below) Dennis Hallinan/FPG / **12.4** Art by Hans & Cassady, Inc. / **12.9** (a) Courtesy of Thomas A. Steitz from *Science*, 246:1135–1142, December 1, 1989 / **12.10** Art by Raychel Ciemma and American Composition & Graphics / **12.12** Nik Kleinberg / **12.13** Art by Raychel Ciemma / **12.14** Brian Matthews, University of Oregon / **12.15** Art by Palay/Beaubois and Hans & Cassady / **12.16** (a) Micrograph C. J. Harrison et al., *Cytogenetics and Cell Genetics*, 35:21–27, copyright 1983 S. Karger A. G., Basel; (b) art by Palay/Beaubois / **12.17** (a) Dr. Karen Dyer Montgomery / **12.18** Jack Carey / **12.19** Frank B.

Salisbury / **12.20** Art by Betsy Palay/Artemis / **12.21** (a) Art by Betsy Palay/Artemis; (b) Lennart Nilsson © Boehringer Ingelheim International GmbH

Chapter 13

13.1 Lewis L. Lainey / **13.2** (a) Stanley N. Cohen/Science Source/Photo Researchers; (b) Dr. Huntington Potter and Dr. David Dressler / **13.4** After Stephen L. Wolfe, *Molecular and Cellular Biology*, Wadsworth, 1993 / **13.6** Cellmark Diagnostics, Abingdon, U.K. / **13.9** Michael Maloney/San Francisco Chronicle / **13.10** (a) W. Merrill; (b) Keith V. Wood; (c), (d) Monsanto Company / **13.11** R. Brinster and R. E. Hammer, School of Veterinary Medicine, University of Pennsylvania

Page 215 S. Stammers/SPL/Photo Researchers

Chapter 14

14.1 Elliot Erwitt/Magnum Photos, Inc. / **14.2** (a) Jen & Des Bartlett/Bruce Coleman Ltd.; (b) Kenneth W. Fink/Photo Researchers; (c) Dave Watts/A.N.T. Photo Library / **14.4** (a) (left) Courtesy George P. Darwin, Darwin Museum, Down House; (right) Heather Angel; (b) Christopher Ralling; (c) photograph Dieter & Mary Plage/Survival Anglia; art by Leonard Morgan / **14.5** (a) Lee Kuhn/FPG; (b) Field Museum of Natural History, Chicago, and the artist, Charles R. Knight (Neg. No. CK21T) / **14.6** (a) Heather Angel; (b) David Cavagnaro; (c) George W. Cox; (d) Alan Root/Bruce Coleman Ltd. / **14.7** (a) Down House and The Royal College of Surgeons of England; (b) John H. Ostrom, Yale University / **14.8** Alan Solem / **14.11** J. A. Bishop and L. M. Cook / **14.13** (a), (b) Warren Abrahamson; (c) Forest W. Buchanan/Visuals Unlimited; (d) Kenneth McCrea and Warren Abrahamson / **14.15** (a) Thomas Bates Smith / **14.16** Bruce Beehler / **14.18** (above) David Neal Parks; (below) W. Carter Johnson / **14.19** Courtesy of Jerry Coyne / **14.20** Photograph David Cavagnaro / **14.21** Kjell Sandved/Visuals Unlimited / **14.22** D. Avon/Ardea, London

Chapter 15

15.1 (a) Photograph Gary Head; snail courtesy of Larry Reed; (b) adapted from R. K. Selander and D. W. Kaufman, *Evolution*, Vol. 29, No. 3, December 31, 1975; / **15.2** Jen & Des Bartlett/Bruce Coleman Ltd. / **15.3** After F. Ayala and J. Valentine, *Evolving*, Benjamin-Cummings, 1979 / **15.4** G. Ziesler/ZEFA / **15.5** (a) Fred McConnaughey/Photo Researchers; (b) Patrice Geisel/Visuals Unlimited / **15.6** After W. Jensen and F. B. Salisbury, *Botany: An Ecological Approach*, Wadsworth, 1972 / **15.7** Photograph Robert C. Simpson/Nature Stock / **15.9** After P. Dodson, *Evolution: Process and Product*, Third edition, Prindle, Weber & Schmidt

Chapter 16

16.1 (left) Vatican Museums; (right) Martin Dohrn/SPL/Photo Researchers / **16.2** (a) Donald Baird, Princeton Museum of Natural History; (b) A. Feduccia, *The Age of Birds*, Harvard University Press, 1980; (c) H. P. Banks; (d) Jonathan Blair / **16.3** David Noble/FPG / **16.4** (a), (b) Gary Head; (c–e) art by Precision Graphics after E. Guerrant, *Evolution*, 36:699–712 / **16.5** (b) From T. Storer et al., *General Zoology*, Sixth edition, McGraw-Hill, 1979. Reproduced by permission of McGraw-Hill, Inc. / **16.6** Art by Victor Royer / **16.7** Art by Raychel Ciemma / **16.8** (top) Douglas P. Wilson/Eric & David Hosking; (center) Superstock, Inc.; (bottom) E. R. Degginger / **16.10** (top) Kjell B. Sandved/Visuals Unlimited; (center) Jeffrey Sylvester/FPG; (bottom) Thomas D. Mangelsen/Images of Nature / **Page 258** (left to right) Larry Lefever/Grant Heilman Inc.; R.I.M. Campbell/Bruce Coleman Ltd.; Runk and Schoenberger/Grant Heilman Inc.; Bruce Coleman Ltd

Page 261 © 1990 Arthur M. Greene

Chapter 17

17.1 Jeff Hester and Paul Scowen (Arizona State University), and NASA / **17.2** Painting by William K. Hartmann / **17.3** (a) Painting by Chesley Bonestell / **17.4** Art by Precision Graphics / **17.5** (a) Sidney W. Fox; (b) W. Hargreaves and D. Deamer / **17.7** Art by Leonard Morgan / **17.8** Maps by Lloyd K. Townsend after Krohn and Sündermann, 1983 / **17.9** Art by Precision Graphics / **17.10** Bill Bachman/Photo Researchers / **17.11** (a) Stanley W. Awramik; (b–f) Andrew H. Knoll, Harvard University / **17.12** P. L. Walne and J. H. Arnott, *Planta*, 77:325–354, 1967 / **17.13** Art by Raychel Ciemma / **17.14** Robert K. Trench / **17.15** (a) Neville Pledge/South Australian Museum; (b) Chip Clark / **17.16** Patricia G. Gensel / **17.17** Art by Raychel Ciemma / **Page 275** Map by Lloyd K. Townsend after Krohn and Sündermann, 1983 / **Page 276** Map by Lloyd K. Townsend after Ziegler, Scotese, and Barrett, 1983 / **17.18** (above) Painting by Megan Rohn courtesy of David Dilcher; (below) art by Precision Graphics / **17.19** © John Gurche 1989 / **17.20** (a) NASA Galileo Imaging Team; (b) art by Raychel Ciemma / **Page 279** Map by Lloyd K. Townsend after Ziegler, Scotese, and Barrett, 1983 / **17.21** (a), (b) Paintings © Ely Kish; (c) Field Museum of Natural History, Chicago, and the artist Charles R. Knight (Neg. No.CK8T) / **17.22** (left) Maps by Lloyd K. Townsend after A. M. Ziegler, C. R. Scotese, and S. F. Barrett, "Mesozoic and Cenozoic Paleogeographic Maps" and J. Krohn and J. Sündermann, "Paleotides Before the Permian" in F. Brosche and J. Sündermann (Eds.), *Tidal Friction and the Earth's Rotation II*, Springer-Verlag, 1983

Chapter 18

18.1 (a–c) Tony Brain and David Parker/SPL/Photo Researchers; (d) Lee D. Simon/Photo Researchers; (e) Gary W. Grimes and Steven L'Hernault **18.2** Art by Raychel Ciemma / **18.3** L. J. LeBeau, University of Illinois Hospital/BPS / **18.4** (a) Stanley Flegler/Visuals Unlimited; (b) CNRI/SPL/Photo Researchers / **18.5** Art by Raychel Ciemma; micrograph L. Santo / **18.7** (a) © 1994 Barrie Rokeach; (b) R. Robinson/Visuals Unlimited / **18.8** (a) John D. Cunningham/Visuals Unlimited; (b) Tony Brain /SPL/Photo Researchers; (c) P. W. Johnson and J. McN. Sieburth, University of Rhode Island/BPS / **18.9** (above) Centers for Disease Control; (below) photograph Edward S. Ross / **18.10** (a), (c) Art by Nadine Sokol; (b) after Stephen L. Wolfe *Molecular Biology of the Cell*, Wadsworth, 1993 / **18.11** (a) George Musil/Visuals Unlimited; (b) K. G. Murti /Visuals Unlimited; (c), (d) Kenneth M. Corbett / **18.12, 18.13** Art by Palay/Beaubois and Precision Graphics / **18.14** (a) Heather Angel; (b) W. Merrill / **18.15** (a) Art by Leonard Morgan; (b) M. Claviez, G. Gerish, and R. Guggenheim; (c) London Scientific Films; (d–f) Carolina Biological Supply Company; (g) photograph courtesy Robert R. Kay from R. R. Kay et al., *Development*, 1989 Supplement, pp. 81–90, © The Company of Biologists Ltd. 1989 / **18.16** Edward S. Ross / **18.17** (a) M. Abbey/Visuals Unlimited; (b) John Clegg/Ardea, London; (c,e) art redrawn from V. & J. Pearse and M. & R. Buchsbaum, *Living Invertebrates*, The Boxwood Press, 1987. Used by permission; (d) G. Shih and R. Kessel/Visuals Unlimited; (e) micrograph T. E. Adams/Visuals Unlimited / **18.18** (a-above) Frieder Sauer/Bruce Coleman Ltd.; (a-below), (b) art by Raychel Ciemma redrawn from V. & J. Pearse and M. & R. Buchsbaum, *Living Invertebrates*, The Boxwood Press, 1987. Used by permission. / **18.19** (a) Jerome Paulin / Visuals Unlimited; (b) David M. Phillips/Visuals Unlimited; (c) John D. Cunningham/Visuals Unlimited / **18.20** Art by Leonard Morgan; micrograph Steven L'Hernault / **18.21** Art by Palay/Beaubois / **18.22** (a) Ronald W. Hoham, Dept. of Biology, Colgate University; (b) Jan Hinsch/SPL/Photo Researchers; (c) Florida Department of Environmental Protection, Florida Marine Research Institute, St. Petersburg; (d) C. C. Lockwood / **18.23** (a) D. P. Wilson/Eric & David Hosking; (b) from Tom Garrison, *Oceanography: An Invitation to Marine Science*, Wadsworth, 1993 / **18.24** Photograph D. J. Patterson/Seaphot Limited: Planet Earth Pictures; art by Raychel Ciemma; (b) Carolina Biological Supply Company

Chapter 19

19.1 (a) Pat & Tom Leeson/Photo Researchers; (b) Dave Schiefelbein; (c) Robert Glenn Ketchum / **19.4** Photograph Jane Burton/Bruce Coleman Ltd.; art by Raychel Ciemma / **19.5** (a) Roger K. Burnard; (b) John D. Cunningham/Visuals Unlimited / **19.6** (above) Field Museum of Natural History, Chicago; (Neg. #7500C); (below) Brian Parker/Tom Stack & Associates / **19.7** (a) Kingsley R. Stern; (b) Edward S. Ross; (c) W. H. Hodge; (d) Edward S. Ross / **19.8** (above) Art by Raychel Ciemma; photograph A. & E. Bomford/Ardea, London; (below) photograph Lee Casebere / **19.9** Photograph Edward S. Ross; art by Raychel Ciemma / **19.10** (a) Ed Reschke; (b) John H. Gerard; (c) Kingsley R. Stern; (d) Edward S. Ross; (e) Edward S. Ross; (inset) F. J. Odendaal, Duke University/BPS / **19.11** (a) Art by Jennifer Wardrip; (b) M.P.L. Fogden/Ardea, London; (c) Heather Angel; (d) Peter F. Zika/Visuals Unlimited / **19.12** Art by Raychel Ciemma / **19.13** (a), (c–e) Robert C. Simpson/Nature Stock; (b) Jane Burton/Bruce Coleman Ltd. / **19.14** Art by Raychel Ciemma; micrograph Garry T. Cole, University of Texas, Austin/BPS / **Page 320** Micrograph J. D. Cunningham/Visuals Unlimited / **19.15** Micrographs Ed Reschke; art by Raychel Ciemma / **19.16** (a) After T. Rost et al., *Botany*, Wiley, 1979; (b) M. Eichelberger/Visuals Unlimited; (c) G. L. Barron, University of Guelph; (d) Garry T. Cole, University of Texas, Austin/BPS / **19.17** N. Allin and G. L. Barron / **19.18** (above) Dr. P. Marazzi/SPL/Photo Researchers; (below) Eric Crichton/Bruce Coleman Ltd. / **19.19** After Raven, Evert, and Eichhorn, *Biology of Plants*, Fourth edition, Worth Publishers, New York, 1986 / **19.20** (a) Mark Mattock/Planet Earth Pictures; (b) Edward S. Ross / **19.21** (a) © 1990 Gary Braasch; (b) F. B. Reeves

Chapter 20

20.1 (a) Courtesy of Department of Library Services, American Museum of Natural History (Neg. No. K10273); (c) Jack Carey and Lisa Starr **20.2** Art by D. & V. Hennings / **20.3** Art by Raychel Ciemma / **20.4** (left) After Laszlo Meszoly in L. Margulis, *Early Life*, Jones and Bartlett Publishers, Inc., Boston, © 1982 / **20.5** Bruce Hall / **20.6** (a–c) Art by Raychel Ciemma; (d) (left) after Bayer and Owre, *The Free-Living Lower Invertebrates*, © 1968 Macmillan; (right) Don W. Fawcett/Visuals Unlimited; (e) Marty Snyderman/Planet Earth Pictures **20.7** (b) Frieder Sauer/Bruce Coleman Ltd.; (c) Kim Taylor/Bruce Coleman Ltd.; (d) art by Raychel Ciemma / **20.8** (a) Douglas Faulkner/Photo Researchers; (b) Christian DellaCorte; (c) Andrew Mounter/Seaphot Limited: Planet Earth Pictures; (d) art by Precision Graphics after T. Storer et al., *General Zoology*, Sixth edition, © 1979 McGraw-Hill / **20.9** Photograph Kim Taylor/Bruce Coleman Ltd.; art by Raychel Ciemma / **20.10** (a) Cath Ellis, University of Hull/SPL/Photo Researchers; (b) Robert & Linda Mitchell / **20.11** Art by Raychel Ciemma / **20.12** J. Solliday/BPS / **20.13** Art by Raychel Ciemma / **20.14** Art by Raychel Ciemma; (d) photograph Carolina Biological Supply Company / **20.15** (a) Lorus J. and Margery Milne; (b) Dianora Niccolini / **Page 337** Art by Palay/Beaubois / **20.16** (b) Alex Kerstitch; (c) Jeff Foott/Tom Stack & Associates; (d) J. Grossauer/ZEFA; (e) Gary Head / **20.17, 20.18** Art by Raychel Ciemma / **20.19** (a) J.A.L. Cooke/Oxford Scientific Films; (b) © Cabisco/Visuals Unlimited; (c) Jon Kenfield/Bruce Coleman Ltd.; (d) (left) from Eugene N. Kozloff, *Invertebrates*, copyright © 1990 by Saunders College Publishing. Reproduced by permission of the publisher. (Right) adapted from Rasmussen, *Ophelia*, Vol. 11 in Eugene N. Kozloff, *Invertebrates*, 1990 / **20.20** Art by Raychel Ciemma / **20.21** Jane Burton/Bruce Coleman Ltd. / **20.22** (above) Angelo Giampiccolo/FPG; (below) Jane Burton/Bruce Coleman Ltd. / **20.23** (a) P. J. Bryant, University of California, Irvine/BPS; (b) Ken Lucas/Seaphot Limited: Planet Earth Pictures; (c) John H. Gerard; (d) redrawn from *Living Invertebrates*, V. & J. Pearse/M. & R. Buchsbaum, The Boxwood Press, 1987. Used by permission. / **20.24** (a) Franz Lanting/Bruce Coleman Ltd.; (b) Hervé Chaumeton; (c) Agence Nature; (d) Runk & Schoenberger/Grant Heilman, Inc.; (e) Fred Bavendam/ Peter Arnold, Inc. / **20.25** After David H. Milne, *Marine Life and the Sea*, Wadsworth 1995 / **20.26** (a) Z. Leszczynski/Animals Animals; (b) Steve Martin/Tom Stack & Associates / **20.27** Art by D. & V. Hennings / **20.28** From Georges Pasteur, "Jean Henri Fabre," *Scientific American*, July

1994. Copyright © 1994 by Scientific American, Inc. All rights reserved. / **20.29** (a–d) Edward S. Ross; (e) C. P. Hickman, Jr.; (f–i) Edward S. Ross; (j) David Maitland/Seaphot Limited: Planet Earth Pictures / **20.30** (a) Chris Huss/The Wildlife Collection; (b) Ian Took/Biofotos; (c) Kjell B. Sandved; (d) John Mason/Ardea, London / **20.31** (a) Art by L. Calver; (b), (c) Hervé Chaumeton/Agence Nature; (d) Jane Burton/Bruce Coleman Ltd. / **20.33** Walter Deas/Seaphot Limited: Planet Earth Pictures

Chapter 21

21.1 (a) Jean Phillipe Varin/Jacana/Photo Researchers; (b) Tom McHugh/Photo Researchers / **21.3** (a) Photograph Rick M. Harbo; (b), (c) sketches redrawn from *Living Invertebrates*, V. & J. Pearse and M. & R. Buchsbaum, The Boxwood Press, 1987. Used by permission. / **21.4** (a) Art by Raychel Ciemma; (b) Runk & Schoenberger/Grant Heilman, Inc. / **21.5** Art by Raychel Ciemma adapted from A. S. Romer and T. S. Parsons, *The Vertebrate Body*, Sixth edition, Saunders College Publishing, 1986; photograph Al Giddings/Images Unlimited / **21.7** (a), (b) After David H. Milne, *Marine Life and the Sea*, Wadsworth, 1995; (c) Heather Angel / **21.8** (a) Erwin Christian/ZEFA; (b) Allan Power/Bruce Coleman Ltd.; (c) Tom McHugh/Photo Researchers / **21.9** (a) Bill Wood/Bruce Coleman Ltd.; (b) art by Raychel Ciemma / **21.10** (a) Art by Raychel Ciemma after C. T. Regan and E. Trewavas, 1932; (b) from Tom Garrison, *Oceanography: An Invitation to Marine Science*, Wadsworth, 1993; (c) Robert & Linda Mitchell; (d) Patrice Ceisel/© 1986 John G. Shedd Aquarium; (e) Peter Scoones/Seaphot Limited: Planet Earth Pictures / **21.11** (a) © Marianne Collins; (b) art by Laszlo Meszoly and D. & V. Hennings / **21.12** (a) Jerry W. Nagel; (b) Stephen Dalton/Photo Researchers; (c) John Serraro/Visuals Unlimited; (d) Juan M. Renjifo/Animals Animals / **Page 361** Art by Leonard Morgan adapted from A. S. Romer and T. S. Parsons, *The Vertebrate Body*, Sixth edition, Saunders College Publishing, 1986, and others / **21.13** (a) Zig Leszczynski/Animals Animals; (b) Kevin Schafer/Tom Stack & Associates; (c) art by D. & V. Hennings; (d) D. Kaleth/Image Bank; (e) Heather Angel; (f) Andrew Dennis/A.N.T. Photo Library; (g) Stephen Dalton/Photo Researchers; art by Raychel Ciemma / **21.14** (a) Gerard Lacz/A.N.T. Photo Library; (b) Rajesh Bedi; (c) Thomas D. Mangelsen/Images of Nature; (d) J.L.G. Grande/Bruce Coleman Ltd. / **21.15** (b) Art by D. & V. Hennings / **21.16** (a) Sandy Roessler/FPG; (b) art by Raychel Ciemma after M. Weiss and A. Mann, *Human Biology and Behavior*, Fifth edition, HarperCollins, 1990 / **21.17** (a) D. & V. Blagden/A.N.T. Photo Library; (b) Jack Dermid; (c) Douglas Faulkner/Photo Researchers; (d) Clem Haagner/Ardea, London; (e) J. Scott Altenbach, University of New Mexico; (f) Christopher Crowley / **21.18** (a) Bruce Coleman Ltd.; (b) Tom McHugh/Photo Researchers; (c) Larry Burrows/Aspect Picture Library / **21.19** Art by D. & V. Hennings / **21.22** © Time Inc. 1965/Larry Burrows Collection / **21.23** (b) Dr. Donald Johanson, Institute of Human Origins; (c) Louise M. Robbins / **21.24** (a) Art by D. & V. Hennings; (b) photographs by John Reader copyright 1981 / **21.26** Douglas Mazonowicz/Gallery of Prehistoric Art

Page 375 Gary Head

Chapter 22

22.1 (a) Roger Werth; (b) © 1980 Gary Braasch; (c) © 1989 Gary Braasch / **22.2** Art by Raychel Ciemma / **22.4** Micrograph James D. Mauseth, *Plant Anatomy*, Benjamin-Cummings, 1988 / **22.5** Art by D. & V. Hennings / **22.6** (a–c) Biophoto Associates / **22.8** George S. Ellmore / **22.9** Art by D. & V. Hennings / **22.10** (a) Robert & Linda Mitchell; (b) E. R. Degginger / **22.11** Art by D. & V. Hennings / **22.12** Art by D. & V. Hennings; (a) (center) Ray F. Evert; (right) James W. Perry; (b) (center) Carolina Biological Supply Company; (right) James W. Perry / **22.13** (a) Art by Raychel Ciemma; (b) C. E. Jeffree et al., *Planta*, 172(1):20–37, 1987. Reprinted by permission of C. E. Jeffree and Springer-Verlag; (c) Jeremy Burgess/SPL/Photo Researchers / **22.14** Heather Angel / **Page 385** Art redrawn by Precision Graphics from Beryl B. Simpson and Molly C. Ogorzaly, *Economic Botany: Plants in Our World*, © 1986 McGraw-Hill Inc. Used by permission of McGraw-Hill, Inc. / **22.15** John E. Hodgin / **22.16** (a) Art by Hans & Cassady, Inc.; (b) Ripon Microslides, Inc. / **22.17** (a) Micrograph Chuck Brown; (b) Carolina Biological Supply Company / **22.18** Ripon Microslides; sketch after T. Rost et al., *Botany: A Brief Introduction to Plant Biology*, Second edition, © 1984, John Wiley & Sons / **22.20** Art by Raychel Ciemma / **22.22** (a) Micrograph Jerry D. Davis; (b) H. A. Core, W. A. Coté, and A. C. Day, *Wood Structure and Identification*, Second edition, Syracuse University Press, 1979 / **Page 390** Edward S. Ross / **Page 391** Micrograph Ripon Microslides

Chapter 23

23.1 (a), (b–left) Robert & Linda Mitchell; (b-right) Robert C. Simpson/Nature Stock; (c) John N.A. Lott, *Scanning Electron Microscope Study of Green Plants*, St. Louis: C. V. Mosby Company, 1976 / **23.2** Micrograph Jean Paul Revel / **23.3** (a) Micrograph Chuck Brown; (b), (c) art by Leonard Morgan / **23.4** (b) Mark E. Dudley and Sharon R. Long; (c) Adrian P. Davies/Bruce Coleman Ltd.; (d) NifTAL Project, University of Hawaii, Maui; art by Jennifer Wardrip / **23.5** (a) James T. Brock; (b) U.S. Department of Agriculture / **23.6** George S. Ellmore / **23.8** T. A. Mansfield / **23.9** Jeremy Burgess/SPL/Photo Researchers / **23.10** Micrographs H. A. Core, W. A. Coté, and A. C. Day, *Wood Structure and Identification*, Second edition, Syracuse University Press, 1979 / **23.11** Art by Raychel Ciemma / **23.12** Art by Hans & Cassady, Inc. / **23.13** Martin Zimmermann, *Science*, 133:73–79, © AAAS 1961 / **23.14** (a–c) Art by Palay/Beaubois / **23.16** W. Thomson, *American Journal of Botany*, 57(3):316, 1970

Chapter 24

24.1 (a) Merlin D. Tuttle, Bat Conservation International; (b) Robert A. Tyrrell / **24.2** Thomas Eisner, Cornell Univeristy / **24.3** (a) Photographs Gary Head / **24.4** John Shaw/Bruce Coleman Ltd. / **24.5** (a), (b) David M. Phillips/Visuals Unlimited; (c) David Scharf/Peter Arnold, Inc. / **24.6** Art by Raychel Ciemma / **24.7** (a), (b) Patricia Schulz; (c), (d) Ray F. Evert; (e), (f) Ripon Microslides, Inc. / **24.8** (a) Janet Jones; (b) B. J. Miller, Fairfax, VA/BPS; (c) R. Carr/Bruce Coleman Ltd.; (d) Richard H. Gross / **24.9** Russell Kaye/© 1993 The Walt Disney Co. Reprinted with permission of Discover Magazine / **24.10** (a) Runk/Schoenberger/Grant Heilman, Inc.; (b) Kingsley R. Stern / **24.11** (a) B. Bracegirdle and P. Miles, *An Atlas of Plant Structure*, Heinemann Educational Books, 1977; (b) photograph Carolina Biological Supply Company; art by Hans & Cassady, Inc.; (c) photograph Hervé Chaumeton/Agence Nature; art by Hans & Cassady, Inc. / **24.13** (a) Kingsley R. Stern / **24.14** Michael A. Keller/FPG / **24.15** (a), (b) micrographs courtesy of Randy Moore from "How Roots Respond to Gravity," M. L. Evans, R. Moore, and K. Hasenstein, *Scientific American*, December 1986; (c) John Digby and Richard Firn / **24.16**(b) Frank B. Salisbury / **24.17** Gary Head / **24.18** Cary Mitchell / **24.20** Frank B. Salisbury / **24.22** (a) Jan Zeevart; (b) Gary Head / **24.23** N. R. Lersten / **24.24** R. J. Downs / **24.25** Diana Starr / **24.26** Grant Heilman Inc.

Page 425 © Kevin Schafer

Chapter 25

25.1 David Macdonald / **25.2** (a) Focus on Sports; (inset) Manfred Kage/Bruce Coleman Ltd.; art by Palay/Beaubois; photographs; (b) (left) Lennart Nilsson from *Behold Man*, © 1974 by Albert Bonniers Forlag and Little, Brown and Company, Boston; (center) Manfred Kage/Bruce Coleman Ltd.; (right) Ed Reschke/Peter Arnold, Inc. / **25.3** Art by Raychel Ciemma / **25.4** Photograph Gregory Dimijian/Photo Researchers; art by Raychel Ciemma adapted from C. P. Hickman, Jr., L. S. Roberts, and A. Larson, *Integrated Principles of Zoology*, Ninth edition, Wm. C. Brown, 1995 / **25.5** Micrographs (a–c), (e), (f) Ed Reschke; (d) Fred Hossler/Visuals Unlimited / **25.6** Photograph Roger K. Burnard; (left) art by Joel Ito; (right) art by L. Calver / **25.7, 25.8** Micrographs Ed Reschke / **25.9** Art by Robert Demarest / **25.10** (a) Lennart Nilsson from *Behold Man*, © 1974 Albert Bonniers Forlag and Little, Brown and Company, Boston; (b) Kim Taylor/Bruce Coleman Ltd. / **25.11** Art by L. Calver / **25.14** Photograph Fred Bruemmer

Chapter 26

26.1 (a) Chaumeton-Lanceau/Agence Nature; (b) Michael Neveux / **26.2** Art by L. Calver / **26.3** (left) Ed Reschke; (right) CNRI/SPL/Photo Researchers; art by Robert Demarest / **26.4** Sketches from *The Life of Birds*, Fourth edition, by Luis Baptista and Joel Carl Welty, copyright © 1988 by Saunders College Publishing. Reproduced by permission of the publisher. / **26.5** Michael Keller/FPG / **26.6** Linda Pitkin/Planet Earth Pictures / **26.7** Photograph Stephen Dalton/Photo Researchers; art by Raychel Ciemma / **26.8** Art by D. & V. Hennings / **26.9** Art by Raychel Ciemma / **26.10** Art by Joel Ito; micrograph Ed Reschke / **26.11** Art by K. Kasnot / **26.12** National Osteoporosis Foundation / **26.13** C. Yokochi and J. Rohen, *Photographic Anatomy of the Human Body*, Second edition, Igaku-Shoin Ltd., 1979 / **26.15** (a) N.H.P.A./A.N.T. Photo Library; (b) art by Robert Demarest / **26.16** Art by Raychel Ciemma / **26.17** (b) Micrograph John D. Cunningham; (c) micrograph D. W. Fawcett, *The Cell*, Philadelphia: W. B. Saunders Co., 1966; art by Robert Demarest / **26.18** Art by Nadine Sokol / **26.19** Art by Robert Demarest

Chapter 27

27.1 (a) Photograph courtesy of The New York Academy of Medicine Library; (b) from A. D. Waller, *Physiology, The Servant of Medicine*, Hitchcock Lectures, University of London Press, 1910 / **27.3** (b) (left) After M. Labarbera and S. Vogel, *American Scientist*, 70:54–60, 1982 / **27.4** Art by Precision Graphics / **27.5** Art by Palay/Beaubois / **27.6** CNRI/SPL/Photo Researchers / **27.7** Art by Raychel Ciemma / **27.8** (a), (b) Lester V. Bergman & Associates, Inc.; (c) after F. Ayala and J. Kiger, *Modern Genetics*, © 1980 Benjamin-Cummings / **27.9** Art by Nadine Sokol after Gerard J. Tortora and Nicholas P. Anagnostakos, *Principles of Anatomy and Physiology*, Sixth edition, Copyright © 1990 by Biological Sciences Textbooks, Inc., A & P Textbooks, Inc. and Elia-Sparta, Inc. Reprinted by permission of HarperCollins Publishers / **27.11** Art by Kevin Somerville / **27.12** (b) C. Yokochi and J. Rohen, *Photographic Anatomy of the Human Body*, Second edition, Igaku-Shoin Ltd, 1979; (a), (c) art by Raychel Ciemma / **27.14** (a) Micrograph Michael Abbey/Photo Researchers; art by Raychel Ciemma / **27.15** (b–e) Art by Robert Demarest based on A. Spence, *Basic Human Anatomy*, Benjamin-Cummings, 1982 / **27.16** Sheila Terry/SPL/Photo Researchers / **27.17** (a) (above) Ed Reschke; (below) F. Sloop and W. Ober/Visuals Unlimited / **27.20** Lennart Nilsson © Boehringer Ingelheim International GmbH / **27.21, 27.22** Art by Raychel Ciemma / **27.23** Lennart Nilsson from *Behold Man*, © 1974 by Albert Bonniers Forlag and Little, Brown and Company, Boston

Chapter 28

28.1 (a) The Granger Collection, New York; (b) Lennart Nilsson © Boehringer Ingelheim International GmbH / **28.2** Art by Nadine Sokol / **28.3** Micrograph Robert R. Dourmashkin, courtesy of Clinical Research Centre, Harrow, England / **28.4** Lennart Nilsson © Boehringer Ingelheim International GmbH / **28.5** Art by Raychel Ciemma / **28.8** Art by Raychel Ciemma / **28.9** Art by Raychel Ciemma; micrograph Morton H. Nielsen and Ole Werdlin, University of Copenhagen / **28.10** Art by Raychel Ciemma; photograph courtesy Don C. Wiley, Harvard University / **28.11** Art by Hans & Cassady, Inc. / **28.13, 28.14** Art by Palay/Beaubois after B. Alberts et al., *Molecular Biology of the Cell*, Garland Publishing Company, 1983 / **28.16** (left) Matt Meadows/Peter Arnold, Inc.; (right) Lowell Georgia/Science Source/Photo Researchers / **28.17** (left) David Scharf/Peter Arnold, Inc.; (right) Kent

Wood/Photo Researchers / **28.18** Ted Thai/Time Magazine / **28.19** After Stephen L. Wolfe, *Molecular Biology of the Cell*, Wadsworth, 1993 / **28.20** Micrographs Z. Salahuddin, National Institutes of Health

Chapter 29

29.1 Galen Rowell/Peter Arnold, Inc. / **29.2** Giorgio Gualco/Bruce Coleman Ltd. / **29.5** (a) Peter Parks/Oxford Scientific Films; (b) Hervé Chaumeton/Agence Nature / **29.6** Micrograph Ed Reschke / **29.7** Art by Raychel Ciemma / **29.8** (a) M.P.L. Fogden/Bruce Coleman Ltd. / **29.9** Micrograph H. R. Duncker, Justus-Liebig University, Giessen, Germany / **29.10** Art by Kevin Somerville / **29.11** Modified from A. Spence and E. Mason, *Human Anatomy and Physiology*, Fourth edition, 1992, West Publishing Company / **29.12** Art by K. Kasnot / **29.14** Art by Leonard Morgan / **29.15** CNRI/SPL/Photo Researchers / **29.16** O. Auerbach/Visuals Unlimited / **29.17** Micrograph Lennart Nilsson from *Behold Man*, © 1974 by Albert Bonniers Forlag and Little, Brown and Company, Boston / **29.18** Steve Lissau/Rainbow

Chapter 30

30.1 Gary Head / **30.3** Art by Raychel Ciemma / **30.4** (a) D. Robert Franz/Planet Earth Pictures; (c) adapted from A. Romer and T. Parsons, *The Verterbrate Body*, Sixth edition, Saunders College Publishing, 1986 / **30.5** Art by Kevin Somerville / **30.7** Art by Robert Demarest; micrograph Omikron/SPL/Photo Researchers / **30.8** After A. Vander et al., *Human Physiology: Mechanisms of Body Function*, Fifth edition, McGraw-Hill, 1990. Used by permission; (c) Redrawn from page 763 of *Human Anatomy and Physiology*, Fourth edition, by A. Spence and E. Mason. Copyright © 1992 by West Publishing Company. All rights reserved. / **30.9** (a) Art by Raychel Ciemma; (b), (c-right) Lennart Nilsson © Boehringer Ingelheim International GmbH; (c-left) Biophoto Associates/ SPL/Photo Researchers; art by Robert Demarest / **30.10** Art by Raychel Ciemma / **Page 519** (in-text art) After A. Vander et al., *Human Physiology: Mechanisms of Body Function*, Fifth edition, McGraw-Hill, 1990. Used by permission. / **30.11** Modified after A. Vander et al., *Human Physiology*, Fourth edition, McGraw-Hill, 1985 / **Page 521** Photograph Ralph Pleasant/FPG / **30.15** Photograph Gary Head / **30.16** Photographs Dr. Douglas Coleman, The Jackson Laboratory

Chapter 31

31.1 (a) Claude Steelman/Tom Stack & Associates; (b) photograph David Noble/FPG / **31.3, 31.4** Art by Robert Demarest / **31.5** (above) Art by Robert Demarest; (below) art by Precision Graphics / **31.7** (a), (b) From Tom Garrison, *Oceanography: An Invitation to Marine Science*, Wadsworth, 1993; (c) Thomas D. Mangelsen/Images of Nature / **31.8** (a) Bob McKeever/Tom Stack & Associates; (b) Colin Monteath, Hedgehog House, New Zealand / **31.9** The Bettmann Archive

Chapter 32

32.1 Comstock, Inc. / **32.2** Kjell B. Sandved / **32.3** Art by Raychel Ciemma / **32.5** (d) Micrograph Manfred Kage/Peter Arnold, Inc. / **32.6, 32.8** Art by Raychel Ciemma / **32.10** (a) Art by Kevin Somerville; (b) Ed Reschke; (c) micrograph Dr. Constantino Sotelo from *International Cell Biology*, page 83, 1977. Used by copyright permission of the Rockefeller University Press / **32.12** Micrograph R. G. Kessel and R. H. Kardon from *Tissues and Organs: A Text-Atlas of Scanning Electron Microscopy*, W. H. Freeman and Company, 1979. Used by permission of R. G. Kessel; art by Robert Demarest / **32.14** (b) Art by Robert Demarest / **32.15** Painting by Sir Charles Bell, 1809, courtesy of Royal College of Surgeons, Edinburgh / **32.16** (below) Art by Raychel Ciemma / **32.17** Art by Kevin Somerville / **32.19** (a) Art by Robert Demarest; (b) micrograph Manfred Kage/Peter Arnold, Inc. / **32.20** (a) Colin Chumbley/Science Source/Photo Researchers); (b) photograph C. Yokochi and J. Rohen, *Photographic Anatomy of the Human Body*, Second edition, Igaku-Shoin Ltd., 1979 / **32.21** (left) Marcus Raichle, Washington University School of Medicine / **32.23** (a), (b) From Edythe D. London et al., *Archives of General Psychiatry*, 47:567–574 (1990); (right) photograph Ogden Gigli/Photo Researchers / **32.24** Eric A. Newman / **32.25** (left) Art by Palay/Beaubois after Penfield and Rasmussen, *The Cerebral Cortex of Man*, copyright © 1950 Macmillan Publishing Company, Inc. Renewed 1978 by Theodore Rasmussen; (right) photograph Colin Chumbley/Science Source/Photo Researchers / **32.26** Micrograph Ed Reschke; art by Ron Ervin / **32.27** Merlin D. Tuttle, Bat Conservation International / **32.28** (a–d) Art by Robert Demarest; (e), (f) Robert E. Preston, courtesy Joseph E. Hawkins, Kresge Hearing Research Institute, University of Michigan Medical School / **32.29** (a–d) Sketches after M. Gardiner, *The Biology of Vertebrates*, McGraw-Hill, 1972; (d) photograph E. R. Degginger; (e) Chase Swift / **32.30** Art by Robert Demarest / **32.31** Art by Kevin Somerville / **32.32** Micrograph Lennart Nilsson © Boehringer Ingelheim International GmbH / **32.34** Art by Palay/Beaubois after S. Kuffler and J. Nicholls, *From Neuron to Brain*, Sinauer, 1977 / **32.35** Gerry Ellis/The Wildlife Collection

Chapter 33

33.1 Hugo van Lawick / **33.2** Art by Kevin Somerville / **33.5, 33.6** Art by Robert Demarest / **33.7** (a) Mitchell Layton; (b) Syndication International (1986) Ltd.; (c) photographs courtesy of Dr. William H. Daughaday, Washington University School of Medicine, from A. I. Mendelhoff and D. E. Smith, eds., *American Journal of Medicine*, 20:133 (1956) / **33.8** Art by Leonard Morgan / **33.9** (a), (b) Art by Raychel Ciemma; (c) The Bettmann Archive / **33.11** Biophoto Associates/SPL/Photo Researchers / **33.12** Art by Leonard Morgan / **33.13** (a) John S. Dunning/Ardea, London; (b) Evan Cerasoli / **33.14** (a), (b) From R. C. Brusca and G. J. Brusca, *Invertebrates*, © 1990 Sinauer Associates. Used by permission; (c) Frans Lanting/Bruce Coleman Ltd.; (d) Robert & Linda Mitchell / **33.15** Photograph Roger K. Burnard

Chapter 34

34.1 (a) Art by Raychel Ciemma; (b) Hans Pfletschinger; (c–f) John H. Gerard / **34.2** Evan Cerasoli / **34.4** Carolina Biological Supply Company / **34.5** (left) Photographs Carolina Biological Supply Company; (far right) Peter Parks/Oxford Scientific Films/Animals Animals / **34.7** (a) Art by Raychel Ciemma after V. E. Foe and B. M. Alberts, *Journal of Cell Science*, 61:32, © The Company of Biologists 1983; (b–e) J. B. Morrill / **34.8** Art by Raychel Ciemma; (right) after B. Burnside, *Developmental Biology*, 26:416–441, 1971. Used by permission of Academic Press. / **34.9** (a) Art by Palay/Beaubois after Robert F. Weaver and Philip W. Hedrick, *Genetics*. Copyright © 1989 Wm. C. Brown Publishers; (b) Carolina Biological Supply Company / **34.10, 34.11** Art by Raychel Ciemma / **34.12** Ed Reschke / **34.15** Art by Raychel Ciemma / **34.16** (above) Art by Robert Demarest; (below) art by Raychel Ciemma / **34.17** Photograph Lennart Nilsson from *A Child Is Born*, © 1966, 1977 Dell Publishing Company, Inc. / **34.18** Art by Robert Demarest / **34.19–34.22** Art by Raychel Ciemma / **34.23** Photographs Lennart Nilsson, *A Child Is Born*, © 1966, 1977 Dell Publishing Company, Inc.; art by Raychel Ciemma / **34.24** (a) Art by Raychel Ciemma modified from Keith L. Moore, *The Developing Human: Clinically Oriented Embryology*, Fourth edition, Philadelphia, W. B. Saunders Co., 1988 / **34.25** (a) From Lennart Nilsson, *A Child Is Born*, © 1966, 1977 Dell Publishing Company, Inc.; (b) James W. Hanson, M.D. / **34.26** Art by Robert Demarest / **34.27** Art by Raychel Ciemma / **34.28** Art by Raychel Ciemma adapted from L. B. Arey, *Developmental Anatomy*, Philadelphia, W. B. Saunders Co., 1965 / **34.30** CNRI/SPL/Photo Researchers / **34.31** (a) John D. Cunningham/Visuals Unlimited; (b) David M. Phillips/Visuals Unlimited

Page 627 Alan and Sandy Carey

Chapter 35

35.1 (a) Gary Head; (b) Antoinette Jongen/FPG / **35.4** Photograph E. Vetter/ZEFA / **35.5** (a) Jonathan Scott/Planet Earth Pictures; (b) Fred Bavendam/Peter Arnold, Inc.; (c) Wisniewski/ZEFA / **35.6** Photographs John A. Endler / **35.7** Photograph NASA / **35.8** After G. T. Miller, Jr., *Environmental Science*, Sixth edition, Wadsworth, 1997 / **35.9** Data from Population Reference Bureau after G. T. Miller, Jr., *Living in the Environment*, Eighth edition, Wadsworth, 1993 / **35.10, 35.11** Photographs United Nations / **35.12, 35.13** After G. T. Miller, Jr., *Environmental Science*, Sixth edition, Wadsworth, 1997

Chapter 36

36.1 (left) Edward S. Ross; (right) Donna Hutchins / **36.2** (a), (c) Harlo H. Hadow; (b) Bob and Miriam Francis/Tom Stack & Associates / **36.3** Clara Calhoun/Bruce Coleman Ltd. / **36.4** After G. Gause, 1934 / **36.5** Stephen G. Tilley / **36.6** After N. Weland and F. Bazazz, *Ecology*, 56:681–188, © 1975 Ecological Society of America / **36.7** Photograph John Dominis, Life Magazine, © Time Inc. / **36.8** Photograph Ed Cesar/Photo Researchers / **36.9** (a) James H. Carmichael; (b) Edward S. Ross; (c) W. M. Laetsch / **36.10** Edward S. Ross / **36.11** (a) Edward S. Ross; (b) Thomas Eisner, Cornell University; (c) Douglas Faulkner/Sally Faulkner Collection / **36.12** (a–f), (i) Roger K. Burnard; (g), (h) E. R. Degginger / **36.13** (a), (c) Jane Burton/Bruce Coleman Ltd.; (b) Heather Angel; (d), (e) Jane Lubchenco, *American Naturalist*, 112:23–29, © 1978 by The University of Chicago Press / **36.14** (a) After M. Kusenov, *Evolution*, 11:298–299, 1957; (b) after T. Dobzhansky, *American Scientist*, 38:209–221, 1950 / **36.15** (above) Photograph Dr. Harold Simon/Tom Stack & Associates; (below) after S. Fridriksson, *Evolution of Life on a Volcanic Island*, Butterworth: London, 1975 / **36.16** (a), (b) After J. M. Diamond, *Proceedings of the National Academy of Sciences*, 69:3199–3201, 1972; (c) David Cavagnaro / **Page 661** R. Slavin/FPG

Chapter 37

37.1 Wolfgang Kaehler / **37.3** Photograph Sharon R. Chester / **37.4** Gene C. Feldman and Compton J. Tucker/NASA, Goddard Space Flight Center / **37.8** Photograph Gerry Ellis/The Wildlife Collection / **37.9** (a) Photograph © 1991 Gary Braasch / **37.10** Art by Raychel Ciemma / **37.11** (a) Photograph by Gene E. Likens from G. E. Likens and F. H. Bormann, *Proceedings First International Congress of Ecology*, pp. 330–335, September 1974, Centre Agric. Publ. Doc. Wagenigen, The Hague, The Netherlands; (b) photograph by Gene E. Likens from G. E. Likens et al., *Ecology Monograph*, 40(1):23–47, 1970; (c) after G. E. Likens and F. H. Bormann, "An Experimental Approach to New England Landscapes" in A. D. Hasler (ed.), *Coupling of Land and Water Systems*, Chapman & Hall, 1975 / **37.12** (b) Photograph John Lawlor/FPG / **37.14** After W. Dansgaard et al., *Nature*, 364:218–220, 15 July 1993; D. Raymond et al., *Science*, 259:926–933, February 1993; W. Post, *American Scientist*, 78:310–326, July–August 1990 / **37.16** Photograph William J. Weber/Visuals Unlimited / **37.17** After E. Schulze, *Science*, 244:776–783, 1989

Chapter 38

38.1 (a) (left) David Noble/FPG; (right) Edward S. Ross; (b) (left) Richard Coomber/Planet Earth Pictures; (right) Edward S. Ross / **38.2** Edward S. Ross / **38.4** (b) Art by L. Calver / **38.6, 38.8** From Tom Garrison, *Essentials of Oceanography*, Wadsworth, 1995 / **38.7** Art by Raychel Ciemma / **38.9** (b) Art by Victor Royer / **38.10** Art by Raychel Ciemma / **38.11** Art by D. & V. Hennings after Whittaker; Bland; and Tilman / **38.12** Harlo H. Hadow / **38.13** (above) John D. Cunningham/Visuals Unlimited; (inset) AP/Wide World Photos; (below) Jack Wilburn/Animals Animals / **38.14** (a) Kenneth W. Fink/Ardea, London; (b) Ray Wagner/Save the Tall Grass Prairie, Inc.; (c) Jonathan Scott/Planet Earth Pictures / **38.15** (left) © 1991 Gary Braasch; (right) Thase Daniel; (below) (left) Adolf Schmidecker/FPG; (right) Edward S. Ross / **38.16** Thomas E. Hemmerly / **38.17** (a) Dennis Brokaw; (b) Jack Carey / **38.18** (a) Fred Bruemmer; (b) Doug Sokell/Visuals Unlimited / **38.19** (a) D. W. MacManiman / **38.20** Modified after Edward S. Deevy, Jr., *Scientific American*, October 1951 / **38.21** D. W. Schindler, *Science*, 184:897–899 / **38.22** David Steinberg / **38.23** Art by

Lloyd K. Townsend / **38.24** (a) Dennis Brokaw; (b) McCutcheon/ZEFA; (c) Robert Hessler; Woods Hole Institution of Oceanography; (d) J. Frederick Grassle, Woods Hole Institution of Oceanography / **38.26** (page 700) (top) Jim Doran; (center inset) Sea Studios/Peter Arnold, Inc.; (below) Douglas Faulkner/Sally Faulkner Collection; (page 701) (moray eel) Jeff Rotman; (chambered nautilus) Alex Kerstitch; all other photographs Douglas Faulkner/Sally Faulkner Collection / **38.27** E. R. Degginger; art by D. & V. Hennings / **38.28** (a) Courtesy of J. L. Sumich, *Biology of Marine Life,* Fifth edition, William C. Brown, 1992; (b) (above) © 1991 Gary Braasch; (below) Phil Degginger / **38.29** Adapted from Tom Garrison, *Essentials of Oceanography,* Wadsworth, 1995 / **38.30** (a), (b) Illustrations adapted from Tom Garrison, *Oceanography: An Invitation to Marine Science,* Wadsworth, 1993; sattelite images courtesy of NOAA/UCAR, Gene Carl Feldman, NASA-GSFC, Otis Brown, University of Miami / **38.31** (a) Fred McKinney/FPG; (b) L. L. Rue III/FPG / **38.32** (a) Jack Carey

Chapter 39

39.1 (above) Gary Head; (below) U. S. Geological Survey / **39.2** Photograph United Nations / **39.3** (a) USDA Forest Service; (b) Heather Angel; art after G. T. Miller, Jr., *Environmental Science: An Introduction,* Wadsworth, 1986, and the Environmental Protection Agency / **39.4** Center for Air Pollution Impact and Trend Analysis (CAPITA), Washington University, St. Louis, Missouri / **39.5** (a–c) NASA; photograph National Science Foundation / **39.6** Data from G. T. Miller, Jr. / **39.7** (above) R. Bieregaard/Photo Researchers; (below) after G. T. Miller, Jr. *Living in the Environment,* Eighth edition, Wadsworth, 1993 / **39.8** NASA / **39.9** Gerry Ellis/The Wildlife Collection / **39.10** Thomas G. Meier/USDA Soil Conservation Service / **39.11** Agency for International Development / **39.12** Dr. Charles Henneghien/Bruce Coleman Ltd. / **39.13** From Water Resources Council / **39.14** Ocean Arks International / **39.15** Data from G. T. Miller, Jr. / **39.16** Photograph AP/Wide World; art by Precision Graphics after Marvin H. Dickerson, "ARAC: Modeling an Ill Wind" in *Energy and Technology Review,* August 1987. Used by permission of University of California, Lawrence Livermore National Laboratory and U.S. Department of Energy / **39.17** Alex MacLean/Landslides / **Page 723** J. McLoughlin/FPG / **39.18** © 1983 Billy Grimes

Chapter 40

40.1 (a) Robert Maier/Animals Animals; (b) Jack Clark/Comstock, Inc; (c) John Bova/Photo Researchers; (d) from L. Clark, *Parasitology Today,* Vol. 6, No. 11, 1990, Elsevier Trends Journals, Cambridge, U.K. / **40.2** (a) Eugene Kozloff; (b), (c) Stevan Arnold / **40.3** Photograph Hans Reinhard/Bruce Coleman Ltd.; (a), (b) from G. Pohl-Apel and R. Sussinka, *Journal for Ornithologie,* 123:211–214 / **40.4** (a) Evan Cerasoli; (b) From A. N. Meltzoff and M. K. Moore, "Imitation of Facial and Manual Gestures by Human Neonate," *Science,* 198:75–78. Copyright 1977 by the AAAS. / **40.5** (a) Eric Hosking; (b) Stephen Dalton/Photo Researchers / **40.6** Nina Leen in *Animal Behavior,* Life Nature Library / **40.7** John Alcock / **40.8** (a) Edward S. Ross; (b) E. Mickleburgh/Ardea, London / **40.9** Art by D. & V. Hennings / **40.10** John Alcock / **40.11** (left) David Fritts/Animals Animals; (right) Ray Richardson/Animals Animals / **40.12** Michael Francis/The Wildlife Collection / **40.13** Frank Lane Agency/Bruce Coleman Inc. / **40.14** A. E. Zuckerman/Tom Stack & Associates / **40.15** Fred Bruemmer / **40.16** John Alcock / **40.17** Timothy Ransom / **40.18** Kenneth Lorenzen / **40.19** Gregory D. Dimijian/Photo Researchers / **40.20** Yvon Le Maho / **Page 743** (left) F. Schutz; (right) Lincoln P. Brower

INDEX

INDEX